EFFECTIVE FIELD THEORY IN PARTICLE PHYSICS AND COSMOLOGY

Lecture Notes of the Les Houches Summer School:
Volume 108, Session CVIII

Effective Field Theory in Particle Physics and Cosmology

Edited by

Sacha Davidson, Paolo Gambino, Mikko Laine,
Matthias Neubert, Christophe Salomon

OXFORD
UNIVERSITY PRESS

OXFORD
UNIVERSITY PRESS

Great Clarendon Street, Oxford, OX2 6DP,
United Kingdom

Oxford University Press is a department of the University of Oxford.
It furthers the University's objective of excellence in research, scholarship,
and education by publishing worldwide. Oxford is a registered trade mark of
Oxford University Press in the UK and in certain other countries

Published in the United States of America by Oxford University Press
198 Madison Avenue, New York, NY 10016, United States of America

British Library Cataloguing in Publication Data

Data available

Library of Congress Control Number: 2020934961

ISBN 978–0–19–885574–3

Printed and bound by
CPI Group (UK) Ltd, Croydon, CR0 4YY

École de Physique des Houches

Service inter-universitaire commun
à l'Université Joseph Fourier de Grenoble
et à l'Institut National Polytechnique de Grenoble

Subventionné par l'Université Joseph Fourier de Grenoble,
le Centre National de la Recherche Scientifique,
le Commissariat à l'Énergie Atomique

Directeur:
Christophe Salomon, Directeur de Recherche CNRS, Laboratoire Kastler Brossel, Ecole Normale Supérieure - Département de Physique, Paris, France.

Directeurs scientifiques de la session:
Sacha Davidson, CNRS Researcher in Astroparticle Physics, Laboratoire del'Univers et des Particules de Montpellier, Montpellier, France.
Paolo Gambino, Professor of Theoretical Physics, National Institute of Nuclear Physics, University of Turin, Torino, Italy.
Mikko Laine, Professor of Theoretical Physics, Albert Einstein Centre, Institute for Theoretical Physics, University of Bern, Switzerland.
Matthias Neubert, Director of the Mainz Institute for Theoretical Physics, Professor of Theoretical Physics, Johannes Gutenberg University Mainz,Germany.

Previous sessions

Publishers

- Session VIII: Dunod, Wiley, Methuen
- Sessions IX and X: Herman, Wiley
- Session XI: Gordon and Breach, Presses Universitaires
- Sessions XII–XXV: Gordon and Breach
- Sessions XXVI–LXVIII: North Holland
- Session LXIX–LXXVIII: EDP Sciences, Springer
- Session LXXIX–LXXXVIII: Elsevier
- Session LXXXIX– : Oxford University Press

Preface

The topic of the CVIII session of the École de physique des Houches, held in July 2017, was Effective Field Theory (EFT) in Particle Physics and Cosmology. In both particle physics and cosmology, our current best understanding is captured by a "Standard Model" (SM) which describes all known phenomena, but which is nevertheless unsatisfactory from a number of points of view. The hope is that eventually experimental deviations from the SM are found, which would then indicate a road to physics Beyond the Standard Model (BSM). Ideally this leads, amongst others, to a verifiable explanation for the origin of Dark Matter.

In order to be sure that a given deviation is really an indication of BSM physics, SM predictions must be robust and accurate. This represents a non-trivial challenge, as many particles, with varying masses and momenta, can participate in the interactions. Their effects, independently of whether the particles appear as real or virtual states, need to be systematically accounted for. As EFTs help disentangling the effects of physics at different scales, they simplify this task considerably. These scales could be masses (e.g. the light masses of the particles of the SM versus the heavy masses of yet undiscovered particles), momenta (e.g. the hard, collinear and soft momenta playing a role in jets produced in high-energy collisions), or length scales (e.g. the lattice spacing appearing in numerical simulations versus the pion Compton wavelength of interest to low-energy hadronic interactions).

The underlying idea of EFT is that at each scale the relevant physics can be parametrised with appropriate variables, which may change with the scale. In order to achieve precise predictions there is often no need to know the underlying exact theory: one can work with simpler field theories, describing only the degrees of freedom relevant at a certain scale, while performing a systematic expansion in one or more small parameters, generally ratios of well-separated scales. Hence EFTs are essential tools both for precision analyses within known multi-scale theories, such as the SM, and for a concise parameterisation of hypothetical BSM models.

The goal of this school was to offer a broad introduction to the foundations and modern applications of Effective Field Theory in many of its incarnations. This is all the more important, as there are preciously few textbooks covering the subject, none of them in a complete way. The lecturers were invited to present the concepts in a pedagogical way such that attendees can adapt some of the latest developments in other fields to their own problems.

The present collection of lecture notes, covering almost all the lectures, will hopefully serve as an introduction to the topic and as a reference manual also to many students and researchers who could not participate in the actual school. The basic foundations are laid out in two lecture series, by Matthias Neubert and Aneesh Manohar, which review the field-theoretic background of renormalization, power counting, and operator classification. Well-established classic implementations of Effective Field Theories, used for treating systems with spontaneously broken global symmetries, or those with heavy quarks, are introduced by Antonio Pich and Thomas Mannel, respectively. One of the modern tools of collider and flavour physics, called Soft Collinear Effective Theory, is explained by Thomas Becher, whereas specific applications in flavour physics are reviewed by Luca Silvestrini. Cosmological applications of EFT are discussed by Cliff Burgess (inflation), Tobias Baldauf (large-scale structure formation), and Junji Hisano (dark matter). In the framework of lattice field theory, theoretical background is provided by Rainer Sommer. To widen the perspective, applications of Effective Field Theory in nuclear physics are presented by Ubirajara van Kolck. (In the school there were additionally lectures on thermal field theory and post-Newtonian gravity, unfortunately unavailable in written form.)

The summer school was attended by more than 50 participants, covering a very broad spectrum of scientific specializations and geographic origins, including the Americas, India, Asia, and Australia. Apart from eagerly attending the lectures and challenging the lecturers with many questions, the students also had a chance to present their own work in short talks and through posters.

During and after the school, a number of students made the effort to prepare and type solutions to many of the problems proposed by some of the lecturers. These solutions have been included as a chapter in this book, and we are very grateful to the students who have devoted time to this enterprise.

As is usual for Les Houches, the scientific program was complemented by vigorous social and sports activities in the Chamonix valley. The wonderful restaurant and the other institute facilities contributed to a very enjoyable atmosphere.

We are very grateful for the assistance of the administrative staff of the Les Houches School, who helped us put together this event and look after its day-to-day evolution: our special thanks to Muriel Gardette, Isabelle Lelièvre and Jeff Aubrun. We also thank the then school director, Leticia Cugliandolo, for assisting us in the first steps of the organization.

Finally, organizing the school would not have been possible without the generous financial support of the Université franco-allemande (UFA/DFH), of the Lyon Institute of Origins (LIO), of the Physics Department of the University of Torino, of the Albert Einstein Center of the University of Bern (AEC), as well as of the PRISMA Cluster of Excellence of the University of Mainz.

Montpellier, Torino, Bern, Mainz, May 2019

Sacha Davidson
Paolo Gambino
Mikko Laine
Matthias Neubert

Contents

List of Participants

Organizers

DAVIDSON Sacha
Institut de Physique Nucléaire de Lyon, Villeurbanne, France

GAMBINO Paolo
University of Torino, Italy

LAINE Mikko
Institute for Theoretical Physics, University of Bern, Switzerland

NEUBERT Matthias
Johannes Gutenberg University, Mainz, Germany

Lecturers

PICH Antonio
Institute of Corpuscular Physics, Paterna, Spain

MANOHAR Aneesh
University of California, San Diego, USA

BURGESS Cliff
Perimeter Institute, Waterloo, Canada

BECHER Thomas
Institute for Theoretical Physics, University of Bern, Switzerland

Vanhove Pierre
Institut de physique théorique, Gif-sur-Yvette, France

VAN KOLCK Ubirajara
Orsay Nuclear Physics Institute, France

SILVESTRINI Luca
National Institute for Nuclear Physics, Rome, Italy

MANNEL Thomas
University of Siegen, Germany

HISANA Junji HISANO
Nagoya University, Japan

SOMMER Rainer
The Deutsches Elektronen-Synchrotron, Zeuthen, Germany

BALDAUF Tobias
DAMTP, University of Cambridge, Cambridge, UK

CARON-HUOT Simon
McGill University, Montreal, Canada

Students

ACHARYA Neramballi Ripunjay
University of Bonn, Germany

BIALAS Gabriela
Université Blaise, Pascal France

BALSIGER Marcel
University of Bern, Switzerland

BOGERS Mark
University of Stavanger, Norway

BOIARSKA, Iryna
University of Kyiv, Ukraine

BOUNAKIS, Marios
Newcastle University, UK

BREWER Jasmine
Massachusetts Institute of Technology, USA

CAHILL Caroline
University of Liverpool, UK

CHIGUSA So
University of Tokyo, Japan

CHWAY Dongjin
Seoul National University, South Korea

CULLEN Jonathan
Durham University, UK

DRISSI Mehdi
Université Paris-Saclay, France

FLETT Chris
University of Liverpool, UK

FLORIO Adrien
Swiss Federal Institute of Technology Lausanne, Switzerland

GARGALIONIS John
University of Melbourne, Australia

GASBARRO Andrew
Yale University, USA

GHOSH Shayan
Indian Institute of Science, India

GUSTAFSON Erik
University of Iowa, USA

HASNER Caspar
Technical University Munich, Germany

HAUKSSON Sigtryggur
McGill University, Canada

HERMANSSON TRUEDSSON, Nils
Lund University, Sweden

HONGO Masaru
RIKEN, Japan

INGOLDBY James
Yale University, USA

JACKSON Gregory
University of Cape Town, South Africa

JAISWAL, Sneha
Indian Institute of Technology Guwahati, India

KADAVY, Tomas
Charles University, Czech Republic

KOZOW Pawel
University of Warsaw, Poland

KVEDARAITE Sandra
University of Sussex, UK

LEAK Matthew
University of Liverpool, UK

LÓPEZ IBÁÑEZ María Luisa
University of Valencia, Spain

MAELGER Jan
École Polytechnique, France

MELIS Aurora
University of Valencia, Spain

MELVILLE Scott
Imperial College London, UK

MORENO Daniel
Universitat Autònoma de Barcelona, Spain

MURGUI GALVEZ Clara
University of Valencia, Spain

OOSTERHOF Femke
University of Groningen, The Netherlands

OVCHYNNIKOV Maksym
University of Kyiv, Ukraine

PANCHENKO Mischa
Ludwig Maximillian University of Munich, Germany

RENDON SUZUKI Jesus Gumaro
University of Arizona, USA

RUSOV Aleksey
University of Siegen, Germany

SAPORTA-DRUESNES Albert
Institut de Physique Nucléaire de Lyon, France

SÄPPI Eeli Matias
University of Helsinki, Finland

SCHICHO Phillip
University of Bern, Switzerland

SCHRÖDER Dennis
Bielefeld University, Germany

SKODRAS Dimitrios
Technical University of Dortmund, Germany

STIRNER Tobias
Max Planck Institute for Physics, Germany

TAMMARO Michele
University of Cincinnati, USA

TENÓRIO MAIA, Natália
Sao Paulo State University, Brazil

TOUATI Selim
Laboratory of Subatomic Physics & Cosmology, France

TROTT Timothy
University of California, USA

YOON Jong Min
Stanford University, USA

ZOPPI Lorenzo
University of Amsterdam, The Netherlands

1

Renormalization Theory and Effective Field Theories

Matthias NEUBERT

PRISMA Cluster of Excellence & Mainz Institute for Theoretical Physics
Johannes Gutenberg University, 55099 Mainz, Germany and
Department of Physics & LEPP, Cornell University, Ithaca, NY 14853, U.S.A.

Matthias NEUBERT

Neubert M., *Renormalization Theory and Effective Field Theories* In: *Effective Field Theory in Particle Physics and Cosmology.*
Edited by: Sacha Davidson, Paolo Gambino, Mikko Laine, Matthias Neubert and Christophe Salomon, Oxford University Press (2020).
© Oxford University Press. DOI: 10.1093/oso/9780198855743.003.0001

Chapter Contents

Preface

This chapter reviews the formalism of renormalization in quantum field theories with special regard to effective quantum field theories. It is well known that quantum field theories are plagued by ultraviolet (UV) divergences at very short-distance scales and infrared (IR) divergences at long distances. Renormalization theory provides a systematic way in which to deal with the UV divergences. Effective field theories deal with the separation of physics on different length or energy scales. The short-distance physics is described by means of Wilson coefficient functions, whereas the long-distance physics is contained in the matrix elements of effective operators built out of the quantum fields for the low-energy effective degrees of freedom of the theory. Renormalization theory is as important for effective field theories as for conventional quantum field theories. Moreover, building on the Wilsonian approach to renormalization, effective field theories provide a framework for a deeper understanding of the physical meaning of renormalization. While the subject of renormalization theory is treated in every textbook on quantum field theory (see e.g. [1]–[7]), more advanced topics such as the renormalization of composite operators, the mixing of such operators under scale evolution, and the resummation of large logarithms of scale ratios are not always treated in as much detail as they deserve. Because of the central importance of this subject to the construction of effective field theories, this chapter summarizes the main concepts and applications in a concise manner. It thus sets the basis for many of the more specialized lecture courses delivered at this school.

This chapter assumes that the reader has taken an in-depth course on quantum field theory at the graduate level, including some exposure to the technicalities of renormalization. The primary focus here lies on the treatment of UV divergences in conventional quantum field theories. Only the final section discusses renormalization in the context of effective field theories. We do not explore the structure of IR divergences in this chapter, since they are of a different origin. (Nonetheless, effective theories have provided powerful new insights into the structure of IR divergences, see e.g. [8]–[11]).

1.1   Renormalization in QED

1.1.1   Introduction

Loop diagrams in quantum field theories are plagued by ultraviolet (UV) divergences. The procedure of *renormalization* is a systematic way of removing these divergences by means of a *finite number* of redefinitions of the parameters of the theory. We will review this formalism first with the example of Quantum Electrodynamics (QED), the theory describing the interaction of electrically charged particles with light. For simplicity, we focus on the simplest version of the theory containing a single charged fermion, i.e. electrons and positrons.

1.1.2   UV Divergences and Renormalized Perturbation Theory

The Lagrangian of QED reads (omitting gauge-fixing terms for simplicity)

$$
\begin{aligned}
\mathcal{L}_{\mathrm{QED}} &= \bar{\psi}_0 \left(i\slashed{D} - m_0 \right) \psi_0 - \frac{1}{4} F_{\mu\nu,0} F_0^{\mu\nu} \\
&= \bar{\psi}_0 \left(i\slashed{\partial} - m_0 \right) \psi_0 - \frac{1}{4} F_{\mu\nu,0} F_0^{\mu\nu} - e_0 \bar{\psi}_0 \gamma_\mu \psi_0 A_0^\mu,
\end{aligned}
\tag{1.1}
$$

where $F_0^{\mu\nu} = \partial^\mu A_0^\nu - \partial^\nu A_0^\mu$ is the field-strength tensor. The Dirac field ψ_0 describes the electron and its anti-particle, the positron, and the vector field A_0^μ describes the photon. The parameters m_0 and e_0 account for the electron mass and its electric charge. We use a subscript '0' to distinguish the 'bare' quantities appearing in the Lagrangian from the corresponding 'physical' parameters – i.e., the observable mass and electric charge of the electron – and fields with proper (canonical) normalization. Renormalization theory yields the relations between the bare parameters and fields and the renormalized ones.

By means of the Lehmann–Symanzik–Zimmermann (LSZ) reduction formula [12], scattering amplitudes in quantum field theories are connected to fully connected, amputated Feynman diagrams. Moreover, we can restrict the following discussion to one-particle irreducible (1PI) graphs. One-particle reducible diagrams are simply products of 1PI graphs. A useful concept to classify the UV divergences of such diagrams is the so-called *superficial degree of divergence D*. For an arbitrary QED Feynman graph, the dependence on internal (unrestricted) momenta arises from the loop integrals and propagators:

$$
\sim \int \frac{d^4 k_1 \dots d^4 k_L}{(\slashed{k}_i - m + i0) \dots (k_j^2 + i0) \dots}
\tag{1.2}
$$

The quantity D is defined as the sum of the powers of loop momenta in the numerator minus those in the denominator. Hence

$$
D = 4L - P_e - 2P_\gamma,
\tag{1.3}
$$

where L is the number of loops, and P_e and P_γ are the numbers of electron and photon propagators. We naively expect that diagrams with $D > 0$ are *power divergent* ($\propto \Lambda_{\mathrm{UV}}^D$, where we denote by Λ_{UV} a generic UV cutoff regularizing the integral in the region of large loop momenta), diagrams with $D = 0$ are *logarithmically divergent* ($\propto \ln \Lambda_{\mathrm{UV}}$), and diagrams with $D < 0$ have no UV divergences. We will see later that in many cases the actual degree of divergence is less than D, as a consequence of gauge invariance or due to some symmetries. However, as long as we consider fully connected, amputated Feynman diagrams, the actual degree of divergence is never larger than D.

The beautiful combinatoric identity (Problem 1.1.6.1)

$$L = I - V + 1 \tag{1.4}$$

relates the number of loops L of any Feynman graph to the number of internal lines I and the number of vertices V. For QED, this identity reads

$$L = P_e + P_\gamma - V + 1. \tag{1.5}$$

The only vertex of QED connects two fermion lines to a photon line, and hence we can express

$$V = 2P_\gamma + N_\gamma = \frac{1}{2}(2P_e + N_e), \tag{1.6}$$

where N_γ and N_e denote the number of external photon and fermion lines, respectively. This equation follows since each propagator connects to two vertices, whereas each external line connects to a single vertex. Combining relations (1.3), (1.4) and (1.6), we obtain

$$\begin{aligned} D &= 4\left(P_e + P_\gamma - V + 1\right) - P_e - 2P_\gamma \\ &= 4 - 4V + 3P_e + 2P_\gamma \\ &= 4 - 4V + 3\left(V - \frac{1}{2}N_3\right) + \left(V - N_\gamma\right) \\ &= 4 - \frac{3}{2}N_e - N_\gamma. \end{aligned} \tag{1.7}$$

This relation is remarkable, since it relates the superficial degree of divergence of a graph to the number of external lines, irrespective of the internal complexity (the number of loops and vertices) of the graph. It follows that only a small number of n-point functions (sets of fully connected, amputated diagrams with n external legs) have $D \geq 0$. It is instructive to look at them one by one (in each case, the blob represents infinite sets of graphs):

These so-called 'vacuum diagrams' have $D = 4$ and are badly divergent, but they give no contribution to S-matrix elements. As long as we ignore gravity, they merely produce an unobservable shift of the vacuum energy. (When gravity is taken into account, these graphs give rise to the infamous *cosmological constant problem*).

The one-photon amplitude has $D = 3$, but it vanishes by Lorentz invariance. To see this, note that the amplitude (with the external polarization vector removed) has a Lorentz index μ, but there is no four-vector which could carry this index.

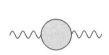

This so-called *photon vacuum polarization* amplitude has a superficial degree of divergence $D = 2$, and hence naively it is quadratically divergent. However, QED is a gauge theory, and gauge invariance requires that the vacuum polarization function has Lorentz structure $\pi^{\mu\nu}(k) = (k^2 g^{\mu\nu} - k^\mu k^\nu)\,\pi(k^2)$, see Section 1.3.2. This means that two powers of loop momenta are replaced by external momenta, and hence the true degree of divergence is $D - 2 = 0$, corresponding to a logarithmic UV divergence.

The three-photon amplitude has $D = 1$, and hence naively it is linearly divergent. In QED, this amplitude vanishes as a result of invariance under C parity (Furry's theorem). The same is true for all $(2n+1)$-photon amplitudes.

The four-photon amplitude has $D = 0$, and hence naively it is logarithmically divergent. Due to gauge invariance, however, the amplitude involves four powers of external momenta (Problem 1.1.6.2). Consequently, the true degree of divergence is $D - 4 = -4$, and so the amplitude is finite.

This so-called 'electron self energy' has $D = 1$, and hence naively it is linearly UV divergent. Chiral symmetry, i.e. the fact that in the limit $m_0 = 0$ left-handed and right-handed spinors transform under different irreducible representations of the Lorentz group, implies that the true degree of divergence is $D - 1 = 0$, corresponding to a logarithmic UV divergence.

The electromagnetic vertex function has $D = 0$ and is logarithmically UV divergent.

All other n-point functions in QED are UV finite.

Note that due to symmetries (Lorentz invariance, C parity, gauge invariance, and the chiral symmetry of massless QED) the true degree of divergence is often less than the superficial degree of divergence D. In fact, the only divergent n-point functions are the two-point functions for the photon and the electron and the electromagnetic vertex function, which captures the quantum corrections to the only vertex of QED. This is a remarkable fact, which allows us to remove these divergences by multiplicative redefinitions of the electron and photon fields, the electron mass and the electric charge. We define the so-called 'renormalized' fields (without subscript '0') by

$$\psi_0 = Z_2^{1/2}\,\psi\,, \qquad A_0^\mu = Z_3^{1/2}\,A^\mu. \tag{1.8}$$

The renormalized fields will be chosen such that their two-point functions (the renormalized propagators) have unit residue (or, depending on the renormalization scheme, at least a finite residue) at $p^2 = m^2$, where m is the physical electron mass. The notation Z_2 and Z_3 for the renormalization factors is historical; it would probably make more sense to call them Z_ψ and Z_A. When the Lagrangian (1.1) is rewritten in terms of renormalized fields, we obtain

$$\mathcal{L}_{\text{QED}} = Z_2\,\bar\psi\,(i\slashed\partial - m_0)\,\psi - \frac{Z_3}{4}\,F_{\mu\nu}\,F^{\mu\nu} - Z_2 Z_3^{1/2}\,e_0\,\bar\psi\gamma_\mu\psi\,A^\mu. \tag{1.9}$$

In the next step, we relate the bare mass and electric charge to the corresponding physical quantities. Let us write the corresponding relations in the form

$$Z_2\,m_0 = Z_m\,m\,, \qquad Z_2 Z_3^{1/2}\,e_0 = \mu^{\frac{4-d}{2}}\,Z_1\,e\,. \tag{1.10}$$

The scale μ enters in the dimensional regularization scheme [13], [14], in which the dimensionality of space-time is analytically continued from 4 to $d < 4$ (see Section 1.1.3). It ensures that the renormalized charge e is a dimensionless parameter. Expressed in terms of renormalized fields and parameters, the QED Lagrangian can now be written as

$$
\begin{aligned}
\mathcal{L}_{\text{QED}} &= Z_2\,\bar\psi\,i\slashed\partial\,\psi - Z_m\,m\,\bar\psi\psi - \frac{Z_3}{4}\,F_{\mu\nu}\,F^{\mu\nu} - \mu^{\frac{4-d}{2}}\,Z_1\,e\,\bar\psi\gamma_\mu\psi\,A^\mu \\
&\equiv \bar\psi\,(i\slashed\partial - m)\,\psi - \frac{1}{4}\,F_{\mu\nu}\,F^{\mu\nu} - \mu^{\frac{4-d}{2}}\,e\,\bar\psi\gamma_\mu\psi\,A^\mu \\
&\quad + \bar\psi\,(\delta_2\,i\slashed\partial - \delta m)\,\psi - \frac{\delta_3}{4}\,F_{\mu\nu}\,F^{\mu\nu} - \mu^{\frac{4-d}{2}}\,\delta_1\,e\,\bar\psi\gamma_\mu\psi\,A^\mu,
\end{aligned}
\tag{1.11}
$$

where we have defined

$$
\begin{aligned}
\delta_2 &= Z_2 - 1\,, & \delta_3 &= Z_3 - 1\,, \\
\delta_1 &= Z_1 - 1\,, & \delta_m &= (Z_m - 1)\,m\,.
\end{aligned}
\tag{1.12}
$$

By construction, scattering amplitudes calculated from this Lagrangian, which are expressed in terms of the physical electron mass m and electric charge e, are free of UV divergences. The first line in (1.11) has the same structure as the original QED Lagrangian (apart from the factor $\mu^{\frac{4-d}{2}}$ in the electromagnetic vertex) and hence gives rise to the usual QED Feynman rules. If that was the entire story, we would still encounter UV-divergent results when computing Feynman graphs. However, the so-called 'counterterms' in the second line give rise to additional Feynman rules, which have the effect of cancelling these UV divergences. The Feynman rules for these counterterms are as follows:

$$= -i\delta_3 \, (p^2 g^{\mu\nu} - p^\mu p^\nu)$$

$$= i \, (\delta_2 \slashed{p} - \delta_m)$$

$$= -i\delta_1 \, \mu^{\frac{4-d}{2}} e\gamma^\mu$$

The Lagrangian (1.11) is the starting point for calculations in 'renormalized perturbation theory' that gives rise to finite scattering amplitudes. The counterterms start at $\mathcal{O}(e^2)$ in perturbation theory and have a perturbative expansion in powers of the renormalized coupling $\alpha = e^2/(4\pi)$. Care must be taken to combine Feynman diagram with elementary vertices and counterterms at the same order in perturbation theory. When this is done consistently, the counterterms remove the UV divergences of Feynman graphs order by order in perturbation theory in α. The proof of this statement is known as the Bogoliubov–Parasiuk–Hepp–Zimmermann (BPHZ) theorem [15]–[17]. It states that all divergences of quantum field theories can be removed by constructing counterterms for the superficially divergent 1PI Feynman graphs. For practical purposes, it is useful to note that renormalization works not only for entire n-point functions, but also for individual Feynman diagrams. Here are two examples:

1.1.3   Calculation of the Renormalization Factors

We now understand that UV divergences only appear in intermediate steps of calculations in quantum field theories. When the counterterms are added to the bare Feynman graphs, these divergences cancel in all predictions for physical observables (e.g. scattering amplitudes). Nevertheless, in order to deal with the UV divergences arising in individual graphs, we must first introduce a regularization scheme. Ideally, the regularization should respect all symmetries of the theory as well as its fundamental properties, such as Lorentz invariance, gauge invariance, chiral symmetry (for $m_0 = 0$), and the analytic structure of scattering amplitudes. Also, the regulator should preserve the freedom to redefine the integration variables (the loop momenta). Initially, the most intuitive regularization scheme, in which we simply cut off loop integrals by means of a hard UV cutoff (such that $k_E^2 < \Lambda_{UV}^2$ after Wick rotation to Euclidean momenta), violates several of these requirements. In fact, the only known regularization scheme which preserves all of them is *dimensional regularization* [13], [14].[1] We have seen in the previous section that the UV divergences of QED n-point functions are at most of logarithmic strength. If we restrict the integrals over the loop momenta to fewer than four space-time dimensions, then these logarithmically divergent integrals become finite. The ingenious idea of dimensional regularization is to take this observation seriously. To this end, we replace the four-dimensional loop integrals by d-dimensional ones:

$$\int \frac{d^4 k}{(2\pi)^4} \to \int \frac{d^d k}{(2\pi)^d} \quad \text{with} \quad d < 4. \tag{1.13}$$

This lowers the degree of divergence of an L-loop diagram by $(d-4)L$, thus rendering logarithmically divergent integrals UV finite. We could now choose $d = 3$ or some smaller integer value, but this would bring us to a lower-dimensional quantum field theory with very different properties than real-world QED. Instead, in dimensional regularization we consider an analytic continuation of space-time to $d = 4 - 2\epsilon$ dimensions, where $\epsilon > 0$ is an infinitesimal parameter. In that way, the regularized theory we consider lives infinitesimally close to the original one.

If you have never been treated to a detailed exposition of dimensional regularization you will feel uncomfortable at this point. You are not alone in having problems imagining a $(4 - 2\epsilon)$-dimensional space-time. The point is that using techniques we review later, loop integrals can be expressed in terms of analytic functions of the space-time dimension d with poles at integer values, reflecting singularities of the integral in d (integer) dimensions. These functions can be analytically continued to the entire complex d-plane (which is more than we will need) in particular, they can be continued to all real values of d. Since we need the dimensional regulator only in intermediate steps of the calculation, it is perfectly fine to work in the immediate vicinity of $d = 4$, even if we cannot imagine what this means geometrically. UV singularities in four space-time dimensions will show up as $1/\epsilon^n$ pole terms. When the counterterms are added to the original Feynman diagrams, these pole terms cancel and we can take the limit $\epsilon \to 0$ in the final result.

When the Lagrangian (1.1) is continued to $d = 4 - 2\epsilon$ space-time dimensions, the canonical dimensions of the fields and parameters change. Using that the action $\int d^d x \mathcal{L}$ is dimensionless (as always in quantum field theory, we work in units where $\hbar = c = 1$), it is straightforward to derive that (the brackets [...] denote the mass dimension of a given quantity)

$$[\psi_0] = \frac{d-1}{2} = \frac{3}{2} - \epsilon, \quad [A_0^\mu] = \frac{d-2}{2} = 1 - \epsilon, \quad [m_0] = 1, \quad [e_0] = \frac{4-d}{2} = \epsilon. \tag{1.14}$$

---

[1] The Pauli–Villars scheme [18] discussed in most textbooks on quantum field theory changes the analytic structure of scattering amplitudes and becomes cumbersome beyond one-loop order.

If we wish to describe the strength of the electromagnetic interaction by means of a dimensionless coupling, we need to extract from the bare coupling e_0 a factor μ^ϵ with some auxiliary mass scale μ, as shown in (1.10), such that

$$e_0 \equiv \mu^\epsilon \, \tilde{e}_0(\mu) = \mu^\epsilon \, Z_e(\mu) \, e(\mu), \qquad Z_e = Z_1 Z_2^{-1} Z_3^{-1/2}, \tag{1.15}$$

where $\tilde{e}_0(\mu)$ is the dimensionless bare coupling and $e(\mu)$ is the renormalized coupling as defined in (1.10). Of course, physical quantities should not depend on the auxiliary scale μ, which we have introduced for convenience only. As discussed in Section 1.4, this fact gives rise to partial differential equations called *renormalization-group equations* (RGEs).

Let me briefly mention a technical complication which will not be of much relevance to these lectures. Since the Clifford algebra $\{\gamma_\mu, \gamma_\nu\} = 2g_{\mu\nu}$ involves the spacetime metric of Minkowski space, it needs to be generalized to d dimensions when the dimensional regularization scheme is employed. It is not difficult to prove the following useful relations (Problem 1.1.6.3):

$$\gamma^\mu \gamma_\mu = d, \qquad \gamma^\mu \gamma_\alpha \gamma_\mu = (2-d) \gamma_\alpha, \qquad \gamma^\mu \gamma_\alpha \gamma_\beta \gamma_\mu = 4g_{\alpha\beta} + (d-4) \gamma_\alpha \gamma_\beta. \tag{1.16}$$

In chiral gauge theories such as the Standard Model (SM) it is also necessary to generalize γ_5 to $d \neq 4$ dimensions. This is a problem full of subtleties, which will not be discussed here (see e.g. [14], [19] for more details).

The evaluation of one-loop integrals (with loop momentum k) in dimensional regularization is a straightforward matter once one has learned a couple of basic techniques, that are taught in any textbook on quantum field theory. We briefly list them here:

1. Combine the denominators of Feynman amplitudes, which contain products of propagators, using Feynman parameters. The general relation reads

$$\frac{1}{A_1 A_2 \cdots A_n} = \Gamma(n) \int_0^1 dx_1 \cdots \int_0^1 dx_n \, \delta\left(1 - \sum_{i=1}^n x_i\right) \frac{1}{(x_1 A_1 + \cdots + x_n A_n)^n}. \tag{1.17}$$

   Taking derivatives with respect to the A_i allows us to derive analogous relations where the factors A_i are raised to integer powers.

2. Introduce a shifted loop momentum

$$\ell^\mu = k^\mu + \sum_i c_i(x_1, \ldots, x_m) \, p_i^\mu, \tag{1.18}$$

   where $\{p_i^\mu\}$ are the external momenta of the diagram and the coefficients c_i are linear functions of Feynman parameters, such that the integral takes on the standard form

$$\int \frac{d^d \ell}{(2\pi)^d} \frac{1}{\left(\ell^2 - \Delta + i0\right)^n} \left(N_0 + N_1 \ell_\mu + N_2 \ell_\mu \ell_\nu + \ldots\right). \tag{1.19}$$

Note the absence of a linear term in ℓ in the denominator. The quantities Δ and N_i depend on the Feynman parameters $\{x_i\}$ and the external momenta $\{p_i^\mu\}$.

3. Use Lorentz invariance to replace

$$\ell_\mu \to 0, \qquad \ell_\mu \ell_\nu \to \frac{g_{\mu\nu}}{d}\ell^2, \text{ etc.} \tag{1.20}$$

under the integral.

4. The remaining integrals are performed using the Wick rotation $\ell^0 \to i\ell_E^0$ (and hence $\ell^2 \to -\ell_E^2$) and using spherical coordinates in d-dimensional Euclidean space. The relevant master formula reads

$$\int \frac{d^d\ell}{(2\pi)^d} \frac{(\ell^2)^\alpha}{(\ell^2 - \Delta + i0)^\beta} = \frac{i(-1)^{\alpha-\beta}}{(4\pi)^{\frac{d}{2}}} (\Delta - i0)^{\alpha-\beta+\frac{d}{2}} \frac{\Gamma(\alpha + \frac{d}{2})\Gamma(\beta - \alpha - \frac{d}{2})}{\Gamma(\beta)\Gamma(\frac{d}{2})}. \tag{1.21}$$

5. Perform the integrals over the Feynman parameters $\{x_i\}$ either in closed form (if possible) or after performing a Laurent expansion about $\epsilon = 0$. At one-loop order, the relevant parameter integrals can all be expressed in terms of logarithms and dilogarithms [20].

Let us now look at the results obtained for the three UV-divergent n-point functions of QED in the dimensional regularization scheme.

1.1.3.1   *Electron Self Energy*

Consider the full electron propagator in momentum space. There are infinitely many diagrams contributing to this object, which we can classify by specifying the number of places in each diagram where the diagram falls apart if we cut a single electron line. Hence, the full propagator can be written as a geometric series of graphs containing more and more insertions of the so-called electron *self energy*, i.e., the infinite set of 1PI graphs with two external fermion legs:

$$= \frac{i}{\not p - m_0 + i0} + \frac{i}{\not p - m_0 + i0}(-i\Sigma)\frac{i}{\not p - m_0 + i0} \tag{1.22}$$

$$+ \frac{i}{\not p - m_0 + i0}(-i\Sigma)\frac{i}{\not p - m_0 + i0}(-i\Sigma)\frac{i}{\not p - m_0 + i0} + \dots$$

$$= \frac{i}{\not p - m_0 - \Sigma + i0}.$$

The self energy $\Sigma \equiv \Sigma(\not{p}, m_0, \alpha_0)$ can be expressed as a function of $\not{p}$ and p^2, as well as of the bare parameters m_0 and $\alpha_0 = \tilde{e}_0^2/(4\pi)$. Since $p^2 = \not{p}\not{p}$, we do not need to list p^2 as an independent variable. The contributions to the self energy arising at one- and two-loop order in perturbation theory are:

The full propagator defined as the Fourier transform of the two-point function of two bare fermion fields has a pole at the position of the physical electron mass m with a residue equal to Z_2, the electron wave-function renormalization constant appearing in (1.8):

$$\overset{p}{\longrightarrow} \bigcirc \longrightarrow \quad \overset{\not{p} \to m}{=} \quad \frac{iZ_2}{\not{p} - m + i0} + \text{ less singular terms.} \tag{1.23}$$

It follows that

$$m = m_0 + \Sigma(\not{p} = m, m_0, \alpha_0),$$

$$Z_2^{-1} = 1 - \frac{d\Sigma(\not{p} = m, m_0, \alpha_0)}{d\not{p}}\bigg|_{\not{p}=m}. \tag{1.24}$$

The action of the derivative operator $d/d\not{p}$ on functions of p^2 is given by $df(p^2)/d\not{p} = 2\not{p}f'(p^2)$. The first relation is an implicit equation for the renormalized mass m in terms of the bare mass parameter m_0. At one-loop order, we find (with Euler's constant $\gamma_E = 0.5772\ldots$)

$$m = m_0\left[1 + \frac{3\alpha_0}{4\pi}\left(\frac{1}{\epsilon} - \gamma_E + \ln 4\pi + \ln\frac{\mu^2}{m_0^2} + \frac{4}{3}\right) + \mathcal{O}(\alpha_0^2)\right],$$

$$Z_2 = 1 - \frac{\alpha_0}{4\pi}\left(\frac{1}{\epsilon} - \gamma_E + \ln 4\pi + \ln\frac{\mu^2}{m_0^2} - 2\ln\frac{m_0^2}{\lambda^2} + 4\right) + \mathcal{O}(\alpha_0^2). \tag{1.25}$$

The derivative of the self energy evaluated at $\not{p} = m$ is IR divergent and gauge dependent. In this section we use the Feynman gauge ($\xi = 1$) in the photon propagator

$$D_F^{\mu\nu}(k) = \frac{-i}{k^2 + i0}\left(g^{\mu\nu} - (1-\xi)\frac{k^\mu k^\nu}{k^2}\right) \tag{1.26}$$

for simplicity. In the above expression for Z_2 we have regularized IR divergences by introducing a fictitious photon mass λ. IR divergences are not our main concern in these lectures, and hence we will not dwell on this issue further.

The first relation in (1.25) appears to suggest that the physical mass m is a quantity which diverges when we take the limit $\epsilon \to 0$. However, we should instead write this equation as a relation for the bare mass parameter m_0 expressed in terms of the renormalized (and thus observable) mass m and the renormalized coupling $\alpha = \alpha_0 + \mathcal{O}(\alpha_0^2)$, such that

$$m_0 = m\left[1 - \frac{3\alpha}{4\pi}\left(\frac{1}{\epsilon} - \gamma_E + \ln 4\pi + \ln\frac{\mu^2}{m^2} + \frac{4}{3}\right) + \mathcal{O}(\alpha^2)\right]. \tag{1.27}$$

Likewise, we can rewrite the result for the wave-function renormalization constant of the electron in the form

$$Z_2 = 1 - \frac{\alpha}{4\pi}\left(\frac{1}{\epsilon} - \gamma_E + \ln 4\pi + \ln\frac{\mu^2}{m^2} - 2\ln\frac{m^2}{\lambda^2} + 4\right) + \mathcal{O}(\alpha^2). \tag{1.28}$$

The parameters m and α on the right-hand side of these equations are measurable quantities. The equations tell us how the bare mass parameter m_0 and the normalization Z_2 of the bare fermion field diverge as the dimensional regulator $\epsilon = (4-d)/2$ is taken to zero. In Section 1.1.3.3, we derive an analogous relation between the bare coupling constant α_0 and the renormalized coupling α.

The definitions (1.24) refer to the so-called on-shell renormalization scheme, in which $m = 0.5109989461(31)\,\text{MeV}$ is the physical mass of the electron [21], given by the pole position in the electron propagator, and in which Z_2 in the relation $\psi_0 = Z_2^{1/2}\,\psi$ is defined such that the renormalized propagator defined as the Fourier transform of the two-point function $\langle\Omega|\,T\{\psi(x)\,\bar\psi(y)\}\,|\Omega\rangle$ has a unit residue at $\not{p} = m$. In the next section we introduce a different renormalization scheme, the so-called modified minimal subtraction $(\overline{\text{MS}})$ scheme, in which the renormalized mass and residue will be defined in a different way.

For completeness, we also quote the renormalization factor of the electron mass as defined in the first relation in (1.10). We obtain

$$Z_m = \frac{Z_2\,m_0}{m} = 1 - \frac{\alpha}{\pi}\left(\frac{1}{\epsilon} - \gamma_E + \ln 4\pi + \ln\frac{\mu^2}{m^2} - \frac{1}{2}\ln\frac{m^2}{\lambda^2} + 2\right) + \mathcal{O}(\alpha^2). \tag{1.29}$$

1.1.3.2 *Photon Vacuum Polarization*

The self energy corrections for the gauge field are traditionally referred to as *vacuum polarization*. Consider the full photon propagator written as a series of contributions with more and more insertions of 1PI diagrams:

Denote by $\pi^{\mu\nu}(k)$ the infinite set of 1PI propagator corrections. Up to two-loop order, the relevant diagrams are:

Gauge invariance implies that $k_\mu \pi^{\mu\nu}(k) = k_\nu \pi^{\mu\nu}(k) = 0$, and hence

$$\pi^{\mu\nu}(k) = \left(g^{\mu\nu} k^2 - k^\mu k^\nu\right) \pi(k^2). \tag{1.30}$$

Performing the sum of the geometric series in an arbitrary covariant gauge, we obtain for the full photon propagator (Problem 1.1.6.4)

$$= \frac{-i}{k^2 \left[1 - \pi(k^2)\right] + i0} \left(g^{\mu\nu} - \frac{k^\mu k^\nu}{k^2 + i0}\right) - i\xi \frac{k^\mu k^\nu}{(k^2 + i0)^2}. \tag{1.31}$$

Here ξ is the gauge parameter ($\xi = 1$ in Feynman gauge). Remarkably, the quantum corrections only affect the first term on the right-hand side, which contains the physical (transverse) polarization states. Also, as long as the function $\pi(k^2)$ is regular at the origin, these corrections do not shift the pole in the propagator. Indeed, the full propagator has a pole at $k^2 = 0$ with residue

$$Z_3 = \frac{1}{1 - \pi(0)}. \tag{1.32}$$

From the relevant one-loop diagram, we obtain

$$\begin{aligned}
Z_3 &= 1 - \frac{\alpha_0}{3\pi} \left(\frac{1}{\epsilon} - \gamma_E + \ln 4\pi + \ln \frac{\mu^2}{m_0^2}\right) + \mathcal{O}(\alpha_0^2) \\
&= 1 - \frac{\alpha}{3\pi} \left(\frac{1}{\epsilon} - \gamma_E + \ln 4\pi + \ln \frac{\mu^2}{m^2}\right) + \mathcal{O}(\alpha^2).
\end{aligned} \tag{1.33}$$

1.1.3.3   *Charge Renormalization*

Besides the electron and photon propagators, our analysis in Section 1.1.2 had indicated that the electromagnetic vertex function coupling a photon to an electron–positron pair contains UV divergences, too. In momentum space, and using the Gordon identity, the vertex function can be written as

$$-i\tilde{e}_0\,\Gamma^\mu(p,p') = -i\tilde{e}_0\left[\gamma^\mu\,\Gamma_1(q^2) + \frac{i\sigma^{\mu\nu}q_\nu}{2m}\,\Gamma_2(q^2)\right], \tag{1.34}$$

where $q^\mu = (p'-p)^\mu$ is the momentum transfer, and $\Gamma^\mu(p,p')$ includes the 1PI vertex-correction graphs, i.e.

$$-i\tilde{e}_0\,\Gamma^\mu(p,p') = \quad \text{[Feynman diagrams]} + \ldots$$

Only the structure $\Gamma_1(q^2)$ is UV divergent. In the on-shell renormalization scheme, we define

$$Z_1 = [\Gamma_1(0)]^{-1}. \tag{1.35}$$

The Ward–Takahashi identity of QED [22], [23]

$$-iq_\mu\,\Gamma^\mu(p,p') = S^{-1}(p') - S^{-1}(p), \tag{1.36}$$

where $S(p)$ denotes the full electron propagator in (1.22), implies that $Z_1 = Z_2$ to all orders of perturbation theory, where we have used (1.23). From (1.10), we then obtain the following relation between the bare and the renormalized electric charges:

$$e_0 = \mu^\epsilon\,Z_1\,Z_2^{-1}Z_3^{-1/2}\,e = \mu^\epsilon\,Z_3^{-1/2}\,e. \tag{1.37}$$

For the coupling $\alpha_0 = \tilde{e}_0^2/(4\pi)$, this relation implies

$$\alpha_0 = \alpha\left[1 + \frac{\alpha}{3\pi}\left(\frac{1}{\epsilon} - \gamma_E + \ln 4\pi + \ln\frac{\mu^2}{m^2}\right) + \mathcal{O}(\alpha^2)\right]. \tag{1.38}$$

Here $\alpha = 1/137.035999139(31)$ is the *fine-structure constant* [21], defined in terms of the photon coupling to the electron at very small momentum transfer ($q^2 \to 0$). It is one of the most precisely known constants of nature. This concludes the calculation of the renormalization constants of QED.

1.1.3.4 Counterterms

Given the above results, it is straightforward to derive the one-loop expressions for the counterterms of QED in the dimensional regularization scheme. We find

$$\delta_1 = \delta_2 = -\frac{\alpha}{4\pi}\left(\frac{1}{\epsilon} - \gamma_E + \ln 4\pi + \ln\frac{\mu^2}{m^2} - 2\ln\frac{m^2}{\lambda^2} + 4\right) + \mathcal{O}(\alpha^2),$$

$$\delta_3 = -\frac{\alpha}{3\pi}\left(\frac{1}{\epsilon} - \gamma_E + \ln 4\pi + \ln\frac{\mu^2}{m^2}\right) + \mathcal{O}(\alpha^2), \tag{1.39}$$

$$\delta_m = -\frac{\alpha}{\pi}\left(\frac{1}{\epsilon} - \gamma_E + \ln 4\pi + \ln\frac{\mu^2}{m^2} - \frac{1}{2}\ln\frac{m^2}{\lambda^2} + 2\right) + \mathcal{O}(\alpha^2).$$

1.1.4 Scale Dependence in the On-shell Renormalization Scheme

So far we have worked in the on-shell renormalization scheme, in which the renormalized mass and coupling constant are related to well-measured physical constants (the electron mass and the fine-structure constant), and in which the renormalized fields are defined such that the renormalized propagators have poles with unit residues at the physical masses. For most calculations in QED (as well as for many calculations in the theory of electroweak interactions) this is the most convenient renormalization scheme.

You might be confused by the following subtlety related to the definition of the renormalized parameters in the on-shell scheme. Clearly, the bare parameters m_0 and e_0 in the QED Lagrangian are independent of the auxiliary scale μ, which we have introduced in (1.10). At first sight, it appears that the renormalized mass and coupling defined in (1.27) and (1.38) must be scale-dependent quantities. But, we saw in previous sections that these parameters have been measured with high precision and thus they are definitely independent of μ. The resolution of this puzzle rests on the fact that the relations between the bare and renormalized quantities are defined in the regularized theory in $d = 4 - 2\epsilon$ space-time dimensions. While in any renormalizable quantum field theory it is possible to take the limit $\epsilon \to 0$ at the end of a calculation of some observable, this limit must not be taken in relations such as (1.27) and (1.38), since the bare parameters m_0 and e_0 would diverge in this limit. Using the fact that in the on-shell scheme both the bare and renormalized parameters are μ independent *after* we take $\epsilon \to 0$, it follows from (1.10) and (1.15) that

$$\mu\frac{d}{d\mu}\Big[\mu^\epsilon Z_e\, e(\mu)\Big] = \left(\mu\frac{d}{d\mu}Z_e\right)\mu^\epsilon e(\mu) + \mu^\epsilon Z_e\left[\epsilon\, e(\mu) + \mu\frac{d}{d\mu}e(\mu)\right] = 0,$$

$$\mu\frac{d}{d\mu}\frac{Z_m}{Z_2} = \left(\frac{\partial}{\partial\ln\mu} + \mu\frac{de(\mu)}{d\mu}\frac{\partial}{\partial e}\right)\frac{Z_m}{Z_2} = 0, \tag{1.40}$$

where $Z_e = Z_3^{-1/2}$. In the on-shell renormalization scheme (but not in other schemes!), the first relation is solved by

$$\mu \frac{d}{d\mu}\, e(\mu) = -\epsilon\, e(\mu), \qquad \mu \frac{d}{d\mu}\, Z_e = \left(\frac{\partial}{\partial \ln \mu} - \epsilon\, e\, \frac{\partial}{\partial e} \right) Z_e = 0. \tag{1.41}$$

In terms of the renormalized coupling $\alpha(\mu)$, this becomes

$$\mu \frac{d}{d\mu}\, \alpha(\mu) = -2\epsilon\, \alpha(\mu), \qquad \mu \frac{d}{d\mu}\, Z_e = \left(\frac{\partial}{\partial \ln \mu} - 2\epsilon\, \alpha\, \frac{\partial}{\partial \alpha} \right) Z_e = 0. \tag{1.42}$$

The first relation states that in the regularized theory in $d = 4 - 2\epsilon$ dimensions the renormalized coupling is indeed scale dependent, but its scale dependence is simply such that $\alpha(\mu) \propto \mu^{-2\epsilon}$. Once we take the limit $\epsilon \to 0$ at the end of a calculation, the renormalized coupling in the on-shell scheme becomes a scale-independent constant, i.e.

$$\lim_{\epsilon \to 0} \alpha(\mu) = \alpha = \frac{1}{137.035999139(31)}. \tag{1.43}$$

Using that $Z_e = Z_3^{-1/2}$ with Z_3 given in (1.33), it is straightforward to check that the second relation in (1.42) is indeed satisfied at one-loop order. Finally, the second relation in (1.40) translates into

$$\left(\frac{\partial}{\partial \ln \mu} - 2\epsilon\, \alpha\, \frac{\partial}{\partial \alpha} \right) \frac{Z_m}{Z_2} = 0. \tag{1.44}$$

It is again easy to see that this relation holds.

1.1.5 Renormalization Schemes

While the on-shell scheme is particularly well motivated physically, it is not the only viable renormalization scheme. The only requirement we really need is that the counterterms remove the UV divergences of Feynman diagrams. The minimal way of doing this is to include *only* the $1/\epsilon^n$ pole terms in the counterterms (where in most cases $n = 1$ at one-loop order) and leave all finite terms out. This is referred to as the minimal subtraction (MS) scheme [24], [25]. In fact, since as we have seen the $1/\epsilon$ poles always come along with an Euler constant and a logarithm of 4π, it is more convenient to remove the poles in

$$\frac{1}{\hat{\epsilon}} \equiv \frac{1}{\epsilon} - \gamma_E + \ln 4\pi. \tag{1.45}$$

The corresponding scheme is called the modified minimal subtraction ($\overline{\text{MS}}$) scheme [26], and it is widely used in perturbative calculations in high-energy physics and in QCD in particular. Let us summarize the QED renormalization factors in the $\overline{\text{MS}}$ scheme. We have

$$Z_1^{\overline{\text{MS}}} = Z_2^{\overline{\text{MS}}} = 1 - \frac{\alpha}{4\pi\hat{\epsilon}} + \mathcal{O}(\alpha^2),$$

$$Z_3^{\overline{\text{MS}}} = 1 - \frac{\alpha}{3\pi\hat{\epsilon}} + \mathcal{O}(\alpha^2), \qquad (1.46)$$

$$Z_m^{\overline{\text{MS}}} = 1 - \frac{\alpha}{\pi\hat{\epsilon}} + \mathcal{O}(\alpha^2).$$

The renormalized electron mass and charge in the $\overline{\text{MS}}$ scheme are free of divergences, but these quantities are *no longer scale independent*. In fact, it is straightforward to relate these parameters to those defined in the on-shell scheme. We obtain

$$m_{\overline{\text{MS}}}(\mu) = m\,\frac{(Z_2/Z_m)^{\overline{\text{MS}}}}{(Z_2/Z_m)^{\text{OS}}} = m\left[1 - \frac{3\alpha}{4\pi}\left(\ln\frac{\mu^2}{m^2} + \frac{4}{3}\right) + \mathcal{O}(\alpha^2)\right],$$

$$\alpha_{\overline{\text{MS}}}(\mu) = \alpha\,\frac{Z_3^{\overline{\text{MS}}}}{Z_3^{\text{OS}}} = \alpha\left(1 + \frac{\alpha}{3\pi}\ln\frac{\mu^2}{m^2} + \mathcal{O}(\alpha^2)\right). \qquad (1.47)$$

While it may appear inconvenient at first sight to express the results of calculations in quantum field theory in terms of such scale-dependent (or 'running') parameters, we will encounter situations where this is indeed very useful. As a rule of thumb, this is always the case when the characteristic energy or mass scale of a process is much larger than the electron mass. The running electron mass $m_{\overline{\text{MS}}}(\mu)$ decreases with increasing μ, while the running coupling $\alpha_{\overline{\text{MS}}}(\mu)$ increases. Note that physical observables such as cross sections for scattering events or decay rates of unstable particles are always scale independent, i.e. the scale dependence of the running parameters is compensated by scale-dependent terms in the perturbative series for these quantities. This will be discussed in more detail in Section 1.3.

1.1.6   Exercises

1.1.6.1   Prove the combinatoric identity (1.4).

1.1.6.2   Gauge invariance requires that the four-photon amplitude $\pi^{\alpha\beta\gamma\delta}(k_1, k_2, k_3, k_4)$ (without external polarization vectors, and with incoming momenta satisfying $k_1 + k_2 + k_3 + k_4 = 0$) vanishes when one of its Lorentz indices is contracted with the corresponding external momentum vector, e.g. $k_{1\alpha}\,\pi^{\alpha\beta\gamma\delta}(k_1, k_2, k_3, k_4) = 0$. Use this fact as well as Bose symmetry to derive the most general form-factor decomposition of this amplitude and show that the amplitude is UV finite.

1.1.6.3 Prove the relations (1.16) for the d-dimensional Dirac matrices using the Clifford algebra $\{\gamma_\mu, \gamma_\nu\} = 2g_{\mu\nu}$.

1.1.6.4 Derive relation (1.31) for the full photon propagator.

1.2   Renormalization in QCD

The Lagrangian of quantum chromodynamics (QCD), the fundamental theory of the strong interactions, is structurally very similar to the QED Lagrangian in (1.1). For the case of a single flavour of quarks, it reads (omitting gauge-fixing terms for simplicity)

$$\mathcal{L}_{\mathrm{QCD}} = \bar{\psi}_{q,0}\left(i\slashed{D} - m_{q,0}\right)\psi_{q,0} - \frac{1}{4}\,G^a_{\mu\nu,0}\,G_0^{\mu\nu,a} + \bar{c}_0^a\,(-\partial^\mu D_\mu^{ab})\,c_0^b. \tag{1.48}$$

Here ψ_q is the Dirac spinor for the quark field, m_q denotes the quark mass, $G^a_{\mu\nu}$ is the field-strength tensor of the gluon fields, and c^a are the Faddeev–Popov ghost fields. As before the subscript '0' is used to indicate 'bare' quantities in the Lagrangian.

In the real world, QCD contains six different types of quark fields with different masses, referred to as 'flavours', which are called up, down, strange, charm, bottom (or beauty), and top (or truth). Strictly speaking, a sum over quark flavours should thus be included in the above Lagrangian.[2]

The main differences between QED and QCD are due to the fact that QCD is a *non-abelian* gauge theory based on the group $SU(N_c)$, where $N_c = 3$ is called the number of colours. While in QED particles carry a single charge (the electric charge $\pm e$ or a fraction thereof), the quarks carry one of three colours $i = 1, 2, 3$. In fact, quarks live in the fundamental representation of the gauge group, and the quark spinor field ψ_q can be thought of as a three-component vector in colour space. The gluons, the counterparts of the photon in QED, live in the adjoint representation of the gauge group, which is $(N_c^2 - 1)$ dimensional. Hence there are eight gluon fields in QCD, labelled by an index $a = 1, \ldots, 8$. When acting on quark fields, the covariant derivative reads

$$iD_\mu = i\partial_\mu + g_s A^a_\mu\, t^a, \tag{1.49}$$

where the eight 3×3 matrices t^a are called the Gell–Mann matrices. They are the generators of colour rotations in the fundamental representation. (When D_μ acts on the ghost fields, the generators t^a must instead be taken in the adjoint representation.) The strong coupling g_s replaces the electromagnetic coupling e in QED. The most important difference results from the form of the QCD field-strength tensor, which reads

$$G^a_{\mu\nu} = \partial_\mu A^a_\nu - \partial_\nu A^a_\mu + g_s f_{abc} A^b_\mu A^c_\nu. \tag{1.50}$$

---

[2] Likewise, in the real world there exist three types of charged leptons, called the electron, the muon, and the tau lepton. In our discussion we ignore the presence of the muon and the tau lepton, which have masses much heavier than the electron.

The quadratic term in the gauge potentials arises since the commutator of two-colour generators is non-vanishing,

$$[t_a, t_b] = i\, f_{abc}\, t_c,\tag{1.51}$$

and it thus reflects the non-abelian nature of the gauge group.

Let us briefly summarize the main differences between QED and QCD, all of which result from the differences between the abelian group $U(1)$ and the non-abelian group $SU(N_c)$:

1. In QED there is a single elementary vertex connecting two electron lines to a photon line. A similar vertex coupling two quark lines to a gluon line also exists in QCD. However, because of the structure of (1.50), there are in addition gluon self-interactions connecting three or four gluons at a single vertex.

2. The gauge-fixing procedure in non-abelian gauge theories gives rise to a non-trivial functional determinant, which is dealt with by introducing Faddeev–Popov ghost fields c^a. These are anti-commuting scalar fields transforming in the adjoint representation, i.e., fields with the wrong spin-statistics relation, which hence cannot appear as external states in scattering amplitudes. Internal ghost fields inside Feynman graphs such as

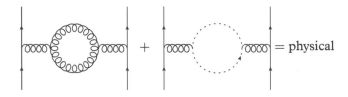

are however needed to cancel the unphysical gluon polarizations in loops. The presence of the ghost fields gives rise to an additional elementary vertex connecting two ghost lines to a gluon line.

3. Unlike in QED, in QCD calculations we encounter non-trivial group-theory factors, the most common ones being

$$C_F = \frac{N_c^2 - 1}{2N_c} = \frac{4}{3}, \qquad C_A = N_c = 3, \qquad T_F = \frac{1}{2}.\tag{1.52}$$

Important relations involving these factors are (summed over repeated indices)

$$t_a\, t_a = C_F \mathbf{1}, \qquad f_{abc} f_{abd} = C_A \delta_{cd}, \qquad \mathrm{Tr}(t_a\, t_b) = T_F \delta_{ab}.\tag{1.53}$$

4. The superficial degree of divergence of a 1PI Feynman diagram in QED has been given in (1.7). For a 1PI graph in QCD, 1 can prove that (Problem 1.2.3.1)

$$D = 4 - \frac{3}{2}N_q - N_g - \frac{3}{2}N_c,$$
(1.54)

where N_q, N_g and N_c are the number of external quark, gluon, and ghost lines. Note that, while scattering amplitudes cannot contain external ghost particles, there do exist UV-divergent 1PI vertex functions involving external ghost fields.

Perhaps the most important difference concerns the phenomenology of the two theories. While QED is a weakly coupled quantum field theory for all relevant energy scales,[3] QCD exhibits strong-coupling behaviour at low energies but weak-coupling behaviour at high energies ('asymptotic freedom'). The strong coupling at low energies gives rise to the phenomenon of *colour confinement*, which is the statement that in the low-energy world quarks and gluons are always locked up inside colourless bound states called hadrons.

1.2.1   Renormalization in QCD

While the on-shell renormalization scheme is useful for many (but not all) calculations in QED, it is *not* a viable renormalization scheme for QCD calculations, for the following reasons:

- Quarks and gluons can never be on-shell because of confinement. In the real world free (isolated) quarks and gluons do not exist, and hence the corresponding two-point functions do not have poles at $p^2 = m_q^2$ or $k^2 = 0$, respectively.

- The strong coupling g_s and the associated parameter $\alpha_s = g_s^2/(4\pi)$ cannot be renormalized at $q^2 = 0$, since QCD is strongly coupled at low or vanishing momentum transfer and hence quarks and gluons are not the relevant degrees of freedom to describe the strong interactions in this regime.

- The masses of the three light quark flavours satisfy $m_q \ll \Lambda_{QCD}$ (for $q = u, d, s$), where Λ_{QCD} is (roughly) the scale at which QCD becomes strongly coupled. It is therefore a good approximation for many purposes to set the light quark masses to zero. We will not consider heavy quarks with masses $m_Q \gg \Lambda_{QCD}$ (for $Q = c, b, t$) in these lectures. The effects of heavy quarks are usually described using some kind of effective field theory. This is discussed in detail in the lecture courses by Mannel, Silvestrini, and Sommer elsewhere in this book.

---

[3] QED would get strongly coupled near the Landau pole of the running coupling $\alpha(\mu)$ in (1.47), which however lies far above the Planck scale.

For all these reasons, we use the $\overline{\text{MS}}$ renormalization scheme for perturbation-theory calculations in QCD.

Let us briefly discuss the structure of UV-divergent vertex functions in QCD. In addition to the analogues of the divergent n-point functions in QED, the following amplitudes which arise only in QCD are UV divergent and require renormalization:

The three-gluon amplitude has $D = 1$, and hence naively it is linearly divergent. Unlike in QED, in QCD this amplitude no longer vanishes (i.e. Furry's theorem does not apply in QCD), but it only contains logarithmic UV divergences due to gauge invariance.

The four-gluon amplitude has $D = 0$ and is logarithmically divergent. The argument holding in QED, stating that gauge invariance renders the four-photon amplitude UV finite, does not apply in QCD, since in a non-abelian gauge theory the elementary four-gluon vertex is part of the gauge-invariant Lagrangian.

The ghost–gluon amplitude has $D = 0$ and is logarithmically divergent.

In analogy with (1.8), we introduce field renormalization constants as

$$\psi_{q,0} = Z_2^{1/2}\,\psi_q, \qquad A_{\mu,0}^a = Z_3^{1/2}\,A_\mu^a, \qquad c_0^a = Z_{2c}^{1/2}\,c^a. \qquad (1.55)$$

Since we neglect the light-quark masses, there is no mass renormalization to consider. We must, however, consider the renormalization of the bare QCD coupling constant $g_{s,0}$. Proceeding as in (1.10), we would define

$$Z_2 Z_3^{1/2}\, g_{s,0} = \mu^{\frac{4-d}{2}} Z_1 g_s, \qquad (1.56)$$

where Z_1 is the renormalization constant associated with the quark–gluon vertex function. However, gauge invariance requires that in QCD *all* interaction vertices are expressed in terms of the same coupling constant, and this feature must be preserved by renormalization. We can thus express the relation between the bare and renormalized couplings in four different ways, using the quark–gluon, three-gluon, four-gluon, and ghost–gluon vertex functions. This yields the following exact relations between renormalization factors:

$$g_{s,0} = \mu^{\frac{4-d}{2}} Z_1 Z_2^{-1} Z_3^{-1/2} g_s$$
$$= \mu^{\frac{4-d}{2}} Z_1^{3g} Z_3^{-3/2} g_s$$
$$= \mu^{\frac{4-d}{2}} \left(Z_1^{4g}\right)^{1/2} Z_3^{-1} g_s \tag{1.57}$$
$$= \mu^{\frac{4-d}{2}} Z_1^{cg} Z_{2c}^{-1} Z_3^{-1/2} g_s,$$

where Z_1^{3g}, Z_1^{4g}, and Z_1^{cg} denote the renormalization constants associated with the three-gluon, four-gluon, and ghost–gluon vertex functions, respectively. The remaining factors arise when the bare fields entering these vertices are expressed in terms of renormalized fields. Note that, unlike in QED, we no longer have the identity $Z_1 = Z_2$ in QCD, since the Ward–Takahashi identity (1.36) must be generalized to the more complicated Slavnov–Taylor identities [27]–[29]. It follows from the above relations that

$$Z_1^{3g} = Z_1 Z_2^{-1} Z_3, \qquad Z_1^{4g} = \left(Z_1 Z_2^{-1}\right)^2 Z_3, \qquad Z_1^{cg} = Z_1 Z_2^{-1} Z_{2c}. \tag{1.58}$$

These are exact relations between renormalization constants, which hold to all orders in perturbation theory.

1.2.2   Calculation of the Renormalization Factors

The calculation of the renormalization factors Z_1, Z_2 and Z_3 proceeds in analogy to the corresponding calculation in QED. We now briefly summarize the results, obtained in a general covariant gauge with gauge parameter ξ.

1.2.2.1   *Quark self energy*

The calculation of the one-loop quark self energy in the limit of vanishing quark mass is a straightforward application of the loop techniques we have reviewed in Section 1.1.3 (Problem 1.2.3.2). The relevant diagram is:

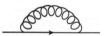

For the field renormalization constant of the quark field in the $\overline{\text{MS}}$ scheme we obtain from (1.24)

$$Z_2 = 1 - \frac{C_F \alpha_s}{4\pi\hat{\epsilon}} \xi + \mathcal{O}(\alpha_s^2). \tag{1.59}$$

Compared with the corresponding relation (1.46) in QED, where we had worked in Feynman gauge $\xi = 1$, we find a simple replacement $\alpha \to C_F \alpha_s$ accounting for the difference in gauge couplings and the colour factor of the one-loop self-energy diagram.

1.2.2.2   *Gluon vacuum polarization*

In addition to the fermion loop graph present in QED, the QCD vacuum polarization function receives several other contributions, which are of genuinely non-abelian origin:

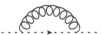

In analogy with (1.30), we decompose the gluon two-point function in the form

$$\pi_{\mu\nu}^{ab}(k) = \delta^{ab}\left(k^2 g_{\mu\nu} - k_\mu k_\nu\right)\pi(k^2). \tag{1.60}$$

The fermion loop graph is obtained from the corresponding diagram in QED by means of the replacement $\alpha \to T_F \alpha_s$, where the factor T_F arises from the trace over colour matrices. The calculation of the remaining diagrams is a bit more involved. Each individual diagram is quadratically UV divergent, and only a consistent regularization scheme such as dimensional regularization allows us to deal with these divergences in such a way that gauge invariance is preserved. After a lengthy calculation, we obtain

$$\pi(k^2) = \frac{\alpha_s}{4\pi}\left\{\left[\left(\frac{13}{6} - \frac{\xi}{2}\right)C_A - \frac{4}{3}T_F n_q\right]\left(\frac{1}{\hat{\epsilon}} - \ln\frac{-k^2 - i0}{\mu^2}\right) + \ldots\right\}, \tag{1.61}$$

where ξ is the gauge parameter and n_q denotes the number of light (approximately massless) quark flavours. For the gluon-field renormalization constant in (1.55) we thus obtain the gauge-dependent expression

$$Z_3 = 1 + \frac{\alpha_s}{4\pi\hat{\epsilon}}\left[\left(\frac{13}{6} - \frac{\xi}{2}\right)C_A - \frac{4}{3}T_F n_q\right] + \mathcal{O}(\alpha_s^2). \tag{1.62}$$

1.2.2.3   *Wave-function renormalization for the ghost field*

At one-loop order, the ghost propagator receives the correction:

From a straightforward calculation of this diagram we can extract the wave-function renormalization constant of the ghost field in a general covariant gauge (Problem 1.2.3.3). The result is

$$Z_{2c} = 1 + \frac{C_A \alpha_s}{4\pi\hat{\epsilon}}\frac{3 - \xi}{4} + \mathcal{O}(\alpha_s^2). \tag{1.63}$$

1.2.2.4   Quark–gluon vertex function

The 1PI one-loop diagrams contributing to the quark–gluon vertex function are:

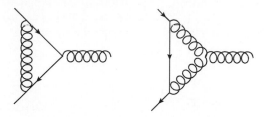

While the first diagram can be obtained from the corresponding QED diagram by the replacement $\alpha \to C_F \alpha_s$, the second graph is of genuinely non-abelian origin. Its calculation requires the colour identity

$$f_{abc}\, t_b\, t_c = \frac{i}{2}\, C_A\, t_a. \tag{1.64}$$

We obtain (Problem 1.2.3.4)

$$Z_1 = 1 - \frac{\alpha_s}{4\pi\hat{\epsilon}} \left(\xi\, C_F + \frac{3+\xi}{4}\, C_A \right) + \mathcal{O}(\alpha_s^2). \tag{1.65}$$

Notice that the 'abelian' part of this result (the term proportional to C_F) is the same as in the expression for Z_2 in (1.59), however the 'non-abelian' part (the term proportional to C_A) violates the identity $Z_1 = Z_2$.

1.2.2.5   Charge renormalization

From the first relation in (1.57), we now obtain for the charge renormalization constant

$$Z_g = Z_1\, Z_2^{-1}\, Z_3^{-1/2} = 1 - \frac{\alpha_s}{4\pi\hat{\epsilon}} \left(\frac{11}{6}\, C_A - \frac{2}{3}\, T_F\, n_q \right) + \mathcal{O}(\alpha_s^2). \tag{1.66}$$

Notice that the dependence on the gauge parameter ξ has disappeared. Compared with the corresponding QED relation (where the factor n_ℓ counts the number of lepton species)

$$Z_e = 1 + \frac{\alpha}{6\pi\hat{\epsilon}}\, n_\ell + \mathcal{O}(\alpha^2), \tag{1.67}$$

we observe that the fermion contributions are identical up to the colour factor $T_F = 1/2$, while the non-abelian contribution proportional to C_A has no counterpart in QED. Crucially, this contribution has the opposite sign of the fermion contribution [30], [31], and this is the reason for the different behaviour of the running coupling constants in QED and QCD.

It is straightforward to calculate the remaining QCD vertex renormalization factors from the relations in (1.58). We find

$$Z_1^{3g} = 1 + \frac{\alpha_s}{4\pi\hat{\epsilon}} \left[\left(\frac{17}{12} - \frac{3\xi}{4} \right) C_A - \frac{4}{3} T_F n_q \right] + \mathcal{O}(\alpha_s^2),$$

$$Z_1^{4g} = 1 + \frac{\alpha_s}{4\pi\hat{\epsilon}} \left[\left(\frac{2}{3} - \xi \right) C_A - \frac{4}{3} T_F n_f \right] + \mathcal{O}(\alpha_s^2), \qquad (1.68)$$

$$Z_1^{cg} = 1 - \frac{\alpha_s}{4\pi\hat{\epsilon}} \frac{\xi}{2} C_A + \mathcal{O}(\alpha_s^2).$$

1.2.3 Exercises

1.2.3.1 Derive relation (1.54) for the superficial degree of divergence of 1PI QCD Feynman graphs.

1.2.3.2 Calculate the one-loop corrections to the quark self-energy in QCD in the limit of vanishing quark mass.

1.2.3.3 Derive the expression (1.63) for the ghost-field wave-function renormalization constant Z_{2c} in a general covariant gauge.

1.2.3.4 Calculate the second diagram contributing at one-loop order to the quark–gluon vertex function (see Section 1.2.2.4) in a general covariant gauge.

1.3 RG Equations and Running Couplings

Now that we have discussed the basics of renormalization in both QED and QCD, we more systematically explore the concept of running couplings and its relevance for multi-scale problems in quantum field theory. A closely related subject is that of the resummation of large logarithmic corrections to all orders of perturbation theory.

Consider a QED observable $\mathcal{O}$ such as a scattering cross section calculated in both the on-shell renormalization scheme and the $\overline{\text{MS}}$ scheme. We have

$$\mathcal{O} = \mathcal{O}_{\text{OS}}\left(\alpha, m, \ln \frac{s}{m^2}, \dots \right) = \mathcal{O}_{\overline{\text{MS}}}\left(\alpha(\mu), m(\mu), \ln \frac{s}{\mu^2}, \dots \right), \qquad (1.69)$$

where $\sqrt{s}$ is the center-of-mass energy, and the dots refer to other kinematic variables such as scattering angles. In the first expression, α is the fine-structure constant defined in the Thomson limit $q^2 \to 0$ and m is the physical mass of the electron, see Section 1.1.3. Both parameters are fundamental physical constants. In the second expression, $\alpha(\mu)$ and $m(\mu)$ are μ-dependent parameters defined in the $\overline{\text{MS}}$ scheme. They are related to the parameters in the on-shell scheme via the relations (1.47). Several comments are in order:

- Both results for the observable $\mathcal{O}$ are equivalent and μ independent (assuming we work to all orders in perturbation theory, otherwise differences arise only beyond the order to which the calculations have been performed).

- Sometimes on-shell renormalization is inconvenient, because it leaves large logarithmic terms in the expression for the observable. For example, if $s \gg m^2$ for a high-energy process, then $\ln(s/m^2)$ is a large logarithm. Typically, these logarithms appear as $[\alpha \ln(s/m^2)]^n$ in higher orders, and they can threaten the convergence of the perturbative expansion if $\alpha \ln(s/m^2) = \mathcal{O}(1)$.[4]

- Choosing the renormalization scale such that $\mu \approx \sqrt{s}$ fixes this problem, giving a well behaved perturbative expansion in terms of the parameters $\alpha(\mu)$ and $m(\mu)$ with $\mu \approx \sqrt{s}$. These are, however, different from the 'physical' electron mass and fine-structure constant. As shown in (1.47), the corresponding relations are

$$
\alpha(\sqrt{s}) = \alpha \left(1 + \frac{\alpha}{3\pi} \ln \frac{s}{m^2} + \dots \right),
$$
$$
m(\sqrt{s}) = m \left[1 - \frac{3\alpha}{4\pi} \left(\ln \frac{s}{m^2} + \frac{4}{3} \right) + \dots \right].
\tag{1.70}
$$

While choosing $\mu \approx \sqrt{s}$ eliminates the large logarithms from the observable itself, it leads to large logarithms in the relations between the parameters in the on-shell scheme and those in the $\overline{\text{MS}}$ scheme. We will discuss in a moment how these logarithms can be resummed.

The μ independence of the observable $\mathcal{O}$ in the $\overline{\text{MS}}$ scheme can be expressed in terms of the partial differential equation

$$
\mu \frac{d}{d\mu} \mathcal{O} = \mu \frac{d\alpha(\mu)}{d\mu} \frac{\partial \mathcal{O}}{\partial \alpha(\mu)} + \mu \frac{dm(\mu)}{d\mu} \frac{\partial \mathcal{O}}{\partial m(\mu)} + \frac{\partial \mathcal{O}}{\partial \ln \mu} = 0.
\tag{1.71}
$$

Equations of this type are referred to as renormalization group (RG) equations. They play a fundamental role in the theory of renormalization. The corresponding equations for Green's functions are called Callan–Symanzik equations [32]–[34]. Above we have assumed that the observable depends on the two running parameters $\alpha(\mu)$ and $m(\mu)$; if it depends on more than two parameters, then (1.71) needs to be generalized accordingly.

To proceed, we define two functions of the coupling $\alpha(\mu)$ via

$$
\mu \frac{d\alpha(\mu)}{d\mu} = \beta\big(\alpha(\mu)\big),
$$
$$
\mu \frac{dm(\mu)}{d\mu} = \gamma_m\big(\alpha(\mu)\big) m(\mu).
\tag{1.72}
$$

---

[4] While for QED this condition would only be satisfied for exceedingly large values of $\sqrt{s}$, large logarithms frequently arise in QCD applications.

The first is referred to as the β-function of QED (admittedly a somewhat dull name), while the second function is called the anomalous dimension of the electron mass. The RG equation (1.71) now takes the form

$$\beta(\alpha)\frac{\partial \mathcal{O}}{\partial \alpha} + \gamma_m(\alpha)\, m\frac{\partial \mathcal{O}}{\partial m} + \frac{\partial \mathcal{O}}{\partial \ln \mu} = 0. \tag{1.73}$$

The strategy for obtaining reliable perturbative results in QED, which are free of large logarithms, is now as follows:

1. Compute the observable to a given order in perturbation theory in terms of renormalized parameters defined in the $\overline{\text{MS}}$ scheme.

2. Eliminate large logarithms in the expansion coefficients by a suitable choice of the renormalization scale μ.

3. Compute the running parameters such as $\alpha(\mu)$ and $m(\mu)$ at that scale by solving the differential equations (1.72). The boundary values in these solutions can be taken as the fine-structure constant α and the physical electron mass m, which are known to excellent accuracy.

The same discussion applies to QCD, where using the $\overline{\text{MS}}$ scheme is the default choice. Setting the light quark masses to zero, we obtain the simpler RG equation

$$\beta(\alpha_s)\frac{\partial \mathcal{O}}{\partial \alpha_s} + \frac{\partial \mathcal{O}}{\partial \ln \mu} = 0, \tag{1.74}$$

where

$$\beta\big(\alpha_s(\mu)\big) = \mu\,\frac{d\alpha_s(\mu)}{d\mu}. \tag{1.75}$$

The running coupling $\alpha_s(\mu)$ is obtained by integrating this equation, using as boundary value the value of α_s at some reference scale, where it is known with high accuracy. A common choice is $\alpha_s(m_Z) = 0.1181(11)$ [21].

1.3.1   Calculation of β-Functions and Anomalous Dimensions

There is an elegant formalism that allows us to extract β-functions and anomalous dimensions from the $1/\epsilon$ poles of the renormalization factors for the various quantities in a quantum field theory. We present the following discussion for the case of QCD with a massive quark of mass m_q, but the same results with obvious replacements apply to QED. From the first equation in (1.57) and the QCD analogue of the first equation in (1.10), we recall the relations between the bare and renormalized colour charge and mass parameter in the form

$$\alpha_{s,0} = \mu^{2\epsilon} \, Z_1^2 \, Z_2^{-2} \, Z_3^{-1} \, \alpha_s(\mu) \equiv \mu^{2\epsilon} \, Z_\alpha(\mu) \, \alpha_s(\mu),$$
$$m_{q,0} = Z_m \, Z_2^{-1} \, m_q(\mu) \equiv Z'_m(\mu) \, m_q(\mu), \tag{1.76}$$

where in the $\overline{\text{MS}}$ scheme (for QED we replace $\alpha_s \to \alpha$, $C_F \to 1$ and $\beta_0 \to -\frac{4}{3} n_\ell$)

$$Z_\alpha(\mu) = 1 - \beta_0 \frac{\alpha_s(\mu)}{4\pi\hat{\epsilon}} + \mathcal{O}(\alpha_s^2); \quad \beta_0 = \frac{11}{3} C_A - \frac{4}{3} T_F n_q,$$
$$Z'_m(\mu) = 1 - 3C_F \frac{\alpha_s(\mu)}{4\pi\hat{\epsilon}} + \mathcal{O}(\alpha_s^2). \tag{1.77}$$

From the fact that the bare parameters are scale independent it follows that

$$\mu \frac{d}{d\mu} \alpha_{s,0} = 0 = \mu^{2\epsilon} \, Z_\alpha(\mu) \, \alpha_s(\mu) \left[2\epsilon + Z_\alpha^{-1} \frac{dZ_\alpha}{d\ln\mu} + \frac{1}{\alpha_s} \frac{d\alpha_s}{d\ln\mu} \right], \tag{1.78}$$

which implies

$$\frac{d\alpha_s}{d\ln\mu} = \alpha_s \left[-2\epsilon - Z_\alpha^{-1} \frac{dZ_\alpha}{d\ln\mu} \right] \equiv \beta(\alpha_s(\mu), \epsilon), \tag{1.79}$$

and

$$\frac{dm_{q,0}}{d\ln\mu} = 0 = Z'_m(\mu) \, m_q(\mu) \left[Z'^{-1}_m \frac{dZ'_m}{d\ln\mu} + \frac{1}{m_q} \frac{dm_q}{d\ln\mu} \right], \tag{1.80}$$

from which it follows that

$$\frac{1}{m_q(\mu)} \frac{dm_q(\mu)}{d\ln\mu} = -Z'^{-1}_m \frac{dZ'_m}{d\ln\mu} \equiv \gamma_m(\alpha_s(\mu)). \tag{1.81}$$

Note that the generalized β-function $\beta(\alpha_s, \epsilon)$ in (1.79) governs the scale dependence of the gauge coupling in the regularized theory at finite ϵ. The limit $\epsilon \to 0$ of this expression will later give us the 'usual' QCD β-function in the renormalized theory.

We now derive some beautiful and very useful relations for the β-function and the anomalous dimension γ_m. For the purposes of this discussion it is convenient to consider for a moment the original MS scheme, in which the Z factors only contain $1/\epsilon^n$ pole terms (with $\epsilon = (4 - d)/2$) and thus depend on μ only through the running coupling $\alpha_s(\mu)$. We can thus write (with $\alpha_s \equiv \alpha_s(\mu)$ throughout)

$$\beta(\alpha_s, \epsilon) = \alpha_s \left[-2\epsilon - \beta(\alpha_s, \epsilon) Z_\alpha^{-1} \frac{dZ_\alpha}{d\alpha_s} \right],$$
$$\gamma_m(\alpha_s) = -\beta(\alpha_s, \epsilon) Z'^{-1}_m \frac{dZ'_m}{d\alpha_s}. \tag{1.82}$$

To solve the first equation we expand

$$\beta(\alpha_s, \epsilon) = \beta(\alpha_s) + \sum_{k=1}^{\infty} \epsilon^k \beta^{[k]}(\alpha_s),$$

$$Z_\alpha = 1 + \sum_{k=1}^{\infty} \frac{1}{\epsilon^k} Z_\alpha^{[k]}(\alpha_s). \tag{1.83}$$

Note that the expansion coefficients $Z_\alpha^{[k]}(\alpha_s)$ start at $\mathcal{O}(\alpha_s^k)$. From the fact that the pole terms $\sim 1/\epsilon^n$ with $n \geq 1$ must cancel in the first relation in (1.82) we can derive an infinite set of relations between $\beta^{[k]}(\alpha_s)$ and $Z_\alpha^{[k]}(\alpha_s)$. The solution to this set of equations is (Problem 1.3.4.1)

$$\beta^{[1]}(\alpha_s) = -2\alpha_s, \qquad \beta^{[k]}(\alpha_s) = 0 \quad \text{for all } k \geq 2,$$

$$\beta(\alpha_s) = 2\alpha_s^2 \frac{dZ_\alpha^{[1]}(\alpha_s)}{d\alpha_s}. \tag{1.84}$$

This yields the exact relation

$$\beta(\alpha_s, \epsilon) = -2\epsilon\,\alpha_s + \beta(\alpha_s) = -2\epsilon\,\alpha_s + 2\alpha_s^2 \frac{dZ_\alpha^{[1]}(\alpha_s)}{d\alpha_s}. \tag{1.85}$$

Likewise, we can show that

$$\gamma_m(\alpha_s) = 2\alpha_s \frac{dZ_m'^{[1]}(\alpha_s)}{d\alpha_s}. \tag{1.86}$$

Also this result is exact. The above relations state that the β-function and anomalous dimension can be computed, to all orders in perturbation theory, in terms of the coefficient of the single $1/\epsilon$ pole in the renormalization factors Z_α and Z_m', respectively. Since the coefficients of the $1/\epsilon$ pole terms in the Z-factors are the same in the MS and $\overline{\text{MS}}$ schemes, these relations also apply to the $\overline{\text{MS}}$ scheme. In the one-loop approximation, we obtain from (1.77)

$$\beta(\alpha_s) = -2\alpha_s \left(\beta_0 \frac{\alpha_s}{4\pi} + \ldots \right), \qquad \gamma_m(\alpha_s) = -6C_F \frac{\alpha_s}{4\pi} + \ldots. \tag{1.87}$$

1.3.2   Leading-order Solutions to the Evolution Equations

In the one-loop approximation for the β-function, equation (1.75) governing the scale dependence (also called the 'running') of the QCD gauge coupling reads

$$\frac{d\alpha_s(\mu)}{d\ln\mu} = -2\beta_0 \frac{\alpha_s^2(\mu)}{4\pi}, \tag{1.88}$$

which using a separation of variables can be rewritten in the form

$$-\frac{d\alpha_s}{\alpha_s^2} = \frac{\beta_0}{4\pi} d\ln\mu^2. \tag{1.89}$$

This can be integrated to obtain

$$-\int_{\alpha_s(Q)}^{\alpha_s(\mu)} \frac{d\alpha_s}{\alpha_s^2} = \frac{1}{\alpha_s(\mu)} - \frac{1}{\alpha_s(Q)} = \frac{\beta_0}{4\pi} \ln\frac{\mu^2}{Q^2}. \tag{1.90}$$

Here Q is some reference scale, at which the value of $\alpha_s(Q)$ is measured with accuracy. A canonical choice is to take Q equal to the mass of the heavy Z boson, $Q = m_Z \approx 91.188\,\mathrm{GeV}$, at which $\alpha_s(Q) = 0.1181(11)$ [21]. Rearranging the above result, we find the familiar form of the running coupling in QCD:

$$\alpha_s(\mu) = \frac{\alpha_s(Q)}{1 + \alpha_s(Q)\frac{\beta_0}{4\pi}\ln\frac{\mu^2}{Q^2}}; \quad \beta_0 = \frac{11}{3}C_A - \frac{4}{3}T_F\,n_q. \tag{1.91}$$

Here n_q is the number of light (massless) quark flavours with masses below the scale μ. The corresponding expression for QED is obtained by replacing $\alpha_s \to \alpha$ and $\beta_0 \to -\frac{4}{3}n_\ell$. Figure 1.1 shows the two couplings as a function of the energy scale. It is not difficult to include higher-order corrections in the calculation of the running couplings of QCD and QED, see Section 1.4.4. These higher-order corrections are included in the figure.

Note that β_0 is *positive* in QCD (since the number of quark generations is fewer than seventeen), while it is negative in QED. As a result, the strong coupling gets weaker at higher energies—a phenomenon referred to as *asymptotic freedom* [30], [31], which was awarded the 2004 Nobel Prize in Physics—while the QED coupling slowly increases with energy. At low energies the QCD coupling grows, and the leading-order expression (1.91) blows up at the scale

$$\mu = Q\exp\left(-\frac{2\pi}{\beta_0\alpha_s(Q)}\right) \approx 0.2\,\mathrm{GeV}. \tag{1.92}$$

QCD becomes strongly coupled at such low scales, and the quarks and gluons are confined inside hadrons. The chiral Lagrangian provides an effective theory for QCD at such low scales. This is discussed in the lectures by Pich [35] in chapter 3 in this book. For QED, the evolution effects of the gauge coupling are more modest but not negligible. At $\mu = m_Z$, the value of $\alpha(m_Z)$ is about 6% larger than the fine-structure constant, which according to (1.47) corresponds to the $\overline{\mathrm{MS}}$ coupling evaluated at the

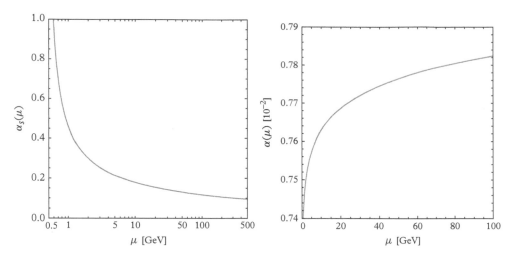

Figure 1.1 *Scale dependence of the running QCD coupling $\alpha_s(\mu)$ (left panel) and the running QED coupling $\alpha(\mu)$ (right panel).*

scale $\mu = m \approx 511$ keV. At very high values of μ, the QED running coupling develops a so-called Landau pole and diverges.

Note that, in the $\overline{\text{MS}}$ scheme, the slope of the running coupling changes whenever μ crosses the mass scale of a fermion. The simple form shown in (1.91) holds only in an interval where the value of n_q is fixed. When μ crosses a quark threshold, the value of β_0 changes abruptly, and the values of $\alpha_s(\mu)$ just above and below the threshold must be matched to each other. For example, at $\mu = m_Z \approx 91.188$ GeV QCD contains five approximately massless quark flavours, while the top quark with mass $m_t \approx 170$ GeV is heavy and is neglected in the running of the coupling. Formula (1.91) can be used to evolve the coupling down to the scale $\mu = m_b(m_b) \approx 4.18$ GeV, below which the mass of the bottom quark can no longer be neglected. In the $\overline{\text{MS}}$ scheme we compute $\alpha_s(m_b)$ from (1.91) using $\beta_0 = \frac{23}{3}$ (corresponding to $n_q = 5$), but for lower scales we replace (1.91) with the analogous relation

$$\alpha_s(\mu) = \frac{\alpha_s(m_b)}{1 + \alpha_s(m_b)\frac{\beta_0}{4\pi}\ln\frac{\mu^2}{m_b^2}}; \quad \mu < m_b, \tag{1.93}$$

where now $\beta_0 = \frac{25}{3}$ (corresponding to $n_q = 4$). The same procedure is repeated when μ falls below the scale of the charm quark ($m_c(m_c) \approx 1.275$ GeV), or when μ is raised above the scale of the top-quark mass. This is explored in more detail in Problem 1.3.4.2.

Let us now study the scale evolution of the running quark masses in QCD. This is important, since free quarks do not exist due to confinement, so unlike in QED the quark masses must always be defined as running parameters. We can rewrite the evolution equation in (1.81) in the form

$$\frac{dm_q(\mu)}{d\ln\mu} = \beta(\alpha_s)\frac{dm_q(\mu)}{d\ln\alpha_s} = m_q(\mu)\,\gamma_m(\alpha_s),\tag{1.94}$$

which using separation of variables can be recast as

$$\frac{dm_q}{m_q} = \frac{\gamma_m(\alpha_s)}{\beta(\alpha_s)}\,d\alpha_s \approx -\frac{\gamma_m^0}{2\beta_0}\frac{d\alpha_s}{\alpha_s}.\tag{1.95}$$

Here we have expanded the anomalous dimension in the perturbative series

$$\gamma_m(\alpha_s) = \gamma_m^0\frac{\alpha_s}{4\pi} + \gamma_m^1\left(\frac{\alpha_s}{4\pi}\right)^2 + \ldots;\quad \gamma_m^0 = -6C_F\tag{1.96}$$

and kept the leading term only. Integrating relation (1.95) in the leading-order approximation yields

$$\ln\frac{m_q(\mu)}{m_q(Q)} = -\frac{\gamma_m^0}{2\beta_0}\ln\frac{\alpha_s(\mu)}{\alpha_s(Q)},\tag{1.97}$$

and hence

$$m_q(\mu) = m_q(Q)\left(\frac{\alpha_s(\mu)}{\alpha_s(Q)}\right)^{-\frac{\gamma_m^0}{2\beta_0}}.\tag{1.98}$$

Since the exponent is positive, it follows that quarks get lighter at higher energies. Again it would not be difficult to include higher-order corrections in this analysis (Problem 1.3.4.3).

As an example of this effect, let us study the evolution of the bottom-quark mass from the scale $\mu = m_b$ to the mass scale of the Higgs boson. The resulting parameter $m_b(m_h)$ governs the effective coupling of the Higgs boson to a pair of b quarks. Starting from $m_b(m_b) \approx 4.18\,\text{GeV}$, we obtain

$$m_b(m_h) \approx m_b(m_b)\left(\frac{\alpha_s(m_h)}{\alpha_s(m_b)}\right)^{12/23} \approx 2.79\,\text{GeV}.\tag{1.99}$$

Obviously, evolution effects have a large impact in this case, and ignoring them would largely overestimate the Higgs bottom coupling at high energies.

1.3.3   Fixed Points of Running Couplings

Now that we have discussed the concept of running couplings and β-functions, we briefly discuss fixed points of RG flows. An interesting possibility is that the β-function $\beta(g)$ for some coupling $g(\mu)$ in a quantum field theory has a zero at some value $g_\star \neq 0$ of the coupling, e.g.

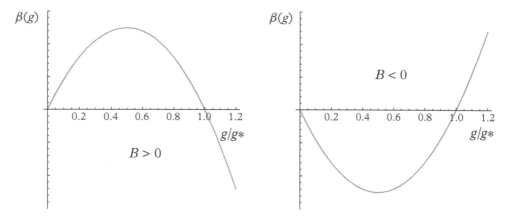

Near such a fixed point we have

$$\beta(g) \approx -B(g - g_\star) = \frac{dg}{d\ln\mu}, \qquad (1.100)$$

and integrating this equation yields

$$g(\mu) \approx g_\star + [g(Q) - g_\star]\left(\frac{\mu}{Q}\right)^{-B}. \qquad (1.101)$$

We can now distinguish two cases:

$$
\begin{array}{llll}
B > 0: & g(\mu) \to g_\star & \text{for } \mu \to \infty & \text{(UV fixed point)} \\
B < 0: & g(\mu) \to g_\star & \text{for } \mu \to 0 & \text{(IR fixed point)}.
\end{array}
\qquad (1.102)
$$

Green's functions obey power-like scaling laws near the fixed point, with *critical exponents* given in terms of anomalous dimensions $\gamma(g_\star)$. Critical phenomena in condensed-matter physics (e.g. phase transitions) are described by anomalous dimensions in simple quantum field theories, such as scalar ϕ^4 theory (see e.g. Chapters 12 and 13 in [3]).

1.3.4   Exercises

1.3.4.1 Prove the relations given in (1.84) and (1.86). The trick is to consider the products $Z_\alpha \beta(\alpha_s, \epsilon)$ and $Z'_m \gamma_m(\alpha_s)$.

1.3.4.2 Section 1.3.2 described the matching procedure, which needs to be applied in the $\overline{\text{MS}}$ scheme whenever the scale μ in the running coupling $\alpha_s(\mu)$ in (1.91) crosses a quark threshold. Using this procedure, computed the values of $\alpha_s(m_t)$, $\alpha_s(m_b)$, and $\alpha_s(m_c)$ starting from $\alpha_s(m_Z) = 0.1181$ and using the masses $m_t(m_t) = 163.4 \,\mathrm{GeV}$, $m_b(m_b) = 4.18 \,\mathrm{GeV}$ and $m_c(m_c) = 1.275 \,\mathrm{GeV}$, where $m_q(m_q)$ are the running quark masses in the $\overline{\mathrm{MS}}$ scheme evaluated at $\mu = m_q$. Also determine the value of μ at which the leading-order formula for the running coupling blows up, see (1.92). Then repeat the same exercise for the running QED coupling. Starting from $\alpha(m_e) = 1/137.036$, compute $\alpha(m_\mu)$, $\alpha(m_\tau)$ and $\alpha(m_Z)$.

1.3.4.3 Integrate the differential equation (1.95) for the running quark mass in QCD keeping the two-loop coefficients β_1 and γ_m^1 in (1.121) and (1.96). Expand the ratio $\gamma_m(\alpha_s)/\beta(\alpha_s)$ in α_s to subleading order and integrate the resulting expression. Find the appropriate generalization of (1.98), which should be of the form

$$m_q(\mu) = m_q(Q) \left(\frac{\alpha_s(\mu)}{\alpha_s(Q)} \right)^{-\frac{\gamma_m^0}{2\beta_0}} \left[1 + c_m \frac{\alpha_s(\mu) - \alpha_s(Q)}{4\pi} + \dots \right]$$

with a constant coefficient c_m that you must determine.

1.4 Effective Field Theories, Composite Operators and the Wilsonian Approach

The basic idea underlying the construction of an effective field theory is that, in a situation where we are faced with a quantum field theory with two (or more) very different energy or length scales, we can construct a simpler theory by performing a systematic expansion in the ratio of these scales. Let us consider an illustrative example.

In view of the fact that the SM of particles physics leaves many questions unanswered, it is plausible that there should exist some 'physics beyond the SM' involving new heavy particles with masses $M \gg v$ much above the scale of electroweak symmetry breaking. While the complete Lagrangian of the UV theory is at present still out of sight, we can construct its low-energy effective theory—the so-called SMEFT—by extending the familiar SM Lagrangian with higher-dimensional local operators built out of SM fields [36]–[40]:

$$\mathcal{L}_{\mathrm{SMEFT}} = \mathcal{L}_{\mathrm{SM}} + \sum_{n \geq 1} \sum_i \frac{C_i^{(n)}}{M^n} \mathcal{O}_i^{(n)}. \tag{1.103}$$

The new operators $\mathcal{O}_i^{(n)}$ with mass dimension $D = 4 + n$ must respect the symmetries of the SM, such as Lorentz invariance and gauge invariance. There is of course an infinite set of such operators, but importantly there exists only a finite set of operators for each dimension D, and the contributions of these operators to any given observable are suppressed by powers of $(v/M)^{D-4}$ relative to the contributions of the operators of the SM. This is discussed in detail in the lectures by Manohar [41] in chapter 2 in this book.

Note that this estimate of the scaling of the higher-order terms assumes that the relevant energies in the process of interest are of order the weak scale v. If we consider high-energy processes characterized by an energy $E \gg v$, then the minimum suppression factor is $(E/M)^{D-4}$ rather than $(v/M)^{D-4}$. An example are transverse-momentum distributions of SM particles produced at the LHC in the region where $p_T \gg v$. If the characteristic energies E are of order the new-physics scale M, then the effective field theory in (1.103) breaks down. Even in this case not all is lost. A different construction based on soft-collinear effective theory [42]–[45]—a non-local effective field theory discussed in the lectures by Becher [46] in chapter 5 in this book—can deal with the case where some kinematical variables in the low-energy theory are parametrically larger than the weak scale [47].

Let me briefly recall how an effective Lagrangian such as (1.103) is derived. 'Integrating out' the heavy degrees of freedom associated with the high scale M from the generating functional of Green's functions we obtain a non-local action functional, which can be expanded in an infinite tower of *local operators* $\mathcal{O}_i^{(n)}$ [48]. For fixed n, the $\{\mathcal{O}_i^{(n)}\}$ form a complete set (a basis) of local, $D = 4 + n$ composite operators built out of the fields of the low-energy theory. These operators are only constrained by the symmetries of the low-energy theory, such as Lorentz invariance, gauge invariance, and global symmetries such as C, P, T, flavour symmetries, etc. The *Wilson coefficients* $C_i^{(n)}$ are dimensionless (this can always be arranged) and contain all information about the short-distance physics which has been integrated out. The above equation is useful only because the infinite sum over n can be truncated at some value $n_{\max}$, since matrix elements of the operators $\mathcal{O}_i^{(n)}$ scale like powers of m, where $m \ll M$ represents the characteristic scale of the low-energy effective theory ($m = v$ in the example of SMEFT), i.e.

$$\langle f| \mathcal{O}_i^{(n)} |i\rangle \sim m^{n+\delta}. \tag{1.104}$$

Here δ is set by the external states. Truncating the sum at $n_{\max}$ creates an error of order $(m/M)^n \ll 1$ relative to the leading term.

1.4.1   Running Couplings and Composite Operators

In essence, in constructing the effective Lagrangian (1.103) we split up the contributions from virtual particles into short- and long-distance modes:

$$\int_0^\infty \frac{d\omega}{\omega} = \int_M^\infty \frac{d\omega}{\omega} + \int_0^M \frac{d\omega}{\omega}, \tag{1.105}$$

where the first term is sensitive to UV physics and is absorbed into the Wilson coefficients $C_i^{(n)}$, while the second term is sensitive to IR physics and is absorbed into the matrix elements $\langle \mathcal{O}_i^{(n)}\rangle$. This is illustrated in panel (a) of Figure 1.2. Now imagine that we are performing a measurement at a characteristic energy scale E, such that $m \ll E < M$. We can then integrate out the high-energy fluctuations of the light SM fields (with frequencies $\omega > E$) from the generating functional, because they will not be needed as source terms for external states. This yields a *different* effective Lagrangian, but one in which the operators $\mathcal{O}_i^{(n)}$ are the same as before (since we have not removed any SM particles). What changes is the split-up of modes, which now reads

$$\int_0^\infty \frac{d\omega}{\omega} = \int_E^\infty \frac{d\omega}{\omega} + \int_0^E \frac{d\omega}{\omega}, \tag{1.106}$$

as shown in panel (b) of Figure 1.2. As a consequence, the values of the Wilson coefficients and operators matrix elements need to be different, i.e.[5]

$$\mathcal{L}_{\text{EFT}} = \sum_{n=0}^\infty \sum_i \frac{C_i^{(n)}(M)}{M^n} \mathcal{O}_i^{(n)}(M) = \sum_{n=0}^\infty \sum_i \frac{C_i^{(n)}(E)}{M^n} \mathcal{O}_i^{(n)}(E). \tag{1.107}$$

We are thus led to study the effective Lagrangian

$$\mathcal{L}_{\text{EFT}} = \sum_{n=0}^\infty \sum_i \frac{C_i^{(n)}(\mu)}{M^n} \mathcal{O}_i^{(n)}(\mu), \tag{1.108}$$

whose matrix elements are, by construction, independent of the arbitrary factorization scale μ (with $m \leq \mu \leq M$), see Figure 1.2(c). Here $\mathcal{O}_i^{(n)}(\mu)$ are renormalized composite operators defined in dimensional regularization and the $\overline{\text{MS}}$ scheme, while $C_i^{(n)}(\mu)$ are the corresponding renormalized Wilson coefficients. These are nothing but the running couplings of the effective theory, in generalization to our discussion of running gauge

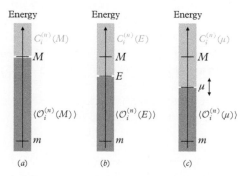

Figure 1.2 *Factorization of an observable into short-distance (light grey) and long-distance (dark grey) contributions, which are accounted for by the Wilson coefficients and operator matrix elements of the effective field theory. The panels differ by the choice of the factorization scale.*

---

[5] The terms with $n = 0$ account for the 'renormalizable' (in an old-fashioned sense) interactions, such as the SM Lagrangian in (1.103).

couplings and running mass parameters in the previous sections. The scale μ serves as the renormalization scale for these quantities, but at the same time it is the factorization scale which separates short-distance (high-energy) from long-distance (low-energy) contributions.

Several comments are in order:

- The terms with $n = 0$ are just the renormalizable Lagrangians of the low-energy theory. As a consequence, parameters such as $\alpha_s(\mu)$ or $m(\mu)$ might, in fact, contain some information about short-distance physics through their scale dependence. For example, the μ dependence of gauge couplings and running mass parameters in supersymmetric extensions of the SM hint at a grand unification of the strong and electroweak forces at a scale $M \sim 10^{16}$ GeV (see e.g. [49]).

- The higher-dimensional operators with $n \geq 1$ are interesting because their coefficients tell us something about the fundamental high-energy scale M. The most prominent example is that of the weak interactions at low energy. These are described by four-fermion operators with mass dimension $D = 6$, whose coefficients are proportional to the Fermi constant $\sqrt{2}\, G_F = 1/v^2$ (see [50] for an excellent review). The numerical value of G_F indicates the fundamental mass scale of electroweak symmetry breaking. Indeed, the masses of the heavy weak gauge bosons $W^\pm$ and Z^0 could be estimated based on observation of weak decays at low energies, long before these particles were discovered.

- The SM also contains the dimension-2 operator $\mu^2 \phi^\dagger \phi$ (here μ^2 is the Higgs mass parameter, not the renormalization scale), which corresponds to $n = -2$. It follows from the assumption of naturalness that $\mu^2 \sim M^2$, and the fact that empirically $|\mu^2|$ is much less than the new-physics scale is known as the *hierarchy problem* of the Standard Model. In a natural theory of fundamental physics, dimension-2 operators should be forbidden by some symmetry, such as supersymmetry (see e.g. [51]).

At any fixed n, the basis $\{\mathcal{O}_i^{(n)}\}$ of composite operators can be renormalized in the standard way, allowing however for the possibility of *operator mixing*. In analogy with the corresponding relations for the field operators in (1.8) and (1.55), we write

$$\mathcal{O}_{i,0}^{(n)} = \sum_j Z_{ij}^{(n)}(\mu)\, \mathcal{O}_j^{(n)}(\mu). \qquad (1.109)$$

The operators on the left-hand side are bare operators (as denoted by the subscript '0'), while the operators on the right-hand side are the renormalized operators. In the presence of operator mixing, in the renormalization of the bare operator $\mathcal{O}_{i,0}^{(n)}$ we need other operators $\mathcal{O}_j^{(n)}(\mu)$ with $j \neq i$ as counterterms. Note that the renormalization constants $Z_{ij}^{(n)}$ contain a wave-function renormalization factor $Z_a^{1/2}$ for each component field contained in the composite operators in addition to renormalization factors absorbing the UV divergences of the 1PI loop corrections to the operator matrix elements. Importantly, in dimensional regularization there is no mixing between operators of different dimension. This fact singles out dimensional regularization as the most convenient regularization scheme. In order not to clutter the notation too much, we henceforth drop the superscript '(n)' on the operators, their Wilson coefficients and the renormalization factors.

There are some important facts about the renormalization properties of composite operators, which are discussed for instance in Section V of [2]. We distinguish three types of composite operators:

- Class-I operators are gauge invariant and do not vanish by virtue of the classical equations of motion.
- Class-II operators are gauge invariant but their matrix elements vanish by virtue of the classical equations of motion.
- Class-III operators are not gauge invariant.

In the renormalization of composite operators it is convenient to use the background-field method [52], which offers an elegant method for renormalizing gauge theories while preserving explicit gauge invariance. Then the following statements hold:

1. The renormalization of class-I operators involves class-I and class-II operators, but not class-III operators as counterterms. In matrix notation

$$\vec{\mathcal{O}}_{\mathrm{I},0} = \boldsymbol{Z}_{\mathrm{I}}\,\vec{\mathcal{O}}_{\mathrm{I}} + \boldsymbol{Z}_{\mathrm{I}\to\mathrm{II}}\,\vec{\mathcal{O}}_{\mathrm{II}}. \qquad (1.110)$$

2. Class-II and class-III operators are renormalized among themselves, i.e.

$$\vec{\mathcal{O}}_{\mathrm{II},0} = \boldsymbol{Z}_{\mathrm{II}}\,\vec{\mathcal{O}}_{\mathrm{II}}, \qquad \vec{\mathcal{O}}_{\mathrm{III},0} = \boldsymbol{Z}_{\mathrm{III}}\,\vec{\mathcal{O}}_{\mathrm{III}}. \qquad (1.111)$$

3. Since on-shell matrix elements of class-II operators vanish by the equations of motion, the contribution of class-II operators in (1.110) has no physical consequences. Furthermore, in background-field gauge class-III operators never arise. Importantly, class-I operators do not appear in (1.111), and hence class-II operators can be ignored for all practical purposes.

Let me add an important comment here. It is often stated that the use of the classical equations of motion to eliminate operators from the basis $\{\mathcal{O}_i^{(n)}\}$ is not justified beyond tree level. This statement is false! Class-II operators can always be removed using *field redefinitions*, which corresponds to a change of variables in the functional integral [53], [54].[6] An explicit proof of this statement is presented in Manohar's lectures [41] in chapter 2 in this book. Special care must be taken when these field redefinitions change the measure of the functional integral. This happens for the case of a fermionic chiral

---

[6] In some cases this generates class-I operators of higher dimension.

transformation and gives rise to the famous chiral anomaly [55]. In any event, the lesson is that at fixed $n \geq 1$ class-II operators can simply be removed from the operator basis.

1.4.2   Anomalous Dimensions of Composite Operators

From the fact that the bare operators on the left-hand side of (1.109) are scale independent, it follows that (a sum over repeated indices is implied)

$$\frac{dZ_{ij}(\mu)}{d\ln\mu}\mathcal{O}_j(\mu) + Z_{ij}(\mu)\frac{d\mathcal{O}_j(\mu)}{d\ln\mu} = 0, \tag{1.112}$$

which can be solved to give

$$\frac{d\mathcal{O}_k(\mu)}{d\ln\mu} = -(Z^{-1})_{ki}(\mu)\frac{dZ_{ij}(\mu)}{d\ln\mu}\mathcal{O}_j(\mu) \equiv -\gamma_{kj}(\mu)\mathcal{O}_j(\mu). \tag{1.113}$$

In matrix notation, this becomes

$$\frac{d\vec{\mathcal{O}}(\mu)}{d\ln\mu} = -\boldsymbol{\gamma}(\mu)\vec{\mathcal{O}}(\mu), \quad \text{with} \quad \boldsymbol{\gamma}(\mu) = Z^{-1}(\mu)\frac{dZ(\mu)}{d\ln\mu}. \tag{1.114}$$

The quantity $\boldsymbol{\gamma}$ is called the *anomalous-dimension matrix* of the composite operators. In analogy with (1.86), this quantity can be obtained from the coefficient of the $1/\epsilon$ pole term in Z via the exact relation (Problem 1.4.5.1)

$$\boldsymbol{\gamma} = -2\alpha_s\frac{\partial Z^{[1]}}{\partial\alpha_s}. \tag{1.115}$$

To calculate the anomalous-dimension matrix we first compute the matrix of renormalization factors Z in (1.109) and then obtain $\boldsymbol{\gamma}$ from the coefficient of the single $1/\epsilon$ pole terms.

1.4.3   RG Evolution Equation for the Wilson Coefficients

The fact that the effective Lagrangian is μ independent by construction implies, for fixed $n \geq 0$, that (a sum over repeated indices is implied)

$$\frac{dC_i(\mu)}{d\ln\mu}\mathcal{O}_i(\mu) + C_i(\mu)\frac{d\mathcal{O}_i(\mu)}{d\ln\mu} = \left[\frac{dC_i(\mu)}{d\ln\mu}\delta_{ij} - C_i(\mu)\gamma_{ij}(\mu)\right]\mathcal{O}_j(\mu) = 0. \tag{1.116}$$

From the linear independence of the basis operators, it follows that

$$\frac{dC_j(\mu)}{d\ln\mu} - C_i(\mu)\,\gamma_{ij}(\mu) = 0 \tag{1.117}$$

for each j, which in matrix notation can be written as

$$\frac{d\vec{C}(\mu)}{d\ln\mu} = \boldsymbol{\gamma}^T(\mu)\,\vec{C}(\mu). \tag{1.118}$$

This matrix differential equation governs the RG evolution of the Wilson coefficients.

In order to solve this equation, we first change variables and express the scale dependence of the various objects via the running QCD coupling $\alpha_s(\mu)$. Using (1.75), this leads to

$$\frac{d\vec{C}(\alpha_s)}{d\alpha_s} = \frac{\boldsymbol{\gamma}^T(\alpha_s)}{\beta(\alpha_s)}\,\vec{C}(\alpha_s). \tag{1.119}$$

Apart from a factor of i on the left-hand side, this equation has the same structure as the time-dependent Schrödinger equation in quantum mechanics, where in our case α_s plays the role of time, $\vec{C}$ corresponds to the Schrödinger wave function, and $\boldsymbol{\gamma}^T/\beta$ plays the role of the Hamiltonian. It follows that the general solution of (1.119) is

$$\vec{C}\big(\alpha_s(\mu)\big) = \mathrm{T}_{\alpha_s}\exp\left[\int\limits_{\alpha_s(M)}^{\alpha_s(\mu)} d\alpha_s\,\frac{\boldsymbol{\gamma}^T(\alpha_s)}{\beta(\alpha_s)}\right]\vec{C}\big(\alpha_s(M)\big), \tag{1.120}$$

where the symbol 'T_{α_s}' implies an ordering of the matrix exponential such that matrices are ordered from left to right according to decreasing α_s values, assuming $\alpha_s(\mu) > \alpha_s(M)$. This is the analogue of the time-ordered exponential in the quantum-mechanical expression for the time-evolution operator. The boundary coefficients $\vec{C}\big(\alpha_s(M)\big)$ correspond to the Wilson coefficients at the high matching scale, which can be computed order by order in QCD perturbation theory. The matrix exponential has the effect of evolving ('running') these coefficients down to a factorization scale $\mu < M$. As we will show in a moment, in this process large logarithms arise (for $\mu \ll M$), which are resummed automatically in the solution (1.120).

At leading order (but not beyond) the ordering symbol becomes irrelevant, and expanding

$$\boldsymbol{\gamma}(\alpha_s) = \boldsymbol{\gamma}_0\,\frac{\alpha_s}{4\pi} + \boldsymbol{\gamma}_1\left(\frac{\alpha_s}{4\pi}\right)^2 + \dots, \qquad \beta(\alpha_s) = -2\alpha_s\left[\beta_0\,\frac{\alpha_s}{4\pi} + \beta_1\left(\frac{\alpha_s}{4\pi}\right)^2 + \dots\right] \tag{1.121}$$

we obtain

$$\vec{C}(\alpha_s(\mu)) \approx \exp\left[-\frac{\pmb{\gamma}_0^T}{2\beta_0}\ln\frac{\alpha_s(\mu)}{\alpha_s(M)}\right]\vec{C}(\alpha_s(M)). \tag{1.122}$$

The matrix exponential can easily be evaluated in Mathematica (Problem 1.4.5.2). For the simplest case of a single operator $\mathcal{O}$, we find

$$C(\alpha_s(\mu)) \approx \left(\frac{\alpha_s(\mu)}{\alpha_s(Q)}\right)^{-\gamma_0/2\beta_0} C(\alpha_s(M)). \tag{1.123}$$

This solution is analogous to that for the running mass in (1.98).

We can use the solution (1.122) to obtain the effective Lagrangian (1.108) at the low-energy scale $\mu = m$, which is characteristic for the mass scale of the low-energy effective theory. At this scale, the matrix elements of the local operators $\mathcal{O}_i(\mu)$ evaluated between physical states can be calculated in fixed-order perturbation theory, since they are free of large logarithms. All potentially large logarithmic corrections are contained in the Wilson coefficients $C_i(\alpha_s(m))$. To see in detail how the large logarithms are resummed, we can substitute from (1.93) the relation

$$\frac{\alpha_s(m)}{\alpha_s(M)} \approx \left[1 - \beta_0 \frac{\alpha_s(M)}{4\pi}\ln\frac{M^2}{m^2}\right]^{-1} \tag{1.124}$$

for the ratio of coupling constants in (1.123), focussing for simplicity on the case of a single operator. This yields

$$\begin{aligned}
C(m) &\approx \left(1 - \beta_0\frac{\alpha_s(M)}{4\pi}\ln\frac{M^2}{m^2}\right)^{\gamma_0/2\beta_0} C(M) \\
&= \left[1 - \frac{\gamma_0}{2}\frac{\alpha_s(M)}{4\pi}\ln\frac{M^2}{m^2} + \frac{\gamma_0(\gamma_0 - 2\beta_0)}{8}\left(\frac{\alpha_s(M)}{4\pi}\ln\frac{M^2}{m^2}\right)^2 + \ldots\right]C(M).
\end{aligned} \tag{1.125}$$

For $\frac{\alpha_s(M)}{4\pi}\ln\frac{M^2}{m^2} = O(1)$ each term in the series contributes at the same order, and resummation is necessary in order to obtain a reliable result.

1.4.4   A Last Remark on the Running QCD Coupling

In order for the above expressions for the Wilson coefficients to make sense, we need to make sure that our formula for $\alpha_s(\mu)$ can be reliably evaluated at any value of μ in the perturbative regime ($\mu \gg \Lambda_{\mathrm{QCD}}$). At leading order we found in (1.91)

$$\alpha_s(\mu) \approx \frac{\alpha_s(Q)}{1 + \beta_0\frac{\alpha_s(Q)}{4\pi}\ln\frac{\mu^2}{Q^2}}. \tag{1.126}$$

We might worry what happens if the logarithm in the denominator becomes large. In other words, we need to demonstrate that higher-order corrections in the β-function do not spoil this formula by introducing additional large logarithms. To see that this does indeed not happen, we keep the next term in the perturbative series for the β-function in (1.121) and study its effect on the solution for the running coupling, which is obtained from (1.75). Separating variables, we obtain

$$-\frac{d\alpha_s}{\alpha_s^2}\frac{1}{1+\frac{\beta_1}{\beta_0}\frac{\alpha_s}{4\pi}+\ldots}=\frac{\beta_0}{4\pi}\,d\ln\mu^2. \tag{1.127}$$

Note that the right-hand side is the single source of logarithms, while no logarithms appear on the left-hand side. As long as we are in the perturbative regime where $\frac{\alpha_s}{4\pi}\ll 1$, we can expand the left-hand side in a perturbative series and obtain, at next-to-leading order,

$$-\frac{d\alpha_s}{\alpha_s^2}\left(1-\frac{\beta_1}{\beta_0}\frac{\alpha_s}{4\pi}+\ldots\right)=\frac{\beta_0}{4\pi}\,d\ln\mu^2. \tag{1.128}$$

Integrating this equation gives

$$\frac{1}{\alpha_s(\mu)}-\frac{1}{\alpha_s(Q)}+\frac{\beta_1}{4\pi\beta_0}\ln\frac{\alpha_s(\mu)}{\alpha_s(Q)}+O\left(\frac{\alpha_s(\mu)-\alpha_s(Q)}{16\pi^2}\right)=\frac{\beta_0}{4\pi}\ln\frac{\mu^2}{Q^2}. \tag{1.129}$$

Multiplying both side with $\alpha_s(Q)$ gives

$$\frac{\alpha_s(Q)}{\alpha_s(\mu)}-\frac{\beta_1}{\beta_0}\frac{\alpha_s(Q)}{4\pi}\ln\frac{\alpha_s(Q)}{\alpha_s(\mu)}+O\left(\frac{\alpha_s(Q)}{4\pi}\frac{\alpha_s(\mu)-\alpha_s(Q)}{4\pi}\right)=1+\beta_0\frac{\alpha_s(Q)}{4\pi}\ln\frac{\mu^2}{Q^2}. \tag{1.130}$$

Once again, the only potentially large logarithm is that on the right-hand side. We can now insert, in an iterative way, the leading-order solution for $\alpha_s(Q)/\alpha_s(\mu)$ in the second term on the left-hand side to obtain

$$\frac{\alpha_s(Q)}{\alpha_s(\mu)}=1+\beta_0\frac{\alpha_s(Q)}{4\pi}\ln\frac{\mu^2}{Q^2}+\frac{\beta_1}{\beta_0}\frac{\alpha_s(Q)}{4\pi}\ln\left(1+\beta_0\frac{\alpha_s(Q)}{4\pi}\ln\frac{\mu^2}{Q^2}\right)+\ldots. \tag{1.131}$$

Even in the 'large-log region', where $\frac{\alpha_s(Q)}{4\pi}\ln\frac{\mu^2}{Q^2}=O(1)$ or larger, the correction proportional to β_1 (the two-loop coefficient of the β-function) is suppressed by at least $\frac{\alpha_s(Q)}{4\pi}\ll 1$ relative to the leading term. The leading-order formula for $\alpha_s(\mu)$ is thus a decent approximation for all values $\mu\gg\Lambda_{\text{QCD}}$.

1.4.5   Exercises

1.4.5.1 Derive relation (1.115), and clarify the origin of the minus sign between this equation and (1.86).

1.4.5.2 The effective weak Lagrangian for the non-leptonic decay $\bar{B}^0 \to \pi^+ D_s^-$ of the neutral B meson contains two dimension-6 four-fermion operators, which differ in their colour structure. Specifically, we find (here i,j are colour indices)

$$\mathcal{L}_{\text{eff}} = -\frac{4 G_F}{\sqrt{2}} V_{cs}^* V_{ub} \left[C_1(\mu) \bar{s}_L^j \gamma_\mu c_L^j \bar{u}_L^i \gamma^\mu b_L^i + C_2(\mu) \bar{s}_L^i \gamma_\mu c_L^j \bar{u}_L^j \gamma^\mu b_L^i \right], \tag{1.132}$$

where $C_1 = 1 + O(\alpha_s)$ and $C_2 = O(\alpha_s)$ follow from tree-level matching of the W-boson exchange diagram onto the effective theory. Using a Fierz rearrangement, the second operator above can also be written as $\bar{u}_L^j \gamma_\mu c_L^j \bar{s}_L^i \gamma^\mu b_L^i$. Note also that

$$\bar{s}_L \gamma_\mu t_a c_L \bar{u}_L \gamma^\mu t_a b_L = \frac{1}{2} \bar{s}_L^i \gamma_\mu c_L^j \bar{u}_L^j \gamma^\mu b_L^i - \frac{1}{2 N_c} \bar{s}_L \gamma_\mu c_L \bar{u}_L \gamma^\mu b_L \tag{1.133}$$

by virtue of a colour Fierz identity, where t_a are the generators of colour $SU(N_c)$. By computing the UV divergences of the two operators in (1.132) at one-loop order (including the effects of wave-function renormalization), show that the anomalous-dimension matrix for the two operators takes the form

$$\gamma = \frac{\alpha_s}{4\pi} \begin{pmatrix} -\frac{6}{N_c} & 6 \\ 6 & -\frac{6}{N_c} \end{pmatrix} + O(\alpha_s^2).$$

Given this result, work out the explicit form of the leading-order solution to the RG equation (1.118), which has been given in (1.122).

Acknowledgements

I thank my colleagues Sacha Davidson, Paolo Gambino, and Mikko Laine for letting me deliver this lecture course despite of being one of the school organizers. Special thanks to Sacha for making sure that the students and lecturers were taken good care of at all times! I am grateful to the students for attending the lectures, asking lots of good questions, and solving homework problems despite the busy schedule of the school. They have made delivering this course a true pleasure. During my stay at Les Houches I enjoyed many interactions with my fellow lecturers, in particular with Thomas Becher, Aneesh

Manohar, and Toni Pich. I also thank my students Stefan Alte, Javier Castellano Ruiz, and Bianka Mecaj for careful proof reading of these lecture notes and suggestions for improvements.

This work has been supported by the Cluster of Excellence *Precision Physics, Fundamental Interactions and Structure of Matter* (PRISMA – EXC 1098) at Johannes Gutenberg University Mainz.

References

[1] C. Itzykson and J. B. Zuber (1980). *Quantum Field Theory*, McGraw-Hill.

[2] P. Pascual and R. Tarrach (1984). QCD: Renormalization for the Practitioner, *Lect. Notes Phys.* **194**, 1.

[3] M. E. Peskin and D. V. Schroeder (1995). *An Introduction to Quantum Field Theory* Addison-Wesley.

[4] J. Collins (2011). Foundations of Perturbative QCD, *Camb. Monogr. Part. Phys. Nucl. Phys. Cosmol.* **32**, 1.

[5] S. Weinberg (2005). *The Quantum Theory of Fields. I: Foundations* Cambridge University Press.

[6] S. Weinberg (2013). *The Quantum Theory of Fields. II: Modern Applications* Cambridge University Press.

[7] M. D. Schwartz (2014). *Quantum Field Theory and the Standard Model* Cambridge University Press.

[8] T. Becher and M. Neubert (2009). *Phys. Rev. Lett.* **102**, 162001. Erratum: [(2013). *Phys. Rev. Lett.* **111**(19): 199905] [arXiv:0901.0722 [hep-ph]].

[9] E. Gardi and L. Magnea (2009). *JHEP* **0903**, 079 [arXiv:0901.1091 [hep-ph]].

[10] T. Becher and M. Neubert (2009). *JHEP* **0906**, 081 Erratum: [(2013). *JHEP* **1311**, 024] [arXiv:0903.1126 [hep-ph]].

[11] T. Becher and M. Neubert (2009). *Phys. Rev. D* **79**, 125004 Erratum: [(2009). *Phys. Rev. D* **80**, 109901] [arXiv:0904.1021 [hep-ph]].

[12] H. Lehmann, K. Symanzik, and W. Zimmermann (1955). Zur Formulierung quantisierter Feldtheorien, *Nuovo Cim.* **1**, 205.

[13] C. G. Bollini and J. J. Giambiagi, (1972). *Nuovo Cim. B* **12**, 20.

[14] G. 't Hooft and M. J. G. Veltman (1972). *Nucl. Phys. B* **44**, 189.

[15] N. N. Bogoliubov and O. S. Parasiuk (1957). *Acta Math.* **97**, 227 [DOI:10.1007/BF02 392399].

[16] K. Hepp (1966). Proof of the Bogoliubov–Parasiuk Theorem on Renormalization. *Commun. Math. Phys.* **2**, 301 (1966).

[17] W. Zimmermann, (1969). Convergence of Bogoliubov's Method of Renormalization in Momentum Space. *Commun. Math. Phys.* **15**, 208 [(2000). *Lect. Notes Phys.* **558**, 217].

[18] W. Pauli and F. Villars (1949). *Rev. Mod. Phys.* **21**, 434 [DOI: 10.1103/RevModPhys.21.434].

[19] S. A. Larin (1993). *Phys. Lett. B* **303**, 113. [hep-ph/9302240].

[20] G. 't Hooft and M. J. G. Veltman (1979). *Nucl. Phys. B* **153**, 365 [DOI: 10.1016/0550-3213(79)90605-9].

[21] M. Tanabashi, *et al.* [Particle Data Group] (2018). *Phys. Rev. D* **98**(3): 030001 [DOI: 10.1103/PhyRevD.98.030001].
[22] J. C. Ward (1950). *Phys. Rev.* **78**, 182 [DOI: 10.1103/PhyRev.78.182].
[23] Y. Takahashi (1957). *Nuovo Cim.* **6**, 371 [DOI: 10.1007/BF02832514].
[24] G. 't Hooft (1973). *Nucl. Phys. B* **61**, 455 [DOI: 10.1016/0550-3213(73)90376-3].
[25] S. Weinberg (1973). *Phys. Rev. D* **8**, 3497.
[26] W. A. Bardeen, A. J. Buras, D. W. Duke, and T. Muta (1978). *Phys. Rev. D* **18**, 3998 [DOI: 10.1103/PhyRevD.18.3998].
[27] G. 't Hooft (1971). *Nucl. Phys. B* **33**, 173 [DOI: 10.1016/0550-3213(71)90395-6].
[28] J. C. Taylor (1971). *Nucl. Phys. B* **33**, 436 [DOI: 10.1016/0550-3213(71)90297-5].
[29] A. A. Slavnov (1972). *Theor. Math. Phys.* **10**, 99 [DOI: 10.1007/BF01090719]. [(1972). *Teor. Mat. Fiz.* **10**, 153].
[30] D. J. Gross and F. Wilczek (1973). *Phys. Rev. Lett.* **30**, 1343 [DOI: 10.1103/PhysRevLett.30.1343].
[31] H. D. Politzer (1973). *Phys. Rev. Lett.* **30**, 1346 [DOI: 10.1103/PhysRevLett.30.1346].
[32] C. G. Callan, Jr. (1970). *Phys. Rev. D* **2**, 1541 [DOI: 10.1103/PhysRevD.2.1541].
[33] K. Symanzik (1970). *Commun. Math. Phys.* **18**, 227 [DOI: 10.1007/BF01649434].
[34] K. Symanzik (1971). *Commun. Math. Phys.* **23**, 49 [DOI: 10.1007/BF01877596].
[35] A. Pich, *Effective Field Theory with Nambu-Goldstone Modes*, arXiv:1804.05664 [hep-ph].
[36] S. Weinberg (1979). *Phys. Rev. Lett.* **43**, 1566 [DOI: 10.1103/PhysRevLett.43.1566].
[37] F. Wilczek and A. Zee (1979). *Phys. Rev. Lett.* **43**, 1571 [DOI: 10.1103/PhysRevLett.43.1571].
[38] W. Buchmüller and D. Wyler (1986). *Nucl. Phys. B* **268**, 621 [DOI: 10.1016/0550-3213(86)90262-2].
[39] C. N. Leung, S. T. Love and S. Rao (1986). *Z. Phys. C* **31**, 433.
[40] B. Grzadkowski, M. Iskrzynski, M. Misiak, and J. Rosiek (2010). *JHEP* **1010**, 085. [arXiv:1008.4884 [hep-ph]].
[41] A. V. Manohar, *Introduction to Effective Field Theories*, arXiv:1804.05863 [hep-ph].
[42] C. W. Bauer, S. Fleming, D. Pirjol, and I. W. Stewart (2001). *Phys. Rev. D* **63**, 114020. [hep-ph/0011336].
[43] C. W. Bauer and I. W. Stewart (2001). *Phys. Lett. B* **516**, 134. [hep-ph/0107001].
[44] C. W. Bauer, D. Pirjol and I. W. Stewart (2002). *Phys. Rev. D* **65**, 054022. [hep-ph/0109045].
[45] M. Beneke, A. P. Chapovsky, M. Diehl and T. Feldmann (2002). *Nucl. Phys. B* **643**, 431. [hep-ph/0206152].
[46] T. Becher, *Les Houches Lectures on Soft-Collinear Effective Theory*, arXiv:1803.04310 [hep-ph].
[47] S. Alte, M. König and M. Neubert (2018). *JHEP* **1808**, 095. [arXiv:1806.01278 [hep-ph]].
[48] J. Polchinski, *Effective Field Theory and the Fermi Surface* [hep-th/9210046].
[49] G. G. Ross (1984). *Grand Unified Theories* (Benjamin/Cummings).
[50] G. Buchalla, A. J. Buras and M. E. Lautenbacher (1996). *Rev. Mod. Phys.* **68**, 1125. [hep-ph/9512380].
[51] J. Wess and J. Bagger (1992). *Supersymmetry and Supergravity* (Princeton University Press).
[52] L. F. Abbott (1981). *Nucl. Phys. B* **185**, 189 [DOI: 10.1016/0550-3213(81)90371-0].
[53] H. D. Politzer (1980). *Nucl. Phys. B* **172**, 349 [DOI: 10.1016/0550-3213(80)90172-8].
[54] H. Georgi (1991). *Nucl. Phys. B* **361**, 339 [DOI: 10.1016/0550-3213(91)90244-R].
[55] K. Fujikawa (1979). *Phys. Rev. Lett.* **42**, 1195 [DOI: 10.1103/PhysRevLett.42.1195].

2

Introduction to Effective Field Theories

Aneesh V. MANOHAR

Department of Physics 0319, University of California at San Diego,
9500 Gilman Drive, La Jolla, CA 92093, USA

Aneesh V. MANOHAR

Manohar, A. V., *Introduction to Effective Field Theories* In: *Effective Field Theory in Particle Physics and Cosmology*. Edited by:
Sacha Davidson, Paolo Gambino, Mikko Laine, Matthias Neubert and Christophe Salomon, Oxford University Press (2020).
© Oxford University Press. DOI: 10.1093/oso/9780198855743.003.0002

Chapter Contents

2.1 Introduction

This chapter compiles an introductory set of lectures on the basic ideas and methods of effective field theories (EFTs). Other lectures in this book go into more detail about the most commonly used effective theories in high energy physics and cosmology. Professor Neubert's lectures [71], delivered concurrently with mine and presented in Chapter 1 of this volume provide an excellent introduction to renormalization in quantum field theory (QFT), the renormalization group equation, operator mixing, and composite operators, and it is assumed the reader already has this knowledge. I also have some twenty-year-old lecture notes from the Schladming school [65] which should be read in conjunction with these lectures. Additional references are [35, 56, 76, 78]. The Les Houches school and this chapter focus on aspects of EFTs as used in high-energy physics and cosmology that are relevant for making contact with experimental observations.

The intuitive idea behind effective theories is that you can calculate without knowing the exact theory. Engineers are able to design and build bridges without any knowledge of strong interactions or quantum gravity. The main inputs in the design are Newton's laws of mechanics and gravitation, the theory of elasticity, and fluid flow. The engineering design depends on parameters measured on macroscopic scales of order meters, such as the elastic modulus of steel. Short-distance properties of Nature, such as the existence of weak interactions, or the mass of the Higgs boson are not needed.

In some sense, the ideas of EFT are 'obvious'. However, implementing them in a mathematically consistent way in an interacting QFT is not so obvious. These lectures provide pedagogical examples of how we actually implement EFT ideas in particle physics calculations of experimentally relevant quantities. Additional details on specific EFT applications are given in other chapters in this volume.

An EFT is a quantum theory in its own right, and like any other QFT, it comes with a regularization and renormalization scheme necessary to obtain finite matrix elements. We can compute S-matrix elements in an EFT from the EFT Lagrangian, with no additional external input, in the same way that we can compute in QED starting from the QED Lagrangian. In many cases, an EFT is the low-energy limit of a more fundamental theory (which might itself be an EFT), often called the 'full theory'.

EFTs allow us to compute an experimentally measurable quantity with some *finite* error. Formally, an EFT has a small expansion parameter δ, known as the power counting parameter. Calculations are done in an expansion to some order n in δ, so that the error is of order δ^{n+1}. Determining the order in δ of a given diagram is done using what is referred to as a *power counting formula*.

A key aspect of EFTs is that we have a systematic expansion, with a well-defined procedure to compute higher-order corrections in δ. Thus we can compute to arbitrarily high order in δ, and make the theoretical error as small as desired, by choosing n

sufficiently large. Such calculations might be extremely difficult in practice because higher-order diagrams are hard to compute, but they are possible in principle. This is very different from modelling, e.g. the non-relativistic quark model provides a good description of hadron properties at the twenty-five per cent level. However, it is *not* the first term in a systematic expansion, and it is not possible to systematically improve the results.

In many examples, there are multiple expansion parameters δ_1, δ_2, etc. For example, in heavy quark effective theory (HQET) [46, 47, 69, 77], b decay rates have an expansion in $\delta_1 = \Lambda_{\text{QCD}}/m_b$ and $\delta_2 = m_b/M_W$. In such cases, we must determine which terms $\delta_1^{n_1}\delta_2^{n_2}$ are retained to reach the desired accuracy goal. Usually, but not always, the expansion parameter is the ratio of a low-energy scale such as the external momentum p, or particle mass m, and a short-distance scale usually denoted by Λ, $\delta = p/\Lambda$. In many examples, we also have a perturbative expansion in a small coupling constant such as $\alpha_s(m_b)$ for HQET.

EFT calculations to order δ^n depend on a finite number of Lagrangian parameters N_n. The number of parameters N_n generally increases as n increases. We get parameter-free predictions in an EFT by calculating more experimentally measured quantities than N_n. For example, HQET computations to order $\Lambda_{\text{QCD}}^2/m_b^2$ depend on two parameters λ_1 and λ_2 of order Λ_{QCD}^2. There are many experimental quantities that can be computed to this order, such as the meson masses, form factors, and decay spectra [69]. Two pieces of data are used to fix λ_1 and λ_2, and then we have parameter-free predictions for all other quantities.

EFTs can be used even when the dynamics are non-perturbative. The most famous example of this type is chiral perturbation theory (χPT), which has an expansion in p/Λ_χ, where $\Lambda_\chi \sim 1\,\text{GeV}$ is the chiral symmetry breaking scale. Systematic computations in powers of p/Λ_χ are in excellent agreement with experiment [33, 74, 73, 81].

The key ingredient used in formulating EFTs is locality, which leads to a separation of scales, i.e. factorization of the field theory amplitudes into short-distance Lagrangian coefficients and long-distance matrix elements. The short-distance coefficients are universal, and independent of the long-distance matrix elements computed [82]. The experimentally measured quantities $\mathcal{O}_i$ are then given as the product of these short-distance coefficients C and long-distance matrix elements. Often, there are multiple coefficients and matrix elements, so that $\mathcal{O}_i = \sum_i C_{ij}M_j$. Sometimes, as in deep-inelastic scattering, C and M depend on a variable x instead of an index i, and the sum becomes a convolution

$$\mathcal{O} = \int_0^1 \frac{dx}{x} C(x)M(x). \tag{2.1}$$

The short-distance coefficient $C(x)$ in this case is called the hard-scattering cross section, and can be computed in QCD perturbation theory. The long-distance matrix elements are the parton distribution functions, which are determined from experiment. The hard-scattering cross section is universal, but the parton distribution functions depend on the hadronic target.

EFTs allow us to organize calculations in an efficient way, and to estimate quantities using the power counting formula in combination with locality and gauge invariance. The tree-level application of EFTs is straightforward; it is simply a series expansion of the scattering amplitude in a small parameter. The true power lies in being able to compute radiative corrections. It is worth repeating that EFTs are fully fledged quantum theories, and we can compute measurable quantities such as S-matrix elements *without any reference or input from a underlying UV theory*. The 1933 Fermi theory of weak interactions [30] was used long before the Standard Model (SM) was invented, or anyone knew about electroweak gauge bosons. Pion-nucleon scattering lengths [79, 80] and $\pi - \pi$ scattering lengths [80] were computed in 1966, without any knowledge of QCD, quarks, or gluons.

Here are some warm-up exercises that will be useful later.

Exercise 2.1 Show that for a *connected* graph, $V - I + L = 1$, where V is the number of vertices, I is the number of internal lines, and L is the number of loops. What is the formula if the graph has n connected components?

Exercise 2.2 Work out the transformation of fermion bilinears $\overline{\psi}(\mathbf{x},t)\,\Gamma\,\chi(\mathbf{x},t)$ under C, P, T, where $\Gamma = P_L, P_R, \gamma^\mu P_L, \gamma^\mu P_R, \sigma^{\mu\nu}P_L, \sigma^{\mu\nu}P_R$. Use your results to find the transformations under CP, CT, PT, and CPT.

Exercise 2.3 Show that for $SU(N)$,

$$[T^A]^\alpha_{\ \beta}\,[T^A]^\lambda_{\ \sigma} = \frac{1}{2}\delta^\alpha_\sigma \delta^\lambda_\beta - \frac{1}{2N}\delta^\alpha_\beta \delta^\lambda_\sigma, \tag{2.2}$$

where the $SU(N)$ generators are normalized to $\operatorname{Tr} T^A T^B = \delta^{AB}/2$. From this, show that

$$\delta^\alpha_\beta \delta^\lambda_\sigma = \frac{1}{N}\delta^\alpha_\sigma \delta^\lambda_\beta + 2[T^A]^\alpha_{\ \sigma}\,[T^A]^\lambda_{\ \beta},$$

$$[T^A]^\alpha_{\ \beta}\,[T^A]^\lambda_{\ \sigma} = \frac{N^2-1}{2N^2}\delta^\alpha_\sigma \delta^\lambda_\beta - \frac{1}{N}[T^A]^\alpha_{\ \sigma}\,[T^A]^\lambda_{\ \beta}. \tag{2.3}$$

Exercise 2.4 Spinor Fierz identities are relations of the form

$$(\overline{A}\,\Gamma_1 B)(\overline{C}\,\Gamma_2 D) = \sum_{ij} c_{ij}(\overline{C}\,\Gamma_i B)(\overline{A}\,\Gamma_j D)$$

where A, B, C, D are fermion fields, and c_{ij} are numbers. They are much simpler if written in terms of chiral fields using $\Gamma_i = P_L, P_R, \gamma^\mu P_L, \gamma^\mu P_R, \sigma^{\mu\nu}P_L, \sigma^{\mu\nu}P_R$, rather than Dirac fields. Work out the Fierz relations for

$$(\overline{A}P_L B)(\overline{C}P_L D), \qquad (\overline{A}\gamma^\mu P_L B)(\overline{C}\gamma_\mu P_L D), \qquad (\overline{A}\sigma^{\mu\nu}P_L B)(\overline{C}\sigma_{\mu\nu}P_L D),$$

$$(\overline{A}P_L B)(\overline{C}P_R D), \qquad (\overline{A}\gamma^\mu P_L B)(\overline{C}\gamma_\mu P_R D), \qquad (\overline{A}\sigma^{\mu\nu}P_L B)(\overline{C}\sigma_{\mu\nu}P_R D).$$

Do not forget the Fermi minus sign. The $P_R \otimes P_R$ identities are obtained from the $P_L \otimes P_L$ identities by using $L \leftrightarrow R$.

2.2   Examples

This section discusses qualitative examples of EFTs illustrating the use of power counting, symmetries such as gauge invariance, and dimensional analysis. Some of the examples are covered in detail in other lectures in this volume.

2.2.1   Hydrogen Atom

A simple familiar example is the computation of the hydrogen atom energy levels, as done in a quantum mechanics class. The Hamiltonian for an electron of mass m_e interacting via a Coulomb potential with a proton treated as an infinitely heavy point particle is

$$\mathscr{H} = \frac{\mathbf{p}^2}{2m_e} - \frac{\alpha}{r}. \tag{2.4}$$

The binding energies, electromagnetic transition rates, etc. are computed from eqn (2.4). The fact that the proton is made up of quarks, weak interactions, neutrino masses, etc. are irrelevant, and we do not need any detailed input from QED or QCD. The only property of the proton we need is that its charge is $+1$; this can be measured at long distances from the Coulomb field.

Corrections to eqn (2.4) can be included in a systematic way. Proton recoil is included by replacing m_e by the reduced mass $\mu = m_e m_p/(m_e + m_p)$, which gives corrections of order m_e/m_p. At this point, we have included one strong-interaction parameter, the mass m_p of the proton, which can be determined from experiments done at low energies, i.e. at energies much below Λ_{QCD}.

The hydrogen fine structure is calculated by including higher-order (relativistic) corrections to the Hamiltonian, and gives corrections of relative order α^2. The hydrogen hyperfine structure (the famous 21 cm line) requires including the spin–spin interaction between the proton and electron, which depends on their magnetic moments. The proton magnetic moment $\mu_p = 2.793\,e\hbar/(2m_p c)$ is the second strong interaction parameter that now enters the calculation, and can be measured in low-energy nuclear magnetic resonance spectroscopy (NMR) experiments. The electron magnetic moment is given by its Dirac value $-e\hbar/(2m_e c)$.

Even more accurate calculations require additional non-perturbative parameters, as well as QED corrections. For example, the proton charge radius r_p, $g - 2$ for the electron, and QED radiative corrections for the Lamb shift all enter to obtain the accuracy required to compare with precision experiments.

For calculations with an accuracy of 10^{-13} eV ~ 50 Hz, it is necessary to include the weak interactions. The weak interactions give a very small shift in the energy levels, and are a tiny correction to the energies. But they are the leading contribution to atomic parity violation effects. The reason is that the strong and electromagnetic interactions conserve parity. Thus the relative size of various higher-order contributions depends on the quantity being computed—there is no universal rule that can be unthinkingly followed in all examples. Even in the simple hydrogen atom example, we have multiple expansion parameters m_e/m_p, α, and m_p/M_W.

2.2.2 Multipole Expansion in Electrostatics

A second familiar example is the multipole expansion from electrostatics,

$$V(\mathbf{r}) = \frac{1}{r} \sum_{l,m} b_{lm} \frac{1}{r^l} Y_{lm}(\Omega), \tag{2.5}$$

which will illustrate a number of useful points. A sample charge configuration with its electric field and equipotential lines is shown in Figure 2.1.

While the following discussion is in the context of the electrostatics example, it holds equally well for other EFT examples. If the typical spacing between charges in Figure 2.1 is of order a, eqn (2.5) can be written as

$$V(\mathbf{r}) = \frac{1}{r} \sum_{l,m} c_{lm} \left(\frac{a}{r}\right)^l Y_{lm}(\Omega), \qquad\qquad b_{lm} \equiv c_{lm} a^l, \tag{2.6}$$

using dimensionless coefficients c_{lm}.

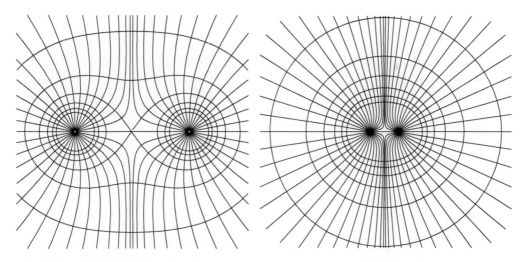

Figure 2.1 *The electric field and potential lines for two point charges of the same sign. The right figure is given by zooming out the left figure.*

- As written, eqn (2.6) has two scales r and a, with $r \gg a$. r is the long-distance, or infrared (IR) scale, and a is the short-distance or ultraviolet (UV) scale. The small expansion parameter is the ratio of the IR and UV scales $\delta = a/r$. The expansion is useful if the two scales are widely separated, so that $\delta \ll 1$. We often work in momentum space, so that the IR scale is $p \sim 1/r$, the UV scale is $\Lambda \sim 1/a$, and $\delta = p/\Lambda$.

- A far away (low-energy) observer measures the potential $V(r)$ as a function of r and $\Omega = (\theta, \phi)$. By Fourier analysis, the observer can determine the short distance coefficients $b_{lm} = c_{lm}a^l \sim c_{lm}/\Lambda^l$. These coefficients are dimensionful, and suppressed by inverse powers of Λ as l increases.

- More accurate values of the potential are given by including more multipoles. The terms in eqn (2.5,2.6) get smaller as l increases. A finite experimental resolution implies that c_{lm} can only be experimentally determined up to a finite maximum value $l_{\max}$ that depends on the resolution. More accurate experiments probe larger $l_{\max}$.

- We can factor out powers of a, as shown in eqn (2.6), and use c_{lm} instead of b_{lm}. Then c_{lm} are order unity. This is dimensional analysis. There is no precise definition of a, and any other choice for a of the same order of magnitude works equally well. a is given from observations by measuring b_{lm} for large values of r, and inferring a by letting $b_{lm} = c_{lm}a^l$, and seeing if some choice of a makes all the c_{lm} of order unity.

- Some c_{lm} can vanish, or be anomalously small due to an (approximate) symmetry of the underlying charge distribution. For example, cubic symmetry implies $c_{lm} = 0$ unless $l \equiv 0 \pmod 2$ and $m \equiv 0 \pmod 4$. Measurements of b_{lm} provide information about the short-distance structure of the charge distribution, and possible underlying symmetries.

- More accurate measurements require higher-order terms in the l expansion. There is only a finite number, $(l_{\max} + 1)^2$, of parameters including all terms up to order $l_{\max}$.

- We can use the l expansion without knowing the underlying short-distance scale a, as can be seen from the first form eqn (2.5). The parameters b_{lm} are determined from the variation of $V(r)$ with respect to the IR scale r. Using $b_{lm} = c_{lm}a^l$ gives us an estimate of the size of the charge distribution. We can determine the short-distance scale a by accurate measurements at the long-distance scale $r \gg a$, or by less-sophisticated measurements at shorter distances r comparable to a.

The above analysis also applies to searches for BSM (beyond SM) physics. Experiments are searching for new interactions at short distances $a \sim 1/\Lambda$, where Λ is larger than the electroweak scale $v \sim 246$ GeV. Two ways of determining the new physics scale are by making more precise measurements at low-energies, as is being done in B physics experiments, or by making measurements at even higher energies, as at the LHC.

Subtleties can arise even in the simple electrostatic problem. Consider the charge configuration shown in Figure 2.2, which is an example of a multiscale problem. The system has two intrinsic scales, a shorter scale d characterizing the individual charge clumps, and a longer scale a characterizing the separation between clumps.

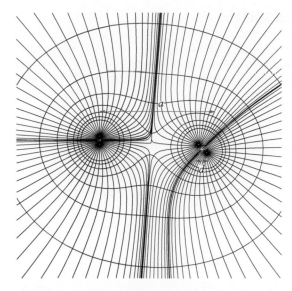

Figure 2.2 *A charge distribution with two intrinsic scales: d, the size of each clump, and a, the distance between clumps.*

Measurements at large values of r determine the scale a. Very accurate measurements of c_{lm} can determine the shorter distance scale d. Discovering d requires noticing patterns in the values of c_{lm}. It is much easier to determine d if we know ahead of time that there is a short-distance scale d that must be extracted from the data. d can be easily determined by making measurements at shorter distances (higher energies) $d \ll r \ll a$, i.e. if we are allowed to measure the electrostatic potential between the two clumps of charges.

Multiscale problems are common in EFT applications. The Standard Model EFT (SMEFT) is an EFT used to characterize BSM physics. The theory has a scale Λ, of order a few TeV, which is the expected scale of BSM physics in the electroweak sector, as well as higher scales Λ_L and Λ_B at which lepton and baryon numbers are broken. χPT has the scales $m_\pi \sim 140$ MeV, $m_K \sim 500$ MeV and the chiral symmetry breaking scale $\Lambda_\chi \sim 1$ GeV. HQET has the scales m_b, m_c and $\Lambda_{\rm QCD}$. EFT methods allow us to separate scales in a multiscale problem, and organize the calculation in a systematic way.

2.2.3 Fermi Theory of Weak Interactions

The Fermi theory of weak interactions [30] is an EFT for weak interactions at energies below the W and Z masses. It is a low-energy EFT constructed from the SM. The EFT power counting parameter is $\delta = p/M_W$, where p is of order the momenta of particles in the weak decay. For example, in μ decay, p is of order the muon mass. In hadronic weak decays, p can be of order the hadron (or quark) masses, or of order $\Lambda_{\rm QCD}$. The theory also has the usual perturbative expansions in $\alpha_s/(4\pi)$ and $\alpha/(4\pi)$. Historically, Fermi's theory was used for weak decay calculations even when the scales M_W and M_Z were not known. We construct the Fermi interaction in Section 2.4.8.

2.2.4   HQET/NRQCD

Heavy quark effective theory (HQET) and non-relativistic QCD (NRQCD [18]) describe the low-energy dynamics of hadrons containing a heavy quark. The theories are applied to hadrons containing b and c quarks. In HQET, the expansion parameter is $\Lambda_{\rm QCD}/m_Q$, where $m_Q = m_b, m_c$ is the mass of the heavy quark. The theory also has an expansion in powers of $\alpha_s(m_Q)/(4\pi)$. The matching from QCD to HQET can be done in perturbation theory, since $\alpha_s(m_Q)/(4\pi)$ is small, $\alpha_s(m_b) \sim 0.22$, $\alpha_s(m_b)/(4\pi) \sim 0.02$. Calculations in HQET contain non-perturbative corrections, which can be included in a systematic way in an expansion in $\Lambda_{\rm QCD}/m_Q$.

NRQCD is similar to HQET, but treats $Q\overline{Q}$ bound states such as the Υ meson. The heavy quarks move non-relativistically, and the expansion parameter is the velocity v of the heavy quarks, which is of order $v \sim \alpha_s(m_Q)$.

HQET and NRQCD are covered in Mannel's lectures in Chapter 9.

2.2.5   Chiral Perturbation Theory

Chiral perturbation theory describes the interactions of pions and nucleons at low momentum transfer. The theory was developed in the 1960s, and the method closest to the modern way of calculating was developed by Weinberg. χPT describes the low-energy dynamics of QCD. In this example, the full theory is known, but it is not possible to analytically compute the matching onto the EFT, since the matching is non-perturbative. Recent progress has been made in computing the matching numerically [6]. The two theories, QCD and χPT, are not written in terms of the same fields. The QCD Lagrangian has quark and gluon fields, whereas χPT has meson and baryon fields. The parameters of the chiral Lagrangian are usually fit to experiment.

Note that computations in χPT, such as Weinberg's calculation of $\pi\pi$ scattering, were done using χPT *before* QCD was even invented. This example shows rather clearly that we can compute in an EFT without knowing the UV origin of the EFT.

The expansion parameter of χPT is p/Λ_χ, where $\Lambda_\chi \sim 1\,{\rm GeV}$ is referred to as the scale of chiral symmetry breaking. χPT can be applied to baryons even though baryon masses are comparable to Λ_χ. The reason is that baryon number is conserved, and so baryons can be treated as heavy particles analogous to heavy quarks in HQET as long as the momentum transfer is smaller than Λ_χ. There is an interesting relation between the large-N_c expansion of QCD and baryon chiral perturbation theory [48, 67].

χPT is covered in Pich's lectures, found in Chapter 3 in this volume.

2.2.6   SCET

Soft-collinear effective theory (SCET [8]–[11]) describes energetic QCD processes where the final states have small invariant mass compared to the centre-of-mass energy of the collision, such as in jet production in high-energy pp collisions. The underlying theory is once again QCD. The expansion parameters of SCET are $\Lambda_{\rm QCD}/Q$, M_J/Q and $\alpha_s(Q)/(4\pi)$, where Q is the centre-of-mass energy of the hard-scattering process, and $M_{\mathcal{J}}$ is the invariant mass of the jet. SCET was originally developed for the decay of B mesons to light particles, such as $B \to X_s \gamma$ and $B \to \pi\pi$.

SCET is covered in Becher's lectures in Chapter 5.

2.2.7 SMEFT

SMEFT is the EFT constructed out of SM fields, and is used to analyse deviations from the SM, and search for BSM physics. The higher-dimension operators in SMEFT are generated at a new physics scale Λ, which is not known. Nevertheless, we can still perform systematic computations in SMEFT, as should be clear from the multipole expansion example in Section 2.2.2. SMEFT is discussed in Section 2.10.

2.2.8 Reasons for using an EFT

We summarize here the many reasons for using an EFT. The points are treated in more detail later in this chapter, and also in the other chapters in the book.

- **Every theory is an EFT:** For example, QED, the first relativistic QFT developed, is an approximation to the SM. It is an EFT obtained from the SM by integrating out all particles other than the photon and electron.

- **EFTs simplify the computation by dealing with only one scale at a time:** For example the B meson decay rate depends on M_W, m_b and Λ_{QCD}, and we can get horribly complicated functions of the ratios of these scales. In an EFT, we deal with only one scale at a time, so there are no functions, only constants. This is done by using a series of theories, e.g. SM $\to$ Fermi theory $\to$ HQET.

- **EFTs make symmetries manifest:** QCD has a spontaneously broken chiral symmetry, which is manifest in the chiral Lagrangian. Heavy quarks have an Isgur–Wise [46] spin-flavour symmetry under which $b \uparrow$, $b \downarrow$, $c \uparrow$, $c \downarrow$ transform as a four-dimensional representation of $SU(4)$. This symmetry is manifest in the HQET Lagrangian [36], which makes it easy to derive the consequences of this symmetry. Symmetries such as spin-flavour symmetry are only true for certain limits of QCD, and so are hidden in the QCD Lagrangian.

- **EFTs include only the relevant interactions:** EFTs have an explicit power counting estimate for the size of various interactions. Thus we can include only the relevant terms in the EFT Lagrangian needed to obtain the required accuracy of the final result.

- **Sum logs of the ratios of scales:** This allows us to use renormalization group improved perturbation theory, which is more accurate, and has a larger range of validity than fixed order perturbation theory. For example, the semileptonic B decay rate depends on powers

$$\left(\frac{\alpha_s}{4\pi} \ln \frac{M_W}{m_b} \right)^n . \tag{2.7}$$

Even though $\alpha_s/(4\pi)$ is small, it is multiplied by a large log, and fixed order perturbation theory can break down. RG improved perturbation theory sums the corrections in eqn (2.7), so that the perturbation expansion is in powers of $\alpha_s/(4\pi)$, without a multiplicative log. The resummation of logs is even more important in SCET, where there are two powers of a log for each α_s, the so-called Sudakov double logarithms.

The leading-log corrections are not small. For example, the strong interaction coupling changes by a factor of two between M_Z and m_b,

$$\alpha_s(M_Z) \sim 0.118, \qquad\qquad \alpha_s(m_b) \sim 0.22.$$

While summing logs might seem like a technical point, it is one of the main reasons why EFTs (or equivalent methods such as factorization formulæ in QCD) are used in practice. In QCD collider processes, resummed cross sections can be dramatically different from fixed order ones.

- **Sum IR logs by converting them to UV logs:** This is related to the previous point. UV logs are summed by the renormalization group equations, since they are related to anomalous dimensions and renormalization counterterms. There is no such summation method for IR logs. However, IR logs in the full theory can be converted to UV logs in the EFT, which can then be summed by integrating the renormalization group equations in the EFT (see Section 2.5.8). QCD leads to a number of different effective theories, HQET, NRQCD, SCET, and χPT. Each is designed to treat a particular IR regime, and sum the corresponding IR logs.

- **Non-perturbative effects can be included in a systematic way:** In HQET, powers of Λ_{QCD} are included through the matrix elements of higher dimension operators, giving the $(\Lambda_{\text{QCD}}/m_b)^n$ expansion.

- **Efficient method to characterize new physics:** EFTs provide an efficient way to characterize new physics, in terms of coefficients of higher-dimension operators. This method includes the constraints of locality, gauge invariance, and Lorentz invariance. All new physics theories can be treated in a unified framework using a few operator coefficients.

2.3   The EFT Lagrangian

2.3.1   Degrees of Freedom

To write down an EFT Lagrangian, we first need to determine the dynamical degrees of freedom, and thus the field content of the Lagrangian. In cases where the EFT is a weakly coupled low-energy version of a UV theory, this is simple—just retain the light fields. However, in many cases, identifying the degrees of freedom in an EFT can be non-trivial.

NRQCD describes $Q\bar{Q}$ bound states, and is an EFT that follows from QCD. One formulation of NRQCD has multiple gluon modes, soft and ultrasoft gluons, which describe different momentum regions contributing to the $Q\bar{Q}$ interaction. SCET describes the interactions of energetic particles, and is applicable to processes such as jet production by $q\bar{q} \to q\bar{q}$ interactions. It has collinear gluon fields for each energetic particle direction, as well as ultrasoft gluon fields.

A famous example which shows that there is no unique 'correct' choice of fields to use in an interacting QFT is the sine–Gordon–Thirring model duality in $1+1$ dimensions [22]. The sine–Gordon model is a bosonic theory of a real scalar field with Lagrangian

$$\mathscr{L} = \frac{1}{2}\partial_\mu \phi\, \partial^\mu \phi + \frac{\alpha}{\beta^2}\cos\beta\phi, \tag{2.8}$$

and the Thirring model is a fermionic theory of a Dirac fermion with Lagrangian

$$\mathscr{L} = \bar{\psi}\left(i\slashed{\partial} - m\right)\psi - \frac{1}{2}g\left(\bar{\psi}\gamma^\mu\psi\right)^2. \tag{2.9}$$

Coleman showed that the two theories were *identical*; they map into each other with the couplings related by

$$\frac{\beta^2}{4\pi} = \frac{1}{1+g/\pi}. \tag{2.10}$$

The fermion in the Thirring model is the soliton of the sine–Gordon model, and the boson of the sine–Gordon model is a fermion–anti-fermion bound state of the Thirring model. The duality exchanges strongly and weakly coupled theories. This example also shows that we cannot distinguish between elementary and composite fields in an interacting QFT.

2.3.2 Renormalization

Here we present a quick summary of renormalization in QCD, to define the notation and procedure we use for EFTs. A detailed discussion is given in Chapter 1.

QCD is a QFT with Lagrangian

$$\mathscr{L} = -\frac{1}{4}F^A_{\mu\nu}F^{A\mu\nu} + \sum_{r=1}^{N_F}\left[\bar{\psi}_r i\slashed{D}\psi_r - m_r\overline{\psi}_r\psi_r\right] + \frac{\theta g^2}{32\pi^2}F^A_{\mu\nu}\tilde{F}^{A\mu\nu}, \tag{2.11}$$

where N_F is the number of flavours. The covariant derivative is $D_\mu = \partial_\mu + igA_\mu$, and the $SU(3)$ gauge field is a matrix $A_\mu = T^A A^A_\mu$, where the generators are normalized to $\mathrm{Tr}\, T^A T^B = \delta^{AB}/2$. Experimental limits on the neutron electric dipole moment give $\theta \lesssim 10^{-10}$, and we will ignore it here.

The basic observables in a QFT are S-matrix elements—on-shell scattering amplitudes for particles with physical polarizations. Green functions of ψ and A_μ, which are the correlation functions of products of fields, are gauge dependent and not experimental observables. The QCD Lagrangian eqn (2.11) is written in terms of fields, but *fields are not particles*. The relation between S-matrix elements of particles and Green functions for fields is through the LSZ reduction formula [60] explained in Section 2.6.1. We can use *any* field $\phi(x)$ to compute the S-matrix element involving a particle state $|p\rangle$ as long as

$$\langle p|\phi(x)|0\rangle \neq 0, \tag{2.12}$$

i.e. the field can create a one-particle state from the vacuum.

Radiative corrections in QCD are infinite, and we need a regularization and renormalization scheme to compute finite S-matrix elements. The regularization and renormalization procedure is part of the *definition* of the theory. The standard method used in modern field theory computations is to use dimensional regularization and the $\overline{\text{MS}}$ subtraction scheme. We use dimensional regularization in $d = 4 - 2\epsilon$ dimensions. A brief summary of the procedure is given here.

The QCD Lagrangian for a single flavour in the CP-conserving limit (so that the θ term is omitted) that gives finite S-matrix elements is

$$\mathscr{L} = -\frac{1}{4}F^A_{0\mu\nu}F_0^{A\mu\nu} + \overline{\psi}_0 i(\slashed{\partial} + ig_0\slashed{A}_0)\psi_0 - m_0\overline{\psi}_0\psi_0 \tag{2.13a}$$

$$= -\frac{1}{4}Z_A F^A_{\mu\nu}F^{A\mu\nu} + Z_\psi \overline{\psi} i(\slashed{\partial} + ig\mu^\epsilon Z_g Z_A^{1/2}\slashed{A})\psi - mZ_m Z_\psi \overline{\psi}\psi \tag{2.13b}$$

where ψ_0, A_0, g_0, and m_0 are the bare fields and parameters, which are related to the renormalized fields and parameters ψ, A, g, and m by

$$\psi_0 = Z_\psi^{1/2}\psi, \qquad A_{0\mu} = Z_A^{1/2}A_\mu, \qquad g_0 = Z_g g\mu^\epsilon, \qquad m_0 = Z_m m. \tag{2.14}$$

The renormalization factors Z_a have an expansion in inverse powers of ϵ,

$$Z_a = 1 + \sum_{k=1}^{\infty}\frac{Z_a^{(k)}}{\epsilon^k}, \qquad\qquad a = \psi, A, g, m, \tag{2.15}$$

with coefficients which have an expansion in powers of $\alpha_s = g^2/(4\pi)$,

$$Z_a^{(k)} = \sum_{r=1}^{\infty}Z_a^{(k,r)}\left(\frac{\alpha_s}{4\pi}\right)^r. \tag{2.16}$$

The renormalized parameters g and m are finite, and depend on μ. The renormalization factors Z_a are chosen to give finite S-matrix elements.

Separating out the 1 from Z_a, the Lagrangian eqn (2.13b) can be written as

$$\mathcal{L} = -\frac{1}{4}F_{\mu\nu}^A F^{A\mu\nu} + \overline{\psi}i(\slashed{\partial} + ig\mu^\epsilon \slashed{A})\psi - m\overline{\psi}\psi + \text{c.t.} \tag{2.17}$$

where c.t. denotes the renormalization counterterms which are pure poles in $1/\epsilon$,

$$\mathcal{L}_{\text{c.t.}} = -\frac{1}{4}(Z_A - 1)F_{\mu\nu}^A F^{A\mu\nu} + (Z_\psi - 1)\overline{\psi}i\slashed{\partial}\psi + \left(Z_\psi Z_g Z_A^{1/2} - 1\right)\overline{\psi}ig\mu^\epsilon \slashed{A}\psi$$
$$- (Z_\psi Z_m - 1)m\overline{\psi}\psi. \tag{2.18}$$

The Lagrangian eqn (2.13a) contains two bare parameters, g_0 and m_0. The Lagrangian eqn (2.13b) contains two renormalized parameters $g(\mu)$, $m(\mu)$ and the renormalization scale μ. As discussed in Chapter 1, the renormalization group equation, which follows from the condition that the theory is μ-independent, implies that there are only two free parameters, for example $g(\mu_r)$ and $m(\mu_r)$ at some chosen reference scale μ_r. The renormalization group equations determine how m and g must vary with μ to keep the observables the same. We will see later how the freedom to vary μ allows us to sum logarithms of the ratio of scales. The variation of renormalized parameters with μ is sometimes referred to as the renormalization group flow.

The bare parameters in the starting Lagrangian eqn (2.13a) are infinite. The infinities cancel with those in loop graphs, so that S-matrix elements computed are finite. Alternatively, we start with the Lagrangian split up into the renormalized Lagrangian with finite parameters plus counterterms, as in eqn (2.17). The infinite parts of loop graphs computed from the renormalized Lagrangian are cancelled by the counterterm contributions, to give finite S-matrix elements. The two methods are equivalent, and give the usual renormalization procedure in the $\overline{\text{MS}}$ scheme. Usually, we compute in perturbation theory in the coupling g, and determine the renormalization factors Z_a order by order in g to ensure finiteness of the S-matrix.

Exercise 2.5 Compute the mass renormalization factor Z_m in QCD at one loop. Use this to determine the one-loop mass anomalous dimension γ_m,

$$\mu\frac{\mathrm{d}m}{\mathrm{d}\mu} = \gamma_m m, \tag{2.19}$$

by differentiating $m_0 = Z_m m$, and noting that m_0 is μ-independent.

2.3.3 Determining the couplings

How do we determine the parameters in the Lagrangian? The bare Lagrangian parameters are infinite, and cannot be measured directly. The renormalized Lagrangian parameters are finite. However, in general, they are scheme dependent, and also not directly measurable. In QCD, the $\overline{\text{MS}}$ quark mass $m_b(\mu)$ is not a measurable quantity. Often, people refer to the quark pole mass m_b^{pole} defined by the location of the pole in the quark propagator in perturbation theory. It is related to the $\overline{\text{MS}}$ mass by

$$m_b^{\text{pole}} = m_b(m_b) \left[1 + \frac{4\alpha_s(m_b)}{3\pi} + \ldots \right]. \tag{2.20}$$

m_b^{pole} is independent of μ, and hence is renormalization group invariant. Nevertheless, m_b^{pole} is not measurable—quarks are confined, and there is no pole in gauge-invariant correlation functions at m_b^{pole}. Instead we determine the B meson mass m_B experimentally. The quark mass $m_b(\mu)$ or m_b^{pole} is fixed by adjusting it till it reproduces the measured meson mass. To actually do this requires a difficult non-perturbative calculation, since m_b^{pole} and m_B differ by order Λ_{QCD} effects. In practice, we use observables that are easier to compute theoretically, such as the electron energy spectrum in inclusive B decays, or the $e^+e^- \to b\bar{b}$ cross section near threshold, to determine the quark mass. Similarly, the gauge coupling $g(\mu)$ is not an observable, and must be determined indirectly.

Even in QED, the Lagrangian parameters are not direct observables. QED has two Lagrangian parameters, and two experimental inputs are used to fix these parameters. We can measure the electron mass m_e^{obs} (which is the pole mass, since electrons are not confined), and the electrostatic potential at large distances, $-\alpha_{\text{QED}}/r$. These two measurements fix the values of the Lagrangian parameters $m_e(\mu)$ and $e(\mu)$. All other observables, such as positronium energy levels, the Bhabha scattering cross section, etc., are then determined, since they are functions of $m_e(\mu)$ and $e(\mu)$.

The number of Lagrangian parameters $N_{\mathscr{L}}$ tells you how many inputs are needed to completely fix the predictions of the theory. In general, we compute a set of observables $\{O_i\}, i = 1, \ldots, N_{\mathscr{L}}$ in terms of the Lagrangian parameters. $N_{\mathscr{L}}$ observables are used to fix the parameters, and the remaining $N_O - N_{\mathscr{L}}$ observables are predictions of the theory:

$$\underbrace{O_1, \ldots, O_{N_{\mathscr{L}}}}_{\text{observables}} \quad \longrightarrow \quad \underbrace{m_i(\mu), g(\mu), \ldots}_{\text{parameters}} \quad \longrightarrow \quad \underbrace{O_{N_{\mathscr{L}}+1}, \ldots}_{\text{predictions}} \tag{2.21}$$

The Lagrangian plays the role of an intermediary, allowing one to relate observables to each other. The S-matrix program of the 1960s avoided any use of the Lagrangian, and related observables directly to each other using analyticity and unitarity.

Given a QFT Lagrangian $\mathscr{L}$, including a renormalization procedure, you can calculate S-matrix elements. No additional outside input is needed, and the calculation is often automated. For example, in QED, it is not necessary to know that the theory is the low-energy limit of the SM, or to consult an oracle to obtain the value of certain loop graphs. All predictions of the theory are encoded in the Lagrangian. A renormalizable theory has only a finite number of terms in the Lagrangian, and hence only a finite number of parameters. We can compute observables to arbitrary accuracy, at least in principle, and obtain parameter-free predictions.

The above discussion applies to EFTs as well, including the last bit about a finite number of parameters, provided that one works to a *finite* accuracy δ^n in the power counting parameter. As an example, consider the Fermi theory of weak interactions, which we discuss in more detail in Section 2.4.8. The EFT Lagrangian in the lepton sector is

$$\mathscr{L} = \mathscr{L}_{\text{QED}} - \frac{4G_F}{\sqrt{2}} (\overline{e}\gamma^\mu P_L \nu_e)(\overline{\nu}_\mu \gamma_\mu P_L \mu) + \ldots, \tag{2.22}$$

where $P_L = (1 - \gamma_5)/2$, and $G_F = 1.166 \times 10^{-5} \text{ GeV}^{-2}$ has dimensions of inverse mass-squared. As in QCD or QED, we can calculate μ-decay directly using eqn (2.22) without using any external input, such as knowing eqn (2.22) was obtained from the low-energy limit of the SM. The theory is renormalized as in eqn (2.13a, 2.13b, 2.14). The main difference is that the Lagrangian eqn (2.22) has an infinite series of operators (only one is shown explicitly), with coefficients that absorb the divergences of loop graphs. The expansion parameter of the theory is $\delta = G_F p^2$. To a fixed order in δ, the theory is just like a regular QFT. However, if we want to work to higher accuracy, more operators must be included in $\mathscr{L}$, so that there are more parameters. If we insist on infinitely precise results, then there are an infinite number of terms and an infinite number of parameters. Thus an EFT is just like a regular QFT, supplemented by a power counting argument that tells us what terms to retain to a given order in δ. The number of experimental inputs used to fix the Lagrangian parameters increases with the order in δ. In the μ-decay example, G_F can be fixed by the muon lifetime. The Fermi theory then gives a parameter-free prediction for the decay distributions, such as the electron energy spectrum, electron polarization, etc.

The parameters of the EFT Lagrangian eqn (2.22) can be obtained from low-energy data. The divergence structure of the EFT is *different* from that of the full theory, of which the EFT is a low-energy limit. This is not a minor technicality, but a fundamental difference. It is crucial in many practical applications, where IR logs can be summed by transitioning to an EFT.

In cases where the EFT is the low-energy limit of a weakly interacting full theory, e.g. the Fermi theory as the low-energy limit of the SM, we construct the EFT Lagrangian to reproduce the same S-matrix as the original theory, a procedure known as matching. The full and effective theory are equivalent; they are different ways of computing the same observables. The change in renormalization properties means that fields in the EFT are not the same as fields in the full theory, even though they are often denoted by the same symbol. Thus the electron field e in eqn (2.22) is not the same as the field e in the SM Lagrangian. The two agree at tree-level, but at higher orders, we must explicitly compute the relation between the two. A given high-energy theory can lead to multiple EFTs, depending on the physical setting. For example, χPT, HQET, NRQCD, and SCET are all EFTs based on QCD.

2.3.4   Inputs

At the beginning I suggested that it was 'obvious' that low-energy dynamics was insensitive to the short-distance properties of the theory. This is true provided the input parameters are obtained from low-energy processes computed using the EFT. QED plus QCD with five flavours of quarks is the low-energy theory of the SM below the electroweak scale. The input couplings can be determined from measurements below 100 GeV.

Now suppose, instead, that the input couplings are fixed at high energies, and their low-energy values are determined by computation. Given the QED coupling $\alpha(\mu_H)$ at a scale $\mu_H > m_t$ above the top-quark mass, for example, we can determine the low-energy value $\alpha(\mu_L)$ for μ_L smaller than m_t. In this case, $\alpha(\mu_L)$ is sensitive to high-energy parameters, such as heavy masses including the top-quark mass. For example, if we vary the top-quark mass, then

$$m_t \frac{d}{dm_t} \left[\frac{1}{\alpha(\mu_L)} \right] = -\frac{1}{3\pi}, \tag{2.23}$$

where $\mu_L < m_t$, and we have kept $\alpha(\mu_H)$ for $\mu_H > m_t$ fixed. Similarly, if we keep the strong coupling $\alpha_s(\mu_H)$ fixed for $\mu_H > m_t$, then the proton mass is sensitive to m_t,

$$m_p \propto m_t^{2/27}. \tag{2.24}$$

A bridge builder would have a hard time designing a bridge if the density of steel depended on the top-quark mass via eqn (2.24). Luckily, knowing about the existence of top-quarks is not necessary. The density of steel is an experimental observable, and its measured value is used in the design. The density is measured in lab experiments at low-energies, on length scales of order meters, not in LHC collisions. How the density depends on m_t or possible BSM physics is irrelevant. There is no sensitivity to high-scale physics if the inputs to low-energy calculations are from low-energy measurements. The short-distance UV parameters are not 'more fundamental' than the long-distance ones. They are just parameters. For example, in QED, is $\alpha(\mu > m_t)$ more fundamental than $\alpha_{\text{QED}} = 1/(137.036)$ given by measuring the Coulomb potential as $r \to \infty$? It is α_{QED}, for example, which is measured in quantum Hall effect experiments.

Combining low-energy EFTs with high-energy inputs mixes different scales, and leads to problems. The natural parameters of the EFT are those measured at low energies. Using high-energy inputs forces the EFT to use inputs that do not fit naturally into the framework of the theory. We will return to this point in the Appendix.

Symmetry restrictions from the high-energy theory feed down to the low-energy theory. QCD (with $\theta = 0$) preserves C, P, and CP, and hence so does χPT. Causality in QFT leads to the spin-statistics theorem. This is a restriction imposed in quantum mechanics, and follows because the quantum theory is the non-relativistic limit of a QFT.

Exercise 2.6 Verify eqn (2.23) and eqn (2.24).

2.3.5 Integrating Out Degrees of Freedom

The old-fashioned view is that EFTs are given by integrating out high-momentum modes of the original theory, and thinning out degrees of freedom as we evolve from the UV to the IR [55, 84, 85]. That is not what happens in the EFTs that are used to describe experimentally observable phenomena, and it is not the correct interpretation of renormalization group evolution in these theories.

In SCET, there are different collinear sectors of the theory labelled by null vectors $n_i = (1, \mathbf{n}_i)$, $\mathbf{n}_i^2 = 1$. Each collinear sector of SCET is the same as the full QCD Lagrangian, so SCET has multiple copies of the original QCD theory, as well as ultrasoft modes that couple the various collinear sectors. The number of degrees of freedom in SCET is much larger than in the original QCD theory. In χPT, the EFT is written in terms of meson and baryon fields, whereas QCD is given in terms of quarks and gluons. Mesons and baryons are created by composite operators of quarks and gluons, but there is no sense in which the EFT is given by integrating out short-distance quarks and gluons.

The renormalization group equations are a consequence of the μ independence of the theory. Thus varying μ changes nothing measurable; S-matrix elements are μ independent. Nothing is being integrated out as μ is varied, and the theory at different values of μ is the same. The degrees of freedom do not change with μ. The main purpose of the renormalization group equations is to sum logs of ratios of scales, as we will see in Section 2.5.10.

It is much better to think of EFTs in terms of the physical problem we are trying to solve, rather than as the limit of some other theory. The EFT is then constructed out of the dynamical degrees of freedom (fields) that are relevant for the problem. The focus should be on what we want, not on what we don't want.

2.4 Power Counting

The EFT functional integral is

$$\int \mathcal{D}\phi \, e^{iS},$$

(2.25)

so that the action S is dimensionless. The EFT action is the integral of a local Lagrangian density

$$S = \int \mathrm{d}^{\mathsf{d}} x \, \mathscr{L}(x),$$

(2.26)

(neglecting topological terms), so that in d space-time dimensions, the Lagrangian density has mass dimension d,

$$[\mathscr{L}(x)] = \mathsf{d},$$

(2.27)

and is the sum

$$\mathscr{L}(x) = \sum_i c_i \, O_i(x),$$

(2.28)

of local, gauge invariant, and Lorentz invariant operators O_i with coefficients c_i. The operator dimension will be denoted by $\mathscr{D}$, and its coefficient has dimension $d - \mathscr{D}$.

The fermion and scalar kinetic terms are

$$S = \int d^d x \, \bar{\psi} \, i\slashed{\partial} \, \psi, \qquad\qquad S = \int d^d x \, \frac{1}{2} \partial_\mu \phi \, \partial^\mu \phi, \qquad (2.29)$$

so that dimensions of fermion and scalar fields are

$$[\psi] = \frac{1}{2}(d - 1), \qquad\qquad [\phi] = \frac{1}{2}(d - 2). \qquad (2.30)$$

The two terms in the covariant derivative $D_\mu = \partial_\mu + igA_\mu$ have the same dimension, so

$$[D_\mu] = 1, \qquad\qquad [gA_\mu] = 1. \qquad (2.31)$$

The gauge field strength $X_{\mu\nu} = \partial_\mu A_\nu - \partial_\nu A_\mu + \dots$ has a single derivative of A_μ, so A_μ has the same dimension as a scalar field. This determines, using eqn (2.31), the dimension of the gauge coupling g,

$$[A_\mu] = \frac{1}{2}(d - 2), \qquad\qquad [g] = \frac{1}{2}(4 - d). \qquad (2.32)$$

In $d = 4$ space-time dimensions,

$$[\phi] = 1, \qquad [\psi] = 3/2, \qquad [A_\mu] = 1, \qquad [D] = 1, \qquad [g] = 0. \qquad (2.33)$$

In $d = 4 - 2\epsilon$ dimensions, $[g] = \epsilon$, so in dimensional regularization, we usually use a dimensionless coupling g and write the coupling in the Lagrangian as $g\mu^\epsilon$, as in eqn (2.14).

The only gauge and Lorentz invariant operators with dimension $\mathscr{D} \leq d = 4$ that can occur in the Lagrangian are

$$\begin{aligned}
\mathscr{D} &= 0: \quad 1 \\
\mathscr{D} &= 1: \quad \phi \\
\mathscr{D} &= 2: \quad \phi^2 \\
\mathscr{D} &= 3: \quad \phi^3, \bar{\psi}\psi \\
\mathscr{D} &= 4: \quad \phi^4, \, \phi\bar{\psi}\psi, \, D_\mu\phi \, D^\mu\phi, \, \bar{\psi} \, i\slashed{D} \, \psi, \, X_{\mu\nu}^2.
\end{aligned} \qquad (2.34)$$

Other operators, such as $D^2\phi$ vanish upon integration over $d^d x$, or are related to operators already included eqn (2.34) by integration by parts. In $d = 4$ space-time dimensions, fermion fields can be split into left-chiral and right-chiral fields, which transform as irreducible representations of the Lorentz group. The projection operators are $P_L = (1 - \gamma_5)/2$ and $P_R = (1 + \gamma_5)/2$. Left-chiral fermions will be denoted $\psi_L = P_L\psi$, etc.

Renormalizable interactions have coefficients with mass dimension ≥ 0, and eqn (2.34) lists the allowed renormalizable interactions in four space-time dimensions. The distinction between renormalizable and non-renormalizable operators should be clear after Section 2.4.2.

In $\mathsf{d} = 2$ space-time dimensions

$$[\phi] = 0, \qquad [\psi] = 1/2, \qquad \left[A_\mu\right] = 0, \qquad [D] = 1, \qquad [g] = 1, \qquad (2.35)$$

so an arbitrary potential $V(\phi)$ is renormalizable, as is the $\left(\bar{\psi}\psi\right)^2$ interaction, so that the sine–Gordon and Thirring models are renormalizable. In $\mathsf{d} = 6$ space-time dimensions,

$$[\phi] = 2, \qquad [\psi] = 5/2, \qquad \left[A_\mu\right] = 2, \qquad [D] = 1, \qquad [g] = -1. \qquad (2.36)$$

The only allowed renormalizable interaction in six dimensions is ϕ^3. There are no renormalizable interactions above six dimensions.[1]

Exercise 2.7 In $\mathsf{d} = 4$ space-time dimensions, work out the field content of Lorentz-invariant operators with dimension $\mathscr{D}$ for $\mathscr{D} = 1, \ldots, 6$. At this point, do not try to work out which operators are independent, just the possible structure of allowed operators. Use the notation ϕ for a scalar, ψ for a fermion, $X_{\mu\nu}$ for a field strength, and D for a derivative. For example, an operator of type $\phi^2 D$ such as $\phi D_\mu \phi$ is not allowed because it is not Lorentz invariant. An operator of type $\phi^2 D^2$ could be either $D_\mu \phi D^\mu \phi$ or $\phi D^2 \phi$, so a $\phi^2 D^2$ operator is allowed, and we will worry later about how many independent $\phi^2 D^2$ operators can be constructed.

Exercise 2.8 For $\mathsf{d} = 2, 3, 4, 5, 6$ dimensions, work out the field content of operators with dimension $\mathscr{D} \leq \mathsf{d}$, i.e. the 'renormalizable' operators.

2.4.1 EFT Expansion

The EFT Lagrangian follows the same rules as the previous section, and has an expansion in powers of the operator dimension

$$\mathscr{L}_{\text{EFT}} = \sum_{\mathscr{D} \geq 0, i} \frac{c_i^{(\mathscr{D})} O_i^{(\mathscr{D})}}{\Lambda^{\mathscr{D} - d}} = \sum_{\mathscr{D} \geq 0} \frac{\mathscr{L}_{\mathscr{D}}}{\Lambda^{\mathscr{D} - d}} \qquad (2.37)$$

where $O_i^{(\mathscr{D})}$ are the allowed operators of dimension $\mathscr{D}$. All operators of dimension $\mathscr{D}$ are combined into the dimension $\mathscr{D}$ Lagrangian $\mathscr{L}_{\mathscr{D}}$. The main difference from the previous discussion is that one does not stop at $\mathscr{D} = \mathsf{d}$, but includes operators of arbitrarily high dimension. A scale Λ has been introduced so that the coefficients $c_i^{(\mathscr{D})}$ are dimensionless.

---

[1] There are exceptions to this in strongly coupled theories where operators can develop large anomalous dimensions.

Λ is the short-distance scale at which new physics occurs, analogous to $1/a$ in the multipole expansion example in Section 2.2.2. As in the multipole example, what is relevant for theoretical calculations and experimental measurements is the product $c_\mathscr{D}\Lambda^{\mathsf{d}-\mathscr{D}}$, not $c_\mathscr{D}$ and $\Lambda^{\mathsf{d}-\mathscr{D}}$ separately. Λ is a convenient device that makes it clear how to organize the EFT expansion.

In $\mathsf{d}=4$,

$$\mathscr{L}_{\mathrm{EFT}} = \mathscr{L}_{\mathscr{D}\leq 4} + \frac{\mathscr{L}_5}{\Lambda} + \frac{\mathscr{L}_6}{\Lambda^2} + \dots \tag{2.38}$$

$\mathscr{L}_{\mathrm{EFT}}$ is given by an infinite series of terms of increasing operator dimension. An important point is that the $\mathscr{L}_{\mathrm{EFT}}$ has to be treated as an expansion in powers of $1/\Lambda$. If you try to sum terms to all orders, you violate the EFT power counting rules, and the EFT breaks down.

2.4.2 Power Counting and Renormalizability

Consider a scattering amplitude $\mathscr{A}$ in d dimensions, normalized to have mass dimension zero. If we work at some typical momentum scale p, then a single insertion of an operator of dimension $\mathscr{D}$ in the scattering graph gives a contribution to the amplitude of order

$$\mathscr{A} \sim \left(\frac{p}{\Lambda}\right)^{\mathscr{D}-\mathsf{d}} \tag{2.39}$$

by dimensional analysis. The operator has a coefficient of mass dimension $1/\Lambda^{\mathscr{D}-\mathsf{d}}$ from eqn (2.37), and the remaining dimensions are produced by kinematic factors such as external momenta to make the overall amplitude dimensionless. An insertion of a set of higher-dimension operators in a tree graph leads to an amplitude

$$\mathscr{A} \sim \left(\frac{p}{\Lambda}\right)^{n} \tag{2.40}$$

with

$$n = \sum_i (\mathscr{D}_i - \mathsf{d}), \qquad n = \sum_i (\mathscr{D}_i - 4) \text{ in } \mathsf{d}=4 \text{ dimensions}, \tag{2.41}$$

where the sum on i is over all the inserted operators. This follows from dimensional analysis, as for a single insertion. Equation (2.41) is known as the EFT power counting formula. It gives the (p/Λ) suppression of a given graph.

The key to understanding EFTs is to understand why eqn (2.41) holds for *any* graph, not just tree graphs. The technical difficulty for loop graphs is that the loop momentum k is integrated over all values of k, $-\infty \leq k \leq \infty$, where the EFT expansion in powers of k/Λ breaks down. Nevertheless, eqn (2.41) still holds. The validity of eqn (2.41) for any graph is explained in Section 2.5.3.

The first example of a power counting formula in an EFT was Weinberg's power counting formula for χPT. This is covered in Chapter 3, and is closely related to eqn (2.41). Weinberg counted powers of p in the numerator, whereas we have counted powers of Λ in the denominator. The two are obviously related.

The power counting formula eqn (2.41) tells us how to organize the calculation. If we want to compute $\mathscr{A}$ to leading order, we only use $\mathscr{L}_{\mathscr{D}\leq\mathsf{d}}$, i.e. the renormalizable Lagrangian. In $\mathsf{d}=4$ dimensions, p/Λ corrections are given by graphs with a single insertion of $\mathscr{L}_5$; $(p/\Lambda)^2$ corrections are given by graphs with a single insertion of $\mathscr{L}_6$, or two insertions of $\mathscr{L}_5$, and so on. As mentioned previously, we do not need to assign a numerical value to Λ to do a systematic calculation. All we are using is eqn (2.41) for a fixed power n.

We can now understand the difference between renormalizable theories and EFTs. In an EFT, there are higher-dimension operators with dimension $\mathscr{D}>\mathsf{d}$. Suppose we have a single dimension-five operator (using the $\mathsf{d}=4$ example). Graphs with two insertions of this operator produce the same amplitude as a dimension-six operator. In general, loop graphs with two insertions of $\mathscr{L}_5$ are divergent, and we need a counterterm, which is an $\mathscr{L}_6$ operator. Even if we set the coefficients of $\mathscr{L}_6$ to zero in the renormalized Lagrangian, we have to add a $\mathscr{L}_6$ counterterm with a $1/\epsilon$ coefficient. Thus the Lagrangian has a coefficient $c_6(\mu)$. $c_6(\mu)$ might vanish at one special value of μ, but in general, it evolves with μ by the renormalization group equations, and so it will be non-zero at a different value of μ. There is nothing special about $c_6=0$ if this condition does not follow from a symmetry. Continuing in this way, we generate the infinite series of terms in eqn (2.37). We can generate operators of arbitrarily high dimension by multiple insertions of operators with $\mathscr{D}-\mathsf{d}>0$.

On the other hand, if we start only with operators in $\mathscr{L}_{\mathscr{D}\leq\mathsf{d}}$, we do not generate any new operators, only those already included in $\mathscr{L}_{\mathscr{D}\leq\mathsf{d}}$. The reason is that $\mathscr{D}-\mathsf{d}\leq0$ in eqn (2.41) so we only generate operators with $\mathscr{D}\leq\mathsf{d}$. Divergences in a QFT are absorbed by local operators, which have $\mathscr{D}\geq0$. Thus new operators generated by loops have $0\leq\mathscr{D}\leq\mathsf{d}$, and have already been included in $\mathscr{L}$. We do not need to add counterterms with negative dimension operators, such as $1/\phi^2(x)$, since there are no divergences of this type. In general, renormalizable terms are those with $0\leq\mathscr{D}\leq d$, i.e. the contribution to n in eqn (2.41) is non-positive.

Renormalizable theories are a special case of EFTs, where we formally take the limit $\Lambda\to\infty$. Then all terms in $\mathscr{L}$ have dimension $\mathscr{D}\leq\mathsf{d}$. Scattering amplitudes can be computed to arbitrary accuracy, as there are no p/Λ corrections. Theories with operators of dimensions $\mathscr{D}>\mathsf{d}$ are referred to as non-renormalizable theories, because an infinite number of higher dimension operators are needed to renormalize the theory. We have seen, however, that as long as we are interested in corrections with some maximum value of n in eqn (2.41), there are only a finite number of operators that contribute, and non-renormalizable theories (i.e. EFTs) are just as good as renormalizable ones.

2.4.3 Photon–Photon Scattering

We now illustrate the use of the EFT power counting formula eqn (2.41) with some simple examples, which show the power of eqn (2.41) when combined with constraints from gauge invariance and Lorentz invariance.

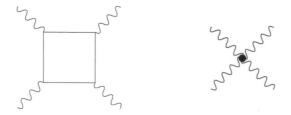

Figure 2.3 *The left figure is the QED contribution to the $\gamma\gamma$ scattering amplitude from an electron loop. The right figure is the low-energy limit of the QED amplitude treated as a local $F_{\mu\nu}^4$ operator in the Euler–Heisenberg Lagrangian.*

Consider $\gamma\gamma$ scattering at energies much lower than the electron mass, $E \ll m_e$. At these low energies, the only dynamical degrees of freedom in the EFT are photons. Classical electromagnetism without charged particles is a free theory, but in QED, photons can interact via electron loops, as shown in Figure 2.3. In the EFT, there are no dynamical electrons, so the 4γ interaction due to electron loops is given by a series of higher-dimension operators involving photon fields. The lowest-dimension interactions that preserve charge conjugation are given by dimension-eight operators, so the EFT Lagrangian has the expansion

$$\mathscr{L} = -\frac{1}{4}F_{\mu\nu}F^{\mu\nu} + \frac{\alpha^2}{m_e^4}\left[c_1\left(F_{\mu\nu}F^{\mu\nu}\right)^2 + c_2\left(F_{\mu\nu}\tilde{F}^{\mu\nu}\right)^2\right] + \ldots. \tag{2.42}$$

This is the Euler–Heisenberg Lagrangian [43]. We can compare eqn (2.42) with the general form eqn (2.37). We have used m_e for the scale Λ, since we know that the higher-dimension operators are generated by the electron loop graph in QED shown in Figure 2.3. Since QED is perturbative, we have included a factor of e^4 from the vertices, and $1/16\pi^2$ from the loop, so that $c_{1,2}$ are pure numbers.

The scattering amplitude computed from eqn (2.42) in the centre-of-mass frame is

$$\mathscr{A} \sim \frac{\alpha^2\omega^4}{m_e^4}, \tag{2.43}$$

where ω is the photon energy. The α^2/m_e^4 factor is from the Lagrangian, and the ω^4 factor is because each field-strength tensor is the gradient of A_μ, and produces a factor of ω. The scattering cross section σ is proportional to $|\mathscr{A}|^2$, and has mass dimension -2. The phase space integral is thus $\propto 1/\omega^2$ to get the correct dimensions, since ω is the only dimensionful parameter in the low-energy problem. The cross section is then

$$\sigma \sim \left(\frac{\alpha^2\omega^4}{m_e^4}\right)^2 \frac{1}{\omega^2}\frac{1}{16\pi} \sim \frac{\alpha^4\omega^6}{16\pi m_e^8}. \tag{2.44}$$

The $1/(16\pi)$ will be explained in Section 2.8. The ω^6 dependence of the cross section follows from the lowest operator being of dimension eight, so that $\mathscr{A} \propto 1/m_e^4$, and $\sigma \propto 1/m_e^8$,

$$A \propto \frac{1}{m_e^4} \Rightarrow \sigma \propto \omega^6. \tag{2.45}$$

If we had assumed (incorrectly) that gauge invariance was not important and written the interaction operator generated by Figure 2.3 as the dimension-four operator

$$\mathscr{L} = c\alpha^2 (A_\mu A^\mu)^2 \tag{2.46}$$

the cross section would be $\sigma \sim \alpha^4/(16\pi\omega^2)$ instead. The ratio of the two estimates is $(\omega/m_e)^8$. For $\omega \sim 1$ eV, the ratio is 10^{48}!
   An explicit computation [28, 27, 43] gives

$$c_1 = \frac{1}{90}, \qquad c_2 = \frac{7}{360}, \tag{2.47}$$

and [61]

$$\sigma = \frac{\alpha^4 \omega^6}{16\pi m_e^8} \frac{15568}{10125}. \tag{2.48}$$

Our estimate eqn (2.44) is quite good (about 50% off), and was obtained with very little work.
   For scalar field scattering, the interaction operator would be ϕ^4, so that $\sigma \sim 1/(16\pi\omega^2)$, whereas Goldstone bosons such as pions have interactions $\Pi^2(\partial\Pi)^2/f^2$, so that $\sigma \sim \omega^4/(16\pi f^4)$. Cross sections can vary by many orders of magnitude (10^{48} between scalars and gauge bosons), so dimensional estimates such as this are very useful to decide whether a cross section is experimentally relevant before starting on a detailed calculation.

2.4.4   Proton Decay

Grand unified theories violate baryon and lepton number. The lowest-dimension operators constructed from SM fields that violate baryon number are dimension-six operators,

$$\mathscr{L} \sim \frac{qqql}{M_G^2}. \tag{2.49}$$

These operators violate baryon number B and lepton number L, but conserve $B-L$. The operator eqn (2.49) leads to the proton decay amplitude $p \rightarrow e^+\pi^0$

$$\mathscr{A} \sim \frac{1}{M_G^2}, \tag{2.50}$$

and the proton decay rate

$$\Gamma \sim \frac{m_p^5}{16\pi M_G^4}. \tag{2.51}$$

In eqn (2.51), we have obtained a decay rate of the correct dimensions using the only scale in the decay rate calculation, the proton mass m_p, and the rule of $1/(16\pi)$ for the final state phase space discussed in Section 2.8. The proton lifetime is

$$\tau = \frac{1}{\Gamma} \sim \left(\frac{M_G}{10^{15}\,\text{GeV}}\right)^4 \times 10^{30}\ \text{years} \tag{2.52}$$

EFT power counting provides a natural explanation for baryon number conservation. In the SM, baryon number is first violated at dimension six, leading to a long proton lifetime.

If baryon number were violated at dimension five (as happens in some supersymmetric models), eqn (2.50) would be replaced by $\mathscr{A} \sim 1/M_G$, and the proton decay rate is

$$\Gamma \sim \frac{m_p^3}{16\pi M_G^2}. \tag{2.53}$$

The proton lifetime is very short,

$$\tau = \frac{1}{\Gamma} \sim \left(\frac{M_G}{10^{15}\,\text{GeV}}\right)^2 \times 1\ \text{years}, \tag{2.54}$$

and is ruled out experimentally.

2.4.5 $n - \overline{n}$ Oscillations

In some theories, baryon number is violated but lepton number is not. Then proton decay is forbidden. The proton is a fermion, and so its decay products must contain a lighter fermion. But the only fermions lighter than the proton carry lepton number, so proton decay is forbidden. These theories do allow for a new experimental phenomenon, namely $n - \overline{n}$ oscillations, which violates only baryon number.

The lowest-dimension operator that leads to $n - \overline{n}$ oscillations, is the $\Delta B = 2$ six-quark operator

$$\mathscr{L} \sim \frac{q^6}{M_G^5}, \tag{2.55}$$

which is dimension nine, and suppressed by five powers of the scale M_G at which the operator is generated. This leads to an oscillation amplitude

$$\mathscr{A} \sim \left(\frac{m_n}{M_G}\right)^5, \tag{2.56}$$

which is strongly suppressed.

2.4.6 Neutrino Masses

The lowest-dimension operator in the SM which gives a neutrino mass is the $\Delta L = 2$ operator of dimension five (see Section 2.10.1.1),

$$\mathscr{L} \sim \frac{(H^\dagger \ell)(H^\dagger \ell)}{M_S}, \tag{2.57}$$

generated at a high scale M_S usually referred to as the seesaw scale. Equation (2.57) gives a Majorana neutrino mass of order

$$m_\nu \sim \frac{v^2}{M_S} \tag{2.58}$$

when $SU(2) \times U(1)$ symmetry is spontaneously broken by $v \sim 246$ GeV. Using $m_\nu \sim 10^{-2}$ eV leads to a seesaw scale $M_S \sim 6 \times 10^{15}$ GeV. Neutrinos are light if the lepton number violating scale M_S is large.

2.4.7 Rayleigh Scattering

The scattering of photons off atoms at low energies also can be analysed using our power counting results. Here low energies means energies small enough that the internal states of the atom are not excited, as they have excitation energies of order electron-Volts.

The atom can be treated as a neutral particle of mass M, interacting with the electromagnetic field. Let $\psi(x)$ denote a field operator that creates an atom at the point x. Then the effective Lagrangian for the atom is

$$\mathscr{L} = \psi^\dagger \left(i\partial_t - \frac{\partial^2}{2M} \right) \psi + \mathscr{L}_{\text{int}}, \tag{2.59}$$

where $\mathscr{L}_{\text{int}}$ is the interaction term. From eqn (2.59), we see that $[\psi] = 3/2$. Since the atom is neutral, covariant derivatives acting on the atom are ordinary derivatives, and do not contain gauge fields. The gauge field interaction term is a function of the electromagnetic field strength $F_{\mu\nu} = (\mathbf{E}, \mathbf{B})$. Gauge invariance forbids terms which depend only on the vector potential A_μ. At low energies, the dominant interaction is one that involves the lowest-dimension operators,

$$\mathscr{L}_{\text{int}} = a_0^3 \, \psi^\dagger \psi \left(c_E \mathbf{E}^2 + c_B \mathbf{B}^2 \right). \tag{2.60}$$

An analogous $\mathbf{E} \cdot \mathbf{B}$ term is forbidden by parity conservation. The operators in eqn (2.60) have $\mathscr{D} = 7$, so we have written their coefficients as dimensionless constants times a_0^3. a_0 is the size of the atom, which controls the interaction of photons with the atom, and $[a_0] = -1$. The photon only interacts with the atom when it can resolve its charged constituents, the electron and nucleus, which are separated by a_0, so a_0 plays the role of $1/\Lambda$ in eqn (2.60).

The interaction eqn (2.60) gives the scattering amplitude

$$\mathscr{A} \sim a_0^3 \omega^2,$$ (2.61)

since the electric and magnetic fields are gradients of the vector potential, so each factor of $\mathbf{E}$ or $\mathbf{B}$ produces a factor of ω. The scattering cross section is proportional to $|\mathscr{A}|^2$. This has the correct dimensions to be a cross section, so the phase space is dimensionless, and

$$\sigma \propto a_0^6 \, \omega^4.$$ (2.62)

Equation (2.62) is the famous ω^4 dependence of the Rayleigh scattering cross section, which explains why the sky is blue—blue light is scattered sixteen times more strongly than red light, since it has twice the frequency.

The argument above also applies to the interaction of low-energy gluons with $Q\bar{Q}$ bound states such as the $\mathcal{J}/\psi$ or Υ. The Lagrangian is eqn (2.60) where $\mathbf{E}^2$ and $\mathbf{B}^2$ are replaced by their QCD analogues, $\mathbf{E}^A \cdot \mathbf{E}^A$ and $\mathbf{B}^A \cdot \mathbf{B}^A$. The scale a_0 is now the radius of the QCD bound state. The Lagrangian can be used to find the interaction energy of the $Q\bar{Q}$ state in nuclear matter. The ψ field is a colour singlet, so the only interaction with nuclear matter is via the the gluon fields. The forward scattering amplitude off a nucleon state is

$$\mathscr{A} = a_0^3 \langle N | c_E \mathbf{E}^A \cdot \mathbf{E}^A + c_B \mathbf{B}^A \cdot \mathbf{B}^A | N \rangle$$ (2.63)

Equation (2.63) is a non-perturbative matrix element of order Λ_{QCD}^2. It turns out that it can evaluated in terms of the nucleon mass and the quark momentum fraction measured in DIS [62]. The binding energy U of the $Q\bar{Q}$ state is related to $\mathscr{A}$ by

$$U = \frac{n\mathscr{A}}{2M_N},$$ (2.64)

where n is the number of nucleons per unit volume in nuclear matter. The $1/(2M_N)$ prefactor is because nucleon states in field theory are normalized to $2M_N$ rather than to 1, as in quantum mechanics. Just using dimensional analysis, with $n \sim \Lambda_{\mathrm{QCD}}^3$, $\mathscr{A} \sim a_0^3 \Lambda_{\mathrm{QCD}}^2$, and neglecting factors of two,

$$U = \frac{a_0^3 \Lambda_{\mathrm{QCD}}^5}{M_N}.$$ (2.65)

With $a_0 \sim 0.2 \times 10^{-15}$ m for the $\mathcal{J}/\psi$, and $\Lambda_{\mathrm{QCD}} \sim 350$ MeV, the binding energy is $U \sim 5$ MeV.

2.4.8 Low Energy Weak Interactions

The classic example of an EFT is the Fermi theory of low-energy weak interactions. The full (UV) theory is the SM, and we can match onto the EFT by transitioning to a theory valid at momenta small compared to $M_{W,Z}$. Since the weak interactions are perturbative, the matching can be done order by order in perturbation theory.

The W boson interacts with quarks and leptons via the weak current:

$$j_W^\mu = V_{ij} \left(\bar{u}_i \gamma^\mu P_L d_j \right) + \left(\bar{\nu}_\ell \gamma^\mu P_L \ell \right), \tag{2.66}$$

where $u_i = u, c, t$ are up-type quarks, $d_j = d, s, b$ are down-type quarks, and V_{ij} is the CKM mixing matrix. There is no mixing matrix in the lepton sector because we are using neutrino flavour eigenstates, and neglecting neutrino masses.

The tree-level amplitude for semileptonic $b \to c$ decay from Figure 2.4 is

$$\mathscr{A} = \left(\frac{-ig}{\sqrt{2}} \right)^2 V_{cb} \left(\bar{c} \gamma^\mu P_L b \right) \left(\bar{\ell} \gamma^\nu P_L \nu_\ell \right) \left(\frac{-ig_{\mu\nu}}{p^2 - M_W^2} \right), \tag{2.67}$$

where $g/\sqrt{2}$ is the W coupling constant. For low-momentum transfers, $p \ll M_W$, we can expand the W propagator,

$$\frac{1}{p^2 - M_W^2} = -\frac{1}{M_W^2} \left(1 + \frac{p^2}{M_W^2} + \frac{p^4}{M_W^4} + \cdots \right), \tag{2.68}$$

giving different orders in the EFT expansion parameter p/M_W. Retaining only the first term gives

$$\mathscr{A} = \frac{i}{M_W^2} \left(\frac{-ig}{\sqrt{2}} \right)^2 V_{cb} \left(\bar{c} \gamma^\mu P_L b \right) \left(\bar{\ell} \gamma_\mu P_L \nu_\ell \right) + \mathcal{O}\left(\frac{1}{M_W^4} \right), \tag{2.69}$$

which is the same amplitude as that produced by the local Lagrangian

$$\mathscr{L} = -\frac{g^2}{2M_W^2} V_{cb} \left(\bar{c} \gamma^\mu P_L b \right) \left(\bar{\ell} \gamma_\mu P_L \nu_\ell \right) + \mathcal{O}\left(\frac{1}{M_W^4} \right). \tag{2.70}$$

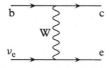

Figure 2.4 *Tree-level diagram for semileptonic $b \to c$ decay.*

Figure 2.5 *$b \to c$ vertex in the Fermi theory.*

Equation (2.70) is the lowest-order Lagrangian for semileptonic $b \to c$ decay in the EFT, and is represented by the vertex in Figure 2.5. It is usually written, for historical reasons, in terms of G_F

$$\frac{G_F}{\sqrt{2}} \equiv \frac{g^2}{8M_W^2} = \frac{1}{2v^2}, \tag{2.71}$$

where $v \sim 246$ GeV is the scale of electroweak symmetry breaking, so

$$\mathscr{L} = -\frac{4G_F}{\sqrt{2}} V_{cb} \left(\bar{c}\gamma^\mu P_L b\right) \left(\bar{\ell}\gamma^\mu P_L \nu_\ell\right). \tag{2.72}$$

Similarly, the μ decay Lagrangian is

$$\mathscr{L} = -\frac{4G_F}{\sqrt{2}} \left(\bar{\nu}_\mu \gamma^\mu P_L \mu\right) \left(\bar{e}\gamma^\mu P_L \nu_e\right). \tag{2.73}$$

The EFT Lagrangian eqn (2.72, 2.73) is the low-energy limit of the SM. The EFT no longer has dynamical W bosons, and the effect of W exchange in the SM has been included via dimension-six four-fermion operators. The procedure used here is referred to as 'integrating out' a heavy particle, the W boson.

The Lagrangian eqn (2.72, 2.73) has been obtained by expanding in p/M_W, i.e. by treating M_W as large compared to the other scales in the problem. Weak decays computed using eqn (2.72, 2.73) still retain the complete dependence on low-energy scales such as m_b, m_c and m_ℓ. Using eqn (2.72) gives the b lifetime,

$$\Gamma(b \to c\ell\bar{\nu}_\ell) = \frac{|V_{cb}|^2 G_F^2 m_b^5}{192\pi^3} f\left(\frac{m_c^2}{m_b^2}\right), \tag{2.74}$$

where we have neglected m_ℓ, and

$$f(\rho) = 1 - 8\rho + 8\rho^3 - \rho^4 - 12\rho^2 \ln\rho, \qquad \rho = \frac{m_c^2}{m_b^2}. \tag{2.75}$$

Equation (2.74) gives the full m_c/m_b dependence of the decay rate, but drops terms of order m_b/M_W and m_c/M_W. The full m_ℓ/m_b dependence can also be included by retaining m_ℓ in the decay rate calculation. The use of the EFT Lagrangian eqn (2.72) simplifies the calculation. We could have achieved the same simplification by computing Figure 2.4 in the SM, and expanding the amplitude using eqn (2.68). The true advantages of EFT show up in higher-order calculations, including radiative corrections from loop graphs, which cannot be computed by simply expanding the SM amplitude.

The Fermi Lagrangian can be used to compute electroweak scattering cross sections such as the neutrino cross section. Here we give a simple dimensional estimate of the cross section,

$$\sigma \sim \frac{1}{16\pi} \left(\frac{4G_F}{\sqrt{2}} \right)^2 E_{\text{CM}}^2 \sim \frac{1}{2\pi} G_F^2 E_{\text{CM}}^2, \tag{2.76}$$

where the G_F factor is from the weak interaction Lagrangian, $1/(16\pi)$ is two-body phase space, and E_{CM} gives σ the dimensions of a cross section. For neutrino scattering off a fixed target, $E_{CM}^2 = 2E_\nu M_T$, so neutrino cross sections grow linearly with the neutrino energy. Neutrino cross sections are weak as long as E_ν is much smaller than the electroweak scale.

Exercise 2.9 Compute the decay rate $\Gamma(b \to ce^-\overline{\nu}_e)$ with the interaction Lagrangian

$$L = -\frac{4G_F}{\sqrt{2}} V_{cb} (\overline{c}\gamma^\mu P_L b)(\overline{\nu}_e \gamma_\mu P_L e)$$

with $m_e \to 0$, $m_\nu \to 0$, but retaining the dependence on $\rho = m_c^2/m_b^2$. It is convenient to write the three-body phase space in terms of the variables $x_1 = 2E_e/m_b$ and $x_2 = 2E_\nu/m_b$.

2.4.9 M_W vs G_F

The weak interactions have two parameters, g and M_W, and the Fermi Lagrangian in eqn (2.72) depends only on the combination G_F in eqn (2.71). Higher-order terms in the expansion eqn (2.68) are of the form

$$-\frac{4G_F}{\sqrt{2}} \left[1 + \frac{p^2}{M_W^2} + \dots \right] = -\frac{2}{v^2} \left[1 + \frac{p^2}{M_W^2} + \dots \right] \tag{2.77}$$

so that the EFT momentum expansion is in powers of $\delta = p/M_W$, even though the first term in eqn (2.77) is $\propto 1/v^2$. The expansion breaks down for $p \sim M_W = gv/2$, which is smaller than $v \sim 246$ GeV.

Despite the theory having multiple scales M_W and v, we can still use our EFT power counting rules of Section 2.4.2. From the μ decay rate computed using eqn (2.73)

$$\Gamma(\mu \to e\nu_\mu \overline{\nu}_e) = \frac{G_F^2 m_\mu^5}{192\pi^3}, \tag{2.78}$$

Figure 2.6 *One-loop correction to $\phi\phi$ scattering from a ϕ^6 interaction.*

and the experimental value of the μ lifetime 2.197×10^{-6} s, we obtain $G_F \sim 1.16 \times 10^{-5}\,\text{GeV}^{-2}$. Using $G_F \sim 1/\Lambda^2$ gives $\Lambda \sim 300\,\text{GeV}$. This indicates that we have an EFT with a scale of order Λ. This is similar to the multipole expansion estimate for a.

We can then use the power counting arguments of Section 2.4.2. They show that the leading terms in the decay amplitude are single insertions of dimension-six operators, the next corrections are two insertions of dimension-six or one insertion of dimension-eight operators, etc. None of these arguments care about the precise value of Λ. They allow us to group terms in the expansion of similar size.

Dimension-eight corrections are p^2/Λ^2 suppressed. In μ-decay, this implies that dimension-eight corrections are suppressed by m_μ^2/Λ^2. The power counting estimate using either $\Lambda \sim M_W$ or $\Lambda \sim v$ shows that they are very small corrections. We can check that these corrections are small from *experiment*. The Lagrangian eqn (2.73) predicts observables such as the phase space distribution of μ decay events over the entire Dalitz plot, the polarization of the final e^-, etc. Comparing these predictions, which neglect dimension-eight contributions, with experiment provides a test that eqn (2.73) gives the correct decay amplitude. Very accurate experiments that are sensitive to deviations from the predictions of eqn (2.73), i.e. have an accuracy $m_\mu^2/M_W^2 \sim 10^{-6}$, can then be used to probe dimension-eight effects, and determine the scale M_W.

Historically, when the SM was developed, G_F was fixed from μ decay, but the values of M_W and M_Z were not known. Their values *were not needed* to apply the Fermi theory to low-energy weak interactions. The value of M_Z was determined by studying the energy dependence of parity violation in electron scattering through $\gamma - Z$ interference effects. This fixed the size of the dimension-eight p^2/M_Z^2 terms in the neutral current analogue of eqn (2.77), and determined the scale at which the EFT had to be replaced by the full SM, with dynamical gauge fields.

2.5 Loops

The real power of EFTs becomes apparent when computing loop corrections. There are several tricky points that must be understood before EFTs can be used at the loop level, which are explained in this section.

For simplicity consider an EFT of a scalar field ϕ, with a higher-dimension operator

$$\mathscr{L} = \mathscr{L}_{\mathscr{D} \leq 4} + \frac{c_6}{\Lambda^2} \frac{1}{6!} \phi^6. \tag{2.79}$$

The dimension-six operator gives a contribution to $\phi - \phi$ scattering from the graph in Figure 2.6,

$$\mathscr{A} = -\frac{c_6}{2\Lambda^2} \int \frac{d^4k}{(2\pi)^4} \frac{1}{k^2 - m_\phi^2}. \tag{2.80}$$

The EFT is valid for $k < \Lambda$, so we can use a momentum-space cutoff $\Lambda_c < \Lambda$. The scalar mass m_ϕ is much smaller than Λ_c, since ϕ is a particle in the EFT. Neglecting m_ϕ, the integral gives

$$\mathscr{A} \approx -\frac{c_6}{2\Lambda^2} \frac{\Lambda_c^2}{16\pi^2}. \tag{2.81}$$

The integral eqn (2.80) is quadratically divergent, which gives the quadratic cutoff dependence in eqn (2.81). Similarly, a dimension eight operator $\phi^4(\partial_\mu\phi)^2$ with coefficient c_8/Λ^4 has an interaction vertex k^2/Λ^4, and gives a contribution

$$\mathscr{A} = -\frac{c_8}{\Lambda^4} \int \frac{d^4k}{(2\pi)^4} \frac{k^2}{k^2 - m_\phi^2} \approx -\frac{c_8}{\Lambda^4} \frac{\Lambda_c^4}{16\pi^2}, \tag{2.82}$$

since the integral is quartically divergent.

The results of eqns (2.81, 2.82) lead to a violation of the power counting formula eqn (2.41), and the EFT expansion in powers of $1/\Lambda$ breaks down, since Λ_c is the same order as Λ. Loops with insertions of higher dimension operators give contributions of leading order in the $1/\Lambda$ expansion, which need to be resummed. We could try using $\Lambda_c \ll \Lambda$, but this turns out not to work. Firstly, Λ_c is an artificial scale that has been introduced, with no connection to any physical scale. In the end, all Λ_c dependence must cancel. For example, the weak interactions would require introducing a cutoff scale $m_b \ll \Lambda_c \ll M_W$ to keep the power divergences in eqns (2.81, 2.82) under control, and this would be an artificial scale that cancels in final results. Furthermore, cutoffs do not allow us to sum large logarithms, which is one of the main reasons why EFTs are used in the first place, since we are restricted to $\Lambda_c \ll \Lambda$. A cutoff has other problems as well, it violates important symmetries such as gauge invariance and chiral symmetry. In fact, no one has successfully performed higher-order log-resummation in EFTs with non-Abelian gauge interactions using a cutoff. Wilson proposed a set of axioms [83] for good regulators which are discussed in [23, Chapter 4].

Often, you will see discussions of EFTs where high momentum modes with $k > \Lambda_c$ are integrated out, and the cutoff is slowly lowered to generate an IR theory. While ideas like this were historically useful, this is not the way to think of an EFT, and it is not the way EFTs are actually used in practice.

Let us go back to a loop graph such as eqn (2.80), and for now, retain the cutoff Λ_c. In addition to the contribution shown in eqn (2.81), the loop graph also contains non-analytic terms in m_ϕ. In more complicated graphs, there would also be non-analytic terms in the external momentum p. Loop graphs have a complicated analytic structure in p and m_ϕ, with branch cuts, etc. The discontinuities across branch cuts from logs in loop graphs are related to the total cross section via the optical theorem. The non-analytic contributions are crucial to the EFT, and are needed to make sure the EFT respects unitarity. The non-analytic part of the integral can be probed by varying m_ϕ and p, and arises from $k \sim m_\phi, p$, i.e. loop momenta of order the physical scales in the EFT. For loop momenta of order Λ_c, $m_\phi, p \ll \Lambda_c$, we can expand in m_ϕ and p, and the integral gives analytic but Λ_c-dependent contributions such as eqn (2.81).

The high-momentum part of the integral is analytic in the IR variables, and has the same structure as amplitudes generated by local operators. This is the concept of locality mentioned in the introduction. Thus the integral has non-analytic pieces we want, plus local pieces that depend on Λ_c. The cutoff integral is an approximation to the actual integral in the full theory. Thus the local pieces computed as in eqn (2.81) are not the correct ones. In fact, in theories such as χPT where the UV theory is not related perturbatively to the EFT, the UV part of the integral is meaningless. Luckily, locality saves the day. The local pieces have the same structure as operators in the EFT Lagrangian, so they can be absorbed into the EFT Lagrangian coefficients. The EFT coefficients are then adjusted to make sure the EFT gives the correct S-matrix, a procedure referred to as 'matching'. The difference in UV structure of the full theory and the EFT is taken care of by the matching procedure. In the end, we only need the EFT to reproduce the non-analytic dependence on IR variables; the analytic dependence is absorbed into Lagrangian coefficients. An explicit calculation is given in Section 2.5.5.

To actually use EFTs in practice, we need a renormalization scheme that automatically implements the procedure above—i.e. it gives the non-analytic IR dependence without any spurious analytic contributions that depend on Λ_c. Such a scheme also maintains the EFT power counting, since no powers of a high scale Λ_c appear in the numerator of loop integrals, and cause the EFT expansion to break down. Dimensional regularization is a regulator that satisfies the required properties. It has the additional advantage that it maintains gauge invariance and chiral symmetry.

2.5.1   Dimensional Regularization

The basic integral we need is

$$
\mu^{2\epsilon} \int \frac{d^d k}{(2\pi)^d} \frac{\left(k^2\right)^a}{\left(k^2 - M^2\right)^b} = \frac{i\mu^{2\epsilon}}{(4\pi)^{d/2}} \frac{(-1)^{a-b}\Gamma(d/2+a)\Gamma(b-a-d/2)}{\Gamma(d/2)\Gamma(b)} \left(M^2\right)^{d/2+a-b}
$$

(2.83)

where $d = 4 - 2\epsilon$. The $\mu^{2\epsilon}$ prefactor arises from μ^ϵ factors in coupling constants, as in eqn (2.14). Equation (2.83) is obtained by analytically continuing the integral from values of a and b where it is convergent. Integrals with several denominators can be converted to eqn (2.83) by combining denominators using Feynman parameters.

The integral eqn (2.83) is then expanded in powers of ϵ. As an example,

$$I = \mu^{2\epsilon} \int \frac{d^d k}{(2\pi)^d} \frac{1}{(k^2 - M^2)^2} = \frac{i\mu^{2\epsilon}}{(4\pi)^{2-\epsilon}} \frac{\Gamma(\epsilon)}{\Gamma(2)} \left(M^2\right)^{-\epsilon},$$

$$= \frac{i}{16\pi^2} \left[\frac{1}{\epsilon} - \gamma + \log \frac{4\pi \mu^2}{M^2} + \mathcal{O}(\epsilon) \right], \tag{2.84}$$

where $\gamma = 0.577$ is Euler's constant. In the $\overline{\text{MS}}$ scheme, we make the replacement

$$\mu^2 = \bar{\mu}^2 \frac{e^\gamma}{4\pi}, \tag{2.85}$$

so that

$$I = \frac{i}{16\pi^2} \left[\frac{1}{\epsilon} + \log \frac{\bar{\mu}^2}{M^2} + \mathcal{O}(\epsilon) \right]. \tag{2.86}$$

The $1/\epsilon$ part, which diverges as $\epsilon \to 0$, is cancelled by a counterterm, leaving the renormalized integral

$$I + \text{c.t.} = \frac{i}{16\pi^2} \log \frac{\bar{\mu}^2}{M^2}. \tag{2.87}$$

The replacement eqn (2.85) removes $\log 4\pi$ and $-\gamma$ pieces in the final result.

There are several important features of dimensional regularization:

- $\bar{\mu}$ only appears as $\log \bar{\mu}$, and there are no powers of $\bar{\mu}$. The only source of $\bar{\mu}$ in the calculation is from powers of μ^ϵ in the coupling constants, and expanding in ϵ shows that only $\log \mu$ (and hence $\log \bar{\mu}$) terms occur.

- Scaleless integrals vanish,

$$\mu^{2\epsilon} \int \frac{d^d k}{(2\pi)^d} \frac{(k^2)^a}{(k^2)^b} = 0. \tag{2.88}$$

This follows using eqn (2.83) and taking the limit $M \to 0$. Since integrals in dimensional regularization are defined by analytic continuation, the limit $M \to 0$ is taken assuming $d/2 + a - b > 0$ so that the limit vanishes. Analytically continuing to $d/2 + a - b \le 0$, the integral remains 0. The vanishing of scaleless integrals plays a very important role in calculations using dimensional regularization.

- There are no power divergences. For example, the quadratically divergent integral

$$\mu^{2\epsilon} \int \frac{d^d k}{(2\pi)^d} \frac{1}{(k^2 - m^2)} = -\frac{i\mu^{2\epsilon}}{(4\pi)^{d/2}} \Gamma(-1+\epsilon) \left(m^2\right)^{1-\epsilon}$$

$$= \frac{i}{16\pi^2} \left[\frac{m^2}{\epsilon} + m^2 \log \frac{\bar{\mu}^2}{m^2} + m^2 + \mathcal{O}(\epsilon) \right], \tag{2.89}$$

Figure 2.7 *One loop correction to the Higgs mass from the* $-\lambda(H^\dagger H)^2$ *interaction.*

depends only on powers of the IR scale m. There is no dependence on any UV scale (such as a cutoff), nor any power-law dependence on $\bar{\mu}$. Similarly, the integral

$$\mu^{2\epsilon}\int\frac{d^d k}{(2\pi)^d}\frac{(k^2)}{(k^2-m^2)}=\frac{i\mu^{2\epsilon}}{(4\pi)^{d/2}}\frac{\Gamma(3-\epsilon)\Gamma(-2+\epsilon)}{\Gamma(2-\epsilon)\Gamma(1)}\left(m^2\right)^{2-\epsilon}$$

$$=\frac{i}{16\pi^2}\left[\frac{m^4}{\epsilon}+m^4\log\frac{\bar{\mu}^2}{m^2}+m^4+\mathcal{O}\left(\epsilon\right)\right],\qquad(2.90)$$

so the quartic divergence of the integral turns into the IR scale m to the fourth power.

The structure of the above integrals is easy to understand. Evaluating integrals using dimensional regularization is basically the same as evaluating integrals using the method of residues. Values of d, a, b are assumed such that the integrand vanishes sufficiently fast as $k \to \infty$ that the contour at infinity can be thrown away. The integrand is then given by the sum of residues at the poles. The location of the poles is controlled by the denominators in the integrand, which only depend on the physical scales in the low-energy theory, such as particle masses and external momenta. Dimensional regularization automatically gives what we want—it keeps all the dependence on the physical parameters, and throws away all unphysical dependence on high-energy scales. It is the simplest physical regulator, and the one used in all higher-order calculations.

2.5.2 No Quadratic Divergences

Let us look at the scalar graph Figure 2.7 which gives a correction to the Higgs mass in the SM,

$$\delta m_H^2 = -12\lambda\mu^{2\epsilon}\int\frac{d^d k}{(2\pi)^d}\frac{1}{(k^2-m_H^2)},\qquad(2.91)$$

where λ is the Higgs self-coupling. You will have heard repeatedly that Figure 2.7 gives a correction

$$\delta m_H^2 \propto \Lambda^2,\qquad(2.92)$$

to the Higgs mass that depends quadratically on the cutoff. This is supposed to lead to a naturalness problem for the SM, because the Higgs is so much lighter than Λ, which is taken to be at the GUT scale or Planck scale. The naturalness problem also goes by the names of hierarchy problem or fine-tuning problem.

The above argument for the naturalness problem is *completely bogus*. The regulator used for the SM is dimensional regularization, which respects gauge invariance. The actual value of the integral is eqn (2.89). Adding the renormalization counterterm cancels the $1/\epsilon$ piece, resulting in a correction to the Higgs mass

$$\delta m_H^2 = -\frac{12\lambda m_H^2}{16\pi^2}\left[\log\frac{m_H^2}{\bar{\mu}^2}+1\right],\qquad(2.93)$$

which is proportional to the Higgs mass. There is no quadratic mass shift proportional to the cutoff; there is no cutoff. The argument eqn (2.92) is based on a regulator that violates gauge invariance and the Wilson axioms, and which is never used for the SM in actual calculations. Bad regulators lead to bad conclusions.

Exercise 2.10 Compute the one-loop scalar graph Figure 2.7 with a scalar of mass m and interaction vertex $-\lambda\phi^4/4!$ in the $\overline{\text{MS}}$ scheme. Verify the answer is of the form eqn (2.93). The overall normalization will be different, because this exercise uses a real scalar field, and H in the SM is a complex scalar field.

2.5.3　Power Counting Formula

We can now extend the power counting formula eqn (2.41) to include loop corrections. If we consider a loop graph with an insertion of EFT vertices with coefficients of order $1/\Lambda^a$, $1/\Lambda^b$, etc. then any amplitude (including loops) will have the Λ dependence

$$\frac{1}{\Lambda^a}\frac{1}{\Lambda^b}\cdots=\frac{1}{\Lambda^{a+b+\dots}}\qquad(2.94)$$

simply from the product of the vertex factors. The discussion of Section 2.5.1 shows that the only scales which can occur in the numerator after doing the loop integrals are from poles in Feynman propagator denominators. These poles are at scales in the EFT, none of which is parametrically of order Λ. Thus there are no compensating factors of Λ in the numerator, i.e. the power of Λ is given by the vertex factors alone, so eqn (2.41), also holds for loop graphs.

Loop graphs in general are infinite, and the infinities ($1/\epsilon$ poles) are cancelled by renormalization counterterms. The EFT must include all operators necessary to absorb these divergences. From $n=\sum_i(\mathcal{D}_i-4)$, we see that if there is an operator with $\mathcal{D}>4$, we will generate operators with arbitrary high dimension. Thus an EFT includes all possible higher dimension operators consistent with the symmetries of the theory. Dimension-six operators are needed to renormalize graphs with two insertions of

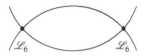

Figure 2.8 *Graph with two insertions of dimension-six operators, which requires a dimension-eight counterterm.*

dimension-five operators; dimension-eight operators are needed to renormalize graphs with two insertions of dimension-six operators (see Figure 2.8), etc. and we have to keep the entire expansion in higher dimension operators

$$\mathscr{L}_{\text{EFT}} = \mathscr{L}_{\mathscr{D} \leq 4} + \frac{\mathscr{L}_5}{\Lambda} + \frac{\mathscr{L}_6}{\Lambda^2} + \dots . \tag{2.95}$$

Even if we focus just on the low-dimension operators, it is understood that the higher dimension operators are still present. It also makes no sense to set their coefficients to zero. Their coefficients depend on $\bar{\mu}$, and on other choices such as the gauge-fixing term, etc. and so setting them to zero is a random unmotivated choice which will no longer hold at a different value of $\bar{\mu}$ unless the operator is forbidden by a symmetry.

2.5.4   An Explicit Computation

We now analyse a simple toy example, and explicitly compute a one-loop amplitude in the full theory, in the EFT, and discuss the matching between the two. The toy example is a two-scale integral that will be evaluated using EFT methods. The entire argument applies almost without change to a practical example, the derivation of the HQET Lagrangian to one-loop [66].

Consider the integral

$$I_F = g^2 \mu^{2\epsilon} \int \frac{d^d k}{(2\pi)^d} \frac{1}{(k^2 - m^2)(k^2 - M^2)} \tag{2.96}$$

where we will take $m \ll M$. M is the UV scale, and m is the IR scale. Integrals such as eqn (2.96) arise in loop calculations of graphs with intermediate heavy and light particles, such as in Figure 2.9. In eqn (2.96), we have set the external momenta to zero to get a simple integral which we can analyse to all orders in m/M.

The integral can be done exactly in $d = 4 - 2\epsilon$ dimensions

$$
\begin{aligned}
I_F &= g^2 \mu^{2\epsilon} \int \frac{d^d k}{(2\pi)^d} \frac{1}{(k^2 - m^2)(k^2 - M^2)} \\
&= \frac{i g^2}{16\pi^2} \left[\frac{1}{\epsilon} - \log \frac{M^2}{\bar{\mu}^2} + \frac{m^2}{M^2 - m^2} \log \frac{m^2}{M^2} + 1 \right],
\end{aligned} \tag{2.97}
$$

Figure 2.9 *A graph that gives a loop integral of the form eqn (2.96). The solid lines are light external fields. The thin dashed line is a light particle with mass m. The thick dashed line is a heavy particle of mass M that is not in the EFT.*

where we have switched to the $\overline{\text{MS}}$ scheme using eqn (2.85). I_F is a relatively simple integral because there are only two mass scales in the denominator. An integral with three denominators with unequal masses gives rise to dilogarithms.

The heavy particle M can be integrated out, as was done for the W boson. The heavy particle propagator is expanded in a power series,

$$\frac{1}{k^2 - M^2} = -\frac{1}{M^2}\left(1 + \frac{k^2}{M^2} + \frac{k^4}{M^4} + \ldots\right). \tag{2.98}$$

The loop graph in the EFT is a series of contributions, one from each term in eqn (2.98),

$$\begin{aligned}
I_{\text{EFT}} &= g^2 \mu^{2\epsilon} \int \frac{d^d k}{(2\pi)^d} \frac{1}{(k^2 - m^2)} \left[-\frac{1}{M^2} - \frac{k^2}{M^4} - \frac{k^4}{M^6} - \cdots \right] \\
&= \frac{ig^2}{16\pi^2 M^2}\left[-\frac{m^2}{\epsilon} + m^2 \log \frac{m^2}{\bar{\mu}^2} - m^2 \right] + \frac{ig^2}{16\pi^2 M^4}\left[-\frac{m^4}{\epsilon} + m^4 \log \frac{m^2}{\bar{\mu}^2} - m^4 \right] \\
&\quad + \frac{ig^2}{16\pi^2 M^6}\left[-\frac{m^6}{\epsilon} + m^6 \log \frac{m^2}{\bar{\mu}^2} - m^6 \right] + \ldots.
\end{aligned} \tag{2.99}$$

The series in eqn (2.99) is sufficiently simple in this example that we can sum it up,

$$I_{\text{EFT}} = \frac{ig^2}{16\pi^2}\left[-\frac{1}{\epsilon} \frac{m^2}{M^2 - m^2} + \frac{m^2}{M^2 - m^2} \log \frac{m^2}{\bar{\mu}^2} - \frac{m^2}{M^2 - m^2} \right], \tag{2.100}$$

to compare with I_F. However, it is best to think of I_{EFT} in the expanded form eqn (2.99), since the EFT is an expansion in powers of $1/M$.

There are several important points to note:

- The two results I_F and I_{EFT} are different. The order of integration and expansion matters.
- The $1/\epsilon$ terms do not agree, they are cancelled by counterterms which differ in the full and EFT. The two theories have *different* counterterms and hence *different* anomalous dimensions. In our example, the $1/\epsilon$ terms in eqn (2.99) give the

anomalous dimensions of the $1/M^2$, $1/M^4$, $1/M^6$, etc. operators. Each operator has its own anomalous dimension.

- The full theory and the EFT are independent theories adjusted to give the same S-matrix. We can use different regulators or gauge-fixing for the two theories.
- The $\log m^2$ terms, which are non-analytic in the IR scale, agree in the two theories. This is the part of I_F which *must* be reproduced in the EFT.
- The $\log M^2$ non-analytic terms in M are not present in the EFT integral. This must be the case, because in the EFT calculation, we integrated an expression which was a power series in $1/M$, and had no non-analytic terms in M.
- The difference between I_F and I_{EFT} is from the UV part of the integral, and is local in the IR mass scale m, so that $I_F - I_{\text{EFT}}$ is local (i.e. analytic) in m. This difference is called the matching contribution to the Lagrangian, and is included in the EFT result by absorbing it into shifts of the EFT Lagrangian coefficients.
- I_F has $\log M^2/m^2$ terms, which involve the ratio of the UV and IR scales. These logs can be summed using the RGE in the EFT.

Exercise 2.11 Compute I_F and I_{EFT} given in eqns (2.97, 2.99) in dimensional regularization in $\mathsf{d} = 4 - 2\epsilon$ dimensions. Both integrals have UV divergences, and the $1/\epsilon$ pieces are cancelled by counterterms. Determine the counterterm contributions $I_{F,\text{ct}}$, $I_{\text{EFT},\text{ct}}$ to the two integrals.

2.5.5 Matching

The infinite parts of I_F and I_{EFT} are cancelled by counterterms in the full theory and the EFT, respectively. The difference of the two renormalized integrals is the matching contribution

$$I_M = \left[I_F + I_{F,\text{c.t.}}\right] - \left[I_{\text{EFT}} + I_{\text{EFT},\text{c.t.}}\right]$$
$$= \frac{ig^2}{16\pi^2}\left[\left(\log\frac{\bar{\mu}^2}{M^2} + 1\right) + \frac{m^2}{M^2}\left(\log\frac{\bar{\mu}^2}{M^2} + 1\right) + \ldots\right]. \qquad (2.101)$$

The terms in parentheses are matching corrections to terms of order 1, order $1/M^2$, etc., from integrating out the heavy particle with mass M. They are analytic in the IR scale m. In our simple toy example, the $(m^2/M^2)^r$ corrections are corrections to the coefficient of the χ^4 operator, where χ is the external field in Figure 2.9. If the mass m is generated from a $\lambda\phi^4/4!$ interaction when a light field ϕ gets a vacuum expectation value $\langle\phi\rangle = v$, $m^2 = \lambda v^2/3$, then we can treat m^2 as $\lambda\phi^2/3$, and the series eqn (2.101) is an expansion in $\chi^4(\phi^2)^r$ operators of increasing dimension. For this reason, we refer to the $1/M$ expansion as being in operators of increasing dimension.

The logarithm of the ratio of IR and UV scales m and M can be written as

$$\log \frac{m^2}{M^2} = \underbrace{-\log \frac{M^2}{\bar{\mu}^2}}_{\text{matching}} + \underbrace{\log \frac{m^2}{\bar{\mu}^2}}_{\text{EFT}}, \tag{2.102}$$

where the scales have been separated using $\bar{\mu}$. The first piece is in the matching condition eqn (2.101), and the second is the EFT result eqn (2.100). We have separated a two-scale calculation into two one-scale calculations. A single scale integral is far easier to compute than a multi-scale integral, so the two-step calculation is much easier to do in practice.

Exercise 2.12 Compute $I_M \equiv \left(I_F + I_{F,\text{ct}}\right) - \left(I_{\text{EFT}} + I_{\text{EFT,ct}}\right)$ and show that it is analytic in m.

2.5.6 Summing Large Logs

The full theory result I_F has $\log M^2/m^2$ terms, depending on the ratio of a UV and an IR scale. At higher orders, we get additional powers of the log,

$$\left[\frac{g^2}{16\pi^2} \log \frac{M^2}{m^2} \right]^n. \tag{2.103}$$

If $M \gg m$, perturbation theory can break down when $g^2/(16\pi^2)\log M^2/m^2 \sim 1$. QCD perturbation theory often breaks down because of such logs, and it is necessary to sum these corrections.

In the EFT approach, I_F has been broken into two pieces, the matching I_M and the EFT result I_{EFT}. I_M only involves the high scale M, and logs in I_M depend on the ratio $M/\bar{\mu}$. These logs are not large if we choose $\bar{\mu} \sim M$. I_M can be computed in perturbation theory with $\bar{\mu} \sim M$, and perturbation theory is valid as long as $g^2/(16\pi^2)$ is small, a much weaker condition than requiring $g^2/(16\pi^2)\log M^2/m^2$ to be small. Similarly, I_{EFT} only involves the scale m, and logs in I_{EFT} are logs of the ratio $m/\bar{\mu}$. The EFT logs are not large if we choose $\bar{\mu} \sim m$. Thus we can compute I_M and I_{EFT} if we use two different $\bar{\mu}$ values. The change in $\bar{\mu}$ is accomplished by using the renormalization group equations in the EFT.

2.5.7 A Better Matching Procedure

While we argued that single-scale integrals were much easier to evaluate than multi-scale ones, the way we computed I_M as the difference $I_F - I_{\text{EFT}}$ still required first computing the multi-scale integral I_F. And if we know I_F, then essentially we have the answer we want anyway. Why bother with constructing an EFT in the first place?

It turns out there is a much simpler way to compute the matching that does not rely on first computing I_F. I_F and I_{EFT} both contain terms non-analytic in the IR scale, but the difference I_M is analytic in m,

$$\underbrace{I_M(m)}_{\text{analytic}} = \underbrace{I_F(m)}_{\text{non-analytic}} - \underbrace{I_{\text{EFT}}(m)}_{\text{non-analytic}} . \tag{2.104}$$

Therefore, we can compute I_M by expanding $I_F - I_{\text{EFT}}$ in an expansion in the IR scale m. This drops the non-analytic pieces, but we know they cancel in $I_F - I_{\text{EFT}}$.

The expansion of I_F is

$$I_F^{(\text{exp})} = g^2 \mu^{2\epsilon} \int \frac{d^d k}{(2\pi)^d} \frac{1}{k^2 - M^2} \left[\frac{1}{k^2} + \frac{m^2}{k^4} + \dots \right]. \tag{2.105}$$

The expansion of I_{EFT} is

$$I_{\text{EFT}}^{(\text{exp})} = g^2 \mu^{2\epsilon} \int \frac{d^d k}{(2\pi)^d} \left[\frac{1}{k^2} + \frac{m^2}{k^4} + \dots \right] \left[-\frac{1}{M^2} - \frac{k^2}{M^4} - \dots \right]. \tag{2.106}$$

Both $I_F^{(\text{exp})}$ and $I_{\text{EFT}}^{(\text{exp})}$ have to be integrated term by term. The expansions $I_F^{(\text{exp})}$ and $I_{\text{EFT}}^{(\text{exp})}$ drop non-analytic terms in m, and do not sum to give I_F and I_{EFT}. However, the non-analytic terms in m cancel in the difference, so $I_F^{(\text{exp})} - I_{\text{EFT}}^{(\text{exp})}$ does sum to give I_M.

Non-analytic terms in dimensional analysis arise from contributions of the form

$$\frac{1}{\epsilon} m^\epsilon = \frac{1}{\epsilon} + \log m + \dots \tag{2.107}$$

in integrals done using dimensional regularization. In eqns (2.105, 2.106), we first expand in the IR scale m, and then expand in ϵ. In this case,

$$\frac{1}{\epsilon} m^\epsilon = \frac{1}{\epsilon} \left[m^\epsilon \Big|_{m=0} + \epsilon m^{\epsilon-1} \Big|_{m=0} + \dots \right]. \tag{2.108}$$

In dimensional regularization, the $m = 0$ limit of all the terms in the square brackets vanishes. Expanding in m sets all non-analytic terms in m to zero.

$I_{\text{EFT}}^{(\text{exp})}$ has to be integrated term by term. Each term is a scaleless integral, and vanishes. For example the first term in the product is

$$g^2 \mu^{2\epsilon} \int \frac{d^d k}{(2\pi)^d} \left[\frac{1}{k^2} \right] \left[-\frac{1}{M^2} \right] = -\frac{1}{M^2} g^2 \mu^{2\epsilon} \int \frac{d^d k}{(2\pi)^d} \frac{1}{k^2} = 0. \tag{2.109}$$

This is not an accident of our particular calculation, but completely general. I_{EFT} was given by expanding the integrand of I_F in inverse powers of the UV scale M. $I_{\text{EFT}}^{(\text{exp})}$ is given by taking the result and expanding the integrand in powers of the IR scale m. The resulting integrand has all scales expanded out, and so is scaleless and vanishes. $I_F^{(\text{exp})}$, on the other hand, now only depends on the UV scale M; the IR scale m has been expanded out. Integrating term by term reproduces eqn (2.101) for the matching integral I_M. Thus the matching is given by evaluating I_F with all IR scales expanded out. This is a much easier way to compute I_M than computing I_F and I_{EFT} and taking the difference.

Exercise 2.13 Compute $I_F^{(\text{exp})}$, i.e. I_F with the IR m scale expanded out

$$I_F^{(\text{exp})} = -i\mu^{2\epsilon} \int \frac{\mathrm{d}^d k}{(2\pi)^d} \frac{1}{(k^2 - M^2)} \left[\frac{1}{k^2} + \frac{m^2}{k^4} + \cdots \right].$$

Note that the first term in the expansion has a $1/\epsilon$ UV divergence, and the remaining terms have $1/\epsilon$ IR divergences.

Exercise 2.14 Compute $I_F^{(\text{exp})} + I_{F,\text{ct}}$ using $I_{F,\text{ct}}$ determined in Exercise 2.11. Show that the UV divergence cancels, and the remaining $1/\epsilon$ IR divergence is the same as the UV counterterm $I_{\text{EFT},ct}$ in the EFT.

Something remarkable has happened. We have taken I_F, and expanded term by term in inverse powers of $1/M$, i.e. by assuming $k \ll M$, to get I_{EFT}. Then we have taken the original I_F and expanded term by term in powers of m, i.e. by assuming $k \gg m$, to get $I_F^{(\text{exp})} = I_M$. The sum of the two contributions is exactly the original integral I_F. Adding two different expansions of the same integrand recovers the original result, not *twice* the original result. The agreement is exact. We might worry that we have counted the region $m \ll k \ll M$ in both integrals. But this double-counting region is precisely $I_{\text{EFT}}^{(\text{exp})}$, and vanishes in dimensional regularization. It does not vanish with other regulators, such as a cutoff. We can understand why the EFT method works by using the analogy of dimensional regularization with integration using the method of residues. The I_F integrand has UV poles at M and IR poles at m. Expanding out in $1/M$ to get I_{EFT} leaves only the IR poles. Expanding out in m leaves only the UV poles in I_M. The sum of the two has all poles, and gives the full result.

Dimensional regularized integrals are evaluated with k set by the physical scales in the problem. There are no artificial scales as in a cutoff regulator that lead to spurious power divergences which have to be carefully subtracted away.

The method of regions [13] is a very clever procedure for evaluating Feynman integrals that is closely related to the above discussion. We find momentum regions which lead to poles in the integrand, expand in a power series in each region, and integrate term-by-term using dimensional regularization. Adding up the contributions of all the momentum regions gives the original integrand. In our example, the two regions were the hard region $k \sim M$, and the soft region $k \sim m$. The method of regions provides useful information to formulate an EFT, but it is not the same as an EFT. In an EFT, we have a Lagrangian, and the EFT amplitudes are given by computing graphs using Feynman rules derived from the Lagrangian. We cannot add or subtract modes depending on which momentum region contributes to an EFT graph. For example, in HQET, graphs get contributions from momenta of order m_b, and of order m_c. Nevertheless, HQET only has a single gluon field, not separate ones for each scaling region. In the method of regions, the contribution of different regions can depend on how loop momenta are routed in a Feynman graph, though the total integral given by summing all regions remains unchanged. In an EFT, the Lagrangian and Feynman rules do not depend on the momentum routing used.

2.5.8   UV and IR Divergences

Let us look in more detail at the $1/\epsilon$ terms. The original integral I_F can have both UV and IR divergences. In our example, it only has a UV divergence. The terms in I_{EFT} are expanded in k^2/M^2 and become more and more UV divergent. The terms in $I_F^{(\text{exp})}$ are expanded in m^2/k^2 and become more and more IR divergent. Dimensional regularization regulates both the UV and IR divergences. It will be useful to separate the divergences into UV and IR, and label them by ϵ_{UV} or ϵ_{IR}. In reality, there is only one $\epsilon = \epsilon_{\text{UV}} = \epsilon_{\text{IR}}$ given by $\epsilon = (4-d)/2$. At higher loops, we must be careful about mixed divergences which are the product of IR and UV divergences.

The log divergent (in $d=4$) scaleless integral vanishes

$$\int \frac{d^d k}{(2\pi)^d} \frac{1}{k^4} = 0. \tag{2.110}$$

It is both UV and IR divergent, and can be split into UV divergent and IR divergent integrals

$$\int \frac{d^d k}{(2\pi)^d} \frac{1}{k^4} = \int \frac{d^d k}{(2\pi)^d} \left[\frac{1}{k^2(k^2-m^2)} - \frac{m^2}{k^4(k^2-m^2)} \right], \tag{2.111}$$

by introducing an arbitrary mass scale m. The first term is UV divergent, and the second is IR divergent. Using $\epsilon_{\text{UV}}, \epsilon_{\text{IR}}$, and evaluating the pieces, eqn (2.111) becomes

$$\int \frac{d^d k}{(2\pi)^d} \frac{1}{k^4} = \frac{i}{16\pi^2} \left[\frac{1}{\epsilon_{\text{UV}}} - \frac{1}{\epsilon_{\text{IR}}} \right] = 0. \tag{2.112}$$

Log divergent scaleless integrals vanish because of the cancellation of $1/\epsilon_{\text{UV}}$ with $1/\epsilon_{\text{IR}}$. Power law divergent scaleless integrals simply vanish, and do not produce $1/\epsilon$ poles, e.g.

$$\int \frac{d^d k}{(2\pi)^d} \frac{1}{k^2} = 0, \qquad\qquad \int \frac{d^d k}{(2\pi)^d} 1 = 0, \tag{2.113}$$

so there are no quadratic or quartic divergences in dimensional regularization.

Let us go back to our matching example. I_F and I_{EFT} have the same IR behaviour, because the EFT reproduces the IR of the full theory. Now consider a particular term in $I_F^{(\text{exp})}$ with coefficient m^r,

$$I_F^{(\text{exp})}(m) = \sum_r m^r I_F^{(r)}. \tag{2.114}$$

We have expanded out the IR scale m, so there can be IR divergences which would otherwise have been regulated by m. The integral is a single scale integral depending only on M, and has the form

$$I_F^{(r)} = \frac{A^{(r)}}{\epsilon_{\text{UV}}} + \frac{B^{(r)}}{\epsilon_{\text{IR}}} + C^{(r)}, \tag{2.115}$$

where $A^{(r)}$ is the UV divergence, $B^{(r)}$ is the IR divergence, and $C^{(r)}$ is the finite part. For example from eqn (2.105)

$$I_F^{(0)} = g^2 \mu^{2\epsilon} \int \frac{\mathrm{d}^d k}{(2\pi)^d} \frac{1}{k^2 - M^2} \frac{1}{k^2} = \frac{ig^2}{16\pi^2} \left[\frac{1}{\epsilon_{\text{UV}}} + \log \frac{\bar{\mu}^2}{M^2} + 1 \right],$$

$$I_F^{(2)} = g^2 \mu^{2\epsilon} \int \frac{\mathrm{d}^d k}{(2\pi)^d} \frac{1}{k^2 - M^2} \frac{1}{k^4} = \frac{ig^2}{16\pi^2} \frac{1}{M^2} \left[\frac{1}{\epsilon_{\text{IR}}} + \log \frac{\bar{\mu}^2}{M^2} + 1 \right], \tag{2.116}$$

so that

$$A^{(0)} = \frac{ig^2}{16\pi^2}, \qquad\qquad A^{(2)} = 0,$$

$$B^{(0)} = 0, \qquad\qquad B^{(2)} = \frac{ig^2}{16\pi^2} \frac{1}{M^2},$$

$$C^{(0)} = \frac{ig^2}{16\pi^2} \left[\log \frac{\bar{\mu}^2}{M^2} + 1 \right], \qquad C^{(2)} = \frac{ig^2}{16\pi^2} \frac{1}{M^2} \left[\log \frac{\bar{\mu}^2}{M^2} + 1 \right]. \tag{2.117}$$

Now look at the terms in $I_{\text{EFT}}^{(\text{exp})}$,

$$I_{\text{EFT}}^{(\text{exp})}(m) = \sum_r m^r I_{\text{EFT}}^{(r)}. \tag{2.118}$$

$I_{\text{EFT}}^{(\text{exp})}$ is a scaleless integral, and vanishes. However, we can still pick out the log divergent terms, and write 0 in the form eqn (2.112). In general, we have

$$I_{\text{EFT}}^{(r)} = -\frac{B^{(r)}}{\epsilon_{\text{UV}}} + \frac{B^{(r)}}{\epsilon_{\text{IR}}} = 0, \tag{2.119}$$

and there is no finite piece, since the integral vanishes. $B^{(r)}$ is the *same* as in eqn (2.115), because the two integrals have the same IR divergence, so the $1/\epsilon_{\text{IR}}$ terms must agree.

In our example, from eqn (2.106),

$$I_{\text{EFT}}^{(0)} = 0 \text{ since there is no } m^0/k^4 \text{ term,}$$

$$I_{\text{EFT}}^{(2)} = -g^2 \frac{1}{M^2} \mu^{2\epsilon} \int \frac{\mathrm{d}^d k}{(2\pi)^d} \frac{1}{k^4} = -\frac{i}{16\pi^2} \frac{1}{M^2} \left[\frac{1}{\epsilon_{\text{UV}}} - \frac{1}{\epsilon_{\text{IR}}} \right], \tag{2.120}$$

so that

$$B^{(0)} = 0, \qquad\qquad B^{(2)} = \frac{ig^2}{16\pi^2} \frac{1}{M^2}, \qquad (2.121)$$

which agree with $B^{(0)}$ and $B^{(2)}$ in eqn (2.117), as expected. The renormalized expression for $I_F^{(r)}$ is given by adding the full theory counterterm $-A^{(r)}/\epsilon_{\rm UV}$,

$$I_F^{(r)} + I_{F,\text{c.t.}}^{(r)} = \frac{B^{(r)}}{\epsilon_{\rm IR}} + C^{(r)}, \qquad (2.122)$$

and the renormalized expression for $I_{\rm EFT}^{(r)}$ by adding the EFT counterterm $B^{(r)}/\epsilon_{\rm UV}$,

$$I_{\rm EFT}^{(r)} + I_{\rm EFT,c.t.}^{(r)} = \frac{B^{(r)}}{\epsilon_{\rm IR}}. \qquad (2.123)$$

Note that one does not cancel IR divergences by counterterms. The difference of eqn (2.122) and eqn (2.123) is

$$I_M^{(r)} = \left[I_F^{(r)} + I_{F,\text{c.t.}}^{(r)} \right] - \left[I_{\rm EFT}^{(r)} + I_{\rm EFT,c.t.}^{(r)} \right] = C^{(r)}. \qquad (2.124)$$

The IR divergences cancel between the two, leaving only the finite part $C^{(r)}$.

Exercise 2.15 Compute $I_{\rm EFT}^{(\exp)}$, i.e. $I_{\rm EFT}$ with the IR m scale expanded out. Show that it is a scaleless integral, which vanishes. Using the known UV divergence from Exercise 2.11, write it in the form

$$I_{\rm EFT}^{(\exp)} = -B \frac{1}{16\pi^2} \left[\frac{1}{\epsilon_{\rm UV}} - \frac{1}{\epsilon_{\rm IR}} \right],$$

and show that the IR divergence agrees with that in $I_F^{(\exp)} + I_{F,ct}$.

Exercise 2.16 Compute $\left(I_F^{(\exp)} + I_{F,\text{ct}} \right) - \left(I_{\rm EFT}^{(\exp)} + I_{\rm EFT,ct} \right)$ and show that all the $1/\epsilon$ divergences (both UV and IR) cancel, and the result is equal to I_M found in Exercise 2.12.

This gives the prescription for the matching condition: Expand I_F in IR scales, and keep only the finite part. However, we have obtained some new information. The anomalous dimension in the full theory is proportional to the UV counterterm $-A$. The anomalous dimension in the EFT is proportional to the EFT counterterm B, which can be different from A. By the argument just given, B is the IR divergence of the full theory. By using an EFT, we have converted IR divergences (i.e. the $\log m$ terms) in the full theory into UV divergences in the EFT. This converts IR logs into UV logs, which can be summed using the renormalization group. In the EFT, $\log M/m$ terms in the full theory are converted to $\log \bar{\mu}/m$ terms, since $M \to \infty$ in the EFT. These are summed by the EFT renormalization group equations.

Exercise 2.17 Make sure you understand why you can compute I_M simply by taking $I_F^{(\exp)}$ and dropping all $1/\epsilon$ terms (both UV and IR).

Finally, if we do the EFT calculation without expanding out the IR scale m, then the EFT calculation is no longer IR divergent and can have a finite part,

$$I_{\mathrm{EFT}}^{(r)} = -\frac{B^{(r)}}{\epsilon_{\mathrm{UV}}} + D^{(r)}, \tag{2.125}$$

where the UV divergence remains the same as before. The finite part of the full amplitude I_F has been split into $C^{(r)} + D^{(r)}$, with $C^{(r)}$ from the matching and $D^{(r)}$ from the EFT. In our example,

$$D^{(0)} = 0,$$

$$D^{(2)} = \frac{ig^2}{16\pi^2} \frac{1}{M^2} \left[\log \frac{m^2}{\mu^2} - 1 \right], \tag{2.126}$$

from eqn (2.99).

2.5.9 Summary

It has taken a while to get to the final answer, but we can now summarize our results. The general procedure is simple to state:

- Compute the full theory graphs expanding in all IR scales. The integrals are single-scale integrals involving only the high scale M. Drop the $1/\epsilon$ terms from both UV and IR divergences. This gives $C^{(r)}(\mu)$. To avoid large logarithms, μ should be chosen to be of order the high scale M. The starting values of the EFT coefficient at the high scale are $C^{(r)}(\mu \sim M)$.
- Evolve the EFT down from $\mu \sim M$ to a low scale $\mu \sim m$ using the renormalization group equations in the EFT. This sums logs of the ratios of scales, $\log M/m$.
- Compute in the EFT using $\mu \sim m$. There are no large logs in the EFT calculation.
- Combine the pieces to get the final result.

One computation has been broken up into several much simpler calculations, each of which involves a single scale.

Exercise 2.18 Compute the QED on-shell electron form factors $F_1(q^2)$ and $F_2(q^2)$ expanded to first order in q^2/m^2 using dimensional regularization to regulate the

IR and UV divergences. This gives the one-loop matching to heavy-electron EFT. Note that it is much simpler to *first* expand and then do the Feynman parameter integrals. A more difficult version of the problem is to compute the on-shell quark form factors in QCD, which gives the one-loop matching to the HQET Lagrangian. For help with the computation, see [66]. Note that in the non-Abelian case, using background field gauge is helpful because the amplitude respects gauge invariance on the external gluon fields.

Exercise 2.19

The SCET matching for the vector current $\overline{\psi}\gamma^\mu\psi$ for the Sudakov form factor is a variant of the previous problem. Compute $F_1(q^2)$ for on-shell massless quarks, in pure dimensional regularization with $Q^2 = -q^2 \neq 0$. Here Q^2 is the big scale, whereas in the previous problem q^2 was the small scale. The spacelike calculation $Q^2 > 0$ avoids having to deal with the $+i0^+$ terms in the Feynman propagator which lead to imaginary parts. The timelike result can then be obtained by analytic continuation.

Exercise 2.20

Compute the SCET matching for time-like q^2, by analytically continuing the previous result. Be careful about the sign of the imaginary parts.

2.5.10   RG Improved Perturbation Theory

We have mentioned several times that renormalization group improved perturbation theory is better than fixed order perturbation theory. To understand the difference, consider an example where an operator coefficient $c(\mu)$ satisfies the one-loop renormalization group equation

$$\mu\frac{\mathrm{d}}{\mathrm{d}\mu}c(\mu) = \left[\gamma_0\frac{g^2(\mu)}{16\pi^2} + \mathcal{O}\left(\frac{g^2(\mu)}{16\pi^2}\right)^2\right]c(\mu), \qquad (2.127)$$

where γ_0 is a constant. The evolution of $g(\mu)$ is given by the β-function equation

$$\mu\frac{\mathrm{d}g(\mu)}{\mathrm{d}\mu} = -b_0\frac{g^3(\mu)}{16\pi^2} + \mathcal{O}\left[\frac{g^5(\mu)}{(16\pi^2)^2}\right]. \qquad (2.128)$$

As long as $g^2(\mu)/(16\pi^2)$ is small, we can integrate the ratio of eqn (2.127) and eqn (2.128) to get

$$\frac{c(\mu_1)}{c(\mu_2)} = \left[\frac{\alpha_s(\mu_1)}{\alpha_s(\mu_2)}\right]^{-\gamma_0/(2b_0)}, \qquad \alpha_s(\mu) = \frac{g^2(\mu)}{4\pi}. \qquad (2.129)$$

Integrating eqns (2.127, 2.128) term by term, or equivalently, expanding eqn (2.129) gives

Figure 2.10 *Graph contributing to the anomalous dimension of O_1 and O_2. We have to sum over gluon exchange between all possible pairs of lines, and also include wavefunction corrections.*

$$\frac{c(\mu_1)}{c(\mu_2)} = 1 + \gamma_0 \frac{\alpha_s(\mu_1)}{4\pi} \log \frac{\mu_1}{\mu_2} - \frac{1}{2}\gamma_0(2b_0 - \gamma_0) \left[\frac{\alpha_s(\mu_1)}{4\pi} \log \frac{\mu_1}{\mu_2}\right]^2$$
$$+ \frac{1}{6}\gamma_0(2b_0 - \gamma_0)(4b_0 - \gamma_0) \left[\frac{\alpha_s(\mu_1)}{4\pi} \log \frac{\mu_1}{\mu_2}\right]^3 + \dots \qquad (2.130)$$

The renormalization group sums the leading log (LL) series $\alpha_s^n \log^n$, as can be seen from eqn (2.130). We can show that the higher-order corrections in eqns (2.127, 2.128) do not contribute to the leading log series, since they are suppressed by $\alpha_s/(4\pi)$ without a log. Including the two-loop terms gives the next-to-leading-log (NLL) series $\alpha_s^n \log^{n-1}$, the three-loop terms give the NNLL series $\alpha_s^n \log^{n-2}$, etc.

The change in $g(\mu)$ and $c(\mu)$ can be very large, even if $\alpha_s(\mu)$ is small. For example, in the strong interactions, $\alpha_s(M_Z) \approx 0.118$ and $\alpha_s(m_b) \approx 0.22$, a ratio of about two. Even though both values of α_s are small, weak decay operator coefficients also change by about a factor of two between M_Z and m_b.

2.5.10.1 *Operator mixing*

Summing logs using the renormalization group equations allows us to include operator mixing effects in a systematic way. This is best illustrated by the simple example of non-leptonic weak $b \to c$ decays via the effective Lagrangian

$$L = -\frac{4G_F}{\sqrt{2}} V_{cb} V_{ud}^* \, (c_1 O_1 + c_2 O_2), \qquad (2.131)$$

where the two operators and their tree-level coefficients at $\mu = M_W$ are

$$O_1 = \left(\bar{c}^\alpha \gamma^\mu P_L b_\alpha\right) \left(\bar{d}^\beta \gamma_\mu P_L u_\beta\right), \qquad c_1 = 1 + \mathcal{O}(\alpha_s), \qquad (2.132)$$
$$O_2 = \left(\bar{c}^\alpha \gamma^\mu P_L b_\beta\right) \left(\bar{d}^\beta \gamma_\mu P_L u_\alpha\right), \qquad c_2 = 0 + \mathcal{O}(\alpha_s), \qquad (2.133)$$

where α and β are color indices. Since the W boson is colour-singlet, only O_1 is produced by the tree-level graph. O_2 is generated by loop graphs involving gluons, which are suppressed by a power of α_s.

The renormalization group equations can be computed from the one-loop graph in Figure 2.10 [38],

$$\mu \frac{d}{d\mu} \begin{bmatrix} c_1 \\ c_2 \end{bmatrix} = \frac{\alpha_s}{4\pi} \begin{bmatrix} -2 & 6 \\ 6 & -2 \end{bmatrix} \begin{bmatrix} c_1 \\ c_2 \end{bmatrix}. \tag{2.134}$$

Exercise 2.21 Compute the anomalous dimension mixing matrix in eqn (2.134). Two other often-used bases are

$$Q_1 = (\bar{b}\gamma^\mu P_L c)(\bar{u}\gamma^\mu P_L d) \qquad\qquad Q_2 = (\bar{b}\gamma^\mu P_L T^A c)(\bar{u}\gamma^\mu P_L T^A d)$$

and

$$O_\pm = O_1 \pm O_2$$

So let

$$\mathcal{L} = c_1 O_1 + c_2 O_2 = d_1 Q_1 + d_2 Q_2 = c_+ O_+ + c_- O_-$$

and work out the transformation between the anomalous dimensions for $d_{1,2}$ and $c_{+,-}$ in terms of those for $c_{1,2}$.

The anomalous dimension matrix is not diagonal, which is referred to as operator mixing. In this simple example, the equations can be integrated by taking the linear combinations $c_\pm = c_1 \pm c_2$,

$$\mu \frac{d}{d\mu} \begin{bmatrix} c_+ \\ c_- \end{bmatrix} = \frac{\alpha_s}{4\pi} \begin{bmatrix} 4 & 0 \\ 0 & -8 \end{bmatrix} \begin{bmatrix} c_+ \\ c_- \end{bmatrix}, \tag{2.135}$$

which decouples the equations. The solution is

$$\frac{c_+(\mu_1)}{c_+(\mu_2)} = \left[\frac{\alpha(\mu_1)}{\alpha(\mu_2)}\right]^{-6/23}, \qquad\qquad \frac{c_-(\mu_1)}{c_-(\mu_2)} = \left[\frac{\alpha(\mu_1)}{\alpha(\mu_2)}\right]^{12/23}, \tag{2.136}$$

using eqn (2.129), with $b_0 = 11 - 2/3 n_f = 23/3$ and $n_f = 5$ dynamical quark flavours. With $\alpha_s(m_b) \sim 0.22$ and $\alpha_s(M_Z) \sim 0.118$,

$$\frac{c_+(m_b)}{c_+(M_W)} = 0.85, \qquad\qquad \frac{c_-(m_b)}{c_-(M_W)} = 1.38, \tag{2.137}$$

so that

$$c_1(m_b) \approx 1.12, \qquad\qquad c_2(m_b) \approx -0.27. \tag{2.138}$$

A substantial c_2 coefficient is obtained at low scales, even though the starting value is $c_2(M_W) = 0$.

Equation (2.130) for the general matrix case is

$$\mathbf{c}(\mu_1) = \left[1 + \gamma_0 \frac{\alpha_s(\mu_1)}{4\pi} \log\frac{\mu_1}{\mu_2} - \frac{1}{2}\gamma_0(2b_0 - \gamma_0)\left[\frac{\alpha_s(\mu_1)}{4\pi}\log\frac{\mu_1}{\mu_2}\right]^2 \right.$$

$$\left. + \frac{1}{6}\gamma_0(2b_0 - \gamma_0)(4b_0 - \gamma_0)\left[\frac{\alpha_s(\mu_1)}{4\pi}\log\frac{\mu_1}{\mu_2}\right]^3 + \dots \right]\mathbf{c}(\mu_2), \qquad (2.139)$$

where γ_0 is a matrix and $\mathbf{c}$ is a column vector. Equation (2.139) shows that $c_2(m_b)$ in eqn (2.138) is a leading-log term, even though it starts at $c_2(M_W) = 0$. In examples with operator mixing, it is difficult to obtain the leading-log series eqn (2.139) by looking at graphs in the full theory. The method used to sum the leading-log series in practice is to integrate anomalous dimensions in the EFT.

The above discussion of renormalization group equations and operator mixing also holds in general EFTs. The EFT Lagrangian is an expansion in higher dimension operators

$$\mathcal{L} = \mathcal{L}_{\mathscr{D} \leq 4} + \frac{1}{\Lambda}c_i^{(5)}O_i^{(5)} + \frac{1}{\Lambda^2}c_i^{(6)}O_i^{(6)} + \dots. \qquad (2.140)$$

The running of the coupling constants in $\mathcal{L}_{\mathscr{D} \leq 4}$ is given by the usual β-functions of the low-energy theory, e.g. by the QCD and QED β-functions. The other terms in $\mathcal{L}$ are higher-dimension operators, and their anomalous dimensions are computed in the same way as eqn (2.134) for the weak interactions. The additional piece of information we have is the EFT power counting formula. This leads to RGE equations of the form

$$\mu\frac{\mathrm{d}}{\mathrm{d}\mu}c_i^{(5)} = \gamma_{ij}^{(5)}c_j^{(5)},$$

$$\mu\frac{\mathrm{d}}{\mathrm{d}\mu}c_i^{(6)} = \gamma_{ij}^{(6)}c_j^{(6)} + \gamma_{ijk}c_j^{(5)}c_k^{(5)}, \qquad (2.141)$$

and in general

$$\mu\frac{\mathrm{d}}{\mathrm{d}\mu}c_i^{(D)} = \gamma_{ij_1j_2\dots j_r}c_{j_1}^{(D_1)}\dots c_{j_r}^{(D_r)}, \qquad (2.142)$$

with $D - 4 = \sum_i(D_i - 4)$, where the anomalous dimensions γ are functions of the coupling constants in $\mathcal{L}_{\mathscr{D} \leq 4}$. The renormalization group equations are *non-linear*. Graphs with two insertions of a dimension-five operator need a dimension-six counterterm, leading to the $c_j^{(5)}c_k^{(5)}$ term in the anomalous dimension for $c_i^{(6)}$, etc. In the presence of mass terms such as m_H^2, we also get mixing to $D - 4 < \sum_r(D_r - 4)$ operators, e.g.

$$\mu\frac{\mathrm{d}}{\mathrm{d}\mu}c_i^{(4)} = m_H^2\gamma_{ij}^{(6\to4)}c_j^{(6)} + \dots. \qquad (2.143)$$

as in SMEFT [53].

2.6   Field redefinitions and equations of motion

2.6.1   LSZ Reduction Formula

Experimentally observable quantities in field theory are S-matrix elements, whereas what we compute from the functional integral are correlation functions of quantum fields. The LSZ reduction formula relates the two. For simplicity, we discuss a theory with a scalar field $\phi(x)$. The momentum space Green's functions are defined by

$$G(q_1,\ldots,q_m;p_1,\ldots,p_n)$$
$$= \prod_{i=1}^{m}\int d^4y_i\, e^{iq_i\cdot y_i}\prod_{j=1}^{n}\int d^4x_j\, e^{-ip_j\cdot x_j}\,\langle 0|\,T\{\phi(y_1)\ldots\phi(y_m)\phi(x_1)\ldots\phi(x_n)\}\,|0\rangle \quad (2.144)$$

where the momenta p_i are incoming, and momenta q_i are outgoing, as shown in Figure 2.11. These Green's functions can be computed in perturbation theory using the usual Feynman diagram expansion. The ϕ propagator in Figure 2.12 is a special case of eqn (2.144),

$$D(p) = \int d^4x\, e^{ip\cdot x}\,\langle 0|\,T\{\phi(x)\phi(0)\}\,|0\rangle. \quad (2.145)$$

If the field $\phi(x)$ can produce a single particle state $|p\rangle$ with invariant mass m from the vacuum,

$$\langle p|\phi(x)|0\rangle \neq 0, \quad (2.146)$$

then the propagator $D(p)$ has a pole at $p^2 = m^2$,

$$D(p) \sim \frac{i\mathcal{R}}{p^2 - m^2 + i\epsilon} + \text{non-pole terms.} \quad (2.147)$$

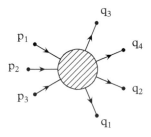

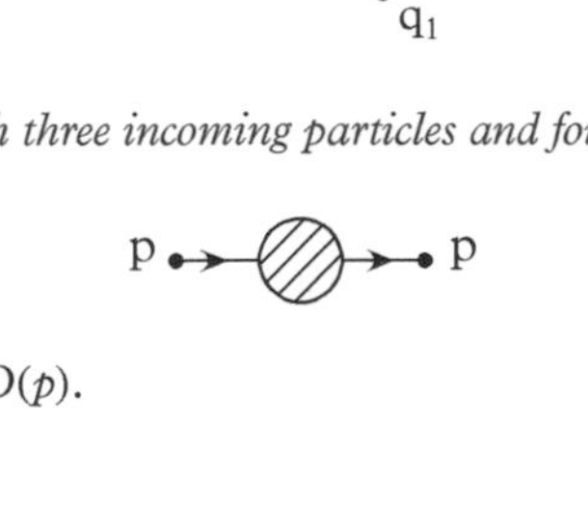

Figure 2.11 *Green's function with three incoming particles and four outgoing particles.*

Figure 2.12 *Two-point function $D(p)$.*

ϕ is called an interpolating field for $|p\rangle$. The wavefunction factor $\mathcal{R}$ is defined by

$$\lim_{\substack{p^2 \to m^2 \\ p^0 > 0}} \left(p^2 - m^2\right) D(p) \equiv i\mathcal{R}. \qquad (2.148)$$

$\mathcal{R}$ is finite, since $D(p)$, the renormalized propagator, is finite.

The S-matrix is computed from the Green's function by picking out the poles for each particle,

$$\lim_{\substack{q_i^2 \to m^2 \\ q_i^0 > 0}} \lim_{\substack{p_j^2 \to m^2 \\ p_j^0 > 0}} \prod_{i=1}^{m} \left(q_i^2 - m^2\right) \prod_{j=1}^{n} \left(p_j^2 - m^2\right) G(q_1, \ldots, q_m; p_1, \ldots, p_n)$$

$$= \prod_{i=1}^{m} \left(i\sqrt{\mathcal{R}_i}\right) \prod_{j=1}^{n} \left(i\sqrt{\mathcal{R}_j}\right) \, {}_{\text{out}}\langle q_1, \ldots, q_m | p_1, \ldots, p_n \rangle_{\text{in}}, \qquad (2.149)$$

i.e. the $n + m$ particle pole of the Green's function gives the S-matrix up to wavefunction normalization factors. Equation (2.149) is called the LSZ reduction formula [60]. The only complication for fermions and gauge bosons is that we have to contract with spinors $u(p, s), v(p, s)$ and polarization vectors $\epsilon_\mu^{(s)}(p)$.

The important feature of eqn (2.149) is that the derivation only depends on eqn (2.146), so that any interpolation field can be used. Particle states are given by the physical spectrum of the theory, and Green's functions are given by correlation functions of fields. S-matrix elements, which are the physical observables, depend on particle states, not fields. Fields and particles are not the same.

2.6.2 Field Redefinitions

It is now easy to see why field redefinitions do not change the S-matrix. The LSZ reduction formula does not care what field is used. To understand this in more detail, consider the functional integral

$$Z[\mathcal{J}] = \int D\phi \, e^{i\int L[\phi] + \mathcal{J}\phi}. \qquad (2.150)$$

The Green's functions

$$\langle 0 | T\{\phi(x_1) \ldots \phi(x_r)\} | 0 \rangle = \frac{\int D\phi \, \phi(x_1) \ldots \phi(x_r) \, e^{iS(\phi)}}{\int D\phi \, e^{iS(\phi)}}, \qquad (2.151)$$

are given by

$$\langle 0 | T\{\phi(x_1) \ldots \phi(x_r)\} | 0 \rangle = \frac{1}{Z[\mathcal{J}]} \frac{\delta}{i\delta\mathcal{J}(x_1)} \cdots \frac{\delta}{i\delta\mathcal{J}(x_r)} Z[\mathcal{J}] \bigg|_{\mathcal{J}=0}. \qquad (2.152)$$

Consider a local field redefinition,

$$\phi(x) = F[\phi'(x)], \tag{2.153}$$

such as

$$\phi(x) = \phi'(x) + c_1 \partial^2 \phi'(x) + c_2 \phi'(x)^3. \tag{2.154}$$

The field redefinition $F[\phi'(x)]$ can involve integer powers of ϕ and a finite number of derivatives. Then L' defined by

$$L[\phi(x)] = L[F[\phi'(x))] = L'[\phi'(x)], \tag{2.155}$$

is the new Lagrangian after the field redefinition eqn (2.153).

The functional integral Z' with the new field $\phi'(x)$ and Lagrangian L'

$$Z'[\mathcal{J}] = \int D\phi' \; e^{i\int L'[\phi'] + \mathcal{J}\phi'} = \int D\phi \; e^{i\int L'[\phi] + \mathcal{J}\phi}, \tag{2.156}$$

gives correlation functions of ϕ' computed using $L'[\phi']$, or equivalently, correlation functions of ϕ computed using $L'[\phi]$, since ϕ is a dummy integration variable and can be replaced by ϕ. The original functional integral eqn (2.150) under the change of variables eqn (2.153) becomes

$$Z[\mathcal{J}] = \int D\phi' \; \left|\frac{\delta F}{\delta \phi'}\right| e^{i\int L'[\phi'] + \mathcal{J}F[\phi']}. \tag{2.157}$$

The Jacobian $|\delta F/\delta \phi'|$ is unity in dimensional regularization, except for the special case of a fermionic chiral transformation, where there is an anomaly [31]. Neglecting anomalies, and dropping primes on the dummy variable ϕ' gives

$$Z[\mathcal{J}] = \int D\phi \; e^{i\int L'[\phi] + \mathcal{J}F[\phi]}. \tag{2.158}$$

Thus $Z[\mathcal{J}]$, which gives the Green's functions of ϕ computed using Lagrangian $L[\phi]$ by eqn (2.150), also gives the Green's functions of $F[\phi]$ computed using Lagrangian $L'[\phi]$. In contrast, $Z'[\mathcal{J}]$ gives the correlation functions of ϕ computed using the new Lagrangian $L'[\phi]$. The two correlation functions are different, so Green's functions change under a field redefinition. However, the S-matrix remains unchanged. $Z[\mathcal{J}]$ computes the S-matrix using Lagrangian $L'[\phi]$ and $F[\phi]$ as the interpolating field, by eqn (2.158). $Z'[\mathcal{J}]$ computes the S-matrix using Lagrangian $L'[\phi]$ and ϕ as the interpolating field, by eqn (2.156). The S-matrix does not care about the choice of interpolating field (i.e. field redefinition) as long as

$$\langle p|F[\phi]|0\rangle \neq 0, \tag{2.159}$$

so a field redefinition leaves the S-matrix unchanged.

In field theory courses, we study renormalizable Lagrangians with terms of dimension $\leqslant 4$. The only field redefinitions allowed are linear transformations,

$$\phi_i' = C_{ij}\,\phi_j. \tag{2.160}$$

These are used to put the kinetic term in canonical form,

$$\frac{1}{2}\partial_\mu \phi_i\,\partial^\mu \phi^i. \tag{2.161}$$

In an EFT, there is much more freedom to make field redefinitions, since the Lagrangian includes higher-dimensional operators. We make field redefinitions that respect the EFT power counting, e.g.

$$\phi \to \phi + \frac{1}{\Lambda^2}\phi^3 + \ldots \tag{2.162}$$

and work order by order in $1/\Lambda$. Field redefinitions are often used to put EFT Lagrangians in canonical form. The EFT Lagrangian is then given by matching from the full theory, followed by a field redefinition, so fields in the EFT are not the same as in the full theory.

2.6.3 Equations of Motion

A special case of field redefinitions is the use of equations of motion [37, 75]. Let $E[\phi]$ be the *classical* equation of motion

$$E[\phi] \equiv \frac{\delta S}{\delta \phi}. \tag{2.163}$$

For example, if

$$\mathscr{L} = \frac{1}{2}\partial_\mu \phi \partial^\mu \phi - \frac{1}{2}m^2 \phi^2 - \frac{1}{4!}\lambda \phi^4, \tag{2.164}$$

$E[\phi]$ is

$$E[\phi] = -\partial^2 \phi(x) - m^2 \phi(x) - \frac{1}{3!}\lambda \phi^3(x). \tag{2.165}$$

Let θ be an operator with a factor of the classical equation of motion,

$$\theta[\phi] = F[\phi]E[\phi] = F[\phi]\frac{\delta S}{\delta \phi}, \tag{2.166}$$

and consider the functional integral

$$Z[\mathcal{J},\widetilde{\mathcal{J}}] = \int D\phi \; e^{i\int L[\phi]+\mathcal{J}\phi+\widetilde{\mathcal{J}}\theta[\phi]}.$$

(2.167)

The correlation function

$$\langle 0|T\{\phi(x_1)\ldots\phi(x_n)\theta(x)\}|0\rangle$$

(2.168)

with one insertion of the equation-of-motion operator θ is given by evaluating

$$\langle 0|T\{\phi(x_1)\ldots\phi(x_n)\theta(x)\}|0\rangle = \frac{1}{Z[\mathcal{J},\widetilde{\mathcal{J}}]} \frac{\delta}{i\delta\mathcal{J}(x_1)} \cdots \frac{\delta}{i\delta\mathcal{J}(x_r)} \frac{\delta}{i\delta\widetilde{\mathcal{J}}(x)} Z[\mathcal{J},\widetilde{\mathcal{J}}]\bigg|_{\mathcal{J}=\widetilde{\mathcal{J}}=0}.$$

(2.169)

Make the change of variables

$$\phi = \phi' - \widetilde{\mathcal{J}}F[\phi']$$

(2.170)

in the functional integral eqn (2.167),

$$Z[\mathcal{J},\widetilde{\mathcal{J}}] = \int D\phi' \left|\frac{\delta\phi}{\delta\phi'}\right| e^{i\int L[\phi']-\frac{\delta S}{\delta\phi}\big|_{\phi'}\widetilde{\mathcal{J}}F[\phi']+\mathcal{J}\phi'-\mathcal{J}\widetilde{\mathcal{J}}F[\phi']+\widetilde{\mathcal{J}}\theta[\phi']+\mathcal{O}(\widetilde{\mathcal{J}})^2},$$

$$= \int D\phi' \left|\frac{\delta\phi}{\delta\phi'}\right| e^{i\int L[\phi']+\mathcal{J}\phi'-\mathcal{J}\widetilde{\mathcal{J}}F[\phi']+\mathcal{O}(\widetilde{\mathcal{J}})^2},$$

(2.171)

by eqn (2.166). The Jacobian

$$\left|\frac{\delta\phi(x)}{\delta\phi'(y)}\right| = \det\left[\delta(x-y) - \widetilde{\mathcal{J}}\frac{\delta F[\phi'(x)]}{\delta\phi'(y)}\right],$$

(2.172)

is unity in dimensional regularization. Relabelling the dummy integration variable as ϕ gives

$$Z[\mathcal{J},\widetilde{\mathcal{J}}] = \int D\phi \; e^{i\int L[\phi]+\mathcal{J}\phi-\mathcal{J}\widetilde{\mathcal{J}}F[\phi]+\mathcal{O}(\widetilde{\mathcal{J}})^2}.$$

(2.173)

Taking the $\widetilde{\mathcal{J}}$ derivative and setting $\widetilde{\mathcal{J}} = 0$ gives, by using the equality of eqn (2.167) and eqn (2.173),

$$\int D\phi \; \theta(x) \; e^{i\int L[\phi]+\mathcal{J}\phi} = -\int D\phi \; \mathcal{J}(x)F[\phi(x)] \; e^{i\int L[\phi]+\mathcal{J}\phi}.$$

(2.174)

Differentiating multiple times with respect to $\mathcal{J}$ gives the equation-of-motion Ward identity

$$\langle 0| T\{\phi(x_1)\dots\phi(x_n)\theta(x)\}|0\rangle$$
$$= i\sum_r \delta(x-x_r)\langle 0| T\{\phi(x_1)\dots\cancel{\phi(x_r)}\dots\phi(x_n)F[\phi(x_r)]\}|0\rangle. \qquad (2.175)$$

The S matrix element with an insertion of θ vanishes,

$$_{\text{out}}\langle q_1,\dots,q_m|\theta|p_1,\dots,p_n\rangle_{\text{in}} = 0, \qquad (2.176)$$

because it is given by picking out the term with $m+n$ poles on the left-hand side of eqn (2.175). But the right-hand side shows that the matrix element of the rth term has no pole in p_r, because of the δ function. Each term in the sum vanishes, leading to eqn (2.176). As a result, equation-of-motion operators can be dropped because they do not contribute to the S-matrix.

Note that eqn (2.176) implies that the *classical* equations of motion can be dropped. The equations of motion have quantum corrections, but the Ward identity eqn (2.176) is for the classical equations of motion without the quantum corrections. The Ward identity holds even for insertions of the equation-of-motion operator in loop graphs, where the particles are off-shell, and do not satisfy the classical equations of motion.

Using the equations of motion is a special case of a field redefinition. Consider the field redefinition (with $\epsilon \ll 1$):

$$\phi(x) = \phi'(x) + \epsilon F[\phi'(x)]. \qquad (2.177)$$

The change in the Lagrangian due to eqn (2.177) is

$$L[\phi] = L[\phi'] + \epsilon F[\phi']\frac{\delta S[\phi']}{\delta\phi'} + \mathcal{O}\left(\epsilon^2\right) = L[\phi'] + \epsilon\theta[\phi'] + + \mathcal{O}\left(\epsilon^2\right). \qquad (2.178)$$

We have already seen that a field redefinition leaves the S-matrix invariant. Thus the S-matrix computed with the new Lagrangian $L'[\phi] = L[\phi] + \epsilon\theta[\phi]$ is the same as that computed with $L[\phi]$.[2] Thus we can shift the Lagrangian by equation of motion terms. The way equations of motion are used in practice is to eliminate operators with derivatives in the EFT Lagrangian.

Exercise 2.22 The classical equation of motion for $\lambda\phi^4$ theory,

$$L = \frac{1}{2}(\partial_\mu\phi)^2 - \frac{1}{2}m^2\phi^2 - \frac{\lambda}{4!}\phi^4,$$

---

[2] Remember ϕ is a dummy variable, so we can use $L'[\phi]$ instead of $L'[\phi']$.

is

$$E[\phi] = (-\partial^2 - m^2)\phi - \frac{\lambda}{3!}\phi^3.$$

The equation of motion Ward identity for $\theta = F[\phi]E[\phi]$ is eqn (2.175). Integrate both sides with

$$\int dx\, e^{-iq\cdot x} \prod_i \int dx_i\, e^{-ip_i\cdot x_i}$$

to get the momentum space version of the Ward identity

$$\langle 0|T\{\tilde\phi(p_1)\ldots\tilde\phi(p_n)\tilde\theta(q)\}|0\rangle = i\sum_{r=1}^{n} \langle 0|T\{\tilde\phi(p_1)\ldots\cancel{\tilde\phi(p_r)}\ldots\tilde\phi(p_n)\tilde F(q+p_r)\}|0\rangle.$$

(a) Consider the equation of motion operator

$$\theta_1 = \phi\, E[\phi] = \phi(-\partial^2 - m^2)\phi - \frac{\lambda}{3!}\phi^4,$$

and verify the Ward identity by explicit calculation at order λ (i.e. tree level) for $\phi\phi$ scattering, i.e. for $\phi\phi \to \phi\phi$.

(b) Take the on-shell limit $p_r^2 \to m^2$ at fixed $q \neq 0$ of

$$\prod_r (-i)(p_r^2 - m^2) \times \text{Ward identity},$$

and verify that both sides of the Ward identity vanish. Note that both sides do not vanish if we first take $q = 0$ and then take the on-shell limit.

(c) Check the Ward identity to one loop for the equation of motion operator

$$\theta_2 = \phi^3\, E[\phi] = \phi^3(-\partial^2 - m^2)\phi - \frac{\lambda}{3!}\phi^6.$$

As an example of the use of the equations of motion, suppose we have an EFT Lagrangian

$$\mathscr{L} = \frac{1}{2}\partial_\mu\phi\partial^\mu\phi - \frac{1}{2}m^2\phi^2 - \frac{1}{4!}\lambda\phi^4 + \frac{c_1}{\Lambda^2}\phi^3\partial^2\phi + \frac{c_6}{\Lambda^2}\phi^6 + \ldots. \tag{2.179}$$

Then making the field redefinition

$$\phi \to \phi + \frac{c_1}{\Lambda^2}\phi^3, \tag{2.180}$$

gives the new Lagrangian

$$\mathcal{L} = \frac{1}{2}\partial_\mu\phi\partial^\mu\phi - \frac{1}{2}m^2\phi^2 - \frac{1}{4!}\lambda\phi^4 + \frac{c_1}{\Lambda^2}\phi^3\partial^2\phi + \frac{c_6}{\Lambda^2}\phi^6$$
$$+ \frac{c_1}{\Lambda^2}\phi^3\left[-\partial^2\phi - m^2\phi - \frac{\lambda}{3!}\phi^3\right] + \dots$$
$$= \frac{1}{2}\partial_\mu\phi\partial^\mu\phi - \frac{1}{2}m^2\phi^2 - \left[\frac{1}{4!}\lambda + \frac{c_1}{\Lambda^2}m^2\right]\phi^4 + \left[\frac{c_6}{\Lambda^2} - \frac{c_1}{\Lambda^2}\frac{\lambda}{3!}\right]\phi^6 + \dots. \qquad (2.181)$$

The two Lagrangians eqn (2.179) and eqn (2.181) give the same S-matrix. In eqn (2.181), we have eliminated the $\phi^3\partial^2\phi$ operator at the cost of redefining the coefficients of the ϕ^4 and ϕ^6 operators. The EFT power counting has been maintained in going from eqn (2.179) to eqn (2.181). It is easier to do computations with eqn (2.181) rather than eqn (2.179), because eqn (2.181) has fewer independent operators. In EFTs, we usually apply the equations of motion to eliminate as many operators with derivatives as possible.

The calculation above only retained terms up to dimension six. If we work to dimension eight, we must retain the terms quadratic in c_1/Λ^2 in the transformed Lagrangian. These terms are second order in the equation of motion. Working to second order in the equations of motion is tricky [50, 66, 68], and it is best to systematically use field redefinitions to eliminate operators to avoid making mistakes.

Using field redefinitions rather than the equations of motion also clears up some subtleties. For example, the fermion kinetic term is

$$\overline{\psi}\,i\slashed{D}\psi. \qquad (2.182)$$

This operator vanishes using the fermion equation of motion $i\slashed{D}\psi = 0$. However, it is not possible to eliminate this term by a field redefinition, so we cannot eliminate the fermion kinetic energy using the equations of motion. We can eliminate higher-order terms such as $\phi^2\overline{\psi}\,i\slashed{D}\psi$. Another interesting example is given in [50].

Exercise 2.23 Write down all possible C-even dimension-six terms in eqn (2.42), and show how they can be eliminated by field redefinitions.

Exercise 2.24 Take the heavy quark Lagrangian

$$\mathcal{L}_v = \bar{Q}_v\left\{iv\cdot D + i\slashed{D}_\perp\frac{1}{2m + iv\cdot D}i\slashed{D}_\perp\right\}Q_v$$
$$= \bar{Q}_v\left\{iv\cdot D - \frac{1}{2m}\slashed{D}_\perp\slashed{D}_\perp + \frac{1}{4m^2}\slashed{D}_\perp(iv\cdot D)\slashed{D}_\perp + \dots\right\}Q_v$$

and use a sequence of field redefinitions to eliminate the $1/m^2$ suppressed $v\cdot D$ term. The equation of motion for the heavy quark field is $(iv\cdot D)Q_v = 0$,

Figure 2.13 *Penguin graph in the weak interactions.*

so this example shows how to eliminate equation of motion operators in HQET. Here v^μ is the velocity vector of the heavy quark with $v \cdot v = 1$, and

$$D_\perp^\mu \equiv D^\mu - (v \cdot D)v^\mu.$$

If you prefer, you can work in the rest frame of the heavy quark, where $v^\mu = (1,0,0,0)$, $v \cdot D = D^0$ and $D_\perp^\mu = (0, \mathbf{D})$. See [66] for help.

In general, there are are many equation of motion operators E_i. Under renormalization, these operators mix among themselves,

$$\mu \frac{\mathrm{d}}{\mathrm{d}\mu} E_i = \gamma_{ij} E_j, \tag{2.183}$$

where γ_{ij} can be gauge dependent. The reason is that the left-hand side vanishes when inserted in an S-matrix element, and this needs to hold for all values of μ. E_i are not observable quantities, and their anomalous dimensions can depend on choice of gauge. For non-equation of motion operators O_i, the anomalous dimensions take the form

$$\mu \frac{\mathrm{d}}{\mathrm{d}\mu} O_i = \gamma_{ij} O_j + \Gamma_{ik} E_k. \tag{2.184}$$

An operator O_i is not an equation of motion operator if O_i contributes to S-matrix elements. Under μ evolution, these operators can mix with $\{E_i\}$, since $\{E_i\}$ have zero contributions to S-matrix elements. Since O_i are observable, γ_{ij} is gauge independent, but Γ_{ik} can be gauge dependent.

A well-known example of the use of equations of motion is for penguin graphs in the weak interactions [38], shown in Figure 2.13. The penguin graph is divergent, and requires the counterterm

$$\mathscr{L} = \frac{4 G_F}{\sqrt{2}} \frac{c_P}{\epsilon} g(\overline{\psi}\gamma^\mu T^A \psi)(D^\nu F_{\mu\nu})^A. \tag{2.185}$$

The penguin counterterm is eliminated from the Lagrangian by making a field redefinition,

$$\mathscr{L} = \frac{4 G_F}{\sqrt{2}} \frac{c_P}{\epsilon} g(\overline{\psi}\gamma^\mu T^A \psi)(D^\nu F_{\mu\nu})^A \rightarrow \frac{4 G_F}{\sqrt{2}} \frac{c_P}{\epsilon} g(\overline{\psi}\gamma^\mu T^A \psi) g(\overline{\psi}\gamma_\mu T^A \psi), \tag{2.186}$$

Figure 2.14 *Penguin and four-quark contribution to $qq \to qq$.*

Figure 2.15 *One-loop contribution to the QED β-function from a fermion of mass m.*

and replacing it by a four-quark operator. The field redefinition needed for eqn (2.186) is

$$A_\mu^A \to A_\mu^A - \frac{4G_F}{\sqrt{2}} \frac{c_P}{\epsilon} g \overline{\psi} \gamma^\mu T^A \psi, \tag{2.187}$$

which is a field redefinition with an infinite coefficient. Green's functions using the redefined Lagrangian eqn (2.186) are infinite, but the S-matrix is finite. There is no counterterm to cancel the penguin graph divergence, but the on-shell four-quark amplitude gets both the penguin and counterterm contributions (Figure 2.14) and is finite.

2.7 Decoupling of heavy particles

Heavy particles do not decouple in a mass-independent subtraction scheme like $\overline{\text{MS}}$. For example, the one-loop QCD β-function coefficient is $b_0 = 11 - 2/3n_f$, where n_f is the number of quark flavours. Thus b_0 has the same value for all μ, independent of the quark masses. We expect that the top quark only contributes to the β-function for $\mu \gg m_t$, and no longer contributes when $\mu \ll m_t$, i.e. heavy particles decouple at low energy.

To understand the decoupling of heavy particles, consider the contribution of a charged lepton of mass m to the one-loop β function in QED. The diagram Figure 2.15 in dimensional regularization gives

$$
\begin{aligned}
&i\frac{e^2}{2\pi^2}\left(p_\mu p_\nu - p^2 g_{\mu\nu}\right)\left[\frac{1}{6\epsilon} - \int_0^1 dx\, x(1-x)\log\frac{m^2 - p^2 x(1-x)}{\overline{\mu}^2}\right] \\
&\equiv i\left(p_\mu p_\nu - p^2 g_{\mu\nu}\right)\Pi(p^2)
\end{aligned}
\tag{2.188}
$$

where p is the external momentum.

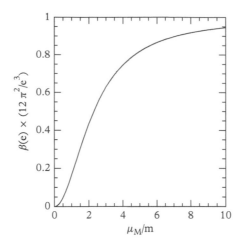

Figure 2.16 *Contribution of a fermion of mass m to the QED β-function. The result is given for the momentum-space subtraction scheme, with renormalization scale μ_M. The β function does not attain its limiting value of $e^3/12\pi^2$ until $\mu_M \gtrsim 10\,m$. The fermion decouples for $\mu_M \ll m$.*

2.7.1   Momentum–Subtraction Scheme

Consider a mass-dependent scheme, the momentum space subtraction scheme, where we subtract the value of the graph at a Euclidean momentum point $p^2 = -\mu_M^2$, to get the renormalized vacuum polarization function,

$$\Pi_{\mathrm{mom}}(p^2, m^2, \mu_M^2) = -\frac{e^2}{2\pi^2}\left[\int_0^1 dx\, x(1-x)\, \log\frac{m^2 - p^2 x(1-x)}{m^2 + \mu_M^2 x(1-x)}\right]. \tag{2.189}$$

The fermion contribution to the QED β-function is obtained by acting on Π with $(e/2)\mu_M\, d/d\mu_M$,

$$
\begin{aligned}
\beta_{\mathrm{mom}}(e) &= -\frac{e}{2}\mu_M \frac{d}{d\mu_M}\frac{e^2}{2\pi^2}\left[\int_0^1 dx\, x(1-x)\, \log\frac{m^2 - p^2 x(1-x)}{m^2 + \mu_M^2 x(1-x)}\right] \\
&= \frac{e^3}{2\pi^2}\int_0^1 dx\, x(1-x)\, \frac{\mu_M^2 x(1-x)}{m^2 + \mu_M^2 x(1-x)}.
\end{aligned}
\tag{2.190}
$$

The fermion contribution to the β-function is plotted in Figure 2.16. When the fermion mass m is small compared with the renormalization point μ_M, $m \ll \mu_M$, the β-function contribution is

$$\beta(e) \approx \frac{e^3}{2\pi^2}\int_0^1 dx\, x(1-x) = \frac{e^3}{12\pi^2}. \tag{2.191}$$

As the renormalization point passes through m, the fermion decouples, and for $\mu_M \ll m$, its contribution to β vanishes as

$$\beta(e) \approx \frac{e^3}{2\pi^2} \int_0^1 dx\, x(1-x) \frac{\mu_M^2 x(1-x)}{m^2} = \frac{e^3}{60\pi^2} \frac{\mu_M^2}{m^2} \to 0. \tag{2.192}$$

Thus in the momentum space scheme, we see the expected behaviour that heavy particles decouple, which is an example of the Appelquist–Carazzone decoupling theorem [7].

2.7.2 The $\overline{\text{MS}}$ Scheme

In the $\overline{\text{MS}}$ scheme, we subtract only the $1/\epsilon$ pole of eqn (2.188), so

$$\Pi_{\overline{\text{MS}}}(p^2, m^2, \overline{\mu}^2) = -\frac{e^2}{2\pi^2} \left[\int_0^1 dx\, x(1-x) \log \frac{m^2 - p^2 x(1-x)}{\overline{\mu}^2} \right]. \tag{2.193}$$

The fermion contribution to the QED β-function is obtained by acting with $(e/2)\overline{\mu}\, d/d\overline{\mu}$ on Π,

$$\begin{aligned}
\beta_{\overline{\text{MS}}}(e) &= -\frac{e}{2}\overline{\mu}\frac{d}{d\overline{\mu}}\frac{e^2}{2\pi^2} \left[\int_0^1 dx\, x(1-x) \log \frac{m^2 - p^2 x(1-x)}{\overline{\mu}^2} \right] \\
&= \frac{e^3}{2\pi^2} \int_0^1 dx\, x(1-x) = \frac{e^3}{12\pi^2},
\end{aligned} \tag{2.194}$$

which is independent of the fermion mass and $\overline{\mu}$.

The fermion contribution to the β-function in the $\overline{\text{MS}}$ scheme does not vanish as $m \gg \overline{\mu}$, so the fermion does not decouple as it should. There is another problem: from eqn (2.193), the finite part of the Feynman graph in the $\overline{\text{MS}}$ scheme at low momentum is

$$\Pi_{\overline{\text{MS}}}(0, m^2, \overline{\mu}^2) = -\frac{e^2}{2\pi^2} \left[\int_0^1 dx\, x(1-x) \log \frac{m^2}{\overline{\mu}^2} \right]. \tag{2.195}$$

For $\overline{\mu} \ll m$ the logarithm becomes large, and perturbation theory breaks down. These two problems are related. The large finite part corrects for the fact that the value of the running coupling used at low energies is 'incorrect', because it was obtained using the 'wrong' β-function.

The two problems can be solved at the same time by integrating out heavy particles. We use a theory including the heavy fermion as a dynamical field when $m < \overline{\mu}$, and a theory without the fermion field when $m > \overline{\mu}$. Effects of the heavy particle in the low-energy theory are included via higher-dimension operators, which are suppressed by inverse powers of the heavy particle mass. The matching condition of the two theories is that S-matrix elements for light particle scattering in the low-energy theory must equal the S-matrix elements for light particle scattering in the high-energy theory. Schematically, we match

$$\mathscr{L}^{(n_l+1)} \to \mathscr{L}^{(n_l)}, \tag{2.196}$$

from a theory with n_l light particles and one heavy particle to a theory with n_l light particles. The effects of the heavy particles are absorbed into changes in the coefficients of $\mathscr{L}$. These are referred to as threshold corrections. Thus at the matching scale, $\mathscr{L}$ changes, both in terms of the field content and the values of the Lagrangian coefficients. However, nothing discontinuous is going on, and the physics (i.e. S-matrix elements) are continuous across the threshold. The description changes, but the resulting S-matrix elements remain the same.

In our example, we can integrate out the heavy lepton at the matching scale $\bar{\mu}$. The effect of the one-loop heavy lepton graph Figure 2.15 can be expanded for $p^2 \ll m^2$ as

$$\begin{aligned}
\Pi_{\overline{\text{MS}}}(p^2, m^2, \bar{\mu}^2) &= -\frac{e^2}{2\pi^2} \int_0^1 dx\, x(1-x) \left\{ \log \frac{m^2}{\bar{\mu}^2} + \log\left[1 - \frac{p^2}{m^2} x(1-x) \right] \right\} \\
&= -\frac{e^2}{2\pi^2} \int_0^1 dx\, x(1-x) \left\{ \log \frac{m^2}{\bar{\mu}^2} - \frac{p^2}{m^2} x(1-x) + \dots \right\} \\
&= -\frac{e^2}{2\pi^2} \left[\frac{1}{6} \log \frac{m^2}{\bar{\mu}^2} - \frac{p^2}{30 m^2} + \mathcal{O}\left(\frac{p^4}{m^4} \right) \right].
\end{aligned} \tag{2.197}$$

The first term is included in $\mathscr{L}^{(n_l)}$ by a shift in the gauge kinetic term. Rescaling the gauge field to restore the kinetic term to its canonical normalization $-F_{\mu\nu}^2/4$ gives a shift in the gauge coupling constant,

$$\frac{1}{e_L^2(\bar{\mu})} = \frac{1}{e_H^2(\bar{\mu})} - \frac{1}{12\pi^2} \log \frac{m^2}{\bar{\mu}^2}. \tag{2.198}$$

where e_L is the gauge coupling in the low-energy theory, and e_H is the gauge coupling in the high-energy theory. The $\bar{\mu}$ dependence of the threshold correction is related to the difference in β-functions of the two theories.

The second term in eqn (2.197) gives a dimension six operator in the low-energy theory,

$$\mathscr{L} = -\frac{e^2}{240\pi^2 m^2} \partial_\alpha F_{\mu\nu} \partial^\alpha F^{\mu\nu}, \tag{2.199}$$

and so on. While the Lagrangian has changed at μ_M, the S-matrix has not. The change in the Lagrangian is exactly the same as the contribution from Figure 2.15, which is present in the high-energy theory but not in the low-energy theory.

Exercise 2.25 Verify that the first term in eqn (2.197) leads to the threshold correction in the gauge coupling given in eqn (2.198). If we match at $\bar{\mu} = m$, then $e_L(\bar{\mu}) =$

$e_H(\bar{\mu})$, and the gauge coupling is continuous at the threshold. Continuity does not hold at higher loops, or when a heavy scalar is integrated out.

Exercise 2.26 Assume the threshold correction is of the form

$$\frac{1}{e_L^2(\bar{\mu})} = \frac{1}{e_H^2(\bar{\mu})} + c \log \frac{m^2}{\bar{\mu}^2}.$$

Find the relation between c and the difference $\beta_H - \beta_L$ of the β-functions in the two theories, and check that this agrees with eqn (2.198).

2.8 Naive dimensional analysis

There is a slightly more sophisticated version of the EFT power counting formula which is referred to as naive dimensional analysis (NDA) [64]. It is a power counting formula that keeps track of the 4π factors from loop graphs. If ϕ, ψ and $X_{\mu\nu}$, g, y, λ denote generic scalar fields, fermion fields, gauge field-strength tensors, gauge couplings, Yukawa couplings, and ϕ^4 couplings, then the NDA formula says that an operator in the EFT should be normalized as

$$\widehat{O} = f^2 \Lambda^2 \left[\frac{\partial}{\Lambda}\right]^{N_p} \left[\frac{\phi}{f}\right]^{N_\phi} \left[\frac{A}{f}\right]^{N_A} \left[\frac{\psi}{f\sqrt{\Lambda}}\right]^{N_\psi} \left[\frac{g}{4\pi}\right]^{N_g} \left[\frac{y}{4\pi}\right]^{N_y} \left[\frac{\lambda}{16\pi^2}\right]^{N_\lambda}. \qquad (2.200)$$

where Λ and f are related by

$$\Lambda = 4\pi f, \qquad (2.201)$$

and Λ is the scale of the EFT derivative expansion. With this normalization, EFT coefficients are expected to be of order unity,

$$\mathcal{L} = \sum \widehat{C}_i \widehat{O}_i, \qquad (2.202)$$

with $\widehat{C}_i \sim 1$. A generalization of NDA to d dimensions can be found in [34]. From eqn (2.200),

$$\frac{D}{\Lambda} = \frac{\partial + igA}{\Lambda} = \frac{\partial}{\Lambda} + i\left[\frac{g}{4\pi}\right]\left[\frac{A}{f}\right] \qquad (2.203)$$

so that both parts of a covariant derivative have the same power counting.

Loop graphs in the EFT maintain the NDA form, i.e. an arbitrary graph with insertions of operators of the form eqn (2.200) generates an operator of the same form. The proof, which relies on counting $1/(16\pi^2)$ factors from each loop and the topological identity for a connected graph $V - I + L = 1$, where V is the number of vertices, I the number of internal lines, and L the number of loops, is left as an exercise.

Exercise 2.27

Show that the power counting formula eqn (2.200) for an EFT Lagrangian is self-consistent, i.e. an arbitrary graph with insertions of vertices of this form generates an interaction that maintains the same form. (See [34] and [64]). Show that eqn (2.200) is equivalent to

$$\widehat{O} \sim \frac{\Lambda^4}{16\pi^2} \left[\frac{\partial}{\Lambda}\right]^{N_p} \left[\frac{4\pi\,\phi}{\Lambda}\right]^{N_\phi} \left[\frac{4\pi\,A}{\Lambda}\right]^{N_A} \left[\frac{4\pi\,\psi}{\Lambda^{3/2}}\right]^{N_\psi} \left[\frac{g}{4\pi}\right]^{N_g} \left[\frac{y}{4\pi}\right]^{N_y} \left[\frac{\lambda}{16\pi^2}\right]^{N_\lambda}.$$

Using the more sophisticated power counting of eqn (2.200) instead of only counting factors of Λ makes a big difference in estimating the coefficients of higher dimension terms in the Lagrangian. For example, the four-quark dimension-six operator is normalized to

$$\widehat{O} = f^2 \Lambda^2 \frac{\left(\overline{\psi}\gamma^\mu\psi\right)^2}{(f\sqrt{\Lambda})^4} = \frac{1}{f^2}\left(\overline{\psi}\gamma^\mu\psi\right)^2 = \frac{16\pi^2}{\Lambda^2}\left(\overline{\psi}\gamma^\mu\psi\right)^2.$$

The extra $16\pi^2$ makes a difference of ~ 150 in the normalization of the operator.

In χPT, the Lagrangian is written in terms of

$$U(x) = e^{2i\Pi(x)/f}, \qquad (2.204)$$

where $\Pi(x)$ is a matrix of pion fields. $U(x)$ satisfies eqn (2.200), since every Π comes with a factor $1/f$. The normalization of the two-derivative term in the chiral Lagrangian is

$$\widehat{O} = \Lambda^2 f^2 \frac{\partial_\mu U}{\Lambda} \frac{\partial^\mu U^\dagger}{\Lambda} = f^2 \partial_\mu U \partial^\mu U^\dagger \qquad (2.205)$$

which is the usual normalization of the kinetic term. The four-derivative term is normalized to

$$\widehat{O} = \Lambda^2 f^2 \frac{\partial_\mu U}{\Lambda} \frac{\partial^\mu U^\dagger}{\Lambda} \frac{\partial_\nu U}{\Lambda} \frac{\partial^\nu U^\dagger}{\Lambda} = \frac{1}{16\pi^2} \partial_\mu U \partial^\mu U^\dagger \partial_\nu U \partial^\nu U^\dagger. \qquad (2.206)$$

The four-derivative coefficients in the chiral Lagrangian are usually denoted by L_i, and eqn (2.206) shows that we expect $L_i \sim 1/(16\pi^2) \sim 4 \times 10^{-3}$, which is true experimentally (see [73]).

The difference between $f = 93\,\text{MeV}$ and $\Lambda = 4\pi f = 1.2\,\text{GeV}$ is very important for χPT. The value of f is fixed from the experimental value of the $\pi \to \mu\bar{\nu}_\mu$ decay rate. If we did not keep track of the 4π factors, this would imply that the scale Λ of χPT is $\Lambda \sim f$, and χPT breaks down for momenta of order f. If this is the case, χPT is not very useful, since the pion mass is around $140\,\mathrm{MeV}$, so χPT breaks down for on-shell pions. Luckily, eqn (2.201) says that Λ_χ, the scale of the χPT derivative expansion is $4\pi f$ [64], which is much larger than f, so that χPT is valid for $\pi - \pi$ scattering at low momentum. Loop corrections in pion χPT are of order $[m_\pi/(4\pi f)]^2 \sim 0.014$, and are a few per cent. χPT for kaons has corrections of order $[m_K/(4\pi f)]^2 \sim 0.2$.

The NDA formula eqn (2.200) implies that if all operators in the Lagrangian are normalized using NDA, then an arbitrary loop graph gives

$$\delta \widehat{C}_i \sim \prod_k \widehat{C}_k, \tag{2.207}$$

where the graph has insertions of Lagrangian terms $\widehat{C}_k \widehat{O}_k$, and produces an amplitude of the form $\widehat{C}_i \widehat{O}_i$. All the 4π factors have disappeared, and we obtain a very simple form for the amplitudes. The results are equally valid for strongly and weakly coupled theories.

The NDA formula eqn (2.207) also shows that in strongly coupled theories $\widehat{C} \lesssim 1$ [64]. The reason is that if $\widehat{C} \gg 1$, then the hierarchy of equations eqn (2.207) is unstable, because higher-order contributions to $\widehat{C}_i$ are much larger than $\widehat{C}_i$. On the other hand, there is no inconsistency if $\widehat{C}_i \ll 1$, since all this implies is that higher-order corrections are small, a sign of a weakly coupled theory. Eqn (2.207) shows that an interaction becomes strongly coupled when $\widehat{C} \sim 1$. For the dimension-four interactions, strong coupling is when gauge couplings are $g \sim 4\pi$, Yukawa couplings are $y \sim 4\pi$ and scalar self-couplings are $\lambda \sim (4\pi)^2$.

We can use NDA for cross sections as well as amplitudes. A cross section is the imaginary part of the forward scattering amplitude, so we can estimate cross sections by using NDA for the forward amplitude, and then multiplying by π, since the imaginary part comes from $\log(-1) = i\pi$. Since two-body final states give a one-loop forward scattering diagram, and n-body final states give a $n - 1$ loop diagram, the 4π counting rules for phase space are: $1/(16\pi)$ for the first two particles, and $1/(16\pi^2)$ for each additional particle. We used this 4π counting rule earlier in these lectures in our estimates of cross sections.

2.9 Invariants

EFT Lagrangians are constructed using gauge and Lorentz invariant operators that are polynomials in the basic fields. Classifying these operators is a fun topic that is extensively studied in the mathematics literature on invariant theory. This section briefly discusses invariant theory. For an elementary summary, see [42, 49].

We start with the simple example of a theory with N_f fermions with mass term

$$\mathscr{L} = -\overline{\psi}_L M \psi_R + \text{h.c.}, \tag{2.208}$$

where M is an $N_f \times N_f$ matrix. We can make a field redefinition (ignoring anomalies),

$$\psi_L \to L\psi_L, \qquad\qquad \psi_R \to R\psi_R, \tag{2.209}$$

under which

$$M \rightarrow LMR^\dagger. \tag{2.210}$$

Under CP, $M \rightarrow M^*$. The S-matrix is invariant under the field redefinition eqn (2.209), and depends only on invariants constructed from M. To eliminate R, define

$$X \equiv MM^\dagger, \qquad\qquad X \rightarrow LXL^\dagger, \tag{2.211}$$

which transforms only under L. Then the invariants are

$$I_{2n} = \langle X^n \rangle, \tag{2.212}$$

where $2n$ is the degree of the invariant in the basic object M, and $\langle \cdot \rangle$ denotes a trace. Suppose $N_f = 1$. Then X is a 1×1 matrix, and

$$\langle X^2 \rangle = I_4 = I_2^2 = \langle X \rangle^2, \qquad\qquad \langle X^3 \rangle = I_6 = I_2^3 = \langle X \rangle^3, \tag{2.213}$$

and there is one independent invariant of every even degree, $I_{2n} = I_2^n = \langle X \rangle^n$.

The Hilbert series is defined as

$$H(q) = \sum_{n=0}^{\infty} N_n q^n \tag{2.214}$$

where N_n is the number of invariants of degree n, and $N_0 = 1$ by convention. In the 1×1 matrix example,

$$H(q) = 1 + q^2 + q^4 + \ldots = \frac{1}{1 - q^2}. \tag{2.215}$$

The denominator of $H(q)$ in eqn (2.215) tells us that there is one generator of degree two, which is $\langle X \rangle$, and that all invariants are given by powers of this generator. Given I_2, we can determine the fermion mass, $m = \sqrt{I_2}$, as a real, non-negative number. The invariant is CP even, since under CP, $X \rightarrow X^*$, and $\langle X \rangle \rightarrow \langle X^* \rangle = \langle X^\dagger \rangle = \langle X \rangle$ since X is Hermitian, and the trace is invariant under transposition of the matrix.

The next case is $N_f = 2$, with invariants

$$\langle X \rangle, \langle X^2 \rangle, \langle X^3 \rangle, \ldots. \tag{2.216}$$

These are not all independent, because the Cayley–Hamilton theorem implies

$$\langle X^3 \rangle = \frac{3}{2} \langle X \rangle \langle X^2 \rangle - \frac{1}{2} \langle X \rangle^3, \tag{2.217}$$

for any 2×2 matrix. This identity eliminates all traces of X^n for $n \geq 3$. There is one invariant of degree 2, $\langle X \rangle$, two of degree four $\langle X \rangle^2$ and $\langle X^2 \rangle$, etc. The Hilbert series is

$$H(q) = 1 + q^2 + 2q^4 + \ldots = \frac{1}{(1-q^2)(1-q^4)}. \tag{2.218}$$

The denominator factors imply that all invariants are generated by products of $\langle X \rangle$ and $\langle X^2 \rangle$. Given $\langle X \rangle$ and $\langle X^2 \rangle$, we can find the two masses by solving

$$\langle X \rangle = m_1^2 + m_2^2, \qquad\qquad \langle X^2 \rangle = m_1^4 + m_2^4. \tag{2.219}$$

For $N_f = 3$, the generators are $\langle X \rangle$, $\langle X^2 \rangle$, $\langle X^3 \rangle$. Higher powers are eliminated by the Cayley–Hamilton theorem,

$$\langle X^4 \rangle = \frac{1}{6} \langle X \rangle^4 - \langle X \rangle^2 \langle X^2 \rangle + \frac{4}{3} \langle X^3 \rangle \langle X \rangle + \frac{1}{2} \langle X^2 \rangle^2, \tag{2.220}$$

and the Hilbert series is

$$H(q) = 1 + q^2 + 2q^4 + \ldots = \frac{1}{(1-q^2)(1-q^4)(1-q^6)}. \tag{2.221}$$

Exercise 2.28 By explicit calculation, show that

$$\left[\frac{1}{2} \langle A \rangle^2 - \frac{1}{2} \langle A^2 \rangle \right] 1 - \langle A \rangle A + A^2 = 0,$$

$$\frac{1}{6} \langle A \rangle^3 - \frac{1}{2} \langle A \rangle \langle A^2 \rangle + \frac{1}{3} \langle A^3 \rangle = 0,$$

for a general 2×2 matrix A and that

$$\langle A \rangle \langle B \rangle \langle C \rangle - \langle A \rangle \langle BC \rangle - \langle B \rangle \langle AC \rangle - \langle C \rangle \langle AB \rangle + \langle ABC \rangle + \langle ACB \rangle = 0.$$

for general 2×2 matrices A, B, C. Identities analogous to this for 3×3 matrices are used in χPT to remove L_0 and to replace it by $L_{1,2,3}$, as discussed in Chapter 3.

Now consider the case of two quark types, u and d, in the SM. There are two mass matrices M_u and M_d which transform as

$$M_u \to L M_u R_u^\dagger, \qquad\qquad M_d \to L M_d R_d^\dagger. \tag{2.222}$$

Equation (2.222) results because the right-handed quarks u_R and d_R are independent fields with independent transformations R_u and R_d in the SM, whereas the left-handed quarks are part of a weak doublet

$$qL = \begin{bmatrix} u_L \\ d_L \end{bmatrix}, \tag{2.223}$$

so $L_u = L_d = L$. To construct invariants, we can eliminate $R_{u,d}$ by constructing

$$X_u = M_u M_u^\dagger, \qquad\qquad X_d = M_d M_d^\dagger, \tag{2.224}$$

which transform as

$$X_u \to L X_u L^\dagger, \qquad\qquad X_d \to L X_d L^\dagger. \tag{2.225}$$

For $N_f = 1$, X_u and X_d are numbers, and the only independent invariants are $\langle X_u \rangle$ and $\langle X_d \rangle$, and the Hilbert series is

$$H(q) = \frac{1}{(1-q^2)^2}. \tag{2.226}$$

For $N_f = 2$, the independent generators are $\langle X_u \rangle$, $\langle X_d \rangle$, $\langle X_u^2 \rangle$, $\langle X_d^2 \rangle$ and $\langle X_u X_d \rangle$, and

$$H(q) = \frac{1}{(1-q^2)^2(1-q^4)^3}. \tag{2.227}$$

$\langle X_u \rangle$ and $\langle X_u^2 \rangle$ determine the two u-quark masses m_u and m_c as in eqn (2.219). $\langle X_d \rangle$ and $\langle X_d^2 \rangle$ determine the two d-quark masses m_d and m_s. $\langle X_u X_d \rangle$ determines the Cabibbo angle,

$$\langle X_u X_d \rangle = (m_u^2 m_d^2 + m_c^2 m_s^2) - (m_c^2 - m_u^2)(m_s^2 - m_d^2)\sin^2\theta. \tag{2.228}$$

If $m_u = m_c$ or if $m_d = m_s$, θ is not defined (or can be rotated away).

All the invariants are CP even, so there is no CP violation in the quark sector for two quark flavours. For example, under CP,

$$\langle X_u X_d \rangle \to \langle X_u^* X_d^* \rangle = \left\langle \left(X_u^* X_d^*\right)^T \right\rangle = \left\langle X_d^\dagger X_u^\dagger \right\rangle = \langle X_d X_u \rangle = \langle X_u X_d \rangle \tag{2.229}$$

since X_u and X_d are Hermitian, and the trace is invariant under transposition and cyclic permutation.

The first non-trivial example is $N_f = 3$. The CP even generators are

$$\langle X_u \rangle, \; \langle X_u^2 \rangle, \; \langle X_u^3 \rangle, \; \langle X_d \rangle, \; \langle X_d^2 \rangle, \; \langle X_d^3 \rangle, \; \langle X_u X_d \rangle, \langle X_u^2 X_d \rangle, \langle X_u X_d^2 \rangle, \langle X_u^2 X_d^2 \rangle. \tag{2.230}$$

They determine the quark masses $m_{u,c,t}$, $m_{d,s,b}$, and the three CKM angles $\theta_{12}, \theta_{13}, \theta_{23}$. However, the terms in eqn (2.230) do not generate all the invariants. We also have the *CP* odd invariant

$$I_- = \left\langle X_u^2 X_d^2 X_u X_d \right\rangle - \left\langle X_d^2 X_u^2 X_d X_u \right\rangle = \frac{1}{3}\left\langle [X_u, X_d]^3 \right\rangle. \tag{2.231}$$

and the *CP* even invariant

$$I_+ = \left\langle X_u^2 X_d^2 X_u X_d \right\rangle + \left\langle X_d^2 X_u^2 X_d X_u \right\rangle. \tag{2.232}$$

I_+ is not independent; it can be written as a linear combination of the lower-order invariants in eqn (2.230).

While I_- is not a linear combination of the invariants in eqn (2.230), it turns out that I_-^2 *is* a linear combination. This is an example of a relation among the invariants. There also can be relations among relations, which are known as syzygies. Thus the independent invariants are arbitrary products of powers of eqn (2.230) plus I_- to at most the first power. This gives the Hilbert series for $N_f = 3$

$$H(q) = \frac{1 + q^{12}}{(1 - q^2)^2(1 - q^4)^3(1 - q^6)^4(1 - q^8)}, \tag{2.233}$$

where the $+q^{12}$ in the numerator is the contribution from I_-. I_- is related to the Jarlskog invariant $\mathcal{J}$,

$$I_- = 2i(m_c^2 - m_u^2)(m_t^2 - m_c^2)(m_t^2 - m_u^2)(m_s^2 - m_d^2)(m_b^2 - m_s^2)(m_b^2 - m_d^2)\mathcal{J}, \tag{2.234}$$

where

$$\mathcal{J} = \text{Im}\left[V_{11}V_{12}^*V_{22}V_{21}^*\right] = c_{12}s_{12}c_{13}s_{13}^2c_{23}s_{23}s_\delta, \tag{2.235}$$

using the CKM matrix convention of the PDG [72].

The *CP*-even invariants in eqn (2.230) determine $\mathcal{J}^2$, and hence $\mathcal{J}$ up to an overall sign. The invariant I_- fixes the sign. This analysis should be familiar from the study of *CP* violation in the SM. By measuring *CP* conserving decay rates, we can determine the lengths of the three sides of the unitarity triangle. This determines the triangle (including the area, which is a measure of *CP* violation) up to an overall reflection, which is fixed by the sign of $\mathcal{J}$. Thus, we can determine if *CP* is violated only from *CP* conserving measurements.

Exercise 2.29 Show that the invariant

$$I_- = \left\langle X_u^2 X_d^2 X_u X_d \right\rangle - \left\langle X_d^2 X_u^2 X_d X_u \right\rangle,$$

is the lowest order *CP*-odd invariant constructed from the quark mass matrices. Show that I_- also can be written in the form

$$I_- = \frac{1}{3}\Big\langle [X_u, X_d]^3 \Big\rangle,$$

and explicitly work out I_- in the SM using the CKM matrix convention of the PDG [72]. Verify eqns (2.234, 2.235).

Exercise 2.30 Compute the Hilbert series for the ring of invariants generated by

(a) x, y (each of dimension 1), and invariant under the transformation $(x, y) \to (-x, -y)$.

(b) x, y, z (each of dimension 1), and invariant under the transformation $(x, y, z) \to (-x, -y, -z)$.

The general structure of $H(q)$ is the ratio of a numerator $N(q)$ and a denominator $D(q)$,

$$H(q) = \frac{N(q)}{D(q)}, \tag{2.236}$$

where the denominator $D(q)$ is a product of the form

$$D(q) = (1 - q^{n_1})^{r_1} (1 - q^{n_2})^{r_2} \ldots \tag{2.237}$$

and the numerator $N(q)$ is a polynomial with non-negative coefficients of degree d_N which is palindromic, i.e.

$$q^{d_N} N(1/q) = N(q). \tag{2.238}$$

The number of denominator factors $\sum r_i$ is the number of parameters [57]. In eqn (2.233) the number of parameters is ten, which are the six masses, three angles, and one phase.

As a non-trivial example, the lepton sector of the seesaw theory for $n_g = 2$ generations has invariants generated by the mass matrices for the charged leptons m_e, neutrinos m_ν and the singlet Majorana neutrino mass matrix M. The Hilbert series is [49]

$$H(q) = \frac{1 + q^6 + 3q^8 + 2q^{10} + 3q^{12} + q^{14} + q^{20}}{(1 - q^2)^3 (1 - q^4)^5 (1 - q^6)(1 - q^{10})}, \tag{2.239}$$

which has a palindromic numerator. The numerator is of degree twenty, and the coefficients are $1, 0, 0, 0, 0, 0, 1, 0, 3, 0, 2, 0, 3, 0, 1, 0, 0, 0, 0, 0, 1$, which is the same string read in either direction.

To construct an EFT, we have basic fields $\psi(x)$, $\phi(x)$, etc., which transform under various symmetries, and we want to construct invariant Lagrangians that are polynomials in the basic fields. This is a problem in invariant theory, with a few additional requirements.

- We can act with covariant derivatives on fields, $D_\mu \phi(x)$, to get an object that transforms the same way as $\phi(x)$ under gauge and flavour symmetries, but adds an extra Lorentz vector index.
- We can drop total derivatives since they vanish when integrated to get the action. Equivalently, we are allowed to integrate by parts.
- We can make field redefinitions or equivalently use the equations of motion to eliminate operators.

Counting invariants including these constraints seems simple, but there is a subtlety. Terms like

$$\partial_\mu(\phi^\dagger \partial^\mu \phi - \partial^\mu \phi^\dagger \phi) \tag{2.240}$$

vanish because they are a total derivative, and also by using the equations of motion. We have to make sure we do not double count the terms eliminated by these two conditions. This is a non-trivial problem that was recently solved in [44] using representations of the conformal group. The HQET/NRQCD dimension-eight operators were recently classified with the help of invariants [59].

2.10 SMEFT

The SMEFT is an EFT constructed using the basic fields of the SM given in Table 2.1. For an extensive recent review, see [15]. The dimension-four terms give the usual SM

Table 2.1 Fields of the SM. The Lorentz group is $SU(2)_L \times SU(2)_R$. The fermions have a generation index $n_g = 1, 2, 3$.

	Lorentz	SU(3)	SU(2)	U(1)
$G_{\mu\nu}$	$(1,0) + (0,1)$	8	1	0
$W_{\mu\nu}$	$(1,0) + (0,1)$	1	3	0
$B_{\mu\nu}$	$(1,0) + (0,1)$	1	1	0
H	$(0,0)$	1	2	$\frac{1}{2}$
q	$(1/2,0)$	3	2	$\frac{1}{6}$
l	$(1/2,0)$	1	2	$-\frac{1}{2}$
u	$(0,1/2)$	3	1	$\frac{2}{3}$
d	$(0,1/2)$	3	1	$-\frac{1}{3}$
e	$(0,1/2)$	1	1	-1

Lagrangian. There is only a single $U(1)$ gauge field in the SM. In theories with multiple Abelian gauge fields, the general kinetic energy for the $U(1)$ gauge fields has the form

$$\mathcal{L} = -\frac{1}{4} C_{ij} F_{\mu\nu}^{(i)} F_{\mu\nu}^{(j)}, \tag{2.241}$$

where C is a real symmetric matrix with positive eigenvalues, which is referred to as kinetic mixing [32, 45].

Constructing the higher dimension operators in SMEFT is not easy. It is useful to note that Lorentz invariance requires that fermion fields come in pairs. The allowed fermion bilinears written in terms of chiral fields are

$$\overline{\psi}_L \gamma^\mu \psi_L, \quad \overline{\psi}_R \gamma^\mu \psi_R, \quad \overline{\psi}_L \psi_R, \quad \overline{\psi}_L \sigma^{\mu\nu} \psi_R, \quad \overline{\psi}_R \psi_L, \quad \overline{\psi}_R \sigma^{\mu\nu} \psi_L. \tag{2.242}$$

We can always replace a right-handed field ψ_R by its charge-conjugate left-handed field ψ_L^c,

$$\psi_R = C \psi_L^{c*}, \tag{2.243}$$

where $C = i\gamma^2$. Thus we can use either a right-handed e_R^- field, or a left-handed e_L^+ field. The SMEFT is usually written using left-handed $SU(2)$ doublet fields, and right-handed $SU(2)$ singlet fields, as shown in Table 2.1.

Mass terms and dipole interactions are written in terms of left-handed field bilinears

$$\overline{\psi}_R \psi_L = \psi_L^{cT} C \psi_L, \qquad \overline{\psi}_R \sigma^{\mu\nu} \psi_L = \psi_L^{cT} C \sigma^{\mu\nu} \psi_L. \tag{2.244}$$

In general, if there are multiple left-handed fields, the mass and dipole operators are

$$\psi_{Lr}^T C \psi_{Ls}, \qquad \psi_{Lr}^T C \sigma^{\mu\nu} \psi_{Ls}, \tag{2.245}$$

where r, s are flavour indices. The mass term is symmetric in rs, and the dipole term is antisymmetric in rs. We must still ensure that the terms in eqn (2.245) respect gauge invariance, so that a mass term $e_L^{+T} C e_L^-$ is allowed, but not $e_L^{-T} C e_L^-$.

Left-handed fields transform as $(1/2, 0)$ under the Lorentz group, so that the fermion bilinear $\chi_L^T C \Gamma \psi_L$ transforms as $(1/2, 0) \otimes (1/2, 0) = (0, 0) \oplus (1, 0)$. The $(0, 0)$ representation is $\chi_L^T C \psi_L$ and the $(1, 0)$ representation is $\chi_L^T C \sigma^{\mu\nu} \psi_L$. The $(1, 0)$ representation is self-dual because of the self-duality condition on $\sigma^{\mu\nu} P_L$,

$$\frac{i}{2} \epsilon^{\alpha\beta\mu\nu} \sigma_{\mu\nu} P_L = \sigma^{\alpha\beta} P_L. \tag{2.246}$$

Similarly, the right-handed matrix satisfies the anti-self-duality condition

$$\frac{i}{2} \epsilon^{\alpha\beta\mu\nu} \sigma_{\mu\nu} P_R = -\sigma^{\alpha\beta} P_R. \tag{2.247}$$

Exercise 2.31 Show that $(\psi_{Lr}^T C \psi_{Ls})$ is symmetric in rs and $(\psi_{Lr}^T C \sigma^{\mu\nu} \psi_{Ls})$ is anti-symmetric in rs.

Exercise 2.32 Prove the duality relations eqns (2.246, 2.247). The sign convention is $\gamma_5 = i\gamma^0\gamma^1\gamma^2\gamma^3$ and $\epsilon_{0123} = +1$.

The lowest dimension term in the SMEFT with $\mathscr{D} > 4$ is the dimension-five term

$$\mathscr{L}^{(5)} = C_5 \underset{rs}{} \epsilon^{ij}\epsilon^{kl}(l_{ir}^T C l_{ks})H_j H_l + \text{h.c.}. \tag{2.248}$$

Here r,s are flavour indices, and i,j,k,l are $SU(2)$ gauge indices. The coefficient C_5 is symmetric in rs, by Exercise 2.31. $\mathscr{L}^{(5)}$ is a $\Delta L = \pm 2$ interaction, and gives a Majorana mass term to the neutrinos when H gets a vacuum expectation value.

Reference [58] shows that invariant operators constructed from SM fields satisfy

$$\frac{1}{2}(\Delta B - \Delta L) \equiv \mathscr{D} \qquad \text{mod } 2. \tag{2.249}$$

Thus a $\mathscr{D} = 5$ operator cannot conserve both baryon and lepton number.

Exercise 2.33

Show that eqn (2.248) is the unique dimension-five term in the SMEFT Lagrangian. How many independent operators are there for n_g generations?

Exercise 2.34 Show that eqn (2.248) generates a Majorana neutrino mass when H gets a vacuum expectation value, and find the neutrino mass matrix M_ν in terms of C_5 and v.

At dimension-six there are eight different operator classes, X^3, H^6, H^4D^2, X^2H^2, ψ^2H^3, ψ^2XH, ψ^2H^2D, and ψ^4, in terms of their field content. Determining the independent operators is a non-trivial task [17, 41]. Here I discuss a few aspects of the analysis.

The four-quark operators ψ^4 can be simplified using Fierz identities. Consider invariants made from two $\bar{l}\Gamma l$ bilinears. Since l is a left-handed field, the only gamma-matrix allowed is $\Gamma = \gamma^\mu$. Bilinears constructed from l can be either $SU(2)$ singlets or $SU(2)$ triplets, so the l^4 invariants are

$$\underset{pr\,st}{Q_{ll}} = (\bar{l}_{ip}\gamma^\mu l^i{}_r)(\bar{l}_{js}\gamma_\mu l^j{}_t),$$

$$\underset{pr\,st}{Q_{ll}^{(3)}} = (\bar{l}_{ip}\gamma^\mu [\tau^a]^i{}_j l^j{}_r)(\bar{l}_{ks}\gamma_\mu [\tau^a]^k{}_m l^m{}_t), \tag{2.250}$$

where p, r, s, t are generation (flavour) indices and i, j, k, m are weak $SU(2)$ indices. Using the $SU(2)$ Fierz identity (Exercise 2.3)

$$[\tau^a]^i{}_j [\tau^a]^k{}_m = 2\delta^i_m \delta^k_j - \delta^i_j \delta^k_m, \qquad (2.251)$$

the second bilinear can be written as

$$Q^{(3)}_{ll \atop pr\,st} = 2(\bar{l}_{ip} \gamma^\mu l^j{}_r)(\bar{l}_{js} \gamma_\mu l^i{}_t) - (\bar{l}_{ip} \gamma^\mu l^i{}_r)(\bar{l}_{js} \gamma_\mu l^j{}_t). \qquad (2.252)$$

Applying the spinor Fierz identity (Exercise 2.4)

$$(\overline{\psi}_1 \gamma^\mu P_L \psi_2)(\overline{\psi}_3 \gamma_\mu P_L \psi_4) = (\overline{\psi}_1 \gamma^\mu P_L \psi_4)(\overline{\psi}_3 \gamma_\mu P_L \psi_2) \qquad (2.253)$$

on the first term of eqn (2.252) gives

$$Q^{(3)}_{ll \atop pr\,st} = 2(\bar{l}_{ip} \gamma^\mu l^i{}_t)(\bar{l}_{js} \gamma_\mu l^j{}_r) - (\bar{l}_{ip} \gamma^\mu l^i{}_r)(\bar{l}_{js} \gamma_\mu l^j{}_t) = 2Q_{ll \atop ptsr} - Q_{ll \atop pr\,st}. \qquad (2.254)$$

Equation (2.254) implies that we do not need to include $Q^{(3)}_{ll \atop pr\,st}$ operators, as they are linear combinations of Q_{ll} operators, so the independent l^4 operators are Q_{ll}.

For lq operators,

$$Q^{(1)}_{lq \atop pr\,st} = (\bar{l}_{ip} \gamma^\mu l^i{}_r)(\bar{q}_{\alpha js} \gamma_\mu q^{\alpha j}{}_t),$$

$$Q^{(3)}_{lq \atop pr\,st} = (\bar{l}_{ip} \gamma^\mu [\tau^a]^i{}_j l^j{}_r)(\bar{q}_{\alpha ks} \gamma_\mu [\tau^a]^k{}_m q^{\alpha m}{}_t), \qquad (2.255)$$

the identity eqn (2.253) cannot be used since it would produce $(\bar{l}q)$ bilinears. Thus both lq operators in eqn (2.255) are independent.

For four-quark operators $(\bar{q}\gamma^\mu q)(\bar{q}\gamma_\mu q)$, there are four possible gauge invariants, written schematically as

$$1 \otimes 1, \quad \tau^a \otimes \tau^a, \quad T^A \otimes T^A, \quad \tau^a T^A \otimes \tau^a T^A, \qquad (2.256)$$

depending on what gauge generators are inserted in each bilinear. The $SU(N)$ version of eqn (2.251) from Exercise 2.3

$$[T^A]^\alpha{}_\beta [T^A]^\lambda{}_\sigma = \frac{1}{2}\delta^\alpha_\sigma \delta^\lambda_\beta - \frac{1}{2N}\delta^\alpha_\beta \delta^\lambda_\sigma, \qquad (2.257)$$

can be used for the colour generators with $N = 3$. We can view the index contractions for the $SU(2)$ and $SU(3)$ generators as either direct or swapped, i.e. in $(\bar{q}_1 \gamma^\mu q_2)(\bar{q}_3 \gamma_\mu q_4)$

contracted between q_1, q_2 and q_3, q_4, or between q_1, q_4 and q_2, q_3. Then the four possible terms in eqn (2.256) are

direct, $SU(2)$ swapped, $SU(3)$ swapped, both $SU(2)$ and $SU(3)$ swapped. (2.258)

The spinor Fierz identity eqn (2.253) exchanges the q fields, so it swaps both the $SU(2)$ and $SU(3)$ indices, and hence converts

direct $\leftrightarrow$ both swapped $\qquad SU(2)$ swapped $\leftrightarrow SU(3)$ swapped. (2.259)

Thus there are only two independent invariants out of the four in eqn (2.256), which are chosen to be $1 \otimes 1$ and $\tau^a \otimes \tau^a$.

For ψ^4 operators involving $\sigma^{\mu\nu}$, the duality relations eqns (2.246, 2.247) can be used to eliminate $\epsilon_{\mu\nu\alpha\beta}$ contracted with $\sigma^{\kappa\lambda}$ matrices. We also has the relation

$$(\overline{A}\sigma^{\mu\nu}P_L B)(\overline{C}\sigma_{\mu\nu}P_R D) = 0 \qquad (2.260)$$

The left-hand side is a Lorentz singlet in the tensor product $(1,0) \otimes (0,1) = (1,1)$, and so must vanish.

Using the above results, we can determine the independent ψ^4 operators.

Exercise 2.35 Prove eqn (2.260).

2.10.1 SMEFT Operators

Since the SMEFT is playing an increasingly important role in current research, I summarize the operators in SMEFT up to dimension six. The number of operators of each type is listed, and their *CP* property is given as a subscript. For non-Hermitian operators $\mathcal{O}$, $\mathcal{O} + \mathcal{O}^\dagger$ is *CP*-even, and $\mathcal{O} - \mathcal{O}^\dagger$ is *CP*-odd. The flavour indices have not been included for notational simplicity. For example, including flavour indices, Q_{eW} is Q_{eW} and Q_{ll} is $Q_{ll}_{pr\,st}$, etc.

Table 2.2 gives a summary of the SMEFT operators up to dimension six. For $n_g = 3$, there are six $\Delta L = 2$ operators plus their Hermitian conjugates, 273 $\Delta B = \Delta L = 1$ operators plus their Hermitian conjugates, and 2,499 Hermitian $\Delta B = \Delta L = 0$ operators [5]. For $n_g = 1$, there are seventy-six Hermitian $\Delta B = \Delta L = 0$ operators. In the literature, you will often see that there are fifty-nine $\Delta B = \Delta L = 0$ operators. This counts the number of operator types listed in the tables in the next subsections. Some of the operators, such as $(H^\dagger H)^3$ are Hermitian, whereas others, such as $(H^\dagger H)(\bar{l}eH)$ are not, and count as two Hermitian operators. Hermitian operators have a real coefficient in the Lagrangian, whereas non-Hermitian operators have a complex coefficient. Counting Hermitian operators is equivalent to counting real Lagrangian parameters.

Table 2.2 Number of operators of each type in the SMEFT up to dimension six.

dim		$n_g = 1$			$n_g = 3$		
		CP-even	CP-odd	Total	CP-even	CP-odd	Total
5	$\Delta L = 2$			1			6
5	$\Delta L = -2$			1			6
6	$\Delta B = \Delta L = 1$			4			273
6	$\Delta B = \Delta L = -1$			4			273
6	X^3	2	2	4	2	2	4
6	H^6	1	0	1	1	0	1
6	$H^4 D^2$	2	0	2	2	0	2
6	$X^2 H^2$	4	4	8	4	4	8
6	$\psi^2 H^3$	3	3	6	27	27	54
6	$\psi^2 X H$	8	8	16	72	72	144
6	$\psi^2 H^2 D$	8	1	9	51	30	81
6	$(\bar{L}L)(\bar{L}L)$	5	0	5	171	126	297
6	$(\bar{R}R)(\bar{R}R)$	7	0	7	255	195	450
6	$(\bar{L}L)(\bar{R}R)$	8	0	8	360	288	648
6	$(\bar{L}R)(\bar{R}L) + \text{h.c.}$	1	1	2	81	81	162
6	$(\bar{L}R)(\bar{L}R) + \text{h.c.}$	4	4	8	324	324	648
	Total $\Delta B = \Delta L = 0$	53	23	76	1350	1149	2499

2.10.1.1 Dimension 5, $\Delta L = 2$

The dimension-five operators Q_5 are $\Delta L = 2$ operators.

$$(LL)HH + \text{h.c.}$$

Q_5	$\frac{1}{2} n_g (n_g + 1)$	$\epsilon^{ij} \epsilon^{kl} (l_{ip}^T C l_{kr}) H_j H_l$
Total	$\frac{1}{2} n_g (n_g + 1) + \text{h.c.}$	

There are $n_g(n_g + 1)/2$ $\Delta L = 2$ operators, and $n_g(n_g + 1)/2$ $\Delta L = -2$ Hermitian conjugate operators. CP exchanges the $\Delta L = \pm 2$ operators. The $\Delta L = \pm 2$ operators give a Majorana neutrino mass when the weak interactions are spontaneously broken. Since neutrino masses are very small, the $\Delta L = \pm 2$ operators are assumed to be generated at a very high scale (which could be the GUT scale).

2.10.1.2 Dimension 6, $\Delta B = \Delta L = 1$

The dimension-six operators can be divided into several groups. The first group are the $\Delta B = \Delta L = 1$ operators and their Hermitian conjugates.

$$\Delta B = \Delta L = 1 + \text{h.c.}$$

Q_{duql}	n_g^4	$\epsilon^{\alpha\beta\gamma}\epsilon^{ij}(d_{\alpha p}^T C u_{\beta r})(q_{\gamma is}^T C l_{jt})$
Q_{qque}	$\frac{1}{2}n_g^3(n_g+1)$	$\epsilon^{\alpha\beta\gamma}\epsilon^{ij}(q_{\alpha ip}^T C q_{\beta jr})(u_{\gamma s}^T C e_t)$
Q_{qqql}	$\frac{1}{3}n_g^2(2n_g^2+1)$	$\epsilon^{\alpha\beta\gamma}\epsilon^{il}\epsilon^{jk}(q_{\alpha ip}^T C q_{\beta jr})(q_{\gamma ks}^T C l_{\ell t})$
Q_{duue}	n_g^4	$\epsilon^{\alpha\beta\gamma}(d_{\alpha p}^T C u_{\beta r})(u_{\gamma s}^T C e_t)$
Total	$\frac{1}{6}n_g^2(19n_g^2+3n_g+2)+\text{h.c.}$	

The $\Delta B = \Delta L = 1$ operators violate baryon number, and lead to proton decay. They are generated in unified theories, and are suppressed by two powers of the GUT scale.

2.10.1.3 Dimension 6, X^3

There are two *CP*-even and two *CP*-odd operators with three field-strength tensors. In this and subsequent tables, the *CP* property is shown as a subscript.

$$X^3$$

Q_G	1_+	$f^{ABC}G_\mu^{A\nu}G_\nu^{B\rho}G_\rho^{C\mu}$
$Q_{\tilde{G}}$	1_-	$f^{ABC}\tilde{G}_\mu^{A\nu}G_\nu^{B\rho}G_\rho^{C\mu}$
Q_W	1_+	$\epsilon^{IJK}W_\mu^{I\nu}W_\nu^{J\rho}W_\rho^{K\mu}$
$Q_{\tilde{W}}$	1_-	$\epsilon^{IJK}\tilde{W}_\mu^{I\nu}W_\nu^{J\rho}W_\rho^{K\mu}$
Total	$2_+ + 2_-$	

2.10.1.4 Dimension 6, H^6

There is a single operator involving six Higgs fields. It adds a h^6 interaction of the physical Higgs particle h to the SMEFT Lagrangian after spontaneous symmetry breaking.

$$H^6$$

Q_H	1_+	$(H^\dagger H)^3$
Total	1_+	

2.10.1.5 Dimension 6, H^4D^2

$$H^4D^2$$

$Q_{H\square}$	1_+	$(H^\dagger H)\square(H^\dagger H)$
Q_{HD}	1_+	$\left(H^\dagger D_\mu H\right)^* \left(H^\dagger D_\mu H\right)$
Total	2_+	

2.10.1.6 Dimension 6, X^2H^2

$$X^2H^2$$

Q_{HG}	1_+	$H^\dagger H \, G^A_{\mu\nu} G^{A\mu\nu}$
$Q_{H\widetilde{G}}$	1_-	$H^\dagger H \, \widetilde{G}^A_{\mu\nu} G^{A\mu\nu}$
Q_{HW}	1_+	$H^\dagger H \, W^I_{\mu\nu} W^{I\mu\nu}$
$Q_{H\widetilde{W}}$	1_-	$H^\dagger H \, \widetilde{W}^I_{\mu\nu} W^{I\mu\nu}$
Q_{HB}	1_+	$H^\dagger H \, B_{\mu\nu} B^{\mu\nu}$
$Q_{H\widetilde{B}}$	1_-	$H^\dagger H \, \widetilde{B}_{\mu\nu} B^{\mu\nu}$
Q_{HWB}	1_+	$H^\dagger \tau^I H \, W^I_{\mu\nu} B^{\mu\nu}$
$Q_{H\widetilde{W}B}$	1_-	$H^\dagger \tau^I H \, \widetilde{W}^I_{\mu\nu} B^{\mu\nu}$
Total	$4_+ + 4_-$	

The X^2H^2 operators are very important phenomenologically. They lead to $gg \to h$ and $h \to \gamma\gamma$ vertices, and contribute to Higgs production and decay. The corresponding SM amplitudes start at one loop, so LHC experiments are sensitive to X^2H^2 operators via interference effects with SM amplitudes [40, 70].

2.10.1.7 Dimension 6, ψ^2H^3

$$(\bar{L}R)H^3 + \text{h.c.}$$

Q_{eH}	n_g^2	$(H^\dagger H)(\bar{l}_p e_r H)$
Q_{uH}	n_g^2	$(H^\dagger H)(\bar{q}_p u_r \widetilde{H})$
Q_{dH}	n_g^2	$(H^\dagger H)(\bar{q}_p d_r H)$
Total	$3n_g^2 + \text{h.c.}$	

These operators are $H^\dagger H$ times the SM Yukawa couplings, and violate the relation that the Higgs boson coupling to fermions is proportional to their mass.

2.10.1.8 Dimension 6, $\psi^2 XH$

$$(\bar{L}R)XH + \text{h.c.}$$

Q_{eW}	n_g^2	$(\bar{l}_p \sigma^{\mu\nu} e_r) \tau^I H W_{\mu\nu}^I$
Q_{eB}	n_g^2	$(\bar{l}_p \sigma^{\mu\nu} e_r) H B_{\mu\nu}$
Q_{uG}	n_g^2	$(\bar{q}_p \sigma^{\mu\nu} T^A u_r) \tilde{H} G_{\mu\nu}^A$
Q_{uW}	n_g^2	$(\bar{q}_p \sigma^{\mu\nu} u_r) \tau^I \tilde{H} W_{\mu\nu}^I$
Q_{uB}	n_g^2	$(\bar{q}_p \sigma^{\mu\nu} u_r) \tilde{H} B_{\mu\nu}$
Q_{dG}	n_g^2	$(\bar{q}_p \sigma^{\mu\nu} T^A d_r) H G_{\mu\nu}^A$
Q_{dW}	n_g^2	$(\bar{q}_p \sigma^{\mu\nu} d_r) \tau^I H W_{\mu\nu}^I$
Q_{dB}	n_g^2	$(\bar{q}_p \sigma^{\mu\nu} d_r) H B_{\mu\nu}$
Total	$8n_g^2 + \text{h.c.}$	

When H gets a VEV, these operators lead to dipole operators for transitions such as $\mu \to e\gamma$, $b \to s\gamma$, and $b \to sg$.

2.10.1.9 Dimension 6, $\psi^2 H^2 D$

$$\psi^2 H^2 D$$

$Q_{Hl}^{(1)}$	$\frac{1}{2} n_g(n_g+1)_+ + \frac{1}{2} n_g(n_g-1)_-$	$(H^\dagger i \overset{\leftrightarrow}{D}_\mu H)(\bar{l}_p \gamma^\mu l_r)$
$Q_{Hl}^{(3)}$	$\frac{1}{2} n_g(n_g+1)_+ + \frac{1}{2} n_g(n_g-1)_-$	$(H^\dagger i \overset{\leftrightarrow}{D}_\mu^I H)(\bar{l}_p \tau^I \gamma^\mu l_r)$
Q_{He}	$\frac{1}{2} n_g(n_g+1)_+ + \frac{1}{2} n_g(n_g-1)_-$	$(H^\dagger i \overset{\leftrightarrow}{D}_\mu H)(\bar{e}_p \gamma^\mu e_r)$
$Q_{Hq}^{(1)}$	$\frac{1}{2} n_g(n_g+1)_+ + \frac{1}{2} n_g(n_g-1)_-$	$(H^\dagger i \overset{\leftrightarrow}{D}_\mu H)(\bar{q}_p \gamma^\mu q_r)$
$Q_{Hq}^{(3)}$	$\frac{1}{2} n_g(n_g+1)_+ + \frac{1}{2} n_g(n_g-1)_-$	$(H^\dagger i \overset{\leftrightarrow}{D}_\mu^I H)(\bar{q}_p \tau^I \gamma^\mu q_r)$
Q_{Hu}	$\frac{1}{2} n_g(n_g+1)_+ + \frac{1}{2} n_g(n_g-1)_-$	$(H^\dagger i \overset{\leftrightarrow}{D}_\mu H)(\bar{u}_p \gamma^\mu u_r)$
Q_{Hd}	$\frac{1}{2} n_g(n_g+1)_+ + \frac{1}{2} n_g(n_g-1)_-$	$(H^\dagger i \overset{\leftrightarrow}{D}_\mu H)(\bar{d}_p \gamma^\mu d_r)$
$Q_{Hud} + \text{h.c.}$	$n_g^2 + \text{h.c.}$	$i(\tilde{H}^\dagger D_\mu H)(\bar{u}_p \gamma^\mu d_r)$
Total	$\frac{1}{2} n_g(9n_g+7)_+ + \frac{1}{2} n_g(9n_g-7)_-$	

The $\psi^2 H^2 D$ operators modify the coupling of electroweak bosons to fermions. $(Q_{Hud} \pm Q_{Hud}^\dagger)$ are *CP*-even/odd combinations, and contribute n_g^2 *CP*-even and n_g^2 *CP*-odd operators to the total.

2.10.1.10 Dimension 6, $(\bar{L}L)(\bar{L}L)$

The ψ^4 operators can be grouped into different sets, depending on the chirality properties of the operators. We have seen earlier why the $(\bar{L}L)(\bar{L}L)$ invariants are the ones listed in the table.

$$(\bar{L}L)(\bar{L}L)$$

Q_{ll}	$\frac{1}{4}n_g^2(n_g^2+3)_+ + \frac{1}{4}n_g^2(n_g^2-1)_-$	$(\bar{l}_p\gamma_\mu l_r)(\bar{l}_s\gamma^\mu l_t)$
$Q_{qq}^{(1)}$	$\frac{1}{4}n_g^2(n_g^2+3)_+ + \frac{1}{4}n_g^2(n_g^2-1)_-$	$(\bar{q}_p\gamma_\mu q_r)(\bar{q}_s\gamma^\mu q_t)$
$Q_{qq}^{(3)}$	$\frac{1}{4}n_g^2(n_g^2+3)_+ + \frac{1}{4}n_g^2(n_g^2-1)_-$	$(\bar{q}_p\gamma_\mu \tau^I q_r)(\bar{q}_s\gamma^\mu \tau^I q_t)$
$Q_{lq}^{(1)}$	$\frac{1}{2}n_g^2(n_g^2+1)_+ + \frac{1}{2}n_g^2(n_g^2-1)_-$	$(\bar{l}_p\gamma_\mu l_r)(\bar{q}_s\gamma^\mu q_t)$
$Q_{lq}^{(3)}$	$\frac{1}{2}n_g^2(n_g^2+1)_+ + \frac{1}{2}n_g^2(n_g^2-1)_-$	$(\bar{l}_p\gamma_\mu \tau^I l_r)(\bar{q}_s\gamma^\mu \tau^I q_t)$
Total	$\frac{1}{4}n_g^2(7n_g^2+13)_+ + \frac{7}{4}n_g^2(n_g^2-1)_-$	

2.10.1.11 Dimension 6, $(\bar{R}R)(\bar{R}R)$

$$(\bar{R}R)(\bar{R}R)$$

Q_{ee}	$\frac{1}{8}n_g(n_g+1)(n_g^2+n_g+2)_+ + \frac{1}{8}(n_g-1)n_g(n_g+1)(n_g+2)_-$	$(\bar{e}_p\gamma_\mu e_r)(\bar{e}_s\gamma^\mu e_t)$
Q_{uu}	$\frac{1}{4}n_g^2(n_g^2+3)_+ + \frac{1}{4}n_g^2(n_g^2-1)_-$	$(\bar{u}_p\gamma_\mu u_r)(\bar{u}_s\gamma^\mu u_t)$
Q_{dd}	$\frac{1}{4}n_g^2(n_g^2+3)_+ + \frac{1}{4}n_g^2(n_g^2-1)_-$	$(\bar{d}_p\gamma_\mu d_r)(\bar{d}_s\gamma^\mu d_t)$
Q_{eu}	$\frac{1}{2}n_g^2(n_g^2+1)_+ + \frac{1}{2}n_g^2(n_g^2-1)_-$	$(\bar{e}_p\gamma_\mu e_r)(\bar{u}_s\gamma^\mu u_t)$
Q_{ed}	$\frac{1}{2}n_g^2(n_g^2+1)_+ + \frac{1}{2}n_g^2(n_g^2-1)_-$	$(\bar{e}_p\gamma_\mu e_r)(\bar{d}_s\gamma^\mu d_t)$
$Q_{ud}^{(1)}$	$\frac{1}{2}n_g^2(n_g^2+1)_+ + \frac{1}{2}n_g^2(n_g^2-1)_-$	$(\bar{u}_p\gamma_\mu u_r)(\bar{d}_s\gamma^\mu d_t)$
$Q_{ud}^{(8)}$	$\frac{1}{2}n_g^2(n_g^2+1)_+ + \frac{1}{2}n_g^2(n_g^2-1)_-$	$(\bar{u}_p\gamma_\mu T^A u_r)(\bar{d}_s\gamma^\mu T^A d_t)$
Total	$\frac{1}{8}n_g(21n_g^3+2n_g^2+31n_g+2)_+ + \frac{1}{8}n_g(n_g^2-1)(21n_g+2)_-$	

2.10.1.12 Dimension 6, $(\bar{L}L)(\bar{R}R)$

$$(\bar{L}L)(\bar{R}R)$$

Q_{le}	$\frac{1}{2}n_g^2(n_g^2+1)_+ + \frac{1}{2}n_g^2(n_g^2-1)_-$	$(\bar{l}_p\gamma_\mu l_r)(\bar{e}_s\gamma^\mu e_t)$
Q_{lu}	$\frac{1}{2}n_g^2(n_g^2+1)_+ + \frac{1}{2}n_g^2(n_g^2-1)_-$	$(\bar{l}_p\gamma_\mu l_r)(\bar{u}_s\gamma^\mu u_t)$
Q_{ld}	$\frac{1}{2}n_g^2(n_g^2+1)_+ + \frac{1}{2}n_g^2(n_g^2-1)_-$	$(\bar{l}_p\gamma_\mu l_r)(\bar{d}_s\gamma^\mu d_t)$
Q_{qe}	$\frac{1}{2}n_g^2(n_g^2+1)_+ + \frac{1}{2}n_g^2(n_g^2-1)_-$	$(\bar{q}_p\gamma_\mu q_r)(\bar{e}_s\gamma^\mu e_t)$
$Q_{qu}^{(1)}$	$\frac{1}{2}n_g^2(n_g^2+1)_+ + \frac{1}{2}n_g^2(n_g^2-1)_-$	$(\bar{q}_p\gamma_\mu q_r)(\bar{u}_s\gamma^\mu u_t)$
$Q_{qu}^{(8)}$	$\frac{1}{2}n_g^2(n_g^2+1)_+ + \frac{1}{2}n_g^2(n_g^2-1)_-$	$(\bar{q}_p\gamma_\mu T^A q_r)(\bar{u}_s\gamma^\mu T^A u_t)$
$Q_{qd}^{(1)}$	$\frac{1}{2}n_g^2(n_g^2+1)_+ + \frac{1}{2}n_g^2(n_g^2-1)_-$	$(\bar{q}_p\gamma_\mu q_r)(\bar{d}_s\gamma^\mu d_t)$
$Q_{qd}^{(8)}$	$\frac{1}{2}n_g^2(n_g^2+1)_+ + \frac{1}{2}n_g^2(n_g^2-1)_-$	$(\bar{q}_p\gamma_\mu T^A q_r)(\bar{d}_s\gamma^\mu T^A d_t)$
Total	$4n_g^2(n_g^2+1)_+ + 4n_g^2(n_g^2-1)_-$	

2.10.1.13 Dimension 6, $(\bar{L}R)(\bar{R}L)$

	$(\bar{L}R)(\bar{R}L)$	+ h.c.
Q_{ledq}	n_g^4	$(\bar{l}_p^j e_r)(\bar{d}_s q_{tj})$
Total	$n_g^4 + \text{h.c.}$	

2.10.1.14 Dimension 6, $(\bar{L}R)(\bar{L}R)$

	$(\bar{L}R)(\bar{L}R)$	+ h.c.
$Q_{quqd}^{(1)}$	n_g^4	$(\bar{q}_p^j u_r)\epsilon_{jk}(\bar{q}_s^k d_t)$
$Q_{quqd}^{(8)}$	n_g^4	$(\bar{q}_p^j T^A u_r)\epsilon_{jk}(\bar{q}_s^k T^A d_t)$
$Q_{lequ}^{(1)}$	n_g^4	$(\bar{l}_p^j e_r)\epsilon_{jk}(\bar{q}_s^k u_t)$
$Q_{lequ}^{(3)}$	n_g^4	$(\bar{l}_p^j \sigma_{\mu\nu} e_r)\epsilon_{jk}(\bar{q}_s^k \sigma^{\mu\nu} u_t)$
Total	$4n_g^4 + \text{h.c.}$	

Exercise 2.36

In the SMEFT for n_g generations, how many operators are there of the following kind (in increasing order of difficulty): (a) Q_{He} (b) Q_{ledq} (c) $Q_{lq}^{(1)}$ (d) $Q_{qq}^{(1)}$ (e) Q_{ll} (f) Q_{uu} (g) Q_{ee} (h) Show that there are a total of 2,499 Hermitian dimension-six $\Delta B = \Delta L = 0$ operators.

The NDA normalization eqn (2.200) for the SMEFT leads to an interesting pattern for the operators [34, 52],

$$
\begin{aligned}
\mathscr{L} \sim\; & \widehat{C}_H \frac{(4\pi)^4}{\Lambda^2} H^6 \\
& + \widehat{C}_{\psi^2 H^3} \frac{(4\pi)^3}{\Lambda^2} \psi^2 H^3 \\
& + \widehat{C}_{H^4 D^2} \frac{(4\pi)^2}{\Lambda^2} H^4 D^2 + \widehat{C}_{\psi^2 H^2 D} \frac{(4\pi)^2}{\Lambda^2} \psi^2 H^2 D + \widehat{C}_{\psi^4} \frac{(4\pi)^2}{\Lambda^2} \psi^4 \\
& + \widehat{C}_{\psi^2 XH} \frac{(4\pi)}{\Lambda^2} g \psi^2 XH \\
& + \widehat{C}_{X^2 H^2} \frac{1}{\Lambda^2} g^2 X^2 H^2 \\
& + \widehat{C}_{X^3} \frac{1}{(4\pi)^2 \Lambda^2} g^3 X^3
\end{aligned}
\tag{2.261}
$$

with 4π factors ranging from $(4\pi)^4$ to $1/(4\pi)^2$, a variation of $\sim 4 \times 10^6$.

The complete renormalization group equations for the SMEFT up to dimension six have been worked out [3, 5, 53, 54]. A very interesting feature of these equations is that they respect holomorphy, reminiscent of what happens in a supersymmetric gauge theory [4]. The renormalization group equations take a simpler form if written using the normalization eqn (2.261).

2.10.2   EFT below M_W

Below the electroweak scale, we can write a low-energy effective theory (LEFT) with quark and lepton fields, and only QCD and QED gauge fields. The operators have been classified in [50, 51]. Since $SU(2)$ gauge invariance is no longer a requirement, there are several new types of operators beyond those in SMEFT.

- There are dimension-three $\nu\nu$ operators, which give a Majorana neutrino mass for left-handed neutrinos.
- There are dimension-five dipole operators. These are the analogue of the $(\bar{L}R)XH$ operators in Section 2.10.1.8, which turn into dimension-five operators when H is replaced by its vacuum expectation value v. There are seventy Hermitian $\Delta B = \Delta L = 0$ dipole operators for $n_g = 3$.
- There are X^3 and ψ^4 operators as in SMEFT, but operators containing H are no longer present.
- There are $\Delta L = 4$ ν^4 operators, and $\Delta L = 2$ $(\bar{\psi}\psi)\nu\nu$ four-fermion operators, as well as four-fermion $\Delta B = -\Delta L$ operators.
- There are 3,631 Hermitian $\Delta B = \Delta L = 0$ dimension-six operators for $n_g = 3$.

The complete renormalization group equations up to dimension six have been worked out for LEFT [50, 51]. Since the theory has dimension-five operators, there are non-linear terms from two insertions of dimension-five operators for the dimension-six operator running. Various pieces of the computation have been studied previously [1, 2, 14, 16, 19–21, 24–26, 29, 39].

Appendix 2.A   Naturalness and the hierarchy problem

In the SM, most Lagrangian terms have dimension four, but there is an operator of dimension two,

$$\mathscr{L} = \lambda v^2 H^\dagger H. \tag{2.A.1}$$

$m_H^2 = 2\lambda v^2$ is the mass of the physical Higgs scalar h. If we assume the SM is an EFT with a power counting scale $\Lambda \gg v$, then blindly applying eqn (2.41) gives

$$\mathscr{L} \sim \Lambda^2 H^\dagger H. \tag{2.A.2}$$

The quadratic Λ^2 dependence in eqn (2.A.2) is the so-called hierarchy problem: that the Higgs mass gets a correction of order Λ. Similarly, the cosmological constant c, the coefficient of the dimension-zero operator **1** is of order Λ^4, whereas we know experimentally that $c \sim (2.8 \times 10^{-3} \text{ eV})^4$.

The power counting argument of eqn (2.41) does *not* imply that $m_H \propto \Lambda$ or $c \propto \Lambda^4$. We have seen in Section 2.5.2 and eqn (2.113) that there are no Λ^2 and Λ^4 contributions from loops in dimensional regularization. By construction, the EFT describes the dynamics of a theory with particles with masses m_H much smaller than Λ. Since H is in our EFT Lagrangian, its mass m_H is a light scale, $m_H \ll \Lambda$. With this starting point, corrections to m_H only depend on other light scales and possible suppression factors of $1/\Lambda$ from higher-dimension terms. There are no positive powers of Λ.

Let us look at the hierarchy problem in more detail. The usual argument is that loop corrections using a cutoff Λ give contributions to m_H^2 of order Λ^2, so that the bare $m_0^2 H^\dagger H$ coupling in the Lagrangian must be fine-tuned to cancel the Λ^2 contribution, leaving a small remainder of order v^2. Furthermore, this cancellation is unnatural, because m_0^2 must be fine-tuned order-by-order in perturbation theory to cancel the Λ^2 terms from higher-order corrections. There are several reasons why this argument is invalid. Firstly, Nature does not use perturbation theory, so what happens order-by-order in perturbation theory is irrelevant. Secondly, in a sensible renormalization scheme that factorizes scales properly, such as dimensional regularization, there are no Λ^2 loop contributions, and no fine-tuning is needed. Explicit computation of the Higgs mass correction in eqn (2.93) shows that the correction is proportional to m_H^2, not Λ^2.

Here is an even better argument—assume there is new BSM physics with particles at a high scale M_G, say the GUT scale. Then loop corrections to the renormalized mass m^2 are proportional to M_G^2, and these must be cancelled order-by-order in perturbation theory to give a Higgs mass m_H much smaller than M_G. The order-by-order problem is irrelevant, as before. However, we still have corrections $m^2 \propto M_G^2$, even if we compute exactly. These terms show the sensitivity of IR physics (the Higgs mass m_H) to UV parameters (M_G). Recall that in the introduction, it was obvious that short- and long-distance physics factorized, and our bridge-builder did not need to know about M_G to design a bridge. The sensitive dependence of m_H on M_G follows because we are computing low-energy observables in terms of high-energy parameters. We have already seen an example of this in Section 2.3.4. The solution is to use parameters defined at the scale of the measurement. Using the EFT ideas discussed so far, it should be clear that if we do this, all M_G effects are either logarithmic, and can be absorbed into running coupling constants, or are *suppressed* by powers of $1/M_G$. There are no corrections with positive powers of M_G.

Naturalness arguments all rely on the sensitivity of low-energy observables to high-energy (short-distance) Lagrangian parameters. But treating this as a fundamental problem is based on attributing an unjustified importance to Lagrangian parameters. For example, the amplitudes program uses on-shell scattering amplitudes as the basic building blocks of the theory, rather than a Lagrangian. Lagrangian parameters are a convenient way of relating physical observables to each other, as discussed in Section 2.3.3. As an example, consider the computation of hadron properties using lattice gauge theory, with Wilson fermions. The bare quark masses m_0 get corrections of order $\Lambda \sim 1/a$, where a is the lattice spacing. m_0 must be adjusted so that the physical pion mass is small (remember that we cannot measure the quark mass), and this is what is done in numerical simulations. Obtaining light Wilson fermions in the continuum limit is a numerical problem, not a fundamental one. The lattice fine-tuning required does not imply that QCD has a naturalness problem. We know this, because there are other ways to calculate in which the fine-tuning is absent. Similarly, in GUTs, there are ways to calculate in which there is no fine-tuning required for the Higgs mass.

We now consider the only version of the hierarchy problem that does not depend on how experimental observables are calculated. Assume we have a theory with two scales M_G, and m_H,

Figure 2.A.1 *Photos of the 21 Aug 2017 solar eclipse.* [Credit: P. Stoffer]

which are widely separated, $m_H \ll M_G$. Here m_H and M_G are *not* Lagrangian parameters, but experimentally measured physical scales. m_H can be obtained by measuring the physical Higgs mass, $m_H \sim 125$ GeV. M_G can be measured, for example, from the proton decay rate (if the proton does decay). The hierarchy problem is simply the statement that two masses, m_H and M_G, are very different. But suppose instead that we had a situation where m_H and M_G were comparable? Then we would have a different naturalness problem—m_H and M_G can differ by many orders of magnitude, so why are they comparable? The only physics problem is to understand why experimentally measurable quantities such as α_{QED}, m_e, m_μ, m_B, etc., have the values they have. Naturalness is not such a problem. The more than forty years of failure in searches for new physics based on naturalness, as well as the non-zero value of the cosmological constant, have shown that Nature does not care about naturalness.

Finally, let me comment on another fine-tuning problem that many of you are excited about. There will be a total solar eclipse on 21 August, 2017, shortly after the Les Houches school ends. The angular diameter of the Sun and Moon as seen from Earth are almost identical—the Moon will cover the Sun, leaving only the solar corona visible (see Figure 2.A.1). The angular diameters of the Sun and Moon are both experimentally measured (unlike in the Higgs problem where the Higgs mass parameter m at the high scale M_G is not measured) and the difference of angular diameters is much smaller than either.[3] Do you want to spend your life solving such problems?

---

[3] The angular diameters are not constant, but change because of the small eccentricity of the orbit. As a result, we can have both total and annular eclipses. Furthermore, the Earth-Sun-Moon system is almost planar; otherwise there would not be an eclipse. These are two additional fine-tunings.

Acknowledgements

I thank Sacha Davidson, Paolo Gambino, Mikko Laine, and Matthias Neubert for organizing a very interesting school, and all the students for listening patiently to the lectures, asking lots of questions, and solving homework problems on weekends instead of hiking. Sacha Davidson, in particular, made sure all the students were well looked after. I also thank Elizabeth Jenkins, Andrew Kobach, John McGreevy, and Peter Stoffer for carefully reading the manuscript, and Peter Stoffer for permission to use the photographs in Figure 2.A.1. This work was supported in part by DOE Grant No. DE-SC0009919.

References

[1] J. Aebischer, A. Crivellin, M. Fael, and C. Greub. (2016). Matching of Gauge-invariant Dimension-six Operators for $b \rightarrow s$ and $b \rightarrow c$ Transitions. *JHEP* 05: 037.

[2] J. Aebischer, M. Fael, C. Greub, and J. Virto (2017). *B* physics Beyond the Standard Model at One Loop: Complete Renormalization Group Evolution below the Electroweak Scale. *JHEP* 09: 158.

[3] R. Alonso, et al. (2014). Renormalization Group Evolution of Dimension-six Baryon Number Violating Operators. *Phys. Lett. B.* 734: 302–307.

[4] R. Alonso, E. E. Jenkins, and A. V. Manohar (2014). Holomorphy without Supersymmetry in the Standard Model Effective Field Theory. *Phys. Lett.* B739: 95–8.

[5] R. Alonso, E. E. Jenkins, A. V. Manohar, and M. Trott (2014). Renormalization Group Evolution of the Standard Model Dimension Six Operators III: Gauge Coupling Dependence and Phenomenology. *JHEP* 04: 159.

[6] S. Aoki, et al. (2017). Review of Lattice Results Concerning Low-energy Particle Physics. *Eur. Phys. J. C.* 77(2): 112.

[7] T. Appelquist and J. Carazzone (1975). Infrared Singularities and Massive Fields. *Phys. Rev. D.* 11: 2856.

[8] C. W. Bauer, S. Fleming, and M. E. Luke (2000). Summing Sudakov Logarithms in $B \rightarrow X_s \gamma$ in Effective Field Theory. *Phys. Rev. D.* 63: 014006.

[9] C. W. Bauer, S. Fleming, D. Pirjol, and I. W. Stewart (2001). An Effective Field Theory for Collinear and Soft Gluons: Heavy to Light Decays. *Phys. Rev. D.* 63: 114020.

[10] C. W. Bauer, D. Pirjol, and I. W. Stewart (2002). Soft Collinear Factorization in Effective Field Theory. *Phys. Rev. D.* D65: 054022.

[11] C. W. Bauer and I. W. Stewart (2001). Invariant Operators in Collinear Effective Theory. *Phys. Lett. B.* 516: 134–42.

[12] T. Becher. Soft Collinear Effective Theory. Chapter 5, this volume.

[13] M. Beneke and V. A. Smirnov (1998). Asymptotic Expansion of Feynman Integrals near Threshold. *Nucl. Phys. B.* 522, 321–44.

[14] T. Bhattacharya, et al. (2015). Dimension-5 CP-odd Operators: QCD Mixing and Renormalization. *Phys. Rev. D.* 92(11): 114026.

[15] I. Brivio and M. Trott (2017). The Standard Model as an Effective Field Theory. Phys. Rep. 793: 1–98.

[16] G. Buchalla, A. J. Buras, and M. E. Lautenbacher (1996). Weak Decays Beyond Leading Logarithms. *Rev. Mod. Phys.* 68: 1125–44.

[17] W. Buchmuller and D. Wyler (1986). Effective Lagrangian Analysis of New Interactions and Flavor Conservation. *Nucl. Phys. B.* 268: 621.

[18] W. E. Caswell and G. P. Lepage (1986). Effective Lagrangians for Bound State Problems in QED, QCD, and Other Field Theories. *Phys. Lett. B.* 167, 437–42.

[19] A. Delis, J. Fuentes Martın, A. Vicente, and J. Virto (2017). DsixTools: The Standard Model Effective Field Theory Toolkit. *Eur. Phys. J. C.* 77(6): 405.

[20] V. Cirigliano, S. Davidson, and Y. Kuno (2017). Spin-dependent $\mu \to e$ Conversion. *Phys. Lett. B.* 771: 242–6.

[21] V. Cirigliano, M. González Alonso, and M. L. Graesser (2013). Non-standard Charged Current Interactions: Beta Decays Versus the LHC. *JHEP* 02: 046.

[22] S. R. Coleman (1975). The Quantum Sine–Gordon Equation as the Massive Thirring Model. *Phys. Rev. D.* 11: 2088.

[23] J. C. Collins (1986). *Renormalization.* Volume 26, Cambridge Monographs on Mathematical Physics, Cambridge University Press.

[24] A. Crivellin, S. Davidson, G. M. Pruna, and A. Signer (2017). Renormalisation-Group Improved Analysis of $\mu \to e$ Processes in a Systematic Effective-field-theory Approach. *JHEP* 05: 117.

[25] S. Davidson (2016). $\mu \to e\gamma$ and Matching at m_W. *Eur. Phys. J. C.,* 76(7): 370.

[26] W. Dekens and J. de Vries (2013). Renormalization Group Running of Dimension-Six Sources of Parity and Time-Reversal Violation. *JHEP* 05: 149.

[27] H. Euler (1936). Über die Streuung von Licht an Licht nach der Diracschen Theorie. *Ann. der Physik* 26: 398.

[28] H. Euler and B. Kockel (1935). Über die Streuung von Licht an Licht nach der Diracschen Theorie. *Naturwiss.* 23: 246.

[29] A. Falkowski, M. González Alonso, and K. Mimouni (2017). Compilation of Low-energy Constraints on 4-fermion Operators in the SMEFT. *JHEP* 08: 123.

[30] E. Fermi (1933). Tentativo di una teoria dell'emissione dei raggi beta. *Ric. Sci.* 4: 491–5.

[31] K. Fujikawa (1979). Path Integral Measure for Gauge Invariant Fermion Theories. *Phys. Rev. Lett.* 42: 1195–8.

[32] P. Galison and A. Manohar (1984). Two Z's or Not Two Z's? *Phys. Lett. B.* 136, 279–83.

[33] J. Gasser and H. Leutwyler (1984). Chiral Perturbation Theory to One Loop. *Annals Phys.* 158: 142.

[34] B. M. Gavela, E. E. Jenkins, A. V. Manohar, and L. Merlo (2016). Analysis of General Power Counting Rules in Effective Field Theory. *Eur. Phys. J. C.* 76(9): 485.

[35] H. Georgi (1984). *Weak Interactions and Modern Particle Theory, Benjamin/Cummings.*

[36] H. Georgi (1990). An Effective Field Theory for Heavy Quarks at Low-Energies. *Phys. Lett. B.* 240, 447–50.

[37] H. Georgi (1991). On-shell Effective Field Theory. *Nucl. Phys. B.* 361: 339–50.

[38] F. J. Gilman and M. B. Wise (1979). Effective Hamiltonian for $\Delta s = 1$ Weak Nonleptonic Decays in the Six Quark Model. *Phys. Rev. D.* 20: 2392.

[39] M. González Alonso, J. Martin Camalich, and K. Mimouni (2017). Renormalization-Group Evolution of New Physics Contributions to (Semi)Leptonic Meson Decays. *Phys. Lett. B.* 772: 777–785.

[40] C. Grojean, E. E. Jenkins, A. V. Manohar, and M. Trott (2013). Renormalization Group Scaling of Higgs Operators and $\Gamma(h \to \gamma\gamma)$. *JHEP* 04: 016.

[41] B. Grzadkowski, M. Iskrzynski, M. Misiak, and J. Rosick (2010). Dimension-Six Terms in the Standard Model Lagrangian. *JHEP* 1010: 085.

[42] A. Hanany, E. E. Jenkins, A. V. Manohar, and G. Torri (2011). Hilbert Series for Flavor Invariants of the Standard Model. *JHEP* 03: 096.

[43] W. Heisenberg and H. Euler (1936). Consequences of Dirac's Theory of Positrons. *Z. Phys.* 98: 714–32.

[44] B. Henning, X. Lu, T. Melia, and H. Murayama (2017). 2, 84, 30, 993, 560, 15456, 11962, 261485, ...: Higher-Dimension Operators in the SM EFT. *JHEP* 08: 016.

[45] B. Holdom (1986). Two U(1)s and Epsilon Charge Shifts. *Phys. Lett. B.* 166: 196–8.

[46] N. Isgur and M. B. Wise (1989). Weak Decays of Heavy Mesons in the Static Quark Approximation. *Phys. Lett. B.* 232: 113–17.

[47] N. Isgur and M. B. Wise (1990). Weak Transition Form-Factors between Heavy Mesons. *Phys. Lett. B.* 237: 527–30.

[48] E. E. Jenkins (1998). Large N_c baryons. *Ann. Rev. Nucl. Part. Sci.* 48: 81–119.

[49] E. E. Jenkins and A. V. Manohar (2009). Algebraic Structure of Lepton and Quark Flavor Invariants and CP Violation. *JHEP* 10: 094.

[50] E. E. Jenkins, A. V. Manohar, and P. Stoffer (2017). Low-Energy Effective Field Theory below the Electroweak Scale: Anomalous Dimensions. JHEP01: 084.

[51] E. E. Jenkins, A. V. Manohar, and P. Stoffer (2017). Low-Energy Effective Field Theory below the Electroweak Scale: Operators and Matching. JHEP03: 016.

[52] E. E. Jenkins, A. V. Manohar, and M. Trott (2013). Naive Dimensional Analysis Counting of Gauge Theory Amplitudes and Anomalous Dimensions. *Phys. Lett. B.* 726: 697–702.

[53] E. E. Jenkins, A. V. Manohar, and M. Trott (2013). Renormalization Group Evolution of the Standard Model Dimension Six Operators I: Formalism and λ Dependence. *JHEP* 10: 087.

[54] E. E. Jenkins, A. V. Manohar, and M. Trott (2014). Renormalization Group Evolution of the Standard Model Dimension Six Operators II: Yukawa Dependence. *JHEP* 01: 035.

[55] L. P. Kadanoff (1966). Scaling Laws for Ising Models Near T_c. *Physics* 2: 263–72.

[56] D. B. Kaplan (1995). Effective Field Theories. In *Beyond the Standard Model 5. Proceedings, 5th Conference.* 29 April–4 May 1997, Balholm, Norway.

[57] F. Knop and P. Littelmann (1987). Der Grad erzeugender Funktionen von Invarianten-ringen. (German) [The Degree of Generating Functions of Rings of Invariants]. *Math. Z.* 196: 211.

[58] A. Kobach (2016). Baryon Number, Lepton Number, and Operator Dimension in the Standard Model. *Phys. Lett. B.* 758: 455–7.

[59] A. Kobach and S. Pal (2017). Hilbert Series and Operator Basis for NRQED and NRQCD/HQET. *Phys. Lett. B.* 772: 225–31.

[60] H. Lehmann, K. Symanzik, and W. Zimmerman (1955). On the Formulation of Quantized Field Theories. *Nuovo Cim.* 1: 205–25.

[61] Y. Liang and A. Czarnecki (2012). Photon–Photon Scattering: A Tutorial. *Can. J. Phys.* 90: 11–26.

[62] M. E. Luke, A. V. Manohar, and M. J. Savage (1992). A QCD Calculation of the Interaction of Quarkonium with Nuclei. *Phys. Lett. B.* 288: 355–9.

[63] T. Mannel. Effective Field Theories for Heavy Quarks. Chapter 9, this volume.

[64] A. V. Manohar and H. Georgi (1984). Chiral Quarks and the Nonrelativistic Quark Model. *Nucl. Phys. B.* 234: 189.

[65] A. V. Manohar (1997). Effective Field Theories. *Springer Lect. Notes Phys.* 479: 311–62.

[66] A. V. Manohar (1997). The HQET / NRQCD Lagrangian to Order α_s/m^3. *Phys. Rev. D.* 56: 230–237.

[67] A. V. Manohar (1998). Large N QCD. In *Probing the Standard Model of Particle Interactions*. Proceedings, Summer School in Theoretical Physics, parts 1–2, 1091–1169. NATO Advanced Study Institute, 68th Session. 28 July–5 September 1997, Les Houches, France.

[68] A. V. Manohar and M. B. Wise (1994). Inclusive Semileptonic B and Polarized Λ_b Decays from QCD. *Phys. Rev. D.* 49: 1310–29.

[69] A. V. Manohar and M. B. Wise (2000). Heavy Quark Physics. *Camb. Monogr. Part. Phys. Nucl. Phys. Cosmol.* 10: 1–191.

[70] A. V. Manohar and M. B. Wise (2006). Modifications to the Properties of the Higgs Boson. *Phys. Lett. B.* 636: 107–13.

[71] M. Neubert. Renormalization and RGEs. Chapter 1, this volume.

[72] C. Patrignani, et al. (2016). Review of Particle Physics. *Chin. Phys. C.* 40(10): 100001.

[73] A. Pich. Chiral Perturbation Theory. Chapter 3, this volume.

[74] A. Pich (1995). Chiral Perturbation Theory. *Rept. Prog. Phys.* 58: 563–610.

[75] H. D. Politzer (1980). Power Corrections at Short Distances. *Nucl. Phys. B.* 172: 349.

[76] I. Z. Rothstein (2003). TASI Lectures on Effective Field Theories.

[77] M. A. Shifman and M. B. Voloshin (1988). On Production of D and D^* Mesons in B Meson Decays. *Sov. J. Nucl. Phys.* 47: 511. [Yad. Fiz.47,801(1988)].

[78] I. W. Stewart (2013). Effective Field Theories. https://ocw.mit.edu/courses/physics/8-851-effective-field-theory-spring-2013/.

[79] Y. Tomozawa (1966). Axial Vector Coupling Renormalization and the Meson Baryon Scattering Lengths. *Nuovo Cim. A.* 46: 707–717.

[80] S. Weinberg (1966). Pion Scattering Lengths. *Phys. Rev. Lett.* 17: 616–621.

[81] S. Weinberg (1979). Phenomenological Lagrangians. *Physica A.* 96: 327–340.

[82] K. G. Wilson (1969). Nonlagrangian Models of Current Algebra. *Phys. Rev.* 179: 1499–1512.

[83] K. G. Wilson (1973). Quantum Field Theory Models in Less Than Four Dimensions. *Phys. Rev. D.* 7: 2911–26.

[84] K. G. Wilson and M. E. Fisher (1972). Critical Exponents in 3.99 Dimensions. *Phys. Rev. Lett.* 28: 240–3.

[85] K. G. Wilson and J. B. Kogut (1974). The Renormalization Group and the Epsilon Expansion. *Phys. Rept.* 12: 75–200.

3

Effective Field Theory with Nambu–Goldstone Modes

Antonio PICH

Departament de Física Teòrica, IFIC, Universitat de València - CSIC
Apt. Correus 22085, E-46071 València, Spain

Antonio PICH

Pich, A., *Effective Field Theory with Nambu–Goldstone Modes* In: *Effective Field Theory in Particle Physics and Cosmology.*
Edited by: Sacha Davidson, Paolo Gambino, Mikko Laine, Matthias Neubert and Christophe Salomon, Oxford University Press (2020).
© Oxford University Press. DOI: 10.1093/oso/9780198855743.003.0003

Chapter Contents

Field theories with spontaneous symmetry breaking (SSB) provide an ideal environment to apply the techniques of effective field theory (EFT). They contain massless Nambu–Goldstone modes, separated from the rest of the spectrum by an energy gap. The low-energy dynamics of the massless fields can then be analysed through an expansion in powers of momenta over some characteristic mass scale. Owing to the Nambu–Goldstone nature of the light fields, the resulting effective theory is highly constrained by the pattern of symmetry breaking.

Quantum chromodynamics (QCD) and the electroweak Standard Model (SM) are two paradigmatic field theories where symmetry breaking plays a critical role. If quark masses are neglected, the QCD Lagrangian has a global chiral symmetry that gets dynamically broken through a non-zero vacuum expectation value of the $\bar{q}q$ operator. With $n_f = 2$ light quark flavours, there are three associated Nambu–Goldstone modes that can be identified with the pion multiplet. The symmetry breaking mechanism is quite different in the SM case, where the electroweak gauge theory is spontaneously broken through a scalar potential with non-trivial minima. Once a given choice of the (non-zero) scalar vacuum expectation value is adopted, the excitations along the flat directions of the potential give rise to three massless Nambu–Goldstone modes, which in the unitary gauge become the longitudinal polarizations of the $W^\pm$ and Z gauge bosons. In spite of the totally different underlying dynamics (non-perturbative dynamical breaking versus perturbative SSB), the low-energy interactions of the Nambu–Goldstone modes are formally identical in the two theories because they share the same pattern of symmetry breaking.

The lectures compiled in this chapter provide an introduction to the EFT description of the Nambu–Goldstone fields, and some important phenomenological applications. A toy model incorporating the relevant symmetry properties is first studied in Section 3.1, and the different symmetry realizations, Wigner–Weyl and Nambu–Goldstone, are discussed in Section 3.2. Section 3.3 analyses the chiral symmetry of massless QCD. The corresponding Nambu–Goldstone EFT is developed next in Section 3.4, using symmetry properties only, while Section 3.5 discusses the explicit breakings of chiral symmetry and presents a detailed description of chiral perturbation theory (χPT), the low-energy effective realization of QCD. A few phenomenological applications are presented in Section 3.6. The quantum chiral anomalies are briefly discussed in Section 3.7, and Sections 3.8 and 3.9 are devoted to the dynamical understanding of the χPT couplings.

The electroweak symmetry breaking is analysed in Section 3.10, which discusses the custodial symmetry of the SM scalar potential and some relevant phenomenological consequences. Section 3.11 presents the electroweak effective theory formalism, while the short-distance information encoded in its couplings is studied in Section 3.12. A few summarizing comments are finally given in Section 3.13.

To prepare these lectures I have made extensive use of my previous reviews on EFT [167], χPT [166, 168, 169] and electroweak symmetry breaking [171]. Complementary information can be found in many excellent reports [33, 34, 44, 64, 78, 87, 98, 145, 146, 191] and books [74, 192], covering related subjects.

3.1   A Toy Lagrangian: The Linear Sigma Model

Let us consider a multiplet of four real scalar fields $\Phi(x)^T \equiv (\vec{\pi}, \sigma)$, described by the Lagrangian

$$\mathcal{L}_\sigma = \frac{1}{2} \partial_\mu \Phi^T \partial^\mu \Phi - \frac{\lambda}{4} \left(\Phi^T \Phi - v^2 \right)^2. \tag{3.1}$$

$\mathcal{L}_\sigma$ remains invariant under (x^μ-independent) $SO(4)$ rotations of the four scalar components. If v^2 were negative, this global symmetry would be realized in the usual Wigner–Weyl way, with four degenerate states of mass $m_\Phi^2 = -\lambda v^2$. However, for $v^2 > 0$, the potential has a continuous set of minima, occurring for all field configurations with $\Phi^T \Phi = v^2$. This is illustrated in Figure 3.1 that shows the analogous three-dimensional potential. These minima correspond to degenerate ground states, which transform into each other under $SO(4)$ rotations. Adopting the vacuum choice

$$\langle 0|\sigma|0 \rangle = v, \qquad\qquad \langle 0|\vec{\pi}|0 \rangle = 0, \tag{3.2}$$

and making the field redefinition $\hat{\sigma} = \sigma - v$, the Lagrangian takes the form

$$\mathcal{L}_\sigma = \frac{1}{2} \left[\partial_\mu \hat{\sigma} \partial^\mu \hat{\sigma} - 2\lambda v^2 \hat{\sigma}^2 + \partial_\mu \vec{\pi} \partial^\mu \vec{\pi} \right] - \lambda v \hat{\sigma} \left(\hat{\sigma}^2 + \vec{\pi}^2 \right) - \frac{\lambda}{4} \left(\hat{\sigma}^2 + \vec{\pi}^2 \right)^2, \tag{3.3}$$

which shows that the three $\vec{\pi}$ fields are massless Nambu–Goldstone modes, corresponding to excitations along the three flat directions of the potential, while $\hat{\sigma}$ acquires a mass $M^2 = 2\lambda v^2$.

The vacuum choice (3.2) is only preserved by $SO(3)$ rotations among the first three field components, leaving the fourth one untouched. Therefore, the potential triggers an $SO(4) \to SO(3)$ SSB, and there are three ($\frac{4 \times 3}{2} - \frac{3 \times 2}{2}$) broken generators that do not respect the adopted vacuum. The three Nambu–Goldstone fields are associated with these broken generators.

To better understand the role of symmetry on the Goldstone dynamics, it is useful to rewrite the sigma-model Lagrangian with different field variables. Using the 2×2 matrix notation

$$\Sigma(x) \equiv \sigma(x) I_2 + i \vec{\tau} \vec{\pi}(x), \tag{3.4}$$

with $\vec{\tau}$ the three Pauli matrices and I_2 the identity matrix, the Lagrangian (3.1) can be compactly written as

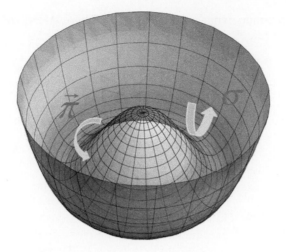

Figure 3.1 *Sigma-model potential.*

$$\mathcal{L}_\sigma = \frac{1}{4} \langle \partial_\mu \Sigma^\dagger \partial^\mu \Sigma \rangle - \frac{\lambda}{16} \left(\langle \Sigma^\dagger \Sigma \rangle - 2v^2 \right)^2, \tag{3.5}$$

where $\langle A \rangle$ denotes the trace of the matrix A. In this notation, $\mathcal{L}_\sigma$ is explicitly invariant under global $G \equiv SU(2)_L \otimes SU(2)_R$ transformations,

$$\Sigma \xrightarrow{G} g_R \Sigma g_L^\dagger, \qquad g_{L,R} \in SU(2)_{L,R}, \tag{3.6}$$

while the vacuum choice $\langle 0 | \Sigma | 0 \rangle = v I_2$ only remains invariant under those transformations satisfying $g_L = g_R$, *i.e.*, under the diagonal subgroup $H \equiv SU(2)_{L+R}$. Therefore, the pattern of symmetry breaking is

$$SU(2)_L \otimes SU(2)_R \longrightarrow SU(2)_{L+R}. \tag{3.7}$$

The change of field variables just makes manifest the equivalences of the groups $SO(4)$ and $SO(3)$ with $SU(2)_L \otimes SU(2)_R$ and $SU(2)_{L+R}$, respectively. The physics content is of course the same.

We can now make the polar decomposition

$$\Sigma(x) = [v + S(x)] \, U(\vec{\phi}), \qquad U(\vec{\phi}) = \exp \left\{ i \frac{\vec{\tau}}{v} \cdot \vec{\phi}(x) \right\}, \tag{3.8}$$

in terms of an Hermitian scalar field $S(x)$ and three pseudoscalar variables $\vec{\phi}(x)$, normalized with the scale v in order to preserve the canonical dimensions. $S(x)$ remains invariant under the symmetry group, while the matrix $U(\vec{\phi})$ inherits the chiral transformation of $\Sigma(x)$:

$$S \xrightarrow{G} S, \qquad\qquad U(\vec{\phi}) \xrightarrow{G} g_R U(\vec{\phi}) g_L^\dagger. \qquad (3.9)$$

Obviously, the fields $\vec{\phi}(x)$ within the exponential transform non-linearly. The sigma-model Lagrangian takes then a very enlightening form:

$$\mathcal{L}_\sigma = \frac{v^2}{4} \left(1 + \frac{S}{v}\right)^2 \langle \partial_\mu U^\dagger \partial^\mu U \rangle + \frac{1}{2}\left(\partial_\mu S \partial^\mu S - M^2 S^2\right) - \frac{M^2}{2v} S^3 - \frac{M^2}{8v^2} S^4. \quad (3.10)$$

This expression shows explicitly the following important properties:

- The massless Nambu–Goldstone bosons $\vec{\phi}$, parametrized through the matrix $U(\vec{\phi})$, have purely derivative couplings. Therefore, their scattering amplitudes vanish at zero momenta. This was not so obvious in eqn (3.3), and implies that this former expression of $\mathcal{L}_\sigma$ gives rise to exact (and not very transparent) cancellations among different momentum-independent contributions. The two functional forms of the Lagrangian should of course give the same physical predictions.

- The potential only depends on the radial variable $S(x)$, which describes a massive field with $M^2 = 2\lambda v^2$. In the limit $\lambda \gg 1$, the scalar field S becomes very heavy and can be integrated out from the Lagrangian. The linear sigma model then reduces to

$$\mathcal{L}_2 = \frac{v^2}{4} \langle \partial_\mu U^\dagger \partial^\mu U \rangle, \qquad (3.11)$$

which contains an infinite number of interactions among the $\vec{\phi}$ fields, owing to the non-linear functional form of $U(\vec{\phi})$. As we will see later, $\mathcal{L}_2$ is a direct consequence of the pattern of SSB in (3.7). It represents a universal (model-independent) interaction of the Nambu–Goldstone modes at very low energies.

- In order to be sensitive to the particular dynamical structure of the potential, and not just to its symmetry properties, we need to test the model-dependent part involving the scalar field S. At low momenta ($p \ll M$), the dominant tree-level corrections originate from S exchange, which generates the four-derivative term

$$\mathcal{L}_4 = \frac{v^2}{8M^2} \langle \partial_\mu U^\dagger \partial^\mu U \rangle^2. \qquad (3.12)$$

The corresponding contributions to the low-energy scattering amplitudes are suppressed by a factor p^2/M^2 with respect to the leading contributions from (3.11).

We can easily identify $\mathcal{L}_\sigma$ with the (non-gauged) scalar Lagrangian of the electroweak SM. However, the non-linear sigma model was originally suggested to describe the low-energy dynamics of the QCD pions [96, 193]. Both theories have the pattern of symmetry breaking displayed in eqn (3.7).

3.2   Symmetry Realizations

Noether's theorem guarantees the existence of conserved quantities associated with any continuous symmetry of the action. If a group G of field transformations leaves the Lagrangian invariant, for each generator of the group T^a, there is a conserved current $j_a^\mu(x)$ such that $\partial_\mu j_a^\mu = 0$ when the fields satisfy the Euler–Lagrangian equations of motion. The space integrals of the time-components $j_a^0(x)$ are then conserved charges, independent of the time coordinate:

$$Q_a = \int d^3x\, j_a^0(x), \qquad\qquad \frac{d}{dt}Q_a = 0. \qquad (3.13)$$

In the quantum theory, the conserved charges become symmetry generators that implement the group of transformations through the unitary operators $U = \exp\{i\theta^a Q_a\}$, being θ^a the continuous parameters characterizing the transformation. These unitary operators commute with the Hamiltonian, *i.e.*, $UHU^\dagger = H$.

In the usual Wigner–Weyl realization of the symmetry, the charges annihilate the vacuum, $Q_a|0\rangle = 0$, so that it remains invariant under the group of transformations: $U|0\rangle = |0\rangle$. This implies the existence of degenerate multiplets in the spectrum. Given a state $|A\rangle = \phi_A^\dagger|0\rangle$, the symmetry transformation $U\phi_A U^\dagger = \phi_B$ generates another state $|B\rangle = \phi_B^\dagger|0\rangle = U|A\rangle$ with the same energy:

$$E_B = \langle B|H|B\rangle = \langle A|U^\dagger H U|A\rangle = \langle A|H|A\rangle = E_A. \qquad (3.14)$$

The previous derivation is no longer valid when the vacuum is not invariant under some group transformations. Actually, if a charge does not annihilate the vacuum, $Q_a|0\rangle$ is not even well defined because

$$\langle 0|Q_a Q_b|0\rangle = \int d^3x\, \langle 0|j_a^0(x)Q_b|0\rangle = \langle 0|j_a^0(0)Q_b|0\rangle \int d^3x = \infty, \qquad (3.15)$$

where we have made use of the invariance under translations of the space-time coordinates, which implies

$$j_a^\mu(x) = e^{iP_\mu x^\mu} j_a^\mu(0)\, e^{-iP_\mu x^\mu}, \qquad (3.16)$$

with P_μ the four-momentum operator that satisfies $[P_\mu, Q_b] = 0$ and $P_\mu|0\rangle = 0$. Thus, one needs to be careful and state the vacuum properties of Q_a in terms of commutation relations that are mathematically well defined. We can easily proof the following important result [100, 101, 150, 151–153].

Nambu–Goldstone theorem: Given a conserved current $j_a^\mu(x)$ and its corresponding conserved charge Q_a, if there exists some operator $\mathcal{O}$ such that $v_a \equiv \langle 0|[Q_a, \mathcal{O}]|0\rangle \neq 0$, then the spectrum of the theory contains a massless state $|\phi_a\rangle$ that couples both to $\mathcal{O}$ and j_a^0, *i.e.*, $\langle 0|\mathcal{O}|\phi_a\rangle \langle\phi_a|j_a^0(0)|0\rangle \neq 0$.

Proof. Using (3.13), (3.16) and the completeness relation $\sum_n |n\rangle\langle n| = 1$, where the sum is over the full spectrum of the theory, the non-zero vacuum expectation value can be written as

$$v_a = \sum_n \int d^3x \left\{ \langle 0|j_a^0(x)|n\rangle\langle n|\mathcal{O}|0\rangle - \langle 0|\mathcal{O}|n\rangle\langle n|j_a^0(x)|0\rangle \right\}$$

$$= \sum_n \int d^3x \left\{ e^{-ip_n \cdot x} \langle 0|j_a^0(0)|n\rangle\langle n|\mathcal{O}|0\rangle - e^{ip_n \cdot x} \langle 0|\mathcal{O}|n\rangle\langle n|j_a^0(0)|0\rangle \right\}$$

$$= (2\pi)^3 \sum_n \delta^{(3)}(\vec{p}_n) \left\{ e^{-iE_n t} \langle 0|j_a^0(0)|n\rangle\langle n|\mathcal{O}|0\rangle - e^{iE_n t} \langle 0|\mathcal{O}|n\rangle\langle n|j_a^0(0)|0\rangle \right\} \neq 0.$$

Since $\mathcal{Q}_a$ is conserved, v_a should be time independent. Therefore, taking a derivative with respect to t,

$$0 = -i(2\pi)^3 \sum_n \delta^{(3)}(\vec{p}_n) E_n \left\{ e^{-iE_n t} \langle 0|j_a^0(0)|n\rangle\langle n|\mathcal{O}|0\rangle + e^{iE_n t} \langle 0|\mathcal{O}|n\rangle\langle n|j_a^0(0)|0\rangle \right\}.$$

The two equations can only be simultaneously true if there exist a state $|n\rangle \equiv |\phi_a\rangle$ such that $\delta^{(3)}(\vec{p}_n) E_n = 0$ (*i.e.*, a massless state) and $\langle 0|\mathcal{O}|n\rangle \langle n|j_a^0(0)|0\rangle \neq 0$.

The vacuum expectation value v_a is called an order parameter of the symmetry breaking. Obviously, when $\mathcal{Q}_a|0\rangle = 0$ the parameter v_a is trivially zero for all operators of the theory. Notice that we have proved the existence of massless Nambu–Goldstone modes without making use of any perturbative expansion. Thus, the theorem applies to any (Poincaré invariant) physical system where a continuous symmetry of the Lagrangian is broken by the vacuum, either spontaneously or dynamically.

3.3 Chiral Symmetry in Massless QCD

Let us consider n_f flavours of massless quarks, collected in a vector field in flavour space: $q^T = (u, d, \dots)$. Colour indices are omitted, for simplicity. The corresponding QCD Lagrangian can be compactly written in the form:

$$\mathcal{L}_{\text{QCD}}^0 = -\frac{1}{4} G_{\mu\nu}^a G_a^{\mu\nu} + i\bar{q}_L \gamma^\mu D_\mu q_L + i\bar{q}_R \gamma^\mu D_\mu q_R, \tag{3.17}$$

with the gluon interactions encoded in the flavour-independent covariant derivative D_μ. In the absence of a quark mass term, the left and right quark chiralities separate into two different sectors that can only communicate through gluon interactions. The

QCD Lagrangian is then invariant under independent global $G \equiv SU(n_f)_L \otimes SU(n_f)_R$ transformations of the left- and right-handed quarks in flavour space:[1]

$$q_L \xrightarrow{G} g_L q_L, \qquad q_R \xrightarrow{G} g_R q_R, \qquad g_{L,R} \in SU(n_f)_{L,R}. \tag{3.18}$$

The Noether currents associated with the chiral group G are:

$$\mathcal{J}_X^{a\mu} = \bar{q}_X \gamma^\mu T^a q_X, \qquad (X = L, R; \quad a = 1, \ldots, n_f^2 - 1), \tag{3.19}$$

where T^a denote the $SU(n_f)$ generators that fulfil the Lie algebra $[T^a, T^b] = i f_{abc} T^c$. The corresponding Noether charges $\mathcal{Q}_X^a$ satisfy the commutation relations

$$[\mathcal{Q}_X^a, \mathcal{Q}_Y^b] = i \delta_{XY} f_{abc} \mathcal{Q}_X^c, \tag{3.20}$$

involving the $SU(n_f)$ structure constants f_{abc}. These algebraic relations were the starting point of the successful Current-Algebra methods of the sixties [5, 67], before the development of QCD.

The chiral symmetry (3.18), which should be approximately good in the light quark sector (u,d,s), is however not seen in the hadronic spectrum. Since parity exchanges left and right, a normal Wigner–Weyl realization of the symmetry would imply degenerate mirror multiplets with opposite chiralities. However, although hadrons can be nicely classified in $SU(3)_V$ representations, degenerate multiplets with the opposite parity do not exist. Moreover, the octet of pseudoscalar mesons is much lighter than all the other hadronic states. These empirical facts clearly indicate that the vacuum is not symmetric under the full chiral group. Only those transformations with $g_R = g_L$ remain a symmetry of the physical QCD vacuum. Thus, the $SU(3)_L \otimes SU(3)_R$ symmetry dynamically breaks down to $SU(3)_{L+R}$.

Since there are eight broken axial generators, $\mathcal{Q}_A^a = \mathcal{Q}_R^a - \mathcal{Q}_L^a$, there should be eight pseudoscalar Nambu–Goldstone states $|\phi^a\rangle$, which we can identify with the eight lightest hadronic states $(\pi^+, \pi^-, \pi^0, \eta, K^+, K^-, K^0$ and $\bar{K}^0)$. Their small masses are generated by the quark-mass matrix, which explicitly breaks the global chiral symmetry of the QCD Lagrangian. The quantum numbers of the Nambu–Goldstone bosons are dictated by those of the broken axial currents $\mathcal{J}_A^{a\mu}$ and the operators $\mathcal{O}^b$ that trigger the needed non-zero vacuum expectation values, because $\langle 0|\mathcal{J}_A^{a0}|\phi^a\rangle \langle \phi^a|\mathcal{O}^b|0\rangle \neq 0$. Therefore $\mathcal{O}^b$ must be pseudoscalar operators. The simplest possibility is $\mathcal{O}^b = \bar{q}\gamma_5 \lambda^b q$, with λ^b the set of eight 3×3 Gell–Mann matrices, which satisfy

$$\langle 0|[\mathcal{Q}_A^a, \bar{q}\gamma_5 \lambda^b q]|0\rangle = -\frac{1}{2} \langle 0|\bar{q}\{\lambda^a, \lambda^b\}q|0\rangle = -\frac{2}{3}\delta_{ab}\langle 0|\bar{q}q|0\rangle. \tag{3.21}$$

---

[1] Actually, the Lagrangian (3.17) has a larger $U(n_f)_L \otimes U(n_f)_R$ global symmetry. However, the $U(1)_A$ part is broken by quantum effects (the $U(1)_A$ anomaly), while the quark-number symmetry $U(1)_V$ is trivially realized in the meson sector.

The quark condensate

$$\langle 0|\bar{u}u|0\rangle = \langle 0|\bar{d}d|0\rangle = \langle 0|\bar{s}s|0\rangle \neq 0 \tag{3.22}$$

is then the natural order parameter of the dynamical chiral symmetry breaking (χSB). The $SU(3)_V$ symmetry of the vacuum guarantees that this order parameter is flavour independent.

With $n_f = 2$, $q^T = (u,d)$, one recovers the pattern of χSB in eqn (3.7). The corresponding three Nambu–Goldstone bosons are obviously identified with the pseudoscalar pion multiplet.

3.4   Nambu–Goldstone Effective Lagrangian

Since there is a mass gap separating the Nambu–Goldstone bosons from the rest of the spectrum, we can build an EFT containing only the massless modes. Their Nambu–Goldstone nature implies strong constraints on their interactions, which can be most easily analysed on the basis of an effective Lagrangian, expanded in powers of momenta over some characteristic scale, with the only assumption of the pattern of symmetry breaking $G \to H$. In order to proceed we need first to choose a good parametrization of the fields.

3.4.1   Coset-Space Coordinates

Let us consider the $O(N)$ sigma model, described by the Lagrangian (3.1), where now $\Phi(x)^T \equiv (\varphi_1, \varphi_2, \cdots, \varphi_N)$ is an N-dimensional vector of real scalar fields. The Lagrangian has a global $O(N)$ symmetry, under which $\Phi(x)$ transforms as an $O(N)$ vector, and a degenerate ground-state manifold composed by all field configurations satisfying $|\Phi|^2 = \sum_i \varphi_i^2 = v^2$. This vacuum manifold is the $N-1$ dimensional sphere S^{N-1}, in the N-dimensional space of scalar fields. Using the $O(N)$ symmetry, we can always rotate the vector $\langle 0|\Phi|0\rangle$ to any given direction, which can be taken to be

$$\Phi_0^T \equiv \langle 0|\Phi|0\rangle^T = (0,0,\cdots,0,v). \tag{3.23}$$

This vacuum choice only remains invariant under the $O(N-1)$ subgroup, acting on the first $N-1$ field coordinates. Thus, there is an $O(N) \to O(N-1)$ SSB. Since $O(N)$ has $N(N-1)/2$ generators, while $O(N-1)$ only has $(N-1)(N-2)/2$, there are $N-1$ broken generators $\widehat{T}^a$. The $N-1$ Nambu–Goldstone bosons parametrize the corresponding rotations of Φ_0 over the vacuum manifold S^{N-1}.

Taking polar coordinates, we can express the N-component field $\Phi(x)$ in the form

$$\Phi(x) = \left(1 + \frac{S(x)}{v}\right) U(x)\,\Phi_0, \tag{3.24}$$

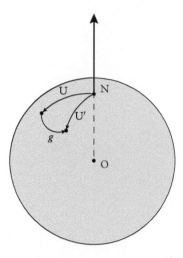

Figure 3.2 *Geometrical representation of the vacuum manifold S^{N-1}. The arrow indicates the chosen vacuum direction Φ_0.*

with $S(x)$ the radial excitation of mass $M^2 = 2\lambda v^2$, and the $N-1$ Nambu–Goldstone fields $\phi_a(x)$ encoded in the matrix

$$U(x) = \exp\left\{ i\, \widehat{T}^a\, \frac{\phi_a(x)}{v} \right\}. \tag{3.25}$$

Figure 3.2 displays a geometrical representation of the vacuum manifold S^{N-1}, with the north pole of the sphere indicating the vacuum choice Φ_0. This direction remains invariant under any transformation $h \in O(N-1)$, *i.e.*, $h\Phi_0 = \Phi_0$, while the broken generators $\widehat{T}^a$ induce rotations over the surface of the sphere. Thus, the matrix $U(x)$ provides a general parametrization of the S^{N-1} vacuum manifold. Under a global symmetry transformation $g \in O(N)$, $U(x)$ is transformed into a new matrix $g\,U(x)$ that corresponds to a different point on S^{N-1}. However, in general, the matrix $g\,U(x)$ is not in the standard form (3.25), the difference being a transformation $h \in O(N-1)$:

$$g\,U(x) = U'(x)\, h(g, U). \tag{3.26}$$

This is easily understood in three dimensions: applying two consecutive rotations $g\,U$ over an object in the north pole N is not equivalent to directly making the rotation U'; an additional rotation around the ON axis is needed to reach the same final result. The two matrices $g\,U$ and U' describe the same Nambu–Goldstone configuration over the sphere S^{N-1}, but they correspond to different choices of the Goldstone coordinates in the coset space $O(N)/O(N-1) \equiv S^{N-1}$. The compensating transformation $h(g, U)$ is non-trivial because the vacuum manifold is curved; it depends on both the transformation g and the original configuration $U(x)$.

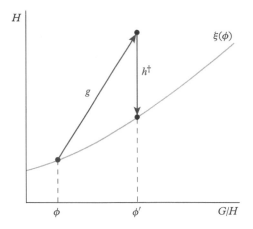

Figure 3.3 *Under the action of $g \in G$, the coset representative $\xi(\vec{\phi})$ transforms into some element of the $\vec{\phi}'$ coset. A compensating $h(\vec{\phi}, g) \in H$ transformation is needed to go back to $\xi(\vec{\phi}')$.*

In a more general situation, we have a symmetry group G and a vacuum manifold that is only invariant under the subgroup $H \subset G$, generating a $G \to H$ SSB with $N = \dim(G) - \dim(H)$ Nambu–Goldstone fields $\vec{\phi} \equiv (\phi_1, \cdots, \phi_N)$, corresponding to the number of broken generators. The action of the symmetry group on these massless fields is given by some mapping

$$\vec{\phi}(x) \xrightarrow{G} \vec{\phi}'(x) = \vec{\mathcal{F}}(g, \vec{\phi}), \tag{3.27}$$

which depends both on the group transformation $g \in G$ and the vector field $\vec{\phi}(x)$. This mapping should satisfy $\vec{\mathcal{F}}(e, \vec{\phi}) = \vec{\phi}$, where e is the identity element of the group, and the group composition law $\vec{\mathcal{F}}(g_1, \vec{\mathcal{F}}(g_2, \vec{\phi})) = \vec{\mathcal{F}}(g_1 g_2, \vec{\phi})$.

Once a vacuum choice $\vec{\phi}_0$ has been adopted, the Nambu–Goldstone fields correspond to quantum excitations along the full vacuum manifold. Since the vacuum is invariant under the unbroken subgroup, $\vec{\mathcal{F}}(h, \vec{\phi}_0) = \vec{\phi}_0$ for all $h \in H$. Therefore,

$$\vec{\phi}(x) = \vec{\mathcal{F}}(g, \vec{\phi}_0) = \vec{\mathcal{F}}(gh, \vec{\phi}_0) \qquad \forall h \in H. \tag{3.28}$$

Thus, the function $\vec{\mathcal{F}}$ represents a mapping between the members of the (left) coset equivalence class $gH = \{gh \mid h \in H\}$ and the corresponding field configuration $\vec{\phi}(x)$. Since the mapping is isomorphic and invertible,[2] the Nambu–Goldstone fields can be then identified with the elements of the coset space G/H.

For each coset, and therefore for each field configuration $\vec{\phi}(x)$, we can choose an arbitrary group element to be the coset representative $\xi(\vec{\phi})$, as shown in Figure 3.3 that

---

[2] If two elements of the coset space $g_1 H$ and $g_2 H$ are mapped into the same field configuration, i.e. $\vec{\mathcal{F}}(g_1, \vec{\phi}_0) = \vec{\mathcal{F}}(g_2, \vec{\phi}_0)$, then $\vec{\mathcal{F}}(g_1^{-1} g_2, \vec{\phi}_0) = \vec{\phi}_0$, implying that $g_1^{-1} g_2 \in H$ and therefore $g_2 \in g_1 H$.

visualizes the partition of the group elements (points in the plane) into cosets (vertical lines). Under a transformation $g \in G$, $\vec{\phi}(x)$ changes as indicated in (3.27), but the group element representing the original field does not get necessarily transformed into the coset representative of the new field configuration $\vec{\phi}'(x)$. In general, one needs a compensating transformation $h(\vec{\phi},g) \in H$ to get back to the chosen coset representative [51, 63]:

$$\xi(\vec{\phi}) \xrightarrow{G} \xi(\vec{\phi}') = g\,\xi(\vec{\phi})\,h^{\dagger}(\vec{\phi},g). \tag{3.29}$$

In the $O(N)$ model, the selected coset representative was the matrix $U(x)$ in (3.25), which only involves the broken generators.

3.4.2 Chiral Symmetry Formalism

Let us now particularize the previous discussion to the χSB

$$G \equiv SU(n_f)_L \otimes SU(n_f)_R \longrightarrow H \equiv SU(n_f)_V, \tag{3.30}$$

with $n_f^2 - 1$ Nambu–Goldstone fields $\phi_a(x)$, and a choice of coset representative $\xi(\vec{\phi}) \equiv (\xi_L(\vec{\phi}),\xi_R(\vec{\phi})) \in G$. Under a chiral transformation $g \equiv (g_L,g_R) \in G$, the change of the field coordinates in the coset space G/H is given by

$$\xi_L(\vec{\phi}) \xrightarrow{G} g_L\xi_L(\vec{\phi})\,h^{\dagger}(\vec{\phi},g), \qquad \xi_R(\vec{\phi}) \xrightarrow{G} g_R\xi_R(\vec{\phi})\,h^{\dagger}(\vec{\phi},g). \tag{3.31}$$

The compensating transformation $h(\vec{\phi},g)$ is the same in the two chiral sectors because they are related by a parity transformation that leaves H invariant.

We can get rid of $h(\vec{\phi},g)$ by combining the two chiral relations in (3.31) into the simpler form

$$U(\vec{\phi}) \equiv \xi_R(\vec{\phi})\xi_L^{\dagger}(\vec{\phi}) \xrightarrow{G} g_R\,U(\vec{\phi})\,g_L^{\dagger}. \tag{3.32}$$

We will also adopt the canonical choice of coset representative $\xi_R(\vec{\phi}) = \xi_L^{\dagger}(\vec{\phi}) \equiv u(\vec{\phi})$, involving only the broken axial generators. The $n_f \times n_f$ unitary matrix

$$U(\vec{\phi}) = u(\vec{\phi})^2 = \exp\left\{i\sqrt{2}\frac{\Phi}{F}\right\}, \qquad \Phi(x) \equiv \sqrt{2}\,\widehat{T}^a\phi_a(x), \tag{3.33}$$

gives a very convenient parametrization of the Nambu–Goldstone modes, with F some characteristic scale that is needed to compensate the dimension of the scalar fields. With $n_f = 3$,

$$\Phi(x) \equiv \frac{\vec{\lambda}}{\sqrt{2}}\vec{\phi} = \begin{pmatrix} \frac{1}{\sqrt{2}}\pi^0 + \frac{1}{\sqrt{6}}\eta_8 & \pi^+ & K^+ \\ \pi^- & -\frac{1}{\sqrt{2}}\pi^0 + \frac{1}{\sqrt{6}}\eta_8 & K^0 \\ K^- & \bar{K}^0 & -\frac{2}{\sqrt{6}}\eta_8 \end{pmatrix}. \qquad (3.34)$$

The corresponding $n_f = 2$ representation, $\Phi(x) \equiv \vec{\tau}\,\vec{\phi}/\sqrt{2}$, reduces to the upper-left 2×2 submatrix, with the pion fields only. The given field labels are of course arbitrary in the fully symmetric theory, but they will correspond (in QCD) to the physical pseudoscalar mass eigenstates, once the symmetry-breaking quark masses will be taken into account. Notice that $U(\vec{\phi})$ transforms linearly under the chiral group, but the induced transformation on the Nambu–Goldstone fields $\vec{\phi}$ is non-linear.

In QCD, we can intuitively visualize the matrix $U(\vec{\phi})_{ij}$ as parametrizing the zero-energy excitations over the quark vacuum condensate $\langle 0|\bar{q}_L^j q_R^i|0\rangle \propto \delta_{ij}$, where i,j denote flavour indices.

3.4.3   Effective Lagrangian

In order to obtain a model-independent description of the Nambu–Goldstone dynamics at low energies, we should write the most general Lagrangian involving the matrix $U(\vec{\phi})$, which is consistent with the chiral symmetry (3.30), *i.e.*, invariant under the transformation (3.32). We can organize the Lagrangian as an expansion in powers of momenta or, equivalently, in terms of an increasing number of derivatives. Owing to parity conservation, the number of derivatives should be even:

$$\mathcal{L}_{\text{eff}}(U) = \sum_n \mathcal{L}_{2n}. \qquad (3.35)$$

The terms with a minimum number of derivatives will dominate at low energies.

The unitarity of the U matrix, $U^\dagger U = I$, implies that all possible invariant operators without derivatives are trivial constants because $\langle (U^\dagger U)^m\rangle = \langle I\rangle = n_f$, where $\langle A\rangle$ denotes the flavour trace of the matrix A. Therefore, we need at least two derivatives to generate a non-trivial interaction. To lowest order (LO), there is only one independent chiral-symmetric structure:

$$\mathcal{L}_2 = \frac{F^2}{4}\,\langle \partial_\mu U^\dagger \partial^\mu U\rangle. \qquad (3.36)$$

This is precisely the operator (3.11) that we found with the toy sigma model discussed in Section 3.1, but now we have derived it without specifying any particular underlying Lagrangian (we have only used chiral symmetry). Therefore, (3.36) is a universal low-energy interaction associated with the χSB (3.30).

Expanding $U(\vec{\phi})$ in powers of Φ, the Lagrangian $\mathcal{L}_2$ gives the kinetic terms plus a tower of interactions involving an increasing number of pseudoscalars. The requirement that the kinetic terms are properly normalized fixes the global coefficient $F^2/4$ in eqn (3.36). All interactions are then predicted in terms of the single coupling F that characterizes the dynamics of the Nambu–Goldstone fields:

$$\mathcal{L}_2 = \frac{1}{2} \langle \partial_\mu \Phi \partial^\mu \Phi \rangle + \frac{1}{12F^2} \langle (\Phi \overset{\leftrightarrow}{\partial_\mu} \Phi)(\Phi \overset{\leftrightarrow}{\partial^\mu} \Phi) \rangle + \mathcal{O}(\Phi^6/F^4), \tag{3.37}$$

where $(\Phi \overset{\leftrightarrow}{\partial_\mu} \Phi) \equiv \Phi(\partial_\mu \Phi) - (\partial_\mu \Phi)\Phi$.

The calculation of scattering amplitudes becomes now a trivial perturbative exercise. For instance, for the $\pi^+ \pi^0$ elastic scattering, we get the tree-level amplitude [207]

$$T(\pi^+ \pi^0 \to \pi^+ \pi^0) = \frac{t}{F^2}, \tag{3.38}$$

in terms of the Mandelstam variable $t \equiv (p'_+ - p_+)^2$ that obviously vanishes at zero momenta. Similar results can be easily obtained for $\pi\pi \to 4\pi, 6\pi, 8\pi \dots$. The non-linearity of the effective Lagrangian relates processes with different numbers of pseudoscalars, allowing for absolute predictions in terms of the scale F.

The derivative nature of the Nambu–Goldstone interactions is a generic feature associated with the SSB mechanism, which is related to the existence of a symmetry under the shift transformation $\phi'_a(x) = \phi_a(x) + c_a$. This constant shift amounts to a global rotation of the whole vacuum manifold that leaves the physics unchanged.

The next-to-leading order (NLO) Lagrangian contains four derivatives:

$$\mathcal{L}_4^{SU(3)} = L_1 \langle \partial_\mu U^\dagger \partial^\mu U \rangle^2 + L_2 \langle \partial_\mu U^\dagger \partial_\nu U \rangle \langle \partial^\mu U^\dagger \partial^\nu U \rangle + L_3 \langle \partial_\mu U^\dagger \partial^\mu U \partial_\nu U^\dagger \partial^\nu U \rangle. \tag{3.39}$$

We have particularized the Lagrangian to $n_f = 3$, where there are three independent chiral-invariant structures.[3] In the general $SU(n_f)_L \otimes SU(n_f)_R$ case, with $n_f > 3$, we must also include the term $\langle \partial_\mu U^\dagger \partial_\nu U \partial^\mu U^\dagger \partial^\nu U \rangle$. However, for $n_f = 3$, this operator can be expressed as a combination of the three chiral structures in (3.39), applying the Cayley–Hamilton relation

$$\langle ABAB \rangle = -2\langle A^2 B^2 \rangle + \frac{1}{2} \langle A^2 \rangle \langle B^2 \rangle + \langle AB \rangle^2, \tag{3.40}$$

which is valid for any pair of traceless, Hermitian 3×3 matrices $A = i(\partial_\mu U^\dagger)U$ and $B = iU^\dagger \partial_\mu U$. For $n_f = 2$, the L_3 term can also be eliminated with the following algebraic relation among arbitrary $SU(2)$ matrices $a, b, c,$ and d:

$$2\langle abcd \rangle = \langle ab \rangle \langle cd \rangle - \langle ac \rangle \langle bd \rangle + \langle ad \rangle \langle bc \rangle. \tag{3.41}$$

---

[3] Terms such as $\langle \Box U^\dagger \Box U \rangle$ or $\langle \partial_\mu \partial_\nu U^\dagger \partial^\mu \partial^\nu U \rangle$ can be eliminated through partial integration and the use of the $\mathcal{O}(p^2)$ equation of motion: $U^\dagger \Box U - (\Box U^\dagger)U = 0$. Since the loop expansion is a systematic expansion around the classical solution, the LO equation of motion can be consistently applied to simplify higher-order terms in the Lagrangian.

Therefore,

$$\mathcal{L}_4^{SU(2)} = \frac{\ell_1}{4} \langle \partial_\mu U^\dagger \partial^\mu U \rangle^2 + \frac{\ell_2}{4} \langle \partial_\mu U^\dagger \partial_\nu U \rangle \langle \partial^\mu U^\dagger \partial^\nu U \rangle. \tag{3.42}$$

While the LO Nambu–Goldstone dynamics is fully determined by symmetry constraints and a unique coupling F, three (two) more free parameters L_i (ℓ_i) appear at NLO for $n_f = 3$ ($n_f = 2$). These couplings encode all the dynamical information about the underlying 'fundamental' theory. The physical predictions of different short-distance Lagrangians, sharing the same pattern of SSB, only differ at long distances in the particular values of these low-energy couplings (LECs). The $SO(4)$ sigma model discussed in Section 3.1, for instance, is characterized at tree level by the couplings $F = v$, $\ell_1 = v^2/(2M^2) = 1/(4\lambda)$ and $\ell_2 = 0$, where v and λ are the parameters of the potential.

3.4.4　Quantum Loops

The effective Lagrangian defines a consistent quantum field theory, involving the corresponding path integral over all Nambu–Goldstone field configurations. The quantum loops contain massless boson propagators and give rise to logarithmic dependences with momenta, with their corresponding cuts, as required by unitarity. Since the loop integrals are homogeneous functions of the external momenta, we can easily determine the power suppression of a given diagram with a straightforward dimensional counting.

Weinberg power counting theorem: Let us consider a connected Feynman diagram Γ with L loops, I internal boson propagators, E external boson lines and N_d vertices of $\mathcal{O}(p^d)$. Γ scales with momenta as p^{d_Γ}, where [211]

$$d_\Gamma = 2L + 2 + \sum_d (d-2) N_d. \tag{3.43}$$

Proof. Each loop integral contributes four powers of momenta, while propagators scale as $1/p^2$. Therefore, $d_\Gamma = 4L - 2I + \sum_d d N_d$. The number of internal lines is related to the total number of vertices in the diagram, $V = \sum_d N_d$, and the number of loops, through the topological identity $L = I + 1 - V$. Therefore, (3.43) follows.

Thus, each loop increases the power suppression by two units. This establishes a crucial power counting that allows us to organize the loop expansion as a low-energy expansion in powers of momenta. The leading $\mathcal{O}(p^2)$ contributions are obtained with $L = 0$ and $N_{d>2} = 0$. Therefore, at LO we must only consider tree-level diagrams with $\mathcal{L}_2$ insertions. At $\mathcal{O}(p^4)$, we must include tree-level contributions with a single insertion of $\mathcal{L}_4$ ($L = 0$, $N_4 = 1$, $N_{d>4} = 0$) plus any number of $\mathcal{L}_2$ vertices, and one-loop graphs with the LO Lagrangian only ($L = 1$, $N_{d>2} = 0$). The $\mathcal{O}(p^6)$ corrections would involve tree-level diagrams with a single insertion of $\mathcal{L}_6$ ($L = 0$, $N_4 = 0$, $N_6 = 1$, $N_{d>6} = 0$), one-loop graphs with one insertion of $\mathcal{L}_4$ ($L = 1$, $N_4 = 1$, $N_{d>4} = 0$), and two-loop contributions from $\mathcal{L}_2$ ($L = 2$, $N_{d>2} = 0$).

The ultraviolet loop divergences need to be renormalized. This can be done order by order in the momentum expansion, thanks to Weinberg's power counting. Adopting a regularization that preserves the symmetries of the Lagrangian, such as dimensional regularization, the divergences generated by the loops have a symmetric local structure and the needed counter-terms necessarily correspond to operators that are already included in the effective Lagrangian, because $\mathcal{L}_{\text{eff}}(U)$ contains by construction all terms permitted by the symmetry. Therefore, the loop divergences can be reabsorbed through a renormalization of the corresponding LECs, appearing at the same order in momentum.

In the usually adopted χPT renormalization scheme [92], we have at $\mathcal{O}(p^4)$

$$L_i = L_i^r(\mu) + \Gamma_i \Delta, \qquad\qquad \ell_i = \ell_i^r(\mu) + \gamma_i \Delta, \qquad\qquad (3.44)$$

where

$$\Delta = \frac{\mu^{D-4}}{32\pi^2} \left\{ \frac{2}{D-4} - \log(4\pi) + \gamma_E - 1 \right\}, \qquad\qquad (3.45)$$

with D the space-time dimension. Notice that the subtraction constant differs from the $\overline{\text{MS}}$ one by a factor -1. The explicit calculation of the one-loop generating functional gives in the $n_f = 3$ theory [93]

$$\Gamma_1 = \frac{3}{32}, \qquad \Gamma_2 = \frac{3}{16}, \qquad \Gamma_3 = 0, \qquad\qquad (3.46)$$

while for $n_f = 2$ we find [92]

$$\gamma_1 = \frac{1}{3}, \qquad \gamma_2 = \frac{2}{3}. \qquad\qquad (3.47)$$

The renormalized couplings $L_i^r(\mu)$ and $\ell_i^r(\mu)$ depend on the arbitrary scale μ of dimensional regularization. Their logarithmic running is dictated by (3.44):

$$L_i^r(\mu_2) = L_i^r(\mu_1) + \frac{\Gamma_i}{(4\pi)^2} \log\left(\frac{\mu_1}{\mu_2}\right), \qquad\qquad \ell_i^r(\mu_2) = \ell_i^r(\mu_1) + \frac{\gamma_i}{(4\pi)^2} \log\left(\frac{\mu_1}{\mu_2}\right).$$
$$(3.48)$$

This renormalization scale dependence cancels exactly with that of the loop amplitude, in all measurable quantities.

A generic $\mathcal{O}(p^4)$ amplitude consists of a non-local (non-polynomial) loop contribution, plus a polynomial in momenta that depends on the unknown constants $L_i^r(\mu)$ or $\ell_i^r(\mu)$. Let us consider, for instance, the elastic scattering of two Nambu–Goldstone particles in the $n_f = 2$ theory:

$$\mathcal{A}(\pi^a \pi^b \to \pi^c \pi^d) = A(s,t,u)\, \delta_{ab}\delta_{cd} + A(t,s,u)\, \delta_{ac}\delta_{bd} + A(u,t,s)\, \delta_{ad}\delta_{bc}. \qquad\qquad (3.49)$$

Figure 3.4 *Feynman diagrams contributing to $\pi^a \pi^b \to \pi^c \pi^d$ at the NLO.*

Owing to crossing symmetry, the same analytic function governs the s, t, and u channels, with the obvious permutation of the three Mandelstam variables. At $\mathcal{O}(p^4)$, we must consider the one-loop Feynman topologies shown in Figure 3.4, with $\mathcal{L}_2$ vertices, plus the tree-level contribution from $\mathcal{L}_4^{SU(2)}$. We obtain the result [92]:

$$
\begin{aligned}
A(s,t,u) = \frac{s}{F^2} &+ \frac{1}{F^4} \left[2\ell_1^r(\mu) s^2 + \ell_2^r(\mu)(t^2 + u^2) \right] \\
&+ \frac{1}{96\pi^2 F^4} \left\{ \frac{4}{3} s^2 + \frac{7}{3}(t^2 + u^2) + \frac{1}{2}(s^2 - 3t^2 - u^2) \log\left(\frac{-t}{\mu^2}\right) \right. \\
&\left. + \frac{1}{2}(s^2 - t^2 - 3u^2) \log\left(\frac{-u}{\mu^2}\right) - 3s^2 \log\left(\frac{-s}{\mu^2}\right) \right\},
\end{aligned}
\tag{3.50}
$$

which also includes the leading $\mathcal{O}(p^2)$ contribution. Using (3.48), it is straightforward to check that the scattering amplitude is independent of the renormalization scale μ, as it should.

The non-local piece contains the so-called chiral logarithms that are fully predicted as a function of the LO coupling F. This chiral structure can be easily understood in terms of dispersion relations. The non-trivial analytic behaviour associated with physical intermediate states (the absorptive contributions) can be calculated with the LO Lagrangian $\mathcal{L}_2$. Analyticity then allows us to reconstruct the full amplitude, through a dispersive integral, up to a subtraction polynomial. The effective theory satisfies unitarity and analyticity; therefore, it generates perturbatively the correct dispersion integrals and organizes the subtraction polynomials in a derivative expansion. In addition, the symmetry embodied in the effective Lagrangian implies very strong constraints that relate the scattering amplitudes of different processes, i.e. all subtraction polynomials are determined in terms of the LECs.

3.5 Chiral Perturbation Theory

So far, we have been discussing an ideal theory of massless Nambu–Goldstone bosons where the symmetry is exact. However, the physical pions have non-zero masses because chiral symmetry is broken explicitly by the quark masses. Moreover, the pion dynamics is sensitive to the electroweak interactions that also break chiral symmetry. In order to incorporate all these sources of explicit symmetry breaking, it is useful to introduce external classical fields coupled to the quark currents.

Let us consider an extended QCD Lagrangian, with the quark currents coupled to external Hermitian matrix-valued fields v_μ, a_μ, s, p:

$$\mathcal{L}_{\text{QCD}} = \mathcal{L}_{\text{QCD}}^0 + \bar{q}\gamma^\mu(v_\mu + \gamma_5 a_\mu)q - \bar{q}(s - i\gamma_5 p)q, \qquad (3.51)$$

where $\mathcal{L}_{\text{QCD}}^0$ is the massless QCD Lagrangian (3.17). The external fields can be used to parametrize the different breakings of chiral symmetry through the identifications

$$r_\mu \equiv v_\mu + a_\mu = -eQA_\mu,$$

$$\ell_\mu \equiv v_\mu - a_\mu = -eQA_\mu - \frac{e}{\sqrt{2}\sin\theta_W}(W_\mu^\dagger T_+ + \text{h.c.}),$$

$$s = \mathcal{M},$$

$$p = 0, \qquad (3.52)$$

with Q and $\mathcal{M}$ the quark charge and mass matrices ($n_f = 3$), respectively,

$$Q = \frac{1}{3}\text{diag}(2, -1, -1), \qquad \mathcal{M} = \text{diag}(m_u, m_d, m_s). \qquad (3.53)$$

The v_μ field contains the electromagnetic interactions, while the scalar source s accounts for the quark masses. The charged-current couplings of the $W^\pm$ bosons, which govern semileptonic weak transitions, are incorporated into ℓ_μ, with the 3×3 matrix

$$T_+ = \begin{pmatrix} 0 & V_{ud} & V_{us} \\ 0 & 0 & 0 \\ 0 & 0 & 0 \end{pmatrix} \qquad (3.54)$$

carrying the relevant quark mixing factors. We could also add the Z couplings into v_μ and a_μ, and the Higgs–Yukawa interaction into s. More exotic quark couplings to other vector, axial, scalar, or pseudoscalar fields, outside the SM framework, could also be easily included in a similar way.

The Lagrangian (3.51) is invariant under local $SU(3)_L \otimes SU(3)_R$ transformations, provided the external fields are enforced to transform in the following way:

$$q_L \longrightarrow g_L q_L, \qquad q_R \longrightarrow g_R q_R, \qquad s + ip \longrightarrow g_R(s + ip)g_L^\dagger,$$

$$\ell_\mu \longrightarrow g_L \ell_\mu g_L^\dagger + ig_L \partial_\mu g_L^\dagger, \qquad r_\mu \longrightarrow g_R r_\mu g_R^\dagger + ig_R \partial_\mu g_R^\dagger. \qquad (3.55)$$

This formal symmetry can be used to build a generalized effective Lagrangian in the presence of external sources. In order to respect the local invariance, the gauge fields v_μ and a_μ can only appear through the covariant derivatives

$$D_\mu U = \partial_\mu U - ir_\mu U + iU\ell_\mu, \qquad D_\mu U^\dagger = \partial_\mu U^\dagger + iU^\dagger r_\mu - i\ell_\mu U^\dagger, \qquad (3.56)$$

and through the field strength tensors

$$F_L^{\mu\nu} = \partial^\mu \ell^\nu - \partial^\nu \ell^\mu - i\,[\ell^\mu, \ell^\nu], \qquad F_R^{\mu\nu} = \partial^\mu r^\nu - \partial^\nu r^\mu - i\,[r^\mu, r^\nu]. \tag{3.57}$$

At LO in derivatives and number of external fields, the most general effective Lagrangian consistent with Lorentz invariance and the local chiral symmetry (3.55) takes the form [93]:

$$\mathcal{L}_2 = \frac{F^2}{4}\, \langle D_\mu U^\dagger D^\mu U + U^\dagger \chi + \chi^\dagger U \rangle, \tag{3.58}$$

with

$$\chi = 2B_0\,(s + ip). \tag{3.59}$$

The first term is just the universal LO Nambu–Goldstone interaction, but now with covariant derivatives that include the external vector and axial-vector sources. The scalar and pseudoscalar fields, incorporated into χ, give rise to a second invariant structure with a coupling B_0, which, like F, cannot be fixed with symmetry requirements alone.

Once the external fields are frozen to the particular values in (3.52), the symmetry is of course explicitly broken. However, the choice of a special direction in the flavour space breaks chiral symmetry in the effective Lagrangian (3.58), in exactly the same way as it does in the fundamental short-distance Lagrangian (3.51). Therefore, (3.58) provides the correct low-energy realization of QCD, including its symmetry breakings.

The external fields provide, in addition, a powerful tool to compute the effective realization of the chiral Noether currents. The Green functions of quark currents are obtained as functional derivatives of the generating functional $Z[v,a,s,p]$, defined via the path-integral formula

$$\exp\{iZ\} = \int \mathcal{D}q\,\mathcal{D}\bar{q}\,\mathcal{D}G_\mu \exp\left\{i \int d^4x\, \mathcal{L}_{\text{QCD}}\right\} = \int \mathcal{D}U \exp\left\{i \int d^4x\, \mathcal{L}_{\text{eff}}\right\}. \tag{3.60}$$

This formal identity provides a link between the fundamental and effective theories. At lowest order in momenta, the generating functional reduces to the classical action $S_2 = \int d^4x\, \mathcal{L}_2$. Therefore, the low-energy realization of the QCD currents can be easily computed by taking the appropriate derivatives with respect to the external fields:

$$\mathcal{J}_L^\mu = \bar{q}_L \gamma^\mu q_L \doteq \frac{\delta S_2}{\delta \ell_\mu} = \frac{i}{2} F^2 D_\mu U^\dagger U = \frac{F}{\sqrt{2}} D_\mu \Phi - \frac{i}{2}\left(\Phi \overleftrightarrow{D^\mu} \Phi\right) + \mathcal{O}(\Phi^3/F),$$

$$\mathcal{J}_R^\mu = \bar{q}_R \gamma^\mu q_R \doteq \frac{\delta S_2}{\delta r_\mu} = \frac{i}{2} F^2 D_\mu U U^\dagger = -\frac{F}{\sqrt{2}} D_\mu \Phi - \frac{i}{2}\left(\Phi \overleftrightarrow{D^\mu} \Phi\right) + \mathcal{O}(\Phi^3/F). \tag{3.61}$$

Thus, at $\mathcal{O}(p^2)$, the fundamental chiral coupling F equals the pion decay constant, $F = F_\pi = 92.2$ MeV, defined as

$$\langle 0|(\mathcal{J}_A^\mu)^{12}|\pi^+\rangle \equiv i\sqrt{2}F_\pi\, p^\mu. \tag{3.62}$$

Taking derivatives with respect to the external scalar and pseudoscalar sources,

$$\bar{q}_L^j q_R^i \doteq -\frac{\delta S_2}{\delta(s-ip)^{ji}} = -\frac{F^2}{2} B_0\, U(\vec{\phi})_{ij},$$

$$\bar{q}_R^j q_L^i \doteq -\frac{\delta S_2}{\delta(s+ip)^{ji}} = -\frac{F^2}{2} B_0\, U^\dagger(\vec{\phi})_{ij}, \tag{3.63}$$

we also find that the coupling B_0 is related to the quark vacuum condensate:

$$\langle 0|\bar{q}^j q^i|0\rangle = -F^2 B_0\, \delta^{ij}. \tag{3.64}$$

The Nambu–Goldstone bosons, parametrized by the matrix $U(\vec{\phi})$, represent indeed the zero-energy excitations over this vacuum condensate that triggers the dynamical breaking of chiral symmetry.

3.5.1   Pseudoscalar Meson Masses at Lowest Order

With $s = \mathcal{M}$ and $p = 0$, the non-derivative piece of the Lagrangian (3.58) generates a quadratic mass term for the pseudoscalar bosons, plus Φ^{2n} interactions proportional to the quark masses. Dropping an irrelevant constant, we get

$$\frac{F^2}{4} 2B_0 \langle \mathcal{M}(U+U^\dagger)\rangle = B_0\left\{-\langle\mathcal{M}\Phi^2\rangle + \frac{1}{6F^2}\langle\mathcal{M}\Phi^4\rangle + \mathcal{O}\left(\frac{\Phi^6}{F^4}\right)\right\}. \tag{3.65}$$

The explicit evaluation of the trace in the quadratic term provides the relation between the masses of the physical mesons and the quark masses:

$$\begin{aligned}
M_{\pi^\pm}^2 &= 2\hat{m}B_0, & M_{\pi^0}^2 &= 2\hat{m}B_0 - \varepsilon + \mathcal{O}(\varepsilon^2),\\
M_{K^\pm}^2 &= (m_u+m_s)B_0, & M_{K^0}^2 &= (m_d+m_s)B_0,\\
M_{\eta_8}^2 &= \frac{2}{3}(\hat{m}+2m_s)B_0 + \varepsilon + \mathcal{O}(\varepsilon^2),
\end{aligned} \tag{3.66}$$

with[4]

$$\hat{m} = \frac{1}{2}(m_u+m_d), \qquad \varepsilon = \frac{B_0}{4}\frac{(m_u-m_d)^2}{(m_s-\hat{m})}. \tag{3.67}$$

---

[4] The $\mathcal{O}(\varepsilon)$ corrections to $M_{\pi^0}^2$ and $M_{\eta_8}^2$ originate from a small mixing term between the ϕ_3 and ϕ_8 fields: $-B_0\langle\mathcal{M}\Phi^2\rangle \longrightarrow -(B_0/\sqrt{3})(m_u-m_d)\phi_3\phi_8$. The diagonalization of the quadratic ϕ_3, ϕ_8 mass matrix, gives the mass eigenstates, $\pi^0 = \cos\delta\,\phi_3 + \sin\delta\,\phi_8$ and $\eta_8 = -\sin\delta\,\phi_3 + \cos\delta\,\phi_8$, where $\tan(2\delta) = \sqrt{3}(m_d-m_u)/(2(m_s-\hat{m}))$.

Owing to chiral symmetry, the meson masses squared are proportional to a single power of the quark masses, the proportionality constant being related to the vacuum quark condensate [97]:

$$F_\pi^2 M_{\pi^\pm}^2 = -\hat{m}\langle 0|\bar{u}u + \bar{d}d|0\rangle. \tag{3.68}$$

Taking out the common proportionality factor B_0, the relations (3.66) imply the old current algebra mass ratios [97, 209],

$$\frac{M_{\pi^\pm}^2}{2\hat{m}} = \frac{M_{K^+}^2}{m_u + m_s} = \frac{M_{K^0}^2}{m_d + m_s} \approx \frac{3M_{\eta_8}^2}{2\hat{m} + 4m_s}, \tag{3.69}$$

and, up to $\mathcal{O}(m_u - m_d)$ corrections, the Gell–Mann–Okubo [95, 155] mass relation,

$$3M_{\eta_8}^2 = 4M_K^2 - M_\pi^2. \tag{3.70}$$

Chiral symmetry alone cannot fix the absolute values of the quark masses, because they are short-distance parameters that depend on QCD renormalization conventions. The renormalization scale and scheme dependence cancels out in the products $m_q \bar{q}q \sim m_q B_0$, which are the relevant combinations governing the pseudoscalar masses. Nevertheless, χPT provides information about quark mass ratios, where the dependence on B_0 drops out (QCD is flavour blind). Neglecting the tiny $\mathcal{O}(\varepsilon)$ corrections, we get the relations

$$\frac{m_d - m_u}{m_d + m_u} = \frac{(M_{K^0}^2 - M_{K^+}^2) - (M_{\pi^0}^2 - M_{\pi^+}^2)}{M_{\pi^0}^2} = 0.29, \tag{3.71}$$

$$\frac{m_s - \hat{m}}{2\hat{m}} = \frac{M_{K^0}^2 - M_{\pi^0}^2}{M_{\pi^0}^2} = 12.6. \tag{3.72}$$

In the first equation, we have subtracted the electromagnetic pion mass-squared difference to account for the virtual photon contribution to the meson self-energies. In the chiral limit ($m_u = m_d = m_s = 0$), this correction is proportional to the square of the meson charge and it is the same for K^+ and π^+.[5] The mass formulae (3.71) and (3.72) imply the quark mass ratios advocated by Weinberg [209]:

$$m_u : m_d : m_s = 0.55 : 1 : 20.3. \tag{3.73}$$

Quark mass corrections are therefore dominated by the strange quark mass m_s, which is much larger than m_u and m_d. The light-quark mass difference $m_d - m_u$ is not small compared with the individual up and down quark masses. In spite of that, isospin turns

---

[5] This result, known as Dashen's theorem [66], can be easily proved using the external sources ℓ_μ and r_μ in (3.52), with formal electromagnetic charge matrices $\mathcal{Q}_L$ and $\mathcal{Q}_R$, respectively, transforming as $\mathcal{Q}_X \to g_X \mathcal{Q}_X g_X^\dagger$. A quark (meson) electromagnetic self-energy involves a virtual photon propagator between two interaction vertices. Since there are no external photons left, the LO chiral-invariant operator with this structure is $e^2 \langle \mathcal{Q}_R U \mathcal{Q}_L U^\dagger \rangle = -2e^2 (\pi^+\pi^- + K^+K^-)/F^2 + \mathcal{O}(\phi^4/F^4)$.

out to be a very good symmetry, because isospin-breaking effects are governed by the small ratio $(m_d - m_u)/m_s$.

The Φ^4 interactions in eqn (3.65) introduce mass corrections to the $\pi\pi$ scattering amplitude (3.38),

$$T(\pi^+\pi^0 \to \pi^+\pi^0) = \frac{t - M_\pi^2}{F^2}, \qquad (3.74)$$

showing that it vanishes at $t = M_\pi^2$ [207]. This result is now an absolute prediction of chiral symmetry, because the scale $F = F_\pi$ has been already fixed from pion decay.

The LO chiral Lagrangian (3.58) encodes in a very compact way all the current algebra results, obtained in the sixties [5, 67]. These successful phenomenological predictions corroborate the pattern of χSB in (3.30) and the explicit breaking incorporated by the QCD quark masses. Besides its elegant simplicity, the EFT formalism provides a powerful technique to estimate higher-order corrections in a systematic way.

3.5.2 Higher-Order Corrections

In order to organize the chiral expansion, we must first establish a well-defined power counting for the external sources. Since $p_\phi^2 = M_\phi^2$, the physical pseudoscalar masses scale in the same way as the external on-shell momenta. This implies that the field combination χ must be counted as $\mathcal{O}(p^2)$, because $B_0 m_q \propto M_\phi^2$. The left and right sources, ℓ_μ and r_μ, are part of the covariant derivatives (3.56) and, therefore, are of $\mathcal{O}(p)$. Finally, the field strength tensors (3.57) are obviously $\mathcal{O}(p^2)$ structures. Thus:

$$U(\vec{\phi}) \sim \mathcal{O}(p^0); \qquad D_\mu, \ell_\mu, r_\mu \sim \mathcal{O}(p^1); \qquad F_L^{\mu\nu}, F_R^{\mu\nu}, \chi \sim \mathcal{O}(p^2). \qquad (3.75)$$

The full LO Lagrangian (3.58) is then of $\mathcal{O}(p^2)$ and, moreover, Weinberg's power counting (3.43) remains valid in the presence of all these symmetry-breaking effects.

At $\mathcal{O}(p^4)$, the most general Lagrangian, invariant under Lorentz symmetry, parity, charge conjugation and the local chiral transformations (3.55), is given by [93]

$$
\begin{aligned}
\mathcal{L}_4 ={}& L_1 \left\langle D_\mu U^\dagger D^\mu U \right\rangle^2 + L_2 \left\langle D_\mu U^\dagger D_\nu U \right\rangle \left\langle D^\mu U^\dagger D^\nu U \right\rangle + L_3 \left\langle D_\mu U^\dagger D^\mu U D_\nu U^\dagger D^\nu U \right\rangle \\
&+ L_4 \left\langle D_\mu U^\dagger D^\mu U \right\rangle \left\langle U^\dagger \chi + \chi^\dagger U \right\rangle + L_5 \left\langle D_\mu U^\dagger D^\mu U \left(U^\dagger \chi + \chi^\dagger U \right) \right\rangle \\
&+ L_6 \left\langle U^\dagger \chi + \chi^\dagger U \right\rangle^2 + L_7 \left\langle U^\dagger \chi - \chi^\dagger U \right\rangle^2 + L_8 \left\langle \chi^\dagger U \chi^\dagger U + U^\dagger \chi U^\dagger \chi \right\rangle \\
&- i L_9 \left\langle F_R^{\mu\nu} D_\mu U D_\nu U^\dagger + F_L^{\mu\nu} D_\mu U^\dagger D_\nu U \right\rangle + L_{10} \left\langle U^\dagger F_R^{\mu\nu} U F_{L\mu\nu} \right\rangle \\
&+ H_1 \left\langle F_{R\mu\nu} F_R^{\mu\nu} + F_{L\mu\nu} F_L^{\mu\nu} \right\rangle + H_2 \left\langle \chi^\dagger \chi \right\rangle.
\end{aligned}
\qquad (3.76)
$$

The first three terms correspond to the $n_f = 3$ Lagrangian (3.39), changing the normal derivatives by covariant ones. The second line contains operators with two covariant derivatives and one insertion of χ, while the operators in the third line involve two powers of χ and no derivatives. The fourth line includes operators with field strength tensors. The last structures proportional to H_1 and H_2 are just needed for renormalization purposes; they only contain external sources and, therefore, do not have any impact on the pseudoscalar meson dynamics.

Thus, at $\mathcal{O}(p^4)$, the low-energy behaviour of the QCD Green functions is determined by ten chiral couplings L_i. They renormalize the one-loop divergences, as indicated in (3.44), and their logarithmic dependence with the renormalization scale is given by eqn (3.48) with [93]

$$\Gamma_1 = \frac{3}{32}, \quad \Gamma_2 = \frac{3}{16}, \quad \Gamma_3 = 0, \quad \Gamma_4 = \frac{1}{8}, \quad \Gamma_5 = \frac{3}{8}, \quad \Gamma_6 = \frac{11}{144},$$

$$\Gamma_7 = 0, \quad \Gamma_8 = \frac{5}{48}, \quad \Gamma_9 = \frac{1}{4}, \quad \Gamma_{10} = -\frac{1}{4}, \quad \widetilde{\Gamma}_1 = -\frac{1}{8}, \quad \widetilde{\Gamma}_2 = \frac{5}{24}, \quad (3.77)$$

where $\widetilde{\Gamma}_1$ and $\widetilde{\Gamma}_2$ are the corresponding quantities for the two unphysical couplings H_1 and H_2.

The structure of the $\mathcal{O}(p^6)$ χPT Lagrangian has been also thoroughly analysed. It contains $90 + 4$ independent chiral structures of even intrinsic parity (without Levi-Civita pseudotensors) [35], the last four containing external sources only, and twenty-three operators of odd intrinsic parity [40, 77]:[6]

$$\mathcal{L}_6 = \sum_{i=1}^{94} C_i\, O_i^{p^6} + \sum_{i=1}^{23} \widetilde{C}_i\, \widetilde{O}_i^{p^6}. \qquad (3.78)$$

The complete renormalization of the χPT generating functional has been already accomplished at the two-loop level [36], which determines the renormalization group equations for the renormalized $\mathcal{O}(p^6)$ LECs.

χPT is an expansion in powers of momenta over some typical hadronic scale Λ_χ, associated with the χSB, which can be expected to be of the order of the (light-quark) resonance masses. The variation of the loop amplitudes under a rescaling of μ, by say e, provides a natural order-of-magnitude estimate of the χSB scale: $\Lambda_\chi \sim 4\pi F_\pi \sim 1.2\,\text{GeV}$ [144, 211].

At $\mathcal{O}(p^2)$, the χPT Lagrangian is able to describe all QCD Green functions with only two parameters, F and B_0, a quite remarkable achievement. However, with $p \lesssim M_K\,(M_\pi)$, we expect $\mathcal{O}(p^4)$ contributions to the LO amplitudes at the level of $p^2/\Lambda_\chi^2 \lesssim 20\%\,(2\%)$. In order to increase the accuracy of the χPT predictions beyond this level, the inclusion of NLO corrections is mandatory, which introduces ten additional unknown LECs. Many more free parameters $(90 + 23)$ are needed to account for $\mathcal{O}(p^6)$ contributions. Thus, increasing the precision reduces the predictive power of the effective theory.

The LECs parametrize our ignorance about the details of the underlying QCD dynamics. They are, in principle, calculable functions of Λ_{QCD} and the heavy-quark masses, which can be analysed with lattice simulations. However, at present, our main source of information about these couplings is still low-energy phenomenology. At

---

[6] The $n_f = 2$ theory contains $7 + 3$ independent operators at $\mathcal{O}(p^4)$ [92], while at $\mathcal{O}(p^6)$ it has $52 + 4$ structures of even parity [35, 111] plus five odd-parity terms (thirteen if a singlet vector source is included) [40, 77].

Table 3.1 Phenomenological determinations of the renormalized couplings $L_i^r(M_\rho)$ from $\mathcal{O}(p^4)$ and $\mathcal{O}(p^6)$ χPT analyses, and from lattice simulations (fourth column). The last two columns show the RχT predictions of Section 3.8, without (column 5) and with (column 6) short-distance information. Values labelled with † have been used as inputs.

i	$L_i^r(M_\rho) \times 10^3$				
	$\mathcal{O}(p^4)$ [38]	$\mathcal{O}(p^6)$ [38]	Lattice [76]	RχT [80]	RχT$_{SD}$ [79, 168]
1	1.0 ± 0.1	0.53 ± 0.06		0.6	0.9
2	1.6 ± 0.2	0.81 ± 0.04		1.2	1.8
3	-3.8 ± 0.3	-3.07 ± 0.20		-2.8	-4.8
4	0.0 ± 0.3	0.3 (fixed)	0.09 ± 0.34	0.0	0.0
5	1.2 ± 0.1	1.01 ± 0.06	1.19 ± 0.25	$1.2^\dagger$	1.1
6	0.0 ± 0.4	0.14 ± 0.05	0.16 ± 0.20	0.0	0.0
7	-0.3 ± 0.2	-0.34 ± 0.09		-0.3	-0.3
8	0.5 ± 0.2	0.47 ± 0.10	0.55 ± 0.15	$0.5^\dagger$	0.4
9	6.9 ± 0.7	5.9 ± 0.4		$6.9^\dagger$	7.1
10	-5.2 ± 0.1	-4.1 ± 0.4		-5.8	-5.3

$\mathcal{O}(p^4)$, the elastic $\pi\pi$ and πK scattering amplitudes are sensitive to $L_{1,2,3}$. The two-derivative couplings $L_{4,5}$ generate mass corrections to the meson decay constants (and mass-dependent wave function renormalizations), while the pseudoscalar meson masses get modified by the non-derivative terms $L_{6,7,8}$. L_9 is mainly responsible for the charged-meson electromagnetic radius and L_{10} only contributes to amplitudes with at least two external vector or axial-vector fields, like the radiative semileptonic decay $\pi \to e\nu\gamma$.

Table 3.1 summarizes our current knowledge on the $\mathcal{O}(p^4)$ constants L_i. The quoted numbers correspond to the renormalized couplings, at a scale $\mu = M_\rho$. The second column shows the LECs extracted from $\mathcal{O}(p^4)$ phenomenological analyses [38], without any estimate of the uncertainties induced by the missing higher-order contributions. In order to assess the possible impact of these corrections, the third column shows the results obtained from a global $\mathcal{O}(p^6)$ fit [38], which incorporates some theoretical priors (prejudices) on the unknown $\mathcal{O}(p^6)$ LECs. In view of the large number of uncontrollable parameters, the $\mathcal{O}(p^6)$ numbers should be taken with caution, but they can give a good idea of the potential uncertainties. The $\mathcal{O}(p^6)$ determination of $L_{10}^r(M_\rho)$ has been directly extracted from hadronic τ decay data [106]. For comparison, the fourth column shows the results of lattice simulations with $2+1+1$ dynamical fermions by the HPQCD colaboration [76]. Similar results with $2+1$ fermions have been obtained by the MILC collaboration [27], although the quoted errors are larger. An analogous compilation of LECs for the $n_f = 2$ theory can be found in [17, 38].

The values quoted in the table are in good agreement with the expected size of the couplings L_i in terms of the scale of χSB:

$$L_i \sim \frac{F_\pi^2/4}{\Lambda_\chi^2} \sim \frac{1}{4(4\pi)^2} \sim 2 \times 10^{-3}. \qquad (3.79)$$

We have just taken as reference values the normalization of $\mathcal{L}_2$ and $\Lambda_\chi \sim 4\pi F_\pi$. Thus, all $\mathcal{O}(p^4)$ couplings have the correct order of magnitude, which implies a good convergence of the momentum expansion below the resonance region, i.e. for $p < M_\rho$. However, Table 3.1 displays, a clear dynamical hierarchy with some couplings being large, while others seem compatible with zero.

χPT allows us to record correctly phenomenological information in terms of some LECs. Once these couplings have been fixed, we can predict other quantities. In addition, the information contained in Table 3.1 is very useful to test QCD-inspired models or non-perturbative approaches. Given any particular theoretical framework aiming to correctly describe QCD at low energies, we no longer need to make an extensive phenomenological analysis to test its reliability; it suffices to calculate the predicted LECs and compare them with their phenomenological values in Table 3.1. For instance, the linear sigma model discussed in Section 3.1 has the right chiral symmetry and, therefore, leads to the universal Nambu–Goldstone Lagrangian (3.36) at LO. However, its dynamical content fails to reproduce the data at NLO because, as shown in eqn (3.12), it only generates a single $\mathcal{O}(p^4)$ LEC, L_1, in complete disagreement with the pattern displayed by the table.

3.6   QCD Phenomenology at Very Low Energies

Current χPT analyses have reached an $\mathcal{O}(p^6)$ precision. This means that the most relevant observables are already known at the two-loop level. Thus, at $\mathcal{O}(p^6)$, the leading double logarithmic corrections are fully known in terms of F and the meson masses, while single chiral logarithms involve the $L_i^r(\mu)$ couplings through one loop corrections with one insertion of $\mathcal{L}_4$. The main pending problem is the large number of unknown LECs, $C_i^r(\mu)$ and $\widetilde{C}_i^r(\mu)$, from tree-level contributions with one $\mathcal{L}_6$ insertion.

To maximize the available information, we make use of lattice simulations (with χPT relations implemented into the lattice analyses), unitarity constraints, crossing, and analyticity, mainly in the form of dispersion relations. The limit of an infinite number of QCD colours turns also to be a very useful tool to estimate the unknown LECs.

An exhaustive description of the chiral phenomenology is beyond the scope of this chapter. Instead, I present a few examples at $\mathcal{O}(p^4)$ to illustrate both the power and limitations of χPT.

3.6.1   Meson Decay Constants

The low-energy expansion of the pseudoscalar-meson decay constants in powers of the light quark masses is known to next-to-next-to-leading order (NNLO) [14]. We

Figure 3.5 *Feynman diagrams contributing to the meson decay constants at the NLO.*

only show here the NLO results [93], which originate from the Feynman topologies displayed in Figure 3.5. The red square indicates an insertion of the axial current, while the black dot is an $\mathcal{L}_2$ vertex. The tree-level diagram involves the NLO expression of the axial current, which is obtained by taking the derivative of $\mathcal{L}_4$ with respect to the axial source a_μ. Obviously, only the L_4 and L_5 operators in (3.76) can contribute to the one-particle matrix elements. The middle topology is a one-loop correction with the LO axial current, *i.e.*, the Φ^3 term in (3.61). The final diagram is a wave function renormalization correction.

In the isospin limit ($m_u = m_d = \hat{m}$), the χPT expressions take the form:

$$F_\pi = F\left\{1 - 2\mu_\pi - \mu_K + \frac{4M_\pi^2}{F^2}L_5^r(\mu) + \frac{8M_K^2 + 4M_\pi^2}{F^2}L_4^r(\mu)\right\},$$

$$F_K = F\left\{1 - \frac{3}{4}\mu_\pi - \frac{3}{2}\mu_K - \frac{3}{4}\mu_{\eta_8} + \frac{4M_K^2}{F^2}L_5^r(\mu) + \frac{8M_K^2 + 4M_\pi^2}{F^2}L_4^r(\mu)\right\},$$

$$F_{\eta_8} = F\left\{1 - 3\mu_K + \frac{4M_{\eta_8}^2}{F^2}L_5^r(\mu) + \frac{8M_K^2 + 4M_\pi^2}{F^2}L_4^r(\mu)\right\}, \tag{3.80}$$

with

$$\mu_P \equiv \frac{M_P^2}{32\pi^2 F^2}\log\left(\frac{M_P^2}{\mu^2}\right). \tag{3.81}$$

Making use of (3.48) and (3.77), we show easily the renormalization scale independence of these results.

The L_4 contribution generates a universal shift of all pseudoscalar-meson decay constants, $\delta F = 8L_4 B_0 \langle \mathcal{M} \rangle$, which can be eliminated taking ratios. Using the most recent lattice average [17]

$$F_K/F_\pi = 1.193 \pm 0.003, \tag{3.82}$$

we can then determine $L_5^r(M_\rho)$; this gives the result quoted in Table 3.1. Moreover, we get the absolute prediction

$$F_{\eta_8}/F_\pi = 1.31 \pm 0.02. \tag{3.83}$$

Figure 3.6 *Feynman diagrams contributing to the vector form factor at the NLO.*

The absolute value of the pion decay constant is usually extracted from the measured $\pi^+ \to \mu^+ \nu_\mu$ decay amplitude, taking $|V_{ud}| = 0.97417 \pm 0.00021$ from superallowed nuclear β decays [114]. We get then $F_\pi = (92.21 \pm 0.14)$ MeV [161]. The direct extraction from lattice simulations gives $F_\pi = (92.1 \pm 0.6)$ MeV [17], without any assumption concerning V_{ud}.

Lattice simulations can be performed at different (unphysical) values of the quark masses. Approaching the massless limit, we can then be sensitive to the chiral scale F. In the $n_f = 2$ theory, we find [17]

$$F_\pi/F = 1.062 \pm 0.007. \tag{3.84}$$

The relation between the fundamental scales of $n_f = 2$ and $n_f = 3$ χPT is easily obtained from the first equation in (3.80):

$$F_{SU(2)} = F_{SU(3)} \left\{ 1 - \bar{\mu}_K + \frac{8\bar{M}_K^2}{F^2} L_4^r(\mu) \right\}, \tag{3.85}$$

where barred quantities refer to the limit $m_u = m_d = 0$ [93].

3.6.2 Electromagnetic Form Factors

At LO, the pseudoscalar bosons have the minimal electromagnetic coupling that is generated through the covariant derivative. Higher-order corrections induce a momentum-dependent form factor, which is already known to NNLO [37, 42]:

$$\langle \pi^+ \pi^- | \mathcal{J}_{em}^\mu | 0 \rangle = (p_+ - p_-)^\mu \, F_V^\pi(s), \tag{3.86}$$

where $\mathcal{J}_{em}^\mu = \frac{2}{3} \bar{u} \gamma^\mu u - \frac{1}{3} \bar{d} \gamma^\mu d - \frac{1}{3} \bar{s} \gamma^\mu s$ is the electromagnetic current carried by the light quarks. The same expression with the $K^+ K^-$ and $K^0 \bar{K}^0$ final states defines the analogous kaon form factors $F_V^{K^+}(s)$ and $F_V^{K^0}(s)$, respectively. Current conservation guarantees that $F_V^\pi(0) = F_V^{K^+}(0) = 1$, while $F_V^{K^0}(0) = 0$. Owing to charge-conjugation, there is no corresponding form factor for π^0 and η.

The topologies contributing at NLO to these form factors are displayed in Figure 3.6. The red box indicates an electromagnetic current insertion, at NLO in the tree-level graph and at LO in the one-loop diagrams, while the black dot is an $\mathcal{L}_2$ vertex. In the isospin limit, we find the result [94]:

$$F_V^\pi(s) = 1 + \frac{2L_9^r(\mu)}{F^2}s - \frac{s}{96\pi^2 F^2}\left[A\left(\frac{M_\pi^2}{s},\frac{M_\pi^2}{\mu^2}\right) + \frac{1}{2}A\left(\frac{M_K^2}{s},\frac{M_K^2}{\mu^2}\right)\right], \quad (3.87)$$

where

$$A\left(\frac{M_P^2}{s},\frac{M_P^2}{\mu^2}\right) = \log\left(\frac{M_P^2}{\mu^2}\right) + \frac{8M_P^2}{s} - \frac{5}{3} + \sigma_P^3\log\left(\frac{\sigma_P+1}{\sigma_P-1}\right), \quad (3.88)$$

with $\sigma_P \equiv \sqrt{1 - 4M_P^2/s}$.

The kaon electromagnetic form factors can also be expressed in terms of the same loop functions:

$$F_V^{K^0}(s) = \frac{s}{192\pi^2 F^2}\left[A\left(\frac{M_\pi^2}{s},\frac{M_\pi^2}{\mu^2}\right) - A\left(\frac{M_K^2}{s},\frac{M_K^2}{\mu^2}\right)\right],$$

$$F_V^{K^+}(s) = F_V^\pi(s) + F_V^{K^0}(s). \quad (3.89)$$

At $\mathcal{O}(p^4)$, there is only one local contribution that originates from the L_9 operator. This LEC can then be extracted from the pion electromagnetic radius, defined through the low-energy expansion

$$F_V^{\phi^+}(s) = 1 + \frac{1}{6}\langle r^2\rangle_V^{\phi^+} s + \mathcal{O}(s^2), \qquad F_V^{K^0}(s) = \frac{1}{6}\langle r^2\rangle_V^{K^0} s + \mathcal{O}(s^2). \quad (3.90)$$

From (3.87), we obtain easily

$$\langle r^2\rangle_V^{\pi^\pm} = \frac{12L_9^r(\mu)}{F^2} - \frac{1}{32\pi^2 F^2}\left\{2\log\left(\frac{M_\pi^2}{\mu^2}\right) + \log\left(\frac{M_K^2}{\mu^2}\right) + 3\right\}, \quad (3.91)$$

while (3.89) implies

$$\langle r^2\rangle_V^{K^0} = -\frac{1}{16\pi^2 F^2}\log\left(\frac{M_K}{M_\pi}\right), \qquad \langle r^2\rangle_V^{K^\pm} = \langle r^2\rangle_V^{\pi^\pm} + \langle r^2\rangle_V^{K^0}. \quad (3.92)$$

In addition to the L_9 contribution, the meson electromagnetic radius $\langle r^2\rangle_V^{\phi^+}$ gets logarithmic loop corrections involving meson masses. The dependence on the renormalization scale μ cancels exactly between the logarithms and $L_9^r(\mu)$. The measured electromagnetic pion radius, $\langle r^2\rangle_V^{\pi^\pm} = (0.439 \pm 0.008)\,\mathrm{fm}^2$ [13], is used as input to estimate the coupling L_9 in Table 3.1. The numerical value of this observable is dominated by the $L_9^r(\mu)$ contribution, for any reasonable value of μ.

Since neutral bosons do not couple to the photon at tree level, $\langle r^2 \rangle_V^{K^0}$ only gets a loop contribution, which is moreover finite (there cannot be any divergence because symmetry forbids the presence of a local operator to renormalize it). The value predicted at $\mathcal{O}(p^4)$, $\langle r^2 \rangle_V^{K^0} = -(0.04 \pm 0.03)\,\mathrm{fm}^2$, is in good agreement with the experimental determination $\langle r^2 \rangle_V^{K^0} = -(0.077 \pm 0.010)\,\mathrm{fm}^2$ [161]. The measured K^+ charge radius, $\langle r^2 \rangle_V^{K^\pm} = (0.34 \pm 0.05)\,\mathrm{fm}^2$ [12], has a much larger experimental uncertainty. Within present errors, it is in agreement with the parameter-free relation in eqn (3.92).

The loop function (3.88) contains the non-trivial logarithmic dependence on the momentum transfer, dictated by unitarity. It generates an absorptive cut above $s = 4M_P^2$, corresponding to the kinematical configuration where the two intermediate pseudoscalars in the middle graph of Figure 3.6 are on-shell. According to the Watson theorem [206], the phase induced by the $\pi\pi$ logarithm coincides with the phase shift of the elastic $\pi\pi$ scattering with $I = \mathcal{J} = 1$, which at LO is given by

$$\delta_1^1(s) = \theta(s - 4M_\pi^2)\,\frac{s}{96\pi F^2}\left(1 - 4M_\pi^2/s\right)^{3/2}. \tag{3.93}$$

3.6.3 $K_{\ell 3}$ Decays

The semileptonic decays $K^+ \to \pi^0 \ell^+ \nu_\ell$ and $K^0 \to \pi^- \ell^+ \nu_\ell$ are governed by the corresponding hadronic matrix elements of the strangeness-changing weak left current. Since the vector and axial components have $\mathcal{J}^P = 1^-$ and 1^+, respectively, the axial piece cannot contribute to these $0^- \to 0^-$ transitions. The relevant vector hadronic matrix element contains two possible Lorentz structures:

$$\langle \pi | \bar{s}\gamma^\mu u | K \rangle = C_{K\pi}\left\{ (p_K + p_\pi)^\mu\, f_+^{K\pi}(t) + (p_K - p_\pi)^\mu\, f_-^{K\pi}(t) \right\}, \tag{3.94}$$

where $t \equiv (p_K - p_\pi)^2$, $C_{K^+\pi^0} = -1/\sqrt{2}$ and $C_{K^0\pi^-} = -1$. At LO, the two form factors reduce to trivial constants: $f_+^{K\pi}(t) = 1$ and $f_-^{K\pi}(t) = 0$. The normalization at $t = 0$ is fixed to all orders by the conservation of the vector current, in the limit of equal quark masses. Owing to the Ademollo–Gatto theorem [2, 29], the deviations from one are of second order in the symmetry-breaking quark mass difference: $f_+^{K^0\pi^-}(0) = 1 + \mathcal{O}[(m_s - m_u)^2]$. However, there is a sizeable correction to $f_+^{K^+\pi^0}(t)$, due to π^0–η_8 mixing, which is proportional to $m_d - m_u$:

$$f_+^{K^+\pi^0}(0) = 1 + \frac{3}{4}\frac{m_d - m_u}{m_s - \hat{m}} = 1.017. \tag{3.95}$$

The $\mathcal{O}(p^4)$ corrections to $f_+^{K\pi}(0)$ can be expressed in a parameter-free manner in terms of the physical meson masses. Including those contributions, we obtain the more precise values [94]

$$f_+^{K^0\pi^-}(0) = 0.977, \qquad\qquad \frac{f_+^{K^+\pi^0}(0)}{f_+^{K^0\pi^-}(0)} = 1.022. \tag{3.96}$$

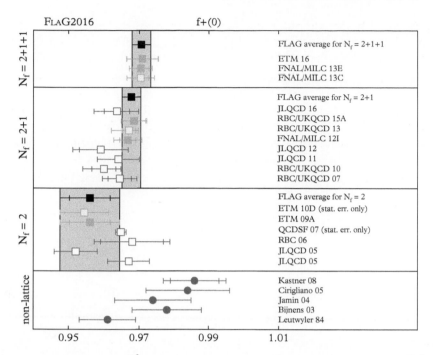

Figure 3.7 *Determinations of $f_+^{K^0\pi^-}(0)$ from lattice simulations and χPT analyses [17].*

From the measured experimental decay rates, we get a very accurate determination of the product [16, 147]

$$|V_{us}|\, f_+^{K^0\pi^-}(0) = 0.2165 \pm 0.0004. \tag{3.97}$$

A theoretical prediction of $f_+^{K^0\pi^-}(0)$ with a similar accuracy is needed in order to profit from this information and extract the most precise value of the Cabibbo–Kobayashi–Maskawa matrix element V_{us}. The present status is displayed in Figure 3.7.

Since 1984, the standard value adopted for $f_+^{K^0\pi^-}(0)$ has been the $\mathcal{O}(p^4)$ chiral prediction, corrected with a quark-model estimate of the $\mathcal{O}(p^6)$ contributions, leading to $f_+^{K^0\pi^-}(0) = 0.961 \pm 0.008$ [141]. However, this is not good enough to match the current experimental precision. The two-loop χPT corrections, computed in 2003 [43], turned out to be larger than expected, increasing the predicted value of $f_+^{K^0\pi^-}(0)$ [43, 59, 125, 131]. The estimated errors did not decrease, unfortunately, owing to the presence of $\mathcal{O}(p^6)$ LECs that need to be estimated in some way. Lattice simulations were in the past compatible with the 1984 reference value, but the most recent and precise determinations [28, 54], done with $2+1+1$ active flavours, exhibit a clear shift to higher values, in agreement with the $\mathcal{O}(p^6)$ χPT expectations. Taking the present $(2+1+1)$ lattice average [17]

$$f_+^{K^0\pi^-}(0) = 0.9706 \pm 0.0027, \tag{3.98}$$

we get

$$|V_{us}| = 0.2231 \pm 0.0007. \tag{3.99}$$

3.6.4   Meson and Quark Masses

The mass relations (3.66) get modified by $\mathcal{O}(p^4)$ contributions that depend on the LECs L_4, L_5, L_6, L_7, and L_8. However, it is possible to obtain one relation between the quark and meson masses, which does not contain any $\mathcal{O}(p^4)$ coupling. The dimensionless ratios

$$Q_1 \equiv \frac{M_K^2}{M_\pi^2}, \qquad\qquad Q_2 \equiv \frac{(M_{K^0}^2 - M_{K^+}^2)_{\mathrm{QCD}}}{M_K^2 - M_\pi^2}, \tag{3.100}$$

get the same $\mathcal{O}(p^4)$ correction [93]:

$$Q_1 = \frac{m_s + \hat{m}}{2\hat{m}} \{1 + \Delta_M\}, \qquad\qquad Q_2 = \frac{m_d - m_u}{m_s - \hat{m}} \{1 + \Delta_M\}, \tag{3.101}$$

where

$$\Delta_M = -\mu_\pi + \mu_{\eta_8} + \frac{8}{F^2} (M_K^2 - M_\pi^2) \left[2L_8^r(\mu) - L_5^r(\mu)\right]. \tag{3.102}$$

Therefore, at this order, the ratio Q_1/Q_2 is just given by the corresponding ratio of quark masses,

$$Q^2 \equiv \frac{Q_1}{Q_2} = \frac{m_s^2 - \hat{m}^2}{m_d^2 - m_u^2}. \tag{3.103}$$

To a good approximation, (3.103) can be written as the equation of an ellipse that relates the quark mass ratios:

$$\left(\frac{m_u}{m_d}\right)^2 + \frac{1}{Q^2}\left(\frac{m_s}{m_d}\right)^2 = 1. \tag{3.104}$$

The numerical value of Q can be directly extracted from the meson mass ratios (3.100), but the resulting uncertainty is dominated by the violations of Dashen's theorem at $\mathcal{O}(e^2 \mathcal{M})$, which have been shown to be sizeable. A more precise determination has been recently obtained from a careful analysis of the $\eta \to 3\pi$ decay amplitudes, which leads to $Q = 22.0 \pm 0.7$ [62].

Obviously, the quark mass ratios (3.73), obtained at $\mathcal{O}(p^2)$, satisfy the elliptic constraint (3.104). However, at $\mathcal{O}(p^4)$, it is not possible to make a separate estimate of m_u/m_d and m_s/m_d without having additional information on some LECs. The determination of the individual quark mass ratios from eqns (3.101) would require to fix first the constant L_8. However, there is no way to find an observable that isolates this coupling. The reason is an accidental symmetry of the effective Lagrangian $\mathcal{L}_2 + \mathcal{L}_4$, which remains invariant under the following simultaneous change of the quark mass matrix and some of the chiral couplings [130]:

$$\mathcal{M}' = \alpha \mathcal{M} + \beta (\mathcal{M}^\dagger)^{-1} \det \mathcal{M}, \qquad\qquad B_0' = B_0/\alpha,$$

$$L_6' = L_6 - \zeta, \qquad L_7' = L_7 - \zeta, \qquad L_8' = L_8 + 2\zeta, \qquad (3.105)$$

with α and β arbitrary constants, and $\zeta = \beta f^2/(32\alpha B_0)$. The only information on the quark mass matrix $\mathcal{M}$ that was used to construct the effective Lagrangian was that it transforms as $\mathcal{M} \to g_R \mathcal{M} g_L^\dagger$. The matrix $\mathcal{M}'$ transforms in the same manner; therefore, symmetry alone does not allow us to distinguish between $\mathcal{M}$ and $\mathcal{M}'$. Since only the product $B_0 \mathcal{M}$ appears in the Lagrangian, α merely changes the value of the constant B_0. The term proportional to β is a correction of $\mathcal{O}(\mathcal{M}^2)$; when inserted in $\mathcal{L}_2$, it generates a contribution to $\mathcal{L}_4$ that gets reabsorbed by the redefinition of the three $\mathcal{O}(p^4)$ couplings. All chiral predictions will be invariant under the transformation (3.105); therefore, it is not possible to separately determine the values of the quark masses and the constants B_0, L_6, L_7, and L_8. We can only fix those combinations of chiral couplings and masses that remain invariant under (3.105).

The ambiguity can be resolved with additional information from outside the pseudoscalar meson Lagrangian framework. For instance, by analysing isospin breaking in the baryon mass spectrum and the ρ–ω mixing, it is possible to fix the ratio [91]

$$R \equiv \frac{m_s - \hat{m}}{m_d - m_u} = 43.7 \pm 2.7. \qquad (3.106)$$

This ratio can also be extracted from lattice simulations, the current average being $R = 35.6 \pm 5.16$ [17] (with $2 + 1 + 1$ dynamical fermions). Inserting this number in (3.103), the two separate quark mass ratios can be obtained. Moreover, we can then determine L_8 from (3.101).

The quark mass ratios can be directly extracted from lattice simulations [17]:

$$\frac{m_s}{\hat{m}} = 27.30 \pm 0.34, \qquad\qquad \frac{m_u}{m_d} = 0.470 \pm 0.056. \qquad (3.107)$$

However, the second ratio includes, some phenomenological information, particularly on electromagnetic corrections. Using only the first ratio and the numerical value of Q extracted from $\eta \to 3\pi$, we would predict $m_u/m_d = 0.44 \pm 0.03$, in excellent agreement with the lattice determination and with a smaller uncertainty [62].

3.7 Quantum Anomalies

Until now, we have assumed that the symmetries of the classical Lagrangian remain valid at the quantum level. However, symmetries with different transformation properties for the left and right fermion chiralities are usually subject to quantum anomalies. Although the Lagrangian is invariant under local chiral transformations, this is no longer true for the associated generating functional because the path-integral measure transforms non-trivially [89, 90]. The anomalies of the fermionic determinant break chiral symmetry at the quantum level [3, 4, 26, 31].

3.7.1 The Chiral Anomaly

Let us consider again the $n_f = 3$ QCD Lagrangian (3.51), with external sources v_μ, a_μ, s, and p, and its local chiral symmetry (3.55). The fermionic determinant can always be defined with the convention that $Z[v, a, s, p]$ is invariant under vector transformations. Under an infinitesimal chiral transformation

$$g_L = 1 + i(\alpha - \beta) + \cdots, \qquad g_R = 1 + i(\alpha + \beta) + \cdots, \qquad (3.108)$$

with $\alpha = \alpha_a T^a$ and $\beta = \beta_a T^a$, the anomalous change of the generating functional is then given by [26]:

$$\delta Z[v, a, s, p] = -\frac{N_C}{16\pi^2} \int d^4x \, \langle \beta(x) \, \Omega(x) \rangle, \qquad (3.109)$$

where $N_C = 3$ is the number of QCD colours,

$$\Omega(x) = \varepsilon^{\mu\nu\sigma\rho} \left[v_{\mu\nu} v_{\sigma\rho} + \frac{4}{3} \nabla_\mu a_\nu \nabla_\sigma a_\rho + \frac{2}{3} i \{ v_{\mu\nu}, a_\sigma a_\rho \} + \frac{8}{3} i a_\sigma v_{\mu\nu} a_\rho + \frac{4}{3} a_\mu a_\nu a_\sigma a_\rho \right] \qquad (3.110)$$

with $\varepsilon_{0123} = 1$, and

$$v_{\mu\nu} = \partial_\mu v_\nu - \partial_\nu v_\mu - i [v_\mu, v_\nu], \qquad \nabla_\mu a_\nu = \partial_\mu a_\nu - i [v_\mu, a_\nu]. \qquad (3.111)$$

Notice that $\Omega(x)$ only depends on the external fields v_μ and a_μ, which have been assumed to be traceless.[7] This anomalous variation of Z is an $\mathcal{O}(p^4)$ effect in the chiral counting.

We have imposed chiral symmetry to construct the χPT Lagrangian. Since this symmetry is explicitly violated by the anomaly at the fundamental QCD level, we need to add to the effective theory a functional Z_A with the property that its change under a chiral gauge transformation reproduces (3.109). Such a functional was first constructed by Wess and Zumino [212], and reformulated in a nice geometrical way by Witten [216]. It has the explicit form:

---

[7] Since $\langle \sigma^a \{ \sigma^b, \sigma^c \} \rangle = 0$, this non-Abelian anomaly vanishes in the $SU(2)_L \otimes SU(2)_R$ theory. However, the singlet currents become anomalous when the electromagnetic interaction is included because the $n_f = 2$ quark charge matrix, $\mathcal{Q} = \text{diag}(\frac{2}{3}, -\frac{1}{3})$, is not traceless [126].

$$S[U, \ell, r]_{\text{WZW}} = -\frac{iN_C}{48\pi^2} \int d^4x\, \varepsilon_{\mu\nu\alpha\beta} \left\{ W(U, \ell, r)^{\mu\nu\alpha\beta} - W(I_3, \ell, r)^{\mu\nu\alpha\beta} \right\}$$

$$-\frac{iN_C}{240\pi^2} \int d\sigma^{ijklm} \left\langle \Sigma_i^L \Sigma_j^L \Sigma_k^L \Sigma_l^L \Sigma_m^L \right\rangle, \tag{3.112}$$

where

$$W(U, \ell, r)_{\mu\nu\alpha\beta} = \langle U\ell_\mu \ell_\nu \ell_\alpha U^\dagger r_\beta \rangle + \frac{1}{4} \langle U\ell_\mu U^\dagger r_\nu U\ell_\alpha U^\dagger r_\beta \rangle + i \langle U\partial_\mu \ell_\nu \ell_\alpha U^\dagger r_\beta \rangle$$

$$+ i \langle \partial_\mu r_\nu U\ell_\alpha U^\dagger r_\beta \rangle - i \langle \Sigma_\mu^L \ell_\nu U^\dagger r_\alpha U\ell_\beta \rangle + \langle \Sigma_\mu^L U^\dagger \partial_\nu r_\alpha U\ell_\beta \rangle - \langle \Sigma_\mu^L \Sigma_\nu^L U^\dagger r_\alpha U\ell_\beta \rangle$$

$$+ \langle \Sigma_\mu^L \ell_\nu \partial_\alpha \ell_\beta \rangle + \langle \Sigma_\mu^L \partial_\nu \ell_\alpha \ell_\beta \rangle - i \langle \Sigma_\mu^L \ell_\nu \ell_\alpha \ell_\beta \rangle + \frac{1}{2} \langle \Sigma_\mu^L \ell_\nu \Sigma_\alpha^L \ell_\beta \rangle - i \langle \Sigma_\mu^L \Sigma_\nu^L \Sigma_\alpha^L \ell_\beta \rangle$$

$$- (L \leftrightarrow R), \tag{3.113}$$

$$\Sigma_\mu^L = U^\dagger \partial_\mu U, \qquad\qquad \Sigma_\mu^R = U \partial_\mu U^\dagger, \tag{3.114}$$

and $(L \leftrightarrow R)$ stands for the interchanges $U \leftrightarrow U^\dagger$, $\ell_\mu \leftrightarrow r_\mu$, and $\Sigma_\mu^L \leftrightarrow \Sigma_\mu^R$. The integration in the second term of (3.112) is over a five-dimensional manifold whose boundary is four-dimensional Minkowski space. The integrand is a surface term; therefore, both the first and the second terms of S_{WZW} are of $\mathcal{O}(p^4)$, according to the chiral counting rules.

The effects induced by the anomaly are completely calculable because they have a short-distance origin. The translation from the fundamental quark-gluon level to the effective chiral level is unaffected by hadronization problems. In spite of its considerable complexity, the anomalous action (3.112) has no free parameters. The most general solution to the anomalous variation (3.109) of the QCD generating functional is given by the Wess–Zumino–Witten (WZW) action (3.112), plus the most general chiral-invariant Lagrangian that we have been constructing before, order by order in the chiral expansion.

The anomaly term does not get renormalized. Quantum loops including insertions of the WZW action generate higher-order divergences that obey the standard Weinberg's power counting and correspond to chiral-invariant structures. They get renormalized by the LECs of the corresponding χPT operators.

The anomaly functional gives rise to interactions with a Levi–Civita pseudotensor that break the intrinsic parity. This type of vertices are absent in the LO and NLO χPT Lagrangians because chiral symmetry only allows for odd-parity invariant structures starting at $\mathcal{O}(p^6)$. Thus, the WZW functional breaks an accidental symmetry of the $\mathcal{O}(p^2)$ and $\mathcal{O}(p^4)$ chiral Lagrangians, giving the leading contributions to processes with an odd number of pseudoscalars. In particular, the five-dimensional surface term in the second line of (3.112) generates interactions among five or more Nambu–Goldstone bosons, such as $K^+ K^- \to \pi^+ \pi^- \pi^0$.

Taking $v_\mu = -eQA_\mu$ in (3.113), the first line in the WZW action is responsible for the decays $\pi^0 \to 2\gamma$ and $\eta \to 2\gamma$, and the interaction vertices $\gamma 3\pi$ and $\gamma \pi^+ \pi^- \eta$. Keeping only the terms with a single pseudoscalar and two photon fields:

$$\mathcal{L}_{\mathrm{WZW}} \doteq -\frac{N_C \alpha}{24\pi F}\, \varepsilon_{\mu\nu\sigma\rho}\, F^{\mu\nu} F^{\sigma\rho} \left(\pi^0 + \frac{1}{\sqrt{3}}\eta_8\right). \tag{3.115}$$

Therefore, the chiral anomaly makes a very strong non-perturbative prediction for the π^0 decay width,

$$\Gamma(\pi^0 \to \gamma\gamma) = \left(\frac{N_C}{3}\right)^2 \frac{\alpha^2 M_\pi^3}{64\pi^3 F^2} = 7.7\ \mathrm{eV}, \tag{3.116}$$

in excellent agreement with the measured experimental value of (7.63 ± 0.16) eV [161].

3.7.2   The U(1)$_A$ Anomaly

With $n_f = 3$, the massless QCD Lagrangian has actually a larger $U(3)_L \otimes U(3)_R$ chiral symmetry. We would then expect nine Nambu–Goldstone bosons associated with the χSB to the diagonal subgroup $U(3)_V$. However, the lightest $SU(3)$-singlet pseudoscalar in the hadronic spectrum corresponds to a quite heavy state: the $\eta'(958)$.

The singlet axial current $\mathcal{J}_5^\mu = \bar{q}\gamma^\mu\gamma_5 q$ turns out to be anomalous [3, 4, 31],

$$\partial_\mu \mathcal{J}_5^\mu = \frac{g_s^2 n_f}{32\pi^2}\, \varepsilon^{\mu\nu\sigma\rho}\, G_{\mu\nu}^a G_{a,\sigma\rho}, \tag{3.117}$$

which explains the absence of a ninth Nambu–Goldstone boson, but brings very subtle phenomenological implications. Although the right-hand side of (3.117) is a total divergence (of a gauge-dependent quantity), the four-dimensional integrals of this term take non-zero values, which have a topological origin and characterize the non-trivial structure of the QCD vacuum [30, 52, 53, 120, 121, 200, 201]. It also implies the existence of an additional term in the QCD Lagrangian that violates P, T, and CP:

$$\mathcal{L}_\theta = -\theta_0 \frac{g^2}{64\pi^2}\, \varepsilon^{\mu\nu\sigma\rho}\, G_{\mu\nu}^a G_{a,\sigma\rho}. \tag{3.118}$$

When diagonalizing the quark mass matrix emerging from the Yukawa couplings of the light quarks, we need to perform a $U(1)_A$ transformation of the quark fields in order to eliminate the global phase $\arg(\det \mathcal{M})$. Owing to the axial anomaly, this transformation generates the Lagrangian $\mathcal{L}_\theta$ with a θ angle equal to the rotated phase. The experimental upper bound on the neutron electric dipole moment puts then a very strong constraint on the effective angle

$$|\theta| \equiv |\theta_0 + \arg(\det \mathcal{M})| \leq 10^{-9}. \tag{3.119}$$

The reasons why this effective angle is so small remain to be understood (strong CP problem). A detailed discussion of strong CP phenomena within χPT can be found in [172].

The $U(1)_A$ anomaly vanishes in the limit of an infinite number of QCD colours with $\alpha_s N_C$ fixed, i.e. choosing the coupling constant g_s to be of $\mathcal{O}(1/\sqrt{N_C})$ [198, 199, 214]. This is a very useful limit because the $SU(N_C)$ gauge theory simplifies considerably at $N_C \to \infty$, while keeping many essential properties of QCD. There exists a systematic expansion in powers of $1/N_C$, around this limit, which for $N_C = 3$ provides a good quantitative approximation scheme to the hadronic world [146] (see Section 3.9).

In the large-N_C limit, we can then consider a $U(3)_L \otimes U(3)_R \to U(3)_V$ chiral symmetry, with nine Nambu–Goldstone excitations that can be conveniently collected in the 3×3 unitary matrix

$$\tilde{U}(\vec{\phi}) \equiv \exp\left\{i\frac{\sqrt{2}}{F}\,\tilde{\Phi}\right\}, \qquad\qquad \tilde{\Phi} \equiv \frac{\eta_1}{\sqrt{3}}I_3 + \frac{\vec{\lambda}}{\sqrt{2}}\vec{\phi}. \qquad (3.120)$$

Under the chiral group, $\tilde{U}(\vec{\phi})$ transforms as $\tilde{U} \to g_R \tilde{U} g_L^\dagger$, with $g_{R,L} \in U(3)_{R,L}$. The LO interactions of the nine pseudoscalar bosons are described by the Lagrangian (3.58) with $\tilde{U}(\vec{\phi})$ instead of $U(\vec{\phi})$. Notice that the η_1 kinetic term in $\langle D_\mu \tilde{U}^\dagger D^\mu \tilde{U}\rangle$ decouples from the $\vec{\phi}$ fields and the η_1 particle becomes stable in the chiral limit.

To lowest non-trivial order in $1/N_C$, the chiral symmetry-breaking effect induced by the $U(1)_A$ anomaly can be taken into account in the effective low-energy theory, through the term [72, 186, 215]

$$\mathcal{L}_{U(1)_A} = -\frac{F^2}{4}\frac{\tilde{a}}{N_C}\left\{\frac{i}{2}\left[\log{(\det\tilde{U})} - \log{(\det\tilde{U}^\dagger)}\right]\right\}^2, \qquad (3.121)$$

which breaks $U(3)_L \otimes U(3)_R$ but preserves $SU(3)_L \otimes SU(3)_R \otimes U(1)_V$. The parameter $\tilde{a}$ has dimensions of mass squared and, with the factor $1/N_C$ pulled out, is booked to be of $\mathcal{O}(1)$ in the large-N_C counting rules. Its value is not fixed by symmetry requirements alone; it depends crucially on the dynamics of instantons. In the presence of the term (3.121), the η_1 field becomes massive even in the chiral limit:

$$M_{\eta_1}^2 = 3\frac{\tilde{a}}{N_C} + \mathcal{O}(\mathcal{M}). \qquad (3.122)$$

Owing to the large mass of the η', the effect of the $U(1)_A$ anomaly cannot be treated as a small perturbation. Rather, we should keep the term (3.121) together with the LO Lagrangian (3.58). It is possible to build a consistent combined expansion in powers of momenta, quark masses, and $1/N_C$ by counting the relative magnitude of these parameters as [140]:

$$\mathcal{M} \sim 1/N_C \sim p^2 \sim \mathcal{O}(\delta). \qquad (3.123)$$

A $U(3)_L \otimes U(3)_R$ description [115, 116, 127] of the pseudoscalar particles, including the singlet η_1 field, allows us to understand many properties of the η and η' mesons in a quite systematic way.

3.8   Massive Fields and Low-Energy Constants

The main limitation of the EFT approach is the proliferation of unknown LECs. At LO, the symmetry constraints severely restrict the allowed operators, allowing us to derive many phenomenological implications in terms of a small number of dynamical parameters. However, higher-order terms in the chiral expansion are much more sensitive to the non-trivial aspects of the underlying QCD dynamics. All LECs are in principle calculable from QCD, but, unfortunately, we are currently unable to perform such a first-principles computation. Although the functional integration over the quark fields in (3.60) can be explicitly done, we do not know how to perform analytically the remaining integral over the gluon field configurations. Numerical simulations in a discretized space-time lattice offer a promising tool to address the problem, but the current techniques are still not good enough to achieve a complete matching between QCD and its low-energy effective theory. On the other hand, a perturbative evaluation of the gluonic contribution would obviously fail in reproducing the correct dynamics of χSB.

A more phenomenological approach consists in analysing the massive states of the hadronic QCD spectrum that, owing to confinement, is a dual asymptotic representation of the quark and gluon degrees of freedom. The QCD resonances encode the most prominent features of the non-perturbative strong dynamics, and it seems rather natural to expect that the lowest-mass states, such as the ρ mesons, should have an important impact on the physics of the pseudoscalar bosons. In particular, the exchange of those resonances should generate sizeable contributions to the chiral couplings. Below the ρ mass scale, the singularity associated with the pole of a resonance propagator is replaced by the corresponding momentum expansion:

$$\frac{1}{s - M_R^2} = -\frac{1}{M_R^2} \sum_{n=0} \left(\frac{s}{M_R^2}\right)^n \qquad (s \ll M_R^2). \qquad (3.124)$$

Therefore, the exchange of virtual resonances generates derivative Nambu–Goldstone couplings proportional to powers of $1/M_R^2$.

3.8.1   Resonance Chiral Theory

A systematic analysis of the role of resonances in the χPT Lagrangian was first performed at $\mathcal{O}(p^4)$ in [79, 80], and extended later to the $\mathcal{O}(p^6)$ LECS [60]. We write a general chiral-invariant Lagrangian $\mathcal{L}(U, V, A, S, P)$, describing the couplings of meson resonances of the type $V(1^{--})$, $A(1^{++})$, $S(0^{++})$, and $P(0^{-+})$ to the Nambu–Goldstone bosons, at LO in derivatives. The coupling constants of this Lagrangian are phenomenologically extracted from physics at the resonance mass scale. Then we have an effective chiral theory defined in the intermediate energy region, with the generating functional (3.60) given by the path-integral formula

$$\exp\{iZ\} = \int \mathcal{D}U\,\mathcal{D}V\,\mathcal{D}A\,\mathcal{D}S\,\mathcal{D}P \, \exp\left\{i \int d^4x\, \mathcal{L}(U, V, A, S, P)\right\}. \qquad (3.125)$$

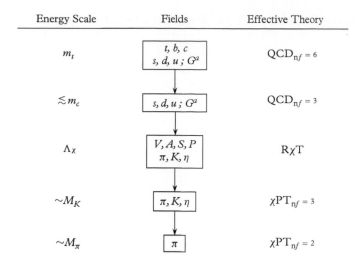

Figure 3.8 *Chain of effective field theories, from m_t to the pion mass scale.*

This resonance chiral theory ($\mathrm{R}\chi\mathrm{T}$) constitutes an interpolating representation between the short-distance QCD description and the low-energy $\chi\mathrm{PT}$ framework, which can be schematically visualized through the chain of effective field theories displayed in Figure 3.8. The functional integration of the heavy fields leads to a low-energy theory with only Nambu–Goldstone bosons. At LO, this integration can be explicitly performed through a perturbative expansion around the classical solution for the resonance fields. Expanding the resulting non-local action in powers of momenta, we get then the local $\chi\mathrm{PT}$ Lagrangian with its LECs predicted in terms of the couplings and masses of the $\mathrm{R}\chi\mathrm{T}$.

The massive states of the hadronic spectrum have definite transformation properties under the vacuum symmetry group $H \equiv SU(3)_V$. In order to couple them to the Nambu–Goldstone modes in a chiral-invariant way, we can take advantage of the compensating transformation $h(\vec{\phi}, g)$ in eqn (3.31), which appears under the action of G on the canonical coset representative $\xi_R(\vec{\phi}) = \xi_L^\dagger(\vec{\phi}) \equiv u(\vec{\phi})$:

$$u(\vec{\phi}) \xrightarrow{G} g_R \, u(\vec{\phi}) \, h^\dagger(\vec{\phi}, g) = h(\vec{\phi}, g) \, u(\vec{\phi}) \, g_L^\dagger. \tag{3.126}$$

A chiral transformation of the quark fields $(g_L, g_R) \in G$ induces a corresponding transformation $h(\vec{\phi}, g) \in H$, acting on the hadronic states.

In practice, we are only interested in resonances transforming as octets or singlets under $SU(3)_V$. Denoting the resonance multiplets generically by $R_8 = \vec{\lambda}\vec{R}/\sqrt{2}$ (octet) and R_1 (singlet), the non-linear realization of G is given by

$$R_8 \xrightarrow{G} h(\vec{\phi}, g) \, R_8 \, h^\dagger(\vec{\phi}, g), \qquad\qquad R_1 \xrightarrow{G} R_1. \tag{3.127}$$

Since the action of G on the octet field R_8 is local, we must define a covariant derivative

$$\nabla_\mu R_8 = \partial_\mu R_8 + [\Gamma_\mu, R_8], \tag{3.128}$$

with the connection

$$\Gamma_\mu = \frac{1}{2}\left\{ u^\dagger(\partial_\mu - ir_\mu)u + u(\partial_\mu - i\ell_\mu)u^\dagger \right\} \tag{3.129}$$

ensuring the proper transformation

$$\nabla_\mu R_8 \xrightarrow{G} h(\vec{\phi},g)\,\nabla_\mu R_8\, h^\dagger(\vec{\phi},g). \tag{3.130}$$

It is useful to define the covariant quantities

$$u_\mu \equiv iu^\dagger(D_\mu U)u^\dagger = u_\mu^\dagger, \qquad \chi_\pm \equiv u^\dagger \chi u^\dagger \pm u\chi^\dagger u, \qquad f_\pm^{\mu\nu} = uF_L^{\mu\nu}u^\dagger \pm u^\dagger F_R^{\mu\nu}u, \tag{3.131}$$

which transform as $SU(3)_V$ octets: $X \xrightarrow{G} h(\vec{\phi},g)\,X\,h^\dagger(\vec{\phi},g)$. Remembering that $U = u^2$, it is a simple exercise to rewrite all χPT operators in terms of these variables. For instance, $\langle D_\mu U^\dagger D^\mu U \rangle = \langle u^\mu u_\mu \rangle$ and $\langle U^\dagger \chi + \chi^\dagger U \rangle = \langle \chi_+ \rangle$. The advantage of the quantities (3.131) is that they can be easily combined with the resonance fields to build chiral-invariant structures.

In the large-N_C limit, the octet and singlet resonances become degenerate in the chiral limit. We can then collect them in a nonet multiplet

$$R \equiv R_8 + \frac{1}{\sqrt{3}}R_0 I_3 = \frac{1}{\sqrt{2}}\vec{\lambda}\vec{R} + \frac{1}{\sqrt{3}}R_0 I_3, \tag{3.132}$$

with a common mass M_R. To determine the resonance-exchange contributions to the $\mathcal{O}(p^4)$ χPT Lagrangian, we only need the LO couplings to the Nambu–Goldstone modes that are linear in the resonance fields. The relevant resonance Lagrangian can be written as [80, 168]

$$\mathcal{L}_{R\chi T} \doteq \sum_R \mathcal{L}_R. \tag{3.133}$$

The spin-0 pieces take the form $(R = S, P)$

$$\mathcal{L}_R = \frac{1}{2}\langle \nabla^\mu R \nabla_\mu R - M_R^2 R^2 \rangle + \langle R\chi_R \rangle. \tag{3.134}$$

Imposing P and C invariance, the corresponding resonance interactions are governed by the $\mathcal{O}(p^2)$ chiral structures

$$\chi_S = c_d\, u_\mu u^\mu + c_m\, \chi_+, \qquad\qquad \chi_P = d_m\, \chi_-. \tag{3.135}$$

At low energies, the solutions of the resonance equations of motion,

$$(\nabla^2 + M_R^2)\, R = \chi_R, \tag{3.136}$$

can be expanded in terms of the local chiral operators that only contain light fields:

$$R = \frac{1}{M_R^2}\, \chi_R + \mathcal{O}\!\left(\frac{p^4}{M_R^4}\right). \tag{3.137}$$

Substituting these expressions back into $\mathcal{L}_R$, in eqn (3.134), we obtain the corresponding LO contributions to the $\mathcal{O}(p^4)$ χPT Lagrangian:

$$\Delta\mathcal{L}_4^R = \sum_{R=S,P} \frac{1}{2M_R^2} \langle \chi_R \chi_R \rangle, \tag{3.138}$$

Rewriting this result in the standard basis of χPT operators in eqn (3.76), we finally get the spin-0 resonance-exchange contributions to the $\mathcal{O}(p^4)$ LECs [80, 168]:

$$L_3^S = \frac{c_d^2}{2M_S^2}, \qquad L_5^S = \frac{c_d c_m}{M_S^2}, \qquad L_8^{S+P} = \frac{c_m^2}{2M_S^2} - \frac{d_m^2}{2M_P^2}. \tag{3.139}$$

Thus, scalar exchange contributes to L_3, L_5, and L_8, while the exchange of pseudoscalar resonances only shows up in L_8.

We must also take into account the presence of the η_1 state, which is the lightest pseudoscalar resonance in the hadronic spectrum. Owing to the $U(1)_A$ anomaly, this singlet state has a much larger mass than the octet of Nambu–Goldstone bosons, and it is integrated out together with the other massive resonances. Its LO coupling can be easily extracted from the $U(3)_L \otimes U(3)_R$ chiral Lagrangian, which incorporates the matrix $\tilde{U}(\vec{\phi})$ that collects the pseudoscalar nonet:

$$\mathcal{L}_2^{U(3)} \doteq \frac{F^2}{4} \langle \tilde{U}^\dagger \chi + \chi^\dagger \tilde{U} \rangle \quad \longrightarrow \quad -i\,\frac{F}{\sqrt{24}}\, \eta_1 \langle \chi_- \rangle. \tag{3.140}$$

The exchange of an η_1 meson generates then the χPT operator $\langle \chi_- \rangle^2$, with a coupling[8]

$$L_7^{\eta_1} = -\frac{F^2}{48 M_{\eta_1}^2}. \tag{3.141}$$

---

[8] As displayed in eqn (3.139), the exchange of a complete nonet of pseudoscalars does not contribute to L_7. The singlet and octet contributions exactly cancel at large N_C.

For technical reasons, the vector and axial-vector mesons are more conveniently described in terms of anti-symmetric tensor fields $V_{\mu\nu}$ and $A_{\mu\nu}$ [80, 92], respectively, instead of the more familiar Proca field formalism.[9] Their corresponding Lagrangians read $(R = V, A)$

$$\mathcal{L}_R = -\frac{1}{2} \langle \nabla^\lambda R_{\lambda\mu} \nabla_\nu R^{\nu\mu} - \frac{M_R^2}{2} R_{\mu\nu} R^{\mu\nu} \rangle + \langle R_{\mu\nu} \chi_R^{\mu\nu} \rangle, \qquad (3.142)$$

with the $\mathcal{O}(p^2)$ chiral structures

$$\chi_V^{\mu\nu} = \frac{F_V}{2\sqrt{2}} f_+^{\mu\nu} + \frac{i G_V}{\sqrt{2}} u^\mu u^\nu, \qquad\qquad \chi_A^{\mu\nu} = \frac{F_A}{2\sqrt{2}} f_-^{\mu\nu}. \qquad (3.143)$$

Proceeding in the same way as with the spin-0 resonances, easily we get the vector and axial-vector contributions to the $\mathcal{O}(p^4)$ χPT LECS [80]:

$$L_1^V = \frac{G_V^2}{8M_V^2}, \qquad\qquad L_2^V = 2L_1^V, \qquad\qquad L_3^V = -6L_1^V,$$

$$L_9^V = \frac{F_V G_V}{2M_V^2}, \qquad\qquad L_{10}^{V+A} = -\frac{F_V^2}{4M_V^2} + \frac{F_A^2}{4M_A^2}. \qquad (3.144)$$

The dynamical origin of these results is graphically displayed in Figure 3.9. Therefore, vector-meson exchange generates contributions to $L_1, L_2, L_3, L_9,$ and L_{10} [75, 80], while A exchange only contributes to L_{10} [80].

The estimated resonance-exchange contributions to the χPT LECS bring a dynamical understanding of the phenomenological values of these constants shown in Table 3.1. The couplings L_4 and L_6, which do not receive any resonance contribution, are much smaller than the other $\mathcal{O}(p^4)$ LECs, and consistent with zero. L_1 is correctly predicted to be positive and the relation $L_2 = 2L_1$ is approximately well satisfied. L_7 is negative, as predicted in (3.141). The absolute magnitude of this parameter can be estimated from the quark-mass expansion of M_η^2 and $M_{\eta'}^2$, which fixes $M_{\eta_1} = 804$ MeV [80]. Taking $F \approx F_\pi = 92.2$ MeV, we predict $L_7 = -0.3 \cdot 10^{-3}$ in perfect agreement with the value given in Table 3.1.

To fix the vector-meson parameters, we take $M_V = M_\rho = 775$ MeV and $|F_V| = 154$ MeV, from $\Gamma(\rho^0 \to e^+ e^-)$. The electromagnetic pion radius determines $F_V G_V > 0$, correctly predicting a positive value for L_9. From this observable we get $|G_V| = 53$ MeV, but this is equivalent to fit L_9. A similar value $|G_V| \sim 69$ MeV, but with a much larger uncertainty, can be extracted from $\Gamma(\rho^0 \to 2\pi)$ [80]. The axial parameters can be

---

[9] The anti-symmetric formulation of spin-1 fields avoids mixings with the Nambu–Goldstone modes and has better ultraviolet properties. The two descriptions are related by a change of variables in the corresponding path integral [41, 128] and give the same physical predictions, once a proper short-distance behaviour is required [79, 175].

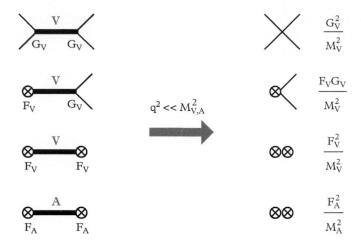

Figure 3.9 *Contributions to the χPT LECs from V and A resonance exchange. The cross denotes the insertion of an external vector or axial-vector current.*

determined using the old Weinberg sum rules [208]: $F_A^2 = F_V^2 - F_\pi^2 = (123\,\text{MeV})^2$ and $M_A^2 = M_V^2 F_V^2 / F_A^2 = (968\,\text{MeV})^2$ (see Section 3.8.2).

The resulting numerical values of the L_i couplings [80] are summarized in the fifth column of Table 3.1. Their comparison with the phenomenologically determined values of $L_i^r(M_\rho)$ clearly establish a chiral version of vector (and axial-vector) meson dominance: whenever they can contribute at all, V and A exchange seem to completely dominate the relevant coupling constants. Since the information on the scalar sector is quite poor, the values of L_5 and L_8 have actually been used to determine c_d/M_S and c_m/M_S (neglecting completely the contribution to L_8 from the much heavier pseudoscalar nonet). Therefore, these results cannot be considered as evidence for scalar dominance, although they provide a quite convincing demonstration of its consistency.

3.8.2 Short-Distance Constraints

Since the RχT is an effective interpolation between QCD and χPT, the short-distance properties of the underlying QCD dynamics provide useful constraints on its parameters [79]:

1. *Vector Form Factor.* At tree-level, the matrix element of the vector current between two Nambu–Goldstone states, is characterized by the form factor

$$F_V(t) = 1 + \frac{F_V\,G_V}{F^2}\,\frac{t}{M_V^2 - t}. \tag{3.145}$$

Since $F_V(t)$ should vanish at infinite momentum transfer t, the resonance couplings should satisfy

$$F_V\,G_V = F^2. \tag{3.146}$$

2. *Axial Form Factor.* The matrix element of the axial current between one Nambu–Goldstone boson and one photon is parametrized by the axial form factor $G_A(t)$. The resonance Lagrangian (3.142) implies

$$G_A(t) = \frac{2 F_V G_V - F_V^2}{M_V^2} + \frac{F_A^2}{M_A^2 - t}, \tag{3.147}$$

which vanishes at $t \to \infty$ provided that

$$2 F_V G_V = F_V^2. \tag{3.148}$$

3. *Weinberg Sum Rules.* In the chiral limit, the two-point function built from a left-handed and a right-handed vector quark currents,

$$2i \int d^4 x \, e^{iqx} \, \langle 0|T[\mathcal{J}_{L,12}^\mu(x)\mathcal{J}_{R,12}^{\nu\dagger}(0)]|0\rangle = (-g^{\mu\nu} q^2 + q^\mu q^\nu) \, \Pi_{LR}(q^2), \tag{3.149}$$

defines the correlator

$$\Pi_{LR}(t) = \frac{F^2}{t} + \frac{F_V^2}{M_V^2 - t} - \frac{F_A^2}{M_A^2 - t}. \tag{3.150}$$

Since gluonic interactions preserve chirality, $\Pi_{LR}(t)$ is identically zero in QCD perturbation theory. At large momenta, its operator product expansion can only get non-zero contributions from operators that break chiral symmetry, which have dimensions $d \geq 6$ when $m_q = 0$. This implies that $\Pi_{LR}(t)$ vanishes faster than $1/t^2$ at $t \to \infty$, [32, 88, 160]. Imposing this condition on (3.150), we get the relations [208]

$$F_V^2 - F_A^2 = F^2, \qquad\qquad M_V^2 F_V^2 - M_A^2 F_A^2 = 0. \tag{3.151}$$

They imply that $F_V > F_A$ and $M_V < M_A$. Moreover,

$$F_V^2 = \frac{F^2 M_A^2}{M_A^2 - M_V^2}, \qquad\qquad F_A^2 = \frac{F^2 M_V^2}{M_A^2 - M_V^2}. \tag{3.152}$$

4. *Scalar Form Factor.* The matrix element of the scalar quark current between one kaon and one pion contains the form factor [123, 124]

$$F_{K\pi}^S(t) = 1 + \frac{4 c_m}{F^2} \left\{ c_d + (c_m - c_d) \frac{M_K^2 - M_\pi^2}{M_S^2} \right\} \frac{t}{M_S^2 - t}. \tag{3.153}$$

Requiring $F_{K\pi}^S(t)$ to vanish at $t \to \infty$, we get the constraints

$$4 c_d c_m = F^2, \qquad\qquad c_m = c_d. \qquad (3.154)$$

5. *SS − PP Sum Rules.* The two-point correlation functions of two scalar or two pseudoscalar currents would be equal if chirality was absolutely preserved. Their difference is easily computed in the RχT:

$$\Pi_{SS-PP}(t) = 16 B_0^2 \left\{ \frac{c_m^2}{M_S^2 - t} - \frac{d_m^2}{M_P^2 - t} + \frac{F^2}{8t} \right\}. \qquad (3.155)$$

For massless quarks, $\Pi_{SS-PP}(t)$ vanishes as $1/t^2$ when $t \to \infty$, with a coefficient proportional to $\alpha_s \langle \bar{q}\Gamma q \bar{q}\Gamma q \rangle$ [122, 195, 196]. The vacuum four-quark condensate provides a non-perturbative breaking of chiral symmetry. In the large-N_C limit, it factorizes as $\alpha_s \langle \bar{q}q \rangle^2 \sim \alpha_s B_0^2$. Imposing this behaviour on (3.155), we get [103]

$$c_m^2 - d_m^2 = \frac{F^2}{8}, \qquad\qquad c_m^2 M_S^2 - d_m^2 M_P^2 = \frac{3\pi\alpha_s}{4} F^4. \qquad (3.156)$$

The relations (3.146), (3.148), and (3.152) determine the vector and axial-vector couplings in terms of M_V and F [79]:

$$F_V = 2 G_V = \sqrt{2} F_A = \sqrt{2} F, \qquad\qquad M_A = \sqrt{2} M_V. \qquad (3.157)$$

The scalar [123, 124] and pseudoscalar parameters are obtained from the analogous constraints (3.154) and (3.156) [168]:

$$c_m = c_d = \sqrt{2} d_m = F/2, \qquad\qquad M_P = \sqrt{2} M_S (1 - \delta)^{1/2}. \qquad (3.158)$$

The last relation involves a small correction $\delta \approx 3\pi\alpha_s F^2/M_S^2 \sim 0.08\alpha_s$ that can be neglected together with the tiny contributions from the light quark masses.

Inserting these values into (3.139) and (3.144), we obtain quite strong predictions for the $\mathcal{O}(p^4)$ LECs in terms of only M_V, M_S and F:

$$2 L_1 = L_2 = \frac{1}{4} L_9 = -\frac{1}{3} L_{10} = \frac{F^2}{8 M_V^2},$$

$$L_3 = -\frac{3 F^2}{8 M_V^2} + \frac{F^2}{8 M_S^2}, \qquad L_5 = \frac{F^2}{4 M_S^2}, \qquad L_8 = \frac{3 F^2}{32 M_S^2}. \qquad (3.159)$$

The last column in Table 3.1 shows the numerical results obtained with $M_V = 0.775$ GeV, $M_S = 1.4$ GeV, and $F = 92.2$ MeV. Also shown is the L_7 prediction in (3.141), taking $M_{\eta_1} = 0.804$ GeV. The excellent agreement with the measured values demonstrates that the lightest resonance multiplets indeed give the dominant contributions to the χPT LECs.

3.9   The Limit of a Very Large Number of QCD Colours

The phenomenological success of resonance exchange can be better understood in the $N_C \to \infty$ limit of QCD [198, 199, 214]. The N_C dependence of the β function determines that the strong coupling scales as $\alpha_s \sim \mathcal{O}(1/N_C)$. Moreover, the fact that there are $N_C^2 - 1 \approx N_C^2$ gluons, while quarks only have N_C colours, implies that the gluon dynamics becomes dominant at large values of N_C.

The counting of colour factors in Feynman diagrams is most easily done considering the gluon fields as $N_C \times N_C$ matrices in colour space, $(G_\mu)^\alpha{}_\beta = G_\mu^a (T^a)^\alpha{}_\beta$, so that the colour flow becomes explicit as in $\bar{q}_\alpha (G_\mu)^\alpha{}_\beta q^\beta$. This can be represented diagrammatically with a double line, indicating the gluon colour-anticolour, as illustrated in Figure 3.10.

Figures 3.11, 3.12, and 3.13 display a selected set of topologies, with their associated colour factors. The combinatorics of Feynman diagrams at large N_C results in simple counting rules, which characterize the $1/N_C$ expansion:

1. Dominance of planar diagrams with an arbitrary number of gluon exchanges (Figure 3.11), and a single quark loop at the edge in the case of quark correlation functions (Figure 3.13).

2. Non-planar diagrams are suppressed by factors of $1/N_C^2$ (last topology in Figure 3.11 and third one in Figure 3.13).

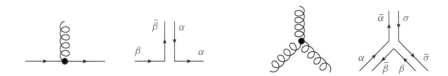

Figure 3.10 *Double-line notation, representing the gluonic colour flow.*

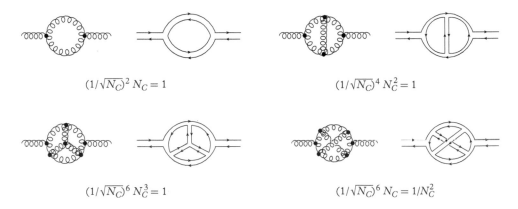

$$(1/\sqrt{N_C})^2 \, N_C = 1 \qquad\qquad (1/\sqrt{N_C})^4 \, N_C^2 = 1$$

$$(1/\sqrt{N_C})^6 \, N_C^3 = 1 \qquad\qquad (1/\sqrt{N_C})^6 \, N_C = 1/N_C^2$$

Figure 3.11 *Large-N_C counting of different gluon topologies.*

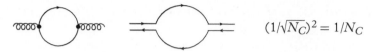

$(1/\sqrt{N_C})^2 = 1/N_C$

Figure 3.12 N_C *suppression of quark loops.*

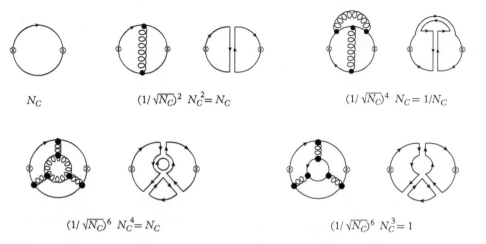

N_C

$(1/\sqrt{N_C})^2\, N_C^2 = N_C$

$(1/\sqrt{N_C})^4\, N_C = 1/N_C$

$(1/\sqrt{N_C})^6\, N_C^4 = N_C$

$(1/\sqrt{N_C})^6\, N_C^3 = 1$

Figure 3.13 *Large-N_C counting of quark correlation functions. The crosses indicate the insertion of a quark current $\bar{q}\Gamma q$.*

3. Internal quark loops are suppressed by factors of $1/N_C$ (Figure 3.12 and last topology in Figure 3.13).

The summation of the leading planar diagrams is a very formidable task that we are still unable to perform. Nevertheless, making the very plausible assumption that colour confinement persists at $N_C \to \infty$, a very successful qualitative picture of the meson world emerges.

Let us consider a generic n-point function of local quark bilinears $\mathcal{J} = \bar{q}\Gamma q$:

$$\langle 0| T[\mathcal{J}_1(x_1)\cdots\mathcal{J}_n(x_n)]|0\rangle \sim \mathcal{O}(N_C). \qquad (3.160)$$

A simple diagrammatic analysis shows that, at large N_C, the only singularities correspond to one-meson poles [214]. This is illustrated in Figure 3.14 with the simplest case of a two-point function. The dashed vertical line indicates an absorptive cut, i.e. an on-shell intermediate state. Clearly, being a planar diagram with quarks only at the edges, the cut can only contain a single $q\bar{q}$ pair. Moreover, the colour flow clearly shows that the intermediate on-shell quarks and gluons form a single colour-singlet state $\bar{q}_\alpha G^\alpha{}_\sigma G^\sigma{}_\beta q^\beta$; no smaller combination of them is separately colour singlet. Therefore, in a confining theory, the intermediate state is a perturbative approximation to a single meson. Analysing other diagrammatic configurations, we can easily check that this is a

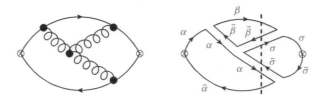

Figure 3.14 *Planar three-gluon correction to a two-point function. The absorptive cut indicated by the dashed line corresponds to the colour-singlet configuration* $\bar{q}_\alpha\, G^\alpha{}_\sigma\, G^\sigma{}_\beta\, q^\beta$.

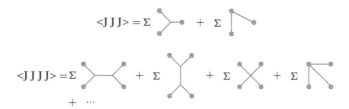

Figure 3.15 *Three-point and four-point correlation functions at large N_C.*

generic feature of planar topologies. Therefore, at $N_C \to \infty$, the two-point function has the following spectral decomposition in momentum space:

$$\langle 0|\mathcal{J}_1(k)\,\mathcal{J}_2(-k)|0\rangle = \sum_n \frac{F_n^2}{k^2 - M_n^2}. \qquad (3.161)$$

From this expression, we can derive the following interesting implications:

1. Since the left-hand side is of $\mathcal{O}(N_C)$, $F_n = \langle 0|\mathcal{J}|n\rangle \sim \mathcal{O}(\sqrt{N_C})$ and $M_n \sim \mathcal{O}(1)$.

2. There are an infinite number of meson states because, in QCD, the correlation function $\langle 0|\mathcal{J}_1(k)\,\mathcal{J}_2(-k)|0\rangle$ behaves logarithmically for large k^2 (the sum on the right-hand side would behave as $1/k^2$ with a finite number of states).

3. The one-particle poles in the sum are on the real axis. Therefore, the meson states are free, stable, and non-interacting at $N_C \to \infty$.

Analysing in a similar way n-point functions, with $n > 2$, confirms that the only singularities of the leading planar diagrams, in any kinematical channel, are one-particle single poles [214]. Therefore, at $N_C \to \infty$, the corresponding amplitudes are given by sums of tree-level diagrams with exchanges of free meson propagators, as indicated in Figure 3.15 for the three-point and four-point correlators. There may be simultaneous poles in several kinematical variables $p_1^2, p_2^2, \ldots p_n^2$. For instance, the three-point function contains contributions with single-poles in three variables, plus topologies with two simultaneous poles. The coefficients of these pole contributions should be non-singular functions of momenta, i.e. polynomials that can be interpreted as local interaction vertices. Thus, the three-pole terms contain an interaction vertex among three mesons, while the two-pole diagrams include a current coupled to two mesons. Similarly, the four-point function contains topologies with four (five) simultaneous single poles and interaction vertices among four (three) mesons, plus three-pole contributions with some currents coupled to two or three mesons.

Since these correlation functions are of $\mathcal{O}(N_C)$, the local interaction vertices among three and four mesons (Figure 3.15) scale as $V_3 \sim \mathcal{O}(N_C^{-1/2})$ and $V_4 \sim \mathcal{O}(N_C^{-1})$, respectively. Moreover, the two-meson current matrix element behaves as $\langle 0|\mathcal{J}|M_1 M_2\rangle \sim \mathcal{O}(1)$, while $\langle 0|\mathcal{J}|M_1 M_2 M_3\rangle \sim \mathcal{O}(N_C^{-1/2})$. In general, the N_C dependence of an effective local interaction vertex among m mesons is $V_m \sim \mathcal{O}(N_C^{1-m/2})$, and a current matrix element with m mesons scales as $\langle 0|\mathcal{J}|M_1 \cdots M_m\rangle \sim \mathcal{O}(N_C^{1-m/2})$. Each additional meson coupled to the current $\mathcal{J}$ or to an interaction vertex brings then a suppression factor $1/\sqrt{N_C}$.

Including gauge-invariant gluon operators, such as $\mathcal{J}_G = \mathrm{Tr}\left(G_{\mu\nu} G^{\mu\nu}\right)$, the diagrammatic analysis can be easily extended to glue states [214]. Since $\langle 0|T\left(\mathcal{J}_{G_1}\cdots\mathcal{J}_{G_n}\right)|0\rangle \sim \mathcal{O}(N_C^2)$, we derive the large-$N_C$ counting rules: $\langle 0|\mathcal{J}_G|G_1\cdots G_m\rangle \sim \mathcal{O}(N_C^{2-m})$ and $V[G_1,\cdots,G_m] \sim \mathcal{O}(N_C^{2-m})$. Therefore, at $N_C \to \infty$, glueballs are also free, stable, non-interacting, and infinite in number. The mixed correlators involving both quark and gluon operators satisfy $\langle 0|T\left(\mathcal{J}_1\cdots\mathcal{J}_n\mathcal{J}_{G_1}\cdots\mathcal{J}_{G_m}|0\rangle\right) \sim \mathcal{O}(N_C)$, which implies $V[M_1,\cdots,M_p;G_1,\cdots,G_q] \sim \mathcal{O}(N_C^{1-q-p/2})$. Therefore, glueballs and mesons decouple at large N_C, their mixing being suppressed by a factor $1/\sqrt{N_C}$.

Many known phenomenological features of the hadronic world are easily understood at LO in the $1/N_C$ expansion: suppression of the $\bar{q}q$ sea (exotics), quark model spectroscopy, Zweig's rule, $SU(3)$ meson nonets, narrow resonances, multiparticle decays dominated by resonant two-body final states, etc. In some cases, the large-N_C limit is in fact the only known theoretical explanation that is sufficiently general. Clearly, the expansion in powers of $1/N_C$ appears to be a sensible physical approximation at $N_C = 3$. Notice that the QED coupling has a similar size $e = \sqrt{4\pi\alpha} \approx 0.3$.

The large-N_C limit provides a weak coupling regime to perform quantitative QCD studies. At LO in $1/N_C$, the scattering amplitudes are given by sums of tree diagrams with physical hadrons exchanged. Crossing and unitarity imply that this sum is the tree approximation to some local effective Lagrangian. Higher-order corrections correspond to hadronic loop diagrams.

3.9.1 χPT at Large N_C

The large-N_C counting rules are obviously well satisfied within χPT. The $U(\vec{\phi})$ matrix that parametrizes the Nambu–Goldstone modes contains the scale $F \sim \mathcal{O}(\sqrt{N_C})$, which compensates the canonical dimension of the fields $\vec{\phi}$. Therefore, each additional pseudoscalar field brings a suppression factor $1/\sqrt{N_C}$. The same scale F governs the single-pseudoscalar coupling to the axial current, which is then of $\mathcal{O}(\sqrt{N_C})$.

The generating functional (3.60) involves classical sources that are coupled to the QCD quark bilinears. Since correlation functions of quark currents are of $\mathcal{O}(N_C)$, the chiral Lagrangian should also scale as $\mathcal{O}(N_C)$, at large N_C. Moreover, the LO terms in $1/N_C$ must involve a single flavour trace because each additional quark loop brings a suppression factor $1/N_C$. The $\mathcal{O}(p^2)$ Lagrangian (3.58) contains indeed a single flavour trace and an overall factor $F^2 \sim \mathcal{O}(N_C)$. The coupling B_0 must then be of $\mathcal{O}(1)$, which is corroborated by (3.64). While $\mathcal{L}_{\text{eff}} \sim \mathcal{O}(N_C)$, chiral loops have a suppression factor $(4\pi F)^{-2} \sim \mathcal{O}(1/N_C)$ for each loop. Thus, the $1/N_C$ expansion is equivalent to a semiclassical expansion.

The $\mathcal{O}(p^4)$ Lagrangian (3.76) contains operators with a single flavour trace that are of $\mathcal{O}(N_C)$, and chiral structures with two traces that should be of $\mathcal{O}(1)$ because of their associated $1/N_C$ suppression. Actually, this is not fully correct due to the algebraic relation (3.40), which has been used to rewrite the LO operator $\langle D_\mu U^\dagger D_\nu U D^\mu U^\dagger D^\nu U \rangle$ as $\frac{1}{2} O^{L_1} + O^{L_2} - 2\, O^{L_3}$. Taking this into account, the large-N_C scaling of the $\mathcal{O}(p^4)$ LECs is given by

$$
\begin{aligned}
\mathcal{O}(N_C): \quad & L_1, L_2, L_3, L_5, L_8, L_9, L_{10}, \\
\mathcal{O}(1): \quad & 2L_1 - L_2, L_4, L_6, L_7.
\end{aligned} \tag{3.162}
$$

This hierarchy of couplings appears clearly manifested in Table 3.1, where the phenomenological (and lattice) determinations of L_4 and L_6 are compatible with zero, and the relation $2L_1 - L_2 = 0$ is well satisfied. The RχT determinations of the LECs are also in perfect agreement with the large-N_C counting rules because they originate from tree-level resonance exchanges that are of LO in $1/N_C$.

A very subtle point arises concerning the coupling L_7. This LEC gets contributions from the exchange of singlet and octet pseudoscalars that, owing to their nonet-symmetry multiplet structure, exactly cancel each other at $N_C \to \infty$, in agreement with (3.162). However, the $U(1)_A$ anomaly, which is an $\mathcal{O}(1/N_C)$ effect, generates a heavy mass for the η_1 that decouples this state from the octet of Nambu–Goldstone pseudoscalars. Therefore, when the η_1 field is integrated out from the low-energy theory, the chiral coupling L_7 receives the η_1-exchange contribution (3.141). Since $M_{\eta_1}^2 \sim \mathcal{O}(1/N_C, \mathcal{M})$, the coupling L_7 could then naively be considered to be of $\mathcal{O}(N_C^2)$ [93]. However, taking the limit of a heavy η_1 mass, while keeping m_s small, amounts to considering N_C small and the large-N_C counting is no longer consistent [163].

3.9.2   RχT Estimates of LECs

The RχT Lagrangian provides an explicit implementation of the large-N_C limit of QCD, which, however, has been truncated to the lowest-mass resonance states. The true hadronic realization of the QCD dynamics at $N_C \to \infty$ corresponds to a (tree-level) local effective Lagrangian with an infinite number of massive resonances. Thus, for each possible choice of quantum numbers, we must include an infinite tower of states with increasing masses. This can be easily implemented in the RχT Lagrangian and taken into account in the determination of the chiral LECs. For the $\mathcal{O}(p^4)$ couplings, the resulting predictions just reproduce the expressions in eqns (3.139) and (3.144), adding to each term the corresponding sum over the tower of states with the given quantum numbers [168]:

$$2L_1 = L_2 = \sum_i \frac{G_{V_i}^2}{4\,M_{V_i}^2}, \qquad L_3 = \sum_i \left\{ -\frac{3\,G_{V_i}^2}{4\,M_{V_i}^2} + \frac{c_{d_i}^2}{2\,M_{S_i}^2} \right\},$$

$$L_5 = \sum_i \frac{c_{d_i}\,c_{m_i}}{M_{S_i}^2}, \qquad L_7 = -\frac{F^2}{48 M_{\eta_1}^2}, \qquad L_8 = \sum_i \left\{ \frac{c_{m_i}^2}{2\,M_{S_i}^2} - \frac{d_{m_i}^2}{2\,M_{P_i}^2} \right\},$$

$$L_9 = \sum_i \frac{F_{V_i}\,G_{V_i}}{2\,M_{V_i}^2}, \qquad L_{10} = \sum_i \left\{ \frac{F_{A_i}^2}{4\,M_{A_i}^2} - \frac{F_{V_i}^2}{4\,M_{V_i}^2} \right\}. \qquad (3.163)$$

Owing to the explicit $1/M_{R_i}^2$ suppression, the largest contributions originate from the exchange of the lightest resonances that we considered in Section 3.8.

The short-distance conditions discussed in Section 3.8.2 must also incorporate the towers of massive states. Since there is an infinite number of resonance couplings involved, we would need to consider an infinite number of constraints through the study of all possible QCD Green functions. Obviously, this is not a feasible task. However, truncating the sums to a few states, we can easily analyse the sensitivity of the results to the inclusion of higher-mass contributions [39, 102, 134, 136]. With a given set of meson states, the RχT Lagrangian provides an effective dynamical description that interpolates between the high-energy QCD behaviour and the very low-energy χPT realization. The short-distance conditions, rather than determining the physical values of the resonance parameters, just fix these couplings so that the interpolation behaves properly. When adding more states, the resonance parameters are readjusted to ensure the best possible interpolation with the available mass spectrum. This explains the amazing success of the simplest determination of LECs with just the lowest-mass resonances [79, 80, 168].

The effects induced by more exotic resonance exchanges with $\mathcal{J}^{PC} = 1^{+-}$ and 2^{++} have been also investigated [82]. The short-distance constraints eliminate any possible contribution to the $\mathcal{O}(p^4)$ LECs from 1^{+-} resonances, and only allow a tiny 2^{++} correction to L_3, $L_3^T = 0.16 \cdot 10^{-3}$, which is negligible compared to the sum of vector and scalar contributions [82]. This small tensor correction had been previously obtained in the $SU(2)$ theory [15, 202].

In order to determine the $\mathcal{O}(p^6)$ (and higher-order) LECs, we must also consider local operators with several massive resonances, as dictated by the large-N_C rules. The relevant RχT couplings can be constrained by studying appropriate sets of three-point (or higher) functions [59, 61, 135, 148, 149, 187]. A very detailed analysis of the ensuing $\mathcal{O}(p^6)$ predictions can be found in [60].

Although the large-N_C limit provides a very successful description of the low-energy dynamics, we are still lacking a systematic procedure to incorporate contributions of NLO in the $1/N_C$ counting. Some relevant subleading effects can be easily pinned down, such as the resonance widths that regulate the poles of the meson propagators [104, 109, 173, 189], or the role of final-state interactions in the physical amplitudes [73, 104, 105, 109, 113, 123, 124, 138, 154, 156]–[159, 173, 189, 203].

More recently, methods to determine the LECs of χPT at the NLO in $1/N_C$ have been developed [58, 176, 177, 184, 185]. This is a very relevant goal because the dependence of the LECs with the chiral renormalization scale is a subleading effect in the $1/N_C$ counting, of the same order than the loop contributions. Since the LO resonance-saturation estimates are performed at $N_C \to \infty$, they cannot control the μ dependence of the LECs. According to Table 3.1, the large-N_C predictions seem to work at $\mu \sim M_\rho$, which appears to be physically reasonable. However, a NLO analysis is mandatory in order to get the right μ dependence of the LECs.

At NLO we need to include quantum loops involving virtual resonance propagators [129, 182, 184]. This constitutes a major technical challenge because their ultraviolet divergences require higher-dimensional counterterms, which could generate a problematic behaviour at large momenta. Therefore, a careful investigation of short-distance QCD constraints, at the NLO in $1/N_C$, becomes necessary to enforce a proper ultraviolet behaviour of RχT and determine the needed renormalized couplings [181]–[184, 188, 190, 217]. Using analyticity and unitarity, it is possible to avoid the technicalities associated with the renormalization procedure, reducing the calculation of one-loop Green functions to tree-level diagrams plus dispersion relations [176, 177, 185]. This allows us to understand the underlying physics in a much more transparent way.

Three $\mathcal{O}(p^4)$ (and three $\mathcal{O}(p^6)$) LECs have been already determined at NLO in $1/N_C$, keeping full control of their μ dependence, with the results [176, 177, 185]:

$$L_8^r(M_\rho) = (0.6 \pm 0.4) \cdot 10^{-3}, \qquad\qquad L_9^r(M_\rho) = (7.9 \pm 0.4) \cdot 10^{-3},$$

$$L_{10}^r(M_\rho) = -(4.4 \pm 0.9) \cdot 10^{-3}. \tag{3.164}$$

These numerical values are in good agreement with the phenomenological determinations in Table 3.1. The result for $L_8^r(M_\rho)$ also agrees with the more precise lattice determination quoted in the table. The NLO RχT predictions for $L_9^r(\mu)$ and $L_{10}^r(\mu)$ are shown in Figure 3.16, as function of the renormalization scale μ (gray bands). The comparison with their LO determinations (dashed red lines) corroborates the numerical success of the $N_C \to \infty$ estimates. At scales $\mu \sim M_\rho$, the differences between the LO and NLO estimates are well within the expected numerical uncertainty of $\mathcal{O}(1/N_C)$.

3.10   Electroweak Symmetry Breaking

The discovery of the Higgs boson represents a major achievement in fundamental physics and has established the SM as the correct description of the electroweak interactions at the experimentally explored energy scales. However, there remain many open questions that the SM is unable to address, such as the nature of the mysterious

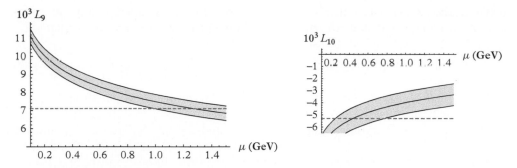

Figure 3.16 $R_\chi T$ *determinations of* $L_9^r(\mu)$ *and* $L_{10}^r(\mu)$, *at NLO in* $1/N_C$ *[176, 177]. The horizontal dashed lines show the LO predictions.*

dark matter that seems to be abundantly present in our Universe, the huge asymmetry between matter and antimatter, and the dynamical origin of the measured hierarchies of fermion masses and mixings, which span many orders of magnitude. Although we are convinced that new physics is needed in order to properly understand these facts, so far, all large hadron collider searches for new phenomena beyond the SM have given negative results. The absence of new massive states at the available energies suggest the presence of an energy gap above the electroweak scale, separating it from the scale where the new dynamics becomes relevant. EFT provides then an adequate framework to parametrize our ignorance about the underlying high-energy dynamics is a model-independent way.

While the measured Higgs properties comply with the SM expectations, the current experimental uncertainties are large enough to accommodate other alternative scenarios of electroweak symmetry breaking (EWSB) [171]. We still ignore whether the discovered Higgs corresponds indeed to the SM scalar, making part of an $SU(2)_L$ doublet together with the three electroweak Nambu–Goldstone fields, or it is a different degree of freedom, decoupled somehow from the Nambu–Goldstone modes. Additional bosons could also exist within an extended scalar sector with much richer phenomenological implications. Another possibility would be a dynamical EWSB similar to what happens in QCD.

To formulate an electroweak EFT, we need to specify its field content (the light degrees of freedom) and symmetry properties. Moreover, a clear power counting must be established in order to organize the effective Lagrangian. The simplest possibility is to build an EFT with the SM gauge symmetries and particle spectrum, assuming that the Higgs boson belongs indeed to the SM scalar doublet. The EWSB is then linearly realized, and the effective Lagrangian is ordered according to the canonical dimensions of all possible gauge-invariant operators:

$$\mathcal{L}_{\text{SMEFT}} = \mathcal{L}_{\text{SM}} + \sum_{d>4}\sum_i \frac{c_i^{(d)}}{\Lambda_{\text{NP}}^{d-4}}\, O_i^{(d)}. \tag{3.165}$$

The SM is just the LO approximation with dimension $d = 4$, while operators with $d > 4$ are suppressed by a factor $1/\Lambda_{\text{NP}}^{d-4}$, with Λ_{NP} representing the new physics scale. The dimensionless couplings $c_i^{(d)}$ contain information on the underlying dynamics. For instance, the exchange of a heavy massive vector boson X_μ, coupled to a quark vector current, would generate a $d = 6$ four-fermion operator in the effective Lagrangian:

$$\mathcal{L}_{\mathrm{NP}} \doteq g_X \left(\bar{q}_L \gamma^\mu q_L \right) X_\mu \quad \longrightarrow \quad -\frac{g_X^2}{M_X^2} \left(\bar{q}_L \gamma^\mu q_L \right) \left(\bar{q}_L \gamma_\mu q_L \right). \tag{3.166}$$

There is only one operator with $d = 5$ (up to Hermitian conjugation and flavour assignments) that violates lepton number by two units [210]. Assuming the separate conservation of the baryon and lepton numbers, and a single SM fermion family, the NLO piece of $\mathcal{L}_{\mathrm{SMEFT}}$ contains fifty-nine independent operators with $d = 6$ [50, 108]. When the three fermion families are included, this number blows up to 1,350 CP-even plus 1,149 CP-odd operators [9]. In addition, there are five $d = 6$ operators that violate B and L [1, 213].

A large number of recent works have studied different aspects of this SM effective field theory (SMEFT), working out the phenomenological constraints on the $d = 6$ Lagrangian. The full anomalous dimension matrix of the $d = 6$ operators is already known [9], allowing us to perform analyses at the one-loop level. An overview of the current status can be found in [44]. Here, we prefer to discuss a more general EFT framework, which does not make any assumption about the nature of the Higgs field. The EWSB is realized non-linearly in terms of its Nambu–Goldstone modes, and the light Higgs is incorporated into the effective Lagrangian as a generic $SU(2)_L$ singlet field with unconstrained couplings to the EWSB sector [45, 48, 174, 175].

3.10.1   Custodial Symmetry

A massless gauge boson has only two polarizations, while a massive spin-1 particle should have three. In order to generate the missing longitudinal polarizations of the $W^\pm$ and Z bosons without breaking gauge invariance, three additional degrees of freedom must be added to the massless $SU(2)_L \otimes U(1)_Y$ gauge theory. The SM incorporates instead a complex scalar doublet $\Phi(x)$ containing four real fields and, therefore, one massive neutral scalar, the Higgs boson, remains in the spectrum after the EWSB. It is convenient to collect the four scalar fields in the 2×2 matrix [18]

$$\Sigma \equiv \left(\Phi^c, \Phi \right) = \begin{pmatrix} \Phi^{0*} & \Phi^+ \\ -\Phi^- & \Phi^0 \end{pmatrix}, \tag{3.167}$$

with $\Phi^c = i\sigma_2 \Phi^*$ the charge-conjugate of the scalar doublet Φ. The SM scalar Lagrangian can then be written in the form [166, 167]

$$\mathcal{L}(\Phi) = \frac{1}{2} \left\langle (D^\mu \Sigma)^\dagger D_\mu \Sigma \right\rangle - \frac{\lambda}{16} \left(\left\langle \Sigma^\dagger \Sigma \right\rangle - v^2 \right)^2, \tag{3.168}$$

where $D_\mu \Sigma \equiv \partial_\mu \Sigma + ig \frac{\vec{\sigma}}{2} \vec{W}_\mu \Sigma - ig' \Sigma \frac{\sigma_3}{2} B_\mu$ is the usual gauge-covariant derivative.

The Lagrangian $\mathcal{L}(\Phi)$ has a global $G \equiv SU(2)_L \otimes SU(2)_R$ symmetry,

$$\Sigma \quad \longrightarrow \quad g_L \Sigma g_R^\dagger, \qquad\qquad g_{L,R} \in SU(2)_{L,R}, \tag{3.169}$$

while the vacuum choice $\langle 0|\Phi^0|0\rangle = v$ is only preserved when $g_L = g_R$, i.e. by the custodial symmetry group $H \equiv SU(2)_{L+R}$ [197]. Thus, the scalar sector of the SM has the same pattern of chiral symmetry breaking (3.30) than QCD with $n_f = 2$ flavours. In fact, $\mathcal{L}(\Phi)$ is formally identical to the linear sigma-model Lagrangian (3.5), up to a trivial normalization of the Σ field and the presence of the electroweak covariant derivative. In the SM, $SU(2)_L$ is promoted to a local gauge symmetry, but only the $U(1)_Y$ subgroup of $SU(2)_R$ is gauged. Therefore, the $U(1)_Y$ interaction in the covariant derivative breaks the $SU(2)_R$ symmetry.

The polar decomposition

$$\Sigma(x) = \frac{1}{\sqrt{2}}\,[v + h(x)]\; U(\vec{\phi}) \tag{3.170}$$

separates the Higgs field $h(x)$, which is a singlet under G transformations, from the three Nambu–Goldstone excitations $\vec{\phi}(x)$, parametrized through the 2×2 matrix

$$U(\vec{\phi}) = \exp\left\{i\vec{\sigma}\,\vec{\phi}/v\right\} \quad \longrightarrow \quad g_L\, U(\vec{\phi})\, g_R^\dagger. \tag{3.171}$$

We can rewrite $\mathcal{L}(\Phi)$ in the form [18, 142, 143]:

$$\mathcal{L}(\Phi) = \frac{v^2}{4}\,\langle D_\mu U^\dagger D^\mu U\rangle + \mathcal{O}(h/v), \tag{3.172}$$

which makes explicit the universal Nambu–Goldstone Lagrangian (3.36) associated with the symmetry breaking (3.30). Up to terms containing the scalar Higgs, the only difference is the presence of the electroweak gauge fields through the covariant derivative $D_\mu U \equiv \partial_\mu U + ig\frac{\vec{\sigma}}{2}\vec{W}_\mu U - ig' U \frac{\sigma_3}{2} B_\mu$. Thus, the same Lagrangian that describes the low-energy pion interactions in two-flavour QCD governs the SM Nambu–Goldstone dynamics, with the trivial change $F \to v$ [167]. The electroweak precision data have confirmed that the pattern of symmetry breaking implemented in the SM is correct, and have determined the fundamental scale $v = (\sqrt{2}\,G_F)^{-1/2} = 246$ GeV [170].

In the unitary gauge, $U(\vec{\phi}) = 1$, the Nambu–Goldstone fields are rotated away and the Lagrangian (3.172) reduces to a quadratic mass term for the electroweak gauge bosons, giving the SM prediction for the $W^\pm$ and Z masses:

$$m_W = m_Z \cos\theta_W = \frac{1}{2}\,vg, \tag{3.173}$$

with $Z^\mu \equiv \cos\theta_W W_3^\mu - \sin\theta_W B^\mu$ and $\tan\theta_W = g'/g$. Thus, these masses are generated by the Nambu–Goldstone modes, not by the Higgs field.[10] The successful mass relation (3.173) is a direct consequence of the pattern of EWSB, providing a clear confirmation of its phenomenological validity. The particular dynamical structure of the SM scalar Lagrangian can only be tested through the Higgs properties. The measured Higgs mass determines the quartic coupling,

---

[10] The QCD pions generate, in addition, a tiny correction $\delta m_W = \delta m_Z \cos\theta_W = F_\pi g/2$, through their coupling to the SM gauge bosons encoded in the χPT Lagrangian (3.58).

$$\lambda = \frac{m_h^2}{2v^2} = 0.13, \tag{3.174}$$

while its gauge couplings are consistent with the SM prediction within the present experimental uncertainties [171].

3.10.2   Equivalence Theorem

The elastic scattering amplitudes of the electroweak Nambu–Goldstone bosons can be obtained from the analogous results for the QCD pions. Thus, they obey the (weak) isospin and crossing relation (3.49), where $A(s, t, u)$ is given in (3.50) with the change $F \rightarrow v$. Actually, eqn (3.50) includes also the one-loop correction with the corresponding $\mathcal{O}(p^4)$ LECs $\ell_{1,2}$ that get different names in the electroweak EFT (discussed later).

The electroweak Nambu–Goldstone modes correspond to the longitudinal polarizations of the gauge bosons. Therefore, the χPT results directly give the scattering amplitudes of longitudinally polarized gauge bosons, up to corrections proportional to their non-zero masses. In the absence of the Higgs field, we get the LO result [65, 139, 205]:

$$T(W_L^+ W_L^- \rightarrow W_L^+ W_L^-) = T(\phi^+ \phi^- \rightarrow \phi^+ \phi^-) + \mathcal{O}\left(\frac{m_W}{\sqrt{s}}\right) = \frac{s+t}{v^2} + \mathcal{O}\left(\frac{m_W}{\sqrt{s}}\right). \tag{3.175}$$

At high energies the amplitude grows as s/v^2, which implies a tree-level violation of unitarity. This is the expected behaviour from the derivative coupling in the Lagrangian (3.172). Making a direct calculation with longitudinal spin-1 bosons, we could naively expect a harder energy dependence, $T(W_L^+ W_L^- \rightarrow W_L^+ W_L^-) \sim g^2 E^4/m_W^4$, because each longitudinal polarization vector grows as $\epsilon_L^\mu(\vec{k}) \sim k^\mu/m_W$ and there are four gauge-boson external legs.[11] Thus, there is a strong gauge cancellation among the first three diagrams in Figure 3.17, which eliminates those contributions with the highest powers of energy.

The SM scalar doublet Φ gives rise to a renormalizable Lagrangian, with good short-distance properties, that obviously satisfies unitarity. The right high-energy behaviour of the scattering amplitude is recovered through the additional contributions from Higgs boson exchange in Figure 3.17, which cancel the unphysical growing:

$$T(W_L^+ W_L^- \rightarrow W_L^+ W_L^-)_{\text{SM}} = \frac{1}{v^2} \left\{ s + t - \frac{s^2}{s - m_h^2} - \frac{t^2}{t - m_h^2} \right\} + \mathcal{O}\left(\frac{m_W}{\sqrt{s}}\right)$$

$$= -\frac{m_h^2}{v^2} \left\{ \frac{s}{s - m_h^2} + \frac{t}{t - m_h^2} \right\} + \mathcal{O}\left(\frac{m_W}{\sqrt{s}}\right). \tag{3.176}$$

---

[11] In the reference frame $k^\mu = (k^0, 0, 0, |\vec{k}|)$, $\epsilon_L^\mu(\vec{k}) = \frac{1}{m_W}(|\vec{k}|, 0, 0, k^0) = \frac{k^\mu}{m_W} + \mathcal{O}(m_W/|\vec{k}|)$.

Figure 3.17 *SM Feynman diagrams contributing to the elastic W^+W^- scattering at LO.*

The SM gauge structure implies an exact cancellation of all terms that grow with energy, changing an E^4 behaviour into E^0. Any small modification of the couplings in Figure 3.17 would spoil this cancellation, generating a phenomenologically unacceptable result. Therefore, large deviations of the gauge couplings from their SM predictions should not be expected. Moreover, small changes in the couplings would require the presence of new contributions to the scattering amplitude in order to maintain the cancellations.

At very large energies, $s \gg m_W^2$, $T(W_L^+ W_L^- \to W_L^+ W_L^-)_{\mathrm{SM}} \approx -2m_h^2/v^2$, which has an S-wave component

$$a_0 \equiv \frac{1}{32\pi} \int_{-1}^{1} d\cos\theta \; T(W_L^+ W_L^- \to W_L^+ W_L^-)_{\mathrm{SM}} \approx -\frac{m_h^2}{8\pi v^2}. \tag{3.177}$$

The elastic unitarity constraint $|a_0| \leq 1$ puts then an upper bound on the Higgs mass. Taking into account the inelastic channels into ZZ and hh, the bound becomes stronger by a factor $\sqrt{2/3}$, leading to the final result $m_h \leq \sqrt{8\pi} v \sqrt{2/3} \approx 1$ TeV [139]. The measured Higgs mass, $m_h = (125.09 \pm 0.24)$ GeV [20], is well below this upper limit. Notice that the bound has been extracted from a perturbative LO calculation. A much heavier Higgs, above 1 TeV, would have indicated the need for large higher-order contributions and a strong-coupling regime with $\lambda > 1$.

QCD provides a good illustration of the role of unitarity at strong coupling. Although the χPT result for the $\pi\pi$ elastic scattering amplitude grows with energy, QCD is a renormalizable theory that satisfies unitarity. The chiral prediction (3.50) is only valid at low energies, below the resonance mass scale. Once higher-mass states are taken into account, as done in RχT, the $\pi\pi$ scattering amplitude recovers its good unitarity properties, provided the proper QCD short-distance behaviour is imposed. The P-wave isovector amplitude ($\mathscr{J} = I = 1$), for instance, gets unitarized by ρ exchange. The unitarization of the S-wave isoscalar amplitude ($\mathscr{J} = I = 0$), with vacuum quantum numbers, proceeds in a much more subtle way because pion loop corrections are very large in this channel; their resummation generates a unitarized amplitude with a pole in the complex plane that corresponds to the controversial σ or $f_0(500)$ meson, a broad resonance structure that is absent at $N_C \to \infty$ [138, 162].

3.11 Electroweak Effective Theory

In order to formulate the electroweak effective theory (EWET) we must consider the most general low-energy Lagrangian that satisfies the SM symmetries and only contains the known light spectrum: the SM gauge bosons and fermions, the electroweak Nambu–Goldstone modes, and the Higgs field h [45, 48, 174, 175]. In addition to the SM gauge symmetries, our main assumption will be the pattern of EWSB:

$$G \equiv SU(2)_L \otimes SU(2)_R \longrightarrow H \equiv SU(2)_{L+R}. \tag{3.178}$$

As we have done before with χPT, we will organize the Lagrangian as an expansion in powers of derivatives and symmetry breakings over the EWSB scale (and/or any new-physics heavy mass scale). The purely Nambu–Goldstone terms are of course formally identical to those present in χPT with $n_f = 2$. However, the EWET contains a richer variety of ingredients, since we must include the SM gauge symmetries, a fermion sector, and a quite different type of symmetry-breaking effects:

$$\mathcal{L}_{\text{EWET}} = \mathcal{L}_2^{\text{EW}} + \mathcal{L}_4^{\text{EW}} + \cdots = \mathcal{L}_{\text{SM}}^{(0)} + \Delta\mathcal{L}_2 + \mathcal{L}_4^{\text{EW}} + \cdots. \tag{3.179}$$

The dots denote the infinite tower of operators with higher chiral dimensions. The LO term $\mathcal{L}_2^{\text{EW}}$ contains the renormalizable massless (unbroken) SM Lagrangian,

$$\mathcal{L}_{\text{SM}}^{(0)} = \mathcal{L}_{\text{YM}} + i \sum_f \bar{f} \gamma^\mu D_\mu f, \tag{3.180}$$

where D_μ is the covariant derivative of the $SU(3)_C \otimes SU(2)_L \otimes U(1)_Y$ gauge group, $\mathcal{L}_{\text{YM}}$ the corresponding Yang–Mills Lagrangian and the sum runs over all SM fermions f. The additional LO piece $\Delta\mathcal{L}_2$ incorporates the EWSB contributions.

We will parametrize the Nambu–Goldstone fields with the coset coordinates (3.31) and the choice of canonical coset representative $\xi_L(\vec{\phi}) = \xi_R^\dagger(\vec{\phi}) \equiv u(\vec{\phi})$. This convention is opposite to (3.126), usually adopted in χPT, but it looks more natural to describe the SM gauge group.[12] Therefore,

$$u(\vec{\phi}) \xrightarrow{G} g_L u(\vec{\phi}) g_h^\dagger = g_h u(\vec{\phi}) g_R^\dagger, \tag{3.181}$$

with $g_h \equiv h(\vec{\phi}, g) \in H$ the compensating transformation needed to get back to the chosen coset representative.

The gauge fields are incorporated in the same way as the left and right sources of χPT, although here we must add the Lagrangian $\mathcal{L}_{\text{YM}}$ because we need to quantize them. Thus, we formally introduce the $SU(2)_L$ and $SU(2)_R$ matrix fields, $\hat{W}_\mu$ and $\hat{B}_\mu$ respectively, transforming as

$$\hat{W}^\mu \longrightarrow g_L \hat{W}^\mu g_L^\dagger + i g_L \partial^\mu g_L^\dagger, \qquad\qquad \hat{B}^\mu \longrightarrow g_R \hat{B}^\mu g_R^\dagger + i g_R \partial^\mu g_R^\dagger, \tag{3.182}$$

---

[12] The two conventions are just related by a permutation of left and right, or equivalently, $u(\vec{\phi}) \leftrightarrow u^\dagger(\vec{\phi})$. Thus, analogous terms in both theories only differ by the change $F \leftrightarrow -v$.

the covariant derivative ($U = u^2 \rightarrow g_L U g_R^\dagger$)

$$D_\mu U = \partial_\mu U - i \hat{W}_\mu U + i U \hat{B}_\mu \quad \longrightarrow \quad g_L D_\mu U g_R^\dagger, \tag{3.183}$$

and the field strength tensors

$$\begin{aligned} \hat{W}_{\mu\nu} &= \partial_\mu \hat{W}_\nu - \partial_\nu \hat{W}_\mu - i[\hat{W}_\mu, \hat{W}_\nu] \quad \longrightarrow \quad g_L \hat{W}_{\mu\nu} g_L^\dagger, \\ \hat{B}_{\mu\nu} &= \partial_\mu \hat{B}_\nu - \partial_\nu \hat{B}_\mu - i[\hat{B}_\mu, \hat{B}_\nu] \quad \longrightarrow \quad g_R \hat{B}_{\mu\nu} g_R^\dagger. \end{aligned} \tag{3.184}$$

The SM gauge fields are recovered through the identification [178]

$$\hat{W}^\mu = -g \frac{\vec{\sigma}}{2} \vec{W}^\mu, \qquad\qquad \hat{B}^\mu = -g' \frac{\sigma_3}{2} B^\mu, \tag{3.185}$$

which explicitly breaks the $SU(2)_R$ symmetry group while preserving the $SU(2)_L \otimes U(1)_Y$ gauge symmetry.

We also define the covariant quantities

$$u_\mu = i u (D_\mu U)^\dagger u = -i u^\dagger D_\mu U u^\dagger = u_\mu^\dagger, \qquad f_\pm^{\mu\nu} = u^\dagger \hat{W}^{\mu\nu} u \pm u \hat{B}^{\mu\nu} u^\dagger, \tag{3.186}$$

which transform as $(u_\mu, f_\pm^{\mu\nu}) \rightarrow g_h (u_\mu, f_\pm^{\mu\nu}) g_h^\dagger$.

The bosonic part of $\Delta\mathcal{L}_2$ can then be written as [175]

$$\Delta\mathcal{L}_2^{\text{Bosonic}} = \frac{1}{2} \partial_\mu h \partial^\mu h - \frac{1}{2} m_h^2 h^2 - V(h/v) + \frac{v^2}{4} \mathcal{F}_u(h/v) \langle u_\mu u^\mu \rangle, \tag{3.187}$$

where $\langle u_\mu u^\mu \rangle$ is the usual LO Nambu–Goldstone operator and

$$V(h/v) = v^4 \sum_{n=3} c_n^{(V)} \left(\frac{h}{v}\right)^n, \qquad \mathcal{F}_u(h/v) = 1 + \sum_{n=1} c_n^{(u)} \left(\frac{h}{v}\right)^n. \tag{3.188}$$

Each chiral-invariant structure can be multiplied with an arbitrary function of h/v because the Higgs field is a singlet under $SU(2)_L \otimes SU(2)_R$ [107]. The electroweak scale v is used to compensate the powers of both the Higgs and the Nambu–Goldstone fields, since they are expected to have a similar underlying origin and, therefore, the coefficients $c_n^{(u)}$ could be conjectured to be of $\mathcal{O}(1)$. The scalar Lagrangian of the SM corresponds to $c_3^{(V)} = \frac{1}{2} m_h^2/v^2$, $c_4^{(V)} = \frac{1}{8} m_h^2/v^2$, $c_{n>4}^{(V)} = 0$, $c_1^{(u)} = 2$, $c_2^{(u)} = 1$ and $c_{n>2}^{(u)} = 0$.

In principle, the quadratic derivative term of the Higgs should also be multiplied with an arbitrary function $\mathcal{F}_h(h/v)$. However, this function can be reabsorbed into a redefinition of the field h [99].

3.11.1   Fermionic Fields

The SM fermion multiplets can be organized into $SU(2)_L$ and $SU(2)_R$ doublets,

$$\psi_L = \begin{pmatrix} t_L \\ b_L \end{pmatrix}, \qquad \psi_R = \begin{pmatrix} t_R \\ b_R \end{pmatrix}, \tag{3.189}$$

and analogous definitions for the other quark and lepton flavours, extending the symmetry group to $\mathcal{G} = SU(2)_L \otimes SU(2)_R \otimes U(1)_X$ with $X = (B-L)/2$, B and L being the baryon and lepton numbers, respectively [119]. They transform under $\mathcal{G}$ as

$$\psi_L \quad \longrightarrow \quad g_X g_L \, \psi_L, \qquad\qquad \psi_R \quad \longrightarrow \quad g_X g_R \, \psi_R, \tag{3.190}$$

with $g_X \in U(1)_X$.

We must introduce the $U(1)_X$ field $\hat{X}_\mu$, transforming like

$$\hat{X}^\mu \quad \longrightarrow \quad \hat{X}^\mu + i g_X \, \partial^\mu g_X^\dagger, \tag{3.191}$$

and its corresponding field strength tensor

$$\hat{X}_{\mu\nu} = \partial_\mu \hat{X}_\nu - \partial_\nu \hat{X}_\mu \tag{3.192}$$

that is a singlet under $\mathcal{G}$. The fermion covariant derivatives take then the form [175]

$$D_\mu^L \psi_L = \left(\partial_\mu - i\hat{W}_\mu - i x_\psi \, \hat{X}_\mu \right) \psi_L, \qquad D_\mu^R \psi_R = \left(\partial_\mu - i\hat{B}_\mu - i x_\psi \, \hat{X}_\mu \right) \psi_R, \tag{3.193}$$

where x_ψ is the corresponding $U(1)_X$ charge of the fermion field.

To recover the SM gauge interactions, the auxiliary fields must be frozen to the values given in (3.185) and

$$\hat{X}_\mu = -g' B_\mu, \tag{3.194}$$

which introduces an explicit breaking of the symmetry group $\mathcal{G}$ to the SM subgroup $SU(2)_L \otimes U(1)_Y$, with $Y = T_{3R} + \frac{1}{2}(B-L)$ [194], i.e.

$$Q = T_{3L} + T_{3R} + \frac{1}{2}(B-L). \tag{3.195}$$

Since bosons have $B = L = 0$, the enlargement of the symmetry group does not modify the bosonic sector.

In order to easily build chiral-invariant structures, it is convenient to introduce covariant fermion doublets [175]

$$\xi_L \equiv u^\dagger \psi_L, \qquad \xi_R \equiv u \psi_R, \qquad d_\mu^L \xi_L \equiv u^\dagger D_\mu^L \psi_L, \qquad d_\mu^R \xi_R \equiv u D_\mu^R \psi_R, \tag{3.196}$$

transforming under $\mathcal{G}$ as

$$\left(\xi_L, \xi_R, d_\mu^L \xi_L, d_\mu^R \xi_R \right) \quad \longrightarrow \quad g_X g_h \left(\xi_L, \xi_R, d_\mu^L \xi_L, d_\mu^R \xi_R \right). \tag{3.197}$$

The combined field $\xi \equiv \xi_L + \xi_R$ and $d_\mu \xi \equiv d_\mu^L \xi_L + d_\mu^R \xi_R$ transform obviously in the same way.

The kinetic fermion Lagrangian can be easily written in covariant form:

$$\mathcal{L}_{\text{Fermionic}}^{(0)} = \sum_\xi i\bar{\xi}\gamma^\mu d_\mu \xi = \sum_\psi \left(i\bar{\psi}_L\gamma^\mu D_\mu^L\psi_L + i\bar{\psi}_R\gamma^\mu D_\mu^R\psi_R \right). \tag{3.198}$$

Similarly, the usual fermion bilinears are compactly expressed as:

$$\mathcal{J}_\Gamma \equiv \bar{\xi}\Gamma\xi = \begin{cases} \bar{\psi}_L\Gamma\psi_L + \bar{\psi}_R\Gamma\psi_R, & \Gamma = \gamma^\mu\,(V), \gamma^\mu\gamma_5\,(A), \\[2mm] \bar{\psi}_L\Gamma\,U(\vec{\phi})\,\psi_R + \bar{\psi}_R\Gamma\,U^\dagger(\vec{\phi})\psi_L, & \Gamma = 1\,(S), i\gamma_5\,(P), \sigma^{\mu\nu}\,(T). \end{cases} \tag{3.199}$$

The Nambu–Goldstone coordinates disappear whenever the left and right sectors decouple; they only remain in those structures that mix the left and right chiralities, such as the scalar (S), pseudoscalar (P), and tensor (T) bilinears. While the spinorial indices get closed within $\mathcal{J}_\Gamma$, the fermion bilinear is an $SU(2)$ tensor, with indices $\mathcal{J}_\Gamma^{mn} = \bar{\xi}_n\Gamma\xi_m$, that transforms covariantly: $\mathcal{J}_\Gamma \to g_h\mathcal{J}_\Gamma g_h^\dagger$.

Fermion masses constitute an explicit breaking of chiral symmetry, which can be incorporated in the effective Lagrangian with a right-handed spurion field $\mathcal{Y}_R$, transforming as

$$\mathcal{Y}_R \longrightarrow g_R\mathcal{Y}_R g_R^\dagger, \qquad \mathcal{Y} \equiv u\mathcal{Y}_R u^\dagger \longrightarrow g_h\mathcal{Y}g_h^\dagger, \tag{3.200}$$

that allows us to add the invariant term

$$\Delta\mathcal{L}_2^{\text{Fermionic}} = -v\,\bar{\xi}_L\mathcal{Y}\xi_R + \text{h.c.} = -v\,\bar{\psi}_L\,U(\vec{\phi})\,\mathcal{Y}_R\psi_R + \text{h.c.} \tag{3.201}$$

The Yukawa interaction is recovered when the spurion field is frozen to the value [19, 23]

$$\mathcal{Y} = \hat{Y}_t(h/v)\,\mathcal{P}_+ + \hat{Y}_b(h/v)\,\mathcal{P}_-, \qquad \mathcal{P}_\pm \equiv \frac{1}{2}\,(I_2 \pm \sigma_3), \tag{3.202}$$

where [48]

$$\hat{Y}_{t,b}(h/v) = \sum_{n=0} \hat{Y}_{t,b}^{(n)}\left(\frac{h}{v}\right)^n. \tag{3.203}$$

In the SM, the $SU(2)$ doublet structure of the Higgs implies $\hat{Y}_{t,b}^{(0)} = \hat{Y}_{t,b}^{(1)} = m_{t,b}/v$, while $\hat{Y}_{t,b}^{(n\geq2)} = 0$.

To account for the flavour dynamics, we promote the fermion doublets $\xi_{L,R}$ to flavour vectors $\xi_{L,R}^A$ with family index A. The spurion field $\mathcal{Y}$ becomes then a 3×3 flavour matrix, with up-type and down-type components $\hat{Y}_u(h/v)$ and $\hat{Y}_d(h/v)$ that parametrize the custodial and flavour symmetry breaking [83]. The corresponding expansion coefficient matrices $\hat{Y}_{u,d}^{(n)}(h/v)$, multiplying the h^n term, could be different for each power n because chiral symmetry does not constrain them. Additional dynamical inputs are needed to determine their flavour structure.

3.11.2 Power Counting

The structure of the LO Lagrangian determines the power counting rules of the EWET, in a quite straightforward way [49, 175]. The Higgs and the Nambu–Goldstone modes do not carry any chiral dimension $\hat{d}$, while their canonical field dimensions are compensated by the electroweak scale v. The external gauge sources $\hat{W}_\mu$, $\hat{B}_\mu$, and $\hat{X}_\mu$ are of $\mathcal{O}(p)$, as they appear in the covariant derivatives, and their corresponding field strength tensors are quantities of $\mathcal{O}(p^2)$. Since any on-shell particle satisfies $p^2 = m^2$, all EWET fields have masses of $\mathcal{O}(p)$. This implies that the gauge couplings are also of $\mathcal{O}(p)$ because $m_W = gv/2$ and $m_Z = \sqrt{g^2 + g'^2}\, v/2$, while $\vec{\hat{W}}_\mu, \hat{B}_\mu \sim \mathcal{O}(p^0)$. All terms in $\mathcal{L}_{\mathrm{YM}}$ and $\Delta\mathcal{L}_2^{\mathrm{Bosonic}}$ are then of $\mathcal{O}(p^2)$, provided the Higgs potential is also assigned this chiral dimension, which is consistent with the SM Higgs self-interactions being proportional to $m_h^2 \sim \mathcal{O}(p^2)$. Thus,

$$u(\vec{\phi}), h, \mathcal{F}_u(h/v), \vec{\hat{W}}_\mu, \hat{B}_\mu \sim \mathcal{O}(p^0), \qquad \hat{W}_{\mu\nu}, \hat{B}_{\mu\nu}, \hat{X}_{\mu\nu}, f_{\pm\mu\nu}, c_n^{(V)} \sim \mathcal{O}(p^2),$$

$$D_\mu U, u_\mu, \partial_\mu, \hat{W}_\mu, \hat{B}_\mu, \hat{X}_\mu, m_h, m_W, m_Z, g, g' \sim \mathcal{O}(,p). \tag{3.204}$$

The assignment of chiral dimensions is slightly more subtle in the fermion sector. The chiral fermion fields must scale as $\xi_{L,R} \sim \mathcal{O}(p^{1/2})$, so that the fermionic component of $\mathcal{L}_{\mathrm{SM}}^{(0)}$ is also of $\mathcal{O}(p^2)$. The spurion $\mathcal{Y}$ is a quantity of $\mathcal{O}(p)$, since the Yukawa interactions must be consistent with the chiral counting of fermion masses, and the fermion mass terms are then also of $\mathcal{O}(p^2)$. Therefore,

$$\xi, \bar{\xi}, \psi, \bar{\psi} \sim \mathcal{O}(p^{1/2}), \qquad m_\psi, \mathcal{Y} \sim \mathcal{O}(,p). \tag{3.205}$$

The chiral dimensions reflect the infrared behaviour at low momenta and lead to a consistent power counting to organize the EWET. In particular, the chiral low-energy expansion preserves gauge invariance order by order, because the kinetic, cubic, and quartic gauge terms have all $\hat{d} = 2$ [119]. With a straight-forward dimensional analysis [144], we can generalize the Weinberg's power counting theorem (3.43) [211]. An arbitrary Feynman diagram Γ scales with momenta as [45, 48, 49, 119, 175]

$$\Gamma \sim p^{\hat{d}_\Gamma}, \qquad \hat{d}_\Gamma = 2 + 2L + \sum_{\hat{d}} (\hat{d} - 2)N_{\hat{d}}, \tag{3.206}$$

with L the number of loops and $N_{\hat{d}}$ the number of vertices with a given value of $\hat{d}$.

As in χPT, quantum loops increase the chiral dimension by two units and their divergences get renormalized by higher-order operators. The loop corrections are suppressed by the usual geometrical factor $1/(4\pi)^2$, giving rise to a series expansion in powers of momenta over the electroweak chiral scale $\Lambda_{\text{EWET}} = 4\pi v \sim 3$ TeV. Short-distance contributions from new physics generate EWET operators, suppressed by the corresponding new physics scale Λ_{NP}. Then we have a combined expansion in powers of p/Λ_{EWET} and p/Λ_{NP}.

Local four-fermion operators, originating in short-distance exchanges of heavier states, carry a corresponding factor $g_{\text{NP}}^2/\Lambda_{\text{NP}}^2$, and similar suppressions apply to operators with a higher number of fermion pairs. Assuming that the SM fermions are weakly coupled, i.e. $g_{\text{NP}} \sim \mathcal{O}(p)$, we must assign an additional $\mathcal{O}(p)$ suppression to the fermion bilinears [49, 175]:

$$(\bar{\eta}\,\Gamma\,\zeta)^n \quad \sim \quad \mathcal{O}\left(p^{2n}\right). \tag{3.207}$$

This has already been taken into account in the Yukawa interactions through the spurion $\mathcal{Y}$. Obviously, the power counting rule (3.207) does not apply to the kinetic term and, moreover, it is consistent with the loop expansion.

When the gauge sources are frozen to the values (3.185) and (3.194), they generate an explicit breaking of custodial symmetry that gets transferred to higher orders by quantum corrections. Something similar happens in χPT with the explicit breaking of chiral symmetry through electromagnetic interactions [80, 81, 204]. This can be easily incorporated into the EWET with the right-handed spurion [175]

$$\mathcal{T}_R \quad \longrightarrow \quad g_R \mathcal{T}_R g_R^\dagger, \qquad\qquad \mathcal{T} = u \mathcal{T}_R u^\dagger \quad \longrightarrow \quad g_h \mathcal{T} g_h^\dagger. \tag{3.208}$$

Chiral-invariant operators with an even number of $\mathcal{T}$ fields account for the custodial symmetry-breaking structures induced through quantum loops with B_μ propagators, provided we make the identification

$$\mathcal{T}_R = -g' \frac{\sigma_3}{2}. \tag{3.209}$$

Being proportional to the coupling g', this spurion has chiral dimension $\hat{d} = 1$ [175]:

$$\mathcal{T}_R \sim \mathcal{T} \sim \mathcal{O}(p). \tag{3.210}$$

3.11.3   NLO Lagrangian

Assuming invariance under CP transformations, the most general $\mathcal{O}(p^4)$ bosonic Lagrangian contains eleven P-even ($\mathcal{O}_i$) and three P-odd ($\widetilde{\mathcal{O}}_i$) operators [174, 175]:

$$\mathcal{L}_4^{\text{Bosonic}} = \sum_{i=1}^{11} \mathcal{F}_i(h/v)\,\mathcal{O}_i + \sum_{i=1}^{3} \widetilde{\mathcal{F}}_i(h/v)\,\widetilde{\mathcal{O}}_i. \tag{3.211}$$

For simplicity, we only consider a single fermion field. The CP-invariant fermionic Lagrangian of $\mathcal{O}(p^4)$ involves then seven P-even ($\mathcal{O}_i^{\psi^2}$) plus three P-odd ($\widetilde{\mathcal{O}}_i^{\psi^2}$) operators

Table 3.2 CP-invariant, P-even operators of the $\mathcal{O}(p^4)$ EWET [174, 175].

i	$\mathcal{O}_i$	$\mathcal{O}_i^{\psi^2}$	$\mathcal{O}_i^{\psi^4}$
1	$\frac{1}{4}\left\langle f_+^{\mu\nu} f_{+\,\mu\nu} - f_-^{\mu\nu} f_{-\,\mu\nu}\right\rangle$	$\langle \mathcal{J}_S\rangle\langle u_\mu u^\mu\rangle$	$\langle \mathcal{J}_S \mathcal{J}_S\rangle$
2	$\frac{1}{2}\left\langle f_+^{\mu\nu} f_{+\,\mu\nu} + f_-^{\mu\nu} f_{-\,\mu\nu}\right\rangle$	$i\left\langle \mathcal{J}_T^{\mu\nu}\left[u_\mu, u_\nu\right]\right\rangle$	$\langle \mathcal{J}_P \mathcal{J}_P\rangle$
3	$\frac{i}{2}\left\langle f_+^{\mu\nu}\left[u_\mu, u_\nu\right]\right\rangle$	$\left\langle \mathcal{J}_T^{\mu\nu} f_{+\,\mu\nu}\right\rangle$	$\langle \mathcal{J}_S\rangle\langle \mathcal{J}_S\rangle$
4	$\left\langle u_\mu u_\nu\right\rangle\left\langle u^\mu u^\nu\right\rangle$	$\hat{X}_{\mu\nu}\left\langle \mathcal{J}_T^{\mu\nu}\right\rangle$	$\langle \mathcal{J}_P\rangle\langle \mathcal{J}_P\rangle$
5	$\left\langle u_\mu u^\mu\right\rangle^2$	$\frac{1}{v}(\partial_\mu h)\left\langle u^\mu \mathcal{J}_P\right\rangle$	$\left\langle \mathcal{J}_V^\mu \mathcal{J}_{V,\mu}\right\rangle$
6	$\frac{1}{v^2}(\partial_\mu h)(\partial^\mu h)\left\langle u_\nu u^\nu\right\rangle$	$\left\langle \mathcal{J}_A^\mu\right\rangle\left\langle u_\mu \mathcal{T}\right\rangle$	$\left\langle \mathcal{J}_A^\mu \mathcal{J}_{A,\mu}\right\rangle$
7	$\frac{1}{v^2}(\partial_\mu h)(\partial_\nu h)\left\langle u^\mu u^\nu\right\rangle$	$\frac{1}{v^2}(\partial_\mu h)(\partial^\mu h)\langle \mathcal{J}_S\rangle$	$\left\langle \mathcal{J}_V^\mu\right\rangle\left\langle \mathcal{J}_{V,\mu}\right\rangle$
8	$\frac{1}{v^4}(\partial_\mu h)(\partial^\mu h)(\partial_\nu h)(\partial^\nu h)$	—	$\left\langle \mathcal{J}_A^\mu\right\rangle\left\langle \mathcal{J}_{A,\mu}\right\rangle$
9	$\frac{1}{v}(\partial_\mu h)\left\langle f_-^{\mu\nu} u_\nu\right\rangle$	—	$\left\langle \mathcal{J}_T^{\mu\nu} \mathcal{J}_{T,\mu\nu}\right\rangle$
10	$\left\langle \mathcal{T} u_\mu\right\rangle^2$	—	$\left\langle \mathcal{J}_T^{\mu\nu}\right\rangle\left\langle \mathcal{J}_{T,\mu\nu}\right\rangle$
11	$\hat{X}_{\mu\nu}\hat{X}^{\mu\nu}$	—	—

with one fermion bilinear, and ten P-even ($\mathcal{O}_i^{\psi^4}$) plus two P-odd ($\tilde{\mathcal{O}}_i^{\psi^4}$) four-fermion operators [175]:

$$
\mathcal{L}_4^{\text{Fermionic}} = \sum_{i=1}^{7} \mathcal{F}_i^{\psi^2}(h/v)\,\mathcal{O}_i^{\psi^2} + \sum_{i=1}^{3} \tilde{\mathcal{F}}_i^{\psi^2}(h/v)\,\tilde{\mathcal{O}}_i^{\psi^2}
$$

$$
+ \sum_{i=1}^{10} \mathcal{F}_i^{\psi^4}(h/v)\,\mathcal{O}_i^{\psi^4} + \sum_{i=1}^{2} \tilde{\mathcal{F}}_i^{\psi^4}(h/v)\,\tilde{\mathcal{O}}_i^{\psi^4}. \tag{3.212}
$$

The basis of independent P-even and P-odd operators is listed in Tables 3.2 and 3.3, respectively.[13] The number of chiral structures has been reduced through field redefinitions, partial integration, equations of motion and algebraic identities.[14] The number of fermionic operators increases dramatically when the quark and lepton flavours are taken into account.

---

[13] A much larger number of operators appears in former EWET analyses, owing to a slightly different chiral counting that handles the breakings of custodial symmetry less efficiently [6, 45, 48].

[14] When QCD interactions are taken into account, the covariant derivatives incorporate the gluon field $\hat{G}_\mu = g_s G_\mu^a T^a$ and two additional P-even operators must be included: $\text{Tr}_C(\hat{G}_{\mu\nu}\hat{G}^{\mu\nu})$ and $\text{Tr}_C(\hat{G}_{\mu\nu}\langle \mathcal{J}_T^{8\,\mu\nu}\rangle)$, where $\hat{G}_{\mu\nu} = \partial_\mu \hat{G}_\nu - \partial_\nu \hat{G}_\mu - i[\hat{G}_\mu, \hat{G}_\nu]$, $\text{Tr}_C(A)$ indicates the colour trace of A and $\mathcal{J}_T^{8\,\mu\nu}$ is the colour-octet tensorial quark bilinear [137].

Table 3.3 CP-invariant, P-odd operators of the $\mathcal{O}(p^4)$ EWET [175].

i	$\tilde{\mathcal{O}}_i$	$\tilde{\mathcal{O}}_i^{\psi^2}$	$\tilde{\mathcal{O}}_i^{\psi^4}$
1	$\frac{i}{2}\langle f_-^{\mu\nu}[u_\mu,u_\nu]\rangle$	$\langle \mathcal{J}_T^{\mu\nu} f_{-\mu\nu}\rangle$	$\langle \mathcal{J}_V^\mu \mathcal{J}_{A,\mu}\rangle$
2	$\langle f_+^{\mu\nu} f_{-\mu\nu}\rangle$	$\frac{1}{v}(\partial_\mu h)\langle u_\nu \mathcal{J}_T^{\mu\nu}\rangle$	$\langle \mathcal{J}_V^\mu\rangle\langle \mathcal{J}_{A,\mu}\rangle$
3	$\frac{1}{v}(\partial_\mu h)\langle f_+^{\mu\nu} u_\nu\rangle$	$\langle \mathcal{J}_V^\mu\rangle\langle u_\mu \mathcal{T}\rangle$	—

All coefficients $\mathcal{F}_i^{(\psi^{2,4})}(z)$ and $\tilde{\mathcal{F}}_i^{(\psi^{2,4})}(z)$ are functions of $z = h/v$; *i.e.*,

$$\mathcal{F}_i^{(\psi^{2,4})}(z) = \sum_{n=0} \mathcal{F}_{i,n}^{(\psi^{2,4})} z^n, \qquad \tilde{\mathcal{F}}_i^{(\psi^{2,4})}(z) = \sum_{n=0} \tilde{\mathcal{F}}_{i,n}^{(\psi^{2,4})} z^n. \qquad (3.213)$$

When the gauge sources are frozen to the values (3.185) and (3.194), the Higgsless term $\mathcal{F}_{2,0}\mathcal{O}_2 + \mathcal{F}_{11,0}\mathcal{O}_{11} + \tilde{\mathcal{F}}_{2,0}\tilde{\mathcal{O}}_2$ becomes a linear combination of the W_μ and B_μ Yang–Mills Lagrangians. Therefore, it could be eliminated through a redefinition of the corresponding gauge couplings.

To compute the NLO amplitudes, we must also include one-loop corrections with the LO Lagrangian $\mathcal{L}_2^{\mathrm{EW}}$. The divergent contributions can be computed with functional methods, integrating all one-loop fluctuations of the fields in the corresponding path integral. The divergences generated by the scalar sector (Higgs and Nambu–Goldstone fluctuations) were first computed in [110], while the full one-loop renormalization has been accomplished recently [10, 46].

According to the EWET power counting, the one-loop contributions are of $\mathcal{O}(p^4)$. Nevertheless, they can give rise to local structures already present in the LO Lagrangian, multiplied by $\mathcal{O}(p^2)$ factors such as m_h^2, $c_n^{(V)}$, or the product of two gauge couplings. In fact, the fluctuations of gauge bosons and fermions belong to the renormalizable sector and do not generate new counterterms [46]. They must be reabsorbed into $\mathcal{O}(p^4)$ redefinitions of the LO Lagrangian parameters and fields. Genuine $\mathcal{O}(p^4)$ structures originate in the scalar fluctuations, owing to the non-trivial geometry of the scalar field manifold [7, 8, 10], and the mixed loops between the renormalizable and non-renormalizable sectors.

In order to avoid lengthy formulae, we only detail here the $\mathcal{O}(p^4)$ divergences, originating from Higgs and Nambu–Goldstone fluctuations, that renormalize the couplings of the bosonic $\mathcal{O}_i$ operators:

$$\mathcal{F}_i(h/v) = \mathcal{F}_i^r(\mu, h/v) + \Gamma_{\mathcal{O}_i}(h/v)\,\Delta_{\overline{\mathrm{MS}}} = \sum_{n=0}\left[\mathcal{F}_{i,n}^r(\mu) + \gamma_{i,n}^{\mathcal{O}}\,\Delta_{\overline{\mathrm{MS}}}\right]\left(\frac{h}{v}\right)^n, \qquad (3.214)$$

with

$$\Delta_{\overline{\mathrm{MS}}} = \frac{\mu^{D-4}}{32\pi^2}\left\{\frac{2}{D-4} - \log(4\pi) + \gamma_E\right\}, \qquad (3.215)$$

Table 3.4 Coefficient functions $\Gamma_{\mathcal{O}_i}(h/v)$, generated by one-loop scalar fluctuations, and their zero-order expansion factors $\gamma_{i,0}^{\mathcal{O}}$ [110].

i	$\Gamma_{\mathcal{O}_i}(h/v)$	$\gamma_{i,0}^{\mathcal{O}}$
1	$\frac{1}{24}(\mathcal{K}^2 - 4)$	$-\frac{1}{6}(1 - a^2)$
2	$-\frac{1}{48}(\mathcal{K}^2 + 4)$	$-\frac{1}{12}(1 + a^2)$
3	$\frac{1}{24}(\mathcal{K}^2 - 4)$	$-\frac{1}{6}(1 - a^2)$
4	$\frac{1}{96}(\mathcal{K}^2 - 4)^2$	$\frac{1}{6}(1 - a^2)^2$
5	$\frac{1}{192}(\mathcal{K}^2 - 4)^2 + \frac{1}{128}\mathcal{F}_u^2\Omega^2$	$\frac{1}{8}(a^2 - b)^2 + \frac{1}{12}(1 - a^2)^2$
6	$\frac{1}{16}\Omega(\mathcal{K}^2 - 4) - \frac{1}{96}\mathcal{F}_u\Omega^2$	$-\frac{1}{6}(a^2 - b)(7a^2 - b - 6)$
7	$\frac{1}{24}\mathcal{F}_u\Omega^2$	$\frac{2}{3}(a^2 - b)^2$
8	$\frac{3}{32}\Omega^2$	$\frac{3}{2}(a^2 - b)^2$
9	$\frac{1}{24}\mathcal{F}_u'\Omega$	$-\frac{1}{3}a(a^2 - b)$

the usual $\overline{\text{MS}}$ subtraction. The computed coefficient functions $\Gamma_{\mathcal{O}_i}(z)$ are given in Table 3.4, in terms of $\mathcal{F}_u(z)$ and [110]

$$\mathcal{K}(z) \equiv \frac{\mathcal{F}_u'(z)}{\mathcal{F}_u^{1/2}(z)}, \qquad \Omega(z) \equiv 2\,\frac{\mathcal{F}_u''(z)}{\mathcal{F}_u(z)} - \left(\frac{\mathcal{F}_u'(z)}{\mathcal{F}_u(z)}\right)^2, \qquad (3.216)$$

where $\mathcal{F}_u'$ and $\mathcal{F}_u''$ indicate the first and second derivative of $\mathcal{F}_u(z)$ with respect to the variable $z = h/v$. The last column of the table shows explicitly the zero-order factors $\gamma_{i,0}^{\mathcal{O}}$ in the expansion of $\Gamma_{\mathcal{O}_i}(h/v)$ in powers of the Higgs field, which only depend on the LO couplings $a \equiv \frac{1}{2}c_1^{(u)}$ and $b \equiv c_2^{(u)}$. The coefficients $\Gamma_{\mathcal{O}_i}(z)$ and $\gamma_{i,n}^{\mathcal{O}}$ govern the running with the renormalization scale of the corresponding $\mathcal{F}_i^r(\mu, z)$ functions and $\mathcal{F}_{i,n}^r(\mu)$ factors, respectively, as dictated in (3.48) for the χPT LECs.

Taking $\mathcal{F}_u(z) = 1$ ($a = b = 0$), the Higgs decouples from the Nambu–Goldstone modes in $\Delta\mathcal{L}_2^{\text{Bosonic}}$, and we recover the known renormalization factors of the Higgsless electroweak chiral Lagrangian [18, 117, 118, 142, 143]. The renormalization factors $\gamma_{4,n}^{\mathcal{O}}$ and $\gamma_{5,n}^{\mathcal{O}}$ reproduce in this case their χPT counterparts $\frac{1}{4}\gamma_2$ and $\frac{1}{4}\gamma_1$ in (3.47). In the SM, $\mathcal{K} = 2$ and $\Omega = 0$ ($a = b = 1$). Thus, all $\mathcal{O}(p^4)$ divergences disappear, as it should, except $\Gamma_{\mathcal{O}_2}(z) = \gamma_{2,0}^{\mathcal{O}} = -\frac{1}{6}$ that is independent of the Higgs field and gets reabsorbed through a renormalization of the gauge couplings in the Yang–Mills Lagrangian.

Since $\frac{1}{4}\Gamma_{\mathcal{O}_1}(z) + \frac{1}{2}\Gamma_{\mathcal{O}_2}(z) = -\frac{1}{12}$ is independent of the Higgs field, the chiral structures $h^n\langle f_+^{\mu\nu} f_{+\,\mu\nu}\rangle$ are not renormalized when $n \neq 0$. Therefore, the interaction vertices $h^n\gamma\gamma$ and $h^n Z\gamma$, with $n \geq 1$, are renormalization group invariant [22, 69, 110].

3.11.4   Scattering of Longitudinal Gauge Bosons at NLO

The one-loop corrections to the elastic scattering of two electroweak Nambu–Goldstone modes (or, equivalently, longitudinal gauge bosons) can be partly taken from the χPT expression (3.50), which only includes the virtual contributions from massless $\vec{\phi}$ propagators. One must add the corrections induced by the Higgs boson, and replace the $\mathcal{O}(p^4)$ χPT LECs by their corresponding EWET counterparts: $\ell_1 \to 4\mathcal{F}_{5,0}, \ell_2 \to 4\mathcal{F}_{4,0}$. In the limit $g = g' = 0$, custodial symmetry becomes exact and the scattering amplitudes follow the weak isospin decomposition (3.49), with the $\mathcal{O}(p^4)$ function [70, 84]

$$
A(s,t,u) = \frac{s}{v^2}\,(1-a^2) + \frac{4}{v^4}\left[\mathcal{F}_{4,0}^r(\mu)\,(t^2+u^2) + 2\mathcal{F}_{5,0}^r(\mu)\,s^2\right]
$$

$$
+ \frac{1}{96\pi^2 v^4}\left\{\frac{2}{3}\,(14\,a^4 - 10\,a^2 - 18\,a^2 b + 9\,b^2 + 5)\,s^2 + \frac{13}{3}\,(1-a^2)^2\,(t^2+u^2)\right.
$$

$$
+ \frac{1}{2}\,(1-a)^2\left[(s^2 - 3t^2 - u^2)\log\left(\frac{-t}{\mu^2}\right) + (s^2 - t^2 - 3u^2)\log\left(\frac{-u}{\mu^2}\right)\right]
$$

$$
\left. - 3\,(2\,a^4 - 2\,a^2 - 2\,a^2 b + b^2 + 1)\,s^2 \log\left(\frac{-s}{\mu^2}\right)\right\}, \tag{3.217}
$$

where $a \equiv \frac{1}{2}c_1^{(u)}$ and $b \equiv c_2^{(u)}$ are the relevant LO couplings of the Higgs to the $\vec{\phi}$ fields in eqn (3.187). All boson masses have been neglected, and the $\mathcal{O}(p^4)$ couplings have been renormalized in the $\overline{\text{MS}}$ scheme, as indicated in (3.214) and (3.215). Taking into account the different renormalization schemes, this result agrees with its corresponding χPT expression (3.50) when $a = b = 0$. In the SM, $a = b = 1$ and $\mathcal{F}_{4,0} = \mathcal{F}_{5,0} = 0$, which implies $A(s,t,u) = 0$. Owing to unitarity, the SM amplitude is of $\mathcal{O}(m_h^2/s)$.

With the same approximations ($g = g' = m_h = 0$) and neglecting in addition the scalar potential, i.e. assuming the Higgs self-interactions to be of $\mathcal{O}(m_h^2)$, the elastic scattering amplitude of two Higgs bosons takes the form [70]

$$
A(hh \to hh) = \frac{2}{v^4}\,\mathcal{F}_{8,0}^r(\mu)\,(s^2+t^2+u^2)
$$

$$
+ \frac{3\,(a^2-b)^2}{32\pi^2 v^4}\left\{2\,(s^2+t^2+u^2) - s^2\log\left(\frac{-s}{\mu^2}\right) - t^2\log\left(\frac{-t}{\mu^2}\right) - u^2\log\left(\frac{-u}{\mu^2}\right)\right\}, \tag{3.218}
$$

which also vanishes in the SM limit ($a = b = 1$, $\mathcal{F}_{8,0} = 0$).

The analogous computation of the $\phi^i \phi^j \to hh$ scattering amplitudes gives [70]:

$$
\mathcal{A}(\phi^i \phi^j \to hh) = \delta_{ij} \left\{ \frac{a^2 - b}{v^2} s + \frac{1}{v^4} \left[2 \mathcal{F}_{6,0}^r(\mu) s^2 + \mathcal{F}_{7,0}^r(\mu) (t^2 + u^2) \right] \right.
$$
$$
+ \frac{a^2 - b}{192 \pi^2 v^4} \left((a^2 - b) t^2 \left[\frac{26}{3} - 3 \log\left(\frac{-t}{\mu^2} \right) - \log\left(\frac{-u}{\mu^2} \right) \right] \right.
$$
$$
+ (a^2 - b) u^2 \left[\frac{26}{3} - \log\left(\frac{-t}{\mu^2} \right) - 3 \log\left(\frac{-u}{\mu^2} \right) \right]
$$
$$
+ (a^2 - b) s^2 \left[\log\left(\frac{-t}{\mu^2} \right) + \log\left(\frac{-u}{\mu^2} \right) \right] + 12 (a^2 - 1) s^2 \log\left(\frac{-s}{\mu^2} \right)
$$
$$
\left. \left. + \frac{1}{3} (72 - 88 a^2 + 16 b) s^2 \right) \right\}. \tag{3.219}
$$

Time-reversal invariance implies that the same expression applies for $\mathcal{A}(hh \to \phi^i \phi^j)$.

With the renormalization factors given in Table 3.4, we can easily check that the scattering amplitudes (3.217), (3.218), and (3.219) are independent of the renormalization scale μ. Similar expressions have been derived for other interesting scattering amplitudes such as $\mathcal{A}(\gamma\gamma \to \phi^i \phi^j)$ [69], $\mathcal{A}(\gamma\gamma \to hh)$ [71], $\mathcal{A}(\phi^i \phi^j \to t\bar{t})$ [56], and $\mathcal{A}(hh \to t\bar{t})$ [56].

3.12 Fingerprints of Heavy Scales

The couplings of the EWET encode all the dynamical information about the underlying ultraviolet dynamics that is accessible at the electroweak scale. Different scenarios of new physics above the energy gap would imply different patterns of LECs, with characteristic correlations that could be uncovered through high-precision measurements. Given a generic strongly coupled theory of EWSB with heavy states above the gap, we aim to identify the imprints that its lightest excitations leave on the effective Lagrangian couplings. To simplify the discussion, we concentrate on the bosonic CP-conserving operators $\mathcal{O}_i$ [174]. A more general analysis, including the fermion sector and gluonic interactions, appears in [137, 175].

Consider an effective Lagrangian containing the SM fields coupled to the lightest scalar, pseudoscalar, vector, and axial-vector colour-singlet resonance multiplets S, P, $V^{\mu\nu}$, and $A^{\mu\nu}$, transforming as $SU(2)_{L+R}$ triplets ($R \to g_h R g_h^\dagger$), and the corresponding singlet states S_1, P_1, $V_1^{\mu\nu}$, and $A_1^{\mu\nu}$ ($R_1 \to R_1$). In order to analyse their implications on the $\mathcal{O}(p^4)$ EWET couplings, we only need to keep those structures with the lowest number of resonances and derivatives. At LO, the most general bosonic interaction with at most one resonance field, which is invariant under the symmetry group G, has the form:

$$
\mathcal{L}_{\mathrm{EWR}\chi\mathrm{T}} \doteq \langle V_{\mu\nu} \chi_V^{\mu\nu} \rangle + \langle A_{\mu\nu} \chi_A^{\mu\nu} \rangle + \langle P \chi_P \rangle + S_1 \chi_{S_1}, \tag{3.220}
$$

with

$$\chi_V^{\mu\nu} = \frac{F_V}{2\sqrt{2}} f_+^{\mu\nu} + \frac{i\,G_V}{2\sqrt{2}} [u^\mu, u^\nu], \qquad \chi_A^{\mu\nu} = \frac{F_A}{2\sqrt{2}} f_-^{\mu\nu} + \sqrt{2}\lambda_1^{hA} \partial^\mu h u^\nu,$$

$$\chi_P = \frac{d_P}{v} (\partial_\mu h)\, u^\mu, \qquad \chi_{S_1} = \frac{c_d}{\sqrt{2}} \langle u_\mu u^\mu \rangle + \lambda_{hS_1} v h^2. \qquad (3.221)$$

The spin-1 fields are described with anti-symmetric tensors and all couplings must be actually understood as functions of h/v. At this chiral order, the singlet vector, axial-vector, and pseudoscalar fields, and the scalar triplet cannot couple to the Nambu–Goldstone modes and gauge bosons.

Integrating out the heavy fields, we recover easily the bosonic $\mathcal{O}(p^4)$ EWET structures. In addition, the exchange of the scalar singlet field S_1 generates a correction to the LO EWET Lagrangian,

$$\Delta \mathcal{L}_{S_1}^{(2)} = \frac{\lambda_{hS_1}}{2M_{S_1}^2} v h^2 \left\{ \lambda_{hS_1} v h^2 + \sqrt{2} c_d \langle u_\mu u^\mu \rangle \right\}, \qquad (3.222)$$

that is suppressed by two powers of the heavy mass scale M_{S_1}. The function $\lambda_{hS_1}(h/v)$ should be assigned a chiral dimension of 2 to get a consistent power counting, so that the two terms in (3.222) have the same chiral order $\mathcal{O}(p^4)$. The term proportional to $\lambda_{hS_1} c_d$ contributes to $\mathcal{F}_u(h/v)$ in (3.187), while the $(\lambda_{hS_1})^2$ term corrects the Higgs potential $V(h/v)$.

The predicted values of the $\mathcal{O}(p^4)$ couplings $\mathcal{F}_i(h/v)$, in terms of resonance parameters, are detailed in Table 3.5. The exchange of triplet vector states contributes to $\mathcal{F}_{1-5}$, while axial-vectors triplets leave their imprints on $\mathcal{F}_{1,2,6,7,9}$. The spin-0 resonances have a much more reduced impact on the LECs: a pseudoscalar triplet only manifests in $\mathcal{F}_7$ and the singlet scalar contributes only to $\mathcal{F}_5$. Possible low-energy contributions from more exotic $\mathcal{J}^{PC} = 1^{+-}$ heavy states are analysed in [57].

The predicted vector and axial-vector contributions are independent of the (anti-symmetric) formalism adopted to describe the spin-1 resonances. The same results are obtained using Proca fields or a hidden-gauge formalism [24, 55], once a proper utraviolet behaviour is required (physical Green functions should not grow at large momenta) [175].

3.12.1 Short-Distance Behaviour

The previous resonance-exchange predictions can be made more precise with mild assumptions about the ultraviolet properties of the underlying fundamental theory. The procedure is similar to that adopted before in RχT. We impose the expected fall-off at large momenta of specific Green functions, and obtain constraints on the resonance parameters that are valid in broad classes of dynamical theories.

Table 3.5 Resonance contributions to the bosonic $\mathcal{O}(p^4)$ LECs. The right column includes short-distance constraints [174].

$$\mathcal{F}_1 \;=\; \frac{F_A^2}{4M_A^2} - \frac{F_V^2}{4M_V^2} \;=\; -\frac{v^2}{4}\left(\frac{1}{M_V^2} + \frac{1}{M_A^2}\right)$$

$$\mathcal{F}_2 \;=\; -\frac{F_A^2}{8M_A^2} - \frac{F_V^2}{8M_V^2} \;=\; -\frac{v^2(M_V^4 + M_A^4)}{8M_V^2 M_A^2 (M_A^2 - M_V^2)}$$

$$\mathcal{F}_3 \;=\; -\frac{F_V G_V}{2M_V^2} \;=\; -\frac{v^2}{2M_V^2}$$

$$\mathcal{F}_4 \;=\; \frac{G_V^2}{4M_V^2} \;=\; \frac{(M_A^2 - M_V^2)v^2}{4M_V^2 M_A^2}$$

$$\mathcal{F}_5 \;=\; \frac{c_d^2}{4M_{S_1}^2} - \frac{G_V^2}{4M_V^2} \;=\; \frac{c_d^2}{4M_{S_1}^2} - \frac{(M_A^2 - M_V^2)v^2}{4M_V^2 M_A^2}$$

$$\mathcal{F}_6 \;=\; -\frac{(\lambda_1^{hA})^2 v^2}{M_A^2} \;=\; -\frac{M_V^2(M_A^2 - M_V^2)v^2}{M_A^6}$$

$$\mathcal{F}_7 \;=\; \frac{d_P^2}{2M_P^2} + \frac{(\lambda_1^{hA})^2 v^2}{M_A^2} \;=\; \frac{d_P^2}{2M_P^2} + \frac{M_V^2(M_A^2 - M_V^2)v^2}{M_A^6}$$

$$\mathcal{F}_8 \;=\; 0$$

$$\mathcal{F}_9 \;=\; -\frac{F_A \lambda_1^{hA} v}{M_A^2} \;=\; -\frac{M_V^2 v^2}{M_A^4}$$

The functional derivatives of the action with respect to the external sources $\hat{W}^\mu$ and $\hat{B}^\mu$ define the corresponding left and right (vector and axial) currents. In complete analogy with QCD, the matrix element of the vector current between two Nambu–Goldstone bosons is characterized by a vector form factor that has the same functional form of eqn (3.145), with F_V and G_V the vector couplings in (3.221) and the electroweak scale v replacing the pion decay constant F. Requiring that this form factor should vanish at infinite momentum transfer implies the constraint

$$F_V G_V = v^2. \tag{3.223}$$

Enforcing a similar condition on the matrix element of the axial current between the Higgs and one Nambu–Goldstone particle, we obtain [179, 180]:

$$F_A \lambda_1^{hA} = \frac{1}{2} c_1^{(u)} v \equiv av, \tag{3.224}$$

where $c_1^{(u)}$ is the $h\,\partial_\mu \vec{\phi}\,\partial^\mu \vec{\phi}$ coupling in (3.187), which for $a \equiv \frac{1}{2} c_1^{(u)} = 1$ reproduces the gauge coupling of the SM Higgs.

The two-point correlation function of one left-handed and one right-handed current is an order parameter of EWSB, formally given by the same expressions as in QCD, eqns (3.149) and (3.150), changing F by v. In asymptotically-free gauge theories, $\Pi_{LR}(t)$ vanishes as $1/t^3$, at large momenta [32], which implies the two Weinberg sum rules [208]:

$$F_V^2 - F_A^2 = v^2, \qquad\qquad F_V^2 M_V^2 - F_A^2 M_A^2 = 0. \tag{3.225}$$

These conditions require $M_V < M_A$ and determine F_V and F_A in terms of the resonance masses and the electroweak scale, as in (3.152). At the one-loop level, and together with (3.223) and (3.224), the super-convergence properties of $\Pi_{LR}(t)$ also impose a relation between the Higgs gauge coupling and the resonance masses [179, 180]:

$$a = M_V^2/M_A^2 < 1. \tag{3.226}$$

Thus, the coupling of the Higgs to the electroweak gauge bosons is predicted to be smaller than the SM value.

With the identities (3.223), (3.224), (3.225), and (3.226), all vector and axial-vector contributions to the $\mathcal{O}(p^4)$ LECs can be written in terms of M_V, M_A and v. The resulting expressions are shown in the last column of Table 3.5, and are valid in dynamical scenarios where the two Weinberg sum rules are fulfilled, as happens in asymptotically free theories. Softer conditions can be obtained imposing only the first sum rule [179, 180], i.e. requiring $\Pi_{LR}(t) \sim 1/t^2$ at large momentum transfer, which is also valid in gauge theories with non-trivial UV fixed points [164].

The expressions derived for the LECs in terms of the vector and axial-vector couplings are generic relations that include the functional dependence on h/v hidden in the couplings. However, this is no longer true for the improved predictions incorporating short-distance constraints, where only constant couplings have been considered. Thus, the expressions on the right-hand side of Table 3.5 refer to $\mathcal{F}_{i,0}$, the $\mathcal{O}(h^0)$ terms in the expansion of the corresponding LECs in powers of h/v.

The numerical predictions for the different LECs $\mathcal{F}_i \equiv \mathcal{F}_{i,0}$, obtained with the short-distance constraints, are shown in Figure 3.18 as functions of M_V. The lightly shaded areas indicate all a priori possible values with $M_A > M_V$. The dashed blue, red, and green lines correspond to $M_V^2/M_A^2 = 0.8, 0.9$, and 0.95, respectively. A single dashed purple curve is shown for $\mathcal{F}_3$, which is independent of M_A. The coupling $\mathcal{F}_2$, which is not displayed in the figure, satisfies the inequality

$$\mathcal{F}_2 \leq -\frac{v^2}{8M_V^2}, \tag{3.227}$$

but its absolute value cannot be bounded with the current information.

The scalar and pseudoscalar contributions can be isolated through the combinations $\mathcal{F}_4 + \mathcal{F}_5$ and $\mathcal{F}_6 + \mathcal{F}_7$ that depend only on the ratios M_{S_1}/c_d and M_P/d_P, respectively. The predicted values are shown in Figure 3.19.

3.12.2 Gauge-Boson Self-Energies

The presence of massive resonance states coupled to the gauge bosons modifies the Z and $W^\pm$ self-energies. The deviations with respect to the SM predictions are characterized through the so-called oblique parameters S and T and U [132, 133, 164, 165] (or equivalently $\varepsilon_1, \varepsilon_2$, and ε_3 [11]), which are bounded by the electroweak precision data to the ranges [112]:

$$S = 0.04 \pm 0.11, \qquad T = 0.09 \pm 0.14, \qquad U = -0.02 \pm 0.11. \tag{3.228}$$

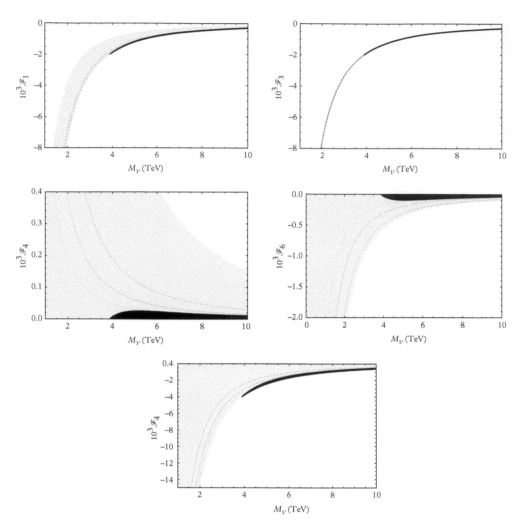

Figure 3.18 *Predicted values of the $\mathcal{O}(p^4)$ LECs in asymptotically free theories, as function of M_V [174]. The lightly shaded regions cover all possible values with $M_A > M_V$, while the blue, red, and green lines correspond to $M_V^2/M_A^2 = 0.8, 0.9,$ and 0.95, respectively. $\mathcal{F}_3$ does not depend on M_A. The S and T constraints restrict the allowed values to the dark areas.*

S measures the difference between the off-diagonal $W_3 B$ correlator and its SM value, while T parametrizes the breaking of custodial symmetry, being related to the difference between the W_3 and $W^\pm$ self-energies. The parameter U is not relevant for our discussion. These electroweak gauge self-energies are tightly constrained by the super-convergence properties of $\Pi_{LR}(t)$ [164].

At LO, the oblique parameter T is identically zero, while S receives tree-level contributions from vector and axial-vector exchanges [165]:

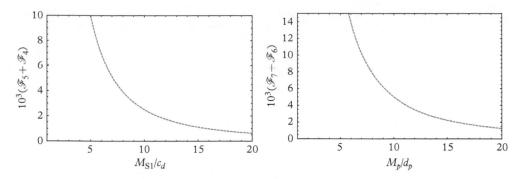

Figure 3.19 *Scalar and pseudoscalar contributions to $\mathcal{F}_5$ and $\mathcal{F}_7$, respectively [174].*

$$S_{\mathrm{LO}} = -16\pi\,\mathcal{F}_1 = 4\pi\left(\frac{F_V^2}{M_V^2} - \frac{F_A^2}{M_A^2}\right) = \frac{4\pi v^2}{M_V^2}\left(1 + \frac{M_V^2}{M_A^2}\right). \tag{3.229}$$

In the EWET, S_{LO} is governed by the LEC $\mathcal{F}_1$. The prediction from $\mathcal{L}_{\mathrm{EWR}\chi\mathrm{T}}$, in terms of vector and axial-vector parameters, reproduces the value of $\mathcal{F}_1$ in Table 3.5. The expression on the right-hand side of (3.229) makes use of the two Weinberg sum rules (3.225), which also imply $M_A > M_V$. Therefore, S_{LO} is bounded to be in the range [178]

$$\frac{4\pi v^2}{M_V^2} < S_{\mathrm{LO}} < \frac{8\pi v^2}{M_V^2}, \tag{3.230}$$

which puts a strong limit on the resonance mass scale: $M_V > 1.9$ TeV (90% CL).

At NLO, S and T receive corrections from the one-loop diagrams displayed in Figure 3.20. The calculation of S can be simplified with the dispersive representation [165]

$$S = \frac{16\pi}{g^2\tan\theta_W}\int_0^\infty \frac{dt}{t}\left[\rho_S(t) - \rho_S(t)^{\mathrm{SM}}\right], \tag{3.231}$$

where $\rho_S(t)$ is the spectral function of the W_3B correlator. Since $\rho_S(t)$ vanishes at short distances, the integral is convergent and, therefore, the dispersive relation does not require any subtractions. The relevant diagrams were computed in [179, 180], at LO in g and g'. The dominant contributions originate from the lightest two-particle cuts, i.e. two Nambu–Goldstone bosons or one Higgs plus one Nambu–Goldstone boson. The corrections generated by $\phi^a V$ and $\phi^a A$ intermediate states are suppressed by their higher mass thresholds.

Up to corrections of $\mathcal{O}(m_W^2/M_R^2)$, the parameter T is related to the difference between the charged and neutral Nambu–Goldstone self-energies [25]. Since the $SU(2)_L$ gauge

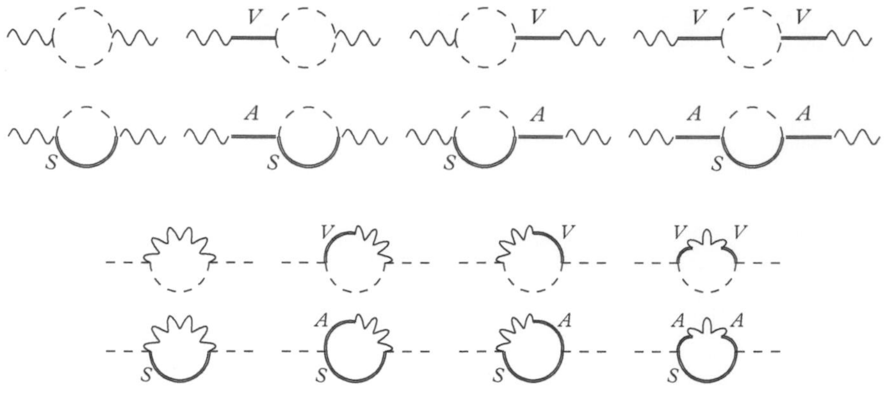

Figure 3.20 *NLO contributions to S (first two lines) and T (last two lines). A dashed (double) line stands for a Nambu–Goldstone (resonance) boson and a wiggled line indicates a gauge boson.*

coupling g does not break custodial symmetry, the one-loop contributions to T must involve the exchange of one B boson. The dominant effects correspond again to the lowest two-particle cuts, i.e. the B boson plus one Nambu–Goldstone or one Higgs boson. Requiring the W_3B correlator to vanish at high energies also implies a good convergence of the Nambu–Goldstone self-energies, at least for the two-particle cuts considered. Therefore, their difference also obeys an unsubtracted dispersion relation, which enables to compute T through the dispersive integral [180]:

$$T = \frac{4\pi}{g'^2 \cos^2 \theta_W} \int_0^\infty \frac{dt}{t^2} \left[\rho_T(t) - \rho_T(t)^{\text{SM}} \right], \tag{3.232}$$

with $\rho_T(t)$ the spectral function of the difference of the neutral and charged Nambu–Goldstone self-energies.

Neglecting terms of $\mathcal{O}(m_h^2/M_{V,A}^2)$ and making use of the short-distance conditions derived previously, the one-loop calculation gives [179, 180]

$$S = 4\pi v^2 \left(\frac{1}{M_V^2} + \frac{1}{M_A^2} \right) + \frac{1}{12\pi} \left\{ \log \frac{M_V^2}{m_h^2} - \frac{11}{6} + \frac{M_V^2}{M_A^2} \log \frac{M_A^2}{M_V^2} \right.$$
$$\left. - \frac{M_V^4}{M_A^4} \left(\log \frac{M_A^2}{m_h^2} - \frac{11}{6} \right) \right\} \tag{3.233}$$

and [180]

$$T = \frac{3}{16\pi \cos^2 \theta_W} \left\{ 1 + \log \frac{m_h^2}{M_V^2} - a^2 \left(1 + \log \frac{m_h^2}{M_A^2} \right) \right\}. \tag{3.234}$$

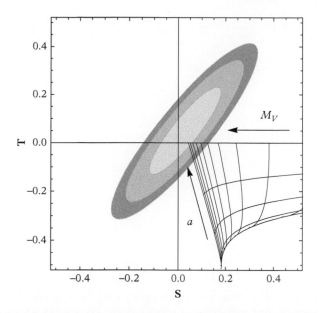

Figure 3.21 *NLO determinations of S and T in asymptotically-free theories [179, 180]. The approximately vertical lines correspond to constant values of M_V, from 1.5 to 6.0 TeV at intervals of 0.5 TeV. The approximately horizontal curves have constant values of a = 0.00, 0.25, 0.50, 0.75, and 1.00. The arrows indicate the directions of growing M_V and a. The ellipses give the experimentally allowed regions at 68% (orange), 95% (green), and 99% (blue) CL.*

These NLO determinations are compared in Figure 3.21 with the experimental constraints. When $a = M_V^2/M_A^2 = 1$, the parameter T vanishes identically, as it should, while the NLO prediction of S reaches the LO upper bound in eqn (3.230). Smaller values of $a = M_V^2/M_A^2 \to 0$ lead to increasingly negative values of T, outside its experimentally allowed range. Thus, the precision electroweak data require that the Higgs-like scalar should have a WW coupling very close to the SM one. At 95% CL, we get [180]

$$0.94 \le a \le 1, \tag{3.235}$$

in nice agreement with the present LHC evidence [21], but much more restrictive. Moreover, the vector and axial-vector states should be very heavy (and quite degenerate) [180]:

$$M_A \ge M_V > 4 \text{ TeV} \quad (95\% \text{ CL}). \tag{3.236}$$

Combining these oblique constraints with the previous resonance-exchange predictions for the LECs of the EWET, we get much stronger constraints on the $\mathcal{F}_i$ couplings. At 95% CL, the allowed regions are reduced to the dark areas shown in Figure 3.18, which imply the limits [180]:

$$-2 \cdot 10^{-3} < \mathcal{F}_1 < 0, \qquad 0 < \mathcal{F}_4 < 2.5 \cdot 10^{-5}, \qquad -2 \cdot 10^{-3} < \mathcal{F}_3 < 0,$$

$$-9 \cdot 10^{-5} < \mathcal{F}_6 < 0, \qquad -4 \cdot 10^{-3} < \mathcal{F}_9 < 0. \qquad (3.237)$$

These constraints are much more restrictive than the current direct bounds obtained from the Higgs signal strengths and anomalous gauge couplings [47, 68, 85, 86]. The limits (3.237) apply to any scenarios of new physics where the two Weinberg sum rules are satisfied, particularly in asymptotically-free theories. More generic limits that remain valid in models where only the first Weinberg sum rule is fulfilled appear in [180].

3.13 Summary

EFT is a very adequate framework to describe the low-energy dynamics of Nambu–Goldstone degrees of freedom. They are massless fields, separated by an energy gap from the massive states of the theory, and their dynamics are highly constrained by the pattern of symmetry breaking. The effective Lagrangian can be organized as an expansion in powers of momenta over some characteristic scale associated with the spontaneous (or dynamical) symmetry breaking. The zero-order term in this expansion vanishes identically because a constant shift of the Nambu–Goldstone coordinates amounts to a global rotation of the whole vacuum manifold that leaves the physics unchanged. Therefore, the Nambu–Goldstone bosons become free particles at zero momenta. Moreover, at the leading non-trivial order in the momentum expansion, the symmetry constraints are so restrictive that all the Nambu–Goldstone interactions are governed by a very low number of parameters, leading to a very predictive theoretical formalism.

This chapter focused on two particular applications of high relevance for fundamental physics: QCD and the electroweak theory. While the dynamical content of these two quantum field theories is very different, they share the same pattern of symmetry breaking (when only two light flavours are considered in QCD). Therefore, the low-energy dynamics of their corresponding Nambu–Goldstone modes is formally identical in the limit where explicit symmetry-breaking contributions are absent.

The low-energy EFT of QCD, χPT, has been thoroughly studied for many years and current phenomenological applications have reached a two-loop accuracy. The main limitation is the large number of LECs that appear at $\mathcal{O}(p^6)$. These LECs encode the short-distance information from the underlying QCD theory and are, in principle, calculable functions of the strong coupling and the heavy quark masses. However, owing to the non-perturbative nature of the strong interaction at low energies, the actual calculation is a formidable task. Nevertheless, a quite good dynamical understanding of these LECs has been already achieved through large-N_C methods and numerical lattice simulations.

The EWET contains a much richer variety of ingredients, such as fermions and gauge symmetries, and a more involved set of explicit symmetry-breaking effects. Therefore, the effective Lagrangian has a more complex structure and a larger amount of LECs.

The number of unknown couplings blows up when the flavour quantum numbers are included, clearly indicating our current ignorance about the fundamental flavour dynamics. The electroweak effective Lagrangian parametrizes the low-energy effects of any short-distance dynamics compatible the SM symmetries and the assumed pattern of EWSB. The crucial difference with QCD is that the true underlying electroweak theory is unknown.

In the absence of direct discoveries of new particles, the only accessible signals of the high-energy dynamics are hidden in the LECs of the EWET. High-precision measurements of scattering amplitudes among the known particles are sensitive to these LECs and could provide indirect evidence for new phenomena. The pattern of LECs emerging from the experimental data could then provide precious hints about the nature of the unknown ultraviolet theory responsible for any observed 'anomalies'.

Acknowledgements

I thank the organizers for the charming atmosphere of this school and all the school participants for their many interesting questions and comments. This work has been supported in part by the Spanish Government and ERDF funds from the EU Commission [Grants FPA2017-84445-P and FPA2014-53631-C2-1-P], by Generalitat Valenciana [Grant Prometeo/2017/053], and by the Spanish Centro de Excelencia Severo Ochoa Programme [Grant SEV-2014-0398].

Exercises

3.1 Prove the current algebra commutation relations in eqn (3.20).

3.2 The quadratic mass term of the $\mathcal{O}(p^2)$ χPT Lagrangian generates a small mixing between the ϕ_3 and ϕ_8 fields, proportional to the quark mass difference $\Delta m \equiv m_d - m_u$ (see footnote 4).

   a) Diagonalize the neutral meson mass matrix and find out the correct π^0 and η_8 mass eigenstates and their masses.

   b) When isospin is conserved, Bose symmetry forbids the transition $\eta \to \pi^0 \pi^+ \pi^-$ (why?). Compute the decay amplitude to first order in Δm.

3.3 a) Compute the axial current at $\mathcal{O}(p^2)$ in χPT and check that $F_\pi = F$ at this order.

   b) Expand the $\mathcal{O}(p^2)$ axial current to $\mathcal{O}(\Phi^3)$ and compute the one-loop corrections to F_π. Remember to include the pion wave function renormalization.

   c) Find the tree-level contribution of the $\mathcal{O}(p^4)$ χPT Lagrangian to the axial current. Renormalize the UV loop divergences with the $\mathcal{O}(p^4)$ LECs and obtain the $\mathcal{O}(p^4)$ expression for F_π in eqn (3.80).

3.4 Compute the elastic scattering amplitude of two electroweak Nambu–Goldstone bosons at LO and obtain the $\mathcal{O}(p^2)$ term in eqn (3.217).

3.5 Consider the two-point correlation function of a left-handed and a right-handed vector currents, either in $R\chi T$ or the EWET with massive states (EWRχT), and compute the LO (tree-level) contributions.

   a) Obtain the result for $\Pi_{LR}(t)$ in eqn (3.150).

   b) Demonstrate that in QCD $\Pi_{LR}(t) \sim 1/t^3$ at $t \to \infty$.

   c) Derive the Weinberg sum rules in eqn (3.151).

3.6 Assume the existence of a hypothetical light Higgs that couples to quarks with the Yukawa interaction

$$\mathcal{L}_{h^0 \bar{q}q} = -\frac{h^0}{v} \sum_q k_q\, m_q\, \bar{q}q.$$

   a) Determine at LO in χPT the effective Lagrangian describing the Higgs coupling to pseudoscalar mesons induced by the light-quark Yukawas.

   b) Determine the effective $h^0 G_a^{\mu\nu} G_{\mu\nu}^a$ coupling induced by heavy quark loops.

   c) The $G_a^{\mu\nu} G_{\mu\nu}^a$ operator can be related to the trace of the energy-momentum tensor, in the three-flavour QCD theory:

$$\Theta_\mu^\mu = \frac{\beta_1 \alpha_s}{4\pi} G_a^{\mu\nu} G_{\mu\nu}^a + \bar{q}\mathcal{M}q,$$

   where $\beta_1 = -\frac{9}{2}$ is the first coefficient of the β function. Using this relation, determine the LO χPT Lagrangian incorporating the Higgs coupling to pseudoscalar mesons induced by the heavy-quark Yukawas.

   d) Compute the decay amplitudes $h^0 \to 2\pi$ and $\eta \to h^0 \pi^0$.

References

[1] L. F. Abbott and M. B. Wise (1980). *Phys. Rev. D.* 22: 2208.

[2] M. Ademollo and R. Gatto (1964). *Phys. Rev. Lett.* 13: 264–5.

[3] S. L. Adler (1969). *Phys. Rev.* 177: 2426–38.

[4] S. L. Adler and W. A. Bardeen (1969). *Phys. Rev.* 182: 1517–36.

[5] S. L. Adler and R. F. Dashen (1968). *Current Algebras and Applications to Particle Physics* Benjamin.

[6] R. Alonso, et al. (2013). *Phys. Lett. B.* 722: 330–5. [Erratum: (2013). *Phys. Lett. B.* 726: 926].

[7] R. Alonso, E. E. Jenkins, and A. V. Manohar (2016). *Phys. Lett. B.* 754: 335–42.

[8] R. Alonso, E. E. Jenkins, and A. V. Manohar (2016). *JHEP* 08: 101.

[9] R. Alonso, E. E. Jenkins, A. V. Manohar, and M. Trott (2014). *JHEP* 04: 159.

[10] R. Alonso, K. Kanshin, and S. Saa (2018). *Phys. Rev. D.* 97(3): 035010.

[11] G. Altarelli and R. Barbieri (1991). *Phys. Lett. B.* 253: 161–7.

[12] S. R. Amendolia, et al. (1986). *Phys. Lett. B.* 178: 435–40.

[13] S. R. Amendolia, et al. (1986). *Nucl. Phys. B.* 277: 168.

[14] G. Amoros, J. Bijnens, and P. Talavera (2000). *Nucl. Phys. B.* 568: 319–63.

[15] B. Ananthanarayan (1998). *Phys. Rev. D.* 58: 036002.

[16] M. Antonelli, et al. (2010). *Eur. Phys. J. C.* 69: 399–424.

[17] S. Aoki, et al. (2017). *Eur. Phys. J. C.* 77(2): 112. http://flag.unibe.ch/.

[18] T. Appelquist and C. E. Bernard (1980). *Phys. Rev. D.* 22: 200.

[19] T. Appelquist, et al. (1985). *Phys. Rev. D.* 31: 1676.

[20] ATLAS and CMS collaborations (2015). *Phys. Rev. Lett.* 114: 191803.

[21] ATLAS and CMS collaborations (2016). *JHEP* 08: 045.

[22] A. Azatov, R. Contino, A. Di Iura, and J. Galloway (2013). *Phys. Rev. D.* 88(7): 075019.

[23] E. Bagan, D Espriu, and J Manzano (1999). *Phys. Rev. D.* 60: 114035.

[24] M. Bando, T. Kugo, and K. Yamawaki (1988). *Phys. Rept.* 164: 217–314.

[25] R. Barbieri, et al. (1993). *Nucl. Phys. B.* 409: 105–27.

[26] W. A. Bardeen (1969). *Phys. Rev.* 184: 1848–57.

[27] A. Bazavov, et al. (2010). *PoS* LATTICE2010, 074.

[28] A. Bazavov, et al. (2014). *Phys. Rev. Lett.* 112(11): 112001.

[29] R. E. Behrends and A. Sirlin (1960). *Phys. Rev. Lett.* 4: 186–7.

[30] A. A. Belavin, A. M. Polyakov, A. S. Schwartz, and Y. S. Tyupkin (1975). *Phys. Lett. B.* 59: 85–7.

[31] J. S. Bell and R. Jackiw (1969). *Nuovo Cim. A.* 60: 47–61.

[32] C. W. Bernard, A. Duncan, J. LoSecco, and S. Weinberg (1975). *Phys. Rev. D.* 12: 792.

[33] V. Bernard and U.-G Meissner (2007). *Ann. Rev. Nucl. Part. Sci.* 57: 33–60.

[34] J. Bijnens (2007). *Prog. Part. Nucl. Phys.* 58: 521–86.

[35] J. Bijnens, G. Colangelo, and G. Ecker (1999). *JHEP* 02: 020.

[36] J. Bijnens, G. Colangelo, and G. Ecker (2000). *Annals Phys.* 280: 100–39.

[37] J. Bijnens, G. Colangelo, and P. Talavera (1998). *JHEP* 05: 014.

[38] J. Bijnens and G. Ecker (2014). *Ann. Rev. Nucl. Part. Sci.* 64: 149–74.

[39] J. Bijnens, E. Gamiz, E. Lipartia, and J. Prades (2003). *JHEP* 04: 055.

[40] J. Bijnens, L. Girlanda, and P. Talavera (2002). *Eur. Phys. J. C.* 23: 539–44.

[41] J. Bijnens and E. Pallante (1996). *Mod. Phys. Lett. A.* 11: 1069–80.

[42] J. Bijnens and P. Talavera (2002). *JHEP* 03: 046.

[43] J. Bijnens and P. Talavera (2003). *Nucl. Phys. B.* 669: 341–62.

[44] I. Brivio and M. Trott (2019). *Phys. Rept.* 793: 1–98.

[45] G. Buchalla and O. Catà (2012). *JHEP* 07: 101.

[46] G. Buchalla. et al. (2018). *Nucl. Phys. B.* 928: 93–106.

[47] G. Buchalla, O. Catà, A. Celis, and C. Krause (2016). *Eur. Phys. J. C.* 76(5): 233.

[48] G. Buchalla, O. Catà, and C. Krause (2014). *Nucl. Phys. B.* 880: 552–73. [Erratum: (2016). *Nucl. Phys. B.* 913: 475].

[49] G. Buchalla, O. Catà, and C. Krause (2014). *Phys. Lett. B.* 731: 80–6.

[50] W. Buchmuller and D. Wyler (1986). *Nucl. Phys. B.* 268: 621–53.

[51] C. G. Callan, Jr., S. R. Coleman, J. Wess, and B. Zumino (1969). *Phys. Rev.* 177: 2247–50.

[52] C. G. Callan, Jr., R. F. Dashen, and D. J. Gross (1976). *Phys. Lett. B.* 63: 334–40.

[53] C. G. Callan, Jr., R. F. Dashen, and D. J. Gross (1978). *Phys. Rev. D.* 17: 2717.

[54] N. Carrasco, et al. (2016). *Phys. Rev. D.* 93(11): 114512.

[55] R. Casalbuoni, et al. (1989). *Int. J. Mod. Phys. A.* 4: 1065.

[56] A. Castillo, R. L. Delgado, A. Dobado, and F. J. Llanes Estrada (2017). *Eur. Phys. J. C.* 77(7): 436.

[57] O. Catà (2014). *Eur. Phys. J. C.* 74(8): 2991.

[58] O. Catà and S. Peris (2002). *Phys. Rev. D.* 65: 056014.

[59] V. Cirigliano, et al. (2005). *JHEP* 04: 006.

[60] V. Cirigliano, et al. (2006). *Nucl. Phys. B.* 753: 139–77.

[61] V. Cirigliano, et al. (2004). *Phys. Lett. B.* 596: 96–106.

[62] G. Colangelo, S. Lanz, H. Leutwyler, and E. Passemar (2017). *Phys. Rev. Lett.* 118(2): 022001.

[63] S. R. Coleman, J. Wess, and B. Zumino (1969). *Phys. Rev.* 177: 2239–2247.

[64] R. Contino (2011). In *Physics of the Large and the Small, TASI 09, Proceedings of the Theoretical Advanced Study Institute in Elementary Particle Physics* 235–306. 1–26 June 2009, Boulder, CO.

[65] J. M. Cornwall, D. N. Levin, and G. Tiktopoulos (1974). *Phys. Rev. D.* 10: 1145. [Erratum: (1975). *Phys. Rev. D.* 11: 972].

[66] R. F. Dashen (1969). *Phys. Rev.* 183: 1245–60.

[67] V. de Alfaro, S. Fubini, G. Furlan, and C. Rossetti (1973). *Currents in Hadron Physics*, North-Holland.

[68] J. de Blas, O. Eberhardt, and C. Krause (2018). JHEP 07: 048.

[69] R. L. Delgado, A. Dobado, M. J. Herrero, and J. J. Sanz Cillero (2014). *JHEP* 07: 149.

[70] R. L. Delgado, A. Dobado, and F. J. Llanes-Estrada (2014). *JHEP* 02: 121.

[71] R. L. Delgado, A. Dobado, and F. J. Llanes-Estrada (2017). *Eur. Phys. J. C.* 77(4): 205.

[72] P. Di Vecchia and G. Veneziano (1980). *Nucl. Phys. B.* 171: 253–72.

[73] A. Dobado, M. J. Herrero, and T. N. Truong (1990). *Phys. Lett. B.* 235: 134–40.

[74] J. F. Donoghue, E. Golowich, and B. R. Holstein (1992). *Dynamics of the Standard Model*, Cambridge Monographs on Particle Physics, Nuclear Physics and Cosmology, Vol. 2 [(2014). Cambridge Monographs on Particle Physics, Nuclear Physics and Cosmology, Vol. 35].

[75] J. F. Donoghue, C. Ramirez, and G. Valencia (1989). *Phys. Rev. D.* 39: 1947.

[76] R. J. Dowdall, C. T. H. Davis, G. P. Lepage, and C. McNeile (2013). *Phys. Rev. D.* 88: 074504.

[77] T. Ebertshauser, H. W. Fearing, and S. Scherer (2002). *Phys. Rev. D.* 65: 054033.

[78] G. Ecker (1995). *Prog. Part. Nucl. Phys.* 35: 1–80.

[79] G. Ecker, et al. (1989). *Phys. Lett. B.* 223: 425–32.

[80] G. Ecker, J. Gasser, A. Pich, and E. de Rafael (1989). *Nucl. Phys. B.* 321: 311–42.

[81] G. Ecker, et al. (2000). *Nucl. Phys. B.* 591: 419–34.

[82] G. Ecker and C. Zauner (2007). *Eur. Phys. J. C.* 52: 315–23.

[83] D. Espriu and J. Manzano (2001). *Phys. Rev. D.* 63: 073008.

[84] D. Espriu, F. Mescia, and B. Yencho (2013). *Phys. Rev. D.* 88: 055002.

[85] A. Falkowski, M. Gonzalez-Alonso, A Greljo, and D. Marzocca (2016). *Phys. Rev. Lett.* 116(1): 011801.

[86] A. Falkowski, et al. (2017). *JHEP* 02: 115.

[87] F. Feruglio (1993). *Int. J. Mod. Phys.* A8: 4937–72.

[88] E. G. Floratos, S. Narison, and E. de Rafael (1979). *Nucl. Phys. B.* 155: 115–49.

[89] K. Fujikawa (1979). *Phys. Rev. Lett.* 42: 1195–98.

[90] K. Fujikawa (1980). *Phys. Rev. D.* 21: 2848. [Erratum: (1980). *Phys. Rev. D* 22: 1499].

[91] J. Gasser and H. Leutwyler (1982). *Phys. Rept.* 87: 77–169.

[92] J. Gasser and H. Leutwyler (1984). *Annals Phys.* 158: 142.

[93] J. Gasser and H. Leutwyler (1985). *Nucl. Phys. B.* 250: 465–516.

[94] J. Gasser and H. Leutwyler (1985). *Nucl. Phys. B.* 250: 517–38.

[95] M. Gell-Mann (1962). *Phys. Rev.* 125: 1067–84.

[96] M. Gell-Mann and M. Levy (1960). *Nuovo Cim.* 16: 705.

[97] M. Gell-Mann, R. J. Oakes, and B. Renner (1968). *Phys. Rev.* 175: 2195–9.

[98] H. Georgi (1993). *Ann. Rev. Nucl. Part. Sci.* 43: 209–52.

[99] G. F. Giudice, C. Grojean, A. Pomarol, and R. Rattazzi (2007). *JHEP* 06: 045.

[100] J. Goldstone (1961). *Nuovo Cim.* 19: 154–64.

[101] J. Goldstone, A. Salam, and S. Weinberg (1962). *Phys. Rev.* 127: 965–70.

[102] M. Golterman, S. Peris, B. Phily, and E. de Rafael (2002). *JHEP* 01: 024.

[103] M. F. L. Golterman and S. Peris (2000). *Phys. Rev. D.* 61: 034018.

[104] D. Gomez Dumm, A. Pich, and J. Portoles (2000). *Phys. Rev. D.* 62: 054014.

[105] A. Gomez Nicola and J. R. Pelaez (2002). *Phys. Rev. D.* 65: 054009.

[106] M. González-Alonso, A. Pich, and A. Rodríguez-Sánchez (2016). *Phys. Rev. D.* 94(1): 014017.

[107] B. Grinstein and M. Trott (2007). *Phys. Rev. D.* 76: 073002.

[108] B. Grzadkowski, M. Iskrzynski, M. Misiak, and J. Rosiek (2010). *JHEP* 10: 085.

[109] F. Guerrero and A. Pich (1997). *Phys. Lett. B.* 412: 382–8.

[110] F.-K. Guo, P. Ruiz-Femenía, and J. J. Sanz Cillero (2015). *Phys. Rev. D.* 92: 074005.

[111] C. Haefeli, M. A. Ivanov, M. Schmid, and G. Ecker (2007). arXiv:0705.0576 [hep-ph].

[112] J. Haller, et al. (2018). *Eur. Phys. J. C.* 78: 675

[113] T. Hannah (1997). *Phys. Rev. D.* 55: 5613–26.

[114] J. C. Hardy and I. S. Towner (2015). *Phys. Rev. C.* 91(2): 025501.

[115] P. Herrera-Siklody, J. I. Latorre, P. Pascual, and J. Taron (1997). *Nucl. Phys. B.* 497: 345–86.

[116] P. Herrera-Siklody, J. I. Latorre, P. Pascual, and J. Taron (1998). *Phys. Lett. B.* 419: 326–32.

[117] M. J. Herrero and E. Ruiz Morales (1994). *Nucl. Phys. B.* 418: 431–55.

[118] M. J. Herrero and E. Ruiz Morales (1995). *Nucl. Phys. B.* 437: 319–55.

[119] J. Hirn and J. Stern (2006). *Phys. Rev. D.* 73: 056001.

[120] R. Jackiw and C. Rebbi (1976). *Phys. Rev. D.* 14: 517.

[121] R. Jackiw and C. Rebbi (1976). *Phys. Rev. Lett.* 37: 172–5.

[122] M. Jamin and M. Munz (1995). *Z. Phys. C.* 66: 633–46.

[123] M. Jamin, J. A. Oller, and A. Pich (2000). *Nucl. Phys. B.* 587: 331–62.

[124] M. Jamin, J. A. Oller, and A. Pich (2002). *Nucl. Phys. B.* 622: 279–308.

[125] M. Jamin, J. A. Oller, and A. Pich (2004). *JHEP* 02: 047.

[126] R. Kaiser (2001). *Phys. Rev. D.* 63: 076010.

[127] R. Kaiser and H. Leutwyler (2000). *Eur. Phys. J. C.* 17: 623–49.

[128] K. Kampf, J. Novotny, and J. Trnka (2007). *Eur. Phys. J. C.* 50: 385–403.

[129] K. Kampf, J. Novotny, and J. Trnka (2010). *Phys. Rev. D.* 81: 116004.

[130] D. B. Kaplan and A. V. Manohar (1986). *Phys. Rev. Lett.* 56: 2004.

[131] A. Kastner and H. Neufeld (2008). *Eur. Phys. J. C.* 57: 541–56.

[132] D. C. Kennedy and P. Langacker (1990). *Phys. Rev. Lett.* 65: 2967–70. [Erratum: (1991). *Phys. Rev. Lett.* 66: 395].

[133] D. C. Kennedy and P. Langacker (1991). *Phys. Rev. D.* 44: 1591–2.

[134] M. Kneckt and E. de Rafael (1998). *Phys. Lett. B.* 424: 335–42.

[135] M. Kneckt and A. Nyffeler (2001). *Eur. Phys. J. C.* 21: 659–78.

[136] M. Kneckt, S. Peris, and E. de Rafael (1998). *Phys. Lett. B.* 443: 255–63.

[137] C. Krause, et al. (2019). JHEP 05: 092

[138] T. Ledwig, et al. (2014). *Phys. Rev. D.* 90(11): 114020.

[139] B. W. Lee, C. Quigg, and H. B. Thacker (1977). *Phys. Rev. D.* 16: 1519.

[140] H. Leutwyler (1996). *Phys. Lett. B.* 374: 163–8.

[141] H. Leutwyler and M. Roos (1984). *Z. Phys. C.* 25: 91.
[142] A. C. Longhitano (1980). *Phys. Rev. D.* 22: 1166.
[143] A. C. Longhitano (1981). *Nucl. Phys. B.* 188: 118–54.
[144] A. V. Manohar and H. Georgi (1984). *Nucl. Phys. B.* 234: 189–212.
[145] A. V. Manohar (1997). *Lect. Notes Phys.* 479: 311–62.
[146] A. V. Manohar (1998). In *Probing the Standard Model of Particle Interactions. Proceedings, Summer School in Theoretical Physics*, parts 1–2, 1091–1169. NATO Advanced Study Institute, 68th Session. 28 July–5 September 1997, Les Houches, France. hep-ph/9802419.
[147] M. Moulson (2014). In *8th International Workshop on the CKM Unitarity Triangle (CKM 2014)*. 8–12 September–2014, Vienna, Austria. arXiv:1411.5252 [hep-ex].
[148] B. Moussallam (1995). *Phys. Rev. D.* 51: 4939–49.
[149] B. Moussallam (1997). *Nucl. Phys. B.* 504: 381–414.
[150] Y. Nambu (1960). *Phys. Rev. Lett.* 4: 380–2.
[151] Y. Nambu (1960). *Phys. Rev.* 117: 648–63.
[152] Y. Nambu and G. Jona-Lasinio (1961). *Phys. Rev.* 122: 345–58.
[153] Y. Nambu and G. Jona-Lasinio (1961). *Phys. Rev.* 124: 246–54.
[154] J. Nieves, A. Pich, and E. Ruiz Arriola (2011). *Phys. Rev. D.* 84: 096002.
[155] S. Okubo (1962). *Prog. Theor. Phys.* 27: 949–66.
[156] J. A. Oller and E. Oset (1999). *Phys. Rev. D.* 60: 074023.
[157] E. Pallante and A Pich (2000). *Phys. Rev. Lett.* 84: 2568–71.
[158] E. Pallante and A Pich (2001). *Nucl. Phys. B.* 592: 294–320.
[159] E. Pallante, A Pich, and I. Scimemi (2001). *Nucl. Phys. B.* 617: 441–74.
[160] P. Pascual and E. de Rafael (1982). *Z. Phys. C.* 12: 127.
[161] C. Patrignani, et al. (2016). *Chin. Phys. C.* 40(10): 100001.
[162] J. R. Palaez (2016). *Phys. Rept.* 658: 1.
[163] S. Peris and E. de Rafael (1995). *Phys. Lett. B.* 348: 539–42.
[164] M. E. Peskin and T. Takeuchi (1990). *Phys. Rev. Lett.* 65: 964–7.
[165] M. E. Peskin and T. Takeuchi (1992). *Phys. Rev. D.* 46: 381–409.
[166] A. Pich (1995). *Rept. Prog. Phys.* 58: 563–610.
[167] A. Pich (1998). In *Probing the Standard Model of Particle Interactions. Proceedings, Summer School in Theoretical Physics*, parts 1–2, 949–1049. NATO Advanced Study Institute, 68th Session. 28 July–5 September 1997, Les Houches, France. hep-ph/9806303.
[168] A. Pich (2002). In *Phenomenology of large N(c) QCD. Proceedings from the Institute for Nuclear Theory*, 239–58. 9–11 January 2002, Tempe, AZ. hep-ph/0205030.
[169] A. Pich (2005). *Int. J. Mod. Phys. A.* 20: 1613–18.
[170] A. Pich (2012). In *Proceedings, High-Energy Physics, 18th European School (ESHEP 2010)*, 1–50. 20 June–3 July 2010, Raseborg, Finland. arXiv:1201.0537 [hep-ph].
[171] A. Pich (2016). *Acta Phys. Polon. B.* 47: 151.
[172] A. Pich and E. de Rafael (1991). *Nucl. Phys. B.* 367: 313–33.
[173] A. Pich and J. Portoles (2001). *Phys. Rev. D.* 63: 093005.
[174] A. Pich, et al. (2016). *Phys. Rev. D.* 93(5): 055041.
[175] A. Pich, I. Rosell, J. Santos, and J. J. Sanz-Cillero (2017). *JHEP* 04: 012.
[176] A. Pich, I. Rosell, and J. J. Sanz-Cillero (2008). *JHEP* 07: 014.
[177] A. Pich, I. Rosell, and J. J. Sanz-Cillero (2011). *JHEP* 02: 109.
[178] A. Pich, I. Rosell, and J. J. Sanz-Cillero (2012). *JHEP* 08: 106.
[179] A. Pich, I. Rosell, and J. J. Sanz-Cillero (2013). *Phys. Rev. Lett.* 110: 181801.
[180] A. Pich, I. Rosell, and J. J. Sanz-Cillero (2014). *JHEP* 01: 157.
[181] J. Portoles, I. Rosell, and P. Ruiz-Femenia (2007). *Phys. Rev. D.* 75: 114011.

[182] I. Rosell, P. Ruiz-Femenia, and J. Portoles (2005). *JHEP* 12: 020.

[183] I. Rosell, P. Ruiz-Femenia, and J. J. Sanz-Cillero (2009). *Phys. Rev. D.* 79: 076009.

[184] I. Rosell, J. J. Sanz-Cillero, and A. Pich (2004). *JHEP* 08: 042.

[185] I. Rosell, J. J. Sanz-Cillero, and A. Pich (2007). *JHEP* 01: 039.

[186] C. Rosenzweig, J. Schechter, and C. G. Trahern (1980). *Phys. Rev. D.* 21: 3388.

[187] P. D. Ruiz-Femeniz, A. Pich, and J. Portoles (2003). *JHEP* 07: 003.

[188] J. J. Sanz-Cillero (2007). *Phys. Lett. B.* 649: 180–5.

[189] J. J. Sanz-Cillero and A. Pich (2003). *Eur. Phys. J. C.* 27: 587–99.

[190] J. J. Sanz-Cillero and J. Trnka (2010). *Phys. Rev. D.* 81: 056005.

[191] S. Scherer (2003). *Adv. Nucl. Phys.* 27: 277.

[192] S. Scherer and M. R. Schindler (2012). *Lect. Notes Phys.* 830: 1–338.

[193] J. S. Schwinger (1957). *Annals Phys.* 2: 407–34.

[194] G. Senjanovic and R. N. Mohapatra (1975). *Phys. Rev. D.* 12: 1502.

[195] M. A. Shifman, A. I. Vainshtein, and V. I. Zakharov (1979). *Nucl. Phys. B.* 147: 448–518.

[196] M. A. Shifman, A. I. Vainshtein, and V. I. Zakharov (1979). *Nucl. Phys. B.* 147: 385–447.

[197] P. Sikivie, L. Susskind, M. B. Voloshin, and V. I. Zakharov (1980). *Nucl. Phys. B.* 173: 189–207.

[198] G. 't Hooft (1974). *Nucl. Phys. B.* 72: 461.

[199] G. 't Hooft (1974). *Nucl. Phys. B.* 75: 461–70.

[200] G. 't Hooft (1976). *Phys. Rev. D.* 14: 3432–50. [Erratum: (1978). *Phys. Rev. D.* 18: 2199 (1978)].

[201] G. 't Hooft (1976). *Phys. Rev. Lett.* 37: 8–11.

[202] D. Toublan (1996). *Phys. Rev. D.* 53: 6602–7. [Erratum: (1998). *Phys. Rev. D.* 57: 4495 (1998)].

[203] T. N. Truong (1988). *Phys. Rev. Lett.* 61: 2526.

[204] R. Urech (1995). *Nucl. Phys. B.* 433: 234–54.

[205] C. E. Vayonakis (1976). *Lett. Nuovo Cim.* 17: 383.

[206] K. M. Watson (1954). *Phys. Rev.* 95: 228–36.

[207] S. Weinberg (1966). *Phys. Rev. Lett.* 17: 616–21.

[208] S. Weinberg (1967). *Phys. Rev. Lett.* 18: 507–9.

[209] S. Weinberg (1977). The Problem of Mass. In L. Motz (Ed), *A Festschrift for I.I. Rabi*, Academy of Sciences, 185.

[210] S. Weinberg (1979). *Phys. Rev. Lett.* 43: 1566–70.

[211] S. Weinberg (1979). *Physica A.* 96: 327–40.

[212] J. Wess and B. Zumino (1971). *Phys. Lett. B.* 37: 95–7.

[213] F. Wilczek and A. Zee (1979). *Phys. Rev. Lett.* 43: 1571–3.

[214] E. Witten (1979). *Nucl. Phys. B.* 160: 57–115.

[215] E. Witten (1980). *Annals Phys.* 128: 363.

[216] E. Witten (1983). *Nucl. Phys. B.* 223: 422–32.

[217] L. Y. Xiao and J. J. Sanz-Cillero (2008). *Phys. Lett. B.* 659: 452–6.

4

Introduction to Effective Field Theories and Inflation

C. P. Burgess

Department of Physics & Astronomy, McMaster University
and Perimeter Institute for Theoretical Physics

C. P. Burgess

Burgess, C. P., *Introduction to Effective Field Theories and Inflation* In: *Effective Field Theory in Particle Physics and Cosmology.*
Edited by: Sacha Davidson, Paolo Gambino, Mikko Laine, Matthias Neubert and Christophe Salomon, Oxford University Press (2020).
© Oxford University Press. DOI: 10.1093/oso/9780198855743.003.0004

Chapter Contents

The lectures compiled in this chapter are meant to provide a brief overview of two topics: the standard (Hot Big Bang, or ΛCDM) model of cosmology and the inflationary universe that presently provides our best understanding of the standard cosmology's peculiar initial conditions. There are several goals here, the first of which is to provide a particle physics audience with some of the tools required by other lecturers at Les Houches. After all, cosmology has become a mainstream topic within particle physics, largely because cosmology provides several of the main pieces of observational evidence for the incompleteness of the Standard Model (SM) of particle physics.

A second goal is to touch on the important role played in cosmology by many of the same methods of effective field theory (EFT) used elsewhere in physics. This second goal is particularly important for the cosmology of the very early universe (such as inflationary or 'bouncing' models) for which a central claim is that quantum fluctuations provide an explanation of the properties of primordial fluctuations presently found writ large across the sky. If true, this claim would imply and that quantum-gravity effects[1] are observable, that their imprint has already been observed cosmologically. Such claims sharpen the need to clarify which parameters control the size of quantum effects in gravity, and along the way more generally to identify the domain of validity of semi-classical methods in cosmology.

The cosmology part of this chapter is divided into two parts: homogeneous, isotropic cosmologies and the fluctuations about them. The first part provides a very brief description of the classic homogeneous and isotropic cosmological models usually encountered in introductory cosmology courses. One goal is to highlight the great success these models have describing the universe as we find it around us. A second goal is to describe the peculiar initial conditions that are required by this observational success. This part then highlights how these puzzling initial conditions suggest the universe once underwent an earlier epoch of accelerated expansion. It closes by describing several simple and representative single-field inflationary models proposed to provide this earlier accelerated epoch.

The cosmology section in this chapter then repeats the same picture, but now for fluctuations about both standard and inflationary cosmologies. It starts by describing the very successful picture of structure formation within the standard ΛCDM model, in which both fluctuations in the cosmic microwave background (CMB) and the distribution of galaxies are attributed to the amplification by gravity of a simple primordial spectrum of small fluctuations. Again the success of standard cosmology proves to rely on a specific choice for the initial spectrum of primordial fluctuations, and again the required initial spectrum can be understood as being produced by

---

[1] Here 'quantum-gravity effects' means quantum phenomena associated with the gravitational field, rather than (the much more difficult) foundational issues about the nature of space-time in a strongly quantum regime.

quantum fluctuations if there were an earlier epoch of accelerated expansion. Accelerated expansion plays double duty: potentially both explaining the initial conditions of the background homogeneous universe and of the primordial spectrum of fluctuations within it.

Because of the important role played by gravitating quantum fluctuations, EFT methods are central to assessing the domain of validity of the entire picture. Consequently the third, non-cosmology, section of this chapter summarizes several of the ways they do so, and how their application can differ in cosmology from those encountered elsewhere in particle physics. This starts by extending standard power-counting arguments to identify the small parameters that control the underlying semiclassical expansion implicitly used in essentially all cosmological models. In my opinion it is the quality of this control over the semiclassical expansion that at present favours inflationary models over their alternatives (such as bouncing cosmologies).[2] Other EFT topics discussed include several issues of principle to do with how to define and use EFTs in explicitly time-dependent situations, and how to quantify the robustness of inflationary predictions to any peculiarities of unknown higher-energy physics.

Parts of these lectures draw on some of my earlier review articles [1]–[3]. Meant as a personal viewpoint about the field rather than a survey of the literature, these lecture notes include references that are not comprehensive (and I apologize in advance to the many friends whose work I inevitably have forgotten to include).

4.1   Cosmology: Background

This section summarizes the standard discussion of background cosmology, both for the ΛCDM model and its inflationary precursors.

4.1.1   Standard ΛCDM cosmology

The starting point is the standard cosmology of the expanding universe revealed to us by astronomical observations. At present an impressively large collection of observations is described very well by the ΛCDM model, in terms of the handful of parameters listed in Figure 4.1. The following sections aim to describe the model, and what some of these parameters mean.

4.1.1.1   FRW geometries

Cosmology became a science once Einstein's discovery of general relativity (GR) related the observed distribution of stress-energy to the measurable geometry of space-time.

---

[2] Of course, although this might explain the current preference among cosmologists for inflationary models, it does not mean that Nature prefers them. Rather, EFT arguments just help set the standard to which formulations of alternative proposals should also aspire to achieve equal credence.

Parameter	[1] *Planck* TT+lowP	[2] *Planck* TE+lowP	[3] *Planck* EE+lowP	[4] *Planck* TT, TE,EE+lowP	$([1]-[4])/\sigma_{[1]}$
$\Omega_b h^2$	0.02222 ± 0.00023	0.02228 ± 0.00025	0.0240 ± 0.0013	0.02225 ± 0.00016	-0.1
$\Omega_c h^2$	0.1197 ± 0.0022	0.1187 ± 0.0021	$0.1150^{+0.0048}_{-0.0055}$	0.1198 ± 0.0015	0.0
$100\theta_{MC}$	1.04085 ± 0.00047	1.04094 ± 0.00051	1.03988 ± 0.00094	1.04077 ± 0.00032	0.2
τ	0.078 ± 0.019	0.053 ± 0.019	$0.059^{+0.022}_{-0.019}$	0.079 ± 0.017	-0.1
$\ln(10^{10}A_s)$	3.089 ± 0.036	3.031 ± 0.041	$3.066^{+0.046}_{-0.041}$	3.094 ± 0.034	-0.1
n_s	0.9655 ± 0.0062	0.965 ± 0.012	0.973 ± 0.016	0.9645 ± 0.0049	0.2
H_0	67.31 ± 0.96	67.73 ± 0.92	70.2 ± 3.0	67.27 ± 0.66	0.0
Ω_m	0.315 ± 0.013	0.300 ± 0.012	$0.286^{+0.027}_{-0.038}$	0.3156 ± 0.0091	0.0
σ_8	0.829 ± 0.014	0.802 ± 0.018	0.796 ± 0.024	0.831 ± 0.013	0.0
$10^9 A_s e^{-2\tau}$	1.880 ± 0.014	1.865 ± 0.019	1.907 ± 0.027	1.882 ± 0.012	-0.1

Figure 4.1 *Summary of cosmological parameters (taken from [4]), as obtained by fits to various combination of data sets (all of which include the most detailed measurements of the properties of the CMB using the Planck satellite). The parameters described in the text are H_0 (present-day Hubble expansion rate, equal to h in units of 100 km/sec/Mpc), Ω_b (baryon abundance), Ω_c (Dark Matter abundance) and $\Omega_m = \Omega_b + \Omega_c$ (non-relativistic matter abundance), which describe the background cosmology, as well as A_s and n_s that describe properties of primordial fluctuations about this background. $\Omega_i := \rho_i/\rho_{crit}$ is defined as the energy density in units of the critical density $\rho_{crit} := 3H_0^2/(8\pi G)$, where G is Newton's gravitational constant. Not discussed are θ_{MC} (the angular size of the sound horizon at last scattering), τ (the optical depth, which measures the amount of ionization in the later universe), and σ_8 (a measure of the amount of gravitational clustering).*

This implies the geometry of the universe as a whole can be tied to the overall distribution of matter at the largest scales. It used to be an article of faith that this geometry should be assumed to be homogeneous and isotropic (the 'cosmological principle'), but these days it is pretty much an experimental fact that the stress-energy of the universe is homogeneous and isotropic on the largest scales visible. One piece of evidence to this effect is the very small—one part in 10^5—temperature fluctuations of the CMB, as shown in Figure 4.2 (and more about which later).

On such large scales the geometry of space-time should therefore be homogeneous and isotropic, and the most general such a geometry in 3+1 dimensions is described by the Friedmann–Robertson–Walker (FRW) metric. The line element for this metric can be written as[3]

$$ds^2 = g_{\mu\nu}\, dx^\mu dx^\nu = -dt^2 + a^2(t)\left[\frac{dr^2}{1-\kappa r^2/R_0^2} + r^2\, d\theta^2 + r^2 \sin^2\theta\, d\phi^2\right] \quad (4.1)$$

$$= -dt^2 + a^2(t)\left[d\ell^2 + r^2(\ell)\, d\theta^2 + r^2(\ell)\sin^2\theta\, d\phi^2\right],$$

[3] For those rusty on what a metric means and perhaps needing a refresher course on GR, an under-graduate-level introduction using the same conventions as used here can be found in *General Relativity: The Notes* at www.physics.mcmaster.ca/~cburgess/Notes/GRNotes.pdf.

Figure 4.2 *Temperature fluctuations in the CMB as a function of direction in the full sky (in galactic coordinates) as measured by the Planck collaboration [5]. The figure subtracts foregrounds due to our galaxy and an order $\delta T/T \sim 10^{-3}$ dipole due to the Earth's motion through the CMB. The fluctuations that remain have a maximum amplitude of order $\delta T/T \sim 10^{-5}$. (Also, notice Stephen Hawking's initials just to the left of centre.)*

where R_0 is a constant with dimension length and κ can take one of the following three values: $\kappa = 1, 0, -1$. The coordinate ℓ is related to r by $d\ell = dr/(1 - \kappa r^2/R_0^2)^{1/2}$, and so

$$r(\ell) = \begin{cases} R_0 \sin(\ell/R_0) & \text{if} \quad \kappa = +1 \\ \ell & \text{if} \quad \kappa = 0 \\ R_0 \sinh(\ell/R_0) & \text{if} \quad \kappa = -1 \end{cases} . \tag{4.2}$$

The quantity $a(t)R_0$ represents the radius of curvature of the spatial slices at fixed t, which are 3-spheres when $\kappa = +1$; 3-hyperbolae for $\kappa = -1$ and are flat for $\kappa = 0$. It is conventional to scale R_0 out of the metric by rescaling the coordinates $\ell \to R_0 \ell$ and $r \to R_0 r$ while at the same time rescaling $a(t) \to a(t)/R_0$. This redefinition makes r and ℓ dimensionless while giving $a(t)$ units of length, and it is often useful to choose cosmological units for which $a(t_0) = 1$ for some t_0 (such as at present). The case $\kappa = 0$ turns out to be of particular interest because all current evidence (coming, for instance, from the measured properties of the CMB) indicates that the spatial slices in the universe are consistent with being flat.

Trajectories along which only t varies are time-like geodesics of this metric and represent the motion of a natural set of static 'co-moving' observers. The co-moving distance, $\Delta\ell$, between two such observers at a fixed time t is related to their physical distance—as measured by the metric (4.1)—by

$$D(\Delta\ell, t) = \Delta\ell\, a(t), \tag{4.3}$$

so the 'scale-factor' $a(t)$ describes the common time-evolution of spatial scales. So long as $a(t)$ is monotonic we can use t or a interchangeably as measures of the passage of time.

The trajectories of photons play a special role in cosmology since until very recently they brought us all of our information about the universe at large. Since they move at the speed of light their trajectories satisfy $ds^2 = 0$ and so

$$g_{\mu\nu}\left(\frac{dx^\nu}{ds}\right)\left(\frac{dx^\nu}{ds}\right) = 0, \tag{4.4}$$

which for radial motion specializes to $dt/ds = \pm a(t)(d\ell/ds)$. Choosing coordinates that place us at the origin means all photons sent to us move along a radial trajectory.

A photon arriving at $t = 0$ from a galaxy situated at fixed co-moving position $\ell = L$ must have departed at time $t = -T$ where

$$L = \int_0^T \frac{dt}{a(t)}. \tag{4.5}$$

Since the universe expands by an amount a_0/a in this time (where $a_0 = a(0)$ is the present-day scale factor and $a = a(-T)$ is its value when the light was emitted), the redshift, z, of the light is given by $z := (\lambda_{\text{obs}} - \lambda_{\text{em}})/\lambda_{\text{em}}$, with $\lambda_{\text{obs}}/\lambda_{\text{em}} = a_0/a$. Consequently z and a are related by

$$1 + z = \frac{a_0}{a}. \tag{4.6}$$

This very usefully ties the universal expansion to the more easily measured redshift of distant objects.[4]

For later purposes, it is worth introducing another useful time coordinate when discussing the evolution of light rays in FRW geometries. Defining 'conformal time', τ, by

$$\tau = \int \frac{dt}{a(t)}, \tag{4.7}$$

allows the metric (4.1) to be written

$$ds^2 = a^2(\tau)\left[-d\tau^2 + d\ell^2 + r^2(\ell)\,d\theta^2 + r^2(\ell)\sin^2\theta\,d\phi^2\right]. \tag{4.8}$$

---

[4] In practice the redshift of any particular object depends also on its 'peculiar' motion relative to co-moving observers, but because peculiar motion effects are smaller than the cosmic redshift for all but relatively nearby galaxies they are ignored in what follows.

The utility of this coordinate system is that the scale-factor $a(\tau)$ completely drops out of the evolution of photons, which simplifies the identification of many of the causal properties of the spacetime (i.e. identifying which events can communicate with each other by exchanging photons).

4.1.1.2 Implications of Einstein's Equations

So far so good, but the story to here is largely just descriptive. The FRW metric, with $a(t)$ specified, says much about how particles move over cosmological distances. But we also need to know how to relate $a(t)$ to the universe's stress-energy content. This connection is made using Einstein's equations,[5]

$$R_{\mu\nu} - \frac{1}{2}Rg_{\mu\nu} = 8\pi G\,T_{\mu\nu}, \tag{4.9}$$

where G is Newton's constant of universal gravitation, $R_{\mu\nu} = R^\alpha{}_{\mu\alpha\nu}$ is the geometry's Ricci tensor (where $R^\alpha{}_{\mu\beta\nu}$ is its Riemann tensor), and $R = g^{\mu\nu}R_{\mu\nu}$.

The twin requirements of homogeneity and isotropy dictate that the most general form for the universe's stress-energy tensor, $T_{\mu\nu}$, is that of a perfect fluid,

$$T_{\mu\nu} = pg_{\mu\nu} + (p+\rho)\,U_\mu U_\nu, \tag{4.10}$$

where p and ρ are respectively the fluid's pressure and energy density, while $U^\mu\partial_\mu = \partial_t$ (or, equivalently, $U_\mu\,dx^\mu = -dt$) is the four-velocity of the co-moving observers.

Specialized to the metric (4.1) the Einstein equations boil down to the following two independent equations:

$$H^2 + \frac{\kappa}{a^2} = \frac{8\pi G}{3}\rho = \frac{\rho}{3M_p^2} \qquad \text{(Friedmann equation)} \tag{4.11}$$

and

$$\dot\rho + 3H(p+\rho) = 0 \qquad \text{(energy conservation)} \tag{4.12}$$

---

[5] These notes use the metric signature $(-+++)$ as well as units with $\hbar = c = k_B = 1$, and conform to the widely used MTW curvature conventions [6] (which differ from Weinberg's conventions [7]—that are often used in my own papers—only by an overall sign for the Riemann curvature tensor, $R^\mu{}_{\nu\lambda\rho}$). The world divides into two camps regarding the metric signature, with most relativists and string theorists using $(-+++)$ and many particle phenomenologists using $(+---)$. As students just forming your own habits now, you should choose one and stick to it. When doing so keep in mind that the $(+---)$ metric becomes the $(----)$ metric in Euclidean signature (such as arises when Wick rotating or for applications at finite temperature), leading to many headaches keeping track of signs because all vector norms become negative: $V^2 = g_{mn}V^m V^n < 0$. Your notation should be your friend, not your adversary.

where G is Newton's gravitational constant and over-dots denote differentiation with respect to t and the Hubble function is defined by $H = \dot{a}/a$. The last equality in eqn (4.11) also defines the 'reduced' Planck mass: $M_p = (8\pi G)^{-1/2} \simeq 10^{18}$ GeV. Differentiating (4.11) and using (4.12) gives a useful formula for the cosmic acceleration

$$\frac{\ddot{a}}{a} = -\frac{1}{6M_p^2}(\rho + 3p). \tag{4.13}$$

Mathematically speaking, finding the evolution of the universe as a function of time requires the integration of eqns (4.11) and (4.12), but in themselves these two equations are inadequate to determine the evolution of the three unknown functions, $a(t)$, $\rho(t)$, and $p(t)$. Another condition is required in order to make the problem well-posed. The missing condition is furnished by the equation of state for the matter in question, which for the present purposes we take to be an expression for the pressure as a function of energy density, $p = p(\rho)$. In particular, the equations of state of interest in ΛCDM cosmology have the general form

$$p = w\rho, \tag{4.14}$$

where w is a t-independent constant.

The first step in solving for $a(t)$ is to determine how p and ρ depend on a, since this is dictated by energy conservation. Using eqn (4.14) in (4.12) allows it to be integrated to obtain

$$\rho = \rho_0 \left(\frac{a_0}{a}\right)^\sigma \quad \text{with} \quad \sigma = 3(1 + w). \tag{4.15}$$

Equation (4.14) implies the pressure also shares this same dependence on a. Similarly using eqn (4.15) to eliminate ρ from (4.11) leads to the following differential equation for $a(t)$:

$$\dot{a}^2 + \kappa = \frac{8\pi G\rho_0 a_0^2}{3}\left(\frac{a_0}{a}\right)^{\sigma-2}. \tag{4.16}$$

When $\kappa = 0$ (and $w \neq -1$) this equation is easily integrated to give

$$a(t) = a_0 \left(\frac{t}{t_0}\right)^\alpha \quad \text{with} \quad \alpha = \frac{2}{\sigma} = \frac{2}{3(1+w)}. \tag{4.17}$$

4.1.1.3   *Equations of state*

In the ΛCDM model of cosmology the total energy density is regarded as the sum of several components, each of which separately satisfies one of the following three basic equations of state.

Non-Relativistic Matter An ideal gas of non-relativistic particles in thermal equilibrium has a pressure and energy density given by

$$p = nT \qquad \text{and} \qquad \rho = nm + \frac{nT}{\gamma - 1}, \tag{4.18}$$

where n is the number of particles per unit volume, m is the particle's rest mass and $\gamma = c_p/c_v$ is its ratio of specific heats, with $\gamma = 5/3$ for a gas of monatomic atoms. For non-relativistic particles the total number of particles is usually also conserved,[6] which implies that

$$\frac{\mathrm{d}}{\mathrm{d}t}\left[na^3\right] = 0. \tag{4.19}$$

Since $m \gg T$ (or else the atoms would be relativistic) the equation of state for this gas may be taken to be

$$\frac{p}{\rho} \sim \frac{T}{m} \ll 1 \qquad \text{and so} \qquad w \simeq 0. \tag{4.20}$$

Since $w \simeq 0$ energy conservation implies $\sigma = 3(1+w) \simeq 3$ and so ρa^3 is a constant. This is appropriate for non-relativistic matter for which the energy density is dominated by the particle rest-masses, $\rho \simeq nm$, because in this case energy conservation is equivalent to conservation of particle number which according to (4.19) implies $n \propto a^{-3}$.

Finally, whenever the total energy density is dominated by non-relativistic matter we know $w = 0$ also implies $\alpha = 2/\sigma = 2/3$ and so if $\kappa = 0$ then the universal scale factor expands like $a \propto t^{2/3}$.

Radiation Thermal equilibrium dictates that a gas of relativistic particles (like photons) must have an energy density and pressure given by

$$\rho = a_B T^4 \qquad \text{and} \qquad p = \frac{1}{3} a_B T^4, \tag{4.21}$$

where $a_B = \pi^2/15 \simeq 0.6580$ is the Stefan–Boltzmann constant (in units where $k_B = c = \hbar = 1$) and T is the temperature. Together, these ensure that ρ and p satisfy the equation of state

---

[6] The total *difference* between the number of non-relativistic particles and their anti-particles can be constrained to be non-zero if they carry a conserved charge (such as baryon number, for protons and neutrons). In the absence of such a charge the density of such particles becomes quite small if they remain in thermal equilibrium since their abundance becomes Boltzmann suppressed, $n \propto e^{-m/T}$, at temperatures $T < m$. This suppression happens because the annihilation of particles and anti-particles is not compensated by their pair-production due to there being insufficient thermal energy.

$$p = \frac{1}{3}\rho \quad \text{and so} \quad w = \frac{1}{3}. \tag{4.22}$$

Equation (4.22) also applies to any other particle whose temperature dominates its rest mass, and so in particular applies to neutrinos for most of the universe's history.

Since $w = 1/3$ it follows that $\sigma = 3(1 + w) = 4$ and so $\rho \propto a^{-4}$. This has a simple physical interpretation for a gas of non-interacting photons, since for these the total number of photons is fixed and so $n_\gamma \propto a^{-3}$. But each photon energy is inversely proportional to its wavelength and so also redshifts like $1/a$ as the universe expands, leading to $\rho_\gamma \propto a^{-4}$.

Whenever radiation dominates the total energy density then $w = 1/3$ implies $\alpha = 2/\sigma = 1/2$, and so if $\kappa = 0$ then $a(t) \propto t^{1/2}$.

The Vacuum If the vacuum is Lorentz invariant—as the success of special relativity seems to indicate—then its stress-energy must satisfy $T_{\mu\nu} \propto g_{\mu\nu}$. This implies the vacuum pressure must satisfy the only possible Lorentz-invariant equation of state:

$$p = -\rho \quad \text{and so} \quad w = -1. \tag{4.23}$$

Furthermore, for $T_{\mu\nu} = -\rho g_{\mu\nu}$ stress-energy conservation, $\nabla^\mu T_{\mu\nu} = 0$, implies ρ must be spacetime-independent (in agreement with (4.15) for $w = -1$). This kind of constant energy density is often called, for historical reasons, a cosmological constant.

Although counter-intuitive, constant energy density can be consistent with energy conservation in an expanding universe. This is because (4.12) implies the total energy satisfies $\mathrm{d}(\rho a^3)/\mathrm{d}t = -p \, \mathrm{d}(a^3)/\mathrm{d}t$. Consequently, the equation of state (4.23) ensures the pressure does precisely the amount of work required to produce the change in total energy required by having constant energy density.

When the vacuum dominates the energy density then $\alpha = 2/\sigma \to \infty$, which shows that the power-law solutions, $a \propto t^\alpha$, are not appropriate. Returning directly to the Friedmann equation, eqn (4.11), shows (when $\kappa = 0$) that $H = \dot{a}/a$ is constant and so the solutions are exponentials: $a \propto \exp[\pm H(t - t_0)]$. Notice that (4.23) implies $\rho + 3p$ is negative if ρ is positive. This furnishes an explicit example of an equation of state for which the universal acceleration, $\ddot{a}/a = -\frac{4}{3}\pi G(\rho + 3p)$, can be positive.

4.1.1.4 *Universal Energy Content*

At present there is direct observational evidence that the universe contains at least four independent types of matter, whose properties are now briefly summarized.

Radiation The universe is known to be awash with photons, and is also believed to contain similar numbers of neutrinos (that until very recently[7] could also be considered to be radiation).

---

[7] Although neutrino masses play an important role in some things (like the formation of galaxies and other structure), I lump them here with radiation because for most of what follows the fact that they very recently likely became non-relativistic does not matter.

Cosmic Photons: The most numerous type of photons found at present in the universe are the photons in the cosmic microwave background (CMB). These are distributed thermally in energy with a temperature that is measured today to be $T_{\gamma 0} = 2.725$ K. The present number density of these CMB photons is determined by their temperature to be

$$n_{\gamma 0} = 4.11 \times 10^8 \text{ m}^{-3}, \tag{4.24}$$

which turns out to be much higher than the average number density of ordinary atoms. Their present energy density (also determined by their temperature) is

$$\rho_{\gamma 0} = 0.261 \text{ MeV m}^{-3} \quad \text{or} \quad \Omega_{\gamma 0} = 5.0 \times 10^{-5}, \tag{4.25}$$

where $\Omega_{\gamma 0} := \rho_{\gamma 0}/\rho_{c0}$ defines the fraction of the total energy density (also the 'critical' density, $\rho_{c0} \simeq 5200$ MeV m$^{-3} \simeq 10^{-29}$ g cm^{-3}) currently residing in CMB photons.

Relict Neutrinos: It is believed on theoretical grounds that there are also as many cosmic relict neutrinos as there are CMB photons running around the universe, although these neutrinos have never been detected. They are expected to have been relativistic until relatively recently in cosmic history, and to be thermally distributed. The neutrinos are expected to have a slightly lower temperature, $T_{\nu 0} = 1.9$ K, and are fermions and so have a slightly different energy-density/temperature relation than do photons.

Their contribution to the present-day cosmological energy budget is not negligible, and if they were massless would be predicted to be

$$\rho_{\nu 0} = 0.18 \text{ MeV m}^{-3} \quad \text{or} \quad \Omega_{\nu 0} = 3.4 \times 10^{-5}, \tag{4.26}$$

leading to a total radiation density, $\Omega_{R0} = \Omega_{\gamma 0} + \Omega_{\nu 0}$, of size

$$\rho_{R0} = 0.44 \text{ MeV m}^{-3} \quad \text{or} \quad \Omega_{r0} = 8.4 \times 10^{-5}. \tag{4.27}$$

Baryons The main constituents of matter we see around us are atoms, made up of protons, neutrons and electrons, and these are predominantly non-relativistic at the present epoch. Although some of this material is now in gaseous form much of it is contained inside larger objects, like planets or stars. But the earlier universe was more homogeneous and at these times atoms and nuclei would have all been uniformly spread around as part of the hot primordial soup. (At least, this is the working hypothesis of the very successful Hot Big Bang picture.)

The absence of anti-particles in the present-day universe indicates that the primordial soup had an over-abundance of baryons (i.e. protons and neutrons) relative to anti-baryons. The same is true of electrons, whose total abundance is also very likely precisely equal to that of protons in order to ensure that the universe carries no net electric charge.

Since protons and neutrons are about 1,840 times more massive than electrons, the energy density in ordinary non-relativistic particles is likely to be well approximated by the total energy in baryons. It turns out it is possible to determine the total number of baryons in the universe (regardless of whether or not they are presently visible), in several independent ways.

One way to determine the baryon density uses measurements of the properties of the CMB, whose understanding depends on things like the speed of sound or on reaction rates—and so also on the density—for the Hydrogen (and some Helium) gas from which the CMB photons last scattered [4]. Another way uses the success of the predictions for the abundances of light elements as nuclei formed during the very early universe, which depends on nuclear reaction rates—again proportional to the total nucleon density. A determination of the baryon abundance as inferred from the primordial He/H abundance ratio (measured from the CMB and from nuclear calculations of primordial element abundances—or Big Bang Nucleosynthesis) is given in Figure 4.3.

These two kinds of inferences are consistent with each other and indicate the total energy density in baryons is[8]

$$\rho_{B0} = 210 \text{ MeV m}^{-3} \qquad \text{or} \qquad \Omega_{B0} = 0.04. \qquad (4.28)$$

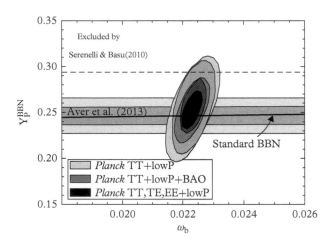

Figure 4.3 *The universal baryon abundance as inferred from CMB measurements and from Big Bang nucleosynthesis (BBN) calculations, taken from [4]. Y_p denotes the primordial He/H abundance while $\omega_b = \Omega_b h^2$ represents the universal baryon fraction (with h as in Figure 4.1).*

---

[8] Ω_{B0} is what is denoted Ω_b in the table in Figure 4.1.

For purposes of comparison, this is about ten times larger than the amount of *luminous* matter, found using the luminosity density for galaxies, $nL = 2 \times 10^8 \ L_\odot \ \mathrm{Mpc}^{-3}$, together with the best estimates of the average mass-to-luminosity ratio of for galactic matter: $M/L \simeq 4M_\odot/L_\odot$.

It should be emphasized that although there is more energy in baryons than in CMB photons, the *number density* of baryons is much smaller, since

$$n_{B0} = \frac{210 \ \mathrm{MeV \ m}^{-3}}{940 \ \mathrm{MeV}} = 0.22 \ \mathrm{m}^{-3} = 5 \times 10^{-10} \ n_{\gamma 0}. \tag{4.29}$$

Dark Matter There are several lines of evidence pointing to the large-scale presence of another form of non-relativistic matter besides baryons, and carrying much more energy. Part of the evidence for this so-called *dark matter* comes from a variety of independent ways of measuring of the total amount of gravitating mass in galaxies and in clusters of galaxies.

The differential rotation rate of numerous galaxies as a function of their radius indicates there is considerably more gravitating mass present than would be inferred by counting the visible luminous matter. Furthermore, the motion of Hydrogen gas clouds and other things orbiting these galaxies indicates this mass is distributed well outside of the radius of the visible stars.

Similarly, the total mass contained within clusters of galaxies appears to be much more than is found when adding up what is visible. This is equally true when galaxy-cluster masses are estimated using the motions of their constituent galaxies, or from the temperature of their hot intergalactic gas or from the amount of gravitational lensing they produce when they are in the foreground of even more distant objects.

Whatever it is, this matter should be non-relativistic since it takes part in the gravitational collapse which gives rise to galaxies and their clusters. (Relativistic matter tends not to cluster in this way as discussed in later sections.)

All of these estimates appear to be consistent with one another, and with several independent ways of measuring energy density in cosmology (more about which later). They indicate a non-relativistic matter density of order

$$\rho_{DM0} = 1350 \ \mathrm{MeV \ m}^{-3} \qquad \mathrm{or} \qquad \Omega_{DM0} = 0.26. \tag{4.30}$$

The errors in this inference of the size of Ω_{DM0} can be seen in Figure 4.1 (where Ω_{DM0} is denoted Ω_c). Provided this has the same equation of state, $p \approx 0$, as have the baryons (as is assumed in the ΛCDM model), this leads to a total energy density in non-relativistic matter, $\Omega_{M0} = \Omega_{B0} + \Omega_{DM0}$, which is of order

$$\rho_{M0} = 1600 \ \mathrm{MeV \ m}^{-3} \qquad \mathrm{or} \qquad \Omega_{m0} = 0.30. \tag{4.31}$$

In Figure 4.1 the quantity Ω_{m0} is denoted Ω_m.

Dark Energy Finally, there are also at least two lines of evidence pointing to a second form of unknown matter in the universe, independent of dark matter. One line is based on the recent observations that the universal expansion is accelerating, and so requires the universe must now be dominated by a form of matter for which $\rho + 3p < 0$. Whatever this is, it cannot be dark matter since the evidence for dark matter shows it to gravitate similarly to non-relativistic matter.

The second line of argument is based on the observational evidence about the spatial geometry of the universe, which favours the universe being spatially flat, $\kappa = 0$. (The evidence for spatial flatness comes from measurements of the angular fluctuations in the temperature of the CMB, since the light we receive from the CMB knows about the geometry of the intervening space through which it passed to get here.) These two lines of evidence are consistent with one another (within sizeable errors) and point to a *dark energy* density that is of order

$$\rho_{DE0} = 3600 \text{ MeV m}^{-3} \qquad \text{or} \qquad \Omega_{DE0} = 0.70. \tag{4.32}$$

The equation of state for the dark energy is only weakly constrained, with observations requiring at present both $\rho_{DE0} \sim 0.7\,\rho_c > 0$ and $w \lesssim -0.7$. The best evidence says w is not changing with time right now, though within large errors. The strength of this evidence is shown in Figure. 4.4, which compares best-fit present-day values for w (called w_0 in the figure) and $w_a = \mathrm{d}w/\mathrm{d}a$.

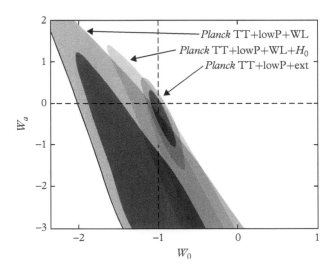

Figure 4.4 *Inferences for the dark energy equation of state parameters, $w_0 = w(a_0)$, as compared with its present-day rate of change, $w_a = (\mathrm{d}w/\mathrm{d}a)_0$, as given in [4]. These show broad agreement with a vacuum-energy equation of state, $w = -1$, independent of a (though within relatively large errors). No well-understood examples of stable matter with $w < -1$ are known, but the broader question of whether they might exist remains controversial.*

If w really is constant it is plausible on theoretical grounds that $w = -1$ and the dark energy is simply the Lorentz-invariant vacuum energy density. Although it is not yet known whether the vacuum need be Lorentz invariant to the precision required to draw cosmological conclusions of sufficient accuracy, in the ΛCDM model it is assumed that the dark energy equation of state is $w = -1$.

4.1.1.5 Earlier Epochs

Given the present-day cosmic ingredients of the previous section, it is possible to extrapolate their relative abundances into the past in order to estimate what can be said about earlier cosmic environments. This evolution can be complicated when the various components of the cosmic fluid significantly interact with one another (e.g. baryons and photons at redshifts larger than about $z \simeq 1100$), but simplifies immensely if the various components of the cosmic fluid do not exchange stress-energy directly with one another. The ΛCDM model assumes there is no such direct energy exchange between other components and the dark matter and dark energy, and that no exchange exists between the two dark components.

When the component fluids do not directly exchange stress-energy things simplify because eqn (4.12) applies separately to each component individually, dictating the dependence $\rho_i(a)$ and $p_i(a)$ for each of them:

- **Radiation:** For photons (and relict neutrinos of sufficiently small mass compared with temperature) we have $w = 1/3$ and so $\rho(a)/\rho_0 = (a_0/a)^4$;
- **Non-Relativistic Matter:** For both ordinary matter (baryons and electrons) and for the dark matter we have $w = 0$ and so $\rho(a)/\rho_0 = (a_0/a)^3$;
- **Vacuum Energy:** Assuming the dark energy has the equation of state $w = -1$ we have $\rho(a) = \rho_0$ for all a.

This implies the total energy density and pressure have the form

$$\rho(a) = \rho_{DE0} + \rho_{M0} \left(\frac{a_0}{a}\right)^3 + \rho_{R0} \left(\frac{a_0}{a}\right)^4$$
$$p(a) = -\rho_{DE0} + \frac{1}{3} \rho_{R0} \left(\frac{a_0}{a}\right)^4, \tag{4.33}$$

showing how the relative contribution of each component within the total cosmic fluid changes as it responds differently to the expansion of the universe (Figure 4.5).

As the universe is run backwards to smaller sizes it is clear that the dark energy becomes less and less important, while relativistic matter becomes more and more important. Although the dark energy is now the dominant contribution to ρ and non-relativistic matter is the next most abundant, when extrapolated backwards they switch roles, so $\rho_M(a) > \rho_{DE}(a)$, relatively recently, at a redshift

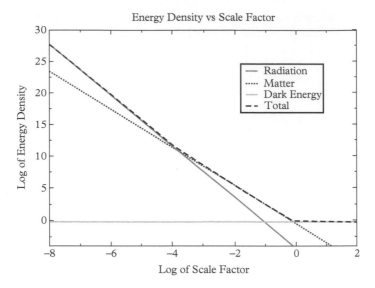

Figure 4.5 *The relative abundance (in the energy density) of radiation, non-relativistic matter and vacuum energy, vs the size of the universe $a/a_0 = (1+z)^{-1}$. The figure assumes negligible direct energy transfer between these fluids, and shows how this implies each type of fluid dominates during particular epochs. The transition from radiation to matter domination (at redshift $z_{\rm eq} \simeq 3600$) plays an important role in the development of structure in the universe.*

$$1 + z = \frac{a_0}{a} > \left(\frac{\Omega_{DE0}}{\Omega_{M0}} \right)^{1/3} \simeq \left(\frac{0.7}{0.3} \right)^{1/3} \simeq 1.3. \tag{4.34}$$

In the absence of dark matter the energy density in baryons alone would become larger than the dark energy density at a slightly earlier epoch

$$1 + z > \left(\frac{\Omega_{DE0}}{\Omega_{B0}} \right)^{1/3} \simeq \left(\frac{0.7}{0.04} \right)^{1/3} \simeq 2.6. \tag{4.35}$$

For times earlier than this the dominant component of the energy density is due to non-relativistic matter, and this remains true back until the epoch when the energy density in radiation became comparable with that in non-relativistic matter. Since $\rho_R \propto a^{-4}$ and $\rho_M \propto a^{-3}$ radiation-matter equality occurs when $z = z_{\rm eq}$ with

$$1 + z_{\rm eq} = \frac{\Omega_{M0}}{\Omega_{R0}} \simeq \frac{0.3}{8.4 \times 10^{-5}} \simeq 3600. \tag{4.36}$$

This crossover would have occurred much later in the absence of dark matter, since the radiation energy density equals the energy density in baryons when

$$1 + z - \frac{\Omega_{B0}}{\Omega_{R0}} \simeq \frac{0.04}{8.4 \times 10^{-5}} \simeq 480. \tag{4.37}$$

Using the dependence of ρ on a in the Friedmann equation then gives H as a function of a

$$H(a) = H_0 \left[\Omega_{DE0} + \Omega_{\kappa 0} \left(\frac{a_0}{a} \right)^2 + \Omega_{M0} \left(\frac{a_0}{a} \right)^3 + \Omega_{R0} \left(\frac{a_0}{a} \right)^4 \right]^{1/2}, \tag{4.38}$$

where (as before) $\Omega_f = \rho_f / \rho_c$ for $f = \text{radiation}(R)$, matter (M) or vacuum (DE). The critical density is defined by $\rho_c := 3H^2 M_p^2$ and the subscript '0' denotes the present epoch. Finally, eqn (4.38) defines the curvature contribution to H as

$$\Omega_\kappa := -\frac{\kappa}{(Ha)^2}. \tag{4.39}$$

As mentioned earlier, observations of the CMB constrain the present-day value for $\Omega_{\kappa 0}$, because they tell us about the overall geometry of space through which photons move on their way to us from where they were last scattered by primordial Hydrogen. These observations indicate the universe is close to spatially flat (i.e. that κ is consistent with 0). Quantitatively, these CMB observations tell us that $\Omega_{\kappa 0}$ is at most of order 10%, and so the Friedmann equation implies $\Omega_0 = \Omega_{DE0} + \Omega_{m0} + \Omega_{r0} \simeq 1$ (and so $\rho_{DE} + \rho_m + \rho_r = \rho_c$). Joint constraints on Ω_κ and $\Omega_b = \Omega_{B0}$ are shown in Figure 4.6.

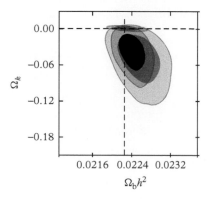

Figure 4.6 *Constraints on the universe's spatial curvature, Ω_κ, and its baryon abundance, Ω_b, obtained from CMB observations [4]. As usual $h = H/(100 km/sec/Mpc) \simeq 0.67$. Different coloured ellipses correspond to fits with different combinations of data sets.*

Because κ/a^2 falls more slowly with increasing a than does either $\rho_m \propto 1/a^3$ or $\rho_r \propto 1/a^4$, the relatively small size of $\Omega_{\kappa 0}$ implies the κ/a^2 term contributes negligibly the further one moves to the remote past. This makes it a very good approximation to take $\kappa = 0$ when discussing the very early universe.

In principle (4.38) can be integrated to obtain $a(t)$. Although in general this dependence must be obtained numerically, many of its features follow on simple analytic grounds because for most epochs there is only a single component of the cosmic fluid that dominates the total energy density. For instance, for redshifts larger than several thousand $a(t) \propto t^{1/2}$ should be a good approximation, as appropriate for the expansion in a universe filled purely by radiation.

Once a_0/a falls below $3,600$ there should be a brief transition to the time-dependence appropriate for a universe dominated by non-relativistic matter and so $a \propto t^{2/3}$. This should apply right up to the very recent past, when a/a_0 is around 0.8, after which there is a transition to vacuum-energy domination, during which the universal expansion accelerates to become exponential with t.

In all likelihood we are currently living in the transition period from matter to vacuum-energy domination. And when $\kappa = 0$ it is also possible to give simple analytic expressions for the time-dependence of a in transition regions like this. Neglecting radiation during the matter/dark-energy transition gives a Friedmann equation of the form

$$\left(\frac{\dot{a}}{a}\right)^2 = H_{DE}^2 \left[1 + \left(\frac{a_e}{a}\right)^3\right], \tag{4.40}$$

where $H_{DE}^2 = 8\pi G \rho_{DE}/3$ is the (constant) Hubble scale during the pure dark-energy epoch and a_e is the value of the scale factor when the energy densities of the matter and dark energy are equal to one another. Integrating this equation (assuming $\dot{a} > 0$), with the boundary condition that $a = 0$ when $t = 0$ then gives the solution

$$a(t) = a_0 \sinh^{2/3}\left(\frac{3H_{DE}t}{2}\right), \tag{4.41}$$

where a_0 is a constant. Notice that this approaches $a/a_0 \propto \exp(H_{de}t)$ if $H_{DE}\,t \gg 1$, as appropriate for dark energy domination, while for $H_{DE}\,t \ll 1$ it instead becomes $a/a_0 \propto t^{2/3}$, as appropriate for a matter-dominated epoch.

4.1.1.6   *Thermal Evolution*

The Hot Big Bang theory of cosmology starts with the idea that the universe was once small and hot enough that it contained just a soup of elementary particles, in order to see if this leads to a later universe that we recognize in cosmological observations. This picture turns out to describe well many of the features we see around us, which are otherwise harder to understand.

This type of hot fluid cools as the universe expands, leading to several types of characteristic events whose late-time signatures provide evidence for the validity of the Hot Big Bang picture. The first type of characteristic event is the departure from equilibrium that every species of particle always experiences eventually once its particle density becomes too low for particles to find one another frequently enough to maintain equilibrium.

The second type of characteristic event is the formation of bound states. At finite temperature the net abundance of bound states (like atoms or nuclei, say) is fixed by detailed balance: the competition between reactions (like $e^- p \to H\gamma$) that form the bound states (in this case Hydrogen) and the inverse reactions (like $H\gamma \to e^- p$) that dissociate them. Once the temperature falls below the binding energy of a bound state the typical collision energy falls below the threshold required for dissociation and so the abundance of the bound state grows until the constituents eventually become sufficiently rare that the formation reactions also effectively turn off the production processes. Once this happens the bound-state abundance freezes and for the purposes of later cosmology these bound states can be regarded as being part of the inventory of 'elementary' particles during later epochs.

There is concrete evidence that the formation of bound states took place at least twice in the early universe. The earliest case happened during the epoch of primordial nucleosynthesis, at redshift $z \simeq 10^{10}$, when temperatures were in the MeV regime and protons and neutrons got cooked into light nuclei. The evidence that this occurred comes from the agreement between the primordial abundances of light nuclear isotopes with the results of precision calculations of their formation rates. This agreement is non-trivial because the total formation rate for each nuclear isotope depends on the density of protons and neutrons at the time, and the same value for the baryon density gives successful agreement between theory and observations for ^{2}H, ^{3}He, ^{4}He, and ^{7}Li. The consistency of these calculations both tells us that this picture of their origins is likely right, and the total density of baryons throughout the universe at this time.

The second important epoch during which bound states formed is the epoch of 'recombination', at redshifts around $z \simeq 1,100$. At this epoch the temperature of the cosmic fluid is around 1,000 K and electrons and nuclei combine to form electrically neutral atoms (like H or He). The evidence that this occurred comes from the existence and properties of the CMB itself. Before neutral atoms formed the charged electrons and protons very efficiently scattered photons, making the universe at that time opaque to light. But this scattering stopped after atoms formed, leaving a transparent universe in which all the photons present in the hot gas remain but no longer scatter very often. Indeed it is this bath of primordial photons, now redshifted down to microwave wavelengths and currently being detected, that we call the CMB.

The distribution of these CMB photons has a beautiful thermal form as a function of the present-day photon wavelength, λ_0, as shown in Figure 4.7. The temperature of this distribution is measured as a function of direction in the sky, $T_\gamma(\theta, \phi)$, and it is the angular average of this measured temperature,

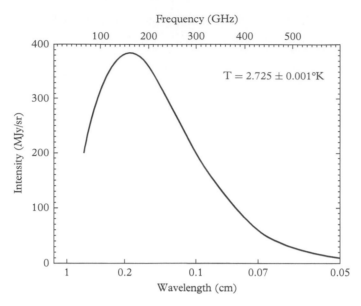

Figure 4.7 *The FIRAS measurement of the thermal distribution of the CMB photons. The experimental points lie on the theoretical curve, with errors that are smaller than the width of the curve.*

$$T_{\gamma 0} = \langle T_\gamma \rangle = \frac{1}{4\pi} \int T_\gamma(\theta,\phi) \sin\theta \; d\theta \; d\phi = 2.2725 \text{ K}, \qquad (4.42)$$

that is used as the present temperature of the relic photons.

The starting point for making such a thermal description precise is a summary of the various types of particles that are believed to be 'elementary' at the temperatures of interest. The highest temperature for which there is direct observational evidence the universe attained in the past is $T \sim 10^{10}$ K, which corresponds to thermal energies of order 1 MeV. The elementary particles which might be expected to be found within a soup having this temperature are:

- **Photons** (γ): bosons that have no electric charge or mass, and can be singly emitted and absorbed by any electrically charged particles.

- **Electrons and Positrons** ($e^\pm$): fermions with charge $\pm e$ and masses equal numerically to $m_e = 0.511$ MeV. Because the positron, e^+, is the anti-particle for the electron, e^-, (and vice versa), these particles can completely annihilate into photons through the reaction

$$e^+ + e^- \leftrightarrow 2\gamma, \qquad (4.43)$$

and do so once the temperature falls below the electron mass.

- **Protons** (p): fermions with charge $+e$ and mass $m_p = 938$ MeV. Unlike all of the other particles described here, the proton and neutron can take part in the strong interactions, e.g. experiencing reactions like

$$p + n \leftrightarrow D + \gamma, \tag{4.44}$$

  in which a proton and neutron combine to produce a deuterium nucleus. The photon that appears in this expression simply carries off any excess energy which is released by the reaction.

- **Neutrons** (n): electrically neutral fermions with mass $m_n = 940$ MeV. Like protons, neutrons participate in the strong interactions. Isolated neutrons are unstable, and left to themselves decay through the weak interactions into a proton, an electron and an electron-antineutrino:

$$n \to p + e^- + \overline{\nu}_e. \tag{4.45}$$

- **Neutrinos and Anti-Neutrinos** ($\nu_e, \overline{\nu}_e, \nu_\mu, \overline{\nu}_\mu, \nu_\tau, \overline{\nu}_\tau$): fermions that are electrically neutral, and have been found to have non-zero masses whose precise values are not known, but that are known to be smaller than 1 eV.

- **Gravitons** (G): electrically neutral bosons that mediate the gravitational force in the same way that photons do for the electromagnetic force. Gravitons only interact with other particles with gravitational strength, which is much weaker that the strength of the other interactions. As a result they turn out never to be in thermal equilibrium for any of the temperatures to which we have observational access in cosmology.

To these must be added whatever makes up the dark matter, provided temperatures and interactions are such that the dark matter can be regarded to be in thermal equilibrium.

How would the temperature of a bath of these particles evolve on thermodynamic grounds as the universe expands? The first step asks how the temperature is related to a (and so also t), in order to quantify the rate with which a hot bath cools due to the universal expansion.

Relativistic Particles The energy density and pressure for a gas of relativistic particles (like photons) when in thermal equilibrium at temperature T_R are given by

$$\rho_R = a_B T_R^4 \qquad \text{and} \qquad p_R = \frac{1}{3} a_B T_R^4, \tag{4.46}$$

where a_B is $g/2$ times the Stefan–Boltzmann constant and g counts the number of internal (spin) states of the particles of interest (and so $g = 2$ for a gas of photons).

Combining this with energy conservation, which says $\rho_R \propto (a_0/a)^4$, shows that the product aT is constant, and so

$$T_R(a) = T_{R0}\left(\frac{a_0}{a}\right) = T_{R0}(1+z). \qquad (4.47)$$

This is equivalent to the statement that the expansion is adiabatic, since the entropy per unit volume of a relativistic gas is $s_R \propto T_R^3$, and so the total entropy in this gas is

$$S_R \propto s_R a^3 \propto (T_R a)^3 = \text{constant}. \qquad (4.48)$$

Although the relation $T \propto a^{-1}$ is derived above assuming thermal equilibrium, it can continue to hold (for relativistic particles) once the particles become insufficiently dense to scatter frequently enough to maintain equilibrium. This is because the thermal distribution functions for relativistic particles are functions of the ratio of particle energy divided by temperature: $f(\epsilon, T) \propto (e^{\epsilon/T} + 1)^{-1}$. Because relativistic particles have energies $\epsilon(\mathbf{p}) = |\mathbf{p}|$, where the physical momentum $\mathbf{p}$ is related to co-moving momentum by $\mathbf{p} = \mathbf{k}/a$, their energies redshift $\epsilon \propto a^{-1}$ with the universal expansion.

This ensures that the distributions remain in the thermal form for all t, provided that their temperature is also regarded as falling like $T \propto a^{-1}$, so that ϵ/T is time-independent. For this reason it makes sense to continue to regard the CMB photon temperature to be falling like $T_R \propto a^{-1}$ even though photons stopped interacting frequently enough to remain in equilibrium once protons and electrons combined into electrically neutral atoms around redshift $z \simeq 1100$.

Non-Relativistic Particles As mentioned earlier, an ideal gas of non-relativistic particles in thermal equilibrium has a pressure and energy density given instead by

$$p_M = n T_M \quad \text{and} \quad \rho_M = nm + \frac{n T_M}{\gamma - 1}, \qquad (4.49)$$

where n is the number density of particles, m is the particle's rest mass and $\gamma = c_p/c_v$ is its ratio of specific heats, with $\gamma = 5/3$ for a gas of monatomic atoms.

In order to repeat the previous arguments using energy conservation to infer how T_M evolves with a we must first determine what n depends on. If the total number of particles is conserved, so

$$\frac{\mathrm{d}}{\mathrm{d}t}\left[n a^3\right] = 0, \qquad (4.50)$$

then consistency of $n \propto a^{-3}$ with energy conservation, eqn (4.12), implies T_M should satisfy

$$\frac{\dot{T}_M}{T_M} + 3(\gamma - 1)\frac{\dot{a}}{a} = 0, \qquad (4.51)$$

and so

$$T_M = T_{M0} \left(\frac{a_0}{a}\right)^{3(\gamma-1)} = T_{M0}(1+z)^{3(\gamma-1)}. \tag{4.52}$$

For example, for a monatomic gas with $\gamma = 5/3$ this implies $T_M \propto (1+z)^2 \propto a^{-2}$, as also would be expected for an adiabatic expansion given that the entropy density for such a fluid varies with T_M like $s_M \propto (mT_M)^{3/2}$.

When a non-relativistic species of particle falls out of equilibrium its energy (because it is non-relativistic) is dominated by its rest-mass: $\epsilon(\mathbf{p}) \simeq m$. Because of this ϵ does not redshift and so the distribution of particles remains frozen at the fixed temperature, T_f, where equilibrium first broke down.

Multi-Component Fluids The previous examples assume negligible energy exchange between these different components, which in particular also precludes them being in thermal equilibrium with one another (allowing their respective temperatures free to evolve independently of one another). But what happens when several components of the fluid *are* in thermal equilibrium with one another? This situation actually happens for $z > 1,100$ when non-relativistic protons and neutrons (or nuclei) are in equilibrium with relativistic photons, electrons and neutrinos.

To see how this works, we now repeat the previous arguments for a fluid that consists of both relativistic and non-relativistic components, coexisting in mutual thermal equilibrium at a common temperature, T. In this case the energy density and pressure are given by

$$p = nT + \frac{1}{3} a_B T^4 \quad \text{and} \quad \rho = nm + \frac{nT}{\gamma-1} + a_B T^4. \tag{4.53}$$

Inserting this into the energy conservation equation, as above, leads to the result

$$\frac{\dot{T}}{T} + \left[\frac{1+s}{s + \frac{1}{3}(\gamma-1)^{-1}}\right] \frac{\dot{a}}{a} = 0, \tag{4.54}$$

where

$$s \equiv \frac{4a_B T^3}{3n} = 74.0 \left[\frac{(T/\deg)^3}{n/\mathrm{cm}^{-3}}\right], \tag{4.55}$$

is the relativistic entropy per non-relativistic gas particle. For example, if the relativistic gas consists of photons, then the number of photons per unit volume is $n_\gamma = [30\zeta(3)/\pi^4] a_B T^3 = 3.7 a_B T^3$, and so $s = 0.37(n_\gamma/n)$.

Equation (4.54) shows how T varies with a, and reduces to the pure radiation result, $Ta = $ constant, when $s \gg 1$ and to the non-relativistic matter result, $Ta^{3(\gamma-1)} = $ constant, when $s \ll 1$. In general, however, this equation has more complicated solutions because $\mathfrak{s}$ need not be a constant. Given that particle conservation implies $n \propto a^{-3}$, we see that the time-dependence of $\mathfrak{s}$ is given by $\mathfrak{s} \propto (T a)^3$.

We are led to the following limiting behaviour. If, initially, $\mathfrak{s} = \mathfrak{s}_0 \gg 1$ then at early times $T \propto a^{-1}$ and so $\mathfrak{s}$ remains approximately constant (and large). For such a gas the common temperature of the relativistic and non-relativistic fluids continues to fall like $T \propto a^{-1}$. In this case the high-entropy relativistic fluid controls the temperature evolution and drags the non-relativistic temperature along with it. Interestingly, it can do so even if $\rho_M \approx nm$ is larger than $\rho_R = a_B T^4$, as can easily happen when $m \gg T$. In practice this happens until the two fluid components fall out of equilibrium with one another, after which their two temperatures begin to evolve separately according to the expressions given previously.

On the other hand if $\mathfrak{s} = \mathfrak{s}_0 \ll 1$ initially, then $T \propto a^{-3(\gamma-1)}$ and so $\mathfrak{s} \propto a^{3(4-3\gamma)}$. This falls as a increases provided $\gamma > 4/3$, and grows otherwise. For instance, the particularly interesting case $\gamma = 5/3$ implies $T \propto a^{-2}$ and so $\mathfrak{s} \propto a^{-3}$. We see that if $\gamma > 4/3$, then an initially small $\mathfrak{s}$ gets even smaller still as the universe expands, implying the temperature of both radiation and matter continues to fall like $T \propto a^{-3(\gamma-1)}$. If, however, $1 < \gamma < 4/3$, an initially small $\mathfrak{s}$ can grow even as the temperature falls, until the fluid eventually crosses over into the relativistic regime for which $T \propto a^{-1}$ and $\mathfrak{s}$ stops evolving.

4.1.2   An Early Accelerated Epoch

This section switches from a general description of the ΛCDM model to a discussion about the peculiar initial conditions on which its success seems to rely. This is followed by a summary of the elements of some simple single-field inflationary models, and why their proposal is motivated as explanations of the initial conditions for the later universe.

4.1.2.1   *Peculiar Initial Conditions*

The ΛCDM model describes well what we see around us, provided that the universe is started off with a very specific set of initial conditions. There are several properties of these initial conditions that seem peculiar, as is now summarized.

Flatness Problem   The first problem concerns the observed spatial flatness of the present-day universe. As described earlier, observations of the CMB indicate that the quantity κ/a^2 of the Friedmann equation, eqn (4.11), is at present consistent with zero. What is odd about this condition is that this curvature term tends to grow in relative importance as the universe expands, so finding it to be small now means that it must have been *extremely* small in the remote past.

More quantitatively, it is useful to divide the Friedmann equation by $H^2(t)$ to give

$$1 + \frac{\kappa}{(aH)^2} = \frac{8\pi G \rho}{3H^2} =: \Omega(a), \tag{4.56}$$

where (as before) the final equality defines $\Omega(a)$. The problem arises because the product aH decreases with time during both matter and radiation domination. For instance, observations indicate that at present $\Omega = \Omega_0$ is unity to within about 10%, and since during the matter-dominated era the product $(aH)^2 \propto a^{-1}$ it follows that at the epoch $z_{eq} \simeq 3600$ of radiation-matter equality we must have had

$$\Omega(z_{eq}) - 1 = \left(\Omega_0 - 1\right)\left(\frac{a}{a_0}\right) = \frac{\Omega_0 - 1}{1 + z_{eq}} \simeq \frac{0.1}{3600} \simeq 2.8 \times 10^{-5}. \tag{4.57}$$

So $\Omega - 1$ had to be smaller than a few tens of a millionth at the time of radiation-matter equality in order to be of order 10% now.

And it only gets worse the further back we go, provided the extrapolation back occurs within a radiation- or matter-dominated era (as seems to be true at least as far back as the epoch of nucleosynthesis). Since during radiation-domination we have $(aH)^2 \propto a^{-2}$ and the redshift of nucleosynthesis is $z_{BBN} \sim 10^{10}$ it follows that at this epoch we must require

$$\Omega(z_{BBN}) - 1 = \left[\Omega(z_{eq}) - 1\right]\left(\frac{1 + z_{eq}}{1 + z_{BBN}}\right)^2 = \frac{0.1}{3600}\left(\frac{3600}{10^{10}}\right)^2 \approx 3.6 \times 10^{-18}, \tag{4.58}$$

requiring Ω to be unity with an accuracy of roughly a part in 10^{18}. The discomfort of having the success of a theory hinge so sensitively on the precise value of an initial condition in this way is known as the Big Bang's *flatness problem*.

Horizon Problem Perhaps a more serious question asks why the initial universe can be so very homogeneous. In particular, the temperature fluctuations of the CMB only arise at the level of one part in 10^5, and the question is how this temperature can be so incredibly uniform across the sky.

Why is this regarded as a problem? It is not uncommon for materials on Earth to have a uniform temperature, and this is usually understood as a consequence of thermal equilibrium. An initially inhomogeneous temperature distribution equilibrates by having heat flow between the hot and cold areas, until everything eventually shares a common temperature.

The same argument is harder to make in cosmology because in the Hot Big Bang model the universe generically expands so quickly that there has not been enough time for light to travel across the entire sky to bring everyone the news as to what the common temperature is supposed to be. This is easiest to see using conformal coordinates, as in (4.8), since in these coordinates it is simple to identify which regions can be connected by light signals. In particular, radially directed light rays travel along lines $d\ell = \pm d\tau$, which can be drawn as straight lines of slope ± 1 in the $\tau - \ell$ plane, as in Figure 4.8. The problem is that $a(\tau)$ reaches zero in a finite conformal time (which we can conventionally choose to happen at $\tau = 0$), since $a(\tau) \propto \tau$ during radiation domination and $a(\tau) \propto \tau^2$ during matter domination. Redshift $z_{rec} \simeq 1,100$ (the epoch of recombination, at which the CMB photons last sampled the temperature of the Hydrogen gas with which they interact) is simply too early for different directions in the sky to have been causally connected in the entire history of the universe up to that point.

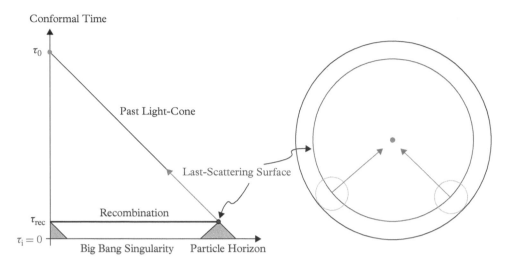

Figure 4.8 *A conformal diagram illustrating how there is inadequate time in a radiation-dominated universe for there to be a causal explanation for the correlation of temperature at different points of the sky in the CMB. (Figure taken from [8].)*

To pin this down quantitatively, let us assume that the universe is radiation dominated for all points earlier than the epoch of radiation-matter equality, t_{eq}, so the complete evolution of $a(t)$ until recombination is

$$a(t) \simeq \begin{cases} a_{\text{eq}}(t/t_{\text{eq}})^{1/2} & \text{for } 0 < t < t_{\text{eq}} \\ a_{\text{eq}}(t/t_{\text{eq}})^{2/3} & \text{for } t_{\text{eq}} < t < t_{\text{rec}} \,. \end{cases} \qquad (4.59)$$

(The real evolution does not have a discontinuous derivative at $t = t_{\text{eq}}$, but this inaccuracy is not important for the argument that follows.) The maximum proper distance, measured at the time of recombination, that a light signal could have travelled by the time of recombination, t_{rec}, then is

$$D_{\text{rec}} = a_{\text{rec}} \left[\int_0^{t_{\text{eq}}} \frac{\mathrm{d}\hat{t}}{a(\hat{t})} + \int_{t_{\text{eq}}}^{t_{\text{rec}}} \frac{\mathrm{d}\hat{t}}{a(\hat{t})} \right] = \frac{a_{\text{rec}} t_{\text{eq}}}{a_{\text{eq}}} \left[3 \left(\frac{t_{\text{rec}}}{t_{\text{eq}}} \right)^{1/3} - 1 \right]$$

$$= \frac{2}{H_{\text{eq}}^+} \left(\frac{a_{\text{rec}}}{a_{\text{eq}}} \right)^{3/2} \left[1 - \frac{1}{3} \left(\frac{a_{\text{eq}}}{a_{\text{rec}}} \right)^{1/2} \right] \simeq \frac{1.6}{H_{\text{rec}}}, \qquad (4.60)$$

where $H_{\text{eq}}^+ = 2/(3t_{\text{eq}})$ denotes the limit of the Hubble scale as $t \to t_{\text{eq}}$ on the matter-dominated side. The approximate equality in this expression uses $H \propto a^{-3/2}$ during matter domination as well as using the redshifts $z_{\text{rec}} \simeq 1,100$ and $z_{\text{eq}} \simeq 3,600$ (as would be true in the ΛCDM model) to obtain $a_{\text{eq}}/a_{\text{rec}} \simeq 1,100/3,600 \simeq 0.31$.

To evaluate this numerically we use the present-day value for the Hubble constant, $H_0 \simeq 70$ km/sec/Mpc—or (keeping in mind our units for which $c = 1$), $H_0^{-1} \simeq 13$ Gyr $\simeq 4$ Gpc. This then gives $H_{\text{rec}}^{-1} \simeq H_0^{-1}(a_{\text{rec}}/a_0)^{3/2} \simeq 3 \times 10^{-5} H_0^{-1} \simeq 0.1$ Mpc, if we use $a_0/a_{\text{rec}} = 1 + z_{\text{rec}} \simeq 1,100$, and so $D_{\text{rec}} \simeq 0.2$ Mpc.

Now CMB photons arriving to us from the surface of last scattering left this surface at a distance from us that is now of order

$$R_0 = a_0 \int_{t_{\text{rec}}}^{t_0} \frac{d\hat{t}}{a(\hat{t})} = 3t_0 - 3t_0^{2/3} t_{\text{rec}}^{1/3} = \frac{2}{H_0}\left[1 - \left(\frac{a_{\text{rec}}}{a_0}\right)^{1/2}\right], \qquad (4.61)$$

again using $a \propto t^{2/3}$ and $H \propto a^{-3/2}$, and so $R_0 \simeq 2/H_0 \simeq 8$ Gpc. So the angle subtended by D_{rec} placed at this distance away (in a spatially flat geometry) is really $\theta \simeq D_{\text{rec}}/R_{\text{rec}}$ where $R_{\text{rec}} = (a_{\text{rec}}/a_0)R_0 \simeq 7$ Mpc is its distance *at the time of last scattering*, leading to[9] $\theta \simeq 0.2/7 \simeq 1°$. Any two directions in the sky separated by more than this angle (about twice the angular size of the Moon, seen from Earth) are so far apart that light had not yet had time to reach one from the other since the Universe's beginning.

How can all the directions we see have known they were all to equilibrate to the same temperature? It is very much as if we were to find a very uniform temperature distribution, *immediately* after the explosion of a very powerful bomb.

Defect Problem Historically, a third problem—called the 'defect' (or 'monopole') problem—is also used to motivate changing the extrapolation of radiation domination into the remote past. A defect problem arises if the physics of the much higher energy scales relevant to the extrapolation involves the production of topological defects, like domain walls, cosmic strings or magnetic monopoles. Such defects are often found in Grand Unified Theories (GUTs); models proposed to unify the strong and electroweak interactions as energies of order 10^{15} GeV.

These kinds of topological defects can be fatal to the success of late-time cosmology, depending on how many of them survive down to the present epoch. For instance if the defects are monopoles, then they typically are extremely massive and so behave like non-relativistic matter. This can cause problems if they are too abundant because they can preclude the existence of a radiation dominated epoch, because their energy density falls more slowly than does radiation as the universe expands.

Defects are typically produced with an abundance of one per Hubble volume, $n_d(a_f) \sim H_f^3$, where $H_f = H(a_f)$ is the Hubble scale at their epoch of formation, at which time $a = a_f$. Once produced, their number is conserved, so their density at later times falls like $n_d(a) = H_f^3(a_f/a)^3$. Consequently, at present the number surviving within a Hubble volume is $n_d(a_0)H_0^{-3} = (H_f a_f/H_0 a_0)^3$.

Because the product aH is a falling function of time, the present-day abundance of defects can easily be so numerous that they come to dominate the universe well before

---

[9] This estimate is related to the quantity θ_{MC} in the table of Figure 4.1.

the nucleosynthesis epoch.[10] This could cause the universe to expand (and so cool) too quickly as nuclei were forming, and so give the wrong abundances of light nuclei. Even if not sufficiently abundant during nucleosynthesis, the energy density in relict defects can be inconsistent with measures of the current energy density.

This is clearly more of a hypothetical problem than are the other two, since whether there is a problem depends on whether the particular theory for the high-energy physics of the very early universe produces these types of defects or not. It can be fairly pressing in GUTs since in these models the production of magnetic monopoles can be fairly generic.

4.1.2.2   Acceleration to the Rescue

The key observation when trying to understand the above initial conditions is that they only seem unreasonable because they are based on extrapolating into the past assuming the universe to be radiation (or matter) dominated (as would naturally be true if the ΛCDM model were the whole story). This section argues that these initial conditions can seem more reasonable if a different type of extrapolation is used; in particular if there were an earlier epoch during which the universal expansion were to accelerate: $\ddot{a} > 0$ [9, 10].

Why should acceleration help? The key point is that the above initial conditions are a problem because the product aH is a falling function as a increases, for both matter and radiation domination. For instance, for the flatness problem the evolution of the curvature term in the Friedmann equation is $\Omega_\kappa \propto (aH)^{-2}$ and this grows as a grows only because aH decreases with a. But if $\ddot{a} > 0$ then $\dot{a} = aH$ *increases* as a increases, and this can help alleviate the problems. For example, finding Ω_κ to be very small in the recent past would be less disturbing if the more-distant past contained a sufficiently long epoch during which aH grew.

How long is long enough? To pin this down suppose there were an earlier epoch during which the universe were to expand in the same way as during dark energy domination, $a(t) \propto e^{Ht}$, for constant H. Then $aH = a_0 H e^{Ht}$ grows exponentially with time and so even if Ht were of order 100 or less it would be possible to explain why Ω_κ could be as small as 10^{-18} or smaller.

Having aH grow also allows a resolution to the horizon problem. One way to see this is to notice that $a(t) \propto e^{Ht}$ implies $\tau = -H^{-1}e^{-Ht}$ plus a constant (with the sign a consequence of having τ increase as t does), and so

$$a(\tau) = -\frac{1}{H(\tau - \tau_0)}, \tag{4.62}$$

with $0 < a < \infty$ corresponding to the range $-\infty < \tau < \tau_0$. Exponentially accelerated expansion allows τ to be extrapolated to arbitrarily negative values, and so allows

---

[10] Whether they do also depends on their dimension, with magnetic monopoles tending to be more dangerous in this regard than, for example, cosmic strings.

Evolution of Scales

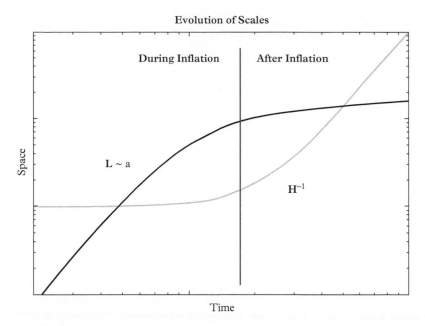

Figure 4.9 *A sketch of the relative growth of physical scales, L(t), and the Hubble length, H^{-1}, during and after inflation. Horizon exit happens during inflation where the curves first cross, and this is eventually followed by horizon re-entry where the curves cross again during the later Hot Big Bang era.*

sufficient time for the two causally disconnected regions of the conformal diagram of Figure 4.8 to have at one point been in causal contact.

Another way to visualize this is to plot physical distance $\lambda(t) \propto a(t)$ and the Hubble radius, H^{-1}, against t, as in Figure 4.9. Focus first on the right-hand side of this figure, which compares these quantities during radiation or matter domination. During these epochs the Hubble length evolves as $H^{-1} \propto t$ while the scale factor satisfies $a(t) \propto t^p$ with $0 < p < 1$. Consequently H^{-1} grows more quickly with t than do physical length scales $\lambda(t)$. During radiation or matter domination systems of any given size eventually get caught by the growth of H^{-1} and so 'come inside the Hubble scale' as the universe expands. Systems involving larger $\lambda(t)$ do so later than those with smaller λ. The largest sizes currently visible have only recently crossed inside of the Hubble length, having spent their entire earlier history larger than H^{-1} (assuming always a radiation- or matter-dominated universe).

Having $\lambda > H^{-1}$ matters because physical quantities tend to freeze when their corresponding length scales satisfy $\lambda(t) > H^{-1}$. This then precludes physical processes from acting over these scales to explain things like the uniform temperature of the CMB. The freezing of super-Hubble scales can be seen, for example, in the evolution of a massless scalar field in an expanding universe, since the field equation $\Box \phi = 0$ becomes in FRW coordinates

$$\ddot{\phi}_k + 3H\dot{\phi}_k + \left(\frac{k}{a}\right)^2 \phi_k = 0, \tag{4.63}$$

where we Fourier expand the field $\phi(x) = \int d^3k\, \phi_k \exp[i\mathbf{k}\cdot\mathbf{x}]$ using co-moving coordinates, $\mathbf{x}$. For modes satisfying $2\pi/\lambda = p = k/a \ll H$ the field equation implies $\dot{\phi}_k \propto a^{-3}$ and so $\phi_k = C_0 + C_1 \int dt/a^3$ is the sum of a constant plus a decaying mode.

Things are very different during exponential expansion, however, as is shown on the left-hand side of Figure 4.9. In this regime $\lambda(t) \propto a(t) \propto e^{Ht}$ grows exponentially with t while H^{-1} remains constant. This means that modes that are initially smaller than the Hubble length get stretched to become larger than the Hubble length, with the transition for a specific mode of length $\lambda(t)$ occurring at the epoch of 'Hubble exit', $t = t_{he}$, defined by $2\pi/\lambda(t_{he}) = p_{he} = k/a(t_{he}) = H$. In this language it is because the criterion for Hubble exit and entry is $k = aH$ that the growth or shrinkage of aH is relevant to the horizon problem.

How much expansion is required to solve the horizon problem? Choosing a mode ϕ_k that is only now crossing the Hubble scale tells us that $k = a_0 H_0$. This same mode would have crossed the horizon during an exponentially expanding epoch when $k = a_{he} H_I$, where H_I is the constant Hubble scale during exponential expansion. So clearly $a_0 H_0 = a_{he} H_I$ where t_{he} is the time of Hubble exit for this particular mode. To determine how much exponential expansion is required solve the following equation for $N_e := \ln(a_{end}/a_{he})$, where a_{end} is the scale factor at the end of the exponentially expanding epoch:

$$1 = \frac{a_{he} H_I}{a_0 H_0} = \left(\frac{a_{he} H_I}{a_{end} H_I}\right)\left(\frac{a_{end} H_I}{a_{eq} H_{eq}}\right)\left(\frac{a_{eq} H_{eq}}{a_0 H_0}\right) = e^{-N_e}\left(\frac{a_{eq}}{a_{end}}\right)\left(\frac{a_0}{a_{eq}}\right)^{1/2}. \tag{4.64}$$

This assumes (for the purposes of argument) that the universe is radiation dominated right from t_{end} until radiation-matter equality, and uses $aH \propto a^{-1}$ during radiation domination and $aH \propto a^{-1/2}$ during matter domination. $N_e = H_I(t_{end} - t_{he})$ is called the number of *e*-foldings of exponential expansion and is proportional to how long exponential expansion lasts.

Using, as above, $(a_{eq} H_{eq})/(a_0 H_0) = (a_0/a_{eq})^{1/2} \simeq 60$, and $(a_{eq} H_{eq})/(a_{end} H_{end}) = a_{end}/a_{eq} = T_{eq}/T_M$ with $T_{eq} \sim 3$ eV, and assuming the energy density of the exponentially expanding phase is transferred perfectly efficiently to produce a photon temperature T_M then leads to the estimate

$$N_e \sim \ln\left[(3 \times 10^{23}) \times 60\right] + \ln\left(\frac{T_M}{10^{15}\text{ GeV}}\right) \approx 58 + \ln\left(\frac{T_M}{10^{15}\text{ GeV}}\right). \tag{4.65}$$

Roughly sixty *e*-foldings of exponential expansion can provide a framework for explaining how causal physics might provide the observed correlations that are observed in the CMB over the largest scales, even if the energy densities involved are as high as 10^{15} GeV. We shall see later that life is even better than this, because in addition to providing a *framework* in which a causal understanding of correlations could be solved, inflation itself can provide the *mechanism* for explaining these correlations (given an inflationary scale of the right size).

4.1.2.3   Inflation or a Bounce?

An early epoch of near-exponential accelerated expansion has come to be known as an 'inflationary' early universe. Acceleration within this framework speeds up an initially expanding universe to a higher expansion rate. However, an attentive reader may notice that although acceleration is key to helping with ΛCDM's initial condition issues, there is no a priori reason why the acceleration must occur in an initially expanding universe, as opposed (say) to one that is initially contracting. Models in which we try to solve the problems of ΛCDM by having an initially contracting universe accelerate to become an expanding one are called 'bouncing' cosmologies.

Since it is really the acceleration that is important, bouncing models should in principle be on a similar footing to inflationary ones. In what follows only inflationary models are considered, for the following reasons:

Validity of the Semiclassical Methods Predictions in essentially all cosmological models are extracted using semiclassical methods: we typically write down the action for some system and then explore its consequences by solving its classical equations of motion. So a key question for all such models is the identification of the small parameter (or parameters) that suppresses quantum effects and so controls the underlying semi-classical approximation. In the absence of such a control parameter classical predictions need not capture what the system really does. Such a breakdown of the semiclassical approximation really means that the 'theory error' in the model's predictions could be arbitrarily large, making comparisons to observations essentially meaningless.

A reason sometimes given for not pinning down the size of quantum corrections when doing cosmology is that gravity plays a central role, and we do not yet know the ultimate theory of quantum gravity. Implicit in this argument is the belief that the size of quantum corrections is incalculable without such an ultimate theory, perhaps because of the well-known divergences in quantum predictions due to the non-renormalizability of general relativity (GR) [11]. But experience with non-renormalizable interactions elsewhere in physics tells us that quantum predictions can sometimes be made, provided they involve an implicit low-energy/long-distance expansion relative to the underlying physical scale set by the dimensionful non-renormalizable couplings. Because of this, the semiclassical expansion parameter in such theories is usually the ratio between this underlying short-distance scale and the distances of interest in cosmology (which, happily enough, aims at understanding the largest distances on offer). EFT provides the general tools for quantifying these low-energy expansions, and this is why EFT methods are so important for any cosmological studies.

As is argued in more detail in Section 4.3, the semiclassical expansion in cosmology is controlled by small quantities like $(\lambda M_p)^{-2}$ where λ is the smallest length scale associated with the geometry of interest. In practice it is often $\lambda \sim H^{-1}$ that provides the relevant scale in cosmology, particularly when all geometrical dimensions are similar in size. So a rule of thumb generically asks the ratio H^2/M_p^2 to be chosen to be small:

$$\frac{H^2}{M_p^2} \propto \frac{\rho}{M_p^4} \ll 1, \qquad (4.66)$$

as a necessary condition[11] for quantum cosmological effects to be suppressed.

For inflationary models H is usually at its largest during the inflationary epoch, with geometrical length scales only increasing thereafter, putting us deeper and deeper into the semiclassical domain. It is a big plus for these models that they can account for observations while wholly remaining within the regime set by (4.66), and this is one of the main reasons why they receive so much attention.

For bouncing cosmologies the situation can be more complicated. The smallest geometrical scale λ usually occurs during the epoch near the bounce, even though H^{-1} itself usually tends to infinity there. In models where λ becomes comparable to M_p^{-1} (or whatever other scale—such as the string length scale, $M_s^{-1} \gg M_p^{-1}$—that governs short-distance gravity), quantum effects during the bounce need not be negligible and the burden on proponents is to justify why semiclassical predictions actually capture what happens during the bounce.

Difficulty of Achieving a Semiclassically Large Bounce Another issue arises even if the scale λ during a bounce does remain much larger than the more microscopic scales of gravity. In this regime the bounce can be understood purely within the low-energy effective theory describing the cosmology, for which GR should be the leading approximation. But (when $\kappa = 0$) the Friedmann equation for FRW geometries in GR states that $H^2 = \rho/3M_p^2$, and so ρ must pass through zero at the instant where the contracting geometry transitions to expansion (since $H = \dot{a}/a$ vanishes at this point). Furthermore, using (4.11) and (4.13), it must also be true that

$$\dot{H} = \frac{\ddot{a}}{a} - H^2 = -\frac{1}{2M_p^2}(\rho + p) > 0, \qquad (4.67)$$

at this point in order for H to change sign there, which means the dominant contributions to the cosmic fluid must satisfy $\rho + p < 0$ during the bounce.[12]

Although there are no definitive no-go theorems, it has proven remarkably difficult to find a convincing physical system that both satisfies the condition $\rho + p < 0$ and does not also have other pathologies, such as uncontrolled runaway instabilities. For instance, within the class of multiple scalar field models for which the Lagrangian density is $\mathcal{L} = \sqrt{-g}\left[\frac{1}{2} G_{ij}(\phi)\, \partial_\mu \phi^i\, \partial^\mu \phi^j + V(\phi)\right]$ we have $\rho + p = G_{ij}(\phi)\, \dot{\phi}^i\, \dot{\phi}^j$ and so $\rho + p < 0$

---

[11] The semiclassical criterion can be stronger than this, although this can often only be quantified within the context of a specific proposal for what quantum gravity is at the shortest scales. For instance, if it is string theory that takes over at the shortest scales then treatment of cosmology using a field theory—rather than fully within string theory—requires (4.66) to be replaced by the stronger condition $H^2/M_s^2 \ll 1$, where $M_s \ll M_p$ is the string scale, set for example by the masses of the lightest string excited states.

[12] This is usually phrased as a violation of the 'null-energy' condition, which states that $T_{\mu\nu} n^\mu n^\nu \geq 0$ for all null vectors n^μ.

requires the matrix of functions $G_{ij}(\phi)$ to have a negative eigenvalue. But if this is true then there is always a combination of fields for which the kinetic energy is negative (what is called a 'ghost'), and so is unstable towards the development of arbitrarily rapid motion. Such a negative eigenvalue also implies the gradient energy $\frac{1}{2} G_{ij} \nabla \phi^i \cdot \nabla \phi^j$ is also unbounded from below, indicating instability towards the development of arbitrarily short-wavelength spatial variations.

Phenomenological Issues In addition to these conceptual issues involving the control of predictions, there are also potential phenomenological issues that bouncing cosmologies must face. Whereas expanding geometries tend to damp out spatially varying fluctuations—such as when gradient energies involve factors like $(k/a)^2$ that tend to zero as $a(t)$ grows—the opposite typically occurs during a contracting epoch for which $a(t)$ shrinks. This implies that inhomogeneities tend to grow during the pre-bounce contraction—even when the gradient energies are bounded from below—and so a mechanism must be provided for why we emerge into the homogeneous and isotropic later universe seen around us in observational cosmology.

It is of course important that bouncing cosmologies be investigated, not least in order to see most fully what might be required to understand the flatness and horizon problems. Furthermore it is essential to know whether there are alternative observational implications to those of inflation that might be used to marshal evidence about what actually occurred in the very early universe. But within the present state of the art inflationary models have one crucial advantage over bouncing cosmologies: they provide concrete semiclassical control over the key epoch of acceleration on which the success of the model ultimately relies. Because of this inflationary models are likely to remain the main paradigm for studying pre-ΛCDM extrapolations, at least until bouncing cosmologies are developed to allow similar control over how primordial conditions get propagated to the later universe through the bounce.

4.1.2.4 Simple Inflationary Models

So far so good, but what kind of physics can provide both an early period of accelerated expansion and a mechanism for ending this expansion to allow for the later emergence of the successful Hot Big Bang cosmology?

Obtaining the benefits of an accelerated expansion requires two things: first, some sort of physics that hangs the universe up for a relatively long period with an accelerating equation of state, $p < -\frac{1}{3} \rho < 0$; then some mechanism for ending this epoch to allow the later appearance of the radiation-dominated epoch within which the usual Big Bang cosmology starts. Although a number of models exist that can do this, none yet seems completely compelling. This section describes some of the very simplest such models.

The central requirement is to have some field temporarily dominate the universe with potential energy, and for the vast majority of models this new physics comes from the dynamics of a scalar field, $\varphi(x)$, called the 'inflaton'. This field can be thought of as an order parameter characterizing the dynamics of the vacuum at the very high energies likely to be relevant to inflationary cosmology. Although the field φ can in principle depend on both position and time, once inflation gets going it rapidly smooths out spatial variations, suggesting the study of homogeneous configurations: $\varphi = \varphi(t)$.

Higgs Field as an Inflaton

No way is known to obtain a viable inflationary model simply using the known particles and interactions, but a minimal model [12] does use the usual scalar Higgs field already present in the SM as the inflaton, provided it is assumed to have a non-minimal coupling to gravity of the form $\delta\mathcal{L} = -\xi \sqrt{-g} \, (\mathcal{H}^\dagger \mathcal{H}) R$, where $\mathcal{H}$ is the usual Higgs doublet and R is the Ricci scalar. Here ξ is a new dimensionless coupling, whose value turns out must be of order 10^4 in order to provide a good description of cosmological observations. Inflation in this case turns out to occur when the Higgs field takes trans-Planckian values, $\mathcal{H}^\dagger \mathcal{H} > M_p^2$, assuming V remains proportional to $(\mathcal{H}^\dagger \mathcal{H})^2$ at such large values.

As argued in [13, 14], although the large values required for both ξ and $\mathcal{H}^\dagger \mathcal{H}$ need not invalidate the validity of the EFT description, they do push the envelope for the boundaries of its domain of validity. In particular, semiclassical expansion during inflation turns out to require the neglect of powers of $\sqrt{\xi} H / M_p$, which during inflation is to be evaluated with $H \sim M_p / \xi$. This means both that the semiclassical expansion is in powers of $1/\sqrt{\xi}$, and that some sort of new physics (or 'UV completion') must intervene at scales $M_p / \sqrt{\xi} \sim \sqrt{\xi} H$, not very far above inflationary energies. Furthermore, it must do so in a way that also explains why the Lagrangian should have the very particular large-field form that is required for inflation. In particular, V must be precisely proportional to the square, f^2, of the coefficient of the non-minimal Ricci coupling, $f(\mathcal{H}^\dagger \mathcal{H})R$, at trans-Planckian field values, since this is ultimately what ensures the potential is flat when expressed in terms of canonically normalized variables in this regime. There are no known proposals for UV completions that satisfy all of the requirements, although conformal or scale invariance seems likely to be relevant [15].

This example raises a more general point that is worth noting in passing: having trans-Planckian fields during inflation need not in itself threaten the existence of a controlled low-energy EFT description. The reason for this–as is elaborated in more detail in Section 4.3–is that the EFT formulation is ultimately a low-energy expansion and so large fields are only dangerous if they also imply large energy densities. Using an EFT to describe trans-Planckian field evolution need not be a problem so long as the evolution satisfies $H \ll M$ at the field values of interest, where $M \lesssim M_p$ is the scale of the physics integrated out to obtain the EFT in question. The condition $H \ll M$ becomes $V \ll M_p^4$ if it happens that $M \sim M_p$. (In any explicit example the precise conditions for validity of EFT methods are obtained using power counting arguments along the lines of those given in Section 4.3.)

New Field as Inflaton

The simplest models instead propose a single new relativistic scalar field, φ, and design its dynamics through choices made for its potential energy, $V(\varphi)$. Taking

$$\mathcal{L} = -\sqrt{-g}\left[\frac{1}{2}\partial_\mu\varphi\,\partial^\mu\varphi + V(\varphi)\right], \tag{4.68}$$

the inflaton field equation becomes $\Box\varphi = V'(\varphi)$, which for homogeneous configurations $\varphi(t)$ reduces in an FRW geometry to

$$\ddot{\varphi} + 3H\dot{\varphi} + V' = 0, \tag{4.69}$$

where $V' = dV/d\varphi$.

The Einstein field equations are as before, but with new φ-dependent contributions to the energy density and pressure: $\rho = \rho_{\text{rad}} + \rho_{\text{m}} + \rho_\varphi$ and $p = \frac{1}{3}\rho_{\text{rad}} + p_\varphi$, where

$$\rho_\varphi = \frac{1}{2}\dot{\varphi}^2 + V(\varphi) \qquad \text{and} \qquad p_\varphi = \frac{1}{2}\dot{\varphi}^2 - V(\varphi). \tag{4.70}$$

The dark energy of the present-day epoch is imagined to arise by choosing V so that its minimum satisfies $\rho_{DE} = V(\varphi_{\text{min}})$. Inflation is imagined to occur when φ evolves slowly through a region where $V(\varphi) \gg V(\varphi_{\text{min}})$ is very large, and ends once φ rolls down towards its minimum.

With these choices energy conservation for the φ field—$\dot{\rho}_\varphi + 3(\dot{a}/a)(\rho_\varphi + p_\varphi) = 0$ follows from the field equation, eqn (4.69). Some couplings must also exist between the φ field and ordinary SM particles in order to provide a channel to transfer energy from the inflaton to ordinary particles, and so reheat the universe as required for the later Hot Big Bang cosmology. But φ is not imagined to be in thermal equilibrium with itself or with the other kinds of matter during inflation or at very late times, and this can be self-consistent if the coupling to other matter is sufficiently weak and if the φ particles are too heavy to be present once the cosmic fluid cools to the MeV energies and below (for which we have direct observations).

Slow-Roll Inflation

To achieve an epoch of near-exponential expansion, we seek a solution to the above classical field equations for $\varphi(t)$ in which the Hubble parameter, H, is approximately constant. This is ensured if the total energy density is dominated by ρ_φ, with ρ_φ also approximately constant. As we have seen, energy conservation implies the pressure must then satisfy $p_\varphi \approx -\rho_\varphi$. Inspection of eqns (4.70) shows that both of these conditions are satisfied if the φ kinetic energy is negligible compared with its potential energy:

$$\frac{1}{2}\dot{\varphi}^2 \ll V(\varphi), \tag{4.71}$$

since then $p_\varphi \simeq -V(\varphi) \simeq -\rho_\varphi$. So long as $V(\varphi)$ is also much larger than any other energy densities, it would dominate the Friedmann equation and $H^2 \simeq V/(3M_p^2)$ would then be approximately constant.

What properties must $V(\varphi)$ satisfy in order to allow (4.71) to hold for a sufficiently long time? This requires a long period of time where φ moves slowly enough to allow *both* the neglect of $\frac{1}{2}\dot{\varphi}^2$ relative to $V(\varphi)$ in the Friedmann equation, (4.11), *and* the neglect of $\ddot{\varphi}$ in the scalar field equation (4.69).

The second of these conditions allows eqn (4.69) to be written in the approximate *slow-roll* form,

$$\dot{\varphi} \approx -\left(\frac{V'}{3H}\right). \tag{4.72}$$

Using this in (4.71) then shows V must satisfy $(V')^2/(9H^2V) \ll 1$, leading to the condition that slow-roll inflation requires φ must lie in a region for which

$$\epsilon := \frac{1}{2}\left(\frac{M_p V'}{V}\right)^2 \ll 1. \tag{4.73}$$

Physically, this condition requires H to be approximately constant over any given Hubble time, inasmuch as $3M_p^2 H^2 \simeq V$ implies $6M_p^2 H\dot{H} \simeq V'\dot{\varphi} \simeq -(V')^2/3H$ and so

$$-\frac{\dot{H}}{H^2} \simeq \frac{(V')^2}{18H^4 M_p^2} \simeq \frac{M_p^2(V')^2}{2V^2} = \epsilon \ll 1. \tag{4.74}$$

Self-consistency also demands that if eqn (4.72) is differentiated to compute $\ddot{\varphi}$ it should be much smaller than $3H\dot{\varphi}$. Performing this differentiation and demanding that $\ddot{\varphi}$ remain small (in absolute value) compared with $3H\dot{\varphi}$, then implies $|\eta| \ll 1$ where

$$\eta := \frac{M_p^2 V''}{V}, \tag{4.75}$$

defines the second slow-roll parameter. The slow-roll parameters ϵ and η are important [16] because (as shown below) the key predictions of single-field slow-roll inflation for density fluctuations can be expressed in terms of the three parameters ϵ, η and the value, H_I, of the Hubble parameter during inflation.

Given an explicit shape for $V(\varphi)$ we can directly predict the amount of inflation that occurs after the epoch of Hubble exit (where currently observable scales become larger than the Hubble length). This is done by relating the amount of expansion directly to the distance φ traverses in field space during this time. To this end, rewriting eqn (4.72) in terms of $\varphi' \equiv d\varphi/da$, leads to

$$\frac{d\varphi}{da} = \frac{\dot{\varphi}}{\dot{a}} = -\frac{V'}{3aH^2} = -\frac{M_p^2 V'}{aV}, \tag{4.76}$$

which when integrated between horizon exit, φ_{he}, and final value, φ_{end}, gives the amount of expansion during inflation as $a_{end}/a_{he} = e^{N_e}$, with

$$N_e = \int_{a_{he}}^{a_{end}} \frac{da}{a} = \int_{\varphi_{end}}^{\varphi_{he}} d\varphi \left(\frac{V}{M_p^2 \, V'} \right) = \frac{1}{M_p} \int_{\varphi_{end}}^{\varphi_{he}} \frac{d\varphi}{\sqrt{2\epsilon}}. \tag{4.77}$$

In these expressions φ_{end} can be defined by the point where the slow-roll parameters are no longer small, such as where $\epsilon \simeq \frac{1}{2}$. Then this last equation can be read as defining $\varphi_{end}(N_e)$, as a function of the desired number of e-foldings between the the epoch of horizon exit and the end of inflation, since this is this quantity constrained to be large by the horizon and flatness problems.

Notice also that if ϵ were approximately constant during inflation, then eqn (4.77) implies that $N_e \approx (\varphi_{he} - \varphi_{end})/(\sqrt{2\epsilon} \, M_p)$. In such a case φ must traverse a range of order $N_e M_p \sqrt{2\epsilon}$ between φ_{he} and φ_{end}. This is larger than order M_p provided only that $1 \gg \epsilon \gtrsim 1/N_e^2$, showing why Planckian fields are often of interest for inflation [17].

It is worth working through what these formulae mean in a few concrete choices for the shape of the scalar potential.

Example I: Quadratic Model

The simplest example of an inflating potential [10, 18] chooses φ to be a free massive field, for which

$$V = \frac{1}{2} m^2 \varphi^2, \tag{4.78}$$

and so $V' = m^2 \varphi$ and $V'' = m^2$, leading to slow-roll parameters of the form

$$\epsilon = \frac{1}{2} \left(\frac{2M_p}{\varphi} \right)^2 \quad \text{and} \quad \eta = \frac{2M_p^2}{\varphi^2}, \tag{4.79}$$

and so $\epsilon = \eta$ in this particular case, and slow roll requires $\varphi \gg M_p$. The scale for inflation in this field range is $V = \frac{1}{2} m^2 \varphi^2$ and so $H_I^2 \simeq m^2 \varphi^2/(6 M_p^2)$. We can ensure $H_I^2/M_p^2 \ll 1$ even if $\varphi \gg M_p$ by choosing m/M_p sufficiently small. Observations will turn out to require $\epsilon \sim \eta \sim 0.01$ and so the regime of interest is $\varphi_{he} \sim 10 M_p$, and so $H_I/M_p \ll 1$ requires $m/M_p \ll 0.1$.

In this large-field regime φ (and so also V and H) evolves only very slowly despite there being no nearby stationary point for V because Hubble friction slows φ's slide down the potential. Since φ evolves towards smaller values, eventually slow roll ends once η and ϵ become $O(1)$. Choosing φ_{end} by the condition $\epsilon(\varphi_{end}) = \eta(\varphi_{end}) = \frac{1}{2}$ implies $\varphi_{end} = 2M_p$. The number of e-foldings between horizon exit and $\varphi_{end} = 2M_p$ is then given by eqn (4.77), which in this instance becomes

$$N_e = \int_{2M_p}^{\varphi_{\text{he}}} d\varphi \left(\frac{\varphi}{2M_p^2} \right) = \left(\frac{\varphi_{\text{he}}}{2M_p} \right)^2 - 1, \qquad (4.80)$$

and so obtaining $N_e \sim 63$ e-foldings (say) requires choosing $\varphi_{\text{he}} \sim 16\,M_p$. In particular $\epsilon_{\text{he}} := \epsilon(\varphi_{\text{he}})$ and $\eta_{\text{he}} := \eta(\varphi_{\text{he}})$ can be expressed directly in terms of N_e, leading to

$$\epsilon_{\text{he}} = \eta_{\text{he}} = \frac{1}{2(N_e + 1)}, \qquad (4.81)$$

which are both of order 10^{-2} for $N_e \simeq 60$. As seen later, the prediction $\epsilon = \eta$ is beginning to be disfavoured by cosmological observations.

Example II: Pseudo-Goldstone Axion

The previous example shows how controlled inflation requires the inflaton mass to be small compared with the scales probed by φ. Small masses arise because the condition $|\eta| \ll 1$ implies the inflaton mass satisfies $m^2 \sim |V''| \sim |\eta\, V/M_p^2| \ll V/M_p^2 \simeq 3H^2$. Consequently m must be very small compared with H, which itself must be Planck suppressed compared with other scales (such as $v \sim V^{1/4}$) during inflation. From the point of view of particle physics, such small masses pose a puzzle because it is fairly uncommon to find interacting systems with very light spinless particles in their low-energy spectrum.[13]

The main exceptions to this statement are Goldstone bosons for the spontaneous breaking of continuous global symmetries since these are guaranteed to be massless by Goldstone's theorem. This makes it natural to suppose the inflaton to be a pseudo-Goldstone boson (i.e. a would-be Goldstone boson for an approximate symmetry, much like the pions of chiral perturbation theory). In this case, Goldstone's theorem ensures the scalar's mass (and other couplings in the scalar potential) must vanish in the limit the symmetry becomes exact, and this 'protects' it from receiving generic UV-scale contributions. For Abelian broken symmetries this shows up in the low-energy EFT as an approximate shift symmetry under which the scalar transforms inhomogeneously: $\varphi \to \varphi + \text{constant}$.

If the approximate symmetry arises as a $U(1)$ phase rotation for some microscopic field, and if this symmetry is broken down to discrete rotations, $Z_N \subset U(1)$, then the inflaton potential is usually trigonometric [19]:

$$V = V_0 + \Lambda^4 \left[1 - \cos\left(\frac{\varphi}{f} \right) \right] = V_0 + 2\Lambda^4 \sin^2\left(\frac{\varphi}{2f} \right), \qquad (4.82)$$

for some scales V_0, Λ and f. Here V_0 is chosen to agree with ρ_{DE} and because ρ_{DE} is so small the parameter V_0 is dropped in what follows. The parameter Λ represents the scale

---

[13] From an EFT perspective having a light scalar requires the coefficients of low-dimension effective interactions like ϕ^2 to have unusually small coefficients like m^2, rather than being as large as the (much larger) UV scales M^2.

associated with the explicit breaking of the underlying $U(1)$ symmetry while f is related to the size of its spontaneous breaking. The statement that the action is approximately invariant under the symmetry is the statement that Λ is small compared with UV scales like f. Expanding about the minimum at $\varphi = 0$ reveals a mass of size $m = \Lambda^2/f \ll \Lambda \ll f$, showing the desired suppression of the scalar mass.

With this choice $V' = (\Lambda^4/f)\sin(\varphi/f)$ and $V'' = (\Lambda^4/f^2)\cos(\varphi/f)$, leading to slow-roll parameters of the form

$$\epsilon = \frac{M_p^2}{2f^2}\cot^2\left(\frac{\varphi}{2f}\right) \quad \text{and} \quad \eta = \frac{M_p^2}{2f^2}\left[\cot^2\left(\frac{\varphi}{2f}\right) - 1\right], \tag{4.83}$$

and so $\eta = \epsilon - (M_p^2/2f^2)$. Notice that in the limit $M_p \lesssim \varphi \ll f$ these go over to the $m^2\varphi^2$ case examined above, with $m = \Lambda^2/f$.

Slow roll in this model typically requires $f \gg M_p$. This can be seen directly from (4.83) for generic $\varphi \simeq f$, but also follows when $\varphi \ll f$ because in this case the potential is close to quadratic and slow roll requires $M_p \ll \varphi \ll f$. The scale for inflation is $V \simeq \Lambda^4$ and so $H_I \sim \Lambda^2/M_p$. This ensures $H_I^2/M_p^2 \ll 1$ follows from the approximate-symmetry limit which requires $\Lambda \ll M_p$. The condition $\epsilon \sim 0.01$ is arranged by choosing $f \sim 10 M_p$, but once this is done the prediction $\epsilon \simeq \eta$ is in tension with recent observations.

The number of e-foldings between horizon exit and φ_{end} is again given by eqn (4.77), so

$$N_e = \frac{2f}{M_p^2}\int_{\varphi_{\text{end}}}^{\varphi_{\text{he}}} d\varphi \, \tan\left(\frac{\varphi}{2f}\right) = \left(\frac{2f}{M_p}\right)^2 \ln\left|\frac{\sin(\varphi_{\text{he}}/2f)}{\sin(\varphi_{\text{end}}/2f)}\right|, \tag{4.84}$$

which is only logarithmically sensitive to φ_{he}, but which can easily be large due to the condition $f \gg M_p$.

While models like this do arise generically from UV completions like string theory [20], axions in string theory typically arise with $f \ll M_p$ [21], making the condition $f \gg M_p$ tricky to arrange [22].

Example III: Pseudo-Goldstone Dilaton

Another case where the inflaton mass is protected by an approximate shift symmetry arises when it is a pseudo-Goldstone boson for a scaling symmetry of the underlying UV theory. Such 'accidental' scale symmetries turn out to be fairly common in explicit examples of UV completions because scale invariances are automatic consequences of higher-dimensional supergravities [23]. Because it is a scaling symmetry the same arguments leading to trigonometric potentials for the compact $U(1)$ rotations instead in this case generically lead to exponential potentials [24].

In this case, the form expected for the scalar potential during the inflationary regime would be

$$V = V_0 - V_1 e^{-\varphi/f} + \cdots, \tag{4.85}$$

for some scales V_0, V_1 and f. Our interest is in the regime $\varphi \gg f$ and in this regime V_0 dominates, and so is chosen as needed for inflationary cosmology, with $H_I^2 \simeq V_0/(3M_p^2)$. Control over the semiclassical limit requires $V_0 \ll M_p^4$.

With this choice the relevant potential derivatives are $V' \simeq (V_1/f)\, e^{-\varphi/f}$ and $V'' \simeq -(V_1/f^2)\, e^{-\varphi/f}$ leading to slow-roll parameters of the form

$$\epsilon \simeq \frac{1}{2}\left(\frac{M_p V_1}{f V_0}\right)^2 e^{-2\varphi/f} \qquad \text{and} \qquad \eta \simeq -\left(\frac{M_p^2 V_1}{f^2 V_0}\right) e^{-\varphi/f}, \tag{4.86}$$

and so

$$\epsilon = \frac{1}{2}\left(\frac{f}{M_p}\right)^2 \eta^2. \tag{4.87}$$

The number of e-foldings between horizon exit and φ_{end} is again given by eqn (4.77), so

$$N_e = \left(\frac{f V_0}{M_p^2 V_1}\right)\int_{\varphi_{\text{end}}}^{\varphi_{\text{he}}} d\varphi\; e^{\varphi/f} = \left(\frac{f^2 V_0}{M_p^2 V_1}\right)\left[e^{\varphi_{\text{he}}/f} - e^{\varphi_{\text{end}}/f}\right], \tag{4.88}$$

which can easily be large as long as $\varphi_{\text{he}} \gg f$ and φ_{end}/f is order unity.

Notice that ϵ and η are generically small whenever $\varphi \gg f$, even if $V_1 \sim V_0$, so there is no need to require f be larger than M_p to ensure a slow roll. Typical examples of underlying UV theories (discussed later) give $f \sim M_p$, in which case $\epsilon \simeq \eta^2$. It turns out that this prediction provides better agreement with experiment than $\epsilon \simeq \eta$ does, and (as seen below) the generic expectation that $\epsilon \sim \eta^2$ has potentially interesting observational consequences for measurements of primordial gravitational waves because it relates the as-yet-unmeasured tensor-to-scalar ratio, $r \lesssim 0.07$, to the observed spectral tilt, $n_s \simeq 0.96$, giving the prediction $r_{\text{th}} \simeq (n_s - 1)^2 \simeq 0.002$.

Interestingly, many successful inflationary models can be recast into this exponential form, usually with specific values predicted for f. The earliest instance using an exponential potential [25] came from a supergravity example with $f = \sqrt{\frac{1}{6}}\,M_p$, with a non-linearly realized $SU(1,1)$ symmetry. Such symmetries are now known to arise fairly commonly when dimensionally reducing higher-dimensional supersymmetric models [23, 24]. This early supergravity example foreshadows the results from a class of explicit higher-dimensional UV completions within string theory [26], which reduce to the above with $f = \sqrt{3}\,M_p$, while the first extra-dimensional examples of this type [27] gave $f = \sqrt{2}\,M_p$.

In fact, the Higgs-inflation model described earlier can also be recast to look like a scalar field with an exponential potential of the form considered here, once it is written with canonically normalized fields. The prediction in this case is $f = \sqrt{\frac{3}{2}}\,M_p$. The same is true for another popular model that obtains inflation using curvature-squared interactions [28], for which again $f = \sqrt{\frac{3}{2}}\, M_p$. Although both of these models are hard to obtain in a controlled way from UV completions directly, their formulation in terms of exponential potentials may provide a way to do so through the back door.

4.2 Cosmology: Fluctuations

This section repeats the previous discussion of ΛCDM cosmology and its peculiar initial conditions, but extends it to the properties of fluctuations about the background cosmology.

4.2.1 Structure formation in ΛCDM

Previous sections show that the universe was very homogeneous at the time of photon last scattering, since the temperature fluctuations observed in the distribution of CMB photons have an amplitude $\delta T/T \sim 10^{-5}$. On the other hand, the universe around us is full of stars and galaxies and so is far from homogeneous. How did the one arise from the other?

The basic mechanism for this in the ΛCDM model is based on gravitational instability: the gravitational force towards an initially overdense region acts to attract even more material towards this region, thereby making it even more dense. This process can feed back on itself until an initially small density perturbation becomes dramatically amplified, such as into a star. This section describes the physics of this instability in the very early universe when the density contrasts are small enough to be analysed perturbatively in the fluctuation amplitude. The discussion follows that of [31, 32].

4.2.1.1 *Non-Relativistic Density Perturbations*

We start with the discussion of gravitational instability for non-relativistic fluids, both for simplicity and since this is the sector that actually displays the instability in practice. The equations found here provide a self-consistent description of how linearized density fluctuations for non-relativistic matter evolve in a matter-dominated universe (which is the main one relevant for structure growth), and also turn out to capture how non-relativistic density fluctuations grow when the total energy density is dominated by radiation or dark energy.

The following equations of motion describe the evolution of a simple non-relativistic fluid with energy density, ρ, pressure, p, entropy density, s, and local fluid velocity $\mathbf{v}$. Each equation expresses a local conservation law,

$$\frac{\partial \rho}{\partial t} + \nabla \cdot (\rho \mathbf{v}) = 0 \qquad \text{(energy conservation)}$$

$$\rho\left[\frac{\partial \mathbf{v}}{\partial t} + (\mathbf{v} \cdot \nabla)\mathbf{v}\right] + \nabla p + \rho \nabla \phi = 0 \qquad \text{(momentum conservation)} \qquad (4.89)$$

$$\frac{\partial s}{\partial t} + \nabla \cdot (s\mathbf{v}) = 0 \qquad \text{(entropy conservation)}$$

$$\nabla^2 \phi - 4\pi G\rho = 0 \qquad \text{(universal gravitation)},$$

and they are imagined supplemented by an equation of state, $p = p(\rho, s)$. Here ϕ denotes the local Newtonian gravitational potential. Because they are non-relativistic these equations are expected to break down for super-Hubble modes, for which $k/a \lesssim H$ and the proscription against motion faster than light plays an important role.

For cosmological applications expand about a homogeneously and radially expanding background fluid configuration. For these purposes consider a fluid background for which $\mathbf{v}_0 = H(t)\mathbf{r}$, where $H(t)$ is assumed a given function of t. In this case $\nabla \cdot \mathbf{v}_0 = 3H(t)$. This flow is motivated by the observation that it corresponds to the proper velocity if particles within the fluid were moving apart from one another according to the law $\mathbf{r}(t) = a(t)\mathbf{y}$, with $\mathbf{y}$ being a time-independent co-moving coordinate. In this case $\mathbf{v}_0 := d\mathbf{r}/dt = \dot{a}\mathbf{y} = H(t)\mathbf{r}(t)$ where $H = \dot{a}/a$. In this sense $H(t)$ describes the Hubble parameter for the expansion of the background fluid.

Background Quantities We now ask what the rest of the background quantities, $\rho_0(t)$, $p_0(t)$, and $\phi_0(t)$ must satisfy in order to be consistent with this flow. The equation of energy conservation implies ρ_0 must satisfy

$$0 = \dot{\rho}_0 + \nabla \cdot (\rho_0 \mathbf{v}_0) = \dot{\rho}_0 + 3H\rho_0, \qquad (4.90)$$

and so, given $H = \dot{a}/a$, it follows that $\rho_0 \propto a^{-3}$. Not surprisingly, the density of a non-relativistic expanding fluid necessarily falls with universal expansion in the same way required by the full relativistic treatment.

Using this density in the law for universal gravitation requires the gravitational potential, ϕ_0, take the form

$$\phi_0 = \frac{2\pi G\rho_0}{3}\mathbf{r}^2, \qquad (4.91)$$

and so $\nabla \phi_0 = \frac{4}{3}\pi G\rho_0 \mathbf{r}$. This describes the radially directed gravitational potential that acts to decelerate the overall universal expansion.

Given this gravitational force, the momentum conservation equation, using $\dot{\mathbf{v}}_0 + (\mathbf{v}_0 \cdot \nabla)\mathbf{v}_0 = [H + \dot{H}/H]\mathbf{v}_0$ and $\mathbf{v}_0 = H\mathbf{r}$, becomes

$$\left[\dot{H} + H^2 + \frac{4\pi G\rho_0}{3}\right]\mathbf{r} = 0. \qquad (4.92)$$

This is equivalent to the Friedmann equation, as is now shown. Notice that if we take $a \propto t^\alpha$ then $H = \alpha/t$ and $\dot{H} = -\alpha/t^2 = -H^2/\alpha$. This, together with $\rho_0 \propto a^{-3} \propto t^{-3\alpha}$, is consistent with eqn (4.92) only if $\alpha = 2/3$, as expected for a matter-dominated universe. Furthermore, with this choice for α we also have $\dot{H} + H^2 = -\frac{1}{2}H^2$, and so eqn (4.92) is equivalent to

$$H^2 = \frac{8\pi G}{3}\rho_0,\tag{4.93}$$

which is the Friedmann equation, as claimed.

When studying perturbations we solve the entropy equation by taking $s_0 = 0$. This is done mostly for simplicity, but it is also true that for many situations the thermal effects described by s_0 play a negligible role.

Perturbations During Matter Domination To study perturbations about this background take $\mathbf{v} = \mathbf{v}_0 + \delta\mathbf{v}$, $\rho = \rho_0 + \delta\rho$, $p = p_0 + \delta p$, $s = \delta s$, and $\phi = \phi_0 + \delta\phi$, and expand the equations of motion to first order in the perturbations. Defining $D_t = \partial/\partial t + \mathbf{v}_0 \cdot \nabla$, the linearized equations in this case become

$$
\begin{aligned}
D_t\,\delta\rho + 3H\,\delta\rho + \rho_0\nabla \cdot \delta\mathbf{v} &= 0\\
\rho_0(D_t\,\delta\mathbf{v} + H\,\delta\mathbf{v}) + \nabla\delta p + \rho_0\nabla\delta\phi &= 0\\
D_t\,\delta s &= 0\\
\nabla^2\delta\phi - 4\pi G\delta\rho &= 0.
\end{aligned}
\tag{4.94}
$$

To obtain this form for the momentum conservation equation requires using the equations of motion for the background quantities.

Our interest is in the evolution of $\delta\rho$, and this can be isolated by taking D_t of the first of eqns (4.94) and the divergence of the second if these equations, and using the results to eliminate $\delta\mathbf{v}$. The remaining equations involve the two basic fluid perturbations, $\delta\rho$ and δs, and imply both $D_t\,\delta s = 0$ and

$$D_t^2\left(\frac{\delta\rho}{\rho_0}\right) + 2HD_t\left(\frac{\delta\rho}{\rho_0}\right) - c_s^2\nabla^2\left(\frac{\delta\rho}{\rho_0}\right) - 4\pi G\rho_0\left(\frac{\delta\rho}{\rho_0}\right) = \frac{\xi}{\rho_0}\delta s,\tag{4.95}$$

where

$$c_s^2 := \left(\frac{\partial p}{\partial\rho}\right)_{s0} \quad\text{and}\quad \xi := \left(\frac{\partial p}{\partial s}\right)_{\rho 0}.\tag{4.96}$$

In order to analyse the solutions to this equation, it is convenient to change variables to a co-moving coordinate, $\mathbf{y}$, defined by $\mathbf{r} = a(t)\,\mathbf{y}$. In this case, for any function $f = f(\mathbf{r}, t)$ we have $(\partial f/\partial t)_{\mathbf{y}} = (\partial f/\partial t)_{\mathbf{r}} + H\mathbf{r} \cdot \nabla f = D_t f$, and $\nabla f = (1/a)\nabla_{\mathbf{y}}f$. Fourier transforming the perturbations in co-moving coordinates, $\delta\rho/\rho_0 = \delta_k(t)\exp[i\mathbf{k} \cdot \mathbf{y}]$, leads to the following master equation governing density perturbations

$$\ddot{\delta}_k + 2H\dot{\delta}_k + \left(\frac{c_s^2\,k^2}{a^2} - 4\pi G\rho_0\right)\delta_k = \left(\frac{\xi}{\rho_0}\right)\delta s,\tag{4.97}$$

where the overdot denotes $\mathrm{d}/\mathrm{d}t$.

These equations have solutions whose character depends on the relative size of k/a and the Jeans wave number,

$$k_{\mathcal{J}}^2(t) = \frac{4\pi\, G\rho_0(t)}{c_s^2(t)} = \frac{3H^2(t)}{2\, c_s^2(t)}, \tag{4.98}$$

with instability occurring once $k/a \ll k_{\mathcal{J}}$. Notice that so long as $c_s \sim O(1)$ the Jeans length is comparable in size to the Hubble length, $\ell_{\mathcal{J}} \sim H^{-1}$. For adiabatic fluctuations ($\delta s_k = 0$) the above equation implies that the short-wavelength fluctuations ($k/a \gg k_{\mathcal{J}}$) undergo damped oscillations of the form

$$\delta_k(t) \propto a^{-1/2} \exp\left[\pm ikc_s \int^t \frac{dt'}{a(t')}\right]. \tag{4.99}$$

The overall prefactor of $a^{-1/2}$ shows how these oscillations are damped due to the universal expansion, or Hubble friction.

Long-wavelength adiabatic oscillations ($k/a \ll k_{\mathcal{J}}$) exhibit an instability, with the unstable mode growing like a power law of t. The approximate solutions are

$$\delta_k(t) \propto t^{2/3} \propto a(t) \quad\text{and}\quad \delta_k(t) \propto t^{-1} \propto a^{-3/2}(t), \tag{4.100}$$

with the $\delta_k(t) \propto a \propto t^{2/3}$ solution describing the unstable mode. The instability has power-law rather than exponential growth because the expansion of space acts to reduce the density, and this effect fights the density increase due to gravitational collapse.

Because both the red-shifted wave-number, k/a, and the Jeans wave number, $k_{\mathcal{J}}$, depend on time, the overall expansion of the background can convert modes from stable to unstable (or vice versa). Whether this conversion is towards stability or instability depends on the the time-dependence of $ak_{\mathcal{J}}$, which is governed by the time-dependence of the combination aH/c_s. If $a \propto t^\alpha$ then $aH \propto t^{\alpha-1} \propto a^{1-1/\alpha}$, and so aH increases with t if $\alpha > 1$ and decreases with t if $\alpha < 1$. Since $\alpha = 2/3$ for the matter-dominated universe of interest here, it follows that $aH \propto t^{-1/3} \propto a^{-1/2}$, and so *decreases* with t. Provided that c_s does not change much, this ensures that in the absence of other influences modes having fixed k pass from being unstable to stable as a increases due to the overall expansion.

Perturbations during Radiation and Vacuum Domination A completely relativistic treatment of density perturbations requires following fluctuations in the matter stress energy as well as in the metric itself (since these are related by Einstein's equations relating geometry and stress-energy). The details of such calculations go beyond the scope of this chapter, although some of the main features are described later. But these considerations suffice to address a result that is an important part of the structure-formation story: the stalling of perturbation growth for non-relativistic matter during radiation- or vacuum-dominated epochs.

To contrast how fluctuations grow during radiation and matter domination it is instructive to examine the transition from radiation to matter domination. To this end we again track the growth of density fluctuations for non-relativistic matter, $\delta\rho_{m0}/\rho_{m0}$, and do so using the same Fourier-transformed equation as before,

$$\ddot{\delta}_{\mathbf{k}} + 2H\dot{\delta}_{\mathbf{k}} + \left(\frac{c_s^2 \mathbf{k}^2}{a^2} - 4\pi G\rho_{m0}\right)\delta_{\mathbf{k}} = 0, \tag{4.101}$$

but with $H^2 = 8\pi G\rho_0/3$ where $\rho_0 = \rho_{m0} + \rho_{r0}$ includes both radiation and matter. In particular, during the transition between radiation and matter domination the Hubble scale satisfies

$$H^2(a) = \frac{8\pi G\rho_0}{3} = \frac{H_{eq}^2}{2}\left[\left(\frac{a_{eq}}{a}\right)^3 + \left(\frac{a_{eq}}{a}\right)^4\right], \tag{4.102}$$

where radiation-matter equality occurs when $a = a_{eq}$, at which point $H(a = a_{eq}) = H_{eq}$.

As described above, any departure from the choice $a(t) \propto t^{2/3}$—such as occurs when radiation is non-negligible in $\rho(a)$—means that the background momentum-conservation equation, eqn (4.92), is no longer satisfied. Instead the expression for H comes from solving the fully relativistic radiation-dominated Friedmann equation, eqn (4.102). But this does not mean that the non-relativistic treatment of the fluctuations, eqns (4.94), must fail, since the important kinematics and gravitational interactions amongst these perturbations remain the Newtonian ones. To first approximation the leading effect of the radiation domination for these fluctuations is simply to change the expansion rate, as parameterized by $H(a)$ in (4.102).

For all modes for which the pressure term, $c_s^2 \mathbf{k}^2/a^2$, is negligible, (4.101) implies $\delta(x)$ satisfies

$$2x(1+x)\delta'' + (3x+2)\delta' - 3\delta = 0, \tag{4.103}$$

where the rescaled scale factor, $x = a/a_{eq}$, is used as a proxy for time and primes denote differentiation with respect to x. As is easily checked, this is solved by $\delta^{(1)} \propto (x + \frac{2}{3})$, and so the growing mode during matter domination (i.e. $x \gg 1$) does not also grow during radiation domination ($x \ll 1$). Furthermore, the solution linearly independent to this one can be found using the Frobenius method, and this behaves for $x \ll 1$ (i.e. deep in the radiation-dominated regime) as $\delta^{(2)} \propto \delta^{(1)} \ln x + $ (analytic), where 'analytic' denotes a simple power series proportional to $1 + c_1 x + \cdots$. These solutions show how density perturbations for non-relativistic matter grow at most logarithmically during the radiation-dominated epoch.

A similar analysis covers the case where dark energy (modelled as a cosmological constant) dominates in an $\Omega = 1$ universe. In this case $4\pi G\rho_{m0} \sim \Omega_m H^2 \ll H^2$ and so the instability term becomes negligible relative to the first two terms of (4.101). This leads to

$$\ddot{\delta} + 2H\dot{\delta} \simeq 0, \qquad (4.104)$$

which has as solution $\dot{\delta} \propto a^{-2}$. Integrating again gives a frozen mode, $\delta \propto a^0$, and a damped mode that falls as $\delta \propto a^{-2}$ when H is constant (as it is when dark energy dominates and $a \propto e^{Ht}$). This shows that non-relativistic density perturbations also stop growing once matter domination ends.

We are now in a position to summarize how inhomogeneities grow in the late universe, assuming the presence of an initial spectrum of very small primordial density fluctuations. The key observation is that three conditions all have to hold in order for there to be appreciable growth of density inhomogeneities:

1. Fluctuations of any type do not grow for super-Hubble modes, for which $k/a \ll H$, regardless of what type of matter dominates the background evolution.

2. Fluctuations in non-relativistic matter can be unstable, growing as $\delta_k \propto a(t)$, but only in a matter-dominated universe and for those modes in the momentum window $H \ll (k/a) \ll H/c_s$.

3. No fluctuations in relativistic matter ever grow appreciably, either inside or outside the Hubble scale. (Although this is not shown explicitly above for relativistic matter, and requires the fully relativistic treatment, the instability window $H \ll k/a \ll H/c_s$ for non-relativistic fluctuations is seen to close as they become relativistic, i.e. as $c_s \to 1$.)

Before pursuing the implications of these conditions for instability, we must first describe what properties of fluctuations are actually measured.

4.2.1.2   *The Power Spectrum*

The presence of unstable density fluctuations implies the universe does not remain precisely homogeneous and isotropic once matter domination begins, and so the view seen by observers like us depends on their locations in the universe relative to the fluctuations. For this reason, when comparing with observations it is less useful to try to track the detailed form of a specific fluctuation and instead better to characterize fluctuations by their statistical properties, since these can be more directly applied to observers without knowing their specific place in the universe. In particular, we imagine there being an ensemble of density fluctuations, whose phases we assume to be uncorrelated and whose amplitudes are taken to be random variables.

On the observation side statistical inferences can be made about the probability distribution governing the distribution of fluctuation amplitudes by measuring statistical properties of the matter distribution observed around us. For instance, a useful statistic measures the mass-mass auto-correlation function

$$\xi(\mathbf{r} - \mathbf{r}') \equiv \frac{\langle \delta\rho(\mathbf{r})\, \delta\rho(\mathbf{r}') \rangle}{\langle \rho \rangle^2}, \qquad (4.105)$$

which might be measured by performing surveys of the positions of large samples of galaxies.[14] When using (4.105) with observations the average $\langle \cdots \rangle$ is interpreted as integration of one of the positions (say, $\mathbf{r}'$) over all directions in the sky.[15]

When making predictions $\langle \cdots \rangle$ instead is regarded as an average over whatever ensemble is thought to govern the statistics of the fluctuations δ_k. Fourier transforming $\delta\rho(\mathbf{r})/\langle\rho\rangle = \int d^3 k \, \delta_k \exp[i\mathbf{k}\cdot\mathbf{r}]$ in comoving coordinates, as before, allows $\xi(\mathbf{r})$ to be related to the following ensemble average over the Fourier mode amplitudes, δ_k.

$$\xi(r) = \int \frac{d^3 k}{(2\pi)^3} \langle |\delta_k|^2 \rangle \exp[i\mathbf{k}\cdot\mathbf{r}] = \frac{1}{2\pi^2} \int_0^\infty \frac{dk}{k} \, k^3 P_\rho(k) \left(\frac{\sin kr}{kr} \right), \qquad (4.106)$$

which defines the density *power spectrum*: $P_\rho(k) := \langle |\delta_k|^2 \rangle$.

For homogeneous and isotropic backgrounds $P_\rho(k)$ depends only on the magnitude $k = |\mathbf{k}|$ and not on direction, and this is used above to perform the angular integrations. The average in these expressions is over the ensemble, and it is this average which collapses the right-hand side down to a single Fourier integral. The last equality motivates the definition

$$\Delta_\rho^2(k) := \frac{k^3}{2\pi^2} P_\rho(k). \qquad (4.107)$$

A variety of observations over the years give the form of $P_\rho(k)$ as inferred from the distribution of structure around us, with results summarized in Figure 4.10. As this figure indicates, inferences about the shape of $P_\rho(k)$ for small k come from measurements of the temperature fluctuations in the CMB; those at intermediate k come from galaxy distributions as obtained through galaxy surveys and those at the largest k come from measurements of the how quasar light is absorbed by intervening Hydrogen gas clouds, the so-called Lyman-α 'forest'. The reasons why different kinds of measurements control different ranges of k are illustrated in Figure 4.11, which shows how the distance accessible to observations is correlated with how far back one looks into the universe: measurements of distant objects in the remote past (e.g. the CMB) determine the shape of $P_\rho(k)$ for small k while measurements of more nearby objects in the more recent past (e.g. the Lyman-α forest) constrain $P_\rho(k)$ for larger k.

---

[14] A practical complication arises because, although galaxies are relatively easy to count, most of the mass density is actually dark matter. Consequently, assumptions are required to relate these to one another, the usual choice being that the galaxy and mass density functions are related to one another through a phenomenologically defined 'bias' factor.

[15] The density correlation function can also be measured using the temperature fluctuations of the CMB, because these fluctuations can be interpreted as redshifts acquired by CMB photons as they climb out of the gravitational potential wells formed by density fluctuations in non-relativistic matter.

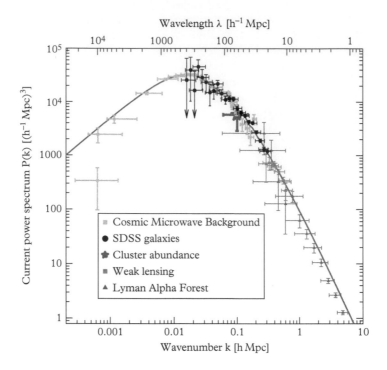

Figure 4.10 *The power spectrum as obtained from measurements of the CMB spectrum, together with the SDSS Galaxy Survey, observations of abundances of galaxy clusters and Lyman-α measurements (taken from [29]).*

The observations summarized in Figure 4.10 are well approximated by the phenomenological formula,

$$P(k) = \frac{A k^{n_s}}{(1 + \alpha k + \beta k^2)^2},$$ (4.108)

where

$$\alpha = 16 \left(\frac{0.5}{\Omega h^2} \right) \text{Mpc} \quad \text{and} \quad \beta = 19 \left(\frac{0.5}{\Omega h^2} \right)^2 \text{Mpc}^2 \quad \text{and} \quad n_s = 0.97.$$ (4.109)

Here $h = H_0/(100 \text{ km/sec/Mpc}) \approx 0.7$, and $\Omega \approx 1$ denotes the present value of ρ/ρ_c. Given that $n_s \approx 1$ the observations suggest the power spectrum is close to linear, $P(k) \propto k$ for $k \ll k_\star \sim 0.07 \text{ Mpc}^{-1}$, and $P(k) \propto k^{-3}$ for $k \gg k_\star$. The value $k_\star$ here is simply defined to be the place where $P_\rho(k)$ turns over and makes the transition from $P_\rho \propto k$ to $P_\rho \propto k^{-3}$.

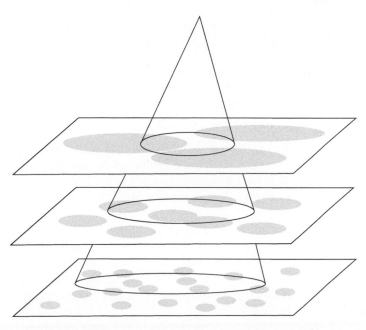

Figure 4.11 *A sketch of several spatial slices intersecting the past light cone of an astronomer on Earth. The empty ovals indicate how the light cone has larger intersections with the spatial slices the further back one looks. The filled ovals indicate regions the size of the Hubble distance on each spatial slice. Correlations outside of these ovals (such as the uniformity of the CMB temperature) represent a puzzle for ΛCDM cosmology. The figure shows how later times (higher slices) have larger Hubble distances, as well as how observations only sample the largest distance scales on the most remote spatial slices. This illustrates why CMB measurements tend to constrain the power spectrum for small k while observations of more nearby objects (like galaxy distributions or the distribution of foreground Lyman-α Hydrogen gas clouds) constrain larger k.*

As described later, there are good reasons to believe that the shape of $P_\rho(k)$ for $k \ll k_\star$ represents the pattern of primordial fluctuations inherited from the very early universe, while the shape for $k > k_\star$ reflects how fluctuations evolve in the later universe. Consequently observations are consistent with primordial fluctuations being close to[16] a *Zel'dovich* spectrum, $P_\rho(k) = A k$, corresponding to $n_s = 1$. As shown later, the parameter n_s is predicted to be close to, but not equal to, unity by inflationary models.

For later purposes it proves more convenient to work with the power spectrum for the Newtonian gravitational potential, $\delta\phi$, that is related to $\delta\rho$ by Poisson's equation—the last of eqns (4.94)—and so $\delta\phi_k \propto \delta_k/k^2$. Because of this relation their power spectra are related by $P_\phi(k) = P_\rho(k)/k^4$ as well as

---

[16] Close to but not equal to. Fits to ΛCDM cosmology establish n_s is significantly different from 1.

$$\Delta_\phi^2(k) := \frac{k^3}{2\pi^2} P_\phi(k) = \frac{P_\rho(k)}{2\pi^2 k} \propto \begin{cases} k^{n_s-1} & \text{if} \quad k \ll k_\star \\ k^{n_s-5} & \text{if} \quad k \gg k_\star \end{cases}. \qquad (4.110)$$

This last expression also clarifies why the choice $n_s = 1$ is called scale invariant. When $n_s = 1$ the primordial ($k \ll k_\star$) spectrum for $\Delta_\phi^2(k)$ becomes k-independent, as would be expected for a scale-invariant process.

4.2.1.3 Late-Time Structure Growth

Before trying to explain the properties of the primordial part of the power spectrum—$\Delta_\phi^2(k) \propto A k^{n_s-1}$—a further digression is in order to explain why the measured distribution has the peculiar hump-shaped form, bending at $k \simeq k_\star$. This shape arises due to the processing of density fluctuations by their evolution in the subsequent universe, as is now described.

The key observations go back to the three criteria, given at the end of Section 4.2.1.1, for when fluctuating modes can grow. These state that the fluctuations that are most important are those involving non-relativistic matter, although these remain frozen unless the universe is matter dominated and the mode number lies within the interval $H \ll k/a \ll H/c_s$. These conditions for growth superimpose a k-dependence on $P_\rho(k)$, for the following reasons.

The important wave-number $k_\star$ corresponds to the wave-number, $k_{\rm eq}$, for which modes satisfy $k/a \sim H$ at the epoch of radiation-matter equality (which occurs at $z_{\rm eq} = 3,600$). Numerically, $k_{\rm eq}$ corresponds to a co-moving wave-number of order $k_{\rm eq} \sim 0.07$ Mpc^{-1}. What is important about this scale is that it divides modes (with $k > k_{\rm eq}$) that re-enter the Hubble scale during radiation domination and those (with $k < k_{\rm eq}$) that re-enter during matter domination.

Because they re-enter during matter domination, all dark matter fluctuation modes with $k < k_{\rm eq}$ are free to begin growing immediately on re-entry and have done so ever since, at least until they either become non-linear—when $\delta_k \sim \mathcal{O}(1)$—or the universe reaches the very recent advent of dark energy domination. So the present-day power spectrum for these modes reflects the primordial one which was frozen into these modes long ago when they left the Hubble scale in the pre-ΛCDM era. It is these modes that reveal the primordial distribution

$$P(k) \propto k^{n_s} \qquad \text{(for } k \ll k_{\rm eq}). \qquad (4.111)$$

By contrast, those modes with $k \gg k_{\rm eq}$ re-enter the Hubble scale during the radiation-dominated epoch that precedes matter-radiation equality. The amplitude of these modes therefore remain frozen at their values at the time of re-entry, because they are unable to grow while the universe is radiation dominated. Consequently they remain stunted in amplitude relative to their longer-wavelength counterparts while waiting for the universe to become matter-dominated, leading to a suppression of $P_\rho(k)$ for $k \gg k_{\rm eq}$.

The *relative* stunting of large-k modes compared to small-k modes can be computed from the information that the unstable modes grow with amplitude $\delta_k(a) \propto a$ during matter domination. For $k < k_{\rm eq}$ this growth applies as soon as they cross the Hubble scale, while for $k > k_{eq}$ the modes cannot grow in this way until the transition from radiation to matter domination. As a result the relative size of two modes, one with $k_0 \ll k_{eq}$ and one with $k \gg k_{eq}$, is

$$\frac{\delta_k(a)}{\delta_{k_0}(a)} \propto \frac{\delta_k(a_k)(a/a_{eq})}{\delta_{k_0}(a_{k_0})(a/a_{k_0})} \propto \frac{\delta_k(a_k)(a/a_k)}{\delta_{k_0}(a_{k_0})(a/a_{k_0})} \left(\frac{k_{eq}}{k}\right)^2, \tag{4.112}$$

where a_k denotes the scale factor at the (k-dependent) epoch of re-entry, defined by $k = a_k H_k$. The first relation in (4.112) uses that modes in the numerator all start growing at the same time (radiation-matter equality), while those in the denominator grow for a k_0-dependent amount a/a_{k_0}. The second relation then makes the k-dependence of the suppression a_k/a_{eq} in the numerator explicit, using the matter-domination evolution $aH \propto a^{-1/2}$ in the re-entry condition to conclude $k = a_k H_k \propto a_k^{-1/2}$ and so $a_k \propto k^{-2}$.

This leads to the expectation that the power spectrum has the form $P(k) = P_{prim}(k)\,\mathcal{T}(k)$, where $P_{prim}(k) = \langle|\delta_k(a)|^2\rangle = \langle|\delta_k(a_k)|^2\rangle (a/a_k)^2$ is the primordial power spectrum and $\mathcal{T}(k)$ is the transfer function that expresses the relative stunting of modes for $k \gg k_{eq}$. Keeping in mind that $P(k) \propto |\delta_k|^2$, this discussion shows we expect $\mathcal{T}(k) \simeq 1$ for $k \ll k_{eq}$ and $\mathcal{T}(k) \simeq (k_{eq}/k)^4$ for $k \gg k_{eq}$. Given a primordial distribution $P_{prim}(k) \simeq Ak^{n_s}$ this leads to

$$P_\rho(k) \propto \begin{cases} k^{n_s} & \text{if } k \ll k_\star \\ k^{n_s-4} & \text{if } k \gg k_\star \end{cases}, \tag{4.113}$$

much as is observed.

It is noteworthy that the success of the above argument contains more evidence for the existence of dark matter. For many modes $\delta_k \simeq \mathcal{O}(1)$ occurs before the present epoch, at which point non-linear gravitational physics is expected to produce the large-scale structure actually seen in galaxy surveys. But the observed isotropy of the CMB implies the amplitude of $\delta_k(a_{rec})$ must have started off very small at the time photons last scattered from the Hydrogen gas at redshift $z_{rec} \sim 1,100$. Given this small start—that modes cannot grow before matter-radiation equality—and given that instability growth is proportional to a, a minimum amount of time is required for fluctuations to become non-linear early enough to account for the observed distribution of non-linear structure (like galaxies). Crucially, if dark matter did not exist then growth could not start until redshift z_{eq}(baryons only) $\simeq 480$—compare with (4.37)—which does not leave enough time. But the presence of dark matter moves back the epoch of radiation-matter equality to $z_{eq} \simeq 3,600$—compare with (4.36)—giving sufficient time for non-linear structure to form at the required scales.

The story of late-time fluctuations is even many richer than this, with detailed comparisons between observations and theory. A spectacular example is provided by the observation of 'baryon acoustic oscillations' (BAO), which are observed correlations between the distribution of galaxies and the distribution of CMB temperature fluctuations. The physical origin of these correlations lies in the coupled late-time evolution of perturbations in the dark matter and baryon-radiation fluid. Once fluctuations in the baryon–photon fluid begin to be free to oscillate the local dark matter evolution acts as a forcing term. This sends out a sound wave in the density of the baryon–photon fluid that initially propagates at a significant fraction of the speed of light, due to the dominance of the photon entropy in this fluid. But the speed of sound for the baryons drops like a rock once the baryons and photons decouple from one another at recombination, causing the sound wave to stall. The resulting correlation has been observed, and its properties again confirm the ΛCDM model with values for the model parameters consistent with other determinations.

4.2.2   Primordial Fluctuations from Inflation

We have discussed that fluctuations in the ΛCDM model provide a successful description of structure in the universe, but only given the initial condition of a primordial spectrum of fluctuations having a specific power-law form: $P_\rho(k) \simeq A_s k^{n_s}$ (or $\Delta_\phi^2(k) \simeq A_s k^{n_s-1}$). It again falls to the earlier universe to explain why primordial fluctuations should have this specific form, and why it should be robust against the many poorly understood details governing the physics of this earlier epoch.

It is remarkable that there is evidence that an earlier period of inflationary expansion can also explain this initial distribution of fluctuations [30]. This section provides a sketch of this evidence. Since the modes of interest start off during ΛCDM outside the Hubble length, $k \ll aH$, and are known to be small, their evolution can be tracked into earlier epochs using linear perturbation theory. Because the modes are super-Hubble in size the treatment must be relativistic, and so involves linearizing the coupled Einstein-matter field equations. The first part of this section sketches how this super-Hubble evolution works, and shows how to relate the primordial fluctuations that re-enter the Hubble scale to those that exit the Hubble scale during the inflationary epoch (see Figure 4.9).

At first sight this just pushes to problem back to an earlier time, requiring an explanation why a particular pattern of fluctuations should exist during inflation. Even worse, within the classical approximation there is good reason to believe there should be no fluctuations at all leaving at horizon exit during inflation. This is because the exponential growth of the scale factor, $a \propto e^{Ht}$, during inflation is absolutely ruthless in ironing out any spacetime wrinkles since momentum-dependent terms like $(k/a)^2$ in the field equations go to zero so quickly.

But the key words here are "within the classical approximation". Quantum fluctuations are *not* ironed away during inflation, and persist at a level proportional to the Hubble scale. Because this Hubble scale is approximately constant, the resulting fluctuations are largely scale-independent, providing a natural explanation for why primordial fluctuations seem to be close to the Zel'dovich spectrum. But H during inflation also cannot be exactly constant since inflation must end eventually. In the explicit models examined earlier the time-dependence of H arises at a level suppressed by the slow-roll parameters ϵ and η and so deviations from scale invariance should arise at the few per cent level. Because of this we shall find that the prediction for n_s in inflationary models is a bit smaller than unity, naturally agreeing with the observed value $n_s \simeq 0.97$.

4.2.2.1 Linear Evolution of Metric-Inflaton Fluctuations

The first task is to evolve fluctuations forward from the epoch of inflationary horizon exit until they re-enter during the later Hot Big Bang era. In particular our focus is on the perturbations of the metric, $\delta g_{\mu\nu}$, since these include perturbations of the Newtonian potential and so also the density fluctuations whose power spectrum is ultimately measured. The discussion here follows that of [31].

The symmetry of the FRW background allows the fluctuations of the metric to be classified by their rotational properties, with fluctuations of different spin not mixing at linear order in the field equations. Fluctuations of the metric come in three such kinds: *scalar*, *vector*, and *tensor*. Specializing to a spatially flat FRW background and transforming to conformal time, $\tau = \int dt/a$, the scalar perturbations may be written

$$\delta_S g_{\mu\nu} = a^2 \begin{pmatrix} 2\phi & \partial_j \mathcal{B} \\ \partial_i \mathcal{B} & 2\psi\, \delta_{ij} + \partial_i \partial_j \mathcal{E} \end{pmatrix}, \tag{4.114}$$

while the vector and tensor ones are

$$\delta_V g_{\mu\nu} = a^2 \begin{pmatrix} 0 & \mathcal{V}_j \\ \mathcal{V}_i & \partial_i \mathcal{W}_j + \partial_j \mathcal{W}_i \end{pmatrix} \quad \text{and} \quad \delta_T g_{\mu\nu} = a^2 \begin{pmatrix} 0 & 0 \\ 0 & h_{ij} \end{pmatrix}. \tag{4.115}$$

Here all vectors are divergence free, as is the tensor (which is also traceless). To these are added the fluctuations in the inflaton field, $\delta\varphi$.

There is great freedom to modify these functions by performing infinitesimal coordinate transformations, so it is useful to define the following combinations that are invariant at linearized order:

$$\Phi = \phi - \frac{1}{a}\Big[a(\mathcal{B} - \mathcal{E}')\Big]', \qquad \Psi = \psi + \frac{a'}{a}(\mathcal{B} - \mathcal{E}') \tag{4.116}$$
$$\delta\chi = \delta\varphi - \varphi'(\mathcal{B} - \mathcal{E}'), \quad V_i = \mathcal{V}_i - \mathcal{W}_i \quad \text{and} \quad h_{ij},$$

in terms of which all physical inferences can be drawn. Here primes denote differentiation with respect to conformal time, τ. Notice that Φ, Ψ, and V_i reduce to ϕ, ψ, and $\mathcal{V}_i$ in the gauge choice where $\mathcal{B} = \mathcal{E} = \mathcal{W}_i = 0$, and so Φ is the relativistic generalization of the Newtonian potential.

These functions are evolved forward in time by linearizing the relevant field equations:

$$\Box\varphi - V'(\varphi) = 0 \quad \text{and} \quad R_{\mu\nu} - \frac{1}{2} R g_{\mu\nu} = \frac{T_{\mu\nu}}{M_p^2}, \tag{4.117}$$

and provided we use the invariant stress-energy perturbations,

$$\delta\mathcal{T}^0{}_0 = \delta T^0{}_0 - \left[t^0{}_0\right]'(\mathcal{B} - \mathcal{E}'),$$

$$\delta\mathcal{T}^0{}_i = \delta T^0{}_i - \left[t^0{}_0 - \frac{1}{3}t^k{}_k\right]\partial_i(\mathcal{B} - \mathcal{E}'), \tag{4.118}$$

$$\delta\mathcal{T}^i{}_j = \delta T^i{}_j - \left[t^i{}_j\right]'(\mathcal{B} - \mathcal{E}'),$$

(where $t^\mu{}_\nu$ denotes the background stress-energy), the results can be expressed purely in terms of the gauge-invariant quantities, eqns (4.116).

The equations which result show that in the absence of vector stress-energy perturbations (i.e. if $\delta T^0{}_i$ is a pure gradient—as would be the case for perturbed inflaton), then vector perturbations, V_i, are not sourced, and decay very rapidly in an expanding universe, allowing them to be henceforth ignored. Similarly, in the absence of off-diagonal stress-energy perturbations (i.e. if $\delta T^i{}_j = \delta p\,\delta^i_j$) it is also generic that $\Psi = \Phi$.

Switching back to FRW time, the equations which govern the evolution of tensor modes then become (after Fourier transforming)

$$\ddot{h}_{ij} + 3H\dot{h}_{ij} + \frac{k^2}{a^2}\,h_{ij} = 0, \tag{4.119}$$

showing that these evolve independent of all other fluctuations. Such primordial tensor fluctuations can be observable if they survive into the later universe, since the differential stretching of space-time that they predict can contribute observably to the polarization of CMB photons that pass through them. The search for evidence for this type of primordial tensor fluctuations is active and ongoing, and (as is shown below) is expected in inflation to be characterized by a near scale-invariant tensor power spectrum,

$$P_h(k) = A_T\,k^{n_T}, \tag{4.120}$$

with n_T close to zero.

The equations evolving the scalar fluctuations are more complicated and similarly reduce to

$$\delta\ddot{\chi} + 3H\delta\dot{\chi} + \frac{k^2}{a^2}\delta\chi + V''(\varphi)\delta\chi - 4\dot{\varphi}\,\dot{\Phi} + 2V'(\varphi)\,\Phi = 0$$

$$\text{and}\quad \dot{\Phi} + H\,\Phi = \frac{\dot{\varphi}}{2M_p^2}\,\delta\chi. \tag{4.121}$$

The homogeneous background fields themselves satisfy the equations

$$\ddot{\varphi} + 3H\dot{\varphi} + V'(\varphi) = 0 \quad\text{and}\quad 3M_p^2 H^2 = \frac{1}{2}\dot{\varphi}^2 + V(\varphi). \tag{4.122}$$

These expressions show that although Φ and $\delta\chi$ would decouple from one another if expanded about a static background (for which $\dot{\varphi} = V' = 0$), they do not when the background is time-dependent.

4.2.2.2 *Slow-Roll Evolution of Scalar Perturbations*

The character of the solutions of these equations depends strongly on the size of k/a relative to H, since this dictates the extent to which the frictional terms can compete with the spatial derivatives. As usual the two independent solutions for $\delta\chi$ that apply when $k/a \gg H$ describe damped oscillations

$$\delta\chi_k \propto \frac{1}{a\sqrt{k}} \exp\left[\pm ik \int^t \frac{dt'}{a(t')}\right]. \tag{4.123}$$

Our interest during inflation is in the limit $k/a \ll H$ in a slow-roll regime for which $\delta\ddot{\chi}$, $\ddot{\varphi}$, and $\dot{\Phi}$ can all be neglected. In this limit the scalar evolution equations simplify to

$$3H\delta\dot{\chi} + V''(\varphi)\delta\chi + 2V'(\varphi)\Phi \simeq 0 \quad \text{and} \quad 2M_p^2 H \Phi \simeq \dot{\varphi}\delta\chi, \tag{4.124}$$

and have approximate solutions (after Fourier transformation) of the form

$$\delta\chi_k \simeq C_k \frac{V'(\varphi)}{V(\varphi)} \quad \text{and} \quad \Phi_k \simeq -\frac{C_k}{2}\left(\frac{V'(\varphi)}{V(\varphi)}\right)^2. \tag{4.125}$$

where C_k is a (potentially k-dependent) constant of integration. Since the background fields satisfy $M_p V'/V = \sqrt{2\epsilon}$ these equations show how the amplitude of $\delta\chi_k$ and Φ_k during inflation track the evolution of the slow-roll parameter, ϵ, for super-Hubble modes, and therefore tend to grow in amplitude as inflation eventually draws to a close.

We have two remaining problems: first, what is the origin of the initial fluctuations at horizon exit; and second, how do we evolve fluctuations from the end of inflation through to the later epoch of horizon re-entry? The latter of these seems particularly vicious since it a priori might be expected to depend on the many details involved in getting the universe from its inflationary epoch to the later Hot Big Bang.

4.2.2.3 *Post-Inflationary Evolution*

For the case of single-field inflation discussed here, post-inflationary evolution of the fluctuation Φ actually turns out to be quite simple. This is because it can be shown that when $k \ll aH$ the quantity

$$\zeta = \Phi + \frac{2}{3}\left(\frac{\Phi + \dot{\Phi}/H}{1 + w}\right) = \frac{1}{3(1 + w)}\left[(5 + 3w)\Phi + \frac{2\dot{\Phi}}{H}\right], \tag{4.126}$$

is *conserved*, inasmuch as $\dot{\zeta} \simeq 0$ for $k \to 0$.

This result follows schematically because the perturbed metric can be written as proportional to $e^\zeta g_{ij}$ and so spatially constant ζ is indistinguishable from the background scale factor, $a(t)$. Conservation has been proven under a wide variety of assumptions [31, 33], but the form used here assumes that the background cosmology satisfies an equation of state $p = w\rho$, but w is *not* assumed to be constant. The same result is known not to be true if there were more than a single scalar field evolving.

Conservation of ζ is a very powerful result because it can be used to evolve fluctuations using $\zeta(t_i) = \zeta(t_f)$, assuming only that they involve a single scalar field, and that the modes in question are well outside the horizon: $k/a \ll H$. Furthermore, although $\dot{\Phi}$ in general becomes non-zero at places where w varies strongly with time, this time-dependence quickly damps due to Hubble friction for modes outside the Hubble scale.

Therefore, for most of the universe's history we also may neglect the dependence of ζ on $\dot{\Phi}$, provided we restrict t_i and t_f to epochs during which w is roughly constant. This allows the expression $\zeta(t_i) = \zeta(t_f)$ to be simplified to

$$\Phi_f = \frac{1 + w_f}{1 + w_i} \left(\frac{5 + 3w_i}{5 + 3w_f} \right) \Phi_i, \tag{4.127}$$

where $w_i = w(t_i)$ and $w_f = w(t_f)$, implying in particular $\Phi_f = \Phi_i$ whenever $w_i = w_f$. Similarly, the values of Φ deep within radiation- and matter-dominated phases are related by $\Phi_{\text{mat}} \simeq \frac{9}{10} \Phi_{\text{rad}}$.

To infer the value of Φ in the later Hot Big Bang era we choose t_i just after horizon exit (where a simple calculation shows $w_i \simeq -1 + \frac{2}{3} \epsilon_{\text{he}}$, with ϵ_{he} the slow-roll parameter at horizon exit). t_f is then chosen in the radiation-dominated universe (where $w_f = \frac{1}{3}$), either just before horizon re-entry for the mode of interest, or just before the transition to matter domination, whichever comes first. Equations (4.125) and (4.127) then imply

$$\Phi_f \simeq \left(\frac{2\Phi}{3\epsilon} \right)_{\text{he}}. \tag{4.128}$$

It remains to grapple with what should be expected for the initial condition for Φ at horizon exit.

4.2.2.4 *Quantum Origin of Fluctuations*

The primordial fluctuation amplitude derived in this way depends on the integration constants C_k, which are themselves set by the initial conditions for the fluctuation at horizon exit, during inflation. But why should this amplitude be non-zero given that all previous evolution is strongly damped, as in eqn (4.123)? The result remains non-zero (and largely independent of the details of earlier evolution) because quantum fluctuations in $\delta\chi$ continually replenish the perturbations long after any initial classical configurations have damped away.

The starting point for the calculation of the amplitude of scalar perturbations is the observation that the inflaton and metric fields whose dynamics we are following are quantum fields, not classical ones. For instance, for spatially flat space-times the linearized inflaton field, $\delta\chi$, is described by the operator

$$\delta\chi(x) = \int \frac{d^3 k}{(2\pi)^3} \Big[c_k u_k(t)\, e^{i\mathbf{k}\cdot\mathbf{r}/a} + c_k^* u_k^*(t)\, e^{-i\mathbf{k}\cdot\mathbf{r}/a} \Big], \tag{4.129}$$

where the expansion is in a basis of eigenmodes of the scalar field equation in the background metric, $u_k(t)\, e^{i\mathbf{k}\cdot\mathbf{x}}$, labelled by the co-moving momentum $\mathbf{k}$. For constant H the time-dependent mode functions are

$$u_k(t) \propto \frac{H}{k^{3/2}} \left(i + \frac{k}{aH} \right) \exp\left(\frac{ik}{aH} \right), \tag{4.130}$$

which reduces to the standard flat-space form, $u_k(t) \propto a^{-1} k^{-1/2} e^{-ik \int dt/a}$, when $k/a \gg H$. [This is perhaps easiest to see using conformal time, for which $\exp(ik/aH) = \exp(-ik\tau)$, or more directly by using $\exp\left(-ik \int dt/a\right) = \exp\left(ik/aH\right)$ when $a \propto e^{Ht}$.] The quantities c_k and their adjoints c_k^* are *annihilation* and *creation operators*, which define the adiabatic vacuum state, $|\Omega\rangle$, through the condition $c_k|\Omega\rangle = 0$ (for all $\mathbf{k}$).

The $\delta\chi$ auto-correlation function in this vacuum, $\langle\delta\chi(x)\delta\chi(x')\rangle$, describes the quantum fluctuations of the field amplitude in the quantum ground state, and the key assumption is that the quantum statistics of the mode leaving the horizon during inflation agrees with the classical fluctuations of the field $\delta\chi$ after evolving outside of the Hubble scale. This assumes the quantum fluctuations to be decohered (for preliminary discussions see [34, 35]) into classical distribution for $\delta\chi$ sometime between horizon exit and horizon re-entry.

It turns out that during inflation interactions with the bath of short-wavelength, sub-Hubble modes is extremely efficient at decohering the quantum fluctuations of long-wavelength, super-Hubble modes [36]. As is usual when a system is decohered through interactions with an environment, the resulting classical distribution is normally defined for the 'pointer basis', that diagonalizes the interactions with the environment. It turns out that the freezing of super-Hubble modes has the effect of making them very classical (WKB-like), and so ensure the fields' canonical momenta become functions of the fields themselves. This ensures that it is always the field basis that diagonalizes any local interactions, and so guarantees that quantum fluctuations become classical fluctuations for the fields (like $\delta\chi$), rather than (say) their canonical momenta.

The upshot is that after several e-foldings, even very weak interactions (like gravitational strength ones) eventually convert quantum fluctuations into classical statistical fluctuations for the classical field, φ, about its spatial mean. For practical purposes, this means in the above calculations we simply can use the initial condition $|\delta\chi_k| \sim [\langle\delta\chi_k\delta\chi_{-k}\rangle]^{1/2} \propto |u_k(t)|$. For observational purposes, what matters is that the classical variance of these statistical fluctuations is well described by the corresponding quantum auto-correlations—a property that relies on the kinds of 'squeezed' quantum states that arise during inflation [31, 40].

Evaluating $\delta\chi_k \sim u_k$ at t_{he} (where $k = aH$) and equating the result to the fluctuation of eqn (4.125) allows the integration constant in this equation to be determined to be

$$C_k = u_k(t_{he}) \left(\frac{V}{V'}\right)_{\varphi_{he}}, \qquad (4.131)$$

where both t_{he} and $\varphi_{he} = \varphi(t_{he})$ implicitly depend on k. Using this to compute Φ_k in eqn (4.125) then gives

$$\Phi_k(t) = -\frac{1}{2}u_k(t_{he}) \left(\frac{V}{V'}\right)_{\varphi_{he}} \left(\frac{V'}{V}\right)^2_{\varphi(t)} = -\epsilon(t) \left(\frac{u_k}{\sqrt{2\epsilon}\,M_p}\right)_{t_{he}}. \qquad (4.132)$$

In particular, evaluating at $t = t_{he}$ then gives

$$\Phi_k(t_{he}) = -\left(\frac{u_k}{M_p}\sqrt{\frac{\epsilon}{2}}\right)_{t_{he}}. \qquad (4.133)$$

4.2.2.5 Predictions for the Scalar Power Spectrum

We are now in a situation to pull everything together and compute in more detail the inflationary prediction for the properties of the primordial fluctuation spectrum. Using (4.133) in (4.128) gives

$$\Phi_k(t_f) \simeq \left(\frac{2\Phi}{3\epsilon}\right)_{he} = -\left(\frac{2u_k}{3\sqrt{2\epsilon}\,M_p}\right)_{t_{he}}. \qquad (4.134)$$

Using this in the definition of the dimensionless power spectrum for Φ, $\Delta_\Phi^2 = k^3 P_\Phi/(2\pi^2)$, then leads to

$$\Delta_\Phi^2(k) = \frac{k^3|\Phi_k(t_f)|^2}{2\pi^2} \propto \frac{k^3|u_k(t_{he})|^2}{\pi^2\epsilon(\varphi_{he})M_p^2}. \qquad (4.135)$$

Once the order-unity factors are included we find

$$\Delta_\Phi^2(k) = \frac{k^3 P_\Phi(k)}{2\pi^2} = \left(\frac{H^2}{8\pi^2 M_p^2\,\epsilon}\right)_{he} = \left(\frac{V}{24\pi^2 M_p^4\,\epsilon}\right)_{he}, \qquad (4.136)$$

It is the quantity V/ϵ evaluated at Hubble exit that controls the amplitude of density fluctuations, and so is to be compared with the observed power spectrum of scalar density fluctuations,

$$\Delta_\Phi^2(k) = \Delta_\Phi^2(\hat{k}) \left(\frac{k}{\hat{k}}\right)^{n_s},$$ (4.137)

where $n_s = 0.968 \pm 0.006$ [4] and

$$\Delta_\Phi^2(\hat{k}) = 2.28 \times 10^{-9},$$ (4.138)

is the amplitude evaluated at the reference 'pivot' point $k = \hat{k} \sim 7.5\, a_0 H_0$. In terms of V this implies

$$\left(\frac{V}{\epsilon}\right)^{1/4}_{\text{pivot}} = 6.6 \times 10^{16} \text{ GeV},$$ (4.139)

for the epoch when the pivot scale underwent Hubble exit. The smaller ϵ becomes, the smaller the required potential energy during inflation. For $\epsilon \sim 0.01$ we have $V \sim 2 \times 10^{16}$ GeV. This is titillatingly close to the scale where the couplings of the three known interactions would unify in GU models, which may indicate a connection between the physics of GU and inflation.[17]

Notice also that the size of $\Delta_\Phi^2(k)$ is set purely by H and ϵ at horizon exit, and these only depend weakly on k (through their weak dependence on time) during near-exponential inflation. This is what ensures the approximate scale-invariance of the primordial power spectrum which inflation predicts for the later universe. To pin down the value of n_s more precisely notice that the power-law form of (4.137) implies

$$n_s - 1 \equiv \left.\frac{\mathrm{d}\ln \Delta_\Phi^2}{\mathrm{d}\ln k}\right|_{\text{he}}.$$ (4.140)

To evaluate this during slow-roll inflation use the condition $k = aH$ (and the approximate constancy of H during inflation) to write $\mathrm{d}\ln k = H\mathrm{d}t$. Since the right-hand side of eqn (4.136) depends on k and t only through its dependence on φ, it is convenient to use the slow-roll equations, eqn (4.72) to further change variables from t to φ: $\mathrm{d}t = \mathrm{d}\varphi/\dot{\varphi} \simeq -(3H/V')\mathrm{d}\varphi$, and so

$$\frac{\mathrm{d}}{\mathrm{d}\ln k} = -M_p^2 \left(\frac{V'}{V}\right) \frac{\mathrm{d}}{\mathrm{d}\varphi} = \sqrt{2\epsilon}\, M_p \frac{\mathrm{d}}{\mathrm{d}\varphi}.$$ (4.141)

Performing the φ derivative using (4.136) finally gives the following relation between n_s and the slow-roll parameters, ϵ and η

---

[17] Of course, V can be much smaller if ϵ is smaller as well, or if primordial fluctuations actually come from another source.

$$n_s - 1 = -6\epsilon + 2\eta, \tag{4.142}$$

where the right-hand side is evaluated at $\varphi = \varphi_{\text{he}}$. For single-field models the right-hand side is negative and typically of order 0.01, agreeing well with the measured value $n_s \simeq 0.97$.

4.2.2.6 Tensor Fluctuations

A similar story applies for the tensor fluctuations, but without the complications involving mixing between $\delta\chi$ and Φ. Tensor modes are also directly generated by quantum fluctuations, in this case where the vacuum is the quantum state of the graviton part of the Hilbert space. Although tensor fluctuations have not yet been observed, they are potentially observable through the polarization effects they produce as CMB photons propagate through them to us from the surface of last scattering.

Just like for scalar fluctuations, for each propagating mode the amplitude of fluctuations in the field h_{ij} is set by $H/(2\pi)$, but because there is no longer a requirement to mix with any other field (unlike Φ, which because it does not describe a propagating particle state has to mix with the fluctuating field $\delta\chi$), the power spectrum for tensor perturbations depends only on H^2 rather than on H^2/ϵ. Repeating the above arguments leads to the following dimensionless tensor power spectrum

$$\Delta_h^2(k) = \frac{8}{M_p^2}\left(\frac{H}{2\pi}\right)^2 = \frac{2V}{3\pi^2 M_p^4}. \tag{4.143}$$

This result is again understood to be evaluated at the epoch when observable modes leave the horizon during inflation, $\varphi = \varphi_{\text{he}}$.

Should both scalar and tensor modes be measured, a comparison of their amplitudes provides a direct measure of the slow-roll parameter ϵ. This is conventionally quantified in terms of a parameter r, defined as a ratio of the scalar and tensor power spectra

$$r := \frac{\Delta_h^2}{\Delta_\Phi^2} = 16\epsilon. \tag{4.144}$$

The absence of evidence for these perturbations to date places an upper limit: $r \lesssim 0.07$ [37] and so $\epsilon \lesssim 0.004$. Because ϵ appears to be so small, the measured value for n_s used with (4.142) permits an inference of how large η can be. Fitting to a global data set gives a less-stringent bound on r [38, 39], leading to

$$\epsilon < 0.012 \text{ (95\% CL)}; \quad \text{and} \quad \eta = -0.0080 \, ^{+0.0080}_{-0.0146} \text{ (68\% CL)}, \tag{4.145}$$

and it is this incipient evidence that $\epsilon \neq \eta$ that drives the tension with some of the model predictions described earlier. This information is given pictorially in Figure 4.12.

The detection of tensor modes in principle also allows a measurement of the k dependence of their power spectrum. This is usually quantified in terms of a tensor spectral index, n_T, defined by eqn (4.120) and so

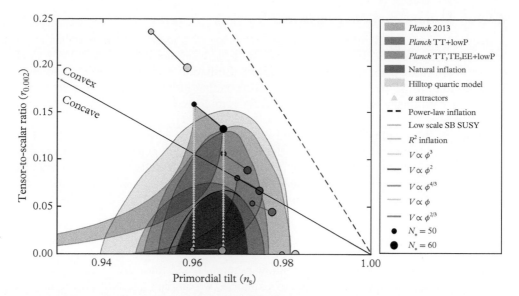

Figure 4.12 *A comparison of a variety of inflationary models to a suite of cosmological measurements (taken from [39]), with the ellipses showing the observationally preferred values for the scalar-to-tensor ratio, r, and primordial 'tilt', n_s. Each model is portrayed as giving a range of values rather than a single point, with N_e chosen for a variety of assumptions about the nature of reheating (compare with eqn (4.65)). The model labelled 'Natural inflation' corresponds to the 'pseudo-Goldstone axion' model described in the text, while the ones called 'α-attractors' represent the 'pseudo-Goldstone dilaton' models.*

$$n_T \equiv \frac{\mathrm{d}\ln \Delta_h^2}{\mathrm{d}\ln k} = -2\epsilon = -\frac{r}{8}, \qquad (4.146)$$

where the second-last equality evaluates the derivative within inflation as before by changing variables from k to φ.

Ultimately single-field models have three parameters: ϵ, η and the Hubble scale during inflation, H_I. But the scalar and tensor fluctuation spectra provide four observables: A_s, A_T, n_s and n_T. The ability to describe four observables using just three parameters implies a predicted relation amongst the observables: $n_T = -r/8$ (as seen from (4.146)). This is a robust prediction shared by all single-field slow-roll inflationary models.

4.2.3 Flies in the Ointment

Although not really the main line of development of this chapter, it would be wrong to leave the impression that inflationary theories must be the last word in early universe cosmology. Indeed they have problems that motivate some to seek out better alternatives [41]. Here are a few of the main complaints.

4.2.3.1 Initial Condition Problems

A major motivation for inflation comes from trying to understand the peculiar initial conditions required for the success of the late-time cosmology of the ΛCDM model. But this cannot be regarded as being a success if inflation itself also requires contrived initial conditions. In particular, there are concerns that inflation might not start unless the universe is initially prepared in a sufficiently homogeneous configuration over several Hubble scales.

Although the fragility of the required initial conditions is in dispute [42], it is true that there are not many explicit calculations done with more generic initial conditions. There are calculations involving more generic random potentials [43] that do indicate that inflation can be a rare occurrence, but it is still being explored how much these calculations depend on the assumptions being made.

4.2.3.2 Fine-Tuning Problems

Slow-roll inflation requires relatively shallow potentials, and these are relatively difficult to obtain within the low-energy limit of explicit UV completions. Not all models are equally bad in this regard, with those based on pseudo-Goldstone bosons being able to arrange shallow potentials in more controlled ways.

But even with these models inflationary predictions are notoriously sensitive to small effects. Because the inflationary effect being sought is gravitational, it is Planck suppressed and so can be threatened even by other Planck-suppressed effective interactions that in any other circumstances would have been regarded as negligible.

4.2.3.3 The Multiverse and the Landscape

Once it gets going, inflation can be hard to stop. And even if it ends in some parts of the universe, if it survives in others these inflating regions expand so quickly that they can come to dominate the volume of the universe. These kinds of effects are made even worse if physical constants are really controlled by the expectation values of fields that can be different in different parts of the universe, as seems to be the case in theories like string theory.

For such theories it becomes difficult to see how to make definite predictions in a traditional way. A great variety of universes might arise within any given framework, and how do we falsify such a theory if the options available become too numerous? One way out is to use anthropic reasoning, but it is not yet clear what the proper rules should be for doing so.

One point that is worth making is that problems with the multiverse (if problems they prove to be) are actually common to pretty much any cosmological framework in theories that admit a complicated landscape of solutions. That is because even if we think early-universe cosmology is described by something besides inflation (such as a bouncing cosmology) all the above predictivity problems in any case arise if inflation nevertheless should unintentionally get going in any remote corner of the landscape.

4.2.3.4   My Two CENTS

My own opinion is to accept that inflationary models are a work in progress, leaving many things to be desired. Regardless, they seem at this point to be in better shape than all of their alternatives, mostly because of the control they allow over all of the approximations being made during their use. This situation could change as alternatives are better explored (as they should certainly be), but the shortage of convincing alternatives shows that inflation already sets a fairly high bar for other theories to pass.

Although it may be premature to speculate about issues of the multiverse that are difficult to compare with observations, inflationary models seem to give a clean answer to the more limited practical question of what kind of extrapolation could be useful into our relatively immediate pre-Big Bang past. Their predictions seem to be under good theoretical control and to agree well with the properties of the primordial fluctuations that have so far been revealed.

4.3   EFT Issues

It may not yet be clear how EFT methods enter into the beautiful story presented thus far, but this section argues EFT methods are actually used throughout (as is also typically true essentially everywhere else in physics). Since these lectures were delivered in a school entirely devoted to EFTs, the logic of this section is not to explain what an EFT is (such as they arise in areas like chiral perturbation theory), but rather to sketch some of the issues that come up when they are applied to gravity- and cosmology-specific problems.

In my opinion the lesson of these applications is twofold. First, there is no evidence (yet) for 'gravitational exceptionalism'—the idea that there is nothing to learn about gravity from experience with other interactions because gravity is fundamentally different. The second lesson is that EFT applications to gravity can sometimes more resemble effective descriptions of particles moving through a medium than they do the traditional uses of EFTs in particle physics. As such, they can be mind broadening to those of us who approach the subject with a particle physics training. Each of the following sections addresses different kinds of examples of this.

4.3.1   Effective Field Theories and Gravity

The most important use of EFT methods in gravity-related problems is the one described here: the justification of the semiclassical approximation that underpins almost all theoretical studies of gravity, including cosmology. Although it is common to think of gravitational interactions as being classical, a question less often asked is why this is true (and, if so, what the small parameter is that suppresses quantum effects).

The claim made here is that the issues for gravitational systems in may ways resemble those arising in non-linear sigma-models,

$$\mathcal{L} = -\frac{f^2}{2}\, G_{ij}(\phi)\, \partial_\mu \phi^i\, \partial^\mu \phi^j, \tag{4.147}$$

such as describe Goldstone (and pseudo-Goldstone) bosons (including those studied in chiral perturbation theory). This similarity arises because both are non-renormalizable, in that their interactions involve inverse powers of a mass scale (f for the sigma-model and M_p for gravity) and both are dominated at low energies by interactions involving only two derivatives but many powers of the interacting fields.

Both of these properties lose their power to paralyse once it is recognized that the action should really also include all possible kinds of higher-derivative interactions, and it is recognized that predictive power is only possible for low-energy observables relative to f (or M_p). For gravity this leads us to regard GR as the leading part of what might be called (in analogy to the Standard Model EFT, or SMEFT) the general relativity EFT, or GREFT.

4.3.1.1   *GREFT*

To see how this works in detail for gravity, we apply to GR the same steps seen in other chapters. (For reviews on GR as an EFT see [2, 44].)

The low-energy degrees of freedom in this case are gravitons, whose field is the metric, $g_{\mu\nu}$, of space-time itself. The low-energy symmetries that constrain the form of the action are general covariance and local Lorentz invariance. Invariance under these symmetries dictate the metric can appear in the action only through curvature invariants built from the Riemann tensor and its contractions and covariant derivatives. The Riemann curvature tensor is defined by

$$R^{\mu}{}_{\nu\rho\lambda} = \partial_\lambda \Gamma^{\mu}_{\nu\rho} + \Gamma^{\mu}_{\lambda\alpha}\Gamma^{\alpha}_{\nu\rho} - (\lambda \leftrightarrow \rho) \tag{4.148}$$

$$\text{with} \quad \Gamma^{\mu}_{\nu\lambda} = \frac{1}{2}g^{\mu\beta}\left(\partial_\nu g_{\beta\lambda} + \partial_\lambda g_{\beta\nu} - \partial_\beta g_{\nu\lambda}\right). \tag{4.149}$$

and its only independent contractions are the Ricci curvature tensor $R_{\mu\nu} = R^{\alpha}{}_{\mu\alpha\nu}$ and its trace $R = g^{\mu\nu}R_{\mu\nu}$, where the inverse metric, $g^{\mu\nu}$, satisfies $g^{\mu\nu}g_{\nu\lambda} = \delta^{\mu}_{\lambda}$. What is important in what follows about these definitions is that, although complicated, the curvature tensors involve precisely two derivatives of the metric.

GREFT is defined (as usual) by writing down a local action involving all possible powers of derivatives of the metric, which general covariance then requires must be built from powers of the curvature tensors and their derivatives. This leads to the following effective Lagrangian:

$$-\frac{\mathcal{L}_{\text{GREFT}}}{\sqrt{-g}} = \lambda + \frac{M_p^2}{2}R$$
$$+ c_{41}R_{\mu\nu}R^{\mu\nu} + c_{42}R^2 + c_{43}R_{\mu\nu\lambda\rho}R^{\mu\nu\lambda\rho} + c_{44}\Box R \tag{4.150}$$
$$+ \frac{c_{61}}{M^2}R^3 + \frac{c_{62}}{M^2}\partial_\mu R \partial^\mu R + \cdots,$$

where $\sqrt{-g} = \sqrt{-\det g_{\mu\nu}}$, as usual. The first line here includes all possible terms involving two or fewer derivatives, and is the Einstein–Hilbert action of GR, with cosmological constant λ. The second line includes all possible terms involving precisely four derivatives, and (for brevity) the third line includes only the first two representative examples of the many possible terms involving six or more derivatives.

The first, cosmological constant, term in eqn (4.150) is the only one with no derivatives. Its appearance complicates power counting arguments (in much the same was as does the appearance of a scalar potential when power counting with a sigma-model—more about this later). Such terms cause problems if their coefficients are too large (such as similar in size as for the two-derivative terms), and the good news is that if λ is regarded as being the dark energy it is measured to be extremely small. The puzzle as to *why* this happens to be true is a well-known, unsolved problem [45]. For simplicity of presentation the cosmological constant term is simply dropped in the power counting argument that follows, although it returns once scalars and their potential are considered in later sections. Once this is done the leading term in the derivative expansion is the Einstein–Hilbert term of GR. Its coefficient defines Newton's constant (and so also the Planck mass, $M_p^{-2} = 8\pi G$).

The constants c_{dn} are dimensionless couplings, with the convention that d counts the number of derivatives of the corresponding effective operator and $n = 1, \cdots, N_d$ runs over the number of such couplings. These couplings are dimensionless because the explicit mass scales, M and M_p, are extracted to ensure this is so. Often we see this action written with only the Planck scale appearing, i.e. with $M \sim M_p$. However, as is usual in an EFT, the scale M is usually of order the lightest particle integrated out to produce this effective theory, leaving only the metric as the variable. Since it is the *smallest* such a mass that dominates, M is generically expected to be much smaller than M_p. (For applications to the solar system M might be the electron mass; for applications to post-nucleosynthesis Big Bang cosmology M might be of order the QCD scale, and so on.) Of course, contributions like $M^2 R$ or R^3/M_p^2 could also exist, but these are completely negligible compared to the terms displayed in eqn (4.150). The central point of EFT methods is that the consequences of (4.150) should be explored as low-energy expansion in powers of q/M and q/M_p, where q is a typical energy/momentum or curvature scale characterizing the observables of interest.

Redundant Interactions

Just as is true in SMEFT, to save needless effort we should eliminate those redundant interactions that can be removed by integrating by parts or performing a field redefinition. The freedom to drop total derivatives allows us to set the coupling c_{44} to zero, as well (in four dimensions) as c_{43}. (For c_{44} this can be done because $\sqrt{-g}\,\Box R$ is a total derivative, and for c_{43} the relevant observation is that the quantity

$$\sqrt{-g}\,X = \sqrt{-g}\Big(R_{\mu\nu\lambda\rho}R^{\mu\nu\lambda\rho} - 4R_{\mu\nu}R^{\mu\nu} + R^2\Big), \tag{4.151}$$

integrates to give a topological invariant in four dimensions, and so is locally also a total derivative. It is therefore always possible to replace, for example, $R_{\mu\nu\lambda\rho}R^{\mu\nu\lambda\rho}$ in the four-derivative effective Lagrangian with the linear combination $4R_{\mu\nu}R^{\mu\nu} - R^2$, with no consequences for any observables, provided these observables are insensitive to the overall topology of space-time (such as are the classical equations, or perturbative particle interactions).

As discussed in other chapters (see also [2]), the freedom to perform field redefinitions allows the dropping of any terms that vanish when evaluated at solutions to the lowest-order equations of motion. This freedom allows the removal of the other two four-derivative terms because (in the absence of other, matter, fields) the lowest order equations of motion are $R_{\mu\nu} = 0$, and the remaining terms vanish when this is imposed. For pure gravity (without a cosmological constant) the first non-trivial effective interaction involves more than four derivatives, such as the term proportional to the cube of the Riemann tensor. This irrelevance of all of the four-derivative terms must be re-examined once matter fields are included, however, since once these are included $R_{\mu\nu}$ need no longer vanish.

4.3.1.2 *Power Counting (Gravity Only)*

In any EFT the central question is which interactions are relevant when computing observables at a specific order in the low-energy expansion in powers of q/M and q/M_p. Because of the similarity in the structure of derivatives appearing in sigma models and GR, power counting for the two types of theories is very similar. This section briefly recaps the result without repeating the details (see however [2]), highlighting those features that differ.

Start by considering the interactions of gravitons propagating in flat space (returning to curved space below). In this case we expand[18] $g_{\mu\nu} = \eta_{\mu\nu} + h_{\mu\nu}$ and identify propagators and interactions for perturbative calculations in the usual way. For the purposes of this power counting all we need to know about the curvatures is that they each involve all possible powers of $h_{\mu\nu}$, but with only precisely two derivatives. Consider an arbitrary graph that contributes at L loops to the amputated[19] E-point graviton-scattering amplitude, $\mathcal{A}_E(q)$, performed with energy q. Suppose also the graph contains V_{id} vertices involving d derivatives and the emission or absorption of i gravitons. Using arguments identical to those used for sigma models elsewhere leads to the following dependence[20] of $\mathcal{A}_E(q)$ on the scales q, M and M_p:

$$\mathcal{A}_E(q) \sim q^2 M_p^2 \left(\frac{1}{M_p}\right)^E \left(\frac{q}{4\pi M_p}\right)^{2L} \prod_i \prod_{d>2} \left[\frac{q^2}{M_p^2}\left(\frac{q}{M}\right)^{(d-4)}\right]^{V_{id}}. \qquad (4.152)$$

---

[18] A factor proportional to $1/M_p$ would appear with $h_{\mu\nu}$ if fluctuations were to be canonical normalized, but this normalization is not required in what follows.

[19] Amputation means that the graphs have no external lines, such as might be encountered when computing the size of coefficients in a low-energy effective action.

[20] Technical point: as is usually the case this power counting result is computed in dimensional regularization, since not including a spurious cutoff scale makes arguments based on dimensional analysis particularly simple.

Notice that since d is even for all of the interactions, the condition $d > 2$ in the product implies there are no negative powers of q in this expression.

Equation (4.152) shows that the weakness of a graviton's coupling (much like the weak couplings of a Goldstone boson) comes purely from the low-energy approximations, $q \ll M_p$ and $q \ll M$. It is also clear that even though the ratio q/M could be much larger than q/M_p, it only arises in $\mathcal{A}_E$ together with a factor of q^2/M_p^2, and only when including interactions with $d \geq 6$ (coming from curvature-cubed interactions or higher).

Furthermore (4.152) shows that the dominant contributions to low-energy graviton scattering amplitudes correspond to graphs with $L = 0$ and $V_{id} = 0$ for all $d > 2$. That is to say, built using only tree graphs constructed purely from the Einstein–Hilbert ($d = 2$) action: it is classical GR that governs the low-energy dynamics of gravitational waves.

But EFTs excel when computing next-to-leading contributions. In this case these are:

- $L = 1$ and $V_{id} = 0$ for any $d \neq 2$ but V_{i2} is arbitrary, or
- $L = 0$, $\sum_i V_{i4} = 1$, V_{i2} is arbitrary, and all other V_{id} vanish.

That is, the next-to-leading contribution is found using one-loop graphs with only the interactions of GR, or by working to tree level and including precisely one insertion of a curvature-squared interaction in addition to any number of interactions from GR. Both of these are suppressed compared to the leading term by a factor of $(q/M_p)^2$. The next-to-leading tree graphs provide precisely the counter-terms required to absorb the UV divergences in the one-loop graphs. And so on.

What this shows is that the small parameter that controls the loop expansion (i.e. the semiclassical expansion) for graviton scattering is the ratio $q^2/(4\pi M_p)^2$; the semiclassical approximation *is* the low-energy approximation.

But the above argument was made specifically for gravitons propagating in flat space. How reliable should these power counting arguments be for drawing conclusions for more general curved environments? Related to this, how important is it to be able to work in momentum space, as is usually done in sigma-model type arguments (and those adapted from them to gravity)?

The issue of momentum space can be put aside, because the arguments for sigma models can equally well be made in position space. The key estimate made to arrive at (4.152) is based on dimensional analysis: all of the factors of M and M_p are tracked by counting how they appear as factors in propagators and vertices, and the remaining dimensions are all filled in as the common low-energy scale q. The analogous argument works also in position space, provided there is also only one scale q that characterizes the observables of interest in the low-energy theory.[21]

---

[21] General EFT arguments still apply when there is more than one scale, but are more complicated. Indeed, many of the complications encountered elsewhere when non-relativistic particles are present can be traced to their having more than a single scale, and the same is true for non-relativistic particles interacting with gravity [47].

Physically, the equivalence of the short-distance position-space and high-energy momentum-space estimates happens because the high-energy contributions arise due to the propagation of modes having very small wavelength, λ. Provided this wavelength is very small compared with the local radius of curvature, r_c, particle propagation behaves just as if it had taken place in flat space. We expect the most singular behaviour to be just as for flat space, with curvature effects appearing in subdominant corrections as powers of λ/r_c. This expectation is borne out explicitly in curved-space calculations using heat-kernel methods [48, 49].

It is often true that the low-energy gravitational system is characterized by a single scale. For cosmological models this scale is often the Hubble scale $q \sim H$. (For black holes it is instead $q \sim r_s^{-1}$ where $r_s = 2GM = M/(4\pi M_p^2)$ is the Schwarzschild radius of a black hole with mass M.) In this case the above power counting arguments imply the semiclassical expansion arises as powers of $H^2/(4\pi M_p)^2$ [or $(4\pi M_p r_s)^{-2} \sim (M_p/M)^2$ in the case of black holes]. We require $H/M_p \ll 1$ (or $M \gg M_p$ for black holes) in order to believe inferences about their properties using semiclassical methods.

4.3.1.3 *Power Counting (Scalar-Tensor Theories)*

So far so good, but for inflationary applications these power counting rules must be extended to include both the metric and inflaton, as is now done following [50]. A little more detail is given because the introduction of the scalar potential changes the reasoning somewhat relative to the case of pure gravity.

To this end add N dimensionless scalar fields, θ^i, expanding the effective Lagrangian to the form

$$-\frac{\mathcal{L}_{\text{eff}}}{\sqrt{-g}} = v^4 V(\theta) + \frac{M_p^2}{2} g^{\mu\nu} \left[W(\theta) R_{\mu\nu} + G_{ij}(\theta) \, \partial_\mu \theta^i \partial_\nu \theta^j \right] \tag{4.153}$$

$$+A(\theta)(\partial\theta)^4 + B(\theta) R^2 + C(\theta) R (\partial\theta)^2 + \frac{E(\theta)}{M^2} (\partial\theta)^6 + \frac{F(\theta)}{M^2} R^3 + \cdots,$$

with terms involving up to two derivatives written explicitly and the rest written schematically, inasmuch as R^3 collectively represents all possible independent curvature invariants involving six derivatives, and so on. The explicit mass scales M_p and M are explicitly written, as before, so that the functions $W(\theta), A(\theta), B(\theta)$, etc, are dimensionless. Also as before, it is natural to assume $M \ll M_p$, where M is the lowest scale integrated out to obtain $\mathcal{L}_{\text{eff}}$. A new scale, v, is also added so that the scalar potential $V(\theta)$ is also dimensionless. The kinetic term for the scalars is chosen to be normalized with the Planck mass as its coefficient, and this has the effect of generically suppressing the couplings of canonically normalized scalars by powers of M_p. With inflationary applications in mind take $H^2 M_p^2 \sim V \sim v^4 \ll M^4 \ll M_p^4$ when $\theta \simeq \mathcal{O}(1)$.

Semiclassical Expansion

Expanding about a classical solution using fields that have canonical dimension, $\theta^i(x) = \vartheta^i(x) + \phi^i(x)/M_p$ and $g_{\mu\nu}(x) = \hat{g}_{\mu\nu}(x) + h_{\mu\nu}(x)/M_p$ allows this Lagrangian to be written

$$\mathcal{L}_{\text{eff}} = \hat{\mathcal{L}}_{\text{eff}} + M^2 M_p^2 \sum_n \frac{c_n}{M^{d_n}} \, \mathcal{O}_n\left(\frac{\phi}{M_p}, \frac{h_{\mu\nu}}{M_p}\right) \tag{4.154}$$

where $\hat{\mathcal{L}}_{\text{eff}} = \mathcal{L}_{\text{eff}}(\vartheta, \hat{g}_{\mu\nu})$ and the interactions, $\mathcal{O}_n$, involve $N_n = N_n^{(\phi)} + N_n^{(h)} \geq 2$ powers of the fields ϕ^i and $h_{\mu\nu}$. Using a parameter d_n to count the number of derivatives appearing in $\mathcal{O}_n$, the coefficients c_n are dimensionless and the prefactor, $M^2 M_p^2$, ensures the kinetic terms (and so also the propagators) are independent of M and M_p.

The Lagrangians of (4.154) and (4.153) make equivalent predictions for physical observables provided an appropriate dependence on M, M_p, and v is assigned to the coefficients c_n. In particular, reproducing the M-dependence of the coefficients of the curvature-cubed and higher terms in (4.153) implies

$$c_n = \left(\frac{M^2}{M_p^2}\right) g_n \qquad (\text{if } d_n > 2), \tag{4.155}$$

where g_n is at most order-unity and independent (up to logarithms) of M and M_p. For terms with no derivatives—i.e. those coming from the scalar potential, $V(\theta)$—we instead find

$$c_n = \left(\frac{v^4}{M^2 M_p^2}\right) \lambda_n \qquad (\text{if } d_n = 0), \tag{4.156}$$

where the dimensionless couplings λ_n are also independent of M_p and M. In terms of the λ_n's the above assumptions mean that the scalar potential has the schematic form

$$V(\phi) = v^4 \left[\lambda_0 + \lambda_2 \left(\frac{\phi}{M_p}\right)^2 + \lambda_4 \left(\frac{\phi}{M_p}\right)^4 + \cdots\right], \tag{4.157}$$

which shows that V ranges through values of order v^4 as ϕ^i ranges through values of order M_p. These choices capture qualitative features of many explicit inflationary models.

For cosmological applications it also proves useful to normalize amplitudes differently than is done in earlier sections, which treated $\mathcal{A}_E$ as appropriate for the sum of amputated Feynman graphs (as would be useful when computing an effective action, say). For cosmology it is more useful to track correlation functions of fields, since our interest is in tracking how quantities like $\langle \phi^2 \rangle$ depend on the scales M, M_p, v, and $q \sim H$. To obtain correlation functions from amputated graphs a propagator is attached to to each external line followed by an integration over the space-time where the external line is attached to the rest of the graph. Using $q \sim H$ for the common low-energy scale, the Feynman amplitude for a correlation function with E external lines scales as $\mathcal{B}_E(H) \simeq \mathcal{A}_E(H) H^{2E-4}$.

Another complication for cosmology is that we separately track dependence on *two* low-energy scales and not just one, since correlators are required as functions of both H and mode momentum k/a. However, this need not alter the power counting dimensional analysis argument if these are the same size (as they are during the epoch of most interest: Hubble exit).

Combining these observations and repeating the usual power counting steps leads to the result

$$\mathcal{B}_E(H) \simeq \frac{M_p^2}{H^2} \left(\frac{H^2}{M_p}\right)^E \left(\frac{H}{4\pi M_p}\right)^{2L} \prod_{d_n=0} \left[\lambda_n \left(\frac{v^4}{H^2 M_p^2}\right)\right]^{V_n}$$

$$\times \left[\prod_{d_n=2} c_n^{V_n}\right] \prod_{d_n \geq 4} \left[g_n \left(\frac{H}{M_p}\right)^2 \left(\frac{H}{M}\right)^{d_n-4}\right]^{V_n}, \tag{4.158}$$

which shows that the presence of vertices with no derivatives ($d_n = 0$) introduces factors where the low-energy scale $q \sim H$ appears in the denominator rather than the numerator. Such terms are potentially dangerous because their repeated insertion threatens to undermine the entire low-energy expansion.

However, these dangerous scalar-potential terms do not pose a problem for the class of inflationary potentials, eqn (4.157), of present interest [13]. That is because once the relationship, $H \simeq v^2/M_p$, connecting the size of H to the scale in the potential is used, the potentially dangerous $d_n = 0$ term becomes

$$\prod_{d_n=0} \left[\lambda_n \left(\frac{v^4}{H^2 M_p^2}\right)\right]^{V_n} \simeq \prod_{d_n=0} \lambda_n^{V_n}, \tag{4.159}$$

leading to the final result

$$\mathcal{B}_E(H) \simeq \frac{M_p^2}{H^2} \left(\frac{H^2}{M_p}\right)^E \left(\frac{H}{4\pi M_p}\right)^{2L} \left[\prod_{d_n=2} c_n^{V_n}\right] \tag{4.160}$$

$$\times \left[\prod_{d_n=0} \lambda_n^{V_n}\right] \prod_{d_n \geq 4} \left[g_n \left(\frac{H}{M_p}\right)^2 \left(\frac{H}{M}\right)^{d_n-4}\right]^{V_n},$$

Equation (4.160) shows why scalar fields do not undermine the validity of the low-energy approximation underlying gravitational semiclassical expansion. The basic loop-counting parameter is small provided $H \ll 4\pi M_p$, and the leading contribution describes classical ($L = 0$) physics, as is assumed in standard treatments of cosmology.

Equation (4.160) also shows why trans-Planckian field values need not be a threat to the validity of the low-energy expansion that underpins EFT methods, even though this expansion demands low energies compared with mass scales that are Planck size or lower. The point is that what would be bad is high energies and this does not necessarily follow from large fields. This fact is baked into the above power counting analysis by the assumption that the scalar potential $V = v^4 U(\theta)$ remained $\mathcal{O}(v^4)$ for generic $\theta = \phi/M_p$ order unity.

This emphasizes the important conceptual difference between expanding in powers of fields, like ϕ, and expanding in powers of derivatives of fields. The derivative expansion is part and parcel of low-energy methods, while the field expansion is only relevant to the exploration of a particular neighbourhood of field space. It can happen that a potential has a known and bounded asymptotic form at large fields rather than small fields—such as perhaps[22] $V(\varphi) \sim V_0 - V_1 e^{-\varphi/f} + \cdots$—while not knowing its behaviour for small φ.

Slow-Roll Suppression

With additional assumptions the previous arguments can be refined to track suppression by slow-roll parameters as well as powers of H/M_p. This is done by recognizing that derivatives of background fields and derivatives of fluctuation fields may be very different sizes. In inflation, derivatives of background fields are suppressed by slow-roll parameters while fluctuations need not be, and suppressed background derivatives can make the effective couplings like g_n or λ_n parametrically small.

There are two ways that slow-roll parameters can enter into these effective couplings. First, they can do so because of the assumed flatness of the inflationary potential. For instance, if all slow-roll parameters are similar in size then the sth derivative of the scalar potential can be written $(\mathrm{d}^s V/\mathrm{d}\varphi^s) \simeq \epsilon^{s/2} V/M_p^s \simeq \epsilon^{s/2} v^4/M_p^s$ and so

$$\lambda_n \simeq \epsilon^{N_n/2} \hat{\lambda}_n, \tag{4.161}$$

where now it is $\hat{\lambda}_n$ that is order unity. Here N_n counts the number of scalar lines that meet at the vertex in question, and having all slow-roll parameters of a similar size means there is a factor of $\sqrt{\epsilon}$ arising for each scalar line that meets in the vertex.

The other way slow-roll parameters enter into eqn (4.160) is through derivatives of the background scalar field, which we assume satisfies

$$\frac{1}{M_p} \frac{\mathrm{d}^n \varphi}{\mathrm{d}t^n} \simeq \epsilon^{n/2} H^n, \tag{4.162}$$

---

[22] This is not a purely hypothetical example, since the energy of an extra-dimensional modulus (such as the radius, r, of an extra-dimensional sphere, say) typically arises as a curvature expansion and so as a series in powers of $1/r$. When regarded as a field in the low-energy 4D EFT this gives Lagrangians of the form $\mathcal{L} \propto (\partial r)^2/r^2 + V(r)$ with $V(r) = V_0 + V_1/r + \cdots$, with the kinetic term coming from the dimensionally reduced Einstein action, generically leading to the exponential form so attractive for inflation [27].

and so in particular satisfies the slow-roll relation $H M_p \dot{\varphi} \simeq V'$. It is straightforward then to track how powers of ϵ appear in any particular graph contributing to $\mathcal{B}_E(H)$, the details of which are in [50].

For single-field slow-roll models these power counting rules imply the standard estimates for the size of the leading few n-point functions. For instance, the leading contributions to metric and inflaton two-point functions correspond to using $E = 2$ and $L = 0$ and taking vertices only from the 2-derivative interactions. The diagonal terms then arise unsuppressed by powers of H/M_p or ϵ, while the leading off-diagonal terms are down by at least one power of $\sqrt{\epsilon}$. Mixed terms are suppressed because they arise at leading order by expanding the scalar kinetic term $\sqrt{-g}\,(\partial\phi)^2 \simeq h\dot{\varphi}\dot{\phi} + \cdots$ with $\dot{\varphi} \propto \sqrt{\epsilon}$.

This leads to the estimates

$$\langle hh \rangle \sim \langle \phi\phi \rangle \sim H^2, \quad \text{while} \quad \langle \phi h \rangle \sim \sqrt{\epsilon}\, H^2. \tag{4.163}$$

Keeping in mind that the standard gauge-invariant variables $\zeta \sim \Phi$ are related to the basic inflaton and metric by

$$\zeta \sim \frac{\phi}{\dot{\varphi}/H} \sim \frac{\phi}{\sqrt{\epsilon}\, M_p} \quad \text{and} \quad t_{\mu\nu} \sim \frac{h_{\mu\nu}}{M_p}. \tag{4.164}$$

then leads to the usual estimates

$$\langle \zeta\zeta \rangle \sim \frac{H^2}{\epsilon\, M_p^2} \quad \text{and} \quad \langle tt \rangle \sim \frac{H^2}{M_p^2}. \tag{4.165}$$

The first of these agrees with the more explicit arguments given earlier for Δ_Φ^2 at Hubble crossing.

The leading powers of H/M_p in the 3-point functions (called 'bispectra' by cosmologists) are similarly obtained by choosing $E = 3$ and $L = 0$ and no vertices used except those with $d_n = 2$. For the quantities $\langle hhh \rangle$ and $\langle h\phi\phi \rangle$ the leading contributions then are

$$\langle hhh \rangle \sim \langle h\phi\phi \rangle \sim \frac{H^4}{M_p}, \tag{4.166}$$

since unsuppressed cubic vertices come from either the Einstein–Hilbert action or the inflaton kinetic term. By contrast, the correlators $\langle hh\phi \rangle$ or $\langle \phi\phi\phi \rangle$ all come suppressed by at least one power of $\sqrt{\epsilon}$, with the leading contribution obtained by inserting a single h-ϕ kinetic mixing into $\langle hhh \rangle$ or $\langle hh\phi \rangle$. This leads to the estimates

$$\langle hh\phi \rangle \sim \langle \phi\phi\phi \rangle \sim \frac{\sqrt{\epsilon}\, H^4}{M_p}. \tag{4.167}$$

Again converting to dimensionless strain and curvature fluctuation using eqn (4.164) then leads to the usual results [51]

$$\langle ttt \rangle \sim \langle tt\zeta \rangle \sim \frac{H^4}{M_p^4} \quad \text{and} \quad \langle t\zeta\zeta \rangle \sim \langle \zeta\zeta\zeta \rangle \sim \frac{H^4}{\epsilon M_p^4}, \tag{4.168}$$

and so on. The last of these, $\langle \zeta\zeta\zeta \rangle$, is a measure of the amount of non-Gaussianity associated with the primordial distribution of density fluctuations, and because CMB measurements tell us $\langle \zeta\zeta \rangle \sim \Delta_\Phi^2 \sim H^2/(\epsilon M_p^2) \sim 10^{-10}$ a simple estimate of the size of this non-Gaussianity in single-field slow-roll models is

$$\langle \zeta\zeta\zeta \rangle \sim \frac{H^4}{\epsilon M_p^4} \sim \epsilon \left(\frac{H^2}{\epsilon M_p^2} \right)^2 \tag{4.169}$$

which is too small to be detected at present. This suppression need not be present for more complicated models, making searches for non-Gaussianity a useful benchmark when testing the single-field hypothesis.

4.3.2   Conceptual Issues for EFTs with Time-Dependent Backgrounds

Besides issues specific to gravity, use of EFTs in cosmology can also involve other complications that are often not seen in particle physics (but do arise in other areas of physics where time-dependent background fields are encountered).

4.3.2.1   *Importance of the Adiabatic Approximation*

An issue specific to cosmology arises due to the appearance there of time-dependent backgrounds. The issue asks: if EFTs are defined by dividing systems into low- and high-energy states how can they be defined in time-dependent problems where energy is not conserved? The short version of this section is that time-dependence (in gravity and elsewhere) imposes additional restrictions on the domain of validity of EFTs, the most important of which is the requirement that the background time-dependence should be adiabatic (i.e. $\dot{\varphi}/\varphi$ should be smaller than the UV scales of interest, for every time-dependent background field φ in the problem.)

Adiabatic motion is important because in essence EFTs organize states according to their energy, and energy is generically not conserved (and so is not useful) in the presence of time-dependent backgrounds. In the special case of adiabatic motion, however, an approximately conserved Hamiltonian, $\mathfrak{H}(t)$, can exist, even though it may drift slowly with time as the background evolves. This allows both the definition of an approximate ground state and an energy in terms of which the low-energy/high-energy split can be defined.

Once the system is partitioned in this way into low-energy and high-energy states, we can ask whether a purely low-energy description of time evolution is possible using only a low-energy, local effective Lagrangian. The main danger is that the time evolution of the system need not keep low-energy states at low energies, or high-energy states at high energies. For instance, this could happen if the background's time-dependence is rapid enough to allow particle production of what were regarded as high-energy states, making the initial ground state unstable. Or, it could happen that the initial gap between high and low energies decreases with time, and so the approximation of expanding in the ratio of these energies becomes a poor approximation. Level crossing is an extreme example of the evolution of gap size, in which the gap eventually vanishes and high- and low-energy states nominally cross one another as time evolves.

A related issue can arise if there is a transfer of states from high energy to low energy as the dividing line between them, $\Lambda(t)$, evolves. For example, this could happen for a charged particle in a decreasing magnetic field if the effective theory is set up so that the dividing energy, $\Lambda(t)$, between low and high energies is not similarly time dependent. In this case, Landau levels continuously enter the low-energy theory as the magnetic field strength wanes. Such a migration of states can also happen in cosmology, such as during an inflationary phase (the so-called trans-Planckian 'problem'). This usually is only a problem for the effective theory formulation if the states that enter in this way are not in their adiabatic ground state when they do so. If they *are* in their adiabatic ground state they do not affect low-energy observables, but if they are *not* they can since then new physical excitations appear at low energies.

What emerges from this is that EFTs can make sense despite the presence of time-dependent backgrounds, provided we can focus on the evolution of low-energy states, $(q < \Lambda(t))$, without worrying about losing probability into high-energy states $(q > \Lambda(t))$. This can often be ensured if the background time evolution is sufficiently adiabatic.

4.3.2.2 *Predicting Background Evolution with EFTs*

There is another issue at stake when using EFTs in cosmology (or other time-dependent settings). Up to now the evolution of the background field is regarded as being given, and the EFT issues of the previous section are to do with understanding how to split the system into low and high energy states relative to an adiabatic energy defined in the presence of this time-dependent background.

But it is often also of interest to know how the background itself responds to events within time-dependent systems. For instance the background might back-react in response to changes in the state of fluctuations with which it interacts. This can also be amenable to EFT analysis, often by solving self-consistently for the background using the field equations of the low-energy theory. Central to this approach is the assumption that solutions to field equations within an EFT actually capture the behaviour of solutions to field equations within the full theory.

Need this always be true? This section—following [3]—argues in general the answer is 'no', although adiabatic motion is a notable exception.

To see why EFTs and UV completions can agree on their solutions to the equations of motion we must hark back to the definitions of the EFT itself. (The EFT formulation used here follows [46].) Consider therefore a theory with high-energy and low-energy fields h and ℓ, with action $S(h,\ell)$. We wish to integrate out h to obtain the effective action,

$S_{\text{eff}}(\ell)$, to examine its equations of motion. For simplicity we do so at the classical level, in which case integrating out h is equivalent to solving its classical field equations as a function of the light field, $h_c(\ell)$ and plugging the result back into the original action:

$$S_{\text{eff}}[\ell] = S[h_c(\ell), \ell], \qquad \text{where} \quad \left(\frac{\delta S}{\delta h} \right)_{h = h_c(\ell)} = 0. \qquad (4.170)$$

(Exercise: verify this statement explicitly by showing that it is equivalent to integrating out h using only tree graphs.)

An immediate consequence of the above derivation seems to be that any solution to the low-energy EFT

$$\left(\frac{\delta S_{EFT}}{\delta \ell} \right)_{\ell = \ell_c} = 0, \qquad (4.171)$$

must also be extrema of the full theory, by virtue of the choice $h = h_c(\ell)$. How can this argument ever fail?

The key step in deriving any EFT, glossed over in the previous paragraphs, is the necessity of expanding to some finite order in powers of the heavy mass scale, $1/M$. It is only after this expansion that an effective action like (4.170) is given by a local Lagrangian density. Because of this we should only trust the equations of motion of any local EFT up to the same order in powers of $1/M$. Solutions of the full theory can differ from those of the effective theory if they are not captured by such a $1/M$ expansion.

It is actually a good thing that the solutions to an EFT are not completely equivalent to solutions to the full theory from which the EFT is derived. One benefit is that EFTs often involve higher time derivatives, and so naively should generically have unstable runaway solutions [52], even if the underlying theory has none.

To see why instabilities might arise within the EFT consider the following toy effective Lagrangian:

$$\frac{L}{v^2} = \frac{1}{2} \dot{\theta}^2 + \frac{1}{2M^2} \ddot{\theta}^2, \qquad (4.172)$$

whose variation $\delta L = 0$ gives the linear equation of motion

$$-\ddot{\theta} + \frac{1}{M^2} \ddddot{\theta} = 0. \qquad (4.173)$$

The general solution to this equation is

$$\theta = A + Bt + Ce^{Mt} + De^{-Mt}, \qquad (4.174)$$

where A, B, C, and D are integration constants.

Now comes the main point. Only the solutions with $C = D = 0$ go over to the solutions to the lowest-order field equation, obtained from the $M \to \infty$ Lagrangian, $L_0 = \frac{1}{2} \dot{\theta}^2$. The others make no sense at any finite order of $1/M$ because for them the $\dot{\theta}^2$ and $\ddot{\theta}^2$ terms are always comparably large. Since a local EFT is only meant to capture the full theory order-by-order in $1/M$ these exponential solutions should not be expected to reproduce the low-energy approximation of the full theory.

4.3.2.3 EFT for Inflationary Fluctuations

The previous sections touch on general issues associated with time-dependent backgrounds. For inflation it is particularly interesting if the time evolution describes a slow roll near de Sitter space-time, since this is the maximally symmetric space-time obtained when solving Einstein's equations with a positive cosmological constant, with equation of state $w = -1$ (corresponding to the case of exponential expansion $a \propto e^{Ht}$ that arises during a potential-dominated epoch).

Slow-roll evolution of a scalar field near a maximally symmetric space like de Sitter spacetime lends itself to symmetry arguments. This is because maximal symmetry in four dimensions implies de Sitter space enjoys a ten-parameter group of isometries (transformations that preserve the form of the metric). For de Sitter space the group is $O(4, 1)$, which has the same number of generators as the Poincaré symmetry group of flat space (though with different commutation relations).

This abundance of symmetries gives the discussion of fluctuations about a near-de Sitter slow roll some fairly universal features, which allow a more general parameterization than might otherwise be possible. These features are described by yet another kind of EFT within the inflationary literature, one that has come to be known as 'the' effective theory of inflation [53] (for a review see [54]). This section provides a telegraphic summary of this specific theory, in order to put it into its context within the broader EFT pantheon.

The EFT of inflation is aimed at single-field inflationary models including, but not restricted to, the simple models considered above. The starting observation of this theory is that homogeneous roll of the single inflaton field, $\varphi(t)$, provides the clock that breaks the symmetries the de Sitter space-time otherwise would have had. There are many equivalent ways to set up spatial slices in maximally symmetric spaces, but the level surfaces of an evolving scalar field pick out a specific preferred frame within which to build spatial slices.

The background φ (and the evolving metric to which it also gives rise) spontaneously breaks the symmetries of de Sitter space and acts as the Goldstone boson for this breaking (and is ultimately eaten by the metric, in the same way that would-be Goldstone bosons for local internal symmetries get eaten by the corresponding gauge bosons). Because of this symmetry representation it is possible to use the non-linearly realized symmetries to classify the kinds of allowed low-energy interactions in a model-independent way, providing a more robust description of the kinds of observables that could potentially arise at low energies for cosmologies based any single-field slow-roll system near de Sitter space.

Concretely, because the slowly rolling field, ϕ, is assumed to be a scalar, by definition under arbitrary space-time motions, $\delta x^\mu = V^\mu(x)$, it transforms as $\delta\phi = V^\mu \partial_\mu \phi$. But the *fluctuation* in such a scalar field about a homogeneous background $\varphi(t)$ transforms differently than this, since if $\phi = \varphi(t) + \hat{\phi}(x,t)$, then although $\delta\hat{\phi}(x,t) = V^i\partial_i\hat{\phi}(x,t)$ remains a scalar under purely spatial motions, $\delta x^i = V^i(x,t)$, at fixed t, it transforms non-linearly under motions $\delta t = V^0(x,t)$, since

$$\delta\hat{\phi}(x,t) = V^0(x,t)\dot{\varphi} + V^0\partial_0\hat{\phi}. \tag{4.175}$$

This kind of inhomogeneous transformation indicates a would-be Goldstone mode.

There are two natural ways to describe the interactions of a field that transforms like (4.175). One way—called 'unitary gauge'—uses this transformation to completely remove the field $\hat{\phi}$ from the problem. A second way trades $\hat{\phi}$ for a 'Stueckelberg field' $\hat{\pi}(x,t) := -V^0(x,t)$, though for brevity's sake this summary sticks with the unitary gauge formulation.[23]

In unitary gauge the physics of $\hat{\phi}$ gets transferred into any other fields that are present and transform under the symmetry. In the present case the only other field is the metric, and in this sense $\hat{\phi}$ gets eaten by the metric. In this representation the remaining symmetries are spatial motions, $\delta x^k = V^k(x)$ and

$$\delta g_{00} = V^k\partial_k g_{00}$$
$$\delta g_{0i} = V^k\partial_k g_{0i} + \partial_i V^k g_{0k} \tag{4.176}$$
$$\text{and}\quad \delta g_{ij} = V^k\partial_k g_{ij} + \partial_i V^k g_{jk} + \partial_j V^k g_{ik}, \tag{4.177}$$

and so respectively transform as a scalar, vector, or tensor. The most general invariant action is built using the spatial scalar g_{00} and the spatial tensors built from the intrinsic and extrinsic curvatures, $R^i{}_{jkl}$ and K_{ij}, of the spatial slices (which—from the Gauss-Codazzi relations—together encode all of the information of the full 4D curvature $R^\mu{}_{\nu\lambda\rho}$). Whereas $R^i{}_{jkl}$ involves two derivatives of the metric, K_{ij} only involves one.

The most general Lagrangian density describing these degrees of freedom is

$$\frac{\mathcal{L}}{\sqrt{-g}} = \frac{M_p^2}{2}R - \alpha(t) - \beta(t)g^{00} + \frac{1}{2}M_2^4(t)(g^{00}+1)^2 + \frac{1}{3!}M_3^4(t)(g^{00}+1)^3 + \cdots, \tag{4.178}$$

where the ellipses include terms with more powers of $g^{00}+1$ and its derivatives, as well as the variation of the extrinsic curvatures, δK_{ij}, computed relative to the extrinsic curvatures evaluated in the background geometry, and so on. The coefficients α, β, M_i, and so on, are in principle arbitrary functions of t, though in a near-de Sitter framework this t-dependence should be suppressed by slow-roll parameters.

An important property of the Lagrangian (4.178) is that the only terms linear in a metric fluctuation are the two terms with coefficients α and β. This is why in all other

---

[23] The formulation involving $\hat{\pi}(x,t)$ has the advantage that for scales of order H many (but not all) quantities can be computed fairly easily using only relatively simple self-interactions of $\hat{\pi}$, without needing the full complications of expanding the metric dependence of the Einstein action.

terms g^{00} is chosen always to appear in the combination $g^{00} + 1$. Although not obvious, this claim relies on the observation that the other linear terms are redundant, in that they are total derivatives or can be removed by a field redefinition. Because of this, only α and β play a role in the evolution of the background metric, $H(t)$. Since the background has the symmetries of an FRW geometry the stress energy computed from $\mathcal{L}$ evaluated at the background is characterized by an energy density and a pressure, ρ and p, and these are related to α and β by $\rho = \alpha + \beta$ and $p = \beta - \alpha$. The background Einstein equations then imply

$$H^2 = \frac{\alpha + \beta}{3M_p^2} \quad \text{and} \quad \frac{\ddot{a}}{a} = \dot{H} + H^2 = -\frac{(2\beta - \alpha)}{3M_p^2}, \tag{4.179}$$

and so (4.178) can be rewritten as

$$\frac{\mathcal{L}}{\sqrt{-g}} = \frac{M_p^2}{2} R - M_p^2 (3H^2 + \dot{H}) + M_p^2 \dot{H} g^{00} \tag{4.180}$$

$$+ \frac{1}{2} M_2^4(t)(g^{00} + 1)^2 + \frac{1}{3!} M_3^4(t)(g^{00} + 1)^3 + \cdots.$$

The Lagrangian obtained by setting all coefficients except for α and β to zero corresponds to the simple single-field slow-roll models considered in the rest of this chapter. This can be seen by taking the scalar Lagrangian, $\mathcal{L}_\phi / \sqrt{-g} := -\frac{1}{2}(\partial\phi)^2 - V(\phi)$ and going to unitary gauge, for which $\phi = \varphi(t)$ and so $\mathcal{L}_\phi / \sqrt{-g} = -\frac{1}{2} \dot{\varphi}^2 g^{00} - V(\varphi)$. Comparing with (4.178) shows $\alpha = V(\varphi)$ and $\beta = \frac{1}{2} \dot{\varphi}^2$. In this case all of the non-linear interactions amongst the fluctuations are contained within the Einstein–Hilbert part of the action.

The remaining coefficients $M_i(t)$ and so on describe deviations from the simplest scalar models. These could correspond to supplementing the basic scalar Lagrangian with higher derivative interactions, like higher powers of the kinetic term—such as $(\partial\phi)^4$ and so on—or through more exotic kinds of choices [55]. Part of the utility of (4.178) is that one does not need to know what these choices might have been in order to use $\mathcal{L}$ to compute how observables can depend on the coefficients M_i. This allows a relatively model-independent survey of what kind of observables are possible at low energies, without having to go through all possible microscopic models beforehand.

4.3.2.4 Open Systems

EFTs applied to gravitational systems can surprise in other ways as well. In particular, during inflation we saw that the main observational consequences are tied up with super-Hubble modes, for which $k/a \ll H$. Since these are the longest-wavelength modes in the system the effective action that describes them has long been sought as the most efficient way to capture inflationary predictions in as model-independent a way as possible. But no such an effective action was ever found.

This doesn't mean that an EFT description for these modes does not exist; rather, it just turns out it need not necessarily be usefully described by a traditional Wilsonian effective action [56]. This unusual situation arises because during inflation the long-wavelength modes are described by the EFT for an *open* system [56]–[59], in that modes are continually moving from sub-Hubble to super-Hubble throughout the inflationary epoch. This should be contrasted with the usual situation with a Wilsonian effective theory, for which high- and low-energy states are forbidden from transitioning into one another by energy conservation.

Because of this mode migration the long- and short-distance sectors can interact in more complicated ways than are normally entertained, such as by entangling and/or decohering with one another. The appropriate language for describing long-wavelength modes in this kind of situation is to use the reduced density matrix, $\varrho_L = \text{Tr}_S \rho$, in which the full system's density matrix is traced over the unwatched (in this case, short-wavelength) sector. It turns out that ϱ_L evolves in time according to a Lindblad-type equation, which need not be writable as a Liouville equation for some choice of effective Hamiltonian.

Using these kinds of arguments it is possible to show that the evolution of the leading effective description of the diagonal parts of the reduced density matrix, $P[\varphi, t] := \langle \varphi | \rho_L | \varphi \rangle$, describing fluctuations amongst super-Hubble modes during inflation, is given by a Fokker–Planck equation. The description of the properties of this equation is called 'stochastic inflation' [60], and the stochastic description of the quantum system is possible because quantum states become very classical, in that they are well-described by the WKB approximation, when outside the Hubble scale. Although a full description goes beyond the scope of this chapter, the evidence now is that this open-system evolution (starting with stochastic inflation) does a good job capturing the late-time evolution of super-Hubble modes. In particular, it resolves problems to do with infrared divergences and the secular breakdown of perturbation theory at late times that are encountered in all but the simplest inflationary models.

A bonus with this picture comes when evolving the off-diagonal components of the reduced density matrix, $\langle \varphi | \rho_L | \psi \rangle$, since this evolution tends towards a diagonal matrix in a basis that diagonalizes the field operators $\phi(x)$. This provides the explanation of why perturbations that leave the Hubble scale as quantum fluctuations eventually re-enter the sub-Hubble regime well-described as a statistical distribution of classical fields [56].

4.4 The Bottom Line

These notes trace four golden threads woven deeply into the great cosmic tapestry.

First, a very beautiful and pragmatic picture is emerging wherein the special initial conditions required of late-time cosmology can be understood in terms of quantum fluctuations during a much earlier accelerating epoch. Although there is debate as to whether the state being accelerated was initially growing—inflation—or shrinking—a bounce—everyone sings from the same quantum scoresheet. If true, quantum effects in gravity are literally written across the entire sky.

Second, as in all other areas of physics, EFT methods provide extremely valuable ways to handle systems with more than one scale. This is especially true for gravitational physics in particular, whose non-renormalizability is a red flag telling us we are dealing with the low-energy limit of something more fundamental. Indeed, EFT ideas provide the bedrock on which all standard semiclassical reasoning is founded. With our present understanding of physics, any departure from an EFT framework when working with gravity always brings loss of control over theoretical error. Cosmologists cannot afford to do so now that cosmology has become a precision science.

Third, at this point there is no uncontroversial evidence in favour of gravitational exceptionalism, inasmuch as the issues encountered applying EFTs to gravitational problems also seem to arise in other areas of physics (for which many powerful tools have been developed). But there are also unsolved puzzles associated with quantum fields interacting gravitationally, and it is always prudent to have one eye out for surprises.

Finally, although the EFTs used in gravity seem to have counterparts elsewhere in physics, these other areas are often not particle physics and so gravity *can* provide surprises to those coming from a particle physics training. These surprises include secular evolution and a generic breakdown of perturbation theory at late times (that arise because the gravitational field never goes away, and so even small secular effects eventually can accumulate to become large). They include stochastic effects where Wilsonian actions need not be the useful way to formulate the problem, because the physics of interest is an open system. For these the effective interactions of particles moving within a larger medium can be better models than intuition based on traditional low-energy Wilsonian EFTs.

But this is a relatively young field and yours is the generation likely to be reaping the rewards of (or discovering) new directions. In situations like this the field is likely to belong to those who bring diverse tools and an open mind: be broad and good luck!

Acknowledgements

I thank the organizers of this school for their kind invitation to present these lectures in such pleasant environs. My thanks also to my many students, collaborators and colleagues—Peter Adshead, Niayesh, Afshordi, Shanta de Alwys, Diego Blas, Michele Cicoli, Ross Diener, Claudia de Rham, Jared Enns, Peter Hayman, Richard Holman, Michael Horbatsch, Matthew Johnson, Sven Krippendorf, Louis Leblond, Hyun-Min Lee, Anshuman Maharana, Leo van Nierop, Matt Williams, Subodh Patil, Fernando Quevedo, Sarah Shandera, Gianmassimo Tasinato, Andrew Tolley and Michael Trott—and especially to my teachers—Willy Fischler, Joe Polchinski and Steven Weinberg—for their help in understanding effective field theories and how they apply in a cosmological context. My research has been supported in part by the Natural Sciences and Engineering Research Council of Canada. Research at the Perimeter Institute is supported in part by the Government of Canada through Industry Canada, and by the Province of Ontario through the Ministry of Research and Information.

References

[1] C. P. Burgess, Lectures on Cosmic Inflation and its Potential Stringy Realizations, *Class. Quant. Grav.* 24: S795 [PoS P **2GC** (2006) 008] [PoS CARGESE **2007** (2007) 003] doi:10.1088/0264-9381/24/21/S04 [arXiv:0708.2865 [hep-th]].

[2] C. P. Burgess, Quantum Gravity in Everyday Life: General Relativity as an Effective Field Theory, *Living Rev. Rel.* 7: 5 doi:10.12942/lrr-2005 [gr-qc/0311082].

[3] C. P. Burgess and M. Williams, Who You Gonna Call? Runaway Ghosts, Higher Derivatives and Time-Dependence in EFTs, *JHEP* 1408: 074 doi:10.1007/JHEP08(2014)074 [arXiv:1404.2236 [gr-qc]].

[4] P. A. R. Ade, et al. [Planck Collaboration], (2016). Planck 2015 Results. XIII. Cosmological Parameters, *Astron. Astrophys.* 594: A13 doi:10.1051/0004-6361/201525830 [arXiv:1502.01589 [astro-ph.CO]].

[5] R. Adam *et al.* [Planck Collaboration], (2016). Planck 2015 Results. I. Overview of Products and Scientific Results, *Astron. Astrophys.* 594: A1 doi:10.1051/0004-6361/201527101 [arXiv:1502.01582 [astro-ph.CO]].

[6] C. W. Misner, J. A. Wheeler, and K. S. Thorne, (1973). *Gravitation,* W. H. Freeman & Company.

[7] S. Weinberg, (1972). *Gravitation and Cosmology: Principles and Applications of the General Theory of Relativity,* Wiley.

[8] D. Baumann, Inflation, doi:10.1142/9789814327183_0010 arXiv:0907.5424 [hep-th].

[9] Alan H. Guth, (1981). The Inflationary Universe: A Possible Solution to the Horizon and Flatness Problems, *Phys. Rev. D* 23: 347-356;
A.D. Linde, (1982). A New Inflationary Universe Scenario: A Possible Solution of the Horizon, Flatness, Homogeneity, Isotropy and Primordial Monopole Problems, *Phys. Lett. B* 108: 389-393;
A. Albrecht and P. J. Steinhardt, (1982). Cosmology for Grand Unified Theories with Radiatively Induced Symmetry Breaking, *Phys. Rev. Lett.* 48: 1220-1223;

[10] A. D. Linde, (1983). Chaotic Inflation, *Phys. Lett. B* 129: 177. doi:10.1016/0370-2693(83)90837-7

[11] R. P. Feynman, (1963). Quantum Theory of Gravitation, *Acta Phys. Polon.* 24: 697;
B. S. DeWitt, (1967). Quantum Theory of Gravity. 1. The Canonical Theory, *Phys. Rev.* 160: 1113. doi:10.1103/PhysRev.160.1113; Quantum Theory of Gravity. 2. The Manifestly Covariant Theory, *Phys. Rev.* 162: 1195. doi:10.1103/PhysRev.162.1195;
S. Mandelstam, (1968). Feynman Rules For The Gravitational Field From The Coordinate Independent Field Theoretic Formalism, *Phys. Rev.* 175: 1604. doi:10.1103/PhysRev.175.1604.

[12] F. L. Bezrukov and M. Shaposhnikov, (2008). The Standard Model Higgs Boson as the Inflaton, *Phys. Lett. B* 659: 703 doi:10.1016/j.physletb.2007.11.072 [arXiv:0710.3755 [hep-th]].

[13] C. P. Burgess, H. M. Lee, and M. Trott, (2009). Power-Counting and the Validity of the Classical Approximation During Inflation, *JHEP* 0909: 103 doi:10.1088/1126-6708/2009/09/103 [arXiv:0902.4465 [hep-ph]].

[14] J. L. F. Barbon and J. R. Espinosa, (2009). On the Naturalness of Higgs Inflation, *Phys. Rev. D* 79: 081302 doi:10.1103/PhysRevD.79.081302 [arXiv:0903.0355 [hep-ph]];

C. P. Burgess, H. M. Lee, and M. Trott, (2010). Comment on Higgs Inflation and Naturalness, *JHEP* 1007: 007 doi:10.1007/JHEP07(2010)007 [arXiv:1002.2730 [hep-ph]].

F. Bezrukov, A. Magnin, M. Shaposhnikov, and S. Sibiryakov, (2011). Higgs Inflation: Consistency and Generalisations, *JHEP* 1101: 016 doi:10.1007/JHEP01(2011)016 [arXiv:1008.5157 [hep-ph]].

[15] R. Kallosh and A. Linde, (2013). Superconformal Generalization of the Chaotic Inflation Model $\frac{\lambda}{4}\phi^4 - \frac{\xi}{2}\phi^2 R$, *JCAP* **1306**: 027 doi:10.1088/1475-7516/2013/06/027 [arXiv:1306.3211 [hep-th]].

[16] A. R. Liddle and D. H. Lyth, (1992). COBE, Gravitational Waves, Inflation and Extended Inflation, *Phys. Lett. B* 291: 391 doi:10.1016/0370-2693(92)91393-N [astro-ph/9208007].

[17] D. H. Lyth, (1997). What Would We Learn by Detecting a Gravitational Wave Signal in the Cosmic Microwave Background Anisotropy?, *Phys. Rev. Lett.* 78: 1861 doi:10.1103/PhysRevLett.78.1861 [hep-ph/9606387].

[18] V. Mukhanov, (2013). Quantum Cosmological Perturbations: Predictions and Observations, *Eur. Phys. J. C* 73: 2486 doi:10.1140/epjc/s10052-013-2486-7 [arXiv:1303.3925 [astro-ph.CO]].

[19] K. Freese, J. A. Frieman, and A. V. Olinto, (1990). Natural Inflation with Pseudo-Nambu–Goldstone bosons, *Phys. Rev. Lett.* 65: 3233. doi:10.1103/PhysRevLett.65.3233

[20] J. J. Blanco-Pillado, et al., (2004). Racetrack Inflation, *JHEP* 0411: 063 doi:10.1088/1126-6708/2004/11/063 [hep-th/0406230]; (2006). Inflating in a Better Racetrack, *JHEP* 0609: 002 doi:10.1088/1126-6708/2006/09/002 [hep-th/0603129].

E. Silverstein and A. Westphal, (2008). Monodromy in the CMB: Gravity Waves and String Inflation, *Phys. Rev. D* 78: 106003 doi:10.1103/PhysRevD.78.106003 [arXiv:0803.3085 [hep-th]].

L. McAllister, E. Silverstein, and A. Westphal, (2010). Gravity Waves and Linear Inflation from Axion Monodromy, *Phys. Rev. D* 82: 046003 doi:10.1103/PhysRevD.82.046003 [arXiv:0808.0706 [hep-th]].

[21] For a recent summary, see: P. Svrcek and E. Witten, (2006). Axions In String Theory, *JHEP* 0606: 051 doi:10.1088/1126-6708/2006/06/051 [hep-th/0605206].

K. S. Choi, H. P. Nilles, S. Ramos-Sanchez, and P. K. S. Vaudrevange, (2009). Accions, *Phys. Lett. B* 675: 381 doi:10.1016/j.physletb.2009.04.028 [arXiv:0902.3070 [hep-th]].

[22] J. E. Kim, H. P. Nilles, and M. Peloso, (2005). Completing Natural Inflation, *JCAP* **0501**: 005 doi:10.1088/1475-7516/2005/01/005 [hep-ph/0409138].

S. Dimopoulos, S. Kachru, J. McGreevy, and J. G. Wacker, (2008). N-flation, *JCAP* 0808: 003 doi:10.1088/1475-7516/2008/08/003 [hep-th/0507205].

[23] E. Witten, (1985). Dimensional Reduction of Superstring Models, *Phys. Lett. B* 155: 151. doi:10.1016/0370-2693(85)90976-1

C. P. Burgess, A. Font, and F. Quevedo, (1986). Low-Energy Effective Action for the Superstring, *Nucl. Phys. B* 272: 661. doi:10.1016/0550-3213(86)90239-7

Y. Aghababaie, et al., (2003). Warped Brane Worlds in Six-Dimensional Supergravity, *JHEP* 0309: 037 doi:10.1088/1126-6708/2003/09/037 [hep-th/0308064].

C. P. Burgess, et al., (2012). On Brane Back-Reaction and de Sitter Solutions in Higher-Dimensional Supergravity, *JHEP* 1204: 018 doi:10.1007/JHEP04(2012)018 [arXiv:1109.0532 [hep-th]].

[24] C. P. Burgess, M. Cicoli, and F. Quevedo, (2013). String Inflation After Planck 2013, *JCAP* 1311: 003 doi:10.1088/1475-7516/2013/11/003 [arXiv:1306.3512 [hep-th]].

R. Kallosh, A. Linde, and D. Roest, (2014). Universal Attractor for Inflation at Strong Coupling, *Phys. Rev. Lett.* 112: no.1, 011303 doi:10.1103/PhysRevLett.112.011303 [arXiv: 1310.3950 [hep-th]].

C. P. Burgess, M. Cicoli, F. Quevedo, and M. Williams, (2014). Inflating with Large Effective Fields, *JCAP* 1411: 045 doi:10.1088/1475-7516/2014/11/045 [arXiv:1404.6236 [hep th]].

C. Csaki, N. Kaloper, J. Serra, and J. Terning, (2014). Inflation from Broken Scale Invariance, *Phys. Rev. Lett.* 113: 161302 doi:10.1103/PhysRevLett.113.161302 [arXiv:1406.5192 [hep-th]].

C. P. Burgess, M. Cicoli, S. de Alwis, and F. Quevedo, (2016). Robust Inflation from Fibrous Strings, *JCAP* 1605(5): no.05, 032 doi:10.1088/1475-7516/2016/05/032 [arXiv:1603.06789 [hep-th]].

[25] A. B. Goncharov and A. D. Linde, (1984). Chaotic Inflation in Supergravity, *Phys. Lett. B* 139: 27. doi:10.1016/0370-2693(84)90027-3

[26] M. Cicoli, C. P. Burgess, and F. Quevedo, (2009). Fibre Inflation: Observable Gravity Waves from IIB String Compactifications, *JCAP* 0903: 013 doi:10.1088/1475-7516/2009/03/013 [arXiv:0808.0691 [hep-th]].

[27] C. P. Burgess, et al., (2002). Brane-Anti-Brane Inflation in Orbifold and Orientifold Models, *JHEP* 0203: 052 doi:10.1088/1126-6708/2002/03/052 [hep-th/0111025].

[28] A. A. Starobinsky, (1980). A New Type of Isotropic Cosmological Models Without Singularity, *Phys. Lett. B* 91: 99. doi:10.1016/0370-2693(80)90670-X

[29] M. Tegmark, et al., (2004). [SDSS Collaboration], *Phys. Rev. D* 69: 103501 doi:10.1103/ PhysRevD.69.103501 [astro-ph/0310723].

[30] V. F. Mukhanov and G. V. Chibisov, (1981). *JETP Lett.* 33: 532 [(1981). Pisma Zh. Eksp. Teor. Fiz. 33: 549];

A. H. Guth and S. Y. Pi, (1982). *Phys. Rev. Lett.* 49: 1110;

A. A. Starobinsky, (1982). *Phys. Lett. B* 117: 175;

S. W. Hawking, (1982). *Phys. Lett. B* 115: 295;

V. N. Lukash, (1980). *Pisma Zh. Eksp. Teor. Fiz.* 31: 631; (1980). *Sov. Phys. JETP* 52: 807 [(1980). *Zh. Eksp. Teor. Fiz.* 79];

W. Press, (1980). *Phys. Scr.* 21: 702;

K. Sato, (1981). *Mon. Not. Roy. Astron. Soc.* 195: 467.

[31] V. Mukhanov, (2005). *Physical Foundations of Cosmology*, Cambridge University Press.

[32] V. F. Mukhanov, H. A. Feldman, and R. H. Brandenberger, (1992). Theory of Cosmological Perturbations. Part 1. Classical Perturbations. Part 2. Quantum Theory of Perturbations. Part 3. Extensions, *Phys. Rept.* 215: 203. doi:10.1016/0370-1573(92)90044-Z

R. H. Brandenberger, (2004). Lectures on the Theory of Cosmological Perturbations, *Lect. Notes Phys.* 646: 127 [arXiv:hep-th/0306071].

[33] V. Assassi, et al., (1983). Spontaneous Creation of Almost Scale-Free Density Perturbations in an Inflationary Universe, *Phys. Rev. D* 28: 679. doi:10.1103/PhysRevD.28.679

D. H. Lyth, (1985). Large Scale Energy Density Perturbations and Inflation, *Phys. Rev. D* 31: 1792. doi:10.1103/PhysRevD.31.1792

D. Wands, K. A. Malik, D. H. Lyth, and A. R. Liddle, (2000). A New Approach to the Evolution of Cosmological Perturbations on Large Scales, *Phys. Rev. D* 62: 043527 doi:10.1103/PhysRevD.62.043527 [astro-ph/0003278].

K. A. Malik, D. Wands, and C. Ungarelli, (2003). Large-scale Curvature and Entropy Perturbations for Multiple Interacting Fluids, *Phys. Rev. D* 67: 063516 doi:10.1103/ PhysRevD.67.063516 [astro-ph/0211602].

D. H. Lyth, K. A. Malik, and M. Sasaki, (2005). A General Proof of the Conservation of the Curvature Perturbation, *JCAP* 0505: 004 doi:10.1088/1475-7516/2005/05/004 [astro-ph/0411220].

(2012). On Soft Limits of Inflationary Correlation Functions, *JCAP* 1211: 047 doi:10.1088/1475-7516/2012/11/047 [arXiv:1204.4207 [hep-th]].

[34] A. H. Guth and S. Y. Pi, (1985). The Quantum Mechanics Of The Scalar Field In The New Inflationary Universe, *Phys. Rev. D* 32: 1899;

M. A. Sakagami, (1988). Evolution From Pure States Into Mixed States In De Sitter Space, *Prog. Theor. Phys.* 79: 442;

L. P. Grishchuk and Y. V. Sidorov, (1989). On The Quantum State Of Relic Gravitons, *Class. Quant. Grav.* 6: L161;

R. H. Brandenberger, R. Laflamme, and M. Mijic, (1990). Classical Perturbations From Decoherence Of Quantum Fluctuations In The Inflationary Universe, *Mod. Phys. Lett. A* 5: 2311 (1990);

E. Calzetta and B. L. Hu, (1995). Quantum Fluctuations, Decoherence of the Mean Field, and Structure Formation in the Early Universe, *Phys. Rev. D* 52: 6770 [gr-qc/9505046];

D. Polarski and A. A. Starobinsky, (1996). Semiclassicality and Decoherence of Cosmological Perturbations, *Class. Quant. Grav.* 13: 377 [gr-qc/9504030];

F. C. Lombardo and D. Lopez Nacir, (2005). Decoherence During Inflation: The Generation of Classical Inhomogeneities, *Phys. Rev. D* 72: 063506 [gr-qc/0506051];

C. P. Burgess, R. Holman, and D. Hoover, (2006). On the Decoherence of Primordial Fluctuations During Inflation, [astro-ph/0601646].

[35] C. Kiefer, D. Polarski, and A. A. Starobinsky, (1998). Quantum-to-Classical Transition for Fluctuations in the Early Universe, *Int. J. Mod. Phys. D* 7: 455 [gr-qc/9802003];

J. Lesgourgues, D. Polarski, and A. A. Starobinsky, (1997). Quantum-to-Classical Transition of Cosmological Perturbations for Non-Vacuum Initial States, *Nucl. Phys. B* 497: 479 [gr-qc/9611019].

[36] C. P. Burgess, R. Holman, G. Tasinato, and M. Williams, (2015). EFT Beyond the Horizon: Stochastic Inflation and How Primordial Quantum Fluctuations Go Classical, *JHEP* 1503: 090 doi:10.1007/JHEP03(2015)090 [arXiv:1408.5002 [hep-th]].

H. Collins, R. Holman, and T. Vardanyan, (2017). The Quantum Fokker–Planck Equation of Stochastic Inflation, arXiv:1706.07805 [hep-th].

[37] P. A. R. Ade, et al., (2016). [BICEP2 and Keck Array Collaborations], Improved Constraints on Cosmology and Foregrounds from BICEP2 and Keck Array Cosmic Microwave Background Data with Inclusion of 95 GHz Band, *Phys. Rev. Lett.* 116: 031302 doi:10.1103/PhysRevLett.116.031302 [arXiv:1510.09217 [astro-ph.CO]].

[38] P. A. R. Ade, et al., (2015). [BICEP2 and Planck Collaborations], Joint Analysis of BICEP2/Keck Array and Planck Data, *Phys. Rev. Lett.* 114: 101301 doi:10.1103/PhysRevLett.114.101301 [arXiv:1502.00612 [astro-ph.CO]].

[39] P. A. R. Ade, et al., (2016). [Planck Collaboration], Planck 2015 Results. XX. Constraints on Inflation, *Astron. Astrophys.* 594: A20 doi:10.1051/0004-6361/201525898 [arXiv:1502.02114 [astro-ph.CO]].

[40] L. P. Grishchuk and Y. V. Sidorov, (1990). Squeezed Quantum States Of Relic Gravitons And Primordial Density Fluctuations, *Phys. Rev. D* 42: 3413;

A. Albrecht, P. Ferreira, M. Joyce, and T. Prokopec, (1994). Inflation and Squeezed Quantum States, *Phys. Rev. D* 50: 4807 [astro-ph/9303001].

[41] R. H. Brandenberger and C. Vafa, (1989). Superstrings in the Early Universe, *Nucl. Phys. B* 316: 391. doi:10.1016/0550-3213(89)90037-0

M. Gasperini and G. Veneziano, (1993). Pre-Big Bang in String Cosmology, *Astropart. Phys.* 1: 317 doi:10.1016/0927-6505(93)90017-8 [hep-th/9211021].

J. Khoury, B. A. Ovrut, P. J. Steinhardt, and N. Turok, (2001). The Ekpyrotic Universe: Colliding Branes and the Origin of the Hot Big Bang, *Phys. Rev. D* 64: 123522 doi:10.1103/PhysRevD.64.123522 [hep-th/0103239].

P. J. Steinhardt and N. Turok, (2002). A Cyclic Model of the Universe, *Science* 296: 1436 doi:10.1126/science.1070462 [hep-th/0111030].

[42] W. E. East, M. Kleban, A. Linde, and L. Senatore, (2016). Beginning Inflation in an Inhomogeneous Universe, *JCAP* 1609(9): 010 doi:10.1088/1475-7516/2016/09/010 [arXiv:1511.05143 [hep-th]].

[43] S. A. Kim and A. R. Liddle, (2007). Nflation: Observable Predictions from the Random Matrix Mass Spectrum, *Phys. Rev. D* 76: 063515 doi:10.1103/PhysRevD.76.063515 [arXiv:0707.1982 [astro-ph]].

M. C. D. Marsh, L. McAllister, E. Pajer, and T. Wrase, (2013). Charting an Inflationary Landscape with Random Matrix Theory, *JCAP* 1311: 040 doi:10.1088/1475-7516/2013/11/040 [arXiv:1307.3559 [hep-th]];

A. Linde, (2017). Random Potentials and Cosmological Attractors, *JCAP* 1702(2): 028 doi:10.1088/1475-7516/2017/02/028 [arXiv:1612.04505 [hep-th]].

[44] J. Donoghue, (2017). Quantum Gravity as a Low Energy Effective Field Theory, *Scholarpedia* 12(4): 32997. doi:10.4249/scholarpedia.32997

J. F. Donoghue and B. R. Holstein, (2015). Low Energy Theorems of Quantum Gravity from Effective Field Theory, *J. Phys. G* 42(10): 103102 doi:10.1088/0954-3899/42/10/103102 [arXiv:1506.00946 [gr-qc]].

[45] See, for example:

S. Weinberg, (1989). *Rev. Mod. Phys.* 61: 1;

E. Witten, (2000). The Cosmological Constant from the Viewpoint of String Theory, [hep-ph/0002297];

J. Polchinski, The Cosmological Constant and the String Landscape, [hep-th/0603249];

C. P. Burgess, (2013). The Cosmological Constant Problem: Why it is Hard to Get Dark Energy from Micro-Physics, in *Proceedings of the Les Houches School Cosmology After Planck*, [arXiv:1309.4133];

T. Banks, (2014). Supersymmetry Breaking and the Cosmological Constant, *Int. J. Mod. Phys. A* 29: 1430010 [arXiv:1402.0828 [hep-th]];

A. Padilla, (2015). Lectures on the Cosmological Constant Problem, [arXiv:1502.05296 [hep-th]].

[46] C. P. Burgess, (2007). Introduction to Effective Field Theory, *Ann. Rev. Nucl. Part. Sci.* 57: 329 [hep-th/0701053].

[47] W. D. Goldberger and I. Z. Rothstein, (2006). An Effective Field Theory of Gravity for Extended Objects, *Phys. Rev. D* 73: 104029 doi:10.1103/PhysRevD.73.104029 [hep-th/0409156];

W. D. Goldberger, A. Ross and I. Z. Rothstein, (2014). Black Hole Mass Dynamics and Renormalization Group Evolution, *Phys. Rev. D* 89(12): 124033 doi:10.1103/PhysRevD.89.124033 [arXiv:1211.6095 [hep-th]].

[48] J. Schwinger, (1951). *Phys. Rev.* 82: 664;

B.S. De Witt, (1965). Dynamical Theory of Groups and Fields. In B. S. De Witt and C. De Witt (Eds), *Relativity, Groups and Topology*, Gordon and Breach;

P. B. Gilkey, (1975). *J. Diff. Geom.* 10: 601 (1975);

D.M. McAvity and H. Osborn, (1991). *Class. Quant. Grav.* 8: 603–638.

[49] For a review of heat kernel techniques see:

A. O. Barvinsky and G. A. Vilkovisky, (1985). *Phys. Rept.* 119: 1;

D. V. Vassilevich, (2003). Heat Kernel Expansion: Users Manual, *Phys. Rept.* 388: 279 [hep-th/0306138].

[50] P. Adshead, C. P. Burgess, R. Holman, and S. Shandera, (2017). Power-Counting during Single-Field Slow-Roll Inflation, arXiv:1708.07443 [hep-th].

[51] J. M. Maldacena, (2003). Non-Gaussian Features of Primordial Fluctuations in Single Field Inflationary Models, *JHEP* 0305: 013 doi:10.1088/1126-6708/2003/05/013 [astro-ph/0210603].

[52] M. V. Ostrogradsky, (1950). *Mem. Acad. St. Petersbourg VI* 4: 385;

A. Pais and G. Uhlenbeck, (1950). *Phys. Rev.* 79: 145.

[53] C. Cheung, et al., (2008). The Effective Field Theory of Inflation, *JHEP* 0803: 014 doi:10.1088/1126-6708/2008/03/014 [arXiv:0709.0293 [hep-th]].

[54] L. Senatore, (2016). Lectures on Inflation, doi:10.1142/9789813149441_0008 arXiv:1609.00716 [hep-th].

[55] C. Armendariz-Picon, T. Damour, and V. F. Mukhanov, (1999). k-Inflation, *Phys. Lett. B* 458: 209 doi:10.1016/S0370-2693(99)00603-6 [hep-th/9904075].

For a particularly nice variant complete with a UV pedigree see:

M. Alishahiha, E. Silverstein, and D. Tong, (2004). DBI in the Sky, *Phys. Rev. D* 70: 123505 doi:10.1103/PhysRevD.70.123505 [hep-th/0404084].

[56] C.P. Burgess, R. Holman, G. Tasinato, and M. Williams, (2015). EFT Beyond the Horizon: Stochastic Inflation and How Primordial Quantum Fluctuations Go Classical, *JHEP* 1503: 090 [arXiv:1408.5002 [hep-th]].

[57] C.P. Burgess, R. Holman, and G. Tasinato, (2015). Open EFTs, IR Effects & Late-Time Resummations: Systematic Corrections in Stochastic Inflation, *JHEP* 1503: 090 [arXiv:1512.00169 [gr-qc]].

[58] E. Braaten, H.-W. Hammer, and G. P. Lepage, (2017). Lindblad Equation for the Inelastic Loss of Ultracold Atoms, *Phys. Rev. A* 95(1): 012708 doi:10.1103/PhysRevA.95.012708 [arXiv:1607.08084 [cond-mat.quant-gas]].

E. Braaten, H.-W. Hammer, and G. P. Lepage, (2016). Open Effective Field Theories from Deeply Inelastic Reactions, *Phys. Rev. D* 94(5): 056006 doi:10.1103/PhysRevD.94.056006 [arXiv:1607.02939 [hep-ph]].

[59] C. Agon, V. Balasubramanian, S. Kasko, and A. Lawrence, (2014). Coarse Grained Quantum Dynamics, arXiv:1412.3148 [hep-th].

C. Agn and A. Lawrence, (2017). Divergences in Open Quantum Systems, arXiv:1709.10095 [hep-th].

[60] A. A. Starobinsky, (1986). Stochastic De Sitter (inflationary) Stage in the Early Universe, *Lect. Notes Phys.* 246: 107;

A. A. Starobinsky and J. Yokoyama, (1994). Equilibrium State of a Self-Interacting Scalar Field in the De Sitter Background, *Phys. Rev. D* 50: 6357 [astro-ph/9407016].

5

Soft-Collinear Effective Theory

Thomas BECHER

Albert Einstein Center for Fundamental Physics, Institut für Theoretische Physik, Universität Bern, Sidlerstrasse 5, CH-3012 Bern, Switzerland

Thomas BECHER

Becher, B., *Soft-Collinear Effective Theory* In: *Effective Field Theory in Particle Physics and Cosmology.* Edited by: Sacha Davidson, Paolo Gambino, Mikko Laine, Matthias Neubert and Christophe Salomon, Oxford University Press (2020). © Oxford University Press. DOI: 10.1093/oso/9780198855743.003.0005

Chapter Contents

Preface

This chapter on soft-collinear effective theory (SCET) [1]–[3] assumes some basic familiarity with effective field theory methods. A reader who is completely new to this area should consult the chapters by Aneesh Manohar and Matthias Neubert in this volume, or one of the growing number of lecture notes and books on effective theory [4]–[11].

Alessandro Broggio, Andrea Ferroglia, and I have written an introductory book on soft-collinear effective theory [12]. The construction of the effective theory presented here proceeds along similar lines and follows the same philosophy as the book (which in turn is guided by [3]). However, to keep things interesting for myself and hopefully useful for the reader, the physics examples are completely disjoint from the ones in [12], which focuses on threshold resummation and transverse momentum resummation, while this chapter's collected lectures discuss jet processes and soft-photon physics. In [12], every single computation is spelled out step by step. These lectures required a somewhat quicker pace to cover the relevant material efficiently, but refers to the detailed derivations in the book in those cases where they are not presented here.

5.1 Introduction

Soft-Collinear Effective Theory (SCET) [1]–[3] is the effective field theory (EFT) for processes with energetic particles such as jet production at high-energy colliders. A typical two-jet process is depicted in Figure 5.1. It involves sprays of energetic particles along two directions with momenta $p_{\mathcal{J}}$ and $p_{\bar{\mathcal{J}}}$, accompanied by soft radiation with momentum p_s. Typically such processes involve a scale hierarchy

$$Q^2 = (p_{\mathcal{J}} + p_{\bar{\mathcal{J}}})^2 \gg p_{\mathcal{J}}^2 \sim p_{\bar{\mathcal{J}}}^2 \gg p_s^2 \,. \tag{5.1}$$

In SCET, the physics associated with the hard scale Q^2 is integrated out and absorbed into Wilson coefficients of effective-theory operators, much in the same way that a heavy particle is integrated out when constructing a low-energy theory for light particles only. SCET involves two different types of fields, collinear and soft fields to describe the physics associated with the two low-energy scales $p_{\mathcal{J}}^2$ and p_s^2.

The result of a SCET analysis of a jet cross section is often a factorization theorem of the schematic form

$$\sigma = H \cdot \mathcal{J} \otimes \bar{\mathcal{J}} \otimes S \,. \tag{5.2}$$

The hard function H encodes the physics at the scale Q^2, while the jet functions $\mathcal{J}$ and $\bar{\mathcal{J}}$ depend on the jet scales $p_{\mathcal{J}}^2$ and $p_{\bar{\mathcal{J}}}^2$, respectively, and the soft function S describes the physics at the soft momentum scale. Depending on the observable under consideration, the jet and soft functions are convoluted or multiplied together, as indicated by the $\otimes$ symbol. We derive several such factorization theorems throughout the chapter.

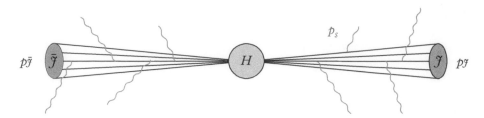

Figure 5.1 *Sketch of a two-jet process.*

The theorem (5.2) is obtained after expanding in the ratios of the scales (5.1) and holds at leading power in the expansion. Its main virtue is scale separation – the property that each of the functions in (5.2) is only sensitive to a single scale. The individual functions furthermore fulfill renormalization group (RG) equations. As was demonstrated in Matthias Neubert's lecture [13], by solving RG equations one can resum the large perturbative logarithms. To do so one evaluates each component function in (5.2) at its natural scale and then evolves them to a common reference scale where they are combined. This resums logarithms such as $\alpha^n \ln^m(Q^2/p_{\bar{\jmath}}^2)$ with $m \leq 2n$, which can spoil the standard perturbative expansion of the cross section. A characteristic feature of jet processes is the presence of double logarithms, $m = 2n$, which are also called Sudakov logarithms. They are the result of an interplay of soft and collinear physics.

In certain cases, the soft or collinear scales can be so low that a perturbative expansion becomes unreliable, even after resumming the large logarithms. Factorization theorems such as (5.2) remain useful also in this situation, because they allow us to separate perturbative from non-perturbative physics. In fact, every high-energy hadron-collider computation involves non-perturbative Parton–Distribution Functions (PDFs). Having a factorization theorem that separates them from the hard scattering process is crucial to be able to make predictions.

Traditionally, factorization theorems were derived purely diagrammatically; see [14]–[17] for an introduction to these methods. SCET is of course based on the same physics as the diagrammatic techniques and there is a close relation between the two approaches. An advantage of SCET is that effective theory provides an operator formulation of the low-energy physics, which simplifies and systematizes the analysis. This is especially important for complicated problems such as the factorization of power suppressed contributions. Via the RG equations, SCET also provides a natural framework to perform resummations.

Compared to traditional EFT such as Fermi theory, SCET involves several complications. First of all, we cannot simply integrate out particles: quarks and gluons are still present in the low-energy theory. Instead, we split the fields into modes

$$\begin{matrix} H & \jmath & \bar{\jmath} & S \\ \phi = \phi_h & + \phi_c & + \phi_{\bar{c}} & + \phi_s \end{matrix} \qquad (5.3)$$

containing the contributions of the different momentum regions of the quark or gluon fields $\phi \in \{q, A^a_\mu\}$. The hard mode ϕ_h, which describes contributions where the particles are off shell by a large amount, is then integrated out, while the low-energy modes become the fields of the effective theory. The term integrating out refers to a path-integral formalism introduced by Wilson, where we would integrate over the field ϕ_h since it does not appear as an external state. This is, however, not how things are done in practice. Instead, we start by writing down the most general effective Lagrangian with the low-energy fields. We then adjusts the couplings of the different terms in the Lagrangian to reproduce the contribution of the hard momentum region. These couplings are also called *Wilson coefficients* and the process of adjusting them to reproduce the full theory result is called *matching*, see e.g. [18]. An important and non-trivial element of the analysis is to identify the relevant momentum modes for the problem at hand, which are the degrees of freedom of the effective theory. This is done by analysing full theory diagrams and provides the starting point of the effective theory construction.

Not only does SCET contain several different fields for each QCD particle, a second complication is that the different momentum components of the fields scale differently. Momentum components transverse to the jet direction are always small, but the components along the jet directions are large. To perform a derivative expansion of the effective Lagrangian, we therefore need to split the momenta into different components. This is done by introducing reference vectors $n^\mu \propto p^\mu_{\bar{\jmath}}$ and $\bar{n}^\mu \propto p^\mu_{\bar{\jmath}}$ in the directions of the two jets. The fact that the momentum components of the collinear particles along the jet are unsuppressed leads to a final complication, namely that we can write down operators with an arbitrary number of such derivatives. One way to take all these operators into account is to make operators non-local along the corresponding light cone directions, as explained later.

5.2 Warm Up: Soft Effective Theory

For the reasons discussed in the introduction, the construction of SCET is technically involved. We address all of the complications, but due to their presence it will take some time to set up the effective Lagrangian. Rather than immediately diving into technicalities, I first discuss a simpler example where only some of the difficulties are present and where we can obtain physics results a bit quicker. Therefore, before turning to jet processes, I begin with electron–electron scattering in QED, as depicted in Figure 5.2.

To keep things simple, I will assume that the electron energies are of the order of the electron mass m_e. The electrons can thus be relativistic but not ultra-relativistic (having them ultra-relativistic would bring us back into the realm of SCET). Instead of SCET, we then deal with soft effective field theory (SET). We will use SET to obtain a classic factorization theorem for soft photon radiation in QED. This seminal result was originally derived by Yennie, Frautschi, and Suura in 1961 using diagrammatic methods [19]. All the steps in our derivation are also valid in QCD and later in the course we will use our results to analyse soft-gluon effects in jet processes.

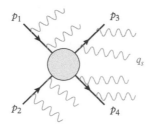

Figure 5.2 *Electron–electron scattering, including soft-photon radiation.*

5.2.1   Soft Photons in Electron–Electron Scattering

Note that we cannot avoid the presence of soft photons in QED processes. No matter how good our detectors are, photons with energies below some threshold will always go unnoticed and since it costs little energy to produce them, any QED final state will always include soft photons. Indeed, trying to compute higher-order corrections to scattering processes without accounting for their presence leads to divergent results for cross sections. The divergence signals that completely exclusive QED cross sections are not physical, which was understood very early on [20]–[22]. When we talk about electron–electron scattering, we really measure the inclusive process

$$e^-(p_1) + e^-(p_2) \rightarrow e^-(p_3) + e^-(p_4) + X_s(q_s), \tag{5.4}$$

where X_s is any state with an arbitrary number of soft photons which carry the total momentum q_s. The constraints imposed on the soft momenta depend of course on the detailed experimental setup, but for our purposes it will be sufficient to assume that the total energy fulfills $E_s \ll m_e$. We will now analyse the process (5.4) up to terms suppressed by powers of the expansion parameter $\lambda = E_\gamma/m_e$. Given the large scale hierarchy, effective-field theory methods should be useful to analyse this situation.

So what is the effective Lagrangian for soft photons with $E_\gamma \ll m_e$ by themselves? As always, it is useful to organize the operators in the effective Lagrangian by their dimension

$$\mathcal{L}_{\mathrm{eff}}^\gamma = \mathcal{L}_4^\gamma + \frac{1}{m_e^2}\mathcal{L}_6^\gamma + \frac{1}{m_e^4}\mathcal{L}_8^\gamma \tag{5.5}$$

because the coefficients of the higher-dimensional operators involve inverse powers of the large scale, which is m_e in our case. The contributions of these operators will therefore be suppressed by powers of λ. The leading Lagrangian only involves a single term

$$\mathcal{L}_4^\gamma = -\frac{1}{4}F_{\mu\nu}F^{\mu\nu}, \tag{5.6}$$

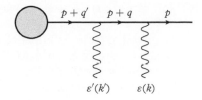

Figure 5.3 *Soft emissions from an outgoing electron. Note that $q' = k + k'$ and $q = k$.*

whose coefficient can be adjusted to the canonical value by rescaling the photon field. The leading-power effective-field-theory Lagrangian is therefore simply the one for free photons. This makes sense, since the effective theory is obtained by integrating out the massive particles which leaves only the photons. Integrating out the electrons does induce higher-power operators which describe photon-photon interactions. While we will not need them, it is an interesting exercise to analyse these higher-power terms; the first non-trivial ones arise at dimension 8; see e.g. [10, 23].

However, $\mathcal{L}_{\text{eff}}^{\gamma}$ is by itself not sufficient. While the energy of the soft radiation is too small to produce additional electron–positron pairs, we do need to include the incoming and outgoing electrons in the effective theory. (Due to fermion number conservation they remain even as $E_\gamma \to 0$.) Consider, as in Figure 5.3, an outgoing electron with momentum $p^\mu = m_e v^\mu$. The momentum on internal fermion lines is $p + q$, where q is a soft photon momentum. We can expand the internal fermion propagators in the small momentum q. Neglecting higher order terms suppressed by q^μ/m_e, we find

$$
\Delta_F(p+q) = i \frac{\not{p} + \not{q} + m_e}{(p+q)^2 - m_e^2 + i0} = i \frac{\not{p} + m_e}{2p \cdot q + i0} = \frac{\not{v} + 1}{2} \frac{i}{v \cdot q + i0}
$$

$$
\equiv P_v \frac{i}{v \cdot q + i0},
$$

(5.7)

where we introduced the projection operator

$$
P_v = \frac{1 + \not{v}}{2},
$$

(5.8)

which has the properties

$$
\not{v} P_v = P_v, \qquad\qquad P_v^2 = P_v, \qquad\qquad P_v \not{\varepsilon} P_v = P_v \varepsilon \cdot v.
$$

(5.9)

Using these, the outgoing-leg part of the diagram in Figure 5.3 simplifies to

$$
\bar{u}(p) P_v \frac{i}{v \cdot q} (-ie\varepsilon \cdot v) P_v \frac{i}{v \cdot q'} (-ie\varepsilon' \cdot v) \dots .
$$

(5.10)

This form of the expanded soft emissions is well known and called the *eikonal approximation*.

Can we obtain the expanded expression from an effective Lagrangian? This should be possible, we just need to view the expanded propagator (5.7) as the propagator in the effective theory and the emissions in the expanded diagram must be resulting from a Feynman rule $-iev^\mu$ for the electron–photon vertex. So we already know the Feynman rules of the effective theory and just need to write down a Lagrangian which produces them! Consider

$$\mathcal{L}_{\text{eff}}^v = \bar{h}_v(x)\, iv \cdot D h_v(x), \qquad (5.11)$$

where h_v is an auxiliary fermion field which fulfills $P_v h_v = \slashed{v} h_v = h_v$. (Such a field can be obtained by multiplying a regular fermion field with P_v.) As usual, the propagator can be obtained by inverting the quadratic part of the Lagrangian in Fourier space and multiplying by i. This indeed yields $i/(v \cdot q)$ as in (5.7). The factor of P_v arises because the external spinor of the auxiliary field h_v includes such a factor due to the property $P_v h_v = h_v$ of the field. Thanks to the projection property $P_v^2 = P_v$ a single power of this matrix on the fermion line is sufficient. Also the photon vertex comes out correctly. Inserting $D_\mu = \partial_\mu + ieA_\mu$ into (5.11), the vertex Feynman rule is $-iev^\mu$ so that the Lagrangian (5.11) correctly reproduces the eikonal expression (5.10) obtained by expanding the original QED diagram. Note that the propagator of the field h_v only has a single pole in the energy corresponding to the fermion. The anti-fermion pole has been lost in the expansion (5.7). This is perfectly fine since the field h_v describes a fermion close to its mass-shell with momentum $m_e v^\mu + q^\mu$, where q^μ is a soft-photon momentum. In this situation anti-fermions cannot arise as external particles and their virtual effects can be absorbed into the Wilson coefficients of the effective theory.

Our construction of (5.11) highlights the close connection between diagrammatic methods and EFT: we constructed the effective Lagrangian in such a way that it reproduces the expansion of the full-theory diagram in (5.10). We will follow the same strategy when setting up the SCET Lagrangian below. The astute reader will have recognized (5.11) as the Lagrangian of heavy quark effective theory (HQET), covered in Thomas Mannel's lectures [29] which contain a path-integral derivation of the same Lagrangian.

The field h_v cannot describe other fermion lines in the process (5.4) which have different velocities. To account for all four fermion lines in the process, we need to include four auxiliary fermion fields so that the full effective Lagrangian takes the form

$$\mathcal{L}_{\text{eff}} = \sum_{i=1}^{4} \bar{h}_{v_i}(x)\, iv_i \cdot D h_{v_i}(x) - \frac{1}{4} F_{\mu\nu} F^{\mu\nu} + \Delta \mathcal{L}_{\text{int}}, \qquad (5.12)$$

where the velocity vectors are given by $v_i^\mu = p_i^\mu / m_e$. This Lagrangian has some features which will also be present in SCET. First of all, $\mathcal{L}_{\text{eff}}$ depends on reference vectors along the direction of the energetic particles. In SCET, we will deal with jets of energetic massless particles and the reference vectors will be light cone vectors along the directions of the jets instead of the velocity vectors. Secondly, we need different fields to represent the electrons along the different directions in the effective theory, while all of these were described by a single field in QED. The same will be true in SCET. In the present case, the different fields are modes of the full-theory fermion field, which live in small momentum regions around the reference momenta $m_e v_i^\mu$.

What remains is to write down the interaction terms. These have the form

$$\Delta \mathcal{L}_{\text{int}} = C_{\alpha\beta\gamma\delta}(v_1, v_2, v_3, v_4, m_e) \, h_{v_1}^\alpha(x) \, h_{v_2}^\beta(x) \, \bar{h}_{v_3}^\gamma(x) \, \bar{h}_{v_4}^\delta(x) \,. \tag{5.13}$$

More elegantly we could write

$$\Delta \mathcal{L}_{\text{int}} = \sum_i C_i(v_1, v_2, v_3, v_4, m_e) \, \bar{h}_{v_3}(x) \Gamma_i h_{v_1}(x) \, \bar{h}_{v_4}(x) \Gamma_i h_{v_2}(x) \,, \tag{5.14}$$

with a basis of Dirac matrices Γ_i since $\Delta \mathcal{L}_{\text{int}}$ has to be a scalar. However, the less elegant form (5.13), in which the Wilson coefficients depend on the Dirac indices α, β, γ, δ of the fields, will be convenient to perform the matching. In principle we could also write down interaction terms involving only two fields such as

$$\Delta \mathcal{L}'_{\text{int}} = C_{\alpha\beta}(v_1, v_3) \, h_{v_1}^\alpha(x) \bar{h}_{v_3}^\beta(x) \,, \tag{5.15}$$

but when performing a matching computation we would find that their Wilson coefficients $C_{\alpha\beta}$ are zero if the velocities are different, since the corresponding operator would describe a process in which a fermion spontaneously changes its velocity, which violates momentum conservation. Adding an additional photon field to the operator (5.15) would allow for very small velocity changes, but $\mathcal{O}(1)$ changes are again impossible, if the only particles in the theory are soft photons. This leaves (5.13) as the simplest non-trivial interaction term. Of course, we could also write interactions terms with covariant derivatives or more fields, but these are higher-dimensional operators, whose contributions are suppressed by powers of the electron mass as in (5.5).

The leading-power effective Lagrangian is thus complete. All that is left is to determine the Wilson coefficients $C_{\alpha\beta\gamma\delta}(v_1, v_2, v_3, v_4)$ in the interaction term. To do so, we should compute the same quantity in QED and in the effective theory and then adjust the Wilson coefficient to reproduce the QED result. The simplest quantity we can use to do the matching is the amputated on-shell Green's function for $e^-(p_1) + e^-(p_2) \to e^-(p_3) + e^-(p_4)$. The relevant QED diagrams are shown in Figure 5.4. In the effective theory, only the interaction Lagrangian (5.13) contributes and the result is directly equal to the Wilson coefficient

$$= C_{\alpha\beta\gamma\delta}(v_1, v_2, v_3, v_4, m_e) \,. \tag{5.16}$$

Figure 5.4 *QED tree-level diagrams for the matching computation needed to extract the Wilson coefficient (5.13).*

To reproduce the QED result, the Wilson coefficient must be set equal to the on-shell QED Green's function (which is the same as the scattering amplitude, up to the external spinors). At the moment, we are only discussing tree-level matching but the same simple relation also holds at loop level in dimensional regularization. The reason is that all loop corrections to the on-shell amplitude vanish in the effective theory because they are given by scaleless integrals. This is the case since all the photon momenta were set to zero and the electron mass is no longer present in the low-energy theory. This makes it very convenient to use such amplitudes for the matching and shows that the Wilson coefficient has a simple interpretation.

Our effective theory factorizes low- and high-energy physics: the hard scattering of the electrons is part of the Wilson coefficient, which depends on the high-energy scale m_e, while the low-energy diagrams in the effective theory only depend on photon-energy scales. We can obtain a very elegant form of the low-energy matrix element by introducing the *Wilson line*

$$S_i(x) = \exp\left[-ie \int_{-\infty}^{0} ds\, v_i \cdot A(x + sv_i)\right]. \tag{5.17}$$

One way to obtain such a Wilson line is to add a point-like source which travels along the path $y^\mu(s) = x^\mu + sv_i^\mu$ to the Lagrangian of electrodynamics. This is indeed how the outgoing electrons behave: since their energy is much larger than the photon energies, they travel without recoiling when emitting photons.

To see that (5.17) reproduces the emission pattern of an incoming electron, let us take the matrix element with a photon in the final state. To obtain it, we can expand the Wilson line in the coupling and since we are taking this matrix element in a free theory (see (5.5)) only the first-order term gives a non-vanishing contribution:

$$\begin{aligned}
\langle \gamma(k)|S_i(0)|0\rangle &= -ie \int_{-\infty}^{0} ds\, v_i^\mu \langle \gamma(k)|A_\mu(sv_i)|0\rangle \\
&= -ie \int_{-\infty}^{0} ds\, v_i \cdot \varepsilon(k) e^{isv_i \cdot k} = e \frac{v_i \cdot \varepsilon(k)}{-v_i \cdot k + i0}.
\end{aligned} \tag{5.18}$$

We indeed reproduce the eikonal structure (5.10) we found expanding the diagram. To ensure the convergence of the integral in the second line of (5.18) at $s = -\infty$, the exponent $v_i \cdot k$ must have a negative imaginary part, which amounts to the $+i0$ prescription in the eikonal propagator. Analogously to (5.17), we can also define a Wilson line describing the radiation of an outgoing particle

$$\bar{S}_i^\dagger(x) = \exp\left[-ie \int_0^\infty ds\, v_i \cdot A(x + sv_i)\right]. \qquad (5.19)$$

We define the dagger of the Wilson line since the outgoing particle is produced by the conjugate field. The matrix element of $\bar{S}_i(x)$ is the same eikonal expression as (5.18), but with a $-i0$ prescription.

What is important in the following is that the Wilson line fulfills the equation

$$v_i \cdot D S_i(x) = 0, \qquad (5.20)$$

and another way of introducing the object (5.17) is to define it as the solution to this differential equation. The reader is invited to check that expression (5.17) indeed fulfills (5.20). An explicit solution to this exercise can be found in Appendix D of [12].

Let us now perform a field redefinition by writing the fermion fields of the incoming fermion fields in the form

$$h_{v_i}(x) = S_i(x)\, h_{v_i}^{(0)}(x), \qquad (5.21)$$

as a Wilson line along the corresponding direction times a new fermion field $h_{v_i}^{(0)}(x)$. The fermion Lagrangian then takes the form

$$
\begin{aligned}
\bar{h}_{v_i}(x)\, iv_i \cdot D h_{v_i}(x) &= \bar{h}_{v_i}^{(0)}(x)\, S_i^\dagger(x)\, iv_i \cdot D S_i(x)\, h_{v_i}^{(0)}(x) \\
&= \bar{h}_{v_i}^{(0)}(x)\, S_i^\dagger(x)\, S_i(x)\, iv_i \cdot \partial\, h_{v_i}^{(0)}(x) \\
&= \bar{h}_{v_i}^{(0)}(x)\, iv_i \cdot \partial\, h_{v_i}^{(0)}(x).
\end{aligned}
\qquad (5.22)
$$

The field $h_{v_i}^{(0)}(x)$ is a free field; we were able to remove the interactions with the soft photons using the *decoupling transformation* (5.21). For the fields describing the outgoing fields, the decoupling is performed using $\bar{S}_i(x)$.[1] The same method is used in SCET to decouple soft gluons [2] as we show in the next section.

---

[1] If we have in- and outgoing fields along the same direction, it is no longer clear how to decouple the soft radiation. Indeed, if we consider nearly forward scattering, the effective theory we formulated is no longer appropriate. We need to include and resum the effects associated with the Coulomb potential between the two electrons in the process. Similar effects arise in QCD and go under the name of Coulomb or Glauber gluons.

While the Wilson lines cancel in the fermion Lagrangian, they are present in the interaction Lagrangian (5.14) which now takes the form

$$\Delta \mathcal{L}_{\text{int}} = \sum_i C_i(v_1, v_2, v_3, v_4)\, \bar{h}_{v_3}^{(0)}\, \bar{S}_3^\dagger\, \Gamma_i\, S_1\, h_{v_1}^{(0)}\; \bar{h}_{v_4}^{(0)}\, \bar{S}_4^\dagger\, \Gamma_i\, S_2\, h_{v_2}^{(0)} \tag{5.23}$$

so that we end up with Wilson lines along the directions of all particles in the scattering process.

As a final step, we now use our effective theory to compute the scattering amplitude for $\mathcal{M}(e^-(p_1) + e^-(p_2) \to e^-(p_3) + e^-(p_4) + X_s(k))$, where the final state contains n photons, $X_s(k) = \gamma(k_1) + \gamma(k_2) + \dots \gamma(k_n)$. Since the photons no longer interact with the fermions after the decoupling, the relevant matrix element factorizes into a fermionic part times a photonic matrix element. Using the form (5.14) of the interaction Lagrangian, the amplitude is given by

$$\mathcal{M} = \sum_i C_i\, \bar{u}(v_3)\, \Gamma_i\, u(v_1)\, \bar{u}(v_4)\, \Gamma_i\, u(v_2)\, \langle X_s(k) | \bar{S}_3^\dagger\, S_1\, \bar{S}_4^\dagger\, S_2 | 0 \rangle$$

$$= \mathcal{M}_{ee}\, \langle X_s(k) | \bar{S}_3^\dagger\, S_1\, \bar{S}_4^\dagger\, S_2 | 0 \rangle\,, \tag{5.24}$$

where we have used in the second line that the Wilson coefficient times the spinors is simply the amplitude $\mathcal{M}_{ee} = \mathcal{M}(e^-(p_1) + e^-(p_2) \to e^-(p_3) + e^-(p_4))$ for the process without soft photons. So we have shown that the amplitude factorizes into an amplitude without soft photons times a matrix element of Wilson lines. Analogous statements hold for soft gluon emissions in QCD, except that the Wilson lines will be matrices in colour space and we have to keep track of the colour indices. We can square our factorized amplitude to obtain the cross section, which takes the form

$$\sigma = \mathcal{H}(m_e, \{\underline{v}\})\, \mathcal{S}(E_s, \{\underline{v}\})\,, \tag{5.25}$$

where the hard function $\mathcal{H}$ is the cross section for the process without soft photons,

$$\mathcal{H}(m_e, \{\underline{v}\}) = \frac{1}{2E_1 2E_2 |\vec{v}_1 - \vec{v}_2|} \frac{d^3 p_3}{(2\pi)^3 2E_3} \frac{d^3 p_4}{(2\pi)^3 2E_4} |\mathcal{M}_{ee}|^2 (2\pi)^4 \delta^{(4)}(p_1 + p_2 - p_3 - p_4)\,, \tag{5.26}$$

while the soft function $\mathcal{S}$ is the Wilson line matrix element squared, together with the phase-space constraints on the soft radiation,

$$\mathcal{S}(E_s, \{\underline{v}\}) = \sum_{X_s}\!\!\!\!\!\int \left| \langle X_s | \bar{S}_3^\dagger\, S_1\, \bar{S}_4^\dagger\, S_2 | 0 \rangle \right|^2 \theta(E_s - E_{X_s})\,. \tag{5.27}$$

The sum and integral symbol indicates that we have to sum over the different multi-photon final states and integrate over their phase space. Note that both the hard and soft functions depend on the directions $\{\underline{v}\} = \{v_1, \dots, v_4\}$ of the electrons. For simplicity, we only constrain the total soft energy; the constraints in real experiments will of course be more complicated. To obtain (5.25) we expanded the small soft momentum out of the momentum conservation δ function

$$\delta^{(4)}(p_1 + p_2 - p_3 - p_4 - k) = \delta^{(4)}(p_1 + p_2 - p_3 - p_4) + \mathcal{O}(\lambda), \qquad (5.28)$$

which is then part of the hard function $\mathcal{H}$ in (5.26).

The Wilson-line matrix elements such as (5.27) which define the soft functions have a very interesting property in QED: they exponentiate,

$$\mathcal{S}(E_s, \{\underline{v}\}) = \exp\left[\frac{\alpha}{4\pi} S^{(1)}(E_s, \{\underline{v}\})\right], \qquad (5.29)$$

so that the all-order result is obtained by exponentiating the first-order result. We will not derive this formula here, but the key ingredient in the derivation is the eikonal identity

$$\frac{1}{v \cdot k_1 \, v \cdot (k_1 + k_2)} + \frac{1}{v \cdot k_2 \, v \cdot (k_1 + k_2)} = \frac{1}{v \cdot k_1} \frac{1}{v \cdot k_2}, \qquad (5.30)$$

which allows us to rewrite sums of diagrams with multiple emissions as products of diagrams with a single one. In non-Abelian gauge theories such as QCD, there are genuine higher-order corrections to soft matrix elements since the different diagrams, and therefore the different terms on the left-hand side of (5.30), have different colour structures and cannot be combined. However, the higher-order corrections only involve certain maximally non-Abelian colour structures [24]–[28].

While the inclusive cross section is finite, the hard and soft functions in (5.25) individually suffer from divergences. The soft function suffers from ultraviolet (UV) divergences, which can be regularized using dimensional regularization. These UV divergences can be absorbed into the Wilson coefficients of the effective theory, which are encoded in the hard function. This renormalization renders the hard function finite, at the expense of introducing a renormalization scale μ, which in our context is often called the factorization scale. After renormalization the theorem takes the form

$$\sigma = \mathcal{H}(m_e, \{\underline{v}\}, \mu) \, \mathcal{S}(E_s, \{\underline{v}\}, \mu), \qquad (5.31)$$

and the μ dependence of the functions fulfills an RG equation. Since the cross section is finite, the hard and soft anomalous dimensions must be equal and opposite. On a more concrete level, we see that the on-shell amplitudes that define the hard function suffer from infrared (IR) divergences which cancel against the UV divergences of the soft function. Since the soft function exponentiates, also the divergences in the hard amplitudes must have this property.

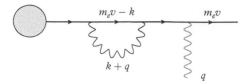

Figure 5.5 *Loop correction on a fermion line.*

5.2.2   Expansion of Loop Integrals and the Method of Regions

When constructing $\mathcal{L}_{\text{eff}}$, we have expanded in the soft-photon momenta. This is fine for tree-level diagrams, but how about loops? Of course, the Taylor expansion does *not* commute with the loop integrations and to correct for this, we have to perform matching computations. We will now see that the part which gets lost in the naive low-energy expansion can be obtained by expanding the loop integrand in the region of large loop momentum. This is an example of a general technique called the *method of regions* [30, 31] to expand loop integrals around various limits. We will use this method when constructing SCET, but it is instructive to discuss it with a simple example integral in QED.

Let us consider the diagram shown in Figure 5.5. For our discussion the numerators of the diagrams will not lead to complications and for simplicity we will therefore consider the associated scalar integral

$$F = \int d^d k \frac{1}{(k+q)^2} \frac{1}{(m_e v - k)^2 - m_e^2}, \tag{5.32}$$

where q is a soft-photon momentum. In the low-energy theory, we assumed that $k^\mu \sim q^\mu \ll m_e$. Expanding the integrand in this region gives integrals of the form

$$F_{\text{low}} = \int d^d k \frac{1}{(k+q)^2} \frac{1}{-2m_e v \cdot k} \left\{ 1 + \frac{k^2}{2m_e v \cdot k} + \dots \right\}, \tag{5.33}$$

The expansion produces exactly the linear propagators $i/v \cdot k$ encountered in our tree-level discussion in the previous section. Of course, the loop momentum is integrated all the way to infinity and the expansion we have performed is no longer valid when $k^\mu \sim m_e$ or larger. We could follow Wilson and work with a hard cutoff to ensure that the loop momentum would never be too large, but we want to use dimensional regularization for the low-energy theory and not restrict the loop momentum. Looking at the integrals (5.33) we see that the expansion has produced ultraviolet divergences which are stronger than the one in the original integral, but the integrals are well-defined in dimensional regularization.

To correct the problems from naively expanding the integrand, we consider the difference

$$F_{\text{high}} \equiv F - F_{\text{low}} \tag{5.34}$$

$$= \int d^d k \frac{1}{(k+q)^2} \left[\frac{1}{(m_e v - k)^2 - m_e^2} - \frac{1}{-2m_e v \cdot k} \left\{ 1 + \frac{k^2}{2mv \cdot k} + \cdots \right\} \right].$$

By construction, the integrand has only support for $k^\mu \gg q^\mu$ since the square bracket tends to zero for $k^\mu \sim q^\mu$. We can therefore expand the integrand around $q^\mu = 0$ by expanding the first propagator denominator

$$F_{\text{high}} = \int d^d k \frac{1}{k^2} \left\{ 1 - \frac{q^2}{2q \cdot k} + \cdots \right\}$$

$$\times \left[\frac{1}{(m_e v - k)^2 - m_e^2} - \frac{1}{-2m_e v \cdot k} \left\{ 1 + \frac{k^2}{2mv \cdot k} + \cdots \right\} \right].$$

Now we can evaluate the integrals one by one to get the high-energy part. What simplifies this task is that integrals of the form

$$I(\alpha, \beta, \gamma) = \int d^d k \left(k^2 \right)^\alpha (v \cdot k)^\beta (q \cdot k)^\gamma = 0 \tag{5.35}$$

all vanish. To show this, rescale $k \to \lambda k$. This yields $I(\alpha, \beta, \gamma) = \lambda^{d+2\alpha+\beta+\gamma} I(\alpha, \beta, \gamma)$ for any $\lambda > 0$. Dropping the scaleless integrals, we get

$$F_{\text{high}} = \int d^d k \frac{1}{k^2} \left\{ 1 - \frac{q^2}{2q \cdot k} + \cdots \right\} \frac{1}{(m_e v - k)^2 - m_e^2}, \tag{5.36}$$

which is the expansion of the integrand for $k^\mu \sim m_e \gg q^\mu$. So we observe that we obtain the full result by performing the expansion of the integrand in two regions (low and high k^μ), integrating each term and adding the two results.

We can summarize the method of regions expansion as follows:

(a) Consider all relevant scalings (regions) of the loop momenta. In our example the scalings are $k_\mu \sim m_h$ (hard region) and $k_\mu \sim q_\mu$ (soft region).

(b) Expand the loop integral in each region.

(c) Integrate each term over the full phase space $\int d^d k$.

(d) Add up the contributions.

This technique provides a general method to expand loop integrals around different limits and can be used in many different kinematical situations [31]. Seeing that we integrate every region over the full phase space, we could be worried that this would lead to a double counting, but this is not the case. As our construction shows, the overlap region is given by scaleless integrals which can be dropped as we did in the last step. For this to be true, it was important that we consistently expanded away small momenta in the low energy region. Because of this, we ended up with single scale integrals, which become scaleless upon further expansion. If this is not done, we will need to eliminate the overlap region using subtractions, which are also called zero-bin subtractions in the context of SCET [32]. A second important ingredient for the method is that we have to ensure that the expanded integrals are properly regularized and in some cases dimensional regularization alone is not sufficient. The reader interested to learn more about these issues can consult [31, 33].

The method of region technique has a close connection to EFTs in that the low-energy regions correspond to degrees of freedom in the EFT and the expanded full theory diagrams are equivalent to effective-theory diagrams, as we have seen in the example of soft effective theory. The contribution from the hard region gets absorbed into the Wilson coefficients. In the next section, we apply the method of regions technique to the Sudakov form factor integral, which allows us to identify the degrees of freedom relevant in this case. In addition to the soft region, we will find contributions from momentum configurations where the loop momentum is collinear to external momentum. As a consequence, the relevant effective theory then contains not only soft, but also collinear, particles.

5.3   Soft-Collinear Effective Theory (SCET)

This section reviews the construction of the effective theory in detail. To do so, we consider the simplest problem in which both soft and collinear particles play a role, namely the Sudakov form factor. By itself this is not a physical quantity, but it arises as a crucial element in many collider processes.

5.3.1   Method of Regions for the Sudakov Form Factor

Figure 5.6 shows the one-loop contribution to the Sudakov form factor. We define $L^2 = -l^2 - i0$, $P^2 = -p^2 - i0$ and $Q^2 = -(l-p)^2 - i0$ and will analyse the form factor in the limit

$$L^2 \sim P^2 \ll Q^2 . \tag{5.37}$$

This is the limit of large momentum transfer and small invariant mass, the same kinematics which is relevant for the jet process depicted in Figure 5.1. Indeed, the corresponding loop correction will also arise in the computation of the jet cross section. The small off-shellness of the external lines arises in this case because of soft and collinear emissions from these lines. If the quantities Q^2, P^2, and L^2 are all positive, the form factor is real and analytic, the results in other regions can be obtained by analytic continuation, taking into account the $i0$ prescriptions specified above (5.37).

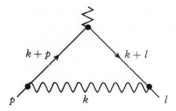

Figure 5.6 *One-loop contribution to the Sudakov form factor.*

We want to find out which momentum modes are relevant in the Sudakov problem and how the different components of the momenta scale compared to the external momenta. Tensor loop integrals involve exactly the same momentum regions as scalar ones since all the complications arise from expanding propagator denominators. For our purposes it is therefore sufficient to study the scalar loop integral

$$I = i\pi^{-d/2}\mu^{4-d}\int d^d k \frac{1}{(k^2+i0)\left[(k+p)^2+i0\right]\left[(k+l)^2+i0\right]}. \tag{5.38}$$

For convenience, we have included a prefactor to make the integral dimensionless; in the computation of the full diagram the associated scale μ arises from the coupling in $d = 4 - 2\varepsilon$ dimensions.

To perform the expansion of the integral around the limit (5.37), it is useful to introduce light-like reference vectors along p^μ and l^μ, in analogy to the vectors v_i^μ we introduced in our discussion of soft photons. To be explicit, we choose our coordinate system such that

$$n^\mu = (1,0,0,1) \approx p^\mu/p^0, \tag{5.39}$$

$$\bar{n}^\mu = (1,0,0,-1) \approx l^\mu/l^0, \tag{5.40}$$

with $n^2 = \bar{n}^2 = 0$ and $n\cdot\bar{n} = 2$. Using these light cone vectors any four vector can be decomposed in the form

$$\begin{aligned}
p^\mu &= n\cdot p\frac{\bar{n}^\mu}{2} + \bar{n}\cdot p\frac{n^\mu}{2} + p_\perp^\mu \\
&= \quad p_+^\mu \quad + \quad p_-^\mu \quad + p_\perp^\mu.
\end{aligned} \tag{5.41}$$

The quantities $(n\cdot p, \bar{n}\cdot p, p_\perp^\mu)$ are the light cone components of the vector and we now discuss how these components scale. To do so, it is useful to define a small expansion parameter

$$\lambda^2 \sim P^2/Q^2 \sim L^2/Q^2 \ll 1. \tag{5.42}$$

Note that

$$p^2 = n \cdot p\, \bar{n} \cdot p + p_\perp^2 .\tag{5.43}$$

and due to $p^2 \sim \lambda^2 Q^2$ and $p^\mu = p_-^\mu + \mathcal{O}(\lambda) = Q/2\, n^\mu + \mathcal{O}(\lambda)$ the components of the external momenta must scale as

$$(n \cdot p, \bar{n} \cdot p,\ p_\perp^\mu)$$
$$p^\mu \sim (\lambda^2,\ 1,\ \lambda)\, Q,$$
$$l^\mu \sim (1,\ \lambda^2,\ \lambda)\, Q.\tag{5.44}$$

In our context the term *scaling as λ^a* means that the given quantity approaches the limit $\lambda \to 0$ as the a-th power of the expansion parameter λ.

We now perform the region expansion of our integral after assigning different scalings to the loop momentum k^μ and expanding the integrand in each region, proceeding exactly as outlined at the end of Section 5.2.2. The following scalings yield non-zero contributions

$$
\begin{array}{llc}
 & & (n \cdot k, \bar{n} \cdot k, k_\perp^\mu) \\
\text{hard} & (h) & (1,\ 1,\ 1)\, Q, \\
\text{collinear to } p^\mu & (c) & (\lambda^2,\ 1,\ \lambda)\, Q, \\
\text{collinear to } l^\mu & (\bar{c}) & (1,\ \lambda^2,\ \lambda)\, Q, \\
\text{soft} & (s) & (\lambda^2,\ \lambda^2,\ \lambda^2)\, Q.
\end{array}
\tag{5.45}
$$

For brevity, we refer to the momenta which are collinear to p^μ simply as *collinear* and to the ones collinear to l^μ *anti-collinear*. All other scaling choices $(\lambda^a, \lambda^b, \lambda^c)$ for the loop momentum lead to scaleless integrals upon performing the expansion—pick one and check!

Since in the soft region all components of the loop momentum scale as λ^2, we have

$$k_s^2 \sim \lambda^4 Q^2 \sim \frac{P^2 L^2}{Q^2} \ll P^2 \sim L^2 .\tag{5.46}$$

This is the hierarchy we advertised in (5.1), and since $k_s^2 \ll k_c^2$, this mode is sometimes also called ultra-soft. For some other observables, the soft mode scales as $(\lambda, \lambda, \lambda)$. The version of SCET for this situation is called SCET$_\mathrm{II}$ to distinguish it from the one relevant for the Sudakov form factor which is also called SCET$_\mathrm{I}$. SCET$_\mathrm{II}$ involves so-called rapidity logarithms which are not present in SCET$_\mathrm{I}$ and there are different formalisms to deal with them. Direct exponentiation based on the *collinear anomaly* [34], or resummation using the *rapidity* RG [35, 36]. For our introduction, we will exclusively work with SCET$_I$. The reader interested to learn more about SCET$_{II}$ can consult the book [12].

Let us now expand the integrand in the different regions to leading power. To get the leading-power integrand in the hard region, we simply set all the suppressed momentum components to zero, which amounts to replacing $p^\mu \to p_-^\mu$ and $l^\mu \to l_+^\mu$. This leads to

$$I_h = i\pi^{-d/2}\mu^{4-d} \int d^d k \frac{1}{\left(k^2 + i0\right)\left(k^2 + 2k_- \cdot l_+ + i0\right)\left(k^2 + 2k_+ \cdot p_- + i0\right)}. \qquad (5.47)$$

The corresponding loop integral is standard; it is simply the on-shell form factor integral. Performing it in the usual way, we obtain

$$I_h = \frac{\Gamma(1+\varepsilon)}{2l_+ \cdot p_-} \frac{\Gamma^2(-\varepsilon)}{\Gamma(1-2\varepsilon)} \left(\frac{\mu^2}{2l_+ \cdot p_-}\right)^\varepsilon. \qquad (5.48)$$

For $\varepsilon \to 0$, we encounter soft and collinear infrared divergences when evaluating the integral.

In the collinear region the integration momentum scales as $k^\mu \sim (\lambda^2, 1, \lambda)\, Q$ and $k^2 \sim \lambda^2 Q^2$. We can therefore expand

$$(k+l)^2 = 2k_- \cdot l_+ + \mathcal{O}(\lambda^2), \qquad (5.49)$$

while the other two propagators stay as they are. We end up with the integral

$$I_c = i\pi^{-d/2}\mu^{4-d} \int d^d k \frac{1}{\left(k^2 + i0\right)\left(2k_- \cdot l_+ + i0\right)\left[(k+p)^2 + i0\right]}. \qquad (5.50)$$

This loop integral involves a propagator which is linear in the loop momentum. To deal with it, we can use a variant of the usual Feynman parametrization,

$$\frac{1}{A^n B^m} = \frac{\Gamma(m+n)}{\Gamma(m)\Gamma(n)} \int_0^\infty d\eta \frac{\eta^{m-1}}{(A+\eta B)^{n+m}}, \qquad (5.51)$$

where B is the linear propagator, while A is quadratic. Using this parametrization, we again end up with a standard loop integral. Evaluating it yields

$$I_c = -\frac{\Gamma(1+\varepsilon)}{2l_+ \cdot p_-} \frac{\Gamma^2(-\varepsilon)}{\Gamma(1-2\varepsilon)} \left(\frac{\mu^2}{P^2}\right)^\varepsilon. \qquad (5.52)$$

While it is some work to evaluate the integral, at least its scaling is easily obtained. I_c can depend on the invariants P^2 and $l_+ \cdot p_-$. Looking at the integrand we observe that $I_c \to I_c/\alpha$ under a rescaling $l_+ \to \alpha\, l_+$. Together with dimensional analysis, this dictates the dependence on P^2 so that the only unknown is the ε-dependent prefactor. The contribution of the $\bar{c}$-region is obtained by replacing $P^2 \to L^2$ in (5.52). This leaves the soft region in which $k^\mu \sim (\lambda^2, \lambda^2, \lambda^2) \, Q$ so that we can approximate

$$(k+l)^2 = 2k_- \cdot l_+ + l^2 + \mathcal{O}(\lambda^3), \qquad (k+p)^2 = 2k_+ \cdot p_- + p^2 + \mathcal{O}(\lambda^3). \qquad (5.53)$$

Dropping the higher order terms, we are now left with two linear propagators, which can be handled by using (5.51) twice. Performing the loop integration and integrating over the Feynman parameters we obtain

$$
\begin{aligned}
I_s &= i\pi^{-d/2} \mu^{4-d} \int d^d k \frac{1}{(k^2+i0)\left(2k_- \cdot l_+ + l^2 + i0\right)\left(2k_+ \cdot p_- + p^2 + i0\right)} \\
&= -\frac{\Gamma(1+\varepsilon)}{2l_+ \cdot p_-} \Gamma(\varepsilon)\Gamma(-\varepsilon) \left(\frac{2l_+ \cdot p_- \mu^2}{L^2 P^2}\right)^\varepsilon.
\end{aligned}
\qquad (5.54)
$$

Having obtained the contributions from all the different momentum regions, we can now add them up and verify whether we reproduce the full integral (5.38). This involves some non-trivial cancellations since the individual integrals are all divergent, while the full integral is finite in $d=4$. To make the divergences explicit, let us expand in ε and list the individual contributions.

$$
\begin{aligned}
I_h &= \frac{\Gamma(1+\varepsilon)}{Q^2} \left(\frac{1}{\varepsilon^2} + \frac{1}{\varepsilon}\ln\frac{\mu^2}{Q^2} + \frac{1}{2}\ln^2\frac{\mu^2}{Q^2} - \frac{\pi^2}{6}\right) \\
I_c &= \frac{\Gamma(1+\varepsilon)}{Q^2} \left(-\frac{1}{\varepsilon^2} - \frac{1}{\varepsilon}\ln\frac{\mu^2}{P^2} - \frac{1}{2}\ln^2\frac{\mu^2}{P^2} + \frac{\pi^2}{6}\right) \\
I_{\bar{c}} &= \frac{\Gamma(1+\varepsilon)}{Q^2} \left(-\frac{1}{\varepsilon^2} - \frac{1}{\varepsilon}\ln\frac{\mu^2}{L^2} - \frac{1}{2}\ln^2\frac{\mu^2}{L^2} + \frac{\pi^2}{6}\right) \\
I_s &= \frac{\Gamma(1+\varepsilon)}{Q^2} \left(\frac{1}{\varepsilon^2} + \frac{1}{\varepsilon}\ln\frac{\mu^2 Q^2}{L^2 P^2} + \frac{1}{2}\ln^2\frac{\mu^2 Q^2}{L^2 P^2} + \frac{\pi^2}{6}\right)
\end{aligned}
\qquad (5.55)
$$

$$I_{\text{tot}} = \frac{1}{Q^2}\left(\ln\frac{Q^2}{L^2}\ln\frac{Q^2}{P^2} + \frac{\pi^2}{3}\right).$$

The sum $I_{\text{tot}} = I_h + I_c + I_{\bar{c}} + I_s$ is indeed finite and reproduces the leading term in the expansion of the full integral I. The cancellation of divergences is especially remarkable for the $1/\varepsilon$ pieces which all involve logarithms of different scales which must add up to zero. The cancellation implies that the infrared divergences of the hard integrals are the same as the ultra-violet divergences of the SCET matrix elements. Similar relations arise for n-point Green's functions and imply strong constraints on the infrared divergences of on-shell amplitudes in QCD and other gauge theories. These were analysed in [37]–[40] using SCET and with diagrammatic methods in [41]–[44]. The constraints fully fix the infrared structure of an arbitrary two-loop n-point amplitudes in terms of quantities which can be extracted from Sudakov form factors. A new structure, which involves four legs, first arises at three loops and was recently computed [45]–[47].

5.3.2 Effective Lagrangian

Using the form factor integral, we have identified the relevant momentum regions for the Sudakov problem. We now proceed similar to Section 5.2 and construct a Lagrangian whose Feynman rules directly yield the expanded diagrams obtained using method of regions expansion. To do so, we introduce EFT fields ϕ_c, $\phi_{\bar{c}}$ and ϕ_s (where ϕ denotes a quark or gluon field) whose momenta scale exactly as appropriate for the relevant momentum region.

The standard approach to constructing an effective theory is to write down the most general Lagrangian for the given fields and to then adjust the Wilson coefficients to reproduce the low-energy behaviour of the full theory. At tree level, we can use a shortcut and simply substitute

$$
\begin{aligned}
\psi &\to \psi_c + \psi_{\bar{c}} + \psi_s, \\
A^\mu &\to A_c^\mu + A_{\bar{c}}^\mu + A_s^\mu,
\end{aligned}
\tag{5.56}
$$

in QCD, expand away the suppressed terms and read off the effective Lagrangian. At loop level, matching corrections arise, which can modify the coefficients of the tree-level operators and induce new operators, but we will find that the matching corrections only affect the terms involving both c and $\bar{c}$ fields. To construct the purely soft and collinear Lagrangians, the tree-level short cut (5.56) is useful and efficient.

Both the fermion and gauge fields have several components, which scale differently with the expansion parameter λ. To read off the scaling of the fields, it is easiest to consider the propagators. Let us start with the gluon-field propagator

$$
\langle 0 | T\left\{ A_\mu^a(x) A_\nu^b(0) \right\} | 0 \rangle = \int \frac{d^4 p}{(2\pi)^4} \frac{i}{p^2 + i0} e^{-ip\cdot x} \left[-g_{\mu\nu} + \xi \frac{p_\mu p_\nu}{p^2} \right] \delta^{ab}.
\tag{5.57}
$$

Here a and b are the colour indices of the fields. In the following, we will usually work with matrix fields $A_\mu(x) = A_\mu^a(x) t^a$, where the t^as are the generators of the gauge group, and keep the colour implicit. The position argument x^μ of the fields is conjugate to the momentum p^μ in the Fourier exponent $p \cdot x \sim \mathcal{O}(\lambda^0)$. The part of the propagator involving the gauge parameter ξ scales like $d^4 p / (p^2)^2 \, p_\mu p_\nu \sim p_\mu p_\nu$. In a generic gauge, the gluon field thus scales exactly like the momentum $A_\mu \sim p_\mu$. This is of course expected since gauge symmetry ties the field to the momentum. For soft and collinear gluons, the field component thus scale as

$$
\begin{aligned}
\left(n \cdot A_s, \bar{n} \cdot A_s, A_{s\perp}^\mu \right) &\sim \left(\lambda^2, \lambda^2, \lambda^2 \right), \\
\left(n \cdot A_c, \bar{n} \cdot A_c, A_{c\perp}^\mu \right) &\sim \left(\lambda^2, 1, \lambda \right).
\end{aligned}
\tag{5.58}
$$

From the scaling we immediately see that in terms involving both soft and collinear gluons, the soft gluons are power suppressed, except for the contribution from the $n \cdot A_s$ component, which is commensurate with its collinear counterpart.

Next, let's consider the fermion propagators. For a soft fermion, we have

$$\langle 0 | T\{\psi_s(x)\bar{\psi}_s(0)\} |0\rangle = \int \frac{d^4 p}{(2\pi)^4} \frac{i\slashed{p}}{p^2 + i0} e^{-ip\cdot x} \sim (\lambda^2)^4 \frac{\lambda^2}{\lambda^4} = \lambda^6, \qquad (5.59)$$

from which we conclude that the soft fermions scale as $\psi_s(x) \sim \lambda^3$. The collinear case is more complicated because the numerator of the propagator must be decomposed into

$$\slashed{p} = n \cdot p \frac{\slashed{\bar{n}}}{2} + \bar{n} \cdot p \frac{\slashed{n}}{2} + \slashed{p}^\perp \qquad (5.60)$$

and the three terms have different scaling, which implies that different parts of the fermion spinor scale differently. To take this into account, we split the fermion field into two parts

$$\psi_c = \xi_c + \eta_c = P_+ \psi_c + P_- \psi_c \qquad (5.61)$$

using the projection operators

$$P_+ = \frac{\slashed{n}\slashed{\bar{n}}}{4}, \qquad\qquad P_- = \frac{\slashed{\bar{n}}\slashed{n}}{4}, \qquad (5.62)$$

which fulfill $P_+ + P_- = 1$ and $P_\pm^2 = P_\pm$. The propagator

$$\langle 0 | T\{\xi_c(x)\bar{\xi}_c(0)\} |0\rangle = \int \frac{d^4 p}{(2\pi)^4} e^{-ip\cdot x} \frac{\slashed{n}\slashed{\bar{n}}}{4} \frac{i\slashed{p}}{p^2 + i0} \frac{\slashed{\bar{n}}\slashed{n}}{4}$$
$$= \int \frac{d^4 p}{(2\pi)^4} e^{-ip\cdot x} \frac{i\bar{n}\cdot p \frac{\slashed{n}}{2}}{p^2 + i0} \sim \lambda^4 \frac{1}{\lambda^2} = \lambda^2, \qquad (5.63)$$

shows that the field scales as $\xi_c \sim \lambda$ and repeating the exercise for η_c we read off that $\eta_c \sim \lambda^2$. We observe that soft fermions are power suppressed with respect to collinear fermions.

Now that we know the scaling of all fields, we can plug the decomposition (5.56) into the QCD action and read off the tree-level EFT Lagrangian,

$$S = S_s + S_c + S_{\bar{c}} + S_{c+s} + S_{\bar{c}+s} + \dots \qquad (5.64)$$

where we collected the terms according to their field content. S_s contains the purely soft terms, S_c the collinear terms, and S_{c+s} describes the soft-collinear interactions. Since the

$S_{\bar{c}}$ and $S_{\bar{c}+s}$ can be obtained from the collinear terms using simple substitution rules, we will suppress them for the moment.

Let us first write the purely soft part, which has exactly the same form as the standard QCD action, except that the field is replaced with the soft field

$$S_s = \int d^4x\, \bar{\psi}_s i\slashed{D}_s \psi_s - \frac{1}{4}(F_s^a)_{\mu\nu}(F_s^a)^{\mu\nu}, \tag{5.65}$$

where the soft covariant derivative is $iD_s^\mu = i\partial^\mu + A_s^\mu$ and the soft field strength tensor $(F_s^a)_{\mu\nu}$ is of course built from this derivative. Note that all the terms in the action are $\mathcal{O}(\lambda^0)$ since the integration measure d^4x counts as $\mathcal{O}(\lambda^{-8})$ because the position vector components scale as $x^\mu \sim 1/p_s^\mu \sim \lambda^{-2}$. That we reproduce the standard QCD action is of course expected. The purely soft quarks and gluons behave exactly as standard quarks and gluons and since everything scales in the same way, there is nothing which can be expanded away. Next, we consider the purely collinear part of the action. This is again a simple copy of the QCD action, but we write out the fermion field in its two components $\psi_c = \xi_c + \eta_c$ and get

$$S_c = \int d^4x\, (\bar{\xi}_c + \bar{\eta}_c)\left[\frac{\slashed{n}}{2} i\bar{n}\cdot D_c + \frac{\slashed{\bar{n}}}{2} in\cdot D_c + i\slashed{D}_{c\perp}\right](\xi_c + \eta_c) - \frac{1}{4}(F_c^a)_{\mu\nu}(F_c^a)^{\mu\nu}$$
$$= \int d^4x\, \bar{\xi}_c \frac{\slashed{\bar{n}}}{2} in\cdot D_c \xi_c + \bar{\xi}_c\, i\slashed{D}_{c\perp}\, \eta_c + \bar{\eta}_c\, i\slashed{D}_{c\perp}\, \xi_c + \bar{\eta}_c \frac{\slashed{n}}{2} i\bar{n}\cdot D_c \eta_c - \frac{1}{4}(F_c^a)_{\mu\nu}(F_c^a)^{\mu\nu}. \tag{5.66}$$

Using the scaling of the fields and taking into account that the integration measure now scales as $d^4x \sim \lambda^{-4}$ because the position is conjugate to a collinear momentum, we see that each term is $\mathcal{O}(\lambda^0)$. This form of the Lagrangian is inconvenient since it involves both the large component $\xi_c \sim \lambda$ and the power suppressed field $\eta_c \sim \lambda^2$ and the two mix into each other. To construct operators, it is simplest to avoid this complication by integrating out the small components of the fermion field, similar to what is done when constructing the Lagrangian in HQET, where the small component has a propagator which scales as $1/m_q$, where m_q is the heavy quark mass. In this sense, the small component is not really a dynamical field and can be removed. In close analogy, the propagator of the small component of the collinear field scales as $1/Q$ and can therefore be integrated out. In practical computations, it can be convenient to keep the small component since we can then use the standard QCD Feynman rules for collinear computations, as was done, e.g. in [48], and some authors have advocated to formulate the theory [49] without integrating out the small components. At any rate, the action (5.66) is quadratic in η_c and we can therefore integrate out the field exactly. To do so, we shift the field

$$\eta_c \to \eta_c - \frac{\slashed{\bar{n}}}{2} \frac{1}{i\bar{n}\cdot D_c} i\slashed{D}_{c\perp} \xi_c \tag{5.67}$$

to complete the square in the action S_c. After the shift, the action takes the form

$$\mathcal{L}_c = \bar{\xi}_c \frac{\bar{\slashed{n}}}{2} \left[in \cdot D_c + i\slashed{D}_{c\perp} \frac{1}{i\bar{n} \cdot D} i\slashed{D}_{c\perp} \right] \xi_c - \frac{1}{4} (F_c^a)_{\mu\nu} (F_c^a)^{\mu\nu} + \bar{\eta}_c \frac{\slashed{n}}{2} i\bar{n} \cdot D_c \eta_c \qquad (5.68)$$

and we can integrate out the field η_c, which leaves a determinant $\det(\frac{\slashed{n}}{2} i\bar{n} \cdot D_c)$. To make the determinant and the inverse derivative in the shift (5.67) well defined, we should adopt an $i0$ prescription for the $i\bar{n} \cdot D_c$ operator. The prescription is without physical consequences since it concerns the region near $\bar{n} \cdot p = 0$, while the effective theory deals with processes with $\bar{n} \cdot p \sim Q$. It turns out that the determinant is trivial, which can be seen by evaluating it in the gauge $\bar{n} \cdot A_c = 0$ and noting that it is gauge invariant, or by realizing that the associated closed-loop diagrams all vanish since for a given $i0$ prescription all propagator poles in the closed η_c loops are on the same side of the integration contour and can be avoided. After dropping the trivial determinant, the collinear SCET Lagrangian is therefore given by (5.68) without the last term.

Next we consider the soft-collinear interaction terms in S_{s+c}. The general construction of the terms is quite involved and was performed in [50] in the position-space formalism we are using here. For simplicity, we will restrict ourselves to leading-power terms which are usually sufficient for collider physics applications. Nevertheless there is currently a lot of effort to analyse power corrections, the interested reader can consult [51]–[57] for more information. Getting the leading-power soft-collinear interactions is quite simple once we remember the power counting of the fields discussed above

- ψ_s is power suppressed compared to the collinear fermion fields, so that no soft quarks are present in leading-power interactions with collinear fields.
- Since $\bar{n} \cdot A_s$ and $A_{s\perp}^{\mu}$ are power suppressed compared to their collinear counterparts, only the component $n \cdot A_s$ arises.

Taken together, this implies that the leading power interaction terms can be obtained by substituting

$$A_c^{\mu} \rightarrow A_c^{\mu} + n \cdot A_s \frac{\bar{n}^{\mu}}{2} \qquad (5.69)$$

in S_c.

The final step is to perform a derivative expansion in the resulting Lagrangian which corresponds to the expansion in small momentum components. To do so, consider the interaction term of a soft gluon with a collinear fermion

$$S_{c+s} = \int d^4x \, \bar{\xi}_c(x) \frac{\bar{\slashed{n}}}{2} n \cdot A_s(x) \xi_c(x), \qquad (5.70)$$

which arises from the substitution (5.69) into (5.68), together with purely gluonic interactions. The momentum is given by a soft and a collinear momentum which scales like a collinear momentum so that x^μ is conjugate to a collinear momentum. Explicitly, this implies the scaling

$$p_c^\mu + p_s^\mu \sim p_c^\mu \sim \left(\lambda^2, 1, \lambda \right),$$

$$x^\mu \sim \left(1, \lambda^{-2}, \lambda^{-1} \right), \tag{5.71}$$

which follows from the fact $p_c \cdot x = 1/2\, n \cdot p_c \bar{n} \cdot x + 1/2\, \bar{n} \cdot p_c n \cdot x + p_c^\perp \cdot x^\perp \sim 1$. When we therefore consider a product of a soft momentum with the position vector

$$p_s \cdot x = \frac{1}{2} \underbrace{n \cdot p_s\, \bar{n} \cdot x}_{\mathcal{O}(1)} + \frac{1}{2} \underbrace{\bar{n} \cdot p_s \bar{n} \cdot x}_{\mathcal{O}(\lambda^2)} + \underbrace{p_s^\perp \cdot x^\perp}_{\mathcal{O}(\lambda)}$$

$$= \quad p_{s+} \cdot x_- \quad + \quad p_{s-} \cdot x_+ \quad + \quad p_\perp \cdot x_\perp \tag{5.72}$$

only the term involving x_- is of leading power. We can therefore expand the interaction term into a Taylor series,

$$S_{c+s} = \int d^4x\, \bar{\xi}_c(x) \frac{\slashed{\bar{n}}}{2} \xi_c(x) \Big[1 + \underbrace{x_\perp \cdot \partial_\perp}_{\mathcal{O}(\lambda)} + \underbrace{x_+ \cdot \partial_-}_{\mathcal{O}(\lambda^2)} + \dots \Big] n \cdot A_s(x) \Big|_{x=x_-}$$

$$= \int d^4x\, \bar{\xi}_c(x) \frac{\slashed{\bar{n}}}{2} \xi_c(x) n \cdot A_s(x_-) + \mathcal{O}(\lambda), \tag{5.73}$$

and the leading power interactions are obtained by replacing $x^\mu \to x_-^\mu$ in the argument of the soft fields. This derivative expansion was called the *multipole expansion* in [3], since it has similarity to what is done when approximating charge distributions at large distances in electrodynamics. Let us note that the original SCET papers [1, 2] and a large fraction of the current SCET literature uses a different method to expand in small momenta. In this approach, we split the momenta into large and small components and treats the large components in Fourier space, while the small ones remain in position space. This is similar to the procedure we used in soft effective theory, where we split the electron momentum into $p^\mu = m_e v^\mu + k^\mu$, where k^μ is the soft residual momentum in which we expand. The position dependence of the field $h_v(x)$ is conjugate to the residual momentum k^μ and the large part $m_e v^\mu$ became a label on the field $h_v(x)$. For this reason, this hybrid momentum-and-position-space formulation of SCET is called the *label formalism*. We will work in position space and not cover the label formulation in more detail. The book [12] contains a comparison of the two formulations. At leading power, it is simple to translate between them.

After performing the multipole expansion, we arrive at the final form of the leading-power SCET Lagrangian

$$\mathcal{L}_{\text{SCET}} = \bar{\psi}_s i\slashed{D}_s \psi_s + \bar{\xi}_c \frac{\slashed{\bar{n}}}{2} \Big[in \cdot D + i\slashed{D}_{c\perp} \frac{1}{i\bar{n} \cdot D_c} i\slashed{D}_{c\perp} \Big] \xi_c - \frac{1}{4} \left(F_{\mu\nu}^{s,a} \right)^2 - \frac{1}{4} \left(F_{\mu\nu}^{c,a} \right)^2. \tag{5.74}$$

This expression involves the collinear and soft covariant derivatives

$$iD^s_\mu = i\partial_\mu + gA^s_\mu(x), \qquad\qquad iD^c_\mu = i\partial_\mu + gA^c_\mu(x), \qquad (5.75)$$

as well as the mixed derivative

$$in \cdot D = in \cdot \partial + gn \cdot A_c(x) + gn \cdot A_s(x_-). \qquad (5.76)$$

The associated field-strength tensors are

$$igF^{s,a}_{\mu\nu} t^a = \left[iD^s_\mu, iD^s_\nu\right], \qquad\qquad igF^{c,a}_{\mu\nu} t^a = \left[iD_\mu, iD_\nu\right], \qquad (5.77)$$

where the derivative appearing in the second commutator in (5.77) is defined as

$$D^\mu = n \cdot D\frac{\bar{n}^\mu}{2} + \bar{n} \cdot D_c\frac{n^\mu}{2} + D^\mu_{c\perp}. \qquad (5.78)$$

As stated above, we omitted the terms involving the $\bar{c}$ fields for brevity in the Lagrangian (5.74). They have the same form as those involving c fields, but with $n \leftrightarrow \bar{n}$ and $x_- \leftrightarrow x_+$.

To finish the discussion of the Lagrangian, let us briefly discuss gauge transformations. Since we split the gauge field into different soft and collinear components, we can consider separate gauge transformations

soft: $$V_s(x) = \exp[i\alpha^a_s(x)t^a], \qquad (5.79)$$

collinear: $$V_c(x) = \exp[i\alpha^a_c(x)t^a], \qquad (5.80)$$

where the gauge transformations scale like the associated fields $\partial_\mu\alpha^a_s(x) \sim \lambda^2\alpha^a_s(x)$ and $\partial_\mu\alpha^a_c(x) \sim (\lambda^2, 1, \lambda)\alpha^a_c(x)$. The soft gauge transformations act on the soft fields in the usual way,

$$\psi_s(x) \to V_s(x)\,\psi_s(x),$$
$$A^\mu_s(x) \to V_s(x)\,A^\mu_s(x)\,V^\dagger_s(x) + \frac{i}{g}V_s(x)(\partial_\mu V^\dagger_s(x)), \qquad (5.81)$$

so that the covariant derivative transforms as $D^\mu_s(x) \to V_s(x)\,D^\mu_s(x)\,V^\dagger_s(x)$ while the collinear fields transform as

$$\psi_c(x) \to V_s(x_-)\,\psi_c(x),$$
$$A^\mu_c(x) \to V_s(x_-)\,A^\mu_c(x)\,V^\dagger_s(x_-). \qquad (5.82)$$

This differs in two important aspects from the transformation of the soft fields. First of all, we have performed the multipole expansion and have replaced $x \to x_-$ in the soft fields. Without this expansion, the gauge transformations would induce a tower of power corrections, which would be inconvenient. Secondly, we observe that the collinear gauge field $A_c^\mu(x)$ transforms as a matter field, i.e. without the inhomogeneous term present in (5.81). This ensures that the mixed derivative transforms in the correct way as

$$D^\mu \to V_s(x_-)\, D^\mu(x)\, V_s^\dagger(x_-)\,. \tag{5.83}$$

Next, we consider collinear transformations. Since these involve a collinear field which carries a large momentum, the soft fields must remain invariant under these transformations, while the collinear ones transform in the usual way

$$
\begin{aligned}
\xi_c(x) &\to V_c(x)\,\xi_c(x) & D^\mu &\to V_c(x)\, D^\mu\, V_c^\dagger(x)\,, \\
\psi_s(x) &\to \psi_s(x) & D_s^\mu &\to D_s^\mu\,.
\end{aligned}
\tag{5.84}
$$

To see what kind of transformation this implies for the field $A_c^\mu(x)$, the reader can consult [12]. Given that the covariant derivatives transform in the expected way, it is straightforward to verify that the Lagrangian (5.74) is invariant under both soft and collinear gauge transformations.

Let us finally come back to the statement that there are no matching corrections to the Lagrangian. For the purely soft Lagrangian this is clear, since it is a copy of the QCD Lagrangian. All purely soft processes are therefore identically reproduced by the effective theory. For this to be true, it is important that we work with dimensional regularization rather than with a hard cutoff. If we would restrict the soft momenta to be below a certain cutoff in the loop diagrams of the low-energy theory, we would not reproduce the higher-energy parts of QCD and would need to correct for this by performing a matching computation. The same statements apply to the collinear Lagrangian. It looks different than the QCD Lagrangian because we integrated out two components, but since we did not perform any approximation, the collinear SCET Lagrangian is equivalent to QCD. We might worry that soft loops would contribute to purely collinear processes and that we would need to remove these contributions, but dimensional regularization is again very efficient in eliminating unnecessary contributions. It turns out that soft loop corrections to purely collinear diagrams are all scaleless and vanish. We can understand this by looking at the soft loop integral (5.54). It involves the scale $\Lambda_s^2 = P^2 L^2/Q^2$, which is non-zero only when both c and $\bar c$ particles are involved. Matching corrections are therefore only present for interactions involving both types of fields, such as the current operators discussed in the next section.

5.3.3 The Vector Current in SCET

We have constructed $\mathcal{L}_s$, $\mathcal{L}_c$ and $\mathcal{L}_{s+c}$ in the previous section and can obtain $\mathcal{L}_{\bar c}$ and $\mathcal{L}_{s+\bar c}$ from simple substitutions in the c terms. What is missing are operators involving both c and $\bar c$ fields. In the Sudakov problem only the electromagnetic current operator connects the two fields. The necessary hard momentum transfer is provided by the virtual photon on the external line. The tree-level diagram in QCD can be reproduced by a SCET operator

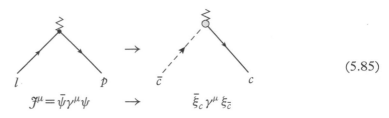

$$\mathcal{J}^\mu = \bar{\psi}\gamma^\mu\psi \qquad \longrightarrow \qquad \bar{\xi}_c\gamma^\mu\xi_{\bar{c}}$$

(5.85)

where we have indicated the $\bar{c}$ fermion with a green dashed line and the c field with a blue line. Due to the projection properties of the two fermion fields, we can simplify the SCET operator a bit further

$$\bar{\xi}_c\gamma^\mu\xi_{\bar{c}} = \bar{\xi}_c\left[n^\mu\frac{\bar{\slashed{n}}}{2} + \bar{n}^\mu\frac{\slashed{n}}{2} + \gamma_\perp^\mu\right]\xi_{\bar{c}} = \bar{\xi}_c\gamma_\perp^\mu\xi_{\bar{c}}.$$

(5.86)

However, beyond tree-level and for processes involving collinear gluons, the above operator is insufficient and it will now take quite some work to write down the most general gauge-invariant leading-power operator.

One problem with the operator in (5.86) is the following. Usually operators with derivatives are power suppressed but in SCET the derivatives corresponding to the large momentum component of the collinear fields are unsuppressed,

$$\bar{n}\cdot\partial\,\phi_c \sim \lambda^0 Q\phi_c.$$

(5.87)

We therefore need to include operators with an arbitrary number of such derivatives!

We could work with infinitely many operators and Wilson coefficients C_n for the operator with n such derivatives, but there is a more elegant way to deal with this complication. To understand it, consider the series

$$\phi_c(x+t\bar{n}) = \sum_{n=0}^{\infty}\frac{t^n}{n!}(\bar{n}\cdot\partial)^n\phi_c(x).$$

(5.88)

Inserting this into a convolution integral we get

$$\int dt\, C(t)\phi_c(x+t\bar{n}) = \sum_{n=0}^{\infty}\frac{C_n}{n!}(\bar{n}\cdot\partial)^n\phi_c(x).$$

(5.89)

where C_n is the nth moment of the coefficient function

$$C_n = \int dt\, C(t)\, t^n. \tag{5.90}$$

Instead of including an arbitrary number of derivatives, we can smear the field ϕ_c along the light cone as in (5.89). The function $C(t)$ then encodes the information about the Wilson coefficients a_n of the higher-derivative operators.

However, with the smearing the current operator becomes non-local and when putting operators at different points in a gauge theory, we need to be careful to maintain gauge invariance. Consider for example an operator with two collinear fermions at different points

$$\bar{\xi}_c(x + t\bar{n})\,[x + t\bar{n}, x]\,\frac{\slashed{\bar{n}}}{2}\,\xi_c(x). \tag{5.91}$$

The proton matrix elements of this operator define the quark PDFs. In order to make it gauge invariant we need to transport the gauge transformation at point x to the point $x + t\bar{n}$. This can be achieved using the Wilson line

$$[x + t\bar{n}, x] = \boldsymbol{P}\exp\left[ig\int_0^t dt'\,\bar{n}\cdot A_c(x + t\bar{n})\right]. \tag{5.92}$$

Since the exponent is a colour matrix, we need an ordering prescription to define it. The symbol $\boldsymbol{P}$ indicates that the matrices at different times should be path ordered, i.e. that those at a later time are to the left of those that are earlier. The Wilson line transforms as

$$[x + t\bar{n}, x] \rightarrow V_c(x + t\bar{n})\,[x + t\bar{n}, x]\,V_c^\dagger(x), \tag{5.93}$$

which renders the bilocal operator (5.91) gauge invariant. The gauge transformation (5.93) and other properties of Wilson lines are derived in Appendix D of [12]. Since the other components of the gluon field are power suppressed, a Wilson line along a non-straight path would differ from (5.92) by power corrections.

In SCET, it is useful to take a detour and define a Wilson line which runs from infinity along $\bar{n}^\mu$ to the point x^μ as follows

$$W_c(x) = [x, x - \infty\bar{n}], \tag{5.94}$$

so that the finite segment can be written as a product

$$[x + t\bar{n}, x] = W_c(x + t\bar{n})\,W_c^\dagger(x). \tag{5.95}$$

In other words, to move from x to $x + t\bar{n}$, we first move from x to infinity and then back $x + t\bar{n}$. The segment we travel in both directions drops out by unitarity of the corresponding matrix, which leaves the finite segment. The advantage of the Wilson line W_c is that it allows us to define the building blocks

$$
\begin{aligned}
\chi_c(x) &\equiv W_c^\dagger(x)\xi_c(x)\,,\\
\mathcal{A}_c^\mu &\equiv W_c^\dagger(D_c^\mu W_c)\,,
\end{aligned}
\tag{5.96}
$$

which are invariant under collinear gauge transformations which vanish at infinity. With these building blocks, we can then easily build gauge invariant SCET operators. In addition to the collinear Wilson line W_c, we later introduce Wilson lines built from soft fields as in (5.17) to decouple soft interactions.

After all this preparation, we are finally ready to write down the most general leading-power SCET current operator. It takes the form

$$
\mathcal{J}^\mu(0) = \int ds \int dt\, C_V(s,t)\, \bar{\chi}_c(t\bar{n})\, \gamma_\perp^\mu\, \chi_{\bar{c}}(sn)\,.
\tag{5.97}
$$

The Wilson coefficient $C_V(s,t)$ (the V stands for vector current) needs to be determined by matching. At tree level it is given by $C_V(s,t) = \delta(t)\delta(s)$ to reproduce the tree-level current (5.85). Since it is related to the large derivatives, the Fourier transform of the coefficient encodes the dependence on the large momentum transfer $Q^2 = n\cdot l\,\bar{n}\cdot p$. To see this, we use the momentum operator to shift the fields to the point $x = 0$,

$$
\phi(x) = e^{iP\cdot x}\phi(0)e^{-iP\cdot x}\,,
\tag{5.98}
$$

and take the matrix element of the current operator between a state with an incoming quark q with momentum l^μ and outgoing one with momentum p^μ. We obtain

$$
\begin{aligned}
\langle q(p)|\mathcal{J}^\mu(0)|q(l)\rangle &= \int ds \int dt\, C_V(s,t)\, e^{-isn\cdot l}\, e^{it\bar{n}\cdot p}\, \bar{u}(p)\, \gamma_\perp^\mu\, u(l)\\
&= \tilde{C}_V(n\cdot l\,\bar{n}\cdot p)\, \bar{u}(p)\, \gamma_\perp^\mu\, u(l)\,.
\end{aligned}
\tag{5.99}
$$

Note that the Fourier transformed coefficient $\tilde{C}_V$ only depends on the product $Q^2 = n\cdot l\,\bar{n}\cdot p$ and not on the individual components. One way to see this is to note that SCET is invariant under a rescaling $n^\mu \to \alpha\, n^\mu$ and $\bar{n}^\mu \to 1/\alpha\,\bar{n}^\mu$. This invariance is manifest already in the decomposition (5.41) and is part of a larger set of reparametrization invariances [58], which express the independence of the physics on the exact choice of the reference vectors used to set up the effective theory.

The Wilson coefficient $\tilde{C}_V$ encodes the dependence on the large momentum scale Q^2, which arises from the hard momentum region. To determine it, we should perform a matching computation. The loop diagrams contributing to the one-loop Sudakov form factor are shown in Figure 5.7. On the EFT side, there are several contributions. In addition to loop diagrams involving collinear fields (blue), anti-collinear fields (green) and soft exchanges (red), there is a contribution from the one-loop correction to the

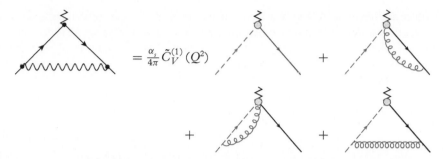

Figure 5.7 *One-loop contributions to the Sudakov form factor in QCD and in SCET. In addition to the diagrams shown, there are external leg corrections which we suppress. The Feynman rules for gluons emitted from the current are obtained after expanding the Wilson lines which are part of the collinear building blocks (5.96) in powers of the gauge coupling.*

Wilson coefficient. The vertices in which an (anti-)collinear gluon is emitted from the current are obtained after expanding the collinear Wilson lines in the building blocks (5.96) in the coupling. The different diagrams are in one-to-one correspondence to the contributions of the different momentum regions computed in Section 5.3.1, except for the fact that we left out the numerators of the diagrams in the region computation. What remains the same is the fact that each contribution involves a different momentum scale. Setting the low-energy scales P^2 and L^2 to zero, both the soft and the collinear loop integrals become scaleless. In this case, the full-theory result becomes equal to the contribution of the hard region and on the effective-theory side only the one-loop correction to the Wilson coefficient remains. The most efficient way to extract the Wilson coefficient is thus to compute the on-shell form factor $P^2 = L^2 = 0$. Performing the on-shell computation, we obtain

$$\tilde{C}_V^{\text{bare}}(\varepsilon, Q^2) = 1 + \frac{\alpha_s(\mu)}{4\pi} C_F \left(-\frac{2}{\varepsilon^2} - \frac{3}{\varepsilon} + \frac{\pi^2}{6} - 8 + \mathcal{O}(\varepsilon) \right) \left(\frac{Q^2}{\mu^2} \right)^{-\varepsilon} + \mathcal{O}\left(\alpha_s^2 \right), \quad (5.100)$$

with colour structure $t^a t^a = C_F \mathbf{1} = (N_c^2 - 1)/(2N_c)\mathbf{1}$. To get this result, we have expressed the bare coupling $\alpha_s^0 = g^2/(4\pi)$ in terms of the $\overline{\text{MS}}$ renormalized coupling constant $\alpha_s(\mu)$ via the relation $Z_\alpha \alpha_s(\mu) \mu^{2\varepsilon} = e^{-\varepsilon \gamma_E} (4\pi)^\varepsilon \alpha_s^0$, where $Z_\alpha = 1 + \mathcal{O}(\alpha_s)$ at our accuracy. We have added a label bare to the Wilson coefficient to indicate that we still need to renormalize it, which is done by absorbing the divergences into a multiplicative Z-factor,

$$\tilde{C}_V(Q^2, \mu) = \lim_{\varepsilon \to 0} Z^{-1}\left(\varepsilon, Q^2, \mu \right) \tilde{C}_V^{\text{bare}}(\varepsilon, Q^2). \quad (5.101)$$

Doing so, leaves us with the finite, renormalized Wilson coefficient

$$\tilde{C}_V(Q^2, \mu) = 1 + \frac{\alpha_s(\mu)}{4\pi} C_F \left(-\ln^2 \frac{Q^2}{\mu^2} + 3\ln \frac{Q^2}{\mu^2} + \frac{\pi^2}{6} - 8 \right) + \mathcal{O}(\alpha_s^2). \quad (5.102)$$

In the renormalized coefficient we have taken the limit $\varepsilon \to 0$, but it depends on the renormalization scale μ. The whole procedure is of course the same as renormalization in standard quantum field theory, up to the fact that we had to deal with $1/\varepsilon^2$ divergences, which arise because we have both soft and collinear divergences. As a consequence, the Wilson coefficient contains double logarithms. Due to the presence of the double logarithms, the anomalous dimension governing the RG equation for the Wilson coefficient has a logarithmic piece,

$$\frac{d}{d\ln\mu}\tilde{C}_V(Q^2,\mu) = \left[C_F \gamma_{\mathrm{cusp}}(\alpha_s)\ln\frac{Q^2}{\mu^2} + \gamma_V(\alpha_s)\right]\tilde{C}_V(Q^2,\mu), \qquad (5.103)$$

where, at order α_s, the functions γ_{cusp} and γ_V are given by

$$\gamma_{\mathrm{cusp}}(\alpha_s) = 4\frac{\alpha_s(\mu)}{4\pi}, \qquad \text{and} \qquad \gamma_V(\alpha_s) = -6C_F\frac{\alpha_s(\mu)}{4\pi}. \qquad (5.104)$$

The on-shell form factor has been computed to three loops [59, 60], and all these ingredients are known to this accuracy. The presence of the logarithm in the anomalous dimension is the distinguishing feature of this RG (and other RG equations in SCET). It is important that only a single logarithm appears so that the expansion of the anomalous dimension is not spoiled by the presence of large logarithms. Otherwise, RG-improved perturbation theory would no longer work. The linearity in the logarithm follows from factorization, which we discuss next.

In soft photon effective theory we were able to decouple the soft radiation from the electron field by the field redefinition (5.21) involving a soft Wilson line. In the same way, soft emissions can be decoupled from the collinear field by considering the Wilson line

$$S_n(x) = \mathbf{P}\exp\left[ig\int_{-\infty}^{0} ds\, n\cdot A_s(x+sn)\right], \qquad (5.105)$$

which fulfills the equation $n\cdot D_s S_n(x) = 0$. Redefining

$$\begin{aligned} \xi_c &= S_n(x_-)\xi_c^{(0)}, \\ A_c^\mu &= S_n(x_-)A_c^{(0)\mu} S_n^\dagger(x_-), \end{aligned} \qquad (5.106)$$

the soft-collinear interaction term becomes

$$\begin{aligned} \mathcal{L}_{c+s} &= \bar{\xi}_c \frac{\slashed{\bar{n}}}{2} in\cdot D\xi_c = \bar{\xi}_c \frac{\slashed{\bar{n}}}{2}\left(in\cdot D_s + n\cdot A_c\right)\xi_c \\ &= \bar{\xi}_c^{(0)}\frac{\slashed{\bar{n}}}{2}\left(in\cdot\partial_s + n\cdot A_c^{(0)}\right)\xi_c^{(0)} = \bar{\xi}_c^{(0)}\frac{\slashed{\bar{n}}}{2}in\cdot D_c^{(0)}\xi_c^{(0)} \end{aligned} \qquad (5.107)$$

so that the decoupling has completely removed the soft-collinear interactions from the leading-power Lagrangian. Performing an analogous decoupling also for the anti-collinear fields, it takes the form

$$\mathcal{L}_{\text{SCET}} = \mathcal{L}_c^{(0)} + \mathcal{L}_{\bar{c}}^{(0)} + \mathcal{L}_s. \tag{5.108}$$

Since there are no longer any interactions, we are dealing with independent theories of soft and collinear particles and also the states separate as

$$|X\rangle = |X_c\rangle \otimes |X_{\bar{c}}\rangle \otimes |X_s\rangle. \tag{5.109}$$

Of course, this does not imply that the soft physics is no longer present. As in the soft-photon case, it manifests itself as soft Wilson lines along the directions of the energetic particles. The vector current operator $\mathcal{J}^\mu$, for example, takes the form

$$\bar{\chi}_c(t\bar{n}) \, \gamma_\perp^\mu \, \chi_{\bar{c}}(sn) = \bar{\chi}_c^{(0)}(t\bar{n}) \, \bar{S}_n^\dagger(0) \, \gamma_\perp^\mu \, S_{\bar{n}}(0) \, \chi_{\bar{c}}^{(0)}(sn) \tag{5.110}$$

after the decoupling. Note that the decoupling for the anti-collinear fields involves an incoming soft Wilson line $S_{\bar{n}}$ along the $\bar{n}$ direction, while we used the outgoing one $\bar{S}_n$ for the collinear field. The soft Wilson lines in the current operator are evaluated at $x = 0$, since the position argument in the current has hard scaling because the operator involves both collinear and anti-collinear fields.

5.3.4 Resummation by RG Evolution

Computing the factorized form factor in the decoupled theory, we obtain a result of the form

$$F(Q^2, L^2, P^2) = \tilde{C}_V(Q^2, \mu) \, \bar{\mathcal{J}}(L^2, \mu) \, \mathcal{J}(P^2, \mu) \, S(\Lambda_s^2, \mu), \tag{5.111}$$

which is shown graphically in Figure 5.8. The collinear and anti-collinear functions $\bar{\mathcal{J}}$ and $\mathcal{J}$ are of course identical, the bar simply indicates to which sector the function belongs. The soft function is given by the matrix element of the Wilson line in (5.110) and its scale is $\Lambda_s^2 = L^2 P^2 / Q^2$. We have indicated that the renormalized effective theory matrix elements and the Wilson coefficient $\tilde{C}_V$ depend on the renormalization scale. This dependence must cancel in the product, which implies that the anomalous dimensions of the ingredients must add up to zero. We gave the RG equation for the Wilson coefficient in (5.103) and those for the collinear and soft factors are

$$\frac{d}{d\ln\mu} \mathcal{J}\left(P^2, \mu^2\right) = -\left[C_F \gamma_{\text{cusp}}(\alpha_s) \ln \frac{P^2}{\mu^2} + \gamma_{\mathcal{J}}(\alpha_s)\right] \mathcal{J}\left(P^2, \mu^2\right),$$

$$\frac{d}{d\ln\mu} S\left(\Lambda_s^2, \mu^2\right) = \left[C_F \gamma_{\text{cusp}}(\alpha_s) \ln \frac{\Lambda_s^2}{\mu^2} + \gamma_S(\alpha_s)\right] S\left(\Lambda_s^2, \mu^2\right). \tag{5.112}$$

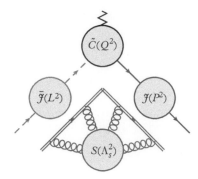

Figure 5.8 *Schematic form of the factorized form factor. The double lines indicate the soft Wilson lines.*

The μ-independence of $F(Q^2, L^2, P^2)$ then requires that

$$
\begin{aligned}
& C_F\, \gamma_{\mathrm{cusp}}(\alpha_s) \ln \frac{Q^2}{\mu^2} + \gamma_V(\alpha_s) \\
& -C_F\, \gamma_{\mathrm{cusp}}(\alpha_s) \left(\ln \frac{L^2}{\mu^2} + \ln \frac{P^2}{\mu^2} \right) - 2\gamma_{\mathcal{J}}(\alpha_s) \\
& +C_F\, \gamma_{\mathrm{cusp}}(\alpha_s) \ln \frac{\Lambda_s^2}{\mu^2} + \gamma_S(\alpha_s) = 0.
\end{aligned}
\tag{5.113}
$$

In order for the logarithms to cancel it is crucial that the anomalous dimensions are linear in the logarithm and all proportional to the same coefficient γ_{cusp}. To prove linearity we can make a more general ansatz for the anomalous dimensions and then show that scale independence of the sum, together with the kinematic dependence of the individual pieces, imply that higher-log terms must be absent. Note that the soft anomalous dimension is given by a closed Wilson loop with a cusp at $x = 0$, where the direction changes from $\bar{n}$ to n. Polyakov [61] and Brandt, Neri, and Sato [62] proved that Wilson lines with cusps require renormalization and the two-loop anomalous dimension for light-like Wilson lines with a cusp was first computed in [63]. At the moment, there is a lot of effort to also compute this cusp anomalous dimension to four loops. While the analytic QCD result is not yet known, some partial analytical and numerical results are available [64]–[68].

To resum the large logarithms in the form factor, we can evaluate each ingredient at its characteristic scale and then use the RG to evolve them to a common scale (Figure 5.9). For the hard function $\tilde{C}_V(Q^2, \mu)$, for example, we chose $\mu_h \sim Q$ as the initial value. For such a choice, the function does not involve any large logarithms and we can evaluate it perturbatively. Solving the RG equation, we then get the hard function also at lower scales as

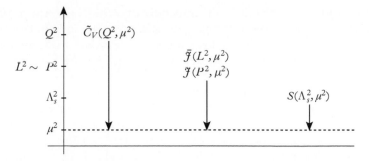

Figure 5.9 *Resummation by RG evolution.*

$$\tilde{C}_V(Q^2,\mu) = \exp\left\{\int_{\mu_h}^{\mu} d\ln\mu'\left[C_F\gamma_{\text{cusp}}(\alpha_s)\ln\frac{Q^2}{\mu'^2}+\gamma_V(\alpha_s)\right]\right\}\tilde{C}_V(Q^2,\mu_h)$$
$$\equiv U(\mu_h,\mu)\,\tilde{C}_V(Q^2,\mu_h).$$

(5.114)

Since the form of the RG of the other ingredients is the same, also their solution can immediately be written down. Using the definition of the β-function

$$\frac{d\alpha_s(\mu)}{d\ln\mu} = \beta(\alpha_s(\mu))$$

(5.115)

and rewriting the logarithm in the exponent of the evolution matrix $U(\mu_h,\mu)$ as

$$\ln\frac{\nu}{\mu} = \int_{\alpha_s(\mu)}^{\alpha_s(\nu)}\frac{d\alpha}{\beta(\alpha)},$$

(5.116)

it can be written in the form

$$U(\mu_h,\mu) = \exp\left[2C_F S(\mu_h,\mu)-A_{\gamma_V}(\mu_h,\mu)\right]\left(\frac{Q^2}{\mu_h^2}\right)^{-C_F A_{\gamma_{\text{cusp}}}(\mu_h,\mu)}.$$

(5.117)

The quantities S and A_γ are defined as

$$S(\nu,\mu) = -\int_{\alpha_s(\nu)}^{\alpha_s(\mu)}d\alpha\frac{\gamma_{\text{cusp}}(\alpha)}{\beta(\alpha)}\int_{\alpha_s(\nu)}^{\alpha}\frac{d\alpha'}{\beta(\alpha')},$$
$$A_{\gamma_i}(\nu,\mu) = -\int_{\alpha_s(\nu)}^{\alpha_s(\mu)}d\alpha\frac{\gamma_i(\alpha)}{\beta(\alpha)},$$

(5.118)

with $i \in \{V, \text{cusp}\}$. They can be computed by expanding the anomalous dimensions and the β-function order by order in perturbation theory and then performing the integrations.

Their explicit form can be found in the appendix of [69]. This form of perturbation theory, which involves couplings at the different scales but is free from large logarithms, is called RG improved perturbation theory and was covered in Matthias Neubert's lectures [13]. It is the standard way in which resummations are performed in SCET and other EFTs.

Let me finish this section with an admission. While all the steps we took to achieve factorization and to perform resummation would be perfectly appropriate for a physical observable, there is a problem for the Sudakov form factor. We stated earlier that the off-shell form factor is unphysical and gauge dependent and that we only consider it because it is the simplest example involving both soft and collinear physics. To achieve factorization, we have used the decoupling transformation, which is a field redefinition. Such redefinitions leave physical quantities invariant, as Aneesh Manohar showed [18], but they do lead to different off-shell Green's functions. In particular, we find that the soft matrix elements change. After the field redefinition, they involve not only logarithms but also IR divergences and are given by scaleless integrals. The anomalous dimensions we discussed above are not directly affected by this problem, but if one wants to compute them after the decoupling, we will need to separate the UV from the IR divergences. Fortunately, we are not really interested in the off-shell form factor and none of these problems will be present for the physical observables studied in the next section.

5.4   Applications in Jet Physics

While the Sudakov form factor discussed in Section 5.3 is by itself not a physical quantity, the effective-theory electromagnetic current operator we have constructed is relevant in a variety of physical processes. Three examples are shown in Figure 5.10. On the left, we show the process $e^- + p \to e^- + X$ in a situation, where the hadronic final state X contains many particles. This is called deep-inelastic scattering (DIS) and, as in the case of the Sudakov form factor, SCET is relevant in the limit where the final state invariant mass M_X is much smaller than the momentum transfer Q mediated by the virtual photon. Introducing the variable x via $M_X^2 = Q^2(1-x)/x$, this corresponds to the limit $x \to 1$. To analyse the process in SCET, it is easiest to work in the Breit frame in which the virtual photon has momentum $q^\mu = (0,0,0,Q)$. We then introduce a reference vector n^μ along the out-going low-mass jet of particles and a vector $\bar{n}^\mu$ for the incoming proton. The vector current of quarks is also relevant at hadron colliders, for example, for $pp \to \gamma^*/Z \to e^+ + e^- + X$. Here SCET is relevant in cases where the outgoing radiation X is either soft or collinear to the beam directions and the reference vectors n^μ and $\bar{n}^\mu$ point along the beams. Finally, and most obviously, SCET can be used in cases where there are low-mass final-state jets such as the two-jet process displayed on the right side of Figure 5.10. Here the reference vectors point along the jet directions. The processes involving hadrons in the initial state involve collinear matrix elements with protons. Such matrix elements are called *beam functions* [70], while the collinear matrix elements with a vacuum initial state are called *jet functions*. In addition to a perturbatively calculable part,

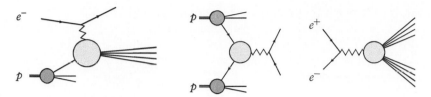

Figure 5.10 *Collider processes induced the vector or axial-vector current $\mathcal{J}^{\mu}$, indicated by the light gray blob. On the left, we have DIS, in the middle the Drell–Yan process, and on the right two-jet production in e^+e^- collisions. The processes with protons in the initial state involve PDFs, indicated in dark gray.*

the beam functions contain the usual PDFs, which arise in all processes with hadrons in the initial state.

It would of course be interesting to discuss jet production processes at the LHC. However, these processes involve energetic particles both along the beam directions and the jet directions. To analyse two-jet production at the LHC, we would therefore introduce light cone reference vectors $n_1, \ldots, n_4$ and conjugate vectors $\bar{n}_1, \ldots, \bar{n}_4$ and one would have four types of collinear fields. To keep things simple, we stick to processes with only two directions for which we can choose $n_1 = n$ and $n_2 = \bar{n}$. Since we want to have jets in the final state, we consider leptonic collisions and study the two jet process depicted in the right panel of Figure 5.10. The reader interested in PDFs and beam functions can consult [12], which discusses examples of hadron collider observables with two energetic directions, namely threshold resummation and transverse momentum for the Drell–Yan process.

So far, we have used the word jet loosely to talk about sprays of energetic particles with low invariant mass. To define jet cross sections, we need to be more concrete. One possibility to define a two-jet process is to demand that all energy except for a small fraction is contained inside two cones. This is the original jet definition proposed by Sterman and Weinberg [71] in 1977. This basic idea has evolved into a variety of modern jet definitions [72]. A simpler set of observables are *event shapes* which characterize the geometry of collider events. Rather than attributing particles to jets, one introduces a quantity which indicates how pencil-like the final state is. The prototypical event shape is the thrust T introduced by Farhi in the same year as the jet definition [73].

The resummation of large logarithms for actual jet processes is complicated and most SCET papers have therefore focused on event shapes. We first analyse the the event shape thrust but then briefly discuss the resummation for actual jet observables.

5.4.1   Factorization for the Event-Shape Variable Thrust

The definition of thrust [73],

$$T = \frac{1}{Q} \max_{\vec{n}_T} \sum_i |\vec{n}_T \cdot \vec{p}_i|, \tag{5.119}$$

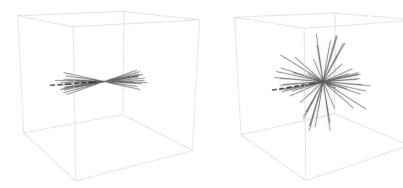

Figure 5.11 *Two sample collider events. On the left side a pencil-like two-jet event with small $\tau = 1 - T \approx 0.002$; on the right side an almost spherical event with large $\tau \approx 0.35$. The thrust axis is shown as a red dashed line.*

needs a few explanations. It involves a sum over all particles in the event and one sums the projections of their momenta along the thrust axis $\vec{n}_T$ which must be chosen to maximize the sum. The momentum flow along $\vec{n}_T$ is then normalized to the centre-of-mass energy, which for massless particles is given by

$$Q = \sum_i |\vec{p}_i|. \qquad (5.120)$$

The thrust T thus measures the fraction of momentum flowing along the thrust axis. For $T = 1$ all momentum must flow along $\vec{n}_T$, i.e. all particles are parallel or anti-parallel to $\vec{n}_T$. Events with large thrust T are thus pencil-like and involve two low-mass jets. It is convenient to define $\tau = 1 - T$ such that the end-point is at $\tau = 0$. The quantity τ defines the SCET expansion parameter which was denoted by λ in the discussion of the Sudakov problem. The maximum value of τ (minimum value of T) is obtained for a completely spherical event. It is a short exercise to verify that such a configuration has $\tau = 1/2$. Figure 5.11 shows two example events, together with their thrust values.

The definition (5.119) is useful to distinguish pencil-like from spherical events, but not suitable to single out configurations with more than two jets. However, minimizing over several reference vectors one can generalize thrust in such a way that it vanishes for events with N massless jets. This generalized quantity is called N-jettiness [74]. At hadron colliders, it is also useful define event shapes in the transverse plane [75]. An analysis of factorization for transverse thrust at a hadron colliders can be found in [76]. As discussed earlier, the relevant effective theory involves four collinear sectors.

Thrust is soft and collinear safe, i.e. its value does not change under exactly collinear splittings or infinitely soft emissions. This property makes it possible to compute it perturbatively. However, for small $\tau \ll 1$ we encounter large logarithms. To analyse this limit, let us choose the SCET reference vectors as $n^\mu = (1, \vec{n}_T)$ and $\bar{n}^\mu = (1, -\vec{n}_T)$. Figure 5.12 shows an event with small τ. It will involve energetic particles collinear to

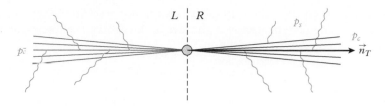

Figure 5.12 *An event with large thrust T (corresponding to small τ). The thrust axis $\vec{n}_T$, shown as an arrow, splits the event into two hemispheres denoted by L and R.*

n^μ and $\bar{n}^\mu$, together with soft large-angle radiation. Performing a region analysis, one will find the same regions (5.45) we identified in the Sudakov form factor, but with expansion parameter $\lambda = \tau$. We can thus separate the sum over particles in (5.119) into individual sums in the soft and collinear sectors and write

$$
\begin{aligned}
\tau Q &= \sum_i |\vec{p}_i| - |\vec{n}_T \cdot \vec{p}_i| \\
&= \sum_i n \cdot p_{ci} + \sum_i \bar{n} \cdot p_{\bar{c}i} + \sum_i n \cdot p_{si}^R + \sum_i \bar{n} \cdot p_{si}^L \\
&= n \cdot p_{X_c} + n \cdot p_{X_s}^R + \bar{n} \cdot p_{X_{\bar{c}}} + \bar{n} \cdot p_{X_s}^L ,
\end{aligned}
\tag{5.121}
$$

where we have split the soft particles into left- and right-moving ones in order to be able write the sums in terms of light cone components. In the last line, we have introduced the total momentum in each category. This result has a simple physical interpretation. Before deriving it, we note that due to the definition of the thrust axis, the total transverse momentum is zero in each hemisphere. As a result, also the *total* collinear transverse momentum $p_{X_c}^\perp$ is zero, up to terms which are of the same order as the soft momentum. Up to power corrections, we therefore write the invariant mass of all particles in the right hemisphere as

$$
\begin{aligned}
M_R^2 &= (p_{X_c} + p_{X_s}^R)^2 \\
&= p_{X_c}^2 + \bar{n} \cdot p_{X_c} n \cdot p_{X_s}^R \\
&= \bar{n} \cdot p_{X_c} n \cdot p_{X_c} + \bar{n} \cdot p_{X_c} n \cdot p_{X_s}^R \\
&= Q(n \cdot p_{X_c} + n \cdot p_{X_s}^R) .
\end{aligned}
\tag{5.122}
$$

The expression in the last line is exactly the contribution of the right-moving particles to (5.121). Up to power corrections, we obtain the equality

$$
\tau Q^2 = M_L^2 + M_R^2 = p_{X_c}^2 + p_{X_{\bar{c}}}^2 + Q\left(n \cdot p_{X_s}^R + \bar{n} \cdot p_{X_s}^L \right) .
\tag{5.123}
$$

For small values, τ is equal to the sum of the invariant masses in the two hemispheres, normalized to Q^2. The fact that thrust is additive in the soft and collinear contributions, will be important to establish factorization.

Let us now compute the cross section

$$\frac{d\sigma}{d\tau} = \frac{1}{2Q^2} \sum_X \hspace{-1.4em}\int \; |\mathcal{M}(e^+e^- \to \gamma^* \to X)|^2 (2\pi)^4 \delta^{(4)}(q - p_X) \delta(\tau - \tau(X)), \qquad (5.124)$$

where τ is the thrust value we prescribe, while $\tau(X)$ is the value obtained with the given final state momenta in the state X, which is equal to the expression in (5.123) at leading power. The momentum $q = q_1 + q_2$ is the total lepton momentum and $q^2 = Q^2$. The amplitude can be written as a product of a leptonic amplitude times a hadronic one. For the squared amplitude, this implies $|\mathcal{M}|^2 = L^{\mu\nu} H_{\mu\nu}$, where the hadronic tensor has the form

$$H_{\mu\nu}(q, \tau) = \sum_X \hspace{-1.4em}\int \; \langle 0| \mathcal{F}_\nu^\dagger(0) |X\rangle \langle X| \mathcal{F}_\mu(0) |0\rangle (2\pi)^4 \delta^{(4)}(q - p_X) \delta(\tau - \tau(X)), \qquad (5.125)$$

and

$$L^{\mu\nu}(q_1, q_2) = \frac{e^4 Q_q^2}{Q^4} \bar{v}(q_2) \gamma^\mu u(q_1) \bar{u}(q_1) \gamma^\nu v(q_2) \qquad (5.126)$$

contains the trivial leptonic part. For convenience, we include the photon propagators as well as the quark charge eQ_q in $L^{\mu\nu}$. Averaging over the spins of the incoming leptons produces a trace of Dirac matrices and the lepton tensor becomes

$$L^{\mu\nu}(q_1, q_2) = \frac{e^4 Q_q^2}{Q^4} \left(q_1^\mu q_2^\nu + q_2^\mu q_1^\nu - q_1 \cdot q_2 g^{\mu\nu} \right). \qquad (5.127)$$

We want the cross section to be differential in the angle θ of the thrust vector with respect to the incoming electron. To factor out the integral over this direction, we now explicitly distinguish the reference vector $\vec{n}$ in SCET from the thrust axis $\vec{n}_T$, which is derived from the particles in a given event and introduce the dummy integration

$$1 = \int d^3\vec{n}\, \delta^{(3)}(\vec{n} - \vec{n}_T). \qquad (5.128)$$

In the effective theory, the thrust axis is given by $\vec{n}_T = \vec{p}_{X_c}/|\vec{p}_{X_c}|$ up to power corrections. Inserting this into the above equation and using the fact that momentum conservation fixes $|\vec{p}_{X_c}| = Q/2$, we can rewrite it in the form

$$\int d^3\vec{n}\, \delta^{(3)}(\vec{n} - \vec{n}_T) = (2\pi) \int d\cos\theta \left(\frac{Q}{2}\right)^2 \delta^{(2)}(p_{X_c}^\perp), \qquad (5.129)$$

where the prefactor (2π) is from the integration over the azimuthal angle. The net effect of these manipulations is that the cross section involves a δ-function which fixes the total transverse momentum of the collinear radiation to be zero, as required by the thrust definition. Combining it with the momentum conservation δ-functions and expanding away small momentum components, we get

$$
\delta^{(4)}(q - p_{X_c} - p_{X_{\bar{c}}} - p_{X_s})\delta^{(2)}(p_{X_c}^\perp)
$$
$$
= 2\delta(\bar{n} \cdot p_{X_c} - Q)\delta(n \cdot p_{X_{\bar{c}}} - Q)\delta^{(2)}(p_{X_c}^\perp)\delta^{(2)}(p_{X_{\bar{c}}}^\perp). \tag{5.130}
$$

The factor of 2 is the Jacobian for converting to light cone components.

After this preparation, we can now plug in the factorized SCET current (5.110) into the hadron tensor (5.125). In the Sudakov form factor, we had an incoming and an outgoing particle, so that the current (5.110) had an incoming soft Wilson line S_n and an an outgoing Wilson line $\bar{S}_{\bar{n}}^\dagger$ after decoupling. In the present case the particle and anti-particle are both outgoing so that the appropriate soft Wilson-lines structure in the current is $\bar{S}_{\bar{n}}^\dagger \bar{S}_n$. Since all QCD particles in this section are outgoing, we drop the bar on the soft Wilson lines in the following.

Using that the states factorize in the form (5.109), the hadronic tensor contains a collinear, an anti-collinear, and a soft matrix elements, tied together by the thrust constraint (5.123). To separate the individual contributions to thrust, we introduce three more integrations

$$
1 = \int dM_c^2\, \delta\left(M_c^2 - p_{X_c}^2\right) \int dM_{\bar{c}}^2\, \delta\left(M_{\bar{c}}^2 - p_{X_{\bar{c}}}^2\right) \int d\omega\, \delta\left(\omega - n \cdot p_{X_s}^R - \bar{n} \cdot p_{X_s}^L\right). \tag{5.131}
$$

Putting all this together, we obtain the cross section in the factorized form

$$
\frac{d\sigma}{d\tau\, d\cos\theta} = \frac{\pi}{2} L_{\mu\nu} |\tilde{C}_V(-Q^2 - i0, \mu)|^2 \int dM_c^2 \int dM_{\bar{c}}^2 \int d\omega\, \delta\left(\tau - \frac{M_c^2 + M_{\bar{c}}^2 + Q\omega}{Q^2}\right)
$$
$$
\times \sum_{X_c} \langle 0| \chi_{c,\delta}^a(0)|X_c\rangle \langle X_c| \bar{\chi}_{c,\alpha}^b |0\rangle\, \delta\left(M_c^2 - p_{X_c}^2\right) \delta^{(2)}\left(p_{X_c}^\perp\right) \delta\left(\bar{n} \cdot p_{X_c} - Q\right)
$$
$$
\times \sum_{X_{\bar{c}}} \langle 0| \bar{\chi}_{\bar{c},\gamma}^d(0)|X_{\bar{c}}\rangle \langle X_{\bar{c}}| \chi_{\bar{c},\beta}^e |0\rangle\, \delta\left(M_{\bar{c}}^2 - p_{X_{\bar{c}}}^2\right) \delta^{(2)}\left(p_{X_{\bar{c}}}^\perp\right) \delta\left(n \cdot p_{X_{\bar{c}}} - Q\right)
$$
$$
\times \sum_{X_s} \langle 0|[S_n^\dagger S_{\bar{n}}]_{da} |X_s\rangle \langle X_s| [S_{\bar{n}}^\dagger S_n]_{be} |0\rangle\, \delta\left(\omega - n \cdot p_{X_s}^R - \bar{n} \cdot p_{X_s}^L\right)
$$
$$
\times (2\pi)^4\, (\gamma_\perp^\mu)_{\alpha\beta}\, (\gamma_\perp^\nu)_{\gamma\delta}. \tag{5.132}
$$

In this result, the Latin letters a, b, d, e denote the quark colours, and the Greek letters their Dirac indices. While it is lengthy, the above expression is simply the result of inserting the different ingredients and then separating the contributions to the different sectors. In line two, we have the collinear matrix element defining the jet function $\mathcal{J}$, line three defines $\bar{\mathcal{J}}$ and line four the soft function S. As they stand, these functions are matrices in colour and Dirac space, but the expressions can still be massaged further. An important property is that the collinear matrix elements are colour diagonal and therefore proportional to $\delta^{ab}\delta^{de}$. This contracts the colour indices in the soft matrix element, and we can define a scalar soft function as

$$S(\omega) = \frac{1}{N_c}\sum_{X_s}\langle 0|\left[S_n^\dagger S_{\bar{n}}\right]_{ab}|X_s\rangle\langle X_s|\left[S_{\bar{n}}^\dagger S_n\right]_{ba}|0\rangle\,\delta(\omega - n\cdot p_{X_s}^R - \bar{n}\cdot p_{X_s}^L)\,. \tag{5.133}$$

The prefactor $1/N_c$ has been added so that $S(\omega) = \delta(\omega)$ at lowest order. Next, we consider the collinear and anti-collinear matrix elements in the second and third lines. They can be written in the form

$$\frac{\delta^{ab}}{2(2\pi)^3}\left[\frac{\not{n}}{2}\right]_{\delta\alpha}\mathcal{J}(M^2) = \sum_{X_c}\langle 0|\,\chi_{c,\delta}^a(0)\,|X_c\rangle\,\langle X_c|\,\bar{\chi}_{c,\alpha}^b(0)\,|0\rangle$$
$$\times\,\delta(M^2 - p_{X_c}^2)\,\delta^{(2)}(p_{X_c}^\perp)\delta(\bar{n}\cdot p_{X_c} - Q)\,, \tag{5.134}$$

$$\frac{\delta^{de}}{2(2\pi)^3}\left[\frac{\not{\bar{n}}}{2}\right]_{\beta\gamma}\mathcal{J}(M^2) = \sum_{X_{\bar{c}}}\langle 0|\,\bar{\chi}_{\bar{c},\gamma}^d(0)\,|X_{\bar{c}}\rangle\,\langle X_{\bar{c}}|\,\chi_{\bar{c},\beta}^e(0)\,|0\rangle$$
$$\times\,\delta(M^2 - p_{X_{\bar{c}}}^2)\,\delta^{(2)}(p_{X_{\bar{c}}}^\perp)\delta(n\cdot p_{X_{\bar{c}}} - Q)\,. \tag{5.135}$$

The final state X_c in the collinear matrix element has the quantum numbers of a quark, while the $X_{\bar{c}}$ is a state with anti-quark quantum numbers. This is the case because we have chosen $\vec{n}_T$ to point along the quark direction. To understand why these matrix elements only involve a single scalar jet function $\mathcal{J}(M^2)$, we note that due to the property $\not{n}\chi_c = 0$, the collinear matrix elements must vanish when multiplied by $\not{n}$ on either side. Because of this constraint the field χ_c is effectively a two-component spinor and the associated Dirac basis has only the four matrices $\not{n}$, $\not{n}\gamma_5$, and $\not{n}\gamma_\perp^\mu$. Due to parity invariance, the second structure cannot arise and since it does not involve a transverse vector which could be dotted into $\gamma_\perp^\mu$, also the third one must be absent, leaving the first one as the only choice. The prefactor multiplying the Dirac structure on the left-hand side of (5.134) and (5.135) was chosen to have $\mathcal{J}(M^2) = \delta(M^2)$ at leading order.

A convenient representation for computing the jet function is to rewrite the matrix element as the imaginary part of the collinear propagator

$$\mathcal{J}(p^2) = \frac{1}{\pi}\,\mathrm{Im}\left[i\mathcal{J}(p^2)\right] \tag{5.136}$$

where

$$\frac{\not n}{2}\,\bar n\cdot p\,\mathcal J(p^2) = \int d^4x\,e^{-ip\cdot x}\,\langle 0|\,\mathrm T\{\chi_c(0)\overline\chi_c(x)\}\,|0\rangle. \tag{5.137}$$

To derive the equality of the two representations, we insert a complete set of states into (5.137), then translate the fields from point x to 0 using (5.98) and use the Fourier representation

$$\theta(x^0) = -\frac{1}{2\pi i}\int d\omega\,\frac{e^{-i\omega x^0}}{\omega + i0} \tag{5.138}$$

for the Heaviside functions in the definition of the time-ordered product. The imaginary parts of the two terms in the time-ordered product then arise from states with quark and anti-quark quantum numbers, as in (5.134) and (5.135) and are identical by the charge conjugation symmetry of QCD. The representation (5.137) makes it clear that the jet function has a simple QCD interpretation. We have stressed earlier that the collinear Lagrangian is equivalent to the QCD Lagrangian and the field $\chi_c = W_c^\dagger P_+ \psi_c$ can be obtained in QCD using the same projection operator and Wilson line. In the light cone gauge $\bar n\cdot A = 0$ the Wilson line is absent and we can thus view (and compute [77]) the jet function as the imaginary part of the quark propagator in this particular gauge.

We can now insert (5.134) and (5.135) into the factorized cross section (5.132) and use the definition (5.133) of the soft function. What remains are some simple manipulations, namely to evaluate the Dirac algebra of the hadron tensor

$$\mathrm{tr}\!\left[\gamma_\mu^\perp\frac{\bar{\not n}}{2}\gamma_\nu^\perp\frac{\not n}{2}\right] = n_\mu\bar n_\nu + n_\nu\bar n_\mu - 2g_{\mu\nu} \equiv -2g_{\mu\nu}^\perp \tag{5.139}$$

and to perform the contraction with the leptonic tensor. The contraction produces scalar products with lepton momenta, which can be written in terms of the scattering angle using $n\cdot q_1 = Q/2\,(1+\cos\theta)$, $\bar n\cdot q_1 = Q/2\,(1-\cos\theta)$ and the same relations with opposite signs in front of the cosines for q_2. Doing so and collecting the prefactors, we arrive at the final result

$$\frac{d\sigma}{d\tau\,d\cos\theta} = \frac{\pi N_c Q_f^2 \alpha^2}{2Q^2}(1+\cos^2\theta)|\tilde C_V(-Q^2-i0,\mu)|^2\int dM_c^2\int dM_{\bar c}^2\int d\omega$$
$$\delta\!\left(\tau - \frac{M_c^2 + M_{\bar c}^2 + Q\omega}{Q^2}\right)\mathcal J(M_c^2,\mu)\,\mathcal J(M_{\bar c}^2,\mu)\,S(\omega,\mu), \tag{5.140}$$

where we give the result for a single quark flavour with charge Q_q and have written it in terms of the fine-structure constant $\alpha = e^2/(4\pi)$. The dependence on θ is trivial and we can easily integrate over it. We have defined the jet and soft functions, such that they reduce to δ-functions at lowest order and $\tilde{C}_V = 1 + \mathcal{O}(\alpha_s)$. Plugging in the lowest-order expressions and integrating over θ and τ we reproduce the correct lowest order $e^+ e^- \to q\bar{q}$ cross section

$$\sigma_0 = \frac{4\pi\alpha^2 N_c Q_f^2}{3Q^2}. \tag{5.141}$$

We have indicated in (5.140) that the hard, jet, and soft functions depend on μ. To resum large logarithms, we can again solve the RG equations and evolve to a common reference scale as we discussed for the Sudakov form factor (Figure 5.9). The only complication is that in the factorization formula (5.140) the soft and jet functions are convolved, while we were dealing with a simple product in the case of the Sudakov form factor. This complication can be avoided by working in Laplace space since the transformation turns the convolution into a product. The Laplace transforms of the jet and soft functions fulfill a RG equation of the same form as $\tilde{C}_V$ and the inversion to momentum space can be performed analytically [78]. Working in Laplace space, the resummation for thrust was performed at next-to-next-to-leading order (NNLO) in RG improved perturbation theory in [79], which corresponds to next-to-next-to-next-to-leading logarithmic (N^{3}LL) accuracy in traditional terminology. For this accuracy we need the two-loop results for the hard, jet, and soft functions, together with three-loop anomalous dimensions. In fact, because of the extra logarithm in the cusp piece, we need four-loop accuracy for the cusp anomalous dimension γ_{cusp}; as discussed earlier, the four-loop piece is not yet fully known but has a very small effect on the result. Figure 5.13, [79] shows the fixed-order expansion to NNLO [80], compared to resummed results at different orders. The plots show the breakdown of fixed-order perturbation theory at small τ, where it diverges, and demonstrate that resummation dramatically improves the convergence of the expansion. We note that the resummed results were matched to the fixed-order predictions at larger τ, i.e. they are constructed in such a way that they reproduce the fixed-order result at higher τ. The different orders in the resummed result shown in the figure correspond to different accuracies in the resummation and in the matching, see [79] for more details. One-loop calculations for the hard, jet, and soft functions can be found in [81].

The resummed result for thrust was fit to the available measurements to extract a value of the strong coupling constant α_s [83]. Since the thrust distribution is affected by non-perturbative hadronization corrections, the authors not only fit for α_s but also for hadronization parameters using data at different centre-of-mass energies. The leading effect of hadronization is a shift of the distribution to the right and we observe that the theoretical prediction without hadronization shown in Figure 5.13 should be shifted to match the experimental result. Fitting for this shift, the authors of [83] find significant hadronization effects which reduce the extracted value of α_s by more than 7%. To extract α_s reliably, good control over the hadronization effects and their uncertainty is important. The extracted value $\alpha_s(m_Z) = 0.1135 \pm (0.0002)_{\text{expt}} \pm (0.0005)_{\text{hadr}} \pm (0.0009)_{\text{pert}}$ from the analysis [83] has small uncertainties and is significantly lower than the world average. The source of this discrepancy is not understood.

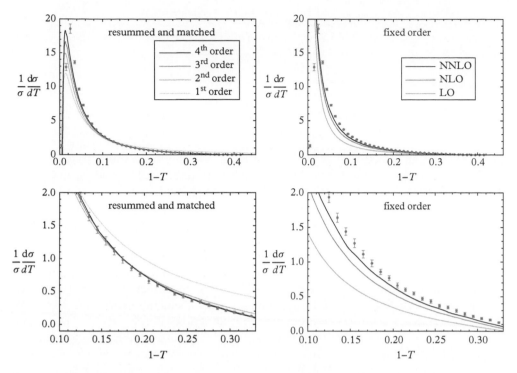

Figure 5.13 *Convergence of resummed and fixed-order distributions. ALEPH data [82] (red) at 91.2 GeV are included for reference. All plots have $\alpha_s(m_Z) = 0.1168$. This figure is taken from [79].*

5.4.2 Factorization and Resummation for Jet Cross Sections

Interestingly, the pattern of logarithms for jet cross sections such as the Sterman–Weinberg cross section is much more complicated than for thrust. Even at the leading-logarithmic level, one finds new colour structures at each loop order rather than a simple exponentiation, as would follow from an RG equation such as (5.103) for $\tilde{C}_V$. This complicated structure of logarithms was discovered by Dasgupta and Salam [84] and arises whenever an observable is insensitive to soft radiation in certain regions of phase space, as is the case for jet cross sections which do not constrain emissions inside the jets. Such observables are called non-global and the additional non-exponentiating terms are often called non-global logarithms (NGLs). Dasgupta and Salam resummed the leading NGLs at large N_c using a parton shower. They can also be computed by solving a non-linear integral equation derived by Banfi, Marchesini, and Smye (BMS) [85]. This approach was extended to finite N_c in [86] and some finite-N_c results are by now available [87, 88]. A resummation of subleading NGLs, on the other hand, has not yet been achieved.

Factorization for such observables was not understood until recently and given that the logarithms obey a non-linear equation, while RG equations are linear, it was even

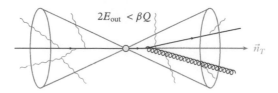

Figure 5.14 *A cone jet with three hard partons inside the jets, together with soft radiation.*

speculated that an effective theory treatment for non-global observables might not be possible. One recent proposal to deal with them is to resum global logarithms in a series of subjet observables with the goal of reducing the numerical size of the remaining non-global logarithms after combining the contributions from different subjet multiplicities [89]. At leading-logarithmic accuracy, the expansion in subjets amounts to an iterative solution of the BMS equation [90]. In a series of recent papers [91]–[94], we have attacked the problem directly and have shown that a resumation of NGLs *can* be obtained using effective-theory methods. We have derived factorization theorems for a number of such observables, starting with the Sterman–Weinberg cross section. Interestingly, the associated RG equations reduce to a parton shower at leading-logarithmic accuracy. Our effective-theory approach reproduces the leading-logarithmic result of Dasgupta and Salam, but also shows what ingredients are needed to resum subleading logarithms. While they are rather different at first sight, our results closely relate to the colour-density matrix formalism of reference [95]. We comment on the precise connection later.

We now analyse the simplest example of a non-global observable, namely the Sterman–Weinberg cross section [71]. For simplicity, we choose the thrust axis $\vec{n}_T$ as the jet axis and then put two oppositely directed cones of half-angle δ on the axis. We will work with a large cone angle δ, and will impose $2E_{\text{out}} < \beta Q$ on the energy outside the jet cones (Figure 5.14). Choosing $\beta \ll 1$ restricts the outside energy to be much smaller than the centre-of-mass energy of the collisions Q, but the cross section will suffer from large logarithms of β, which we will resum in the following. The case of small angle δ was analysed in [91, 92], but is more complicated because also collinear logarithms associated with the small opening angle are present and need to be resummed. As always, we should first identify which momentum regions are relevant for the problem at hand. This is particularly simple in our setup. Since the opening angle is large, $\delta \sim 1$, there is not really a direction which is singled out. Of course, each event has a thrust axis $\vec{n}_T$, which we use as the jet axis, but since we do not impose that the thrust is large, the particles will in general not be collinear to this axis. So we only have two momentum regions

$$
\begin{aligned}
\text{hard} &\quad (h) \quad k^\mu \sim Q, \\
\text{soft} &\quad (s) \quad k^\mu \sim \beta Q.
\end{aligned}
\tag{5.142}
$$

The hard momentum region is relevant for the partons inside the jet. These partons cannot be outside otherwise they would violate the constraint $2E_{\text{out}} < \beta Q$. Only the soft partons are allowed to be outside the jet. We should now analyse the factorization of the soft partons from the hard ones. Fortunately, and not entirely by coincidence, we have analysed exactly this situation in Section 5.1, which discussed soft effective theory. We derived that the soft radiation in QED is given by Wilson lines along the energetic particles, see (5.24). The same is true in QCD, except that the soft Wilson lines are colour matrices, which act on the colours of the hard amplitude. It is convenient to suppress the colour indices and use the colour-space notation of Catani and Seymour [96, 97], in which the factorization for an amplitude with m hard partons takes the form

$$S_1(n_1)\, S_2(n_2) \ldots S_m(n_m)\, |\mathcal{M}_m(\{\underline{p}\})\rangle\,, \tag{5.143}$$

where $\{\underline{p}\} = \{p_1, \ldots, p_m\}$ are the particle momenta and $\{\underline{n}\} = \{n_1, \ldots, n_m\}$ are light-cone vectors along their directions. The amplitude $|\mathcal{M}_m\rangle$ is a ket vector in colour space and the Wilson-line matrix $S_i(n_i)$ acts on the colour index of particle i. Note that the matrices acting on different partons trivially commute since they act on different indices. To obtain the cross section, we need to square (5.143), which will give rise to a hard function times a soft matrix element as in (5.25). However, in Section 5.2 we considered soft radiation for an exclusive process, while we deal with an inclusive cross section in the present case. To get the inclusive result, we need to sum over the number of partons in the final state and integrate over their directions. The cross section then takes the form [91, 92]

$$\sigma(Q,\beta) = \sum_{m=k}^{\infty} \langle \mathcal{H}_m(\{\underline{n}\}, Q, \mu) \otimes \boldsymbol{S}_m(\{\underline{n}\}, \beta Q, \mu)\rangle. \tag{5.144}$$

The symbol $\otimes$ indicates the integral over the directions and $\langle \ldots \rangle$ denotes the colour trace, which is taken after multiplying the two functions. The soft function is

$$\boldsymbol{S}_m(\{\underline{n}\}, \beta Q, \mu) =$$
$$\sum_{X_s} \langle 0| S_1^{\dagger}(n_1) \ldots S_m^{\dagger}(n_m) |X_s\rangle \langle X_s| S_1(n_1) \ldots S_m(n_m) |0\rangle\, \theta(\beta Q - 2E_{\text{out}}), \tag{5.145}$$

which can be compared to (5.27). The hard function is

$$\mathcal{H}_m(\{\underline{n}\}, Q, \mu) = \frac{1}{2Q^2} \sum_{\text{spins}} \prod_{i=1}^{m} \int \frac{dE_i\, E_i^{d-3}}{(2\pi)^{d-2}}\, |\mathcal{M}_m(\{\underline{p}\})\rangle \langle \mathcal{M}_m(\{\underline{p}\})|$$
$$\times (2\pi)^d\, \delta\Big(Q - \sum_{i=1}^{m} E_i\Big)\, \delta^{(d-1)}(\vec{p}_{\text{tot}})\, \Theta_{\text{in}}(\{\underline{p}\}). \tag{5.146}$$

The phase space constraint $\Theta_{\text{in}}(\{\underline{p}\})$ restricts the hard partons to the inside of the jets. As in our earlier examples, the factorization theorem (5.144) achieves scale separation. The hard function only depends on the large scale Q, while the soft function depends on the soft scale $Q\beta$. As usual, these functions involve divergences, which can be renormalized after which both the Wilson coefficients $\mathcal{H}_m$ and the soft functions $\mathcal{S}_m$ depend on the renormalization scale. For experts, let us briefly explain how the above results relate to the formalism of reference [95] in which a colour-density matrix U is used to track hard partons. The hard functions (5.146) are obtained by expanding the colour-density functional to the m-th power in U and the low-energy Wilson lines matrix elements 5.145 arise when performing a low-energy average over U.

With formula (5.144) the resummation of large logarithms (global and non-global) becomes standard, at least in principle. All we need to do is to solve the RG evolution for the hard functions

$$\frac{d}{d\ln\mu}\,\mathcal{H}_m(\{\underline{n}\},Q,\mu) = -\sum_{l=k}^{m}\mathcal{H}_l(\{\underline{n}\},Q,\mu)\,\boldsymbol{\Gamma}^H_{lm}(\{\underline{n}\},Q,\mu) \qquad (5.147)$$

and evolve them from an initial scale $\mu_h \sim Q$, where they are free of large logarithms, down to a lower scale $\mu_s \sim Q\beta$ at which the soft functions are free from large corrections. The only difficulty is that there are infinitely many hard functions $\mathcal{H}_l$ that mix under renormalization because the anomalous dimension $\boldsymbol{\Gamma}^H_{lm}$ is a matrix not only in colour space but also in the multiplicity of the partons. At one loop this matrix has the simple structure

$$\boldsymbol{\Gamma} = \frac{\alpha_s}{4\pi}\begin{pmatrix} V_2 & R_2 & 0 & 0 & \cdots \\ 0 & V_3 & R_3 & 0 & \cdots \\ 0 & 0 & V_3 & R_3 & \cdots \\ 0 & 0 & 0 & V_4 & \cdots \\ \vdots & \vdots & \vdots & \vdots & \ddots \end{pmatrix} + \mathcal{O}(\alpha_s^2), \qquad (5.148)$$

because the one-loop corrections either are purely virtual in which case they do not change the multiplicity, or involve a single real emission. This structure then imprints itself onto the $\boldsymbol{Z}$-factor and the anomalous dimension $\boldsymbol{\Gamma}$. That such a mixing must be present is familiar from fixed-order perturbative computations. Amplitudes with fixed multiplicities are not finite and to get a finite result, we need to combine them with lower multiplicity terms. For example, in a NLO computation, the divergences of the real-emission diagrams cancel against the virtual ones.

The anomalous dimension encodes UV divergences of the soft functions, which are in one-to-one correspondence to soft divergences in the amplitudes. These can be extracted by considering soft limits of amplitudes after which we obtain the explicit results [92]

$$V_m = 2\sum_{(ij)}(\boldsymbol{T}_{i,L}\cdot\boldsymbol{T}_{j,L} + \boldsymbol{T}_{i,R}\cdot\boldsymbol{T}_{j,R})\int\frac{d\Omega(n_l)}{4\pi}\,W^l_{ij}, \qquad (5.149)$$

$$R_m = -4\sum_{(ij)}\boldsymbol{T}_{i,L}\cdot\boldsymbol{T}_{j,R}\,W^{m+1}_{ij}\,\Theta_{\text{in}}(n_{m+1}),$$

where the dipole radiator is defined as

$$W_{ij}^l = \frac{n_i \cdot n_j}{n_i \cdot n_l \, n_j \cdot n_l}. \tag{5.150}$$

This dipole is the combination of two eikonal factors as in (5.10), from a soft exchange between legs i and j.

We now consider the resummation of the leading logarithms, which corresponds to evolving with the anomalous dimension at $\mathcal{O}(\alpha_s)$ and evaluating the hard and soft functions at $\mathcal{O}(\alpha_s^0)$. Since the anomalous dimension matrix is very sparse, it is convenient to write out the one-loop RG equation explicitly

$$\frac{d}{dt} \mathcal{H}_m(t) = \mathcal{H}_m(t) \, V_m + \mathcal{H}_{m-1}(t) \, R_{m-1}, \tag{5.151}$$

where we have traded the dependence on μ for the variable

$$t = \frac{1}{2\beta_0} \ln \frac{\alpha(\mu)}{\alpha(\mu_h)} = \frac{\alpha_s}{4\pi} \ln \frac{\mu_h}{\mu} + \mathcal{O}(\alpha_s^2), \tag{5.152}$$

which is zero for $\mu = \mu_h$ and increases as we evolve to lower scales. For $\mu = \mu_s$ the logarithm becomes large and compensates the suppression by the coupling constant. For this reason t is treated as a quantity of $\mathcal{O}(1)$ in RG-improved perturbation theory. Solving the homogeneous equation and using variation of the constant, equation (5.152) can also be written as

$$\mathcal{H}_m(t) = \mathcal{H}_m(t_0) \, e^{(t-t_0)V_m} + \int_{t_0}^{t} dt' \, \mathcal{H}_{m-1}(t') \, R_{m-1} \, e^{(t-t')V_m}. \tag{5.153}$$

This form is usually written for parton showers, which involve an evolution time t. One can evolve from t_0 to t either without any additional emissions (first term in (5.153)) or by adding an emission to the lower multiplicity cross section (second term). The connection to parton showers becomes even more clear when we consider the initial condition. At the high scale $\mu = \mu_h$, corresponding to $t = 0$, only the hard function $\mathcal{H}_2$ is present since the higher multiplicity functions involve powers of α_s and are free of large logarithms for this scale choice. Starting with the function $\mathcal{H}_2$, we can then iteratively generate the higher functions as

$$\mathcal{H}_2(t) = \mathcal{H}_2(0) \, e^{tV_2},$$

$$\mathcal{H}_3(t) = \int_0^t dt' \, \mathcal{H}_2(t') \, R_2 \, e^{(t-t')V_3},$$

$$\mathcal{H}_4(t) = \int_0^t dt' \, \mathcal{H}_3(t') \, R_3 \, e^{(t-t')V_4}, \tag{5.154}$$

$$\mathcal{H}_5(t) = \ldots,$$

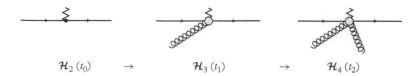

$$\mathcal{H}_2\,(t_0) \qquad \rightarrow \qquad \mathcal{H}_3\,(t_1) \qquad \rightarrow \qquad \mathcal{H}_4\,(t_2)$$

Figure 5.15 *Solution of the RG equation using Monte Carlo methods.*

see Figure 5.15. To get the resummed result, we evolve to the appropriate value of t, which is set by the scales μ_h and μ_s in (5.152). The leading-logarithmic cross section is obtained from the sum

$$\sigma_{\mathrm{LL}}(Q,\beta) = \sum_{m=2}^{\infty} \langle \mathcal{H}_m(t) \,\hat{\otimes}\, 1 \rangle$$

$$= \langle \mathcal{H}_2(t) + \int \frac{d\Omega_1}{4\pi} \mathcal{H}_3(t) + \int \frac{d\Omega_1}{4\pi} \int \frac{d\Omega_2}{4\pi} \mathcal{H}_4(t) + \ldots \rangle, \qquad (5.155)$$

where we have used that the soft functions at the scale μ_s are trivial at leading order $\mathcal{S}_m = 1 + \mathcal{O}(\alpha_s)$. The integrals over t as well as the angular integrals for the cross section can be computed using Monte Carlo methods as is done in all parton shower programs. To do so, we generate a time step Δt_1, evaluates $\mathcal{H}_2(t_1)$ at $t_1 = \Delta t_1$ using (5.154). This is then used as the initial condition for computing $\mathcal{H}_3(t_2)$ after the next time step Δt_2 which yields $t_2 = \Delta t_1 + \Delta t_2$ after randomly choosing the direction of the extra parton, etc. Repeating these steps we obtain an ensemble of hard functions that yield the cross section (5.155). The only stumbling block for an implementation is that the hard functions and anomalous dimensions are matrices in colour space. Also note that the shower acts on the level of the amplitudes in (5.146), not on the cross section. In fact, our shower is within the general class of showers described in [98]. To avoid the difficulty with colour, we can take the large N_c limit which renders the colour structure trivial [92]. An efficient implementation of the resulting parton shower was given by Dasgupta and Salam in [84].

It is remarkable that we recover a parton shower as the solution of a RG equation in SCET. Of course it is not a general purpose shower, but simply computes the logarithms for a class of observables and nothing more, but also nothing less. We are guaranteed that we obtain the correct logarithmic structure from this shower, since it is based on a factorization theorem in QCD. The NGL shower differs in some important aspects from standard Monte Carlo showers. For example, these codes take great care to conserve momentum in the emissions, while our shower does not do this, since small momenta are systematically expanded away. We only generate angles, not full momenta, because the evolution generates the logarithms of the energy. However, since our shower is based on RG equations, we know which elements are needed to extend the resummation to higher logarithms, in contrast to standard showers. Next-to-leading logarithmic resummation corresponds to NLO in RG improved perturbation theory. We thus need to include the hard functions and soft functions to $\mathcal{O}(\alpha_s)$. Concretely, this implies that the one-loop corrections to $\mathcal{H}_2(0)$ and the tree-level $\mathcal{H}_3(0)$, together with the one-loop correction to all functions $\mathcal{S}_m$ need to be included, as well as the two-loop anomalous dimension. This anomalous dimension, based on soft limits, was computed in the colour-density matrix formalism in [95] (for $\mathcal{N} = 4$ even three-loop corrections were obtained [99]). It should therefore be possible to set up a shower that also performs subleading resummation for non-global observables.

The connection to parton showers is also interesting from the point of view of automating resummations. For global observables automated resummation was pioneered in [100] at NLL and extended to NNLL for leptonic collisions in [101]. Also in the framework of SCET some resummations have been automated. These include NNLL resummation for processes involving a jet-veto [102], NLL resummation for two-jet event shapes [103] and the NNLO computation of soft functions [104, 105]. Such automation will be important to make higher-logarithmic computations more available and to extend them to more complicated observables, similarly to what has been achieved by automating fixed-order computations up to NLO.

Given that our field of research is high-energy physics, an effective theory for energetic particles has obviously many applications, beyond the few examples discussed in these lectures. An overview over the areas in which this EFT has been used can be found in the last chapter of [12]. In addition to collider-QCD applications, there has been a lot of work on heavy quark physics (in particular on B-meson decays), work on electroweak resummation for collider processes and dark matter annihilation, work on low-energy QED processes, applications in heavy-ion collisions, as well as more exotic topics such as soft-collinear gravity [106, 107] and supersymmetry [108]. An important area of recent progress concerns Glauber gluons. A version of SCET that includes their effects has been put forward [109] and been used to study factorization violation [110, 111]. Another important recent application of SCET has been to use it not just for resummations, but to simplify fixed-order computations. Using a method called N-jettiness subtraction [112, 113], a generalization of q_T-subtraction [114], we are able to trade an NNLO computation in QCD for an NNLO computation in SCET, together with an NLO computation in QCD. This technique has successfully been used to perform many NNLO computations. Jet substructure (see [115] for a recent review) is another area where there has been a lot of progress using SCET. Factorization theorems for jet substructure observables have been derived [116, 117] and substructure techniques have been used to improve observables such that they can be predicted with better accuracy [89, 118, 119]. Finally, we mentioned earlier that the analysis of power corrections has progressed a lot over the past year [51]–[57]. These examples illustrate that, while SCET has become a a standard method to analyse factorization and perform resummation, the development of these methods continues to be a very active area of research.

Acknowledgements

I thank the school organizers Sacha Davidson, Paolo Gambino, Mikko Laine, and Matthias Neubert for the opportunity to present these lectures. I thank the students for their attention, for interesting discussions, and of course the wine. I attended a Les Houches summer school as a PhD student and it was a great pleasure to be back (and slightly shocking to discover how many years passed since my first visit). Two of the students, Marcel Balsiger and Philipp Schicho, were kind enough to read through a draft version of these lecture notes (they have the bad fortune to do their PhD here in Bern so that I could easily solicit their help). I thank them for pointing out some of the more egregious typos; please let me know if you find more! I thank Samuel Favrod, Dingyu Shao, and Rudi Rahn for discussions, and Andrea Ferroglia and Alessandro Broggio for letting me recycle some equations from our book.

References

[1] C. W. Bauer, S. Fleming, D. Pirjol, and I. W. Stewart, *Phys. Rev. D* 63: 114020 [hep-ph/0011336].

[2] C. W. Bauer, D. Pirjol, and I. W. Stewart, *Phys. Rev. D* 65: 054022 [hep-ph/0109045].

[3] M. Beneke, A. P. Chapovsky, M. Diehl, and T. Feldmann, *Nucl. Phys. B* 643: 431 [hep-ph/0206152].

[4] J. F. Donoghue, E. Golowich, and B. R. Holstein, (1992). *Dynamics of the Standard Model*, Cambridge Monographs on Particle Physics, Nuclear Physics and Cosmology, Vol. 2 [(2014). Cambridge Monographs on Particle Physics, Nuclear Physics and Cosmology, Vol. 35].

[5] H. Georgi, *Ann. Rev. Nucl. Part. Sci.* 43: 209.

[6] A. Pich, hep-ph/9806303.

[7] I. Z. Rothstein, hep-ph/0308266.

[8] G. Ecker, hep-ph/0507056.

[9] D. B. Kaplan, nucl-th/0510023.

[10] A. G. Grozin, arXiv:0908.4392 [hep-ph].

[11] A. A. Petrov and A. E. Blechman, *Effective Field Theories*, World Scientific.

[12] T. Becher, A. Broggio, and A. Ferroglia, *Lect. Notes Phys.* 896: 1 [arXiv:1410.1892 [hep-ph]].

[13] M. Neubert, Les Houches Lectures on Renormalization Theory and Effective Field Theories. In *Proceedings of the 2017 Les Houches Summer School on Effective Field Theory in Particle Physics and Cosmology* 108th Session. July 31–August 25, 2017, Les Houches, France. [arXiv:1901.06573 [hep-ph]].

[14] J. C. Collins, D. E. Soper, and G. F. Sterman, *Adv. Ser. Direct. High Energy Phys.* 5, 1 [hep-ph/0409313].

[15] G. F. Sterman, hep-ph/9606312.

[16] G. F. Sterman, *Lect. Notes Phys.* 479: 209.

[17] J. Collins, (2011). *Foundations of perturbative QCD*, Cambridge Monographs on Particle Physics, Nuclear Physics and Cosmology, Vol. 32.

[18] A. V. Manohar, Introduction to Effective Field Theories. In *Proceedings of the 2017 Les Houches Summer School on Effective Field Theory in Particle Physics and Cosmology* 108th Session. July 31–August 25, 2017, Les Houches, France. [arXiv:1804.05863 [hep-ph]].

[19] D. R. Yennie, S. C. Frautschi, and H. Suura, *Annals Phys.* 13: 379.

[20] F. Bloch and A. Nordsieck, *Phys. Rev.* 52: 54.

[21] T. Kinoshita, *J. Math. Phys.* 3: 650.

[22] T. D. Lee and M. Nauenberg, *Phys. Rev. B.* 133: 1549.

[23] C. P. Burgess, *Ann. Rev. Nucl. Part. Sci.* 57: 329 [hep-th/0701053].

[24] J. G. M. Gatheral, *Phys. Lett. B*. 133: 90.

[25] J. Frenkel and J. C. Taylor, *Nucl. Phys. B* 246: 231.

[26] A. Mitov, G. Sterman, and I. Sung, *Phys. Rev. D* 82: 096010 [arXiv:1008.0099 [hep-ph]].

[27] E. Gardi, E. Laenen, G. Stavenga, and C. D. White, *JHEP* 1011: 155 [arXiv:1008.0098 [hep-ph]].

[28] E. Gardi, J. M. Smillie, and C. D. White, *JHEP* 1306: 088 [arXiv:1304.7040 [hep-ph]].

[29] T. Mannel, Effective Field Theories for Heavy Quarks. In *Proceedings of the 2017 Les Houches Summer School on Effective Field Theory in Particle Physics and Cosmology* 108th Session. July 31–August 25, 2017, Les Houches, France.

[30] M. Beneke and V. A. Smirnov, *Nucl. Phys. B* 522: 321 [hep-ph/9711391].

[31] V. A. Smirnov, (2002). *Applied Asymptotic Expansions in Momenta and Masses*, Springer Tracts in Modern Physics, Vol. 177.

[32] A. V. Manohar and I. W. Stewart, *Phys. Rev. D* 76, 074002 [hep-ph/0605001].

[33] B. Jantzen, *JHEP* 1112: 076 [arXiv:1111.2589 [hep-ph]].

[34] T. Becher and M. Neubert, *Eur. Phys. J. C.* 71: 1665 [arXiv:1007.4005 [hep-ph]].

[35] J. Y. Chiu, A. Jain, D. Neill and I. Z. Rothstein, *Phys. Rev. Lett.* 108: 151601 [arXiv:1104.0881 [hep-ph]].

[36] J. Y. Chiu, A. Jain, D. Neill, and I. Z. Rothstein, *JHEP* 1205: 084 [arXiv:1202.0814 [hep-ph]].

[37] T. Becher and M. Neubert, *Phys. Rev. Lett.* 102: 162001 Erratum: [*Phys. Rev. Lett.* 111(19): 199905] [arXiv:0901.0722 [hep-ph]].

[38] T. Becher and M. Neubert, *JHEP* 0906: 081 Erratum: [*JHEP* 1311: 024] [arXiv:0903.1126 [hep-ph]].

[39] T. Becher and M. Neubert, *Phys. Rev. D.* 79: 125004 Erratum: [*Phys. Rev. D.* 80: 109901] [arXiv:0904.1021 [hep-ph]].

[40] V. Ahrens, M. Neubert, and L. Vernazza, *JHEP* 1209: 138 [arXiv:1208.4847 [hep-ph]].

[41] E. Gardi and L. Magnea, *JHEP* 0903: 079 [arXiv:0901.1091 [hep-ph]].

[42] L. J. Dixon, E. Gardi, and L. Magnea, *JHEP* 1002: 081 [arXiv:0910.3653 [hep-ph]].

[43] V. Del Duca, et al., *Phys. Rev. D.* 85: 071104 [arXiv:1108.5947 [hep-ph]].

[44] V. Del Duca, C. Duhr, E. Gardi, L. Magnea and C. D. White, *JHEP* 1112: 021 [arXiv:1109.3581 [hep-ph]].

[45] Ø. Almelid, C. Duhr, and E. Gardi, *Phys. Rev. Lett.* 117(17): 172002 [arXiv:1507.00047 [hep-ph]].

[46] J. M. Henn and B. Mistlberger, *Phys. Rev. Lett.* 117(17): 171601 [arXiv:1608.00850 [hep-th]].

[47] Ø. Almelid, et al., *JHEP* 1709: 073 [arXiv:1706.10162 [hep-ph]].

[48] T. Becher and M. Neubert, *Phys. Lett. B.* 637: 251 [hep-ph/0603140].

[49] S. M. Freedman and M. Luke, *Phys. Rev. D.* 85: 014003 [arXiv:1107.5823 [hep-ph]].

[50] M. Beneke and T. Feldmann, *Phys. Lett. B.* 553: 267 [hep-ph/0211358].

[51] I. Moult, et al., *Phys. Rev. D.* 95(7): 074023 [arXiv:1612.00450 [hep-ph]].

[52] R. Boughezal, X. Liu, and F. Petriello, *JHEP* 1703: 160 [arXiv:1612.02911 [hep-ph]].

[53] D. Bonocore, et al., *JHEP* 1612: 121 [arXiv:1610.06842 [hep-ph]].

[54] I. Feige, D. W. Kolodrubetz, I. Moult, and I. W. Stewart, *JHEP* 1711: 142 [arXiv:1703.03411 [hep-ph]].

[55] I. Moult, et al., *Phys. Rev. D.* 97(1): 014013 [arXiv:1710.03227 [hep-ph]].

[56] R. Goerke and M. Inglis-Whalen, *JHEP* 1805: 023 [arXiv:1711.09147 [hep-ph]].

[57] M. Beneke, M. Garny, R. Szafron, and J. Wang, *JHEP* 1803: 001 [arXiv:1712.04416 [hep-ph]].

[58] A. V. Manohar, T. Mehen, D. Pirjol, and I. W. Stewart, *Phys. Lett. B.* 539: 59 [hep-ph/0204229].

[59] P. A. Baikov, et al., *Phys. Rev. Lett.* 102: 212002 [arXiv:0902.3519 [hep-ph]].

[60] T. Gehrmann, et al., *JHEP* 1011: 102 [arXiv:1010.4478 [hep-ph]].

[61] A. M. Polyakov, *Nucl. Phys. B.* 164: 171.

[62] R. A. Brandt, F. Neri, and M. A. Sato, *Phys. Rev. D.* 24: 879.

[63] I. A. Korchemskaya and G. P. Korchemsky, *Phys. Lett. B.* 287: 169.

[64] J. Henn, et al., *JHEP* 1703: 139 [arXiv:1612.04389 [hep-ph]].

[65] S. Moch, et al., *JHEP* 1710: 041 [arXiv:1707.08315 [hep-ph]].

[66] R. H. Boels, T. Huber, and G. Yang, *Phys. Rev. Lett.* 119(20): 201601 [arXiv:1705.03444 [hep-th]].

[67] R. H. Boels, T. Huber, and G. Yang, *JHEP* 1801: 153 [arXiv:1711.08449 [hep-th]].

[68] L. J. Dixon, *JHEP* 1801: 075 [arXiv:1712.07274 [hep-th]].

[69] T. Becher, M. Neubert, and B. D. Pecjak, *JHEP* 0701: 076 [hep-ph/0607228].

[70] I. W. Stewart, F. J. Tackmann, and W. J. Waalewijn, *Phys. Rev. D.* 81: 094035 [arXiv:0910.0467 [hep-ph]].

[71] G. F. Sterman and S. Weinberg, *Phys. Rev. Lett.* 39: 1436.

[72] G. P. Salam, *Eur. Phys. J. C.* 67: 637 [arXiv:0906.1833 [hep-ph]].

[73] E. Farhi, *Phys. Rev. Lett.* 39: 1587.

[74] I. W. Stewart, F. J. Tackmann, and W. J. Waalewijn, *Phys. Rev. Lett.* 105: 092002 [arXiv:1004.2489 [hep-ph]].

[75] A. Banfi, G. P. Salam, and G. Zanderighi, *JHEP* 1006: 038 [arXiv:1001.4082 [hep-ph]].

[76] T. Becher and X. Garcia i Tormo, *JHEP* 1506: 071 [arXiv:1502.04136 [hep-ph]].

[77] T. Becher and G. Bell, *Phys. Lett. B.* 695: 252 [arXiv:1008.1936 [hep-ph]].

[78] T. Becher and M. Neubert, *Phys. Rev. Lett.* 97: 082001 [hep-ph/0605050].

[79] T. Becher and M. D. Schwartz, *JHEP* 0807: 034 [arXiv:0803.0342 [hep-ph]].

[80] A. Gehrmann-De Ridder, T. Gehrmann, E. W. N. Glover, and G. Heinrich, *JHEP* 0711: 058 [arXiv:0710.0346 [hep-ph]].

[81] M. D. Schwartz, (2014). *Quantum Field Theory and the Standard Model*, Cambridge University Press.

[82] A. Heister et al. [ALEPH Collaboration], *Eur. Phys. J. C.* 35: 457.

[83] R. Abbate, et al., *Phys. Rev. D.* 83: 074021 [arXiv:1006.3080 [hep-ph]].

[84] M. Dasgupta and G. P. Salam, *Phys. Lett. B.* 512: 323 [hep-ph/0104277].

[85] A. Banfi, G. Marchesini, and G. Smye, *JHEP* 0208: 006 [hep-ph/0206076].

[86] H. Weigert, *Nucl. Phys. B.* 685: 321 [hep-ph/0312050].

[87] Y. Hatta and T. Ueda, *Nucl. Phys. B.* 874: 808 [arXiv:1304.6930 [hep-ph]].

[88] Y. Hagiwara, Y. Hatta, and T. Ueda, *Phys. Lett. B.* 756: 254 [arXiv:1507.07641 [hep-ph]].

[89] A. J. Larkoski, I. Moult, and D. Neill, *JHEP* 1509: 143 [arXiv:1501.04596 [hep-ph]].

[90] A. J. Larkoski, I. Moult, and D. Neill, *JHEP* 1611: 089 [arXiv:1609.04011 [hep-ph]].

[91] T. Becher, M. Neubert, L. Rothen, and D. Y. Shao, *Phys. Rev. Lett.* 116(19): 192001 [arXiv:1508.06645 [hep-ph]].

[92] T. Becher, M. Neubert, L. Rothen, and D. Y. Shao, *JHEP* 1611: 019 Erratum: [*JHEP* 1705: 154] [arXiv:1605.02737 [hep-ph]].

[93] T. Becher, B. D. Pecjak, and D. Y. Shao, *JHEP* 1612: 018 [arXiv:1610.01608 [hep-ph]].

[94] T. Becher, R. Rahn, and D. Y. Shao, *JHEP* 1710: 030 [arXiv:1708.04516 [hep-ph]].

[95] S. Caron-Huot, *JHEP* 1803: 036 [arXiv:1501.03754 [hep-ph]].

[96] S. Catani and M. H. Seymour, *Phys. Lett. B.* 378: 287 [hep-ph/9602277].

[97] S. Catani and M. H. Seymour, *Nucl. Phys. B.* 485: 291 Erratum: [*Nucl. Phys. B.* 510: 503] [hep-ph/9605323].

[98] Z. Nagy and D. E. Soper, *JHEP* 0709: 114 [arXiv:0706.0017 [hep-ph]].

[99] S. Caron-Huot and M. Herranen, *JHEP* 1802: 058 [arXiv:1604.07417 [hep-ph]].

[100] A. Banfi, G. P. Salam, and G. Zanderighi, *JHEP* 0503: 073 [hep-ph/0407286].

[101] A. Banfi, H. McAslan, P. F. Monni, and G. Zanderighi, *JHEP* 1505: 102 [arXiv:1412.2126 [hep-ph]].

[102] T. Becher, R. Frederix, M. Neubert, and L. Rothen, *Eur. Phys. J. C.* 75(4): 154 [arXiv:1412.8408 [hep-ph]].

[103] D. Farhi, I. Feige, M. Freytsis, and M. D. Schwartz, *JHEP* 1608: 112 [arXiv:1507.06315 [hep-ph]].

[104] G. Bell, R. Rahn, and J. Talbert, (2016). *PoS (RADCOR 2015)* 235: 052 [arXiv:1512.06100 [hep-ph]].

[105] G. Bell, R. Rahn, and J. Talbert, (2018). *PoS (RADCOR 2017)* 290: 047 [arXiv:1801.04877 [hep-ph]].

[106] M. Beneke and G. Kirilin, *JHEP* 1209: 066 [arXiv:1207.4926 [hep-ph]].

[107] T. Okui and A. Yunesi, *Phys. Rev. D.* 97(6): 066011 [arXiv:1710.07685 [hep-th]].

[108] T. Cohen, G. Elor, and A. J. Larkoski, *JHEP* 1703: 017 [arXiv:1609.04430 [hep-th]].

[109] I. Z. Rothstein and I. W. Stewart, *JHEP* 1608: 025 [arXiv:1601.04695 [hep-ph]].

[110] M. D. Schwartz, K. Yan, and H. X. Zhu, *Phys. Rev. D.* 96(5): 056005 [arXiv:1703.08572 [hep-ph]].

[111] M. D. Schwartz, K. Yan, and H. X. Zhu, *Phys. Rev. D.* 97(9): 096017 [arXiv:1801.01138 [hep-ph]].

[112] R. Boughezal, C. Focke, X. Liu, and F. Petriello, *Phys. Rev. Lett.* 115(6): 062002 [arXiv:1504.02131 [hep-ph]].

[113] J. Gaunt, M. Stahlhofen, F. J. Tackmann, and J. R. Walsh, *JHEP* 1509: 058 [arXiv:1505.04794 [hep-ph]].

[114] S. Catani and M. Grazzini, *Phys. Rev. Lett.* 98: 222002 [hep-ph/0703012].

[115] A. J. Larkoski, I. Moult, and B. Nachman, arXiv:1709.04464 [hep-ph].

[116] A. J. Larkoski, I. Moult, and D. Neill, *JHEP* 1605: 117 [arXiv:1507.03018 [hep-ph]].

[117] A. J. Larkoski, I. Moult, and D. Neill, *JHEP* 1802: 144 [arXiv:1710.00014 [hep-ph]].

[118] C. Frye, A. J. Larkoski, M. D. Schwartz, and K. Yan, *JHEP* 1607: 064 [arXiv:1603.09338 [hep-ph]].

[119] A. H. Hoang, S. Mantry, A. Pathak, and I. W. Stewart, arXiv:1708.02586 [hep-ph].

6

Effective Field Theories for Nuclear and (Some) Atomic Physics

U. van KOLCK

Institut de Physique Nucléaire, CNRS/IN2P3,
Université Paris-Sud, Université Paris-Saclay,
91406 Orsay, France
and
Department of Physics, University of Arizona,
Tucson, AZ 85721, USA

U. van KOLCK

van Kolck, U., *Effective Field Theories for Nuclear and (Some) Atomic Physics* In: *Effective Field Theory in Particle Physics and Cosmology*.
Edited by: Sacha Davidson, Paolo Gambino, Mikko Laine, Matthias Neubert and Christophe Salomon, Oxford University Press (2020).
© Oxford University Press. DOI: 10.1093/oso/9780198855743.003.0006

Chapter Contents

Dedicated to the memory of Professor Cécile DeWitt-Morette

Professor DeWitt-Morette founded the Les Houches School and was a towering figure in mathematical physics. I had the privilege of her guidance while studying stochastic systems in my early graduate student days at the University of Texas. Like other great theorists, she did not look at physics from the perspective of a specific field, but strove to build a consistent view of nature. Besides an accomplished scientist, she was passionate about science, and she was kind and supportive. Near the end of my PhD she recommended me to come to Les Houches, but it took me a whole quarter of a century to get here. It is a sad twist of fate that, when writing this dedication, I found out I had to add 'the memory of' to it.

6.1   Introduction

For humans, who (unlike frogs) are born basically as smaller versions of ourselves when we have babies, one of the most natural transformations of the physical world is that of scale, where all distances change by a common numerical factor. Despite its familiarity, scale invariance is usually not directly useful in physics—we do not particularly resemble stars or molecules. As we embark on an exploration of a physical problem, the first task is always to identify its relevant scales.[1]

The existence of characteristic scales means that our resolution when probing a system is important. Tracking changes in resolution is the task of the renormalization group (RG). Once a resolution is chosen, we have at least two momentum scales to contend with, that of the physics of interest (call it M_{lo}) and that of the physics at much shorter distances (call it $M_{\mathrm{hi}} \gg M_{\mathrm{lo}}$). Effective field theory (EFT) is the formalism to exploit this separation of scales and expand observable quantities in powers of $M_{\mathrm{lo}}/M_{\mathrm{hi}}$ without making assumptions about the short-range dynamics, other than its symmetries [1]. Schematically, the T matrix for momentum $Q \sim M_{\mathrm{lo}}$ (from which the S matrix and other observables can be obtained) can be written as

$$T(Q) = \sum_{\nu=0}^{\infty} \left(\frac{M_{\mathrm{lo}}}{M_{\mathrm{hi}}}\right)^{\nu} F_{\nu}\left(\frac{Q}{M_{\mathrm{lo}}}; \left\{\gamma^{(\nu)}\right\}\right) \equiv \sum_{\nu=0}^{\infty} T^{(\nu)}(Q), \qquad (6.1)$$

where each F_{ν} is a calculable function parametrized by a finite set of 'low-energy constants' (LECs) or 'Wilson coefficients' $\{\gamma^{(\nu)}\}$. Clearly this is a paradigm to tackle any physical system, as the variety of topics in this school attests to. As far as we know, nature is but a pile of EFTs.

---

[1] And, for convenience, simple units. Since I am interested in problems that are rooted in quantum mechanics and relativity, I use throughout these lectures natural units where Planck's constant and the speed of light are $\hbar = c = 1$. Then distance and time have the same units as inverse momentum, energy and mass. (To convert to more conventional units, use $\hbar c \simeq 200$ MeV fm.)

Combined in this chapter, these lectures show how the paradigm fares when facing nuclear physics, a very traditional field, where it is hard to come up with something that has not been tried before, usually without success. The strong interactions encapsulated in QCD produce non-trivial structures at the most fundamental level we can study today. Understanding how hadrons and their bound states—nuclei—arise remains an open problem in the Standard Model (SM) of particle physics, which hampers our ability to make predictions about processes involving new physics and astrophysical reactions.

It has been around twenty-five years since Weinberg [2, 3] and Rho [4] proposed that EFT could reproduce much of what was known in low-energy nuclear physics, while at the same time explaining some of its mysteries. Weinberg had earlier articulated the EFT paradigm [5] and was interested in general ways to set up the electroweak symmetry-breaking sector of the SM [1, 6]. Nuclear forces from pion exchange naturally come to mind when pondering about how Goldstone bosons couple to matter. The hope was that one would be able to formulate a renormalizable theory of nuclear interactions, overcoming the obstacles faced since the 1950s. 'Chiral potentials' constructed according to Weinberg's suggestion have now become the favourite input to '*ab initio*' methods to calculate nuclear structure and reactions. In traditional nuclear physics, the *initio* is a non-relativistic potential among nucleons, yet the remarkable recent progress in Lattice QCD [7] means that soon the starting point will be QCD itself [8, 9].

Unfortunately these chiral potentials produce scattering amplitudes that do not respect RG invariance. In the process of discovering and fixing this problem, other nuclear EFTs have been formulated, which apply to various energy regimes. Weinberg's basic insights have survived but building nuclear EFT turned out to be much more interesting than he anticipated. Some of the issues and advantages of explicit pion exchange are mentioned at the end of these notes, but I cannot possibly cover all the twists and turns of this story, not even all nuclear EFTs. I focus instead on the simplest one, Pionless EFT, and refer to others only occasionally. Pionless EFT is so simple, in fact, that with small changes it can be applied to systems of cold, neutral atoms as well. The lectures compiled here serve also as a (somewhat idiosyncratic) introduction to some of the physics that reinvigorated the atomic field over roughly the same time frame. I make no attempt to provide an extensive coverage of the literature; only papers that are best suited to make specific points are cited.

Pionless EFT is sufficient for a sample taste of nuclear EFTs. In contrast to many of the other lectures in the school, here we shall deal with a situation where the leading order (LO) of the M_{lo}/M_{hi} expansion must be non-perturbative in order to generate poles of the S matrix: the bound states and resonances that we identify as nuclei (or molecules). The combination of non-perturbative LO and perturbative corrections— relative $\mathcal{O}(M_{lo}/M_{hi})$ at next-to-leading order (NLO), relative $\mathcal{O}(M_{lo}^2/M_{hi}^2)$ at N^{2}LO, etc.—is at the core of the beauty of the nuclear and atomic applications of EFT. Pionless EFT is the 'poster EFT' to describe such combination. There is much regularity in the properties of nuclear and atomic bound states and resonances, and Pionless EFT captures a class of these regularities that sometimes goes by the name of 'Efimov physics'. In a magical paper almost half-a-century ago [10], Efimov showed that for certain non-relativistic systems, if a two-body bound state lives on the verge of non-existence, then a geometric tower of three-body bound states exists, with the ground state potentially quite deep. We now know that this phenomenon is not limited to the three-body system, but reverberates through larger clusters and even 'infinite' matter. As we will see, scale invariance is the key to understand this sort of structure.

6.2   Some Nuclear and Atomic Scales

A cursory look at the Particle Data Book [11] shows a bewildering variety of hadrons, which nevertheless fall into isospin multiplets containing various charge states with approximately the same mass. When made out of light quarks, they have masses in the 1, 2 GeV range, such as the proton and the neutron that can be paired in an isospin doublet of mass $m_N \simeq 940$ MeV. We can infer that QCD has a characteristic scale $M_{\text{QCD}} \sim 1$ GeV associated with its non-perturbative dynamics.

The one clear exception is the isospin triplet of light pions, with mass $m_\pi \simeq 140$ MeV $\ll M_{\text{QCD}}$. This low mass has long been understood as the result of the spontaneous breaking of an approximate $\text{SU}(2)_L \times \text{SU}(2)_R \sim \text{SO}(4)$ chiral symmetry of independent rotations in the space of two flavours for left- and right-handed quarks. Because the diagonal subgroup $\text{SU}(2)_{L+R} \sim \text{SO}(3)$ of isospin rotations remains unbroken, the three Goldstone bosons in the coset space $\text{SO}(4)/\text{SO}(3) \sim S^3$ can be identified with the three pions. Their interactions are governed by a dimensionful parameter, the pion decay constant $f_\pi \simeq 92$ MeV $\sim M_{\text{QCD}}/(4\pi)$, which is the radius of this 'chiral sphere'. Pions are not exactly massless because chiral symmetry is explicitly broken by the quark masses. From perturbation theory one expects $m_\pi^2 \sim M_{\text{QCD}}\bar{m}$, which is in the right ballpark if the average quark mass $\bar{m}$ is a few MeV. Isospin is broken by the down-up quark mass difference $\varepsilon \bar{m}$, where $\varepsilon \sim 1/3$ [12], which gives rise to small splittings among isospin multiplets. Of course, isospin is also broken explicitly by electromagnetic interactions. All this has been understood for quite some time and is discussed in some detail in Pich's lectures [6].

What might be somewhat surprising is that once quarks hug tight into nucleons and pions with intrinsic sizes $\sim M_{\text{QCD}}^{-1} \simeq 0.2$ fm, nucleons form nuclei of much larger size and feebler binding. One proton and one neutron bind into an isospin singlet—the simplest nucleus, the deuteron (^{2}H)—with total spin $S = 1$ and a binding energy $B_d \simeq 2.2$ MeV. When $S = 0$, proton and neutron are part, with two neutrons and two protons, of an isospin-triplet virtual state at $B_{d^*} \simeq 0.08$ MeV. (A bound state is a pole of the S matrix with positive imaginary momentum, while a virtual state is a pole with negative imaginary momentum—both have negative energy $-B$. A resonance consists of a pair of poles in the lower half of the complex-momentum plane with opposite, non-zero real parts.) As shown in the leftmost column of Table 6.1, the binding energy per particle B_A/A increases for the trinucleon isodoublet with $S = 1/2$—triton (^{3}H) and a slightly less-bound helion (^{3}He)—and even more for the isospin-singlet alpha particle (^{4}He) with $S = 0$. The exclusion principle makes it harder for more nucleons to be together

Table 6.1 Ground-state (first-excited) binding energies per particle B_A/A (B_{A^*}/A) of selected light nuclei and ^{4}He atomic clusters, in units of the three-body binding energy per particle, where $B_3 = 8.48$ MeV and $B_3 = 0.1265$ K, respectively. (To convert between K and eV, use $k = 8.6 \cdot 10^{-5}$ eV K^{-1}.) In the nuclear case, an entry corresponds to the deepest isobar state with the respective nucleon number A, a parenthesis indicating a virtual state. Entries in the left column are experimental. Entries in the middle and right columns are from Pionless EFT at LO: from [15, 16, 9, 17] away from unitarity and from [18] at unitarity. For ^{4}He atomic clusters, the left column shows the results [19, 20, 21] of calculations with phenomenological ^{4}He-^{4}He potentials. Entries in the middle and right columns are from Pionless EFT at LO: from [22, 23, 24] away from unitarity and from [22, 25, 26] at unitarity. For simplicity I do not indicate quantities used in the construction of the interactions nor show error estimates, which can be found in the original references. For some entries similar numbers exist from other references.

$3B_A/(AB_3)$	nucleons			^{4}He atoms		
A	experiment	LO EFT	unitarity	potential	LO EFT	unitarity
2	0.39	0.39	0	0.0156	0.0152	0
3	1	1	1	1	1	1
3*	(0.17)	(0.19)	0.0019	0.0180	0.0175	0.0019
4	2.50	2.6	3.5	3.3	3.2	3.46
4*	0.71	0.90	0.75	0.755	0.759	0.752
5	1.94	?	?	6.2	5.7	6.3
6	1.89	1.4	?	9.2	8.2	8.9
$\vdots$						
16	2.82	2.5	?	?	?	27.4
$\vdots$						
$\to \infty$	5.7	?	?	180	?	90

and B_A/A first decreases, but then increases again (on average) till it reaches a maximum of about 8 MeV for $A = 56$, decreasing slowly beyond that. Table 6.1 also shows the value of the binding energy per particle for nuclear matter, $\lim_{A\to\infty}(B_A/A) \equiv b_\infty$. Nuclear matter is an idealized system without surface or electroweak interactions, defined by extrapolation from heavy nuclei via the 'liquid-drop' relation $B_A/A - b_\infty \propto A^{-1/3}$. Consonant with such 'saturation', the typical nucleus size is $\sim r_{\text{nuc}}A^{1/3}$, with $r_{\text{nuc}} \simeq 1.2$ fm—but it can be much larger for more loosely bound nuclei, such as light nuclei and 'halo nuclei', which have a cloud of loose nucleons orbiting a more tightly bound core. (The typical example is ^{6}He, which is thought to be essentially two neutrons around a ^{4}He core.) Halo EFT [13, 14] provides a description of this type of state similar to Pionless EFT, but with the core as an additional degree of freedom.

This disparity between QCD and nuclear scales tells us several things. First, nucleons are relatively far apart inside nuclei and retain their identity. Second, they move slowly, that is, are approximately non-relativistic. Third, their interactions must come from the exchange of the lightest colour singlets—the pions, giving rise to a force of range $\sim m_\pi^{-1} \simeq$ 1.4 fm—plus complicated mechanisms of shorter range $\sim M_{\mathrm{QCD}}^{-1}$. Nuclei should thus be described by an EFT with non-relativistic nucleons and pions (and possibly the lightest nucleon excitations) subject to approximate chiral symmetry—this is Chiral EFT, whose restriction to $A = 0, 1$ is Chiral Perturbation Theory (ChPT) [5, 6]. Moreover, in the long distances relevant for light nuclei, even pion exchange can be considered a short-range interaction. This is the regime of Pionless EFT, where the only explicit degrees of freedom are non-relativistic nucleons.

The situation is not totally dissimilar for some (neutral) *atoms*, like the boson ^{4}He. Instead of QCD binding quarks into hadrons, QED binds a nucleus and Z electrons into atoms with energies $E_{\mathrm{at}} \sim -(Z\alpha)^2 m_e$, where $\alpha \simeq 1/137$ is the fine-structure constant and m_e the electron mass. The atoms themselves form much more loosely bound molecules, through the exchange of (at least) two photons and shorter-range interactions. The former gives rise to the Van der Waals potential $\sim -l_{\mathrm{vdW}}^4/(2\mu r^6)$, where μ is the reduced mass and l_{vdW} is the 'Van der Waals length', which depends on what kind of atom we are considering. For a nice compilation of values, as well as a discussion of long-range interactions, see [27]. For certain types of atoms, clusters have sizes that are significantly larger than l_{vdW}; in these cases, the short-range interactions dominate and we can, up to a point, treat the system as one with short-range forces only. This Contact EFT is analogous to Pionless EFT, just with a field for the atom substituted for the nucleon's.

^{4}He is a particularly interesting atom because a macroscopic sample remains liquid at zero temperature and exhibits the remarkable property of superfluidity. The ^{4}He dimer has been measured to have an average separation $\langle r \rangle = 52(4)$ Å [28], which is an order of magnitude larger than the corresponding $l_{\mathrm{vdW}} \simeq 5.4$ Å [29, 30]. Experimental numbers exist for the dimer [31, 32] and excited-trimer [33] binding energies, which confirm that they are loosely bound. Sophisticated ^{4}He-^{4}He potentials have been developed over the years, which are consistent with a variety of experimental data and allow for the prediction of the energies of larger clusters. The results for one of these potentials— dubbed the 'LM2M2 potential' [34]—are shown in the left ^{4}He column of Table 6.1. Other potentials give similar results. Apart from a huge difference in overall scale, the numbers for $A \leq 4$ atoms have some qualitative similarity to those for nucleons. The most obvious difference is the lack of exclusion principle, which translates into a monotonic increase in binding energy, which starts approximately as $(N - 2)^2 B_3$ [24]. Just as for nucleons, though, the interaction saturates to a constant binding energy per particle, the value of which was calculated with another potential—the 'HFDHE2 potential' [35]—in [19, 36]. Similarly to nuclei, cluster size first decreases, then starts to increase till it settles into an $\sim r_{\mathrm{at}} A^{1/3}$ behaviour, with $r_{\mathrm{at}} \simeq 2.2$ Å [36].

These two families of systems—nuclei and ^{4}He atomic clusters—are peculiar for their large sizes compared to the interaction range R. A rough estimate of the characteristic momentum Q_A of the particles in the bound state is obtained by assuming that every one of the A particles of mass m contributes the same energy $Q_A^2/(2m)$ to B_A:

$$Q_A \sim \sqrt{2mB_A/A}. \tag{6.2}$$

(This estimate reflects the correct location of the bound state in relative momentum for $A = 2$ and gives a finite Q_A in the limit $A \to \infty$.) For nucleons, we find $Q_3 \sim 70$ MeV, while for ^{4}He atoms $Q_3 \sim (12 \text{ Å})^{-1}$, each one about half of the corresponding R^{-1}. That is, particles in the three-body bound state are separated by a distance about twice as large as the range of the interaction. How can this be? Classically, the size of an orbit is given by the range of the force. Thus, these are intrinsically quantum-mechanical systems, which hold a few surprises in store for us.

6.3   EFT of Short-Range Forces

In the class of systems sometimes referred to as 'quantum halos', we are interested in the S matrix for processes with a typical external momentum $Q \ll M_{\text{hi}}$, where the EFT breakdown scale M_{hi} is related to the inverse of the force range, R^{-1}—m_π or l_{vdW}^{-1}, as the case may be. The few-body structures we want to describe are characterized by momenta $Q \sim Q_3$, so I will use Q_3 as a proxy for M_{lo}. The idea is to construct an expansion of the form (6.1) from the most general Lagrangian (density) allowed by the symmetries supported by the relevant degrees of freedom, exploiting the 'folk theorem' that the resulting quantum field theory will then generate the most general S matrix allowed by the same symmetries [5].

6.3.1   Degrees of Freedom

In systems whose sizes are larger than the range R of the force, the constituent particles are not able to resolve details of the potential. They feel the interaction as a contact: the potential can be represented by a Dirac delta function and its derivatives. The only degrees of freedom we need to consider are the constituent particles themselves—there is no need to account explicitly for other particles whose exchanges are responsible for the specific form of the potential.

Since in the cases of interest here the particle mass $m \gtrsim M_{\text{hi}}$, a non-relativistic expansion must hold, which yields a much simpler field theory than in the relativistic case. Because it takes $\gtrsim 2m$ in energy to create a particle–anti-particle pair, there is no need to include anti-particles—unless we are interested in processes that include anti-particles to start with, or in processes that involve particle- (baryon- and/or lepton-) number violation. We can exploit this simplification by a convenient choice of field. (I remind you that fields are not directly observable and in an EFT all choices that represent the same degrees of freedom are equivalent [37, 38].) It is sufficient to employ a field $\psi_a(x)$ that only annihilates particles,

$$\psi_a(x) \equiv \int \frac{d^3p}{(2\pi)^3} \frac{e^{-ip\cdot x}}{2p^0} u_s(\vec{p}) a_{\vec{p}} \tag{6.3}$$

where $a_{\vec{p}}$ is the annihilation operator for a particle of momentum $\vec{p} \equiv \vec{k}$, $u_s(\vec{p})$ carries information about the spin, and $p^0 = \sqrt{\vec{p}^2 + m^2} = m + \vec{p}^2/(2m) - \vec{p}^4/(8m^3) + \ldots \equiv m + k^0$ (I am sorry, I am using the 'wrong', West Coast metric here: $p \cdot x = p^0 t - \vec{p} \cdot \vec{r}$.) Propagation is only forward in time, and it is represented in Feynman diagrams by a line going up (when time is represented as increasing upwards, as I do here). The absence of pair creation means that, if particle number is conserved, these lines go through the diagram. The theory breaks into separate sectors, each with a given number of particles A. Life is made tremendously easier by the fact that we can deal in turn with sectors of increasing A without simultaneously having to worry about the feedback from larger number of particles.

To produce an $M_{\text{lo}}/M_{\text{hi}}$ expansion, we expect some form of derivative expansion in the action. Yet derivatives of $\psi_a(x)$ contain the large m coming from the trivial factor $\exp(-imt)$ in the evolution of the field. In the particle's rest frame, where the particle's four-velocity v^μ, $v^2 = 1$, is $(1, \vec{0})$ it is convenient to remove this factor by defining a new 'heavy field' [39]

$$\psi_h(x) \equiv e^{imv \cdot x} \psi_a(x). \tag{6.4}$$

The evolution of the new field is governed by the kinetic energy $k^0 \ll m$, since particles will exchange small three-momenta. Instead of the $-ip_\mu$ that contains the large m, $\partial_\mu \psi_h(x)$ gives $-ik_\mu$ in momentum space.

For cases where anti-particles are present, a conjugate field can be introduced similarly. Of course the possibility of pair annihilation when both particles and anti-particles are present allows for large momenta, and the applicability of a Q/m expansion is limited to times prior to annihilation.

6.3.2 Symmetries

So far, no sign has been seen of violation of Lorentz invariance or the product CPT of charge conjugation (C), parity (P), and time reversal (T). For simplicity I will also neglect the small effects that arise from the violation of P, T, and baryon and lepton numbers, although they are part of the EFT. All these symmetries restrict the form of the terms allowed in the action. For example, the transformation associated with particle number is an arbitrary phase change

$$\psi_h(x) \rightarrow e^{i\alpha} \psi_h(x), \tag{6.5}$$

and particle-number conservation implies that ψ_h enters the Lagrangian in combination with $\psi_h^\dagger$.

For a heavy field, the symmetry whose implementation is least obvious is Lorentz invariance. To start with, note that the definition of ψ_h can be made in other frames where v^μ is not necessarily $(1,\vec{0})$. We introduce the residual momentum k^μ through $p^\mu \equiv mv^\mu + k^\mu$. The residual momentum is only constrained by $k^2 = -2mv \cdot k \ll m^2$. There is freedom to consider a different residual momentum $k^\mu - q^\mu$ with $q^2 = -2mv \cdot q \ll m^2$, and a simultaneous relabeling of the velocity, $v^\mu \rightarrow v^\mu + q^\mu/m$. (You can check that $(v + q/m)^2 = 1$). Thus the theory must be invariant under this 'reparametrization invariance' (RPI) [40], when the field transforms as

$$\psi_h(x) \rightarrow e^{iq \cdot x} D(v + q/m, v) \, \psi_h(x), \tag{6.6}$$

where D is determined by the representation of the Lorentz group encoded in u_s. Because of the phase, $\partial_\mu \psi_h(x)$ is not covariant under reparametrization. As standard in such cases, we can introduce a 'reparametrization covariant derivative'

$$\mathcal{D}_\mu \psi_h(x) \equiv \left[\partial_\mu - imv_\mu \right] \psi_h(x) \rightarrow e^{iq \cdot x} D(v + q/m, v) \, \mathcal{D}_\mu \psi_h(x). \tag{6.7}$$

We account for Lorentz invariance by properly contracting Lorentz indices in the Lagrangian as usual, but using the reparametrization covariant derivative.

As an example, take the kinetic terms of a scalar field, whose RPI form is

$$\mathcal{L}_{\text{kin}} = (\mathcal{D}^\mu \psi_h)^\dagger \mathcal{D}_\mu \psi_h - m^2 \psi_h^\dagger \psi_h = 2m \psi_h^\dagger \left[iv \cdot \partial - \partial^2/(2m) \right] \psi_h. \tag{6.8}$$

The large mass term has disappeared, as desired, except for an overall normalization. The remaining terms resemble the inverse of the usual non-relativistic propagator, but contain also $\partial_0^2 \psi$. To bring the propagator to the usual form, we can define yet another field, which is canonically normalized:

$$\psi(x) \equiv \frac{1}{\sqrt{2m}} \left[1 - \frac{i}{4m} v \cdot \partial + \dots \right] \psi_h(x). \tag{6.9}$$

Substitution into eqn (6.8) leads to

$$\mathcal{L}_{\text{kin}} = \psi^\dagger \left\{ iv \cdot \partial + \frac{1}{2m} \left[(v \cdot \partial)^2 - \partial^2 \right] + \ldots \right\} \psi. \tag{6.10}$$

Now in the rest frame the term in square brackets reduces to the usual $\vec{\nabla}^2$.

Exercise 6.1 Consider the higher orders in $1/m$ in the '...' of eqn (6.9). Show that an appropriate field choice makes the next term in the '...' of eqn (6.10) equal to $\vec{\nabla}^4/(8m^3)$ in the particle's rest frame, in line with $k^0 = \vec{k}^2/(2m) - \vec{k}^4/(8m^3) + \ldots$

The same procedure can be followed for interaction terms. As the number of derivatives increase, we have to contend with two or more time derivatives. In this case, there is a non-trivial relation between the time derivative of the field and the field's conjugate momentum. As a result the interaction Hamiltonian is not simply minus the interaction Lagrangian, but contains additional, Lorentz non-covariant terms. It is no problem to include these interactions, as other non-covariant pieces arise in covariant perturbation theory from contractions involving derivatives, and in a time-ordered formalism from its inherent non-covariance. We can show explicitly [41] that the sum of diagrams contributing to any given process is indeed covariant. Alternatively, we can integrate over the field's momentum in the path integral arriving at an effective Lagrangian, which is covariant but has additional terms compared to the classical Lagrangian we started from [42]. For heavy fields these complications can be avoided altogether by including in the '...' of eqn (6.9) terms with additional fields, designed to remove $\partial_0 \psi$ from interactions as well.

Accounting for a non-zero spin manifest in the $D(v + q/m, v)$ in eqn (6.6) is a bit more complicated. To represent only particles, the fields obey constraints. For example, of the four components of a Dirac spinor only two are associated with the two spin states of a particle with $S = 1/2$. In the rest frame we can project onto the two upper (in the Dirac representation) components with the projector $(1 + \gamma^0)/2$ and thereby use a Pauli spinor, the gamma matrices reducing to the spin operator $(0, \vec{\sigma}/2)$. In a generic frame the two 'large' components can be selected by the constraint $(1 - \slashed{v})\psi = 0$. Spin properties are encoded in $S^\mu = i\gamma_5 \sigma^{\mu\nu} v_\nu/2$, which satisfies $S \cdot v = 0$ and $S^2 = -3/4$, as a little gamma-matrix algebra shows. The 'reparametrization covariant spin' that appears in interactions is then $\Sigma^\mu = -\gamma_5 \sigma^{\mu\nu} \mathcal{D}_\nu/(2m)$, which satisfies $\psi^\dagger \Sigma \cdot \mathcal{D} \psi = 0$ and $\psi^\dagger \Sigma^2 \psi = 3\psi^\dagger \mathcal{D}^2 \psi/(4m^2)$.

At the end of the day, we simply end up with a Lagrangian that complies with Lorentz invariance in a Q/m expansion. Of course other, perhaps simpler, ways exist to implement the constraints of Lorentz invariance, for example an explicit reduction of a fully relativistic theory. Heavy fields with RPI are, however, more EFT-like, because they do not rely on detailed knowledge of the theory at $Q \sim m$. This is particularly relevant in nuclear physics, where, if fully relativistic nucleons are to be considered, we need also to include mesons heavier than the pion, for which no systematic Q/M_{hi} expansion is known. RPI is thus well adapted to the expansion in Q/M_{hi} we seek. More details about the heavy field formalism can be found in Chapter 9 in this volume [43]. In the following, for simplicity, I will drop the interactions that are explicitly spin-dependent and work in the rest frame.

6.3.3   The Action

The most general action for non-relativistic particles of mass m interacting under these symmetries through short-range forces can therefore be written as

$$
\mathcal{S} = \int d^4x\, \mathcal{L} = \int \frac{dt}{2m} \int d^3r \left\{ \psi^\dagger \left(2im\frac{\partial}{\partial t} + \vec{\nabla}^2 + \dots \right) \psi \right.
$$
$$
- 4\pi \left[C_0 \left(\psi^\dagger \psi \right)^2 + C_2 \left(\psi^\dagger \psi \right) \left(\psi^\dagger \vec{\nabla}^2 \psi + \mathrm{H.c.} \right) + \dots \right]
$$
$$
\left. - \frac{(4\pi)^2}{3} D_0 \left(\psi^\dagger \psi \right)^3 + \dots \right\},
\tag{6.11}
$$

where $C_{0,2}$, D_0, etc. are the LECs and '...' include terms with more derivatives and/or fields. Beware that I chose a normalization of the LECs that is not normally used in the literature, but is very convenient for the subsequent discussion of orders of magnitude and scale invariance. With this choice, the mass appears in the combination t/m, and in observables together with the energy E as $mE \equiv k^2 + \dots$. As a consequence, binding energies are all $\propto 1/m$. This is at the root of a choice of units frequently made in atomic physics: $m = 1$. I prefer to keep m explicit instead.

In writing eqn (6.11) I neglected spin projections in the interaction terms. For fermions, we can use Fierz reordering, which encodes the exclusion principle, to reduce the number of independent terms. For a two-state fermion (e.g. a neutron with spin up or down, in the absence of protons), the C_0 interaction operates only when the two particles are in different states. Similarly, the D_0 interaction, which requires three particles at the same point, vanishes. For more-state fermions, more than one C_0- and/or D_0-type interaction is possible, depending on the symmetries. For example, because a nucleon consists of four states—isospin up (proton p) and down (neutron n), each with two spin states—four C_0-type interactions exist, corresponding to the four S-wave channels: the isospin-singlet pn channel with spin $S = 1$—the deuteron channel—and the pp, pn, and nn interactions in the isospin-triplet channels with $S = 0$. In situations where isospin is a good approximate symmetry, these interactions are reduced in a first approximation to two, the one operating in the $^3S_1$ channel, and a single one in $^1S_0$. (We commonly use the spectroscopic notation $^{2S+1}L_{\mathcal{J}}$ where S is the spin, L is the orbital angular momentum, and $\mathcal{J}$ the total angular momentum.) Similarly, there is a single D_0-type interaction, operating in the nucleon-deuteron (Nd) $^2S_{1/2}$ wave—the triton/helion channel. (In the $^4S_{3/2}$ channel there are at least two protons or two neutrons in the same spin state.)

Exercise 6.2 Show that Fermi statistics and isospin symmetry imply that two nucleons interact in two S waves: *i)* $^3S_1$, which is symmetric in spin and anti-symmetric in isospin, and thus involves an np pair; and *ii)* $^1S_0$, which is anti-symmetric in spin and symmetric in isospin, and is split by isospin-violating interactions into pp, pn, and nn. Analyse the two S waves for the Nd system, $^4S_{3/2}$ and $^2S_{1/2}$, similarly.

I have not considered electromagnetic interactions explicitly in eqn (6.11). The associated $U(1)_e$ gauge symmetry can be introduced in the usual way, that is, by requiring that all interactions be built out of the electromagnetic covariant derivative and the field strength. In the neutral-atom case, the covariant derivative is just the usual derivative, but electromagnetic interactions still proceed through the coupling of the atom to the field strength. The longest-range effects arise from two-photon exchange, the most important being the Van der Waals force. By remaining at momenta below l_{vdW}^{-1}, dimensional analysis shows that the non-analytic contributions from two-photon exchange enter only at $\mathcal{O}(Q^3)$, so at lower orders we can pretend to have only short-range interactions. In the nuclear case, where the proton is charged, one-photon exchange leads to long-range interactions, the most important being the well-known Coulomb force. There is further isospin violation from shorter distances stemming from 'hard' photons and from the quark mass difference. The associated isospin violation splits various charge states. For nuclear ground states one can argue [44, 45] that isospin-breaking effects are subleading and bring no fundamental changes to the discussion below.

To keep the notation lean, I ignore spin-isospin complications in the formulas that follow, and pretend that only one short-range interaction of each type is important. I simply remark on the changes that take place when $S > 0$.

6.3.4   Renormalization Group Invariance and Power Counting

As in any EFT, the interactions in the action (eqn 6.11) are local and require regularization, that is, the introduction of a method to suppress explicit high-momentum, or equivalently short-distance, modes. There is an infinite number of ways of doing this. What is common to all methods is the presence of a parameter with dimensions of mass, which I will denote Λ. In perturbation theory, frequently the cleanest method is dimensional regularization because it keeps no terms that go as negative powers of Λ. (In most subtractions it keeps also no positive powers.) However, dimensional regularization is difficult (impossible?) to apply in generic non-perturbative contexts, where we are restricted (at least for now) to momentum or distance regulators. The simplest examples of these regulators are, respectively, a sharp momentum cut-off Λ and a minimum length, or lattice spacing, Λ^{-1}. More generally, one can use a function of the momentum $\vec{p}$ which vanishes smoothly as $|\vec{p}|$ increases beyond Λ, say a Gaussian function of $|\vec{p}|/\Lambda$ with unit coefficient, or a smooth representation of a delta function in coordinate space $\vec{r}$, say a Gaussian function of $|\vec{r}|\Lambda/2$ with a coefficient that scales as Λ^3. From here on I talk mostly about a momentum cut-off, with only occasional remarks about dimensional regularization.

The effects of the high-momentum modes must, of course, reappear in the LECs, which are thus dependent on Λ. Renormalization is the process of ensuring that this dependence is such that observables are independent of the arbitrary regulator, or

$$\frac{dT(Q)}{d\Lambda} = 0. \tag{6.12}$$

That this must be possible is a consequence of the uncertainty principle coupled to our accounting of *all* interactions allowed by symmetries: modes of momentum $\gtrsim \Lambda$ can be absorbed in interactions of range $\lesssim \Lambda^{-1}$ with the same symmetries. If $\Lambda \gtrsim M_{hi}$, this adds no extra limitation to the EFT. If $\Lambda \lesssim M_{hi}$, we cannot apply the EFT all the way to its physical breakdown scale.

A central role is played in EFT by 'power counting', that is, a rule that organizes the infinite sequence of interactions according to their relevance to observables by relating the counting index ν in eqn (6.1) to the properties of the various terms in the action (number of derivatives, fields, etc.). To preserve the insensitivity to the regulating procedure, we want each truncation of eqn (6.1),

$$T^{[\bar{\nu}]}(Q) \equiv \sum_{\nu=0}^{\bar{\nu}} T^{(\nu)}(Q) \tag{6.13}$$

to satisfy

$$\frac{\Lambda}{T^{[\bar{\nu}]}(Q)} \frac{dT^{[\bar{\nu}]}(Q)}{d\Lambda} = \mathcal{O}\left(\frac{Q}{\Lambda}\right). \tag{6.14}$$

In this way, the regulator does not increase the truncation error, as long as $\Lambda \gtrsim M_{hi}$. This places a constraint on the power counting: that it contains enough interactions at each order to remove non-negative powers of Λ from observables. If we do not include all interactions needed to ensure eqn (6.14), observables are sensitive to the arbitrary regulator—not only its dimensionful parameter but also its form—and there is no guarantee that results reflect the low-energy limit of the underlying theory. The canonical dimension of a LEC suggests a lower bound on its magnitude, when we use M_{hi} to make it dimensionless and $\mathcal{O}(1)$. However, as we are going to see, our problem involves significant departures from this simple dimensional analysis, in which case RG invariance in the form of eqn (6.14) offers particularly useful guidance regarding the LEC sizes. Regularization schemes—such as dimensional regularization with minimal subtraction—that kill positive powers of Λ are not the most useful in this context, because they might lead you to overlook the need for a LEC at the order under consideration. (A physically relevant example is given in [13].)

This approach is a generalization of the old concept of 'renormalizable theory', where renormalization is achieved by a finite number of parameters and at the end $\Lambda \to \infty$ is taken. As Weinberg is fond of saying (see, e.g. [46]), 'non-renormalizable theories' are just as renormalizable as 'renormalizable theories'. Now we only need a finite number of parameters *at each order* and $\Lambda \gtrsim M_{hi}$. Each of the LECs obeys an RG equation that tells us how it depends on Λ. A generic observable will look like

$$O^{[\bar{\nu}]}(\Lambda) = O^{[\bar{\nu}]}(\infty) \left[1 + \alpha_O^{[\bar{\nu}]} \frac{M_{lo}}{\Lambda} + \beta_O^{[\bar{\nu}]} \frac{M_{lo}^2}{\Lambda^2} + \dots \right], \tag{6.15}$$

with $\alpha_O^{[\bar{v}]}, \beta_O^{[\bar{v}]}$, etc., numbers of $\mathcal{O}(1)$. Values for a finite number of observables are used to determine the LECs—for these observables $\alpha_O^{[\bar{v}]} = \beta_O^{[\bar{v}]} = \ldots = 0$. For other observables, a non-zero $\alpha_O^{(\bar{v})}$ indicates the existence of at least one interaction at next order, since it generates a term of size $\mathcal{O}(M_{\mathrm{lo}}/M_{\mathrm{hi}})$ when $\Lambda \sim M_{\mathrm{hi}}$. There is no need to take $\Lambda \to \infty$ for the theory contains in any case errors of $\mathcal{O}(M_{\mathrm{lo}}/M_{\mathrm{hi}})$ from the truncation in the action. But regulator cut-off variation from $\sim M_{\mathrm{hi}}$ to much larger values does usually provide an estimate of the truncation error. Regularization schemes that kill *negative* powers of Λ, such as dimensional regularization, deprive you of this tool.

This newer view of renormalization arose from the combination of the traditional view with the more intuitive 'Wilsonian RG'. The latter is usually applied to a (e.g. condensed-matter) system where the underlying theory is known, and the effective theory is constructed by explicitly reducing Λ starting from M_{hi}. In this process all interactions allowed by symmetries are generated, and of course depend on Λ by construction. For example, even if there is an underlying potential that is mostly two-body, its expansion at large distances will contain higher-body components, stemming from successive two-body encounters at unresolved distances and times. (In most situations these higher-body forces are relatively small, but not always, as we will see below.) In the Wilsonian RG, Λ marks the highest on-shell momentum to which we can apply the EFT. As Λ is decreased it eventually crosses a physical scale M'_{hi} where the EFT needs to be reorganized, for example, due to the emergence of new degrees of freedom (say, a Goldstone boson or another type of low-energy collective effect). If we extend this new EFT to Λ *above* M'_{hi}, we are in the situation described earlier: real momenta up to M'_{hi} can be considered as Λ cuts off virtual momenta only. Thus, while intuitive, there is no need, in fact no advantage, in keeping the regulator parameter Λ below the physical breakdown scale we are interested in. This is particularly true when the underlying theory is unknown (as for the SM) or known but hard to solve explicitly (as for QCD at low energies).

Beware that applications of this more general EFT implementation of the RG have been muddled by the multiple uses of the term 'cut-off'. When the Wilsonian RG is favoured, typically 'cut-off' is used for the regulator, which also limits the range applicability of the EFT. In contrast, in modern particle physics where dimensional regularization is almost exclusively employed, 'cut-off' is often used for the physical breakdown scale. I will try to consistently distinguish the regulator cut-off Λ, which is not physical, and the physical breakdown scale M_{hi}.

6.4 Two-Body System

Consider how this EFT works in the simplest case: the two-body system. Were we considering photons explicitly, we could look at the electromagnetic properties of one particle, such as form factors (accessible through electron scattering) or polarizabilities (through Compton scattering). But the corresponding amplitudes are purely perturbative (except possibly around special kinematic points), and my goal here is to illustrate the more challenging issue of building a power counting when a subset of interactions has to be treated non-perturbatively. That is the case when the T matrix has low-energy poles. In particular, I want to tackle the situation relevant to the systems discussed in Section 6.2: a shallow two-body bound or virtual state.

6.4.1 Amplitude

Because the EFT splits into sectors of different A, we can focus on $A = 2$ without worrying, as we do in relativistic theories, about interactions with more than four fields in the action (eqn 6.11). Of the one-body terms, all we need is the propagator for a particle of four-momentum l,

$$iD_1(l^0, \vec{l}) = iD_1^{(0)}(l^0, \vec{l}) + iD_1^{(0)}(l^0, \vec{l}) \frac{i\vec{l}^4}{8m^3} iD_1^{(0)}(l^0, \vec{l}) + \ldots, \qquad (6.16)$$

where

$$iD_1^{(0)}(l^0, \vec{l}) = \frac{i}{l^0 - \vec{l}^2/(2m) + i\epsilon}, \qquad (6.17)$$

with $\epsilon > 0$, is the LO propagator. Because they are suppressed by $\mathcal{O}(Q^2/m^2)$, relativistic corrections come at N^2LO or higher depending on how we decide to count m relative to M_{hi}. There is not much consensus about the most efficient scheme to do this, but the issue does not appear up to NLO, which is sufficient for these lectures. Likewise, for an on-shell particle of momentum $\vec{p}$, to this order we need only the first term in the expansion of the energy, $E = k^2/(2m) - k^4/(8m^3) + \ldots$, where $k = |\vec{p}|$. Note that I did keep the recoil term $\propto 1/m$ in the denominator of eqn (6.17). Were we considering light probes (such as photons) with momenta $\mathcal{O}(Q)$, they would deposit energies $\mathcal{O}(Q)$ onto the virtual particles. In this case, we could also expand the propagator (6.17) in powers of Q/m, the leading term being the static propagator $i/(l^0 + i\epsilon)$. Instead, here particles start off with $E = \mathcal{O}(Q^2/m)$ and remain nearly on-shell: $l^0 = \mathcal{O}(Q^2/m)$ and thus comparable to $\vec{l}^2/(2m)$. The propagator at LO is *not* static. This is the first indication that the power counting for processes involving two or more heavy particles is different from that for the simpler one-body processes.

The simplest of the four-field (or in non-relativistic parlance, two-body) interactions in the action (eqn 6.11) is the C_0 term, which is represented by a vertex with four legs. It gives a momentum-independent tree-level contribution to the two-body T matrix,

$$T_2^{(0,0)}(\Lambda) = -\frac{4\pi}{m} C_0^{(0)}(\Lambda) \equiv -V_2^{(0)}(\Lambda), \qquad (6.18)$$

where $C_0^{(0)}(\Lambda)$ denotes the LO part of C_0. Things are more interesting at one-loop level, where two C_0 vertices are connected by two non-relativistic propagators (eqn 6.17). In the centre-of-mass system, where one incoming particle has four-momentum $(p^0, \vec{p})$ and the other $(p^0, -\vec{p})$,

$$T_2^{(0,1)}(k;\Lambda) = i\left(\frac{4\pi i}{m}C_0^{(0)}(\Lambda)\right)^2 \int \frac{d^3l}{(2\pi)^3}\int\frac{dl^0}{2\pi}\, D_1^{(0)}(l^0+p^0,\vec{l}+\vec{p})\, D_1^{(0)}(-l^0+p^0,-\vec{l}-\vec{p})$$

$$= \left(\frac{4\pi}{m}C_0^{(0)}(\Lambda)\right)^2 \int \frac{d^3l}{(2\pi)^3}\,\frac{m}{\vec{l}^2-k^2-i\epsilon}. \qquad (6.19)$$

Here, I first integrated over the 0th component of the loop momentum, as it is standard in non-relativistic theories. We can do this by contour integration: closing the contour in the upper plane, we pick a contribution from the residue of the pole at $p^0 - (\vec{l}+\vec{p})^2/(2m) + i\epsilon$; closing on the lower plane, a contribution from the other pole, which gives, of course, the same result. As first noticed in this context by Weinberg [2, 3], had we neglected recoil in the one-body propagator (eqn 6.17), we would have faced a pinched singularity at the origin. The reflection of this is the appearance of the large m in the numerator of eqn (6.19), where I additionally relabelled $\vec{l} \to \vec{l} - \vec{p}$. This form of the one-loop amplitude should come as no surprise: we simply have an integration over the virtual three-momentum of the standard Schrödinger propagator.

In the form of eqn (6.19), it is clear that the integral would diverge without regularization. The most intuitive regularization consists of a 'non-local' regulator that depends separately on the incoming and outgoing nucleon momenta: a function $F(l/\Lambda)$ with the properties that $F(0) = 1$ (to preserve the physics at low momentum) and $F(x \to \infty) = 0$ (to kill high momenta):

$$\int \frac{d^3l}{(2\pi)^3}\,\frac{m}{\vec{l}^2-k^2-i\epsilon} \equiv \frac{m}{2\pi^2}\int_0^\infty dl\,\frac{l^2}{l^2-k^2-i\epsilon}F(l/\Lambda) \equiv I_1(k;\Lambda). \qquad (6.20)$$

The simplest choice is a step function, $F(x) = \theta(1-x)$:

$$I_1(k;\Lambda) = \frac{m}{2\pi^2}\int_0^\Lambda dl\,\frac{l^2}{l^2-k^2-i\epsilon}$$

$$= \frac{m}{2\pi^2}\left(\Lambda + \frac{k^2}{2}\int_{-\Lambda}^\Lambda dl\,\frac{1}{l+k+i\epsilon}\frac{1}{l-k-i\epsilon}\right)$$

$$= \frac{m}{4\pi}\left(\frac{2}{\pi}\Lambda + ik - \frac{2k^2}{\pi\Lambda} + \mathcal{O}(k^4/\Lambda^3)\right), \qquad (6.21)$$

where I redefined $\epsilon \to 2k\epsilon > 0$ and again used contour integration, now in the complex-l plane. More generally,

$$I_1(k;\Lambda) = \frac{m}{4\pi}\left(\theta_1\Lambda + ik + \theta_{-1}\frac{k^2}{\Lambda} + \dots\right), \qquad (6.22)$$

where θ_{1-2n}, $n = 0, 1, \dots$, are numbers that depend on the specific form of the regulator [47].

Exercise 6.3 Show that $\theta_{1-2n} = 0$ in dimensional regularization with minimal subtraction. While a subtraction of the pole in two spatial dimensions ('power divergence subtraction', or PDS) [48, 49] retains the linear divergence, subtracting instead all poles leads to a result identical to that of a momentum-cut-off regulator [50].

We then arrive at

$$T_2^{(0,1)}(k; \Lambda) = \frac{4\pi}{m} C_0^{(0)2}(\Lambda) \left(\theta_1 \Lambda + ik + \theta_{-1} \frac{k^2}{\Lambda} + \ldots \right). \tag{6.23}$$

As we will show shortly, the regulator-dependent terms can be eliminated. As in any EFT, the meaningful loop term is the non-analytic (in energy) ik, the 'unitarity term'. Relative to the tree-level eqn (6.18), it is $\mathcal{O}(C_0 Q)$. This is to be contrasted with the analogous situation in ChPT [6], where the one-loop diagrams are down with respect to the tree by $Q^2/(4\pi f_\pi)^2$ [5, 51]. The difference is that, whereas the relativistic loop gives $Q^2/(4\pi)^2$, the non-relativistic loop gives $mQ/(4\pi)$. There are *two* enhancements for heavy particle processes: an infrared enhancement by m [2, 3] and an 'angular' enhancement of 4π [52, 48, 49, 47]. We can think of $mQ/(4\pi)$ as arising from the simple rules:

$$\text{heavy particle propagator} \to m/Q^2,$$
$$\text{loop integral} \to Q^5/(4\pi m),$$
$$\text{derivative} \to Q. \tag{6.24}$$

If we make the dimensional guess $C_0^{(0)} = \mathcal{O}(M_{\mathrm{hi}}^{-1})$, we see that the loop is suppressed by one order of the expansion parameter, $\mathcal{O}(Q/M_{\mathrm{hi}})$. The theory is perturbative and any pole of T_2 can only have binding momentum of $\mathcal{O}(M_{\mathrm{hi}})$ or higher, which is outside the EFT. In order to accommodate a shallow pole with $Q_2 = \mathcal{O}(M_{\mathrm{lo}})$, we must assume instead that $C_0^{(0)} = \mathcal{O}(M_{\mathrm{lo}}^{-1})$. I discuss this assumption in Section 6.5; for now, let us see what it implies.

Under this assumption the one-loop diagram is $\mathcal{O}(Q/M_{\mathrm{lo}})$ compared to the tree diagram. It is not difficult to see that the n-loop diagram $T_2^{(0,n)}$ is proportional to the nth power of one-loop diagram and its magnitude is $\mathcal{O}(Q^n/M_{\mathrm{lo}}^n)$ compared to the tree. For $Q \gtrsim M_{\mathrm{lo}}$, the whole geometric series must be resummed,

$$T_2^{(0)}(k; \Lambda) \equiv \sum_{n=0}^{\infty} T_2^{(0,n)}(k; \Lambda) = -\frac{4\pi}{m} \left(\frac{1}{C_0^{(0)}(\Lambda)} + \theta_1 \Lambda + ik + \theta_{-1} \frac{k^2}{\Lambda} + \ldots \right)^{-1}. \tag{6.25}$$

This LO amplitude is shown in Figure 6.1. Even though we derived it by explicitly summing up the individual diagrams, it can be obtained directly from an integral equation, the Lippmann–Schwinger (LS) equation, also shown in Figure 6.1. Quite generally, in a way that also applies to other EFTs for heavy particles, we can define the potential V as the sum of diagrams that cannot be split by cutting only heavy particle

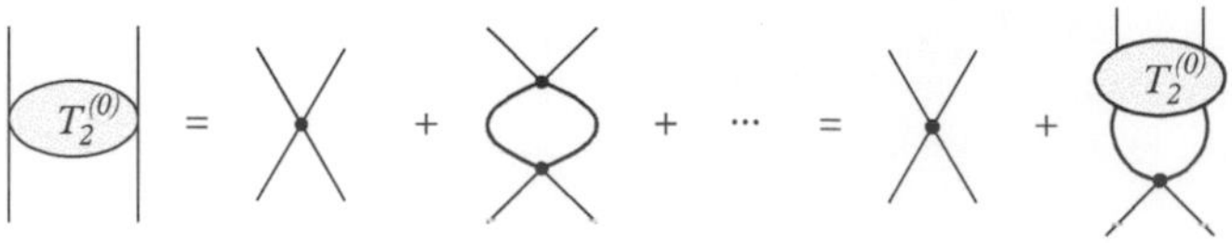

Figure 6.1 *Two-body T matrix in Pionless EFT at LO, $T_2^{(0)}$. Solid lines denote particle propagation up, eqn (6.17). The dotted vertex stands for the $C_0^{(0)}$ contact interaction, eqn (6.18). The first equality represents eqn (6.25); the second, eqn (6.26).*

lines [2, 3]. In Pionless EFT V reduces to the sum of all tree diagrams, but in Chiral EFT it includes also loop diagrams with pions (Section 6.9). Apart from a sign, the potential gives the T matrix in first Born approximation, one-loop amplitude diagrams represent the second Born approximation, and so on. Here, the LO potential is given by eqn (6.18) and the LS equation is, after integrating over the 0th component of the loop momentum,

$$T_2^{(0)}(k, \Lambda) = -V_2^{(0)}(\Lambda) - \int \frac{d^3 l}{(2\pi)^3} \, T_2^{(0)}(k, \Lambda) \frac{m}{l^2 - k^2 - i\epsilon} \, V_2^{(0)}(\Lambda). \tag{6.26}$$

More generally the T matrix depends on the incoming $\vec{p}$ and outgoing $\vec{p}'$ relative momenta, as well as the energy. The LS equation then involves the 'half-off-shell' T matrix, since one of the momenta is integrated over. In contrast, our LS eqn (6.26) with a non-local or 'separable' regulator can be solved easily with the *Ansatz* that $T_2^{(0)}$ depends only on the energy: by taking $T_2^{(0)}(k, \Lambda)$ and $V_2^{(0)}(\Lambda)$ out of the integral in the right-hand side and combining this term with the left-hand side, we obtain eqn (6.25) directly from eqn (6.26) and eqn (6.18). When the regulator is chosen to be 'local' or 'non-separable'— a function solely of the momentum transfer $\vec{p}' - \vec{p}$, which translates into a function of the spatial coordinate $\vec{r}$—the loops do not form a simple geometric series and we are usually forced to solve eqn (6.26) numerically.

Equation (6.25) still contains explicit Λ dependence which, if not controlled, will lead to regulator dependence in observables. For example, there is a pole in eqn (6.25) at imaginary momentum, or equivalently negative energy; if $C_0^{(0)}$ were Λ independent, then the corresponding binding energy would be $\propto \Lambda^2/m$. To avoid such a disaster, the Λ dependence of $C_0^{(0)}(\Lambda)$ must cancel the term linear in Λ,

$$\frac{1}{C_0^{(0)}(\Lambda)} = -\theta_1 \Lambda + \frac{1}{C_{0R}^{(0)}} \left(1 + \frac{\zeta_{-1}}{C_{0R}^{(0)} \Lambda} + \ldots \right), \tag{6.27}$$

with $C_{0R}^{(0)}$—the renormalized LEC—a constant to be determined from experimental data or from matching to the underlying theory, and ζ_{-1-2n}, $n = 0, 1, \ldots$, a set of arbitrary numbers related to the choice of input. In terms of the renormalized LEC the LO amplitude becomes

$$T_2^{(0)}(k;\Lambda) = -\frac{4\pi}{m}\left[\frac{1}{C_{0R}^{(0)}} + ik + \frac{1}{\Lambda}\left(\frac{\zeta_{-1}}{C_{0R}^{(0)2}} + \theta_{-1}k^2\right) + \dots\right]^{-1}. \tag{6.28}$$

Now the LO amplitude has a physical pole at $k = i\kappa^{(0)}$, where the binding momentum $\kappa^{(0)} = 1/C_{0R}^{(0)} + \dots$. This is a bound state if $C_{0R}^{(0)} > 0$ and a virtual state if $C_{0R}^{(0)} < 0$. In either case the binding energy is

$$B_2^{(0)}(\Lambda) = \frac{1}{mC_{0R}^{(0)2}}\left[1 + \frac{2}{C_{0R}^{(0)}\Lambda}(\zeta_{-1} - \theta_{-1}) + \dots\right]. \tag{6.29}$$

The renormalized LEC is obtained from one datum. For example, we could use the value of the amplitude at a chosen momentum $k = \mu$: in order to have $T_2^{(0)}(\mu;\Lambda) = T_2(\mu)$ Λ independent, we take $\zeta_{-1} = -\theta_{-1}C_{0R}^{(0)2}\mu^2$ etc., and the renormalized LEC is fixed by $C_{0R}^{(0)} = -[4\pi/(mT_2(\mu)) + i\mu]^{-1}$. If we choose $\mu = i\sqrt{mB_2}$, then $B_2^{(0)}(\Lambda) = B_2$ is Λ independent, but other observables, such as the amplitude at zero momentum, will have residual regulator dependence. For other choices of μ, the binding energy contains non-vanishing Λ^{-1} terms. We might as well choose to fit $C_{0R}^{(0)}$ to several low-energy data simultaneously, each with a certain weight, in which case $C_{0R}^{(0)}$ does not necessarily depend on a single fixed momentum μ.

Regardless of the choice of input observable(s), we have eliminated the dangerous regulator dependence: the amplitude (eqn 6.28) satisfies eqn (6.14), only negative powers of Λ appear in observables, and they can be made arbitrarily small by increasing Λ. The bare parameter $C_0^{(0)}(\Lambda)$ has disappeared as well. Its size does not matter; it is of course the surviving, renormalized LEC that has size $C_{0R}^{(0)} = \mathcal{O}(M_{\text{lo}}^{-1})$, giving $T_2^{(0)}(k;\Lambda) = \mathcal{O}(4\pi/(mM_{\text{lo}}))$ and $\kappa^{(0)} = \mathcal{O}(M_{\text{lo}}) \ll M_{\text{hi}}$. Equation (6.29), for example, is exactly of the generic form (eqn 6.15). Renormalization at LO has been completed: the simple contact interaction is renormalizable at the two-body level.

Now we can describe the physics of the shallow state in a controlled expansion. Let us see how this works at NLO. We expect non-vanishing NLO corrections on the basis that the residual-regulator effects in observables such as eqn (6.29) can be as large as $\mathcal{O}(M_{\text{lo}}/M_{\text{hi}})$. Indeed, we can write the LO amplitude as

$$T_2^{(0)}(k;\Lambda) = -\frac{4\pi}{m}\left(\frac{1}{C_{0R}^{(0)}} + ik\right)^{-1} + \delta T_2^{(1)}(k;\Lambda) + \dots, \tag{6.30}$$

where

$$\delta T_2^{(1)}(k;\Lambda) \equiv \frac{m}{4\pi\Lambda}\left[T_2^{(0)}(k;\infty)\right]^2\left(\frac{\zeta_{-1}}{C_{0R}^{(0)2}} + \theta_{-1}k^2\right) \tag{6.31}$$

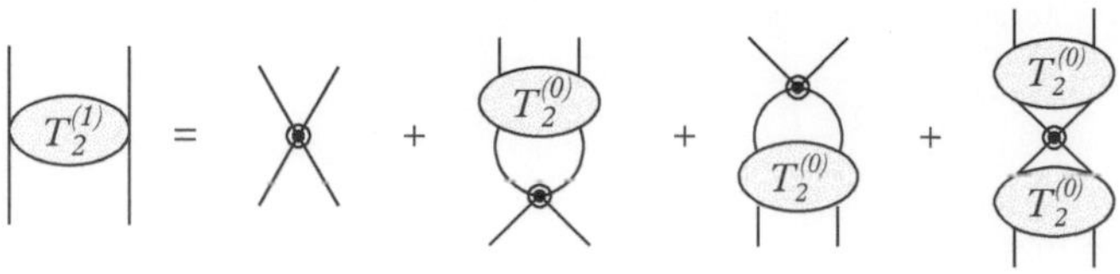

Figure 6.2 *Two-body T matrix in Pionless EFT at NLO, $T_2^{(1)}$. The circled vertex stands for the $C_0^{(1)}$ and $C_2^{(1)}$ contact interactions, eqn (6.32).*

has the size of an NLO term: $\delta T_2^{(1)}(k; M_{\mathrm{hi}}) = \mathcal{O}(4\pi/(mM_{\mathrm{hi}}))$. The form of this correction suggests that the NLO interaction that removes such a residual regulator dependence is the C_2 term in eqn (6.11), plus a perturbative correction to C_0. At tree level,

$$T_2^{(1,0)}(\vec{p}',\vec{p};\Lambda) = -\frac{4\pi}{m}\left(C_0^{(1)}(\Lambda) - C_2^{(1)}(\Lambda)\frac{\vec{p}'^2 + \vec{p}^2}{2}\right) = -V_2^{(1)}(\vec{p}',\vec{p};\Lambda), \qquad (6.32)$$

where I denoted the NLO parts of $C_{0,2}$ by $C_{0,2}^{(1)}(\Lambda)$. We expect $T_2^{(1,0)}(\vec{p}',\vec{p};M_{\mathrm{hi}}) \sim \delta T_2^{(1)}(k; M_{\mathrm{hi}})$, which is a perturbative correction to the LO amplitude.

However, adding a loop connecting $V_2^{(1)}$ to $T_2^{(0)}$ through a Schrödinger propagator gives a contribution of the same size. Thus, at NLO we have only one vertex representing NLO interactions, but it is dressed in all possible ways by LO interactions (Figure 6.2). In the language of the LS equation, eqn (6.32) is an NLO correction to the potential. The subleading potentials are solved in what is called *distorted-wave* Born approximation, which differs from the ordinary Born approximation in that the LO is not just a plane wave.

Exercise 6.4 Show that when the external particles are on-shell, $|\vec{p}'| = |\vec{p}| = k$, the amplitude in Figure 6.2 is [47]

$$T_2^{(1)}(k;\Lambda) = -\frac{m}{4\pi}\left[\frac{T_2^{(0)}(k;\Lambda)}{C_0^{(0)}(\Lambda)}\right]^2\left[C_0^{(1)}(\Lambda) - \left(-\frac{4}{\pi}\theta_3\Lambda^3\,C_0^{(0)}(\Lambda) + k^2\right)C_2^{(1)}(\Lambda) + \ldots\right],$$

$$(6.33)$$

where θ_3 is a regulator-dependent number.

Combining eqn (6.33) and eqn (6.31), we can write the full NLO amplitude as

$$T_2^{(1)}(k;\Lambda) + \delta T_2^{(1)}(k;\Lambda) = -\frac{m}{4\pi}\left[\frac{T_2^{(0)}(k;\infty)}{C_{0R}^{(0)}}\right]^2\left(C_{0R}^{(1)} - C_{2R}^{(1)}k^2\right) + \ldots, \qquad (6.34)$$

if we impose that the NLO LECs be given by

$$\frac{C_2^{(1)}(\Lambda)}{C_0^{(0)2}(\Lambda)} = \frac{C_{2R}^{(1)}}{C_{0R}^{(0)2}} - \frac{\theta_{-1}}{\Lambda} + \dots, \tag{6.35}$$

$$\frac{C_0^{(1)}(\Lambda)}{C_0^{(0)2}(\Lambda)} = -\frac{4}{\pi}\theta_3 \Lambda^3 \frac{C_2^{(1)}(\Lambda)}{C_0^{(1)}(\Lambda)} + \frac{C_{0R}^{(1)}}{C_{0R}^{(0)2}} + \frac{\zeta_{-1}}{C_{0R}^{(0)2}\Lambda} + \dots, \tag{6.36}$$

in terms of the renormalized LECs $C_{0R}^{(1)} = \mathcal{O}(M_{\text{hi}}^{-1})$, $C_{2R}^{(1)} = \mathcal{O}((M_{\text{lo}}^2 M_{\text{hi}})^{-1})$. We have succeeded in renormalizing the NLO amplitude.

There is only one new physical parameter at NLO, $C_{2R}^{(1)}$, which is related to the energy dependence of the amplitude. The LEC $C_{0R}^{(1)}$ is introduced for convenience only, so as to allow us to keep the observable used to fix $C_0^{(0)}(\Lambda)$ unchanged. For example, if we want to keep the energy-independent part of the amplitude unchanged, we take $C_{0R}^{(1)} = 0$, but more generally $C_{0R}^{(1)}$ depends on $C_{0R}^{(0)}$ and $C_{2R}^{(1)}$. In any case, the leading regulator dependence from eqn (6.30) gets replaced by the physical effect from $C_{2R}^{(1)}$. In particular, ζ_{-1} disappeared, showing that the different choices of input observable at LO are an NLO effect. Similarly, the freedom we have now to fix $C_{2R}^{(1)}$ is a higher-order effect.

To see how $C_{2R}^{(1)}$ affects the bound or virtual state it is convenient to write the amplitude up to NLO as

$$T_2^{[1]}(k;\Lambda) = T_2^{(0)}(k;\infty)\left[1 - \frac{m}{4\pi}\frac{T_2^{(0)}(k;\infty)}{C_{0R}^{(0)2}}\left(C_{0R}^{(1)} - C_{2R}^{(1)}k^2\right) + \dots\right]$$

$$= -\frac{4\pi}{m}\left(\frac{1}{C_{0R}^{(0)}} + ik - \frac{C_{0R}^{(1)}}{C_{0R}^{(0)2}} + \frac{C_{2R}^{(1)}}{C_{0R}^{(0)2}}k^2\right)^{-1}\left[1 + \mathcal{O}\left(\frac{M_{\text{lo}}^2}{M_{\text{hi}}^2}, \frac{M_{\text{lo}}^2}{\Lambda M_{\text{hi}}}\right)\right] \tag{6.37}$$

Now the binding energy becomes

$$B_2^{[1]}(\Lambda) = \frac{1}{mC_{0R}^{(0)2}}\left[1 - 2\left(\frac{C_{0R}^{(1)}}{C_{0R}^{(0)}} + \frac{C_{2R}^{(1)}}{C_{0R}^{(0)3}}\right) + \dots\right]. \tag{6.38}$$

If we choose to fit B_2 at LO and want it unchanged at NLO, then we choose $C_{0R}^{(1)} = -C_{2R}^{(1)}/C_{0R}^{(0)2}$. If we choose to fit two other observables at NLO—say the energy-independent part of the amplitude and its linear dependence on the energy—the binding energy will have a physical shift $\mathcal{O}(M_{\text{lo}}/M_{\text{hi}})$ instead of the regulator dependence $\mathcal{O}(M_{\text{lo}}/\Lambda)$ in eqn (6.29). The remaining regulator dependence in the '...' is $\mathcal{O}(M_{\text{lo}}^2/(\Lambda M_{\text{hi}}))$ and no larger than N^{2}LO as long as $\Lambda \gtrsim M_{\text{hi}}$: the binding energy is predicted to a better precision.

The procedure can be generalized to higher orders [47, 53]. At N^2LO, for example, we have to consider two insertions of C_2 and one of a four-derivative, four-field operator, both acting on S waves only. At the same order in the two-nucleon case there is also a tensor operator that mixes ${}^3S_1$ and ${}^3D_1$ [48, 49, 53]. The first contributions to higher-wave phase shifts enter at N^3LO via two-derivative four-field operators. The simple rule [52, 48, 49, 47, 53] is that—under the assumption that M_{lo} does not contaminate other waves—an operator gets an $(M_{hi}/M_{lo})^n$ enhancement over dimensional analysis, where $n = 1, 2$ is the number of S waves it connects. For example, since C_2 connects S to S waves, $C_{2R}^{(1)} = \mathcal{O}((M_{lo}^2 M_{hi})^{-1})$ instead of $C_{2R}^{(1)} = \mathcal{O}(M_{hi}^{-3})$.

6.4.2 Connection to the Effective-Range Expansion

The form of eqn (6.37) should be familiar from scattering in quantum mechanics. Recall that S waves should dominate at low energies for their lack of centrifugal barrier, and that the corresponding phase shift $\delta_2(k)$ is defined in terms of the S matrix as $S_2(k) = \exp(2i\delta_2(k))$. Thus,

$$T_2(k) = -\frac{2\pi i}{mk} [S_2(k) - 1] = -\frac{4\pi}{m} [-k\cot\delta_2(k) + ik]^{-1}. \qquad (6.39)$$

At very low energies, the effective-range expansion (ERE) [54] is known to hold,

$$k\cot\delta_2(k) = -1/a_2 + r_2 k^2/2 - P_2^3 k^4/4 + \dots, \qquad (6.40)$$

where a_2, r_2, P_2, ... are real parameters—respectively, the scattering length, effective range, shape parameter, *etc.*, all with dimensions of distance. Comparison with eqn (6.37) gives

$$a_2 = C_{0R}^{(0)} + C_{0R}^{(1)} + \dots = \mathcal{O}(M_{lo}^{-1}), \qquad r_2 = -2\frac{C_{2R}^{(1)}}{C_{0R}^{(0)2}} + \dots = \mathcal{O}(M_{hi}^{-1}), \qquad \dots \quad (6.41)$$

The scattering length is just the amplitude at zero energy and fixes $C_{0R}^{(0)}$ at LO if we choose the renormalization scale $\mu = 0$ in the discussion below eqn (6.29). The effective range provides a simple, explicit example of the usefulness of a high regulator cut-off. Because we chose a regulator for which we could iterate $C_0^{(0)}$ analytically, we could separate $\delta T_2^{(1)}(k; \Lambda)$ from $T_2^{(0)}(k; \infty)$ in eqn (6.30). Since Λ is not physical, we might as well take the amplitude at a given order to be its $\Lambda \to \infty$ limit, as in old times. Alas, in a numerical calculation—which is necessary for this EFT with a local regulator that depends on the transferred momentum, or for $A \geq 3$, or for other EFTs—this cannot be done. An LO calculation will have subleading pieces buried in them. In the case considered here, an effective range $-\theta_1 \Lambda^{-1}$ is induced. For the error of the LO calculation not to exceed its truncation error, which is $\pm\mathcal{O}(M_{hi}^{-1})$ for r_2, we have to take $\Lambda \gtrsim M_{hi}$. Even though the theory is renormalizable at LO, a 'Wilsonian cut-off' $\Lambda \lesssim M_{hi}$

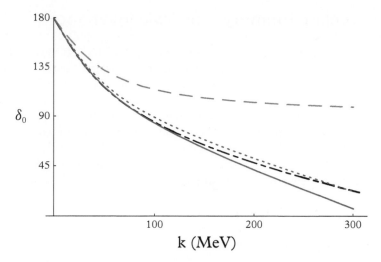

Figure 6.3 $^3S_1$ *two-nucleon phase shift* δ_0 *(in degrees) as function of the centre-of-mass momentum k (in MeV) in Pionless EFT: LO, (magenta) dashed line; NLO, (red) dotted line; and N^{2}LO, (blue) solid line. The Nijmegen potential result [55] is given by the (black) dash-dotted line. Reprinted from [53] with permission from Elsevier.*

would lead, in the absence of an explicit integration of modes in a known underlying theory, to excessive errors. Moreover, the magnitude of the change in the LO amplitude upon variation in Λ from M_{hi}, when $|\theta_1|\Lambda^{-1} = \mathcal{O}(M_{\text{hi}}^{-1})$, to much larger values, when $\theta_1\Lambda^{-1} \simeq 0$, gives an estimate of the actual size of r_2 and thus of the LO error.

Given the direct relation between renormalized LECs and ERE parameters in eqn (6.41), the latter are often used as input data in Pionless EFT. Once the LECs present at each order are determined from an equal number of data points, the full phase shifts can be predicted within the truncation error, as long as $\Lambda \gtrsim M_{\text{hi}}$. As an example, figure 6.3 shows the phase shifts in the $^3S_1$ two-nucleon system at lowest orders obtained in [53].

You might be disappointed that, after so much work, we found that Pionless EFT in the two-body system is essentially just the ERE, known for seventy years. In fact, we can show [47] that this EFT is also equivalent to even older approaches: short-distance boundary conditions [56, 57] and Fermi's pseudopotential [58]. The EFT can be seen as a derivation of these older results, once EFT's general framework is deployed to this particular class of non-relativistic systems with short-range interactions. As such, it is the proverbial cannon to kill a fly. However, a cannon can kill more. The calculation above can be generalized (numerically) to Chiral EFT, where it provides a guide for much more complicated renormalization. Moreover, Pionless EFT applies to systems that are outside the scope of the ERE. For example, in the nuclear case we can look at the electromagnetic properties of the deuteron [53], thanks to the inclusion of consistent one- and two-nucleon electromagnetic operators. I instead describe how Pionless EFT allows us to generalize the ERE to more than two particles, where amazing new structures arise.

6.5 Fine Tuning, Unitarity, and Scale Invariance

Before we consider more bodies, a few remarks about the EFT in the two-body system.

We would naively expect all ERE parameters to be comparable to the force range R and indeed, for most parameter values of most finite-range potentials, that is what one finds, with bound states or resonances that are not particularly shallow. In this case, for $k \ll R^{-1}$ each term in the ERE is larger than the next and the corresponding EFT is purely perturbative. But by dialling one or more potential parameters we can make $|a_2| \gg R$ with other ERE parameters still of $\mathcal{O}(R)$. For $k \ll |a_2|^{-1}$, the EFT is still perturbative but as we have just seen, if we want to continue to describe physics of momentum up to R^{-1}, we have to resum (only!) $C_0^{(0)}$. By *fine tuning* we can reach the extreme point, the 'unitarity limit' where $|a_2| \to \infty$ and $T_2(k)$ takes (up to range corrections) its maximum value $\propto k^{-1}$.

Exercise 6.5 Consider a three-dimensional spherical well with dimensionless depth α,

$$V_2(\vec{r}) = -\frac{\alpha^2}{mR^2}\, \theta(R - r). \tag{6.42}$$

Solve the Schrödinger equation for the S wave in the usual way, i.e., by matching inside and outside solutions. Obtain $\delta_2(k)$ from the outside solution (asymptotically $\psi_2(r) \propto [\sin(kr + \delta_2(k))]/r$) and show that

$$T_2(k) = -\frac{2\pi i}{mk}\left[e^{-2ikR}\frac{\sqrt{\alpha^2 + (kR)^2}\cot\sqrt{\alpha^2 + (kR)^2} + ikR}{\sqrt{\alpha^2 + (kR)^2}\cot\sqrt{\alpha^2 + (kR)^2} - ikR} - 1\right], \tag{6.43}$$

and thus

$$a_2 = R\left(1 - \frac{\tan\alpha}{\alpha}\right), \quad r_2 = R\left(1 - \frac{R}{\alpha^2 a_2} - \frac{R^2}{3a_2^2}\right), \quad \ldots \tag{6.44}$$

For generic values of α, we see that $a_2 \sim r_2 \sim \ldots \sim R$. However, when $\alpha \simeq (2n + 1)\pi/2 \equiv \alpha_c$, while still $r_2 \sim \ldots \sim R$, $a_2 \simeq R/[\alpha_c(\alpha - \alpha_c)] \gg R$. For α just below α_c, there is a shallow virtual state; as the attraction increases past α_c, a shallow bound state appears. In this example, $M_{\text{hi}} \sim 1/R$ and $M_{\text{lo}} \sim \alpha_c|\alpha - \alpha_c|/R$ [47].

As we saw in the previous section, we incorporate such fine tuning in the EFT by allowing the LECs to scale with the small M_{lo}. Although I used the methods of quantum field theory (the proverbial cannon), we can also use standard quantum mechanics, supplemented by regularization and renormalization. The LO interaction at tree level, eqn (6.18), is just a constant in momentum space, which in coordinate space is a delta function. The appearance of the Schrödinger propagator at one loop, eqn (6.19), reveals that the iteration of this interaction is equivalent to solving the Schrödinger equation with a delta-function potential. In fact, we can show on general grounds that the LS equation is equivalent to the Schrödinger equation. Even if we prefer to solve the Schrödinger equation, regularization and renormalization are still necessary. For example, the LO potential in coordinate space can be written as

$$V_2^{(0)}(\vec{r}; \Lambda) = \frac{4\pi}{m} C_0^{(0)}(\Lambda) \delta_\Lambda^{(3)}(\vec{r}), \tag{6.45}$$

where $\delta_\Lambda^{(3)}(\vec{r})$ is a regularization of the three-dimensional delta function, that is, a smearing over distances $r \lesssim \Lambda^{-1}$ with $\lim_{\Lambda \to \infty} \delta_\Lambda^{(3)}(\vec{r}) = \delta^{(3)}(\vec{r})$. Whatever the underlying potential is, we can use, say, the spherical well (6.42) as such regularization. In this guise, the range of the well functions as a regulator parameter, $\Lambda \equiv 1/R$, and $\alpha = \alpha(R)$ can be adjusted so that a_2 in eqn (6.44) is R independent and reproduces a given experimental value. Another popular regularization for its analytical simplicity is a delta-shell potential. For many-body calculations a Gaussian regularization is particularly convenient for its smoothness.

Exercise 6.6 Solve the Schrödinger equation for the potential (6.45). *Hint*: Fourier transform to momentum space and choose a sharp momentum regulator. Show that the LO results of the previous section are reproduced and, in addition, the negative-energy wave function is

$$\psi_2^{(0)}(r) \propto \frac{1}{r} \exp(-r/a_2) \tag{6.46}$$

as $\Lambda \to \infty$. Thus a_2, which is basically the scattering amplitude at $k = 0$, is a measure of the size of the system. A real bound (virtual) state has a (non-) normalizable wave function and corresponds to $a_2 > 0$ ($a_2 < 0$).

The bare LECs change with the regulator parameter Λ. At LO, renormalization requires eqn (6.27). This is the solution of the RG equation

$$\mu \frac{d}{d\mu} \left(\mu C_0^{(0)}(\mu) \right) = \mu C_0^{(0)}(\mu) \left(1 + \mu C_0^{(0)}(\mu) \right), \tag{6.47}$$

with $\mu \equiv \theta_1 \Lambda$. This equation admits two fixed points: a 'trivial' $\mu C_0^{(0)}(\mu) = 0$ and the 'non-trivial' $\mu C_0^{(0)}(\mu) = -1$ [3]. When $\mu \ll |a_2|^{-1}$, we are near the trivial point where perturbation theory holds. On the other hand, for $\mu \gg |a_2|^{-1}$,

$$C_0^{(0)}(\Lambda; C_{0R}^{(0)}) = -\frac{1}{\theta_1 \Lambda} \left[1 + \frac{1}{\theta_1 \Lambda C_{0R}^{(0)}} + \mathcal{O}\left(\frac{1}{\Lambda^2 C_{0R}^{(0)2}} \right) \right], \tag{6.48}$$

the flow is close to the non-trivial fixed point, and all diagrams containing only this vertex should be resummed. The unitarity limit corresponds to the non-trivial fixed point,

$$C_0^{(0)}(\Lambda; \infty) = -\frac{1}{\theta_1 \Lambda}. \tag{6.49}$$

The fine tuning needed to produce the unitarity limit can be carried out experimentally for cold atoms through the mechanism of Feshbach resonances [59]—this was one of the reasons for the explosion of interest in these systems. The mechanism works when the system has two coupled channels with different spins and thresholds—'open' and 'closed' channels—and the relative position of a bound state in the closed channel can be changed by an external magnetic field. In the open channel, the scattering length varies and diverges as the energy of the bound state crosses the open threshold. A short-range EFT for this situation is discussed in [60]. In contrast, the ^{4}He dimer just happens to be close to the unitarity limit even in the absence of a magnetic field. The scattering length and effective range calculated with the LM2M2 potential are [61] $a_2 \simeq 100$ Å $\simeq 18 l_{\text{vdW}}$ and $r_2 \simeq 7.3$ Å $\simeq 1.3 l_{\text{vdW}}$, with similar values for other sophisticated potentials—even though, of course, a_2 is very sensitive to potential details because of fine tuning.

Nucleons are not as close to unitarity: for np in the $^3S_1$ (deuteron) channel, $a_{2,S=1} \simeq 5.4$ fm $\simeq 3.8 m_\pi^{-1}$ and $r_{2,S=1} \simeq 1.8$ fm $\simeq 1.3 m_\pi^{-1}$. However, in the $^1S_0$ channel, where there is a shallow virtual state, the relative magnitudes of np parameters are not very different from atomic ^{4}He, $a_{2,S=0} \simeq -23.7$ fm $\simeq -17 m_\pi^{-1}$ and $r_{2,S=0} \simeq 2.7$ fm $\simeq 1.9 m_\pi^{-1}$. In QCD, the only free parameters are the quark masses, and we can imagine alternative worlds where the interactions are fundamentally unchanged but explicit chiral symmetry breaking is larger and the range of the pion-exchange force is, consequently, smaller. As the quark masses change, so do nuclear binding energies. Because heavier quarks are easier to evolve in imaginary time in a four-dimensional space-time lattice, Lattice QCD has provided so far only 'alternative facts' about light nuclei [7]. The situation is still in flux, with different methods of signal extraction leading to contradictory results, but in the majority of calculations it seems that nuclei at larger $\bar{m}$ are more bound versions of their counterparts in our world [62]. At large quark masses, where there is no Chiral EFT, Pionless EFT offers the only viable description of these nuclei. We can take few-body observables calculated in Lattice QCD as input to Pionless EFT, thus bypassing the need for experimental data, and use Pionless EFT to calculate the structure of heavier nuclei [8, 9]. A possible scenario [63] for quark-mass variation is one in which the deuteron and the $^1S_0$ virtual state become, respectively, unbound below and bound above, but near, the physical point. If this is the case, then the mechanism of fine tuning in QCD is parallel to that of Feshbach resonances for atoms, with the magnetic field replaced by the quark masses.

Even when $|a_2| \gg R$ is finite, as for nucleons at physical quark masses and ^{4}He atoms, the unitarity limit is useful: in the 'unitarity window' $|a_2|^{-1} \ll k \ll R^{-1}$, $T_2(k)$ is close to the maximum value $\propto k^{-1}$:

$$T_2(|a_2|^{-1} \ll k \ll R^{-1}) = \frac{4\pi}{m} (ik)^{-1} \left[1 + \mathcal{O}\left(k M_{\text{hi}}^{-1}, k\Lambda^{-1}, (ka_2)^{-1} \right) \right]. \quad (6.50)$$

When we retain only the first term, there is no dimensionful parameter other than k itself. The vanishing of the binding energy then is a reflection of scale invariance. Under a change of scales [64] with parameter $\alpha > 0$,

$$r \to \alpha r, \qquad t/m \to \alpha^2 t/m, \qquad \Lambda \to \alpha^{-1}\Lambda, \qquad \psi \to \alpha^{-3/2}\psi, \quad (6.51)$$

the first two terms in eqn (6.11)—the nucleon bilinear and the $C_0^{(0)}$ contact interaction—are invariant on account of eqn (6.49). Under a scale change, $mE \to \alpha^{-2}mE$, but in the unitarity limit there is no scale, so $B_2 = 0$. In this limit the $A = 2$ system is also conformally invariant [65].

Away from the unitarity limit, scale symmetry is explicitly broken by the dimensionful parameter $C_{0R}^{(0)} = a_2$ in eqn (6.48). At subleading orders scale symmetry is also broken by M_{hi} in the form of the higher ERE parameters. The dependence of B_2 on dimensionful parameters may be determined with the 'spurion field' method [66], which is designed to exploit the consequences of an approximate symmetry. The idea is that *if* under scale invariance these parameters changed according to their canonical dimension, then the system would remain invariant. For example, if a_2 changed to αa_2, then the first two terms in eqn (6.11) would still be invariant. In that case, the energy after the transformation should equal the transformed energy: $B_2(\alpha a_2) = \alpha^{-2} B_2(a_2)$. This implies $B_2(a_2) \propto (ma_2^2)^{-1}$, see eqn (6.29). Now, since a_2 is actually fixed, $B_2(a_2)$ reflects the specific way in which a_2 breaks scale invariance. In this particular case the spurion method is just dimensional analysis, since by allowing a_2 to vary we are changing all dimensionful quantities appearing to this order according to their (inverse mass) dimension. And, of course, this relation was obtained earlier directly from eqn (6.28), but the spurion method illustrates how considerations of symmetry underlie dynamical results.

The message is that we are dealing with fine-tuned systems, where for $A = 2$ we are close to the non-trivial fixed point associated with unitarity and scale invariance. There are significant departures from naive dimensional analysis, but renormalization provides a useful guide to infer the corresponding enhancements.

6.6 Three-Body System

Whatever the reason for fine tuning, we can ask what structures it produces. The first surprise comes when $A = 3$.

Many-body forces are not forbidden by any symmetry, and yet we are used to think of them as small. That is a consequence of their high dimensionality. From dimensional analysis we might expect them to be highly suppressed, e.g. $D_0 = \mathcal{O}(M_{\text{hi}}^{-4})$. If this is the case, the properties of many-body systems are determined, to a very good approximation, by two-body interactions. The simplest connected diagram for three particles consists of an LO interaction between two particles (say 1 and 2), followed by propagation of one particle (say 2), and its interaction with the third particle. In the next simplest diagram, either particle 2 or 3 further propagates till it interacts with particle 1, giving rise to a loop. Using the power counting rules (6.24), the expected size of the latter diagram relative to the former is

$$\frac{\mathcal{O}\left((4\pi C_0^{(0)}/m)^3 \, (Q_3^5/(4\pi m)) \, (m/Q_3^2)^3 \right)}{\mathcal{O}\left((4\pi C_0^{(0)}/m)^2 \, (m/Q_3^2) \right)} = \mathcal{O}(C_0^{(0)} Q_3). \qquad (6.52)$$

This counting extends straightforwardly to diagrams with more loops. For $Q_3 \gtrsim C_{0R}^{(0)-1} = \mathcal{O}(M_{\text{lo}})$, the arguments of Section 6.4 apply to any of the three two-body subsystems, meaning the LO interactions $C_0^{(0)}$ must resummed into the LO two-body T matrix $T_2^{(0)}$. Subleading corrections are treated perturbatively. This argument applies also to the scattering of one particle on a two-body bound state, such as nd scattering or particle-dimer scattering in the atomic lingo. From the corresponding three-body T matrix, T_3, we can find the three-body bound states. The issue now is, is T_3 properly renormalized up to N^4LO, when we naively expect the appearance of the first three-body force?

6.6.1 Auxiliary Field

In the systems we want to describe, where there are shallow T-matrix poles, it is often times convenient to introduce auxiliary fields with the quantum numbers of these poles. They can be thought of as 'composite' fields for the corresponding states, which are not essential but do simplify the description of the larger systems, particularly reactions involving the bound state.

Most useful is a field for the dimer, the 'dimeron', denoted by d, with the quantum number of the $A = 2$ pole—first introduced in this context in [67]. From the evolution of this field, whose mass is defined as $2m - \Delta$, $2m$ is removed by a field redefinition. The corresponding action is obtained by replacing the Lagrangian $\mathcal{L}$ in eqn (6.11) by

$$\mathcal{L}_d = \psi^\dagger \left(i \frac{\partial}{\partial t} + \frac{\vec{\nabla}^2}{2m} + \dots \right) \psi + d^\dagger \left[\Delta + \sigma \left(i \frac{\partial}{\partial t} + \frac{\vec{\nabla}^2}{4m} + \dots \right) \right] d$$
$$- \frac{g_0}{\sqrt{2}} \left(d^\dagger \psi \psi + \psi^\dagger \psi^\dagger d \right) - h_0 \, d^\dagger d \psi^\dagger \psi + \dots, \tag{6.53}$$

where $\sigma = \pm 1$ and $g_0, h_0, \dots$ are LECs. In particle-dimer scattering with a relative momentum $Q \lesssim Q_2$, when the dimer cannot be broken up, a lower-energy halo EFT can be constructed where d is an 'elementary' field. For that, take $g_0 = 0$ and $\sigma = +1$, with h_0 being the leading contact interaction between particle and dimer analogous to C_0 in eqn (6.11). Here, where $Q \gtrsim Q_2$, the coupling $g_0 \neq 0$ to two particles ensures the composite nature of the dimeron field. The particle–particle interaction proceeds through the dimeron propagator, and h_0 represents a three-body force. Integrating out the d field brings back eqn (6.11).

The power counting of Section 6.4 is reproduced if $\Delta = \mathcal{O}(M_{\text{lo}})$ and $g_0 = \mathcal{O}(\sqrt{4\pi/m})$. In this case the kinetic terms of the dimeron are subleading. The full dimeron propagator at LO is the sum of bubbles,

$$iD_2^{(0)}(p^0, \vec{p}; \Lambda) = \frac{i}{\Delta^{(0)} + i\epsilon} \left[1 - \frac{g_0^2}{\Delta^{(0)} + i\epsilon} I_1(k; \Lambda) + \left(\frac{g_0^2}{\Delta^{(0)} + i\epsilon} I_1(k; \Lambda) \right)^2 + \dots \right]$$

$$= i \frac{4\pi}{mg_0^2} \left[\frac{4\pi}{m} \left(\frac{\Delta^{(0)}}{g_0^2} + I_1(k; \Lambda) \right) + i\epsilon \right]^{-1}, \tag{6.54}$$

and the NLO correction is

$$iD_2^{(1)}(p^0, \vec{p}; \Lambda) = i \left(\Delta^{(1)} + \frac{\sigma^{(1)} k^2}{m} \right) \left(iD_2^{(0)}(p^0, \vec{p}; \Lambda) \right)^2, \tag{6.55}$$

where in these expressions $k \equiv \sqrt{mp^0 - \vec{p}^2/4 + i\epsilon}$. The dimeron can be thought to represent the bound- or virtual-state propagator once we multiply numerator and denominator by $\Delta^{(0)}/g_0^2 - I_1(k; \Lambda)$ in order to remove the square root of the energy from the denominator. Expanding around the pole we can obtain the residue, that is, the wave function renormalization,

$$Z_2^{-1} = \frac{\partial}{\partial p^0} \left(D_2(p^0, \vec{p}) \right)^{-1} \Bigg|_{p^0 = -B_2}. \tag{6.56}$$

Attaching two external particle legs to the dimeron propagator multiplies it by $-g_0^2$ and gives iT_2. Equation (6.54) shows that there is only one parameter, $\Delta^{(0)}/g_0^2$, at LO. This redundancy is frequently eliminated with a redefinition of the auxiliary field to make [68]

$$g_0 \equiv \sqrt{\frac{4\pi}{m}}, \tag{6.57}$$

which elevates σ to a full-blown LEC $\sigma(\Lambda)$ rather than a just sign. With this choice and the renormalization scale $\mu = 0$ for simplicity, the T matrix has the forms of eqn (6.30) and eqn (6.34) with

$$\Delta_R^{(0)} \equiv \Delta^{(0)}(\Lambda) + \theta_1 \Lambda = \frac{1}{C_{0R}^{(0)}} = \frac{1}{a_2}, \qquad \Delta_R^{(1)} \equiv \Delta^{(1)}(\Lambda) = 0, \tag{6.58}$$

and

$$\sigma_R^{(1)} \equiv \sigma^{(1)}(\Lambda) + \theta_{-1} \frac{m}{\Lambda} = \frac{m C_{2R}^{(1)}}{C_{0R}^{(0)2}} = -\frac{m r_2}{2}. \tag{6.59}$$

Note that $r_2 > 0$, as in most situations, requires $\sigma < 0$, that is, d is a ghost field—and yet the two-body amplitude is perfectly fine. Renormalization is different than before, however, because the dimeron induces energy-dependent corrections instead of the momentum-dependent C_2 corrections. Thus, no Λ^3 divergence appears at NLO. In this case we can resum the NLO corrections without running into problems with the RG [47], contrary to the case of momentum-dependent corrections [69].

6.6.2   Amplitude

In terms of the auxiliary field, we can represent the scattering of a particle on a dimer at LO through the 'one-particle' exchange diagrams shown in Figure 6.4. The whole series of 'pinball' diagrams with multiple such exchanges needs to be included on account of eqn (6.52), giving rise to an integral equation known as the Skorniakov–Ter–Martirosian equation. In these diagrams the dimeron propagator is the LO propagator (eqn 6.54). At NLO, we include one insertion of the NLO propagator (eqn 6.55) in all possible ways, and analogously for higher orders.

Let us work again in the centre-of-mass frame, where at LO the incoming (outgoing) dimer has energy $k^2/(4m) - B_2^{(0)}$ $(k^2/(4m) - B_2^{(0)} + E')$ and momentum $\vec{p}$ $(\vec{p}')$, and the incoming (outgoing) particle has energy $k^2/(2m)$ $(k^2/(2m) - E')$ and momentum $-\vec{p}$ $(-\vec{p}')$. The total energy is $E = 3k^2/(4m) - B_2^{(0)}$. We can take the initial state to be on-shell, $|\vec{p}| = k$. For simplicity, we take the $\Lambda \to \infty$ limit in the dimer propagator. The integration over the 0th component of the loop momentum is similar to the one done in Section 6.4.1: we pick a pole from, say, the particle propagator, and are left with a three-dimensional integral involving the dimeron propagator. At this point we can set $E' = (k^2 - \vec{p}'^2)/(2m)$, which holds when the final state is on-shell. With the choice (eqn 6.57), we find for the half-off-shell amputated amplitude

$$t_3(\vec{p}',\vec{p}) = -v_3(\vec{p}',\vec{p}) - \lambda \int \frac{d^3 l}{(2\pi)^3} \frac{t_3(\vec{p}',\vec{l})\,v_3(\vec{l},\vec{p})}{-1/a_2 + \sqrt{3\vec{l}^2/4 - mE}}, \qquad (6.60)$$

where

$$v_3(\vec{p}',\vec{p}) = \frac{8\pi}{mE - \vec{p}'^2 - \vec{p}^2 - \vec{p}'\cdot\vec{p}}. \qquad (6.61)$$

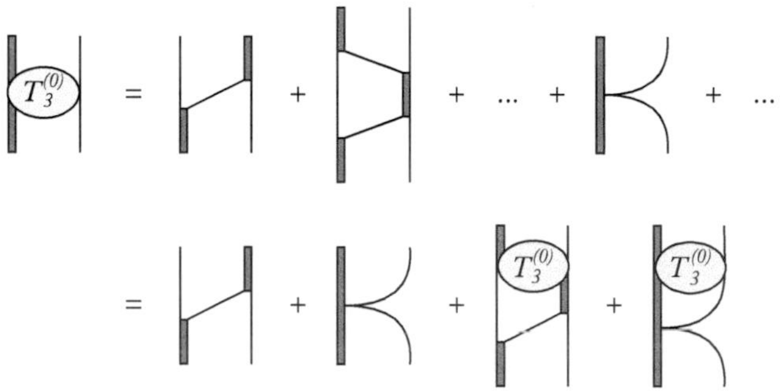

Figure 6.4 *Three-body T matrix in Pionless EFT at LO, $T_3^{(0)}$. The filled double line represents the full dimeron propagator (eqn 6.54), when internal, or the corresponding wave-function renormalization (eqn 6.56), when external. The two-particle–dimeron vertex stands for the g_0 interaction, eqn (6.57), while the particle-dimeron contact is the three-body force $h_0^{(0)}$, eqn (6.68).*

The above equation with $\lambda = 1$ is derived from the EFT for bosons [70, 22]. For three nucleons with total spin $S = 3/2$, the equation takes the same form but with $\lambda = -1/2$ [71, 72]. Instead, when $S = 1/2$ one finds a pair of coupled integral equations. In the ultraviolet (UV) limit where scattering length and binding energy can be discarded, these equations decouple [15] into a pair of equations like eqn (6.60), one with $\lambda = 1$, the other with $\lambda = -1/2$. Thus, even though I consider here the single eqn (6.60), the lessons learned from different values of λ can be applied to nucleons in the triton channel as well.

Now, for simplicity we focus on the most important, S wave. We can project on it by integrating over the angle between $\vec{p}'$ and $\vec{p}$. Performing the integration also over the angles in the loop integral, the equation simplifies to

$$t_{3,0}(p',k) = -v_{3,0}(p',k) - \frac{\lambda}{2\pi^2} \int_0^\Lambda dl\, l^2\, \frac{t_{3,0}(p',l)\, v_{3,0}(l,k)}{-1/a_2 + \sqrt{3l^2/4 - mE}}, \tag{6.62}$$

where I took a sharp (three-body) cut-off for definiteness and

$$v_{3,0}(p',k) = \frac{4\pi}{p'k} \ln\left(\frac{p'^2 - p'k + k^2 - mE}{p'^2 + p'k + k^2 - mE}\right). \tag{6.63}$$

The on-shell scattering amplitude is obtained by making $p' = k$ and accounting for wave-function renormalization,

$$T_{3,0}(k) = \sqrt{Z_2^{(0)}}\, t_{3,0}(k,k)\, \sqrt{Z_2^{(0)}}. \tag{6.64}$$

As in the two-body case, the first step to solve the integral equation is to look at the UV region, $p' \gg k \gtrsim 1/a_2$, where the equation reduces to

$$t_{3,0}(p' \gg k) = \frac{4\lambda}{\sqrt{3}\pi} \int_0^\Lambda \frac{dl}{p'} \ln\left(\frac{p'^2 + p'l + l^2}{p'^2 - p'l + l^2}\right) t_{3,0}(l \gg k). \tag{6.65}$$

This equation is homogeneous so it cannot fix the overall normalization of $t_{3,0}(p' \gg k)$, but it does determine the dependence on p' in the region $\Lambda > p' \gg k$. Scale invariance (eqn 6.51) suggests the *ansatz* $t_{3,0}(p' \gg k) \propto p'^{-(s+1)}$, which works if s obeys

$$\frac{8\lambda}{\sqrt{3}s} \frac{\sin(\pi s/6)}{\cos(\pi s/2)} = 1. \tag{6.66}$$

This relation is analysed in detail in [73]. Because of the additional inversion symmetry $p' \to 1/p'$, the roots come in pairs. For $-1/2 \le \lambda \le \lambda_c \equiv 3\sqrt{3}/(4\pi) \simeq 0.4135$, the roots are real. The root with $\Re(s) > -1$ ensures that $t_{3,0}(p' \gg k)$ goes to zero, in which case the amplitude has no essential sensitivity to the regulator and predictions about the three-body system can be made at LO. In particular, for three nucleons with $S = 3/2$, when

$\lambda = -1/2$, we find that $t_{3,0}(p' \gg k) \propto p'^{-3.17}$, which is softer than the p'^{-2} behaviour expected in perturbation theory from $v_{3,0}(p' \gg k) \propto p'^{-2}$. The numerical solution of eqn (6.62) gives a low-energy amplitude in good agreement with phenomenology, which improves at subleading orders [71, 72]. Because of the good UV behaviour of the LO amplitude, we can resum higher-order terms to make calculations easier without jeopardizing RG invariance. As an example, the $S = 3/2$ nd scattering length $a_{nd,S=3/2} = 5.09 + 0.89 + 0.35 + \ldots$ fm $= 6.33 \pm 0.05$ fm [71], to be compared with the experimental value 6.35 ± 0.02 fm [74].

In contrast, for other λ values the solutions are complex, and for $\lambda > \lambda_c \equiv 3\sqrt{3}/(4\pi)$ the roots are imaginary. In particular, for the bosonic case $\lambda = 1$ there is a pair of imaginary solutions $s = \pm i s_0$, with $s_0 \simeq 1.00624$. The two solutions are equally acceptable (or actually unacceptable...) and lead to an asymptotic behavior of the half-off-shell amplitude of the form

$$t_{3,0}(p' \gg k) \propto \cos\left(s_0 \ln(p'/\Lambda) + \delta\right), \tag{6.67}$$

where δ is a dimensionless, p'-independent number. A numerical solution of eqn (6.62) confirms this oscillatory behaviour with $\delta = 0.76 \pm 0.01$ [70, 22]. Small changes in Λ propagate to lower momenta and lead to dramatic changes in the observable $t_{3,0}(k,k)$, but the changes are periodic. One can show that this solution, first found in [75], supports a sequence of bound states that appear with the same periodicity as Λ increases, with the binding energy of each state growing as Λ^2/m [76]. This solution is obviously unacceptable: the first two terms in eqn (6.11) are not renormalizable beyond $A = 2$.

How can we maintain RG invariance? The only possibility is a three-body force, and the one provided by h_0 is the simplest. For this fix to work, this force has to be assumed to be LO, but even then it is not obvious that it can remove the regulator dependence when iterated. Upon including

$$h_0^{(0)}(\Lambda) \equiv 8\pi \frac{H(\Lambda)}{\Lambda^2}, \tag{6.68}$$

where $H(\Lambda)$ is dimensionless, we have additional diagrams, also shown in Figure 6.4. The LO amplitudes $t_3^{(0)}(\vec{p}',\vec{p})$ and $t_{3,0}^{(0)}(p',k)$ still satisfy eqn (6.60) and eqn (6.62) but with, respectively,

$$v_3(\vec{p}',\vec{p}) \rightarrow v_3(\vec{p}',\vec{p}) + h_0^{(0)}(\Lambda) \equiv v_3^{(0)}(\vec{p}',\vec{p};\Lambda), \tag{6.69}$$

$$v_{3,0}(p',k) \rightarrow v_{3,0}(p',k) + h_0^{(0)}(\Lambda) \equiv v_{3,0}^{(0)}(p',k;\Lambda). \tag{6.70}$$

The asymptotic equation (eqn 6.65) now becomes, for the physically relevant $\lambda = 1$,

$$t_{3,0}^{(0)}(p' \gg k) = \frac{4}{\sqrt{3}\pi} \int_0^\Lambda \frac{dl}{p'} \left[\ln\left(\frac{p'^2 + p'l + l^2}{p'^2 - p'l + l^2}\right) - 2\frac{p'l}{\Lambda^2} H(\Lambda) \right] t_{3,0}^{(0)}(l \gg k). \tag{6.71}$$

Only for $p' \sim \Lambda$ is the three-body force important. In the region $p' \ll \Lambda$, the behaviour (eqn 6.67) still holds, but now δ is determined by $H(\Lambda)$. We can define the physical, dimensionful parameter $\Lambda_\star$ through

$$\delta(H(\Lambda)) = s_0 \ln(\Lambda/\Lambda_\star). \tag{6.72}$$

Since

$$t_{3,0}^{(0)}(p' \gg k) \propto \cos\left(s_0 \ln(p'/\Lambda_\star)\right) \tag{6.73}$$

is now essentially cut-off independent, so will the low-energy on-shell amplitude. $\Lambda_\star$ can then be determined from low-energy data or matching to the underlying theory. Again, numerical experimentation shows [70, 22] that this can indeed be achieved. We can also show that bound states now accrete periodically from below (that is, from very large binding energies) as the regulator cut-off becomes large enough to accommodate them. As Λ increases their binding energies approach constants. With the addition of the three-body force, the EFT is renormalizable at LO for $A = 3$.

An approximate form can be obtained for $H(\Lambda)$ by going back to eqn (6.62) for two values of the regulator cut-off, Λ and $\Lambda' > \Lambda$. Imposing that the two equations agree in the region $\Lambda > p' \gg k$, and making the approximation (eqn 6.73) for the Λ' solution also when $p' \sim \Lambda'$, we find [70, 22]

$$H(\Lambda) \simeq c \frac{\sin\left(s_0 \ln(\Lambda/\Lambda_\star) - \tan^{-1}(s_0^{-1})\right)}{\sin\left(s_0 \ln(\Lambda/\Lambda_\star) + \tan^{-1}(s_0^{-1})\right)}, \tag{6.74}$$

where $c \simeq 1$. $H(\Lambda)$ can also be extracted purely numerically by demanding that one low-energy datum (for example, the particle–dimer scattering length a_3) be reproduced at any value of Λ. The agreement between approximate and numerical results is shown in Figure 6.5. The best fit gives $c = 0.879$ [77].

At tree level, the h_0 particle–dimeron interaction generates the three-particle force

$$D_0^{(0)}(\Lambda) = \frac{3 h_0^{(0)}(\Lambda)}{4\pi \Delta^{(0)2}(\Lambda)} \propto \frac{1}{\Lambda^4} \frac{\sin\left(s_0 \ln(\Lambda/\Lambda_\star) - \tan^{-1}(s_0^{-1})\right)}{\sin\left(s_0 \ln(\Lambda/\Lambda_\star) + \tan^{-1}(s_0^{-1})\right)} \left[1 + \mathcal{O}\left((C_{0R}^{(0)}\Lambda)^{-1}\right)\right]. \tag{6.75}$$

In coordinate space, the corresponding potential is

$$V_3^{(0)}(\vec{r}_{12}, \vec{r}_{23}; \Lambda) = \frac{(4\pi)^2}{m} D_0^{(0)}(\Lambda)\, \delta_\Lambda^{(3)}(\vec{r}_{12}) \delta_\Lambda^{(3)}(\vec{r}_{23}). \tag{6.76}$$

where $\vec{r}_{ij}$ is the position of particle i with respect to particle j.

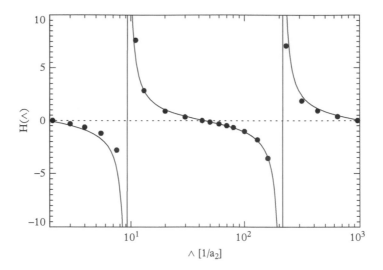

Figure 6.5 *Dimensionless three-body force H in Pionless EFT at LO, eqn (6.68), as a function of the regulator cut-off Λ (in units of a_2^{-1}): numerical solution for $a_3 = 1.56\,a_2$, dots; and eqn (6.74) with $\Lambda_\star = 19.5\,a_2^{-1}$, solid line. Reprinted from [22] with permission from Elsevier.*

The argument above applies directly to bosons and indirectly to nucleons with $S = 1/2$. Therefore a three-nucleon force is needed for RG invariance [15], consistently with the fact that the D_0 force has a non-vanishing projection onto the $^2S_{1/2}$ channel. In channels with angular momentum $l > 0$ similar equations are obtained with the logarithm replaced by Legendre polynomials of the second kind [73]. An equation for s analogous to eqn (6.66) is obtained, involving a hypergeometric function. For both $\lambda = 1$ and $\lambda = -1/2$, $s \simeq l+1$ in fair agreement with the expectation from perturbation theory, which can be shown from the Legendre polynomials to be $t_{3,l}(p' \gg k) \propto p^{-(l+2)}$. There is therefore no need for additional three-body forces at LO.

The perturbative NLO correction that accounts for two-body range effects induces a finite change in the three-body system and does not require an additional three-body force for RG invariance [22, 78]; a correction $D_0^{(1)}(\Lambda)$ is sufficient. A two-derivative three-body force does enter, however, at N²LO [79]. Thus, while the LO three-body force is enhanced by $(M_{hi}/M_{lo})^4$ over simple dimensional analysis, three-body force corrections, which are amenable to perturbation theory, seem to be suppressed by the expected relative factors of M_{lo}/M_{hi}.

6.6.3 Bound States and Correlations

The EFT produces a series of discrete bound states whose spacing depends on the two-body scattering length. The three-body binding momenta quickly exceed M_{hi}, so that only a finite number of states ($\sim \ln(|a_2|/R)/\pi$ for an underlying potential of range R

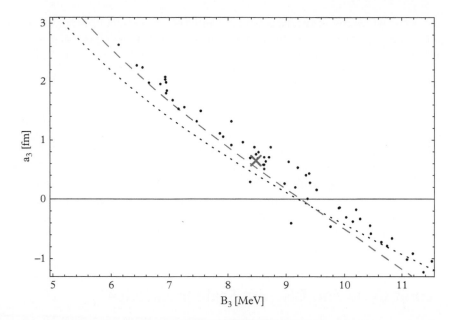

Figure 6.6 *Phillips line in the plane of doublet nd scattering length $a_3 \equiv a_{nd,S=1/2}$ (in fm) and triton binding energy $B_3 \equiv B_t$ (in MeV), in Pionless EFT: LO, (black) dotted line; NLO, (red) dashed line. The dots represent a variety of nuclear potentials with the same two-nucleon scattering lengths and effective ranges [84]. The cross is the experimental result. Reprinted from [79] with permission from Elsevier.*

[80, 15]) are within the range of applicability of the EFT. For atomic ^{4}He, for example, both the ground [81] and first-excited [33] states have been detected, with a ratio of binding energies of about sixty (Table 6.1). For nucleons only the triton (and helion, separated from triton only by small isospin-breaking effects) is observed, but there is a *virtual nd* state with a binding energy about six times smaller.

Because a single parameter emerges in the three-body force up to NLO, we expect correlations among these binding energies and phase shifts in channels not affected by the exclusion principle. The classic example is the Phillips line [83]: a line in the plane spanned by the triton binding energy B_t and the $S = 1/2$ *nd* scattering length $a_{nd,S=1/2}$. This correlation was first discovered empirically, as a line formed by points representing various phenomenological potentials, which describe two-nucleon data up to relatively high momenta. From the potential model perspective, this line is a mystery: we would expect results to form an amorphous cluster around the experimental point. From the EFT point of view, instead, this line is indication that these potentials differ by one relevant parameter not determined by two-body physics. As $\Lambda_\star$ is varied, the LO EFT also produces a line [22, 15], which lies close not only to the experimental point but also to the phenomenological line. At NLO the line position changes [79], approaching models and experiment (Figure 6.6). Taking the EFT error into account, the line is actually a band. This generalizes an earlier (regulator-dependent) explanation [84].

As expected [22], this property is generic and ^{4}He potentials also fall on a Phillips line [85]. Other correlations can be understood similarly. This means that the various phenomenological potentials, with their many parameters and varied forms, are basically equivalent to the same EFT with different values of $\Lambda_\star$. For $A \geq 3$, Pionless EFT is definitely not just the ERE.

The proximity of the EFT Phillips line to the experimental point means that once one datum is used to determine $\Lambda_\star$ at LO and NLO, other three-body data can be predicted or postdicted in agreement with experiment. For example, if we fit $a_{nd,S=1/2}$, the triton binding energy $B_t = 8.0 + 0.8 + \ldots$ MeV [15, 78], compared with the measured 8.48 MeV. Alternatively, we can use the experimental value of B_t as input. Agreement with phase shifts is good already at LO [15], improves at NLO [78], and improves further still at N^{2}LO [79], where a second three-body parameter is needed as input. The three-body amplitude and various observables have been calculated fully perturbatively up to N^{2}LO for bosons in [86, 87] and for nucleons in [88]. Reviews of the state-of-the-art three-body calculations in Pionless EFT can be found in [89, 90].

6.7   Limit Cycle and Discrete Scale Invariance

The three-body force (eqn 6.75) has a surprising cut-off dependence. $H(\Lambda)$ is the solution of an unusual RG equation,

$$\mu \frac{d}{d\mu} H(\mu) \simeq 2 \left(1 + H^2(\mu)\right) \tag{6.77}$$

and is log-periodic, taking the same value for Λ and $\alpha_l \Lambda$, with

$$\alpha_l = e^{l\pi/s_0} \simeq (22.7)^l, \tag{6.78}$$

l an integer. This is an RG limit cycle. The possibility of such a cycle in QCD had been conjectured [91], and the three-body system provided the first example in a field theory [92]. Not many such examples exist—for a short review, see [93].

This force appears at LO not only for small $|a_2^{-1}|$ but also in the unitarity limit. In this limit there is no $A = 2$ dimensionful parameter at LO, yet renormalization for $A = 3$ forces on us a dimensionful parameter $\Lambda_\star$. This is an example of 'dimensional transmutation': the scale invariance present in the unitarity limit is 'anomalously' broken. The limit cycle signals a remaining discrete scale invariance (DSI) [70, 22, 15]. Because of the characteristic dependence on Λ in eqn (6.75), the three-body term in eqn (6.11) is invariant under the transformation (eqn 6.51), but only for the discrete values in eqn (6.78). Other examples of the anomalous breaking of scale invariance and of DSI with its characteristic log-periodicity are discussed in [94] and [95], respectively.

The first consequence of the breaking of scale invariance is that $\Lambda_\star$ offers a dimensionful scale for binding energies. By dimensional analysis, the three-body binding energies can be written as

$$B_{3;n} = \frac{\kappa_\star^2}{m} \beta_{3;n}\left((a_2\kappa_\star)^{-1}\right) = \frac{\kappa_\star^2}{m}\left[\beta_{3;n}(0) + \frac{\beta'_{3;n}(0)}{a_2\kappa_\star} + \dots\right], \quad \beta_{3;0}(0) = 1, \quad (6.79)$$

where $n \geq 0$ ($n = 0$ denoting the deepest state within the EFT), $\beta_{3;n}((a_2\kappa_\star)^{-1})$ are universal, dimensionless functions, and $\beta_{3;n}(0)$, $\beta'_{3;n}(0)$, etc. are pure numbers arising from an expansion in $(a_2\kappa_\star)^{-1}$. Because $\Lambda_\star$ is only defined up to a factor $\exp(n_\star\pi/s_0)$, with $n_\star$ an integer, it was traded above by a fixed scale $\kappa_\star$ defined from the ground-state energy at unitarity:

$$\ln(\kappa_\star) = \ln(\beta\,\Lambda_\star) \quad \mathrm{mod}\ \pi/s_0, \quad (6.80)$$

with $\beta \simeq 0.383$ [96].

DSI manifests itself in the spectrum. The energy of a bound state after a discrete scale transformation should equal the transformed energy but not necessarily of the same level, so that

$$\beta_{3;n+l}(0) = \alpha_l^{-2}\beta_{3;n}(0) \quad \Leftrightarrow \quad \beta_{3;n}(0) = e^{-2n\pi/s_0}\beta_{3;0}(0). \quad (6.81)$$

Thus discrete scale invariance leads to a geometric tower of states extending up to threshold, with successive states having a ratio of binding energies

$$B_{3;n+1}/B_{3;n} = \exp(-2n\pi/s_0) \simeq 1/515. \quad (6.82)$$

This amazing structure was first predicted by Efimov [10] and its signals have now been seen in cold-atom systems around Feshbach resonances [97, 98].

Away from unitarity DSI is only an approximate symmetry, even at LO. Although the deep spectrum might be little affected, a finite a_2^{-1} distorts the spectrum in the infrared (IR). Using the spurion field method, the deviation from unitarity due to the two-body scattering length can be followed,

$$\beta_{3;n+l}\left((\alpha_l a_2\kappa_\star)^{-1}\right) = \alpha_l^{-2}\beta_{3;n}\left((a_2\kappa_\star)^{-1}\right). \quad (6.83)$$

This relation gives information about how Efimov's tower evolves as $|a_2^{-1}|$ grows. For example, taking a derivative and expanding in $(a_2\kappa_\star)^{-1}$, we see the leading effect of tower deformation:

$$\beta'_{3;n+l}(0) = \alpha_l^{-1}\beta'_{3;n}(0) \quad \Leftrightarrow \quad \beta'_{3;n}(0) = e^{-n\pi/s_0}\beta'_{3;0}(0), \quad (6.84)$$

where $\beta'_{3;0}(0) \simeq 2.11$ [96]. Note that here the spurion method is not simply dimensional analysis because $\Lambda_\star$ is kept fixed. It instead tracks how the two-body scattering length explicitly breaks DSI. Equation (6.83) is only valid to the extent that the three-body force retains DSI except for $(a_2 \Lambda)^{-1}$ terms—that is, as long as eqn (6.75) contains no $a_2 \Lambda_\star$ dependence, which would require in the spurion method that we scaled $\Lambda_\star$ as well.

As $a_2^{-1} > 0$ grows the two-body bound state moves away from threshold quadratically in a_2^{-1}. Progressively more excited Efimov states fail to have energy below the particle-dimer threshold, disappearing as virtual states [82, 17]. We have the counter-intuitive situation where *fewer* three-body states survive as the two-body attraction increases. For $a_2^{-1} < 0$, three-body bound states exist even though there are no two-body bound states—the system is said to be Borromean in reference to the coat of arms of the Borromeo family, which displays three rings with the property that, when one is removed, the other two are free. The properties of the Efimov spectrum are discussed in detail in [96].

NLO corrections from the two-body effective range can be handled similarly. Since eqn (6.75) contains no $r_2 \Lambda_\star$ dependence, the coefficients of linear corrections should scale with α_l^{-3} [99], as can be easily verified with the spurion method. (However, an explicit calculation [99] indicates that these coefficients vanish.) The generalization to higher orders is obvious.

In the nuclear case, the existence of a single three-body force at LO leads to an additional approximate symmetry, which is exact in the unitarity window: independent rotations in spin and isospin, which form the SU(4)$_W$ group proposed by Wigner [100] to explain some of the properties of heavier nuclei. Away from unitarity, the symmetry is broken by the difference in inverse scattering lengths between $^3S_1$ and $^1S_0$ channels, in ranges at NLO, etc. The approximate SU(4)$_W$ symmetry of this EFT was elaborated upon in [101, 102].

6.8 More-Body Systems

Let us summarize the EFT so far. At LO, the action is given by one-body kinetic, two-body C_0 and three-body D_0 terms in eqn (6.11), and at NLO by the two-body C_2 term. Other interactions need to be accounted for at N^{2}LO—including another three-body force and, for nucleons, a two-body tensor force—and higher orders. At two-body unitarity, there is a single scale at LO, $\Lambda_\star$ (or equivalently the $\kappa_\star$ of eqn (6.80)). The crucial issue now is whether higher-body forces appear at LO. If they do, new scales will be introduced in an essential, non-perturbative way. If they do not, all low-energy properties for $A \geq 4$ can be predicted at LO, and the issue becomes whether any of these higher-body forces show up at NLO, or at another relatively low order that causes sizable distortions to the LO predictions.

A difficulty we face in answering these questions is the complexity of $A \geq 4$ calculations. In perturbative EFTs, where there is no fine tuning to dramatically enhance LO, the size of interactions can be guessed by naive dimensional analysis [51]. This rule is inferred by looking at the regulator dependence of loops in perturbation theory. In our case, we need instead to look at the regulator dependence of an integral equation at LO and of the distorted Born approximation in subleading orders, as we have just done for $A = 3$.

All $A \geq 4$ Pionless EFT calculations I am aware of are based on the numerical solution of (some version of) the many-body Schrödinger equation. At LO this is done with the potential

$$V^{(0)} = \sum_{\{ij\}} V_2^{(0)}(\vec{r}_{ij}; \Lambda) + \sum_{\{ijk\}} V_3^{(0)}(\vec{r}_{ij}, \vec{r}_{jk}; \Lambda), \qquad (6.85)$$

where $\{ij\}$ and $\{ijk\}$ denote doublets and triplets, respectively, while $V_2^{(0)}$ and $V_3^{(0)}$ are given by eqn (6.45) and eqn (6.76). If a many-body force is missing, many-body observables will not be renormalized properly, i.e. they will fail to converge as the regulator cut-off Λ increases. If there is one regulator for which lack of convergence is seen, renormalizability is disproved.

A hand-waving argument suggests that higher-body forces are *not* needed at LO for RG invariance. The two-body potential (eqn 6.45) is singular but $C_0^{(0)}(\Lambda)$ in eqn (6.48) has an overall Λ^{-1} so that the potential scales at high momentum just as the kinetic repulsion. Smaller terms $\propto (C_{0R}^{(0)}\Lambda)^{-1}$ are adjusted to give enough attraction for the two-body state to be slightly bound, or slightly virtual. When we embed the two-body potential in an A-body system, the number $A(A-1)/2$ of doublets grows faster than the number $A-1$ of kinetic terms (one term goes into the kinetic energy of the centre of mass), so the system collapses [76]. An effectively repulsive three-body force—eqn (6.76), which at high-momentum scales just as the two-body potential thanks to eqn (6.75)—is needed to keep $A = 3$ stable. Because the number $A(A-1)(A-2)/6$ of triplets grows even faster than doublets, $A \geq 4$ systems should not collapse but have instead well-defined limits for $\Lambda \gtrsim M_{\text{hi}}$.

This argument, simplistic as it is, seems to be borne out by explicit calculations. The pioneering $A = 4$ calculations in [23, 103, 104] have found convergence—at least in the range of cut-off values examined—in the binding energies of the ground state for bosons and nucleons, as well as of the first excited state for bosons. This result has been confirmed several times afterwards with various regulators [105, 106, 24, 18, 9]. Similarly, the ground-state energies of $A = 6, 16$ nucleons converge without many-nucleon forces [16, 9]. The four spin–isospin nucleon states require five- and more-body forces to include derivatives, which should suppress them. This exclusion–principle suppression is absent for bosons, but calculations of $A = 5, 6$ ground-state energies [24] showed no evidence of the need for those forces, either. In fact, binding energies show just the behaviour (eqn 6.15) expected from a properly renormalized order. Discussions found in the literature regarding this issue are summarized in [107].

The absence of LO higher-body forces has fundamental implications for the physics of $A \geq 4$ systems. One is that there are correlations among low-energy many-body observables through $\Lambda_\star$, similar to the Phillips line. The simplest example is the Tjon line [108] in the plane of the four- and three-body ground-state energies, $B_{4;0}$ and $B_{3;0}$. As with the Phillips line, this correlation was discovered empirically by plotting results of phenomenological nuclear potentials. It also exists for ^{4}He atoms [109]. It materializes in EFT as a variation of $\Lambda_\star$ at fixed two-body input [23, 103, 104]. The EFT line at LO again is close to both the phenomenological line and the experimental point, suggesting the EFT might converge for $A = 4$ as well. There is at least a very large class of potentials that do not seem to have an extra essential parameter introduced by a four-body force. The absence of higher-body forces at LO further implies the existence of 'generalized Tjon lines' in the planes spanned by other ground-state energies, for example [24], $B_{5;0}$ or $B_{6;0}$ for $A = 5, 6$, and $B_{3;0}$. Again, such correlations had been discovered earlier in the context of potential models [110, 111].

Table 6.1 summarizes existing results for binding energies of nuclei and ^{4}He atoms at LO in EFT. For nuclei [15, 16, 9, 17], we can see agreement at the level expected from an expansion where one of the parameters is $r_2/a_2 \sim 30\%$ (in the $^3S_1$ channel). Results for atomic ^{4}He obtained with potential models [19, 20, 21] and with LO EFT [22, 23, 24] are also given in Table 6.1. Here the discrepancy is no larger than $\simeq 10\%$ for $A \le 6$, which is surprising because an estimate of the binding momentum, eqn (6.2), suggests that $Q_6 \, l_{\text{vdW}} \sim 1.3$. Perhaps eqn (6.2) for Q_A is an overestimate.

Until recently, calculations for $A \ge 4$ that went beyond LO included a resummation of the NLO two-body interaction. Although they show improved results over LO for $A = 4, 16, 40$ nuclei [105, 112, 113], they are limited to cut-off values $\Lambda \lesssim M_{\text{hi}}$. A test of RG invariance requires a perturbative treatment of subleading corrections, which was carried out for bosons at NLO in [114]. Surprisingly, a four-body force is necessary and sufficient for renormalization of the $A \ge 4$ energies. Once it is fixed to $B_{4;0}$, $B_{5;0}^{[1]}$ and $B_{6;0}^{[1]}$ come out well, strengthening the case that Pionless EFT converges better than expected. The limit of validity of Pionless EFT remains an open question.

Implications of the absence of other LO forces are even stronger at unitarity, where DSI is expected to hold: except for small corrections, all states within the validity of the EFT, i.e. those states that are insensitive to the details of physics at distances $r \lesssim R$, are fixed by a *single* parameter $\Lambda_\star$. Hammer and Platter [104] discovered that for bosons an $A = 3$ state spawns two $A = 4$ states, one very close to the $A = 3$ threshold and one about four times deeper. According to the accurate calculation of [25], for the two lowest $A = 4$ states at unitarity, $B_{4;0;0}/B_{3;0} \simeq 4.611$ and $B_{4;0;1}/B_{3;0} \simeq 1.002$. These states have been spotted in atomic systems [115].

Remarkably, potential-model calculations show that this doubling process continues with increasing number of *bosons* [116, 117, 118, 119], so that for a given A there are $2^{(A-3)}$ 'interlocking' towers of states. Generalizing eqn (6.79),

$$B_{A;n;\{i\}} = \frac{\kappa_\star^2}{m} \beta_{A;n;\{i\}} \left((a_2 \kappa_\star)^{-1} \right), \tag{6.86}$$

by labelling each state with the $A = 3$ ancestor state (n) and a set $\{i\} = \{i_1, i_2, ..., i_{A-3}\}$ tracking the doubling, with $i_j = 0$ (1) denoting the lower (higher) state in the jth doubling. Just as before, the dimensionless functions $\beta_{A;n;\{i\}}$ of $(a_2 \kappa_\star)^{-1}$ reduce at unitarity to pure numbers $\beta_{A;n;\{i\}}(0)$, which obey

$$\beta_{A;n+l;\{i\}}(0) = \alpha_l^{-2} \beta_{A;n;\{i\}}(0) \quad \Leftrightarrow \quad \beta_{A;n;\{i\}}(0) = e^{-2n\pi/s_0} \beta_{A;0;\{i\}}(0). \tag{6.87}$$

Again the spurion method gives

$$\beta_{A;n+l;\{i\}}\left((\alpha_l a_2 \kappa_\star)^{-1}\right) = \alpha_l^{-2} \beta_{A;n;\{i\}}\left((a_2\kappa_\star)^{-1}\right), \tag{6.88}$$

with similar implications as for $A = 3$.

I am not aware of an explanation for the doubling, which has a topological interpretation [120], but the replicating towers are a reflection of the surviving DSI. For $A \geq 4$, all but the two lower states appear as resonances in the scattering of a particle on the $(A - 1)$-particle ground state. Because of the tower structure, we can focus on these two lower states. The higher one is near the ground state of the system with one less particle,

$$\beta_{A;0;0,\ldots,0,1}(0) \simeq \beta_{A-1;0;0,\ldots,0}(0), \tag{6.89}$$

and thus can be thought of as a two-body system of a particle and an $(A - 1)$ cluster. This state and its cousins up the tower are examples of 'halo states', such as observed in halo nuclei—states that have a clusterized structure in which a certain number of 'valence' particles orbits a tight cluster of the remaining particles.

The ground states close to unitarity get deeper and deeper as A increases. We can write

$$\frac{B_A}{A} \equiv \frac{B_{A;0;\{0\}}}{A} \simeq \frac{3}{A}\beta_{A;0;\{0\}}(0)\frac{B_{3;0}}{3} \equiv \kappa_A \frac{B_3}{3}. \tag{6.90}$$

where the set of numbers κ_A encapsulates the dynamical information about the ground states at unitarity. Relation (6.90) expresses the generalized Tjon lines [24] at unitarity.

The κ_A for $A \leq 60$ bosons have been calculated recently using Monte Carlo techniques to solve the Schrödinger equation [26], with selected results shown in Table 6.1. At small A, κ_A is approximately linear in A, as obtained earlier [116, 118, 121, 122, 123, 24], but eventually saturation sets in, where the growth tapers off (Figure 6.7). This change in behaviour is fitted well by a 'liquid-drop' formula,

$$\kappa_A = \kappa\left(1 + \eta A^{-1/3} + \ldots\right), \tag{6.91}$$

with $\kappa = 90 \pm 10$ and $\eta = -1.7 \pm 0.3$, respectively, the dimensionless 'volume' and 'surface' terms. The factor of $\simeq 90$ is large but still well below the $\simeq 515$ that provides an upper bound for the EFT breakdown scale.

Atomic ^{4}He is close enough to unitarity to sustain two trimers, an excited tetramer not too far from the ground-state trimer, and a ground-state tetramer about 4.4 deeper than the ground-state trimer. As shown in Table 6.1, $A = 5, 6$ systems have energies close to unitarity values as well. And an equation of the type of eqn (6.91) also describes calculated ground-state energies [36], yielding $\kappa \simeq 180$ and $\eta \simeq -2.7$. The energy of the bulk is thus ~ 2 away from unitarity. It is possibly beyond an EFT approach [124].

For $A > 4$ *multistate fermions* at unitarity the spectral pattern is not clear, as no calculations are available, but towers must also exist. For the ground states, eqn (6.90)

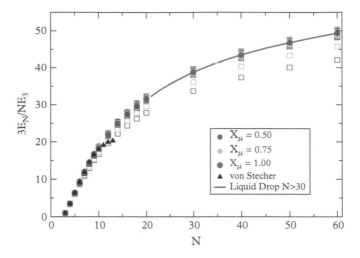

Figure 6.7 *Binding energy per particle at unitarity scaled by the three-body binding energy per particle, $\kappa_N = 3E_N/(NE_3)$, as function of the number N of bosons in a cluster. Pionless EFT results at LO for various regulator cut-offs are shown as open and filled (coloured) symbols. Black triangles show potential-model results up to $N = 15$ from [116]. The (blue) solid line shows a fit to the $N \geq 30$ (blue) points, corresponding to the largest cut-off values, using the liquid-drop formula (6.91). Reprinted figure with permission from [26]. Copyright 2017 by the American Physical Society.*

still holds, but with a different set of numbers κ_A. The four-state fermion system, for example, reduces to a bosonic system [125], similarly to the three-nucleon system [15]. κ_4 is then the same in both cases, but $\kappa_{A \geq 5}$ must differ on account of the exclusion principle and shell structure.

Although it can be seen from Table 6.1 that the two-nucleon system is not as close to unitarity as two ^{4}He atoms, the first excited state of the triton is almost bound, the alpha particle has an excited level close to the triton, and the alpha-particle ground state is about three times deeper. We argued recently [18] that nuclei are close enough to unitarity for a perturbative expansion in $(a_2\kappa_\star)^{-1}$ to converge. Reference [18] shows, for example, that the Tjon line can be obtained in perturbation theory from the Tjon line at unitarity. It would be extremely gratifying if it turns out that one can indeed devise a theory of nuclear physics based on a single parameter, plus perturbative corrections. For more speculation along these lines, see [107].

6.9 Long-Range Forces

Pionless EFT acquired a life of its own. But it was born to address the non-perturbative renormalization issues befalling Chiral EFT. Despite Pionless EFT's successes, many feel Chiral EFT should be better suited to nuclear physics, where traditionally the pion has been thought to be an important ingredient. And indeed, the binding momenta estimated with eqn (6.2) are not far below the pion mass. (How seriously one should take factors of 2 or 3 in eqn (6.2), which are important in this comparison, remains unclear.) So I conclude with a glimpse at the basic issues that confront Chiral EFT. These challenges are more basic than pion exchange, as they arise from the renormalization of singular interactions and affect also Van der Waals forces. But, since little has been done for the latter, my main focus here is on pion exchange. Much more complete accounts of Chiral EFT can be found in various reviews, e.g. [126, 127, 128].

Chiral EFT is an EFT for momenta $Q \sim m_\pi \simeq 140$ MeV, when pions appear as explicit degrees of freedom. (We might want to include also the lowest-lying nucleon excitations such as the Delta isobar, which is separated in mass from the nucleon by only $m_\Delta - m_N \simeq 300$ MeV.) As pseudo-Goldstone bosons of chiral symmetry, pions can be included through a non-linear realization of SO(4), when the isospin doublet of nucleon fields N transforms as under the SO(3) subgroup of isospin, but with 'parameters' that depend on the pion fields. The isospin triplet of pion fields π itself parametrizes the coset space S^3 of radius f_π, and always appear in the combination π/f_π. We define chiral covariant derivatives of pion and nucleon (and maybe nucleon-excitation) fields, and constructs all interactions that are isospin symmetric, which then will be automatically chiral invariant. Quark masses break chiral symmetry (including isospin) as components of certain SO(4) vectors, so we add to the action all interactions that break the symmetry as tensor products of the corresponding vectors. This procedure ensures that chiral symmetry is broken in Chiral EFT just as in QCD. Details can be found in [129].

For our purposes here, the most important terms are a subset of the leading Lagrangian: pion kinetic and mass terms, and the dominant pion–nucleon interaction, that is,

$$\Delta\mathcal{L}_\pi = \frac{1}{2}\left(\partial_\mu \boldsymbol{\pi}\, \partial^\mu \boldsymbol{\pi} - m_\pi^2 \boldsymbol{\pi}^2\right) + \frac{g_A}{f_\pi}\, N^\dagger \left(\vec{S}\cdot\boldsymbol{\tau}\cdot\vec{\nabla}\boldsymbol{\pi}\right)N + \ldots, \qquad (6.92)$$

where $\boldsymbol{\tau}$ are the Pauli matrices in isospin space and $g_A \simeq 1.27$ is a LEC, the axial-vector coupling constant. I relegate to '...' interactions with more fields, derivatives and powers of quark masses, including the chiral partners of the terms shown explicitly. The Chiral EFT Lagrangian consists of the Lagrangian in eqn (6.11) (with $\psi \to N$) supplemented by its chiral partners and the additional terms in eqn (6.92). However, since this is a different theory, the LECs of short-range interactions in eqn (6.11) have new values: they run differently and their renormalized values are, in general, also different—we have to repeat the renormalization procedure to relate them to data in the presence of pions.

The EFT still splits into sectors of different nucleon number A, but now there is more interesting physics at $A = 0, 1$ (the domain of ChPT), some of which is covered in Chapter 3 in this volume. Still, the sectors with $A \geq 2$ are richer because they involve *two* types of loops [2, 3]: the 'reducible' loops we have seen already in the context of Pionless EFT, where we can separate a diagram in at least two parts by cutting horizontally through heavy particle lines only; and 'irreducible' loops, where we must cut also through one or more pion lines. In the first type of loop, we pick a pole from one heavy particle propagator, the typical energy is $\sim Q^2/m$, and the magnitude of the contribution can be estimated by the rule (6.24). In the second type of loop, the pion propagators give energies $\sim Q$, so that nucleon propagators are in a first approximation static—recoil is suppressed by a relative $\mathcal{O}(Q/m)$ and can be included perturbatively. For irreducible loops the power counting rules are, instead, those used in ChPT [5],

$$
\begin{aligned}
&\text{pion propagator} \to Q^{-2}, \\
&\text{heavy particle propagator} \to Q^{-1}, \\
&\text{loop integral} \to Q^4/(4\pi)^2, \\
&\text{derivative, pion mass} \to Q.
\end{aligned}
\tag{6.93}
$$

Only the second type of loop is present for $A = 0, 1$. An extra irreducible loop amounts to $\mathcal{O}(Q^2/(4\pi f_\pi)^2)$ [5, 51], so an expansion is possible for $Q \sim m_\pi \ll 4\pi f_\pi$. By demanding RG invariance, one can infer the scaling of LECs known as NDA [51]. For consistency, it requires $4\pi f_\pi = \mathcal{O}(M_{\mathrm{QCD}}) = m_N$, so that both the loop and non-relativistic expansions are part of the m_π/M_{QCD} expansion.

To isolate irreducible loops when $A \geq 2$, we define the potential as the sum of these irreducible loops. In addition to the contact interactions of Pionless EFT, we have also pion exchange. The power counting (eqn 6.93) indicates that one-pion exchange (OPE) between two nucleons is the dominant long-range potential, with corrections starting *two* orders down the $M_{\mathrm{lo}}/M_{\mathrm{hi}}$ expansion [130]. A crucial difference with respect to $A = 0, 1$ is that the sum of irreducible loops does not generate the full T matrix, is not directly observable, and is *not* RG invariant in itself. The logic we follow for T is the same as for $A = 0, 1$, but we now have to infer the sizes of LECs taking into account the reducible loops, as we do in the absence of pions. Since these loops generate regulator dependence, the potential itself cannot be regulator independent.

Challenges start already at the level of OPE: in momentum space

$$
\Delta V_\pi(\vec{p}', \vec{p}) = -\frac{4\pi}{m_N M_{NN}} \, \tau_1 \cdot \tau_2 \, \frac{\vec{\sigma}_1 \cdot \vec{q} \, \vec{\sigma}_2 \cdot \vec{q}}{\vec{q}^2 + m_\pi^2} + \dots,
\tag{6.94}
$$

and in coordinate space

$$
\begin{aligned}
\Delta V_\pi(\vec{r}) = \frac{\tau_1 \cdot \tau_2}{m_N M_{NN}} \Bigg\{ & S_{12}(\hat{r}) \left[1 + m_\pi r + \frac{(m_\pi r)^2}{3} \right] \frac{e^{-m_\pi r}}{r^3} \\
& + \frac{\vec{\sigma}_1 \cdot \vec{\sigma}_2}{3} \left(-4\pi \delta^{(3)}(\vec{r}) + \frac{m_\pi^2}{r} e^{-m_\pi r} \right) \Bigg\} + \dots
\end{aligned}
\tag{6.95}
$$

Here the subscripts $_{1,2}$ label nucleons, $\vec{q} \equiv \vec{p} - \vec{p}'$ is the transferred three-momentum and $q^{02} \ll \vec{q}^2$ was neglected in the pion propagator since it is a higher-order effect. The tensor operator, defined as $S_{12}(\hat{r}) \equiv 3\vec{\sigma}_1 \cdot \hat{r} \vec{\sigma}_2 \cdot \hat{r} - \vec{\sigma}_1 \cdot \vec{\sigma}_2$, can be shown to vanish for total spin $S = 0$. For $S = 1$ it mixes waves of $l = j \pm 1$, where it has one positive and one negative eigenvalue, except for $^3P_0$ where it is diagonal with a negative eigenvalue.

It also acts on states with $l = j$, where it has a positive eigenvalue. The 'tensor force', which is highly singular, is attractive in some uncoupled waves like $^3P_0$ and $^3D_2$, and in one of the eigenchannels of each coupled wave. In contrast, $\vec{\sigma}_1 \cdot \vec{\sigma}_2$ takes the value 3 or -1 when $S = 0$ or $S = 1$, respectively. (An analogous remark holds for $\tau_1 \cdot \tau_2$ and total isospin $I = 0, 1$.) This 'central force' has two components: *i*) a contact term that can be absorbed in the two $C_0^{(0)}$ contact interactions; and *ii*) a long-range component with Yukawa form, which is attractive in isovector (isoscalar) channels for $S = 0$ ($S = 1$). In the '…' we find higher orders, which are more singular still.

In order to bring the numerical factor in eqn (6.94) to the same form used for contact interactions, we use

$$M_{NN} \equiv \frac{16\pi f_\pi^2}{g_A^2 m_N}, \tag{6.96}$$

following [48, 49]. From the rule (eqn 6.24) we expect

$$\frac{\mathcal{O}((4\pi/(m_N M_{NN}))^2 \, (m_N Q/(4\pi)))}{\mathcal{O}(4\pi/(m_N M_{NN}))} = \mathcal{O}(Q/M_{NN}) \tag{6.97}$$

for the magnitude of the ratio between the non-analytic part of once-iterated OPE and single OPE. By NDA, $M_{NN} = \mathcal{O}(f_\pi)$, but numerically $M_{NN} \simeq 290$ MeV, or about three times larger. There are also other numbers of $\mathcal{O}(1)$ floating around. If we take $M_{NN} = \mathcal{O}(M_{\text{hi}})$, then OPE is an NLO effect in an m_π / M_{NN} expansion [48, 49]. In this case LO would be formally identical to LO in Pionless EFT, perhaps explaining why the latter seems to work well even for the alpha particle where the binding momentum (eqn 6.2) is not very small compared to m_π. More generally, we can show [48, 49] that in this case the same power counting of Pionless EFT applies, with the pion mass counting as a derivative. All pion exchanges are perturbative. Results at NLO show the expected improvement over LO, but unfortunately at N^{2}LO, where the first iteration of OPE appears, all hell breaks loose: in channels where the tensor force is attractive, no convergence is found for momenta $Q \gtrsim 100$ MeV [131].

The inference is that the purely numerical factors do not in general help convergence and the NDA estimate $M_{NN} = \mathcal{O}(f_\pi)$ is realistic. For quark masses such that $m_\pi \lesssim M_{NN}$, pions are perturbative for $Q \sim m_\pi$, but for larger masses we must consider OPE as an LO potential. This has the virtue of providing a scale for binding momenta which is related to chiral symmetry: the amplitude is a series of the schematic form [126]

$$T_2^{(0)}(Q) \sim \frac{4\pi}{m_N M_{NN}} \left[1 + \frac{Q}{M_{NN}} + \left(\frac{Q}{M_{NN}} \right)^2 + \dots \right] \sim \frac{4\pi}{m_N} \frac{1}{M_{NN} - Q}, \tag{6.98}$$

allowing for a pole at $Q \sim M_{NN} = \mathcal{O}(f_\pi) \ll M_{\text{QCD}}$. The existence of shallow nuclei, except perhaps the very lightest where binding momenta are even smaller, is then tied to spontaneous chiral symmetry breaking.

Taking OPE as LO means that we need to deal with its singular nature, as for delta functions in Pionless EFT. For large momentum, the potential in eqn (6.94) is not more singular than in eqn (6.18) and, on the surface, does not seem to offer further challenges. However, the more intricate angular dependence leads to an r^{-3} behaviour at short distances instead of $\delta^{(3)}(\vec{r})$, with contributions also to waves higher than S. Stronger singularities such as r^{-5} appear at higher orders. Things remain similar to atomic systems, in that we need to deal with the r^{-6} singularity of the Van der Waals interaction if we are interested in momenta $Q \sim l_{\text{vdW}}^{-1}$. And smaller but more singular components exist in this context as well [27].

The quantum mechanics of singular interactions has a long history [132], when it was more or less agreed that an attractive singular interaction in itself is not sufficient to define the solution of the Schrödinger equation. The reason is that for a potential of the type $-\alpha^2/(2\mu r^n)$, with μ the reduced mass and $n \geq 2$, the two allowed solutions both oscillate with decreasing amplitude as the distance decreases, in contrast with a regular potential for which there are two clearly distinct solutions—regular and irregular. As a consequence, there is an undetermined phase, just like in eqn (6.67): the zero-energy $l = 0$ wave function, for example, can be shown to be given at small r by

$$\psi(r) \propto r^{n/4-1} \cos\left(\frac{\alpha}{n/2-1} r^{1-n/2} + \delta_n\right) + \dots \qquad (6.99)$$

for $n > 2$. For $n = 2$, $\alpha r^{1-n/2}/(n/2-1) \to \sqrt{\alpha^2 - 1/4}\ln(r/R)$. Equation (6.99) is parallel to eqn (6.67). (In fact, Efimov [10] first arrived at his geometric spectrum by consideration of the three-particle equation in coordinate space.) The phase δ_n can be fixed by a single counterterm regardless of the value of n [133]. For example, with a spherical-well regularization of the corresponding delta function (see Section 6.5), progressively more unphysical bound states cross threshold as the regulator distance R decreases—similarly to the Thomas collapse in the three-body system—unless the depth is adjusted to produce a value of the wave function at R that gives the phase we want. The phase in turn determines the low-energy properties of the scattering amplitude. Just like in the three-body case earlier, the LEC oscillates as $\Lambda \sim R^{-1}$ increases. However, only for $n = 2$, when the classical system is scale invariant, is the dependence periodic in $\ln \Lambda$. The RG analysis of singular potentials is discussed in [134]. From the EFT point of view, the quantum mechanics of singular potentials is just the renormalization of the LO amplitude.

For $l > 0$, the centrifugal barrier effectively suppresses the effect of the long-range potential on the amplitude by factors of l^{-1}. Only for lower waves does the long-range potential need to be iterated in the low-energy region where the EFT applies [135, 136, 137]. At subleading orders where the potential gets more singular, renormalization can still be carried out with further contact interactions [138], at least as long as corrections are treated in perturbation theory as done for Pionless EFT (Section 6.4).

In the nuclear case there are complications arising from the tensor and spin operators:

- The $^3S_1$ wave gets mixed with $^3D_1$ by the OPE tensor force. The tensor force has one attractive eigenvector, which is finite in the chiral limit, and the C_0 LEC in this channel, expected on the basis of NDA, is sufficient for renormalization at LO [139, 63].

- In spin-triplet channels where the tensor force is repulsive, OPE can be iterated without RG problems [135]. In contrast, when it is attractive, RG invariance is destroyed by iteration [135, 140]. We expect an angular-momentum suppression similar to the one seen for a central force, so that beyond a critical angular momentum l_{cr} OPE can be considered subleading. However, in lower waves like $^3S_1$-$^3D_1$ and $^3P_0$, one can argue [135, 136, 137] that OPE needs to be iterated in the low-energy region, in agreement with the findings of [131]. In these waves, additional contact interactions with derivatives, which would be expected by NDA only at higher orders, are necessary and sufficient for renormalization at LO [135].

- In the simplest spin-singlet channel, $^1S_0$, OPE takes the form of an attractive Yukawa interaction $\propto m_\pi^2$, which by itself generates a finite amplitude but is far from providing enough binding to explain the virtual state. A contact C_0 interaction must still be present at LO. However, the interference between the two interactions gives rise to a $m_\pi^2 \ln \Lambda$ term—in perturbation theory, it comes from a diagram where OPE takes place between two contact interactions, but the same regulator dependence is seen non-perturbatively [141]. Renormalization requires a chiral symmetry-breaking contact interaction with LEC $m_\pi^2 D_2$ at LO [141]. Thus this contact interaction is also enhanced with respect to NDA, and there is no straightforward chiral expansion of the contact interactions.

- In all other spin-singlet channels, the absence of a contact interaction in LO means OPE can be iterated without RG problems [135]. However, as the angular momentum l increases, factors of l^{-1} suppress its contribution, and OPE is really perturbative and thus subleading [142].

- Residual cut-off dependence indicates the need at NLO for a single two-nucleon correction from the $^1S_0$ C_2 contact interaction, treated in perturbation theory [143]—in the same way as in Pionless EFT (Section 6.4).

- Since OPE changes the asymptotic behaviour of the two-nucleon amplitude, it is not immediately obvious whether a three-body force is needed for renormalization at LO or perhaps NLO. Explicit calculations [135, 144] show it is not.

During the period while nuclear EFTs were being formulated, rapid progress was achieved in the development of methods to solve the Schrödinger equation for increasingly higher A with a given potential. While they first used purely phenomenological potentials, eventually most calculations became based on Weinberg's original suggestion [2, 3] to use 'chiral potentials', where: i) contact interactions are assumed to have sizes given by NDA; ii) the expansion of irreducible diagrams is truncated at a certain order; and iii) the truncated potential is treated exactly. The resulting amplitudes are not renormalizable and much work goes into finding the 'best' regulator to fit data with. In contrast, a properly renormalized EFT has only been explored beyond NLO in the two-nucleon sector where it has given promising results [145, 146, 147, 148, 143]. There is still much to learn about implementing corrections in perturbation theory. Despite its age, this is a field with plenty of open problems.

6.10   Summary

EFT is not only a tool for inferring new degrees of freedom and symmetries, but also for understanding the emergence of new structures. I hope to have given you a flavour of this latter aspect of EFT's power in the context of nuclear and atomic physics. Through pionless EFT, I described how renormalization in a non-perturbative setting can be very different from perturbation theory, yet sufficiently tractable for us to observe the emergence of new phenomena: the non-trivial fixed point of two-body unitarity, the limit cycle of three-body physics, and the description of larger structures from a single essential parameter. The remaining discrete scale invariance allows for many-body spectra reminiscent of Russian dolls, for ground states that saturate as the number of particles grows very large, and for a (quantum) liquid. How such a picture can be matched with Chiral EFT, where equally bizarre renormalization takes place, is a question for you to tackle.

Acknowledgments

I am grateful to Cécile DeWitt-Morette for teaching me, among other things, the value of passion and determination in science. I first met Paolo Gambino when I attended the Cargèse school 25 years ago upon her recommendation. I thank him for the invitation to Les Houches. I greatly appreciated the hospitality of the directors of the school, in particular Sacha Davidson, who gently oversaw everything, including, with extraordinary diligence and patience, the completion of these lecture notes. This work was supported in part by the U.S. Department of Energy, Office of Science, Office of Nuclear Physics, under award DE-FG02-04ER41338 and by the European Union Research and Innovation program Horizon 2020 under grant No. 654002.

References

[1] A.V. Manohar, *Introduction to Effective Field Theories*, Chapter 2 in this volume. Also at arXiv:1804.05863 [hep-ph].
[2] S. Weinberg, (1990). *Phys. Lett. B* 251: 288.
[3] S. Weinberg, (1991). *Nucl. Phys. B* 363: 3.
[4] M. Rho, (1991). *Phys. Rev. Lett.* 66: 1275.
[5] S. Weinberg, (1979). *Physica A* 96: 327.
[6] A. Pich, *Effective Field Theory with Nambu-Goldstone Modes*, Chapter 3, this volume. See also arXiv:1804.05664 [hep-ph].
[7] S. R. Beane, W. Detmold, K. Orginos, and M. J. Savage, (2011). *Prog. Part. Nucl. Phys.* 66: 1 [arXiv:1004.2935 [hep-lat]].

[8] N. Barnea, et al., (2015). *Phys. Rev. Lett.* 114: 052501 [arXiv:1311.4966 [nucl-th]].

[9] L. Contessi, et al., (2017). *Phys. Lett. B* 772: 839 [arXiv:1701.06516 [nucl-th]].

[10] V. Efimov, (1970). *Phys. Lett. B* 33: 563.

[11] M. Tanabashi et al., [Particle Data Group], (1991). *Phys. Rev. D* 98: 030001.

[12] S. Weinberg, (1977). *Trans. New York Acad. Sci.* 38: 185.

[13] C. A. Bertulani, H.-W. Hammer, and U. van Kolck, (2002). *Nucl. Phys. A* 712: 37 [nucl-th/0205063].

[14] P. F. Bedaque, H.-W. Hammer, and U. van Kolck, (2003). *Phys. Lett. B* 569: 159 [nucl-th/0304007].

[15] P. F. Bedaque, H.-W. Hammer, and U. van Kolck, (2000). *Nucl. Phys. A* 676: 357 [arXiv:nucl-th/9906032].

[16] I. Stetcu, B. R. Barrett, and U. van Kolck, (2007). *Phys. Lett. B* 653: 358 [nucl-th/0609023].

[17] G. Rupak, A. Vaghani, R. Higa, and U. van Kolck, (2019). *Phys. Lett. B* 791: 414 [arXiv:1806.01999 [nucl-th]].

[18] S. König, H. W. Grießhammer, H.-W. Hammer, and U. van Kolck, (2017). *Phys. Rev. Lett.* 118: 202501, [arXiv:1607.04623 [nucl-th]].

[19] M. H. Kalos, M. A. Lee, P. A. Whitlock, and G. V. Chester, (1981). *Phys. Rev. B* 24: 115.

[20] D. Blume and C. H. Greene, (2000). *J. Chem. Phys.* 112: 8053.

[21] E. Hiyama and M. Kamimura, (2012). *Phys. Rev. A* 85: 022502 [arXiv:1111.4370 [physics.atom-ph]].

[22] P. F. Bedaque, H.-W. Hammer, and U. van Kolck, (1999). *Nucl. Phys. A* 646: 444 [nucl-th/9811046].

[23] L. Platter, H.-W. Hammer, and U.-G. Meißner, (2004). *Phys. Rev. A* 70: 052101 [cond-mat/0404313].

[24] B. Bazak, M. Eliyahu, and U. van Kolck, (2016). *Phys. Rev. A* 94: 052502 [arXiv:1607.01509 [cond-mat.quant-gas]].

[25] A. Deltuva, (2010). *Phys. Rev. A* 82: 040701 [arXiv:1009.1295 [physics.atm-clus]].

[26] J. Carlson, S. Gandolfi, U. van Kolck, and S. A. Vitiello, (2017). *Phys. Rev. Lett.* 119: 223002 [arXiv:1707.08546 [cond-mat.quant-gas]].

[27] A. Calle Cordón and E. Ruiz Arriola, (2010). *Phys. Rev. A* 81: 044701 [arXiv:0912.1714 [cond-mat.other]].

[28] R. E. Grisenti et al., (2000). *Phys. Rev. Lett.* 85: 2284.

[29] Z.-C. Yan, J. F. Babb, A. Dalgarno, and G. W. F. Drake, (1996). *Phys. Rev. A* 54: 2824 [atom-ph/9607002].

[30] J.-Y. Zhang, et al., (2006). *Phys. Rev. A* 74: 014704 [physics/0603232].

[31] W. Cencek, et al., (2012). *J. Chem. Phys.* 136: 224303.

[32] S. Zeller, et al., (2016). *Proc. Natl. Acad. Sci. U.S.A.* 113: 14651 [arXiv:1601.03247 [physics.atom-ph]].

[33] M. Kunitski, et al., (2006). *Science* 348: 551 [arXiv:1512.02036 [physics.atm-clus]].

[34] R. A. Aziz and M. J. Slaman, (1991). *J. Chem. Phys.* 94: 8047.

[35] R. A. Aziz, et al., (1979). *J. Chem. Phys.* 70: 4330.

[36] V. R. Pandharipande, et al., (1983). *Phys. Rev. Lett.* 50: 1676.

[37] J. S. R. Chisholm, (1961). *Nucl. Phys.* 26: 469.

[38] S. Kamefuchi, L. O'Raifeartaigh, and A. Salam, (1961). *Nucl. Phys.* 28: 529.

[39] H. Georgi, (1990). *Phys. Lett. B* 240: 447.

[40] M. E. Luke and A. V. Manohar, (1992). *Phys. Lett. B* 286: 348 [hep-ph/9205228].

[41] T. D. Lee and C. N. Yang, (1962). *Phys. Rev.* 128: 885.

[42] A. Salam and J. A. Strathdee, (1970). *Phys. Rev. D* 2: 2869.

[43] T. Mannel, *Effective Field Theories for Heavy Quarks*, Chapter 9, this volume.

[44] S. König, H. W. Grießhammer, H.-W. Hammer, and U. van Kolck, (2016). *J. Phys. G* 43: 055106 [arXiv:1508.05085 [nucl-th]].

[45] S. König, (2017). *J. Phys. G* 44: 064007 [arXiv:1609.03163 [nucl-th]].

[46] S. Weinberg, (2018). *Phys. Rev. Lett.* 121: 220001.

[47] U. van Kolck, (1999). *Nucl. Phys. A* 645: 273 [arXiv:nucl-th/9808007].

[48] D. B. Kaplan, M. J. Savage, and M. B. Wise, (1998). *Phys. Lett. B* 424: 390 [nucl-th/9801034].

[49] D. B. Kaplan, M. J. Savage, and M. B. Wise, (1998). *Nucl. Phys. B* 534: 329 [nucl-th/9802075].

[50] D. R. Phillips, S. R. Beane, and M. C. Birse, (1999). *J. Phys. A* 32: 3397 [hep-th/9810049].

[51] A. Manohar and H. Georgi, (1984). *Nucl. Phys. B* 234: 189.

[52] U. van Kolck, (1998). *Lect. Notes Phys.* 513: 62 [hep-ph/9711222].

[53] J. W. Chen, G. Rupak, and M. J. Savage, (1999). *Nucl. Phys. A* 653: 386 [nucl-th/9902056].

[54] H. A. Bethe, (1949). *Phys. Rev.* 76: 38.

[55] V. G. J. Stoks, R. A. M. Klomp, C. P. F. Terheggen, and J. J. de Swart, (1994). *Phys. Rev. C* 49: 2950 [nucl-th/9406039].

[56] H. A. Bethe and R. Peierls, (1935). *Proc. Roy. Soc. Lond. A* 148: 146.

[57] H. A. Bethe and R. Peierls, (1935). *Proc. Roy. Soc. Lond. A* 149: 176.

[58] E. Fermi, (1936). *Ric. Scientifica* 7: 13.

[59] T. Köhler, K. Goral, and P. S. Julienne, (2006). *Rev. Mod. Phys.* 78: 1311 [cond-mat/0601420].

[60] T. D. Cohen, B. A. Gelman, and U. van Kolck, (2004). *Phys. Lett. B* 588: 57 [nucl-th/0402054].

[61] A. R. Janzen and R. A. Aziz, (1995). *J. Chem. Phys.* 103: 8626.

[62] M. L. Wagman, et al., (2017). *Phys. Rev. D* 96: 114510 [arXiv:1706.06550 [hep-lat]].

[63] S. R. Beane, P. F. Bedaque, M. J. Savage, and U. van Kolck, (2002). *Nucl. Phys. A* 700: 377 [nucl-th/0104030].

[64] C. R. Hagen, (1972). *Phys. Rev. D* 5: 377.

[65] T. Mehen, I. W. Stewart, and M. B. Wise, (2000). *Phys. Lett. B* 474: 145 [hep-th/9910025].

[66] B. d'Espagnat and J. Prentki, (1956). *Nuovo Cim.* 3: 1045.

[67] D. B. Kaplan, (1997). *Nucl. Phys. B* 494: 471 [nucl-th/9610052].

[68] H. W. Grießhammer, (2004). *Nucl. Phys. A* 744: 192 [nucl-th/0404073].

[69] S. R. Beane, T. D. Cohen, and D. R. Phillips, (1998). *Nucl. Phys. A* 632: 445 [nucl-th/9709062].

[70] P. F. Bedaque, H.-W. Hammer, and U. van Kolck, (1999). *Phys. Rev. Lett.* 82: 463 [nucl-th/9809025].

[71] P. F. Bedaque and U. van Kolck, (1998). *Phys. Lett. B* 428: 221 [nucl-th/9710073].

[72] P. F. Bedaque, H.-W. Hammer, and U. van Kolck, (1998). *Phys. Rev. C* 58: R641 [nucl-th/9802057].

[73] H. W. Grießhammer, (2005). *Nucl. Phys. A* 760: 110 [nucl-th/0502039].

[74] W. Dilg, L. Koester, and W. Nistler, (1971). *Phys. Lett. B* 36: 208.

[75] G. S. Danilov, (1961). *Sov. Phys. JETP* 13: 349.

[76] L. H. Thomas, (1935). *Phys. Rev.* 47: 903.

[77] E. Braaten, D. Kang, and L. Platter, (2011). *Phys. Rev. Lett.* 106: 153005 [arXiv:1101.2854 [cond-mat.quant-gas]].

[78] H.-W. Hammer and T. Mehen, (2001). *Phys. Lett. B* 516: 353 [nucl-th/0105072].

[79] P. F. Bedaque, G. Rupak, H. W. Grießhammer, and H.-W. Hammer, (2003). *Nucl. Phys. A* 714: 589 [nucl-th/0207034].

[80] R. D. Amado and J. V. Noble, (1972). *Phys. Rev. D* 5: 1992.

[81] W. Schöllkopf and J. P. Toennies, (1996). *J. Chem. Phys.* 104: 1155.

[82] S. K. Adhikari and L. Tomio, (1982). *Phys. Rev. C* 26: 83.

[83] A. C. Phillips, (1968). *Nucl. Phys. A* 107: 209.

[84] V. Efimov and E. G. Tkachenko, (1988). *Few-Body Syst.* 4: 71.

[85] V. Roudnev and M. Cavagnero, (2012). *Phys. Rev. Lett.* 108: 110402 [arXiv:1109.4656 [physics.atm-clus]].

[86] C. Ji, D. R. Phillips, and L. Platter, (2012). *Annals Phys.* 327: 1803 [arXiv:1106.3837 [nucl-th]].

[87] C. Ji and D. R. Phillips, (2013). *Few-Body Syst.* 54: 2317 [arXiv:1212.1845 [nucl-th]].

[88] J. Vanasse, (2013). *Phys. Rev. C* 88: 044001 [arXiv:1305.0283 [nucl-th]].

[89] C. Ji, (2016). *Int. J. Mod. Phys. E* 25: 1641003 [arXiv:1512.06114 [nucl-th]].

[90] J. Vanasse, (2016). *Int. J. Mod. Phys. E* 25: 1641002 [arXiv:1609.03086 [nucl-th]].

[91] K.G. Wilson, (1971). *Phys. Rev. D* 3: 1818.

[92] K.G. Wilson, (2005). *Nucl. Phys. Proc. Suppl.* 140: 3 [hep-lat/0412043].

[93] K. M. Bulycheva and A. S. Gorsky, (2014). *Phys. Usp.* 57: 171 [arXiv:1402.2431 [hep-th]].

[94] H. E. Camblong and C. R. Ordóñez, (2003). *Phys. Rev. D* 68: 125013 [hep-th/0303166].

[95] D. Sornette, (1998). *Phys. Rept.* 297: 239 [cond-mat/9707012 [cond-mat.stat-mech]].

[96] E. Braaten and H.-W. Hammer, (2006). *Phys. Rept.* 428: 259 [cond-mat/0410417].

[97] T. Kraemer et al., (2006). *Nature* 440: 315 [cond-mat/0512394].

[98] B. Huang et al., (2014). *Phys. Rev. Lett.* 112: 190401 [arXiv:1402.6161 [cond-mat.quant-gas]].

[99] L. Platter, C. Ji, and D. R. Phillips, (2009). *Phys. Rev. A* 79: 022702 [arXiv:0808.1230 [cond-mat.other]].

[100] E. Wigner, (1937). *Phys. Rev.* 51: 106.

[101] T. Mehen, I. W. Stewart, and M. B. Wise, (1999). *Phys. Rev. Lett.* 83: 931 [hep-ph/9902370].

[102] J. Vanasse and D. R. Phillips, (2017). *Few-Body Syst.* 58: 26 [arXiv:1607.08585 [nucl-th]].

[103] L. Platter, H.-W. Hammer, and U.-G. Meißner, (2005). *Phys. Lett. B* 607: 254 [nucl-th/0409040].

[104] H.-W. Hammer and L. Platter, (2007). *Eur. Phys. J. A* 32: 113 [nucl-th/0610105].

[105] J. Kirscher, H. W. Grießhammer, D. Shukla, and H. M. Hofmann, (2010). *Eur. Phys. J. A* 44: 239 [arXiv:0903.5538 [nucl-th]].

[106] J. Kirscher, et al., (2015). *Phys. Rev. C* 92: 054002 [arXiv:1506.09048 [nucl-th]].

[107] U. van Kolck, (2017). *Few-Body Syst.* 58: 112.

[108] J. A. Tjon, (1975). *Phys. Lett. B* 56: 217.

[109] S. Nakaichi, Y. Akaishi, H. Tanaka, and T. K. Lim, (1978). *Phys. Lett. A* 68: 36.

[110] S. Nakaichi, T. K. Lim, Y. Akaishi, and H. Tanaka, (1979). *J. Chem. Phys.* 71: 4430.

[111] T. K. Lim, S. Nakaichi, Y. Akaishi, and H. Tanaka, (1980). *Phys. Rev. A* 22: 28.

[112] V. Lensky, M. C. Birse, and N. R. Walet, (2016). *Phys. Rev. C* 94: 034003 [arXiv:1605.03898 [nucl-th]].

[113] A. Bansal, et al., (2018). *Phys. Rev. C* 98: 054301 [arXiv:1712.10246 [nucl-th]].

[114] B. Bazak, et al., (2019) *Phys. Rev. Lett.* 122: 143001 [arXiv:1812.00387 [cond-mat.quant-gas]].

[115] F. Ferlaino, et al., (2009). *Phys. Rev. Lett.* 102: 140401 [arXiv:0903.1276 [cond-mat.other]].

[116] J. von Stecher, (2010). *J. Phys. B* 43: 101002 [arXiv:0909.4056 [cond-mat.quant-gas]].

[117] M. Gattobigio, A. Kievsky, and M. Viviani, (2011). *Phys. Rev. A* 84: 052503 [arXiv:1106.3853 [physics.atm-clus]].

[118] J. von Stecher, (2011). *Phys. Rev. Lett.* 107: 200402 [arXiv:1106.2319 [cond-mat.quant-gas]].

[119] M. Gattobigio, A. Kievsky, and M. Viviani, (2012). *Phys. Rev. A* 86: 042513 [arXiv:1206.0854 [physics.atm-clus]].

[120] Y. Horinouchi and M. Ueda, (2016). *Phys. Rev. A* 94: 050702 [arXiv:1603.05328 [cond-mat.quant-gas]].

[121] A. N. Nicholson, (2012). *Phys. Rev. Lett.* 109: 073003 [arXiv:1202.4402 [cond-mat.quant-gas]].

[122] A. Kievsky, N. K. Timofeyuk, and M. Gattobigio, (2014). *Phys. Rev. A* 90: 032504 [arXiv:1405.2371 [cond-mat.quant-gas]].

[123] Y. Yan and D. Blume, (2015). *Phys. Rev. A* 92: 033626 [arXiv:1508.00081 [cond-mat.quant-gas]].

[124] A. Kievsky, A. Polls, B. Juliá-Díaz, and N. K. Timofeyuk, (2017). *Phys. Rev. A* 96: 040501 [arXiv:1707.05628 [cond-mat.quant-gas]].

[125] L. Platter, (2005). PhD dissertation, University of Bonn.

[126] P. F. Bedaque and U. van Kolck, (2002). *Ann. Rev. Nucl. Part. Sci.* 52: 339 [nucl-th/0203055].

[127] E. Epelbaum, H.-W. Hammer, and U.-G. Meißner, (2009). *Rev. Mod. Phys.* 81: 1773 [arXiv:0811.1338 [nucl-th]].

[128] R. Machleidt and D. R. Entem, (2011). *Phys. Rept.* 503: 1 [arXiv:1105.2919 [nucl-th]].

[129] S. Weinberg, (2013). *The Quantum Theory of Fields: Modern applications* Vol. 2, Cambridge University Press.

[130] C. Ordóñez and U. van Kolck, (1992). *Phys. Lett. B* 291: 459.

[131] S. Fleming, T. Mehen, and I. W. Stewart, (2000). *Nucl. Phys. A* 677: 313 [nucl-th/9911001].

[132] W. Frank, D. J. Land, and R. M. Spector, (1971). *Rev. Mod. Phys.* 43: 36.

[133] S. R. Beane, et al., (2001). *Phys. Rev. A* 64: 042103 [quant-ph/0010073].

[134] M. Pavón Valderrama and E. Ruiz Arriola, (2008). *Annals Phys.* 323: 1037 [arXiv:0705.2952 [nucl-th]].

[135] A. Nogga, R. G. E. Timmermans, and U. van Kolck, (2005). *Phys. Rev. C* 72: 054006 [nucl-th/0506005].

[136] M. C. Birse, (2006). *Phys. Rev. C* 74: 014003 [nucl-th/0507077].

[137] S. Wu and B. Long, arXiv:1807.04407 [nucl-th].

[138] B. Long and U. van Kolck, (2008). *Annals Phys.* 323: 1304 [arXiv:0707.4325 [quant-ph]].

[139] T. Frederico, V. S. Timóteo, and L. Tomio, (1999). *Nucl. Phys. A* 653: 209 [nucl-th/9902052].

[140] M. Pavón Valderrama and E. Ruiz Arriola, (2006). *Phys. Rev. C* 74: 064004 [Erratum: Phys. Rev. C 75: 059905] [nucl-th/0507075].

[141] D. B. Kaplan, M. J. Savage, and M. B. Wise, (1996). *Nucl. Phys. B* 478: 629 [nucl-th/9605002].

[142] M. Pavón Valderrama, et al., (2017). *Phys. Rev. C* 95: 054001 [arXiv:1611.10175 [nucl-th]].

[143] B. Long and C.-J. Yang, (2012). *Phys. Rev. C* 86: 024001 [arXiv:1202.4053 [nucl-th]].

[144] Y.-H. Song, R. Lazauskas, and U. van Kolck, (2017). *Phys. Rev. C* 96: 024002 [arXiv:1612.09090 [nucl-th]].

[145] M. Pavón Valderrama, (2011). *Phys. Rev. C* 83: 024003 [arXiv:0912.0699 [nucl-th]].

[146] M. Pavón Valderrama, (2011). *Phys. Rev. C* 84: 064002 [arXiv:1108.0872 [nucl-th]].

[147] B. Long and C.-J. Yang, (2011). *Phys. Rev. C* 84: 057001 [arXiv:1108.0985 [nucl-th]].

[148] B. Long and C.-J. Yang, (2012). *Phys. Rev. C* 85: 034002 [arXiv:1111.3993 [nucl-th]].

7

Effective Field Theory of Large-Scale Structure

Tobias BALDAUF

Department for Applied Mathematics and Theoretical Physics
Centre for Theoretical Cosmology
University of Cambridge

Tobias BALDAUF

Baldauf, T., *Effective Field Theory of Large-Scale Structure* In: *Effective Field Theory in Particle Physics and Cosmology*. Edited by:
Sacha Davidson, Paolo Gambino, Mikko Laine, Matthias Neubert and Christophe Salomon, Oxford University Press (2020).
© Oxford University Press. DOI: 10.1093/oso/9780198855743.003.0007

Chapter Contents

7.1   Introduction

So far, the observed state of the Universe is compatible with the concordance ΛCDM model, where around 70% of the energy content of the Universe is in the form of dark energy, 25 % in the form of dark matter, and only 5% in the form of baryonic matter. The fluctuations follow a spectrum resembling the one predicted by single-field, slow-roll inflation and are very Gaussian. However, there are still many open questions, which we would like to answer:

- What is causing the current accelerated expansion of the Universe? Is it simply a cosmological constant, is it a scalar field or do we need to modify Einstein's gravity?
- What is the nature of dark matter?
- What are the dynamics and field content of inflation?
- What is the mass of the neutrinos, which of the two hierarchies applies?
- How did the rich structures in our Universe arise from the small initial perturbations and how can we understand the process analytically?

In contrast to particle physics, where a particle collision can be repeated many times, we have only one Universe at our disposal and need to extract as much information as possible from the observables that are available to us:

- The cosmic microwave background (CMB)–temperature fluctuations, polarization (lensing and primordial B-modes) and spectral distortions
- Large-scale structure (LSS)–galaxy positions and weak gravitational lensing of the photons emitted by background galaxies or the CMB
- 21 cm intensity mapping–measuring the density of neutral hydrogen using the 21 cm hyperfine transition
- Ly-alpha forest–absorbtion of redshifted Ly-alpha photons by gas along the line of sight

In these lectures we will be considering the evolution of fluctuations in the Universe after recombination. The growth of perturbations leads to the breakdown of linear evolution equations and we will need to consider a perturbative approach to describe the evolution of densities, displacements and velocities. The basics of clustering statistics and perturbation theory are nicely summarized in the review [11]. In the end of the chapter we will go beyond the dynamically dominating dark matter component and consider models that allow us to describe observable tracers of LSS, such as galaxies. The relation between the underlying matter distribution and the tracer distribution is encoded in bias models, which were recently reviewed by [16].

Why would we want to go through this trouble? The analysis of the mostly linear CMB fluctuations has tightly constrained the six parameters of the minimal ΛCDM model and further improvements are to be expected from small-scale experiments, lensing, and

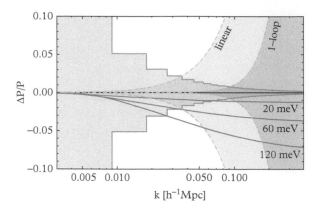

Figure 7.1 *Ratio of various corrections to the non-linear power spectrum and the linear power spectrum. On large scales statistical inference is hindered by the small number of modes available and leads to the statistical error bar, cosmic variance, shown by the gray shaded region. On small scales the analysis of LSS needs to take non-linear clustering into account, such that the tree level calculation has errors shown by the light red region and the one-loop calculation has uncertainties shown by the darker red region. Theoretically interesting signals, such as the suppression of power due to massive neutrinos, can only be detected if they are outside both statistical and theoretical error regions.*

polarization measurements. However, many of the constraints are limited by the number of available modes ($N_{CMB} \approx l_{max}^2 \approx 10^6$) in the CMB and tighter constraints will require the exploration of additional independent modes. The number of independent Fourier modes is relevant as we are essentially trying to measure the variance of fluctuations given a finite sample and the error of that measurement scales as $N^{-1/2}$. LSS, the distribution of matter and galaxies in the low redshift universe ($z < 6$), can provide additional information, especially if we are able to increase survey volume and the maximum wave-number up to which we can reliably analyze the data. The number of modes then scales as $N_{LSS} \approx (k_{max}/k_{min})^3 \approx 10^8$ up to $z = 2.5$ and $k_{max} = 0.3\ h\mathrm{Mpc}^{-1}$. The relative statistical and theoretical errors as well as the imprint of interesting physics of massive neutrinos on the two-point function in Fourier space are shown in Figure 7.1. While numerical simulations can help to model the LSS statistics, they are expensive and by themselves too slow for Markov chain Monte Carlo sampling of the cosmological parameter space. In these notes, we thus describe a theoretical model that allows for a perturbative, model independent, fast calculation of LSS statistics while accounting for theoretical uncertainties.

7.2 Random fields in 3D Space

The currently favoured model for creating the fluctuations in the Universe is cosmological inflation, a quantum mechanical process. Due due its quantum nature, it is impossible to predict the exact fluctuation pattern realized in our Universe, but only its statistical properties. There are many different realizations of the Universe that are statistically equivalent. To compare theory and observation, we need to measure statistical properties of the Universe and compare the result to the same statistics evaluated in the model of the Universe under consideration. The statistical approach to the matter distribution in the late time Universe is described in great detail in [24].

The cosmological principle asserts that on large enough scales the Universe is isotropic and homogeneous. We demand that the statistics are:

- invariant under rotations ⇒ *statistical isotropy*,
- invariant translations ⇒ *statistical homogeneity*.

We start to consider random fields in three dimensional space. As these fields describe deviations from or fluctuations around the background Universe, we only consider mean zero fields $\langle f \rangle = 0$. For instance, we can consider the temperature fluctuations in the CMB

$$\theta = \frac{T}{\bar{T}} - 1 = \frac{\Delta T}{T} \tag{7.1}$$

or the the *overdensity*

$$\delta = \frac{\rho}{\bar{\rho}} - 1 = \frac{n}{\bar{n}} - 1 \tag{7.2}$$

where ρ is a density and $n = N/V$ is a number density as appropriate for discrete objects.

7.2.1 Power Spectrum

Correlators are expectation values of products of field values at different spatial locations (or different Fourier modes). The most prominent two-point statistic is the *power spectrum*[1]

$$\langle \delta(\mathbf{k})\delta(\mathbf{k}') \rangle = (2\pi)^3 \delta^{(\mathrm{D})}(\mathbf{k} + \mathbf{k}') P(|\mathbf{k}|). \tag{7.3}$$

By statistical isotropy the power spectrum may depend only on the magnitude of $\mathbf{k}$. The Dirac delta function ensures that the power spectrum is invariant under spatial translations. We will show this in more detail below. Distinct Fourier modes are uncorrelated and thus statistically independent. The above definition gives the power per unit Fourier space volume $\mathrm{d}^3 k/(2\pi)^3$; sometimes we define the power in a shell or *dimensionless power spectrum*

---

[1] The prefactor of $(2\pi)^3$ is due to the Fourier convention used here, for other conventions it might be absent or arise with an exponen of 3/2 instead of 3.

$$\Delta^2(k) = \frac{k^3 P(k)}{2\pi^2}, \tag{7.4}$$

such that $\Delta^2(k)\mathrm{d}\ln k = P(k)\mathrm{d}^3k/(2\pi)^3$ after integrating out the azimuthal angle. For real valued configuration space fields we have $\delta(k) = \delta^*(-k)$, and thus $\langle\delta(k)\delta^*(k')\rangle = \langle\delta(k)\delta(-k')\rangle$, which is why we might sometimes encounter a different definition of the power spectrum

$$\langle\delta(k)\delta^*(k')\rangle = (2\pi)^3\delta^{(D)}(k - k')P(|k|). \tag{7.5}$$

7.2.2   Correlation Function

The real space equivalent of the power spectrum is the *correlation function*

$$\langle\delta(x)\delta(x+r)\rangle = \xi(|r|). \tag{7.6}$$

By statistical homogeneity the correlation function can only depend on the difference of the positions $x+r$ and x and statistical isotropy enforces dependence on the magnitude only. Correlation function and power spectrum are related by a simple Fourier transformation and thus contain the same information. However, they do have their respective advantages and disadvantages, both from a theoretical point of view regarding how apparent certain features are but also in terms of estimating the statistic.

An alternative interpretation of the correlation function defined above can be found in terms of the multipoint probability distribution functions [24]. We consider the background density field to be traced by a certain species with number density $\bar{n}$ and consider infinitesimally small volumes δV, which either host or don't host one tracer particle. This setup is sketched in Figure 7.2. The one-point probability for finding a particle in the small volume δV_1 is $\mathbb{P}_{1\mathrm{pt}}(1) = \bar{n}\delta V_1$. If we had a purely random field the joint or two point probability of finding particles both in volumes δV_1 and δV_2 separated by $r_{12} = |x_1 - x_2|$ would be given by the product of the independent probabilities

$$\mathbb{P}_{2\mathrm{pt}}(1,2) = \mathbb{P}_{1\mathrm{pt}}(1)\mathbb{P}_{1\mathrm{pt}}(2) = \bar{n}^2\delta V_1\delta V_2. \tag{7.7}$$

For a correlated sample the probabilities will no longer be independent and the correlation function can now be defined as the excess over random probability of finding two particles in volumes δV_1 and δV_2 separated by r_{12}

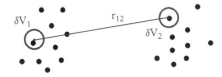

Figure 7.2 *Clustering in a point distribution with a preferred distance, i.e. correlation length.*

$$\mathbb{P}_{2pt}(1,2) = \bar{n}^2 \left[1 + \xi(r_{12})\right] \delta V_1 \delta V_2. \tag{7.8}$$

Since the probability of having a particle in δV_1 is given by $\bar{n}\delta V_1$, we can write the conditional probability to find a particle in δV_2 given there is one in δV_1

$$\mathbb{P}_{1pt}(2|1) = \frac{\mathbb{P}_{2pt}(1,2)}{\mathbb{P}_{1pt}(1)} = \bar{n}[1 + \xi(r_{12})]\delta V_2, \tag{7.9}$$

where we used Bayes theorem for the conditional probability. So we see that for correlated samples ($\xi(r_{12}) > 0$) the probability of finding a second particle is enhanced over random, whereas it is suppressed over random for the anti-correlated case ($\xi(r_{12}) < 0$). We can straightforwardly deduce a correlation function estimator from the above equation: count the number of neighbours in a shell of volume V around a given particle and compare to the expected number of pairs in a random field $\bar{n}V$.

We can also consider moments of the fields, which are products of the field at the same spatial location, for instance, the *variance* of the field

$$\sigma_R^2 = \left\langle \delta_R^2 \right\rangle = \xi_R(r = 0). \tag{7.10}$$

Here the subscript R symbolizes a smoothing on a spatial scale R. We relate the variance to the power spectrum shortly but note here that it is an important cosmological parameter describing the typical amplitude of fluctuations, which by convention is quantified in spheres of radius $R = 8\ h^{-1}$Mpc and amounts to about $\sigma_8 \approx 0.8$.

We understand the Universe to be a realization of an ensemble of universes and thus expectation values are understood as averages over many realizations. As we have only one Universe at our disposal to perform measurements we need to make use of *ergodic theorem* to replace ensemble averages by spatial averages. That the two agree is non-trivial and only the case if the correlation of field values vanishes in the large distance limit.

7.2.3   Higher-Order Correlators

In the simple case of linearly evolved fluctuations from single field inflation, the power spectrum or two-point function is sufficient to describe the statistics. However, deviations from the simple inflationary models, non-linear evolution, or the consideration of tracers of the cosmic density distribution will require higher-order correlators. The simplest of those is the *bispectrum*

$$\langle \delta(\boldsymbol{k}_1)\delta(\boldsymbol{k}_2)\delta(\boldsymbol{k}_3) \rangle = (2\pi)^3 \delta^{(D)}(\boldsymbol{k}_1 + \boldsymbol{k}_2 + \boldsymbol{k}_3)B(k_1, k_2, k_3),. \tag{7.11}$$

Statistical homogeneity leads to the delta function that forces the three wave vectors to form a triangle and due to statistical isotropy the triangle can be fully described by three lengths or two lengths and one enclosed angle. The Fourier transform of the bispectrum is the three-point correlation function $\zeta = \mathrm{FT}(B)$, which in the above probabilistic interpretation can be defined as

$$\mathbb{P}_{3\text{pt}}(1,2,3) = \bar{n}^3 \left[1 + \xi(r_{12}) + \xi(r_{23}) + \xi(r_{31}) + \zeta(r_{12}, r_{23}, r_{31}) \right] \delta V_1 \delta V_2 \delta V_3. \quad (7.12)$$

Adding yet another field we obtain the *trispectrum*

$$\langle \delta(\mathbf{k}_1) \delta(\mathbf{k}_2) \delta(\mathbf{k}_3) \delta(\mathbf{k}_4) \rangle = (2\pi)^3 \delta^{(\text{D})}(\mathbf{k}_1 + \mathbf{k}_2 + \mathbf{k}_3 + \mathbf{k}_4) T(k_1, k_2, k_3, k_4, |\mathbf{k}_1 + \mathbf{k}_2|,$$
$$|\mathbf{k}_2 + \mathbf{k}_3|), \quad (7.13)$$

which we encounter again when discussing cosmic variance later.

7.3 Fourier Space Conventions

It will prove convenient to build up the actual density field from a superposition of modes that describe the behaviour on a certain scale.

We introduce the following Fourier convention:

$$\delta(\mathbf{k}) = \int \mathrm{d}^3 r \exp\left[\mathrm{i}\mathbf{k} \cdot \mathbf{r}\right] \delta(\mathbf{r}), \quad (7.14)$$

$$\delta(\mathbf{r}) = \int \frac{\mathrm{d}^3 k}{(2\pi)^3} \exp\left[-\mathrm{i}\mathbf{k} \cdot \mathbf{r}\right] \delta(\mathbf{k}). \quad (7.15)$$

The k-space representation of the nabla operator is given by $\nabla \rightarrow -\mathrm{i}\mathbf{k}$. More often than not, the configuration space fields will be real, leading to $\delta^*(\mathbf{k}) = \delta(-\mathbf{k})$. Under a spatial shift $\mathbf{x} \rightarrow \mathbf{x} + \mathbf{\Delta x}$ the Fourier modes transform as

$$\delta(\mathbf{k}) \rightarrow \exp\left[\mathrm{i}\mathbf{k} \cdot \mathbf{\Delta x}\right] \delta(\mathbf{k}) \quad (7.16)$$

and the power spectrum thus transforms as

$$\delta(\mathbf{k}) \delta(\mathbf{k}') \rightarrow \delta(\mathbf{k}) \delta(\mathbf{k}') \exp\left[\mathrm{i}(\mathbf{k} + \mathbf{k}') \cdot \mathbf{\Delta x}\right], \quad (7.17)$$

thus invariance under translations obviously requires $\mathbf{k}' = -\mathbf{k}$ and thus $\delta^{(\text{D})}(\mathbf{k} + \mathbf{k}')$. An equivalent argument can be made for higher n-point functions.

The Dirac delta function is thus given by

$$\delta^{(\text{D})}(\mathbf{x} + \mathbf{x}') = \int \frac{\mathrm{d}^3 q}{(2\pi)^3} \exp\left[\mathrm{i}(\mathbf{x} + \mathbf{x}')\mathbf{q}\right]. \quad (7.18)$$

In particular, for finite volumes this leads to

$$\delta^{(\text{D})}(\mathbf{k} - \mathbf{k}') = \frac{V}{(2\pi)^3} \delta^{(\text{K})}_{\mathbf{k}, \mathbf{k}'}. \quad (7.19)$$

An important advantage of working in Fourier space is that convolutions in real space become simple multiplications in k-space

$$f(x) = \int \mathrm{d}^3 y\, g(y) h(x-y) \Rightarrow f(k) = g(k) h(k). \tag{7.20}$$

This is of particular advantage when *smoothing* operations are considered.

$$f_R(x) = \int \mathrm{d}^3 y\, W_R(y) f(x-y) \Rightarrow f_R(k) = f(k) W_R(k). \tag{7.21}$$

For the variance of spheres in real space we consider a spatial *top-hat filter* $W_{\mathrm{TH},R}(r) = 3\theta(R-r)/4\pi R^3$ leading to

$$W_{\mathrm{TH},R}(k) = 3\,\frac{\sin(kR) - (kR)\cos(kR)}{(kR)^3} \tag{7.22}$$

In case of spherical symmetry we can perform the angular integration in the definition of the Fourier transform

$$f(r) = \frac{1}{2\pi^2} \int \mathrm{d}k\, k^2 \frac{\sin kr}{kr} f(k) = \frac{1}{2\pi^2} \int \mathrm{d}k\, k^2 j_0(kr) f(k) \tag{7.23}$$

where j_0 is the spherical Bessel function. For the inverse transform this yields

$$f(k) = 4\pi \int \mathrm{d}r\, r^2 j_0(kr) f(r). \tag{7.24}$$

The correlation function can now be expressed as

$$
\begin{aligned}
\xi(r) &= \int \frac{\mathrm{d}^3 k}{(2\pi)^3} \int \frac{\mathrm{d}^3 k'}{(2\pi)^3} \left\langle \delta(k)\delta(k') \right\rangle \exp\left[-i k \cdot x\right] \exp\left[-i k' \cdot (x+r)\right] \\
&= \int \frac{\mathrm{d}^3 q}{(2\pi)^3} P(k) \exp\left[-i k' \cdot r\right] = \frac{1}{2\pi^2} \int \mathrm{d}k\, k^2 P(k) j_0(kr).
\end{aligned}
\tag{7.25}
$$

In the other direction we have

$$
\begin{aligned}
\left\langle \delta(k)\delta(k') \right\rangle &= \int \mathrm{d}^3 x \int \mathrm{d}^3 x'\, \exp\left[i k \cdot x\right] \exp\left[i k' \cdot x\right] \left\langle \delta(x)\delta(x') \right\rangle \\
&= (2\pi)^3 \delta^{(\mathrm{D})}(k+k') \int \mathrm{d}^3 r\, \exp\left[i k' \cdot r\right] \xi(r).
\end{aligned}
\tag{7.26}
$$

Since the last line has the same form as the definition of the power spectrum, the power spectrum is in turn related to the correlation function by

$$P(k) = \int d^3 r \xi(r) \exp\left[i\boldsymbol{k} \cdot \boldsymbol{r}\right] = 4\pi \int dr r^2 \xi(r) j_0(kr). \tag{7.27}$$

Let us check that the transformations actually close.

$$P(k) = 4\pi \int \frac{dk k^2}{2\pi^2} P(k') \int dr r^2 j_0(kr) j_0(k'r) = P(k), \tag{7.28}$$

where we used the closure relation for spherical Bessel functions

$$\int_0^\infty dx x^2 j_\alpha(ux) j_\alpha(vx) = \frac{\pi}{2u^2} \delta^{(D)}(u-v). \tag{7.29}$$

Let us finally consider the *variance* of the smoothed density field

$$\sigma_R^2 = \left\langle \delta_{\mathrm{TH},R}^2 \right\rangle = \frac{1}{2\pi^2} \int dk k^2 P(k) W_{\mathrm{TH},R}^2(k). \tag{7.30}$$

7.4   Shape of the Matter Power Spectrum

The shape of the matter power spectrum is given by linear transformations of the power spectrum generated by inflation. It can be shown that the power spectrum of the metric perturbations scales as

$$P_{\zeta\zeta}(k) \propto k^{n_{\mathrm{s}}-4} \tag{7.31}$$

where n_{s} is proportional to the slow-roll parameters and observed to be close to unity $n_{\mathrm{s}} \approx 0.967$. The modes seeded during inflation are subject to different total growth rates depending on whether they re-entered the horizon before or after matter-radiation equality. To account for this fact it is convenient to introduce the *transfer function* $T(k)$, which can be implicitly defined as a relation between the late matter power spectrum after recombination and the seed fluctuations produced by inflation

$$P_{\delta\delta}(k) \propto T^2(k) k^4 P_{\zeta\zeta}(k) \propto T^2(k) k^{n_{\mathrm{s}}}, \tag{7.32}$$

where we used that inside the horizon, the Poisson equation yields $k^2\zeta \propto k^2\phi \propto \delta$. Depending on the wavelength of the fluctuations we can thus distinguish two qualitatively different regimes:

- outside the horizon, potential fluctuations are constant $\Rightarrow$ transfer function constant on large scales

- modes that enter during radiation domination have a supressed growth $\Rightarrow$ transfer function is supressed on small scales

The transfer function can be parametrized as

$$T^2(k) \propto \begin{cases} 1 & k < k_{\text{eq}} \\ \frac{k_{\text{eq}}^4}{k^4}\left(1+\ln\left(\frac{k}{k_{\text{eq}}}\right)\right)^2 & k > k_{\text{eq}}. \end{cases} \tag{7.33}$$

The expansion factor at equality is given by

$$a_{\text{eq}} = \frac{\Omega_{\text{r},0}}{\Omega_{\text{m},0}} \tag{7.34}$$

and we can use $\rho_\text{r} = \frac{\pi^2}{15} T_{\text{CMB}}^4$ to calculate the radiation density parameter $\Omega_{\text{r},0} \approx 8 \times 10^{-5}$ yielding $a_{\text{eq}} \approx 3 \times 10^{-4}$. The horizon wave number at equality is then given by $k_{\text{eq}}/\mathcal{H}_{\text{eq}} = 1$, giving $k_{\text{eq}} \approx 0.015 h^{-1}\text{Mpc}$. An example of the power spectrum as well as its low- and high-k limits is given in Figure 7.3. The power spectrum thus contains information about the initial conditions, which are unperturbed on large scales as well as information on the various components affecting the suppression of growth on small scales. For example, the mass of neutrinos directly affects the matter transfer function and allows us to put constraints on neutrino mass that are close to the minimal mass

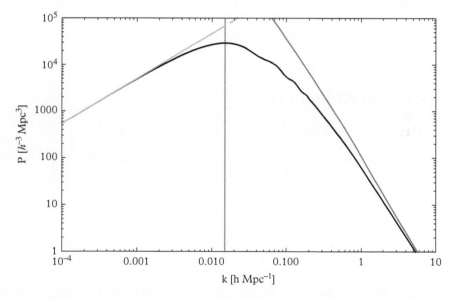

Figure 7.3 *Linear power spectrum of the ΛCDM model and its low- and high-k approximations.*

required from neutrino oscillation experiments. The full shape of the transfer function can be obtained using Boltzmann solvers such as CMBfast, CAMB, or CLASS.

We now focus on the impact and relevance of the *baryon acoustic oscillation* (BAO) feature in the correlation function and power spectrum. In the correlation function the BAO shows up as a distinct feature at $r_{BAO} \approx 100\ h^{-1}$Mpc. Let us for simplicity consider the BAO to be a Dirac delta function in one spatial dimension and calculate the corresponding power spectrum

$$P_{1D}(k) = \int dx \left[\delta^{(D)}(x - r_{BAO}) + \delta^{(D)}(x + r_{BAO}) \right] \exp[ikx] = 2\cos(kr_{BAO}). \quad (7.35)$$

We thus see that the wiggles in the power spectrum correspond to a spatially localized feature in the correlation function. In reality the BAO gets broadended at a scale w in the correlation function and the corresponding wiggles in the power spectrum are suppressed at high wave numbers

$$
\begin{aligned}
P(k) &= \int dx\, \frac{1}{\sqrt{2\pi}\, w} \left\{ \exp\left[-\frac{(x - r_{BAO})^2}{2w^2} \right] + \exp\left[-\frac{(x + r_{BAO})^2}{2w^2} \right] \right\} \exp[ikx] \\
&= 2\cos(kr_{BAO}) \exp\left[-\frac{1}{2} k^2 w^2 \right].
\end{aligned}
\quad (7.36)
$$

The BAO scale is very well known from CMB observations and thus forms a standard ruler in LSS. When observing this scale in transverse galaxy clustering we are probing the angular diameter distance and when observing it along the line of sight, we are probing the Hubble rate. Both of these quantities depend on the expansion history of the Universe and thus on the dark energy equation of state $w = p/\rho$. Constraining the dark energy equation of state at the 10% level requires 1% precision in the measurement of distances or $H(z)$.

7.5 Gaussian Random Fields

7.5.1 Probability and Characteristic Function

A vector $f = [f_1, \ldots, f_N]$ of random variables is called Gaussian if the joint probability density function (PDF) is a multivariate Gaussian

$$\mathbb{P}(f) = \frac{1}{\sqrt{(2\pi)^N |C|}} \exp\left[-\frac{1}{2} f_i C_{ij}^{-1} f_j \right], \quad (7.37)$$

where the positive definite, symmetric $N \times N$-matrix $C_{ij} = \langle f_i f_j \rangle$ is called the *covariance matrix*.

A random field $f : \mathbb{R}^3 \to \mathbb{R}$ is a *Gaussian random field* (GRF) if for arbitrary collections of field points $(x_1, \ldots, x_n)$ the variables $[f(x_1), \ldots, f(x_N)]$ are joint Gaussian variables.

For the GRF, the PDF can be expressed as a Gaussian functional of f but for practical purposes we will often work with a finite set of tracer points or pixels and denote $f_i = f(x_i)$.

For a GRF, by statistical homogeneity C can only depend on the separation $x_i - x_j$ and by statistical isotropy only on the magnitude $|x_i - x_j|$. As $f(k)$ is linear in $f(x)$, the PDF of the Fourier modes $f(k)$ is Gaussian as well. As you can show yourself, irrelevant degrees of freedom can be integrated out from the PDF

$$\mathbb{P}(f_1,\ldots,f_{N-M}) = \left\{ \prod_{i=N-M+1}^{n} \int df_i \right\} \mathbb{P}(f) \tag{7.38}$$

The simplest inflationary models predict Gaussian primordial fluctuations, but there are distinct models that can predict non-Gaussian features. This property is conserved by linear evolution. As the (primary) CMB is very nearly linear in the initial fluctuations the fluctuations look very Gaussian, there are however small non-Gaussian deviations imprinted by gravitational lensing of CMB photons by the large-scale structure of the Universe. Non-linear structure formation at late times leads to strong deviations from Gaussianity that will be discussed in the last part of this chapter.

Let us first note that the multivariate Gaussian is appropriately normalized, i.e. that the probability integrates to unity $\int d^n f \mathbb{P}(f) = 1$

$$\int d^n f \; \mathbb{P}(f) = \prod_i \left\{ \int_{-\infty}^{\infty} df_i \right\} \frac{1}{\sqrt{(2\pi)^N |C|}} \exp\left[-\frac{1}{2} f_i C_{ij}^{-1} f_j \right] \tag{7.39}$$

Let us orthogonalize $C^{-1} = O^{-1} D O$ (where $O^{-1} = O^T$) and $y = O^{-1} f$, i.e. we are performing a rotation in the Euclidean space over which we integrate. With $D = \text{diag}\{1/\sigma_i^2\}$ we have

$$\prod_i \left\{ \int_{-\infty}^{\infty} dy_i \frac{1}{\sqrt{2\pi\sigma_i^2}} \exp\left[-\frac{1}{2} \frac{y_i^2}{\sigma_i^2} \right] \right\} = 1. \tag{7.40}$$

For the density field at one point we have thus

$$\mathbb{P}(\delta) = \frac{1}{\sqrt{2\pi}\sigma} \exp\left[-\frac{1}{2} \frac{\delta^2}{\sigma^2} \right]. \tag{7.41}$$

Let us first show that the correlation function is indeed $\langle f_i f_j \rangle = C_{ij}$. For this purpose we calculate the *characteristic function* of the Gaussian PDF

$$\mathcal{M}(i\mathcal{J}) = \langle \exp\left[i\mathbf{J}\cdot\mathbf{f}\right]\rangle = \int d^N f \frac{1}{\sqrt{(2\pi)^N|C|}} \exp\left[-\frac{1}{2}f_i C_{ij}^{-1}f_j + i\mathcal{J}_i f_i\right] = \exp\left[-\frac{1}{2}\mathcal{J}_i C_{ij}\mathcal{J}_j\right].$$
(7.42)

This result can be easily obtained by completing the squares. Generic moments can now be generated from derivatives of the characteristic function

$$\langle f_{i_1}\dots f_{i_n}\rangle = (-i)^n \frac{\partial}{\partial\mathcal{J}_{i_1}} \cdots \frac{\partial}{\partial\mathcal{J}_{i_n}} \langle \exp\left[i\mathbf{J}\cdot\mathbf{f}\right]\rangle\Big|_{\mathcal{J}=0}.$$
(7.43)

For $n = 2$ this yields

$$\langle f_m f_n\rangle = (-i)^2 \frac{\partial}{\partial\mathcal{J}_m}\frac{\partial}{\partial\mathcal{J}_n} \langle \exp\left[i\mathbf{J}\cdot\mathbf{f}\right]\rangle\Big|_{\mathcal{J}=0}$$

$$= (-i)^2 \frac{\partial}{\partial\mathcal{J}_m}\left(\mathcal{J}_j C_{jm}\exp\left[-\frac{1}{2}\mathcal{J}_i C_{ij}\mathcal{J}_j\right]\right)\Big|_{\mathcal{J}=0} = C_{mn}.$$
(7.44)

7.5.2   Wick Theorem

In the Wick theorem,[2] for a mean zero Gaussian random field the reduced correlation functions of order higher than two either vanish or are expressible in terms of products of two-point functions summed over all possible pairings.

The odd case obviously vanishes; for the even case we take eqn (7.43)

$$\langle f_i,\dots,f_{2N+1}\rangle = 0,$$
(7.45)

$$\langle f_1,\dots,f_{2N}\rangle = \sum_{\text{ordered pairings }\mathcal{P}_i} \prod_{\text{pairs }(i,j)\text{ in the pairing }\mathcal{P}_i} C_{ij}.$$
(7.46)

Alternatively we can also consider all $(2N)!$ possible permutations of $(i_1,\dots,i_{2N})$, cut each of them up in subsequent pairs and remove the redundancies by dividing by appropriate factors

$$\langle f_{i_1},\dots,f_{i_{2N}}\rangle = \frac{1}{2^N N!} \underbrace{(C_{i_1 i_2} C_{i_3 i_4}\dots C_{i_{2N-1} i_{2N}} + \text{perm})}_{2N!\text{ permutations}}.$$
(7.47)

The factor 2^N counts the redundant terms arising from exchanges of indices $C_{ij}\leftrightarrow C_{ji}$ which leaves the correlator unaffected due to the symmetry of the covariance matrix. The factor $N!$ counts possible reorderings of whole pairs $C_{ij}C_{mn}\leftrightarrow C_{mn}C_{ij}$, which leave

---

[2] In cosmology this theorem was introduced by Isserlis, but is often associated with the QFT version introduced by Wick.

the correlator unaffected as well. The number of products of correlation functions in the correlator of $2N$ fields is thus $(2N)!/(2^N N!) = (2N-1)!!$. You can convince yourself that this is the case by calculating the four-point function explicitly and by induction going from $2N$ to $2N+2$.

In summary, the procedure for calculating correlators of GRF is thus:

1. We first generate all $(2N-1)!!$ possible ordered pairings of indices from $(i_1,\ldots,i_{2N})$, i.e., we generate $\mathcal{P} = \{[(i_1,i_2),\ldots,(i_{2N-1},i_{2N})],\ldots,[(i_1,i_2N),\ldots,(i_2,i_{2N-1})]]\}$. For the sake of definiteness we choose $i < j$ for the pairings (i,j).
2. For each of these pairings in $\mathcal{P}$ calculate the product of N correlators, e.g. for $\mathcal{P}_1$ evaluate $C_{i_1 i_2} \ldots C_{i_{2N-1} i_{2N}}$
3. Sum over all of these products of correlators.

Let us consider the example of the four-point function. We expect that there will be $3!! = 3$ contributing terms. Indeed, there are three different fields with which f_1 can be correlated. Once this partner for f_1 has been chosen, the remaining fixed pair is correlated as well:

$$\langle f_1 f_2 f_3 f_4\rangle = \langle f_1 f_2 f_3 f_4\rangle + \langle f_1 f_2 f_3 f_4\rangle + \langle f_1 f_2 f_3 f_4\rangle \tag{7.48}$$
$$= C_{12} C_{34} + C_{13} C_{24} + C_{14} C_{23}.$$

In Fourier space this leads to

$$\langle \delta(\boldsymbol{k}_1),\ldots,\delta(\boldsymbol{k}_{2N+1})\rangle = 0, \tag{7.49}$$
$$\langle \delta(\boldsymbol{k}_1),\ldots,\delta(\boldsymbol{k}_{2N})\rangle = \sum_{\text{pairs } P\{(i,j)\}} \prod \langle \delta(\boldsymbol{k}_i),\delta(\boldsymbol{k}_j)\rangle. \tag{7.50}$$

7.5.3 Weakly Non-Gaussian Fields

Let us consider the *moment generating function*, which is closely related to the characteristic function discussed above

$$\mathcal{M}(\mathcal{J}) = \sum_{p=0}^{\infty} \frac{\langle \delta^p\rangle}{p!} \mathcal{J}^p = \langle \exp[\mathcal{J}\delta]\rangle = \int d\delta \, \mathbb{P}(\delta) \exp[\mathcal{J}\delta] \tag{7.51}$$

For a Gaussian field we obviously have

$$\mathcal{M}(\mathcal{J}) = \exp\left[\frac{1}{2}\mathcal{J}^2 \sigma^2\right] = 1 + \frac{\sigma^2}{2!}\mathcal{J}^2 + \frac{3\sigma^2}{4!}\mathcal{J}^4 + \ldots. \tag{7.52}$$

The moment generating function is the Laplace transform of the PDF and thus the PDF can be written as the inverse Laplace transform of the moment generating function

$$\mathbb{P}(\delta) = \int_{-i\infty}^{i\infty} \frac{\mathrm{d}\mathcal{J}}{2\pi i} \exp[\delta\mathcal{J}]\mathcal{M}(\mathcal{J}) \tag{7.53}$$

For a Gaussian PDF we can replace $\lambda = -i\mathcal{J}$ and perform the Gaussian integral

$$\mathbb{P}(\delta) = \int_{-\infty}^{\infty} \frac{\mathrm{d}y}{2\pi} \exp\left[-\frac{1}{2}\sigma^2 y^2 - iy\delta\right] = \frac{1}{\sqrt{2\pi}\sigma} \exp\left[-\frac{1}{2}\frac{\delta^2}{\sigma^2}\right] \tag{7.54}$$

Let us now consider a field that has a non-vanishing cubic moment $\langle\delta^3\rangle = S_3\sigma^2$ with $S_3 \ll 1$. Writing the moment generating function as

$$\mathcal{M}(\mathcal{J}) = \exp\left[\frac{1}{2}\mathcal{J}^2\sigma^2\right]\left(1 + \frac{\sigma^2 S_3}{3!}\mathcal{J}^3\right) \tag{7.55}$$

$$\mathbb{P}(\delta) = \frac{1}{\sqrt{2\pi}\sigma} \exp\left[-\frac{1}{2}\frac{\delta^2}{\sigma^2}\right]\left(1 + \frac{S_3(\delta^3 - 3\delta\sigma^2)}{3!\sigma^4}\right). \tag{7.56}$$

This is the leading correction to the Gaussian PDF, a more general form is known as the Edgeworth or Gram–Charlier expansion of the PDF

$$\mathbb{P}(\delta) = \frac{1}{\sqrt{2\pi\sigma^2}} \exp\left[-\frac{1}{2}\nu^2\right]\left[1 + \sigma\frac{S_3 H_3(\nu)}{6} + \sigma^2\left(\frac{S_4 H_4(\nu)}{24} + \frac{S_3^2 H_6(\nu)}{72}\right) + \cdots\right] \tag{7.57}$$

where $\nu = \delta/\sigma$ and $H_n(\nu)$ are the Hermite polynomials.

7.5.4   The Simplest Form of Non-Gaussianity

Non-Gaussian fields can be straightforwardly generated from Gaussian fields by non-linear transformations. Let us consider a very simple model for relating the fluctuations in the galaxy distribution δ_g to the underlying matter Gaussian matter distribution δ, the so-called local bias model, which we discuss in much more detail at the end of this chapter:

$$\delta_g(x) = b_1\delta(x) + \frac{b_2}{2}(\delta^2(x) - \sigma^2) + \mathcal{O}(\delta^3). \tag{7.58}$$

Here the subtraction of the variance in the squared term ensures that the galaxy overdensity averages to zero. Let us now consider the above model in Fourier space; as we have seen before, squaring in real space corresponds to convolutions in Fourier space

$$\delta_g(k) = b_1\delta(k) + \frac{b_2}{2}\int\frac{\mathrm{d}^3q}{(2\pi)^2}\delta(q)\delta(k-q) = \delta_g^{(1)}(k) + \delta_g^{(2)}(k) \tag{7.59}$$

We can now write down the definition of the galaxy bispectrum and write down the contributions, which will start at fourth order through a correlator of a second-order contribution (quadratic in the fields in the above equation) with two linear contributions.

$$(2\pi)^3 \delta^{(\mathrm{D})}(\boldsymbol{k}_1 + \boldsymbol{k}_2 + \boldsymbol{k}_3) B_{\mathrm{g}}(\boldsymbol{k}_1, \boldsymbol{k}_2, \boldsymbol{k}_3) = \langle \delta_{\mathrm{g}}(\boldsymbol{k}_1) \delta_{\mathrm{g}}(\boldsymbol{k}_2) \delta_{\mathrm{g}}(\boldsymbol{k}_3) \rangle$$

$$= \langle \delta_{\mathrm{g}}^{(2)}(\boldsymbol{k}_1) \delta_{\mathrm{g}}^{(1)}(\boldsymbol{k}_2) \delta_{\mathrm{g}}^{(1)}(\boldsymbol{k}_3) \rangle + 2 \text{ cyc.} \tag{7.60}$$

Using the implicit definition of the linear and quadratic bias contributions in eqn (7.59) we obtain

$$\langle \delta_{\mathrm{g}}^{(2)}(\boldsymbol{k}_1) \delta_{\mathrm{g}}^{(1)}(\boldsymbol{k}_2) \delta_{\mathrm{g}}^{(1)}(\boldsymbol{k}_3) \rangle = \frac{1}{2} b_1^2 b_2 \left\langle \int \frac{d^3 q}{(2\pi)^3} \delta(\boldsymbol{q}) \delta(\boldsymbol{k}_1 - \boldsymbol{q}) \delta(\boldsymbol{k}_2) \delta(\boldsymbol{k}_3) \right\rangle. \tag{7.61}$$

Wick's theorem would allow for three different pairings; however, the $\langle \delta(\boldsymbol{q}) \delta(\boldsymbol{k}_1 - \boldsymbol{q}) \rangle$ correlator only contributes to the homogenous $\boldsymbol{k}_1 = 0$ mode that is irrelevant for our purposes. The remaining two correlators link the $\boldsymbol{q}$-mode with either $\boldsymbol{k}_2$ or $\boldsymbol{k}_3$. Both of these give the same result as we can exchange $\boldsymbol{q} \leftrightarrow \boldsymbol{k} - \boldsymbol{q}$ in the momentum integral. We obtain

$$\langle \delta_{\mathrm{g}}^{(2)}(\boldsymbol{k}_1) \delta_{\mathrm{g}}^{(1)}(\boldsymbol{k}_2) \delta_{\mathrm{g}}^{(1)}(\boldsymbol{k}_3) \rangle = b_1^2 b_2 \int \frac{d^3 q}{(2\pi)^3} (2\pi)^3 \delta^{(\mathrm{D})}(\boldsymbol{q} + \boldsymbol{k}_2) P(k_2) (2\pi)^3$$

$$\delta^{(\mathrm{D})}(\boldsymbol{k}_1 - \boldsymbol{q} + \boldsymbol{k}_3) P(k_3)$$

$$= b_1^2 b_2 (2\pi)^3 \delta^{(\mathrm{D})}(\boldsymbol{k}_1 + \boldsymbol{k}_2 + \boldsymbol{k}_3) P(k_2) P(k_3). \tag{7.62}$$

One of the delta functions collapsed the momentum integral and the remaining momentum-conserving delta function just reproduces that required in the definition of the bispectrum in eqn (7.60). Thus, we finally obtain for the bispectrum in the simple quadratic bias model

$$B_{\mathrm{g}}(k_1, k_2, k_3) = b_1^2 b_2 \left[P(k_1) P(k_2) + 2 \text{ cyc.} \right]. \tag{7.63}$$

We discuss a prescription to derive this result from Feynman diagrams later in this chapter.

7.6 Estimators and Cosmic Variance

Given a dataset, i.e. a realization of the underlying statistical ensemble, we need to estimate the relevant statistics and the uncertainty of that estimate.

In observations and numerical simulations the volume is limited, which leaves us with finite Fourier modes, the smallest of them given by the *fundamental mode* $k_{\mathrm{f}} = 2\pi / L$

and the corresponding volume of the *fundamental cell* is $V_f = (2\pi)^3/V$. The Dirac Delta function rewritten for discrete k as

$$\delta^{(D)}(\mathbf{k}_i - \mathbf{k}_j) = \delta^{(D)}((i-j)k_f) = \frac{1}{k_f^3}\delta^{(D)}(i-j) - \frac{V}{(2\pi)^3}\delta_{i,j}^{(K)} \qquad (7.64)$$

All the Fourier modes can now be expressed as $\mathbf{k}_i = ik_f$, where i is an integer vector.

7.6.1   Power Spectrum Estimator and Variance

The power spectrum for discrete cells is thus given by

$$V\delta_{\mathbf{k},-\mathbf{k}'}^{(K)}P(|\mathbf{k}|) = \langle\delta(\mathbf{k})\delta(\mathbf{k}')\rangle \qquad (7.65)$$

We estimate the power spectrum in bins, i.e. spherical shells corresponding to an intervall in wave vector magnitude $[k\pm] = [k_-, k_+)$ centred at k as depicted in Figure 7.4 averaging all possible directions for the wave vector (making use of statistical isotropy)

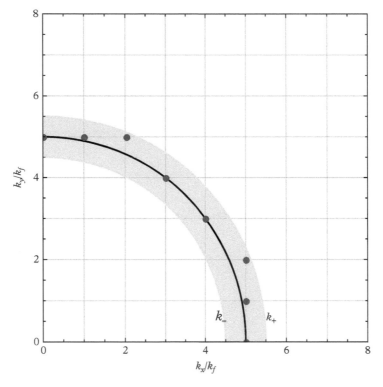

Figure 7.4 *Discrete Fourier grid and modes (red points) contributing to a wave number bin (gray shaded region) centred around k.*

$$\hat{P}(k) = \frac{1}{N_k V} \sum_{k_i \in [k_-, k_+)} \delta(k_i)\delta(-k_i), \tag{7.66}$$

where N_k is the number of cells in the k-bin. Effectively, we are estimating the variance of δ in a shell using N_k observations. Note that the estimator is unbiased since $\langle \hat{P} \rangle = P$. Obviously there is some freedom in choosing the configuration of the bins. Basically, the above estimator gives the mean power in the shell $[k_\pm]$, and thus it is advisable to average the theory calculations over the same bins for a fair comparison. The number of grid cells in the bin is given by the shell volume

$$V_s = \int_{[k_\pm]} d^3 q \int d^3 q' \delta^{(D)}(q + q') \tag{7.67}$$

divided by the volume of the fundamental cell

$$N_k = \frac{V_s}{V_f} = \frac{4\pi k^3 d\ln k}{V_f}. \tag{7.68}$$

Let us now calculate the variance of the power spectrum estimator

$$\left\langle \hat{P}^2(k) \right\rangle - \left\langle \hat{P}(k) \right\rangle^2 = \frac{1}{N_k^2 V^2} \sum_{k_i, jk_f \in [k_\pm]} \left\langle \delta(k_i)\delta(-k_i)\delta(k_j)\delta(-k_j) \right\rangle - P^2(k)$$

$$= \frac{1}{N_k^2} \sum_{k_i, jk_f \in [k_\pm]} P(k_i)P(k_j) + \frac{2}{N_k^2} \sum_{k_i \in [k_\pm]} P^2(k_i) - P^2(k) \tag{7.69}$$

$$= \frac{2}{N_k} P^2(k) = \frac{2V_f}{4\pi k^3 d\ln k} P^2(k).$$

Here we assumed Gaussianity of the underlying field, in which case the covariance matrix is diagonal, i.e. power spectrum estimates for distinct wave numbers are independent. In the more general case of non-Gaussian fields, the connected trispectrum contributes as well. Note also that the covariance matrix of power spectra of non-overlapping bins is diagonal, i.e. estimates of power in non-overlapping bins are independent.

7.6.2 Bispectrum

The bispectrum estimator for a fixed configuration $\{k_1, k_2, \mu = k_1 \cdot k_2\}$ can be estimated as

$$\hat{B}(k_1, k_2, \mu) = \frac{1}{N_{tr} V} \sum_{k_i \in [k_\pm]} \sum_{jk_f \in [k_\pm, \mu_\pm]} \delta(k_i)\delta(k_j)\delta(-k_i - k_i), \tag{7.70}$$

and the estimator is unbiased since $\langle \hat{B} \rangle = B$. The number of triangles in the bin is given by the shell volume divided by the volume of the fundamental cell squared

$$N_{\mathrm{tr}} = \frac{V_{123}}{V_{\mathrm{f}}^2} = \frac{8\pi^2 k_1 k_2 k_3 (\mathrm{d}k)^3}{V_{\mathrm{f}}^2}. \tag{7.71}$$

For the variance of the bispectrum estimator we have

$$\left\langle \hat{B}^2(k_1, k_2, \mu) \right\rangle - \left\langle \hat{B}(k_1, k_2, \mu) \right\rangle^2 = \frac{1}{N_{\mathrm{tr}}^2 V^2} \sum_{i,j,l,m} \langle \delta(\mathbf{k}_i) \delta(\mathbf{k}_j) \delta(-\mathbf{k}_i - \mathbf{k}_j) \delta(\mathbf{k}_l) \delta(\mathbf{k}_m)$$

$$\delta(-\mathbf{k}_l - \mathbf{k}_m)\rangle - B^2(k_1, k_2, \mu)$$

$$= s_{123} \frac{V}{N_{\mathrm{tr}}^2} \sum_{i,j} P(\mathbf{k}_i) P(\mathbf{k}_j) P(-\mathbf{k}_i - \mathbf{k}_j)$$

$$= s_{123} \frac{V}{N_{\mathrm{tr}}} P(k_1) P(k_2) P(k_3)$$

$$= s_{123} \frac{(2\pi)^3 V_{\mathrm{f}}}{8\pi^2 k_1 k_2 k_3 (\mathrm{d}k)^3 \mathrm{d}\mu} P(k_1) P(k_2) P(k_3). \tag{7.72}$$

The factor s_{123} takes on values of $6, 2, 1$ for general, isosceles, and equilateral triangles. This is a simple consequence of the fact that for equilateral triangles the k-modes are indistinguishable. We again assumed Gaussianity, for which $B = 0$.

7.6.3   Generation of Gaussian Random Fields

Note that the real and imaginary part of the complex density $\delta(\mathbf{k}) = a(\mathbf{k}) + ib(\mathbf{k})$ field are independent Gaussian random fields with variance $P/2$

$$\langle a(\mathbf{k}) a(\mathbf{k}') \rangle = \frac{1}{4} \langle (\delta + \delta^*)(\delta + \delta^*) \rangle = (2\pi)^3 \delta^{(\mathrm{D})}(\mathbf{k} - \mathbf{k}') \frac{P}{2} + (\mathbf{k}' \to -\mathbf{k}') \tag{7.73}$$

$$\langle b(\mathbf{k}) b(\mathbf{k}') \rangle = -\frac{1}{4} \langle (\delta - \delta^*)(\delta - \delta^*) \rangle = (2\pi)^3 \delta^{(\mathrm{D})}(\mathbf{k} - \mathbf{k}') \frac{P}{2} - (\mathbf{k}' \to -\mathbf{k}') \tag{7.74}$$

$$\langle a(\mathbf{k}) b(\mathbf{k}') \rangle = \frac{1}{4i} \langle (\delta + \delta^*)(\delta - \delta^*) \rangle = 0. \tag{7.75}$$

Thus realizations can be generated by drawing real and imaginary parts from independent Gaussian distributions with mean zero and variance $P/2$. While the real and imaginary parts are independent Gaussian distributed, the magnitude of δ follows a Rayleigh distribution. This can be seen as follows: both real and imaginary parts are Gaussian distributed with variance $P/2$

$$\mathbb{P}_a(x) = \mathbb{P}_b(x) = \frac{1}{\sqrt{\pi P}} \exp\left[-\frac{x^2}{P}\right]. \tag{7.76}$$

The probability has to be invariant under a reparametrization and thus going to polar coordinates $\delta(\mathbf{k}) = a(\mathbf{k}) + ib(\mathbf{k}) = r \exp[i\phi]$ yields

$$\mathbb{P}_{|\delta|}(r)\mathbb{P}_\varphi(\varphi)\mathrm{d}r\mathrm{d}\varphi = \mathbb{P}_a(a)\mathbb{P}_a(b)\mathrm{d}a\mathrm{d}b = \frac{r}{\pi P} \exp\left[-\frac{r^2}{P}\right]\mathrm{d}r\mathrm{d}\varphi. \tag{7.77}$$

Thus, the probability is uniform for the phase of the mode and Rayleigh for the magnitude

$$\mathbb{P}_{|\delta|}(r) = \frac{2r}{P} \exp\left[-\frac{r^2}{P}\right], \qquad\qquad \mathbb{P}_\varphi(\varphi) = \frac{1}{2\pi}. \tag{7.78}$$

7.7 Dynamics in the Newtonian Regime

7.7.1 Equations of Motion

Let us now consider the equations governing the cosmological fluid in the Newtonian limit, i.e. for small distances $x \ll H^{-1}$ and small velocities $v \ll 1$. The equation of motion for a particle at physical position $\mathbf{r}$ is

$$\ddot{\mathbf{r}} = -\nabla_r\Phi. \tag{7.79}$$

Defining comoving coordinates as $\mathbf{r} = a\mathbf{x}$ we have $\nabla_x = a\nabla_r$. From now on we use the gradient with respect to comoving coordinates.[3] Let us take the derivative of the physical coordinate with respect to physical time and rewrite in terms of the comoving position

$$\dot{\mathbf{r}} = \mathcal{H}\mathbf{x} + \mathbf{x}'. \tag{7.80}$$

Likewise we have for the second derivative

$$\ddot{\mathbf{r}} = \frac{1}{a}\left(\mathcal{H}'\mathbf{x} + \mathcal{H}\mathbf{x}' + \mathbf{x}''\right) = -\frac{1}{a}\nabla\Phi. \tag{7.81}$$

the term proportional to the position is peculiar, since it leads to a spatial dependence of the particle acceleration. This term arises from the comoving coordinates and accounts

---

[3] For an arbitrary function $f(t)$ we have $a\dot{f} = f'$ $a^2\ddot{f} = f'' - \mathcal{H}f'$ Unless otherwise quoted we refer to dots as derivatives with respect to coordinate time and dashes as derivatives with respect to conformal time.

for the expansion of space-time. We can thus bring it to the right-hand side of the above equation and define the peculiar potential[4]

$$\Phi = -\frac{1}{2}\mathcal{H}'x^2 + \phi. \tag{7.82}$$

Thus we have for the equations of motion in terms of physical and conformal time

$$\ddot{x} + 2H\dot{x} = -\frac{\nabla\phi}{a^2}, \qquad\qquad x'' + \mathcal{H}x' = -\nabla\phi. \tag{7.83}$$

The peculiar potential is solely seeded by the energy density fluctuations in the Universe. Since we are assuming that dark energy is homogeneous, at late times these energy density fluctuations are dominated by the fluctuations in the matter density. Hence, the Poisson equation for the peculiar potential can be expressed as

$$\nabla_x^2\phi = \frac{3}{2}\mathcal{H}^2\Omega_m(a)\delta = \frac{3}{2}\Omega_{m,0}H_0^2\frac{\delta}{a}. \tag{7.84}$$

Defining the *canonical momentum*

$$p = amu, \tag{7.85}$$

where $u = x'$ is the comoving velocity we have for the equation of motion

$$p' = -am\nabla_x\phi. \tag{7.86}$$

Let us finally stress again that these equations are only true in the Newtonian regime.

7.7.2   The Fluid Equations

The particle distribution in phase space is conveniently described by the distribution function $f(x,p,\tau)$. The number of particles in a infinitesimal phase space volume $d^3x d^3p$ is thus given by $dN = f(x,p,\tau)d^3x d^3p$. Louiville theorem asserts the conservation of the phase space density. For collisionless dark matter, this conservation of phase space density then yields the *collisionless Boltzmann equation*, also known as *Vlasov equation*

$$\frac{df}{d\tau} = \frac{\partial f}{\partial\tau} + \frac{dx}{d\tau}\cdot\frac{\partial f}{\partial x} + \frac{dp}{d\tau}\frac{\partial f}{\partial p} \tag{7.87}$$

$$= \frac{\partial f}{\partial\tau} + \frac{p}{ma}\cdot\frac{\partial f}{\partial x} - am\nabla\phi\cdot\frac{\partial f}{\partial p} = 0 \tag{7.88}$$

---

[4] In the background we have $\mathcal{H}'x = -\nabla\phi_b$ with $\nabla^2\phi_b = 4\pi G\bar\rho$.

where we have used the equation of motion (7.86) in the last line.

We rarely are interested in the full phase space distribution and thus there is no need to solve this non-linear seven-dimensional differential equation. Instead we will be mostly concerned with fluid properties such as density, mean streaming velocity, and velocity dispersion, which are readily obtained as moments of the distribution function

$$\rho(x,\tau) = \frac{m}{a^3} \int d^3p\, f(x,p,\tau),$$

(7.89)

$$v_i(x,\tau) = \int d^3p\, \frac{p_i}{am} f(x,p,\tau) \Big/ \int d^3p\, f(x,p,\tau),$$

(7.90)

$$\sigma_{ij}(x,\tau) = \int d^3p\, \frac{p_i}{am}\frac{p_j}{am} f(x,p,\tau) \Big/ \int d^3p\, f(x,p,\tau) - v_i(x)v_j(x).$$

(7.91)

The velocity dispersion is sometimes also referred to as anisotropic stress and describes the deviation from a single coherent flow as is obvious in eqn (7.91).

The equations of motion for these quantities can now be obtained by taking moments of the Vlasov equation (7.88). The 0th moment of the Vlaslov equation yields the *continuity equation*. Upon integrating over the momentum, we have to integrate the last term by parts and use that the potential is independent of the momentum

$$\delta' + \nabla \cdot \big[v(1+\delta) \big] = 0$$

(7.92)

Taking the the first moment and using the continuity equation yields the *Euler equation*

$$v_i' + \mathcal{H}v_i + v \cdot \nabla v_i = -\nabla_i \phi - \frac{1}{\rho}\nabla_i(\rho\sigma_{ij})$$

(7.93)

or conservation of momentum. In principle we could have continued to hierarchy of equations, which couples the equation of motion for n-th moment of the Vlaslov equation to the $n+1$th moment. To close the hierarchy we postulate that all moments beyond the velocity are vanishing, an assumption that is denoted the pressureless perfect fluid. This assumption is reasonable in the linear regime but needs to be validated numerically at late times, when structures collapse, virialize, and shell crossing occurs. The fluid velocity can be decomposed into a scalar and a vector part $v = v_{\parallel} + v_{\perp}$, where $\nabla \times v_{\parallel} = 0$ and $\nabla \cdot v_{\perp} = 0$. The velocity field can thus be described by its vorticity $w = \nabla \times v$ and its divergence $\theta = \nabla \cdot v$.

7.7.3 Linearized Equations

Let us neglect all the quadratic terms in the continuity and Euler equation and assume that the velocity dispersion vanishes

$$\delta' + \theta = 0$$

(7.94)

$$v' + \mathcal{H}v = -\nabla\phi.$$

(7.95)

The system can be solved straightforwardly after rewriting the Euler equation in terms of velocity, vorticity, and divergence

$$\theta' + \mathcal{H}\theta = -\Delta\phi \tag{7.96}$$

$$\boldsymbol{w}' + \mathcal{H}\boldsymbol{w} = 0. \tag{7.97}$$

The solution of the vorticity equation is simply $\boldsymbol{w} \propto a^{-1}$, i.e. any initially present vorticity decays at linear level. To solve the scalar equation, we take the time derivative of eqn (7.94) and replace θ' with eqn (7.96). In the resulting equation, we can replace θ using eqn (7.96) and $\Delta\phi$ using the Poisson eqn (7.84). We obtain the *linear growth equation*

$$\delta''(\boldsymbol{k}, \tau) + \mathcal{H}(\tau)\delta'(\boldsymbol{k}, \tau) - \frac{3}{2}\Omega_{\mathrm{m}}(\tau)\mathcal{H}^2(\tau)\delta(\boldsymbol{k}, \tau) = 0. \tag{7.98}$$

It is sometimes convenient to rewrite this in terms of derivatives with respect to scale factor a

$$-a^2\mathcal{H}^2\partial_a^2\delta + \frac{3}{2}\mathcal{H}^2\big[\Omega_{\mathrm{m}}(a) - 2\big]a\partial_a\delta + \frac{3}{2}\Omega_{\mathrm{m}}\mathcal{H}^2\delta = 0. \tag{7.99}$$

We easily can confirm the above differential equation has a growing and a decaying mode solution $\delta(\boldsymbol{k}, \tau) = D_+(\tau)\delta_{+,0}(\boldsymbol{k}) + D_-(\tau)\delta_{-,0}(\boldsymbol{k})$, where $D_-(\tau) = D_{-,0}H = \mathcal{H}/a$. The growing mode solution can then be obtained as

$$D_+(\tau) = D_{+,0}H(\tau) \int_0^{a(\tau)} \frac{\mathrm{d}a'}{\mathcal{H}^3(a')}, \tag{7.100}$$

where $D_{+,0}$ is a normalization factor used to achieve $D_+(a = 1) = 1$. Let us first discuss the solution in a Einstein–de Sitter (EdS) matter-only Universe. Since $a \propto t^{2/3}$ we have $H = a^{-3/2}$ and thus $D_+ = a$ and $D_- = a^{-3/2}$. This special case and the solution for our fiducial ΛCDM model are shown in the left panel of Figure 7.5. We see that the growth in the ΛCDM Universe stalls at late times when the cosmological constant starts to dominate. In what follows we concentrate on the growing mode solutions and use $D \equiv D_+$ unless otherwise stated.

7.7.4 Velocities

The previous section showed that the vorticity decays at linear level in the absence of anisotropic stress. At non-linear level we have

$$\boldsymbol{w}' + \mathcal{H}\boldsymbol{w} + \nabla \times (\boldsymbol{v} \times \boldsymbol{w}) = 0. \tag{7.101}$$

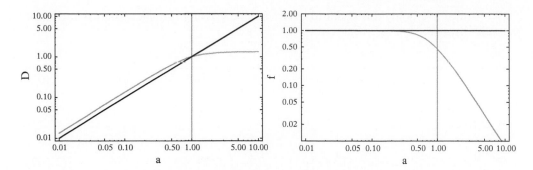

Figure 7.5 *Linear growth of structure for our fiducial* Λ*CDM (blue) and a matter only EdS Universe (red).* Left panel: *Linear growth factor D.* Right panel: *Logarithmic growth factor f.*

This equation tells us that, even at non-linear level, if there is no initial vorticity, evolution will not generate it. Together with the knowledge that vorticity decays at early times when the fluctuations are still linear we can conclude that vorticity will be negligible throughout evolution in absence of anisotropic stress. However, there is evidence from simulations that at late times velocity dispersion and thus vorticity is generated in high-density regions. In the following, we present standard perturbation theory, ignoring the possible vorticity. This assumption restricts the validity of the solutions to large scales.

In case of vanishing curl we have

$$v(k) = \mathrm{i}\frac{k}{k^2}\theta(k). \tag{7.102}$$

From the linearized continuity equation in Fourier space we have

$$\theta(k,\tau) = -\mathrm{i}k \cdot v(k,\tau) = -\delta'(k,\tau) = -\mathcal{H}f\delta(k,\tau), \tag{7.103}$$

where we defined the logarithmic growth factor

$$f = \frac{\mathrm{d}\ln D}{\mathrm{d}\ln a}. \tag{7.104}$$

For the linear growing mode defined above in eqn (7.100), we have

$$f(a) = \frac{\mathrm{d}\ln H}{\mathrm{d}\ln a} + \frac{a}{(aH)^3}\frac{1}{\int_0^a \mathrm{d}a'\,[a'\,H(a')]^3}, \tag{7.105}$$

which is unity for EdS. The right panel of Figure 7.5 shows the logarithmic growth factor for a EdS and our fiducial ΛCDM cosmology. The late time dominance of the cosmological constant is even more apparent in this plot where f decays to $f(a=1) \approx 0.48$ at present time.

We can now employ this result to derive simple velocity statistics, such as the linear velocity dispersion

$$\kappa_{ij} = \langle v_i v_j \rangle - \langle v_i \rangle \langle v_j \rangle = \langle v_i v_j \rangle = \mathcal{H}^2 f^2 \int \frac{\mathrm{d}^3 q}{(2\pi)^3} \frac{q_i q_j}{q^4} P_{\mathrm{lin}}(q), \qquad (7.106)$$

and its trace

$$\sigma_v^2 = \frac{1}{3} \mathrm{Tr}\left(\kappa_{ii}^2\right) = \frac{\mathcal{H}^2 f^2}{6\pi^2} \int \mathrm{d}q \, P_{\mathrm{lin}}(q). \qquad (7.107)$$

In the following we frequently consider the displacement dispersion $\sigma_d = \sigma_v / \mathcal{H}f$, for which we obtain in our fiducial cosmology $\sigma_d \approx 6 \, h^{-1}\mathrm{Mpc}.$[5] Even though this number is calculated in linear theory, it turns out to be a fairly accurate description of what is measured in N-body simulations.

7.7.5 Fluid Equations in Fourier Space

After having obtained some intuition on the solutions in the linear regime, where the quadratic terms are negligible, we now return to the full equations. To facilitate the analysis, we work in Fourier space, where the Euler and continuity equations read as

$$\delta'(\boldsymbol{k}) + \theta(\boldsymbol{k}) = -\int \frac{\mathrm{d}^3 q}{(2\pi)^3} \int \frac{\mathrm{d}^3 q}{(2\pi)^3} (2\pi)^3$$
$$\delta^{(\mathrm{D})}(\boldsymbol{k} - \boldsymbol{q} - \boldsymbol{q}') \alpha(\boldsymbol{q}, \boldsymbol{q}') \theta(\boldsymbol{q}) \delta(\boldsymbol{q}'), \qquad (7.108)$$

$$\theta'(\boldsymbol{k}) + \mathcal{H}\theta(\boldsymbol{k}) + \frac{3}{2}\Omega_{\mathrm{m}}(a)\mathcal{H}^2 \delta(\boldsymbol{k}) = -\int \frac{\mathrm{d}^3 q}{(2\pi)^3} \int \frac{\mathrm{d}^3 q'}{(2\pi)^3} (2\pi)^3$$
$$\delta^{(\mathrm{D})}(\boldsymbol{k} - \boldsymbol{q} - \boldsymbol{q}') \beta(\boldsymbol{q}, \boldsymbol{q}') \theta(\boldsymbol{q}) \theta(\boldsymbol{q}'). \qquad (7.109)$$

The coupling kernels on the right-hand side are defined as

$$\alpha(\boldsymbol{k}_1, \boldsymbol{k}_2) = \frac{\boldsymbol{k}_1 \cdot (\boldsymbol{k}_1 + \boldsymbol{k}_2)}{k_1^2} \qquad (7.110)$$

---

[5] Here we have used that the displacement $\boldsymbol{\Psi}$ is given by

$$\boldsymbol{\Psi}(\boldsymbol{k}, \tau) = \int \mathrm{d}\tau \, \boldsymbol{v}(\boldsymbol{k}, \tau) = \frac{i\boldsymbol{k}}{k^2} \int \mathrm{d}\tau \, \theta(\boldsymbol{k}, \tau) = \frac{-i\boldsymbol{k}}{k^2} \int \mathrm{d}\tau \, \delta'(\boldsymbol{k}, \tau) = -\frac{i\boldsymbol{k}}{k^2}\delta(\boldsymbol{k}, \tau),$$

such that $\sigma_v = \mathcal{H}f\sigma_d$.

$$\beta(k_1, k_2) = \frac{1}{2}(k_1 + k_2)^2 \frac{k_1 \cdot k_2}{k_1^2 k_2^2} = \frac{1}{2} \frac{k_1 \cdot k_2}{k_1 k_2} \left(\frac{k_2}{k_1} + \frac{k_1}{k_2} \right) + \frac{(k_1 \cdot k_2)^2}{k_1^2 k_2^2}. \tag{7.111}$$

Note that $\alpha(k_1, k_2)$ is not symmetric in its arguments but $\beta(k_1, k_2)$ is. The fluid eqn (7.108) and eqn (7.109) are non-linear coupled differential equations for the density and velocity divergence. A closed-form solution does in general not exist. However, we can try to solve them perturbatively in the regime, where $\delta \ll 1$ and $\theta \ll 1$. We discuss the perturbative solutions in much more detail in the next section.

7.8 Perturbative Solution of the Fluid Equations

The continuity and Euler equations can be rewritten as second-order differential equations

$$\mathcal{H}^2 \left\{ -a^2 \partial_a^2 + \frac{3}{2} (\Omega_m(a) - 2) a \partial_a + \frac{3}{2} \Omega_m(a) \right\} \delta = \mathcal{S}_\beta - \mathcal{H} \partial_a \left(a \mathcal{S}_\alpha \right),$$

$$\mathcal{H} \left\{ a^2 \partial_a^2 + \left(4 - \frac{3}{2} \Omega_m(a) \right) a \partial_a + (2 - 3\Omega_m) \right\} \theta = \partial_a \left(a \mathcal{S}_\beta \right) - \frac{3}{2} \Omega_m(a) \mathcal{H} \mathcal{S}_\alpha, \tag{7.112}$$

with source terms given by

$$\mathcal{S}_\alpha(k, \tau) = - \int \frac{d^3 q}{(2\pi)^3} \alpha(q, k - q) \theta(q, \tau) \delta(k - q, \tau),$$

$$\mathcal{S}_\beta(k, \tau) = - \int \frac{d^3 q}{(2\pi)^3} \beta(q, k - q) \theta(q, \tau) \theta(k - q, \tau). \tag{7.113}$$

The corresponding Green's functions for δ and θ in ΛCDM are given by

$$G_\delta(a, a') = \Theta(a - a') \frac{2}{5} \frac{1}{\mathcal{H}_0^2 \Omega_m^0} \frac{D_+(a')}{a'} \left\{ \frac{D_-(a)}{D_-(a')} - \frac{D_+(a)}{D_1(a')} \right\},$$

$$G_\theta(a, a') = -\mathcal{H} f(a) G_\delta(a, a'). \tag{7.114}$$

We could solve these equations numerically order by order. We can however gain more insights into the structure of the solutions by studying a power series ansatz in a matter-only EdS Universe.

7.8.1 Series Ansatz and Coupling Kernels

SPT aims to solve the fluid equations perturbatively using a power law ansatz. This approach simplifies significantly in an EdS Universe, where $D = a$. Hence we study this case first and discuss the generalization to ΛCDM later. Furthermore, we neglect the decaying mode. The power law ansatz reads

$$\delta(k,\tau) = \sum_{i=1}^{\infty} a^i(\tau)\delta^{(i)}(k), \qquad \theta(k,\tau) = -\mathcal{H}(\tau)\sum_{i=1}^{\infty} a^i(\tau)\tilde{\theta}^{(i)}(k). \qquad (7.115)$$

The expansion is in powers of the linear density field discussed above, i.e. $^{(i)}\delta = \mathcal{O}(^{(1)}\delta^i)$. We can now write the nth order solutions as convolutions of linear density fields

$$\delta^{(n)}(k) = \prod_{m=1}^{n} \left\{ \int \frac{\mathrm{d}^3 q_m}{(2\pi)^3} \delta^{(1)}(q_m) \right\} F_n(q_1,\ldots,q_n)(2\pi)^3 \delta^{(\mathrm{D})}(k - q|_1^n) \qquad (7.116)$$

$$\tilde{\theta}^{(n)}(k) = \prod_{m=1}^{n} \left\{ \int \frac{\mathrm{d}^3 q_m}{(2\pi)^3} \delta^{(1)}(q_m) \right\} G_n(q_1,\ldots,q_n)(2\pi)^3 \delta^{(\mathrm{D})}(k - q|_1^n) \qquad (7.117)$$

where $q|_i^j = \sum_{m=i}^{j} q_l$. A diagrammatic representation of this expansion is shown in Figure 7.6. We can derive the following recursion relations for the convolution kernels

$$F_n(q_1,\ldots,q_n) = \sum_{m=1}^{n-1} \frac{G_m(q_1,\ldots,q_m)}{(2n+3)(n-1)} \left[(2n+1)\alpha\left(q|_1^m, q|_{m+1}^n\right) F_{n-m}\left(q_{m+1},\ldots,q_n\right) \right.$$
$$\left. + 2\beta\left(q|_1^m, q|_{m+1}^n\right) G_{n-m}\left(q_{m+1},\ldots,q_n\right) \right] \qquad (7.118)$$

$$G_n(q_1,\ldots,q_n) = \sum_{m=1}^{n-1} \frac{G_m(q_1,\ldots,q_m)}{(2n+3)(n-1)} \left[3\alpha\left(q|_1^m, q|_{m+1}^n\right) F_{n-m}\left(q_{m+1},\ldots,q_n\right) \right.$$
$$\left. + 2n\beta\left(q|_1^m, q|_{m+1}^n\right) G_{n-m}\left(q_{m+1},\ldots,q_n\right) \right]. \qquad (7.119)$$

For general ΛCDM the series ansatz in eqn (7.115) can be generalized to

$$\delta(k,\tau) = \sum_{i=1}^{\infty} D^i(\tau)\delta^{(i)}(k) \qquad \theta(k,\tau) = -\mathcal{H}(\tau)f(\tau)\sum_{i=1}^{\infty} D^i(\tau)\tilde{\theta}^{(i)}(k). \qquad (7.120)$$

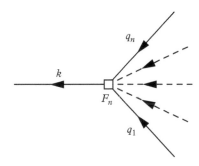

Figure 7.6 *Diagrammatic representation of the series expansion of the density field in Eq. (7.116). The points on the right side are initial density fields and the points on the left side are n-th order density fields.*

The exact solution deviates from the above solution and doesn't allow for a separation of time and space dependence as in eqn (7.120). The differences are discussed in detail in [28] (and shown in Figure 7.10) and are generally at the sub-percent level. We thus stick to the approximation eqn (7.120), which is sufficiently accurate for our purpose. Let us come back to the kernels and evaluate them explicitly at second and third order. The second-order density kernel is given by

$$
\begin{aligned}
F_2(\boldsymbol{k}_1, \boldsymbol{k}_2) &= \frac{5}{7}\alpha(\boldsymbol{k}_1, \boldsymbol{k}_2) + \frac{2}{7}\beta(\boldsymbol{k}_1, \boldsymbol{k}_2) \\
&= \frac{5}{7} + \frac{1}{2}\frac{\boldsymbol{k}_1 \cdot \boldsymbol{k}_2}{k_1 k_2}\left(\frac{k_2}{k_1} + \frac{k_1}{k_2}\right) + \frac{2}{7}\frac{(\boldsymbol{k}_1 \cdot \boldsymbol{k}_2)^2}{k_1^2 k_2^2}.
\end{aligned}
\tag{7.121}
$$

Here we defined $\boldsymbol{k}_1 \cdot \boldsymbol{k}_2 = k_1 k_2 \mu_{12}$ and symmetrized eqn (7.116) over the momenta. The latter can be rewritten as

$$
F_2(\boldsymbol{k}_1, \boldsymbol{k}_2) = \frac{17}{21} + \frac{1}{2}\frac{\boldsymbol{k}_1 \cdot \boldsymbol{k}_2}{k_1 k_2}\left(\frac{k_2}{k_1} + \frac{k_1}{k_2}\right) + \frac{2}{7}\left[\frac{(\boldsymbol{k}_1 \cdot \boldsymbol{k}_2)^2}{k_1^2 k_2^2} - \frac{1}{3}\right]
\tag{7.122}
$$

where the quadratic density term, the shift term, and the anisotropic stress term are more obvious. At the same time, the angular structure of monopole, dipole, and quadrupole is more obvious. In configuration space the second-order density field is thus given as a sum of a growth term, an advection term, and a tidal term

$$
\delta^{(2)}(\boldsymbol{x}) = \frac{17}{21}\delta^2(\boldsymbol{x}) - \boldsymbol{\Psi}(\boldsymbol{x}) \cdot \nabla\delta(\boldsymbol{x}) + \frac{2}{7}s^2(\boldsymbol{x}).
\tag{7.123}
$$

The second-order velocity kernel reads

$$
\begin{aligned}
G_2(\boldsymbol{k}_1, \boldsymbol{k}_2) &= \frac{3}{7}\alpha(\boldsymbol{k}_1, \boldsymbol{k}_2) + \frac{4}{7}\beta(\boldsymbol{k}_1, \boldsymbol{k}_2), \\
&= \frac{3}{7} + \frac{1}{2}\frac{\boldsymbol{k}_1 \cdot \boldsymbol{k}_2}{k_1 k_2}\left(\frac{k_2}{k_1} + \frac{k_1}{k_2}\right) + \frac{4}{7}\frac{(\boldsymbol{k}_1 \cdot \boldsymbol{k}_2)^2}{k_1^2 k_2^2}.
\end{aligned}
\tag{7.124}
$$

The explicit expressions for F_3 and G_3 are

$$
\begin{aligned}
F_3(\boldsymbol{q}_1, \boldsymbol{q}_2, \boldsymbol{q}_3) &= \frac{1}{18}\Big[7\alpha(\boldsymbol{q}_1, \boldsymbol{q}_2 + \boldsymbol{q}_3)F_2(\boldsymbol{q}_2, \boldsymbol{q}_3) + 2\beta(\boldsymbol{q}_1, \boldsymbol{q}_2 + \boldsymbol{q}_3)G_2(\boldsymbol{q}_2, \boldsymbol{q}_3)\Big] \\
&\quad + \frac{G_2(\boldsymbol{q}_1, \boldsymbol{q}_2)}{18}\Big[7\alpha(\boldsymbol{q}_1 + \boldsymbol{q}_2, \boldsymbol{q}_3) + 2\beta(\boldsymbol{q}_1 + \boldsymbol{q}_2, \boldsymbol{q}_3)\Big],
\end{aligned}
\tag{7.125}
$$

$$
\begin{aligned}
G_3(\boldsymbol{q}_1, \boldsymbol{q}_2, \boldsymbol{q}_3) &= \frac{1}{18}\Big[3\alpha(\boldsymbol{q}_1, \boldsymbol{q}_2 + \boldsymbol{q}_3)F_2(\boldsymbol{q}_2, \boldsymbol{q}_3) + 6\beta(\boldsymbol{q}_1, \boldsymbol{q}_2 + \boldsymbol{q}_3)G_2(\boldsymbol{q}_2, \boldsymbol{q}_3)\Big] \\
&\quad + \frac{G_2(\boldsymbol{q}_1, \boldsymbol{q}_2)}{18}\Big[3\alpha(\boldsymbol{q}_1 + \boldsymbol{q}_2, \boldsymbol{q}_3) + 6\beta(\boldsymbol{q}_1 + \boldsymbol{q}_2, \boldsymbol{q}_3)\Big].
\end{aligned}
\tag{7.126}
$$

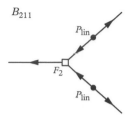

Figure 7.7 *Tree-level bispectrum: A second-order density field is contracted with two linear density fields to yield the leading-order contribution to the matter bispectrum.*

The above formulae are not symmetrized over the arguments yet. Upon integration over three equivalent density fields $\delta(\boldsymbol{q}_1)\delta(\boldsymbol{q}_2)\delta(\boldsymbol{q}_3)$ we have to symmetrize, accounting both for the cyclic and odd permutations of the arguments in the kernels.

For the following discussion, it is important to know how the kernels scale if one of the momenta becomes very large. The kernels obey scaling laws such as

$$\lim_{q\to\infty} F_n(\boldsymbol{k}_1,\dots,\boldsymbol{k}_{n-2},\boldsymbol{q},-\boldsymbol{q}) \propto \frac{k^2}{q^2}, \tag{7.127}$$

where $\boldsymbol{k}=\boldsymbol{k}_1+\dots\boldsymbol{k}_{n-2}$. For us it turns out to be important that also for F_2 and F_3 a similar scaling holds when the sum of the arguments remains finite while one of the momenta goes to infinity, i.e.

$$\lim_{q\to\infty} F_2(\boldsymbol{q},\boldsymbol{k}-\boldsymbol{q}) \propto \lim_{q\to\infty} F_2(\boldsymbol{k}_1-\boldsymbol{q},\boldsymbol{k}_2+\boldsymbol{q}) \propto \frac{k^2}{q^2},$$
$$\lim_{q\to\infty} F_3(\boldsymbol{q},\boldsymbol{k}_1-\boldsymbol{q},\boldsymbol{k}_2) \propto \frac{k^2}{q^2}, \tag{7.128}$$

where we assumed that the momenta $k_1 \sim k_2 \sim k$ are of the same order.

7.8.2   Diagrams and Feynman Rules for the n-Spectra

Even though the diagrams can be straightforwardly translated into the corresponding mathematical expressions, we write down the Feynman rules explicitly for the sake of definiteness. For the calculation of the ith order contribution to the n-spectrum the steps are:

1. Draw all distinct connected diagrams with n external lines up to the desired order i in $\delta^{(1)}$

   i) For each vertex V with ingoing momenta $\boldsymbol{q}_i$ and outgoing momentum $\boldsymbol{p}$ write a delta function $(2\pi)^3\delta^{(\mathrm{D})}\left(\boldsymbol{p}-\sum_i \boldsymbol{q}_i\right)$ and a coupling kernel $V(\boldsymbol{q}_1,\dots,\boldsymbol{q}_n)$ (for instance $F_n(\boldsymbol{q}_1,\dots,\boldsymbol{q}_n)$)

ii) Assign a linear power spectrum $(2\pi)^3 \delta^{(D)}(\boldsymbol{q} + \boldsymbol{q}')P_{\mathrm{lin}}(q)$ to each of the dots with outgoing momenta $\boldsymbol{q}$ and $\boldsymbol{q}'$

iii) Integrate over all inner momenta $\int \mathrm{d}^3 q_i / (2\pi)^3$

iv) Multiply with the symmetry factor

v) Sum over all distinct labelings of the external lines.

2. Add up the resulting expressions from all diagrams.

7.8.3   Power Spectrum and Bispectrum

The discussion in the previous section allowed us to express non-linear density and velocity fields as a sum of products of linear density fields. If we are interested in *n*-spectra of the fields, we have to correlate two of these non-linear fields with each other. This will again lead to a sum of correlators. Secction 7.5 showed that, only even correlators of linear density fields contribute and these correlators can be expressed as products of linear power spectra, whose form is given by the process seeding the fluctuations and the subsequent linear growth. Thus we were able to reduce the problem of calculating spectra of non-linear fields to convolutions of linear power spectra.

In Section 7.5 we mentioned that Gaussian random fields have a vanishing bispectrum and that their trispectrum can be written as a product of power spectra. The gravitational evolution changes this behaviour and leads to a non-vanishing bispectrum that arises from expanding one of the fields in the correlator of three Fourier modes to second order

$$\langle \delta(\boldsymbol{k}_1, \delta(\boldsymbol{k}_2)\delta(\boldsymbol{k}_3)\rangle = \left\langle \delta^{(2)}(\boldsymbol{k}_1)\delta^{(1)}(\boldsymbol{k}_2)\delta^{(1)}(\boldsymbol{k}_3)\right\rangle + 2 \text{ cyc.} \tag{7.129}$$

We thus have (see Section 7.5.4 for details of the calculation)

$$B(\boldsymbol{k}_1, \boldsymbol{k}_2, \boldsymbol{k}_3) = 2F_2(\boldsymbol{k}_1, \boldsymbol{k}_2)P_{\mathrm{lin}}(k_1)P_{\mathrm{lin}}(k_2) + 2 \text{ cyc.} \tag{7.130}$$

Here, cyc. stands for a cyclic permutation of the three *k*-vectors in the arguments of the power spectra and coupling kernels.

As this calculation becomes more and more tedious order by order, we usually truncate the calculation at next-to-leading order or next-to-next-to-leading order. In this context it is also useful to introduce the notion of *loops*. The leading order contribution to the power spectrum is second order in the fields and doesn't involve any momentum integrations and is thus formally a zero-loop result. The next-to-leading order has to be of fourth order in the fields by Wick theorem. There are two possibilities to achieve this: by correlating two second-order density fields or by correlating a linear density field with a second-order density field

$$\langle \delta(\boldsymbol{k})\delta(\boldsymbol{k}')\rangle = \left\langle \delta^{(1)}(\boldsymbol{k})\delta^{(1)}(\boldsymbol{k}')\right\rangle + 2\left\langle \delta^{(1)}(\boldsymbol{k})\delta^{(3)}(\boldsymbol{k}')\right\rangle + \left\langle \delta^{(2)}(\boldsymbol{k})\delta^{(2)}(\boldsymbol{k}')\right\rangle \tag{7.131}$$

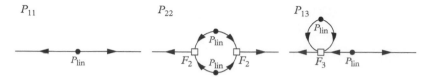

Figure 7.8 *Diagrammatic representation of the one loop matter power spectrum in eqn (7.132) where* $P_{11} = P_{lin}$.

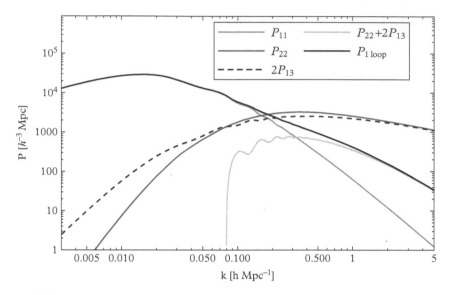

Figure 7.9 *Matter power spectrum for the currently favoured ΛCDM cosmology. We show the linear power spectrum (gray), the two contributions of the one-loop power spectrum (green and blue), the total one-loop corrections (gray), and the full one-loop power spectrum (black). Dashed parts are absolute values of negative contributions.*

$$P_{1\text{-loop}}(k) = P_{\text{lin}}(k) + 2P_{13}(k) + P_{22}(k). \tag{7.132}$$

A diagrammatic representation of the above power spectrum is given in Figure 7.9. The coupling of two second-order density fields leads to the mode coupling power spectrum P_{22}

$$P_{22}(k) = 2 \int \frac{d^3q}{(2\pi)^3} P(q) P(|\boldsymbol{k} - \boldsymbol{q}|) |F_2(\boldsymbol{q}, \boldsymbol{k} - \boldsymbol{q})|^2. \tag{7.133}$$

For explicit evaluations it is convenient to parametrize the momenta as $\boldsymbol{k} = (0, 0, k)$ and $\boldsymbol{q} = rk(\sqrt{1 - \mu^2}, 0, \mu)$

$$P_{22} = \frac{k^3}{2\pi^2} \int \mathrm{d}r\, r^2 \int \mathrm{d}\mu\; P(rk)P(\psi(r,\mu)k)\left|F_{2,\mathrm{d}}(q,\mu)\right|^2, \tag{7.134}$$

where $\psi(r,\mu) = \sqrt{1+r^2-2r\mu}$ and

$$F_{2,\mathrm{d}}(r,\mu) = \frac{7\mu + (3-10\mu^2)\,r}{14r\,(r^2-2\mu r+1)}. \tag{7.135}$$

The correlation of the linear field with a cubic field leads to the propagator P_{13}

$$P_{13}(k) = 3P_{\mathrm{lin}}(k) \int \frac{\mathrm{d}^3 q}{(2\pi)^3} P_{\mathrm{lin}}(q) F_3(\boldsymbol{k},\boldsymbol{q},-\boldsymbol{q}). \tag{7.136}$$

We realize immediately that this term is a product of a linear power spectrum and a k-dependent correction. Thus, this term is usually interpreted as the gravitational modification of the initial linear power spectrum and thus denoted propagator. The integral over the angular part can be analytically calculated leading to

$$P_{13}(k) = \frac{k^3}{252(2\pi)^2} P_{\mathrm{lin}}(k) \int \mathrm{d}r\, r^2 P_{\mathrm{lin}}(kr) \times$$
$$\times \left[\frac{12}{r^4} - \frac{158}{r^2} + 100 - 42r^2 + \frac{3}{r^5}\left(7r^2+2\right)\left(r^2-1\right)^3 \log\left(\frac{r+1}{r-1}\right) \right]. \tag{7.137}$$

As we show in detail below, the next-to-leading order calculations involve one momentum integral and are thus one-loop terms. Up to sixth order in the fields, i.e. two loops, we have

$$\langle \delta(\boldsymbol{k})\delta(\boldsymbol{k}') \rangle = \langle \delta^{(1)}(\boldsymbol{k})\delta^{(1)}(\boldsymbol{k}') \rangle + 2\langle \delta^{(1)}(\boldsymbol{k})\delta^{(3)}(\boldsymbol{k}') \rangle + 2\langle \delta^{(1)}(\boldsymbol{k})\delta^{(5)}(\boldsymbol{k}') \rangle$$
$$+ \langle \delta^{(2)}(\boldsymbol{k})\delta^{(2)}(\boldsymbol{k}') \rangle + \langle \delta^{(3)}(\boldsymbol{k})\delta^{(3)}(\boldsymbol{k}') \rangle + 2\langle \delta^{(2)}(\boldsymbol{k})\delta^{(4)}(\boldsymbol{k}') \rangle. \tag{7.138}$$

Here we separated the terms correlating non-linear density field and linear density field (propagator terms) in the first line and the mode coupling terms in the second line.

7.8.4   Properties of the One-Loop Power Spectrum

In is interesting to understand the behaviour of the mode coupling term in the limit of very small and very large external momenta. We are first considering the high-k limit $k \gg q$

$$P_{22}(k) \xrightarrow{k \gg q} \left[\frac{569}{735} P_{\text{lin}}(k) - \frac{47}{105} k \frac{dP}{dk} + \frac{1}{10} k^2 \frac{d^2P}{dk^2} \right] \int \frac{d^3q}{(2\pi)^3} P_{\text{lin}}(q) + \frac{1}{3} k^2 P_{\text{lin}}(k)$$

$$\int \frac{d^3q}{(2\pi)^3} \frac{P_{\text{lin}}(q)}{q^2} = \left[\frac{569}{735} P_{\text{lin}}(k) - \frac{47}{105} k \frac{dP}{dk} + \frac{1}{10} k^2 \frac{d^2P}{dk^2} \right] \sigma^2 + k^2 P_{\text{lin}}(k) \sigma_d^2.$$

$$(7.139)$$

In the last equality, we used the definition of the displacement dispersion σ_d^2 defined below eqn (7.107).

The low-k limit yields

$$P_{22}(k) \xrightarrow{k \ll q} \frac{9}{98} k^4 \int \frac{d^3q}{(2\pi)^3} \frac{P^2(q)}{q^4}. \tag{7.140}$$

While the exact numerical prefactor needs to be calculated from taking the limit of the F_2 kernel, the k^4 scaling can be inferred from the general asymptotic properties of the F_n kernels discussed in eqn (7.128).

In the high-k limit ($k \gg q$) P_{13} asymptotes to

$$P_{13}(k) = -\frac{1}{3} k^2 P(k) \int \frac{d^3q}{(2\pi)^3} \frac{P(q)}{q^2} \left(1 - \frac{116}{105} \frac{q^2}{k^2} + \frac{188}{245} \frac{q^4}{k^4} + \cdots \right) = -\frac{1}{2} k^2 P(k) \sigma_d^2(k).$$

$$(7.141)$$

We immediately see that the σ_d^2 contribution in $2P_{13}$ exactly cancels with the corresponding positive contribution in P_{22} above.

In the low k-limit we have instead

$$P_{13}(k) = -\frac{1}{3} k^2 P(k) \int \frac{d^3q}{(2\pi)^3} \frac{P(q)}{q^2} \left(\frac{61}{210} - \frac{2}{35} \frac{k^2}{q^2} + \cdots \right) = -\frac{61}{210} k^2 P(k) \sigma_d^2(k). \tag{7.142}$$

The convergence properties of P_{13}, P_{22} and the full one-loop result are summarized in Table 7.1.

Table 7.1 Convergence properties of the one-loop power spectrum and its components P_{22} and P_{13}.

	UV-divergent	IR-divergent
P_{13}	$n \geq -1$	$n \leq -1$
P_{22}	$n \geq 1/2$	$n \leq -1$
$P_{1-\text{loop}}$	$n \geq -1$	$n \leq -3$

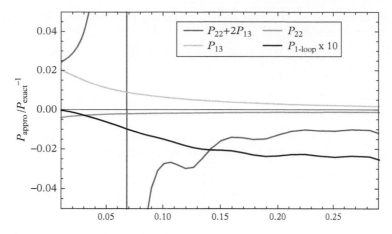

Figure 7.10 *Comparison of contributions to the one-loop matter power spectrum in the EdS approximation and the exact ΛCDM solution. Note that the fractional contribution for the full one-loop result has been multiplied by a factor of ten.*

7.9 Effective Field Theory Approach

Let us consider what we did so far—we wrote down a power series in density perturbations assuming that they are small. Let us calculate their typical size. The expectation value of δ itself vanishes, but we can calculate the variance. Let us be a bit more specific and consider the variance of the modes with wave numbers below some cut-off $k < \Lambda$. This corresponds to the variance of a field smoothed on a scale $R = 1/\Lambda$

$$\sigma_\Lambda^2 = \frac{1}{2\pi^2} \int_0^\Lambda \mathrm{d}\ln q \; q^3 P_{\mathrm{lin}}(q) \approx \frac{\Lambda^3 P_{\mathrm{lin}}(\Lambda)}{2\pi^2} \approx \Delta^2(\Lambda). \tag{7.143}$$

A power law expansion is clearly not warranted once the typical size of fluctuations exceeds unity. The variance is a growing function of Λ and the wave number at which it crosses unity defines the *non-linear wave number* k_{NL}. For the currently favoured ΛCDM model this happens at $k_{\mathrm{NL}} \approx 0.3 \; h\mathrm{Mpc}^{-1}$.

7.9.1 Problems of the Standard Treatment

Let us summarize the shortcomings of the SPT treatment that we have been using so far:

- **No well-defined expansion parameter:** On small scales the density is large, thus the power series expansion is not convergent.
- **Deviations from PPF:** On small scales, shell crossing leads to multistreaming and thus deviations from the pressureless perfect fluid model.

- **Divergences:** Table 7.1 shows that for certain power law initial power spectra the loop integrals are divergent.
- **Performance:** Figure 7.11 shows there is no notion of convergence towards the true answer.

We will see below how these issues can be addressed in the Effective Field Theory of Large-Scale Structure (EFTofLSS), pioneered by [10, 13].

7.9.2 Coarse-Grained Equations of Motion

As demonstrated above, the density is not a good variable to expand in as it can be large on small scales. We will thus integrate out the small scales by spatially smoothing the distribution function on a scale $1/\Lambda$ with $\Lambda < k_{NL}$ following [10, 13]

$$f_l(x,p,\tau) = \int d^3x'\, W_\Lambda(x-x')f(x',p,\tau). \qquad (7.144)$$

The coarse-graining procedure defines the long wavelength part of fluctuations of arbitrary fields as

$$X_l = [X]_\Lambda(x) = \int d^3x'\, W_\Lambda(|x-x'|)X(x') \qquad (7.145)$$

and we define the short wavelength part through $X_s = X - X_l$. While the exact functional form of the smoothing function is not extremely important, a Gaussian is very convenient.

$$W_\Lambda(x) = \left(\frac{\Lambda}{\sqrt{2\pi}}\right)^3 \exp\left[-\frac{1}{2}\Lambda^2 x^2\right], \qquad W_\Lambda(k) = \exp\left[-\frac{1}{2}\frac{k^2}{\Lambda^2}\right].$$

The coarse-grained density and momentum are then given by

$$\rho_l = \int d^3p\, f_l(x,p,\tau), \qquad (7.146)$$

$$\pi_l = \rho_l v_l = \frac{m}{a^3} \int d^3p\, \frac{p}{ma} f_l(x,p,\tau). \qquad (7.147)$$

Note that the long wavelength velocity is defined through $v_l = \pi_l/\rho_l$.[6] The coarse-grained fluid equations thus read

---

[6] Defining velocity in this fashion makes the velocity a contact or composite operator; thus we require counterterms for correlators involving this bare velocity.

$$\delta_1' + \partial_j \left[(1 + \delta_1) v_{1,j} \right] = 0,$$

$$v_{1,i}' + \mathcal{H} v_{1,i} + \partial_i \phi_1 + v_{1,j} \partial_j v_{1,i} = -\frac{1}{\rho_1} \partial_j \left[\tau_{ij} \right]_\Lambda.$$

The fact that we generated a non-vanishing stress tensor arises from the fact that coarse-grained products of fluctuations lead to products of long fluctuations plus corrections, for instance, the coarse-grained product of short modes (see [10] for a derivation)

$$[fg]_\Lambda = f_1 g_1 + [f_s g_s]_\Lambda + \frac{1}{\Lambda^2} \nabla f_1 \nabla g_1 + \cdots. \tag{7.148}$$

The derivative corrections in the last term will be negligible in the end as we only consider the theory for $k \ll \Lambda$. The stress tensor then arises from the microscopic stress tensor discussed above plus corrections that arise from the coarse-grained products of short fluctuations [10, 13]

$$\tau_{ij} = \rho \sigma_{ij} + \rho v_i^s v_j^s - \frac{\phi_{,k}^s \phi_{,k}^s \delta_{ij}^{(K)} - 2\phi_{,i}^s \phi_{,j}^s}{8\pi G}. \tag{7.149}$$

This implies that, even in absence of microscopic velocity dispersion σ_{ij}, the coarse-graining procedure produces an effective stress tensor or velocity dispersion.

The short scales are strongly coupled and thus cannot be treated in the EFT, but we are not really interested in their explicit behaviour. We can thus take expectation values over the short wavelength fluctuations. In QFT this procedure is known as *integrating out the UV degrees of freedom*. The result of the expectation value depends on the local amplitude of the long modes, for instances, through tidal effects. We thus split them into an expectation value over the long modes, a response and fluctuations around the mean and response

$$[f_s g_s]_\Lambda = \langle f_s g_s \rangle_\Lambda |_{\delta_1=0} + \left. \frac{\partial \langle f_s g_s \rangle}{\partial \delta_\Lambda} \right|_{\delta_1=0} \delta_1 + [f_s g_s]_\Lambda |_{\delta_1=0} - \langle f_s g_s \rangle_\Lambda + \ldots \tag{7.150}$$

This yields the *effective stress tensor* of an imperfect fluid, that at linear order in the fluctuations [13]

$$[\tau_{ij}]_\Lambda = p \delta_{ij}^{(K)} + \bar{\rho} \tilde{c}_s^2 \delta_{ij}^{(K)} \delta_1 - \bar{\rho} \frac{\tilde{c}_{v,b}^2}{\mathcal{H}} \delta_{ij}^{(K)} \partial_m v_{1,m}$$
$$- \frac{3}{4} \bar{\rho} \frac{\tilde{c}_{v,s}^2}{\mathcal{H}} \left[\partial_i v_{1,j} + \partial_j v_{1,i} - \frac{2}{3} \delta_{ij}^{(K)} \partial_m v_{1,m} \right] + \Delta \tau_{ij}. \tag{7.151}$$

Here we introduced the background pressure, the speed of sound, and the bulk- and shear-viscosity. Finally, $\Delta \tau_{ij}$ describes the deviations between expectation values and the actual realization that does not correlate with the long modes. We thus refer to this term as the *stochastic term*. Taking two spatial derivatives of the effective stress tensor, we obtain the source term of the Euler equation for the velocity divergence

$$\tau_\theta = \partial_i\partial_j\tau_{ij} = \bar{\rho}\left[\tilde{c}_s^2\partial^2\delta_1 - \frac{\tilde{c}_{v,b}^2}{\mathcal{H}}\partial^2\theta_1 - \frac{3}{4}\frac{\tilde{c}_{v,s}^2}{\mathcal{H}}\partial^2\theta_1\right] + \Delta\mathcal{J},$$

$$= \bar{\rho}\left[\tilde{c}_s^2\partial^2\delta_1 - \frac{\tilde{c}_v^2}{\mathcal{H}}\partial^2\theta_1\right] + \Delta\mathcal{J},$$

(7.152)

where $\Delta\mathcal{J} = \partial_i\partial_j\Delta\tau_{ij}$. The effective stress tensor can in principle also be motivated from combinations of second derivatives of the gravitational potential due to equivalence principle. For instance, at first order we have[7]

$$\tau_{ij} = l_1\partial_i\partial_j\phi + l_2\delta_{ij}^{(K)}\partial^2\phi$$

(7.153)

leading to $\partial_i\partial_j\tau_{ij} = \tilde{c}_s^2\partial^2\delta^{(1)}$. At second order (as relevant for the bispectrum) we would have

$$\tau_{ij} = e_1\partial_i\partial_l\phi\partial_l\partial_j\phi + e_2\delta_{ij}^{(K)}\partial^2\phi\partial^2\phi + e_3\partial_i\partial_j\phi\partial^2\phi.$$

(7.154)

Note that the second-order stress tensor has these products of density fields plus explicit second-order fields in the linear stress tensor.

Having calculated the stress tensor, we now need to solve the equations again in presence of this correction. For this purpose it is good to develop a notion of the size of various terms. As we will motivate later, we consider $c_s^2 = \mathcal{O}([\delta^{(1)}]^2)$ and $\Delta\mathcal{J} = \mathcal{O}([\delta^{(1)}]^2)$.

The inclusion of the source terms is best performed using the Greens function method for the coupled system

$$\mathcal{H}^2\left\{-a^2\partial_a^2 + \frac{3}{2}(\Omega_m(a) - 2)a\partial_a + \frac{3}{2}\Omega_m(a)\right\}\delta = \mathcal{S}_\beta - \mathcal{H}\partial_a(a\mathcal{S}_\alpha),$$

$$\mathcal{H}\left\{a^2\partial_a^2 + \left(4 - \frac{3}{2}\Omega_m(a)\right)a\partial_a + (2 - 3\Omega_m)\right\}\theta = \partial_a(a\mathcal{S}_\beta) - \frac{3}{2}\Omega_m(a)\mathcal{H}\mathcal{S}_\alpha.$$

(7.155)

The source terms now have a new contribution from the effective stress tensor

$$\mathcal{S}_\alpha(\boldsymbol{k},\tau) = -\int\frac{d^3q}{(2\pi)^3}\,\alpha(\boldsymbol{q},\boldsymbol{k}-\boldsymbol{q})\theta(\boldsymbol{q},\tau)\delta(\boldsymbol{k}-\boldsymbol{q},\tau),$$

$$\mathcal{S}_\beta(\boldsymbol{k},\tau) = -\int\frac{d^3q}{(2\pi)^3}\,\beta(\boldsymbol{q},\boldsymbol{k}-\boldsymbol{q})\theta(\boldsymbol{q},\tau)\theta(\boldsymbol{k}-\boldsymbol{q},\tau) + \tau_\theta(\boldsymbol{k},\tau).$$

(7.156)

---

[7] We often consider a gravitational potential defined by $\nabla^2\phi = \delta$, i.e. absorb $4\pi G\bar{\rho}$ into the definition of ϕ.

The solution for δ can now be obtained using the Green's function

$$c_s^2 k^2 \delta^{(1)}(k) = \delta_{c_s^2} = \int \mathrm{d}a' \, G_\delta(a, a') k^2 \left[\tilde{c}_s^2(a') + \tilde{c}_v^2(a') \right] \delta^{(1)}(k, a') \tag{7.157}$$

or with a power law ansatz $\tilde{c}_s^2 \propto a^\gamma$ leading to recursion relations for the EFT correction terms.

Finally, we obtain for the third-order density field in the EFTofLSS

$$\delta(k, \tau) = \delta^{(1)}(k, \tau) + \delta^{(2)}(k, \tau) + \delta^{(3)}(k, \tau) - k^2 c_s^2(\tau) \delta^{(1)}(k, \tau) + \delta_{\mathcal{J}}(k, \tau), \tag{7.158}$$

where we have removed the tilde on c_s^2 to account for the time integration.

7.9.3   EFT and the Power Spectrum

We are now in the position to calculate the power spectrum including the counterterms

$$P(k) = P_{\mathrm{lin}}(k) + P_{22,\Lambda}(k) + 2P_{13,\Lambda}(k) - 2c_{s,\Lambda}^2 k^2 P_{\mathrm{lin}}(k) + P_{\mathcal{J}\mathcal{J},\Lambda}(k) \tag{7.159}$$

Here we have neglected the explicit dependence of P_{11} on the cut-off Λ as it is negligible for $k \ll \Lambda$. Equation (7.159) looks like it explicitly depends on the arbitrary cut-off scale Λ that was introduced to regularize the theory. In fact this dependence is only apparent, since the speed-of-sound and stochastic counterterms have the correct functional form to capture the cut-off dependence of P_{13} and P_{22}, respectively.

Let us first consider the effect of splitting the P_{13} integral at Λ

$$P_{13,\infty}(k) = 3P_{\mathrm{lin}}(k) \int_0^\Lambda \frac{\mathrm{d}^3 q}{(2\pi)^3} F_{3,s}(k, q, -q) P_{\mathrm{lin}}(q) + 3P_{\mathrm{lin}}(k) \int_\Lambda^\infty \frac{\mathrm{d}^3 q}{(2\pi)^3} F_{3,s}(k, q, -q) P_{\mathrm{lin}}(q)$$

$$= P_{13,\Lambda}(k) - k^2 P_{\mathrm{lin}}(k) \frac{61}{210} \frac{1}{6\pi^2} \int_\Lambda^\infty \mathrm{d}q P_{\mathrm{lin}}(q). \tag{7.160}$$

Here we have used that the effect of the smoothing function $W_\Lambda(k)$ is roughly equivalent to cutting off the integral at Λ and use the low-k limit of the integral for the contributions coming from $q > \Lambda \gg k$.

The counterterms per se depend on the cut-off in a way to capture the cut-off dependence of the SPT integrals, for instance c_s^2

$$c_{s,\infty}^2 = c_{s,\Lambda}^2 - \frac{61}{210} \frac{1}{6\pi^2} \int_\Lambda^\infty \mathrm{d}q \, P_{\mathrm{lin}}(q) \tag{7.161}$$

We can now combine the cut-off dependence of $P_{13,\Lambda}$ and the running of $c_{s,\Lambda}^2$ to obtain the cut-off independent result:

$$P_{13,\Lambda}(k) - c_{s,\Lambda}^2 k^2 P_{\text{lin}}(k) = P_{13,\infty}(k) - c_{s,\infty}^2 k^2 P_{\text{lin}}(k). \qquad (7.162)$$

The residual $c_{s,\infty}^2 k^2 P_{\text{lin}}(k)$ is a correction to the standard SPT result that remains even after undoing the smoothing operation. It is a free parameter of the effective theory, a so-called low-energy constant that has to be fitted to the data. There are two ways to do this: by fitting the one-loop power spectrum, to a non-linear power spectrum or by a measurement of the effective stress tensor for a fixed smoothing scale Λ and extrapolation to $c_{s,\infty}^2$ using eqn (7.161). As we will motivate Section 7.9.5 shows that the EFT is an expansion in powers of k/k_{NL} for $k < k_{\text{NL}}$. In such an expansion contributions with a shallower slope are obviously more important. This argument motivates that the P_{13} correction from c_s^2 is more relevant than the correction of P_{22}.

We now discuss the remaining stochastic contribution. So far we have not said anything about its behaviour; only that is arises from fluctuations in the short modes. These fluctuations can be understood as a local reshuffling of material, for instance, from the original linear configuration δ to a highly non-linear configuration $\tilde{\delta}$. The mere fact that the local non-linear reshuffling has to obey mass and momentum conservation can inform us about the behaviour of the stochastic term [23, 24]:

$$
\begin{aligned}
\delta_{\mathcal{J}}(\mathbf{k}) &= \int_R \mathrm{d}^3 x \exp\left[i\mathbf{k}\cdot(\mathbf{x}_0 + \mathbf{x})\right]\left[\delta(\mathbf{x}_0 + \mathbf{x}) - \tilde{\delta}(\mathbf{x}_0 + \mathbf{x})\right] \\
&= \exp\left[i\mathbf{k}\cdot\mathbf{x}_0\right]\int_R \mathrm{d}^3 x\left[1 + i\mathbf{k}\cdot\mathbf{x} - \frac{1}{2}(\mathbf{k}\cdot\mathbf{x})^2\right]\left[\delta(\mathbf{x}_0 + \mathbf{x}) - \tilde{\delta}(\mathbf{x}_0 + \mathbf{x})\right] \qquad (7.163) \\
&= \mathcal{O}(k^2).
\end{aligned}
$$

Mass conservation ensures that the integral over $\delta - \tilde{\delta}$ vanishes and momentum conservation ensures that the same is true for the change of centre-of-mass $\mathbf{k}\cdot\mathbf{x}(\delta - \tilde{\delta})$. We can thus infer that the power spectrum of the stochastic term has to scale as

$$P_{\mathcal{J}\mathcal{J}}(k) \propto k^4. \qquad (7.164)$$

This is great news, as it corresponds to the momentum dependence of P_{22} for small wave numbers discussed in eqn (7.140). In particular it allows us to set up the running or Λ-dependence of $P_{\mathcal{J}\mathcal{J}}(k)$ in such a way that it cancels the corresponding cut-off dependence of P_{22}

$$P_{22,\Lambda}(k) + P_{\mathcal{J}\mathcal{J},\Lambda}(k) = P_{22,\infty}(k) + P_{\mathcal{J}\mathcal{J},\infty}(k). \qquad (7.165)$$

We can thus send the cut-off to infinity, i.e. calculate the usual SPT loop integrals. In the observationally favoured ΛCDM model these integrals are formally convergent, but the fact that they are running over non-perturbative wave numbers requires a counterterm

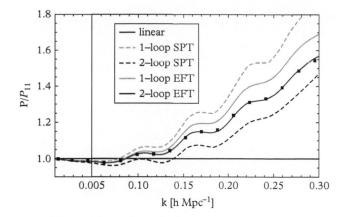

Figure 7.11 *Ratio of the non-linear and linear matter power spectrum. We show one-loop (red) and two-loop (blue) results for SPT (dashed) and EFT (solid). Clearly in SPT there is no notion of convergence as we go to higher loops. EFT fares much better, providing a good fit to the data up to $k = 0.3\ h\mathrm{Mpc}^{-1}$.*

to capture the unphysical contributions. This counterterm is conveniently provided by the residual non-vanishing correction $c_{s,\infty}^2$:

$$P(k) = P_{11}(k) + P_{22}(k) + 2P_{13}(k) - 2c_{s,\infty}^2 k^2 P_{11}(k) + P_{\mathcal{JJ},\infty}(k). \qquad (7.166)$$

Should one desire to evaluate the SPT integrals for power law initial conditions for which they would be formally divergent, the functional form of the counterterms $c_{s,\infty}^2 k^2 P_{\mathrm{lin}}$ and $P_{\mathcal{JJ},\infty}(k) \propto k^4$ are able to absorb these divergencies such that the result remains finite.

Based on a large-scale fit to the amplitude of $c_{s,\infty}^2$ we can predict the one- and two-loop power spectrum in the EFT and compare to non-linear data. As we show in Figure 7.11, the EFT predictions greatly improve the agreement with N-body simulations for $k < 0.1\ h\mathrm{Mpc}^{-1}$ (one-loop) and $k < 0.3\ h\mathrm{Mpc}^{-1}$ (two-loop) [5].

7.9.4   EFT and the Bispectrum

As mentioned in Section 7.2.3, the bispectrum as the leading non-Gaussian statistics is an important diagnostic for dynamics and field content of inflation. Uncovering the primordial bispectrum will require a precise understanding of the late time non-linear matter bispectrum. Furthermore, even for Gaussian initial conditions, clustering distributes information from the two-point function to all higher point functions and using the full potential of current and upcoming LSS surveys will require a joint analysis of the power spectrum and at least the bispectrum.

The one-loop matter bispectrum is given by

$$B = B_{112} + B_{123-I} + B_{123-II} + B_{222-I} + B_{114} \qquad (7.167)$$

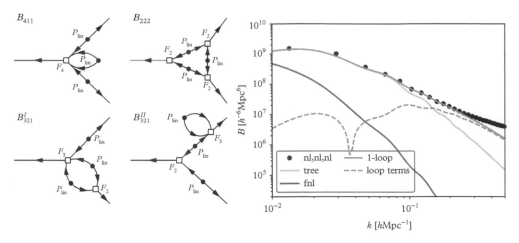

Figure 7.12 Left panel: *Diagrammatic representation of the one-loop contributions to the matter bispectrum in eqn (7.167).* Right panel: *Equilateral bispectrum $B(k,k,k)$ measured in N-body simulations, tree-level, and one-loop predictions.*

with

$$B_{222} = 8 \int_q F_2(-q, q+k_1) F_2(q+k_1, -q+k_2) F_2(k_2-q, q)$$
$$P_{\text{lin}}(q) P_{\text{lin}}(|q+k_1|) P_{\text{lin}}(|q-k_2|), \tag{7.168}$$

$$B_{321}^I = 6 P_{\text{lin}}(k_3) \int_q F_3(-q, q-k_2, -k_3) F_2(q, k_2-q) \, P_{\text{lin}}(q) P_{\text{lin}}(|q-k_2|)$$
$$+ 5 \text{ perm.}, \tag{7.169}$$

$$B_{321}^{II} = 6 F_2(k_2, k_3) \, P_{\text{lin}}(k_2) P_{\text{lin}}(k_3) \int_q F_3(k_3, q, -q) \, P_{\text{lin}}(q) + 5 \text{ perm.},$$
$$= F_2(k_2, k_3) \, P_{\text{lin}}(k_2) P_{13}(k_3) + 5 \text{ perm.}, \tag{7.170}$$

$$B_{411} = 12 P_{\text{lin}}(k_2) P_{\text{lin}}(k_3) \int_q F_4(q, -q, -k_2, -k_3) \, P_{\text{lin}}(q) + 2 \text{ cyc.}. \tag{7.171}$$

While the B_{123-II}-diagram is regularized by the c_s^2-counterterm introduced above, the regularization of the B_{411}-diagram requires a new counterterm that can be derived from the second order stress tensor in eqn (7.154) [2, 4]. Figure 7.12 shows the diagrams contributing to the one-loop bispectrum as well as the performance of tree-level and one-loop perturbation theory in reproducing the matter bispectrum measured in a N-body simulation.

7.9.5   Importance of Terms in a Scaling Universe

The understanding of the importance of various terms and the cancellation of divergencies is rather difficult in the observationally favoured ΛCDM model with the somewhat complex power spectrum discussed in Section 7.4. To gain more insight, let us consider a matter-only EdS universe with a power law initial power spectrum. These initial conditions are also denoted scale-invariant initial conditions as there is no specific scale like the matter-radiation equality that would lead to the turnover of the linear power spectrum. However, there is still the non-linear wave number k_{NL}, where the variance of the linear modes crosses unity.[8] In such a scale-invariant universe, the non-linear wave number is the only scale, and thus the linear power spectrum can be written as

$$\Delta^2_{\mathrm{lin}} = \left(\frac{k}{k_{\mathrm{NL}}} \right)^{3+n}. \tag{7.172}$$

The finite part of the loop-integrals counts the power spectra, for instance, at one-loop we have two power spectra and more generally at l-loops [23]:

$$\Delta^2_{l-\mathrm{loop}} = \left(\frac{k}{k_{\mathrm{NL}}} \right)^{(3+n)(1+l)}. \tag{7.173}$$

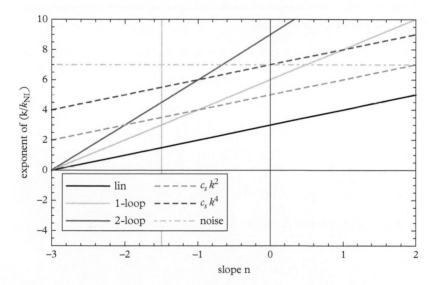

Figure 7.13 *Effective slopes of the contributions to the power spectrum in a scaling universe with initial slope n. The vertical line at n = −3/2 marks the effective slope of the power spectrum in the currently favoured ΛCDM model at the non-linear wave number k_{NL}. At this wave number the ordering of relative importance is: linear power spectrum, one-loop power spectrum, speed-of-sound corrections, two-loop power spectrum. The stochastic term is sub-leading.*

---

[8] We consider initial power spectra $P_{\mathrm{lin}}(k) = k^n$ with $n > -3$ such that $\Delta^2(k)$ is a growing function of k.

Finally, we saw that the speed of sound leads to $k^2 P_{\text{lin}}$ and the stochastic term to $P_{\tilde{J}\tilde{J}} \propto k^4$. Translating these contributions to dimensionless power spectra $\Delta^2 = k^3 P/2\pi^2$ we obtain

$$\Delta^2_{\text{stoch}} = \left(\frac{k}{k_{\text{NL}}}\right)^7, \qquad\qquad \Delta^2_{c_s^2} = \left(\frac{k}{k_{\text{NL}}}\right)^{(5+n)}. \qquad (7.174)$$

Figure 7.13 shows the amplitude of the various terms as a function of the power law slope n. Now we ask, which slope might be relevant for our ΛCDM universe. It turns out a meaningful approximation is to consider the effective slope of the linear power spectrum at the non-linear wave number k_{NL}, which is roughly $n = -3/2$. Note that a smaller slope means a bigger contribution for $k < k_{\text{NL}}$. Considering this, the ordering of importance would be: linear power > one loop SPT > one-loop counterterm c_s^2 > two loop SPT > two-loop counterterm > stochastic term. For practical purposes it is thus safe to ignore $P_{\tilde{J}\tilde{J}}$ in eqn (7.166).

7.10 Lagrangian Perturbation Theory

7.10.1 Equation of Motion

Lagrangian perturbation theory (LPT) starts from the initial unperturbed Lagrangian position q of the particles and follows their trajectory to their final position x. The Lagrangian and Eulerian coordinates are related via the mapping Ψ

$$x(q,\tau) = q + \Psi(q,\tau). \qquad (7.175)$$

The corresponding change of the density is given by the continuity relation

$$[1 + \delta(x)]\,\mathrm{d}^3 x = \mathrm{d}^3 q, \qquad (7.176)$$

where the volume distortion is described by the Jacobian

$$\mathcal{J} = \left|\frac{\mathrm{d}^3 x}{\mathrm{d}^3 q}\right| = \det\left[\delta^{(K)}_{ij} + \frac{\partial \Psi_i}{\partial q_j}\right] = \det\left[\delta^{(K)}_{ij} + \Psi_{i,j}\right]. \qquad (7.177)$$

Thus we have $\delta(x) = 1/\mathcal{J} - 1$ and consequently $\delta^{(1)}(x) = -\mathcal{J}^{(1)} = -\Psi^{(1)}_{i,i}$. Taking the divergence of the particle equation of motion eqn (7.83) and the Poisson equation we obtain

$$\mathcal{J}\,\nabla_x \cdot \left[\frac{\mathrm{d}^2 x}{\mathrm{d}\tau^2} + \mathcal{H}\frac{\mathrm{d}x}{\mathrm{d}\tau}\right] = \frac{3}{2}\Omega_{\text{m}}(a)\mathcal{H}^2(\mathcal{J} - 1). \qquad (7.178)$$

This equation is not yet completely expressed in terms of Lagrangian variables since $\nabla_{x_i} = (\delta_{ij}^{(K)} + \Psi_{i,j})^{-1}\nabla_{q_j}$. The equation can be solved perturbatively with the ansatz

$$\Psi = \Psi^{(1)} + \Psi^{(2)} + \Psi^{(3)} + \cdots \tag{7.179}$$

The determinant is equivalently expanded as

$$\mathcal{J} = 1 + \mathcal{J}^{(1)} + \mathcal{J}^{(2)} + \mathcal{J}^{(3)} + \cdots \tag{7.180}$$

expressed in terms of the displacement field as [12]

$$\mathcal{J}^{(1)} = \mathcal{L}^{(1)} = \sum_i \Psi_{i,i}^{(1)} \tag{7.181}$$

$$\mathcal{J}^{(2)} = \mathcal{L}^{(2)} + \mathcal{K}^{(2)} = \sum_i \Psi_{i,i}^{(2)} + \frac{1}{2}\sum_{i\neq j}\left\{\Psi_{i,i}^{(1)}\Psi_{j,j}^{(1)} - \Psi_{i,j}^{(1)}\Psi_{j,i}^{(1)}\right\} \tag{7.182}$$

$$\mathcal{J}^{(3)} = \mathcal{L}^{(3)} + \mathcal{K}^{(3)} + \mathcal{M}^{(3)} = \sum_i \Psi_{i,i}^{(3)} + \sum_{i\neq j}\left\{\Psi_{i,i}^{(2)}\Psi_{j,j}^{(1)} - \Psi_{i,j}^{(2)}\Psi_{j,i}^{(1)}\right\} + \det\Psi_{i,j}^{(1)} \tag{7.183}$$

where $\mathcal{L}, \mathcal{K},$ and $\mathcal{M}$ are invariant scalars of the deformation tensor. At first order we have for the equation of motion

$$\left[\frac{d^2}{d\tau^2} + \mathcal{H}\frac{d}{d\tau} - \frac{3}{2}\Omega_{\rm m}(a)\mathcal{H}^2\right]\Psi_{i,i}^{(1)} = 0. \tag{7.184}$$

This is just the linear growth equation we have seen before eqn (7.98). At first order we thus have

$$\Psi^{(1)}(\mathbf{k},\tau) = -i\frac{\mathbf{k}}{k^2}\delta^{(1)}(\mathbf{k})D(\tau). \tag{7.185}$$

Furthermore, for EdS we have with $\Psi^{(n)} \propto a^n$ we have $d^2\Psi^{(n)}/da^2 = n(n-1)\Psi^{(n)}/a^2$ and $d\Psi^{(n)}/da = n\Psi^{(n)}/a$

$$\mathcal{J}(\delta_{ij}^{(K)} + \Psi_{i,j})^{-1}\left(n^2 + \frac{n}{2}\right)\mathcal{H}^2\Psi_{i,j}^{(n)} = \frac{3}{2}\mathcal{H}^2(\mathcal{J} - 1). \tag{7.186}$$

Up to second order we have

$$(\delta_{ij}^{(K)} + \Psi_{i,j})^{-1} = \delta_{ij}^{(K)} - \Psi_{i,j} + \Psi_{i,l}\Psi_{l,j}. \tag{7.187}$$

We can now collect all the second-order terms (intrinsically second-order terms and products of first-order terms)

$$\frac{3}{2}\mathcal{J}^{(1)}\Psi_{i,i}^{(1)} - \frac{3}{2}\Psi_{i,j}^{(1)}\Psi_{i,j}^{(1)} + 5\Psi_{i,i}^{(2)} = \frac{3}{2}\mathcal{J}^{(2)} \qquad (7.188)$$

Using the expansion of the determinant we finally have

$$\frac{3}{2}\left[\Psi_{i,i}^{(1)}\Psi_{j,j}^{(1)} - \Psi_{i,j}^{(1)}\Psi_{i,j}^{(1)}\right] + 5\Psi_{i,i}^{(2)} = \frac{3}{2}\Psi_{i,i}^{(2)} + \frac{3}{4}\left[\Psi_{i,i}^{(1)}\Psi_{j,j}^{(1)} - \Psi_{i,j}^{(1)}\Psi_{i,j}^{(1)}\right]. \qquad (7.189)$$

Collecting the divergence of the second-order displacement field we obtain

$$\mathcal{K}^{(2)} = \Psi_{i,i}^{(2)} = -\frac{3}{14}\sum_{i,j}\left[\Psi_{i,i}^{(1)}\Psi_{j,j}^{(1)} - \Psi_{i,j}^{(1)}\Psi_{i,j}^{(1)}\right] = -\frac{3}{7}\mathcal{L}^{(2)}. \qquad (7.190)$$

7.10.2 Perturbative Solutions

In Fourier space we can express the nth order displacement field as a convolution of n linear density fields using the coupling kernels L_n

$$\boldsymbol{\Psi}^{(n)}(\boldsymbol{k}) = -\frac{i}{n!}\prod_{i=1}^{n}\left\{\int\frac{d^3q_i}{(2\pi)^3}\delta^{(1)}(\boldsymbol{q}_i)\right\}L_n(\boldsymbol{q}_1,\dots,\boldsymbol{q}_n)(2\pi)^3\delta^{(D)}(\boldsymbol{k}-\boldsymbol{q}_1\dots\boldsymbol{q}_n). \qquad (7.191)$$

The first order kernel is obviously $L_1(\boldsymbol{k}) = 1$. For the second-order solution we have with eqn (7.190)

$$L_2 = \frac{3}{7}\frac{k}{k^2}\left(1 - \frac{(\boldsymbol{q}_1\cdot\boldsymbol{q}_2)^2}{q_1^2 q_2^2}\right) \qquad (7.192)$$

and finally for the third-order solution [12, 20]

$$L_3 = \frac{5}{7}\frac{k}{k^2}\left[1 - \left(\frac{\boldsymbol{q}_1\cdot\boldsymbol{q}_2}{q_1 q_2}\right)^2\right]\left[1 - \left(\frac{(\boldsymbol{q}_1+\boldsymbol{q}_2)\cdot\boldsymbol{q}_3}{|\boldsymbol{q}_1+\boldsymbol{q}_2|q_3}\right)^2\right]$$
$$-\frac{1}{3}\frac{k}{k^2}\left[1 - 3\left(\frac{\boldsymbol{q}_1\cdot\boldsymbol{q}_2}{q_1 q_2}\right)^2 + 2\frac{\boldsymbol{q}_1\cdot\boldsymbol{q}_2\boldsymbol{q}_2\cdot\boldsymbol{q}_3\boldsymbol{q}_3\cdot\boldsymbol{q}_1}{q_1^2 q_2^2 q_3^2}\right]. \qquad (7.193)$$

At third order there is also a vector component, which, however, only enters two-point functions at sixth order, i.e. two-loops.

7.10.3 Calculating the Density Field

$$\delta(\mathbf{k}) = \int \mathrm{d}^3 x \exp\left[i\mathbf{k} \cdot \mathbf{x}\right] \delta(\mathbf{x}) = \int \mathrm{d}^3 x \exp\left[i\mathbf{k} \cdot \mathbf{x}\right] \left[1 + \delta(\mathbf{x})\right] - \int \mathrm{d}^3 q \exp\left[i\mathbf{k} \cdot \mathbf{q}\right]$$

$$= \int \mathrm{d}^3 q \exp\left[i\mathbf{k} \cdot \mathbf{q}\right] \left(\exp\left[i\mathbf{k} \cdot \mathbf{\Psi}(\mathbf{q})\right] - 1\right). \tag{7.194}$$

Where we used the continuity equation $[1 + \delta(\mathbf{x})]\,\mathrm{d}^3 x = \mathrm{d}^3 q$. Expanding this equation in the density field we can see that order by order LPT and SPT are equivalent.

The power spectrum is given by

$$P(k) = \int \mathrm{d}^3 r \exp\left[i\mathbf{k} \cdot \mathbf{r}\right] \langle \exp\left[i\mathbf{k} \cdot \Delta\mathbf{\Psi}\right] - 1\rangle, \tag{7.195}$$

where $\mathbf{r} = \mathbf{q} - \mathbf{q}'$ and $\Delta\mathbf{\Psi}(r) = \mathbf{\Psi}(\mathbf{q}) - \mathbf{\Psi}(\mathbf{q}')$ and that this quantity can only depend on the distance between the points by translation invariance. We can use the cumulant expansion theorem to bring the expectation value into the exponential[9] $\langle \exp\left[i\mathbf{k} \cdot \Delta\mathbf{\Psi}\right]\rangle = \exp\left[-1/2\langle(\mathbf{k} \cdot \Delta\mathbf{\Psi})^2\rangle\right]$. While LPT is correctly accounting for the effect of long wavelength motions on the BAO peak in the correlation function, per se it leads to a worse agreement for the broadband power. The reason for this is that unphysical and potentially large small-scale displacements are kept in the exponent. To retain the benefits of the resummation of long modes and account for the renormalization of the non-perturbative small-scale fluctuations, the concept of *IR resummation* was introduced [27, 6].

7.11 Biased Tracers

Apart from gravitational lensing, we have no direct way to see the dark matter distribution. What we can observe, however, are galaxies, which form at particular locations in the density field. In particular, galaxies tend to form in virialized accumulations of dark matter that are referred to as *dark matter haloes*. We can identify these dark matter haloes in N-body simulations and study their clustering statistics, but we want to develop some analytical understanding and extend the perturbation theory techniques developed to the observable tracers of LSS. Figure 7.14 shows an overview of the modelling approaches that allow us to describe the clustering of galaxies and their host haloes in terms of a given Gaussian or weakly non-Gaussian initial density fluctuation.

---

[9] For generic non-Gaussian fields we have

$$\langle \exp\left[X\right]\rangle = \exp\left[\sum_{n=1}^{\infty} \frac{1}{n!} \langle X^N\rangle_c\right]$$

where the $\langle X^N\rangle_c$ are the connected moments or cumulants.

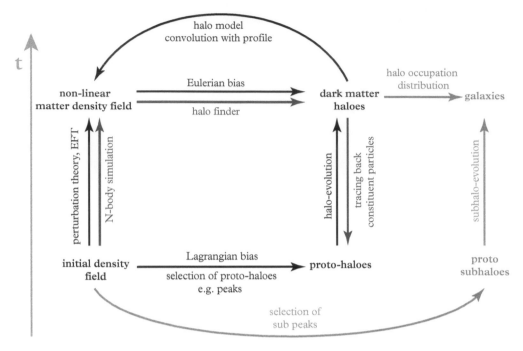

Figure 7.14 *Overview of the methods that can be employed to understand and model the distribution of matter, haloes, and galaxies starting from the close-to-Gaussian initial conditions. As described in the main text, halo biasing can be understood as the selection of regions of the smoothed Gaussian density field that exceed a certain density threshold. Those regions then need to be propagated to Eulerian space. However, the Eulerian bias treatment links the halo density to the evolved non-linear matter field. Galaxies can then be added into the framework using their halo occupation statistics.*

7.11.1   Spherical Collapse

Consider a matter-only universe of mean density ρ_b. We now consider a ball of radius $R_{b,0}$ and compress its contents into a smaller ball of radius R_0. Due to the Birkhoff theorem, the background Universe expands uniformly with scale factor $a_b = R_b/R_{b,0} = (3/2H_{b,0}t)^{2/3}$. The evolution of the overdense inner ball is given by the Friedmann equation

$$H^2 = \frac{8\pi G\rho}{3a^3} - \frac{K}{a^2} = H_0^2\left[\Omega_{m,0}a^{-3} + \left(1 - \Omega_{m,0}\right)a^{-2}\right], \tag{7.196}$$

where $a = R/R_0$ and mass conservation yields $R_0^3\rho = R_{b,0}^3\rho_b$. The time evolution of the growth factor can be parametrized by the cycloid solution

$$a = \frac{R}{R_0} = A(1 - \cos\theta), \qquad\qquad t = B(\theta - \sin\theta), \tag{7.197}$$

where

$$A = \frac{4\pi G \rho}{3K} = \frac{\Omega_{m,0}}{2(\Omega_{m,0} - 1)}, \qquad B = \frac{4\pi G \rho}{3K^{3/2}} = \frac{\Omega_{m,0}}{2H_0(\Omega_{m,0} - 1)^{3/2}}. \qquad (7.198)$$

Time runs from 0 to $t_{\text{coll}} = 2\pi B$. We expand this to obtain

$$a = A\frac{\theta^2}{2}\left(1 - \frac{\theta^2}{12}\right), \qquad\qquad t = B\frac{\theta^3}{6}\left(1 - \frac{\theta^2}{20}\right). \qquad (7.199)$$

The leading order solution is just $t = B\theta^3/6$ and $a = (3/2H_{b,0}t)^{2/3}\Omega_{m,0}^{1/3} \propto a_b$ and the correction to the background evolution is given by

$$a \approx \frac{A}{2}\left(\frac{6t}{B}\right)^{2/3}\left[1 - \frac{1}{20}\left(\frac{6t}{B}\right)^{2/3}\right]. \qquad (7.200)$$

Figure 7.15 shows the time dependence of the background solution as well as the linear and non-linear evolution of the spherical overdensity. The linear density contrast is

$$\delta_{\text{lin}} = \frac{R_b^3}{R_{\text{lin}}^3} - 1 = \frac{3}{20}\left(\frac{6t}{B}\right)^{2/3} \propto a_b. \qquad (7.201)$$

At collapse $\theta = 2\pi$ we have $t = 2\pi B$ and thus $\delta_{\text{lin}} = \frac{3}{20}(12\pi)^{2/3} \approx 1.686$. For haloes that are collapsing today, we thus expect the present day linear overdensity to equal the critical collapse threshold $\delta_c = 1.686$. The object will not collapse completely, but instead virialize halfway between turnaround and collapse. The present day, non-linear overdensity of such an object is then $1 + \delta = a_b^3(2\pi)/a^3(3\pi/2) \approx 178$.

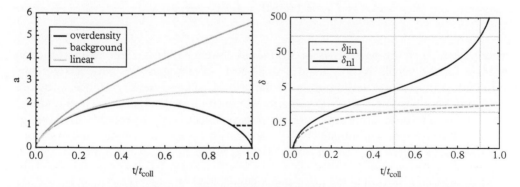

Figure 7.15 *Spherical collapse solution.* Left panel: *Scale factors of the background expansion, non-linear spherical-collapse solution and its linear approximation. Note that overdensity does not collapse all the way but instead virializes at* $\theta \approx 3\pi/2$. Right panel: *Evolution of the spherical overdensity.*

7.11.2   The Mass Function

The spherical collapse model tells us that a spherical region of size R whose present-day linear overdensity is δ_c will collapse into a virialized object of mass

$$M = \frac{4\pi}{3} R^3 \bar{\rho}. \tag{7.202}$$

Here we neglected the overdensity, since it is negligible compared to the mean density in Lagrangian space. The above one-to-one relation between mass and radius allows us to use the two interchangeably in the following. First we estimate the probability of finding a spherical region with a given density in a Gaussian random field. For this purpose we first calculate the rms amplitude of fluctuations at the scale R or mass M, smoothing the density field with a top-hat window of size R

$$\sigma^2(M) = \int \frac{d^3 q}{(2\pi)^3} P(q) W_{\mathrm{TH}}^2[qR(M)]. \tag{7.203}$$

Then the probability of finding a region that has an overdensity of δ_c is given by the probability density function of a Gaussian random variable

$$\mathbb{P}(\delta|M) = \frac{1}{\sqrt{2\pi\sigma^2(M)}} \exp\left[-\frac{1}{2}\frac{\delta^2}{\sigma^2(M)}\right]. \tag{7.204}$$

The probability for a region to exceed the density threshold δ_c is given by

$$\mathbb{P}(> \delta_c|M) = \int_{\delta_c}^{\infty} d\delta \, \mathbb{P}(\delta|M) = \int_{\nu}^{\infty} dx \, \frac{1}{\sqrt{2\pi}} \exp\left[-\frac{x^2}{2}\right], \tag{7.205}$$

where $\nu = \delta_c/\sigma$ is the *peak height* or *significance*. Since $\sigma(M)$ is a decreasing function of mass for the vanilla ΛCDM model, small-scale inhomogeneities have a larger rms amplitude and are thus the first to cross the critical collapse density. Hence, structure formation happens in a *bottom-up* scenario, where small-scale objects collapse first and merge to form more and more increasingly massive objects as time proceeds. In universes whose matter content is dominated by hot dark matter, structure formation follows a top-down scenario, where large objects form first and subsequently disintegrate into smaller objects.

Since we are interested in the regions that form a halo of mass M, we need to account for the fact that a high overdensity of mass M might be part of a larger region containing mass $M + dM$ that also exceeds the critical collapse density and could thus form a more massive halo. Thus, the fraction of regions that form haloes of mass M is given by

$$\mathbb{P}(> \delta_c | [M, M + dM]) = |\mathbb{P}(> \delta_c | M + dM) - \mathbb{P}(> \delta_c | M)| \approx -\frac{d\mathbb{P}}{dM} \tag{7.206}$$

$$= -\frac{d\mathbb{P}}{d\nu}\frac{d\nu}{d\sigma}\frac{d\sigma}{dM} = -\frac{1}{\sqrt{2\pi}}\frac{\delta_c}{\sigma^2}\exp\left[-\frac{\delta_c^2}{2\sigma^2}\right]\frac{d\sigma}{dM}. \tag{7.207}$$

The above argument accounts only for half of the mass in the Universe, since underdense regions never collapse. This can be formally seen by integrating the above formula over σ or ν, yielding $1/2$. To correct for this problem, PS introduced an ad hoc factor of 2, which can be explained in the language of uncorrelated random walks. To obtain the abundance of haloes of mass M, i.e. their *mass function*, we need to multiply this number by the maximum number of regions of mass M in a certain volume V containing mass M_{tot}, which is given by $N_{max} = M_{tot}/M$. Thus, the maximum number density is $n_{max} = M_{tot}/M/V = \bar{\rho}/M$, which is independent of the norm volume.

$$n(M) = n_{max}\mathbb{P}(> \delta_c | [M, M + dM]) = -\sqrt{\frac{2}{\pi}}\frac{\bar{\rho}}{M}\frac{\delta_c}{\sigma^2}\exp\left[-\frac{\delta_c^2}{2\sigma^2}\right]\frac{d\sigma}{dM} \tag{7.208}$$

$$= \sqrt{\frac{2}{\pi}}\frac{\bar{\rho}}{M}\exp\left[-\frac{\nu^2}{2}\right]\frac{d\nu}{dM} \tag{7.209}$$

The Press–Schechter mass function is derived by smoothing the density field on a scale $R \propto M^{1/3}$ and then looking for the biggest such region that exceeds the spherical collapse density threshold δ_c. The resulting number density of objects can be written as

$$n(M) = -\sqrt{\frac{2}{\pi}}\frac{\bar{\rho}}{M}\frac{\nu}{\sigma}\exp\left[-\frac{\nu^2}{2}\right]\frac{d\sigma}{dM} = -\sqrt{\frac{2}{\pi}}\frac{\bar{\rho}}{M^2}\nu\exp\left[-\frac{\nu^2}{2}\right]\frac{d\ln\sigma}{d\ln M}. \tag{7.210}$$

It is important to note that the above calculation was performed in the (rescaled) Gaussian initial conditions, rather than the final Eulerian distribution of matter, where the non-linear density halo patch is roughly 180 times the background density.

7.11.3 Lagrangian Bias

Biasing can be understood in the framework of the peak-background split. Similar to what we did for the coarse graining in the EFT, we define long wavelength fluctuations by smoothing the linear density field on a scale R. In contrast to the EFT above, where the smoothing scale was an ad hoc regularization scale, here the scale is a physical scale related to the mass of the object. However, as before we are interested in the long wavelength behaviour, i.e. the large-scale clustering of the haloes on scales $k < 1/R$. The scales smaller than R will determine the internal dynamics of the halo, which are somewhat irrelevant for the calculation of halo clustering statistics.

The short wavelength fluctuations with $k > 1/R$ dominate the variance, whereas the long wavelength fluctuations lead to a overall modulation of the fluctuations. Figure 7.16 shows that in a crest of the long wavelength mode δ_l it is easier to cross the collapse

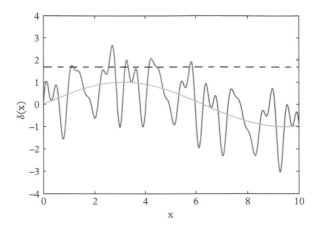

Figure 7.16 *The presence of a long wavelength mode δ_l (red) facilitates the crossing of the collapse threshold δ_c (black dashed) for a short mode (gray). This effect can be accounted for by considering an effective collapse threshold $\delta_c - \delta_l$.*

threshold, and the effective threshold thus becomes $\delta_c \to \delta_c - \delta_l$. Thus the peak height becomes

$$\nu = \frac{\delta_c}{\sigma} \to \tilde{\nu} = \frac{\delta_c - \delta_l}{\sigma}. \tag{7.211}$$

We can thus expand the mass function (local abundance of haloes) in powers of the long wavelength density fluctuation around its background value

$$n(\tilde{\nu}) = \bar{n}(\nu) + \frac{\partial n(\tilde{\nu})}{\partial \delta_l}\delta_l + \frac{1}{2}\frac{\partial^2 n(\tilde{\nu})}{\partial \delta_l^2}\delta_l^2 + \cdots. \tag{7.212}$$

Usually we are not interested in the number density per se, but the fluctuations around the mean number density, the *galaxy overdensity*

$$\delta_g(\boldsymbol{q}) = \frac{n}{\bar{n}} - 1 = b_1^{(L)}\delta_1(\boldsymbol{q}) + \frac{b_2^{(L)}}{2}\delta_1^2(\boldsymbol{q}) + \frac{b_3^{(L)}}{3!}\delta_1^3(\boldsymbol{q}) + \cdots, \tag{7.213}$$

where obviously

$$b_i^{(L)} = \frac{1}{\bar{n}}\frac{\partial^i n}{\partial \delta_1^i}$$
$$b_1^{(L)} = \frac{1}{\bar{n}}\frac{\partial n}{\partial \nu}\frac{\partial \tilde{\nu}}{\partial \delta_1} = -\frac{1}{\bar{n}}\frac{\partial n}{\partial \nu}\frac{\partial \nu}{\partial \delta_c}. \tag{7.214}$$

Here we introduced the superscript (L) to highlight that this expansion is performed in Lagrangian space, i.e. before the non-linear gravitational evolution sets in. It turns out that a power series expansion has some odd effects, starting with $\langle \delta_g \rangle \neq 0$, so we rather expand in Hermite polynomials

$$\delta_g(q) = \sum \frac{1}{i!} b_i^{(L)} H_i(\delta_1/\sigma) \sigma^i \qquad (7.215)$$

where $H_1(x) = x$, $H_2(x) = x^2 - 1$ and $H_3(x) = x^3 - 3x$.

For the Press–Schechter mass function we have

$$b_1^{(L)} = \frac{\nu^2 - 1}{\delta_c}, \qquad\qquad b_2^{(L)} = \frac{\nu^2 (\nu^2 - 3)}{\delta_c^2}. \qquad (7.216)$$

Before we discuss biasing in the Eulerian, i.e. observational late time setting, let us get the two-point function of biased tracers in Lagrangian space directly from our considerations about n-point functions of Gaussian random fields.

7.11.4 Threshold Bias

The Lagrangian matter density field is a Gaussian random field whose statistical properties are fully described by its two-point function. This remains to be true after smoothing the density field at the mass scale of the halo sample under consideration. Hence, we can calculate the exact correlation function of regions above threshold following [18].

The one- and two-point probability density functions (PDFs) of a Gaussian random field are given by

$$\mathbb{P}_{1pt}(\delta) = \frac{1}{\sqrt{2\pi\sigma^2}} \exp\left[-\frac{1}{2} \frac{\delta^2}{\sigma^2} \right] \quad \mathbb{P}_{2pt}(Y) = \frac{1}{\sqrt{(2\pi)^2 \det C}} \exp\left[-\frac{1}{2} Y C^{-1} Y^T \right] \qquad (7.217)$$

where $Y = (\delta(x_1), \delta(x_2)) = (\delta_1, \delta_2)$ and C is the covariance matrix of density fluctuations at positions x_1 and x_2. The covariance matrix is given by

$$C = \langle \delta(x_1)\delta(x_2) \rangle = \begin{pmatrix} \sigma^2 & \xi(r) \\ \xi(r) & \sigma^2 \end{pmatrix}, \qquad (7.218)$$

where $r = |x_1 - x_2|$ and ξ is the usual correlation function of the smoothed linear density field. Here we omitted the subscript R, but the density fields, correlation functions, and variance should be understood as smoothed on the scale R using a top hat filter. The inverse covariance matrix is readily written as

$$C^{-1} = \frac{1}{\sigma^4 - \xi^2(r)} \begin{pmatrix} \sigma^2 & -\xi(r) \\ -\xi(r) & \sigma^2 \end{pmatrix}. \qquad (7.219)$$

Introducing $\delta = \sigma\nu$ the two point PDF can be written as

$$\mathbb{P}_{2\text{pt}}(\nu_1, \nu_2) = \frac{1}{(2\pi)^3 \sigma^2 \sqrt{1 - \xi^2(r)/\sigma^4}} \exp\left[-\frac{1}{2} \frac{\nu_1^2 + \nu_2^2 - 2\nu_1\nu_2\xi(r)/\sigma^2}{1 - \xi^2(r)/\sigma^4}\right]. \qquad (7.220)$$

Integrating the two point PDF over all peak heights, we recover the underlying two-point function. If we are interested in the distribution of protohaloes, we would rather like to consider the regions above threshold in a Gaussian field smoothed on the mass scale of the halo sample under consideration. The probability for finding the overdensity in a certain regime of peak heights can be found by integrating the above PDFs between two limiting peak heights or from a threshold peak height to infinity. The latter case will select all haloes above a certain mass threshold and be dominated by peak heights close to the threshold due to the steeply decreasing mass function.

Section 7.2.2 shows that the correlation function is defined as the excess over random probability of finding two objects separated by a distance r

$$\mathbb{P}_{2\text{pt}}(r) = \mathbb{P}_{1\text{pt}}^2 \left[1 + \xi_{\text{tr}}(r)\right]. \qquad (7.221)$$

Thus we have for the correlation function of the thresholded regions

$$1 + \xi_{\text{tr}}(r) = \frac{\int d\nu_1 \int d\nu_2 p_{2\text{pt}}(\nu_1, \nu_2)}{\int d\nu_1 p_{1\text{pt}}(\nu_1) \int d\nu_2 p_{1\text{pt}}(\nu_2)}. \qquad (7.222)$$

The one-point function or abundance in the denominator reminds us of the PS results discussed earlier. The difference is that we are not considering the differential abundance between masses M and $M + dM$. The above equation is a non-perturbative expression for the correlation of thresholded regions in a Gaussian random field. To find the relation between the correlation function of thresholded regions and the underlying matter correlation function we can expand in the large-distance, small-correlation limit and obtain

$$\xi_{\text{tr}}(r) = \sum_{i=1}^{\infty} \frac{b_i^2}{i!} \xi^i(r). \qquad (7.223)$$

We see that at leading order in this expansion, the correlation function of thresholded regions is a linearly biased version of the smoothed correlation function. The bias is given by

$$b_1 = \frac{1}{\sigma} \frac{\int d\nu \, \nu \exp\left[-\nu^2/2\right]}{\int d\nu \exp\left[-\nu^2/2\right]} \approx \frac{\nu}{\sigma} \approx \frac{\nu^2}{\delta_c} \qquad (7.224)$$

The higher-order biases are given as

$$b_i = \frac{\int \mathrm{d}\nu f_i(\nu) \exp\left[-\nu^2/2\right]}{\int \mathrm{d}\nu \exp\left[-\nu^2/2\right]} \approx \frac{\nu^i}{\sigma^i} \approx b_1^i, \tag{7.225}$$

where the integral kernels can be expressed in terms of the probabilists Hermite polynomials H_n as

$$f_n = \frac{1}{n!\sigma^2} H_n(\nu). \tag{7.226}$$

7.11.5 Eulerian Bias

Traditionally, the galaxy density was written as a functional of the halo density using a local approximation

$$\delta_{\mathrm{g}}^{(\mathrm{E})}(x) = \delta_{\mathrm{g}}\left[\delta(x')\right] = b_1^{(\mathrm{E})}\delta(x) + \frac{1}{2}b_2^{(\mathrm{E})}\left(\delta(x) - \langle\delta^2\rangle\right) + \cdots \tag{7.227}$$

It turns out that this model is too restrictive. It does not provide a consistent fit to various halo statistics in simulations and it turns out that it is incompatible with time evolution even if the Lagrangian initial bias turns out to be local.

7.11.5.1 Dynamical Prediction

The Eulerian bias model by itself is not predictive in the sense that the Eulerian bias parameters in eqn (7.227) are free parameters that would have to be measured from the data. However, under some simplifying assumptions we can gain some insight on their amplitude by considering mapping the properties of the halo formation sites in Lagrangian space eqn (7.213) to the final evolved haloes. In Section 7.10.1 we saw that the mapping between the Lagrangian and Eulerian volume elements can be expressed as

$$[1 + \delta(x)]\,\mathrm{d}^3 x = \mathrm{d}^3 q. \tag{7.228}$$

Assuming that the haloes are comoving with the matter and that the number of haloes is conserved, we can write down a similar equation relating the halo overdensities in Lagrangian and Eulerian space

$$\left[1 + \delta_{\mathrm{h}}^{(\mathrm{E})}(x)\right]\mathrm{d}^3 x = \left[1 + \delta_{\mathrm{h}}^{(\mathrm{L})}(q)\right]\mathrm{d}^3 q. \tag{7.229}$$

Combining the two equations yields the mapping from the Lagrangian to Eulerian halo overdensity

$$\delta_{\mathrm{h}}^{(\mathrm{E})}(x) = \delta_{\mathrm{h}}^{(\mathrm{L})}(q) + \delta(x)\delta_{\mathrm{h}}^{(\mathrm{L})}(q) + \delta(x). \tag{7.230}$$

At first order we thus have

$$\delta_{\rm h}^{(\rm E,1)}(x) = \delta_{\rm h}^{(\rm L,1)}(q) + \delta^{(1)}(x) = \left[b_1^{(\rm L)} + 1\right]\delta^{(1)}(x), \qquad (7.231)$$

where we used that at first order $\delta^{(1)}(q) = \delta^{(1)}(x)$. We thus found the well-known relation $b_1^{(\rm E)} = 1 + b_1^{(\rm L)}$. At second order we have [8, 14]

$$\delta_{\rm h}^{(\rm E,2)}(x) = \left(b_1^{(\rm L)} + 1\right)\delta^{(2)}(x) + \frac{1}{2}\left(b_2^{(\rm L)} + \frac{8}{21}b_1^{(\rm L)}\right)\left[\delta^{(1)}(x)\right]^2 - \frac{2}{7}b_1^{(\rm L)}s^2(x), \quad (7.232)$$

where $s^2 = s_{ij}s_{ij}$ and the tidal tensor is defined as

$$s_{ij} = \phi_{,ij} - \frac{1}{3}\nabla^2\phi\delta_{ij}^{(\rm K)} = \left(\frac{\nabla_i\nabla_j}{\nabla^2} - \frac{1}{3}\delta_{ij}^{(\rm K)}\right)\delta. \qquad (7.233)$$

This calculation can be extended to cubic order bias parameters, which are relevant for the one-loop galaxy or halo power spectrum [25].

7.11.5.2 *Symmetry based*

We can also argue that the galaxy or halo overdensity can be written in a basis that is constructed out of all non-degenerate terms that can be constructed from scalar combinations of second derivatives of the gravitational potential (here defined as $\nabla^2\phi = \delta$) and velocity potential ($\nabla^2\phi_v = -\theta/\mathcal{H}$). At first order there is obviously only $\delta = \nabla^2\phi$ since the velocity and gravitational potential agree at this order

$$\delta_{\rm g}^{(\rm E,1)}(x) = b_1^{(\rm E)}\delta^{(1)}(x). \qquad (7.234)$$

At second order we can consider the square of the overdensity $(\nabla^2\phi)^2 = \delta^2$ but also $\phi_{,ij}\phi_{,ij}$. Subtracting the trace from $\phi_{,ij}\phi_{,ij}$, we recover the tidal field eqn (7.233) and $s^2 = s_{ij}s_{ij}$. Sometimes this operator is also expressed in terms of the second-order Galileon

$$\mathcal{G}_2(\phi) = (\nabla_i\nabla_j\phi)^2 - (\nabla^2\phi)^2, \qquad (7.235)$$

where the difference with respect to s^2 can be absorbed by the prefactor of δ^2. The difference between the velocity potential and gravitational potential is a second-order quantity, but can be expressed as a linear combination of δ^2 and s^2 as

$$\nabla^2\phi_v - \nabla^2\phi = \delta^{(2)} - \theta^{(2)} = -\frac{4}{21}\delta^2 + \frac{2}{7}s^2. \qquad (7.236)$$

Thus, the velocity potential does not explicitly appear at second order. In summary, beyond the square of the density field the general bias expansion at second order thus only needs to contain the square of the tidal tensor [21]

$$\delta_g^{(E,2)}(x) = b_1^{(E)} \delta^{(2)}(x) + \frac{b_2^{(E)}}{2} \delta^2(x) + b_{s^2}^{(E)} s^2(x). \tag{7.237}$$

Note that the functional form agrees exactly with what we predicted from the dynamics in eqn (7.232). The symmetry or EFT, based approach treats the bias parameters as nuisance parameters that need to be marginalized over for cosmological analyses. This approach is more conservative, in that it requires fewer assumptions, at the price of a larger set of nuisance parameters. A full treatment of the one-loop galaxy power spectrum requires the consideration of all bias parameters to cubic order [21, 3, 22]. There is obviously the cube of the density field δ^3. The first non-trivial new term at cubic order is the third-order Galileon operator

$$\mathcal{G}_3(\phi_g) = -\frac{1}{2}\left[(\nabla^2\phi_g)^3 + 2\nabla_i\nabla_j\phi_g\nabla^j\nabla^k\phi_g\nabla_k\nabla_i\phi_g - 3(\nabla_i\nabla_j\phi_g)^2\nabla^2\phi_g\right]. \tag{7.238}$$

Furthermore, there is now an explicit dependence on the difference of gravitational potential and velocity potential through

$$\Gamma_3(\phi_v, \phi_g) = \mathcal{G}_2^{(3)}(\phi_g) - \mathcal{G}_2^{(3)}(\phi_v), \tag{7.239}$$

with $\mathcal{G}_2^{(3)}$ is given by

$$\mathcal{G}_2^{(3)}(\phi) = 2\left(\nabla_i\nabla_j\phi^{(1)}\nabla^i\nabla^j\phi^{(2)} - \nabla^2\phi^{(1)}\nabla^2\phi^{(2)}\right). \tag{7.240}$$

Finally, there is a product of the second-order Galileon and the matter density field $\mathcal{G}_2\delta$. The cubic terms contribute to the tree-level trispectrum, which is a complicated statistic to extract from data or simulations due to the number of configurations. These operators were thus recently detected and quantified using cross-power spectra of cubic fields with the halo density field [1, 19].

7.11.6   Derivative Bias

The non-local corrections to the bias model discussed above arose from dynamical effects. While the smoothing scale $R \propto 1/\Lambda$ in the effective field theory of large-scale structure was an arbitrary and unphysical regulator that must not affect the results, our considerations in Section 7.11.1 indicate that for haloes there is a physical scale related to their mass $R \propto M^{1/3}$. All of the fields on the right-hand side of eqn (7.227) should thus be interpreted as being smoothed on the halo scale. In Fourier space this smoothing corresponds to multiplication with the Fourier transform of the smoothing function, which on large scales can be expanded as $W_R(k) \approx 1 - R^2 k^2/2$. This motivates that even at linear level in the density field there should be an additional derivative bias operator

$$\delta_g^{(E)}(x) = b_1^{(E)}\delta(x) - b_{\nabla^2\delta}\nabla^2\delta(x), \tag{7.241}$$

where $b_{\nabla^2\delta} \sim b_1^{(E)}R^2$. In Fourier space this obviously corresponds to a $k^2\delta^{(1)}(k)$ correction, which is exactly the functional form encountered for the impact of the effective stress tensor on the matter overdensity in eqn (7.158). When using the matter perturbation theory in the halo model the two effects are degenerate and can be absorbed into a new combined $b_{\nabla^2\delta}$ coefficient. There are additional considerations, for instance, in the peak model [9, 15, 16] that lead to similar $k^2\delta$ corrections, which are of particular importance around the BAO scale.

7.11.7   Stochasticity

Let us consider a finite number of tracers N such as galaxies at positions x_i in a finite volume V. Their Fourier space density field is then given by

$$\delta_g(k) = \frac{1}{\bar{n}}\sum_i \exp\left[ikx_i\right], \tag{7.242}$$

where $\bar{n} = N/V$. The power spectrum of the discrete tracers in the finite volume can then be computed as

$$
\begin{aligned}
P_{gg}(k) &= \frac{1}{V}\langle\delta(k)\delta(-k)\rangle, \\
&= \frac{V}{N^2}\sum_{i=j}\langle\exp\left[ik(x_i - x_j)\right]\rangle + \frac{V}{N^2}\sum_{i\neq j}\langle\exp\left[ik(x_i - x_j)\right]\rangle, \\
&= \frac{1}{\bar{n}} + P_{gg,\mathrm{cont}}(k).
\end{aligned}
\tag{7.243}
$$

Here, the constant $1/\bar{n}$ is denoted the *shot noise* term or *stochasticity* and have identified the non-zero lag expectation value with the continuous part of the discrete tracer power spectrum $P_{g,\mathrm{cont}}(k)$. In the local bias model at linear order we have $P_{g,\mathrm{cont}}(k) = b_1^2 P(k)$. This is clearly just an approximation for the continuous part of the tracer correlation function. The shot noise leads to a constant k^0 contribution to the observed halo or galaxy power spectrum. While the above model relates its amplitude to the inverse number density, the true stochasticity receives corrections from non-perturbative exclusion effects as well as all higher orders of bias. In the spirit of the EFT the amplitude of the k^0 term should thus be considered a free parameter, the stochasticity ϵ_2. It should be noted that stochasticity only appears in the auto power spectrum of the tracer and not in cross correlations, which are, for instance, measured in the cross correlation between galaxy clustering and CMB lensing or galaxy weak lensing.

Let us now consider the bispectrum. Following the same steps that lead to the power spectrum above, we obtain

$$B_{\mathrm{ggg}} = \frac{1}{V} \langle \delta(\boldsymbol{k}_1) \delta(\boldsymbol{k}_2) \delta(-\boldsymbol{k}_1 - \boldsymbol{k}_2) \rangle \,,$$

$$= \frac{V^2}{N^3} \sum_{i=j=l} \left\langle \exp\left[i \boldsymbol{k}_1 (\boldsymbol{x}_i - \boldsymbol{x}_l) + i \boldsymbol{k}_2 (\boldsymbol{x}_j - \boldsymbol{x}_l) \right] \right\rangle$$

$$+ 3 \frac{V^2}{N^3} \sum_{i=l \neq j} \left\langle \exp\left[i \boldsymbol{k}_1 (\boldsymbol{x}_i - \boldsymbol{x}_l) + i \boldsymbol{k}_2 (\boldsymbol{x}_j - \boldsymbol{x}_l) \right] \right\rangle \qquad (7.244)$$

$$+ \frac{V^2}{N^3} \sum_{i \neq j \neq l} \left\langle \exp\left[i \boldsymbol{k}_1 (\boldsymbol{x}_i - \boldsymbol{x}_l) + i \boldsymbol{k}_2 (\boldsymbol{x}_j - \boldsymbol{x}_l) \right] \right\rangle \,,$$

$$= \frac{1}{\bar{n}^2} + \frac{1}{\bar{n}} \left[P_{\mathrm{g,cont}}(k_1) + 2 \text{ perm} \right] + B_{\mathrm{g,cont}}(k_1, k_2, k_3) \,.$$

Again, the non-zero lag correlators are identified with the continuous power spectrum and bispectrum of the tracer field. We see that two different stochasticity corrections arise: a $1/\bar{n}^2$ constant shot noise term and a product of the shot noise and the continuous power spectrum. As above for the power spectrum, the new stochasticity term in the bispectrum needs to be regularized by an effective stochasticity ϵ_3.

7.11.8   Summary: Bias Parameters

In summary, the *n*-point functions of the galaxy/halo density field depend on the following bias parameters:

- *Tree-level power spectrum*: The galaxy/halo auto power depends on the linear bias parameter b_1 multiplying the linear power spectrum and the stochasticity ϵ_2.

- *Tree-level bispectrum*: The tree level bispectrum has a linear bias parameter multiplying the matter bispectrum as well as terms arising from the explicit quadratic couplings in the galaxy/halo density field proportional to b_2, b_{s^2} in eqn (7.237). The bispectrum shot noise needs to be parametrized by an effective stochasticity ϵ_3.

- *One-loop power spectrum*: The linear bias b_1 multiplies the one-loop matter power spectrum in eqn (7.166) including c_s^2. Beyond the second-order bias parameters b_2, b_{s^2} only the cubic bias parameter b_{Γ_3} explicitly contributes [3, 1]. The derivative bias $b_{\nabla^2 \delta}$ has to be considered at this order and can absorb all of the speed-of-sound corrections in the matter power spectrum. Finally, the stochasticity needs to be accounted for by ϵ_2. The latter can differ from the tree-level calculation as it gets renormalized by b_2 and b_{s^2}. The final expression for the galaxy–galaxy power spectrum is

$$P_{\mathrm{gg}}(k) = b_1^2 P_{\delta,\delta}^{(\mathrm{loop})}(k) + \frac{1}{4} b_2^2 P_{\delta^2,\delta^2}(k) + b_2 b_{s^2} P_{\delta^2,s^2}(k) + 2 b_{s^2}^2 P_{s^2,s^2}(k)$$

$$+ 2 b_1 \left(b_{s^2} + \frac{2}{5} b_{\Gamma_3} \right) \mathcal{I}_{s^2 \delta^{(1)}}(k) + b_1 b_2 \mathcal{I}_{\delta^{(2)} \delta^2}(k) + 2 b_1 b_{s^2} \mathcal{I}_{\delta^{(2)} s^2}(k) \qquad (7.245)$$

$$- 2 b_1 b_{\nabla^2 \delta} k^2 P_{\delta,\delta}^{(\mathrm{tree})}(k) + \epsilon_2,$$

where

$$\mathcal{I}_{s^2\delta}(k) = 4P_{\text{lin}}(k) \int_q \left(S_2(q, k-q)F_2(k, -q) - \frac{34}{63} \right) P_{\text{lin}}(q), \tag{7.246}$$

$$\mathcal{I}_{\delta^{(2)}\delta^2}(k) = 2 \int_q F_2(k-q, q)P_{\text{lin}}(q)P_{\text{lin}}(|k-q|), \tag{7.247}$$

$$\mathcal{I}_{\delta^{(2)}s^2}(k) = 2 \int_q F_2(k-q, q)S_2(k-q, q)P_{\text{lin}}(q)P_{\text{lin}}(|k-q|). \tag{7.248}$$

This model captures the weakly non-linear regime of the halo power spectrum [1, 25].

7.12   Redshift Space Distortions

While the angular positions of galaxies can be directly observed, the radial distance needs to be inferred from the redshift of the spectral lines. If the galaxies were comoving with the Hubble flow, this would indeed provide an accurate distance. However, as we have seen before, inhomogeneities in the density distribution lead to velocities beyond the Hubble flow. These peculiar velocities lead to a shift between the observed and true radial position, referred to as redshift space distortions. We will only provide a very brief description of the consequences for the observed galaxy power spectrum and refer the reader to [17] for a more in-depth discussion.

The position in redshift space s is given by the position in configuration space plus a correction proportional to the line of sight velocity

$$s = x + \frac{v \cdot \hat{n}}{\mathcal{H}} \hat{n}. \tag{7.249}$$

As the objects are simply displaced rather then created, we can employ a continuity equation between configuration and redshift space

$$[1 + \delta_s(s)] \, \mathrm{d}^3 s = [1 + \delta(x)] \, \mathrm{d}^3 x. \tag{7.250}$$

In the limit of far-away galaxies with a transverse spread small compared to the distance, we can work in the plane-parallel approximation and take the line of sight to agree with the z-axis. For the density field we thus have in analogy with the above transformations between Lagrangian and Eulerian space

$$\begin{aligned} \delta_s(k) &= \int \mathrm{d}^3 x \, \exp\left[ik \cdot s\right] \delta_s(s) \\ &= \int \mathrm{d}^3 x \, \exp\left[ik \cdot (x + v_z \hat{z}/\mathcal{H})\right](1 + \delta(x)) - \int \mathrm{d}^3 x \exp\left[ik \cdot x\right] \\ &= \delta(k) + \int \mathrm{d}^3 x \, \exp\left[ik \cdot x\right] \left[\exp\left[ik \cdot \hat{z} v_z/\mathcal{H}\right] - 1\right]\left[1 + \delta(x)\right]. \end{aligned} \tag{7.251}$$

At leading order in density and velocity we obtain the *Kaiser formula*

$$\delta_s(\boldsymbol{k}) = \delta(\boldsymbol{k}) + \mathrm{i}k \cdot \hat{\boldsymbol{n}} \frac{\hat{\boldsymbol{n}} \cdot \boldsymbol{v}(\boldsymbol{k})}{\mathcal{H}} = \left(1 + f\mu^2\right)\delta(\boldsymbol{k}), \tag{7.252}$$

where $\mu = k_{\parallel}/k$. Due to equivalence principle, on large scales biased tracers are comoving with the dark matter. Thus only the density is biased with respect to the dark matter such that we obtain for the galaxy density field in redshift space

$$\delta_s(\boldsymbol{k}) = \delta_{\mathrm{g}}(\boldsymbol{k}) + \mathrm{i}k \cdot \hat{\boldsymbol{n}} \frac{\hat{\boldsymbol{n}} \cdot \boldsymbol{v}(\boldsymbol{k})}{\mathcal{H}} = \left(1 + \beta\mu^2\right)b_1\delta(\boldsymbol{k}), \tag{7.253}$$

where $\beta = f/b_1$. The power spectrum is thus given by $P_{\mathrm{s}}(k,\mu) = b_1^2\left(1 + \beta\mu^2\right)^2 P_{\mathrm{lin}}(k)$. It is immediately obvious that redshift space distortions break statistical isotropy and that the power spectrum now has a monopole, quadrupole, and hexadecupole in terms of the direction cosine between k-mode and line of sight. As expectd, modes transverse to the line of sight ($\mu = 0$) agree with the underlying matter or tracer distribution.

Up to third order we have from eqn (7.251)

$$\begin{aligned}
\delta_s(\boldsymbol{k}) = \delta(\boldsymbol{k}) &+ \mathrm{i}\frac{k_{\parallel}}{\mathcal{H}}v_{\parallel}(\boldsymbol{k}) + \frac{\mathrm{i}^2}{2!}\left(\frac{k_{\parallel}}{\mathcal{H}}\right)^2 [v_{\parallel} \star v_{\parallel}](\boldsymbol{k}) + \frac{\mathrm{i}^3}{3!}\left(\frac{k_{\parallel}}{\mathcal{H}}\right)^3 [v_{\parallel} \star v_{\parallel} \star v_{\parallel}](\boldsymbol{k}) \\
&+ \mathrm{i}\frac{k_{\parallel}}{\mathcal{H}}[\delta \star v_{\parallel}](\boldsymbol{k}) + \frac{\mathrm{i}^2}{2!}\left(\frac{k_{\parallel}}{\mathcal{H}}\right)^2 [\delta \star v_{\parallel} \star v_{\parallel}](\boldsymbol{k}).
\end{aligned} \tag{7.254}$$

This expression contains products of density and velocity fields in configuration space, these so-called contact operators thus contain products of long and short modes, which need to be regularized by appropriate counterterms. These corrections are discussed in detail in [26] and can be included by replacing the bare correlators by a sum of the bare correlator and the appropriate counterterms

$$[v_{\parallel} \star v_{\parallel}](\boldsymbol{k}) \to [v_{\parallel} \star v_{\parallel}](\boldsymbol{k}) + c_1 + c_2\delta(\boldsymbol{k}) + c_3\frac{k_{\parallel}^2}{k^2}\delta(\boldsymbol{k}), \tag{7.255}$$

$$[v_{\parallel} \star v_{\parallel} \star v_{\parallel}](\boldsymbol{k}) \to [v_{\parallel} \star v_{\parallel} \star v_{\parallel}](\boldsymbol{k}) + 3c_1 v_{\parallel}(\boldsymbol{k}), \tag{7.256}$$

$$[\delta \star v_{\parallel} \star v_{\parallel}](\boldsymbol{k}) \to [\delta \star v_{\parallel} \star v_{\parallel}](\boldsymbol{k}) + c_1\delta(\boldsymbol{k}). \tag{7.257}$$

Identifying the momentum as $\pi_{\parallel}(\boldsymbol{k}) = v_{\parallel}(\boldsymbol{k}) + [\delta \star v_{\parallel}](\boldsymbol{k})$ we have

$$\begin{aligned}
\delta_s(\boldsymbol{k}) = \delta(\boldsymbol{k}) &+ \frac{\mathrm{i}k_{\parallel}}{\mathcal{H}}\pi_{\parallel}(\boldsymbol{k}) - \frac{1}{2}\left(\frac{k_{\parallel}}{\mathcal{H}}\right)^2 \left\{[v_{\parallel} \star v_{\parallel}](\boldsymbol{k}) + [\delta \star v_{\parallel} \star v_{\parallel}](\boldsymbol{k})\right\} \\
&- \frac{\mathrm{i}}{6}\left(\frac{k_{\parallel}}{\mathcal{H}}\right)^3 [v_{\parallel} \star v_{\parallel} \star v_{\parallel}](\boldsymbol{k}) - \frac{1}{2}(c_1 + c_2)\left(\frac{k_{\parallel}}{\mathcal{H}}\right)^2 \delta(\boldsymbol{k}) \\
&- \frac{1}{2}c_3\frac{k_{\parallel}^4}{k^2\mathcal{H}^2}\delta(\boldsymbol{k}) - \frac{\mathrm{i}}{2}c_1\left(\frac{k_{\parallel}}{\mathcal{H}}\right)^3 v_{\parallel}(\boldsymbol{k}).
\end{aligned} \tag{7.258}$$

In the redshift space power spectrum, the new counterterms appear as $(1 + f\mu^2)k^2\mu^2 P$ and $(1 + f\mu^2)k^2\mu^4 P$. The final expression is given by [26]

$$
\begin{aligned}
P_{\mathrm{s}}(k\mu) = {}& P_{\delta,\delta}^{(\mathrm{loop})}(k) + 2\frac{\mu^2}{\mathcal{H}}P_{\delta,\delta'}^{(\mathrm{loop})}(k) + \frac{\mu^4}{\mathcal{H}^2}P_{\delta',\delta'}^{(\mathrm{loop})}(k) \\
& + \frac{1}{4}\frac{k^4\mu^4}{\mathcal{H}^4}P_{[v^2][v^2]}^{(\mathrm{loop})}(k) - \frac{k^2\mu^2}{\mathcal{H}^2}P_{\delta,[v^2]}^{(\mathrm{loop})}(k) - \frac{k^2\mu^4}{\mathcal{H}^3}P_{\delta',[v^2]}^{(\mathrm{loop})}(k) \\
& - \left(1 + f\mu^2\right)\frac{k^2\mu^2}{\mathcal{H}^2}P_{\delta,[\delta v^2]}^{(\mathrm{tree})}(k) - \frac{i}{3}\left(1 + f\mu^2\right)\frac{k^3\mu^3}{\mathcal{H}^3}P_{\delta,[v^3]}^{(\mathrm{tree})}(k) \\
& + \left(1 + f\mu^2\right)\left[c_A\mu^2 + c_B\mu^4\right]k^2 P_{\delta,\delta}^{(\mathrm{tree})}(k).
\end{aligned}
\tag{7.259}
$$

Here we used that at leading order $\delta' = f\delta$. Note that $P_{\delta,\delta}^{(\mathrm{loop})}$, $P_{\delta',\delta}^{(\mathrm{loop})}$, and $P_{\delta',\delta'}^{(\mathrm{loop})}$ contain the speed-of-sound counterterm and its time derivatives and that due to $v \sim \delta/k$, only the terms in the last line are derivative suppressed by multiplicative k^2.

7.13 Summary

This chapter has explored the perturbative modelling of Large-Scale Structure starting from the statistical description of the fluctuations in terms of n-point functions via the dynamics of structure formation to the the large-scale (low-energy) effective theory that parametrizes our ignorance about the strongly coupled, non-perturbative small-scale clustering. We found that at the level of the one-loop matter power spectrum we need to include one new parameter: the speed-of-sound of the effective fluid. The matter power spectrum is relevant for the analysis of weak lensing or CMB lensing auto power spectra. It also enters in the clustering statistics for galaxies and their host haloes. For these biased tracers of LSS, we need to introduce additional operators and their corresponding amplitudes to account for the fact that halo formation is a generic non-linear function of the underlying matter distribution. Finally, galaxies are observed in redshift space, the perturbative description of which requires additional counterterms.

In summary, we end up with a theory that contains a number of free parameters but that provides a model-independent description of the LSS clustering statistics. We can then add physically relevant effects, such as the suppression of power from massive neutrinos shown in Figure 7.1, modified gravity effects, or primordial non-Gaussianity. Marginalizing over the EFT parameters, we can then constrain the physics parameters, as long as their imprints are non-degenerate with the functional forms of the EFT operators [7].

References

[1] M. M. Abidi and T. Baldauf, (2018). Cubic Halo Bias in Eulerian and Lagrangian Space. *JCAP* 1807(07): 029.

[2] Angulo, Raul E., Foreman, Simon, Schmittfull, Marcel, and Senatore, Leonardo (2015). The One-Loop Matter Bispectrum in the Effective Field Theory of Large-Scale Structures. *JCAP* 1510(10): 039.

[3] C. Assassi, D. Baumann, D. Green, and M. Zaldarriaga, (2014). Renormalized Halo Bias. *JCAP* 1408: 056.

[4] T. Baldauf, L. Mercolli, M. Mirbabayi, and E. Pajer, (2015). The Bispectrum in the Effective Field Theory of Large Scale Structure. *JCAP* 1505(05): 007.

[5] T. Baldauf, L. Mercolli, and M. Zaldarriaga, (2015). Effective Field Theory of Large Scale Structure at Two Loops: The Apparent Scale Dependence of the Speed of Sound. *Phys. Rev. D.* 92(12): 123007.

[6] T. Baldauf, M. Mirbabayi, M. Simonović, and M. Zaldarriaga (2015). Equivalence Principle and the Baryon Acoustic Peak. *Phys. Rev. D.* 92(4): 043514.

[7] T. Baldauf, M. Mirbabayi, M. Simonović, and M. Zaldarriaga (2016). LSS Constraints with Controlled Theoretical Uncertainties. [arXiv:1602.00674].

[8] T. Baldauf, U. Seljak, V. Desjacques, and P. McDonald, (2012). Evidence for Quadratic Tidal Tensor Bias from the Halo Bispectrum. *Phys. Rev. D.* 86: 083540.

[9] J. M. Bardeen, J. R. Bond, N. Kaiser, and A. S. Szalay, (1986). The Statistics of Peaks of Gaussian Random Fields. *Astrophys. J.* 304: 15–61.

[10] D. Baumann, A. Nicolis, L. Senatore, and M. Zaldarriaga, (2012). Cosmological Non-Linearities as an Effective Fluid. *JCAP*, 1207: 051.

[11] F. Bernardeau, S. Colombi, E. Gaztanaga, and P. Scoccimarro, (2002). Large Scale Structure of the Universe and Cosmological Perturbation Theory. *Phys. Rept.* 367: 1–248.

[12] F. R. Bouchet, S. Colombi, E. Hivon, and R. Juszkiewicz, (1995). Perturbative Lagrangian Approach to Gravitational Instability. *Astron. Astrophys.* 296: 575.

[13] Carrasco, John Joseph M., Hertzberg, Mark P., and Senatore, Leonardo (2012). The Effective Field Theory of Cosmological Large Scale Structures. *JHEP* 09: 082.

[14] K. C. Chan, R. Scoccimarro, and R. K. Sheth, (2012, April). Gravity and large-scale nonlocal bias. *Phys. Rev.* 85(8): 083509.

[15] V. Desjacques, (2010). The Large-Scale Clustering of Massive Dark Matter Haloes. [arXiv:1005.1105].

[16] V. Desjacques, D. Jeong, and F. Schmidt, (2018). Large-Scale Galaxy Bias. *Phys. Rept.* 733: 1–193.

[17] A. J. S. Hamilton, (1997). Linear Redshift Distortions: A Review. In *Ringberg Workshop on Large Scale Structure*. 23–28 September 1996, Ringberg, Germany.

[18] N. Kaiser, (1984, September). On the Spatial Correlations of Abell Clusters. *Astrophys. J. Let.* 284: L9–12.

[19] T. Lazeyras and F. Schmidt, (2018). Beyond LIMD Bias: A Measurement of the Complete Set of Third-Order Halo Bias Parameters. *JCAP* 1809(09): 008.

[20] T. Matsubara, (2008). Resumming Cosmological Perturbations via the Lagrangian Picture: One-Loop Results in Real Space and in Redshift Space. *Phys. Rev. D.* 77: 063530.

[21] P. McDonald and A. Roy, (2009). Clustering of Dark Matter Tracers: Generalizing Bias for the Coming Era of Precision LSS. *JCAP* 0908: 020.

[22] M. Mirbabayi, F. Schmidt, and M. Zaldarriaga, (2015). Biased Tracers and Time Evolution. *JCAP* 1507(07): 030.

[23] E. Pajer and M. Zaldarriaga, (2013). On the Renormalization of the Effective Field Theory of Large Scale Structures. *JCAP* 1308: 037.

[24] P. J. E. Peebles, (1980). *The Large-Scale Structure of the Universe*, Princeton University Press.

[25] S. Saito, et al., (2014). Understanding Higher-Order Nonlocal Halo Bias at Large Scales by Combining the Power Spectrum with the Bispectrum. *Phys. Rev. D.* 90(12): 123522.

[26] L. Senatore and M. Zaldarriaga, (2014). Redshift Space Distortions in the Effective Field Theory of Large Scale Structures. [arXiv:1409.1225].

[27] L. Senatore and M. Zaldarriaga, (2015). The IR-resummed Effective Field Theory of Large Scale Structures. *JCAP* 1502(02): 013.

[28] R. Takahashi, (2008). Third Order Density Perturbation and One-Loop Power Spectrum in a Dark Energy Dominated Universe. *Prog. Theor. Phys.* 120: 549–59.

8

Effective Theories for Quark Flavour Physics

Luca SILVESTRINI

National Institute for Nuclear Physics, Rome, Italy and CERN, Geneva, Switzerland

Luca SILVESTRINI

Silvestrini, L., *Effective Theories for Quark Flavour Physics* In: *Effective Field Theory in Particle Physics and Cosmology.* Edited by: Sacha Davidson, Paolo Gambino, Mikko Laine, Matthias Neubert and Christophe Salomon, Oxford University Press (2020). © Oxford University Press. DOI: 10.1093/oso/9780198855743.003.0008

Chapter Contents

8.1 Introduction

Quark flavour physics is among the most powerful probes of new physics (NP) beyond the standard model (SM) of electroweak and strong interactions. The sensitivity to NP in the flavour sector stems from a few peculiarities of the SM: first of all, the absence of flavour changing neutral currents (FCNC) at the tree level, which makes FCNC processes finite and therefore predictable; second, the Glashow–Iliopoulos–Maiani (GIM) suppression at the loop level [1]; third, the hierarchical structure of quark masses and mixing angles, resulting in the smallness of Jarlskog commutator [2]. Thanks to these suppression factors, NP contributions to FCNC processes generated by the exchange of heavy new particles can compete with SM amplitudes, leading to stringent bounds on the NP mass scale. As an example, in Figure 8.1 we report the bounds on the NP scale Λ obtained from $\Delta F = 2$ processes (i.e. FCNC $\bar{q}_i q_j \leftrightarrows q_i \bar{q}_j$ transitions), assuming NP contributes at tree level with coupling equal to one in all possible chiral structures. We will return to this plot at the end of these lectures, after working out the basic ingredients of the phenomenological analysis leading to these results; we can however already see that, under the above assumptions, scales up to $\mathcal{O}(10^5)$ TeV can be probed, demonstrating the extraordinary NP sensitivity of FCNC processes.

We have stated above that FCNC processes are calculable in the SM, in the sense that a prediction can be obtained (at least in principle) once all the parameters in the SM are known. However, in practice the computation of FCNC processes in the quark sector is in general a very complicated problem, for several reasons. First of all, what can be measured are transitions between a hadronic initial state (for example a K, $D_{(s)}$, $B_{(s)}$ meson or a baryon) and a leptonic, semileptonic, or non-leptonic final state. Thus, non-perturbative QCD effects connected to quark confinement are always involved, at least in the form of meson decay constants, form factors or other hadronic matrix elements. Furthermore, for non-leptonic final states we must include final state interactions, another very difficult task. Finally, the energy scales involved span several orders of magnitude, from the strong interaction scale Λ_{QCD} to the weak interaction scale M_W to even larger energies if NP is involved. Effective theories are then the best tool to cope with such multiscale processes, allowing for a systematic expansion in small ratios of widely different scales, and providing the scale separation needed to disentangle perturbative and non-perturbative strong interaction effects.

The absence of tree-level FCNC in the SM implies that NP contributions to FCNC transitions must appear as higher-dimensional operators suppressed by the NP scale Λ. If the NP scale is much larger than the weak scale, then these higher-dimensional operators will be invariant under the SM gauge group, leading to the so-called standard model effective theory (SMEFT) (see Chapter 2 [3] Chapter 3 [4] in this book for a detailed discussion of the SMEFT). Indeed, the bounds

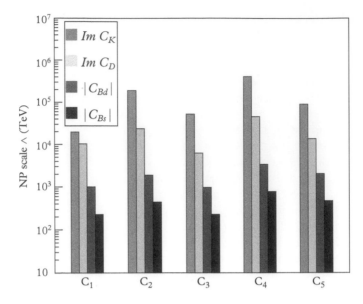

Figure 8.1 *Summary of the 95% probability lower bound on the NP scale* Λ. *See the text for details.*

presented in Figure 8.1 can be interpreted as bounds on the coefficients of SMEFT operators [5].

The impressive bounds on the NP scale reported in Figure 8.1 correspond to a generic flavour structure. This suggests that any NP close to the EW scale must have a flavour structure either identical or very similar to the SM one. This observation leads to the formulation of effective theories based on the hypothesis of minimal flavour violation (MFV) [6] or on approximate flavour symmetries [7]–[11], which can be viable at scales within the LHC reach.

The goal of these lectures is to give the reader a basic idea of how the stringent bounds in Figure 8.1 are obtained. After giving a concise review of the flavour structure of the SM in Section 8.2, Section 8.3 introduces the effective Hamiltonian for quark weak decays. Section 8.4 then considers $\Delta F = 2$ processes, generalizing to the case of NP. Section 8.5 discusses how meson–anti-meson mixing and CP violation can be described in terms of the effective Hamiltonians introduced in the previous sections. Finally, we put everything together in the context of the Unitarity Triangle Analysis in Section 8.6. Section 8.7 contains suggestions for further reading. A few useful formulæ are collected in Appendix ??.

8.2 The Flavour Structure of the Standard Model

To set the stage for our discussion, and to fix the notation, let us quickly review the flavour structure of the SM. The SM is described by the most general renormalizable

$SU(3)_c \otimes SU(2)_L \otimes U(1)_Y$ gauge-invariant Lagrangian involving three generations of leptons and quarks and one Higgs doublet:[1]

$$\mathcal{L}_{\text{SM}} = \mathcal{L}_{\text{gauge}} + \mathcal{L}_{\text{fermionic}} + \mathcal{L}_{\text{Higgs}} + \mathcal{L}_{\text{Yukawa}}, \tag{8.1}$$

$$\mathcal{L}_{\text{gauge}} = \frac{1}{4} G^a_{\mu\nu} G^{a\mu\nu} + \frac{1}{4} W^\alpha_{\mu\nu} W^{\alpha\mu\nu} + \frac{1}{4} B_{\mu\nu} B_{\mu\nu}, \tag{8.2}$$

$$\mathcal{L}_{\text{fermionic}} = \sum_f \overline{\psi}_f i D^\mu \gamma_\mu \psi_f, \tag{8.3}$$

$$\mathcal{L}_{\text{Higgs}} = \left(D_\mu \phi\right)^\dagger \left(D^\mu \phi\right) + \mu^2 \phi^\dagger \phi - \frac{\lambda}{4} \left(\phi^\dagger \phi\right)^2, \tag{8.4}$$

$$\mathcal{L}_{\text{Yukawa}} = Y^u_{ij} \overline{Q}^i_L u^j_R \phi + Y^d_{ij} \overline{Q}^i_L d^j_R \tilde{\phi} + \text{H.c.} + \dots \tag{8.5}$$

with $a = 1,\dots,8$ and $\alpha = 1,2,3$ indices in the adjoint representation of $SU(3)_c$ and $SU(2)_L$ respectively, $f = \{Q^i_L, u^i_R, d^i_R, L^i_L, \ell^i_R\}$, i and j generation indices and the ellipse in the last equation denotes lepton Yukawa couplings. $Q^i_L, u^i_R, d^i_R, L^i_L$ and ℓ^i_R represent left-handed $SU(2)_L$ quark doublets, right-handed up- and down-quarks, left-handed lepton $SU(2)_L$ doublets and right-handed charged leptons, respectively. ϕ denotes the Higgs boson doublet, with $\tilde{\phi}^i = \epsilon^{ij} \phi^*_j$. $G_{\mu\nu}$, $W_{\mu\nu}$ and $B_{\mu\nu}$ represent the field strength tensors for $SU(3)_c, SU(2)_L$, and $U(1)_Y$, respectively.

Let us now focus on the flavour quantum numbers. The first three terms in eqn (8.1) are invariant under global $U(3)_{Q_L} \otimes U(3)_{u_R} \otimes U(3)_{d_R} \otimes U(3)_{L_L} \otimes U(3)_{\ell_R}$ transformations acting on generation indices. From now on, we concentrate on quarks. The Yukawa couplings break the $U(3)_{Q_L} \otimes U(3)_{u_R} \otimes U(3)_{d_R}$ symmetry to $U(1)_B$, corresponding to baryon number conservation, an accidental symmetry of the SM. The top Yukawa coupling provides an $\mathcal{O}(1)$ breaking of the $U(3)^3$ flavour symmetry in the quark sector, while an approximate $U(2)^3$ symmetry remains valid up to terms of $\mathcal{O}(Y_c) \sim 10^{-2}$. Due to the $U(3)^3$ invariance of $\mathcal{L}_{\text{gauge}} + \mathcal{L}_{\text{fermionic}} + \mathcal{L}_{\text{Higgs}}$, the SM Yukawa couplings are defined up to an $SU(3)^3 \otimes U(1)^2$ transformation (they are invariant under $U(1)_B$ transformations), which allows to eliminate nine real parameters and seventeen phases from $Y^{u,d}$, leaving us with nine observable real parameters and one phase. Since the Lagrangian in eqn (8.1) is CP-invariant only for real Yukawa couplings, we see that the observable phase in the Yukawa couplings is responsible for CP violation in weak interactions. For two generations of fermions, the Yukawa couplings would contain $8 - (9 - 1) = 0$ observable phases, leading to CP conservation. Thus, the presence of three generations of fermions is crucial to allow for CP violation in weak interactions [12]. This strongly restricts the number of processes in which we can observe CP violation in weak interactions: CP violation can occur only in processes where all the three generations are involved, either as interfering real states or as virtual ones.

---

[1] We neglect the QCD θ term since it is irrelevant for our discussion.

8.2.1 The Cabibbo–Kobayashi–Maskawa Mixing Matrix

Let us now take into account electroweak symmetry breaking induced by the vacuum expectation value (vev) of the neutral component of the Higgs doublet ϕ:

$$\langle\phi\rangle = \frac{1}{\sqrt{2}}\begin{pmatrix} v \\ 0 \end{pmatrix}. \tag{8.6}$$

In the SM, the Higgs vev generates masses for the $W^\pm$ and Z^0 bosons through electroweak interactions as well as for the fermions through Yukawa interactions. For the latter we obtain

$$\mathcal{L}_m = m_{ij}^u \bar{u}_L^i u_R^j + m_{ij}^d \bar{d}_L^i d_R^j + H.c. \tag{8.7}$$

with $m_{ij}^{u,d} \equiv Y_{ij}^{u,d} v/\sqrt{2}$. The complex mass matrices $m_{ij}^{u,d}$ can be brought to diagonal form via a biunitary transformation:

$$U_{u_L} m^u m^{u\dagger} U_{u_L}^\dagger = U_{u_R} m^{u\dagger} m^u U_{u_R}^\dagger = (m_D^u)^2, \tag{8.8}$$

$$U_{u_L} m^u U_{u_R}^\dagger = m_D^u, \tag{8.9}$$

$$U_{d_L} m^d m^{d\dagger} U_{d_L}^\dagger = U_{d_R} m^{d\dagger} m^d U_{d_R}^\dagger = (m_D^d)^2, \tag{8.10}$$

$$U_{d_L} m^d U_{d_R}^\dagger = m_D^d, \tag{8.11}$$

with m_D^d (m_D^u) a diagonal matrix with the masses of down, strange and bottom (up, charm and top) quarks on the diagonal. We can go to the mass eigenstate basis for quarks defining

$$u_{L,R}' = U_{u_{L,R}} u_{L,R}, \qquad d_{L,R}' = U_{d_{L,R}} d_{L,R}. \tag{8.12}$$

Given the $U(3)_{u_R} \otimes U(3)_{d_R}$ invariance of $\mathcal{L}_{\text{gauge}} + \mathcal{L}_{\text{fermionic}} + \mathcal{L}_{\text{Higgs}}$, switching from unprimed to primed right-handed quarks has no effect. For left-handed fermions, it is convenient to rewrite the transformations in eqn (8.12) in the form of a transformation on Q_L followed by an additional transformation on u_L:

$$d_L' = U_{d_L} d_L, \qquad u_L' = V U_{d_L} u_L, \tag{8.13}$$

where $V \equiv U_{u_L} U_{d_L}^\dagger$ is the Cabibbo–Kobayashi–Maskawa (CKM) matrix [12, 13].

The CKM matrix can be parameterized in terms of three angles and one phase, as in the so-called 'standard' parameterization:

$$V = \begin{pmatrix} c_{12}c_{13} & s_{12}c_{13} & s_{13}e^{-i\delta} \\ -s_{12}c_{23} - c_{12}s_{23}s_{13}e^{i\delta} & c_{12}c_{23} - s_{12}s_{23}s_{13}e^{i\delta} & s_{23}c_{13} \\ s_{12}s_{23} - c_{12}c_{23}s_{13}e^{i\delta} & -c_{12}s_{23} - s_{12}c_{23}s_{13}e^{i\delta} & c_{23}c_{13} \end{pmatrix}, \tag{8.14}$$

where we have introduced the shorthand notation $s_{ij} = \sin(\theta_{ij})$, $c_{ij} = \cos(\theta_{ij})$.

Given that $s_{13} \ll s_{23} \ll s_{12} \ll 1$, a perturbative expansion in powers of the sine of the Cabibbo angle s_{12} can be performed [14], defining

$$\lambda \equiv s_{12}, \qquad A \equiv s_{23}/\lambda^2, \qquad (\rho + i\eta) \equiv s_{13}e^{i\delta}/(A\lambda^3) \tag{8.15}$$

and imposing the unitarity constraint at the desired order [15]. In particular, expanding all matrix elements up to $\mathcal{O}(\lambda^5)$, one obtains

$$V = \begin{pmatrix} 1 - \frac{\lambda^2}{2} - \frac{\lambda^4}{8} & \lambda & A\lambda^3(\rho - i\eta) \\ -\lambda + A^2\lambda^5\left(\frac{1}{2} - \rho - i\eta\right) & 1 - \frac{\lambda^2}{2} - \frac{\lambda^4(1+4A^2)}{8} & A\lambda^2 \\ A\lambda^3(1 - \overline{\rho} - i\overline{\eta}) & -A\lambda^2\left(1 - \frac{\lambda^2}{2}\right) - A\lambda^4(\rho + i\eta) & 1 - \frac{A^2\lambda^4}{2} \end{pmatrix},$$

with

$$\overline{\rho} = \rho\left(1 - \frac{\lambda^2}{2}\right), \qquad \overline{\eta} = \eta\left(1 - \frac{\lambda^2}{2}\right). \tag{8.16}$$

The unitarity of the CKM matrix implies triangular relations, which however involve sides of very different lengths, except for the ones corresponding to transitions between the first and third families, namely:

$$V_{ud}V_{ub}^* + V_{cd}V_{cb}^* + V_{td}V_{tb}^* = 0, \tag{8.17}$$
$$V_{ud}V_{td}^* + V_{us}V_{ts}^* + V_{ub}V_{tb}^* = 0, \tag{8.18}$$

where all sides are of $\mathcal{O}(\lambda^3)$. Let us focus on the relation in eqn (8.17) and divide it by the last term, defining the so-called Unitarity Triangle (UT):

$$-\frac{V_{ud}V_{ub}^*}{V_{cd}V_{cb}^*} - \frac{V_{td}V_{tb}^*}{V_{cd}V_{cb}^*} = R_b e^{i\gamma} + R_t e^{-i\beta} = 1 \simeq (\overline{\rho} + i\overline{\eta}) + (1 - \overline{\rho} - i\overline{\eta}), \tag{8.19}$$

where

$$R_b \equiv \left|\frac{V_{ud}V_{ub}^*}{V_{cd}V_{cb}^*}\right|, \quad R_t \equiv \left|\frac{V_{td}V_{tb}^*}{V_{cd}V_{cb}^*}\right|, \quad \gamma \equiv \arg\left(-\frac{V_{ud}V_{ub}^*}{V_{cd}V_{cb}^*}\right), \quad \beta \equiv \arg\left(-\frac{V_{cd}V_{cb}^*}{V_{td}V_{tb}^*}\right). \tag{8.20}$$

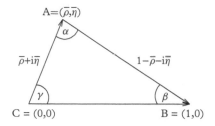

Figure 8.2 *The Unitarity Triangle.*

The UT can then be represented as a triangle in the complex $(\bar{\rho}, \bar{\eta})$ plane, see Figure 8.2. It is useful to define also

$$\alpha \equiv \arg\left(-\frac{V_{td}V_{tb}^*}{V_{ud}V_{ub}^*}\right), \qquad \beta_s \equiv \arg\left(-\frac{V_{ts}V_{tb}^*}{V_{cs}V_{cb}^*}\right). \tag{8.21}$$

The latter angle enters the 'squashed' UT corresponding to the (b,s) unitarity relation.

8.2.2   Weak Interactions Below the EW scale

After electroweak symmetry breaking, going to the quark mass eigenstate basis through the transformations in eqn (8.12) and dropping primes for simplicity, we are left with the following couplings of gauge bosons with fermionic currents:

$$\mathcal{L}_{\text{int}} = -\frac{g_2}{\sqrt{2}}\left(W_\mu^+ \mathcal{J}_{\text{ch}}^\mu + W_\mu^- \mathcal{J}_{\text{ch}}^{\mu\dagger}\right) - g_1 \cos\theta_W A_\mu \mathcal{J}_{\text{em}}^\mu - \frac{g_2}{\cos\theta_W} Z_\mu \mathcal{J}_Z^\mu, \tag{8.22}$$

$$\mathcal{J}_{\text{em}}^\mu = \sum_{f=\ell,u,d} Q_f \bar{f}_i \gamma^\mu f_i, \tag{8.23}$$

$$\mathcal{J}_Z^\mu = \sum_{f=\ell,u,d} \left(I_f^3 - Q_f \sin^2\theta_W\right) \bar{f}_L^i \gamma^\mu f_L^i - Q_f \sin^2\theta_W \bar{f}_R^i \gamma^\mu f_R^i, \tag{8.24}$$

$$\mathcal{J}_{\text{ch}}^\mu = \bar{u}_L^i V^{ij} \gamma^\mu d_L^j + \bar{\nu}_L^i \gamma^\mu \ell_L^i, \tag{8.25}$$

with g_1 and g_2 the $U(1)_Y$ and $SU(2)_L$ gauge couplings respectively, A_μ the photon field, Z_μ the Z^0 field, ν_L left-handed neutrinos, θ_W the weak mixing angle, $e = g_1 \cos\theta_W = g_2 \sin\theta_W$, $Q_\ell = -1$, $Q_u = 2/3$, $Q_d = -1/3$ and I^3 the third component of weak isospin. In the 't-Hooft–Feynman gauge, we have the following Feynman rules:

$$W^\mu \xrightarrow{\;k\;} W^\nu = \frac{-ig^{\mu\nu}}{k^2 - M_W^2 + i\epsilon}, \qquad \phi \xrightarrow{\;k\;} \phi = \frac{i}{k^2 - M_W^2 + i\epsilon}, \tag{8.26}$$

$$= \frac{ig_2}{\sqrt{2}} \gamma_\mu P_L V_{u_i d_j}^* , \qquad = \frac{ig_2}{\sqrt{2}} \gamma_\mu P_L V_{u_i d_j} , \qquad (8.27)$$

$$\phi ---- = \frac{-ig_2}{\sqrt{2}M_W} [m_{d_j} P_L - m_{u_i} P_R] \, V_{u_i d_j}^* , \qquad (8.28)$$

$$\phi ---- = \frac{-ig_2}{\sqrt{2}M_W} [m_{d_j} P_R - m_{u_i} P_L] \, V_{u_i d_j} , \qquad (8.29)$$

with $P_{L,R} = (1 \mp \gamma_5)/2$ the left- and right-handed chiral projectors.

Having fixed the notation, in the next sections we introduce the effective Hamiltonians relevant for quark flavour physics.

8.3   Effective Hamiltonians for Quark Weak Decays

Let us start by considering the amplitude for the $u\bar{d} \to \nu_\ell \bar{\ell}$ transition, which is generated at lowest order by the following Feynman diagrams:

$$(8.30)$$

The Goldstone boson exchange can be neglected here since its couplings are proportional to light fermion masses. The amplitude mediated by the W reads

$$i\mathcal{A}_W = \left(\frac{ig_2}{\sqrt{2}}\right)^2 V_{ud}^* \left(\bar{u}_{\nu_\ell}(p_{\nu_\ell})\gamma_\nu P_L v_\ell(p_\ell)\right) \left(\bar{v}_d(p_d)\gamma_\mu P_L u_u(p_u)\right) \frac{-ig^{\mu\nu}}{k^2 - M_W^2 + i\epsilon}, \qquad (8.31)$$

with $k = (p_u + p_d) = (p_\ell + p_\nu)$. Now, if we are interested in low-energy processes such as pion leptonic or semileptonic decays, we should consider external momenta of the order of the pion mass, therefore much lower than M_W. Thus, we can perform an expansion of the W propagator in powers of the momentum k, leading to

$$
i\mathcal{A}_W = -i\frac{V_{ud}^* g_2^2}{2M_W^2} \left(\bar{u}_{\nu_\ell}(p_{\nu_\ell})\gamma^\mu P_L v_\ell(p_\ell)\right) \left(\bar{v}_d(p_d)\gamma_\mu P_L u_u(p_u)\right) \sum_{n=0}^{\infty} \left(\frac{k^2}{M_W^2}\right)^n
$$

$$
\simeq -i\frac{4G_F}{\sqrt{2}} V_{ud}^* \left(\bar{u}_{\nu_\ell}(p_{\nu_\ell})\gamma^\mu P_L v_\ell(p_\ell)\right) \left(\bar{v}_d(p_d)\gamma_\mu P_L u_u(p_u)\right) + \mathcal{O}\left(\frac{k^2}{M_W^2}\right),
$$

(8.32)

where we have introduced the Fermi constant

$$
\frac{G_F}{\sqrt{2}} \equiv \frac{g_2^2}{8M_W^2},
$$

(8.33)

with $G_F = 1.1663787(6) \cdot 10^{-5}$ GeV^{-2} [16]. The dominant term in eqn (8.32) corresponds to the matrix element of the following local operator:

$$
Q^{\bar{d}u\bar{\nu}_\ell \ell} \equiv \bar{d}_L \gamma^\mu u_L \bar{\nu}_{\ell L} \gamma_\mu \ell_L,
$$

(8.34)

while the terms of order $n > 0$ in the expansion in powers of k^2/M_W^2 correspond to the matrix elements of higher dimensional local operators containing $2n$ derivatives. Keeping only the dimension six operator in this operator product expansion (OPE), we obtain

$$
\mathcal{A}_W = \langle -\mathcal{H}_{\text{eff}} \rangle + \mathcal{O}\left(\frac{k^2}{M_W^2}\right), \qquad \mathcal{H}_{\text{eff}} = \frac{G_F}{\sqrt{2}} V_{ud}^* Q^{\bar{d}u\bar{\nu}_\ell \ell},
$$

(8.35)

where we have introduced the effective Hamiltonian for $\bar{d}u \to \bar{\ell}\nu_\ell$ transitions. The effects of the exchange of the heavy W boson are encoded in the so-called Wilson coefficient, i.e. the coefficient in front of the local operator $Q^{\bar{d}u\bar{\nu}_\ell \ell}$ in $\mathcal{H}_{\text{eff}}$. External momenta are irrelevant in the matching between the full and effective theory performed in eqn (8.35), since the dynamics at scales much lower than M_W is identical in the full and effective theory, up to the desired order in the OPE.

Of course, we should now worry about the effects of strong interactions. Given the low scale at which pion decays occur, we cannot invoke any argument to suppress strong corrections to the diagram in (8.30) such as the one in the first row of Figure 8.3. However, such corrections are identical in the full theory and in the effective one, i.e. the diagrams in the first and second row of Figure 8.3 are identical. Therefore, in this example we do not need to take strong corrections into account in the matching; all strong interactions will be captured by the matrix element of $Q^{\bar{d}u\bar{\nu}_\ell \ell}$ between the relevant initial and final states.

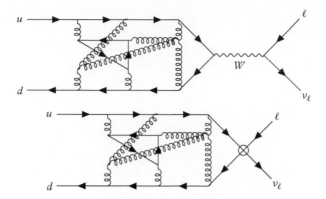

Figure 8.3 *An example of strong interaction corrections to the diagram in (8.30) (top) and its counterpart in the effective theory (bottom). The blob denotes the insertion of the four-fermion operator in eqn (8.34).*

8.3.1　Four-Quark Current–Current Operators

The situation changes dramatically if we now turn to nonleptonic decays. Consider for example $c\bar{s} \to u\bar{d}$ transitions. Neglecting Goldstone boson exchange and QCD corrections, in the SM these are described by diagram (a) in Figure 8.4. Just as in the case of $u\bar{d} \to \nu_{\ell}\bar{\ell}$ transitions discussed above, since the energy scale at which charm decays take place is much lower than M_W, the W boson will propagate over very short distances, so we can perform an OPE and consider dimension six operators only. In this case, the amplitude we obtain expanding diagram (a) at the lowest order in k^2/M_W^2 is proportional to the one generated by diagram (e) with the insertion of operator

$$Q_1^{\bar{s}c\bar{u}d} \equiv \bar{s}_L \gamma^\mu c_L \bar{u}_L \gamma_\mu d_L. \tag{8.36}$$

Imposing that the two amplitudes be equal,

$$\mathcal{A}_W = \langle -\mathcal{H}_{\text{eff}} \rangle + \mathcal{O}\left(\frac{k^2}{M_W^2}\right), \tag{8.37}$$

we obtain the corresponding Wilson coefficient:

$$\mathcal{H}_{\text{eff}} = \frac{4G_F}{\sqrt{2}} V_{ud} V_{cs}^* C_1 Q_1^{\bar{s}c\bar{d}u}, \qquad C_1 = 1. \tag{8.38}$$

As in the case of leptonic and semileptonic decays, the corrections to the SM amplitude generated by the exchange of gluons between the c and s quarks, such as the one in Figure 8.4(b), are identical in the full and in the effective theory, represented in Figure 8.4(f), so they do not enter the matching. They will be taken into account in the evaluation of the relevant hadronic matrix element of the effective Hamiltonian. The

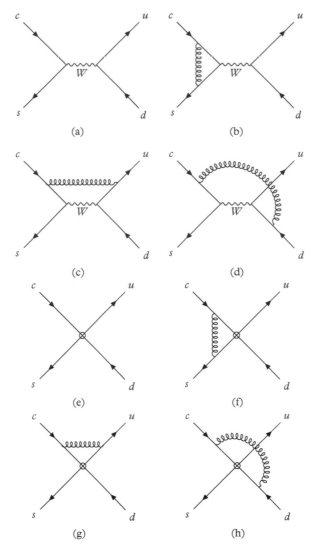

Figure 8.4 *Diagrams relevant for $c \to su\bar{d}$ transitions in the full and effective theory, including leading-order QCD corrections. See the text for details.*

same argument applies to the exchange of gluons between u and d quarks, which has not been explicitly reported in Figure 8.4.

The situation is however totally different for the exchange of gluons between the two currents coupled to the W boson (Figure 8.4(c)–(d)). Here, the W propagator has momentum $k - \ell$, where ℓ is the loop momentum; in the region $(k - \ell)^2 \sim M_W^2$ the W propagator opens up and it falls as ℓ^{-2} for $\ell^2 \gg M_W^2$, making the loop integral convergent. Thus, M_W acts as an ultraviolet (UV) regulator. Indeed, evaluating the amplitude explicitly, putting the quarks off-shell with $p^2 < 0$ to avoid infrared divergences, we find a term proportional to $\alpha_s \log\left(\frac{M_W^2}{-p^2}\right)$. Taken at face value, such term implies the breakdown of perturbation theory, since the effective expansion parameter $\alpha_s \log\left(\frac{M_W^2}{-p^2}\right)$ becomes $\mathcal{O}(1)$ for quark momenta of $\mathcal{O}(\Lambda_{\text{QCD}})$, due to the large logarithm. Fortunately, the effective theory can save us from this disaster.

General considerations

The effective theory counterpart of Figure 8.4(c)–(d) appears in Figure 8.4(g)–(h). Having removed the W propagator in the effective theory, the latter diagrams are divergent, so their UV behaviour is very different from the corresponding SM diagrams. This is no surprise, since we worked out the effective theory as an OPE expanding in powers of k^2/M_W^2, so we expect it to be valid up to a cut-off Λ of $\mathcal{O}(M_W)$; above this cut-off the contribution of all higher dimensional operators becomes unsuppressed and the expansion breaks down. Regulating diagrams (g) and (h) with the introduction of a cut-off Λ, one would obtain terms proportional to $\alpha_s \log\left(\frac{\Lambda^2}{-p^2}\right)$. Since we have seen that M_W acts as a regulator in the SM amplitude, and since the infrared logs of external momenta must be identical in the full and effective theory, it is clear that the coefficients of the log terms in the SM and in the effective theory are equal.

From the technical point of view, rather than introducing an explicit cut-off, it is much more convenient to subtract the divergences and work in the renormalized theory. In this case, the cut-off is removed but a renormalization scale μ is introduced, so that after renormalization $\log\left(\frac{\Lambda^2}{-p^2}\right)$ terms are replaced by $\log\left(\frac{\mu^2}{-p^2}\right)$.

We can match the amplitudes obtained from diagrams (c) and (d) with the ones obtained from diagrams (g) and (h) after subtracting the divergence in the effective theory via a renormalization constant Z. The infrared logs cancel and we are left with terms proportional to

$$\alpha_s \log\left(\frac{M_W^2}{-p^2}\right) - \alpha_s \log\left(\frac{\mu^2}{-p^2}\right) = \alpha_s \log\left(\frac{M_W^2}{\mu^2}\right). \tag{8.39}$$

Choosing a renormalization scale $\mu_W \sim M_W$, we can therefore get rid of large logs in the matching procedure. In this way, we can go from the full theory to the effective one using ordinary perturbation theory. The Wilson coefficient obtained from the matching now carries an explicit dependence on the renormalization scale μ, which cancels against the renormalization scale dependence of the matrix element of the renormalized operator, since the amplitude in the full theory does not depend on μ:

$$\mu\frac{\mathrm{d}}{\mathrm{d}\mu}\mathcal{A}_{\text{full}} = 0 = \mu\frac{\mathrm{d}}{\mathrm{d}\mu}\left(C(\mu)\langle Q^{\text{ren}}(\mu)\rangle\right) = \mu\frac{\mathrm{d}}{\mathrm{d}\mu}C(\mu)\langle Q^{\text{ren}}(\mu)\rangle + C(\mu)\mu\frac{\mathrm{d}}{\mathrm{d}\mu}\langle Q^{\text{ren}}(\mu)\rangle. \tag{8.40}$$

Performing the matching at $\mu_W \sim M_W$ we got rid of large logs in the matching procedure, but we actually just shifted them into the effective theory. Computing the matrix element in the effective theory, large logs of $(\mu_W^2/-p^2)$ would arise again, bringing us back into trouble. However, as discussed in detail in Chapter 1, the renormalization scale dependence of a renormalized operator is governed by the renormalization group equations (RGE) in terms of its anomalous dimension γ_Q (which is nothing else but the coefficient of the divergent terms, i.e. the coefficient of the $\log\left(\frac{M_W^2}{-p^2}\right)$ terms in the full theory):

$$\mu\frac{\mathrm{d}}{\mathrm{d}\mu}Q^{\mathrm{ren}} = -\gamma_Q Q^{\mathrm{ren}}, \qquad \gamma_Q = \frac{\mathrm{d}\log Z}{\mathrm{d}\log \mu}. \tag{8.41}$$

Combining eqns (8.40) and (8.41) we obtain the RGE for the renormalization scale dependence of the Wilson coefficient:

$$\mu\frac{\mathrm{d}}{\mathrm{d}\mu}C(\mu) = \gamma_Q C(\mu), \tag{8.42}$$

which allows us to obtain the Wilson coefficient for any μ, starting from its value at μ_W:

$$C(\mu) = U(\mu, \mu_W)C(\mu_W), \qquad U(\mu, \mu_W) = e^{\int_{g_s(\mu_W)}^{g_s(\mu)} \mathrm{d}g_s' \frac{\gamma_Q(g_s')}{\beta(g_s')}}, \tag{8.43}$$

where the β function governs the running of the strong coupling constant with the renormalization scale:

$$\beta(g_s) = -g_s\frac{\mathrm{d}\log Z_{g_s}}{\mathrm{d}\log \mu}, \tag{8.44}$$

with Z_{g_s} the renormalization constant of the $SU(3)_c$ coupling g_s. We can now run down from $\mu_W \sim M_W$ to a low renormalization scale μ_h close to the physical scale at which the process we are interested in computing occurs, and then compute the relevant matrix element (between an initial state i with momenta p_i and a final state f with momenta p_f) without encountering large logs, since $\mu_h \sim p_i \sim p_f$:

$$\langle f(p_f)|\mathcal{H}_{\mathrm{eff}}|i(p_i)\rangle = C(\mu_h)\langle f(p_f)|Q(\mu_h)|i(p_i)\rangle. \tag{8.45}$$

Where have the large logs gone? They have been resummed via the renormalization group evolution! Thus, the effective theory allows us to perform the matching using perturbation theory in the strong interactions and to resum large logs using the RGE. The evaluation of the relevant matrix elements of the local operators in $\mathcal{H}_{\mathrm{eff}}$ can then be performed, if necessary (and possible), with a nonperturbative method such as Lattice QCD.

The calculation of the β function and of the anomalous dimensions is particularly simple in mass-independent renormalization schemes such as modified minimal subtraction ($\overline{\mathrm{MS}}$) [17, 18]. In dimensional regularization, logarithmic divergences appear as singularities as the number of space-time dimensions tends to four:

$$\log\left(\frac{\Lambda^2}{-p^2}\right) \Leftrightarrow \frac{1}{\bar{\epsilon}} + \log\left(\frac{\mu^2}{-p^2}\right), \tag{8.46}$$

where

$$\frac{1}{\bar{\epsilon}} = \frac{2}{4-D} - \gamma_E + \log(4\pi) \tag{8.47}$$

and μ is the renormalization scale. In the $\overline{\text{MS}}$ scheme we renormalize the operator by subtracting the $\frac{1}{\bar{\epsilon}}$ divergence. Dropping the bar for simplicity, and writing the renormalization constant Z as a series in inverse powers of ϵ,

$$Z = 1 + \sum_k \frac{1}{\epsilon^k} Z_k(g_s), \tag{8.48}$$

and

$$\beta(g_s, \epsilon) = \frac{dg_s(\mu)}{d\log\mu} = -\epsilon g_s + \beta(g_s), \tag{8.49}$$

we obtain from eqn (8.41)

$$\gamma_Q\left(1 + \frac{1}{\epsilon} Z_1(g_s) + \ldots\right) = \frac{1}{\epsilon}\frac{dZ_1}{d\log\mu} + \ldots = \frac{1}{\epsilon}\frac{dZ_1}{dg_s}\frac{dg_s}{d\log\mu} + \ldots = \frac{1}{\epsilon}\frac{dZ_1}{dg_s}(-\epsilon g_s + \beta(g_s)) + \ldots,$$

$$\tag{8.50}$$

where the ellipses denote higher terms in the $1/\epsilon$ expansion. The finiteness of γ_Q implies

$$\gamma_Q = -2\alpha_s \frac{dZ_1}{d\alpha_s}, \tag{8.51}$$

so the anomalous dimension is directly obtained from the $\frac{1}{\epsilon}$ terms in the renormalization constant of the operator Q.

If we are interested in the dominant, log-enhanced QCD corrections, we can drop gluonic corrections to the matching (since no large logs arise at $\mu_W \sim M_W$) and compute the anomalous dimension of the operators in $\mathcal{H}_{\text{eff}}$ at the first order in α_s. In general, if we expand the anomalous dimension matrix and the Wilson coefficients in a series in α_s,

$$C(\mu) = \sum_{n=0}^{n}\left(\frac{\alpha_s}{4\pi}\right)^n C^{(n)}(\mu), \tag{8.52}$$

$$\gamma = \sum_{n=0}^{n}\left(\frac{\alpha_s}{4\pi}\right)^{(n+1)} \gamma_n, \tag{8.53}$$

we can classify the accuracy of the expansion in α_s as follows:

A leading order (LO) calculation resums all terms of $\mathcal{O}\left(\alpha_s \log\left(\frac{M_W^2}{-p^2}\right)\right)^n$, by computing the anomalous dimensions at $\mathcal{O}(\alpha_s)$ and the matching and matrix elements neglecting α_s corrections.

Expanding eqns (8.43), (8.44), and (8.45) we obtain

$$\mathcal{A}_{\rm LO} = C^{(0)}(\mu_h)\langle Q(\mu_h)\rangle^{(0)}, \tag{8.54}$$

where $\langle Q(\mu_h)\rangle^{(n)}$ denotes a matrix element computed at nth order in strong interactions and

$$C^{(0)}(\mu_h) = U_0(\mu_h,\mu_W)C^{(0)}(\mu_W), \qquad U_0 = \left(\frac{\alpha_s(\mu_W)}{\alpha_s(\mu_h)}\right)^{\frac{\gamma_0}{2\beta_0}}. \tag{8.55}$$

The LO evolutor U_0 resums all large logs. In general, as discussed in Chapter 1, QCD corrections induce mixing among different operators, so that in general $\mathcal{H}_{\rm eff}$ comprises several operators. The equations above still apply, provided we consider C and Q as vectors and γ as a matrix; in this case, $U_0 = \left(\frac{\alpha_s(\mu_W)}{\alpha_s(\mu_h)}\right)^{\frac{\gamma_0^T}{2\beta_0}}$.

A (next-to-)^m leading order (N^mLO) calculation resums all terms of $\mathcal{O}\left(\alpha_s^{n+m}\log\left(\frac{M_W^2}{-p^2}\right)\right)^n$, by computing the anomalous dimensions at $\mathcal{O}(\alpha_s^{m+1})$ and the matching and matrix elements at $\mathcal{O}(\alpha_s^m)$.

For example, at NLO we resum all terms of $\mathcal{O}\left(\alpha_s^{n+1}\log^n\left(\frac{M_W^2}{-p^2}\right)\right)$, by computing the anomalous dimension at $\mathcal{O}(\alpha_s^2)$ and the matching and matrix elements at $\mathcal{O}(\alpha_s)$. Explicitly, we have

$$\mathcal{A}_{\rm NLO} = C^{(0)}(\mu_h)\langle Q(\mu_h)\rangle^{(1)} + \frac{\alpha_s(\mu_h)}{4\pi}C^{(1)}(\mu_h)\langle Q(\mu_h)\rangle^{(0)}, \tag{8.56}$$

where

$$C^{(1)}(\mu_h) = U_0(\mu_h,\mu_W)C^{(1)}(\mu_W) + \left(\mathcal{J}U_0(\mu_h,\mu_W) + \frac{\alpha_s(M_W)}{\alpha_s(\mu)}U_0(\mu_h,\mu_W)\mathcal{J}\right)C^{(0)}(\mu_W), \tag{8.57}$$

where the matrix $\mathcal{J}$ is obtained from γ_1 and β_1 as explained by Buras in his renowned lectures [19], where a complete pedagogical introduction to the subtleties of NLO calculations is presented. Although here we confine ourselves to LO calculations, the importance of computing weak Hamiltonians to NLO (or above) cannot be overemphasized.

We emphasize that the $\overline{\rm MS}$ scheme, although very convenient for perturbative calculations, is not the only option. For example, matrix elements computed in Lattice

QCD (LQCD) potentially take into account strong interactions to all orders in the non-perturbative regime; it is therefore possible (and desirable) to perform non-perturbative renormalization, subtracting divergences to all orders in perturbation theory [20]. To achieve this result, it is convenient to use the so-called regularization-independent renormalization schemes, that are defined by fixing the value of a given number of renormalized Green functions. For example, instead of defining the renormalized four-quark operator by subtracting the $1/\epsilon$ poles in dimensional regularization at a given perturbative order, we could define it by imposing that its matrix element on given initial and final states be equal to a given number, for example to the tree-level matrix element of the same operator. This renormalization condition can be implemented in any regularization at any perturbative order, making it possible to match the perturbative calculation of the Wilson coefficient with the non-perturbative calculation of the hadronic matrix element.

Current-current operators at LO

As discussed, if we are interested in capturing the dominant, log-enhanced QCD corrections only, we just need to start at μ_W with the effective Hamiltonian obtained from tree-level matching, eqn (8.38), and run it down to μ_h using eqn (8.42) with the anomalous dimension computed at $\mathcal{O}(\alpha_s)$. To compute the latter, we need to identify the $1/\epsilon$ terms generated by Figure 8.4(f)–(h), plus the 'mirror' ones reported in Figure 8.5. We can actually skip diagrams (f) and (i) since they cancel against the renormalization constants for the quark fields due to the Ward identity that protects the conserved weak current. Let us therefore start from diagram (g). Assigning momentum p to the incoming c and outgoing u quarks, and loop momentum k to the fermions in the loop, we obtain the following amplitude:

$$i\mathcal{A}_{(g)} = \frac{4G_F V_{ud} V_{cs}^*}{\sqrt{2}} \int \frac{d^D k}{(2\pi)^D} \overline{u}_i^u \left(ig_s\gamma_\mu T_{ij}^A\right) \frac{i}{\slashed{k}}\gamma^\rho P_L v_j^d \, \overline{v}_k^s \gamma_\rho P_L \frac{i}{\slashed{k}}\left(ig_s\gamma_\nu T_{kl}^B\right) u_l^c \frac{-ig^{\mu\nu}\delta^{AB}}{(k-p)^2}$$

where we have used the following Feynman rules for QCD in the Feynman gauge:

$$g_A^\mu \, \text{eeee} \, g_B^v = \frac{-ig^{\mu v}\,\delta^{AB}}{k^2 + i\epsilon}, \qquad g_\mu^A \, \text{wwww} = ig_s\,\gamma_\mu T_{ij}^A,$$

(8.58)

with A, B colour indices in the adjoint representation, i, j, k, l colour indices in the fundamental representation and T the $SU(3)$ generators for the fundamental representation.

Pulling out the Dirac structure we can rewrite the amplitude as

$$i\mathcal{A}_{(g)} = -i\frac{4G_F}{\sqrt{2}} V_{ud} V_{cs}^* g_s^2 \, T_{ij}^A T_{kl}^A \overline{u}_i^u \gamma_\mu\gamma_\alpha\gamma^\rho P_L v_j^d \, \overline{v}_k^s \gamma_\rho P_L \gamma_\beta \gamma^\mu u_l^c \mathcal{I}^{\alpha\beta}, \qquad (8.59)$$

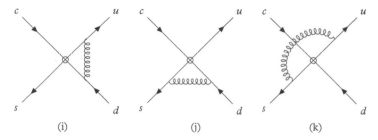

Figure 8.5 *'Mirror' diagrams relevant for $c \to s u \bar{d}$ transitions in the effective theory. See the text for details.*

with

$$
\begin{aligned}
\mathcal{I}^{\alpha\beta} &= \int \frac{d^D k}{(2\pi)^D} \frac{k^\alpha k^\beta}{(k^2)^2 (k-p)^2} \\
&= \int_0^1 dx\, 2(1-x) \int \frac{d^D k}{(2\pi)^D} \frac{k^\alpha k^\beta}{[k^2(1-x) + (k-p)^2 x]^3} \\
&= \int_0^1 dx\, 2(1-x) \int \frac{d^D \ell}{(2\pi)^D} \frac{(\ell+px)^\alpha (\ell+px)^\beta}{[\ell^2 + p^2 x(1-x)]^3} \\
&= \int_0^1 dx\, 2(1-x) \int \frac{d^D \ell}{(2\pi)^D} \frac{\ell^\alpha \ell^\beta}{[\ell^2 + p^2 x(1-x)]^3} + \text{f.t.},
\end{aligned}
\tag{8.60}
$$

where we have introduced the Feynman parameter x using

$$
\frac{1}{a^n b} = n \int_0^1 dx \frac{x^{n-1}}{[(1-x)b + xa]^{n+1}}
\tag{8.61}
$$

(a particular case of the general parameterization in Appendix 8.A.1), and the last equality holds up to non-divergent terms.

Computing the integral on ℓ using the formulae in Appendix 8.A.2, we obtain

$$
\begin{aligned}
\mathcal{I}^{\alpha\beta} &= \int_0^1 dx\, 2(1-x) \frac{g^{\alpha\beta}}{2} \frac{i}{(4\pi)^{D/2}} \frac{\Gamma(3 - D/2 - 1)}{\Gamma(3)} \left(\frac{\mu^2}{-p^2 x(1-x)} \right)^{3 - D/2 - 1} \\
&= \frac{i}{16\pi^2} \frac{g^{\alpha\beta}}{4} \frac{1}{\epsilon} + \text{f.t.},
\end{aligned}
\tag{8.62}
$$

again up to irrelevant finite terms.

Since we are only interested in the divergent terms, we can also perform the Dirac algebra in four dimensions. This greatly simplifies things since we are then authorized to use the so-called Fierz identities [21]. Indeed, in four dimensions we can identify a complete basis for objects carrying two spinor indices. For example, we might choose

$$\mathbb{1}, \; \gamma_5, \; \gamma^\mu, \; \gamma^\mu \gamma_5, \; \sigma^{\mu\nu}, \tag{8.63}$$

but for our purposes it is more convenient to work in a chiral basis such as

$$P_L, \; P_R, \; \gamma^\mu P_L, \; \gamma^\mu P_R, \; \sigma^{\mu\nu}. \tag{8.64}$$

Thus, any object carrying two spinor indices can be projected on this basis. Now, looking at the Dirac string in eqn (8.59), we see several Dirac matrices with Lorentz indices contracted across the two different fermionic lines. Simplifying the Dirac structure would be simple if those Dirac matrices were in the same fermionic line. We can bring all those matrices together using a Fierz transformation,[2] by projecting

$$\left(P_L v_j^d \bar{v}_k^s P_R \right)_{\alpha\beta} \tag{8.65}$$

on the basis in eqn (8.64). Since in eqn (8.65) we have left-handed chirality on the left and right-handed chirality on the right, it is clear that the only structure we can project on is $\gamma^\mu P_R$. To find out the coefficient, we can act on eqn (8.65) with the operator $\frac{1}{2} \mathrm{Tr} \, \gamma^\mu P_L$, which is a projector on $\gamma^\mu P_R$, since

$$\frac{1}{2} \mathrm{Tr} \, \gamma^\mu P_L P_L = 0, \qquad \frac{1}{2} \mathrm{Tr} \, \gamma^\mu P_L P_R = 0, \qquad \frac{1}{2} \mathrm{Tr} \, \gamma^\mu P_L \gamma^\nu P_L = 0, \tag{8.66}$$

$$\frac{1}{2} \mathrm{Tr} \, \gamma^\mu P_L \gamma^\nu P_R = g^{\mu\nu}, \qquad \frac{1}{2} \mathrm{Tr} \, \gamma^\mu P_L \sigma^{\nu\rho} = 0. \tag{8.67}$$

We obtain

$$\frac{1}{2} \mathrm{Tr} \, \gamma^\mu P_L P_L v_j^d \bar{v}_k^s P_R = -\frac{1}{2} \bar{v}_k^s \gamma^\mu P_L v_j^d, \tag{8.68}$$

where we have added a minus sign since the v are anti-commuting spinors. We have thus obtained that

$$\left(P_L v_j^d \bar{v}_k^s P_R \right)_{\alpha\beta} = -\frac{1}{2} \bar{v}_k^s \gamma^\mu P_L v_j^d \left(\gamma_\mu P_R \right)_{\alpha\beta}. \tag{8.69}$$

Substituting eqn (8.69) in eqn (8.59), and keeping into account that the divergent part of $\mathcal{I}^{\alpha\beta}$ is proportional to $g^{\alpha\beta}$, the Dirac string becomes

$$-\frac{1}{2} \bar{v}_k^s \gamma^\nu P_L v_j^d \, \bar{u}_i^u \gamma_\mu \gamma_\alpha \gamma_\rho \gamma_\nu P_R \gamma^\rho \gamma^\alpha \gamma^\mu u_l^c. \tag{8.70}$$

Using the four-dimensional Dirac algebra rules, we move the chiral projector to the right and apply thrice the identity $\gamma^\alpha \gamma^\mu \gamma_\alpha = -2\gamma^\mu$ to obtain

---

[2] I am indebted to R. K. Ellis for pointing this trick out to me in an exercise session at the Parma school of theoretical physics in September 2001.

$$4\bar{v}_k^s \gamma^\nu P_L v_j^d \, \bar{u}_i^u \gamma_\nu P_L u_l^c. \tag{8.71}$$

Notice that the ordering of the spinors here is different with respect to the tree-level amplitude generated by $Q_1^{\bar{s}\bar{c}du}$; thus, to identify the relevant counterterms, let us use the Fierz trick again to exchange the v_j^d and u_l^c spinors. This is easily achieved:

$$\bar{v}_k^s \gamma^\nu P_L v_j^d \, \bar{u}_i^u \gamma_\nu P_L u_l^c = -\frac{1}{2} \bar{u}_i^u \gamma_\mu P_L v_j^d \, \bar{v}_k^s \gamma^\nu \gamma^\mu P_R \gamma_\nu u_l^c = \bar{u}_i^u \gamma_\mu P_L v_j^d \, \bar{v}_k^s \gamma^\mu P_L u_l^c \tag{8.72}$$

where in the first step we used again eqn (8.69). Before putting all the pieces together, let us look at the colour factor in eqn (8.59). We can simplify it using the $SU(N_c)$ identity

$$T_{ij}^A T_{kl}^A = \frac{1}{2}\delta_{il}\delta_{kj} - \frac{1}{2N_c}\delta_{ij}\delta_{kl}. \tag{8.73}$$

Putting everything together, we obtain the amplitude generated by diagram (g):

$$i\mathcal{A}_{(g)} = \frac{4G_F}{\sqrt{2}} V_{ud} V_{cs}^* \frac{\alpha_s}{4\pi} \frac{1}{2\epsilon} \left(\bar{u}_i^u \gamma_\mu P_L v_j^d \, \bar{v}_j^s \gamma^\mu P_L u_i^c - \frac{1}{N_c} \bar{u}^u \gamma_\mu P_L v^d \, \bar{v}^s \gamma^\mu P_L u^c \right). \tag{8.74}$$

We see that this diagram generates a divergence proportional to the matrix element of a new operator,

$$Q_2^{\bar{s}\bar{c}ud} \equiv \bar{s}_L^\alpha \gamma^\mu c_L^\beta \, \bar{u}_L^\beta \gamma_\mu d_L^\alpha. \tag{8.75}$$

Thus, the divergence we obtain in the effective theory (or, equivalently, the large log present in the full theory) is proportional not only to the operator generated omitting QCD corrections, but also to a new operator. This forces us to promote the anomalous dimension to a matrix and the operator basis to a vector.

Let us now consider Figure 8.4(h). For convenience, let us assign momentum $-p$ to the $\bar{u}d$ quark line, so that we end up with exactly the same integral as in diagram (g). Since we are interested in the UV divergence, the choice of external momenta is irrelevant. We obtain:

$$i\mathcal{A}_{(h)} = -i\frac{4G_F}{\sqrt{2}} V_{ud} V_{cs}^* g_s^2 \, T_{ij}^A T_{kl}^A \bar{u}_i^u \gamma^\rho P_L \gamma_\alpha \gamma_\mu v_j^d \, \bar{v}_k^s \gamma_\rho P_L \gamma_\beta \gamma^\mu u_l^c (-\mathcal{I}^{\alpha\beta}) \tag{8.76}$$

$$= \frac{4G_F}{\sqrt{2}} V_{ud} V_{cs}^* T_{ij}^A T_{kl}^A \frac{\alpha_s}{4\pi} \frac{1}{8} \frac{1}{\epsilon} \bar{v}_k^s \gamma^\nu P_L v_j^d \, \bar{u}_i^u \gamma_\rho \gamma_\alpha \gamma_\mu \gamma_\nu P_R \gamma^\rho \gamma^\alpha \gamma^\mu u_l^c$$

$$= \frac{4G_F}{\sqrt{2}} V_{ud} V_{cs}^* T_{ij}^A T_{kl}^A \frac{\alpha_s}{4\pi} (-4) \frac{1}{\epsilon} \bar{v}_k^s \gamma^\mu P_L v_j^d \, \bar{u}_i^u \gamma_\mu P_L u_l^c$$

$$= \frac{4G_F}{\sqrt{2}} V_{ud} V_{cs}^* \frac{\alpha_s}{4\pi} (-4) \frac{1}{\epsilon} \left(\frac{1}{2} \bar{v}_i^s \gamma^\mu P_L u_i^c \, \bar{u}_j^u \gamma_\mu P_L v_i^d - \frac{1}{2N_c} \bar{v}^s \gamma^\mu P_L u^c \, \bar{u}^u \gamma_\mu P_L v^d \right),$$

where, in addition to eqns (8.62), (8.69), (8.72), and (8.73), we have used the identity $\gamma^\mu \gamma_\alpha \gamma_\beta \gamma_\gamma \gamma_\mu = -2\gamma_\gamma \gamma_\beta \gamma_\alpha$, always performing Dirac algebra in four dimensions.

Looking at Figure 8.5, we see that (i) cancels against quark wave function renormalization, while diagrams (j) and (k) give the same result as diagrams (g) and (h), respectively.

Putting everything together we obtain

$$ i\mathcal{A} = \frac{4G_F}{\sqrt{2}} V_{ud} V_{cs}^* \frac{\alpha_s}{4\pi} \frac{-3}{\epsilon} \left(\bar{u}_i^u \gamma_\mu P_L v_j^d \, \bar{v}_j^s \gamma^\mu P_L u_i^c - \frac{1}{N_c} \bar{u}^u \gamma_\mu P_L v^d \, \bar{v}^s \gamma^\mu P_L u^c \right). \quad (8.77) $$

In order to obtain the two-by-two anomalous dimension matrix we need to compute the one-loop renormalization of $Q_2^{\bar{s}c\bar{d}u}$, by inserting it in diagrams (f) to (k). The only difference in the calculation is given by the colour factors, which now read

$$ T_{il}^A T_{kj}^A = \frac{1}{2}\delta_{ij}\delta_{kl} - \frac{1}{2N_c}\delta_{il}\delta_{kj}. \quad (8.78) $$

Defining

$$ Q_i^{\text{bare}} = Z_{ij} Q_j^{\text{ren}} \quad (8.79) $$

we thus obtain in $\overline{\text{MS}}$

$$ Z = 1 + \frac{\alpha_s}{4\pi}\frac{1}{\epsilon} Z_1 = 1 + \frac{\alpha_s}{4\pi}\frac{1}{\epsilon} \begin{pmatrix} \frac{3}{N_c} & -3 \\ -3 & \frac{3}{N_c} \end{pmatrix}. \quad (8.80) $$

From this we obtain, using eqn (8.51),

$$ \gamma_0 = \begin{pmatrix} -\frac{6}{N_c} & 6 \\ 6 & -\frac{6}{N_c} \end{pmatrix}. \quad (8.81) $$

The corresponding evolutor can be obtained from eqn (8.42). In practice, it can be evaluated by going to the basis in which γ_0 is diagonal, defining

$$ Q_\pm = \frac{Q_1 \pm Q_2}{2}, \quad C_\pm = C_1 \pm C_2, \quad \gamma_0^\pm = \pm 6\frac{N_c \mp 1}{N_c}, \quad U_0^\pm = \left(\frac{\alpha_s(\mu_W)}{\alpha_s(\mu_h)}\right)^{\frac{\gamma_0^\pm}{2\beta_0}}. \quad (8.82) $$

Notice that β_0 depends on the number of active flavours, so if we want to compute the Wilson coefficients at $\mu_h = 2$ GeV we need to take into account the bottom quark threshold at a scale $\mu_b \sim m_b$:

$$ C_\pm(2\text{GeV}) = \left(\frac{\alpha_s(\mu_b)}{\alpha_s(2\text{GeV})}\right)^{\frac{\gamma_0^\pm}{2\beta_0(4)}} \left(\frac{\alpha_s(\mu_W)}{\alpha_s(\mu_b)}\right)^{\frac{\gamma_0^\pm}{2\beta_0(5)}} C_\pm(\mu_W). \quad (8.83) $$

At LO, we have $C_\pm(\mu_W) = 1$ and the evolution decreases C_+ and increases C_-, since γ_0^+ is positive and γ_0^- is negative.

8.3.2   Penguin Operators

Let us now change the flavour content of our current-current operators and consider W exchange between a $\bar{s}u$ and a $\bar{u}d$ current. Following exactly the same considerations as in Section 8.3.1, we obtain the following effective Hamiltonian:

$$\mathcal{H}_{\text{eff}}^{\bar{s}\to\bar{d}} = \frac{4G_F}{\sqrt{2}} V_{ud} V_{us}^* \left(C_1 Q_1^{\bar{s}u\bar{u}d} + C_2 Q_2^{\bar{s}u\bar{u}d} \right). \tag{8.84}$$

However, when computing the renormalization of the operators in $\mathcal{H}_{\text{eff}}^{\bar{s}\to\bar{d}}$, additional diagrams [22] (called penguin diagrams[3]) arise by contracting the u and $\bar{u}$ fields in $Q_{1,2}$ and attaching a gluon, as in Figure 8.6(a).

Let us think for a second on the possible structure of the amplitude induced by these diagrams. It is clearly a FCNC $\bar{s}_i \Gamma^\mu T_{ij}^A d_j$ coupling, which as we have seen cannot arise in the SM Lagrangian. Thus, the amplitude must be proportional to the matrix element of some higher dimensional operator. Indeed, gauge invariance forces us to impose $q^\mu \bar{s} \Gamma^\mu T_{ij}^A d = 0$, so we can identify two possible structures:

$$\bar{s}\left(q^2 \gamma_\mu - q_\mu \slashed{q} \right) T^A d \quad \text{and} \quad \bar{s}\sigma^{\mu\nu} q_\nu T^A d. \tag{8.85}$$

The second one connects quarks of different helicity, so for massless quarks it cannot be generated. The first structure corresponds to the matrix element of operator $\bar{s}\gamma^\mu T^A d D^\nu G_{\mu\nu}^A$. Since the equations of motion imply

$$D^\nu G_{\mu\nu}^A = g_s \sum_f \bar{q}_f \gamma_\mu T^A q_f, \tag{8.86}$$

with f any active quark flavour, this is equivalent to operator

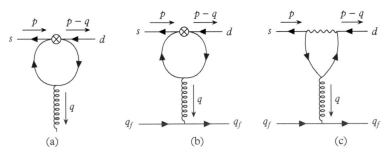

Figure 8.6 *'Penguin' diagrams relevant for $\bar{s} \to \bar{d}$ transitions. See the text for details.*

---

[3] For an instructive recollection of how the term 'penguin diagram' made its way in particle physics see [23].

$$\bar{s}\gamma_\mu T^A d \sum_f \bar{q}_f \gamma^\mu T^A q_f. \tag{8.87}$$

A diagrammatic representation of the equation of motion can be obtained by attaching the gluon line to a quark line of flavour f, as in Figure 8.6(b). When contracted with the gluon propagator, the q^μ term in the FCNC vertex acts on the fermionic current giving zero for gauge invariance. The q^2 term, instead, cancels the pole of the gluon propagator, yielding precisely the local amplitude $\bar{v}_s \gamma_\mu T^a v_d \sum_f \bar{u}_f \gamma^\mu T^a u_f$, i.e. the matrix element of the local operator in eqn (8.87). By power counting, diagrams (a) or (b) are logarithmically divergent, so we will find a $1/\epsilon$ term which must be subtracted, forcing us to enlarge the operator basis again. Since the new operator in eqn (8.87) must be renormalized itself, we must insert it in all the crosses in the diagrams in Figure 8.4, Figure 8.5, and Figure 8.6. As in the case of the insertion of operator Q_1, gluonic corrections will change the original colour structure; moreover, the Dirac algebra will be different for the left- and right-handed components of the q_f current, so we will get different renormalization constants for the two components. All in all, we need to add four 'penguin' operators:

$$Q_3^{\bar{s}d} = \bar{s}_L \gamma_\mu d_L \sum_f \bar{q}_{fL} \gamma^\mu q_{fL}, \tag{8.88}$$

$$Q_4^{\bar{s}d} = \bar{s}_L^\alpha \gamma_\mu d_L^\beta \sum_f \bar{q}_{fL}^\beta \gamma^\mu q_{fL}^\alpha, \tag{8.89}$$

$$Q_5^{\bar{s}d} = \bar{s}_L \gamma_\mu d_L \sum_f \bar{q}_{fR} \gamma^\mu q_{fR}, \tag{8.90}$$

$$Q_6^{\bar{s}d} = \bar{s}_L^\alpha \gamma_\mu d_L^\beta \sum_f \bar{q}_{fR}^\beta \gamma^\mu q_{fR}^\alpha. \tag{8.91}$$

When we insert operators Q_{3-6} in the diagram in Figure 8.6, we should remember that we are using $\overline{MS}$ which is a mass-independent renormalization scheme, so we should manually decouple quark flavours at a threshold $\mu_f \sim m_f$. The anomalous dimensions (as well as the β function) depend on the number of active flavours $n_f(\mu)$, which also determines the summation range in operators Q_{3-6}.

As an explicit example of how penguin operators are generated, let us evaluate Figure 8.6(b) with the insertion of operator $Q_1^{\bar{s}u\bar{u}d}$. We get, omitting the prefactor $\frac{4G_F}{\sqrt{2}} V_{ud} V_{us}^*$ and setting for simplicity $p = 0$,

$$\begin{aligned}
\mathcal{A}_{(b)} &= \int \frac{d^D k}{(2\pi)^D} \bar{v}_d^i \gamma^\mu P_L \frac{i}{\not{k} - \not{q}} (ig_s \gamma^\nu T_{ij}^a) \frac{i}{\not{k}} \gamma_\mu P_L v_s^j \frac{-i}{q^2} \bar{u}_f^k (ig_s \gamma_\nu T_{kl}^a) u_f^l \\
&= -i\frac{g_s^2}{q^2} \int \frac{d^D k}{(2\pi)^D} \bar{v}_d^i \gamma^\mu P_L \frac{\not{k} - \not{q}}{(k-q)^2} \gamma^\nu T_{ij}^a \frac{\not{k}}{k^2} \gamma_\mu P_L v_s^j \bar{u}_f^k \gamma_\nu T_{kl}^a u_f^l \tag{8.92} \\
&= -i\frac{g_s^2}{q^2} I_{\alpha\beta} \bar{v}_d^i \gamma^\mu \gamma^\alpha \gamma^\nu T_{ij}^a \gamma^\beta \gamma_\mu P_L v_s^j \bar{u}_f^k \gamma_\nu T_{kl}^a u_f^l,
\end{aligned}$$

with

$$
\begin{aligned}
I_{\alpha\beta} &= \int \frac{d^D k}{(2\pi)^D} \frac{(k-q)^\alpha k^\beta}{(k-q)^2 k^2} \\
&= \int_0^1 dx \int \frac{d^D k}{(2\pi)^D} \frac{(k-q)^\alpha k^\beta}{\left[(k-q)^2 x + k^2(1-x)\right]^2} \\
&= \int_0^1 dx \int \frac{d^D k}{(2\pi)^D} \frac{(k-q)^\alpha k^\beta}{\left[k^2 - 2q \cdot kx + q^2 x\right]^2} \\
&= \int_0^1 dx \int \frac{d^D l}{(2\pi)^D} \frac{(l - q(1-x))^\alpha (l + qx)^\beta}{\left[l^2 + q^2 x(1-x)\right]^2} \\
&= \int_0^1 dx \int \frac{d^D l}{(2\pi)^D} \frac{l^\alpha l^\beta - q^\alpha q^\beta x(1-x)}{\left[l^2 + q^2 x(1-x)\right]^2} \\
&= \frac{-i}{16\pi^2} \int_0^1 dx \left(\frac{g^{\alpha\beta}}{2} q^2 x(1-x) + q^\alpha q^\beta x(1-x) \right) \left(\frac{1}{\epsilon} + \text{f.t.} \right) \\
&= \frac{-i}{16\pi^2} \left(\frac{g^{\alpha\beta}}{2} q^2 + q^\alpha q^\beta \right) \left(\frac{1}{6\epsilon} \right),
\end{aligned}
\tag{8.93}
$$

where we have used the Feynman parameterization and loop integrals reported in Appendices 8.A.1 and 8.A.2. Substituting eqn (8.93) in eqn (8.92) we obtain

$$
\begin{aligned}
&\frac{-g_s^2}{16\pi^2} \frac{1}{12} \frac{1}{\epsilon} \bar{v}_d^i T_{ij}^a \gamma^\mu \left(\gamma^\alpha \gamma^\nu \gamma_\alpha + 2\frac{q\!\!\!/\gamma^\nu q\!\!\!/}{q^2} \right) \gamma_\mu P_L v_s^j \bar{u}_f^k \gamma_\nu T_{kl}^a u_f^l \\
&= \frac{\alpha_s}{4\pi} \frac{1}{6} \frac{1}{\epsilon} \bar{v}_d^i T_{ij}^a \left(\gamma_\alpha \gamma^\nu \gamma^\alpha + 2\frac{q\!\!\!/\gamma^\nu q\!\!\!/}{q^2} \right) P_L v_s^j \bar{u}_f^k (q) \gamma_\nu T_{kl}^a u_f^l \\
&= -\frac{\alpha_s}{4\pi} \frac{1}{3} \frac{1}{\epsilon} \bar{v}_d^i T_{ij}^a \left(\gamma^\nu - \frac{q\!\!\!/(2q^\nu - q\!\!\!/\gamma^\nu)}{q^2} \right) P_L v_s^j \bar{u}_f^k \gamma_\nu T_{kl}^a u_f^l \\
&= -\frac{\alpha_s}{4\pi} \frac{2}{3} \frac{1}{\epsilon} \bar{v}_d^i T_{ij}^a \left(\gamma^\nu - \frac{q^\nu q\!\!\!/}{q^2} \right) P_L v_s^j \bar{u}_f^k \gamma_\nu T_{kl}^a u_f^l \\
&= -\frac{\alpha_s}{4\pi} \frac{2}{3} \frac{1}{\epsilon} \bar{v}_d^i T_{ij}^a \gamma^\nu P_L v_s^j \bar{u}_f^k \gamma_\nu T_{kl}^a u_f^l.
\end{aligned}
\tag{8.94}
$$

From eqn (8.94) we see explicitly that the FCNC gluon vertex is proportional to $q^2 \gamma^\nu - q^\nu q\!\!\!/$, and that we obtain in the end a local four-quark operator as implied by the equations of motion.

Let us now notice that also current-current operators with charm quarks can mix into penguin operators via the diagrams in Figure 8.6. Thus, we should add to the effective Hamiltonian in eqn (8.84) the corresponding operators with charm, leading to

$$
\mathcal{H}_{\text{eff}}^{\bar{s} \to \bar{d}} = \frac{4 G_F}{\sqrt{2}} \left[V_{ud} V_{us}^* \left(C_1 Q_1^{\bar{s}u\bar{u}d} + C_2 Q_2^{\bar{s}u\bar{u}d} \right) + V_{cd} V_{cs}^* \left(C_1 Q_1^{\bar{s}c\bar{c}d} + C_2 Q_2^{\bar{s}c\bar{c}d} \right) \right].
\tag{8.95}
$$

Now, when inserted in penguin diagrams, operators $Q_{1,2}^{\bar{s}u\bar{u}d}$ and $Q_{1,2}^{\bar{s}c\bar{c}d}$ give exactly the same divergent part, since the divergence is independent on the mass of the quarks running in the loop. Thus, the penguin operators will be generated with a coefficient proportional to $V_{ud}V_{us}^* + V_{cd}V_{cs}^* = -V_{td}V_{ts}^*$. Taking everything into account, we end up with the following effective Hamiltonian:

$$
\begin{aligned}
\mathcal{H}_{\text{eff}}^{\bar{s}\to\bar{d}} &= \frac{4G_F}{\sqrt{2}} \left\{ V_{ud}V_{us}^* \left(C_1 Q_1^{\bar{s}u\bar{u}d} + C_2 Q_2^{\bar{s}u\bar{u}d} \right) + V_{cd}V_{cs}^* \left(C_1 Q_1^{\bar{s}c\bar{c}d} + C_2 Q_2^{\bar{s}c\bar{c}d} \right) \right. \\
&\quad \left. - V_{td}V_{ts}^* \sum_{i=3}^{6} C_i Q_i^{\bar{s}d} \right\} \\
&= \frac{G_F}{\sqrt{2}} \left\{ V_{ud}V_{us}^* \left[C_1 \left(Q_1^{\bar{s}u\bar{u}d} - Q_1^{\bar{s}c\bar{c}d} \right) + C_2 \left(Q_2^{\bar{s}u\bar{u}d} - Q_2^{\bar{s}c\bar{c}d} \right) \right] \right. \\
&\quad \left. - V_{td}V_{ts}^* \left[C_1 Q_1^{\bar{s}c\bar{c}d} + C_2 Q_2^{\bar{s}c\bar{c}d} + \sum_{i=3}^{6} C_i Q_i^{\bar{s}d} \right] \right\},
\end{aligned}
\tag{8.96}
$$

where we have used CKM unitarity to eliminate the $V_{cd}V_{cs}^*$ term. Now, the $V_{ud}V_{us}^*$ part contains current-current operators appearing in the GIM-suppressed difference of up and charm, which does not mix into penguin operators since the divergent part of the diagrams in Figure 8.6 does not depend on the mass of the quark running in the loop. Penguin operators Q_{3-6} are instead generated by the RG running with the top CKM factor, due to the insertion in the diagrams in Figure 8.6 of the operators $Q_{1,2}^{\bar{s}c\bar{c}d}$ in the last line of eqn (8.96). Their anomalous dimension also gets contributions from the insertion of Q_{3-6} in the diagrams of Figures 8.4, 8.5, and 8.6.

Had we performed the matching at $\mathcal{O}(\alpha_s)$, we would have encountered Figure 8.6(c) in the full theory, with u, c, and t quarks running in the loop. Diagrams (b) and (c) are of course not identical, since diagram (b) is logarithmically divergent while diagram (c) is finite (keep in mind the two powers of external momenta required by gauge invariance; see eqn (8.85)). However, if we differentiate with respect to external momenta or quark masses, then also diagram (b) becomes finite, and therefore the loop integration is dominated by momenta of the order of the external momenta and quark masses. After differentiating, we are thus allowed to replace the W propagator in diagram (c) with $1/M_W^2$. In this way, we obtain the following relation between the amplitudes generated by diagrams (b) and (c) with quark i running in the loop:

$$
\mathcal{A}_{(c)}^{(i)} = \mathcal{A}_{(b)}^{(i)} + \mathcal{O}\left(\frac{p^2, m_i^2}{M_W^2} \right) + K,
\tag{8.97}
$$

where p denotes external momenta and K is a constant term, independent on quark masses or momenta, proportional to the matrix element of operators Q_{3-6}. Equation (8.97) implies that the contribution of u and c quarks cancels in the matching up to the constant K and to negligible corrections of $\mathcal{O}\left(\frac{p^2, m_i^2}{M_W^2} \right)$, while the top quark contribution generates a nontrivial contribution to $C_{3-6}^1(\mu_W)$ since in the effective theory we do not have diagram (b) with top quarks running in the loop [24, 25].

8.3.3   Electroweak Penguins

We may wonder what happens if we replace in the diagrams of Figure 8.4–8.6 gluon exchange with photon exchange. Electromagnetic corrections will also get a logarithmic enhancement, making them comparable to NLO QCD corrections, since $\alpha \log(\mu_W^2/\mu_h^2) \sim \alpha_s$. While we do not need to resum these logarithmic terms, we need to include them when working at NLO in QCD [26, 27]. QED corrections bring a novelty: the operator basis must be enlarged, due to the electroweak penguin operators generated by the diagrams in Figure 8.6 replacing the gluon with a photon [28]. While the FCNC photon coupling emerging from Figure 8.6(a) is equivalent to the gluonic one, the equation of motion introduces a quark charge dependence, giving rise to operators with flavour structure $\bar{s}d \sum e_q \bar{q}q$, with e_q the electric charge of flavour q. As in the case of QCD penguins, strong interaction corrections will generate a new colour structure, and the left- and right-handed components of the quark current will renormalize differently, so we need to add four more operators to the basis:

$$Q_7^{\bar{s}d} = \frac{3}{2}\bar{s}_L\gamma_\mu d_L \sum_f e_q \bar{q}_{fR}\gamma^\mu q_{fR}, \tag{8.98}$$

$$Q_8^{\bar{s}d} = \frac{3}{2}\bar{s}_L^\alpha\gamma_\mu d_L^\beta \sum_f e_q \bar{q}_{fR}^\beta\gamma^\mu q_{fR}^\alpha, \tag{8.99}$$

$$Q_9^{\bar{s}d} = \frac{3}{2}\bar{s}_L\gamma_\mu d_L \sum_f e_q \bar{q}_{fL}\gamma^\mu q_{fL}, \tag{8.100}$$

$$Q_{10}^{\bar{s}d} = \frac{3}{2}\bar{s}_L^\alpha\gamma_\mu d_L^\beta \sum_f e_q \bar{q}_{fL}^\beta\gamma^\mu q_{fL}^\alpha. \tag{8.101}$$

Let us now briefly discuss the matching for operators Q_{7-10}. Also in this case, these operators get a contribution from the top loop in the matching from Figure 8.6(c) with the exchange of a photon. However, this is not the only contribution. Indeed, one should also consider diagram (c) with the exchange of a Z^0 instead of a photon, and box diagrams with the exchange of two W bosons. We do not dwell into the details of the matching, but there is an important point we would like to stress. It is instructive to consider diagram (c) with the exchange of a Z^0 in two steps: first, the evaluation of the FC Z coupling from the loop integration, and then the evaluation of the Z exchange. While $SU(3)_c \otimes U(1)_{em}$ gauge invariance forbids dimension four FCNC gluon or photon couplings, this is not the case for FCNC Z couplings, which can indeed arise at dimension four: a $Z_\mu\bar{s}_L\gamma^\mu d_L$ coupling can be generated once $SU(2)_L \otimes U(1)_Y$ is broken. However, for this to happen the diagram must 'feel' the EW symmetry breaking, so the FC Z coupling must vanish linearly with m_q^2/M_W^2 for small m_q. Thus, the loop is dominated by the top quark [24].[4] Having obtained a top-induced FC Z vertex from the loop

---

[4] This m_q^2/M_W^2 suppression is sometimes called 'hard GIM', as opposed to the logarithmic dependence on quark masses which arises for example when matching on the dimension six FC gluon and photon couplings, the so-called 'soft GIM'. We will discuss more in detail GIM suppression in Section 8.4.

integration, we consider the Z exchange in the lower part of diagram (c). Expanding the Z propagator for small momenta in the same way as for the W propagator in eqn (8.32) gives rise to local four-fermion operators, which can be identified with the (electroweak) penguins discussed above.

8.3.4 Anomalous Dimension

We conclude this Section reporting the LO anomalous dimension for the full set of four-quark $\Delta F = 1$ operators listed above [29]–[33]:

$$
\gamma_0 =
\begin{pmatrix}
-\frac{6}{N_c} & 6 & -\frac{2}{3N_c} & \frac{2}{3} & -\frac{2}{3N_c} & \frac{2}{3} & 0 & 0 & 0 & 0 \\
6 & -\frac{6}{N_c} & 0 & 0 & 0 & 0 & 0 & 0 & 0 & 0 \\
0 & 0 & -\frac{22}{3N_c} & \frac{22}{3} & -\frac{4}{3N_c} & \frac{4}{3} & 0 & 0 & 0 & 0 \\
0 & 0 & 6-\frac{2n_f}{3N_c} & -\frac{6}{N_c}+\frac{2n_f}{3} & -\frac{2n_f}{3N_c} & \frac{2n_f}{3} & 0 & 0 & 0 & 0 \\
0 & 0 & 0 & 0 & \frac{6}{N_c} & -6 & 0 & 0 & 0 & 0 \\
0 & 0 & -\frac{2n_f}{3N_c} & \frac{2n_f}{3} & -\frac{2n_f}{3N_c} & 6\frac{1-N_c^2}{N_c}+\frac{2n_f}{3} & 0 & 0 & 0 & 0 \\
0 & 0 & 0 & 0 & 0 & 0 & \frac{6}{N_c} & -6 & 0 & 0 \\
0 & 0 & \frac{-2(n_u-n_d/2)}{3N_c} & \frac{2(n_u-n_d/2)}{3} & \frac{-2(n_u-n_d/2)}{3N_c} & \frac{2(n_u-n_d/2)}{3} & 0 & 6\frac{1-N_c^2}{N_c} & 0 & 0 \\
0 & 0 & \frac{2}{3N_c} & -\frac{2}{3} & \frac{2}{3N_c} & -\frac{2}{3} & 0 & 0 & -\frac{6}{N_c} & 6 \\
0 & 0 & \frac{-2(n_u-n_d/2)}{3N_c} & \frac{2(n_u-n_d/2)}{3} & \frac{-2(n_u-n_d/2)}{3N_c} & \frac{2(n_u-n_d/2)}{3} & 0 & 0 & 6 & -\frac{6}{N_c}
\end{pmatrix},
$$
(8.102)

where $n_{f,u,d} = n_{f,u,d}(\mu)$ is the number of active (up- or down-type) quarks at the scale μ.

8.3.5 The $\Delta I = 1/2$ Rule

As an example of the applications of the $\Delta S = 1$ effective Hamiltonian, let us consider the $\Delta I = 1/2$ rule. We start by writing down the decay amplitudes for a Kaon (anti-Kaon) to decay in two pions in terms of final states with different isospin. The two-pion state in S-wave must have a symmetric isospin wave function, so it can only be in $I = 0$ or $I = 2$ states. Denoting by $|I, I_3\rangle$ a state with isospin I and third component I_3, we can write

$$\langle 2,1| = \frac{1}{\sqrt{2}} \left(\langle \pi^+\pi^0| + \langle \pi_0\pi^+| \right),$$
(8.103)

$$\langle 2,0| = \frac{1}{\sqrt{6}} \left(\langle \pi^+\pi^-| + \langle \pi^-\pi^+| + 2\langle \pi^0\pi^0| \right),$$
(8.104)

$$\langle 0,0| = \frac{1}{\sqrt{3}} \left(\langle \pi^+\pi^-| + \langle \pi^-\pi^+| - \langle \pi^0\pi^0| \right).$$
(8.105)

The initial state Kaon is a doublet, and the $\Delta S = 1$ effective Hamiltonian has $I = 1/2$ and $I = 3/2$ components. Coupling the initial state and the effective Hamiltonian through the Wigner-Eckart theorem, we have:

$$\mathcal{H}_{\text{eff}}|K^+\rangle = (\mathcal{H}_{3/2,1/2} + \mathcal{H}_{1/2,1/2})|1/2,1/2\rangle \tag{8.106}$$

$$= \sqrt{\frac{3}{4}}A_{3/2}|2,1\rangle - \frac{1}{2}A_{3/2}|1,1\rangle + A_{1/2}|1,1\rangle,$$

$$]\mathcal{H}_{\text{eff}}|K^0\rangle = (\mathcal{H}_{3/2,1/2} + \mathcal{H}_{1/2,1/2})|1/2,-1/2\rangle \tag{8.107}$$

$$= \frac{1}{\sqrt{2}}A_{3/2}|2,0\rangle + \frac{1}{\sqrt{2}}A_{3/2}|1,0\rangle + \frac{1}{\sqrt{2}}A_{1/2}|1,0\rangle + \frac{1}{\sqrt{2}}A_{1/2}|0,0\rangle.$$

Finally, from eqns (8.103)–(8.107) we obtain

$$A(K^+ \to \pi^+\pi^0) = \frac{\sqrt{3}}{2\sqrt{2}}A_{3/2}, \tag{8.108}$$

$$A(K^0 \to \pi^+\pi^-) = \frac{1}{2\sqrt{3}}\left(A_{3/2} + \sqrt{3}A_{1/2}\right),$$

$$A(K^0 \to \pi^0\pi^0) = \frac{1}{\sqrt{6}}\left(\sqrt{2}A_{3/2} - A_{1/2}\right).$$

It is convenient to define $A_{0,2} = 1/\sqrt{6}A_{1/2,3/2}$ and to extract, without loss of generality, a CP-invariant phase $\delta_{0,2}$ from each isospin amplitude, so that $A_{0,2} \to A_{0,2}^*$ under CP:

$$A(K^0 \to \pi^+\pi^-) = A_0 e^{i\delta_0} + \frac{A_2}{\sqrt{2}}e^{i\delta_2},$$

$$A(K^0 \to \pi^0\pi^0) = -A_0 e^{i\delta_0} + \sqrt{2}A_2 e^{i\delta_2},$$

$$A(K^+ \to \pi^+\pi^0) = \frac{3}{2}A_2 e^{i\delta_2}. \tag{8.109}$$

We now write the relevant S matrix as:

$$S = \begin{pmatrix} K \to K & K \to (\pi\pi)_0 & K \to (\pi\pi)_2 \\ (\pi\pi)_0 \to K & (\pi\pi)_0 \to (\pi\pi)_0 & (\pi\pi)_0 \to (\pi\pi)_2 \\ (\pi\pi)_2 \to K & (\pi\pi)_2 \to (\pi\pi)_0 & (\pi\pi)_2 \to (\pi\pi)_2 \end{pmatrix} \simeq \begin{pmatrix} 1 & -iT_0 & -iT_2 \\ -iT(T_0) & e^{i\Delta_0} & 0 \\ -iT(T_2) & 0 & e^{i\Delta_2} \end{pmatrix},$$
$$\tag{8.110}$$

where we have assumed for simplicity elastic, isospin-conserving $\pi\pi$ strong interaction scattering, represented by the phases $\Delta_{0,2}$ for $I = 0,2$ $\pi\pi$ states, and we are working at lowest order in weak interactions. $\mathcal{T}$ denotes time reversal. Unitarity of the S matrix implies:

$$(S^\dagger S)_{12} = 0 = -iT_0 + e^{i\Delta}i\mathcal{T}(T_0)^*. \tag{8.111}$$

Writing, as we did in eqn (8.109),

$$T_i = A_i e^{i\delta_i}, \qquad \mathcal{T}(T_i) = CP(T_i) = A_i^* e^{i\delta_i}, \tag{8.112}$$

we obtain from eqn (8.111)

$$-iA_0 e^{i\delta_0} + e^{i\Delta_0}i(A_0^* e^{i\delta_0})^* = 0 \quad \Rightarrow \delta_0 = \Delta_0/2, \tag{8.113}$$

so the CP-even phase of the weak decay amplitude is just half the phase describing strong-interaction scattering of the final state. This is known as Watson theorem [34], and can be generalized to the case of multi-channel strong-interactions unitarity (see [35] for an example of application of Watson theorem to D decays).

Experimentally, $\mathrm{Re}A_2/\mathrm{Re}A_0 \sim 1/22$, so $\Delta I = 1/2$ transitions happen at a much higher rate than $\Delta I = 3/2$. This is commonly denoted as the $\Delta I = 1/2$ rule. One of the most difficult problems in the study of weak decays, still lacking a complete solution, is in fact the theoretical prediction of the ratio A_2/A_0. Let us start from the $\Delta S = 1$ Hamiltonian in eqn (8.84). Considering the lowering operator for the third component of isospin, I_-, we have $I_- u = d$, $I_- \bar{d} = -\bar{u}$, and $I_- d = I_- \bar{u} = 0$. Then the action of I_- on the operator Q_- of eqn (8.82) is given by

$$2I_- Q_- = \bar{s}_L \gamma^\mu (I_- u_L) \bar{u}_L \gamma_\mu d_L + \bar{s}_L \gamma^\mu u_L (I_- \bar{u}_L) \gamma_\mu d_L + \bar{s}_L \gamma^\mu u_L \bar{u}_L \gamma_\mu (I_- d_L)$$
$$- \bar{s}_L^\alpha \gamma^\mu (I_- u_L^\beta) \bar{u}_L^\beta \gamma_\mu d_L^\alpha - \bar{s}_L^\alpha \gamma^\mu u_L^\beta (I_- \bar{u}_L^\beta) \gamma_\mu d_L^\alpha - \bar{s}_L^\alpha \gamma^\mu u_L^\beta \bar{u}_L^\beta \gamma_\mu (I_- d_L^\alpha)$$
$$= \bar{s}_L \gamma^\mu (d_L) \bar{u}_L \gamma_\mu d_L - \bar{s}_L^\alpha \gamma^\mu (d_L^\beta) \bar{u}_L^\beta \gamma_\mu d_L^\alpha = 0, \tag{8.114}$$

where in the last step we have fierzed the Dirac structure. Thus, Q_- is the lower component of an isospin doublet. Doing the same exercise on Q_+ shows instead that Q_+ is an admixture of $I = 1/2$ and $I = 3/2$. Therefore, the enhancement of C_- over C_+ due to RG evolution goes in the right direction to explain the $\Delta I = 1/2$ rule [30], although it can only account for about a factor of two in the amplitude ratio.

Another contribution to the $\Delta I = 1/2$ rule comes from QCD penguin operators $Q_{3...6}$ in eqns (8.88)–(8.91) [22], since these operators are isospin doublets. Still, the effect must largely come from the matrix elements, since perturbative RG effects cannot bring the amplitude ratio close to the experimental value. Unfortunately, computing the relevant matrix elements from first principles with Lattice QCD is a tremendous task. Indeed, this calculation poses all the most difficult challenges to lattice QCD calculations: final state interactions, chiral symmetry breaking, power divergences, disconnected diagrams, etc. Thus, it comes as no surprise that only very recently a pioneering lattice calculation of the matrix elements relevant for the $\Delta I = 1/2$ rule has been achieved [36]. According to this calculation, there is a large deviation from the Vacuum Insertion Approximation (VIA) in the matrix elements of current-current operators, causing a negative interference, and thus a large cancellation, in $\Delta I = 3/2$ matrix elements, which are therefore suppressed with respect to $\Delta I = 1/2$ ones. Such deviation from the VIA, with the corresponding negative interference, is also seen in $\Delta S = 2$ matrix elements [37]. However, the same calculation failed to reproduce the phase of the $\Delta I = 1/2$ amplitude, casting some doubts on the robustness of the estimate of final state interactions. Fortunately, with increased statistics and an improved treatment of the two-pion state, a much better agreement with the experimental value of δ_0 was very recently obtained [38]. We are looking forward to the corresponding update of the results on the $\Delta I = 1/2$ rule, and hopefully to an independent confirmation from another lattice collaboration in the future.

8.4   Effective Hamiltonians for $\Delta F = 2$ Processes

Let us now turn to the transitions that give the very stringent bounds reported in Figure 8.1: $\Delta F = 2$ processes. In particular, let us consider $\bar{s}d \rightarrow \bar{d}s$ transitions. Such FCNC processes cannot arise at the tree level in the SM, so we must consider one-loop contributions. These contributions must be finite, since renormalizability of the SM implies that no counterterm for FCNC amplitudes can arise. In the 't Hooft–Feynman gauge we have the diagrams in Figure 8.7.

Let us start by computing diagram (a). Neglecting external momenta, the amplitude reads

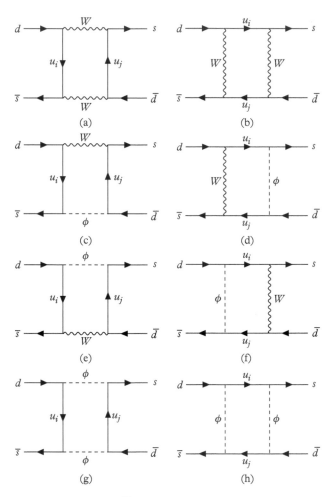

Figure 8.7 *Feynman diagrams for $\bar{s}d \rightarrow \bar{d}s$ transitions in the SM.*

$$i\mathcal{A}^{(a)} = \int \overline{u}_s \left(\frac{ig_2}{\sqrt{2}} \right) \gamma_\mu P_L V_{u_j s}^* \frac{i}{\slashed{q} - m_{u_j}} \left(\frac{ig_2}{\sqrt{2}} \right) \gamma_\nu P_L V_{u_j d} v_d \tag{8.115}$$

$$\times \, \overline{v}_s \left(\frac{ig_2}{\sqrt{2}} \right) \gamma^\nu P_L V_{u_i s}^* \frac{i}{\slashed{q} - m_{u_i}} \left(\frac{ig_2}{\sqrt{2}} \right) \gamma^\mu P_L V_{u_i d} u_d \left(\frac{-i}{q^2 - M_W^2} \right)^2 \frac{\mathrm{d}^4 q}{(2\pi)^4}.$$

Left-handed projectors kill the quark mass terms in the numerator of quark propagators, so we obtain

$$i\mathcal{A}^{(a)} = \frac{g_2^4}{4} V_{u_i s}^* V_{u_i d} V_{u_j s}^* V_{u_j d} \, \overline{u}_s \gamma_\mu \gamma^\alpha \gamma_\nu P_L v_d \, \overline{v}_s \gamma^\nu \gamma^\beta \gamma^\mu P_L u_d I_{\alpha\beta}^{ij}, \tag{8.116}$$

with

$$I_{\alpha\beta}^{ij} \equiv \int \frac{q_\alpha q_\beta}{(q^2 - M_W^2)^2 (q^2 - m_{u_i}^2)(q^2 - m_{u_j}^2)} \frac{\mathrm{d}^4 q}{(2\pi)^4}. \tag{8.117}$$

We can simplify the integral using partial fractioning in the form

$$\frac{m_{u_i}^2 - m_{u_j}^2}{(q^2 - m_{u_i}^2)(q^2 - m_{u_j}^2)} = \frac{1}{q^2 - m_{u_i}^2} - \frac{1}{q^2 - m_{u_j}^2}, \tag{8.118}$$

obtaining

$$I_{\alpha\beta}^{ij} = \frac{I_{\alpha\beta}^i - I_{\alpha\beta}^j}{m_{u_i}^2 - m_{u_j}^2} \tag{8.119}$$

with

$$\begin{aligned} I_{\alpha\beta}^i &= \int \frac{q_\alpha q_\beta}{\left(q^2 - M_W^2 \right)^2 \left(q^2 - m_{u_i}^2 \right)} \frac{\mathrm{d}^4 q}{(2\pi)^4} \\ &= \frac{g_{\alpha\beta}}{4} \int \frac{q^2 + (-m_{u_i}^2 + m_{u_i}^2)}{\left(q^2 - M_W^2 \right)^2 \left(q^2 - m_{u_i}^2 \right)} \frac{\mathrm{d}^4 q}{(2\pi)^4} \\ &= \frac{g_{\alpha\beta}}{4} m_{u_i}^2 \int \frac{1}{\left(q^2 - M_W^2 \right)^2 \left(q^2 - m_{u_i}^2 \right)} \frac{\mathrm{d}^4 q}{(2\pi)^4} + K, \end{aligned} \tag{8.120}$$

where K represents terms independent on $m_{u_i}^2$ which drop in $I_{\alpha\beta}^{ij}$. Introducing Feynman parameters as in eqn (8.61), we obtain

$$\int \frac{1}{(q^2 - M_W^2)^2 (q^2 - m_{u_i}^2)} \frac{\mathrm{d}^4 q}{(2\pi)^4}$$

$$= 2 \int_0^1 \mathrm{d}x \int \frac{\mathrm{d}^4 q}{(2\pi)^4} \frac{x}{[(q^2 - M_W^2)x + (q^2 - m_{u_i}^2)(1-x)]^3}$$

$$= 2 \int_0^1 dx \int \frac{d^4q}{(2\pi)^4} \frac{1}{[q^2 - M_W^2 x - m_{u_i}^2 (1-x)]^3}$$

$$= -\frac{i}{16\pi^2} \int_0^1 dx \frac{x}{M_W^2 x + m_{u_i}^2 (1-x)}$$

$$= -\frac{i}{16\pi^2 M_W^2} \int_0^1 dx \frac{x}{x + x_i (1-x)} \tag{8.121}$$

$$= -\frac{i}{16\pi^2 M_W^2} \int_0^1 dx \frac{x}{x_i + x(1-x_i)}$$

$$= -\frac{i}{16\pi^2 M_W^2} \int_0^1 \frac{dx}{1-x_i} \frac{(1-x_i)x + x_i - x_i}{x_i + x(1-x_i)}$$

$$= -\frac{i}{16\pi^2 M_W^2} \left(\frac{-x_i}{1-x_i} \int_0^1 \frac{dx}{(1-x_i)x + x_i} + \frac{1}{1-x_i} \right)$$

$$= -\frac{i}{16\pi^2 M_W^2} \left(\frac{1}{1-x_i} + \frac{x_i \log x_i}{(1-x_i)^2} \right),$$

where $x_i = m_{u_i}^2 / M_W^2$. Thus, up to terms that do not depend on $m_{u_i}^2$, we have

$$I_{\alpha\beta}^i = -\frac{g_{\alpha\beta}}{4} \frac{i}{16\pi^2} \mathcal{J}(x_i), \tag{8.122}$$

with

$$\mathcal{J}(x_i) = \frac{x_i}{1-x_i} + \frac{x_i^2 \log x_i}{(1-x_i)^2}, \tag{8.123}$$

and therefore

$$I_{\alpha\beta}^{ij} = -\frac{g_{\alpha\beta}}{4M_W^2} \frac{i}{16\pi^2} A(x_i, x_j), \tag{8.124}$$

with

$$A(x_i, x_j) = \frac{\mathcal{J}(x_i) - \mathcal{J}(x_j)}{x_i - x_j}. \tag{8.125}$$

We now turn to the Dirac structure

$$\bar{u}_s \gamma_\mu \gamma_\alpha \gamma_\nu P_L v_d \bar{v}_s \gamma^\nu \gamma^\beta \gamma^\mu P_L u_d \tag{8.126}$$

and use again the Fierz identity in eqn (8.69) to obtain

$$-\frac{1}{2} \bar{v}_s \gamma^\rho P_L v_d \bar{u}_s \gamma_\mu \gamma_\alpha \gamma_\nu \gamma_\rho P_R \gamma^\nu \gamma^\alpha \gamma^\mu u_d = 4 \bar{v}_s \gamma^\mu P_L v_d \bar{u}_s \gamma_\mu P_L u_d. \tag{8.127}$$

Putting everything together we obtain the amplitude generated by diagram (a) as

$$
\begin{aligned}
i\mathcal{A}^{(a)} &= -\frac{i}{16\pi^2}\frac{4g^4}{16M_W^2}\sum_{i,j=u,c,t} v_{is}^* V_{id} V_{js}^* V_{jd} A(x_i,x_j)\bar{v}_s\gamma^\mu P_L v_d \bar{u}_s\gamma_\mu P_L u_d \\
&= -\frac{iG_F^2 M_W^2}{2\pi^2}\sum_{i,j}\lambda_{sd}^i\lambda_{sd}^j A(x_i,x_j)\bar{v}_s\gamma^\mu P_L v_d \bar{u}_s\gamma_\mu P_L u_d,
\end{aligned}
\tag{8.128}
$$

with $\lambda_{sd}^i = V_{is}^* V_{id}$.

Figure 8.7(b) is identical to (a) if we exchange an incoming s anti-quark with an outgoing s quark, and vice versa:

$$
i\mathcal{A}^{(b)} = \frac{iG_F^2 M_W^2}{2\pi^2}\sum_{i,j}\lambda_{sd}^i\lambda_{sd}^j A(x_i,x_j)\bar{u}_s\gamma^\mu P_L u_d \bar{v}_s\gamma_\mu P_L v_d.
\tag{8.129}
$$

We now notice that the amplitudes generated by Figure 8.7(a)–(b) can be written as the matrix element of a local operator, so we can introduce the following $\Delta S = 2$ effective Hamiltonian:

$$
\mathcal{H}_{\text{eff}}^{\Delta S=2} = C\bar{s}\gamma^\mu P_L d\,\bar{s}\gamma_\mu P_L d,
\tag{8.130}
$$

with C a Wilson coefficient with mass dimension -2. This effective Hamiltonian generates the following amplitude:

$$
\begin{aligned}
iT^{\mathcal{H}}(\bar{s}d \to \bar{d}s) &= -iC\langle \bar{d}s|\bar{s}\gamma^\mu P_L d\,\bar{s}\gamma_\mu P_L d|\bar{s}d\rangle \\
&= -2iC\left(\bar{u}_s\gamma^\mu P_L v_d\bar{v}_s\gamma_\mu P_L u_d - \bar{u}_s\gamma^\mu P_L u_d\bar{v}_s\gamma_\mu P_L v_d\right).
\end{aligned}
\tag{8.131}
$$

Matching it with the amplitude in the full theory $i\mathcal{A}^{(a)+(b)}$ we obtain

$$
C^{(a)+(b)} = \frac{G_F^2 M_W^2}{4\pi^2}\sum_{i,j}\lambda_{sd}^i\lambda_{sd}^j A(x_i,x_j).
\tag{8.132}
$$

Evaluating Figures 8.7(c)–(h) we obtain

$$
\begin{aligned}
C^{(c)+(d)} &= C^{(e)+(f)} \\
&= -\frac{G_F^2 M_W^2}{4\pi^2}\sum_{i,j}\lambda_{sd}^i\lambda_{sd}^j A'(x_i,x_j)x_i x_j,
\end{aligned}
\tag{8.133}
$$

$$
C^{(g)+(h)} = \frac{1}{4}\frac{G_F^2 M_W^2}{4\pi^2}\sum_{i,j}\lambda_{sd}^i\lambda_{sd}^j A(x_i,x_j)x_i x_j,
\tag{8.134}
$$

with

$$A'(x_i, x_j) = \frac{\mathscr{J}'(x_i) - \mathscr{J}'(x_j)}{x_i - x_j}, \qquad \mathscr{J}'(x) = \frac{1}{1-x} + \frac{x \log x}{(1-x)^2}. \tag{8.135}$$

Putting everything together we obtain

$$C = \frac{G_F^2 M_W^2}{4\pi^2} \sum_{i,j} \lambda_{sd}^i \lambda_{sd}^j \overline{A}(x_i, x_j), \tag{8.136}$$

where

$$\overline{A}(x_i, x_j) = A(x_i, x_j) - x_i x_j A'(x_i, x_j) + \frac{1}{4} x_i x_j A(x_i, x_j). \tag{8.137}$$

Next, we use CKM unitarity in the form

$$\lambda_{sd}^u = -\lambda_{sd}^c - \lambda_{sd}^t \tag{8.138}$$

to eliminate λ_{sd}^u, and we obtain

$$\sum_{i,j} \lambda_{sd}^i \lambda_{sd}^j \overline{A}(x_i, x_j) = \left(\lambda_{sd}^c + \lambda_{sd}^t\right)^2 \overline{A}(x_u, x_u) + 2\lambda_{sd}^c \lambda_{sd}^t \overline{A}(x_c, x_t)$$

$$+ \left(\lambda_{sd}^c\right)^2 \overline{A}(x_c, x_c) + \left(\lambda_{sd}^t\right)^2 \overline{A}(x_t, x_t)$$

$$- 2\lambda_{sd}^t \left(\lambda_{sd}^c + \lambda_{sd}^t\right) \overline{A}(x_u, x_t) - 2\lambda_{sd}^c \left(\lambda_{sd}^c + \lambda_{sd}^t\right) \overline{A}(x_c, x_u)$$

$$= \left(\lambda_{sd}^t\right)^2 S_0(x_t) + \left(\lambda_{sd}^c\right)^2 S_0(x_c) + 2\lambda_{sd}^t \lambda_{sd}^c S_0(x_c, x_t), \tag{8.139}$$

with

$$S_0(x_t) = \overline{A}(x_t, x_t) + \overline{A}(x_u, x_u) - 2\overline{A}(x_u, x_t), \tag{8.140}$$

$$S_0(x_c) = \overline{A}(x_c, x_c) + \overline{A}(x_u, x_u) - 2\overline{A}(x_u, x_c), \tag{8.141}$$

$$S_0(x_c, x_t) = \overline{A}(x_c, x_t) + \overline{A}(x_u, x_u) - \overline{A}(x_u, x_c) - \overline{A}(x_u, x_t). \tag{8.142}$$

We finally obtain

$$\mathcal{H}_{\text{eff}}^{\Delta S=2} = \frac{G_F^2 M_W^2}{4\pi^2} \left[\left(\lambda_{sd}^t\right)^2 S_0(x_t) + \left(\lambda_{sd}^c\right)^2 S_0(x_c) + 2\lambda_{sd}^t \lambda_{sd}^c S_0(x_c, x_t) \right] \bar{s}\gamma^\mu P_L d\, \bar{s}\gamma_\mu P_L d. \tag{8.143}$$

Notice that the S_0 functions are differences of $\overline{A}$ functions with different arguments. If we Taylor-expand $\overline{A}$ in powers of quark masses, the zeroth-order term cancels in S_0. Thus, for massless quarks no FCNC vertices arise, and the latter are suppressed by the GIM cancellation mechanism [1]. For small x the loop function S_0 vanishes linearly.

Neglecting the contribution of the third family, the FCNC coupling in eqn (8.130) is proportional to

$$G_F^2 M_W^2 \lambda^2 x_c = G_F^2 m_c^2 \lambda^2, \tag{8.144}$$

where λ is the Wolfenstein parameter introduced in eqn (8.15). In other words, the (hard) GIM mechanism converts the effective $\Delta S = 2$ coupling from an $\mathcal{O}(1/M_W^2)$ effect to an $\mathcal{O}(m_c^2/M_W^4)$ one. Notice also that SM fermions do not decouple, since S_0 grows linearly for large x; this non-decoupling explains the relevance of the top quark even in low-energy FCNC processes, and can be easily understood since the coupling to the would-be Goldstone bosons is proportional to fermion masses (or, more precisely, to Yukawa couplings).

8.4.1 Locality and Higher-Dimensional Operators

The matching calculation we performed above might look as an academic exercise: what can we say on K-$\overline{K}$ mixing from a matrix element between zero-momentum quarks with no strong interactions? Indeed, neglecting external momenta a local operator is generated by construction, but this is a reasonable approximation only if the dependence on external momenta is negligible. Let us now discuss this problem in some detail. First of all, we notice that diagrams containing up quarks only are in fact non-local contributions, which cannot be estimated by matching onto a local effective Hamiltonian. Indeed, up-quark contributions cancel in the matching against the matrix element of two $\Delta S = 1$ effective Hamiltonians. The latter represents the long-distance contribution to K-$\overline{K}$ mixing which must be evaluated using some non-perturbative method. This point can be explicitly checked using the same argument discussed in Section 8.3.2. The diagrams in Figure 8.7 become divergent when substituting the W propagator with a local interaction, which corresponds to the matrix element of two $\Delta S = 1$ effective Hamiltonians. However, differentiating thrice with respect to quark masses and/or momenta, the diagram becomes convergent even when the W propagator becomes local, allowing us to identify the diagrams in Figure 8.7 with the matrix element of two $\Delta S = 1$ effective Hamiltonians up to a constant term, to a term proportional to p^2/M_W^2 and to a term proportional to m_i^2/M_W^2, where p represents external momenta and m_i the mass of the quark running in the loop.

However, if we look at the CP-odd part of the effective Hamiltonian we can drop the up-quark contribution since we can always choose a phase convention such that $\mathrm{Im}\lambda_{sd}^u = 0$. In this convention, we have $\mathrm{Im}\lambda_{sd}^c = -\,\mathrm{Im}\lambda_{sd}^t$, so that

$$\mathrm{Im}\left(\lambda_{sd}^t\right)^2 = 2\mathrm{Im}\lambda_{sd}^t \mathrm{Re}\,\lambda_{sd}^t, \tag{8.145}$$

$$\mathrm{Im}\left(\lambda_{sd}^c\right)^2 = -2\mathrm{Im}\lambda_{sd}^t \mathrm{Re}\,\lambda_{sd}^c, \tag{8.146}$$

$$\mathrm{Im}\lambda_{sd}^c\lambda_{sd}^t = \mathrm{Im}\lambda_{sd}^t\left(\mathrm{Re}\,\lambda_{sd}^c - \mathrm{Re}\,\lambda_{sd}^t\right), \tag{8.147}$$

leading to the following Wilson coefficient:

$$\frac{G_F^2 M_W^2}{2\pi^2} \operatorname{Im}\lambda_{sd}^t \left[\operatorname{Re}\lambda_{sd}^t \left(S_0(x_t) - S_0(x_t, x_c) \right) - \operatorname{Re}\lambda_{sd}^c \left(S_0(x_c) - S_0(x_t, x_c) \right) \right]. \qquad (8.148)$$

We can indeed check that loops of up quarks drop in the differences of S_0 functions in eqn (8.148), leaving us with the following expressions:

$$S_0(x_t) - S_0(x_c, x_t) = \overline{A}(x_t, x_t) - \overline{A}(x_t, x_c) - \overline{A}(x_t, x_u) + \overline{A}(x_c, x_u), \qquad (8.149)$$

$$S_0(x_c) - S_0(x_c, x_t) = \overline{A}(x_t, x_u) - \overline{A}(x_t, x_c) - \overline{A}(x_c, x_u) + \overline{A}(x_c, x_c). \qquad (8.150)$$

Now,

$$S_0(x_t) - S_0(x_c, x_t) = \frac{m_t^2}{M_W^2} S_t(x_t, x_c), \qquad (8.151)$$

$$S_0(x_c) - S_0(x_c, x_t) = \frac{m_c^2}{M_W^2} S_c(x_t, x_c), \qquad (8.152)$$

with $S_{c,t}(x_t, x_c)$ non-vanishing in the limit $x_c \to 0$. Had we kept the dependence on external momenta p in the evaluation of the loop functions, we would have obtained terms of $\mathcal{O}(p^2/M_W^2)$ (or higher) in S_0, corresponding to a correction of $\mathcal{O}(p^2/m_t^2)$ to eqn (8.151) and to an $\mathcal{O}(p^2/m_c^2)$ correction to eqn (8.152). The first is fully negligible, but the second is potentially relevant, since $m_K^2/m_c^2 \sim 10\%$. Fortunately, we have a systematic way to keep these corrections into account, since a contribution of $\mathcal{O}(p^2/M_W^2)$ to the amplitude can be described by the matrix element of a local operator of dimension eight. To perform the matching of the full amplitude onto the effective theory including dimension eight operators, we need to expand the diagrams we computed above at $\mathcal{O}(p^2/M_W^2)$. However, this is not enough since at dimension eight the operator basis includes an operator involving the commutator of two covariant derivatives,

$$g_s \bar{s} \gamma_\mu P_L \tilde{G}^{\mu\nu} d \bar{s} \gamma_\nu d, \qquad (8.153)$$

which has vanishing matrix element on four-quark states. Therefore, we need to consider external states with four quarks and a gluon to complete the matching at dimension eight. In this way we can estimate the corrections of $\mathcal{O}(p^2/m_c^2)$, which turn out to be at the few percent level [39, 40].

To summarize, the expansion in local operators is safe and systematically improvable by going to dimension eight operators for the CP violating part of the Hamiltonian, while the CP conserving one is dominated by long distance contributions, which must be evaluated as a long distance matrix element of two $\Delta S = 1$ effective Hamiltonians.

8.4.2   QCD Corrections

The inclusion of LO QCD corrections goes exactly along the lines of Section 8.3.1, the only difference being that in the $\Delta F = 2$ case we do not need to introduce the second operator with a different colour structure since we can Fierz Dirac indices to fall back on the original operator in eqn (8.143):

$$\bar{s}^{\alpha} \gamma^{\mu} P_L d^{\beta} \bar{s}^{\beta} \gamma_{\mu} P_L d^{\alpha} = \bar{s} \gamma^{\mu} P_L d \bar{s} \gamma_{\mu} P_L d. \tag{8.154}$$

The relevant anomalous dimension can then be obtained by a straightforward combination of the results in Section 8.3.1, yielding the same result as for Q_+ in eqn (8.82), namely

$$\gamma_0 = 6 \frac{N_c - 1}{N_c}, \tag{8.155}$$

leading to a suppression of $C(\mu_h)$ with respect to $C(M_W)$.

The calculation of NLO (and of NNLO) QCD corrections is more involved and goes beyond the scope of these lectures; the interested reader can find all the details in [19, 41]–[47].

8.4.3 $\Delta B = 2$ Effective Hamiltonian

In the previous paragraphs we introduced the $\Delta S = 2$ effective Hamiltonian. If we consider instead $\bar{b}q \to \bar{q}b$ transitions, with $q = d, s$, we see that in this case at the scale $\mu_h \sim m_b$ up and charm quarks remain dynamical and thus their contribution cancels in the matching, leaving us with the top-quark contribution only. Recalling the relative size of the relevant CKM factors,

$$\lambda_{bs}^t \sim \lambda_{bs}^c \gg \lambda_{bs}^u, \qquad \lambda_{bd}^t \sim \lambda_{bd}^c \sim \lambda_{bd}^u, \tag{8.156}$$

and the relative size of the loop functions $S_0(x_t)$ and $S_0(x_c, x_t)$, we immediately realize that the top-charm contribution enters at $\mathcal{O}(m_c^2/m_t^2)$ and is therefore fully negligible, leaving us with the effective Hamiltonian

$$\mathcal{H}_{\mathrm{eff}}^{\Delta B = 2} = \frac{G_F^2 M_W^2}{4\pi^2} \left(\lambda_{bq}^t \right)^2 S_0(x_t) \bar{b} \gamma^{\mu} P_L q \bar{b} \gamma_{\mu} P_L q. \tag{8.157}$$

QCD corrections can be included up to NLO following the same line as for the top-top contribution to $\Delta S = 2$. Electroweak corrections have been computed in ref. [48].

8.4.4 $\Delta C = 2$ Effective Hamiltonian

One could think of applying the same procedure as for $\Delta S = 2$ processes to obtain an effective Hamiltonian for $\Delta C = 2$ transitions. However, in this case the role played by the charm in $\Delta S = 2$ goes to the strange quark, which is still dynamical at the charm scale. One would then be in a situation similar to $\Delta B = 2$, except that for $\Delta C = 2$ one has

$$\lambda_{cu}^d \sim \lambda_{cu}^s \gg \lambda_{cu}^b, \tag{8.158}$$

and $\lambda_{cu}^b/\lambda_{cu}^s \ll m_b/m_s$,[5] so that the process is dominated by the matrix element of two $\Delta C = 1$ effective Hamiltonians. Indeed, to an excellent approximation GIM cancellation in $\Delta C = 2$ processes coincides with flavour $SU(3)$.

8.4.5 $\Delta F = 2$ Hamiltonians Beyond the SM

While generalizing $\Delta F = 1$ Hamiltonians beyond the SM, i.e. writing down the most general $\Delta F = 1$ Hamiltonian including all dimension six, $SU(3) \otimes U(1)_{em}$ gauge-invariant operators, increases the number of operators up to ~ 120 [49, 50], the number of independent operators that may arise is much smaller for $\Delta F = 2$ transitions, so let us discuss this as an illustrative example of going beyond the SM.

There is a large degree of arbitrariness in the choice of the operator basis, since Fierz transformations can be used to get rid of a Dirac and colour structure in favour of a different one. As an example, let us choose the basis in ref. [51]:

$$
\begin{aligned}
Q_1^{sdsd} &= \bar{s}_L \gamma^\mu d_L \bar{s}_L \gamma^\mu d_L & \tilde{Q}_1^{sdsd} &= \bar{s}_R \gamma^\mu d_R \bar{s}_R \gamma^\mu d_R \\
Q_2^{sdsd} &= \bar{s}_L d_R \bar{s}_L d_R & \tilde{Q}_2^{sdsd} &= \bar{s}_R d_L \bar{s}_R d_L \\
Q_3^{sdsd} &= \bar{s}_L^\alpha d_R^\beta \bar{s}_L^\beta d_R^\alpha & \tilde{Q}_3^{sdsd} &= \bar{s}_R^\alpha d_L^\beta \bar{s}_R^\beta d_L^\alpha \\
Q_4^{sdsd} &= \bar{s}_L d_R \bar{s}_R d_L \\
Q_5^{sdsd} &= \bar{s}_L^\alpha d_R^\beta \bar{s}_R^\beta d_L^\alpha
\end{aligned}
\tag{8.159}
$$

With respect to the SM, where only Q_1^{sdsd} is present, we need to add two new Dirac structures, each one with two different colour structures, plus the operators obtained by the $L \leftrightarrow R$ transformation. As originally pointed out in ref. [52] and confirmed at NLO in refs. [49, 53, 54], the additional operators have large anomalous dimensions (especially Q_4^{sdsd}) which strongly enhance their coefficients at the hadronic scale with respect to the high scale, making them very important in phenomenological studies of $\Delta F = 2$ processes beyond the SM.

8.4.6 $\Delta F = 2$ Matrix Elements in the VIA

Before closing this Section on $\Delta F = 2$ effective Hamiltonians, let us briefly discuss how their matrix elements can be estimated in the VIA. VIA matrix elements are useful not only because they give a first (and not too rough) estimate of the matrix elements, but also because it is often easier and more accurate to compute the ratio of the full matrix element normalized to the VIA than the absolute matrix element. For this reason, matrix elements are often expressed in terms of VIA results times the so-called B-parameters, which in fact parameterize the ratio of the full matrix element with respect to the VIA one.

---

[5] The strange quark mass in the denominator should actually be replaced by a suitable hadronic scale, making the ratio even smaller.

For the sake of concreteness, let us consider $\Delta S = 2$ processes. The SM effective Hamiltonian only contains Q_1^{sdsd}, whose VIA matrix element is given by

$$\langle \overline{K}^0 | \bar{s}\gamma^\mu P_L d\bar{s}\gamma_\mu P_L d | K^0 \rangle_{\text{VIA}} = 2(\langle \overline{K}^0 | \bar{s}\gamma^\mu P_L d | 0 \rangle \langle 0 | \bar{s}\gamma_\mu P_L d | K^0 \rangle \tag{8.160}$$
$$+ \langle \overline{K}^0 | \bar{s}^\alpha \gamma^\mu P_L d^\beta | 0 \rangle \langle 0 | \bar{s}^\beta \gamma_\mu P_L d^\alpha | K^0 \rangle),$$

where the second term corresponds to Fierzed contractions with respect to the first term. Using

$$\langle 0 | \bar{s}\gamma^\mu \gamma_5 d | K^0 \rangle = -i p_K^\mu \frac{F_K}{\sqrt{2m_K}} \tag{8.161}$$

we obtain for the first term

$$\frac{1}{4} \frac{F_K^2 m_K^2}{2m_K} = \frac{1}{8} m_K F_K^2. \tag{8.162}$$

For the second term we perform a colour Fierz transformation:

$$\delta_{\alpha\beta}\delta_{\gamma\delta} = 2T_{\alpha\delta}^a T_{\gamma\beta}^a + \frac{1}{3}\delta_{\alpha\delta}\delta_{\gamma\beta} \tag{8.163}$$

getting

$$\langle \overline{K}^0 | \bar{s}^\alpha \gamma^\mu P_L d^\beta | 0 \rangle \langle 0 | \bar{s}^\beta \gamma_\mu P_L d^\alpha | K^0 \rangle = \frac{1}{3} \langle \overline{K}^0 | \bar{s}\gamma^\mu P_L d | 0 \rangle \langle 0 | \bar{s}\gamma_\mu P_L d | K^0 \rangle \tag{8.164}$$
$$+ 2 \langle \overline{K}^0 | \bar{s}T^a\gamma^\mu P_L d^\beta | 0 \rangle \langle 0 | \bar{s}T^a\gamma_\mu P_L d | K^0 \rangle.$$

The second term vanishes and the first one reduces to eqn (8.162), so that in the end we obtain

$$\langle \overline{K}^0 | \bar{s}\gamma^\mu P_L d\bar{s}\gamma_\mu P_L d | K^0 \rangle_{\text{VIA}} = 2(1 + \frac{1}{3})\frac{1}{8}F_K^2 m_K = \frac{1}{3}F_K^2 m_K. \tag{8.165}$$

In general we can therefore write

$$\langle \overline{K}^0 | \bar{s}\gamma^\mu P_L d\bar{s}\gamma_\mu P_L d | K^0 \rangle = \langle \overline{K}^0 | \bar{s}\gamma^\mu P_L d\bar{s}\gamma_\mu P_L d | K^0 \rangle_{\text{VIA}} B_K$$
$$= \frac{1}{3}F_K^2 m_K B_K. \tag{8.166}$$

It is now interesting to look at the VIA matrix elements of the additional operators that arise beyond the SM. From eqn (8.159) we see that the operators Q_i^{sdsd} for $i > 1$ are built by products of scalar/pseudoscalar densities. We have

$$\partial_\mu \langle 0 | \bar{s}\gamma^\mu \gamma_5 d | K^0 \rangle = \begin{cases} -i(m_s + m_d)\langle 0 | \bar{s}\gamma_5 d | K^0 \rangle \\ \frac{m_K^2 F_K}{\sqrt{2m_K}} \end{cases} \tag{8.167}$$

where in the first case we applied the derivative to the quark bilinear while in the second case we applied it to the whole matrix element. Using eqn (8.167) we obtain

$$\langle \overline{K}^0 | \bar{s}\gamma_5 d | 0 \rangle \langle 0 | \bar{s}\gamma_5 d | K^0 \rangle = -\left(\frac{m_K}{m_s + m_d}\right)^2 \frac{m_K F_K^2}{2}, \tag{8.168}$$

which for Kaons is chirally enhanced by one order of magnitude since $m_K \sim 3(m_s + m_d)$. Combining this chiral enhancement with the RG enhancement one sees that these operators play a crucial role in $\Delta S = 2$ processes beyond the SM. We will return to this point later. For completeness, we write down the matrix elements for all the operators in eqn (8.159) in the VIA:

$$\langle \overline{K}^0 | Q_1 | K^0 \rangle = \frac{1}{3} m_K F_K^2, \tag{8.169}$$

$$\langle \overline{K}^0 | Q_2 | K^0 \rangle = -\frac{5}{24}\left(\frac{m_K}{m_s + m_d}\right)^2 m_K F_K^2, \tag{8.170}$$

$$\langle \overline{K}^0 | Q_3 | K^0 \rangle = \frac{1}{24}\left(\frac{m_K}{m_s + m_d}\right)^2 m_K F_K^2, \tag{8.171}$$

$$\langle \overline{K}^0 | Q_4 | K^0 \rangle = \left[\frac{1}{24} + \frac{1}{4}\left(\frac{m_K}{m_s + m_d}\right)^2\right] m_K F_K^2, \tag{8.172}$$

$$\langle \overline{K}^0 | Q_5 | K^0 \rangle = \left[\frac{1}{8} + \frac{1}{12}\left(\frac{m_K}{m_s + m_d}\right)^2\right] m_K F_K^2. \tag{8.173}$$

A word of caution is necessary at this point. Operators $Q_{4,5}$ have VIA matrix elements that contain two contributions, one from pseudoscalar density matrix elements and one from axial vector currents. To define the corresponding B-parameters it is convenient to choose as normalization just the pseudoscalar density contributions. However, this corresponds to having $B \neq 1$ in the VIA [55].

8.5 Effective Hamiltonians at Work: Meson–Anti-meson Mixing and CP Violation

Having discussed the basics of $\Delta F = 1$ and $\Delta F = 2$ effective Hamiltonians, let us now use them to study meson–anti-meson mixing and CP violation.

8.5.1 Meson–Anti-meson Mixing

There are four neutral mesons which differ from their antiparticles just because of their flavour quantum numbers: K^0, D^0, B_d and B_s mesons. While strong and electromagnetic interactions preserve flavour, the full Hamiltonian does not, due to flavour-changing weak interactions. Therefore, its eigenstates will be superpositions of mesons and anti-mesons, giving rise to the phenomenon of meson–anti-meson oscillations, which entails a difference of mass and width of the two eigenstates [56]. Let us first write down the formalism for a generic neutral meson, which we denote by M^0, and then specialize to the four cases above, in which different simplifying assumptions can be made.

Notice that a CP transformation takes a neutral meson into its anti-particle with an arbitrary phase shift ξ:

$$CP|M^0\rangle = e^{i\xi}|\overline{M}^0\rangle, \tag{8.174}$$
$$CP|\overline{M}^0\rangle = e^{-i\xi}|M^0\rangle.$$

The matrix elements of the full Hamiltonian between M^0 and $\overline{M}^0$ states can be written as a two-by-two complex matrix:

$$\hat{H} = \begin{pmatrix} H_{11} & H_{12} \\ H_{21} & H_{22} \end{pmatrix} \equiv \begin{pmatrix} \langle M^0|H|M^0\rangle & \langle M^0|H|\overline{M}^0\rangle \\ \langle \overline{M}^0|H|M^0\rangle & \langle \overline{M}^0|H|\overline{M}^0\rangle \end{pmatrix} \equiv \hat{M} - \frac{i}{2}\hat{\Gamma}, \tag{8.175}$$

where in the last equality we have split the complex matrix $\hat{H}$ in its Hermitian ($\hat{M}$) and anti-Hermitian ($-i/2\hat{\Gamma}$) parts.

CPT invariance requires $M_{11} = M_{22}$ and $\Gamma_{11} = \Gamma_{22}$, while it does not constrain the off-diagonal matrix elements. CP invariance instead connects off-diagonal elements among themselves:

$$H_{21} = \langle \overline{M}^0|H|M^0\rangle \xrightarrow{CP} e^{i\xi}\langle M^0|H^{CP}|\overline{M}^0\rangle e^{i\xi}, \tag{8.176}$$

so that CP conservation ($H^{CP} = H$) implies

$$H_{21} = e^{2i\xi}H_{12} \Rightarrow |H_{21}| = |H_{12}| \Rightarrow \text{Im}\left(M_{12}^*\Gamma_{12}\right) = 0 \Rightarrow \text{Im}\left(\frac{\Gamma_{12}}{M_{12}}\right) = 0. \tag{8.177}$$

The eigenvalue equation reads, assuming CPT invariance,

$$\det\left(\hat{H} - \lambda\mathbb{1}\right) = 0 = (H_{11} - \lambda)^2 - H_{12}H_{21} \Rightarrow \lambda = H_{11} \pm \sqrt{H_{12}H_{21}}. \tag{8.178}$$

Defining

$$\lambda_{1,2} = m_{1,2} - i/2\Gamma_{1,2}, \quad m = \frac{m_1 + m_2}{2}, \quad \Gamma = \frac{\Gamma_1 + \Gamma_2}{2}, \tag{8.179}$$
$$\Delta m = m_1 - m_2, \quad \Delta\Gamma = \Gamma_1 - \Gamma_2, \quad x = \frac{\Delta m}{\Gamma}, \quad y = \frac{\Delta\Gamma}{2\Gamma},$$

we can alternatively label the two eigenstates by their mass, i.e. defining $\Delta m = m_H - m_L$ to be positive (here H stands for heavy and L for light), or by their width, i.e. defining $\Delta\Gamma = \Gamma_S - \Gamma_L$ to be positive (here S stands for short-lived and L for long-lived).

We have

$$\Delta\lambda = \lambda_1 - \lambda_2 = \Delta m - \frac{i}{2}\Delta\Gamma = 2\sqrt{H_{12}H_{21}}, \tag{8.180}$$

$$(\Delta m)^2 - \frac{1}{4}(\Delta\Gamma)^2 - i\Delta m\Delta\Gamma = 4H_{12}H_{21} \tag{8.181}$$

$$= 4\left(|M_{12}^2| - \frac{1}{4}|\Gamma_{12}|^2\right) - 4i\mathrm{Re}\left(M_{12}\Gamma_{12}^*\right).$$

Taking real and imaginary parts we obtain

$$(\Delta m)^2 - \frac{1}{4}(\Delta\Gamma)^2 = 4\left(|M_{12}|^2 - \frac{1}{4}|\Gamma_{12}|^2\right), \tag{8.182}$$

$$\Delta m\Delta\Gamma = 4\mathrm{Re}\left(M_{12}\Gamma_{12}^*\right). \tag{8.183}$$

Notice that

$$(\Delta m)^2 = 4(\mathrm{Re}\sqrt{H_{12}H_{21}})^2 = 2|H_{12}H_{21}| + 2\mathrm{Re}H_{12}H_{21}, \tag{8.184}$$

so that

$$\begin{aligned}
(|H_{12}| + |H_{21}|)^2 &= |H_{12}|^2 + |H_{21}|^2 + 2|H_{12}H_{21}| \\
&= |H_{12}|^2 + |H_{21}|^2 - 2\mathrm{Re}(H_{12}H_{21}) + (\Delta m)^2 \\
&= |H_{12} - H_{21}^*|^2 + (\Delta m)^2 \\
&= |\Gamma_{12}|^2 + (\Delta m)^2.
\end{aligned} \tag{8.185}$$

Let us write the eigenstates as

$$|M_{1,2}\rangle = p|M^0\rangle \pm q|\overline{M}^0\rangle, \text{ with } |p^2| + |q^2| = 1. \tag{8.186}$$

Then we have

$$H_{11}p \pm H_{12}q = \lambda_{1,2}p \Rightarrow H_{11} \pm \frac{q}{p}H_{12} = H_{11} \pm \sqrt{H_{12}H_{21}}$$

$$\Rightarrow \frac{q}{p} = \sqrt{\frac{H_{21}}{H_{12}}} = \frac{2M_{12}^* - i\Gamma_{12}^*}{\Delta m - \frac{i}{2}\Delta\Gamma} = \frac{\Delta m - \frac{i}{2}\Delta\Gamma}{2M_{12} - i\Gamma_{12}}. \tag{8.187}$$

Using eqn (8.177) we see that CP conservation implies

$$\frac{q}{p} = \sqrt{\frac{H_{21}}{H_{12}}} = \sqrt{\frac{H_{12}e^{2i\xi}}{H_{12}}} = e^{i\xi}, \quad \text{so that} \quad \left|\frac{q}{p}\right| = 1. \tag{8.188}$$

Thus,

$$\left|\frac{q}{p}\right| \neq 1 \text{ implies CP violation,} \tag{8.189}$$

usually denoted as *CP violation in mixing*.

It is useful to define

$$x_{12} = 2\frac{|M_{12}|}{\Gamma}, \quad y_{12} = \frac{|\Gamma_{12}|}{\Gamma}, \quad and \quad \Phi_{12} = \arg\left(\frac{\Gamma_{12}}{M_{12}}\right), \tag{8.190}$$

so that

$$\left|\frac{q}{p}\right| = \frac{|2M_{12}^* - i\Gamma_{12}^*|}{\Gamma|x - iy|} = \frac{\sqrt{x_{12}^2 + y_{12}^2 + 2x_{12}y_{12}\sin\Phi_{12}}}{\sqrt{x^2 + y^2}}, \tag{8.191}$$

implying that

$$\left|\frac{q}{p}\right| \neq 1 \Leftrightarrow \sin\Phi_{12} \neq 0. \tag{8.192}$$

Finally, CP violation in meson–anti-meson mixing can also be expressed in terms of the parameter δ defined as

$$\delta \equiv \frac{|H_{12}| - |H_{21}|}{|H_{12}| + |H_{21}|} = \langle M_1 | M_2 \rangle = |p|^2 - |q|^2 = \frac{1 - \left|\frac{q}{p}\right|^2}{1 + \left|\frac{q}{p}\right|^2}. \tag{8.193}$$

We have

$$1 + \delta^2 = 1 + \frac{|H_{12}|^2 + |H_{21}|^2 - 2|H_{12}||H_{21}|}{|H_{12}|^2 + |H_{21}|^2 + 2|H_{12}||H_{21}|} = 2\frac{|H_{12}|^2 + |H_{21}|^2}{(|H_{12}| + |H_{21}|)^2}, \tag{8.194}$$

$$|H_{12}|^2 + |H_{21}|^2 = \frac{4|M_{12}|^2 + |\Gamma_{12}|^2}{2}, \tag{8.195}$$

so that

$$\frac{\delta}{1 + \delta^2} = \frac{1}{2}\frac{|H_{12}|^2 - |H_{21}|^2}{|H_{12}|^2 + |H_{21}|^2} = \frac{2|M_{12}\Gamma_{12}|\sin\Phi_{12}}{4|M_{12}|^2 + |\Gamma_{12}|^2} \tag{8.196}$$

and

$$\delta = \frac{2|M_{12}\Gamma_{12}|\sin\Phi_{12}}{(\Delta m)^2 + |\Gamma_{12}|^2}. \tag{8.197}$$

Let us also write down the expressions for $|M_{12}|$, $|\Gamma_{12}|$ and Φ_{12} in terms of Δm, $\Delta\Gamma$, and δ:

$$|M_{12}| = \sqrt{4\frac{(\Delta m)^2 + \delta^2(\Delta\Gamma)^2}{16(1-\delta^2)}} \sim \frac{\Delta m}{2} + \mathcal{O}(\delta^2), \tag{8.198}$$

$$|\Gamma_{12}| = \sqrt{\frac{(\Delta\Gamma)^2 + 4\delta^2(\Delta m)^2}{4(1-\delta^2)}} \sim \frac{\Delta\Gamma}{2} + \mathcal{O}(\delta^2), \tag{8.199}$$

$$\sin\Phi_{12} = \frac{4|\Gamma_{12}|^2 + 16|M_{12}|^2 - (4(\Delta m)^2 + (\Delta\Gamma)^2)|q/p|^2}{16|M_{12}\Gamma_{12}|} \tag{8.200}$$

$$\sim \frac{4(\Delta m)^2 + (\Delta\Gamma)^2}{2\Delta m\Delta\Gamma}\delta + \mathcal{O}(\delta^2),$$

and vice versa:

$$(\Delta m)^2 = \frac{4|M_{12}|^2 - \delta^2|\Gamma_{12}|^2}{1+\delta^2} \sim 4|M_{12}|^2 + \mathcal{O}(\delta^2), \tag{8.201}$$

$$(\Delta\Gamma)^2 = \frac{4|\Gamma_{12}|^2 - 16\delta^2|M_{12}|^2}{1+\delta^2} \sim 4|\Gamma_{12}|^2 + \mathcal{O}(\delta^2). \tag{8.202}$$

Notice that the transformation in eq. (8.186) is unitary only if $|p^2| - |q^2| = \delta = 0$, i.e. if CP is conserved in the mixing. Therefore, if CP is violated, one must be careful in defining outgoing $M_{1,2}$ states using the so-called reciprocal basis:

$$\langle \tilde{M}_{1,2}| = \frac{q\langle M^0| \pm p\langle \overline{M}^0|}{2pq}.$$

8.5.2   Time Evolution of Mixed Meson States

Having obtained the eigenstates of the Hamiltonian in eqns (8.178) and (8.179), we can write down the time evolution of a state initially produced as an M^0 or as an $\overline{M}^0$. We start from the time evolution of the eigenstates,

$$|M_{1,2}(t)\rangle = e^{-i\lambda_{1,2}t}|M_{1,2}(0)\rangle, \tag{8.203}$$

and use eqn (8.186) to rotate back to the $\overline{M}^0$:

$$|M^0(t)\rangle = \frac{1}{2p}(M_1(t) + M_2(t)) = g_+(t)|M^0\rangle + \frac{q}{p}g_-(t)|\overline{M}^0\rangle, \tag{8.204}$$

$$|\overline{M}^0(t)\rangle = \frac{1}{2q}(M_1(t) - M_2(t)) = \frac{p}{q}g_-(t)|M^0\rangle + g_+(t)|\overline{M}^0\rangle, \tag{8.205}$$

with

$$g_\pm(t) = \frac{e^{-i\lambda_1 t} \pm e^{-i\lambda_2 t}}{2}. \tag{8.206}$$

The probability that a meson initially produced as a $\overline{M}^0$ remains such at time t is given by $|g_+(t)|^2$, while the probabilities of an M^0 becoming an $\overline{M}^0$, and vice-versa, are not equal to each other if CP is violated:

$$P(M^0(0) \to \overline{M}^0(t))) = \left|\frac{q}{p}\right|^2 |g_-(t)|^2, \tag{8.207}$$

$$P(\overline{M}^0(0) \to M^0(t))) = \left|\frac{p}{q}\right|^2 |g_-(t)|^2. \tag{8.208}$$

We have

$$|g_\pm(t)|^2 = \frac{e^{-\Gamma_1 t} + e^{-\Gamma_2 t} \pm 2e^{-\Gamma t}\cos(\Delta m t)}{4} \tag{8.209}$$

$$= \frac{e^{-\Gamma t}}{2}\left(\cosh(\Delta\Gamma t/2) \pm \cos(\Delta m t)\right),$$

$$g_+(t)g_-^*(t) = \frac{e^{-\Gamma_1 t} - e^{-\Gamma_2 t} - 2e^{-\Gamma t}\sin(\Delta m t)}{4} \tag{8.210}$$

$$= -\frac{e^{-\Gamma t}}{2}\left(\sinh(\Delta\Gamma t/2) - i\sin(\Delta m t)\right).$$

8.5.3 Observables for Meson–Anti-meson Mixing

Since the M^0 mesons are unstable, we must consider their weak decays in building observables related to meson mixing. The probabilities in eqns (8.207) and (8.208) can be directly accessed using semileptonic decays, since for decays of down-type quarks one has $M^0 \not\to \ell\overline{\nu}_\ell X$ and $\overline{M}^0 \not\to \ell\nu_\ell X$, where X represents an unspecified hadronic final state. Therefore, those decays can only happen through mixing, and we can define the semileptonic CP asymmetry as the difference of the number of semileptonic decays to wrong sign leptons in $\overline{M}^0$ decays normalized to the total number of such decays:

$$a_{\mathrm{SL}} \equiv \frac{N(\overline{M}^0 \to \ell\nu_\ell X) - N(M^0 \to \ell\overline{\nu}_\ell X)}{N(\overline{M}^0 \to \ell\nu_\ell X) + N(M^0 \to \ell\overline{\nu}_\ell X)}. \tag{8.211}$$

Assuming that M^0 and $\overline{M}^0$ are produced in equal number N_0 and CP invariance of the semileptonic decay amplitude A, one has

$$N(M^0 \to \ell\overline{\nu}_\ell X) = N_0|A|^2 \left|\frac{q}{p}\right|^2 \int_0^\infty |g_-(t)|^2 \mathrm{d}t, \tag{8.212}$$

$$N(\overline{M}^0 \to \ell\nu_\ell X) = N_0|A|^2 \left|\frac{p}{q}\right|^2 \int_0^\infty |g_-(t)|^2 \mathrm{d}t. \tag{8.213}$$

All factors except for the mixing parameters drop in the ratio, leading to

$$a_{\mathrm{SL}} = \frac{\left|\frac{p}{q}\right|^2 - \left|\frac{q}{p}\right|^2}{\left|\frac{p}{q}\right|^2 + \left|\frac{q}{p}\right|^2} = \frac{1 - \left|\frac{q}{p}\right|^4}{1 + \left|\frac{q}{p}\right|^4}. \tag{8.214}$$

In general, the decay amplitude into a given final state $|f\rangle$ and its CP-conjugate $|\bar{f}\rangle = e^{-i\xi_f}\mathcal{CP}|f\rangle$ in the SM can be always written in the following form:[6]

$$A(M \to f) \equiv \mathcal{A}_f = A_f e^{i\phi_f} e^{i\delta_f}(1 + r_f e^{i\phi_{r_f}} e^{i\delta_{r_f}}), \qquad (8.215)$$

$$A(\overline{M} \to \bar{f}) \equiv \overline{\mathcal{A}}_{\bar{f}} = e^{-i\xi} e^{i\xi_f} A_f e^{-i\phi_f} e^{i\delta_f}(1 + r_f e^{-i\phi_{r_f}} e^{i\delta_{r_f}}), \qquad (8.216)$$

$$A(M \to \bar{f}) \equiv \mathcal{A}_{\bar{f}} = A_{\bar{f}} e^{i\phi_{\bar{f}}} e^{i\delta_{\bar{f}}}(1 + r_{\bar{f}} e^{i\phi_{r_{\bar{f}}}} e^{i\delta_{r_{\bar{f}}}}), \qquad (8.217)$$

$$A(\overline{M} \to f) \equiv \overline{\mathcal{A}}_f = e^{-i\xi} e^{-i\xi_f} A_{\bar{f}} e^{-i\phi_{\bar{f}}} e^{i\delta_{\bar{f}}}(1 + r_{\bar{f}} e^{-i\phi_{r_{\bar{f}}}} e^{i\delta_{r_{\bar{f}}}}), \qquad (8.218)$$

with A_f, $A_{\bar{f}}$, r_f, and $r_{\bar{f}}$ real, and $\mathcal{CP}|M\rangle = e^{i\xi}|\overline{M}\rangle$. Indeed, using CKM unitarity where needed we can have at most two independent CKM factors in each decay amplitude, corresponding for example in eqn (8.215) to A_f and $A_f r_f$. For later convenience, we have written the second amplitude as a multiplicative factor, since in several cases one has $|r_f| \ll 1$ and an expansion in r_f can be performed. We have written explicitly the CP-odd weak phases ϕ_i and the CP-even strong phases δ_i.

We see that

$$\phi_{r_f} \neq 0 \text{ and } \delta_{r_f} \neq 0 \Leftrightarrow \mathcal{A}_f \neq \overline{\mathcal{A}}_{\bar{f}}. \qquad (8.219)$$

This is usually denoted as *'direct' CP violation* or *CP violation in the decay*. The corresponding CP asymmetry can be written as

$$A_{\text{CP}}^{\text{dir}}(f) \equiv \frac{|A(\overline{M} \to \bar{f})|^2 - |A(M \to f)|^2}{|A(\overline{M} \to \bar{f})|^2 + |A(M \to f)|^2} = \frac{2r_f \sin\phi_{r_f} \sin\delta_{r_f}}{1 + r_f^2 + 2r_f \cos\phi_{r_f} \cos\delta_{r_f}}. \qquad (8.220)$$

For neutral meson decays, we can consider the case of a final state which is a CP eingenstate with eigenvalue η_f:

$$A(M \to f_{\text{CP}}) = A_f e^{i\phi_f} e^{i\delta_f}(1 + r_f e^{i\phi_{r_f}} e^{i\delta_{r_f}}), \qquad (8.221)$$

$$A(\overline{M} \to f_{\text{CP}}) = e^{-i\xi} \eta_f A_f e^{-i\phi_f} e^{i\delta_f}(1 + r_f e^{-i\phi_{r_f}} e^{i\delta_{r_f}}). \qquad (8.222)$$

It is useful to introduce

$$\lambda_f = \frac{q}{p}\frac{\overline{\mathcal{A}}_f}{\mathcal{A}_f} = \frac{q}{p}\frac{e^{-i\xi} e^{-i\xi_f} A_{\bar{f}} e^{-i\phi_{\bar{f}}} e^{i\delta_{\bar{f}}}(1 + r_{\bar{f}} e^{-i\phi_{r_{\bar{f}}}} e^{i\delta_{r_{\bar{f}}}})}{A_f e^{i\phi_f} e^{i\delta_f}(1 + r_f e^{i\phi_{r_f}} e^{i\delta_{r_f}})}, \qquad (8.223)$$

which is manifestly rephasing invariant since for $|M^0\rangle \to e^{i\Xi}|M^0\rangle$ and $|\overline{M}^0\rangle \to e^{i\overline{\Xi}}|\overline{M}^0\rangle$ we have

$$\frac{q}{p} \to e^{i(\Xi - \overline{\Xi})}\frac{q}{p}, \qquad \frac{\overline{\mathcal{A}}_f}{\mathcal{A}_f} \to e^{i(\overline{\Xi} - \Xi)}\frac{\overline{\mathcal{A}}_f}{\mathcal{A}_f} \qquad (8.224)$$

(or equivalently $\xi \to \xi + \Xi - \overline{\Xi}$), so that $\lambda_f \to \lambda_f$.

---

[6] We prefer to keep the equations in a symmetric form, keeping in mind that an overall phase could be dropped since it is physically irrelevant.

For decays to a CP eigenstate this simplifies to

$$\lambda_{f_{\mathrm{CP}}} = \frac{q}{p} e^{-i\xi} \eta_f e^{-2i\phi_f} \frac{(1 + r_f e^{-i\phi_{r_f}} e^{i\delta_{r_f}})}{(1 + r_f e^{i\phi_{r_f}} e^{i\delta_{r_f}})} \tag{8.225}$$

$$= \frac{q}{p} e^{-i\xi} \eta_f e^{-2i\phi_f} (1 - 2i r_f e^{i\delta_{r_f}} \sin\phi_{r_f} + \mathcal{O}(r_f^2)).$$

CP conservation implies $q/p = e^{i\xi}$ (see eq. (8.177)) and $\phi_{f,r_f} = 0$, i.e. Im $\lambda_{f_{\mathrm{CP}}} = 0$. Therefore,

$$\mathrm{Im}\lambda_{f_{\mathrm{CP}}} \neq 0 \text{ implies CP violation.} \tag{8.226}$$

This form of CP violation requires the *interference between mixing and decay*. Indeed, $\lambda_{f_{\mathrm{CP}}}$ determines the time evolution of $\overline{M}^0$ decays into f_{CP}. Using eqns (8.215), (8.218), and (8.204)–(8.205) we can write

$$\Gamma(M^0(t) \to f_{\mathrm{CP}}) = |g_+(t)|^2 |A_{f_{\mathrm{CP}}}|^2 + \left|\frac{q}{p}\right|^2 |g_-(t)|^2 |\overline{A}_{f_{\mathrm{CP}}}|^2 \tag{8.227}$$

$$+ 2\mathrm{Re}\left(g_+^*(t)g_-(t)\frac{q}{p}A_{f_{\mathrm{CP}}}^*\overline{A}_{f_{\mathrm{CP}}}\right)$$

$$= |A_{f_{\mathrm{CP}}}|^2 \left[|g_+(t)|^2 + |\lambda_{f_{\mathrm{CP}}}|^2 |g_-(t)|^2 + 2\mathrm{Re}\left(g_+^*(t)g_-(t)\lambda_{f_{\mathrm{CP}}}\right)\right]$$

$$= |A_{f_{\mathrm{CP}}}|^2 \frac{e^{-\Gamma t}}{2}\left[\left(1 + |\lambda_{f_{\mathrm{CP}}}|^2\right)\cosh(\Delta\Gamma t/2) + \left(1 - |\lambda_{f_{\mathrm{CP}}}|^2\right)\cos(\Delta m t)\right.$$

$$\left. - 2\mathrm{Re}\lambda_{f_{\mathrm{CP}}}\sinh(\Delta\Gamma t/2) + 2\mathrm{Im}\lambda_{f_{\mathrm{CP}}}\sin(\Delta m t)\right],$$

$$\Gamma(\overline{M}^0(t) \to f_{\mathrm{CP}}) = \left|\frac{p}{q}\right|^2 |g_-(t)|^2 |A_{f_{\mathrm{CP}}}|^2 + |g_+(t)|^2 |\overline{A}_{f_{\mathrm{CP}}}|^2$$

$$+ 2\mathrm{Re}\left(g_+^*(t)g_-(t)\frac{p}{q}A_{f_{\mathrm{CP}}}\overline{A}_{f_{\mathrm{CP}}}^*\right)$$

$$= |\overline{A}_{f_{\mathrm{CP}}}|^2 \left[|g_+(t)|^2 + |\lambda_{f_{\mathrm{CP}}}|^{-2} |g_-(t)|^2 + 2\mathrm{Re}\left(g_+(t)^*g_-(t)\lambda_{f_{\mathrm{CP}}}^{-1}\right)\right]$$

$$= |\overline{A}_{f_{\mathrm{CP}}}|^2 \frac{e^{-\Gamma t}}{2}\left[\left(1 + |\lambda_{f_{\mathrm{CP}}}|^{-2}\right)\cosh(\Delta\Gamma t/2) + \left(1 - |\lambda_{f_{\mathrm{CP}}}|^{-2}\right)\cos(\Delta m t)\right.$$

$$\left. - 2\mathrm{Re}\lambda_{f_{\mathrm{CP}}}^{-1}\sinh(\Delta\Gamma t/2) + 2\mathrm{Im}\lambda_{f_{\mathrm{CP}}}^{-1}\sin(\Delta m t)\right],$$

where in the last step we have used eqns (8.209) and (8.210).

Using the expressions above we can find an explicit form for the so-called 'time-dependent CP asymmetry', defined as follows:

$$\mathcal{A}_{CP}(t) \equiv \frac{\Gamma(M^0(t) \to f_{CP}) - \Gamma(\overline{M}^0(t) \to f_{CP})}{\Gamma(M^0(t) \to f_{CP}) + \Gamma(\overline{M}^0(t) \to f_{CP})}$$

$$= \frac{\left(1 - |\lambda_{f_{CP}}|^2\right)\cos(\Delta m t) - 2\mathrm{Im}\lambda_{f_{CP}}\sin(\Delta m t)}{\left(1 + |\lambda_{f_{CP}}|^2\right)\cosh(\Delta\Gamma t/2) - 2\mathrm{Re}\lambda_{f_{CP}}\sinh(\Delta\Gamma t/2)} \qquad (8.228)$$

$$+ \mathcal{O}\left(1 - \left|\frac{p}{q}\right|^2\right).$$

Neglecting CP violation in mixing, i.e. assuming, according to eqn (8.189), $|q/p| = 1$, the coefficient of the $\cos(\Delta m t)$ term is non-vanishing in the presence of direct CP violation in the $M \to f_{CP}$ decay (eqn (8.219)), while the coefficient of the $\sin(\Delta m t)$ term signals CP violation in the interference between mixing and decay (eqn (8.226)).

Let us now discuss in turn K, D, B_d, and B_s mixing. As we shall see, different simplifying assumptions can be made in each sector.

8.5.4   Kaon Mixing and ϵ_K

If CP were conserved, the CP-odd eigenstate would not decay in a two-pion final state, resulting in a much longer lifetime. Allowing for small CP violation, it remains true that one eigenstate has a much longer lifetime, so it is convenient to label the eigenstate by the lifetime as long- and short-lived. Thus, we have

$$K_{S,L} = p_K|K^0\rangle \pm q_K|\overline{K}^0\rangle. \qquad (8.229)$$

We can simplify the general expressions in eqns (8.180)–(8.187) using two peculiarities of the Kaon system. First of all, the $\Delta I = 1/2$ rule implies that

$$\Gamma_{12} \approx A_0^* \overline{A}_0. \qquad (8.230)$$

Furthermore, one has

$$\Delta\Gamma_K \sim -2\Delta m_K. \qquad (8.231)$$

From eqn (8.197), using eqns (8.199), (8.231), and (8.230), we obtain

$$\delta = \frac{2\mathrm{Im}\left(M_{12}^*\Gamma_{12}\right)}{(\Delta m)^2 + |\Gamma_{12}|^2} \simeq \frac{2\mathrm{Im}\left(M_{12}^*\Gamma_{12}\right)}{-2(\Delta m)|\Gamma_{12}|} \simeq \frac{\mathrm{Im}\left(M_{12}A_0\overline{A}_0^*\right)}{(\Delta m)|A_0\overline{A}_0^*|}$$

$$= \frac{1}{\Delta m}\mathrm{Im}\left(M_{12}\left[1 + 2i\frac{\mathrm{Im}A_0}{\mathrm{Re}A_0} + \mathcal{O}\left(\left(\frac{\mathrm{Im}A_0}{\mathrm{Re}A_0}\right)^2\right)\right]\right) \qquad (8.232)$$

$$\simeq \frac{1}{\Delta m}\left(\mathrm{Im}M_{12} + 2\frac{\mathrm{Im}A_0}{\mathrm{Re}A_0}\mathrm{Re}M_{12}\right) \simeq \frac{1}{\Delta m}\mathrm{Im}M_{12} + \frac{\mathrm{Im}A_0}{\mathrm{Re}A_0},$$

where we have taken into account that in the standard phase convention

$$\text{Im}A_0 \ll \text{Re}A_0. \tag{8.233}$$

Neglecting $\text{Im}A_0$, dimension eight operators in $\text{Im}M_{12}$ and non-local matrix elements of two insertions of $\Delta S = 1$ Hamiltonians, one has

$$\delta \approx \frac{\text{Im}M_{12}^{\text{SD};D=6}}{\Delta m_K}. \tag{8.234}$$

This approximation has an accuracy of $\sim 5\%$; going beyond it requires evaluating $\text{Im}A_0$, long-distance contributions to $\text{Im}M_{12}$ and the contribution of dimension-eight operators to $\text{Im}M_{12}$, a formidable task [39, 40, 57].

To evaluate eqn (8.234) we make use of the results obtained in Section 8.4, and in particular of eqns (8.148) and (8.166):

$$\text{Im}M_{12}^{\text{SD};D=6} = \frac{G_F^2 M_W^2}{6\pi^2} F_K^2 m_K B_K(\mu) \text{Im}\lambda_{sd}^t \left[\text{Re}\lambda_{sd}^c (\eta_c(\mu)S_0(x_c) - \eta_{tc}(\mu)S_0(x_t,x_c)) \right.$$
$$\left. - \text{Re}\lambda_{sd}^t (\eta_t(\mu)S_0(x_t) - \eta_{tc}(\mu)S_0(x_t,x_c)) \right], \tag{8.235}$$

where the QCD corrections from the matching and from the RG evolution have been lumped in the factors $\eta_{t,c,tc}(\mu)$. Notice that in the literature it is customary to define the scale-invariant parameters $\hat{B}_K$ and $\eta_{1,2,3}$ by combining the μ-dependent part of $\eta_{t,c,tc}(\mu)$ with $B_K(\mu)$, thereby cancelling explicitly the μ dependence at the given order [58].

Phenomenology of CP violation in the Kaon system

In the CP-invariant case, $K_{S,L}$ would correspond to CP eigenstates and K_L decays to two pions would be forbidden. To test CP conservation, we can therefore measure

$$\eta_{00} = \frac{\langle \pi^0 \pi^0 | H | K_L \rangle}{\langle \pi^0 \pi^0 | H | K_S \rangle} \quad \text{and} \quad \eta_{+-} = \frac{\langle \pi^+ \pi^- | H | K_L \rangle}{\langle \pi^+ \pi^- | H | K_S \rangle}. \tag{8.236}$$

Defining

$$A_f = \langle f | H | K^0 \rangle, \quad \overline{A}_f = \langle f | H | \overline{K}^0 \rangle \text{ and } \lambda_f = \left(\frac{q}{p}\right)_K \frac{\overline{A}_f}{A_f}, \tag{8.237}$$

we have

$$\eta_f = \frac{1 - \lambda_f}{1 + \lambda_f}. \tag{8.238}$$

Writing $\pi\pi$ decay amplitudes in terms of final states with fixed isospin as in eqn (8.109), we take the combination

$$\epsilon_K = \frac{1}{3}(\eta_{00} + 2\eta_{+-}) = \frac{1-\lambda_0}{1+\lambda_0} + \mathcal{O}\left(\frac{A_2^2}{A_0^2}\right), \tag{8.239}$$

selecting a pure $I = 0$ state up to 2%. Since there is only one final state strong phase, the conditions of eqn (8.219) are not met and there is no direct CP violation in ϵ_K. We have

$$\mathrm{Re}\,\epsilon_K = \frac{1-|\lambda_0|^2}{1+2\mathrm{Re}\,\lambda_0 + |\lambda_0|^2}, \tag{8.240}$$

so that

$$\mathrm{Re}\,\epsilon_K \neq 0 \quad \Rightarrow \quad |\lambda_0| \neq 1 \quad \Rightarrow \quad \left|\frac{q_K}{p_K}\right| \neq 1 \tag{8.241}$$

implies CP violation in $K - \bar{K}$ mixing (eqn (8.189)), while

$$\mathrm{Im}\,\epsilon_K = \frac{-2\mathrm{Im}\,\lambda_0}{1+2\mathrm{Re}\,\lambda_0 + |\lambda_0|^2}, \tag{8.242}$$

so that

$$\mathrm{Im}\,\epsilon_K \neq 0 \quad \Rightarrow \quad \mathrm{Im}\,\lambda_0 \neq 0 \tag{8.243}$$

implies CP violation in the interference between mixing and decay (eqn (8.226)). Experimentally, $\arg \epsilon_K \approx \pi/4$, so the two CP-violating effects are comparable.

From eqns (8.240), (8.237), (8.193), (8.234), and (8.235) we obtain

$$\mathrm{Re}\,\epsilon_K \simeq \frac{1}{2}\frac{1-|\lambda_0|^2}{1+|\lambda_0|^2} = \frac{1}{2}\frac{1-|\frac{q_K}{p_K}|^2}{1+|\frac{q_K}{p_K}|^2} = \frac{\delta}{2}$$

$$= \frac{G_F^2 M_W^2}{12\Delta m_K \pi^2}F_K^2 m_K B_K(\mu)\mathrm{Im}\,\lambda_{sd}^t\left[\mathrm{Re}\,\lambda_{sd}^c\left(\eta_c(\mu)S_0(x_c) - \eta_{tc}(\mu)S_0(x_t,x_c)\right)\right.$$
$$\left. - \mathrm{Re}\,\lambda_{sd}^t\left(\eta_t(\mu)S_0(x_t) - \eta_{tc}(\mu)S_0(x_t,x_c)\right)\right]. \tag{8.244}$$

The expression above is valid up to corrections from dimension eight operators, from nonlocal matrix elements of two $\Delta S = 1$ effective Hamiltonians and from the deviation from $\pi/4$ of the phase of ϵ_K. These effects have been partially estimated [57], leading to a correction factor of 0.94 ± 0.02. At NLO, the SM prediction from [59]

$$|\epsilon_K| = (1.97 \pm 0.18) \cdot 10^{-3} \tag{8.245}$$

compares very well with the experimental value

$$|\epsilon_K| = (2.228 \pm 0.011) \cdot 10^{-3}. \tag{8.246}$$

We will come back again to ϵ_K when discussing the UT analysis in the SM and beyond in Section 8.6.

We can form another interesting combination of η_{+-} and η_{00}:

$$
\begin{aligned}
\epsilon' &\equiv \frac{1}{3}\left(\eta_{+-} - \eta_{00}\right) \\
&\simeq \frac{\langle(\pi\pi)_{I=0}|H|K_L\rangle\langle(\pi\pi)_{I=2}|H|K_S\rangle - \langle(\pi\pi)_{I=0}|H|K_S\rangle\langle(\pi\pi)_{I=2}|H|K_L\rangle}{\sqrt{2}\langle(\pi\pi)_{I=0}|H|K_S\rangle^2} \\
&\simeq \frac{ie^{i(\delta_2 - \delta_0)}}{\sqrt{2}}\,\mathrm{Im}\left(\frac{A_2}{A_0}\right) \simeq \frac{ie^{i(\delta_2 - \delta_0)}}{\sqrt{2}}\left(\frac{\mathrm{Im}\,A_2}{\mathrm{Re}\,A_0} - \omega\frac{\mathrm{Im}\,A_0}{\mathrm{Re}\,A_0}\right),
\end{aligned}
\tag{8.247}
$$

where $\omega = \mathrm{Re}\,A_2/\mathrm{Re}\,A_0$ and the equalities are valid up to corrections of relative order $\mathcal{O}(\omega, \epsilon_K, \mathrm{Im}\,A_0/\mathrm{Re}\,A_0)$. For $\delta_2 \neq \delta_0$ and $\mathrm{Im}\,(A_2/A_0) \neq 0$ the conditions for CP violation in the decay are satisfied and we have $\mathrm{Re}\,\epsilon' \neq 0$.

Obtaining a solid estimate of ϵ' is an extremely difficult task: it contains all the difficulties of the $\Delta I = 1/2$ rule and it is also affected by the cancellation between the two terms in the right-hand side of eqn (8.247). Indeed, in the SM the CP-violating effects from QCD penguins in A_0 and from electroweak penguins in A_2 cancel to a large extent, leading typically to predictions for Re ϵ'/ϵ in the 10^{-4} range [60]–[62], below the world average of Re $\epsilon'/\epsilon = (16.6 \pm 2.3)10^{-4}$ [63]–[65]. Very recently, a first estimate of Re ϵ'/ϵ in Lattice QCD has been obtained in the same framework of the first estimate of the $\Delta I = 1/2$ rule, pointing to a value in the low 10^{-4} range, but with a large uncertainty [66, 67]. This result has triggered a reanalysis of the SM prediction combining lattice QCD results with phenomenological considerations and/or arguments based on Dual QCD [68, 69], leading to a claimed discrepancy of $\sim 3\sigma$ with the experimental value. On the other hand, the lattice calculation underestimates the $I = 0$ strong interaction phase, and underestimating final state interactions could bring to an underestimate of ϵ'/ϵ, as noted in [70]–[76] and more recently stressed in [77]–[79]. Further progress in the evaluation of the relevant matrix elements is needed to assess the compatibility of the SM prediction with the experimental value, keeping in mind that ϵ'/ϵ is one of the observables with higher sensitivity to NP.

8.5.5   D – D̄ mixing and CP Violation

In complete analogy with $\Delta S = 2$ transitions, M_{12} and Γ_{12} for $D - \bar{D}$ mixing have the following structure:

$$(\lambda_{cu}^s)^2(f_{dd} + f_{ss} - 2f_{ds}) + 2\lambda_{cu}^s\lambda_{cu}^b(f_{dd} - f_{ds} - f_{db} + f_{sb}) + (\lambda_{cu}^b)^2(f_{dd} + f_{bb} - 2f_{db}), \tag{8.248}$$

where $\lambda_{cu}^q = V_{cq}V_{uq}^*$, $f_{q_iq_j}$ represents an intermediate state with flavours q_i and q_j, and intermediate states containing a b quark only appear in M_{12}. We see that the third generation here plays a very minor role with respect to $K - \bar{K}$ mixing, since its contribution is suppressed by m_b^2/m_t^2 with respect to $\Delta S = 2$ amplitudes. Indeed, we can safely neglect the term proportional to $(\lambda_{cu}^b)^2$. Then, the GIM mechanism essentially coincides with the U-spin subgroup of the flavour $SU(3)$ symmetry of strong interactions. Repeating the arguments of Section 8.4.1 we see that in this case the mixing amplitudes are dominated by non-local contributions, making even a rough estimate of M_{12} and Γ_{12} a tremendous task. While we may hope that in the future the pioneering studies of Δm_K on the lattice [80] may be extended to $D - \bar{D}$ mixing, it turns out that CP violation in $D - \bar{D}$ mixing is already today a very powerful probe of NP. Indeed, the approximate decoupling of the third generation implies a strong suppression of CP-violating effects. We can quantify this suppression by looking at the relevant combination of CKM elements:

$$r = \mathrm{Im}\, \frac{\lambda_{cu}^b}{\lambda_{cu}^s} \simeq 6.5 \cdot 10^{-4}. \tag{8.249}$$

The long-distance contributions to M_{12} and Γ_{12} can be parameterized in terms of their U-spin quantum numbers:

$$(\lambda_{cu}^s)^2 (\Delta U = 2) + 2\lambda_{cu}^s \lambda_{cu}^b (\Delta U = 1 + \Delta U = 2) + \mathcal{O}((\lambda_{cu}^b)^2) \approx (\lambda_{cu}^s)^2 \epsilon^2 + 2\lambda_{cu}^s \lambda_{cu}^b \epsilon, \tag{8.250}$$

so that we expect CP violation to arise at the level of $r\epsilon/\epsilon^2 \approx 2 \cdot 10^{-3} \approx 0.1°$ for an U-spin breaking of the order of 30%. Given the current experimental errors, it is therefore adequate to assume all SM amplitudes to be real, and interpret the (non)-observation of CP violation in $D - \bar{D}$ mixing as an effect of (a constraint on) NP. In fact, heavy NP could generate a short-distance contribution to $\mathrm{Im}\, M_{12}$, which could be observable either via $|q_D/p_D| \neq 1$ or equivalently via $\phi \equiv \arg(q/p)_D \neq 0$ (the two are not independent if all decay amplitudes are real [81]). Allowing for NP-induced CP violation in M_{12} only, and keeping all decay amplitudes real, a global combination of D-mixing related decays can be performed, leading to stringent constraints on NP. For example, the Summer 2018 update of the analysis of [82] finds the distributions for $|q_D/p_D|$, ϕ, $|M_{12}|$ and its phase Φ_{12} reported in Figure 8.8, corresponding to a bound on $|\Phi_{12}| < 3.5°$@95% probability.

8.5.6 $\mathbf{B_d - \bar{B}_d}$ Mixing

Let us consider the structure of M_{12} and Γ_{12} for $B_d - \bar{B}_d$ mixing:

$$(\lambda_{bd}^{c*})^2 (f_{uu} + f_{cc} - 2f_{uc}) + 2\lambda_{bd}^{c*} \lambda_{bd}^{t*} (f_{uu} - f_{uc} - f_{ut} + f_{ct}) + (\lambda_{bd}^{t*})^2 (f_{uu} + f_{tt} - 2f_{ut}), \tag{8.251}$$

where $\lambda_{bd}^q = V_{qb}^* V_{qd}$, $f_{q_i q_j}$ represents an intermediate state with flavours q_i and q_j, and again intermediate states containing a t quark only appear in M_{12}. A few remarks are in order:

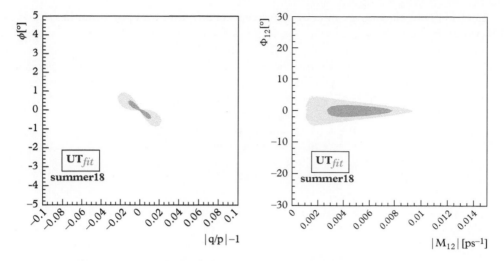

Figure 8.8 *Probability density functions for ϕ vs $|q_D/p_D| - 1$ (left panel) and for Φ_{12} vs $|M_{12}|$ (right panel). The darker (lighter) regions correspond to 68% (95%) probability.*

- contrary to the case of $\Delta S = 2$ transitions, $\lambda^c_{bd} \sim \lambda^t_{bd}$, so there is no CKM enhancement of light quark contributions and M_{12} is dominated by top quark exchange, i.e.

$$M_{12} \simeq (\lambda^{t*}_{bd})^2 f_{tt}. \tag{8.252}$$

Following the reasoning in Section 8.4.1 we see that corrections to the leading contribution from nonlocal matrix elements of two $\Delta B = 1$ effective Hamiltonians and from higher dimensional operators arise at $\mathcal{O}(m_b^2/M_W^2, m_b^2/m_t^2)$ and are thus fully negligible;

- also at variance with $K - \bar{K}$ mixing, Γ_{12} is suppressed with respect to M_{12} since the top quark does not contribute there, so that

$$\left| \frac{\Gamma_{12}}{M_{12}} \right| \sim \mathcal{O}\left(\frac{m_b^2}{m_t^2} \right) \ll 1; \tag{8.253}$$

- CP violation in mixing is even further suppressed, since the dominant contribution to both M_{12} and Γ_{12} is proportional to $(\lambda^{t*}_{bd})^2$, so that the CKM phase drops in the ratio Γ_{12}/M_{12}. CP violation is then induced solely by the other GIM-suppressed contributions to Γ_{12};

- last but not least, since the number of channels contributing to Γ_{12} is large, and the momentum of intermediate states is of $\mathcal{O}(m_b)$, we can advocate quark-hadron duality and perform an operator product expansion for Γ_{12} as well. While a detailed discussion of this subject goes well beyond the scope of these lectures, the interested reader will find all the details in refs. [83]–[85].

Let us now work out the expressions of Section 8.5.1 with the approximation $|\Gamma_{12}| \ll |M_{12}|$:

$$\Delta m_{B_d} = 2|M_{12}|, \tag{8.254}$$

$$\frac{\Delta \Gamma_{B_d}}{\Delta m_{B_d}} = \mathrm{Re}\frac{\Gamma_{12}}{M_{12}}, \tag{8.255}$$

$$\left(\frac{q}{p}\right)_{B_d} = \frac{M_{12}^*}{|M_{12}|}\left(1 - \frac{1}{2}\mathrm{Im}\frac{\Gamma_{12}}{M_{12}}\right), \tag{8.256}$$

$$\left|\frac{q}{p}\right|_{B_d} - 1 = -\frac{1}{2}\mathrm{Im}\frac{\Gamma_{12}}{M_{12}}. \tag{8.257}$$

The mass difference is obtained taking the matrix element of the $\Delta B = 2$ effective Hamiltonian as

$$\Delta m_{B_d} = \frac{G_F^2 M_W^2}{2\pi^2}\left|V_{tb}V_{td}\right|^2 S_0(x_t)\eta_b m_{B_d} f_{B_d}^2 B_{B_d}, \tag{8.258}$$

where the QCD corrections [26, 27, 86] have been absorbed in η_b and B_{B_d} is the B-parameter computed in the same scheme and at the same scale as η_b. The Summer 2018 SM prediction by the UTfit collaboration is

$$\Delta m_{B_d}^{\mathrm{SM}} = (0.54 \pm 0.03)\ \mathrm{ps}^{-1} \tag{8.259}$$

which compares very well with the experimental average

$$\Delta m_{B_d}^{\mathrm{exp}} = (0.5064 \pm 0.0019)\mathrm{ps}^{-1}. \tag{8.260}$$

The experimental sensitivity to $\Delta\Gamma_{B_d}$ is still well above the SM prediction, and the same is true for the semileptonic asymmetry $A_{B_d}^{\mathrm{SL}}$ defined in eqn (8.211), which measures CP violation in mixing.

From the phenomenological point of view, B_d mesons have three peculiarities that make them a golden system to study meson–anti-meson oscillations and CP violation [87, 88]:

- since CKM angles involving the third generation are small, the B_d lifetime is of $\mathcal{O}(\mathrm{ps}^{-1})$, so that a relatively small boost is enough to allow for a B_d meson to fly a measurable distance before it decays;

- the $B_d - \bar{B}_d$ mass difference is comparable to the B_d lifetime, opening the possibility to measure the time dependence of the oscillations;

- the time-dependent CP asymmetry defined in eqn (8.228) allows to measure the CP-violating $\mathrm{Im}\lambda_f$ for a variety of final states f, allowing for an extensive test of the CKM mechanism and of possible NP contributions.

For these reasons, the idea of an asymmetric B-factory, where entangled pairs of $B_d - \bar{B}_d$ mesons could be produced with a boost sufficient to observe the time oscillation, was put forward [89] and developed, leading to the extraordinary success of the BaBar and Belle experiments at SLAC and KEK [90].

Time-dependent CP asymmetry in $B_d \to J/\Psi K_S$

Let us now discuss the time-dependent CP asymmetries for a series of final states, starting from the famous 'golden channel' $B_d \to J/\Psi K_S$. The underlying weak decay is $\bar{b} \to \bar{c}c\bar{s}$, which is generated by the following piece of the $\Delta B = 1$ effective Hamiltonian (see eqns (8.96) and (8.98)–(8.101)):

$$
\mathcal{H}_{\text{eff}}^{\bar{b} \to \bar{c}c\bar{s}} = \frac{4G_F}{\sqrt{2}} \left\{ \lambda_{bs}^c \left(C_1 Q_1^{\bar{b}c\bar{c}s} + C_2 Q_2^{\bar{b}c\bar{c}s} + \sum_{i=3}^{10} C_i Q_i^{\bar{b}s} \right) \right.
$$
$$
\left. + \lambda_{bs}^u \left(C_1 Q_1^{\bar{b}u\bar{u}s} + C_2 Q_2^{\bar{b}u\bar{u}s} + \sum_{i=3}^{10} C_i Q_i^{\bar{b}s} \right) \right\}. \tag{8.261}
$$

To obtain the $\langle J/\Psi K^0 | H_{\text{eff}}^{\bar{b} \to \bar{c}c\bar{s}} | B_d \rangle$ matrix element we need to consider all possible Wick contractions of the fields in $H_{\text{eff}}^{\bar{b} \to \bar{c}c\bar{s}}$ with the initial and final states. Following refs. [91, 92], where the interested reader can find all details, we can classify the different Wick contraction topologies as in Figures 8.9 and 8.10, where the left (right) panels contain 'disconnected' ('connected') topologies. In the infinite m_b limit, the case in which the 'emitted' meson (i.e. M_1 in Figure 8.9) is light becomes computable in terms of form factors, decay constants and perturbative QCD corrections, as argued in ref. [93] and carefully demonstrated in refs. [94, 95]. The basic idea is that the emitted light meson flies away too fast for soft gluons to be exchanged with the B meson and with the other final state meson, the so-called 'colour transparency' argument. In spite of this tremendous theoretical progress, however, a full-fledged computation of the diagrams in Figures 8.9 and 8.10 for realistic values of the b-quark mass remains well beyond our capabilities. Indeed, long-distance contributions and rescattering effects arising at $O(\Lambda/m_b)$ are not systematically computable and have a strong phenomenological impact in two-body nonleptonic B decays [91, 96, 97]. A particularly dangerous class of long-distance contributions are the so-called 'charming penguins', namely penguin matrix elements as in Figure 8.10 with a charm quark running in the loop, which are affected by $D_{(s)}^{(*)} - \bar{D}_{(s)}^{(*)}$ rescattering into light mesons.

To be able to obtain robust phenomenological results, one must therefore seek observables where the dangerous long-distance contributions are either absent or strongly suppressed. To this aim, it is convenient to consider renormalization-group invariant combinations of Wilson coefficients times Wick contractions [92], and to express the decay amplitudes in terms of these parameters. In the case of $B_d \to J/\Psi K_S$ we obtain, using eq. (8.203) for the K_S in the final state:

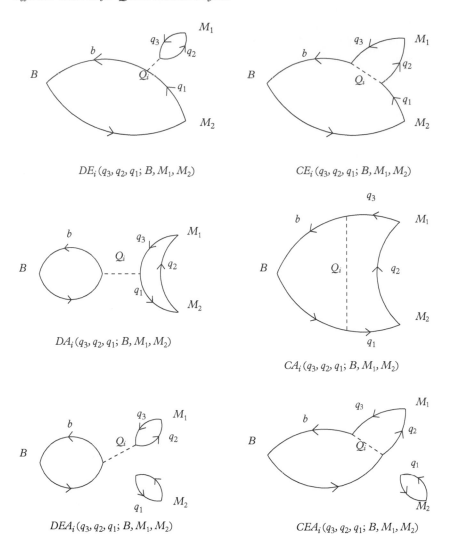

$DE_i(q_3, q_2, q_1; B, M_1, M_2)$ $CE_i(q_3, q_2, q_1; B, M_1, M_2)$

$DA_i(q_3, q_2, q_1; B, M_1, M_2)$

$CA_i(q_3, q_2, q_1; B, M_1, M_2)$

$DEA_i(q_3, q_2, q_1; B, M_1, M_2)$ $CEA_i(q_3, q_2, q_1; B, M_1, M_2)$

Figure 8.9 *Emission, annihilation, and emission–annihilation topologies of Wick contractions in the matrix elements of operators* Q_i *[92].*

$$A_{B_d \to \mathcal{J}/\Psi K_S} = \left[\lambda_{bs}^c (E_2 + P_2) + \lambda_{bs}^u (P_2 - P_2^{\text{GIM}}) \right]/(2p_K), \qquad (8.262)$$

where E_2 contains emission matrix elements of $Q_{1,2}^{\bar{b}c\bar{c}d}$ in the colour-suppressed combination $C_1 CE + C_2 DE$; P_2 contains penguin-emission matrix elements of $Q_{1,2}^{\bar{b}c\bar{c}d}$ together with emission, emission-annihilation and penguin emission matrix elements of $Q_{3-10}^{\bar{b}d}$; P_2^{GIM} contains penguin-emission matrix elements of the GIM-suppressed combinations

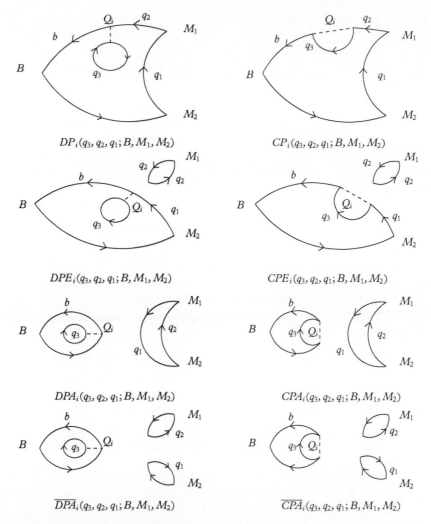

Figure 8.10 *Penguin, penguin–emission, penguin–annihilation, and double-penguin-annihilation topologies of Wick contractions in the matrix elements of operators Q_i [92].*

$Q_{1,2}^{\bar{b}c\bar{c}d} - Q_{1,2}^{\bar{b}u\bar{u}d}$; p_K appears to project the K^0 onto the K_S final state. We expect the dominant contribution to come from E_2, since P_2 is suppressed either by small Wilson coefficients or by penguin matrix elements, and P_2^{GIM} is suppressed by penguin matrix elements and by the GIM mechanism.

Thus, we can use the expansion of eqns (8.221)–(8.222) with

$$r_{J/\Psi K_S} = \left| \frac{\lambda_{bs}^u}{\lambda_{bs}^c} \right| \left| \frac{P_2 - P_2^{\text{GIM}}}{E_2 + P_2} \right| \lesssim \left| \frac{V_{ub}}{V_{cb}} \right| \sim \mathcal{O}(10^{-2}). \tag{8.263}$$

Let us first be bold and put $r_{J/\Psi K_S}$ to zero. Then we obtain from eqns (8.225) and (8.262)

$$
\begin{aligned}
\lambda_{J/\Psi K_s} &= \left(\frac{q}{p}\right)_{B_d} \frac{\lambda_{bs}^{c*}}{\lambda_{bs}^{c}} \left(\frac{p}{q}\right)_K = \frac{(\lambda_{bd}^{t})^2}{(\lambda_{bd}^{t*})^2} \frac{\lambda_{bs}^{c*}}{\lambda_{bs}^{c}} \frac{\lambda_{sd}^{c*}}{\lambda_{sd}^{c}} \\
&= \frac{V_{tb}^* V_{td}}{V_{tb} V_{td}^*} \frac{V_{cb} V_{cs}^*}{V_{cb}^* V_{cs}} \frac{V_{cs} V_{cd}^*}{V_{cs}^* V_{cd}} = \frac{V_{cb} V_{cd}^*}{V_{tb} V_{td}^*} \frac{V_{tb}^* V_{td}}{V_{cb}^* V_{cd}} = e^{-2i\beta},
\end{aligned} \tag{8.264}
$$

where the angle β of the Unitarity Triangle is defined in eqn (8.20). Plugging eqn (8.264) in eqn (8.228) we obtain

$$
\mathcal{A}_{\mathrm{CP}}^{B_d \to J/\Psi K_S}(t) = -\sin 2\beta \sin(\Delta m_{B_d} t), \tag{8.265}
$$

taking into account that the final state is CP odd.

Thus, in the approximation $r_{J/\Psi K_S} = 0$, measuring the time-dependent asymmetry in this channel we should find a vanishing coefficient of the $\cos(\Delta m_{B_d} t)$ term, and the coefficient of the $\sin(\Delta m_{B_d} t)$ measures $\sin 2\beta$. The current world average of $\mathcal{A}_{\mathrm{CP}}^{B_d \to J/\Psi K_{S,L}}(t)$ gives [98]–[102]

$$
\sin 2\beta = 0.690 \pm 0.018, \tag{8.266}
$$

corresponding to $\beta \approx 21.8°$.

Let us now go back to the assumption $r_{J/\Psi K_S} = 0$, under which we obtained eqn (8.265), and investigate if we can get any theoretical or experimental handle on the actual value of $r_{J/\Psi K_S}$, or at least an upper bound on its value. A theoretical calculation of $r_{J/\Psi K_S}$ from first principles is currently impossible even in the infinite m_b limit, since the emitted meson is heavy. The direct CP asymmetry, i.e. the coefficient of the $\cos(\Delta m_{B_d} t)$ term in the time-dependent asymmetry, according to eqn (8.220) is sensitive to

$$
r_{J/\Psi K_S} \sin\phi_{r_{J/\Psi K_S}} \sin\delta_{r_{J/\Psi K_S}} \approx \lambda^2 R_b \left| \frac{P_2 - P_2^{\mathrm{GIM}}}{E_2 + P_2} \right| \sin\gamma \sin\arg\left(\frac{P_2 - P_2^{\mathrm{GIM}}}{E_2 + P_2}\right), \tag{8.267}
$$

with the UT parameters λ, R_b, and γ defined in eqns (8.15), (8.20) and (8.21). The last term in eqn (8.267), i.e. the sine of the strong phase difference between the two amplitudes, prevents us from using directly the direct CP asymmetry as a bound on $r_{J/\Psi K_S}$. Even if we ignore this problem, bounding $r_{J/\Psi K_S}$ at the level of the direct CP asymmetry would anyway give a theoretical error comparable to the experimental one. A way out can be found using the SU(3)-related decay channel $B_d \to J/\Psi \pi^0$ [103, 104]. The effective Hamiltonian governing this decay is given by

$$\mathcal{H}_{\text{eff}}^{\bar{b}\to\bar{c}c\bar{d}} = \frac{4G_F}{\sqrt{2}} \left\{ \lambda_{bd}^c \left(C_1 Q_1^{\bar{b}c\bar{c}d} + C_2 Q_2^{\bar{b}c\bar{c}d} + \sum_{i=3}^{10} C_i Q_i^{\bar{b}d} \right) \right.$$

$$\left. + \lambda_{bd}^u \left(C_1 Q_1^{\bar{b}u\bar{u}d} + C_2 Q_2^{\bar{b}u\bar{u}d} + \sum_{i=3}^{10} C_i Q_i^{bd} \right) \right\}, \tag{8.268}$$

and the decay amplitude is given in the SU(3) limit by

$$A_{B_d\to\mathcal{J}/\Psi\pi^0} = \lambda_{bd}^c(E_2 + P_2) + \lambda_{bd}^u(P_2 - P_2^{\text{GIM}}), \tag{8.269}$$

neglecting a small, colour suppressed emission-annihilation contribution. In eqn (8.269) the second term is not doubly-Cabibbo-suppressed anymore, so that this channel is much more sensitive to $P_2^{\text{GIM}} - P_2$. Using the information from $B_d \to \mathcal{J}/\Psi\pi^0$ we can constrain the theoretical error in the extraction of $\sin 2\beta$ from $B_d \to \mathcal{J}/\Psi K_S$ to be subdominant even allowing for an SU(3) breaking of 100%. It is however crucial that in the future the experimental progress on $B_d \to \mathcal{J}/\Psi\pi^0$ parallels the one on $B_d \to \mathcal{J}/\Psi K_S$, so that the theory uncertainty remains subdominant.

Time-dependent CP asymmetry in $B \to \pi\pi$

Thanks to isospin symmetry, $B \to \pi\pi$ decays have the unique property that all decay amplitudes can be determined experimentally, allowing for a measurement of the CKM angle α with essentially no theoretical input other than isospin [105]. Effects of isospin breaking due to electromagnetic interactions and to quark masses are negligible with respect to current experimental uncertainties, so we will not discuss them here [106].

Using the isospin decomposition of eqn (8.109) for B decays, we see that the independent parameters are the relative strong phase of $I = 0$ and $I = 2$ amplitudes, the weak phases of $I = 0$ and $I = 2$ amplitudes, and their absolute values, so five independent parameters. If we consider time-dependent CP asymmetries, we should add $(q/p)_{B_d}$; neglecting CP violation in the mixing, this amounts to another parameter, $\arg(q/p)_{B_d}$. The observables are three CP-averaged branching ratios ($\mathcal{B}_{+-}$, $\mathcal{B}_{+0}$ and $\mathcal{B}_{00}$) and four CP asymmetries (the coefficients $S_{+-,00}$ of $\sin\Delta m_{B_d}t$ and $C_{+-,00}$ of $\cos\Delta m_{B_d}t$ terms in $\mathcal{A}_{CP}^{B_d\to\pi^{+,0}\pi^{-,0}}$), so the system is overdetermined. In practice, however, the measurement of the time-dependent CP asymmetry in $B_d \to \pi^0\pi^0$ is very difficult, but there are enough observables to determine all parameters even if only C_{00} is used.

It is convenient to write the decay amplitudes separating terms with different weak phases rather than with different strong phases as in eq. (8.109). In particular, using CKM unitarity, we separate the amplitudes in terms proportional to λ_{bd}^u and λ_{bd}^t. Taking into account that $(q/p)_{B_d} \simeq (\lambda_{bd}^t/\lambda_{bd}^{t*})^2$ and that $\lambda_f = q/p\bar{A}_f/A_f$, it is convenient to absorb a factor of λ_{bd}^{t*} (λ_{bd}^t) in A_f ($\bar{A}_f$). In this way we obtain

$$A(B_d \to \pi^+ \pi^-) = e^{-i\alpha} T^{+-} + P, \tag{8.270}$$

$$A(B_d \to \pi^0 \pi^0) = \left(e^{-i\alpha} T^{00} - P \right), \tag{8.271}$$

$$A(B^- \to \pi^- \pi^0) = \frac{1}{\sqrt{2}} e^{-i\alpha} \left(T^{+-} + T^{00} \right). \tag{8.272}$$

We can then extract α, together with T^{+-}, T^{00}, P, and their relative phases, up to an eight-fold ambiguity (explicit formulæ can be found in [107, 108]). It is however clear that the degeneracies in α correspond to different values of the parameters T^{+-}, T^{00} and P. We can then follow the same argument used in Sec. 8.5.6 and relate $B \to \pi^+ \pi^-$ decays to $B_s \to K^+ K^-$ via a U-spin transformation. Since $B_s \to K^+ K^-$ is a $\bar{b} \to \bar{s} u \bar{u}$ transition, the T and P terms in the amplitudes are weighted by a different CKM factor, breaking the degeneracy between different solutions of the $B \to \pi \pi$ system. Thus, the isospin analysis of $B \to \pi \pi$ supplemented by $B_s \to K^+ K^-$ is more efficient [109].

Notice that the isospin analysis of $B \to \pi \pi$ can be generalized beyond the SM as long as new physics does not enhance electroweak penguins by orders of magnitude, and as long as it does not contribute sizeably to current-current operators. Then, we can still extract α even allowing for a NP weak phase to be present in P [110, 111], although with a slightly larger uncertainty.

Finally, the same analysis presented for $B \to \pi \pi$ can be carried out for each polarization of the $B \to \rho \rho$ decays; it turns out that the latter profits from larger branching ratios, making it more sensitive than the $\pi \pi$ channel.

Extracting α from $B \to \rho \pi$ decays

In general, decays to final states including vector mesons can be analyzed with a very powerful tool, the Dalitz plot, which allows in principle to extract the absolute values of all amplitudes contributing to a given final state, and all their relative phases, provided that they interfere among each other in a non-negligible region of phase space. Although the isospin structure of $B \to \rho \pi$ decays is richer than the one of $\pi \pi$, since the final state can also have isospin one, this just turns the triangular relation for $B \to \pi \pi$, $A(B_d \to \pi^+ \pi^-) + A(B_d \to \pi^0 \pi^0) = \sqrt{2} A(B^+ \to \pi^+ \pi^0)$, into a pentagonal relation, $A(B_d \to \pi^+ \rho^-) + A(B_d \to \pi^- \rho^+) + 2A(B_d \to \pi^0 \rho^0) = \sqrt{2} \left(A(B^+ \to \rho^+ \pi^0) + A(B^+ \to \rho^0 \pi^+) \right)$. Again, this allows us to determine the relative phase of the $I = 3/2$ amplitudes for B and $\bar{B}$ decays, which corresponds to 2α [110, 112, 113]. While a detailed discussion of Dalitz analyses of three-body heavy meson decays goes beyond the scope of these lectures, we refer the interested reader to chapter 13 of [90] for a review of several Dalitz analysis techniques.

8.5.7 $B_s - \bar{B}_s$ Mixing

The structure of M_{12} and Γ_{12} for $B_s - \bar{B}_s$ mixing is analogous to the one for $B_d - \bar{B}_d$ mixing given in (8.251), with the substitution $\lambda_{bd}^f \to \lambda_{bs}^f$. However, while in the case of $B_d - \bar{B}_d$ mixing one has $|\lambda_{bd}^{u,c,t}| \sim \lambda^3$, so all three factors arise at third order in the CKM parameter λ, for $b \to s$ transitions the relative weight of the three CKM factors is instead hierarchical:

$$|\lambda_{bs}^{t,c}| \simeq \lambda^2 \gg \lambda_{bs}^u \simeq \lambda^4. \tag{8.273}$$

This has three very important phenomenological consequences:

1. CP violation in $B_s - \bar{B}_s$ mixing is tiny, since the $\mathcal{O}(\lambda^2)$ decoupling of the first generation is reflected in the smallness of the angle $\beta_s \sim \mathcal{O}(\lambda^2)$ defined in eqn (8.21). This suppression acts on top of the mechanism already discussed for $B_d - \bar{B}_d$ mixing, leading to $\mathrm{Im}(\Gamma_{12}/M_{12}) \sim \mathcal{O}(10^{-5})$;

2. since $\Delta m_{B_s}/\Delta m_{B_d}$ goes approximately like the ratio $V_{ts}/V_{td} \sim 1/\lambda$ while $\Gamma_{B_s} \sim \Gamma_{B_d}$, one has $\Delta m_{B_s}/\Gamma_{B_s} \sim 25$, making it much more difficult to resolve experimentally the time-dependence of the mixing;

3. the enhancement factor $\Delta m_{B_s}/\Gamma_{B_s}$ brings $\Delta\Gamma_{B_s}/\Gamma_{B_s} \sim 25\Delta\Gamma_{B_s}/\Delta m_{B_s}$ to the observable level of $\mathcal{O}(10\%)$.

Therefore, in studying $B_s - \bar{B}_s$ mixing we should keep the terms proportional to $\Delta\Gamma_{B_s}$ in the expressions of Section 8.5.3, in particular in eqn (8.228).

The Summer 2018 prediction for Δm_{B_s} in the SM by the UTfit collaboration is

$$\Delta m_{B_s}^{\mathrm{SM}} = (17.25 \pm 0.85)\mathrm{ps}^{-1}, \tag{8.274}$$

which compares very well with the experimental average

$$\Delta m_{B_s}^{\mathrm{exp}} = (17.757 \pm 0.0021)\mathrm{ps}^{-1}, \tag{8.275}$$

while the prediction for $\Delta\Gamma_{B_s}$ yields

$$(\Delta\Gamma_{B_s}/\Gamma_{B_s})^{\mathrm{SM}} = 0.15 \pm 0.01, \tag{8.276}$$

well compatible with the experimental average

$$(\Delta\Gamma_{B_s}/\Gamma_{B_s})^{\mathrm{exp}} = 0.132 \pm 0.008. \tag{8.277}$$

Time-dependent CP asymmetry in $B_s \to J/\Psi\phi$

If we apply the same arguments presented in Section 8.5.6 and consider a $\bar{b} \to \bar{c}c\bar{s}$ transition for B_s decays, we are led to $B_s \to J/\Psi\phi$ as the golden channel for the measurement of the CKM angle β_s:

$$\begin{aligned}
\lambda_{J/\Psi\phi} &= \left(\frac{q}{p}\right)_{B_s} \frac{\lambda_{bs}^{c*}}{\lambda_{bs}^c} = \frac{(\lambda_{bs}^t)^2}{(\lambda_{bs}^{t*})^2}\frac{\lambda_{bs}^{c*}}{\lambda_{bs}^c} \\
&= \frac{V_{tb}^* V_{ts}}{V_{tb} V_{ts}^*}\frac{V_{cb} V_{cs}^*}{V_{cb}^* V_{cs}} = \frac{V_{tb}^* V_{ts}}{V_{cb}^* V_{cs}}\frac{V_{cb} V_{cs}^*}{V_{tb} V_{ts}^*} = e^{2i\beta_s},
\end{aligned} \tag{8.278}$$

where we have assumed $r_{\mathcal{J}/\Psi\phi} = 0$ and for simplicity we have omitted the CP parity of the final state, to be determined with an angular analysis of the decay products of the $\mathcal{J}/\Psi\phi$ intermediate state. In the case of the B_s meson, we cannot neglect the terms proportional to $\Delta\Gamma_{B_s}$ in eqn (8.228), so the result of the measurement is a combined fit of $\Delta\Gamma_{B_s}$ and $\mathrm{Im}\lambda_{\mathcal{J}/\Psi\phi}$.

However, if we now allow for a nonvanishing $r_{\mathcal{J}/\Psi\phi}$, which again we can estimate, following eqn (8.263), as

$$r_{\mathcal{J}/\Psi\phi} = \left|\frac{\lambda^u_{bs}}{\lambda^c_{bs}}\right| \left|\frac{P_2 - P_2^{\mathrm{GIM}}}{E_2 + P_2}\right| \lesssim \left|\frac{V_{ub}}{V_{cb}}\right| \sim \mathcal{O}(10^{-2}), \qquad (8.279)$$

we immediately see that the correction to $\mathrm{Im}\lambda_{\mathcal{J}/\Psi\phi}$ is of the same order of $\sin 2\beta_s$:

$$\mathrm{Im}\lambda_{\mathcal{J}/\Psi\phi} = \sin 2\beta_s - 2r_{\mathcal{J}/\Psi\phi} \sin\gamma \cos\delta_{r_{\mathcal{J}/\Psi\phi}} + \mathcal{O}(r^2_{\mathcal{J}/\Psi\phi}, r_{\mathcal{J}/\Psi\phi}\lambda^2). \qquad (8.280)$$

In other words, both $B_d \to \mathcal{J}/\Psi K_S$ and $B_s \to \mathcal{J}/\Psi\phi$ suffer from doubly Cabibbo suppressed corrections, but the leading term is of $\mathcal{O}(1)$ for B_d and doubly Cabibbo suppressed for B_s. Still, the time-dependent CP asymmetry in $B_s \to \mathcal{J}/\Psi\phi$ remains a most precious tool to constrain possible NP contributions to CP violation in B_s mixing, at least down to the level of $r_{\mathcal{J}/\Psi\phi}$. We could of course envisage a strategy to keep the corrections due to $r_{\mathcal{J}/\Psi\phi}$ under control, using $SU(3)$ as was discussed in Section 8.5.6. However, this approach is complicated by the mixed singlet-octet flavour structure of the ϕ meson, requiring a detailed analysis of several final states. We refer the interested reader to the discussion in [114].

8.6   The Unitarity Triangle Analysis in the SM and Beyond

Let us now very quickly review how we can combine a large amount of theoretical and experimental information using the Unitarity Triangle introduced in Section 8.2.1. Since the CKM matrix is governing all flavour and CP violation in weak interactions, we can translate virtually any flavour- or CP-violating process into a constraint on the UT. Let us start from charged-current processes arising at the tree level in the SM, before turning to FCNC transitions.

8.6.1   The UT from Tree-Level Decays

The CKM matrix elements $|V_{ud}|$ and $|V_{us}|$ can be measured from super-allowed β decays [115, 116], and from semileptonic/leptonic kaon decays [16, 117, 118] respectively, providing an accurate determination of the sine of the Cabibbo angle. Similarly, $|V_{cb}|$ and $|V_{ub}|$ can be determined using (semi-)leptonic B decays. In this case, we can use either exclusive or inclusive decays, which have different theoretical and experimental systematic errors. For $b \to c$ transitions, the analysis of inclusive semileptonic decays relies on heavy quark symmetry and on global quark-hadron duality, while the study of inclusive semileptonic $b \to u$ transitions requires local quark-hadron duality, as well as some model-dependent regularization of singularities that are absent in $b \to c$ decays. In exclusive decays, an estimate of the relevant form factors, as well as of their momentum dependence, is needed to extract CKM factors. Unfortunately, determinations of $|V_{cb}|$ and $|V_{ub}|$ from inclusive and exclusive semileptonic B decays have been displaying a $\sim 3\sigma$ discrepancy for quite a while [16], although it was recently noticed that for $|V_{cb}|$ the situation improves considerably if one relaxes some assumptions on the momentum dependence of the form factors based on the heavy quark limit [119]–[121]. Hopefully more precise data and improved lattice calculations will bring to a resolution of this long-standing puzzle.

The measurements discussed above provide us with the normalization of the UT and with the length of one of the non-unit sides, R_b. Fortunately, we can complete the determination of the UT using only tree-level decays by measuring the angle γ, defined in eqn (8.20). The measurement of γ can be achieved by exploiting the interference between $\bar{b} \to \bar{c}u\bar{q} \to f$ and $\bar{b} \to \bar{u}c\bar{q} \to f$ transitions, where $q = d,s$ and f is a generic final state accessible through both decay chains [122]–[125]. The theoretical uncertainty in the extraction of γ can be always kept subdominant [126], so future experimental progress will have a strong impact on the UT analysis.

Figure 8.11 shows the current status of the UT determined through tree level decays only. Notice that two regions in the $\bar{\rho} - \bar{\eta}$ planes are selected, since we can determine γ only up to $\pm 180°$. We will discuss below how this ambiguity can be lifted using measurements of CP violation in $B - \bar{B}$ mixing [127].

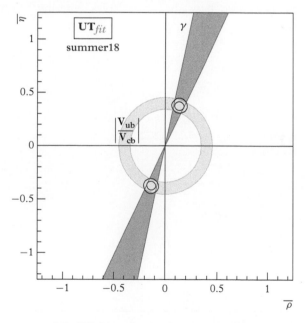

Figure 8.11 *Current status of the UT determination from tree-level decays, from the UTfit Collaboration.*

8.6.2   Adding FCNC to the UT Analysis in the Standard Model and Beyond

We are now ready to add to the processes used in Section 8.6.1 meson-antimeson mixing in K, B_d and B_s sectors, using eqn (8.244) for ϵ_K, eqn (8.258) for $\Delta m_{B_{d,s}}$, eqn (8.265) for $\sin 2\beta$ and the results of Section 8.5.6 for α. This allows to break the degeneracy between the first and third quadrant. The global fit displays a very good consistency of all observables within the SM, as can be seen from Figure 8.12.

The consistency of the SM fit can be translated into constraints on NP contributions to meson-antimeson mixing. Let us proceed in two steps. First, we generalize the UT analysis by parameterizing the relevant NP contributions. Second, we translate the constraints on NP contributions into bounds on the scale of NP.

Following [111, 128], we introduce the following parameters to account for possible NP contributions to meson–anti-meson mixing:

$$C_{B_q} e^{2i\phi_{B_q}} = \frac{M_{12}^{B_q,\text{full}}}{M_{12}^{B_q,\text{SM}}}, \qquad (q = d, s) \tag{8.281}$$

$$C_{\epsilon_K} = \frac{\text{Im}\, M_{12}^{K,\text{full}}}{\text{Im}\, M_{12}^{K,\text{SM}}}. \tag{8.282}$$

We can then immediately see how the observables entering the UT analysis are affected:

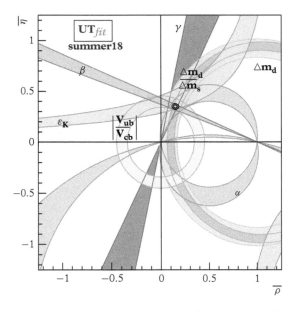

Figure 8.12 *Current status of the UT determination in the SM, from the UTfit Collaboration.*

$$\Delta m_{B_q} = C_{B_q}(\Delta m_{B_q})^{\text{SM}} \tag{8.283}$$

$$\lambda_{\mathcal{J}/\Psi K_S} = e^{-2i(\beta + \phi_{B_d})}, \tag{8.284}$$

$$\lambda_{\mathcal{J}/\Psi \phi} = e^{2i(\beta_s - \phi_{B_s})}, \tag{8.285}$$

$$\alpha_{\text{exp}} = \alpha - \phi_{B_d}, \tag{8.286}$$

where α_{exp} denotes the value of α extracted from $B_d \to \pi\pi$, $\rho\pi$ and $\rho\rho$ decays.[7]

As pointed out in [127], the presence of ϕ_{B_q} can have a large impact on the semileptonic asymmetries defined in eqn (8.211). As we have seen in Section 8.5.6, in the SM the dominant contributions to M_{12} and Γ_{12} have the same CKM phase which drops in the ratio Γ_{12}/M_{12}, so that $\text{Im}(\Gamma_{12}/M_{12})$ only arises from subdominant GIM-suppressed contributions to Γ_{12}. However, if the mixing amplitude is affected by NP so that it gets an additional phase ϕ_{B_q}, the phase cancellation between M_{12} and Γ_{12} is spoiled and one gets a contribution to $\text{Im}(\Gamma_{12}/M_{12})$ from the dominant term, proportional to $\text{Re}(\Gamma_{12}/M_{12})^{\text{SM}}/C_{B_q} \cos 2\phi_{B_q}$. It is then evident that the region in the third quadrant in Figure 8.11, allowed at the tree-level, requires a large value of ϕ_{B_d} which is ruled out at more than 95% probability by the experimental value of $A_{\text{SL}}^{B_d}$.[8]

We can therefore perform a simultaneous determination of the UT and of the NP parameters introduced in eqns (8.281) and (8.282). The Summer 18 update from the UTfit collaboration is reported in Figure 8.13. It is instructive to extract from the C_{B_q} and ϕ_{B_q} parameters the absolute value and phase of the NP contributions relative to the SM:

$$C_{B_q} e^{2i\phi_{B_q}} = 1 + \frac{A_q^{\text{NP}} e^{2i\phi_q^{\text{NP}}}}{A_q^{\text{SM}}}. \tag{8.287}$$

The current constraints on A_q^{NP} and ϕ_q^{NP} are reported in Figure 8.14. We see that our knowledge of the UT in the presence of NP is roughly a factor of two worse than in the SM, and that NP contributions to SM mixing amplitudes at the level of $\approx 30-40\%$ are still allowed at 95% probability, especially if their phase does not differ too much from the SM one. This shows that ample room is left for improvements, both from the experimental and theoretical point of view, until we will be sensitive to NP contributions in the flavour sector at the percent or sub-percent level. However, given the combined loop and GIM suppression of these observables in the SM, and given the hierarchical structure of quark masses and mixings, already this relatively rough sensitivity to NP contributions is able to provide us with the most stringent constraints on the NP scale.

---

[7] In the presence of NP contributions to loop-mediated SM processes, in the isospin or amplitude analysis we should allow the penguin contribution to have a phase different from the SM one [128].

[8] Also in this case when allowing for NP to be present in loop-mediated SM processes we should allow for penguin contributions to Γ_{12} to have a phase different from the SM one [128].

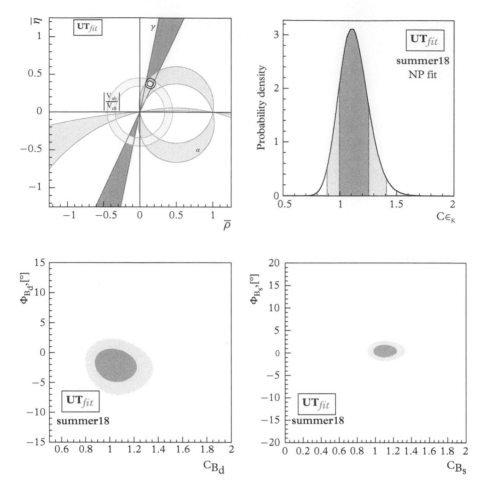

Figure 8.13 *From left to right and from top to bottom: probability density functions for* $(\bar{\rho}, \bar{\eta})$, C_{ϵ_K}, (C_{B_d}, ϕ_{B_d}), (C_{B_s}, ϕ_{B_s}). *Darker (lighter) regions correspond to smallest 68% (95%) probability regions.*

8.6.3   Constraining the NP Scale with $\Delta F = 2$ Amplitudes

We now combine the results on the $\Delta F = 2$ effective Hamiltonian beyond the SM in Section 8.4.5 and 8.4.6 with the constraints on NP contributions obtained in Section 8.5.5 and 8.6.2 to learn more on NP.

Assuming, as we did above, that NP has a negligible impact on processes that arise at the tree-level in the SM, we write for meson-antimeson mixing in all sectors

$$M_{12} = M_{12}^{\text{SM}} + \frac{F_i L_i}{\Lambda^2} \langle M^0 | Q_i | \overline{M}^0 \rangle, \qquad \Gamma_{12} \simeq \Gamma_{12}^{\text{SM}}, \qquad (8.288)$$

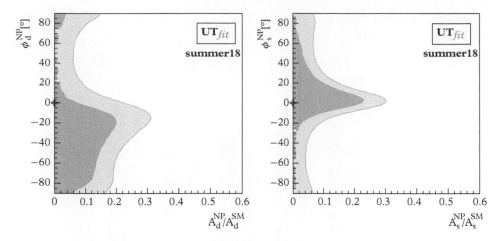

Figure 8.14 *Probability density functions for A_q^{NP}, ϕ_q^{NP}. Darker (lighter) regions correspond to smallest 68% (95%) probability regions.*

where F_i is a function of the (complex) NP flavour couplings, L_i is a loop factor that is present in models with no tree-level flavour changing neutral currents (FCNC), and Λ is the scale of NP, i.e. the typical mass of the new particles mediating the $\Delta F = 2$ transition. For a generic strongly-interacting theory with arbitrary flavour structure, we expect $F_i \sim L_i \sim \mathcal{O}(1)$ so that the allowed range for each of the NP contributions can be immediately translated into a lower bound on Λ. Specific assumptions on the flavour structure of NP, for example minimal or next-to-minimal flavour violation (MFV or NMFV), correspond to particular choices of the F_i functions, as detailed below. Notice that in eqn (8.288) the SM contribution M_{12}^{SM} should be computed using for the CKM parameters the results of the UT analysis in the presence of NP.

Switching on one operator at a time, assuming that $F_i \sim L_i \sim \mathcal{O}(1)$, running its coefficient down from the NP scale Λ to the hadronic scale μ at which the relevant matrix elements have been computed ([129]–[140] show computations of the matrix elements for the full set of relevant operators), computing its contribution according to eqn 8.288 and comparing it to the results presented in Sections 8.5.5 and 8.6.2, we obtain the 95% probability lower bounds on Λ presented in Figure 8.1. The bounds are dominated by CP violation in $K - \bar{K}$ and $D - \bar{D}$ mixing, as expected from the extreme suppression of these processes in the SM, and by the contributions of the chirality-violating operators, which are enhanced both by the RG evolution and by the matrix elements. These bounds are clearly beyond the reach of any direct detection experiment, and strongly suggest us that any NP close to the EW scale must have a hierarchical flavour structure analogous to the SM one. We can then envisage the so-called NMFV scenario, in which one has $F_i \simeq F^{SM}$, where F^{SM} is the CKM factor of the relevant SM amplitude. The bounds on the NP scale in NMFV for $L_i = 1$ are reported in Figure 8.15. We see that the chiral and RG enhancement of Q_4 pushes the NP scale to $\mathcal{O}(100)$ TeV; to keep Λ below 10 TeV

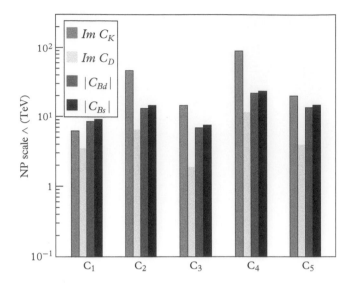

Figure 8.15 *Summary of the 95% probability lower bound on the NP scale Λ for NMFV. See the text for details.*

we must not only enforce the same flavour structure of the SM, but also the same chiral structure.

Requiring the same flavour and chiral structure of the SM corresponds to the so-called MFV framework, initially formulated as the requirement of NP contributions to FCNC observables being just a redefinition of the loop function associated to the top-quark contribution, also known as Constrained MFV (CMFV) [141, 142]. The CMFV hypothesis allows for an improved determination of the UT in the presence of NP and for several tests of consistency, since it implies the independence on NP of ratios of observables in which the top-mediated loop function drops, such as for example $\Delta m_{B_d}/\Delta m_{B_s}$ [143, 144]. More generally, we can observe that the requirement that NP has the same flavour and chiral structure of the SM can be formulated in terms of the flavour symmetry of the SM Lagrangian when the Yukawa couplings are put to zero: the MFV hypothesis then amounts to the requirement that Yukawa couplings are the only source of violation of the flavour symmetry [6]. This automatically leads to small deviations from the SM for NP scales close to the EW one, provided that Yukawa couplings are close to their 'SM' value (i.e. to the value they would take in the SM), while for example in two Higgs doublets models with a large v_2/v_1 ratio of the two vacuum expectation values one could face larger deviations enhanced by this ratio.

While assuming that MFV holds exactly amounts to assuming that Yukawa couplings are fundamental, thus giving up the hope of finding a dynamical explanation of their hierarchical structure, it is certainly true that a NP scale close to the EW one implies a flavour structure close to MFV. This is clearly possible if the NP responsible for the origin of the Yukawa couplings structure is much heavier than the EW scale.

8.7 Summary and Further Reading

The goal of this chapter is to allow the reader to get a first idea of how we can probe NP with flavour observables. The level of refinement of current forefront analyses in this field is clearly way beyond the few basic elements here presented. Fortunately, several excellent reviews are available on most of the topics sketched in the previous sections. First of all, there are several other lectures on the same topics which are more detailed, more general and more inspired than ours, starting from the classic Les Houches lectures by A.J. Buras [19], SLAC and Trieste lectures by Y. Nir [145], and from the excellent book by Branco, Lavoura, and Silva [146], continuing to more recent lectures [144, 147]–[167]. Review articles include the NLO classic [58], while among the many NP-oriented reviews I find [168, 169] particularly inspiring. Ref. [170] contains a remarkably complete discussion of meson–anti-meson mixing in the charm and bottom sectors. Finally, while it was impossible to collect here all original references for the topics we discussed, the reader is strongly encouraged to read the original papers where all details can be found.

Appendix 8.A   Formulæ for Loop Calculations

We collect here a few useful formulæ for loop calculations in dimensional regularization.

8.A.1   Feynman Parameters

Feynman parameters are useful to group denominators in loop amplitudes. The basic formula is the following:

$$\frac{1}{AB} = \int_0^1 dxdy \frac{\delta(x+y-1)}{[xA+yB]^2} = \int_0^1 dx \frac{1}{[xA+(1-x)B]^2}. \tag{8.289}$$

It can be easily verified explicitly:

$$\int_0^1 dx \frac{1}{[xA+(1-x)B]^2} = -\frac{1}{A-B} \left[\frac{1}{xA+(1-x)B} \right]_0^1 = \frac{1}{A-B} \left(\frac{1}{A} - \frac{1}{B} \right) = \frac{1}{AB}. \tag{8.290}$$

We can raise the powers in the denominator by differentiating:

$$\frac{1}{AB^n} = \frac{(-1)^{n-1}}{(n-1)!} \frac{\partial^{n-1}}{\partial B^{n-1}} \frac{1}{AB} = \frac{(-1)^{n-1}}{(n-1)!} \int_0^1 dxdy \frac{\delta(x+y-1)n!y^{n-1}(-1)^{n-1}}{[xA+yB]^{n+1}}$$
$$= \int_0^1 dxdy \frac{\delta(x+y-1)ny^{n-1}}{[xA+yB]^{n+1}}. \tag{8.291}$$

We can add further terms in the denominator by iterating with eqns (8.289) and (8.291):

$$\frac{1}{ABC} = \frac{1}{AB}\frac{1}{C} = \frac{1}{C}\int_0^1 dxdy\frac{\delta(x+y-1)}{[xA+yB]^2}$$

$$= \int_0^1 dw\,dz\,dx\,dy\,2w\frac{\delta(w+z-1)\delta(x+y-1)}{[zC+w(xA+yB)]^3} \begin{array}{c} x'=wx \\ y'=wy \end{array} \int_0^1 dw\,dz\,\delta(w+z-1)$$

$$\int_0^w dx'\,dy'\,2\frac{\delta(x'+y'-w)}{[zC+x'A+y'B]^3}$$

$$= \int_0^1 dz\int_0^{1-z} dx'\,dy'\,2\frac{\delta(x'+y'+z-1)}{[zC+x'A+y'B]^3} = \int_0^1 dxdydz\frac{2\delta(x+y+z-1)}{[xA+yB+zC]^3}. \qquad (8.292)$$

We thus obtain the general formula

$$\frac{1}{A_1^{m_1}A_2^{m_2}\ldots A_n^{M_n}} = \int_0^1 dx_1\,dx_2\ldots dx_n\delta\left(\sum_i x_i - 1\right)\frac{\prod_i x_i^{m_i-1}}{\left[\sum_i x_iA_i\right]^{\sum_i m_i}}\frac{\Gamma(\sum_i m_i)}{\prod_i \Gamma(m_i)}. \qquad (8.293)$$

8.A.2 Loop Integrals

Momentum Shift

After grouping the denominators with Feynman parameters using eqn (8.293), the denominator will contain in general not only the square of the loop momentum and constant terms, but also terms linear in the loop momentum (from dot products with external momenta). We get rid of linear terms in the denominator by performing a shift of the loop momentum, which brings us to the general form

$$\int \frac{d^d k}{(2\pi)^d}\frac{k^{\mu_1}\ldots k^{\mu_n}}{(k^2-D+i\epsilon)^m}. \qquad (8.294)$$

Then integrals with odd powers of k in the numerator vanish by symmetry, and we are left with even powers of k only.

Wick Rotation

The $i\epsilon$ term in eqn (8.294) is there to remind us that we should be careful about the poles of the propagators entering the diagram we are calculating. We can form a closed contour in the complex K^0 plane by going from $-\infty$ to $+\infty$ on the real axis, from $+\infty$ to $-\infty$ on the imaginary axis, and closing the contour with two arcs at infinity from the real to the imaginary axis. Noting that the $i\epsilon$ prescription moves the poles to the second and fourth quadrant, so that they are not inside the contour, and neglecting the contribution at infinity, we see that the integral on the real axis is equal to the integral on the imaginary axis. We then go to the Euclidean with $k^0 = ik_E^0$, and obtain

$$\int \frac{d^d k}{(2\pi)^d}\frac{1}{(k^2-D)^m} = i(-1)^m\int\frac{d^d k_E}{(2\pi)^d}\frac{1}{(k_E^2+D)^m} \qquad (8.295)$$

Angular integration

Let us now split the integration pulling out the angular one:

$$\int \frac{d^d k_E}{(2\pi)^d} \frac{1}{(k_E^2 + D)^m} = \int \frac{d\Omega_d}{(2\pi)^d} \int k_E^{d-1} dk_E \frac{1}{(k_E^2 + D)^m}. \tag{8.296}$$

We can obtain the angular term using a Gaussian integral:

$$(\sqrt{\pi})^d = \left(\int_{-\infty}^{\infty} dx e^{-x^2} \right)^d = \int_{-\infty}^{\infty} d[d] x e^{-x^2} = \int d\Omega_d \int_0^{\infty} dx x^{d-1} e^{-x^2}$$

$$= \int d\Omega_d \frac{1}{2} \int_0^{\infty} dx^2 (x^2)^{\frac{d}{2}-1} e^{-x^2} = \int d\Omega_d \frac{1}{2} \Gamma\left(\frac{d}{2}\right) \tag{8.297}$$

so that

$$\int d\Omega_d = \frac{2\pi^{\frac{d}{2}}}{\Gamma\left(\frac{d}{2}\right)}. \tag{8.298}$$

Momentum Integration

We now turn to the integral over the absolute value of the Euclidean momentum:

$$\int_0^{\infty} k_E^{d-1} dk_E \frac{1}{(k_E^2 + D)^m} = \frac{1}{2} \int_0^{\infty} (k_E^2)^{\frac{d}{2}-1} dk_E^2 \frac{1}{(k_E^2 + D)^m} \underset{x = \frac{D}{k_E^2 + D}}{=\!=\!=\!=\!=}$$

$$- \frac{1}{2} \int_1^0 Dx^{-2} dx \frac{(D/x)^{\frac{d}{2}-1} (1-x)^{\frac{d}{2}-1}}{(D/x)^m}$$

$$= \frac{1}{2} D^{\frac{d}{2}-m} \int_0^1 dx x^{m-\frac{d}{2}-1} (1-x)^{\frac{d}{2}-1} \tag{8.299}$$

$$= \frac{1}{2} \left(\frac{1}{D}\right)^{m-\frac{d}{2}} B\left(m - \frac{d}{2}, \frac{d}{2}\right),$$

where

$$\int_0^1 dx x^{\alpha-1} (1-x)^{\beta-1} = B(\alpha, \beta) = \frac{\Gamma(\alpha)\Gamma(\beta)}{\Gamma(\alpha+\beta)}. \tag{8.300}$$

Expansion for $d = 4 - 2\epsilon$

We now put together the results in eqns (8.295), (8.296), (8.298), and (8.299):

$$\int \frac{d^d k}{(2\pi)^d} \frac{1}{(k^2 - D)^m} = \frac{i}{(4\pi)^{\frac{d}{2}}} (-1)^m \left(\frac{1}{D}\right)^{m-\frac{d}{2}} \frac{\Gamma(m - \frac{d}{2})}{\Gamma(m)} \tag{8.301}$$

and expand for d close to four in the small parameter ϵ:

$$\int \frac{d^d k}{(2\pi)^d} \frac{1}{(k^2 - D)^m} = \frac{i}{16\pi^2} \left(-\frac{1}{D}\right)^{m-2} \left(\frac{1}{4\pi D}\right)^{\epsilon} \frac{\Gamma(m - 2 + \epsilon)}{\Gamma(m)}. \tag{8.302}$$

Using the expansion of Euler Γ

$$\Gamma(\epsilon) = \frac{1}{\epsilon} - \gamma_E + \mathcal{O}(\epsilon), \tag{8.303}$$

$$\Gamma(-n+\epsilon) = \frac{(-1)^n}{n!} \left(\frac{1}{\epsilon} - \gamma_E + 1 + \ldots + \frac{1}{n} + \mathcal{O}(\epsilon)\right),$$

one obtains for example

$$\int \frac{d^d k}{(2\pi)^d} \frac{1}{(k^2 - D)^2} = \frac{i}{16\pi^2} \left(\frac{1}{\epsilon} - \gamma_E - \ln 4\pi - \ln D + \mathcal{O}(\epsilon) \right)$$
$$\equiv \frac{i}{16\pi^2} \left(\frac{1}{\bar{\epsilon}} - \ln D + \mathcal{O}(\bar{\epsilon}) \right), \qquad (8.304)$$

where we introduced for convenience the parameter $\bar{\epsilon}$ defined in eqn (8.47).

Some Useful Integrals

The reader may find the following list of integrals useful:

$$\int \frac{d^d k}{(2\pi)^d} \frac{1}{(k^2 - D)} = \frac{i}{(4\pi)^{\frac{d}{2}}} (-1)^m \left(\frac{1}{D} \right)^{m-\frac{d}{2}} \frac{\Gamma(m - \frac{d}{2})}{\Gamma(m)}, \qquad (8.305)$$

$$\int \frac{d^d k}{(2\pi)^d} \frac{k^2}{(k^2 - D)} = \frac{i}{(4\pi)^{\frac{d}{2}}} (-1)^{m-1} \frac{d}{2} \left(\frac{1}{D} \right)^{m-\frac{d}{2}-1} \frac{\Gamma(m - \frac{d}{2} - 1)}{\Gamma(m)}, \qquad (8.306)$$

$$\int \frac{d^d k}{(2\pi)^d} \frac{k^\mu k^\nu}{(k^2 - D)} = \frac{i}{(4\pi)^{\frac{d}{2}}} (-1)^{m-1} \frac{g^{\mu\nu}}{2} \left(\frac{1}{D} \right)^{m-\frac{d}{2}-1} \frac{\Gamma(m - \frac{d}{2} - 1)}{\Gamma(m)}, \qquad (8.307)$$

$$\int \frac{d^d k}{(2\pi)^d} \frac{(k^2)^2}{(k^2 - D)} = \frac{i}{(4\pi)^{\frac{d}{2}}} (-1)^m \frac{d(d+2)}{4} \left(\frac{1}{D} \right)^{m-\frac{d}{2}-2} \frac{\Gamma(m - \frac{d}{2} - 2)}{\Gamma(m)}, \qquad (8.308)$$

$$\int \frac{d^d k}{(2\pi)^d} \frac{k^\mu k^\nu k^\rho k^\sigma}{(k^2 - D)} = \frac{i}{(4\pi)^{\frac{d}{2}}} (-1)^m \frac{g^{\mu\nu} g^{\rho\sigma} + g^{\mu\rho} g^{\nu\sigma} + g^{\mu\sigma} g^{\nu\rho}}{4} \left(\frac{1}{D} \right)^{m-\frac{d}{2}-2} \frac{\Gamma(m - \frac{d}{2} - 2)}{\Gamma(m)}. \qquad (8.309)$$

Acknowledgements

It is a great pleasure to thank the organizers for providing such a pleasant and stimulating environment, and the students for following my lectures with undeserved interest. Any meaningful statement you may find in this writeup originally comes from the teaching I received from G. Martinelli, M. Ciuchini, E. Franco, L. Reina, A. Masiero, R. Petronzio, A. J. Buras, and A. Romanino (in chronological order, in the form of lectures, notes, and countless discussions). All confusing, misleading or plainly wrong statements are instead my own contribution. I am indebted to M. Ciuchini and to M. Valli for carefully reading this manuscript, and to S. Davidson for her patience and for her comments on the manuscript. This work has received funding from the European Research Council (ERC) under the European Union's Horizon 2020 research and innovation program (grant agreement n° 772369).

References

[1] S. L. Glashow, J. Iliopoulos, and L. Maiani, (1970) Weak Interactions with Lepton-Hadron Symmetry *Phys. Rev. D.* 2: 1285–92.
[2] C. Jarlskog, Commutator of the Quark Mass Matrices in the Standard Electroweak Model and a Measure of Maximal CP Violation, (1985), *Phys. Rev. Lett.* 55: 1039.

[3] A. V. Manohar, (2018). *Introduction to Effective Field Theories*, in *Les Houches Summer School on EFT in Particle Physics and Cosmology*. 3–28 July 2017, Les Houches, France. arXiv:1804.05863.

[4] A. Pich, (2018). *Effective Field Theory with Nambu–Goldstone Modes*, in *Les Houches Summer School on EFT in Particle Physics and Cosmology*. 3–28 July 2017, Les Houches, France. arXiv:1804.05664.

[5] L. Silvestrini and M. Valli, (2019). Model-Independent Bounds on the Standard Model Effective Theory from Flavour Physics, *Phys. Lett. B* 799: 135062. arXiv:1812.10913.

[6] G. D'Ambrosio, G. F. Giudice, G. Isidori, and A. Strumia, (2002). Minimal Flavor Violation: An Effective Field Theory Approach, *Nucl. Phys. B*. 645: 155–87, [hep-ph/0207036].

[7] R. Barbieri, G. R. Dvali, and L. J. Hall, (1996). Predictions from a U(2) Flavor Symmetry in Supersymmetric Theories, *Phys. Lett. B*. 377: 76–82, [hep-ph/9512388].

[8] R. Barbieri, L. J. Hall, and A. Romanino, (1997). Consequences of a U(2) Flavor Symmetry, *Phys. Lett. B*. 401: 47–53, [hep-ph/9702315].

[9] R. G. Roberts, A. Romanino, G. G. Ross, and L. Velasco-Sevilla, (2001). Precision Test of a Fermion Mass Texture, *Nucl. Phys. B*. 615: 358–84, [hep-ph/0104088].

[10] E. Dudas, G. von Gersdorff, S. Pokorski, and R. Ziegler, (2014). Linking Natural Super-symmetry to Flavour Physics, *JHEP* 01: 117, [arXiv:1308.1090].

[11] M. Linster and R. Ziegler, (2018). A Realistic $U(2)$ Model of Flavor, *JHEP* 08: 058, [arXiv:1805.07341].

[12] M. Kobayashi and T. Maskawa, (1973). CP Violation in the Renormalizable Theory of Weak Interaction, *Prog. Theor. Phys.* 49: 652–7.

[13] N. Cabibbo, (1963). Unitary Symmetry and Leptonic Decays, *Phys. Rev. Lett.* 10: 531–3.

[14] L. Wolfenstein, (1983). Parametrization of the Kobayashi–Maskawa Matrix, *Phys. Rev. Lett.* 51: 1945.

[15] A. J. Buras, M. E. Lautenbacher, and G. Ostermaier, (1994). Waiting for the Top Quark Mass, $K^+ \to pi^+ \nu\bar{\nu}$, B(s)0 - Anti-B(s)0 Mixing and CP Asymmetries in B Decays, *Phys. Rev. D*. 50: 3433–46, [hep-ph/9403384].

[16] Particle Data Group Collaboration, M. Tanabashi et al., (2018). Review of Particle Physics, *Phys. Rev. D*. 98(3): 030001.

[17] G. 't Hooft, (1973). Dimensional Regularization and the Renormalization Group, *Nucl. Phys. B*. 61: 455–68.

[18] W. A. Bardeen, A. J. Buras, D. W. Duke, and T. Muta, (1978). Deep Inelastic Scattering Beyond the Leading Order in Asymptotically Free Gauge Theories, *Phys. Rev. D*. 18: 3998.

[19] A. J. Buras, (1998). Weak Hamiltonian, CP Violation and Rare Decays, in *Probing the Standard Model of Particle Interactions. Proceedings, Summer School in Theoretical Physics*, parts 1–2, 281–539. NATO Advanced Study Institute, 68th Session. 28 July–5 September 1997, Les Houches, France. hep-ph/9806471.

[20] G. Martinelli, et al., (1995). A General Method for Nonperturbative Renormalization of Lattice Operators, *Nucl. Phys. B*. 445: 81–108, [hep-lat/9411010].

[21] M. Fierz, (1937). Zur fermischen theorie des β-zerfalls, *Zeitschrift für Physik* 104: 553–65.

[22] A. I. Vainshtein, V. I. Zakharov, and M. A. Shifman, (1975). A Possible Mechanism for the Delta T = 1/2 Rule in Nonleptonic Decays of Strange Particles, *JETP Lett.* 22: 55–6. [(1975). Pisma Zh. Eksp. Teor. Fiz. 22: 123(1975)].

[23] M. A. Shifman, (1999). Foreword, In M. Shifman (Ed), *ITEP Lectures in Particle Physics and Field Theory*, vol. 1, World Scientific. hep-ph/9510397.

[24] T. Inami and C. S. Lim, (1981). Effects of Superheavy Quarks and Leptons in Low-Energy Weak Processes $k_L \to \mu$ Anti-μ, $K^+ \to pi^+$ Neutrino Anti-Neutrino and $K^0 \leftrightarrow$ Anti-K^0, *Prog. Theor. Phys.* 65: 297. [Erratum: (1981). Prog. Theor. Phys. 65: 1772].

[25] A. J. Buras, M. Jamin, M. E. Lautenbacher, and P. H. Weisz, (1992). Effective Hamiltonians for $\Delta S = 1$ and $\Delta B = 1$ Nonleptonic Decays Beyond the Leading Logarithmic Approximation, *Nucl. Phys. B.* 370: 69–104. [Addendum: *Nucl. Phys. B* 375: 501].

[26] A. J. Buras, M. Jamin, and M. E. Lautenbacher, (1993). Two Loop Anomalous Dimension Matrix for Delta S = 1 Weak Nonleptonic Decays. 2. O(alpha-alpha-s), *Nucl.Phys. B.* 400: 75–102, [hep-ph/9211321].

[27] M. Ciuchini, E. Franco, G. Martinelli, and L. Reina, (1994). The Delta S = 1 effective Hamiltonian Including Next-to-Leading Order QCD and QED Corrections, *Nucl. Phys. B.* 415: 403–62, [hep-ph/9304257].

[28] M. Lusignoli, (1989). Electromagnetic Corrections to the Effective Hamiltonian for Strangeness Changing Decays and ϵ'/ϵ, *Nucl. Phys. B.* 325: 33–61.

[29] M. K. Gaillard and B. W. Lee, (1974). Δ I = 1/2 Rule for Nonleptonic Decays in Asymptotically Free Field Theories, *Phys. Rev. Lett.* 33: 108.

[30] G. Altarelli and L. Maiani, (1974), Octet Enhancement of Nonleptonic Weak Interactions in Asymptotically Free Gauge Theories, *Phys. Lett. B.* 52: 351–4.

[31] F. J. Gilman and M. B. Wise, (1979). Effective Hamiltonian for Delta $s = 1$ Weak Nonleptonic Decays in the Six Quark Model, *Phys. Rev. D.* 20: 2392.

[32] B. Guberina and R. D. Peccei, (1980). Quantum Chromodynamic Effects and CP Violation in the Kobayashi–Maskawa Model, *Nucl. Phys. B.* 163: 289–311.

[33] J. Bijnens and M. B. Wise, (1984). Electromagnetic Contribution to Epsilon-Prime/Epsilon, *Phys. Lett. B.* 137: 245–50.

[34] K. M. Watson, (1954). Some General Relations Between the Photoproduction and Scattering of pi Mesons, *Phys. Rev.* 95: 228–236.

[35] E. Franco, S. Mishima, and L. Silvestrini, (2012). The Standard Model Confronts *CP* Violation in $D^0 \to \pi^+\pi^-$ and $D^0 \to K^+K^-$, *JHEP* 05: 140, [arXiv:1203.3131].

[36] RBC, UKQCD Collaboration, (2013). P. A. Boyle et al., Emerging Understanding of the $\Delta I = 1/2$ Rule from Lattice QCD, *Phys. Rev. Lett.* 110: 152001, [arXiv:1212.1474].

[37] ETM Collaboration, N. Carrasco, V. Lubicz, and L. Silvestrini, (2014). Vacuum Insertion Approximation and the $\Delta I = 1/2$ Rule: A Lattice QCD Test of the Naïve Factorization Hypothesis for K , D , B and Static Mesons, *Phys. Lett. B.* 736: 174–9, [arXiv:1312.6691].

[38] RBC, UKQCD Collaboration, (2019). C. Sachrajda, Precision Kaon Physics and Lattice QCD.   Presented at *Towards the Ultimate Precision in Flavour Physics*. 2–4 April 2019, Durham, UK. https://indico.cern.ch/event/760368/contributions/3316194/ attachments/1823201/2982902/Sachrajda.pdf.

[39] O. Cata and S. Peris, (2003). Long Distance Dimension Eight Operators in B(K), *JHEP* 03: 060, [hep-ph/0303162].

[40] M. Ciuchini, et al., Power Corrections to CP Violation in K^0–$\bar{K}^0$ Mixing, *in preparation*.

[41] A. J. Buras, M. Jamin, and P. H. Weisz, (1990). Leading and Next-to-Leading QCD Corrections to ϵ Parameter and $B^0 - \bar{B}^0$ Mixing in the Presence of a Heavy Top Quark, *Nucl. Phys. B.* 347: 491–536.

[42] S. Herrlich and U. Nierste, (1994). Enhancement of the $K(L)$ - $K(S)$ Mass Difference by Short Distance QCD Corrections Beyond Leading Logarithms, *Nucl. Phys. B.* 419: 292–322, [hep-ph/9310311].

[43] S. Herrlich and U. Nierste, (1995). Indirect CP Violation in the Neutral Kaon System Beyond Leading Logarithms, *Phys. Rev. D.* 52: 6505–18, [hep-ph/9507262].

[44] U. Nierste, (1995). Indirect CP Violation in the Neutral Kaon System Beyond Leading Logarithms and Related Topics. PhD thesis, Munich, Tech. U., 1995. hep-ph/9510323.

[45] S. Herrlich and U. Nierste, (1996). The Complete $|\Delta S| = 2$ Hamiltonian in the Next-to-Leading Order, *Nucl. Phys. B*. 476: 27–88, [hep-ph/9604330].

[46] J. Brod and M. Gorbahn, (2010). ϵ_K at Next-to-Next-to-Leading Order: The Charm-Top-Quark Contribution, *Phys. Rev. B*. 82: 094026, [arXiv:1007.0684].

[47] J. Brod and M. Gorbahn, (2012). Next-to-Next-to-Leading-Order Charm-Quark Contribution to the CP Violation Parameter ϵ_K and ΔM_K, *Phys. Rev. Lett.* 108: 121801, [arXiv:1108.2036].

[48] P. Gambino, A. Kwiatkowski, and N. Pott, (1999). Electroweak Effects in the B0 - Anti-B0 Mixing, *Nucl. Phys. B*. 544: 532–56, [hep-ph/9810400].

[49] A. J. Buras, M. Misiak, and J. Urban, (2000). Two Loop QCD Anomalous Dimensions of Flavor Changing Four Quark Operators Within and Beyond the Standard Model, *Nucl. Phys. B*. 586: 397–426, [hep-ph/0005183].

[50] J. Aebischer, M. Fael, C. Greub, and J. Virto, (2017). B Physics Beyond the Standard Model at One Loop: Complete Renormalization Group Evolution Below the Electroweak Scale, *JHEP* 09: 158, [arXiv:1704.06639].

[51] F. Gabbiani, E. Gabrielli, A. Masiero, and L. Silvestrini, (1996). A Complete Analysis of FCNC and CP Constraints in General SUSY Extensions of the Standard Model, *Nucl. Phys. B*. 477: 321–52, [hep-ph/9604387].

[52] J. A. Bagger, K. T. Matchev, and R.-J. Zhang, (1997). QCD Corrections to Flavor Changing Neutral Currents in the Supersymmetric Standard Model, *Phys. Lett. B*. 412: 77–85, [hep-ph/9707225].

[53] M. Ciuchini, et al., (1998). Next-to-Leading Order QCD Corrections to Delta $F = 2$ Effective Hamiltonians, *Nucl. Phys. B*. 523: 501–25, [hep-ph/9711402].

[54] M. Ciuchini, et al., (2006). Next-to-Leading Order Strong Interaction Corrections to the Delta $F = 2$ Effective Hamiltonian in the MSSM, *JHEP* 09: 013, [hep-ph/0606197].

[55] C. R. Allton, et al., (1999). B Parameters for Delta $S = 2$ Supersymmetric Operators, *Phys. Lett. B*. 453: 30–9, [hep-lat/9806016].

[56] M. Gell-Mann and A. Pais, (1955). Behavior of Neutral Particles Under Charge Conjugation, *Phys. Rev.* 97: 1387–9.

[57] A. J. Buras, D. Guadagnoli, and G. Isidori, (2010). On ϵ_K Beyond Lowest Order in the Operator Product Expansion, *Phys. Lett. B*. 688: 309–13, [arXiv:1002.3612].

[58] G. Buchalla, A. J. Buras, and M. E. Lautenbacher, (1996). Weak Decays Beyond Leading Logarithms, *Rev. Mod. Phys.* 68: 1125–44, [hep-ph/9512380].

[59] UTfit Collaboration, M. Bona et al., (2005). The 2004 UTfit Collaboration Report on the Status of the Unitarity Triangle in the Standard Model, *JHEP* 07: 028, [hep-ph/0501199]. Online updates at http://www.utfit.org.

[60] A. J. Buras, M. Jamin, and M. E. Lautenbacher, (1993). The Anatomy of ϵ'/ϵ Beyond Leading Logarithms with Improved Hadronic Matrix Elements, *Nucl. Phys. B*. 408: 209–85, [hep-ph/9303284].

[61] M. Ciuchini, et al., (1995). An Upgraded Analysis of Epsilon-Prime Epsilon at the Next-to-Leading Order, *Z. Phys. C*. 68: 239–56, [hep-ph/9501265].

[62] S. Bosch, et al., (2000). Standard Model Confronting New Results for Epsilon-Prime / Epsilon, *Nucl. Phys. B*. 565: 3–37, [hep-ph/9904408].

[63] NA48 Collaboration, J. R. Batley et al., (2002). A Precision Measurement of Direct CP Violation in the Decay of Neutral Kaons into Two Pions, *Phys. Lett. B*. 544: 97–112, [hep-ex/0208009].

[64] KTeV Collaboration, A. Alavi-Harati et al., (2003). Measurements of Direct CP Violation, CPT Symmetry, and Other Parameters in the Neutral Kaon System, *Phys. Rev. D.* 67: 012005, [hep-ex/0208007]. [Erratum: (2004). *Phys. Rev. D* 70: 079904].

[65] KTeV Collaboration, E. Abouzaid et al., (2011). Precise Measurements of Direct CP Violation, CPT Symmetry, and Other Parameters in the Neutral Kaon System, *Phys. Rev. D.* 83: 092001, [arXiv:1011.0127].

[66] RBC, UKQCD Collaboration, Z. Bai et al., (2015). Standard Model Prediction for Direct *CP* Violation in $K \to \pi\pi$ Decay, *Phys. Rev. Lett.* 115(21): 212001, [arXiv:1505.07863].

[67] T. Blum et al., (2015). $K \to \pi\pi$ $\Delta I = 3/2$ *Decay Amplitude in the Continuum Limit*, *Phys. Rev. D.* 91(7): 074502, [arXiv:1502.00263].

[68] A. J. Buras, M. Gorbahn, S. Jäger, and M. Jamin, (2015). Improved Anatomy of ε'/ε in the Standard Model, *JHEP* 11: 202, [arXiv:1507.06345].

[69] T. Kitahara, U. Nierste, and P. Tremper, (2016). Singularity-Free Next-to-Leading Order $\Delta S = 1$ Renormalization Group Evolution and ϵ'_K/ϵ_K in the Standard Model and beyond, *JHEP* 12: 078, [arXiv:1607.06727].

[70] V. Antonelli, S. Bertolini, M. Fabbrichesi, and E. I. Lashin, (1996). The Delta $I = 1/2$ Selection Rule, *Nucl. Phys. B.* 469: 181–201, [hep-ph/9511341].

[71] S. Bertolini, J. O. Eeg, and M. Fabbrichesi, (1996). A New Estimate of Epsilon-Prime / Epsilon, *Nucl. Phys. B.* 476: 225–54, [hep-ph/9512356].

[72] E. Pallante and A. Pich, (2000). Strong Enhancement of Epsilon-Prime / Epsilon Through Final State Interactions, *Phys. Rev. Lett.* 84: 2568–71, [hep-ph/9911233].

[73] E. Pallante and A. Pich, (2001). *Final State Interactions in Kaon Decays*, *Nucl. Phys. B.* 592: 294–320, [hep-ph/0007208].

[74] M. Buchler, G. Colangelo, J. Kambor, and F. Orellana, (2001). A Note on the Dispersive Treatment of $K \to \pi\pi$ with the Kaon Off-Shell, *Phys. Lett. B.* 521: 29–32, [hep-ph/0102289].

[75] M. Buchler, G. Colangelo, J. Kambor, and F. Orellana, (2001). Dispersion Relations and Soft Pion Theorems for $K \to \pi\pi$, *Phys. Lett. B.* 521: 22–8, [hep-ph/0102287].

[76] E. Pallante, A. Pich, and I. Scimemi, (2001). The Standard Model Prediction for Epsilon-Prime / Epsilon, *Nucl. Phys. B.* 617: 441–74, [hep-ph/0105011].

[77] H. Gisbert and A. Pich, (2018). Direct CP Violation in $K^0 \to \pi\pi$: Standard Model Status, *Rept. Prog. Phys.* 81(7): no. 7 076201, [arXiv:1712.06147].

[78] H. Gisbert and A. Pich, (2018). Updated Standard Model Prediction for ε'/ε, in *21st High-Energy Physics International Conference in Quantum Chromodynamics (QCD 18)*. 2–6 July 2018, Montpellier, France. arXiv:1810.04904.

[79] H. Gisbert, (2018). Current Status of ε'/ε in the Standard Model, in *13th Conference on Quark Confinement and the Hadron Spectrum (Confinement XIII)*. 31 July–6 August 2018, Maynooth, Ireland. 2018. arXiv:1811.12206.

[80] RBC, UKQCD Collaboration, N. Christ, et al., (2014). Calculating the $K_L - K_S$ Mass Difference and ϵ_K to Sub-Percent Accuracy, Proceedings of the 31st International Symposium on Lattice Field Theory (Lattice 2013). 29 July–3 August 2013, Mainz, Germany [arXiv:1402.2577].

[81] M. Ciuchini, et al., (2007). D - $\bar{D}$ Mixing and New Physics: General Considerations and Constraints on the MSSM, *Phys. Lett. B.* 655: 162–6, [hep-ph/0703204].

[82] UTfit Collaboration, A. J. Bevan et al., (2014). The UTfit Collaboration Average of D Meson Mixing Data: Winter 2014, *JHEP* 03: 123, [arXiv:1402.1664].

[83] M. Beneke, G. Buchalla, and I. Dunietz, (1996). Width Difference in the $B_s - \bar{B}_s$ System, *Phys. Rev. D.* 54: 4419–31, [hep-ph/9605259].

[84] M. Beneke, et al., (1999). Next-to-Leading-Order QCD Corrections to the Lifetime Difference of $B(s)$ Mesons, *Phys.Lett. B.* 459: 631–40, [hep-ph/9808385].

[85] M. Ciuchini, et al., (2003). Lifetime Differences and CP Violation Parameters of Neutral *B* Mesons at the Next-to-Leading Order in QCD, *JHEP* 08: 031, [hep-ph/0308029].

[86] A. J. Buras, M. Jamin, M. E. Lautenbacher, and P. H. Weisz, Two Loop Anomalous Dimension Matrix for Delta $S = 1$ Weak Nonleptonic Decays. 1. *O(alpha-s**2)*, *Nucl.Phys. B.* 400: 37–74, [hep-ph/9211304].

[87] A. B. Carter and A. I. Sanda, (1981). CP Violation in *B* Meson Decays, *Phys. Rev. D.* 23: 1567.

[88] I. I. Y. Bigi and A. I. Sanda, (1981). Notes on the Observability of CP Violations in *B* Decays, *Nucl. Phys. B.* 193: 85–108.

[89] P. Oddone, (1987). Detector Considerations. Presented at *Workshop on Conceptual Design of a Test Linear Collider: Possibilities for a B Anti-B Factory.* 26–30 Jan 1987, Los Angeles, California *eConf* C870126: 423–46.

[90] BaBar, Belle Collaboration, A. J. Bevan et al., (2014). The Physics of the B Factories, *Eur. Phys. J. C.* 74: 3026, [arXiv:1406.6311].

[91] M. Ciuchini, E. Franco, G. Martinelli, and L. Silvestrini, (1997). Charming Penguins in B Decays, *Nucl. Phys. B* 501: 271–96, [hep-ph/9703353].

[92] A. J. Buras and L. Silvestrini, (2000). Nonleptonic Two-Body *B* Decays Beyond Factorization, *Nucl. Phys. B.* 569: 3–52, [hep-ph/9812392].

[93] J. D. Bjorken, Topics in B Physics, (1989). *Nucl. Phys. Proc. Suppl.* 11: 325–41.

[94] M. Beneke, G. Buchalla, M. Neubert, and C. T. Sachrajda, (1999). QCD Factorization for $B \to \pi\pi$ Decays: Strong Phases and CP Violation in the Heavy Quark Limit, *Phys. Rev. Lett.* 83: 1914–17, [hep-ph/9905312].

[95] M. Beneke, G. Buchalla, M. Neubert, and C. T. Sachrajda, (2000). QCD Factorization for Exclusive, Nonleptonic B Meson Decays: General Arguments and the Case of Heavy Light Final States, *Nucl. Phys. B.* 591: 313–418, [hep-ph/0006124].

[96] M. Ciuchini, et al., (2001). Charming Penguins Strike Back, *Phys. Lett. B.* 515: 33–41, [hep-ph/0104126].

[97] M. Beneke, G. Buchalla, M. Neubert, and C. T. Sachrajda, (2001). QCD Factorization in $B \to \pi K, \pi\pi$ Decays and Extraction of Wolfenstein Parameters, *Nucl. Phys. B.* 606: 245–321, [hep-ph/0104110].

[98] BaBar Collaboration, B. Aubert et al., (2009). Measurement of Time-Dependent CP Asymmetry in $B^0 \to c\bar{c}K^{(*)0}$ Decays, *Phys. Rev. D.* 79: 072009, [arXiv:0902.1708].

[99] I. Adachi et al., (2012). Precise Measurement of the CP Violation Parameter $\sin 2\phi_1$ in $B^0 \to (c\bar{c})K^0$ Decays, *Phys. Rev. Lett.* 108: 171802, [arXiv:1201.4643].

[100] LHCb Collaboration, R. Aaij et al., (2015). Measurement of *CP* Violation in $B^0 \to J/\psi K_S^0$ Decays, *Phys. Rev. Lett.* 115(3): 031601, [arXiv:1503.07089].

[101] LHCb Collaboration, R. Aaij et al., (2017). Measurement of *CP* Violation in $B^0 \to J/\psi K_S^0$ and $B^0 \to \psi(2S)K_S^0$ Decays, *JHEP* 11: 170, [arXiv:1709.03944].

[102] Heavy Flavor Averaging Group Collaboration, Y. Amhis et al., (2017). Averages of *b*-Hadron, *c*-Hadron, and τ-Lepton Properties as of Summer 2016, *Eur. Phys. J. C.* 77: 895, [arXiv:1612.07233]; updated results and plots available at https://hflav.web.cern.ch.

[103] M. Ciuchini, M. Pierini, and L. Silvestrini, (2005). The Effect of Penguins in the $B_d \to J/\Psi K^0$ CP Asymmetry, *Phys. Rev. Lett.* 95: 221804, [hep-ph/0507290].

[104] M. Ciuchini, M. Pierini, and L. Silvestrini, (2011). Theoretical Uncertainty in sin 2β: An Update. In *CKM unitarity triangle, Proceedings, 6th International Workshop, CKM 2010*. 6–10 September 2010, Warwick, UK. arXiv:1102.0392.

[105] M. Gronau and D. London, (1990). Isospin Analysis of CP Asymmetries in B Decays, *Phys. Rev. Lett.* 65: 3381–4.

[106] M. Gronau and J. Zupan, (2005). Isospin-Breaking Effects on Alpha Extracted in $B \to \pi\pi, \rho\rho, \rho\pi$, *Phys. Rev. D.* 71: 074017, [hep-ph/0502139].

[107] J. Charles, (1999). Taming the Penguin in the $B_d^0(t) \to \pi + \pi-$ CP Asymmetry: Observables and Minimal Theoretical Input, *Phys. Rev. D.* 59: 054007, [hep-ph/9806468].

[108] UTfit Collaboration, M. Bona et al., (2007). Improved Determination of the CKM Angle Alpha from B to pi pi Decays, *Phys. Rev. D.* 76: 014015, [hep-ph/0701204].

[109] M. Ciuchini, E. Franco, S. Mishima, and L. Silvestrini, (2012). Testing the Standard Model and Searching for New Physics with $B_d \to \pi\pi$ and $B_s \to KK$ Decays, *JHEP* 10: 029, [arXiv:1205.4948].

[110] H. J. Lipkin, Y. Nir, H. R. Quinn, and A. Snyder, (1991). Penguin Trapping with Isospin Analysis and CP Asymmetries in B Decays, *Phys. Rev. D.* 44: 1454–60.

[111] UTfit Collaboration, M. Bona et al., (2008). Model-independent Constraints on $\Delta F = 2$ Operators and the Scale of New Physics, *JHEP* 03: 049, [arXiv:0707.0636].

[112] A. E. Snyder and H. R. Quinn, (1993). *Measuring CP Asymmetry in $B \to \rho\pi$ Decays Without Ambiguities*, *Phys. Rev. D.* 48: 2139–44.

[113] H. R. Quinn and J. P. Silva, (2000). The Use of Early Data on $B \to \rho\pi$ Decays, *Phys. Rev. D.* 62: 054002, [hep-ph/0001290].

[114] K. De Bruyn and R. Fleischer, (2015). A Roadmap to Control Penguin Effects in $B_d^0 \to J/\psi K_S^0$ and $B_s^0 \to J/\psi\phi$, *JHEP* 03: 145, [arXiv:1412.6834].

[115] I. S. Towner and J. C. Hardy, (2010). *The Evaluation of V(ud) and its Impact on the Unitarity of the Cabibbo–Kobayashi–Maskawa Quark-Mixing Matrix*, *Rept. Prog. Phys.* 73: 046301.

[116] J. Hardy and I. S. Towner, (2016). $|V_{ud}|$ from Nuclear β Decays In 9th International Workshop on the CKM Unitarity Triangle. 28 November–3 December 2016, Mumbai, India.

[117] M. Moulson, (2017). *Experimental Determination of V_{us} from Kaon Decays*. In 9th International Workshop on the CKM Unitarity Triangle. 28 November–3 December 2016, Mumbai, India [arXiv:1704.04104].

[118] Flavour Lattice Averaging Group Collaboration, S. Aoki et al., (2019). FLAG Review 2019, arXiv:1902.08191.

[119] B. Grinstein and A. Kobach, (2017). Model-Independent Extraction of $|V_{cb}|$ from $\bar{B} \to D^*\ell\bar{\nu}$, *Phys. Lett. B.* 771: 359–64, [arXiv:1703.08170].

[120] D. Bigi, P. Gambino, and S. Schacht, (2017). A Fresh Look at the Determination of $|V_{cb}|$ from $B \to D^*\ell\nu$, *Phys. Lett. B.* 769: 441–5, [arXiv:1703.06124].

[121] F. U. Bernlochner, Z. Ligeti, and D. J. Robinson, (2019). N = 5, 6, 7, 8: Nested Hypothesis Tests and Truncation Dependence of $|V_{cb}|$, *Phys. Rev. D.* 100: 013005. arXiv:1902.09553.

[122] M. Gronau and D. London, (1991). *How to Determine all the Angles of the Unitarity Triangle from $B_d^0 \to DK_s$ and $B_s^0 \to D^0$*, *Phys. Lett. B.* 253: 483–8.

[123] D. Atwood, I. Dunietz, and A. Soni, (1997). Enhanced CP Violation with $B \to KD^0(\bar{D}^0)$ Modes and Extraction of the CKM Angle Gamma, *Phys. Rev. Lett.* 78: 3257–60, [hep-ph/9612433].

[124] D. Atwood, I. Dunietz, and A. Soni, (2001). Improved Methods for Observing CP Violation in $B^{+-} \to KD$ and Measuring the CKM Phase Gamma, *Phys. Rev. D.* 63: 036005, [hep-ph/0008090].

[125] A. Giri, Y. Grossman, A. Soffer, and J. Zupan, (2003). Determining Gamma Using $B^{+-} \to DK^{+-}$ with Multibody D Decays, *Phys. Rev. D.* 68: 054018, [hep-ph/0303187].

[126] J. Brod and J. Zupan, (2014). The Ultimate Theoretical Error on γ from $B \to DK$ Decays, *JHEP* 01: 051, [arXiv:1308.5663].

[127] S. Laplace, Z. Ligeti, Y. Nir, and G. Perez, (2002). Implications of the CP Asymmetry in Semileptonic B Decay, *Phys. Rev. D.* 65: 094040, [hep-ph/0202010].

[128] UTfit Collaboration, M. Bona et al., The UTfit Collaboration Report on the Status of the Unitarity Triangle Beyond the Standard Model. I. Model-Independent Analysis and Minimal Flavor Violation, *JHEP* 03: 080, [hep-ph/0509219].

[129] C. M. Bouchard, et al., (2011). Neutral B Mixing from $2 + 1$ Flavor Lattice-QCD: the Standard Model and Beyond. In Proceedings of the XXIX International Symposium on Lattice Field Theory (Lattice 2011) - Weak Decays. July 10–16 2011, Lake Tahoe, CA [arXiv:1112.5642].

[130] ETM Collaboration, V. Bertone et al., (2013). Kaon Mixing Beyond the SM from $N_f = 2$ tmQCD and Model Independent Constraints from the UTA, *JHEP* 03: 089, [arXiv:1207.1287]. [Erratum: (2013). *JHEP* 7, 143].

[131] RBC, UKQCD Collaboration, P. A. Boyle, N. Garron, and R. J. Hudspith, (2012). *Neutral Kaon Mixing Beyond the Standard Model with $n_f = 2 + 1$ Chiral Fermions*, *Phys. Rev. D.* 86: 054028, [arXiv:1206.5737].

[132] SWME Collaboration, T. Bae et al., (2013). Neutral Kaon Mixing from New Physics: Matrix Elements in $N_f = 2 + 1$ Lattice QCD, *Phys. Rev. D.* 88(7): 071503, [arXiv:1309.2040].

[133] ETM Collaboration, N. Carrasco et al., (2014). B-Physics from $N_f = 2$ *tmQCD*: the Standard Model and Beyond, *JHEP* 03: 016, [arXiv:1308.1851].

[134] N. Carrasco et al., (2014). $D^0 - \bar{D}^0$ Mixing in the Standard Model and Beyond from $N_f = 2$ Twisted Mass QCD, *Phys. Rev. D.* 90(1): 014502, [arXiv:1403.7302].

[135] SWME Collaboration, J. Leem et al., (2014). Calculation of BSM Kaon B-Parameters Using Staggered Quarks. In Proceedings of the 32nd International Symposium on Lattice Field Theory (Lattice 2014). 23–28 June 2014, New York, NY [arXiv:1411.1501].

[136] SWME Collaboration, B. J. Choi et al., (2016). Kaon BSM B-Parameters Using Improved Staggered Fermions from $N_f = 2 + 1$ Unquenched QCD, *Phys. Rev. D.* 93(1): 014511, [arXiv:1509.00592].

[137] ETM Collaboration, N. Carrasco, et al., (2015). $\Delta S = 2$ and $\Delta C = 2$ Bag Parameters in the Standard Model and Beyond from N_f=2+1+1 Twisted-Mass Lattice QCD, *Phys. Rev. D.* 92(3): 034516, [arXiv:1505.06639].

[138] RBC/UKQCD Collaboration, N. Garron, R. J. Hudspith, and A. T. Lytle, (2016). Neutral Kaon Mixing Beyond the Standard Model with $n_f = 2 + 1$ Chiral Fermions Part 1: Bare Matrix Elements and Physical Results, *JHEP* 11: 001, [arXiv:1609.03334].

[139] Fermilab Lattice, MILC Collaboration, A. Bazavov et al., (2016). $B^0_{(s)}$-Mixing Matrix Elements from Lattice QCD for the Standard Model and Beyond, *Phys. Rev. D.* 93(11): 113016, [arXiv:1602.03560].

[140] RBC, UKQCD Collaboration, P. A. Boyle, et al., (2017). Neutral Kaon Mixing Beyond the Standard Model with $n_f = 2 + 1$ Chiral Fermions. Part 2: Non Perturbative Renormalisation of the $\Delta F = 2$ Four-Quark Operators, *JHEP* 10: 054, [arXiv:1708.03552].

[141] E. Gabrielli and G. F. Giudice, (1995). Supersymmetric Corrections to Epsilon Prime / Epsilon at the Leading Order in QCD and QED, *Nucl. Phys. B.* 433: 3–25, [hep-lat/9407029]. [Erratum: *Nucl. Phys. B.* 507: 549].

[142] A. J. Buras, P. Gambino, M. Gorbahn, S. Jager, and L. Silvestrini, (2001). Epsilon-Prime / Epsilon and rare K and B decays in the MSSM, *Nucl. Phys. B.* 592: 55–91, [hep-ph/0007313].

[143] A. J. Buras, et al., (2001). Universal Unitarity Triangle and Physics Beyond the Standard Model, *Phys. Lett. B.* 500: 161–67, [hep-ph/0007085].

[144] A. J. Buras, *Minimal flavor violation,* (2003). Acta Phys. Polon. B34: 5615–68, [hep-ph/0310208].

[145] Y. Nir, (1999). CP Violation in and Beyond the Standard Model, in *Proceedings, 27th SLAC Summer Institute on Particle Physics: CP Violation In and Beyond the Standard Model (SSI 99)* 165–243. 7–16 July 1999, Stanford, CT. hep-ph/9911321.

[146] G. C. Branco, L. Lavoura, and J. P. Silva, (1999). *CP Violation,* Oxford University Press.

[147] J. L. Rosner, (2000). CP Violation in *B* Decays, in *Flavor Physics for the Millennium. Proceedings, Theoretical Advanced Study Institute in Elementary Particle Physics, TASI 2000* 431–80. 4–30 June 2000, Boulder, CO. hep-ph/0011355.

[148] A. J. Buras, (2002). Flavor Dynamics: CP Violation and Rare Decays, *Subnucl. Ser.* 38: 200–337, [hep-ph/0101336].

[149] Y. Nir, (2001). CP Violation: A New era, in *Heavy Flavor Physics: Theory and Experimental Results in Heavy Quark Physics and CP Violation. Proceedings, 55th Scottish Universities Summer School in Physics, SUSSP 2001* 147–200. 7–23 August 2001, St. Andrews, Scotland. hep-ph/0109090.

[150] A. J. Buras, (2004). CP Violation in B and K Decays: 2003, *Lect. Notes Phys.* 629: 85–135, [hep-ph/0307203].

[151] R. Fleischer, (2004). Flavor Physics and CP Violation, in *High-energy physics. Proceedings, European School* 81–150. 24 August–6 September 2003, Tsakhkadzor, Armenia. hep-ph/0405091.

[152] A. J. Buras, (2005). *Flavor Physics and CP Violation,* in *2004 European School of High-Energy Physics* 95–168. 30 May–12 June 2004, Sant Feliu de Guixols, Spain. hep-ph/0505175.

[153] R. Fleischer, (2006). Flavour Physics and CP Violation, in *High-energy physics. Proceedings, European School* 71–153. 21 August–3 September 2005, Kitzbuhel, Austria. hep-ph/0608010.

[154] I. I. Bigi, (2007). Flavour Dynamics and CP Violation in the Standard Model: A Crucial Past and an Essential Future, in *High-energy physics. Proceedings, European School* 115–96. 18 June–1 July 2006, Aronsborg, Sweden. hep-ph/0701273.

[155] T. Mannel, (2008). Aspects of Flavour Physics, in *Proceedings, 2007 European School of High-Energy Physics (ESHEP 2007)* 209–76. 19 August–1 September 2007, Trest, Czech Republic.

[156] R. Fleischer, (2008). Flavour Physics and CP Violation: Expecting the LHC, in *High-energy physics. Proceedings, 4th Latin American CERN-CLAF School* 105–57. 18 February–3 March 2008, Vina del Mar, Chile. arXiv:0802.2882.

[157] U. Nierste, (2009). Three Lectures on Meson Mixing and CKM Phenomenology, in *Heavy quark physics. Proceedings, Helmholtz International School, HQP08* 1–38. 11–21 August 2008, Dubna, Russia. arXiv:0904.1869.

[158] Y. Nir, (2010). Flavour Physics and CP Violation, in *High-energy physics. Proceedings, 5th CERN-Latin-American School.* 15–28 March 2009, Recinto Quirama, Columbia. arXiv:1010.2666.

[159] G. Isidori, (2014). Flavor Physics and CP Violation, in *Proceedings, 2012 European School of High-Energy Physics (ESHEP 2012)* 69–105. 6–19 June 2012, La Pommeraye, Anjou, France. arXiv:1302.0661.

[160] G. Castelo-Branco and D. Emmanuel-Costa, (2015). Flavour physics and CP Violation in the Standard Model and Beyond, *Springer Proc. Phys.* 161: 145–86, [arXiv:1402.4068].

[161] E. Kou, (2014). Flavour Physics and CP Violation (CERN-2014-001), in *Proceedings, 1st Asia-Europe-Pacific School of High-Energy Physics (AEPSHEP)* 151–76. 14–27 October 2012, Fukuoka, Japam. arXiv:1406.7700.

[162] B. Grinstein, (2015). TASI-2013 Lectures on Flavor Physics, in *Theoretical Advanced Study Institute in Elementary Particle Physics: Particle Physics: The Higgs Boson and Beyond (TASI 2013)*. 3–28 June 2013, Boulder, CO. arXiv:1501.05283.

[163] J. F. Kamenik, (2016). Flavour Physics and CP Violation, in *Proceedings, 2014 European School of High-Energy Physics (ESHEP 2014)* 79–94. 18 June–1 July 2014, Garderen, The Netherlands. arXiv:1708.00771.

[164] M. Blanke, (2017). Introduction to Flavour Physics and CP Violation, *CERN Yellow Rep. School Proc.* 1705: 71–100, [arXiv:1704.03753].

[165] Y. Grossman and P. Tanedo, (2016). Just a Taste: Lectures on Flavor Physics, in *Proceedings, Theoretical Advanced Study Institute in Elementary Particle Physics : Anticipating the Next Discoveries in Particle Physics (TASI 2016)* 109–295. 6 June–1 July 2016, Boulder, CO. arXiv:1711.03624.

[166] A. Pich, (2018). Flavour Dynamics and Violations of the CP Symmetry, *CERN Yellow Rep. School Proc.* 4: 65, [arXiv:1805.08597].

[167] J. Zupan, (2019). Introduction to Flavour Physics, in *2018 European School of High-Energy Physics (ESHEP2018)*. 20 June–3 July 2018, Maratea, Italy. arXiv:1903.05062.

[168] Y. Grossman, Y. Nir, and R. Rattazzi, (1998). CP Violation Beyond the Standard Model, *Adv. Ser. Direct. High Energy Phys.* 15: 755–94, [hep-ph/9701231].

[169] G. Isidori, Y. Nir, and G. Perez, (2010). Flavor Physics Constraints for Physics Beyond the Standard Model, *Ann. Rev. Nucl. Part. Sci.* 60: 355, [arXiv:1002.0900].

[170] A. L. Kagan and M. D. Sokoloff, (2009). On Indirect CP Violation and Implications for $D0$ - Anti-$D0$ and $B(s)$ - Anti-$B(s)$ Mixing, *Phys. Rev. D.* 80: 076008, [arXiv:0907.3917].

9

Effective Field Theories for Heavy Quarks: Heavy Quark Effective Theory and Heavy Quark Expansion

Thomas MANNEL

Theoretische Elementarteilchenphysik, Naturwissensch.–techn. Fakultät,
Universität Siegen, 57068 Siegen, Germany

Thomas MANNEL

Mannel, T., *Effective Field Theories for Heavy Quarks: Heavy Quark Effective Theory and Heavy Quark Expansion* In: *Effective Field Theory in Particle Physics and Cosmology.* Edited by: Sacha Davidson, Paolo Gambino, Mikko Laine, Matthias Neubert and Christophe Salomon, Oxford University Press (2020). © Oxford University Press. DOI: 10.1093/oso/9780198855743.003.0009

Chapter Contents

9.1   Introduction

Effective Field Theory (EFT) and Renormalization Group (RG) methods have developed into universal tools that can be applied in various fields of physics. Most efficient use of EFT methods can be made in systems, in which vastly different mass scales appear and appropriate ratios of these mass scales define small parameters one aims to expand in.

In nuclear and particle physics, the obvious scales are the masses of the fundamental constituents. Here we do not discuss the Higgs mechanism, which is assumed to give masses to the quarks and leptons; rather, we assume that the masses are fundamental parameters. Another relevant scale in nuclear and particle physics is the scale $\Lambda_{QCD} \sim 300$ MeV, which is generated by 'dimensional transmutation' in QCD; this scale is the typical scale of the masses of light hadrons and also governs the running (i.e. the dependence on the renormalization scale μ) of the strong coupling 'constant' $\alpha_s(\mu)$.

When considering weak interactions, the typical scale is set by the W boson mass M_W which at low energies manifests itself in the Fermi coupling constant $G_F \sim 1/M_W^2$ relevant for the four-fermion coupling in the EFT for weak interactions. When studying a weak decay of a bottom hadron, the typical scale is set by the b-quark mass m_b. Due to confinement of QCD, this b quark is bound in a hadron, and the relevant scale for this binding in Λ_{QCD}.

The elementary interaction for weak processes is expressed in terms of quark currents, however, the observed states are hadrons. Therefore we have to deal with the effects of strong interactions, which are described in QCD. One important feature of QCD is its *asymptotic freedom*, which implies that its running coupling constant $\alpha_s(\mu) \to 0$ as $\mu \to \infty$. In practical terms this means that $\alpha_s(M_W)$ as well as $\alpha_s(m_b)$ is a small parameter which allows us to perform a perturbative expansion.

A hadronic matrix element of a quark current evaluated at the scale $\mu \sim m_b$ still contains perturbatively computable pieces, which can be extracted by switching to an EFT description, which, for the cases here, is the heavy quark effective theory (HQET). By applying HQET, the hadronic matrix elements are expressed as a combination of perturbatively computable coefficients and new, suitably defined matrix elements, which contain the 'real' non-perturbative contributions.

For some cases it is convenient to also treat the mass of the charm quark m_c as a perturbative scale, which requires also describing the c quark in HQET. However, once we arrive at scales $\mu \sim \Lambda_{QCD}$, the strong coupling $\alpha_s(\mu)$ becomes order one or bigger, indicating that perturbation theory becomes useless.

Thus weak decays involve a sequence of vastly different mass scales. Assuming that the Standard Model (SM) itself is an EFT and that we have physics beyond the SM (BSM) at some high scale Λ_{NP}, we have $\Lambda_{NP} \gg M_W \gg m_b \gg m_c \gg \Lambda_{QCD}$ where - except for the last step - the QCD effects can be treated perturbatively by a tower of suitably constructed EFTs, until finally the non-perturbative QCD effects remain as matrix elements depending solely on the small scale Λ_{QCD}.

Up to that point this describes the ubiquitous machinery of EFT in general. However, EFTs for heavy quarks are quite special in one aspect. In weak interactions at low scales all effects of, for example, the W boson appear only in the couplings (like the Fermi coupling), and at low scales this is the only remnant of the heavy W boson. However, consider now a bottom quark in QCD. In pure QCD, the bottom number is a conserved quantity, and this statement is independent of the scale. Thus a hadron with one unit of bottom quantum number will at low scales (i.e. below the bottom quark mass) still have the bottom quark inside, however, *this bottom quark will behave like a static source of a colour field*. This is in full analogy to the hydrogen atom: Although it is a two particle problem, it is a very good approximation to treat the proton inside the H atom as a static source of a Coulomb field, in which the electron moves; any corrections to this picture will be of order $m_{\text{electron}}/m_{\text{Proton}}$. The simplest type of such a theory is the already mentioned HQET which describes systems with a single heavy quark, where all light degrees of freedom are 'soft', i.e. all components of their momenta are of the order Λ_{QCD}. The first part of the chapter discusses such systems.

The second part of the chapter is devoted to inclusive processes. Using the Operator Product Expansion (OPE), a standard method in quantum firld theory, we will set up an expansion (the heavy quark expansion, or HQE), which has become the basis of many precision calculations in heavy quark physics.

Heavy quarks can decay weakly into light quarks, and hence there is also the kinematic situation, where light quarks acquire energies (in the rest frame of the decaying heavy quark) which scale with the heavy quark mass. For these situations an EFT has been developed, which is called *Soft Collinear Effective Theory* (SCET). This theory has also many applications in high-energy collider physics and will be covered by a different lecture at this school.

A second class are systems with a heavy quark and a heavy antiquark forming a bound system such as a charmonium, a bottomonium and also a B_c. Such systems require to set up yet a different kind of EFT, which is called non-relativistic QCD (NRQCD). However, due to space and time limitations, this type of theory cannot be covered here.

The original work on HQET dates back to Eichten and Hill [1, 2], Grinstein [3], and Georgi [4]. The main impact of HQET are the additional symmetries emerging in the infinite mass limit; this has been noticed first by Shifman and Voloshin [5] and Isgur and Wise [6, 7]. The OPE-based method for inclusive processes has been first set up by Chay, Georgi, and Grinstein [8], by Bigi and colleagues [9, 10], by Manohar and Wise [11], and by Mannel [12].

There are many reviews on this subject, which are too numerous to be listed here; a subjective selection is [13]–[16]. Finally, there are a few textbooks [17]–[20] that elaborate on the subject.

9.2   Heavy Quark Effective Theory

We start with the simplest heavy quark expansion, which is the heavy quark effective theory for systems with a single heavy quark. We first construct its Lagrangian by integrating out the heavy degrees of freedom. The remarkable and for phenomenology very relevant features of HQET are the heavy quark symmetries (HQS) which eventually yield constraints on the non-perturbative matrix elements at low scales that are not evident in full QCD. Since $\alpha_s(m_b)$ is a perturbative scale, we will compute the one-loop matching of full QCD to HQET, which will give us some insight into the anatomy of HQET. Finally we will collect a few results that are used in current phenomenology.

9.2.1   Construction of the HQET Lagrangian

There are two ways to construct the Lagrangian of HQET. We follow directly the idea of EFT by identifying the heavy degrees of freedom and integrating them out from the functional integral [21]. This approach is quite instructive, since it can be explicitly performed at tree level and also at one loop; it also leads to a closed form for the HQET Lagrangian, at least at tree level.

A second approach follows usual non-relativistic reduction of the Dirac equation [22], finally leading to a recursive construction of the terms of higher order. This approach has the disadvantage that it does not explicitly involve the typical steps of the construction of an EFT. We will not discuss this alternative approach here.

The two approaches seem to have different results, since the Lagrangians derived in the two cases look different. However, it has been shown that the two approaches are related by a field redefinition, and that the results for physical quantities are the same in both cases.

We will first consider the derivation of the HQET Lagrangian from the usual machinery of EFT, following [21]. The starting point is the Lagrangian of QCD with a single heavy quark Q written as

$$\mathcal{L}_{\text{QCD}} = \bar{Q}(i\slashed{D} - m_Q)Q + \mathcal{L}_{\text{light}} \tag{9.1}$$

where m_Q is the mass of the heavy quark, $D_\mu = \partial_\mu + igA_\mu$ is the usual QCD covariant derivative including the interaction with the gluon A_μ, and $\mathcal{L}_{\text{light}}$ is the Lagrangian for the light quarks and gluons[1].

To obtain the Green functions of the corresponding quantum field theory one may gather them in a generating functional, which is expressed as a functional integral over the field variables. Thus we write

$$Z(\eta, \bar{\eta}, \lambda) = \int [dQ][d\bar{Q}][d\phi_\lambda] \exp\left\{ i \int d^4x \mathcal{L}_{\text{QCD}} + i \int d^4x (\bar{\eta}Q + \bar{Q}\eta + \phi_\lambda \lambda) \right\}, \tag{9.2}$$

---

[1] $\mathcal{L}_{\text{light}}$ also contains all terms relevant for the gauge fixing and possibly ghost fields needed for the quantization of QCD.

where $\phi_\lambda = q$, A_μ^a denotes the light degrees of freedom (light quarks q and gluons A_μ). Functional differentiation with respect to the source terms $\eta, \bar{\eta}$ and λ and subsequently setting the sources to zero yields the Green functions of QCD[2], e.g.

$$\frac{\delta}{\delta\eta(x)} \frac{\delta}{\delta\bar{\eta}(y)} Z(\eta, \bar{\eta}, \lambda)|_{\eta=0,\bar{\eta}=0,\lambda=0} = \langle 0| T[Q(y)\bar{Q}(x)]|0\rangle \tag{9.3}$$

In order to derive the HQET Lagrangian we consider a system with a single heavy quark which is bound in a heavy hadron. This hadron has a mass m_H and moves with a certain momentum p_H. In case the hadron contains only a single heavy quark, its mass will scale with the heavy quark mass, likewise its momentum will scale with the heavy quark mass. To this end, it is convenient to define a four velocity

$$v = \frac{p_H}{m_H}, \quad v^2 = 1, \quad v_0 > 0. \tag{9.4}$$

which is independent of the heavy quark mass. This vector defines a specific frame, e.g. $v = (1,0,0,0)$ is the rest frame of the heavy hadron.

Eventually we want to consider the heavy quark inside the heavy hadron; since most of the hadron mass is given by the quark mass, the heavy quark moves with almost the same velocity as the heavy hadron. Thus the momentum of the heavy quark my be written as $p_Q = m_Q v + k$ where k is a small 'residual' momentum satisfying $k \ll m_Q$.

To implement this idea on the technical side, we use this 'external' velocity vector v to decompose the heavy-quark field Q into an 'upper' (or 'large') component ϕ and a 'lower' (or 'small') component χ

$$\phi_v = \frac{1}{2}(1+\slashed{v})Q \equiv P_+ Q, \quad \slashed{v}\phi_v = \phi, \tag{9.5}$$

$$\chi_v = \frac{1}{2}(1-\slashed{v})Q \equiv P_- Q \quad \slashed{v}\chi_v = -\chi, \tag{9.6}$$

and to define a decomposition of the covariant derivative into a 'time' and a 'spatial' ($\perp$) part

$$D_\mu = v_\mu(v \cdot D) + D_\mu^\perp, \quad D_\mu^\perp = (g_{\mu\nu} - v_\mu v_\nu)D^\nu, \quad \{\slashed{D}^\perp, \slashed{v}\} = 2(v \cdot D^\perp) = 0. \tag{9.7}$$

In terms of these new fields (9.5, 9.6) and using (9.7), the Lagrangian of the heavy quark field (i.e. the first term of (9.1)) takes the form

$$\mathcal{L}_{\text{heavy}} = \bar{\phi}_v(i(v \cdot D) - m_Q)\phi_v - \bar{\chi}_v(i(v \cdot D) + m_Q)\chi_v + \bar{\phi}_v i\slashed{D}^\perp \chi_v + \bar{\chi}_v i\slashed{D}^\perp \phi_v \tag{9.8}$$

---

[2] We note that the funtional integral (9.2) is mathematically ill defined. For our purposes we take the practitioner's point of view and look at (9.2) as a short-hand notation for perturbation theory, which results from expanding in the strong coupling g_s.

To proceed further, we now implement the decomposition of the heavy quark momentum into a 'large' and a residual piece. This is achieved by multiplying the heavy quark field by a phase

$$\phi_v = e^{-im_Q(v\cdot x)} h_v, \qquad \chi_v = e^{-im_Q(v\cdot x)} H_v. \tag{9.9}$$

Note that the momentum of a field is the derivative acting on the field, i.e.

$$p_Q^\mu \sim i\partial^\mu Q(x) \quad \text{hence} \quad i\partial^\mu \phi_v(x) = e^{-im_Q(v\cdot x)}(m_Q v^\mu + i\partial^\mu) h_v(x)$$

which means that the derivative acting on the field h_v reproduces the residual momentum introduced above. This observation provides us with the power counting of HQET: once we have reformulated the theory in terms of h_v, we aim at an expansion in the residual momentum, i.e. in iD_μ/m_Q.

We express the Lagrangian of the heavy quark in term of the fields h_v and H_v and obtain

$$\mathcal{L}_{\text{heavy}} = \bar{h}_v i(v \cdot D) h_v - \bar{H}_v (i(v \cdot D) + 2m_Q) H_v + \bar{h}_v i\slashed{D}^\perp H_v + \bar{H}_v i\slashed{D}^\perp h_v. \tag{9.10}$$

With this form of the Lagrangian we can now easily identify the degrees of freedom. The field h_v does not have a mass term, while the field H_v has acquired a mass term $2m_Q$; the remaining terms are couplings between h_v and H_v. Thus in the sense of EFT the field H_v is the heavy degree of freedom, while the field h_v is light.

To construct the EFT, we have to 'integrate out' the heavy degree of freedom, which is H_v. In the language of functional integrals this means that we have to integrate over the field H_v in the generating functional (9.2). It is interesting to note, that in the case at hand this functional integration can be explicitly performed, at least for the tree-level Lagrangian (9.10), since there are only quadratic dependences on the relevant field, hence the functional integral over the field H_v is a Gaussian integral.

In order to integrate over the heavy field H_v, we first split the source terms in (9.2) according to

$$\int d^4x (\bar{\eta} Q + \bar{Q}\eta) = \int d^4x (\bar{\rho}_v h_v + \bar{h}_v \rho_v + \bar{R}_v H_v + \bar{H}_v R_v), \tag{9.11}$$

where ρ_v and R_v are now source terms for the upper-component field h_v and the lower-component part H_v, respectively. When studying processes at scales well below the scale $2m_Q$, no Green function involving the heavy field H_v will be relevant, hence we can put the corresponding sources to zero. Performing the Gaussian integral over the field H_v we obtain

$$Z(\rho_v, \bar{\rho}_v, \lambda) = \int [dh_v][d\bar{h}_v][d\lambda]\, \Delta$$

$$\times \exp\left\{ iS + S_\lambda + i\int d^4x (\bar{\rho}_v h_v + \bar{h}_v \rho_v + \phi_\lambda \lambda) \right\}, \tag{9.12}$$

where now the action functional for the heavy quark becomes a non-local object

$$S = \int d^4x \left[\bar{h}_v i(v \cdot D) h_v - \bar{h}_v \slashed{D}^{\perp} \left(\frac{1}{i(v \cdot D) + 2m_Q - i\epsilon} \right) \slashed{D}^{\perp} h_v \right]. \tag{9.13}$$

depending solely on the field h_v and (via the covariant derivatives) on gluon fields.

The quantity Δ is the determinant resulting form the Gaussian integration, which may formally be written as

$$\Delta = \exp\left(\frac{1}{2} \ln\left[i(v \cdot D) + 2m_Q \right] \right) \tag{9.14}$$

$$= \text{const} \exp\left(\frac{1}{2} \ln\left[1 + \frac{1}{i(v \cdot \partial) - 2m_Q + i\epsilon} g_s(v \cdot A) \right] \right).$$

However, unlike in other quantum field theories, this determinant is a constant (i.e. independent of the gluon fields). This can bee seen by either chosing the gauge $v \cdot A = 0$ or by expanding the logarithm which leads to an expression which looks like the fermion bubble diagrams in ordinary QCD; however, here the particles propagate only in forward time-like directions, since the propagator in configuration space is

$$\frac{1}{i(v \cdot \partial) - 2m_Q + i\epsilon} = \delta^3(x^{\perp}) \theta(v \cdot x) e^{i2m(v \cdot x)}. \tag{9.15}$$

Hence a closed loop always yields a zero result.

In general, integrating our degrees of freedom yields non-local action functionals such as (9.13). However, if the degree of freedom that has been integrated out is heavy, it is in general possible to expand the result in inverse powers of the mass of the heavy scale. In our case this is quite evident, since we have $(v \cdot D) \ll 2m_Q$ because $(v \cdot D)$ is related to the residual momentum of the heavy quark. Consequently we may expand to get

$$\frac{1}{i(v \cdot D) + 2m_Q - i\epsilon} = \frac{1}{2m_Q} \sum_{n=0}^{\infty} \left(\frac{-i(v \cdot D)}{2m_Q} \right)^n \tag{9.16}$$

which expresses the non-local distribution on the left-hand side as a series of local distributions.

Truncating at some order N yields a local action functional, and hence we get as the Lagranian

$$\mathcal{L}_{1/m_Q-\text{Expansion}} = \bar{h}_v i(v \cdot D) h_v - \frac{1}{2m_Q} \bar{h}_v \slashed{D}^{\perp} \sum_{n=0}^{N} \left(\frac{-i(v \cdot D)}{2m_Q} \right)^n \slashed{D}^{\perp} h_v. \tag{9.17}$$

This expression is the expansion of the QCD Lagrangian up to the order $1/m_Q^{N+1}$. The leading term

$$\mathcal{L}_{\text{HQET}} = \bar{h}_v i(v \cdot D) h_v \tag{9.18}$$

is the Lagrangian for a static heavy quark moving with the four velocity v, i.e. the Lagrangian of heavy quark effective theory (HQET).

By itself this Lagrangian is not very useful; by choosing an axial gauge $v \cdot A = 0$ the coupling to the gluons can even be made to vanish. However, this Lagrangain becomes useful as soon as additional interactions are implemented which are 'hard', meaning that these interactions change the velocity of the heavy quark. In most applications we consider, these are typically electroweak interactions which inject a large momentum transfer into the system.

To illustrate this in some more detail, let us consider the semileptonic decay $B \to D\ell\bar{\nu}$. The relevant hadronic matrix element is

$$\langle B(p)|\bar{b}\gamma_\mu c|D(p')\rangle$$

which may be obtained from inserting the weak transition current into the generating functional (9.2)

$$Z^{(b \to c)}(\eta_b, \bar{\eta}_b, \eta_c, \bar{\eta}_c, \lambda) = \int [db][d\bar{b}][dc][d\bar{c}][d\phi_\lambda]\,\bar{b}(0)\gamma_\mu c(0)$$

$$\times \exp\left\{ i \int d^4x \mathcal{L}_{\mathrm{QCD}} + i \int d^4x (\bar{\eta}_b b + \bar{b}\eta_b + \bar{\eta}_c c + \bar{c}\eta_c + \phi_\lambda\lambda) \right\}, \qquad (9.19)$$

corresponding to an insertion of the weak $b \to c$ current into the QCD Green functions.

At scales below m_c we may use the static limit for both the b and the c quark, however, the two mesons have different velocities $v = p/M_b$ and $v' = p'/M_D$, so we need to introduce two static quarks b_v and c_v with different velocities. Going through the same steps as before, now for two heavy quarks with different velocities, we get

$$\mathcal{L}_{\mathrm{HQET}}^{b \to c} = \bar{b}_v i(v \cdot D)b_v + \bar{c}_{v'} i(v' \cdot D)c_{v'}. \qquad (9.20)$$

Although this looks like a Lagrangian with two heavy quarks, the weak current ensures that for $x_0 \leq 0$ we have only the bottom quark (moving with velocity v) which at $x_0 = 0$ decays into a charm quark, moving with velocity v'.

This kind of approximation is well known since almost one century. It has been used already in the context of the infrared problem of QED, which for soft photons becomes 'heavy electron effective theory' [23, 24]. Furthermore, once there are two velocities v and v' in the game, there is no possibility to trivialize the theory by a choice of gauge.

Once we consider operator insertions in the Greens functions as in (9.19) we also need to rewrite the fields appearing in the current, which amounts to re-expressing the full QCD field by the static field h_v. We get

$$Q(x) = e^{-im_Q vx} [h_v + H_v] = e^{-im_Q vx} \left[1 + \left(\frac{1}{2m + ivD} \right) i\slashed{D}_\perp \right] h_v$$

$$= e^{-im_Q vx} \left[1 + \frac{1}{2m_Q}\slashed{D}_\perp + \left(\frac{1}{2m_Q} \right)^2 (-ivD)\slashed{D}_\perp + \cdots \right] h_v. \qquad (9.21)$$

Note that this expression is inserted in the functional integral (9.19), so we inter-grate out the heavy field(s) H_v, resulting in the replacement of H_v in the second step of (9.21).

The Hamiltonian which can be derived from (9.17) has the unusual property that it contains time derivatives (i.e. terms involving $(iv\partial)Q_v$). However, these can be removed by field redefinitions, resulting in an Hamiltonian without time derivatives. This Hamiltonian can also be constructed from the start by performing a transformation (a so-called Foldy–Wouthuysen transformation), which decouples the large and the small components of the spinor Q. This yields a Lagrangian which looks different from the one derived above, starting at $1/m_Q^3$; likewise, the expansion of the fields in terms if the static field also looks different.

For physical matrix elements both approaches eventually yield the same answer. To see how this works, we consider a matrix element with a heavy-to-light current of the form $\bar{q}\Gamma Q$, with a heavy meson in the initial state $|M(v)\rangle$ and some final state $|A\rangle$. Computing to order $1/m_Q$ we get

$$\langle A|\bar{q}\Gamma Q|M(v)\rangle = \langle A|\bar{q}\Gamma h_v|H(v)\rangle + \frac{1}{2m_Q}\langle A|\bar{q}\Gamma P_- i\slashed{D} h_v|H(v)\rangle$$

$$- i\int d^4x \langle A|T\{L_1(x)\bar{q}\Gamma h_v\}|H(v)\rangle + \mathcal{O}(1/m^2)\,, \tag{9.22}$$

where L_1 are the $1/m$ corrections to the Lagrangian as given in (9.17). In addition, $|M(v)\rangle$ is the state of the heavy meson in full QCD, including all of its mass dependence, while $|H(v)\rangle$ is the corresponding state in the infinite-mass limit.

A contribution to L_1 with a time derivative will become—upon insertion into the T product—a local operator, which in turn means that it could as well be absorbed into the first term by a field redefinition. Using the Hamiltonian without time derivative (such as the one derived from the Foldy–Wouthuysen transformation) will not have any local contributions in the second term, while the closed expression (9.17) and (9.21) will generate such terms, which in the other approach will be contained in the first term. In this way the result for the the physical matrix element will be the same.

9.2.2 Symmetries of HQET

Probably the most important property of HQET for phenomenology are the heavy quark symmetries (HQS) [5]–[7]. These appear in the infinite-mass limit and are not present in full QCD. These symmetries have very simple physical origins and are already manifest in the Lagrangians derived in the last section. In addition, there is another symmetry related to the fact that the construction of HQET requires the introduction of a four-velocity vector, which is not present in full QCD. Thus a change of this velocity vector by an amount of the order $1/m_Q$ may not change the physics. This so-called reparametrization invariance [25]–[28] has interesting consequences, since it relates different orders in the $1/m_Q$ expansion.

9.2.2.1   Flavour Symmetry

The QCD Lagrangian is known to have flavour symmetries in the case where quarks become mass-degenerate: the approximate degeneracy of the up and the down quark leads to the isospin symmetry, in case all quarks are assumed to be massless, QCD has a chiral symmetry, of which the flavour $SU(3)$ symmetry is manifest. The underlying reason is that the interaction of the quarks with the gluons does not depend on the mass, it depends only on the colour charge of the quarks which is defined by putting all quarks into the fundamental representation of colour $SU(3)$.

This still remains true in the infinite mass limit. Once a heavy quark becomes a static source of colour, its flavour becomes irrelevant. To make this explicit, we consider the $b \to c$ HQET Hamiltonian (9.20) for the case of two equal velocities

$$\mathcal{L}^{b,c}_{\mathrm{HQET}} = \bar{b}_v i(v \cdot D) b_v + \bar{c}_v i(v \cdot D) c_v = (\bar{b}_v, \bar{c}_v) \begin{pmatrix} i(v \cdot D) & 0 \\ 0 & i(v \cdot D) \end{pmatrix} \begin{pmatrix} b_v \\ c_v \end{pmatrix} \tag{9.23}$$

which as a manifest $SU(2)$ symmetry: for any unitary 2×2 matrix U we define the transformation

$$\begin{pmatrix} b_v \\ c_v \end{pmatrix}' = U \begin{pmatrix} b_v \\ c_v \end{pmatrix},$$

under which the Lagrangian (9.23) remains invariant. Note that this symmetry relates only heavy quarks moving with the same velocity v.

As a practical application, consider a semileptonic decay of a B meson into a D meson. Assuming both b and c to be heavy, we may look into the point of maximal momentum transfer to the leptons, which is $q^2_{\max} = (m_B - m_D)^2 \approx (m_b - m_c)^2$. Looking at this decay in the rest frame of the B meson (which is also the rest frame of the b quark as $m_b \to \infty$), the final state D meson (as well as the c quark as $m_c \to \infty$) remains at rest at this kinematic point, while the two leptons carry away the energy difference $m_B - m_D \approx m_b - m_c$ in a back-to-back momentum configuration. As a consequence of heavy flavour symmetry, the light degrees of freedom (the light quark(s) and gluons forming the meson) cannot be affected by this transition (at this special kinematic point), which means that their state did not change! We return to this example when discussing weak transition form factors.

9.2.2.2   Spin Symmetry

The second HQS is the so-called heavy quark spin symmetry. It originates from the fact that in gauge theories like QED and QCD the interaction of the spin of a particle is always of the form $\vec{\sigma} \cdot \vec{B}$, where $\vec{B}$ is the corresponding (chromo)magnetic field. However, this is a dimension-five operator, and its coupling constant is $g/(2m_Q)$, which is the QCD analogue of the Bohr magneton of the particle. As a consequence, the spin of a particle decouples in QCD and hence the rotations of the particle's spin become a symmetry.

To make this explicit, we look at the HQET Lagrangian and decompose the heavy quark field into the two spin components. This is achieved by introducing a spin vector s with $s \cdot v = 0$ and $s^2 = -1$ such that we can define the projections

$$h_v^{\pm s} = \frac{1}{2}(1 \pm \gamma_5 \slashed{s})h_v \qquad h_v = h_v^{+s} + h_v^{-s}. \tag{9.24}$$

In terms of these projections we have

$$\mathcal{L} = \bar{h}_v^{+s}(ivD)h_v^{+s} + \bar{h}_v^{-s}(ivD)h_v^{-s} = (\bar{h}_v^{+s}, \bar{h}_v^{-s}) \begin{pmatrix} i(v \cdot D) & 0 \\ 0 & i(v \cdot D) \end{pmatrix} \begin{pmatrix} h_v^{+s} \\ h_v^{-s} \end{pmatrix}. \tag{9.25}$$

Then, similarly as before we have an $SU(2)$ symmetry: for any unitary 2×2 matrix U we define the transformation

$$\begin{pmatrix} h_v^{+s} \\ h_v^{-s} \end{pmatrix}' = U \begin{pmatrix} h_v^{+s} \\ h_v^{-s} \end{pmatrix},$$

under which the HQET Lagrangian remains invariant. Note that this symmetry relates again only heavy quarks moving with the same velocity v.

9.2.2.3 *Consequences of Heavy Quark Symmetries*

These symmetries have a few interesting consequences which are important to make HQET a useful tool, since they constrain the non-perturbative matrix elements of HQET.

The spin symmetry of the heavy quark has the consequence that all the heavy-hadron states moving with the velocity v fall into spin-symmetry doublets as $m_Q \to \infty$. In Hilbert space, this symmetry is generated by operators $S_v(\epsilon)$ as

$$[h_v, S_v(\epsilon)] = i \slashed{\epsilon} \slashed{v} \gamma_5 h_v, \tag{9.26}$$

where ϵ, with $\epsilon^2 = -1$, is the rotation axis. The simplest spin-symmetry doublet in the mesonic case consists of the pseudoscalar meson $H(v)$ and the corresponding vector meson $H^*(v, \epsilon)$, since a spin rotation yields

$$\exp\left(iS_v(\epsilon)\frac{\pi}{2}\right)|H(v)\rangle = (-i)|H^*(v, \epsilon)\rangle, \tag{9.27}$$

where we have chosen an arbitrary phase to be $(-i)$.

Thus the pseudoscalar ground state meson forms a spin-symmetry doublet with the vector ground state meson; assuming that the bottom is heavy we have the doublets

$$\begin{aligned}
|(b\bar{u})_{\mathcal{J}=0}\rangle &= |B^-\rangle & \longleftrightarrow && |(b\bar{u})_{\mathcal{J}=1}\rangle &= |B^{*-}\rangle \\
|(b\bar{d})_{\mathcal{J}=0}\rangle &= |\overline{B}^0\rangle & \longleftrightarrow && |(b\bar{d})_{\mathcal{J}=1}\rangle &= |\overline{B}^{*0}\rangle \\
|(b\bar{s})_{\mathcal{J}=0}\rangle &= |\overline{B}_s\rangle & \longleftrightarrow && |(b\bar{s})_{\mathcal{J}=1}\rangle &= |\overline{B}_s^*\rangle
\end{aligned} \tag{9.28}$$

which become degenerate in the infinite-mass limit.

For baryons, the situation is more complicated, since the two light quarks can have either spin 0 or spin 1. The doublets with u and d quarks are

$$\big|[(ud)_0 Q]_{1/2}\big\rangle = |\Lambda_Q\rangle \qquad |\Lambda_Q \Uparrow\rangle \longleftrightarrow |\Lambda_Q \Downarrow\rangle \tag{9.29}$$

$$\left|[(uu)_1 Q]_{1/2}\right\rangle, \left|[(ud)_1 Q]_{1/2}\right\rangle, \left|[(dd)_1 Q]_{1/2}\right\rangle = \left|\Sigma_Q\right\rangle \tag{9.30}$$

$$\left|[(uu)_1 Q]_{3/2}\right\rangle, \left|[(ud)_1 Q]_{3/2}\right\rangle, \left|[(dd)_1 Q]_{3/2}\right\rangle = \left|\Sigma_Q^*\right\rangle \qquad \left|\Sigma_Q\right\rangle \longleftrightarrow \left|\Sigma_Q^*\right\rangle \tag{9.31}$$

and similar relation for the strange baryons Ξ_b and Ω_b. Note that for the Λ_b baryon, the two spin directions of the Λ_b are the spin-symmetry doublet, since the light degrees of freedom are in a spinless state and thus the spin of the baryon is the heavy-quark spin, at least to leading order in $1/m_b$.

To leading order, the mass of a heavy Q hadron is the mass of the quark m_Q. However, we may expand the hadron mass in terms of the quark mass, which reads for the mesonic ground states

$$m_H = m_Q + \bar{\Lambda} - \frac{\lambda_1 - 3\lambda_2}{2m_Q} + \dots \tag{9.32}$$

$$m_{H^*} = m_Q + \bar{\Lambda} - \frac{\lambda_1 + \lambda_2}{2m_Q} + \dots \tag{9.33}$$

where we have introduced new parameters $\bar{\Lambda}$, λ_1 and λ_2. $\bar{\Lambda}$ is the binding-energy parameter for the heavy hadron

$$\bar{\Lambda} = \frac{\langle 0|q \overleftarrow{ivD} \gamma_5 h_v|\tilde{H}(v)\rangle}{\langle 0|q\gamma_5 h_v|\tilde{H}(v)\rangle}, \tag{9.34}$$

while λ_1 and λ_2 are defined by the HQET matrix elements

$$2m_H\lambda_1 = \langle \tilde{H}(v)|\bar{h}_v(iD^\perp)^2 h_v|\tilde{H}(v)\rangle \tag{9.35}$$

$$2m_H\lambda_2 = \langle \tilde{H}(v)|\bar{h}_v(iD_\mu^\perp)(iD_\nu^\perp)(i\sigma^{\mu\nu})h_v|\tilde{H}(v)\rangle \tag{9.36}$$

where $|\tilde{H}(v)\rangle$ is the pseudoscalar Q meson ground state in the infinite-mass limit.

These parameters have a simple physical interpretation: λ_1 is the kinetic energy induced by the residual motion of the heavy quark, λ_2 corresponds to the interaction of the chromomagnetic moment of the heavy quark induced by the interaction with the chromomagnetic field $\vec{\sigma} \cdot \vec{B}$ produced by the light degrees of freedom. This implies in particular that (taking the b and the c quark to be heavy)

$$m_{B^*}^2 - m_B^2 = m_{D^*}^2 - m_D^2 = 4\lambda_2 + \mathcal{O}(1/m_Q) \tag{9.37}$$

and so we get $\lambda_2 \approx 0.12$ GeV^2 is indeed of the order of Λ_{QCD}^2. Similar relations can be written for the Q baryons [31, 32]; in particular, the chromomagnetic parameter λ_2 vanishes for Λ_Q baryons, since the light degrees are in a spin-0 state and hence cannot induce a chromomagnetic field.

HQS also constrain hadronic matrix elements. In order to extract the corresponding relations, it is useful to write down a representation for the spins in the ground state mesons. Introducing a spinor $v(v,\pm)$ with spin direction $\pm$ for the light anti-quarks and $u(v,\pm)$ for the heavy quark, we may couple the spins to get the total spin of the meson

$$|(b\bar{u})_{\mathcal{J}=0}(v)\rangle \to \frac{1}{\sqrt{2}}\left[u_\alpha(v,+)\bar{v}_\beta(v,-) - u_\alpha(v,-)\bar{v}_\beta(v,+)\right] \propto \left(\gamma_5\frac{\displaystyle\not{v}-1}{2}\right)_{\alpha\beta}$$

$$|(b\bar{u})_{\mathcal{J}=1,M=0}(v)\rangle \to \frac{1}{\sqrt{2}}\left[u_\alpha(v,+)\bar{v}_\beta(v,-) + u_\alpha(v,-)\bar{v}_\beta(v,+)\right] \propto \left(\not{\epsilon}_{\text{long}}\frac{\displaystyle\not{v}-1}{2}\right)_{\alpha\beta}$$

where α and β are spinor indices. Including the proper normalization of the states, we define the representation matrices for these states

$$H(v) = \frac{1}{2}\sqrt{m_H}\gamma_5(\not{v}-1) \qquad \text{for the pseudoscalar meson,} \tag{9.38}$$

$$H^*(v,\epsilon) = \frac{1}{2}\sqrt{m_H}\not{\epsilon}(\not{v}-1) \qquad \text{for the vector meson,} \tag{9.39}$$

where the two indices of the matrices correspond to the indices of the heavy quark and the light anti-quark, respectively, and ϵ is the polarization vector of the vector meson.

We may now use these representation matrices to exploit the consequences of spin symmetry in a very simple fashion. We look at a transition current of the form $\bar{h}_{v'}\Gamma h_v$ induced, e.g. by a weak transition (such as a $b \to c$ semileptonic process). Spin symmetry implies that the spin of the heavy quark in the current is the same as the one of the quark inside the meson, which means that the heavy-quark index of the representation matrix has to hook directly to the Dirac matrix Γ in the current. Thus for a $0^- \to 0^-$ transition we have

$$\langle M(v')|\bar{h}_{v'}\Gamma h_v|M(v)\rangle = \text{Tr}\left[\overline{H}(v')\Gamma H(v)\mathcal{M}(v,v')\right] \tag{9.40}$$

where the two light-quark indices of $\overline{H}(v')\Gamma H(v)$ will be contracted with a Dirac-matrix valued function $\mathcal{M}(v,v')$ of v and v' which describes the dynamics of the light quarks in the transition. This matrix can be decomposed into the basis of the sixteen Dirac matrices, so we can write (note that due to parity conservation in strong interactions there are no contributions with γ_5 and $\gamma_\mu\gamma_5$)

$$\mathcal{M}(v,v') = \mathbf{1}\,\xi_1(v \cdot v') + \not{v}\,\xi_2(v \cdot v') + \not{v}'\,\xi_3(v \cdot v') + \not{v}\not{v}'\,\xi_4(v \cdot v') \tag{9.41}$$

with scalar functions ξ_i. Inserting this into (9.40) we see that for any Γ this collapses into

$$\langle M(v')|\bar{h}_{v'}\Gamma h_v|M(v)\rangle = \text{Tr}\left[\overline{H}(v')\Gamma H(v)\right]\xi(v \cdot v') \tag{9.42}$$

with $\xi(v \cdot v') = \xi_1(v \cdot v') + \xi_2(v \cdot v') - \xi_3(v \cdot v') - \xi_4(v \cdot v')$

Likewise we can discuss the transitions $0^- \to 1^-$ and $1^- \to 1^-$ between ground state mesons. Spin symmetry tells us that the function $\mathcal{M}(v,v')$ for the light degrees of freedom is the same in all cases, and hence:

> *Any transition within the ground-state spin flavour multiplet $\mathcal{H}(v)$ to the ground-state multiplet $\mathcal{H}(v')$, where $\mathcal{H}(v)$ denotes either $H(v)$ or $H^*(v,\epsilon)$ is described by a single non-perturbative function $\xi(v \cdot v')$.*

The function ξ is called the Isgur–Wise (IW) function, and relation (9.42) is one of the Wigner–Eckart Theorems of spin symmetry and can be written as

$$\langle \mathcal{H}(v') | \bar{h}_{v'} \Gamma h_v | \mathcal{H}(v) \rangle = \xi(v \cdot v') \, \mathrm{Tr} \left\{ \overline{\mathcal{H}}(v') \Gamma \mathcal{H}(v) \right\}. \tag{9.43}$$

This relation has remarkable consequences. Assuming that both b and c are heavy, we find that the six form factors describing the semileptonic transitions $B \to D$ and $B \to D^*$ in the infinite-mass limit for both b and c quark reduce to a single one, the IW function. Furthermore, the current $\bar{b} \gamma_\mu b$ is conserved in pure QCD, which translates into a normalization statement for the IW function

$$\xi(v \cdot v' = 1) = 1 \tag{9.44}$$

where the physical argument for this normalization has been given in the last paragraph of sec. 9.2.2.1. Note that $v \cdot v' = 1$ corresponds exactly to the point q^2_{max} of maximal recoil to the leptons also discussed there.

For phenomenological applications these symmetries are very useful, however, only once the corrections to the symmetry limit can be somehow handled. There are two sources of corrections, which are on the one hand the radiative corrections through hard gluons, and on the other, those induced by subleading terms in the $1/m_Q$ expansion.

We will discuss the latter and consider the $1/m_Q$ corrections to the normalization statement (9.44) that originated from the conservation of the heavy quark current, which in turn is related to HQS. In this case we can apply a general theorem originally derived by Ademollo and Gatto [29] to the case of HQS. The theorem, derived by Luke [30] in the context of HQS, states:

> *In the presence of explicit symmetry breaking, the matrix elements of the currents that generate the symmetry are normalized up to terms which are second-order in the symmetry-breaking interaction.*

For the case of HQS, the argument can be outlined in a simple way, taking as an example the $b \to c$ case. The relevant symmetry is the heavy-flavour symmetry between b and c in the case $m_{b,c} \to \infty$. This symmetry is an $SU(2)$ symmetry and is generated by three operators $Q_\pm$ and Q_3, where[3]

$$Q_+ = \int d^3x \, \bar{b}_v(x) \gamma_0 c_v(x), \quad Q_- = \int d^3x \, \bar{c}_v(x) \gamma_0 b_v(x),$$

$$Q_3 = \int d^3x \, (\bar{b}_v(x) \gamma_0 b_v(x) - \bar{c}_v(x) \gamma_0 c_v(x)),$$

$$[Q_+, Q_-] = Q_3, \qquad [Q_+, Q_3] = -2Q_+, \qquad (Q_+)^\dagger = Q_-. \tag{9.45}$$

---

[3] We set $x_0 = 0$ in the following relations, such that we have to compute equal-time (anti)commutators. In case the symmetry is exact, the operators do not depend on x_0 anyway, but once the symmetry is broken, there will be a small dependence on x_0.

Let us denote the ground-state flavour symmetry multiplet by $|B\rangle$ and $|D\rangle$. The operators then act in the following way:

$$Q_3|B\rangle = |B\rangle, \qquad Q_3|D\rangle = -|D\rangle,$$
$$Q_+|D\rangle = |B\rangle, \qquad Q_-|B\rangle = |D\rangle,$$
$$Q_+|B\rangle = Q_-|D\rangle = 0. \tag{9.46}$$

The Hamiltonian of this system has a $1/m_Q$ expansion, which is decomposed into a symmetric and a symmetry breaking part

$$\begin{aligned} H &= H_0^{(b)} + H_0^{(c)} + \frac{1}{2m_b}H_1^{(b)} + \frac{1}{2m_c}H_1^{(c)} + \cdots \\ &= H_0^{(b)} + H_0^{(c)} + \frac{1}{2}\left(\frac{1}{2m_b} + \frac{1}{2m_c}\right)(H_1^{(b)} + H_1^{(c)}) \\ &\quad + \frac{1}{2}\left(\frac{1}{2m_b} - \frac{1}{2m_c}\right)(H_1^{(b)} - H_1^{(c)}) + \cdots \\ &= H_{symm} + H_{break}. \end{aligned} \tag{9.47}$$

Note that the symmetry breaking term does not commute any more with Q_+ but it still commutes with Q_3 (which only means that we can still distinguish B and D). Thus to order $1/m_Q$ we still have common eigenstates of H and Q_3, which we denote by $|\tilde{B}\rangle$ and $|\tilde{D}\rangle$. Sandwiching the commutation relation, we obtain

$$\begin{aligned} 1 &= \langle\tilde{B}|Q_3|\tilde{B}\rangle = \langle\tilde{B}|[Q_+, Q_-]|\tilde{B}\rangle \\ &= \sum_n \left[\langle\tilde{B}|Q_+|\tilde{n}\rangle\langle\tilde{n}|Q_-|\tilde{B}\rangle - \langle\tilde{B}|Q_-|\tilde{n}\rangle\langle\tilde{n}|Q_+|\tilde{B}\rangle\right] \\ &= \sum_n \left[|\langle\tilde{B}|Q_+|\tilde{n}\rangle|^2 - |\langle\tilde{B}|Q_-|\tilde{n}\rangle|^2\right], \end{aligned} \tag{9.48}$$

where the $|\tilde{n}\rangle$ form a complete set of states of the Hamiltonian $H_{symm} + H_{break}$. The matrix elements may be written as

$$\langle\tilde{B}|Q_\pm|\tilde{n}\rangle = \frac{1}{E_B - E_n}\langle\tilde{B}|[H_{break}, Q_\pm]|\tilde{n}\rangle, \tag{9.49}$$

where E_B and E_n are the energies of the states $|\tilde{B}\rangle$ and $|\tilde{n}\rangle$, respectively. In the case $|\tilde{n}\rangle = |\tilde{D}\rangle$ the matrix element on the left-hand side will be of order unity, since both the numerator and the energy difference in the denominator are of the order of the symmetry breaking. For all other states, the energy difference in the denominator is non-vanishing in the symmetry limit, and hence this difference is of order unity; thus the matrix element for these states will be of the order of the symmetry breaking. From this we conclude that

$$\langle\tilde{B}|Q_+|\tilde{D}\rangle = 1 + \mathcal{O}\left[\left(\frac{1}{2m_b} - \frac{1}{2m_c}\right)^2\right]. \tag{9.50}$$

For simplicity we have used states normalized to unity.

In order to relate this to the form factor normalization, we observe that the generators are obtained from integrating over the time components of the current; the general expression for the matrix elements reads ($q = p - p'$)

$$\langle B(p)|\bar{b}\gamma_\mu c|D(p)\rangle = \frac{1}{\sqrt{4v_0 v_0'}}\left((v_\mu + v'_\mu)f_+(q^2) + (v_\mu - v'_\mu)f_-(q^2)\right). \tag{9.51}$$

Switching to HQET for b and c, taking the time component and integrating over $\vec{x}$ yields

$$\int d^3\vec{x}\,\langle B(p)|\bar{b}_v(x)\gamma_0 c_v(x)|D(p)\rangle = (2\pi)^3\delta^3(\vec{p} - \vec{p}')f_+(q^2_{\text{max}}) \tag{9.52}$$

where the δ function appears because we are using momentum eigenstates, thus it corresponds to the above normalization to unity. Furthermore, $q^2_{\text{max}} = (m_B - m_D)^2$ is the maximal momentum transfer in the $B \to D$ transition, corresponding to the point $v = v'$. Comparing to (9.50) we find

$$f_+(q^2_{\text{max}}) = 1 + \mathcal{O}\left[\left(\frac{1}{2m_b} - \frac{1}{2m_c}\right)^2\right]. \tag{9.53}$$

Note that the statement on the corrections only holds for the form factors that are normalized due to the symmetry; we also have $f_-(q^2) = 0$ from HQS, however, including corrections this means

$$f_-(q^2_{\text{max}}) = \mathcal{O}\left[\frac{1}{2m_b} - \frac{1}{2m_c}\right]. \tag{9.54}$$

9.2.2.4 *Reparametrization Invariance*

Finally, there is another symmetry in HQET called reparametrization invariance (RPI). It originates from the fact that our starting point was full QCD, which is a Lorentz invariant theory. Clearly, when introducing the velocity vector v we explicitly break Lorentz invariance by fixing a time, like direction which, to some extent, is arbitrary and could be also varied sightly. To this end, an HQET constructed with v and an HQET constructed with $v' = v + \delta v$ should give the same physical results [25]–[28].

In order to study the consequences of this simple fact, we write down the variation δ_{RPI} of the relevant quantities under a small change in the velocity

$$v \to v + \delta v\,, \qquad (v + \delta v)^2 = 1 \text{ and thus } v \cdot \delta v = 0\,,$$
$$h_v \to h_v + \frac{\delta\slashed{v}}{2}\left(1 + P_-\frac{1}{2m_Q + ivD}i\slashed{D}\right)h_v\,,$$
$$iD \to iD - m_Q\delta v\,. \tag{9.55}$$

In particular the last relation, which originates from the splitting of the heavy-quark momentum, leads to the observation that the transformation (9.55) relates different orders in the $1/m_Q$ expansion.

This can be easily illustrated using the Lagrangian as an example. We start from the expression (9.13) for the action of the heavy quark after integrating out the the small-component field H_v. This (non-local) expression is invariant under (9.55). Expanding (9.13) in local operators according to (9.17) shows that (9.55) actually relates terms of subsequent orders such that

$$\delta_{\mathrm{RPI}} \mathcal{L}_{1/m_Q - \mathrm{Expansion}} = \mathcal{O}(1/m_Q^{N+2}) \tag{9.56}$$

since (9.17) includes all terms up to and of order $1/m_Q^{N+1}$. Looking at the leading term we find

$$\delta_{\mathrm{RPI}} \bar{h}_v (ivD) h_v = \bar{h}_v (i\delta v D) h_v,$$

which is exactly cancelled by the variation of the first subleading term

$$\delta_{\mathrm{RPI}} \frac{1}{2m_Q} \bar{h}_v (i\slashed{D}^\perp)^2 h_v = -\bar{h}_v (i\delta v D) h_v$$

and as a consequence we have[4]

$$\delta_{\mathrm{RPI}} \left(\bar{h}_v (ivD) h_v + \frac{1}{2m_Q} \bar{h}_v (iD^\perp)^2 h_v \right) = \mathcal{O}(1/m_Q^2)$$

These relations are all tree-level relations; however, RPI has to hold even when including QCD corrections in HQET, which means the relations derived from RPI should hold to all order in α_s. For the Lagangain this means that we may derive relations between the renormalization constants of the operators appearing in (9.17) which are true to any oder in α_s. In particular it means for renormalization constants of the first few terms

$$Z_h \bar{h}_v (ivD) h_v + (Z_h c_1) \frac{1}{2m_Q} \bar{h}_v (iD^\perp)^2 h_v$$

$$= Z_h \left(\bar{h}_v (ivD) h_v + c_1 \frac{1}{2m_Q} \bar{h}_v (iD^\perp)^2 h_v \right) \tag{9.57}$$

where Z_h is the renormalization constant of the the static heavy-quark field. Thus RPI fixes the renormalization constant of the kinetic energy term to be $c_1 \equiv 1$.

9.2.3 HQET at One Loop

Up to now, all discussions refer to the tree-level expressions. We have set up an expansion in $\Lambda_{\mathrm{QCD}}/m_Q$ which is, however, only one of the small parameters we can expand in. To become a useful tool, the perturbative QCD corrections must also be taken into account.

---

[4] The antisymmetric combination $(i\sigma^{\mu\nu})(iD_\mu^\perp)(iD_\nu^\perp)$ is reparametrization invariant.

The strong coupling constant $\alpha_s(m_Q)$ taken at the scale $\mu \geq 1$ GeV constitutes another small parameter that may serve as an expansion parameter. In particular, the heavy quark-mass scale $\mu = m_Q$ is large enough to warrant a perturbative expansion. This has the advantage that many contributions can be computed perturbatively, in particular, the matching between HQET and full QCD. The next section discusses the underlying technology and studies the one-loop diagrams.

Readers interested in the technical aspects of perturbative calculations in HQET will find all details in Grozin [17], which includes the relevant master integrals even up to two loops.

9.2.3.1   *The Feynman Rules of HQET*

This chapter assumes the reader is to some extent familiar with the Feynman rules of QCD, including the discussion of gauge fixing, so we will not repeat the standard technology of calculations within QCD are not repeated here.

However, to compute within HQET, we need to set up the Feynman rules. These are derived from the HQET Lagrangian (9.18)

$$\mathcal{L}_{\text{HQET}} = \bar{h}_v i(v \cdot D) h_v = \bar{h}_v i(v \cdot \partial) h_v + ig\bar{h}_v i(v \cdot A) h_v \qquad (9.58)$$

The propagator can be read off from the first term, while the heavy quark–gluon coupling is encoded in the second term.

The recipe to obtain the propagator from the first term is to invert the distribution appearing between the two fields according to

$$(v \cdot \partial) P(x) = \delta^4(x). \qquad (9.59)$$

Fourier transforming this relation yields

$$P(x) = \int \frac{d^4k}{(2\pi)^4} \frac{1}{v \cdot k + i\epsilon} e^{-ikx}$$

where we have already fixed the boundary conditions by adding a small imaginary part $i\epsilon$, ensuring that particles propagate into forward time direction. The interpretation of this propagator becomes evident by performing the k integration in the rest frame $v = (1,0,0,0)$; we get

$$P(x) = \theta(x_0)\delta^3(\vec{x}), \qquad (9.60)$$

which is the propagator of a static quark sitting at the origin. In order to insert this into a general Feynman diagram we still have to multiply this by the projector P_+ defined in (9.5).

The second term in (9.58) yields the coupling of a static quark to the gluon field. The resulting Feynman rule has the same form as the usual one for the quark-gluon coupling, with the matrix γ_μ replaced by v_μ, which reflects the heavy quark spin symmetry,

Fig. 9.1 shows the two resulting additional Feynman rules; here k denotes the residual momentum of the heavy quark moving with velocity v.

9.2.3.2 One-Loop Diagrams 1: Quark Self Energy

We are now ready to compute Feynman diagrams. As in full QCD there is a set of divergent diagrams, and the handling of these divergencies requires renormalization. We discuss this here for a few examples at the one-loop level.

We start with a sample calculation of the self energy; Figure 9.2(a) is the self energy in full QCD, while Figure 9.2(b) shows the corresponding diagram in HQET. The expression in full QCD (fig. 9.2 (a)) is well known and reads

$$\Sigma_{\mathrm{QCD}}(p) = -ig^2 T^a T^a \mu^{4-D} \int \frac{d^D l}{(2\pi)^D} \frac{1}{(l^2+i\epsilon)} \frac{\gamma_\mu(\not{p}+\not{l}+m_Q)\gamma^\mu}{(p+l)^2-m_Q^2+i\epsilon}. \tag{9.61}$$

Making use of the Feynman rules we get the expression corresponding to (b)

$$\Sigma(v\cdot k) = -ig^2 T^a T^a \mu^{4-D} \int \frac{d^D l}{(2\pi)^D} \frac{1}{(l^2+i\epsilon)(v\cdot k+v\cdot l+i\epsilon)} P_+, \tag{9.62}$$

where we have anticipated a divergence in $D=4$ and regularize, the diagrams by dimensional regularization. As usual, the factor μ^{4-D} is introduced to keep the dimension of Σ fixed as D varies.

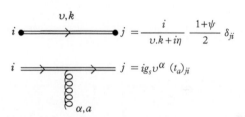

Figure 9.1 *Feynman rules of HQET. All other elements are the same as in full QCD. i and j are color indices, k is the residual momentum of the heavy quark moving with the velocity v, and $g_S = g$ is the strong coupling constant.*

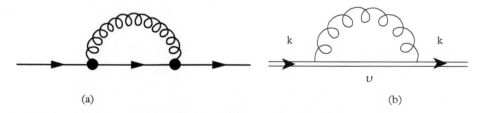

Figure 9.2 *One-loop self energy diagram of a light and a heavy quark.*

In order to evaluate (9.62), we quote a useful relation which we use to combine denominators of propagators

$$\frac{1}{A^n B^m} = 2^m \frac{\Gamma(m+n)}{\Gamma(n)\Gamma(m)} \int\limits_0^\infty d\lambda \frac{\lambda^{m-1}}{(A+2\lambda B)^{m+n}}. \tag{9.63}$$

This relation also holds for non-integer m and n. Using this we can combine the denominators in (9.62) into

$$\Sigma(v \cdot k) = -2ig^2 \mu^{4-D} C_F \int\limits_0^\infty d\lambda \int \frac{d^D l}{(2\pi)^D} \frac{1}{(l^2 + 2\lambda[v \cdot k + v \cdot l] + i\epsilon)^2} \tag{9.64}$$

where we inserted $T^a T^a = C_F \mathbf{I}$ where $C_F = 4/3$. In order to apply the one-loop master formula of dimensional regularization

$$\int \frac{d^D l}{(2\pi)^D} \frac{(l^2)^\alpha}{(l^2 - M^2)^\beta} = (-1)^{\alpha+\beta} \frac{i}{2^D \pi^{D/2}} (M^2)^{\alpha-\beta+D/2} \frac{\Gamma(\alpha+D/2)\Gamma(\beta-\alpha-D/2)}{\Gamma(D/2)\Gamma(\beta)} \tag{9.65}$$

we need to shift the integration variable $l \to l - \lambda v$ which removes the term linear in l in the denominator, leaving us with

$$\Sigma(v \cdot k) = -ig^2 \mu^{2\varepsilon} \frac{8}{3} \int\limits_0^\infty d\lambda \int \frac{d^D l}{(2\pi)^D} \frac{1}{(l^2 - \lambda^2 + \lambda v \cdot k + i\epsilon)^2}, \tag{9.66}$$

where we have defined $D = 4 - 2\varepsilon$. Performing the integration over the loop momentum with the help of (9.65) we find

$$\Sigma(v \cdot k) = C_F \frac{\alpha_s}{2\pi} \Gamma(\varepsilon) \int\limits_0^\infty d\lambda \left(\frac{4\pi \mu^2}{\lambda^2 - 2\lambda v \cdot k}\right)^\varepsilon. \tag{9.67}$$

For the renormalization, we are interested in the divergence as $D \to 4$ (or $\varepsilon \to 0$), which manifests itself as a simple pole

$$\Sigma(v \cdot k) = C_F \frac{\alpha_s}{2\pi} \frac{1}{\varepsilon} (v \cdot k) + \text{finite terms}. \tag{9.68}$$

Renormalization proceeds in the usal way. We insert the self energy into the heavy-quark propagator and get

$$S^{(1)}_{\mathrm{HQET}}(v \cdot k) = \frac{i}{(v \cdot k)} + \frac{i}{(v \cdot k)} (-i\Sigma(v \cdot k)) \frac{i}{(v \cdot k)} + \cdots \tag{9.69}$$

$$= \left(1 + C_F \frac{\alpha_s}{2\pi} \frac{1}{\varepsilon}\right) \frac{i}{(v \cdot k)} + \cdots$$

from which we can read off the wave function renormalization constant of HQET (in the $\overline{MS}$ scheme)

$$Z_{\mathrm{HQET}} = 1 + C_F \frac{\alpha_s}{2\pi} \frac{1}{\varepsilon}. \tag{9.70}$$

This can be compared to the result in full QCD. A similar calculation yields for the wave function renormalization of the quark field in full QCD

$$Z_Q = 1 - C_F \frac{\alpha_s}{4\pi} \frac{1}{\varepsilon}. \tag{9.71}$$

We note that the two renormalization constants are different, which is not a surprise, since the UV behaviour of the two theories is different. In fact, the divergencies in HQET are related to logarithmic mass dependencies of full QCD, which become divergent as $m_Q \to \infty$. For our case this can be made explicit by looking at the result in full QCD obtained from (9.61), which reads (for $p^2 < m^2$)

$$\Sigma(p) = \not{p} \left[-\frac{\alpha_s}{4\pi} C_F \frac{1}{\bar{\varepsilon}} - \frac{\alpha_s}{4\pi} C_F \frac{(m_Q^2 + p^2)(m_Q^2 - p^2)}{(p^2)^2} \ln\left(\frac{m^2 - p^2}{m^2} \right) \right] + \cdots \tag{9.72}$$

where we have used

$$\frac{1}{\bar{\varepsilon}} = \frac{1}{\varepsilon} - \gamma + \ln 4\pi,$$

and the ellipses denote non-logarithmic terms and contributions proportional to m_Q. Inserting $p = mv + k$ and keeping only the leading terms yields

$$\Sigma(v) = \not{v} \left[-\frac{\alpha_s}{4\pi} m_Q C_F \frac{1}{\bar{\varepsilon}} + \frac{\alpha_s}{2\pi} C_F (v \cdot k) \ln\left(\frac{m_Q}{v \cdot k} \right) \right] + \cdots . \tag{9.73}$$

The first term is taken care of by the renormalization of full QCD, i.e. the pole term defines the field renormalization of the full quark field. However, (9.73) shows that the finite term in (9.72) develops in the heavy quark limit a logarithmic divergence, the prefactor of which defines the renormalization constant of the static heavy quark field.

9.2.3.3 *One-Loop Diagrams 2: The $b \to c$ Current*

Next we discuss the $b \to c$ vector current and consider the QCD radiative corrections at one loop. Starting at a high scale above the b quark mass, we compute the one-loop diagrams shown in Figure 9.3(a), together with the corresponding diagrams with self-energy insertions in the external legs. Adding the three contributions and including the proper renormalization, the one-loop result is UV finite, in other words, the current $\bar{b}\gamma_\mu c$ does not have an anomalous dimension. This is actually true at all orders, since this current is conserved in the limit of vanishing masses.

Since the anomalous dimension of this current vanishes, we do not encounter any large logarithms of the form $(\alpha_s/\pi)\ln(M_W^2/m_b^2)$ when we run down to the bottom mass

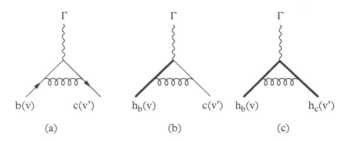

Figure 9.3 *Feynman diagrams for the $b \to c$ vertex corrections in full QCD (a), in the theory with a static b quark (b), and in a theory with static b and c quarks (c). Thick lines denote quarks in HQET.*

scale. At $\mu = m_b$ we have to match the vector current $V_\mu^{(b \to c)}$ of full QCD to operators in a theory where we use HQET for the b quark. We have schematically

$$V_\alpha^{(b \to c)} = \sum_i C_i^{(0)}(\mu) \mathcal{J}_{i,\alpha}^{(b \to c)} + \frac{1}{2m_b} \sum_k C_k^{(1)}(\mu) O_{k,\alpha}^{(b \to c)} + \cdots \qquad (9.74)$$

where $\mathcal{J}_{i,\mu}^{(b \to c)}$ and $O_{k,\mu}^{(b \to c)}$ are local operators and $C_i^{(0)}$ and $C_k^{(1)}(\mu)$ are (computable) Wilson coefficients, and the ellipses denote even higher orders in the $1/m_b$ expansion. In our example we consider only the leading term, which is expressed in terms of two operators

$$\mathcal{J}_{1,\mu}^{(b \to c)} = \bar{c}\gamma_\mu h_v, \qquad (9.75)$$

$$\mathcal{J}_{2,\mu}^{(b \to c)} = \bar{c} h_v \, v_\mu, \qquad (9.76)$$

The relations (9.74) are operator relations, and in order to compute the matching we may use any states we prefer. In this case we want to compute the perturbative corrections, and thus it is convenient to use on-shell states for the b and c quarks. Furthermore, since we are interested in scales well above the charm quark mass, we compute with a massless charm quark.

Computing the one-loop diagrams in full QCD shown in Figure 9.3(a) (together with the diagrams with self-energy insertions in the external legs) and expanding in the result in $1/m_b$ yields the result

$$\langle V_\mu^{(b \to c)} \rangle = \left(1 + \frac{\alpha_s}{2\pi} \left[\ln \frac{m_b^2}{\lambda^2} - \frac{11}{6} \right] \right) \gamma_\mu + \frac{2\alpha_s}{3\pi} v_\mu. \qquad (9.77)$$

As stated, the result is UV finite, but we had to introduce an infrared regulator λ which is a small gluon mass. This is due to the fact that we are using on-shell 'free' quark states in the calculation; if we could compute the matrix element with hadronic states, these IR singularities would be absent.

The next step in the matching procedure is to compute the corresponding diagrams in an HQET where the b quark is replaced by a static quark. Computing the one-loop contribution shown in Figure 9.3(b) (together with the diagrams with self-energy insertions in the external legs) and performing the proper renormalization of the heavy and light quark fields we obtain (using again a small gluon mass to regulate the infrared divergence of the amplitudes)

$$\langle \mathcal{F}_{1,\mu}^{(b \to c)} \rangle = \left(1 + \frac{\alpha_s}{2\pi}\left[\frac{1}{\bar{\varepsilon}} + \ln\frac{\mu^2}{\lambda^2} - \frac{5}{6}\right]\right)\gamma_\mu. \tag{9.78}$$

Note that this result is UV divergent which is expected, since we changed the high-energy behaviour of the theory by switching to a static b quark. Consequently we need to renormalize the current operator; in the $\overline{MS}$ scheme. This just amounts to removing the $1/\bar{\varepsilon}$ pole.

We can now read off the coefficients in (9.74) by taking the corresponding matrix elements of (9.74); we obtain

$$C_1^{(0)}(\mu) = 1 + \frac{\alpha_s}{2\pi}\left[\ln\frac{m_b^2}{\mu^2} - \frac{8}{3}\right] \tag{9.79}$$

$$C_2^{(0)}(\mu) = \frac{2\alpha_s}{3\pi}. \tag{9.80}$$

Note that the IR regulator λ has dropped out which is a general feature of a matching calculation. This is because the IR behaviour in the full and the effective theory have to be the same, once the effective theory is properly constructed.

The UV divergence in the effective theory is related to the mass dependence in the full theory, which can be seen by comparing (9.77) to (9.78), since the prefactor of the $\ln m_b^2$ term in (9.77) is the same as the coefficient in front of the $1/\bar{\varepsilon}$ pole in (9.78). This fact allows us to make use of RG methods in order to resum logarithms of the mass m_b. However, these logarithms are of the form $\ln(m_b^2/m_c^2)$ once we scale down to the charm quark mass m_c, and the RG methods allow us to perform a resummation of terms of order $(\alpha_s/\pi)^n\ln^n(m_b^2/m_c^2)$, which makes sense as soon as the log is so large that it overwhelms the α_s supression. We assume this as we go on, although the constant term $-3/8$ in (9.79) is numerically comparable to the term with the logarithm.

In order to obtain the RG equation, we note that the left-hand side of (9.74) is independent of the scale μ. The μ dependence on the right-hand side originates from the fact that we decided to shift the contributions of scales between m_b and $\mu \leq m_b$ into the Wilson coefficient, while the pieces from scales below μ are still contained in the matrix element of the operator $\mathcal{F}_{1,\mu}^{(b \to c)}$. This observation leads to

$$0 = \mu\frac{d}{d\mu}\langle V_\alpha^{(b \to c)}\rangle = \left(\mu\frac{d}{d\mu}C_1^{(0)}(\mu)\right)\langle \mathcal{F}_{1,\alpha}^{(b \to c)}\rangle_\mu + C_1^{(0)}(\mu)\left(\mu\frac{d}{d\mu}\langle \mathcal{F}_{1,\alpha}^{(b \to c)}\rangle_\mu\right). \tag{9.81}$$

The logarithmic derivative acting on the matrix element of the current is the anomalous dimension, which is computed from the divergence occurring in (9.78). In our one-loop case the anomalous dimension is given by

$$\left(\mu \frac{d}{d\mu} \langle \mathcal{F}_{1,\alpha}^{(b \to c)} \rangle_\mu \right) = -\gamma_{h_b \to c} \langle \mathcal{F}_{1,\alpha}^{(b \to c)} \rangle_\mu = \frac{\alpha_s}{\pi} \langle \mathcal{F}_{1,\alpha}^{(b \to c)} \rangle_\mu \qquad (9.82)$$

This translates into an evolution equation for the coeffiient $C_1^{(0)}(\mu)$

$$\left(\mu \frac{d}{d\mu} - \gamma_{h_b \to c} \right) C_1^{(0)}(\mu) = 0. \qquad (9.83)$$

The coefficient $C_1^{(0)}(\mu)$ is evaluated as a power series in α_s, and hence the μ dependence has actually two sources: aside from the explicit dependence (see (9.79)) there is also the μ dependence of α_s. To make this explicit, we write

$$C_1^{(0)}(\mu) = C_1^{(0)}(\alpha_s, \mu),$$

and write the total derivative as

$$\mu \frac{d}{d\mu} C_1^{(0)}(\mu) = \left(\mu \frac{\partial}{\partial \mu} + \beta(\alpha_s) \frac{\partial}{\partial \alpha_s} \right) C_1^{(0)}(\alpha_s, \mu)$$

where we have introduced the QCD β function as

$$\mu \frac{d}{d\mu} \alpha_s(\mu) = \beta(\alpha_s(\mu)) = -2\alpha_s(\mu) \frac{\alpha_s(\mu)}{4\pi} \left(11 - \frac{2}{3} n_f \right) + \cdots, \qquad (9.84)$$

where n_f is the number of active flavours. Inserting all this yields

$$\left(\mu \frac{\partial}{\partial \mu} + \beta(\alpha_s) \frac{\partial}{\partial \alpha_s} - \gamma_{h_b \to c} \right) C_1^{(0)}(\alpha_s, \mu) = 0. \qquad (9.85)$$

The general solution of this equation is given by

$$C_1^{(0)}(\alpha_s(\mu), \mu) = \exp\left(-\int_{\alpha_s(\mu)}^{\alpha_s(m_b)} \frac{\gamma_{h_b \to c}(a)}{\beta(a)} da \right) C_1^{(0)}(\alpha_s(m_b), m_b), \qquad (9.86)$$

which gives the coefficient $C_1^{(0)}(\alpha_s(\mu), \mu)$ in terms of the initial value $C_1^{(0)}(\alpha_s(m_b), m_b)$ obtained from the matching calculation.

In order to obtain the leading log result, we insert the one-loop results for $\gamma_{h_b \to c}$ and β and use the tree-level value $C_1^{(0)}(\alpha_s(m_b), m_b) = 1$ for the matching coefficient, which yields finally

$$C_1^{(0)}(\alpha_s(\mu), \mu) = \left(\frac{\alpha_s(m_b)}{\alpha_s(\mu)} \right)^{-6/25}. \qquad (9.87)$$

Expanding this result in $\alpha_s(\mu)$ using the one-loop result for the running coupling reproduces the logarithmic term in (9.74).

Scaling further down we eventually arrive at the charm quark mass m_c. Assuming that we can also treat the charm quark as a heavy quark, we may again replace the charm quark by a static quark. This allows us to scale further down to scales below m_c, however, at some point we arrive at $\mu = \Lambda_{QCD}$ where we cannot compute perturbatively any more.

We again look at the vector current, although, since we now compute in HQET only, due to spin symmetry the results will also hold for other currents. At one loop, we need to compute Figure 9.3(c) and match it to the result obtained in the theory where only the b quark is taken to be static.

Figure 9.3(c) contains also an UV divergence, which is related to the logarithmic m_c dependence of (b), if we had included the charm mass in the calculation. Thus we need to include a renormalization of the heavy-to-heavy current, which due to this renormalization has an anomalous dimension, for which we find at one loop

$$\gamma_{h_b \to h_c}(v \cdot v') = \frac{4\alpha_s}{3\pi}[(v \cdot v')r(v \cdot v') - 1] \tag{9.88}$$

with

$$r(x) = \frac{1}{\sqrt{x^2 - 1}}\ln\left(x + \sqrt{x^2 - 1}\right).$$

This result is remarkable, since usually anomalous dimensions do not depend on kinematic variables. However, the velocities in HQET are external variables and thus this is not a problem. We also note that at $v = v'$ the anomalous dimension vanishes, which is necessary, since this current is a generator of HQS at this kinematic point and thus cannot have an anomalous dimension.

The running below m_c is governed by the RGE

$$\left(\mu\frac{\partial}{\partial\mu} + \beta(\alpha_s)\frac{\partial}{\partial\alpha_s} - \gamma_{h_b \to h_c}(v \cdot v')\right)\tilde{C}^{(0)}(\alpha_s, \mu) = 0, \tag{9.89}$$

where the number of active flavours is now three. In the effective theory with both b and c as heavy quarks the matrix element of the current is (up to trivial factors) the IW function, and thus we can write our result as a renormalization of this function

$$\xi(v \cdot v') = \zeta(v \cdot v', m_b, m_c, \mu)\xi_0(v \cdot v', \mu) \tag{9.90}$$

where $\xi_0(v \cdot v', \mu)$ is the bare IW function and

$$\zeta(v \cdot v', m_b, m_c, \mu) = \left(\frac{\alpha_s(m_b)}{\alpha_s(m_c)}\right)^{-6/25}\left(\frac{\alpha_s(m_c)}{\alpha_s(\mu)}\right)^{(8/27)[(v \cdot v')r(v \cdot v') - 1]} \tag{9.91}$$

where the first factor originates from the running from m_b to m_c, while the second comes from the running from m_c to some small scale μ. This result has been derived first in [33].

In full QCD, the amplitude for a $b \to c$ transition via the $\bar{c}\gamma_\mu b$ current can be analytically continued to values of $q^2 \geq (m_B + m_D)^2$ which correspond to the creation of a B and a D meson by the current. In terms of the velocities this is the region where $v \cdot v' \leq -1$, in which case the anomalous dimension (9.88) picks up an imaginary part [34]. At first glance this is puzzling however, it is related to the coulombic phases which appear once the two particles are both in the final state and can re-scatter through soft gluons.

9.2.4 Applications to Phenomenology

This section discusses a few phenomenological results obtained from HQET. The most prominent result is the fact that due to HQS all transitions between ground-state heavy mesons mediated by a bilinear quark current are given in terms of a single form factor, the IW function introduced in (9.43). Assuming that both b and c quarks are heavy, we can consider the decays $B \to D\ell\bar{\nu}$ and $B \to D^*\ell\bar{\nu}$ for which the hadronic matrix element of the current $\bar{c}\gamma_\mu(1 - \gamma_5)b$ is exactly of the form of (9.43).

For heavy quarks it is convenient to use the four velocities of the hadrons v and v' as kinematic variables, which are at leading order the same as the velocities of the heavy quarks. The general parametrization of the matrix elements requires in total six form factors, which can be defined as

$$\langle D(v')|\bar{c}\gamma_\mu b|B(v)\rangle = \sqrt{m_B m_D}\left[\xi_+(y)(v_\mu + v'_\mu) + \xi_-(y)(v_\mu - v'_\mu)\right], \qquad (9.92)$$

$$\langle D^*(v',\epsilon)|\bar{c}\gamma_\mu b|B(v)\rangle = i\sqrt{m_B m_{D^*}}\xi_V(y)\varepsilon_{\mu\alpha\beta\rho}\epsilon^{*\alpha}v'^\beta v^\rho, \qquad (9.93)$$

$$\langle D^*(v',\epsilon)|\bar{c}\gamma_\mu\gamma_5 b|B(v)\rangle = \sqrt{m_B m_{D^*}}\left[\xi_{A1}(y)(vv' + 1)\epsilon^*_\mu - \xi_{A2}(y)(\epsilon^*v)v_\mu\right.$$
$$\left. - \xi_{A2}(y)(\epsilon^*v)v'_\mu\right], \qquad (9.94)$$

where ϵ is the polarization of the charmed vector meson and $y = v \cdot v'$. Applying now (9.43) to (9.92–9.94) we find five relations among the form factors ξ_i

$$\xi_i(y) = \xi(y) \quad \text{for } i = +, V, A1, A3, \qquad \xi_i(y) = 0 \quad \text{for } i = -, A2. \qquad (9.95)$$

which eventually reduces the number of independent form factors to only one.

In addition, we make use of Luke's theorem derived in section 9.2.2.3 which yields a statement about the size of the corrections; we find

$$\xi_i(1) = 1 + \mathcal{O}\left(\left[\frac{1}{2m_c} - \frac{1}{2m_b}\right]^2\right) \quad \text{for } i = +, V, A1, A3,$$

$$\xi_i(1) = \mathcal{O}\left(\frac{1}{2m_c} - \frac{1}{2m_b}\right) \quad \text{for } i = -, A2. \qquad (9.96)$$

This has interesting phenomenological applications. Computing the rates for the exclusive decays $B \to D\ell\bar{\nu}$ and $B \to D^*\ell\bar{\nu}$ in terms of the form factors ξ_i we get

$$\frac{d\Gamma}{dy}(B \to D\ell\nu_\ell) = \frac{G_F^2}{48\pi^3}|V_{cb}|^2(m_B + m_D)^2 \left(m_D\sqrt{y^2-1}\right)^3$$

$$\times \left|\xi_+(y) - \frac{m_B - m_D}{m_B + m_D}\xi_-(y)\right|^2 \qquad (9.97)$$

$$\frac{d\Gamma}{dy}(B \to D^*\ell\nu_\ell) = \frac{G_F^2}{48\pi^3}|V_{cb}|^2(m_B - m_{D^*})^2 m_{D^*}^2 \left(m_{D^*}\sqrt{y^2-1}\right)$$

$$\times (y+1)^2 |\xi_{A1}(y)|^2 \sum_{i=0,\pm} |H_i(y)|^2 \qquad (9.98)$$

with the squared helicity amplitudes

$$|H_\pm(y)|^2 = \frac{m_B^2 - m_{D^*}^2 - 2ym_Bm_{D^*}}{(m_B - m_{D^*})^2}\left[1 \mp \sqrt{\frac{y-1}{y+1}}R_1(y)\right]^2, \qquad (9.99)$$

$$|H_0(y)|^2 = \left(1 + \frac{m_B(y-1)}{m_B - m_{D^*}}[1 - R_2(y)]\right)^2. \qquad (9.100)$$

Here we have defined the form factor ratios

$$R_1(y) = \frac{\xi_V(y)}{\xi_{A1}(y)}, \quad R_2(y) = \frac{\xi_{A3}(y) + \frac{m_B}{m_{D^*}}\xi_{A2}(y)}{\xi_{A1}(y)}. \qquad (9.101)$$

These expressions collapse in the limit $m_b, m_c \to \infty$ into

$$\frac{d\Gamma}{dy}(B \to D\ell\nu_\ell) \to \frac{G_F^2}{48\pi^3}|V_{cb}|^2(m_B + m_D)^2 \left(m_D\sqrt{y^2-1}\right)^3 |\xi(y)|^2, \qquad (9.102)$$

$$\frac{d\Gamma}{dy}(B \to D^*\ell\nu_\ell) \to \frac{G_F^2}{48\pi^3}|V_{cb}|^2(m_B - m_{D^*})^2 m_{D^*}^2 \left(m_{D^*}\sqrt{y^2-1}\right)(y+1)^2$$

$$\times \left[1 + \frac{4y}{y+1}\frac{m_B^2 - m_{D^*}^2 - 2ym_Bm_{D^*}}{(m_B - m_{D^*})^2}\right]|\xi(y)|^2. \qquad (9.103)$$

The impact of these relations is that the absolute normalization of the form factor is given by HQS, and hence a model independent extraction of the CKM matrix element V_{cb} becomes possible by extrapolating the measured differential rates to the kinematic point $y = 1$.

For the decay $B \to D^*\ell\bar\nu$ we find

$$\lim_{y\to 1}\frac{1}{\sqrt{y^2-1}}\frac{d\Gamma}{dy}(B \to D^*\ell\nu_\ell) = \frac{G_F^2}{4\pi^3}(m_B - m_{D^*})^2 m_{D^*}^3|V_{cb}|^2|\xi_{A1}(1)|^2 \qquad (9.104)$$

where the form factor $\xi_{A1}(1) = \xi(1) = 1$ is normalized by HQS. In fact, ξ_{A1} is also protected against linear corrections in $1/m_c$ due to (9.96), and hence we expect a determination of V_{cb} from (9.104) with an uncertainty of about ten percent.

We can also use the process $B \to D\ell\bar{\nu}$ for a determination of V_{cb}, although there is an additional factor of $y^2 - 1$ makes the extrapolation more difficult; furthermore, due to the presence of $\xi_-(1) \sim 1/m_c$ we expect this to be not as precise as for $B \to D^*\ell\bar{\nu}$.

The state of the art is nowadays far more advanced. First of all, QCD corrections have been computed for both the vector and the axial vector current [35]

$$\langle c(v)|\bar{c}\gamma^\mu b|b(v)\rangle = 1 + \frac{2\alpha_s}{3\pi}\left[\frac{3m_b^2 + 2m_c m_b + 3m_c^2}{2(m_b^2 - m_c^2)}\ln\left(\frac{m_b}{m_c}\right) - 2\right] \qquad (9.105)$$

$$\langle c(v)|\bar{c}\gamma^\mu\gamma_5 b|b(v)\rangle = 1 - \frac{\alpha_s}{\pi}\left[\frac{m_b + m_c}{m_b - m_c}\ln\left(\frac{m_c}{m_b}\right) + \frac{8}{3}\right]. \qquad (9.106)$$

Numerically (including also the known α_s^2 corrections) we find [36]:

$$\langle c(v)|\bar{c}\gamma^\mu b|b(v)\rangle = \eta_V = 1.022 \pm 0.004, \qquad (9.107)$$

$$\langle c(v)|\bar{c}\gamma^\mu\gamma_5 b|b(v)\rangle = \eta_A = 0.960 \pm 0.007. \qquad (9.108)$$

Furthermore, QED corrections have been compute as well and amount to an enhancement of the rates by a factor $\eta_{\text{ew}} = 1.007$. In addition the recoil corrections have been estimated by using QCD sum rules which indicate a further decrease of the matrix element of the axial current by another 10%.

More recently, lattice calculations of the form factors have become available at the non-recoil point as well as for $y \neq 1$, even for finite values of the quark masses. All this yields a quite consistent picture giving us a quite reliable value for V_{cb}; a recent analysis [37] yields

$$|V_{cb}| = \left(41.9^{+2.0}_{-1.9}\right) \times 10^{-3}. \qquad (9.109)$$

9.3   Heavy Quark Expansion

In inclusive processes we often make use of the so-called operator product expansion (OPE), a standard tool in quantum field theory. In fact the OPE lies at the heart of the EFT approach, since it is actually this tool that allows us to separate scales.

The most prominent example is deep inelastic scattering $e + p \to e' + X$ (DIS), which is an inclusive process governed by a large scale set by the momentum transfer Q^2 of the electron. Clearly the amplitudes will contain pieces related to this large scale Q^2, which we expect to be computable in perturbation theory, since $\alpha_s(Q^2)$ is small. The non-perturbative parts are eventually the parton distributions of the quarks inside the proton, which are determined by the binding effects of the quarks inside the proton. The expansion set up here in powers of $\Lambda^2_{\text{QCD}}/Q^2$ is very small, such that usually only the leading term is considered.

Here we proceed along the same lines as in DIS. The expansion, the heavy quark expansion (HQE) is in this case in powers of Λ_{QCD}/m_Q; however, in our case we also take subleading terms into account.

9.3.1 Inclusive Decays

Inclusive decays are all processes where a summation over final states is performed. In case we sum over all possible decay channels, we obtain the most inclusive quantity, which is the total decay width of a particle.

However, there are cases where we can single out certain final states. In particular, in weak decays we may also produce leptons, and we may want to sum over the final-state hadrons

$$\Gamma(B \to X\ell\bar{\nu}) = \sum_f \Gamma(B \to f\ell\bar{\nu}).$$

Likewise, we may also consider radiative processes, where photons are emitted.

In both cases we are interested in kinematic distributions such as the energy spectrum of the charged lepton or the photon or the invariant mass spectrum of the leptons

$$\frac{d\Gamma(B \to X\ell\bar{\nu})}{dE_\ell} = \sum_f \frac{d\Gamma(B \to f\ell\bar{\nu})}{dE_\ell}.$$

In the chapter we discuss how to set up an expansion in inverse powers of the b quark mass for inclusive processes.

9.3.2 Operator Product Expansion (OPE)

We start with the total decay rate of a heavy hadron $H(p_H)$. Assuming that H is a ground-state hadron, it can only decay by a weak decay, which is mediatied by an effective Hamiltonian density $\mathcal{H}_{eff}(x)$. To leading order in the weak interaction we obtain—up to trivial factors—for the total rate[5]

$$\Gamma \propto \sum_X (2\pi)^4 \delta^4(p_H - p_X)|\langle X|\mathcal{H}_{eff}(0)|H(p_H)\rangle|^2, \qquad (9.110)$$

where X is the final state with momentum p_X, and we sum over all final states taking into account four-momentum conservation. We use the relation

$$\mathcal{H}_{eff}(x) = e^{-i\hat{P}x}\mathcal{H}_{eff}(0)e^{i\hat{P}x},$$

---

[5] In the following we often drop the argument of field operators or of other space time dependent operators $\mathcal{O}(x)$, we define $\mathcal{O} \equiv \mathcal{O}(0)$. Likewise we write $\partial_\mu \mathcal{O} \equiv (\partial_\mu \mathcal{O}(x))|_{x=0}$.

where $\hat{P}_\mu$ is the (four) momentum operator, and write

$$\sum_X (2\pi)^4 \delta^4(p_H - p_X) |\langle X|\mathcal{H}_{eff}(0)|H(p_H)\rangle|^2$$

$$= \sum_X \int d^4y \exp\left(i(p_H - p_X)y\right) \langle H(p_H)|\mathcal{H}_{eff}(0)|X\rangle\langle X|\mathcal{H}_{eff}(0)|H(p_H)\rangle$$

$$= \sum_X \int d^4y \langle H(p_H)|\mathcal{H}_{eff}(y)|X\rangle\langle X|\mathcal{H}_{eff}(0)|H(p_H)\rangle$$

$$= \int d^4y \langle H(p_H)|\mathcal{H}_{eff}(y)\mathcal{H}_{eff}(0)|H(p_H)\rangle, \tag{9.111}$$

where in the final step we made use of the fact, that

$$\sum_X |X\rangle\langle X| = 1$$

is the unit operator, when summing over all states in the Hilbert space.

Finally we use the optical theorem to relate the matrix element of the product of the Hamiltonian to the time-ordered product

$$\int d^4y \langle H(p_H)|\mathcal{H}_{eff}(y)\mathcal{H}_{eff}(0)|H(p_H)\rangle$$

$$= 2 \operatorname{Im} \int d^4y \langle H(p_H)|T\{\mathcal{H}_{eff}(y)\mathcal{H}_{eff}(0)\}|H(p_H)\rangle. \tag{9.112}$$

This relation is the starting point of all further considerations. The matrix element in (9.112) still contains the heavy quark mass m_Q and our goal is to set up an expansion in inverse powers of this mass. Since $\mathcal{H}_{eff}$ induces a decay of the heavy quark, we expect it to be of the form

$$\mathcal{H}_{eff} = \bar{Q}R + \text{h.c.}, \tag{9.113}$$

where R consists of light(er) quarks and possibly gluons. In order to make the dependence on the heavy mass explicit, we use (9.9) and write[6]

$$Q(x) = \exp(-im_Q(v \cdot x))Q_v(x), \quad v = \frac{p_H}{m_H} \tag{9.114}$$

corresponding to the splitting of the heavy-quark momentum into the large part $m_Q v$ and a residual part related to the derivative acting on Q_v. Note that we do not use the

---

[6] We note that $Q_v(0) = Q(0)$; however, once a derivative is acting on the field Q_v it corresponds to the residual momentum $i\partial_\mu Q_v(0) \sim k_\mu$.

static field introduced above; rather, $Q_v(x)$ is still the field of full QCD, up to the above phase redefinition.

With this phase redefinition we get

$$\int d^4y \langle H(p_H)| T\{\mathcal{H}_{eff}(y)\mathcal{H}_{eff}^{\dagger}(0)\}|H(p_H)\rangle$$

$$= \int d^4y \exp(im_Q(v \cdot y)) \langle H(p_H)| T\{\tilde{\mathcal{H}}_{eff}(y)\tilde{\mathcal{H}}_{eff}^{\dagger}(0)\}|H(p_H)\rangle \qquad (9.115)$$

where $\tilde{\mathcal{H}}_{eff}$ is obtained from $\mathcal{H}_{eff}$ by the replacement (9.114)

$$\tilde{\mathcal{H}}_{eff} = \bar{Q}_v R + \text{h.c.}.$$

Expression (9.115) is the starting point of an OPE, which is a standard method in quantum field theory (for a textbook presentation, see [38]). Without going into details, the main relation is

$$\int d^4y\, e^{-iqx}\, T[O_1(x)O_2(0)] = \sum_n C_n(q)\mathcal{O}_n(0), \qquad (9.116)$$

where O_1 and O_2 are renormalized local operators and $\mathcal{O}_n$ are renormalized local operators which can be ordered by increasing dimension, and $C_n(q)$ are coefficients depending on the momentum transfer q. Note that each term on the right-hand side must have the same dimension, so the increasing dimension of the operators will be compensated by inverse powers of q. Thus, for sufficiently large momentum transfer q we can truncate the series on the right-hand side, and obtain an approximation scheme in terms of powers of $1/q$.

Applying the OPE in the context of QCD, we see that at large q QCD becomes perturbative. This means in particular that we can compute the coefficients in QCD perturbation theory, while the matrix elements of the operators contain the non-perturbative information. This scheme is at the heart of all applications of QCD-based EFTs and has been used in many different contexts such as weak interactions and in DIS.

Inclusive differential rates can be computed for processes with leptons and/or photons in the final state. These rates are inclusive with respect to the final-state hadrons, but we may consider the kinematic distributions of the final state photons and leptons. To be explicit, let us study a semileptonic transition based on the quark decay $Q \rightarrow q + \ell + \bar{\nu}$. The effective Hamiltonian can be written as

$$\mathcal{H}_{eff} = \frac{4G_F V_{\text{CKM}}}{\sqrt{2}} \mathcal{J}_{\mu} L^{\mu} \qquad (9.117)$$

where $\mathcal{J}_{\mu} = \bar{Q}_L \gamma_{\mu} q_L$ is the left-handed hadronic current, $L_{\mu} = \bar{\ell}_L \gamma_{\mu} \nu_L$ is the leptonic current, and G_F is the Fermi-coupling constant. Inserting this into (9.111), we get

$$8G_F^2|V_{CKM}|^2 \sum_{X,\ell\bar{\nu}} (2\pi)^4 \delta^4(p_H - p_X - k - k') |\langle X\ell\bar{\nu}|\mathcal{J}_\mu L^\mu|H(p_H)\rangle|^2 \qquad (9.118)$$

$$= 8G_F^2|V_{CKM}|^2 \sum_X \widetilde{dk}\widetilde{dk}' (2\pi)^4 \delta^4(p_H - p_X - k - k') |\langle X|\mathcal{J}_\mu|H(p_H)\rangle \langle \ell(k)\bar{\nu}(k')|L^\mu|0\rangle|^2,$$

where $\widetilde{dk}$ and $\widetilde{dk}'$ denote the phase space integrations over the leptons. Since the leptons do not have any strong interaction, we can decompose this expression into an hadronic and a leptonic part. We get

$$\sum_X \widetilde{dk}\widetilde{dk}' (2\pi)^4 \delta^4(p_H - p_X - k - k') |\langle X|\mathcal{J}_\mu|H(p_H)\rangle \langle \ell(k)\bar{\nu}(k')|L^\mu|0\rangle|^2 \qquad (9.119)$$

$$= \int \frac{d^4q}{(2\pi)^4} \sum_X (2\pi)^4 \delta^4(p_H - p_X - q)\langle H(p_H)|\mathcal{J}_\alpha^\dagger|X\rangle\langle X|\mathcal{J}_\beta|H(p_H)\rangle$$

$$\times \int \widetilde{dk}\widetilde{dk}' (2\pi)^4 \delta^4(q - k - k')\langle 0|L^{\alpha\dagger}|\ell(k)\bar{\nu}(k')\rangle\langle \ell(k)\bar{\nu}(k')|L^\beta|0\rangle.$$

The leptonic part can be evaluated separately and usually is taken to lowest order in perturbation theory; the hadronic part is encoded in the hadronic tensor, which can be decomposed into scalar functions W_i, $i = 1,\dots,5$

$$W^{\alpha\beta}(q) = \sum_X (2\pi)^4 \delta^4(p_H - p_X - q)\langle H(p_H)|\mathcal{J}^{\alpha\dagger}|X\rangle\langle X|\mathcal{J}^\beta|H(p_H)\rangle \qquad (9.120)$$

$$= -g^{\alpha\beta} W_1 + v^\alpha v^\beta W_2 - i\epsilon^{\alpha\beta\mu\nu} v_\mu q_\nu W_3 + q^\alpha q^\beta W_4 + (v^\alpha q^\beta + v^\beta q^\alpha) W_5,$$

where we introduced $p_H = m_H v$. These scalar functions depend on the two invariants q^2 and $v \cdot q$; in terms of which we get, for example, for the triply differential rate ($E_\ell = v \cdot k$, $E_\nu = v \cdot k'$)

$$\frac{d\Gamma}{dq^2\,dE_\ell\,dE_\nu} = \frac{G_F^2|V_{CKM}|^2}{2\pi^3}\left[W_1 q^2 + W_2\left(2E_\ell E_\nu - \frac{1}{2}q^2 \right) + W_3 q^2(E_\ell - E_\nu) \right] \quad (9.121)$$

where the phase space is restricted by $4E_\ell E_\nu - q^2 \geq 0$.

With the hadronic tensor we can go through the same steps (9.110, ... ,9.112), but we have to insert the phase factor $\exp(-iqy)$ into the y integration:

$$\int d^4y \exp(-iqy)\,\langle H(p_H)|\mathcal{J}_\mu^\dagger(y)\mathcal{J}_\nu(0)|H(p_H)\rangle$$

$$= 2\,\mathrm{Im}\int d^4y \exp(-iqy)\,\langle H(p_H)|T\{\mathcal{J}_\mu^\dagger(y)\mathcal{J}_\nu(0)\}|H(p_H)\rangle. \quad (9.122)$$

Performing the replacement (9.114) we end up with

$$\int d^4 y \exp(-iqy) \langle H(p_H)| T\{\mathcal{J}_\mu^\dagger(y)\mathcal{J}_\nu(0)\}|H(p_H)\rangle \tag{9.123}$$

$$= \int d^4 y \exp(iy(m_Q v - q)) \langle H(p_H)| T\{\widetilde{\mathcal{J}}_\mu^\dagger(y)\widetilde{\mathcal{J}}_\nu(0)\}|H(p_H)\rangle.$$

The time-ordered product of the two hadronic currents has the same decomposition (9.120) as the hadronic tensor with scalar functions T_i, $i = 1, ..., 5$. These functions have an analytic structure as depicted in Figure 9.4: for a fixed value of q^2 we have $p_H - q = p_X$ where p_X is the momentum of the final hadronic state, thus $m_H^2 + q^2 - 2m_H(v \cdot q) = m_X^2$. Thus the maximal value of $v \cdot q$ is given by

$$(v \cdot q)_{\max} = \frac{1}{2m_H}(m_H^2 + q^2 - m_{X\min}^2),$$

where $m_{X\min}$ is the mass of the lightest hadronic state with the correct quantum numbers. Thus for the states with a q quark in the final state, the T_i exhibit a cut

$$-\infty \le (v \cdot q) \le \frac{1}{2m_H}(m_H^2 + q^2 - m_{X\min}^2). \tag{9.124}$$

However, there can also be intermediate states with two Q quarks and a q anti-quark, which yield a branch cut

$$\frac{1}{2m_H}(m_H^2 + q^2 - m_{X(QQ\bar{c})\min}^2) \le (v \cdot q) \le \infty \tag{9.125}$$

where $m_{X(QQ\bar{c})\min}$ denotes the mass of the lightest state with the quark content $QQ\bar{c}$. The relevant W_i are given by the discontinuity T_i of the left-hand cut according to (9.122).

To compute a doubly differential rate we need to integrate over one of the variables, which in the present case is the neutrino energy. Using (9.122) this integration can be replaced by the contour integration depicted in Figure 9.4. Note that there is a gap between the two cuts such that the contour does not get close to the singularity, which indicates that a perturbative calculation is possible for sufficiently 'smeared' quantities.

Before closing the general set-up we need to point out some subtleties. The proof that an OPE exists can strictly only be performed in the deep euclidean region, i.e. for

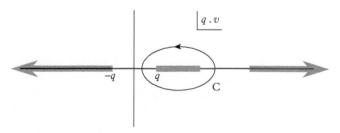

Figure 9.4 *Sketch of the analytic structure of the T_i in the $v \cdot q$ plane for fixed q^2 (Figure taken from [11]).*

$q^2 \to -\infty$ in (9.116). However, in all applications in heavy quark physics we are actually in the Minkowskian region; the momentum in (9.115) is $m_Q v$, which is time-like, as well as the momentum $m_Q v - q$ for the differential rate. This innocent-looking point of analytically continuing from the Euclidean to the Minkowskian region is, however, quite subtle; strictly speaking, the OPE in the Minkowskian region is not proven.

Another related issue is duality, i.e. the question, to what extent partonic results are 'dual' to the real hadronic results. Originally this question has been raised in the context of $e^+e^- \to$ hadrons: how can the differential cross sections obtained from the calculation of $e^+e^- \to$ partons (quarks and gluons) be related to $e^+e^- \to$ hadrons? It has been argued [39] that such a comparison becomes possible once as suitable 'smearing' (i.e. a convolution with smooth weight functions) has been performed.

In the context of heavy-quark physics this has been made more precise in [40], where the question of duality has been connected to the HQE. As we show, the leading term of the HQE for inclusive decays is the parton model, i.e. the decay of a 'free' quark. From a naïve notion of quark-hadron duality one would assume this to be a reasonable approxmation, even for suitably 'smeared' differential quantities. A more quantitative definition of duality as in [40] links this to the HQE, which means on the one hand the convergence of the expansion itself as well as the absence (or at least smallness) of non-analytic terms in the expansion parameters.

9.3.3 Tree-Level Results

To be specific, we start out by constructing the OPE for an inclusive semileptonic $b \to c$ decay. The starting point is the expression (9.123) in the form (now we have $\mathcal{J}_\mu = \bar{c}\gamma_\mu(1 - \gamma_5)b$)

$$\int d^4 y \exp(-iqy)\, T\{\mathcal{J}_\mu^\dagger(y)\mathcal{J}_\nu(0)\} = \int d^4 y \exp(iy(m_Q v - q))\, T\{\widetilde{\mathcal{J}}_\mu^\dagger(y)\widetilde{\mathcal{J}}_\nu(0)\} \qquad (9.126)$$

for which we want to perform an OPE according to (9.116)

$$\int d^4 y \exp(iy(m_Q v - q))\, T\{\widetilde{\mathcal{J}}_\mu^\dagger(y)\widetilde{\mathcal{J}}_\nu(0)\} = \sum_n C_{\mu\nu}^{(n)}\mathcal{O}_n. \qquad (9.127)$$

The key point of the OPE is that it is an operator relation, which means that we can take any matrix element of this relation to compute the coefficients $C_{\mu\nu}^{(n)}$. Hence the simplest way to proceed (having discussed above that we may compute the coefficients in perturbation theory) is to take a matrix element with free quark and gluon states.

We start with a free b quark with momentum $p_b = m_b v + k$ corresponding to the splitting of the quarks momentum into a large part $m_b v$ and a residual part k. The leading tree-level diagram is simply given by the propagator of the free charm quark, leaving us with

$$R^{(0)}_{\mu\nu} = \bar{u}(p_b)\gamma_\mu(1-\gamma_5)\left[\frac{1}{\slashed{Q}+\slashed{k}-m_c}\right]\gamma_\nu(1-\gamma_5)u(p_b) \qquad (9.128)$$

where we introduce $Q = m_b v - q$. At tree level, the construction of the OPE proceeds by expanding in the residual momentum k which is assumed to be small in all its components. For the propagator we get

$$\frac{1}{\slashed{Q}+\slashed{k}-m_c} = \frac{1}{\slashed{Q}-m_c} - \frac{1}{\slashed{Q}-m_c}\slashed{k}\frac{1}{\slashed{Q}-m_c} + \frac{1}{\slashed{Q}-m_c}\slashed{k}\frac{1}{\slashed{Q}-m_c}\slashed{k}\frac{1}{\slashed{Q}-m_c} + \cdots. \qquad (9.129)$$

Starting with the leading term, we get

$$R^{(0,0)}_{\mu\nu} = \frac{1}{Q^2 - m_c^2}\bar{u}(p_b)\gamma_\mu(1-\gamma_5)(\slashed{Q}+m_c)\gamma_\nu(1-\gamma_5)u(p)$$

$$= \frac{2}{Q^2 - m_c^2}\bar{u}(p)\gamma_\mu\slashed{Q}\gamma_\nu(1-\gamma_5)u(p_b). \qquad (9.130)$$

Before we continue, a subtlety should be mentioned. We did not expand the spinors $u(p_b)$ in powers of k, which looks a bit inconsistent on first sight. The matching procedure we employ is to compare the matrix element between fixed states of the right-hand and the left-hand side of (9.127). Consequently, we would also need to compute the matrix element between free quark states with the same momentum $p_b = m_b v + k$ on both sides of the equation. Thus the same spinors $u(p_b)$ will appear on both sides, and the expansion of the spinors would cancel. Therefore we can also drop the expansion of the spinors in both the left- and the right-hand side.

Integrating over the leptonic phase space, neglecting the lepton mass, we find that we have to contract this expression with the tensor

$$L^{\mu\nu} = q^2 g^{\mu\nu} - q^\mu q^\nu. \qquad (9.131)$$

For illustrative reasons we only discuss the first term (even without the factor q^2, so we contract only with the metric tensor) and leave it as an exercise to do the full calculation. We thus get

$$R^{(0,0)} = \frac{2}{Q^2 - m_c^2}\bar{u}(p_b)\gamma_\mu\slashed{Q}\gamma^\mu(1-\gamma_5)u(p_b)$$

$$= \frac{-4}{Q^2 - m_c^2}\bar{u}(p_b)\slashed{Q}(1-\gamma_5)u(p_b). \qquad (9.132)$$

At leading order (with this particular contraction of the indices) we see that the leading order expression for the OPE is

$$\int d^4 y \exp(iyQ)\, T\{\widetilde{\mathcal{F}}^\dagger_\mu(y)\widetilde{\mathcal{F}}^\mu(0)\} = \frac{-4}{Q^2 - m_c^2}\,(b_v(0)\slashed{Q}(1-\gamma_5)b_v(0)) + \mathcal{O}(k). \qquad (9.133)$$

This is the simplest example of a matching calculation, i.e. the comparison of the right-hand side of (9.127) to the expansion of the left-hand side. The leading terms turn out to be a dimension-three operators; it turns out that, the higher-order terms involve higher-dimensional operators.

The next step is to take the matrix elements with the real B meson states. In our mini-example we need to discuss the matrix element of a dimension-three operator

$$\langle B(v)|\bar{b}_v(0)\gamma_\lambda(1-\gamma_5)b_v(0)|B(v)\rangle = \langle B(v)|\bar{b}(0)\gamma_\lambda(1-\gamma_5)b(0)|B(v)\rangle = 2m_B v_\lambda, \quad (9.134)$$

which does not contain any unknown parameter, since the vector current $\bar{b}_v(0)\gamma_\lambda b_v(0)$ is conserved; hence its forward matrix elements between B meson states are normalized, while the corresponding matrix element with the axial current vanishes.

In fact, in other applications different dimension-three matrix elements can appear, which differ from the case at hand only by the Dirac matrix between the heavy quark operators Q_v,

$$\bar{Q}_v \Gamma Q_v = \bar{Q}\Gamma Q : \quad \text{general dimension three operator} \quad (9.135)$$

where Γ is an arbitrary Dirac matrix. However, taking a forward matrix element between the pseudoscalar ground state meson, only $\Gamma = 1$ and $\Gamma = \gamma_\mu$ are non-vanishing.

As pointed out previously, the vector current of the heavy quark Q is conserved, with the consequence that it does not induce an unknown hadronic matrix element:

$$\langle H(p_H)|\bar{Q}_v\gamma_\lambda Q_v|H(p_H)\rangle = \langle H(p_H)|\bar{Q}\gamma_\lambda Q|H(p_H)\rangle = 2p_{H\lambda} \quad (9.136)$$

The matrix element of $\bar{Q}_v Q_v = \bar{Q}Q$ can also be related to the vector current by the equations of motion

$$\not{v}Q_v = Q_v - \frac{i\not{D}}{m_Q}Q_v \quad (9.137)$$

$$(ivD)Q_v = -\frac{1}{2m_Q}(i\not{D})(i\not{D})Q_v. \quad (9.138)$$

Using (9.137) we get

$$\langle H(p_H)|\bar{Q}_v Q_v|H(p_H)\rangle = \langle H(p_H)|\bar{Q}_v\not{v}Q_v|H(p_H)\rangle + \frac{1}{m_Q}\langle H(p_H)|\bar{Q}_v(i\not{D})Q_v|H(p_H)\rangle$$

$$= 2m_H + \frac{1}{m_Q}\langle H(p_H)|\bar{Q}_v(i\not{D})Q_v|H(p_H)\rangle \quad (9.139)$$

which means that to leading order in the HQE no unknown hadronic parameter is induced in general. Furthermore, it is easy to see that (9.139) has no contribution of order $1/m_Q$: starting from the equation of motions (9.137) we get the relation

$$\bar{Q}_v\gamma_\alpha Q_v = \bar{Q}_v\gamma_\alpha\not{v}Q_v + \frac{1}{m_Q}\bar{Q}_v\gamma_\alpha(i\not{D})Q_v. \quad (9.140)$$

Taking the conjugate of (9.137) and multiplying from the right with Q_v we get

$$\bar{Q}_v \gamma_\alpha Q_v = \bar{Q}_v \slashed{v} \gamma_\alpha Q_v + \frac{1}{m_Q} \bar{Q}_v (i\slashed{D}) \gamma_\alpha Q_v + \text{total derivative} \qquad (9.141)$$

where we do not need to take into account the total derivative, since it will not contribute to the forward matrix elements. Averaging (9.140) and (9.141) and taking the forward matrix element ($\langle ... \rangle = \langle H(p_H)|...|H(p_H)\rangle$) yields

$$\langle \bar{Q}_v \gamma_\alpha Q_v \rangle = v_\alpha \langle \bar{Q}_v Q_v \rangle + \frac{1}{m_Q} \langle \bar{Q}_v (iD_\alpha) Q_v \rangle. \qquad (9.142)$$

Contracting this with v^α yields the relation

$$\langle \bar{Q}_v \slashed{v} Q_v \rangle = \langle \bar{Q}_v Q_v \rangle + \frac{1}{m_Q} \langle \bar{Q}_v (ivD) Q_v \rangle. \qquad (9.143)$$

Comparing this to (9.139) we find that

$$\langle \bar{Q}_v (i\slashed{D}) Q_v \rangle = -\langle \bar{Q}_v (ivD) Q_v \rangle = \frac{1}{2m_Q} \langle \bar{Q}_v (i\slashed{D})(i\slashed{D}) Q_v \rangle, \qquad (9.144)$$

where we have used (9.138). Inserting this into (9.139) we finally get

$$\langle \bar{Q}_v Q_v \rangle = 2m_H + \frac{1}{2m_Q^2} \langle \bar{Q}_v (i\slashed{D})(i\slashed{D}) Q_v \rangle \qquad (9.145)$$

which proofs that the corrections are $\mathcal{O}(1/m_Q^2)$.

In order to obtain the total rate, we have to take the imaginary part of (9.133). To this end, we re-install the $i\epsilon$ prescription into the propagator and use the relation

$$2\,\mathrm{Im}\,\frac{1}{x + i\epsilon} = (2\pi)\delta(x) \qquad (9.146)$$

from which we finally obtain

$$\Gamma \sim 2\,\mathrm{Im}\,\int d^4 y \, \exp(iyQ)\, \langle B(v)| T\{\tilde{\mathcal{J}}_\mu^\dagger(y)\tilde{\mathcal{J}}^\mu(0)\}|B(v)\rangle \qquad (9.147)$$

$$= -8(2\pi)\delta(Q^2 - m_c^2)m_B(v \cdot Q) + \cdots \qquad (9.148)$$

This result, in combination with (9.139, 9.145), is a general statement:

> The leading term in the HQE is the partonic result, i.e. the decay of a 'free' heavy quark.

Next we look at the first term in the k expansion

$$R^{(0,1)} = 2\left(\frac{1}{Q^2 - m_c^2}\right)^2 \bar{u}(p_b)\gamma_\mu \slashed{Q}\slashed{k}\slashed{Q}\gamma^\mu(1-\gamma_5)u(p_b)$$

$$= -4\left(\frac{1}{Q^2 - m_c^2}\right)^2 \bar{u}(p_b)[Q^2\slashed{k} - 2(Q\cdot k)\slashed{Q}](1-\gamma_5)u(p_b). \qquad (9.149)$$

Comparing this to the OPE (9.127) we find ($k_\mu \to iD_\mu$)

$$\int d^4y \exp(iyQ)\, T\{\tilde{\mathcal{F}}_\mu^\dagger(y)\tilde{\mathcal{F}}^\mu(0)\} = \text{leading term} \qquad (9.150)$$

$$+ 2\left(\frac{1}{Q^2 - m_c^2}\right)^2 \left[Q^2(b_v(0)(i\slashed{D})(1-\gamma_5)b_v(0)) - 2Q^\mu Q^\nu(b_v(0)(iD_\mu)\gamma_\nu(1-\gamma_5)b_v(0))\right].$$

Again we have to take the forward matrix element of this expression. We use the equations of motion (9.137) and (9.138) and get for the general case (the contribution with γ_5 vanish due to parity)

$$\langle H(p_H)|\bar{Q}_v(iD_\mu)\Gamma Q_v|H(p_H)\rangle = \mathcal{O}\left(\frac{1}{m_Q}\right), \qquad (9.151)$$

where the explicit relation for the first term in (9.150) is given by (9.144). This also shows that these contributions are actually at least of order $1/m_Q^2$. This is in fact a general statement:

> There are no contributions of order $1/m_Q$ in the HQE .

This does not mean that the first term in the k expansion vanishes. Rather the dimension-four matrix elements are $1/m_Q$ suppressed.

In order to obtain the corresponding contribution to the rate, we have to take the imaginary part by re-inserting the $i\epsilon$ prescription into the propagator. Using

$$2\,\mathrm{Im}\left(\frac{1}{x+i\epsilon}\right)^2 = -2\,\mathrm{Im}\frac{d}{dx}\left(\frac{1}{x+i\epsilon}\right) = -(2\pi)\delta'(x), \qquad (9.152)$$

we obtain a contribution to the differential rate proportional to the derivative of the 'on-shell' δ function, which, however is of order $1/m_b^2$.

In order to obtain the full $1/m^2$ contributions, we need to expand $R_{\mu\nu}$ to second order in k_μ, which yields for our toy example

$$R^{(0,2)} = 2\left(\frac{1}{Q^2 - m_c^2}\right)^3 \bar{u}(p_b)\gamma_\mu \slashed{Q}\slashed{k}\slashed{Q}\slashed{k}\slashed{Q}\gamma^\mu(1-\gamma_5)u(p_b), \qquad (9.153)$$

which eventually matches on operators with two derivatives:

$$\bar{Q}_v \Gamma (iD_\mu)(iD_\nu) Q_v : \quad \text{General dimension-five operator.} \tag{9.154}$$

However, at that order an obvious problem arises: while $k_\mu k_\nu$ is clearly a symmetric tensor, the product of two covariant derivatives contains an anti-symmetric part. Since the covariant derivatives do not commute, their commutator is the field-strength tensor

$$\bar{Q}_v \Gamma [(iD_\mu),(iD_\nu)] Q_v = -ig_s \bar{Q}_v \Gamma G_{\mu\nu} Q_v. \tag{9.155}$$

Obviously the expansion in k cannot give us this anti-symmetric piece. However, the anti-symmetric part is related to the field strength, i.e. to the emission of a gluon. In order to pin this down we thus have to compute a matrix element of (9.127) between a quark state and a state with a quark and a gluon. For the left-hand side of (9.115) the leading order result is

$$S_{\mu\nu}^{(0)} = \bar{u}(p_b) \gamma_\mu (1 - \gamma_5) \left[\frac{1}{\slashed{Q} + \slashed{k} - m_c} \right] T^a \slashed{\epsilon}(q) \left[\frac{1}{\slashed{Q} + \slashed{k} + \slashed{q} - m_c} \right] \gamma_\nu (1 - \gamma_5) u(p_b), \tag{9.156}$$

where q is the momentum of the gluon with colour a and polarization ϵ. Note that also the momentum of the gluon is soft, so we have to perform a combined expansion in k and q. Furthermore, the gluon appears as part of the covariant derivative, so also the polarization ϵ counts as one power in the $1/m_b$ expansion; this means that in order to arrive at the second order, we have to expand (9.156) only to first order in k and q.

To obtain the coefficient of the anti-symmetric combination (9.155) we thus have to find the coefficient in front of the combination

$$G_{\alpha\beta} \longleftrightarrow q_\alpha \epsilon_\beta - q_\beta \epsilon_\alpha.$$

This concludes the sketch of the practical aspects of the matching procedure to obtain the coefficients in (9.127). The procedure remains the same even once α_s corrections are included, which means that the expansions in k and gluon momenta and polarization must be performed for the expression including α_s corrections. By comparing the two sides of (9.115) we thus obtain the perturbative expansion of the coefficients.

Finally, it is worthwhile to highlight, that the tree-level expressions for the case of semileptonic decays can be obtained systematically [41], since the ordering of the covariant derivatives can be traced by using an external field propagator of the form for the charm propagator.

$$\left[\frac{1}{\slashed{Q} + i\slashed{D} - m_c} \right].$$

Expanding this under the assumption that the components of iD do not commute yields formally the same expression as (9.129)

$$\frac{1}{\slashed{Q} + i\slashed{D} - m_c} = \frac{1}{\slashed{Q} - m_c} - \frac{1}{\slashed{Q} - m_c} i\slashed{D} \frac{1}{\slashed{Q} - m_c} + \frac{1}{\slashed{Q} - m_c} i\slashed{D} \frac{1}{\slashed{Q} - m_c} i\slashed{D} \frac{1}{\slashed{Q} - m_c} + \cdots, \tag{9.157}$$

but now the ordering of the covariant derivatives is the correct one, i.e. for the second-order term we have

$$\frac{1}{\not{Q}-m_c}i\not{D}\frac{1}{\not{Q}-m_c}i\not{D}\frac{1}{\not{Q}-m_c} = (iD_\mu)(iD_\nu)\left[\frac{1}{\not{Q}-m_c}\gamma^\mu\frac{1}{\not{Q}-m_c}\gamma^\nu\frac{1}{\not{Q}-m_c}\right]$$

where the term in the bracket has the correct anti-symmetric piece in the indices μ and ν. However, this unfortunately only works at tree level.

9.3.4 HQE Parameters

The non-perturbative input in the HQE is given in terms of the hadronic matrix elements of operators, which have the generic form[7]

$$\langle H(p_H)|\bar{Q}_v(iD_{\mu_1})(iD_{\mu_2})\cdots(iD_{\mu_n})\Gamma Q_v|H(p_H)\rangle,$$

where Γ is some Dirac matrix. Note that these operators have dimension $n+3$ and are defined in full QCD, which implies that they still depend on the mass. In principle we can perform an expansion in $1/m_Q$ and the leading term will be just

$$\langle \tilde{H}(v)|\bar{h}_v(iD_{\mu_1})(iD_{\mu_2})\cdots(iD_{\mu_n})\Gamma h_v|\tilde{H}(v)\rangle,$$

with the static field h_v and the meson state $|\tilde{H}(v)\rangle$ in the infinite mass limit.

We have already discussed the dimension-three operators and have shown that there is no unknown matrix element at dimension three, since all matrix elements can be related to the conserved Q-quark vector current, up to terms of order $1/m_Q^2$. We also saw already that all the matrix elements of dimension-four operators are suppressed by one power of $1/m_Q$, and thus the first non-trivial contribution appears at dimension five, i.e. for $n=2$.

Before going into the technicalities a historic remark is in order. The idea that the decay of a ground-state hadron with a single heavy quark can be approximately described by the decay of the 'free' quark inside the hadron is quite old [42]. However, the HQE proves this to be the leading term of a systematic expansion, where the leading non-perturbative corrections turn out to be of the order Λ_{QCD}^2/m_Q^2. In the early days of the HQE this was seen as an embarrassment: as a consequence the lifetimes of all ground-state hadrons of a specific heavy flavour should be identical to leading order, the corrections should be of order Λ_{QCD}^2/m_Q^2, and, as we show later, the lifetime differences should even be of the order Λ_{QCD}^3/m_Q^3. Before the precise measurement of bottom-hadron lifetimes, the lifetimes of charmed hadrons were available; with $m_D \approx m_c \sim 1.8$ GeV and $\Lambda_{QCD} \sim 0.3$ GeV we naively expect lifetime differences to be below 1%. However, the lifetimes of ground-state charmed hadrons vary by a factor of five, which is hard to explain as a Λ_{QCD}^3/m_Q^3 effect. Nevertheless, in the meantime we have a qualitative understanding how such large lifetime differences can emerge.

---

[7] In fact, this does not cover all possible operators; there can also be operators with light quarks, which need to be discussed separately.

At each order in the $1/m_Q$ expansion we need to identify, how many independent parameters actually appear. At dimension five the two independent parameters can be defined as

$$-2m_H\hat{\mu}_\pi^2 = \langle H(p_H)|\bar{Q}_v(iD)^2 Q_v|H(p_H)\rangle \tag{9.158}$$

$$-2m_H\hat{\mu}_G^2 = \langle H(p_H)|\bar{Q}_v(iD_\mu)(iD_\nu)(i\sigma^{\mu\nu})Q_v|H(p_H)\rangle \tag{9.159}$$

and any general matrix element can be related to these two through the 'trace formula'

$$\langle H(p_H)|\bar{Q}_v(iD_\mu)(iD_\nu)\Gamma Q_v|H(p_H)\rangle = -2m_H\frac{\hat{\mu}_\pi^2}{6}\operatorname{Tr}\left(\frac{1+\slashed{v}}{2}\Gamma\right)(g_{\mu\nu}-v_\mu v_\nu)$$

$$-2m_H\frac{\hat{\mu}_G^2}{12}\operatorname{Tr}\left(\frac{1+\slashed{v}}{2}(-i\sigma_{\mu\nu})\frac{1+\slashed{v}}{2}\Gamma\right)+\mathcal{O}\left(\frac{1}{m_Q}\right). \tag{9.160}$$

From this we get our third statement on the HQE:

> The first subleading corrections in the HQE are given by $\hat{\mu}_\pi$ and $\hat{\mu}_G$.

The parameter $\hat{\mu}_\pi^2$ is called the kinetic energy parameter, since it is related to the term $\vec{p}^2/(2m_Q)$ appearing in the Schrödinger equation; the parameter $\hat{\mu}_G$ is called the chromomagnetic moment, since it describes the coupling $\vec{\sigma}\cdot\vec{B}$ of the heavy-quark spin to the chromomagnetic field $\vec{B}$.

The values of these parameters have to be taken from either experimental data or calculations in lattice QCD or in a model. As an example, $\hat{\mu}_G$ can be obtained from hadron spectroscopy. Looking at the expansions of heavy hadron masses in inverse powers of the quark mass (9.141, 9.142), we infer that we may use (9.146) to fix the value of the chomomagnetic moment, while the kinetic energy parameter cannot be obtained from spectroscopy.

Before we continue to higher orders, we point out a few details. In many applications it turns out to be useful to split the covariant derivative into a 'time derivative' and a 'spatial' part according to (9.7). To this end, we may as well use the definitions

$$-2m_H\mu_\pi^2 = \langle H(p_H)|\bar{Q}_v(iD^\perp)^2 Q_v|H(p_H)\rangle \tag{9.161}$$

$$-2m_H\mu_G^2 = \langle H(p_H)|\bar{Q}_v(iD_\mu^\perp)(iD_\nu^\perp)(i\sigma^{\mu\nu})Q_v|H(p_H)\rangle \tag{9.162}$$

and we have $\mu_\pi = \hat{\mu}_\pi + \mathcal{O}(1/m_Q^2)$ and $\mu_G = \hat{\mu}_G + \mathcal{O}(1/m_Q)$. Furthermore, all these parameters still depend on the heavy quark mass; furthermore, expanding this mass dependence yields mass independent parameters λ_1 and λ_2 defined in HQET by

$$2m_H\lambda_1 = \langle \tilde{H}(v)|\bar{h}_v(iD^\perp)^2 h_v|\tilde{H}(v)\rangle \tag{9.163}$$

$$2m_H\lambda_2 = \langle \tilde{H}(v)|\bar{h}_v(iD_\mu^\perp)(iD_\nu^\perp)(i\sigma^{\mu\nu})h_v|\tilde{H}(v)\rangle. \tag{9.164}$$

The advantage of expanding any mass dependence and to define 'static' quantities λ_1 and λ_2 is that these will be the same for *any* heavy quark; thus we might compare inclusive bottom with inclusive charm decays. The advantage of using μ_π and μ_G, or $\hat{\mu}_\pi$ and $\hat{\mu}_G$ becomes clear only when going to higher orders: starting at $1/m_Q$ we also need to take into account the expansion of the state, since we have $|H(p_H)\rangle = |\tilde{H}(v)\rangle + \mathcal{O}(1/m_Q)$, which in general leads to non-local matrix elements involving the subleading terms in the Lagrangian (9.17).

At dimension six we will have three derivatives

$$\langle H(p_H)|\bar{Q}_v(iD_\mu)(iD_\alpha)(iD_\nu)Q_v|H(p_H)\rangle$$

together with four-quark operators. At tree level, we can express all dimension-six matrix elements in terms of two parameters which are given by

$$2m_H\hat{\rho}_D^3 = \langle H(p_H)|\bar{Q}_v(iD_\mu)(ivD)(iD^\mu)Q_v|H(p_H)\rangle, \tag{9.165}$$

$$2m_H\hat{\rho}_{LS}^3 = \langle H(p_H)|\bar{Q}_v(iD_\mu)(ivD)(iD_\nu)(-i\sigma^{\mu\nu})Q_v|H(p_H)\rangle, \tag{9.166}$$

where we have again used the covariant definitions. For these parameters, the same remarks apply as for $\hat{\mu}_\pi$ vs μ_π vs λ_1, etc., which will have differences appearing as terms of subleading order in the $1/m_Q$ expansion.

In a similar fashion as for the terms of dimension five we can write a trace formula, which here reads

$$\langle H(p_H)|\bar{Q}_v(iD_\mu)(iD_\alpha)(iD_\nu)\Gamma Q_v|H(p_H)\rangle = 2m_H\frac{\hat{\rho}_D^3}{6}\text{Tr}\left(\frac{1+\slashed{v}}{2}\Gamma\right)(g_{\mu\nu} - v_\mu v_\nu)v_\alpha$$

$$+2m_H\frac{\hat{\rho}_{LS}^3}{12}\text{Tr}\left(\frac{1+\slashed{v}}{2}(-i\sigma_{\mu\nu})\frac{1+\slashed{v}}{2}\Gamma\right)v_\alpha + \mathcal{O}\left(\frac{1}{m_Q}\right). \tag{9.167}$$

Note that a consistent calculation of higher-order terms requires to also take into account the subleading terms in the trace formula (9.160).

We may continue in the same fashion to higher orders [43], although the number of independent parameters will grow strongly as we proceed to orders higher than $1/m_Q^3$. At order $1/m_Q^4$ there is a total of eleven independent parameters, and at $1/m_Q^5$ there are already twenty-five new parameters. While the four parameters up to $1/m_Q^3$ can be extracted from the data, the large number of parameters appearing at even higher orders have to be modelled or may one day be taken from lattice calculations.

9.3.5 QCD Corrections

The HQE has the potential to compute total and specific differential rates with extremely high precision. However, as discussed previously, the leading term is always the decay of the heavy quark inside the heavy hadron, where the result is the same as if we were discussing a 'free' quark. If we ignore for a moment the mass of the final state particles, the decay width is

$$d\Gamma \sim G_F^2 |V_{\text{CKM}}|^2 m_Q^5, \tag{9.168}$$

which induces an enormously strong dependence on a heavy quark mass. In the early days of the HQE this was considered to be problem, since the heavy quark mass is not a straightforward observable. Unlike for an electron, this mass cannot be just measured as a pole in the propagator, since there are no asymptotic states of outgoing quarks. The quark mass is thus just a parameter in the QCD Lagrangian and, in fact, depends on the scheme one chooses to define it.

It seems that any ambiguity or uncertainty related to the heavy quark mass enters into the predictions of HQET enhanced by a factor of five, although this problem can be controlled and is related to a suitable choice of a scheme in which the mass is actually defined.

9.3.5.1 *Why Do We Need a Mass Scheme?*

When computing Feynman diagrams we insert a quark mass into the quark propagators. This mass is defined by the location of the pole of the propagator, which is the usual definition of what is called the pole mass m_Q^{Pole}. When constructing HQET we redefine the heavy quark momentum by $p_Q = m_Q v + k$, using some mass definition, which we choose to be also the pole mass. However, due to

$$m_H = m_Q + \bar{\Lambda} + \mathcal{O}(1/m_Q)$$

we may compensate any redefinition of the mass by a corresponding shift in the parameter $\bar{\Lambda}$.

The mass renormalization is related to the quark propagator. Including the (one particle irreducible) self-energy contributions $\Sigma(p)$, the renormalized quark propagator becomes

$$S(p) = \frac{-i Z_2^{\text{OS}}}{\not{p} - m_0 + \Sigma(p, m_Q^{\text{Pole}})} \longrightarrow \frac{-i}{\not{p} - m_Q^{\text{Pole}}} \quad \text{as } p^2 \to (m_Q^{\text{Pole}})^2 \tag{9.169}$$

where the pole mass is related to the bare mass by a formal (perturbative) series

$$m_0 = Z_m^{\text{OS}} m_Q^{\text{Pole}} = \left(1 + \sum_{n=1}^{\infty} c_n \left(\frac{\alpha_s}{\pi} \right)^n \right) m_Q^{\text{Pole}}. \tag{9.170}$$

The coefficients c_n are divergent and need to be regularized. The standard way to regularize QCD is dimensional regularization (DimReg) where the loop integrals over momenta are computed in $D = 4 - 2\epsilon$ space-time dimensions. At one loop we obtain

$$c_1 = -C_F \left(\left[\frac{1}{\epsilon} + \gamma_E - 4\pi \right] \frac{3}{4} + 1 + \frac{3}{4} \ln \frac{\mu^2}{(m_Q^{\text{Pole}})^2} + \mathcal{O}(\epsilon) \right)$$

where in the case of the pole mass the scale μ is fixed by the on-shell condition (9.169) and $C_F = 4/3$ is the value of the $SU(3)$ Casimir operator in the fundamental representation.

Alternatively, we can use another definition of the quark mass, e.g. the $\overline{\text{MS}}$ definition, for which we have a relation similar to (9.170)

$$m_0 = Z_m^{\overline{\text{MS}}} m_Q^{\overline{\text{MS}}} = \left(1 + \sum_{n=1}^{\infty} b_n \left(\frac{\alpha_s}{\pi} \right)^n \right) m_Q^{\overline{\text{MS}}} \tag{9.171}$$

where the $\overline{\text{MS}}$ scheme is defined by removing only the $1/\epsilon + \gamma_E - 4\pi$ term, which means at one-loop order

$$b_1 = -C_F \left[\frac{1}{\epsilon} + \gamma_E - 4\pi \right] \frac{3}{4}. \tag{9.172}$$

Note that the $\overline{\text{MS}}$ mass depends on the scale μ and is a running parameter.

The key point relevant for our discussion is that different mass definitions can be related by perturbation theory with finite coefficients. We have

$$m_Q^{\text{Pole}} = z^{\text{Pole} \to \overline{\text{MS}}} m_Q^{\overline{\text{MS}}} = \frac{Z_m^{\overline{\text{MS}}}}{Z_m^{\text{OS}}} m_Q^{\overline{\text{MS}}} \tag{9.173}$$

and

$$z^{\text{Pole} \to \overline{\text{MS}}} = 1 + \sum_{n=1}^{\infty} a_n \left(\frac{\alpha_s}{\pi} \right)^n \quad \text{and} \quad a_1 = -C_F \left(\frac{3}{4} \ln \frac{\mu^2}{(m_Q^{\text{Pole}})^2} + 1 \right). \tag{9.174}$$

Consider now a rate of the form (9.168) and assume that we have fixed the mass scheme to the pole mass. Computing radiative corrections to (9.168) takes the schematic form

$$d\Gamma \sim G_F^2 |V_{\text{CKM}}|^2 (m_Q^{\text{Pole}})^5 \left(1 + \frac{\alpha_s}{\pi} r_1 + \left(\frac{\alpha_s}{\pi} \right)^2 r_2 + \cdots \right) \tag{9.175}$$

with (after proper renormalization) finite coefficients r_i. Switching now to another mass definition, e.g. the $\overline{\text{MS}}$ scheme, we find

$$d\Gamma \sim G_F^2 |V_{\text{CKM}}|^2 (m_Q^{\overline{\text{MS}}})^5 (z^{\text{Pole} \to \overline{\text{MS}}})^5 \left(1 + \frac{\alpha_s}{\pi} r_1 + \left(\frac{\alpha_s}{\pi} \right)^2 r_2 + \cdots \right) \tag{9.176}$$

$$= G_F^2 |V_{\text{CKM}}|^2 (m_Q^{\overline{\text{MS}}})^5 \left(1 + \frac{\alpha_s}{\pi} (r_1 + 5a_1) + \cdots \right).$$

Thus we conclude that the choice of a mass scheme determines the size of the radiative corrections. In other words, with a clever choice of the mass definition we can absorb radiative corrections into the definition of the mass. Clearly such a mass definition must also allow us to obtain the numerical value for the mass from independent data as precisely as possible, since the dependence on the fifth power is still present.

It turns out that the pole mass is a particularly bad choice for a mass scheme, since the coefficients r_1 are large and do not seem to converge well [44]–[46]. Related to the bad convergence is another problem with the pole mass, since it has an intrinsic uncertainty of the order of Λ_{QCD} due to an infrared renormalon. Better definitions are so-called short-distance masses (e.g. the $\overline{\text{MS}}$ mass), which do not have this problem and can thus be determined in principle with arbitrary precision. For most of these short-distance masses the QCD corrections converge much better; there are even mass definitions especially designed for the HQE.

9.3.5.2 *Short-Distance Masses*

We use again the pole mass as a starting point. In terms of the $\overline{\text{MS}}$ mass we have

$$m_Q^{\text{Pole}} = z^{\text{Pole}\to\overline{\text{MS}}}\, m_Q^{\overline{\text{MS}}} = \left(1 + \sum_{n=1}^{\infty} a_n \left(\frac{\alpha_s}{\pi}\right)^n\right) m_Q^{\overline{\text{MS}}}. \tag{9.177}$$

The main point of the following discussion is the fact that this perturbative relation is not converging, rather it is an asymptotic series. This is due to factorially growing contributions in the coefficients $a_n \sim n!$. In fact, we can show that the asymptotic behavior of the perturbative series is [44]

$$z^{\text{Pole}\to\overline{\text{MS}}} = 1 + \frac{C_F e^{5/6}}{\pi} \frac{\mu}{m_Q^{\overline{\text{MS}}}} \alpha_s \sum_n (-2\hat{\beta}_0 \alpha_s)^n\, n! \tag{9.178}$$

where

$$\hat{\beta}_0 = \frac{1}{4\pi}\left(11 - \frac{2n_f}{3}\right)$$

is the leading term of the $\hat{\beta}$ function of QCD and n_f is the number of active flavours.

In order to consider the consequences of this observation, we study the Borel transform of the perturbative series, defined by

$$z(\alpha) = \sum_{n=0}^{\infty} a_n \alpha^{n+1} \quad \longrightarrow \quad B[z](t) = \sum_{n=0}^{\infty} \frac{a_n}{n!} t^n. \tag{9.179}$$

If both series for z and $B[z]$ were convergent, we could define the reverse operation by

$$z(\alpha) = \int_0^{\infty} dt\, \exp\left(-\frac{t}{\alpha}\right) B[z](t), \tag{9.180}$$

which indeed has the same series expansion as the original z. However, the terms shown in (9.178) lead to poles on the positive real axis in the Borel transform, which are called renormalons. Here, the leading term originates from a singularity at $t = 1/2$ and hence the integral in (9.180) cannot be computed without a prescription of how to avoid this pole. This leads to an ambiguity that can be expressed by shifting the singularity in the complex t plane by a small amount ϵ either upwards or downwards; hence we use

$$\frac{1}{t - 1/2 + i\epsilon} - \frac{1}{t - 1/2 - i\epsilon} = 2\pi\delta(t - 1/2)$$

leaving us with an ambiguity of the form

$$\Delta z(\alpha) \propto \frac{\mu}{m_Q^{\overline{\text{MS}}}} \exp\left(-\frac{1}{2\alpha(\mu)}\right) \sim \frac{\Lambda_{\text{QCD}}}{m_Q^{\overline{\text{MS}}}} \tag{9.181}$$

where we have inserted the running coupling of QCD in terms of Λ_{QCD}.

Although these arguments can still be made more stringent, we have at least seen the essence of the reasoning which leads to the conclusion that the pole mass has an intrinsic uncertainty of the order of Λ_{QCD}, related to infrared contributions, which can be related to the coulombic self interactions of a heavy quark.[8]

To this end, it means that the pole mass cannot be used for precise predictions. In particular, inserting the pole mass into (9.175) yields large QCD corrections that are mainly due to this particular choice of the mass. In other words, a more clever choice of the mass definition can minimize the size of the QCD corrections and lead to a much better convergence.

Another problem becomes apparent once power corrections are included. Given an intrinsic uncertainty of the order Λ_{QCD} in the mass in (9.175) renders the power corrections, which are by themselves $\mathcal{O}(\Lambda_{\text{QCD}}^2/m_Q^2)$, completely meaningless.

Thus it is obvious that we have to switch to a 'short-distance' mass such as the $\overline{\text{MS}}$ mass. This mass depends on the scale μ which is usually taken to be $\mu \geq m_Q^{\overline{\text{MS}}}$; below this scale one should switch to HQET, and thus it becomes clear that we should use mass definitions that are 'designed' to go to scales as low as 1 GeV. There are two mass schemes that are frequently used in the context of the HQE and that are the kinetic mass scheme and the $1S$ mass scheme.

9.3.5.3　*Kinetic Mass Scheme*

As discussed, the pole mass contains a renormalon ambiguity of the order of Λ_{QCD}. In order to avoid this problem, we look at the expansion of the heavy hadron mass of the pseudoscalar ground state meson (9.32). We have a physical quantity (the hadron mass) on the left-hand side, which cannot suffer from such an ambiguity. However, on the right-hand side, we have not yet specified, what mass definition is used. If we use the

---

[8] We note that the mass $m_Q^{\overline{\text{MS}}}$ does not suffer from this problem. This can be seen from its relation to the bare mass, where only the ultraviolet $1/\epsilon$ poles are removed.

pole mass, we find that this ambiguity has to cancel between m_Q^{Pole} and the binding-energy parameter $\bar{\Lambda}$ defined in (9.34).

Clearly $\bar{\Lambda}$ as well as the parameters in the HQE of the heavy hadron mass are non-perturbative. However, we may write down a QCD sum rule, called a 'small-velocity' sum rule [47], which allows us to estimate these parameters. This sum rules also allows us to compute the perturbative contribution to these parameters in a hard cut-off scheme. As we pointed out in the discussion of the HQE for the heavy quark mass, all ambiguities in the heavy quark mass have to cancel against the ambiguities in the HQE parameters $\bar{\Lambda}$ and μ_π^2, such that the meson mass is a well defined physical quantity. Writing down an expression similar as for the hadron mass up to the kinetic energy term and inserting the perturbative expressions for $\bar{\Lambda}$ and μ_π^2 yields the definition of the quark mass in the 'kinetic' scheme.

$$m_Q^{\text{kin}}(\mu) = m_Q^{\text{Pole}} - [\bar{\Lambda}(\mu)]_{\text{pert}} - \frac{1}{2m_Q^{\text{kin}}(\mu)}[\mu_\pi^2(\mu)]_{\text{pert}}, \qquad (9.182)$$

which is a short-distance mass like the $\overline{\text{MS}}$ mass, since the renormalon ambiguities in the pole mass cancel against those in $\bar{\Lambda}$ and μ_π^2. Here, the leading order expression for $\bar{\Lambda}$ and μ_π^2 read [47]

$$[\bar{\Lambda}(\mu)]_{\text{pert}} = \frac{16}{9}\frac{\alpha_s(\mu)}{\pi}\mu + \mathcal{O}(\alpha_s^2) \qquad (9.183)$$

$$[\mu_\pi^2(\mu)]_{\text{pert}} = \frac{4}{3}\frac{\alpha_s(\mu)}{\pi}\mu^2 + \mathcal{O}(\alpha_s^2), \qquad (9.184)$$

where μ is the hard cut-off.

The kinetic mass is a short distance mass and can be extracted, e.g. from the thresholds in $e^+e^- \to$ hadrons with a very high precision; currently the uncertainty is about 50 MeV [48].

9.3.5.4 1S Mass Scheme

As discussed in [44, 45], the infrared contributions of the pole mass can be attributed to the colour Coulomb field of the heavy quark. In a non-relativistic picture, this Coulomb field is the main contribution to the binding of a system with heavy quarks, and hence it is suggestive to try to find a definition of the b quark mass in terms of a bottomonium state. If we choose the ground state of the bottomonium, we may compute this in terms of the mass parameter in the Lagrangian in a non-relativistic picture, i.e. in NRQCD (see the last part of the chapter).

To this end, we consider the mass of the lowest lying $\mathcal{J}^{PC} = 1^{--}$, $^3S_1$ bottomonium state, called Υ, for which precise measurements exist. The (perturbative) relation between the Υ mass and the pole mass reads schematically

$$m_\Upsilon = 2m_b^{\text{pole}} \left(1 - \frac{(\alpha_s C_F)^2}{8} \left\{ 1 + \frac{\alpha_s \beta_0}{\pi}(\ell+1) + \left(\frac{\alpha_s \beta_0}{\pi}\right)^2 \left(\frac{1}{2}\ell^2 + \ell + 1\right) + \cdots + \right.\right.$$
$$\left.\left. + \left(\frac{\alpha_s \beta_0}{\pi}\right)^n \left(\frac{1}{n!}\ell^n + \frac{1}{(n-1)!}\ell^{n-1} + \dots + \ell + 1\right) + \cdots \right\} \right) \qquad (9.185)$$

where

$$\ell = \ln\left[\frac{\mu}{m_b \alpha_s C_F}\right], \quad \beta_0 = 4\pi\hat\beta_0 = 11 - \frac{2n_f}{3} \quad \text{and} \quad C_F = \frac{4}{3}.$$

We note that there is a mismatch concerning the orders in α_s, since the non-relativistic binding energy is α_s^2 and hence the series expansion is in powers of $\{\alpha_s^2, \alpha_s^3 \beta_0, \dots, \alpha_s^{n+2}\beta_0^n, \dots\}$. This is in contrast to the usual perturbative expansion (e.g. the relation of the $\overline{\text{MS}}$ mass to the pole mass) which is in terms of $\{\alpha_s, \alpha_s^2 \beta_0, \dots, \alpha_s^{n+1}\beta_0^n, \dots\}$. Hoang, Ligeti and Manohar [49, 50] note that this mismatch disappears in high orders, since we have for large n

$$\left(\frac{1}{n!}\ell^n + \frac{1}{(n-1)!}\ell^{n-1} + \dots + \ell + 1\right) \approx \exp(\ell) = \frac{\mu}{m_b \alpha_s C_F} \qquad (9.186)$$

which fixes the mismatch in the series expansion. In fact, this exponentiation also ensures the cancellation of the renormalons of the pole mass [49].

Thus at low orders we must compare different orders in α_s, which can be made explicit by introducing a counting parameter $\epsilon = 1$ such that

$$m_\Upsilon = 2m_b^{\text{pole}} \left(1 - \frac{(\alpha_s C_F)^2}{8} \left\{ \epsilon + \frac{\alpha_s \beta_0}{\pi}(\ell+1)\epsilon^2 + \left(\frac{\alpha_s \beta_0}{\pi}\right)^2 \left(\frac{1}{2}\ell^2 + \ell + 1\right)\epsilon^3 + \right.\right.$$
$$\left.\left. + \cdots + \left(\frac{\alpha_s \beta_0}{\pi}\right)^n \left(\frac{1}{n!}\ell^n + \frac{1}{(n-1)!}\ell^{n-1} + \dots + \ell + 1\right)\epsilon^{n+1} + \cdots \right\}\right), \quad (9.187)$$

while for the decay rate for a B decay we write for the leading order (schematically)

$$\Gamma = \frac{G_F^2 |V_{\text{CKM}}|^2}{192\pi^3}(m_b^{\text{Pole}})^5 \left[1 + c_1 \frac{\alpha_s}{\pi}\epsilon + c_2 \frac{\alpha_s^2}{\pi^2}\beta_0 \epsilon^2 + \dots + c_n \frac{\alpha_s^n}{\pi^n}\beta_0^{n-1}\epsilon^n + \dots\right]. \quad (9.188)$$

Replacing the pole mass in this relation by the $\Upsilon(1S)$ mass (9.187) and combining the corresponding orders in ϵ yields

$$\Gamma = \frac{G_F^2 |V_{\text{CKM}}|^2}{192\pi^3}\left(\frac{m_\Upsilon}{2}\right)^5 \left[1 + \hat c_1 \epsilon + \hat c_2 \epsilon^2 + \dots + \hat c_n \epsilon^n + \dots\right], \qquad (9.189)$$

which exhibits a quick convergence. In combination with a precise measurement of m_Υ we can get precise predictions for semileptonic decays as well as for moments of spectra.

9.3.6 End-Point Regions

When studying the spectra of photons and leptons within the HQE we find in some regions of phase space a pathological behaviour which prevents us to interpret the spectra point by point. These regions are related to endpoints of the spectra where the HQE breaks down. As an example, let us consider the endpoint of the electron spectrum in semileptonic B decays. The maximal lepton energy is given by

$$E_{\max} = \frac{m_B^2 - m^2}{2m_B}$$

where m is the mass of the lightest final state that can be produced. Close to this energy the possible final states are very few, in the extreme case only the single state with mass m contributes. Clearly we cannot expect an inclusive calculation to be correct here; in other words, the HQE breaks down in this region.

Neglecting the mass of the final-state quark (which we expect to be a good approximation for the $b \to u$ case) already the partonic result behaves pathologically in the endpoint region, since it is a θ function. In fact, we find for the charged lepton energy spectrum up to $1/m_b^2$

$$\frac{d\Gamma}{dy} = \frac{G_F^2 m_b^5 |V_{ub}|^2}{192\pi^3} \left[\theta(2E - m_b)y \left\{ (3 - 2y)y - \frac{5y^2}{3}\frac{\mu_\pi^2}{m_b^2} + \frac{y}{3}(6 + 5y)\frac{\mu_G^2}{m_b^2} \right\} \right.$$
$$\left. + \frac{\mu_\pi^2 - 11\mu_G^2}{6m_b^2}\delta(1 - y) + \frac{\mu_\pi^2}{6m_b^2}\delta'(1 - y) \right] \tag{9.190}$$

with $y = 2E/m_b$. Nevertheless, the integrated inclusive rate exists and can be computed in a $1/m_b$ expansion as shown above; here we get

$$\Gamma = \frac{G_F^2 m_b^5 |V_{ub}|^2}{192\pi^3} \left[1 - \frac{\mu_\pi^2 + 3\mu_G^2}{2m_b^2} \right]. \tag{9.191}$$

In addition, we can show with the same steps as for the total rate that also moments of the spectra can be computed in the HQE.

Obviously the spectrum cannot be interpreted point by point, in particular, close to the endpoint, since the true expansion parameter is $\Lambda_{\rm QCD}/(m_b - 2E)$, which becomes large there. Very close to the endpoint we have a region dominated by single states or resonances, where a description in terms of (a sum over a few) exclusive states is appropriate, and this region is defined by $0 \leq (m_b - 2E)^2 \leq \Lambda_{\rm QCD}^2$.

This particularly means that such a fine 'resolution' of the spectrum in the endpoint region is impossible within the HQE. However, if we look at the structure of the terms of the HQE, we see that the 'most singular' term (i.e. the term with the highest derivative of the δ-function) is the last term in (9.190); in fact, proceeding to $1/m_b^3$ exhibits a term with $\delta''(1 - x)$ etc. These terms can be summed by a technique analogous to to what is done in DIS, leading to non-perturbative functions instead of non-perturbative parameters.

These techniques have been set up in [51]–[53] and put the model suggested in [54] on a firm theoretical basis.

In order to illustrate this technique, we (instead of $B \to X_u \ell \bar{\nu}$) consider $B \to X_s \gamma$. The leading contribution to this process is mediated by the operator

$$O_7 = \frac{e^2}{16\pi^2} m_b \, \bar{s}_L \sigma^{\mu\nu} b_R \, F_{\mu\nu} \qquad H_{\text{eff}} = \frac{4 G_F}{\sqrt{2}} V_{tb} V_{ts}^* C_7(\mu) O_7(\mu). \qquad (9.192)$$

Computing the inclusive rate for $B \to X_s \gamma$ using only this operator yields up to order $1/m_b^2$

$$\Gamma(B \to X_s \gamma) = \frac{\alpha \, G_F^2 m_b^5}{16\pi^4} |V_{tb} V_{ts}^*|^2 |C_7(m_b)|^2 \left[1 - \frac{\mu_\pi^2 + 3\mu_G^2}{2m_b^2} \right]. \qquad (9.193)$$

However, we may also compute the photon spectrum for this decay, which at tree level and to leading order is a δ function, fixing the photon energy to the value $E_\gamma = m_b/2$ determined by the two-particle kinematics of the partonic process. This behaviour persists also for the tree-level expressions at higher orders in the HQE, leading to derivatives of δ functions.

Up to terms of order $1/m_b^2$ we find for the spectrum

$$\frac{d\Gamma}{dy} = \frac{\alpha \, G_F^2 m_b^5}{32\pi^4} |V_{tb} V_{ts}^*|^2 |C_7(m_b)|^2 \qquad (9.194)$$

$$\times \left(\delta(1-y) - \frac{\mu_\pi^2 + 3\mu_G^2}{2m_b^2} \delta(1-y) + \frac{\mu_\pi^2 - \mu_G^2}{2m_b^2} \delta'(1-y) + \frac{\mu_\pi^2}{6m_b^2} \delta''(1-y) \right).$$

Gluon emission will eventually lead to a non-trivial spectrum, although a perturbative calculation is possible only in the region where the gluon and the final-state strange quark have a sizable invariant mass to warrant a perturbative treatment. Close to the endpoint we face the same situation as in $B \to X_u \ell \bar{\nu}$: the spectrum computed for the HQE cannot be interpreted point by point.

However, instead of studying the spectrum point by point, we may take moments of the spectrum. In fact, we may interpret the result (9.194) in terms of an expansion in singular functions, i.e. a moment expansion of the form [55]

$$\frac{1}{\Gamma} \frac{d\Gamma}{dy} = \sum_{n=0}^{\infty} \frac{M_n}{n!} \delta^{(n)}(1-y), \qquad (9.195)$$

where $\delta^{(n)}$ denotes the nth derivative of the δ function, and the moments are defined as

$$M_n = \int_0^{\infty} (y-1)^n \left(\frac{1}{\Gamma} \frac{d\Gamma}{dy} \right). \qquad (9.196)$$

From the structure of the HQE we infer that the moments M_n have a $1/m_b$ expansion, the leading term of which is of order $1/m_b^n$. For the case of $B \to X_s \gamma$ we get

$$M_1 = \mathcal{O}(1/m_b^2) = \frac{\mu_\pi^2 - \mu_G^2}{2m_b^2} \tag{9.197}$$

$$M_2 = \frac{\mu_\pi^2}{6m_b^2} + \mathcal{O}(1/m_b^3)$$

$$M_3 = -\frac{\rho_D^3}{18m_b^3} + \mathcal{O}(1/m_b^4).$$

From this structure it is evident that a resummation scheme would be desirable in which the leading contribution to each moment is resummed. In order to set this up we take a look at the tree-level calculation of $B \to X_s \gamma$. Taking the time ordered product of two effective Hamiltonians from (9.192) and using the external field propagator as in (9.129) (in this case of the massless s quark) we have

$$\frac{1}{\slashed{Q} + i\slashed{D}} = \frac{\slashed{Q} + i\slashed{D}}{Q^2 + 2(Q \cdot iD) + (i\slashed{D})^2} \tag{9.198}$$

where $Q = m_b v - q$, and q is the photon momentum. In the case where Q^2 is large compared to the terms with the covariant derivatives, we obtain the usual power counting and can perform the expansion as in (9.129) with $m_c \to 0$. However, we have $Q^2 = m_b^2(1 - y)$ and thus this quantity is not large compared to the other terms in the denominator, in which case we cannot expand as in (9.129). Instead we are in the kinematic region, where Q^2 is small and $v \cdot Q$ is of the order m_b.

The region we are interested in is that where Q^2 and $(Q \cdot iD)$ are of the same order: is $m_b \Lambda_{\text{QCD}}$. Note that this is not the resonance region, where—as discussed above—Q^2 is actually of order Λ_{QCD}^2. Thus in the endpoint region $m_b(1 - y) \sim \Lambda_{\text{QCD}}$ we can re-sum the leading contributions to the moments by approximating

$$\frac{1}{\slashed{Q} + i\slashed{D}} = \frac{\slashed{Q}}{Q^2 + 2(Q \cdot iD)} + \cdots . \tag{9.199}$$

Since Q is (almost) a light-like vector, it is convenient to introduce light cone vectors n and $\bar{n}$ according to

$$n^2 = 0 = \bar{n}^2 \quad n \cdot \bar{n} = 2 \quad v = \frac{n + \bar{n}}{2} \quad \text{and} \quad Q \cdot iD \approx (v \cdot Q)(n \cdot iD). \tag{9.200}$$

Introducing the shape function (or light come distribution function) f according to

$$2M_B f(\omega) = \langle B(v) | \bar{b}_v \delta(\omega + i(n \cdot D)) | B(v) \rangle$$

allows us now to write this as a convolution, so we get (e.g. for $\Gamma = 1$)

$$\langle B(v) | \bar{b}_v \frac{\slashed{Q}}{Q^2 + 2(Q \cdot iD)} b_v | B(v) \rangle = \int d\omega f(\omega) \frac{v \cdot Q}{Q^2 + 2\omega(v \cdot Q)}. \tag{9.201}$$

The shape function is a non-perturbative function, where the moments of f are given in terms of HQE parameters

$$f(\omega) = \delta(\omega) + \frac{\mu_\pi^2}{6}\delta''(\omega) - \frac{\rho_D^3}{18}\delta'''(\omega) + \cdots . \tag{9.202}$$

Using this function and using (9.199) we obtain for the spectrum of $B \to X_s\gamma$

$$\frac{d\Gamma}{dy} = \frac{\alpha G_F^2 m_b^6}{32\pi^4}|V_{tb}V_{ts}^*|^2|C_7(m_b)|^2 f(m_b(y-1)) \tag{9.203}$$

$$= \frac{\alpha G_F^2 m_b^5}{32\pi^4}|V_{tb}V_{ts}^*|^2|C_7(m_b)|^2 \left(\delta(1-y) + \frac{\mu_\pi^2}{6m_b^2}\delta''(1-y) - \frac{\rho_D^3}{18m_b^3}\delta'''(1-y) + \cdots\right)$$

showing that the shape function indeed resums the most singular terms, i.e. the terms with the highest derivatives of δ functions.

The shape function f plays the same role as the parton distributions of DIS. It is genuinely non-perturbative; however, it is also universal. For the case of B decays, this means that this shape function appears in the end point regions $m_b(1-y) \sim \Lambda_{QCD}$ of any inclusive heavy-to-light transition. In other words, it also appears in the description of the end-point region of $B \to X_u\ell\bar{\nu}$. This leads to a relation between this decay and $B \to X_s\gamma$, which is exploited in phenomenological analyses.

The shape function has a few interesting properties. First of all, we note that the first moment vanishes, i.e. the term with the first derivative of the δ function is absent. This is a consequence of the equations of motion. The second moment is non-vanishing; since this moment is taken with respect to the partonic end point (defined by the quark mass), this means that the shape function has to extend beyond the partonic endpoint $y = 1$ corresponding to the photon energy $E_\gamma = m_b/2$ and $\omega = 0$. The shape function is non-vanishing for $-\infty \le \omega \le \bar{\Lambda}$, where the region $0 \le \omega \le \bar{\Lambda}$ is entirely non-perturbative. The parameter $\bar{\Lambda}$ is exactly the same as that appearing in the expansion of the heavy hadron masses (9.32, 9.33), since the true phase space (ignoring the masses of the final-state hadrons) has a maximal photon energy $E_\gamma^{max} = m_B/2$. Thus the shape function ensures the correct phase space boundary, which is given in terms of the B meson mass.

In fact, the vanishing of the first moment corresponds to a definition of the quark mass. A measurement of the photon spectrum of $B \to X_s\gamma$ yields directly the shape function, and the reference point for which the first moment vanishes yields a measurement of $\bar{\Lambda}$ and hence a definition of the quark mass.

All further discussion, including the way to include radiative corrections, requires more heavy machinery. Since the end point region in heavy-to-light decays is related to (in the restframe of the B meson) energetic light degrees of freedom, the proper tool here is SCET (see Chapter 5, this volume).

9.4 Summary

This chapter outlined the physical foundations of heavy quark methods, with a focus on technical issues. There are many aspects of heavy quark theory that could not be covered:

- Heavy quark expansions (HQET as well as HQE) have an enormous impact on particle physics phenomenology. Until the development of these methods, the hadronic matrix elements had to be modelled, introducing an uncontrollable systematic uncertainty into the theoretical predictions. Heavy quark methods have not made models fully obsolete; however, the use of a model is often necessary only for an estimate of subleading terms, for which we cannot make use of a QCD lattice calculation. To this end, many constraints on physics beyond the Standard Model coming from heavy-flavour physics could be made much more stringent on the basis of heavy quark methods.

- There are still other types of heavy quark methods, which could not be covered by the lectures. As an example, in an exclusive non-leptonic decay like $B \to \pi\pi$, the two pions have (in the rest frame of the B meson) energies of the order of the B meson mass; thus there are light quarks and gluons that have large momenta. Since the settings we have discussed in these lectures are such that the light degrees of feedom have 'small' momenta, it becomes clear that a different effective theory needs to be used in such cases. The relevant theory here is SCET, which has been invented in the context of B decays, but has many applications also in collider physics.

- Another class of heavy quark systems are hadrons with two heavy quarks, such as the quarkonia $b\bar{b}$ and $c\bar{c}$ as well as the $B_c = (b\bar{c})$ or doubly heavy baryons that contain two or even three heavy quarks. Also in this case we must set up a slightly modified effective theory, since the static approximation turns out to be insufficient. Rather we need to include the kinetic energy $\vec{p}^2/(2m)$ into the leading Lagrangian. This leads to non-relativistic QCD (NRQCD), which corresponds to the leading term of a systematic expansion in the relative velocity v between the heavy constituents. The structure of NRQCD turns out to be also more complicated than HQET, since the dynamics of binding generates mass scales, which are the inverse Bohr radius $m_Q v$ and the binding energy $m_Q v^2$. For small v this generates a hierarchy of mass scales, and—depending on the sizes of the dynamically generated mass scales relative to Λ_{QCD}—requires the set up of different EFTs.

Overall, heavy quark methods have put the flavour physics of heavy hadrons onto a solid basis, allowing us to perform in many cases precision calculations including the control over uncertainties. Starting from the original idea encoded in HQET, larger numbers of applications have been discovered and elaborated; it seems that these ideas even have a potential which is not yet fully explored.

Acknowledgements

I thank the organizers of the school for the invitation; it was a real pleasure to lecture for highly motivated students in such an inspiring environment. I also thank Keri Vos for a careful reading of the write-up of the lectures.

References

[1] E. Eichten and B. R. Hill, (1990) *Phys. Lett. B* **234**: 511. doi:10.1016/0370-2693(90)92049-O

[2] E. Eichten and B. R. Hill, (1990). *Phys. Lett. B* **243**: 427. doi:10.1016/0370-2693(90)91408-4

[3] B. Grinstein, (1990). *Nucl. Phys. B* **339**: 253. doi:10.1016/0550-3213(90)90349-I

[4] H. Georgi, (1990). *Phys. Lett. B* **240**: 447. doi:10.1016/0370-2693(90)91128-X

[5] M. A. Shifman and M. B. Voloshin, (1988). *Sov. J. Nucl. Phys.* **47**: 511. [*Yad. Fiz.* **47**: 801].

[6] N. Isgur and M. B. Wise, (1989). *Phys. Lett. B* **232**: 113. doi:10.1016/0370-2693(89)90566-2

[7] N. Isgur and M. B. Wise, (1990). *Phys. Lett. B* **237**: 527. doi:10.1016/0370-2693(90)91219-2

[8] J. Chay, H. Georgi and B. Grinstein, (1990). *Phys. Lett. B* **247**: 399. doi:10.1016/0370-2693(90)90916-T

[9] I. I. Y. Bigi, N. G. Uraltsev and A. I. Vainshtein, (1992). *Phys. Lett. B* **293**: 430. Erratum: [*Phys. Lett. B* **297**: 477] doi:10.1016/0370-2693(92)90908-M, 10.1016/0370-2693(92)91287-J [hep-ph/9207214].

[10] I. I. Y. Bigi, M. A. Shifman, N. G. Uraltsev and A. I. Vainshtein, (199.). *Phys. Rev. Lett.* **71**: 496. doi:10.1103/PhysRevLett.71.496 [hep-ph/9304225].

[11] A. V. Manohar and M. B. Wise, (1994). *Phys. Rev. D* **49**: 1310. doi:10.1103/PhysRevD.49.1310 [hep-ph/9308246].

[12] T. Mannel, (1994). *Nucl. Phys. B* **413**: 396. doi:10.1016/0550-3213(94)90625-4 [hep-ph/9308262].

[13] M. Neubert, (1994). *Phys. Rept.* **245**: 259. doi:10.1016/0370-1573(94)90091-4 [hep-ph/9306320].

[14] I. I. Y. Bigi, M. A. Shifman and N. Uraltsev, (1997). *Ann. Rev. Nucl. Part. Sci.* **47**: 591. doi:10.1146/annurev.nucl.47.1.591 [hep-ph/9703290].

[15] B. Grinstein, (1994). *Proceedings of the Theoretical Advanced Study Institute in Elementary Particle Physics (TASI–94).* 29 May–24 June 1994, Boulder, CO [arXiv:hep-ph/9411275].

[16] T. Mannel, (1997). *Rept. Prog. Phys.* **60**: 1113. doi:10.1088/0034-4885/60/10/003

[17] A. G. Grozin, (2004). *Heavy Quark Effective Theory,* Springer. doi:10.1007/b79301

[18] T. Mannel, (2004). *Effective Field Theories in Flavour Physics,* Springer. doi:10.1007/b79301

[19] A. V. Manohar and M. B. Wise, (2000). *Heavy Quark Physics,* Cambridge University Press

[20] A. A. Petrov and A. E. Blechman, (2016). *Effective Field Theories,* World Scientific

[21] T. Mannel, W. Roberts and Z. Ryzak, (1992). *Nucl. Phys. B* **368**: 204. doi:10.1016/0550-3213(92)90204-O

[22] S. Balk, J. G. Korner and D. Pirjol, (1994) *Nucl. Phys. B* **428**: 499. doi:10.1016/0550-3213(94)90211-9 [hep-ph/9307230].

[23] F. Bloch and A. Nordsieck, (1937). *Phys. Rev.* **52**: 54. doi:10.1103/PhysRev.52.54

[24] D. R. Yennie, S. C. Frautschi and H. Suura, (1961). *Annals Phys.* **13**: 379. doi:10.1016/0003-4916(61)90151-8

[25] M. J. Dugan, M. Golden and B. Grinstein, (1992). *Phys. Lett. B* **282**: 142. doi:10.1016/0370-2693(92)90493-N

[26] M. E. Luke and A. V. Manohar, (1992). *Phys. Lett. B* **286**: 348. doi:10.1016/0370-2693(92)91786-9 [hep-ph/9205228].

[27] Y. Q. Chen, (1993). *Phys. Lett. B* **317**: 421. doi:10.1016/0370-2693(93)91018-I

[28] J. Heinonen, R. J. Hill and M. P. Solon, (2012). *Phys. Rev. D* **86**: 094020. doi:10.1103/PhysRevD.86.094020 [arXiv:1208.0601 [hep-ph]].

[29] M. Ademollo and R. Gatto, (1964). *Phys. Rev. Lett.* **13**: 264. doi:10.1103/PhysRevLett.13.264

[30] M. E. Luke, (1990). *Phys. Lett. B* **252**: 447. doi:10.1016/0370-2693(90)90568-Q

[31] A. F. Falk, (1992). *Nucl. Phys. B* **378**: 79. doi:10.1016/0550-3213(92)90004-U

[32] M. Neubert and C. T. Sachrajda, (1997). *Nucl. Phys. B* **483**: 339. doi:10.1016/S0550-3213(96)00559-7 [hep-ph/9603202].

[33] A. F. Falk, H. Georgi, B. Grinstein and M. B. Wise, (1990). *Nucl. Phys. B* **343**: 1. doi:10.1016/0550-3213(90)90591-Z

[34] W. Kilian, T. Mannel and T. Ohl, (1993). *Phys. Lett. B* **304**: 311. doi:10.1016/0370-2693(93)90301-W [hep-ph/9303224].

[35] M. Neubert, (1991). *Phys. Lett. B* **264**: 455. doi:10.1016/0370-2693(91)90377-3

[36] K. Melnikov, (2008). *Phys. Lett. B* **666**: 336. doi:10.1016/j.physletb.2008.07.089 [arXiv:0803.0951 [hep-ph]].

[37] B. Grinstein and A. Kobach, (2017). *Phys. Lett. B* **771**: 359. doi:10.1016/j.physletb.2017.05.078 [arXiv:1703.08170 [hep-ph]].

[38] J. C. Collins, (2008). *Renormalization : An Introduction to Renormalization, The Renormalization Group, and the Operator Product Expansion,* Cambridge University Press.

[39] E. C. Poggio, H. R. Quinn and S. Weinberg, (1976) *Phys. Rev. D* **13**: 1958. doi:10.1103/PhysRevD.13.1958

[40] M. A. Shifman, doi:10.1142/9789812810458-0032 hep-ph/0009131.

[41] B. M. Dassinger, T. Mannel and S. Turczyk, (2007). *JHEP* **0703**: 087. [doi:10.1088/1126-6708/2007/03/087 [hep-ph/0611168].

[42] B. Guberina, S. Nussinov, R. D. Peccei and R. Ruckl, (1979). *Phys. Lett. B* **89**: 111. doi:10.1016/0370-2693(79)90086-8

[43] T. Mannel, S. Turczyk and N. Uraltsev, (2010). *JHEP* **1011**: 109. doi:10.1007/JHEP11(2010)109 [arXiv:1009.4622 [hep-ph]].

[44] M. Beneke and V. M. Braun, (1994). *Nucl. Phys. B* **426**: 301. doi:10.1016/0550-3213(94)90314-X [hep-ph/9402364].

[45] I. I. Y. Bigi, M. A. Shifman, N. G. Uraltsev and A. I. Vainshtein, (1994). *Phys. Rev. D* **50**: 2234. doi:10.1103/PhysRevD.50.2234 [hep-ph/9402360].

[46] M. Neubert and C. T. Sachrajda, (1995). *Nucl. Phys. B* **438**: 235. doi:10.1016/0550-3213(95)00032-N [hep-ph/9407394].

[47] I. I. Y. Bigi, M. A. Shifman, N. G. Uraltsev and A. I. Vainshtein, (1995). *Phys. Rev. D* **52**: 196. doi:10.1103/PhysRevD.52.196 [hep-ph/9405410].

[48] M. Antonelli et al., (2010). *Phys. Rept.* **494**: 197. doi:10.1016/j.physrep.2010.05.003 [arXiv:0907.5386 [hep-ph]].

[49] A. H. Hoang, Z. Ligeti and A. V. Manohar, (1999). *Phys. Rev. Lett.* **82**: 277. doi:10.1103/PhysRevLett.82.277 [hep-ph/9809423].

[50] A. H. Hoang, Z. Ligeti and A. V. Manohar, (1999). *Phys. Rev. D* **59**: 074017. doi:10.1103/PhysRevD.59.074017 [hep-ph/9811239].

[51] I. I. Y. Bigi, M. A. Shifman, N. G. Uraltsev and A. I. Vainshtein, (1994). *Int. J. Mod. Phys. A* **9**: 2467. doi:10.1142/S0217751X94000996 [hep-ph/9312359].

[52] M. Neubert, (1994). *Phys. Rev. D* **49**: 3392. doi:10.1103/PhysRevD.49.3392 [hep-ph/9311325].

[53] T. Mannel and M. Neubert, (1994). *Phys. Rev. D* **50**: 2037. doi:10.1103/PhysRevD.50.2037 [hep-ph/9402288].

[54] G. Altarelli, et al., (1982). *Nucl. Phys. B* **208**: 365. doi:10.1016/0550-3213(82)90226-7

[55] M. Neubert, (1994). *Phys. Rev. D* **49**: 4623. doi:10.1103/PhysRevD.49.4623 [hep-ph/9312311].

10

Heavy Quark Effective Theory: A Predictive EFT on the Lattice

Rainer SOMMER

John von Neumann Institute for Computing (NIC), DESY, Platanenallee 6, 15738 Zeuthen, Germany

and

Institut für Physik, Humboldt-Universität zu Berlin, Newtonstr. 15, 12489 Berlin, Germany

Rainer SOMMER

Sommer, R., *Heavy Quark Effective Theory: A Predictive EFT on the Lattice* In: *Effective Field Theory in Particle Physics and Cosmology.*
Edited by: Sacha Davidson, Paolo Gambino, Mikko Laine, Matthias Neubert and Christophe Salomon, Oxford University Press (2020).
© Oxford University Press. DOI: 10.1093/oso/9780198855743.003.0010

Chapter Contents

Preface

This chapter describes heavy quark effective theory as an example of a non-perturbative implementation of an effective field theory (EFT). Our emphasis is on the concepts needed for a non-perturbative formulation beyond Feynman graphs rather than on the application. Given that few of the students at the Les Houches school session from where these lectures come work in lattice field theory, it includes a very short description of its essentials.

It also provides a glimpse of Symanzik's effective field theory which describes how lattice field theories approach the continuum limit in Section 10.5.2

Section 10.1 explains why and how EFTs are important for lattice field theory and how lattice field theory is important for EFTs. Section 10.2 is a reminder of (continuum) HQET and introduces our notation; we focus on the theory at zero velocity. Section 10.3 sets the stage for an EFT explaining the relevance of universality and renormalizability at the non-perturbative level. Section 10.4 contains a discussion of the renormalization of HQET and its matching to QCD. At the leading order in $1/m_b$ (Section 10.4.1) we go into some details because we also want to show the limitations in perturbative matching for B-physics. If the reader is not interested in the actual application, but rather in the general EFT aspects, the reader can skip most of this part and study just the case of next to leading order in $1/m_b$ (Section 10.4.2). Finally in Section 10.5 we show how simple it is to put the EFT on the lattice and Section 10.6.2 explains the essential strategy for matching EFT and fundamental theory in practical lattice computations. Section 10.7 provides a summary of the most important messages.

10.1 Lattice for EFTs, EFTs for the Lattice, and EFT on the Lattice

10.1.1 Overview

For concreteness we consider 3+1 dimensions. Lattice field theories then approximate the 4-d (Euclidean) space by the discrete set of points of a lattice. A hyper-cubic one,

$$\Lambda = a\mathbb{Z}^4 = \{x_\mu = an_\mu \mid n_\mu \in \mathbb{Z}, \ \mu = 0,1,2,3\}, \tag{10.1}$$

has enhanced symmetry which is important for renormalization. For numerical computations one considers a finite lattice,

$$\Lambda = \{x_\mu = an_\mu \mid n_\mu = 0,1,\ldots L_\mu/a - 1\} \tag{10.2}$$

Table 10.1 Typical scales in large volume lattice QCD. Quark masses are often varied. Restrictions are due to limits in computer power ($L/a = 32 - 128$) and the need to avoid finite size effects ($m_\pi L \gtrsim 4$).

energy scale		typical range	remark
mass gap	m_π	130 MeV – 500 MeV	infrared scale $L \gg m_\pi^{-1}$
momentum cutoff	$1/a$	1.5 GeV – 6 GeV	
b-quark mass	m_b	$\approx 5\,\text{GeV}$	

with mostly

$$L_0 = T, \; L_1 = L_2 = L_3 = L. \tag{10.3}$$

The action which appears in the path integral weight has the form

$$S = \int d^4x \, \mathscr{L}_{\text{cont}}(x) \rightarrow a^4 \sum_{x \in \Lambda} \mathscr{L}_{\text{lat}}(x) \tag{10.4}$$

$$\equiv a^4 \sum_{n_0,n_1,n_2,n_3} \mathscr{L}_{\text{lat}}((an_0, an_1, an_2, an_3)). \tag{10.5}$$

We discuss Lagrangians $\mathscr{L}_{\text{lat}}$ later.

Euclidean two-point functions,

$$G_2^\Phi(x,y) = \langle \Phi^\star(x)\Phi(y) \rangle, \tag{10.6}$$

allow for the extraction of energies E_n and matrix elements $\mathcal{M}_n = \langle n|\Phi(\mathbf{x})|0 \rangle$ through

$$\frac{1}{L^3} a^6 \sum_{\mathbf{x}\mathbf{y}} G_2^\Phi(x,y) = \sum_{n \geq 1} |\mathcal{M}_n|^2 \, e^{-|x_0-y_0|E_n} \tag{10.7}$$

$$[\text{ e.g. } \Phi = \Phi_\pi = i\bar{u}\gamma_5 d, \; \Phi^* = i\bar{d}\gamma_5 u]. \tag{10.8}$$

At large $x_0 - y_0$, the small n terms dominate and one can determine the low lying energies and matrix elements. In particular when the two-point function is projected to space momentum zero, $\mathbf{p} = 0$, as done here by $a^3 \sum_{\mathbf{x}} f(\mathbf{x}) = a^3 \sum_{\mathbf{x}} e^{i\mathbf{p}\cdot\mathbf{x}} f(\mathbf{x})\big|_{\mathbf{p}=0}$, and when there is a bound state in the channel excited by Φ (as opposed to just resonances), then

$$E_1 = m(a) \tag{10.9}$$

is a particle mass and $\mathcal{M}_1$ is a vacuum-to-one-particle matrix element. The physical mass is given by the continuum limit

$$m_{\text{cont}} = \lim_{a \to 0} m(a). \tag{10.10}$$

The big point is that the Euclidean, latticised, path integral can be evaluated by Monte Carlo simulations [36, 46] and thus truncated expansions in coupling or quark mass are *not* needed. Thus a lattice field theory gives us access to the non-perturbative spectrum and selected matrix elements. It provides the non-perturbative (NP) definition of the QFT and through Monte Carlo simulations also a means for its numerical solution. A feeling for relevant scales and limitations in numerical lattice QCD is provided by Table 10.1.

10.1.2   Why EFT on the Lattice?

The question arises, why then consider EFTs in the context of lattice field theories? Roughly speaking, there are three reasons.

1. Lattice QCD can provide observables that add to experimental ones in order to determine the parameters of EFTs, e.g. we may take quark masses or the electromagnetic coupling to be different from the one in Nature. In this way we can help determine the free parameters of EFTs, e.g. for chiral PT.

2. EFTs can help extrapolations of numerical results of lattice QCD. Prominent examples are the extrapolations $L \to \infty$ and to physical quark masses (chiral PT) and to the continuum limit (Symanzik EFT). In some cases there are parameter-free asymptotic formulae; more generally, the EFT dictates the functional form of extrapolations, which involves a (hopefully small) number of free parameters, usually called low-energy constants. We refer to them just as the parameters of the EFT.

3. Third, there are EFTs that are *not* solvable analytically because they are strongly interacting, such as HQET, which describes heavy quarks interacting non-perturbatively with the other QCD fields. Here the EFT itself is discretized and simulated on a lattice.

Table 10.2 contains examples, the relevant scales, and the relations between the scales that are necessary for the EFT (and lattice QCD) to apply.

Apart from the theoretical interest in describing EFTs non-perturbatively, the advantage of using HQET and not just QCD with the heavy b-quark as a relativistic Dirac field is that when the b-quark is a relativistic Dirac field, its mass (just like any mass) and energy that is relevant needs to be far below the cut-off, $m_b \ll a^{-1}$. In contrast,

$$m_b > \Lambda_{\text{cut}} = a^{-1}, \tag{10.11}$$

is perfectly fine in HQET. Thus b-quarks become treatable without extrapolations or other tricks. Table 10.1 shows the relevant scales.

Table 10.2 Examples for the interplay of EFT and lattice QCD. Special considerations beyond $E \ll a^{-1}$ are marked in the last double-column.

rôle	EFT	applicability range	
		of EFT	of lattice QCD
1. Lattice for EFT			
determine LECs	Chiral PT	low-energy QCD	
2. EFT for Lattice			
discretization effects	Symanzik EFT (Section 10.5.2)	$E \ll a^{-1}$	
finite volume effects	Chiral PT [14, 22]	$L^{-1} \ll m_\pi, \Lambda_{QCD}$	
quark mass effects	Chiral PT [21]	$m_u, m_d \ll \Lambda_{QCD}$	
	Heavy Meson	$m_b \gg \Lambda_{QCD}$	$m_b \ll a^{-1}$
	Chiral PT [24, 31, 45]	$m_u, m_d \ll \Lambda_{QCD}$	
combined effects	HMrsChPT [1]	$m_u, m_d \ll \Lambda_{QCD}$	
		$m_b \gg \Lambda_{QCD}$,	$m_b \ll a^{-1}$
3. EFT on the Lattice			
NP EFT	QCD[(3)]	$E \ll m_c, m_b, m_t$	
NP EFT	HQET	$E, \Lambda_{QCD} \ll m_b$	
NP EFT	NRQCD	$E, \Lambda_{QCD} \ll a^{-1} \ll m_c, m_b$	
NP EFT	Nuclear EFT	see the literature	

This chapter focuses on HQET but emphasizes general features of non-perturbative EFTs and the question how the parameters of the EFT can be determined without losing predictions. It also gives a flavour of Symanzik EFT, the basis for understanding how lattice field theories approach the continuum.

10.2 Heavy Quark Effective Theory at Zero Velocity

The theory was introduced by T. Mannel at this school [38]. Further details concerning the basics as well as the phenomenology can be found in [25, 39, 40]. Our notation is

$$\psi_h = \phi_0|_{\text{Mannel}}, \tag{10.12}$$

$$\psi_{\bar{h}} = \chi_0|_{\text{Mannel}}. \tag{10.13}$$

We repeat some of the definitions for the special case of zero velocity in Euclidean time and in our notation. Until Section 10.5 we work with the formal (no regularization) continuum theory.

10.2.1 Lagrangian and Propagator

The Lagrangian is

$$\mathscr{L} = \mathscr{L}_h^{\text{stat}} + \mathscr{L}_h^{(1)} + \mathscr{L}_{\bar{h}}^{\text{stat}} + \mathscr{L}_{\bar{h}}^{(1)} + \mathrm{O}(\tfrac{1}{m_h^2}) \tag{10.14}$$

$$\mathscr{L}_h^{\text{stat}} = \overline{\psi}_h(m_h + D_0)\psi_h, \qquad \mathscr{L}_{\bar{h}}^{\text{stat}} = \overline{\psi}_{\bar{h}}(m_h - D_0)\psi_{\bar{h}}, \tag{10.15}$$

$$\mathscr{L}_h^{(1)} = -\tfrac{1}{2m_h}(\mathscr{O}_{\text{kin}} + \mathscr{O}_{\text{spin}}). \tag{10.16}$$

The heavy quark fields formally have four components, but due to the constraints

$$\psi_h = P_+\psi_h, \quad \psi_{\bar{h}} = P_-\psi_{\bar{h}}, \tag{10.17}$$

with

$$P_\pm = \tfrac{1}{2}(1 \pm \gamma_0), \quad P_\pm^2 = 1, \quad P_+P_- = 0, \tag{10.18}$$

they only have two degrees of freedom. The mass-dimension five fields composing $\mathscr{L}_h^{(1)}$ are

$$\mathscr{O}_{\text{kin}}(x) = \overline{\psi}_h(x)\,\mathbf{D}^2\,\psi_h(x), \tag{10.19}$$

$$\mathscr{O}_{\text{spin}}(x) = \overline{\psi}_h(x)\,\boldsymbol{\sigma} \cdot \mathbf{B}(x)\,\psi_h(x), \tag{10.20}$$

with

$$\sigma_k = \tfrac{1}{2}\epsilon_{ijk}\sigma_{ij}, \quad B_k = i\tfrac{1}{2}\epsilon_{ijk}[D_i, D_j], \tag{10.21}$$

given in terms of the gauge covariant derivative D_μ and $\sigma_{\mu\nu} = \tfrac{i}{2}[\gamma_\mu, \gamma_\nu]$. Analogous expressions for $\mathscr{L}_{\bar{h}}^{(1)}$ are not needed in the following.

Consider the lowest order $\mathscr{L} = \mathscr{L}_h^{\text{stat}}$. The propagator, G^{stat}, of the field ψ_h is defined by

$$(D_0 + m)\,G^{\text{stat}}(x, y; A) = (\partial_0 + A_0(x) + m)\,G^{\text{stat}}(x, y; A) = \delta(x - y)\,P_+. \tag{10.22}$$

This is the propagator in the presence of a gauge-field $A_\mu(x) = -A_\mu^\dagger(x)$ that we later integrate over in the path-integral. It is easy to write it down explicitly:

$$G^{\text{stat}}(x,y;A) = \theta(x_0 - y_0)\,\delta(\mathbf{x} - \mathbf{y})\,P_+ \exp(-m(x_0 - y_0))\,\mathcal{P}(x \leftarrow y), \quad (10.23)$$

$$\mathcal{P}(x \leftarrow y) = \mathbf{P}_{\text{ord}} \exp\left\{-\int_{y_0}^{x_0} dz_0 A_0(z_0, \mathbf{x})\right\}. \quad (10.24)$$

Here $\mathbf{P}_{\text{ord}}$ denotes path ordering. Similarly we write the propagator of the light quarks in a background gauge field as $G^{\text{light}}(y,x;A_\mu)$. Now consider a two-point function with a heavy quark, for example

$$C_{AA}^{\text{stat}}(x,y;m) = \left\langle A_0^{\text{stat}}(x)(A_0^{\text{stat}})^\dagger(y)\right\rangle_{\text{stat}}, \quad A_0^{\text{stat}} = \bar\psi\gamma_0\gamma_5\psi_h \quad (10.25)$$

$$= -a^3 \sum_{\mathbf{x}} \frac{1}{Z}\int D[A] \quad (10.26)$$

$$\text{tr}\left\{G^{\text{stat}}(x,y;A_\mu)\gamma_5\gamma_0 G^{\text{light}}(y,x;A_\mu)\gamma_0\gamma_5\right\} e^{-S_{\text{eff}}[A]}.$$

All such two-point functions satisfy

$$C^{\text{stat}}(x,y;m) = C^{\text{stat}}(x,y;0)\,\exp(-m(x_0 - y_0)). \quad (10.27)$$

Therefore the mass term in the Lagrangian may be removed (strictly speaking we need $m \to \epsilon > 0$ and then consider the limit $\epsilon \to 0$) and then all energies shifted by it:

$$E_n = E_n|_{m=0} + m. \quad (10.28)$$

Note that we did not use perturbation theory in this simple derivation. Equations (10.27) and (10.28) are valid beyond this expansion.

10.2.2 Symmetries

10.2.2.1 Flavour

If there are F heavy quarks, we just add a corresponding flavour index and use a notation

$$\psi_h \to \psi_h = (\psi_{h1},\ldots,\psi_{hF})^T, \quad \bar\psi_h \to \bar\psi_h = (\bar\psi_{h1},\ldots,\bar\psi_{hF}) \quad (10.29)$$

$$\mathcal{L}_h^{\text{stat}} = \bar\psi_h(D_0 + \epsilon)\psi_h. \quad (10.30)$$

Then we obviously have the symmetry

$$\psi_h(x) \to V\psi_h(x), \quad \bar\psi_h(x) \to \bar\psi_h(x)V^\dagger, \quad V \in SU(F) \quad (10.31)$$

and the same for the anti-quarks. Note that this symmetry emerges in the large mass limit irrespective of how the limit is taken. For example we may take ($F = 2$ with the first heavy flavour identified with charm and the second with beauty)

$$m_b - m_c = c \times \Lambda_{QCD}, \quad \text{or} \quad m_b/m_c = c', \quad m_b \to \infty \tag{10.32}$$

with either c or c' fixed when taking $m_b \to \infty$.

10.2.2.2 Spin

We further note that for each field there are also the two spin components but the Lagrangian contains no spin-dependent interaction. The associated SU(2) rotations are generated by the spin matrices (10.21) (remember that ψ_h, $\overline{\psi}_h$ are kept as four-component fields with two components vanishing)

$$\sigma_k = \frac{1}{2}\epsilon_{ijk}\sigma_{ij} = \begin{pmatrix} \sigma_k & 0 \\ 0 & \sigma_k \end{pmatrix}, \tag{10.33}$$

where the symbol σ_k is used at the same time for the Pauli matrices and the 4×4 matrix. This is in the Dirac representation where

$$\gamma_0 = \begin{pmatrix} 1 & 0 \\ 0 & -1 \end{pmatrix}, P_+ = \begin{pmatrix} 1 & 0 \\ 0 & 0 \end{pmatrix}, P_- = \begin{pmatrix} 0 & 0 \\ 0 & 1 \end{pmatrix}. \tag{10.34}$$

The spin rotation is then

$$\psi_h(x) \to e^{i\alpha_k\sigma_k}\psi_h(x), \quad \overline{\psi}_h(x) \to \overline{\psi}_h(x)e^{-i\alpha_k\sigma_k}, \tag{10.35}$$

with arbitrary real parameters α_k. It acts on each flavour component of the field. Obviously, the symmetry is even bigger. We can take $V \in SU(2F)$ in (10.31). This plays a role in heavy meson ChPT [10, 24, 56].

10.2.2.3 Local Flavour Number

The static Lagrangian contains no space derivative. The transformation

$$\psi_h(x) \to e^{i\eta(x)}\psi_h(x), \quad \overline{\psi}_h(x) \to \overline{\psi}_h(x)e^{-i\eta(x)}, \tag{10.36}$$

is therefore a symmetry for any local phase $\eta(\mathbf{x})$. For every point $\mathbf{x}$ there is a corresponding Noether charge

$$Q_h(x) = \overline{\psi}_h(x)\psi_h(x)\,[= \overline{\psi}_h(x)\gamma_0\psi_h(x)] \tag{10.37}$$

which we call local flavour number. It is conserved,

$$\partial_0 Q_h(x) = 0 \ \forall x. \tag{10.38}$$

We can take the field content together with these symmetries to be the defining properties of the (lowest-order) EFT.

10.3   Non-Perturbative Formulation of EFT

For a short while we abstract from HQET and discuss the general concept and formulation of an EFT. The special features of HQET are discussed in the following subsection.

We consider processes in a fundamental theory (for example, QCD or the Standard Model of particle physics—the important feature is the renormalizability of the theory) at low energy. In particular we first focus on processes (scattering, decays, etc.) of particles that have masses of the order of this low energy or below it (in HQET also the large mass particles *are* involved as is evident by our Lagrangian above and discussed by T. Mannel). In this situation, vacuum fluctuations involving much heavier particles (with masses denoted by m_b in view of our later application) are suppressed and a true creation of the heavier particles is energetically forbidden. We therefore expect to be able to describe the physics of these low energy processes by an EFT containing only the fields of the light particles [53].[1] The leading-order Lagrangian of the theory is formed from the free field theory Lagrangians and all the renormalizable interactions. Restricting to just renormalizable interactions at the lowest order is not always possible. We return to this in Section 10.3.2. First we consider universal EFTs where the lowest-order theory is renormalizable.

10.3.1   Universal EFTs

Assuming the usual power counting, all local composite fields with mass dimension smaller or equal to four are allowed.[2] Let us denote the Lagrangian by $\mathscr{L}_{\mathrm{LO}}$ and the Euclidean action is $S^{\mathrm{LO}} = \int \mathrm{d}^4 x \mathscr{L}_{\mathrm{LO}}(x)$. Observables such as correlation functions are then defined by the standard path integral,

$$\Phi^{\mathrm{LO}} = \langle O \rangle_{\mathrm{LO}} = \frac{1}{Z_{\mathrm{LO}}} \int_{\mathrm{fields}} \mathrm{e}^{-S^{\mathrm{LO}}} O, \quad S^{\mathrm{LO}} = \int \mathrm{d}^4 x \mathscr{L}_{\mathrm{LO}}(x), \qquad (10.39)$$

$$\mathscr{L}_{\mathrm{LO}}(x) = \sum_i \omega_i^{\mathrm{LO}} \mathscr{O}_i^{\mathrm{LO}}(x), \quad [\mathscr{O}_i^{\mathrm{LO}}] \le 4 \quad [\omega_i^{\mathrm{LO}}] \ge 0, \qquad (10.40)$$

($[\mathscr{O}]$ denotes the mass dimension of a field $\mathscr{O}$) with $\langle 1 \rangle_{\mathrm{LO}} = 1$ and O some multilocal product of fields, for example

$$O = \varphi(x)\varphi(y). \qquad (10.41)$$

---

[1] While this statement, referred to as the Weinberg theorem, was originally a conjecture, it has since been tested in many cases and no contradiction has been found.

[2] This is a naive argument. For more details see [48].

In this way we start at LO with a *renormalizable* theory. For a lattice formulation this means that the *continuum limit* of the theory exists when a finite number of bare parameters is varied as a function of the lattice spacing, a, with renormalized parameters kept fixed.

Higher-order terms in the expansion of physical amplitudes (or correlation functions) in $1/m_b$ are given by including fields with higher mass dimension, which is compensated by the appropriate factor of the large mass in the denominator,

$$\mathscr{L}_{\mathrm{NLO}} = \sum_j \omega_j \mathcal{O}_j, \quad \omega_j = \frac{1}{m_{\mathrm{h}}} \hat{\omega}_j, \quad [\mathcal{O}_j] = 5, \quad [\hat{\omega}_j] = 0, \tag{10.42}$$

where the parameters $\hat{\omega}_i$ are dimensionless. In his lectures, M. Neubert referred to these parameters as Wilson coefficients [41]. The fields contained in the (multilocal) O are expanded in the same way as the action,

$$O_{\mathrm{eff}} = O_{\mathrm{LO}} + O_{\mathrm{NLO}} + \dots.$$

We now have to deal with interactions in (10.42) which are *not* renormalizable (by power counting). However, we are only interested in the expansion $\Phi = \Phi_{\mathrm{eff}}^{\mathrm{LO}} + \Phi_{\mathrm{eff}}^{\mathrm{NLO}} + \dots$ of observables Φ in powers of m_{h}^{-1}. It is therefore sufficient to define the theory with the weight in the path integral expanded ($S^{\mathrm{NLO}} = \int \mathrm{d}^4 x \, \mathscr{L}_{\mathrm{NLO}}(x)$)

$$e^{-S} \to e^{-S^{\mathrm{LO}}} \{ 1 - S^{\mathrm{NLO}} + \dots \}, \tag{10.43}$$

in

$$\Phi = \langle O \rangle = \frac{\int_{\mathrm{fields}} e^{-S} O}{\int_{\mathrm{fields}} e^{-S}}. \tag{10.44}$$

At NLO accuracy the expansion is then given by

$$\Phi_{\mathrm{eff}}^{\mathrm{LO}} = \langle O^{\mathrm{LO}} \rangle_{\mathrm{LO}} \tag{10.45}$$

$$\Phi_{\mathrm{eff}}^{\mathrm{NLO}} = \langle O^{\mathrm{NLO}} \rangle_{\mathrm{LO}} - \left(\langle O^{\mathrm{LO}} S^{\mathrm{NLO}} \rangle_{\mathrm{LO}} - \langle O^{\mathrm{LO}} \rangle_{\mathrm{LO}} \langle S^{\mathrm{NLO}} \rangle_{\mathrm{LO}} \right).$$

An important property is that this expansion is renormalizable. Namely, each term, $\Phi_{\mathrm{eff}}^{\mathrm{LO}}, \Phi_{\mathrm{eff}}^{\mathrm{NLO}}, \dots$, is renormalizable with a finite number of counter terms. For convenience we just add the EFT parameters in the fields making up O to the list of ω_i; the index i then runs over the terms in the action and in the composite fields. Adding the counter terms is then equivalent to renormalizing the parameters ω_i.

We demonstrate why it is sufficient to renormalize the ω_i to make the observables finite as the cut-off is removed. Consider O of (10.41) with the positions x, y fixed. Then

$$\varphi^{\text{LO}}(x) \to \varphi_{\text{R}}^{\text{LO}}(x) \tag{10.46}$$

$$\varphi^{\text{NLO}}(x) \to \varphi_{\text{R}}^{\text{NLO}}(x) = \sum_{\{i:\, [\varphi_i] \le [\varphi^{\text{NLO}}]\}} Z_i^{\varphi} \varphi_i(x), \qquad Z_i^{\varphi} = Z_i^{\varphi}(\Lambda_{\text{cut}}, g_0) \tag{10.47}$$

renormalizes the terms $\langle O^{\text{LO}} \rangle_{\text{LO}}$, $\langle O^{\text{NLO}} \rangle_{\text{LO}}$. Assuming for simplicity a single φ^{NLO} without mixing we can write for the second term in $\Phi_{\text{eff}}^{\text{NLO}}$

$$\langle O^{\text{LO}} S^{\text{NLO}} \rangle_{\text{LO}} = \int \mathrm{d}^4 z \, \langle O^{\text{LO}} \sum_j \omega_j \mathscr{O}_j(z) \rangle_{\text{LO}} \quad \to \tag{10.48}$$

$$\langle O^{\text{LO}} S^{\text{NLO}} \rangle_{\text{LO}} \Big|_{\text{R}} = \int \mathrm{d}^4 z \, \langle O^{\text{LO}} \sum_{i,j} \omega_i^{\text{R}} Z_{ij} \mathscr{O}_j(z) \rangle_{\text{LO}} \tag{10.49}$$

$$+ \langle \sum_i w_i O_i \rangle_{\text{LO}}, \qquad Z_{ij} = Z_{ij}(\Lambda_{\text{cut}}, g_0), \ w_i = w_i(\Lambda_{\text{cut}}, g_0)$$

in terms of finite coefficients ω_i^{R} (i running over the terms in the action) and with renormalization constants w_i (i running over the terms in the fields) and Z_{ij}, which absorb divergencies. The terms in the second line of (10.49) are necessary to compensate contact terms of (10.48), which arise when $z \to x$, $z \to y$. The operator product expansion dictates that the contact terms have this form.

The structure of (10.49) is such that we must make all EFT parameters ω_i general functions of the cut-off and coupling g_0 in order to obtain a fully renormalized expression. Starting with a general linear combination of all fields allowed by the symmetries of the action and of φ, this is all that is to be done.

In other words,

$$\omega_i = \omega_i(\Lambda_{\text{cut}}, g_0) \quad \to \quad \text{NLO EFT is finite}$$
$$i \in \text{terms in action} \cup \text{terms in fields } \varphi. \tag{10.50}$$

The finite parts of the coefficients can be chosen to match to the fundamental theory up to $(1/m_b)^2$.

$$\omega_i = \omega_i(\Lambda_{\text{cut}}, m_b, g_0) \quad \to \quad \text{NLO EFT is QCD up to O}((1/m_b)^2). \tag{10.51}$$

Renormalizability is particularly important for a non-perturbative evaluation of the path integral—in practice in a lattice formulation. The continuum limit of an effective theory only exists when we treat the higher-dimensional interactions as insertions in correlation functions in the form of (10.45). The continuum limit is then also expected to be universal, i.e. independent of the specific discretization. There are two important consequences.

The first is that, given a renormalizable lowest order Lagrangian, it is irrelevant that higher-order terms have mass dimension greater than d ($= 4$). The only consequence is that their coefficients have to be determined, by either matching to experiment (phenomenological approach) or by matching to the fundamental theory. Whether coefficients are finite (have a limit as the cut-off is removed) or not is irrelevant.

The second is that the result of the predictions of the EFT are entirely universal in the following sense: they do not depend on the regularization, i.e. on the way the theory was discretized, if we use a lattice. We should not misunderstand this universality to mean a uniqueness of the indicated $O(1/m_b)^2$ corrections. While those do not depend on the regularization, they *do* depend on how the EFT is matched to the fundamental theory.

10.3.2   EFTs with an Intrinsic Cut-off

There are cases where physics dictates that the lowest-order Lagrangian has to contain non-renormalizable terms. One example is the physics of quarkonia. Here NRQCD, not HQET, needs to be used (see [38]). The only difference between NRQCD and HQET is

$$\text{HQET:} \quad \mathscr{L}_{\text{LO}} = \mathscr{L}_{\text{h}}^{\text{stat}}, \quad \mathscr{L}_{\text{NLO}} = -\tfrac{1}{2m_{\text{h}}}(\mathcal{O}_{\text{kin}} + \mathcal{O}_{\text{spin}}) \tag{10.52}$$

$$\text{NRQCD:} \quad \mathscr{L}_{\text{LO}} = \mathscr{L}_{\text{h}}^{\text{stat}} - \tfrac{1}{2m_{\text{h}}}\mathcal{O}_{\text{kin}}, \quad \mathscr{L}_{\text{NLO}} = -\tfrac{1}{2m_{\text{h}}}\mathcal{O}_{\text{spin}}. \tag{10.53}$$

However, this reordering of the expansion means that NRQCD is non-renormalizable. Removing the cut-off requires adding more and more (eventually infinitely many) interaction (or counter-) terms at any given order in the expansion, already at LO. Since infinitely many coefficients cannot be determined, we have to live with the cut-off and discuss its influence.

In a lattice theory the cut-off is the inverse of the lattice spacing, $\Lambda_{\text{cut}} = a^{-1}$. With a fixed number of terms in the Lagrangian we then have cut-off-effects a^k of both positive $k > 0$ (discretization errors) and negative $k < 0$ (divergences). A window,

$$E \ll a^{-1} \ll m_{\text{h}}, \tag{10.54}$$

has to be present for both types of terms to be negligible. Clearly this is a difficult situation. NRQCD on the lattice has been used frequently in the past, but less so nowadays.

10.4   Renormalization of HQET

Before coming to a general discussion, it is instructive to look at the simplest case of renormalization and matching in the static effective theory. Perturbation theory will be our guide, but mostly to understand NP treatment.

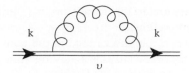

Figure 10.1 *One-loop self energy graph of a static quark. Graph from Mannel's lectures.*

10.4.1　Renormalization of HQET at Leading Order in $1/m_b$

10.4.1.1　Action

Consider the self-energy of the static quark in perturbation theory, namely the diagram Figure 10.1. This diagram behaves like

$$\Sigma \sim g_0^2 \int \mathrm{d}^4 l \, \frac{1}{l^2(l_0 + k_0 + i\epsilon)} \sim g_0^2 \Lambda_{\text{cut}} = \frac{g_0^2}{a} \tag{10.55}$$

in terms of the bare coupling, g_0, and the momentum cut-off, Λ_{cut}. This power divergence has to be compensated by a mass counter-term

$$\delta m \sim \frac{s_1 g_0^2 + s_2 g_0^4 + \cdots}{a}, \tag{10.56}$$

which we expect on the basis of dimensional counting.

The Lagrangian therefore has to contain the counter-term and reads

$$\mathscr{L}_{\text{h}}^{\text{stat}} = \overline{\psi}_{\text{h}}(m_{\text{bare}} + D_0)\psi_{\text{h}}, \quad m_{\text{bare}} = \delta m + m_{\text{finite}}, \quad \delta m \sim \Lambda_{\text{cut}} \sim \frac{1}{a} \tag{10.57}$$

with a single parameter, m_{bare}. Fields such as ψ_{h} are the (bare) fields integrated over in the regularized path integral. We note that the split into a finite mass (different for different heavy flavours f if they are present) and the divergent piece δm is arbitrary. What counts is m_{bare}.

It is important to note that asymptotic freedom tells us that

$$a = \Lambda_{\text{lat}}^{-1} e^{-1/(2b_0 g_0^2)} (b_0 g_0^2)^{-b_1/(2b_0^2)} \times (1 + \mathrm{O}(g_0^2)), \tag{10.58}$$

where Λ_{lat} is the Λ-parameter in the lattice minimal subtraction scheme, see e.g. [54]. The continuum limit is at bare coupling $g_0^2 \to 0$ and we have

$$\delta m \overset{a \to 0}{\sim} [s_1 g_0^2 + s_2' g_0^4 + \cdots] \, e^{1/(2b_0 g_0^2)} (b_0 g_0^2)^{b_1/(2b_0^2)} \Lambda_{\text{lat}}. \tag{10.59}$$

A truncation of the series $[s_1 g_0^2 + \cdots]$ at an order g_0^{2n} leaves a remainder $g_0^{2n+1} e^{1/(2b_0 g_0^2)}$ which diverges exponentially as $g_0 \to 0$, i.e. in the continuum limit. A finite number of terms in the series is not sufficient to determine δm, eq. (10.56), such that energies are finite, i.e. they have a continuum limit.

The counter-term δm, or better immediately the full combination m_{bare}, needs to be determined non-perturbatively. In fact, this is a general property of renormalization: the coefficients of power divergences have to be determined non-perturbatively.

In leading order HQET, we are lucky in that the explicit form of the heavy quark propagator, (10.86), shows that m_{bare} drops out of all observables except for the relation between the QCD quark mass and one energy level in the static theory, e.g. the mass of the B-meson. All energy differences and all properly normalized[3] matrix elements, $\mathcal{M}_n$, are independent of m_{bare},

$$E_n(m_{\text{bare}}) - E_m(m_{\text{bare}}) = E_n(0) - E_m(0), \tag{10.60}$$

$$\mathcal{M}_n^{\text{bare}}(m_{\text{bare}}) = \mathcal{M}_n^{\text{bare}}(0). \tag{10.61}$$

While the statement for the energies is clear from our previous discussions, we accept (10.61) for now and see that it is true after studying (10.104).

10.4.1.2 *Composite Fields*

As we have seen, the renormalization of the leading-order Lagrangian is rather simple, in principle and in practice. However, interesting physics is in transition matrix elements of operators originating from the electroweak Hamiltonian. In the path integral the electroweak Hamiltonian is represented by local composite fields. They need interesting, non-trivial, renormalization (and matching). As an example we choose the time component of the axial current,

$$\text{QCD:} \qquad A_0^{\text{R}}(x) = Z_{\text{A}} \overline{\psi}_{\text{u}}(x) \gamma_5 \gamma_0 \psi_{\text{b}}(x). \tag{10.62}$$

We label the heavy quark in QCD by b in order to distinguish it from the HQET field. The normalization factor is

$$Z_{\text{A}} = 1 + Z_{\text{A}}^{(1)} g_0^2 + \ldots \tag{10.63}$$

with a pure number (no renormalization scale dependence) $Z_{\text{A}}^{(1)}$, which can be chosen (in any regularization) such that the chiral Ward identities hold. The matrix element,

$$\langle 0 | A_0^{\text{R}}(0) | B(\mathbf{p} = 0) \rangle = m_{\text{B}}^{1/2} F_{\text{B}}, \tag{10.64}$$

of A_0^{R} defines the decay constant F_{B}, the single hadronic parameter needed to determine the decay rate $B \to \ell \nu$ (we do not distinguish here between the field in the path integral and the Hilbert space operator).

---

[3] A proper mass-independent non-relativistic normalization has to be chosen. For example, the standard one for B-meson states is $\langle B(\mathbf{p}') | B(\mathbf{p}) \rangle = 2(2\pi)^3 \delta(\mathbf{p} - \mathbf{p}')$.

Now we consider HQET at LO, the static theory. For the moment we just write down the structure of the EFT expression at one-loop order,

$$\mathcal{M}_{\text{QCD}}(m_b) = C^{\text{Wils}}(m_b) \left(\frac{2b_0 \bar{g}^2(\mu)}{2b_0 \bar{g}^2(m_b)} \right)^{-\gamma_0/(2b_0)} \mathcal{M}_{\text{stat}}(\mu) \qquad (10.65)$$

$$\times \left(1 + O(\bar{g}^2) + O(|\mathbf{p}|/m_b) \right), \qquad \mathcal{M} = \langle \alpha | A_0^R | \beta \rangle$$

for any matrix element $\mathcal{M}$ of the current. This is the form used by Neubert and Mannel in their lectures at this Les Houches school, where they discuss that $\gamma_0 \bar{g}^2$ is the leading order anomalous dimension of the field in the static theory,

$$\text{HQET:} \qquad (A_R^{\text{stat}})_0(x) = Z_A^{\text{stat}}(\mu) \overline{\psi}_u(x) \gamma_5 \gamma_0 \psi_h(x). \qquad (10.66)$$

We rewrite this as (to be precise, choose the mass such that $m_b = \overline{m}_{\overline{\text{MS}}}(m_b)$)

$$\mathcal{M}_{\text{QCD}}(m_b) = C^{\text{RGI}}(m_b) \, \mathcal{M}_{\text{stat}}^{\text{RGI}} \times (1 + O(|\mathbf{p}|/m_b)), \qquad (10.67)$$

with

$$C^{\text{RGI}}(m_b) = C^{\text{Wils}}(m_b)/\varphi(\bar{g}(m_b)), \qquad \mathcal{M}_{\text{stat}}^{\text{RGI}} = \varphi(\bar{g}(\mu)) \mathcal{M}_{\text{stat}}(\mu) \qquad (10.68)$$

and

$$\varphi(\bar{g}) = \left[2b_0 \bar{g}^2 \right]^{-\gamma_0/2b_0} \underbrace{\exp\left\{ -\int_0^{\bar{g}} dx \left[\frac{\gamma(x)}{\beta(x)} - \frac{\gamma_0}{b_0 x} \right] \right\}}_{1+O(\bar{g}^2)}. \qquad (10.69)$$

Note that $\mathcal{M}_{\text{stat}}^{\text{RGI}}$ does *not* depend on μ or m_b. Nor does it depend on a renormalization scheme.

Exercise 10.1 Show that $\mathcal{M}_{\text{stat}}^{\text{RGI}}$ does not depend on μ and is scheme independent.

$\mathcal{M}_{\text{stat}}^{\text{RGI}}$ is a pure number, a NP property of the EFT. There are no corrections to (10.69); corrections appear only when β, γ are approximated by perturbation theory at a certain order.[4] There are lattice methods to compute $\mathcal{M}_{\text{stat}}^{\text{RGI}}$ with negligible perturbative truncation error [17]. They involve so-called step scaling strategies.

---

[4] When β is inserted in $n+1$-loop approximation and the other functions in n-loop approximation, the right-hand side of (10.69) is correct up to $O(\alpha(\mu)^n)$ truncation errors.

A complete (but here irrelevant) transition to RGIs is given by expressing C^{Wils}/φ as a function

$$C_{\text{PS}}(M_{\text{b}}/\Lambda) = C^{\text{Wils}}(m_{\text{b}})/\varphi(\bar{g}(m_{\text{b}})), \qquad (10.70)$$

with M_{b} the renormalization group invariant mass and Λ the Λ-parameter of QCD.

10.4.1.3 *Mass Scaling*

Apart from the power corrections, the mass dependence in (10.67) is contained entirely in the $\bar{g}(m_{\text{b}})$-dependence of φ. Just like scale dependence is described by the β-function and anomalous dimensions, it is natural to describe the mass dependence by a mass scaling function γ_{match},[5]

$$\gamma_{\text{match}}(\bar{g}) \equiv \frac{m_{\text{b}}}{\mathcal{M}_{\text{QCD}}} \frac{\partial \mathcal{M}_{\text{QCD}}}{\partial m_{\text{b}}} \overset{\bar{g}\to 0}{\sim} -\gamma_0 \bar{g}^2 - \gamma_1^{\text{match}} \bar{g}^4 + \dots . \qquad (10.71)$$

Upon integration of the first-order term, we obtain the asymptotic mass dependence of the decay constant of a heavy-light pseudo-scalar (e.g. B):[6]

$$F_{\text{B}} \overset{M\to\infty}{\sim} \frac{[\ln(M/\Lambda)]^{\gamma_0/2b_0}}{\sqrt{m_{\text{B}}}} \mathcal{M}_{\text{stat}}^{\text{RGI}} \times [1 + \text{O}([\ln(M/\Lambda)]^{-1})]. \qquad (10.72)$$

Higher perturbative orders yield the indicated logarithmic corrections. Historically, this scaling has been arrived at in a different way in [43, 46]. We have gone through the above for two reasons. One is that the representation in terms of RGIs seems not much appreciated in EFTs and here there is an opportunity to learn about it. Second, the function γ_{match} is very convenient to use when looking at the limitations of perturbative matching.

10.4.1.4 *On the Accuracy of Perturbation Theory*

When we evaluate functions such as C_{PS} in a given order of perturbation theory, various quantities enter such as the β-function and the quark mass anomalous dimension. Apart from γ_{match}, these all have a well behaved perturbative expansion in the $\overline{\text{MS}}$ scheme as seen in Table 10.3 reproduced from Appendix A.2.2 of [47].

Keeping this in mind, we focus on γ_{match}. In the left graph in Figure 10.2 we plot different orders of γ_{match}. For the larger values $\bar{g}^2$ in the plot we may worry about

---

[5] Note that γ_{match} is just the anomalous dimension of the axial current A_0^{stat} in a special renormalization scheme. In this scheme, at scale $\mu = \bar{m}(m_{\text{b}})$, its matrix elements are equal to the QCD ones up to order $1/m_{\text{b}}$,

$$\mathcal{M}_{\text{QCD}} = \mathcal{M}_{\text{match}}(m_{\text{b}}) + \text{O}(1/m_{\text{b}}).$$

We therefore refer to it as the matching scheme.

[6] Note the slow, logarithmic, decrease of the corrections in (10.72). The discussion of Figure 10.2 shows that the perturbative evaluation of $C_{\text{PS}}(M_{\text{b}}/\Lambda)$ is somewhat problematic.

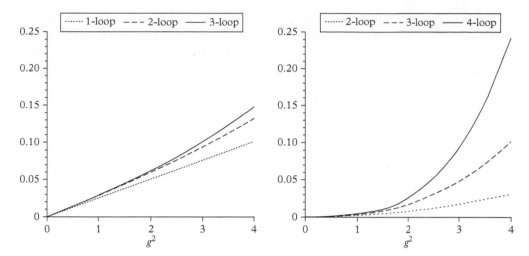

Figure 10.2 *The function $\gamma_{match}(g)$ as a function of $g^2 = \bar{g}^2(m_b)$ for $N_f = 3$ flavours in the $\overline{MS}$ scheme. On the left we show $\gamma_{match} = \gamma_{match}^{A_0}$ for the time component of the axial current, on the right we show the difference $\gamma_{match}^{V_k} - \gamma_{match}^{A_0}$. At one-loop order the latter vanishes.*

Table 10.3 Coefficients of β-function (b_i), anomalous dimension of the mass (d_i), anomalous dimension of the static-light current, (γ_i) and the mass scaling function γ_i^{match} for various currents for three-flavour QCD. The first three refer to the $\overline{MS}$ scheme and also γ_i^{match} are expansion coefficients of the $\overline{MS}$-coupling, but the currents are defined such that they match QCD, i.e. are physical. We use the high order perturbative information accumulated over the years [8, 9, 11, 12, 14, 22, 31, 43, 46, 51, 52]. The normalization (4π-factors) of the numbers is such that these are coefficients of a series in α^i.

coefficient	$i = 1$	$i = 2$	$i = 3$	$i = 4$
$(4\pi)^i b_{i-1}$	0.71620	0.40529	0.32445	0.47367
$(4\pi)^i d_{i-1}$	0.63662	0.76835	0.80114	0.90881
$(4\pi)^i \gamma_{stat,i-1}$	-0.31831	-0.26613	-0.25917	
$(4\pi)^i \gamma_{match,i-1}^{\gamma_0\gamma_5}$	-0.31831	-0.57010	-0.94645	
$(4\pi)^i \gamma_{match,i-1}^{\gamma_k}$	-0.31831	-0.87406	-3.12585	
$(4\pi)^i [\gamma_{match,i-1}^{\gamma_0\gamma_5} - \gamma_{match,i-1}^{\gamma_k}]$	0	0.30396	2.17939	14.803

neglecting higher-order terms. Note that $\bar{g}^2(m_b)$ is around 2.5 for the b-quark and it is out of the range of the graph for the charm quark.

However, a more serious reason for concern derives from the right-hand side graph, which shows the difference of the anomalous dimensions for V_k and A_0. For such differences perturbation theory is known to one loop higher [4] and the perturbative coefficients do grow further. Asymptotic convergence seems to be useful only for rather small couplings or equivalently for masses far above the b-quark mass. At the b-quark mass, every known perturbative order contributes about an equal amount. Such a behaviour raises concern about using perturbation theory for the matching functions. How do we estimate their uncertainty?

The bad behaviour is easily traced back to the function C_{match} [4]. We tried earlier [47] to rearrange the perturbative series in order to find a more stable perturbative prediction, but did not succeed. Possibly the concept of 'renormalon subtraction' [42] helps, but to our knowledge this has not been shown for the case at hand.

10.4.1.5   Summary

The most important facts about renormalization and matching at leading order in $1/m_{\mathrm{b}}$ are:

- The Lagrangian has a single parameter, the ***quark mass***, which needs to be renormalized. It is ***linearly divergent*** and can therefore not be computed in PT. However, it only enters the relation $m_{\mathrm{b}} \leftrightarrow m_{\mathrm{B}}$. All energy differences are independent of m_{bare}, as are matrix elements such as F_{B}.

- Asymptotic freedom allows, in principle, to fully renormalize and match the electroweak currents in PT. One may split the renormalization and matching transparently into the definition of RGI operators in the EFT and a matching function. In practice, looking at three non-trivial orders, higher orders become smaller than the lower ones only when the quark mass is significantly above $m = 5\,\mathrm{GeV}$. For HQET applied to the b-quark there are relevant cases where successive terms in the perturbative expansion are equally large. It is therefore necessary to match the currents non-perturbatively.

10.4.2   Renormalization and Matching of HQET at Higher Orders in $1/m_{\mathrm{b}}$

For quantitative phenomenological results we must also compute $1/m_{\mathrm{b}}$ corrections in HQET. Is it consistent to match perturbatively as we discussed in the previous sections for the static theory? We have seen that the uncertainty due to a truncation of the perturbative matching expressions at n-loop order corresponds to a relative *error*

$$\frac{\delta_{n-\mathrm{loops}}(\mathcal{M}_{\mathrm{QCD}})}{\mathcal{M}_{\mathrm{QCD}}} = \frac{\delta_{n-\mathrm{loops}}(C_{\mathrm{PS}})}{C_{\mathrm{PS}}} \propto [\bar{g}^2(m_{\mathrm{b}})]^n \sim \left[\frac{1}{2b_0 \ln(m_{\mathrm{b}}/\Lambda)}\right]^n,$$

for example, (10.65) and (10.69). As m_{b} is made large, this perturbative error decreases only logarithmically. It becomes dominant over the power correction which we want to include by pushing the HQET expansion to NLO,

$$\frac{\delta_{n-\mathrm{loops}}(\mathcal{M}_{\mathrm{QCD}})}{\mathcal{M}_{\mathrm{QCD}}} = \frac{\delta_{n-\mathrm{loops}}(C_{\mathrm{PS}})}{C_{\mathrm{PS}}} \overset{m_{\mathrm{b}} \gg \Lambda}{\gg} \frac{\Lambda}{m_{\mathrm{b}}}. \tag{10.73}$$

We learn that with a perturbative matching function (or Wilson coefficient), we do not perform a consistent NLO expansion such that errors decrease as $(\Lambda/m_b)^2$.

A practically even more serious issue is that at NLO we must deal with the mixing of operators with lower dimensional ones. For example $\mathcal{O}_{kin} = \overline{\psi}_h \mathbf{D}^2 \psi_h$ mixes with $\overline{\psi}_h D_0 \psi_h$ and $\overline{\psi}_h \psi_h$. In this situation, mixing coefficients are power divergent $\sim a^{-n}$ with $n = 1, 2$.

This is an example of a general property of renormalization: subtracting power divergences in perturbation theory and then computing the matrix elements non-perturbatively always leaves a divergent remainder. Equivalently, from the perspective of a lattice computation we emphasize that matrix elements of perturbatively subtracted operators do not have a non-perturbative continuum limit.

We have to conclude that it is necessary to perform matching and renormalization non-perturbatively. The only alternative is to supplement the theory by assumptions. Namely one frequently assumes that at the lattice spacings available in practice, power divergences of the form $g_0^{2l}/(m_b a)^n$ are small since $m_b a > 1$. This then has to be combined with the assumption that the b-quark mass is not large enough to be in the asymptotic region of (10.73).

10.5 The Lattice Formulation

We have now gathered enough motivation for a treatment of HQET on the lattice. So we return to the formulation of the theory on a hyper-cubic Euclidean lattice.

10.5.1 QCD on the Lattice

Fermion fields $\psi(x), \overline{\psi}(x)$ are associated to the points of the lattice. In order to define discrete gauge-covariant derivatives, we must parallel-transport the fermion fields from one site to the other,

$$P(x \leftarrow x + a\hat{\mu}) = U(x, \mu) = \quad\underleftarrow{\qquad}\quad \in SU(3), \qquad (10.74)$$

$$P(x + a\hat{\mu} \leftarrow x) = U^{\dagger}(x, \mu) = \quad\underrightarrow{\qquad}\quad \in SU(3). \qquad (10.75)$$

These parallel transporters, called links, transform under gauge transformations, $\psi(x) \rightarrow \Omega(x)\psi(x)$, as

$$U(x, \mu) \rightarrow \Omega(x)U(x, \mu)\Omega(x + a\hat{\mu})^{\dagger} \qquad (10.76)$$

such that the (finite difference, backward) derivative,

$$(\nabla_{\mu}^* \psi)(x) = \frac{1}{a}[\psi(x) - U^{\dagger}(x - a\hat{\mu}, \mu)\psi(x - a\hat{\mu})], \qquad (10.77)$$

is gauge covariant. The lattice Dirac operator is formulated in terms of these covariant derivatives. The field of elementary parallel transporters, $U(x, \mu)$, is the lattice gauge field. It is in the gauge group, not in its algebra.

The gauge action has to be local, gauge invariant, lattice (i.e. 90 degree) rotational invariant, and formed in terms of the gauge field U. The trace of the parallel-transporter around a plaquette (elementary square),

$$O_{\mu\nu}(x) = \mathrm{tr}\, U(x,\mu)\, U\,(x + a\,\hat{\mu},\nu)\, U^{-1}(x + a\hat{\nu},\, \mu)\, U^{-1}(x,\nu) = \qquad (10.78)$$

is the most local object and summing over all x, μ, ν leads to the most natural action, the Wilson gauge action [55]. Assuming the lattice gauge fields U to be constructed from a smooth continuum gauge field, the lattice action reproduces the continuum one up to effects of order a^2,

$$a^4 \sum_{\mu,\nu,x} (1 - \tfrac{1}{N_c} O_{\mu\nu}(x)) \sim \int \mathrm{d}^4 x\, \mathrm{tr}\, F_{\mu\nu}(x) F_{\mu\nu}(x) + \mathrm{O}(a^2). \qquad (10.79)$$

Since in the path integral gauge fields are not smooth, this is called the *naive* continuum limit of the action; (10.79) means that we have a theory with the correct classical continuum limit.

Of course, much more should be explained about lattice QCD, for example, the measure in the path integral, and we refer the interested reader to [28].

10.5.2   HQET on the Lattice

10.5.2.1   Static Lagrangian

We simply use [19]

$$D_0\, \psi_h(x) = \nabla_0^* \psi_h(x) \qquad (10.80)$$

and, for later convenience, insert a specific normalization factor defining the static lattice Lagrangian

$$\mathscr{L}_h = \frac{1}{1 + a\delta m} \overline{\psi}_h(x) [\nabla_0^* + \delta m] \psi_h(x). \qquad (10.81)$$

The following points are worth noting, at least for those with some prior knowledge in lattice field theory:

- There are no doubler modes (it is not easy to write down a Dirac operator on the lattice with chiral symmetry, describing a single fermion. However, there is no chiral symmetry for a static quark).
- There is a positive Hermitian transfer matrix and thus a Hermitian Hamiltonian.
- The lattice action preserves all the continuum heavy quark symmetries discussed in the continuum HQET section. Those formulae hold without any change.

10.5.2.2 Propagator

From the Lagrangian (10.81) we have the defining equation for the propagator

$$\frac{1}{1+a\,\delta m}(\nabla_0^* + \delta m)\,G_h(x,y) = \delta(x-y)P_+ \equiv a^{-4}\prod_\mu \delta_{\frac{x_\mu}{a}\frac{y_\mu}{a}}P_+\,. \qquad (10.82)$$

Obviously $G_h(x,y)$ is proportional to $\delta(\mathbf{x}-\mathbf{y})$. Writing $G_h(x,y) = g(n_0,k_0;\mathbf{x})\delta(\mathbf{x}-\mathbf{y})P_+$ with $x_0 = an_0$, $y_0 = ak_0$, the equation yields a simple recursion for $g(n_0+1,k_0;\mathbf{x})$ in terms of $g(n_0,k_0;\mathbf{x})$, which is solved by

$$g(n_0,k_0;\mathbf{x}) = \theta(n_0-k_0)(1+a\delta m)^{-(n_0-k_0)}\mathcal{P}(x\leftarrow y)\,, \qquad (10.83)$$

$$\mathcal{P}(x,x) = 1\,,\quad \mathcal{P}(x\leftarrow y+a\hat{0}) = \mathcal{P}(x\leftarrow y)\,U(y,0)\,, \qquad (10.84)$$

where

$$\theta(n_0-k_0) = \begin{cases} 0 & n_0 < k_0 \\ 1 & n_0 \geq k_0\,. \end{cases} \qquad (10.85)$$

The static propagator then reads

$$G_h(x,y) = \theta(x_0-y_0)\,\delta(\mathbf{x}-\mathbf{y})\,\exp\left(-\widehat{\delta m}\,(x_0-y_0)\right)\mathcal{P}(x\leftarrow y)\,P_+\,, \qquad (10.86)$$

$$\widehat{\delta m} = \tfrac{1}{a}\ln(1+a\delta m)\,. \qquad (10.87)$$

$\mathcal{P}(x,y)$ is just the lattice parallel transporter. Note that the derivation fixes $\theta(0) = 1$ for the lattice θ-function. As in the continuum, the mass counter term, δm, just yields an energy shift; now, on the lattice, the shift is

$$E_{h/\bar{h}}^{\text{QCD}} = E_{h/\bar{h}}^{\text{stat}}\Big|_{\delta m = 0} + m_{\text{bare}}\,,\quad m_{\text{bare}} = \widehat{\delta m} + m\,. \qquad (10.88)$$

It is valid for all energies of states with a single heavy quark or anti-quark. As in the continuum the split between δm and the finite m is convention dependent.

10.5.2.3 Heavy Quark Symmetries

All HQET symmetries are preserved on the lattice. The symmetry transformations can literally be carried over from the continuum, e.g. the local flavour number (10.36). We just replace the continuum fields with the lattice ones.

Note that these HQET symmetries are defined in terms of transformations of the heavy quark fields while the light quark fields and gauge fields do not change. Integrating out just the quark fields in the path integral while leaving the integral over the gauge fields (see (10.25)), they thus yield identities for the integrand or one may say for 'correlation functions in any fixed gauge background field'.

10.5.2.4 Symanzik EFT

According to the—by now well tested—Symanzik conjecture, the cut-off (=discretisation) effects of a lattice theory can be described in terms of an effective *continuum* theory [36, 49, 50]. Once the terms in Symanzik's effective Lagrangian are known, the cut-off effects can be cancelled by adding terms of the same form to the lattice action, resulting in an improved action.

For a static quark, Symanzik's effective action is [34]

$$S_{\text{eff}} = S_0 + a S_1 + \ldots, \quad S_i = \int d^4 x \, \mathcal{L}_i(x) \tag{10.89}$$

where $\mathcal{L}_0(x) = \mathcal{L}_h^{\text{stat}}(x)$ is the continuum static Lagrangian of (10.15) and

$$\mathcal{L}_1(x) = \sum_{i=3}^{5} c_i \mathcal{O}_i(x), \tag{10.90}$$

is given in terms of local fields with mass dimension $[\mathcal{O}_i(x)] = 5$. Their coefficients c_i are functions of the bare gauge coupling. Assuming for simplicity mass-degenerate light quarks with a mass m_l, the set of possible dimension-five fields, which share the symmetries of the lattice theory, is

$$\mathcal{O}_3 = \overline{\psi}_h D_0 D_0 \psi_h, \quad \mathcal{O}_4 = m_l \overline{\psi}_h D_0 \psi_h, \quad \mathcal{O}_5 = m_l^2 \overline{\psi}_h \psi_h. \tag{10.91}$$

Note that $P_+ \sigma_{0j} P_+ = 0$ means there is no term $\overline{\psi}_h \sigma_{0j} F_{0j} \psi_h$, and $\overline{\psi}_h D_j D_j \psi_h$ cannot occur because it violates the local phase invariance (10.36). Finally, $\overline{\psi}_h \sigma_{jk} F_{jk} \psi_h$ is not invariant under the spin rotations (10.35).

Furthermore, we are only interested in on-shell correlation functions and energies. For this class of observables, $\mathcal{O}_3$ and $\mathcal{O}_4$ do not contribute [36, 37] because they vanish by the equation of motion,

$$D_0 \psi_h = 0. \tag{10.92}$$

The only remaining term, $\mathcal{O}_5$, induces a redefinition of the mass counter-term δm, which therefore depends explicitly on the light quark mass in the form of an $a m_l^2$ term.

As noted, for almost all applications δm is explicitly cancelled in the relation between physical observables and thus we have so-called automatic on-shell O(a) improvement for the static action. No parameter has to be tuned to guarantee this property. Still, the improvement of matrix elements and correlation functions requires considering composite fields in the effective theory. For example one finds one O(a) operator for the axial current. A more detailed description is found in [34, 47].

10.5.2.5 Terms at Order $1/m_b$

Valid discretizations for these terms are easily written down:

$$\mathcal{O}_{\text{kin}}(x) = \overline{\psi}_h(x) \, \nabla_k^* \nabla_k \, \psi_h(x), \tag{10.93}$$

$$\mathcal{O}_{\text{spin}}(x) = \overline{\psi}_h(x) \, \boldsymbol{\sigma} \cdot \mathbf{B}(x) \, \psi_h(x). \tag{10.94}$$

10.6   Non-Perturbative HQET

This section provides a simple example of how the different steps of an EFT formulated non-perturbatively work. It is easy to generalize the example. Section 10.6.1 explains the $1/m_b$ expansion of Euclidean correlation functions, assuming the parameters of the HQET Lagrangian and composite fields are known. We also discuss how one determines the physical (Minkowski space) matrix elements and energy levels. Section 10.6.2 turns to the non-trivial question of how we determine the parameters in practice and with controlled uncertainties. The strategy described can in principle be applied to other low-energy EFTs, but this has not been done yet.

10.6.1 $1/m_b$-Expansion of Correlation Functions, Masses, and Matrix Elements

We consider correlation functions of just one composite field, for which we again choose

$$Z_A^{HQET}\,(A_0^{stat}+\omega_A A_0^{(1)}),\quad A_0^{(1)}=\overline{\psi}_1\overleftarrow{D}_j\gamma_j\gamma_5\psi_h. \tag{10.95}$$

In principle there is a second $1/m_b$ term, but it does not contribute when we consider the field at $\mathbf{p}=0$; we just consider that kinematics here. For now we assume that the coefficients

$$O(1):\; m_{bare},\,Z_A^{HQET},$$
$$O(1/m_b):\; \omega_{kin},\,\omega_{spin},\,\omega_A, \tag{10.96}$$

are known as a function of the bare coupling g_0 and the quark mass m. Their non-perturbative determination will be discussed later.

The rules of the $1/m_b$ expansion are illustrated on the QCD correlation function

$$C_{AA,R}^{QCD}(x_0)=a^3\sum_{\mathbf{x}}\Big\langle A_0^R(x)(A_0^R)^\dagger(0)\Big\rangle. \tag{10.97}$$

We use (10.95) and expand the expectation value consistently in $1/m_b$, counting powers of $1/m_b$ as in (10.96). At order $1/m_b$, terms proportional to $\omega_{kin}\times\omega_A$, etc., are to be dropped. They must be dropped to preserve renormalizability.

Exercise 10.2 Think about why this is so.

As a last step, we must take the energy shift between HQET and QCD into account. Therefore, correlation functions with a time separation x_0 obtain an extra factor

$\exp(-x_0\,m)$, where the scheme dependence of m is compensated by a corresponding one in δm.

Dropping all terms $\mathrm{O}(1/m_{\mathrm{b}}^2)$ without further notice, we arrive at the expansion

$$
\begin{aligned}
C_{\mathrm{AA}}^{\mathrm{QCD}}(x_0) &= \mathrm{e}^{-mx_0}(Z_A^{\mathrm{HQET}})^2\Big[C_{\mathrm{AA}}^{\mathrm{stat}}(x_0) + \omega_A\, C_{\delta\mathrm{AA}}^{\mathrm{stat}}(x_0) \\
&\quad + \omega_{\mathrm{kin}}\, C_{\mathrm{AA}}^{\mathrm{kin}}(x_0) + \omega_{\mathrm{spin}}\, C_{\mathrm{AA}}^{\mathrm{spin}}(x_0)\Big]
\end{aligned}
\tag{10.98}
$$

$$
\begin{aligned}
&\equiv \mathrm{e}^{-mx_0}(Z_A^{\mathrm{HQET}})^2\, C_{\mathrm{AA}}^{\mathrm{stat}}(x_0)\Big[1 + \omega_A\, R_{\delta A}^{\mathrm{stat}}(x_0) \\
&\quad + \omega_{\mathrm{kin}}\, R_{\mathrm{AA}}^{\mathrm{kin}}(x_0) + \omega_{\mathrm{spin}}\, R_{\mathrm{AA}}^{\mathrm{spin}}(x_0)\Big]
\end{aligned}
\tag{10.99}
$$

with

$$
C_{\delta\mathrm{AA}}^{\mathrm{stat}}(x_0) = a^3\sum_{\mathbf{x}}\langle A_0^{\mathrm{stat}}(x)(A_0^{(1)}(0))^\dagger\rangle_{\mathrm{stat}} + a^3\sum_{\mathbf{x}}\langle A_0^{(1)}(x)(A_0^{\mathrm{stat}}(0))^\dagger\rangle_{\mathrm{stat}},
$$

$$
C_{\mathrm{AA}}^{\mathrm{kin}}(x_0) = a^3\sum_{\mathbf{x}}\langle A_0^{\mathrm{stat}}(x)(A_0^{\mathrm{stat}}(0))^\dagger a^4\sum_z \mathcal{O}_{\mathrm{kin}}(z)\rangle_{\mathrm{stat}}
$$

$$
C_{\mathrm{AA}}^{\mathrm{spin}}(x_0) = a^3\sum_{\mathbf{x}}\langle A_0^{\mathrm{stat}}(x)(A_0^{\mathrm{stat}}(0))^\dagger a^4\sum_z \mathcal{O}_{\mathrm{spin}}(z)\rangle_{\mathrm{stat}}.
$$

It is now a straightforward exercise to obtain the expansion of the B-meson mass,

$$
m_{\mathrm{B}} = -\lim_{x_0\to\infty}\tilde{\partial}_0\ln C_{\mathrm{AA}}^{\mathrm{QCD}}(x_0)
\tag{10.100}
$$

$$
= m_{\mathrm{bare}} + E^{\mathrm{stat}} + \omega_{\mathrm{kin}}E^{\mathrm{kin}} + \omega_{\mathrm{spin}}E^{\mathrm{spin}} + \omega_A\times 0,
\tag{10.101}
$$

$$
E^{\mathrm{stat}} = -\lim_{x_0\to\infty}\tilde{\partial}_0\ln C_{\mathrm{AA}}^{\mathrm{stat}}(x_0)\Big|_{\delta m=0},
\tag{10.102}
$$

$$
E^{\mathrm{kin}} = -\lim_{x_0\to\infty}\tilde{\partial}_0 R_{\mathrm{AA}}^{\mathrm{kin}}(x_0), \quad E^{\mathrm{spin}} = -\lim_{x_0\to\infty}\tilde{\partial}_0 R_{\mathrm{AA}}^{\mathrm{spin}}(x_0).
$$

The ratios R_{AA}^x are defined by (10.99). They do not depend on δm. The quantities E^{kin}, E^{spin} become energy contributions after multiplying with the dimension-full ω_{kin}, ω_{spin}; they therefore have mass dimension two. We have made the dependence on δm explicit through $m_{\mathrm{bare}} = m_{\mathrm{b}} + \widehat{\delta m}$ and then all quantities are defined in the theory with $\delta m = 0$.

Furthermore, we have already anticipated (10.109). These equations tell us also that the exponential decay of

$$
C_{\mathrm{AA}}^{\mathrm{kin}}(x_0) \overset{x_0\Lambda_{\mathrm{QCD}}\gg 1}{\sim} \text{constant} \times x_0\,\mathrm{e}^{-E^{\mathrm{stat}}x_0}
\tag{10.103}
$$

is accompanied by a factor x_0, just like $C_{\mathrm{AA}}^{\mathrm{spin}}$.

The expansion for the decay constant is

$$F_B\sqrt{m_B} = \lim_{x_0\to\infty} \left\{2\exp(m_B x_0)\, C_{AA}^{QCD}(x_0)\right\}^{1/2} \tag{10.104}$$

$$= Z_A^{HQET}\, \Phi^{stat}\{1+\rho\} \tag{10.105}$$

$$\Phi^{stat} = \lim_{x_0\to\infty} \left\{2\exp(E^{stat} x_0)\, C_{AA}^{stat}(x_0)\right\}^{1/2}$$

$$\rho = \tfrac{1}{2}\lim_{x_0\to\infty}\Big[\,\omega_{kin}\,(x_0 E^{kin} + R_{AA}^{kin}(x_0)) + \omega_{spin}(x_0 E^{spin} + R_{AA}^{spin}(x_0))$$

$$+ \omega_A R_{\delta A}^{stat}(x_0)\,\Big]. \tag{10.106}$$

Inserting $\mathbb{1} = \sum_n |B,n\rangle\langle B,n|$, with finite volume normalization $\langle B,n|B,n\rangle = 2L^3$, and n labelling the eigenstates of the Hamiltonian, we further observe that

$$E^{kin} = -\frac{1}{2L^3}\langle B|a^3\sum_{\mathbf{z}}\mathcal{O}_{kin}(0,\mathbf{z})|B\rangle_{stat} = -\frac{1}{2}\langle B|\mathcal{O}_{kin}(0)|B\rangle_{stat} \tag{10.107}$$

$$E^{spin} = -\frac{1}{2}\langle B|\mathcal{O}_{spin}(0)|B\rangle_{stat}, \tag{10.108}$$

$$0 = \lim_{x_0\to\infty} \tilde{\partial}_0 R_{\delta A}^{stat}(x_0), \tag{10.109}$$

with $|B\rangle \equiv |B,0\rangle$.

Exercise 10.3 Derive (10.107–10.109).

As expected, only the parameters of the action are relevant in the expansion of hadron masses. Note that in lattice regularization there is a so-called transfer matrix, which allows us to define the Hamiltonian and the Hilbert space rigorously; thus $\mathbb{1} = \sum_n |B,n\rangle\langle B,n|$ is rigorous for any spacing a.

A correct split of the terms in equations (10.101) and (10.106) into leading-order and next-to-leading-order pieces, which are separately renormalized and which hence *separately have a continuum limit* and physical meaning beyond a particular regularization, requires more thought on the renormalization of the $1/m_b$-expansion. We postpone that to Sect. 10.6.3.

First, we discuss how the HQET-parameters can be determined such that the effective theory yields the $1/m_b$ expansion of the QCD observables.

10.6.2   Strategy for Non-Perturbative Matching

How can we non-perturbatively match HQET to QCD? Consider the action as well as A_0 (just at $\mathbf{p}=0$) and denote the free parameters of the effective theory by ω_i, $i=1\dots N_{HQET}$.

In static approximation we then have the parameter vector

$$\omega^{\text{stat}} = (m_{\text{bare}}^{\text{stat}}, [\ln(Z_{\text{A}})]^{\text{stat}})^t, \quad N_{\text{HQET}} = 2 \qquad (10.110)$$

and including the first-order terms in $1/m$ together with the static ones, the HQET parameters are

$$\omega^{\text{HQET}} = (m_{\text{bare}}, \ln(Z_{\text{A}}^{\text{HQET}}), \omega_{\text{A}}, \omega_{\text{kin}}, \omega_{\text{spin}})^t, \quad N_{\text{HQET}} = 5. \qquad (10.111)$$

With suitable observables, ($M_{\text{b}} = $ RGI mass of the heavy quark)

$$\Phi_i(L, M_{\text{b}}, a), \ i = 1 \ldots N_{\text{HQET}},$$

in a finite volume with $L = T = L_1 \approx 0.5\,\text{fm}$, we require matching[7]

$$\Phi_i(L, M_{\text{b}}, a) = \Phi_i^{\text{QCD}}(L, M_{\text{b}}, 0), \ i = 1 \ldots N_{\text{HQET}}. \qquad (10.112)$$

Here the restriction to a small volume $L \approx 0.5\,\text{fm}$ allows for very small lattice spacings and therefore a continuum limit can be taken in QCD. This is a crucial step. On the other hand, in HQET we want to extract the bare parameters of the theory from the matching equation and thus we have a finite value of a.

It is convenient to pick observables with HQET expansions linear in ω_i (e.g. $\Phi = \log(C_{\text{AA}})$)

$$\Phi(L, M_{\text{b}}, a) = \eta(L, a) + \varphi(L, a)\, \omega(M_{\text{b}}, a), \qquad (10.113)$$

in terms of a $N_{\text{HQET}} \times N_{\text{HQET}}$ coefficient matrix φ. A natural choice for the first two observables is to choose them as finite volume observables that converge (exponentially) to interesting physical observables that we want to predict,

$$\Phi_1 \equiv Lm_{\text{B}}(L) \stackrel{L\to\infty}{\sim} Lm_{\text{B}} \qquad (10.114)$$

$$\Phi_2 \equiv \ln(\langle 0, L | A_0^{\text{R}} | B, L \rangle) = \log(L^{3/2} F_{\text{B}}(L)\sqrt{m_{\text{B}}(L)/2})$$

$$\stackrel{L\to\infty}{\sim} \log(L^{3/2} F_{\text{B}}\sqrt{m_{\text{B}}/2}). \qquad (10.115)$$

[7] Recall that observables without a superscript refer to HQET.

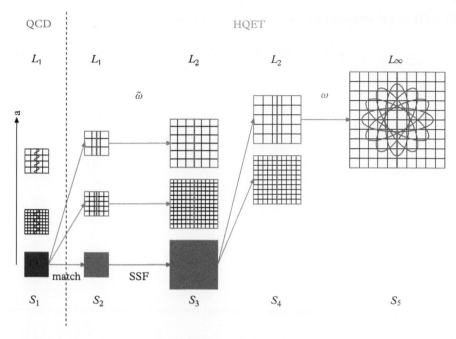

Figure 10.3 *Strategy for non-perturbative HQET [7]. Note that in the realistic implementation[7] finer resolutions are used.*

In static approximation these determine directly ω_1 and ω_2. We do not discuss choices for the other Φ_i, here and of course the above equations are not sufficient to fix Φ_1, Φ_2 completely. For the choices made so far[6], the explicit form of η, φ is

$$\eta = \begin{pmatrix} \Gamma^{\text{stat}} \\ \zeta_A \\ \dots \end{pmatrix}, \quad \varphi = \begin{pmatrix} L & 0 & \dots \\ 0 & 1 & \dots \\ & \dots & \end{pmatrix} \tag{10.116}$$

with

$$\Gamma^{\text{stat}} = -L\, m_{\text{B}}^{\text{stat}}(L)\big|_{m_{\text{bare}}=0}, \quad \zeta_A = \ln\left(L^{3/2} F_{\text{B}}(L)\sqrt{m_{\text{B}}(L)/2}\right)^{\text{stat}}_{m_{\text{bare}}=0}. \tag{10.117}$$

In static approximation, the structure of the matrix φ is perfect: one observable determines one parameter. This is possible since there is no (non-trivial) mixing at that order.

Having specified the matching conditions, the HQET parameters $\omega_i(M_{\text{b}}, a)$ can be obtained from (10.112, 10.113), but only for rather small lattice spacings since a reasonable suppression of lattice artifacts requires $L/a \geq 10$ (in HQET) and thus $a \leq 0.05\,$fm. Larger lattice spacings are needed in large volume. These can be reached by a step scaling strategy, illustrated in Figure 10.3. We go through the four steps of the full strategy. As indicated, the initial value of L is set to $L = L_1$.

(1) Take the continuum limit

$$S1: \quad \Phi_i^{QCD}(L_1, M_b, 0) = \lim_{a/L_1 \to 0} \Phi_i^{QCD}(L_1, M_b, a). \qquad (10.118)$$

This requires $L_1/a = 20 \ldots 40$, or $a = 0.025\,\text{fm} \ldots 0.012\,\text{fm}$.

(2a) Set the HQET observables equal to the QCD ones (10.112) and extract the parameters

$$S2: \quad \tilde{\omega}(M_b, a) \equiv \varphi^{-1}(L_1, a)\,[\Phi(L_1, M_b, 0) - \eta(L_1, a)]. \qquad (10.119)$$

The only restriction here is $L_1/a \gg 1$. We can use $L_1/a = 10 \ldots 20$, which means $a = 0.05\,\text{fm} \ldots 0.025\,\text{fm}$.

(2b.) Choose, e.g. $L_2 = 2L_1$ and insert $\tilde{\omega}$ into $\Phi(L_2, M_b, a)$:

$$S3: \quad \Phi(L_2, M_b, 0) = \lim_{a/L_2 \to 0} \{\eta(L_2, a) + \varphi(L_2, a)\,\tilde{\omega}(M_b, a)\}. \qquad (10.120)$$

(3) Repeat (2a) with the change $L_1 \to L_2$:

$$S4: \quad \omega(M_b, a) \equiv \varphi^{-1}(L_2, a)\,[\Phi(L_2, M_b, 0) - \eta(L_2, a)]. \qquad (10.121)$$

With the same resolutions $L_2/a = 10 \ldots 20$ as for $L = L_1$, we have now reached $a = 0.1\,\text{fm} \ldots 0.05\,\text{fm}$.

(4) Finally insert ω into the expansion of large volume observables, e.g.

$$S5: \quad m_B = \lim_{a \to 0} [\omega_1(M_b, a) + E^{stat}(a)]. \qquad (10.122)$$

In the chosen example the result is the relation between the RGI b-quark mass and the B-meson mass m_B. We can determine the b-quark mass in this way.

It is illustrative to put the different steps into one equation for this simple case:

$$m_B = \qquad\qquad\qquad\qquad\qquad\qquad\qquad\qquad\qquad\qquad\qquad (10.123)$$

$$\lim_{a \to 0} [E^{stat}(a) - \Gamma^{stat}(L_2, a)] \qquad a = 0.1\,\text{fm} \ldots 0.05\,\text{fm} \qquad [S_4, S_5]$$

$$+ \lim_{a \to 0} [\Gamma^{stat}(L_2, a) - \Gamma^{stat}(L_1, a)] \qquad a = 0.05\,\text{fm} \ldots 0.025\,\text{fm} \qquad [S_2, S_3]$$

$$+ \frac{1}{L_1} \lim_{a \to 0} \Phi_1(L_1, M_b, a) \qquad a = 0.025\,\text{fm} \ldots 0.012\,\text{fm} \qquad [S_1].$$

We have indicated the lattices drawn in Figure 10.3 and the typical lattice spacings of these lattices. The explicit expression for the decay constant in static approximation is not more complicated.

Exercise 10.4 Write down the full equation for the decay constant in static approximation.

Including $1/m_b$ terms and also other components of the weak flavour currents follows the same principle. A relevant point is that the matching observables Φ_i can be chosen such that φ, η have a block structure; *not* everything depends on everything.

10.6.3 Separating Orders

Due to renormalization we must be careful in separating different orders in the $1/m_b$ expansion. Still, a proper definition of the HQET parameters at different orders is easily given. The LO ones, ω^{stat}, are defined by dropping $1/m_b$ terms everywhere and the ones that we called ω^{HQET} yield the full LO+NLO results. Since all observables are chosen to be linear in ω, we can then define pure NLO parameters as

$$\omega^{(1/m)} = \omega^{\text{HQET}} - \omega^{\text{stat}}. \tag{10.124}$$

Note that all components $\omega_i^{(1/m)}$ are non-zero, including ω_1, ω_2, which are already present at static order. In fact, our discussion of renormalization of the $1/m_b$ terms shows that $\omega_1^{(1/m)} = m_{\text{bare}}^{(1/m)}$ diverges as $1/(a^2 m)$. In comparison, $\omega_1^{\text{stat}} \sim 1/a$. Of course, physical observables computed with $\omega = \omega^{\text{stat}}$ or $\omega = \omega^{(1/m)}$ are finite.

10.6.4 Important Issues in Practical Applications

We have concentrated here on the principle. In practical lattice field theory simulations of HQET, including matching, various issues had to be studied and problems had to be overcome. A detailed recent account is given in [48]. Here we just give a flavour, pointing to some milestones:

- The statistical precision of HQET correlation functions had to be improved by a change of the discretization [15, 18].
- Suitable finite volume quantities Φ_i had to be found for the simple case discussed above [8, 26, 27] and generalized for the full set of flavour currents [16, 29, 32, 33].
- The feasibility of carrying out all steps had to be demonstrated for QCD with dynamical fermions [5] and for the evaluation of more complicated observables, e.g. form factors [2].

After these steps, the time has come for computations of form factors that are relevant to phenomenology, e.g. for the decay B$\to \pi \ell \nu$. It is further worth thinking about other problems, e.g. inclusive B-decays.

10.7 The Most Important Messages

We finish by summarizing the main results of our discussion and elaborating one point (dependence on matching conditions) a little further.

- An effective theory such as HQET can be implemented non-perturbatively. The result is universal in the following sense: it does not depend on the (lattice) regularization.

- So far, results exist for $N_f = 0, 2$ (light quarks) QCD. The pattern found is that (with the matching as described) NLO = O(10%) LO. No significant deviations of NLO results from results using other methods are known. Therefore all indications are that NLO HQET has a precision of around 1% for b-quarks (with momenta $\sim 500\text{MeV}$).

  Unfortunately, the set of observables that have been evaluated is limited so far: B-leptonic decays and the mass of b-quarks are known at LO and NLO [48]. B_s semileptonic decays at fixed, large, q^2 are known at LO [2] and the methodology has been developed to extend this to NLO [3].

- Our clean, non-perturbative discussion of the matching of fundamental theory and effective theory shows the following important intrinsic limitations in expansions, such as the $1/m_b$-expansion in QFTs. These limitations have nothing to do with the lattice regularization.

  Results at each order in the expansion are ambiguous by terms of the size of the next order. Ambiguous means that they depend on the matching condition imposed. As an example consider the often-written formal HQET-expansion

$$m_B^{av} \equiv \frac{1}{4}[m_B + 3m_{B*}] \tag{10.125}$$

$$= m_b + \bar{\Lambda} + \frac{1}{2m_b}\lambda_1 + O(1/m_b^2) \tag{10.126}$$

with (ignoring renormalization)

$$\lambda_1 = \langle B|\mathcal{O}_{kin}|B\rangle. \tag{10.127}$$

The quantity $\bar{\Lambda}$ is referred to as 'static binding energy' and λ_1 as the kinetic energy of the b-quark inside the B-meson. Depending on how we choose the matching conditions, we change λ_1 by a term of order Λ_{QCD}^2. In fact we could set

$$\Phi_1 = m_B^{av}(L) \text{ for large } L. \tag{10.128}$$

Then we have at LO

$$\text{LO: } m_b + \bar{\Lambda} = m_B^{av} \tag{10.129}$$

and at NLO

$$\text{NLO: } m_b + \bar{\Lambda} + \frac{1}{2m_b}\lambda_1 = m_B^{av}. \tag{10.130}$$

Together it follows that

$$\lambda_1 = 0 \tag{10.131}$$

for this matching condition. If we take L of order 0.5 fm and parametrize m_B^{av} like (10.126) we will have $\lambda_1 = c\Lambda_{QCD}^2 \neq 0$. It is therefore ambiguous. We remark that this is similar to the case of the gluon condensate. A convenient non-perturbative definition is to set λ_1 and the latter to zero.

Acknowledgements

I thank the organizers of this Les Houches school, in particular Sacha Davidson and Mikko Laine, as well as the staff for putting together a very interesting program and for a perfectly smooth organization. I am grateful to Mikko for comments on the manuscript. My special thanks go to the students who asked interesting questions. Finally I thank all those who invaluably contributed to the development of non-perturbative HQET on the lattice—from principle to practice.

References

[1] C. Aubin and C. Bernard (2006). Staggered Chiral Perturbation Theory for Heavy-Light Mesons. *Phys. Rev. D* **73**: 014515.
[2] F. Bahr, et al. (2016). Continuum Limit of the Leading-Order HQET Form Factor in $B_s \rightarrow K\ell\nu$ Decays. *Phys. Lett. B* **757**: 473–9.
[3] D. Banerjee, M. Koren, H. Simma, and R. Sommer (2017). HQET Form Factors for $B_s \rightarrow K\ell\nu$ Decays Beyond Leading Order. In *Proceedings of the 35th International Symposium on Lattice Field Theory (Lattice 2017)*. 18–24 June 2017, Granada, Spain.
[4] S. Bekavac, et al. (2010). Matching QCD and HQET Heavy-Light Currents at Three Loops. *Nucl. Phys. B* **833**: 46–63.
[5] F. Bernardoni, et al. (2014). The b-Quark Mass from Non-Perturbative $N_f = 2$ Heavy Quark Effective Theory at $O(1/m_h)$. *Phys. Lett. B* **730**: 171–7.
[6] B. Blossier, et al. (2012). Parameters of Heavy Quark Effective Theory from $N_f = 2$ Lattice QCD. *JHEP* **1209**: 132.
[7] B. Blossier, M. della Morte, N. Garron, and R. Sommer (2010). HQET at order $1/m$: I. Non-Perturbative Parameters in the Quenched Approximation. *JHEP* **1006**: 002.
[8] D. J. Broadhurst and A. G. Grozin (1991). Two-Loop Renormalization of the Effective Field Theory of a Static Quark. *Phys. Lett. B* **267**: 105–10.
[9] D. J. Broadhurst and A. G. Grozin (1995). Matching QCD and HQET Heavy-Light Currents at Two Loops and Beyond. *Phys. Rev. D* **52**: 4082–98.
[10] G. Burdman and J. F. Donoghue (1992). Union of Chiral and Heavy Quark Symmetries. *Phys. Lett. B* **280**: 287–91.

[11] K. G. Chetyrkin (1997). Quark Mass Anomalous Dimension to O (alpha-s**4). *Phys. Lett. B* **404**: 161–5.

[12] K. G. Chetyrkin and A. G. Grozin (2003). Three-Loop Anomalous Dimension of the Heavy-Light Quark Current in HQET. *Nucl. Phys. B* **666**: 289–302.

[13] G. Colangelo, S. Durr, and C. Haefeli (2005). Finite Volume Effects for Meson Masses and Decay Constants. *Nucl. Phys. B* **721**: 136–74.

[14] M. Czakon (2005). The Four-Loop QCD Beta-Function and Anomalous Dimensions. *Nucl. Phys. B* **710**: 485–98.

[15] M. Della Morte, et al. (2004). Lattice HQET with Exponentially Improved Statistical Precision. *Phys. Lett. B* **581**: 93–8.

[16] M. Della Morte, et al. (2014). Matching of Heavy-Light Flavor Currents Between HQET at Order $1/m$ and QCD: I. Strategy and Tree-Level Study. *JHEP* **1405**: 060.

[17] M. Della Morte, P. Fritzsch, and J. Heitger (2007). Non-Perturbative Renormalization of the Static Axial Current in Two-Flavour QCD. *JHEP* **0702**: 079.

[18] M. Della Morte, A. Shindler, and R. Sommer (2005). On Lattice Actions for Static Quarks. *JHEP* **08**: 051.

[19] E. Eichten, and B. R Hill (1990). An Effective Field Theory for the Calculation of Matrix Elements Involving Heavy Quarks. *Phys. Lett. B* **234**: 511.

[20] J. Gasser, and H. Leutwyler (1984). Chiral Perturbation Theory to One Loop. *Ann. Phys.* **158**: 142.

[21] J. Gasser, and H. Leutwyler (1987). Light Quarks at Low Temperatures. *Phys. Lett. B* **184**: 83.

[22] V. Gimenez (1992). Two Loop Calculation of the Anomalous Dimension of the Axial Current with Static Heavy Quarks. *Nucl. Phys. B* **375**: 582–624.

[23] J. L. Goity (1992). Chiral Perturbation Theory for SU(3) Breaking in Heavy Meson Systems. *Phys. Rev. D* **46**: 3929–36.

[24] B. Grinstein, et al. (1992). Chiral Perturbation Theory for f d(s) / f d and b b(s) / b b. *Nucl. Phys. B* **380**: 369–76.

[25] A. G. Grozin (2004). *Heavy Quark Effective Theory*, Springer.

[26] J. Heitger, M. Kurth, and R. Sommer (2003). Nonperturbative Renormalization of the Static Axial Current in Quenched QCD. *Nucl. Phys. B* **669**: 173–206.

[27] J. Heitger, and R. Sommer (2004). Nonperturbative Heavy Quark Effective Theory. *JHEP* **0402**: 022.

[28] P. Hernandez (2009). Introduction to Lattice QCD. In *Proceedings of the 2009 Les Houches Summer School on Modern Perspectives in Lattice QCD: Quantum Field Theory and High-Performance Computing*, 93rd Session. 3–28 August 2009, Les Houches, France.

[29] D. Hesse, and R. Sommer (2013). A one-Loop Study of Matching Conditions for Static-Light Flavor Currents. *JHEP* **1302**: 115.

[30] E. E. Jenkins (1994). Heavy Meson Masses in Chiral Perturbation Theory With Heavy Quark Symmetry. *Nucl. Phys. B* **412**: 181–200.

[31] X. Ji, and M. J. Musolf (1991). Subleading Logarithmic Mass Dependence in Heavy Meson Form-Factors. *Phys. Lett. B* **257**: 409.

[32] P. Korcyl (2013). Fixing the Parameters of Lattice HQET Including $1/m_B$ Terms. In *Proceedings of the 14th International Conference on B-Physics at Hadron Machines*. 8–12 April, Bologna, Italy, 2013.

[33] P. Korcyl (2013). On One-Loop Corrections to Matching Conditions of lattice HQET Including $1/m_b$ Terms. In *Proceedings of the 31st International Symposium on Lattice Field Theory (LATTICE 2013)*. 29 July–3 August, Mainz, Germany, 2013.

[34] M. Kurth, and R. Sommer (2001). Renormalization and O(a)-Improvement of the Static Axial Current. *Nucl. Phys. B* **597**: 488–518.

[35] M. Lüscher (2010). Computational Strategies in Lattice QCD. In *Proceedings of the 2009 Les Houches Summer School on Modern Perspectives in Lattice QCD: Quantum Field Theory and High-Performance Computing*, 93rd Session. 3–28 August 2009, Les Houches, France.

[36] M. Lüscher, S. Sint, R. Sommer, and P. Weisz (1996). Chiral Symmetry and O(a) Improvement in Lattice QCD. *Nucl. Phys. B* **478**: 365–400.

[37] M. Lüscher, and P. Weisz (1985). On-Shell Improved Lattice Gauge Theories. *Commun. Math. Phys.* **97**: 59.

[38] T. Mannel (2017). Effective Field Theories for Heavy Quarks. Renormalization Theory and Effective Field Theories. Chapter 9, this volume.

[39] A. V. Manohar and M. B. Wise (2000).

[40] M. Neubert (1994). Heavy Quark Symmetry. *Phys. Rept.* **245**: 259–396.

[41] M. Neubert (2017). Renormalization Theory and Effective Field Theories. Chapter 1, this volume.

[42] A. Pineda (2001). Determination of the Bottom Quark Mass from the Upsilon(1S) System. *JHEP* **06**: 022.

[43] H. D. Politzer and M. B. Wise (1988). *Phys. Lett. B* **206**: 681.

[44] J. L. Rosner and M. B Wise (1993). Meson Masses from SU(3) and Heavy Quark Symmetry. *Phys. Rev. D* **47**: 343–5.

[45] S. Schaefer (2009). Simulations with the Hybrid Monte Carlo Algorithm: Implementation and Data Analysis. In *Proceedings of the 2009 Les Houches Summer School on Modern Perspectives in Lattice QCD: Quantum Field Theory and High-Performance Computing*, 93rd Session. 3–28 August 2009, Les Houches, France.

[46] M. A. Shifman and M. B. Voloshin (1987). On Annihilation of Mesons Built from Heavy and Light Quark and Anti-$b_0 \leftrightarrow b_0$ Oscillations. *Sov. J. Nucl. Phys.* **45**: 292.

[47] R. Sommer (2011). Introduction to Non-Perturbative Heavy Quark Effective Theory. In L. Lellouch, et al. (Eds), *Modern Perspectives in Lattice QCD*, Oxford University Press.

[48] R. Sommer (2015). Non-Perturbative Heavy Quark Effective Theory: Introduction and Status. *Nucl. Part. Phys. Proc.* **261-262**: 338–67.

[49] K. Symanzik (1983). Continuum Limit and Improved Action in Lattice Theories. 1. Principles and ϕ^4 Theory. *Nucl. Phys. B* **226**: 187.

[50] K. Symanzik. (1983). Continuum Limit and Improved Action in Lattice Theories. 2. O(N) Nonlinear Sigma Model in Perturbation Theory. *Nucl. Phys. B* **226**: 205.

[51] T. van Ritbergen, J. A. M. Vermaseren, and S. A. Larin (1997). The Four-Loop Beta Function in Quantum Chromodynamics. *Phys. Lett. B* **400**, 379–84.

[52] J. A. M. Vermaseren, S. A. Larin, and T. van Ritbergen (1997). The Four-Loop Quark Mass Anomalous Dimension and the Invariant Quark Mass. *Phys. Lett. B* **405**: 327–33.

[53] S. Weinberg (1979). Phenomenological Lagrangians. *Physa A* **96**: 327–40.

[54] P. Weisz (2011). Renormalization and Lattice Artifacts. In L. Lellouch, et al. (Eds), *Modern Perspectives in Lattice QCD*, Oxford University Press.

[55] K. G. Wilson (1974). Confinement of Quarks. *Phys. Rev. D* **10**: 2445–59.

[56] M. B. Wise (1992). Chiral Perturbation Theory for Hadrons Containing a Heavy Quark. *Phys. Rev. D* **45**: 2188–91.

11
Effective Theory Approach to Direct Detection of Dark Matter

Junji HISANO

Kobayashi–Maskawa Institute for the Origin of Particles and the Universe (KMI),
Nagoya University, Nagoya 464–8602, Japan
Department of Physics, Nagoya University, Nagoya 464–8602, Japan
Kavli Institute for the Physics and Mathematics of the Universe (WPI), Todai Institutes
for Advanced Study, University of Tokyo, Kashiwa 277–8568, Japan

Junji HISANO

Hisano, J., *Effective Theory Approach to Direct Detection of Dark Matter* In: *Effective Field Theory in Particle Physics and Cosmology.*
Edited by: Sacha Davidson, Paolo Gambino, Mikko Laine, Matthias Neubert and Christophe Salomon, Oxford University Press (2020).
© Oxford University Press. DOI: 10.1093/oso/9780198855743.003.0011

Chapter Contents

11.1   Introduction

It is now certain that dark matter (DM) exists in the universe. However, we do not know the nature of DM, since our knowledge about it is limited to the gravitational aspect. We have no DM candidates in the Standard Model (SM) of particle physics, nor in astronomy, and DM is now at the forefront of physics. The idea that DM may be unknown particles produced in the early universe is fascinating, and many candidates for the DM have been proposed [1]. Weakly-interacting massive particles (WIMPs) are one of the leading candidates, and are assumed to be produced in the thermal bath in the early universe. The typical WIMP mass scale is $O(100)$ GeV to $O(1)$ TeV under this assumption. We also expect new physics beyond the SM at the TeV scale from the naturalness point of view. This coincidence is called the WIMP miracle. Many experiments now search for WIMP DM. Direct detection of WIMP DM on Earth is one of the methods. WIMPs are assumed to pass through us; about one million WIMPs may exist in a room. Their interactions are very weak, but there is a small probability that they may collide with nuclei. Direct detection experiments observe the recoiled nuclei, and many such experiments are currently in progress or proposed. Marrodn Undagoitia and Rauch [2] provide recent review of direct detection experiments.

This original lecture evaluated the WIMP DM detection rate from UV theories at the TeV scale. In this evaluation, the effective theory approach works well. UV theories provide the interactions of WIMPs with partons. On the other hand, we need to know the effective interactions of WIMPs with nuclei. We have to derive effective theories at the parton, nucleon, and nuclei levels. We note the quantum chromodynamics (QCD) aspects in the evaluation of WIMP interactions with nucleons. We show that we can handle QCD corrections to the Wilson coefficients in the effective interactions at the parton level well. We then evaluate the next-leading order contribution of α_s, and show the strategy for the evaluation.

The chapter first introduces the WIMP DM. After discussing the effective interactions of WIMPs with nuclei and nucleons in Section 11.3, it briefly reviews the direct detection experiments in Section 11.4. Then, it shows how to evaluate the effective interactions of WIMPs with nucleons from UV theories in Section 11.5. Due to the nucleon matrix elements of the parton-level effective operators, the power counting of α_s in calculating the direct detection rate is not the same as in conventional ones. Furthermore, we do not necessarily need to evaluate the Wilson coefficients for parton-level effective operators at the hadronic scale with renormalization-group (RG) equations, in contrast to the hadronic observables in flavour physics. These topics are all covered in Section 1.5. Section 11.6 shows some results for three UV models as examples: 1) gauge singlet WIMPs coupled with the Higgs boson; 2) gauge singlet WIMPs coupled with coloured scalars and quarks; and 3) $SU(2)_L$ non-singlet WIMPs. This chapter assumes that the WIMPs are Majorana fermions. The application to other WIMPs is straightforward. Finally, Section 11.7 discusses the strategy to evaluate the direct detection rate including the $O(\alpha_s)$ correction, and Section 11.8 concludes the chapter. The Appendix introduces

Fock–Schwinger gauge fixing, which is quite useful in evaluating the Wilson coefficients of the effective operators including gluon field strengths. This chapter is mainly based on the author's recent works [3, 4].

11.2   WIMP DM

If the DM is composed of unknown particles, they are electrically neutral. They are stable, or have a longer lifetime than the age of the universe. They are massive so that they are 'cold' in the structure formation era of the universe, i.e. the free streaming length after production in the early universe is shorter than the size of protogalaxies so that the small-scale structure in the universe is not erased. The cold DM abundance is precisely determined from cosmic microwave background (CMB) power spectrum measurements. The DM particles are non-relativistic in the current universe, and the energy density ρ_X is given by $M_X n_X$, with M_X and n_X the DM particle mass and number density, respectively. It has been found from the CMB measurements that the DM energy density normalized by the critical density in the universe, $\Omega_X \equiv \rho_X/\rho_{\text{critical}}$, is about 27% [5]. The critical density is $\rho_{\text{critical}} \simeq 10^{-5} \text{GeV/cm}^2$. The abundance and the free streaming length depend on the production mechanism in the early universe. In WIMP scenarios, the WIMPs are assumed be in thermal equilibrium in the early hot universe.

One of the representative models for WIMPs is the supersymmetric SM (SUSY SM) [6]. In this model, a Z_2 symmetry called the R-parity is introduced in order to stabilize protons. As a result, the lightest SUSY particle is stable. The neutral components of the fermionic superpartners of the gauge and Higgs bosons, called gauginos and Higgsinos, are WIMP candidates in the SUSY SM. Another representative model is the universal extra dimension (UED) model [7]. In this model, we can impose a parity symmetry in extra-dimensional space so that the lightest Kaluza–Klein particle is stable [8]. The candidate in the minimal model is the Kaluza–Klein photon. The SUSY SM and the UED model are motivated by the naturalness problem in the Higgs boson mass term in the SM so that their energy scale is expected to be at the TeV scale. As will be explained, the WIMPs have masses of about $O(100)$ GeV-$O(1)$ TeV if they were produced in the thermal bath in the early universe. These two observations support the assertion that new physics will appear at the TeV scale. Many models have been proposed in order to explain the naturalness and the WIMP DM.

Here, we evaluate the WIMP DM abundance in the universe. In the WIMP scenarios, the WIMPs have interactions with the SM particles so that the WIMPs are thermalized in the early hot universe. The stability of the WIMPs comes from global symmetries, as given in the previous examples. In the early universe, where the temperature (T) is much higher than the WIMP mass (M_X), they are in thermal equilibrium, and the number density is comparable to those for the SM particles. When the temperature decreases to below the WIMP mass, WIMP pair production by SM particle collisions is suppressed in the thermal bath, so that the WIMP number density deviates from that in thermal equilibrium. The WIMP pair annihilation is frozen when the WIMPs do not find partners for their pair annihilation within a Hubble time, and the number density is only diluted by the expansion of the universe. Thus, the current abundance of WIMPs is determined by the WIMP pair annihilation cross section.

The WIMP abundance is more precisely evaluated with the Boltzmann equation [9],

$$\frac{dn_X}{dt} + 3H(T)n_X = -\langle\sigma|v|\rangle\left[n_X^2 - \left(n_X^{\text{EQ}}\right)^2\right]. \tag{11.1}$$

The second term is for dilution due to the expansion of the universe. The Hubble parameter $H(T)$ in the radiation-dominated (RD) era is given by

$$H(T) = \sqrt{\frac{8\pi}{3M_{\text{pl}}}}\rho \simeq \sqrt{\frac{4\pi}{45}}g_\star^{1/2}\frac{T^2}{M_{\text{pl}}} \tag{11.2}$$

where M_{pl} is the Planck mass and ρ is the energy density ($\rho = (\pi^2/30)g_\star T^4$ with $g_\star = \sum_{\text{boson}} 1 + \sum_{\text{fermion}} 7/8$). If the collision term in the right-hand side of the Boltzmann equation is zero, n_X/s (s is the entropy density, $s = (2\pi^2/45)g_\star T^3$) is constant since s is also diluted by the expansion of the universe. In the collision term, $\langle\sigma|v|\rangle$ is the thermal-averaged WIMP pair annihilation cross section and n_X^{EQ} is the WIMP number density in thermal equilibrium. The collision term is proportional to the square of the WIMP number density since it comes from the WIMP pair annihilation and production.

The Boltzmann equation is rewritten as

$$\frac{x}{Y_{\text{EQ}}}\frac{dY}{dx} = -\frac{\Gamma_A}{H}\left[\left(\frac{Y}{Y_{\text{EQ}}}\right)^2 - 1\right] \tag{11.3}$$

by defining Y and x as $Y \equiv n_X/s$ and $x = M_X/T$, respectively. Here, Γ_A is the probability of annihilation per unit time for a WIMP,

$$\Gamma_A = n_X^{\text{EQ}}\langle\sigma|v|\rangle. \tag{11.4}$$

When $x \gtrsim 1$, the WIMP pair production is kinematically suppressed, and they start to be decoupled from the thermal bath so that $Y/Y_{\text{EQ}} \gtrsim 1$. When $H \gg \Gamma_A$, the annihilation is frozen and Y becomes constant. The freeze-out temperature (T_F) and the WIMP number density at T_F (n_X^F) are approximately determined by $H(T_F) = \Gamma_X$, and it is found that T_F is about $M_X/20$ and $n_X^F \simeq H(T_F)/\langle\sigma|v|\rangle$. Thus, Y and Ω_X are approximately evaluated as

$$Y \simeq \sqrt{\frac{45}{\pi}}g_\star^{-1/2}\frac{1}{T_F M_{\text{pl}}}\frac{1}{\langle\sigma|v|\rangle},$$

$$\Omega_X = \frac{s^{\text{now}}}{\rho_{\text{critical}}}M_X Y \simeq 0.4 \times \left(\frac{x_F \equiv M_X/T_F}{20}\right)\left(\frac{\langle\sigma|v|\rangle}{10^{-9}\text{GeV}^{-2}}\right)^{-1}, \tag{11.5}$$

where s^{now} is the entropy density in the current universe ($s^{\text{now}} \simeq 3000\text{cm}^{-3}$).

Let us discuss some typical cases. When the WIMPs are $SU(2)_L$ singlet fermions, the annihilation cross section into SM fermions is given by $\sigma v \sim \pi\alpha^2 M_X^2/M_S^4$, with M_S the mediator scalar mass. Assuming that the mediator coupling constant α is the same as that of the $U(1)_Y$ gauge interaction, $\sigma v \sim 3 \times 10^{-9} \mathrm{GeV}^{-2}$ for $M_X = M_S = 300 \mathrm{GeV}$. If the mediator mass is heavier than the WIMP mass, the cross section is suppressed by $(M_X/M_S)^4$. If the WIMPs are Majorana fermions, the annihilation into SM fermions suffers from p-wave suppression so that the thermally averaged cross section is more suppressed by $T_F/M_X \sim 1/20$ or the square of the masses of the SM fermions in the final states. Thus, if the WIMPs are $SU(2)_L$ singlet Majorama fermions, the WIMP mass is around 100 GeV. Binos (the fermionic superpartners of the $U(1)_Y$ gauge boson in the SUSY SM) are an example. They are $SU(2)_L$ singlet Majorana fermions. This situation may be changed when some new particles are degenerate with the WIMPs in mass so that co-annihilation occurs [10]. For example, if staus, which are the bosonic superpartners of the tau lepton, are degenerate with the binos in mass, heavier binos are predicted due to their co-annihilation.

On the other hand, if the WIMPs are the neutral component of the $SU(2)_L$ multiplet(s), they annihilate into two weak gauge bosons. The annihilation cross section is approximately given by $\sigma v \sim \pi\alpha_2^2/M_X^2 = 3.5 \times 10^{-9}\ \mathrm{GeV}^{-2} \times (M_X/1\ \mathrm{TeV})^{-2}$. Thus, the WIMP mass is expected to be at the TeV scale. Higgsinos, which are the fermionic superpartners of Higgs bosons in the SUSY SM, are $SU(2)_L$ doublets, and winos, which are those of $SU(2)_L$ gauge bosons, are $SU(2)_L$ triplets. Detailed calculations show that the $SU(2)_L$ doublet and triplet fermion masses are about 1 TeV and 3 TeV, respectively, if they are in thermal equilibrium in the early hot universe [11].

Many kinds of experiments are currently searching for WIMPs in addition to signature of new physics at the TeV scale. The first are direct searches for WIMPs in collider experiments, such as in the LHC. (See [*I*] in Figure 11.1.) While WIMPs do not leave any signatures in detectors in the experiments, their momenta appear missing in their events. LHC experiments are now searching for events with missing transverse momenta. The production cross sections of coloured particles are larger than the those for WIMPs at the LHC, and WIMPs are mainly produced from the decay of the coloured particles if they have interactions.

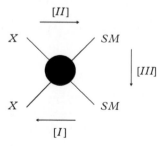

Figure 11.1 *Three approaches to WIMP searches.*

The second approach is indirect detection of WIMP DM in cosmic rays ([*II*] in Figure 11.1). WIMPs are gravitationally accumulated in massive astrophysical objects, such as stars, galaxies, and galactic clusters. The WIMP pair annihilation is increased around these objects since the annihilation rate is proportional to the square of the WIMP density. The final states in the annihilation include gamma rays, positrons, anti-protons, neutrinos, and so on, and they contribute to cosmic rays. The Fermi satellite is observing gamma rays, and it gives constraints on WIMP annihilation from observations of gamma rays from the galactic centre [12] or dwarf galaxies [13]. The atmospheric Cherenkov telescopes, such as HESS and MAGIC, are also searching for gamma rays from WIMP annihilation [14]. AMS-02 on the International Space Station is searching for anti-particles [15]. The ICECUBE [16] and Super-Kamiokande experiments [17] observe solar neutrinos to place constraints on WIMP annihilation in the Sun. If non-standard signatures are found in cosmic rays, they may be interpreted as WIMP annihilation. The indirect detection is sometimes challenging due to the astrophysical backgrounds, though it may be sensitive to heavy WIMPs with masses above the TeV scale. In particular, the annihilation cross sections of $SU(2)_L$ non-singlet WIMPs are enhanced by the attractive force in weak interactions, called the Sommerfeld effect [18].

The third is direct detection of WIMP DM trapped within our galaxy ([*III*] in Figure 11.1). In these experiments, recoiled nuclei are observed in elastic scattering with WIMPs, $X + N \rightarrow X + N$. The typical recoil energy is $\sim m_r^2/m_T v^2$ ($m_r (= m_T M_X/(m_T + M_X)$: reduced mass, m_T: target nucleus mass, and v: DM velocity in the lab flame). The typical DM velocity around the Earth is about $10^{-3}c$ (c: speed of light). Thus, the recoil energy is (1–100) keV. Currently, one ton-class detectors, such as XENON1T [19] and PandaX-II [20], are searching for WIMP DM. We review the experiments after showing the effective interaction of the nucleus/nucleon with WIMPs.

11.3 Effective WIMP-Nucleus/Nucleon Interactions

This section discusses the WIMP-nucleus effective interactions in the non-relativistic limit for elastic scattering processes,

$$X(\vec{p}) + N(\vec{k}) \rightarrow X(\vec{p}') + N(\vec{k}'), \tag{11.6}$$

where $\vec{p}$ and $\vec{p}'$ ($\vec{k}$ and $\vec{k}'$) are the incoming and outgoing WIMP (nucleus, N) three-momenta, respectively. We assume that the WIMPs have spin S_X. This discussion also applies to scalar and vector WIMPs. The effective interactions are given by operators that are constructed with four vectors, the momentum transfer $\vec{q} (\equiv \vec{k} - \vec{k}')$, the incoming WIMP velocity $\vec{v}$, the nucleus spin $\vec{S}_N$, and the WIMP spin $\vec{S}_X$. For convenience, we introduce $\vec{P} (\equiv \vec{p} + \vec{p}' = 2\vec{p} + \vec{q})$ as an alternative to $\vec{v}$. In elastic scattering, $|\vec{P}|$ and $|\vec{q}|$ are proportional to v ($v \equiv |\vec{v}|$).

The effective operators are classified into two categories: (nucleus) spin-independent (SI) and spin-dependent (SD) operators. There are four SI operators [21]:

$$O_1^{(++)} = 1, \qquad O_2^{(-+)} = \vec{S}_X \cdot i\vec{q},$$
$$O_3^{(--)} = \vec{S}_X \cdot \vec{P}, \qquad O_4^{(++)} = \vec{S}_X \cdot (\vec{P} \times i\vec{q}). \qquad (11.7)$$

Here, the superscripts $(\pm\pm)$ are for the transformation properties under parity (P) and charge conjugation (C). The SD operators are the following:

$$
\begin{aligned}
O_5^{(++)} &= \vec{S}_X \cdot \vec{S}_N, & O_6^{(-+)} &= \vec{S}_N \cdot i\vec{q}, \\
O_7^{(--)} &= \vec{S}_N \cdot \vec{P}, & O_8^{(--)} &= (\vec{S}_X \times \vec{S}_N) \cdot i\vec{q}, \\
O_9^{(-+)} &= (\vec{S}_X \times \vec{S}_N) \cdot \vec{P}, & O_{10}^{(++)} &= \vec{S}_N \cdot (\vec{P} \times i\vec{q}), \\
O_{11}^{(++)} &= (\vec{S}_N \cdot \vec{q})(\vec{S}_X \cdot \vec{q}), & O_{12}^{(++)} &= (\vec{S}_N \cdot \vec{P})(\vec{S}_X \cdot \vec{P}), \\
O_{13}^{(+-)} &= (\vec{S}_N \cdot i\vec{q})(\vec{S}_X \cdot \vec{P}) + (\vec{S}_N \cdot \vec{P})(\vec{S}_X \cdot i\vec{q}) \\
O_{14}^{(+-)} &= (\vec{S}_N \cdot i\vec{q})(\vec{S}_X \cdot \vec{P}) - (\vec{S}_N \cdot \vec{P})(\vec{S}_X \cdot i\vec{q}) \\
O_{15}^{(-+)} &= [\vec{S}_N \cdot (\vec{P} \times \vec{q})](\vec{S}_X \cdot \vec{q}) + [\vec{S}_X \cdot (\vec{P} \times \vec{q})](\vec{S}_N \cdot \vec{q}) \\
O_{16}^{(--)} &= [\vec{S}_N \cdot (\vec{P} \times i\vec{q})](\vec{S}_X \cdot \vec{P}) + [\vec{S}_X \cdot (\vec{P} \times i\vec{q})](\vec{S}_N \cdot \vec{P}).
\end{aligned}
$$

The Wilson coefficients for these sixteen operators depend on $|\vec{q}|$. The effective interactions depending on $\vec{P}$ and/or $\vec{q}$ are suppressed by v. Two operators, $O_1^{(++)}$ (SI) and $O_5^{(++)}$ (SD), are dominant in the elastic scattering unless they are accidentally suppressed. In the following, we concentrate on these cases.

The effective interactions of WIMPs with nuclei are derived from those with nucleons. When the WIMPs are Majorana fermions (X), the effective interactions with nucleons ($N = n, p$) are simple, given by[1]

$$L_{\text{eff}} = \sum_{N=n,p} \left(f_N \bar{X} X \bar{N} N + a_N \bar{X} \gamma^\mu \gamma_5 X \bar{N} \gamma_\mu \gamma_5 N \right). \qquad (11.8)$$

The first term is the SI interaction in the non-relativistic limit, while the second is the SD one. Neglecting nucleus form factors, the elastic-scattering cross sections of WIMPs with nuclei are given by

$$\sigma = \sigma_{\text{SI}} + \sigma_{\text{SD}} \qquad (11.9)$$

---

[1] While X is a Majorana fermion, we omit '1/2' in the coefficients of the interactions due to the convention given in Drees and Nojiri [22].

with

$$\sigma_{SI} = \frac{4}{\pi} m_r^2 |Z f_p + (A - Z) f_n|^2,$$

$$\sigma_{SD} = \frac{16}{\pi} m_r^2 \frac{\mathcal{J} + 1}{\mathcal{J}} |a_p \langle S_p \rangle + a_n \langle S_n \rangle|^2. \tag{11.10}$$

Here, Z and A are the atomic and mass numbers, respectively, and $\mathcal{J}$ and $\langle S_N \rangle$ are the total spin of target nucleus and the expectation value of the nucleon spin in the nucleus, respectively.

While nuclei with non-zero spin are sensitive to the SD interactions, nuclei with larger atomic or mass numbers are more sensitive to the SI interactions. The nucleon scalar operators in Eqn. 11.8 become proton and neutron density operators in the non-relativistic limit. Thus, the amplitude is proportional to Z or $A - Z$.

Earlier, we ignored the nucleus form factors. However, they are not negligible in practice. Let us consider the SI cross section. The momentum transfer $q = |\vec{q}|$ is given by $(2m_T E_R)^{1/2}$, where the recoil energy E_R is $E_{th} \leq E_R \leq 2(m_r^2/m_T)v^2$. The nuclear radius r_N is typically $A^{1/3}$ fm, and then, $r_N q \simeq 7 \times 10^{-3} A^{5/6} (E_R(\text{keV}))^{1/2}$. While the SI cross sections are larger for larger nuclei, they suffer from suppression for large E_R and/or large A due to the form factors. Furthermore, the form factors are important to predict the precise spectrum of the nucleus recoil energy in direct detection. The nucleus form factors are reviewed in Lewin and Smith [23].

11.4   Brief Review of Direct Detection Experiments

The DM energy density around the Earth is determined by the stellar motions in our galaxy. However, due to our limited knowledge of the DM density distribution in the Milky Way, the energy distribution is still evaluated with a large uncertainty as [24]

$$\rho_X = (0.2 - 0.6) \text{GeV/cm}^3. \tag{11.11}$$

The DM velocity distribution is more uncertain since we have no way to measure it directly. A Maxwellian distribution $f(\vec{v})$ is assumed in the Galactic Centre coordinate [23]

$$f(\vec{v}) \simeq \frac{1}{(\pi v_0^2)^{3/2}} e^{-(\vec{v} + \vec{\oplus})^2 / v_0^2}, \tag{11.12}$$

where $v_0 \simeq 230 \text{km/sec}$ and $\vec{v}_\oplus$ is the Earth's velocity. This distribution is supported by the N-body simulations. The Earth's velocity is

$$|\vec{v}_\oplus| \simeq 244 + 15 \sin(2\pi y) \ (\text{km/sec}), \tag{11.13}$$

with y the elapsed time from March 2nd. The first term comes from the motion of the Solar System, and the second is from revolution of the Earth around the Sun. If WIMP wind relative to the Earth is detected by direction-sensitive detectors, it would be strong evidence for the DM, although it would still not be conclusive. Annual modulation of the event rate is also expected, and it is at most 3%.

Thus, the signals in the direct detection experiments are only from recoiled nuclei. The signals are observed by detecting phonons/heat, ionization, or light generated by the recoiled nuclei in the experiments. They have to maintain a low background in the experiments. The dominant backgrounds are gamma rays and electrons from the beta and gamma decays. The experiments are performed underground, and the coincidence of the signals, such as ionization and light in the XENON [19] and PandaX experiments [20], are observed.

The event rate per unit mass of the target R is given as

$$dR = \frac{N_0}{A}\sigma v \, dn_X, \tag{11.14}$$

(N_0 is the Avogadro number). Assuming a zero-momentum transfer cross section $\sigma \simeq$ const $= \sigma_0$, the event rate is evaluated as

$$
\begin{aligned}
R_0 &= \frac{2}{\pi}\frac{N_0}{A}\frac{\rho_X}{M_X}\sigma_0 v_0 \\
&\simeq \frac{540}{AM_X}\left(\frac{\sigma_0}{10^{-36}\mathrm{cm}^2}\right)\left(\frac{\rho_0}{0.4\mathrm{GeV/cm}^3}\right)\left(\frac{v_0}{230\mathrm{km/sec}}\right) \text{ event/kg/day,} \tag{11.15}
\end{aligned}
$$

where M_X is in GeV.

The event spectrum as a function of the recoil energy E_R is approximately given by [23]

$$\frac{dR}{dE_R} \simeq \frac{R_0}{E_0}\mathrm{e}^{-E_R/E_0} \tag{11.16}$$

where $E_0 = 2(m_r^2/M_T)v^2$. Thus, if the energy threshold in the experiment is lowered, the event rate is increased and the experiments are more sensitive to WIMPs with lighter masses.

As explained in Section 1.3, the SI cross section of WIMPs with nuclei is enhanced when the target nucleus has large atomic or mass numbers. Thus, many experiments use heavy nuclei as the targets. If the SI interaction is isosinglet, the zero-momentum transfer SI cross sections with nuclei are proportional to the SI cross section with the proton σ_{SI}^p as

$$\sigma_0 = A^2\frac{m_r^2(A)}{m_r^2(1)}\sigma_{\mathrm{SI}}^p \tag{11.17}$$

with $m_r^2(A)$ ($m_r^2(1)$) the reduced mass of the target nucleus (proton) and the WIMP. The sensitivity and exclusion curves of the direct detection experiments are shown on a plane of σ_{SI}^p and M_X.

The current limits on the SI cross section of protons are derived from LUX [25], XENON1T [19], and PandaX-II [20]. They exclude up to $\sigma_{SI}^p \sim 10^{-46}$ cm for $M_X \simeq 50$ GeV. The limits for heavier WIMPs are scaled by $1/M_X$ since the DM number density is ρ_X/M_X. The second-generation experiments, XENONnT, LZ, and PandaX-xT, whose fiducial volumes are $O(1)$ ton, will start in a few years [2]. These experiments will aim for 10^{-48} cm^2.

Neutrino coherent scattering off nuclei is a serious background source in direct detection experiments, called 'the neutrino floor' [26]. It is quite difficult to remove it except in direct detection experiments with directional sensitivity. The solar neutrinos hide the signals for light DM ($M_X \lesssim O(1)$ GeV) if $\sigma_{SI}^p \lesssim 10^{-44}$ cm^2. The atmospheric neutrinos also hide the heavier DM signals if $\sigma_{SI}^p \lesssim 10^{-49}$ (10^{-48}) cm^2 for $M_X \simeq 100$ GeV (1 TeV). The third-generation experiments aim to reach the neutrino floor [2].

11.5 Evaluation of Effective Interaction of WIMPs with Nucleons

We now evaluate the effective interactions of WIMPs with nucleons from UV theories. For this purpose, we first construct effective theories of WIMPs and quarks/gluons for direct detection experiments. The Wilson coefficients $C_i(\mu_{UV})$ ($i = 1, 2, \cdots$) of the effective operators $O_i(\mu_{UV})$ at the UV scale μ_{UV} are derived by integrating out heavy particles in the UV theory. In the normal procedure for hadronic observables in flavour physics, the Wilson coefficients at the hadronic scale $\mu_H \sim 1$ GeV are derived from those at the UV scale using the RG equations. This is the case where our knowledge about the matrix elements of the effective operators is limited to those at the hadronic scale. However, some matrix elements relevant to DM direct detection are available at any μ so that we do not need to evaluate the Wilson coefficients at the hadronic scale ourselves, especially at the leading order of α_s.

Here, we assume that the WIMPs are Majorana fermions, though the derivation in this section is applicable to the scalar and vector WIMPs, as given in Hisano, Nagai and Nagata [3].

11.5.1 Effective Theories at the Parton Level

When the WIMPs are Majorana fermions, the parton-level effective interactions relevant to DM direct detection at the hadronic scale are the following:

$$L_{\text{eff}} = \sum_{p=q,g} C_S^p O_S^p + \sum_{i=1,2} \sum_{p=q,g} C_{T_i}^p O_{T_i}^p + \sum_q C_{AV}^q O_{AV}^q, \qquad (11.18)$$

with

$$O_S^q \equiv \bar{X}X \, m_q \bar{q}q,$$

$$O_S^g \equiv \bar{X}X \, \frac{\alpha_s}{\pi} G_{\mu\nu}^A G^{A\mu\nu},$$

$$O_{T1}^p \equiv \frac{1}{M_X} \bar{X} i \partial^\mu \gamma^\nu X \, O_{\mu\nu}^p,$$

$$O_{T2}^p \equiv \frac{1}{M_X^2} \bar{X} i \partial^\mu i \partial^\nu X \, O_{\mu\nu}^p,$$

$$O_{AV}^q \equiv \bar{X} \gamma_\mu \gamma_5 X \, \bar{q} \gamma^\mu \gamma_5 q, \tag{11.19}$$

up to the equations of motions and the integration by parts. Here, q and $G_{\mu\nu}^A$ denote the light quarks ($q = u, d, s$) and the field strength tensor of the gluon field, respectively; m_q are the masses of the quarks; $\alpha_s \equiv g_s^2/(4\pi)$ is the strong coupling constant, $O_{\mu\nu}^q$ and $O_{\mu\nu}^g$ are called the spin-2 twist-2 operators of the quarks and the gluon, respectively,

$$O_{\mu\nu}^q \equiv \frac{1}{2} \bar{q} i \left(D_\mu \gamma_\nu + D_\nu \gamma_\mu - \frac{1}{2} g_{\mu\nu} \not{D} \right) q,$$

$$O_{\mu\nu}^g \equiv G_\mu^{A\rho} G_{\nu\rho}^A - \frac{1}{4} g_{\mu\nu} G_{\rho\sigma}^A G^{A\rho\sigma}, \tag{11.20}$$

with D_μ as the covariant derivatives. The quark or gluon field strength bilinear operators are up to dimension 4 in Eqn. 11.18. The operators O_{AV}^q contribute to the SD interactions, while the other operators contribute to those that are SI.

The quark/gluon scalar operators in O_S^q and O_S^g are multiplied by m_q and α_s/π, respectively. The reasons are the following. The quark scalar operators are chiral symmetry-breaking so that they are multiplied by m_q. The nucleon matrix elements of $G_{\mu\nu}^A G^{A\mu\nu}$ are multiplied by π/α_s, compared with those of $m_q \bar{q}q$, as we show later. This leads to a change in the power counting of α_s in the perturbation when evaluating the DM direct detection rate. Then, we multiply O_S^g by α_s/π. In addition, O_S^q and O_S^g are RG invariant at all orders and at the one-loop level, respectively, in QCD. These are explained in Section 11.5.2 in more detail.

Twist is defined as the mass dimension minus the spin of the operators. Higher spin or higher twist operators have more mass dimensions than the spin-2 twist-2 operators so they are negligible in DM direct detection [22]. The operators in O_{T1}^p and O_{T2}^p ($p = q, g$) have mass dimensions 8 and 9, respectively, higher than the others in Eqn. 11.19. However, in the non-relativistic limit for WIMPs, $O_{T1}^p \simeq X^\dagger X O_{00}^p$ and $O_{T2}^p \simeq \bar{X} X O_{00}^p$, which behave as dimension-7 operators, since the operators in O_{T1}^p and O_{T2}^p are multiplied by $1/M_X$ and $1/M_X^2$, respectively [22].[2] Unless the mediator particles

---

[2] This is more transparent in heavy particle effective theories, where the WIMPs are treated as non-relativistic fields [27].

generating those operators have masses much larger than M_X, the Wilson coefficients for O^p_{T1} and O^p_{T2} are not suppressed. Thus, they may contribute to the SI interactions, comparable to O^q_S and O^g_S.

11.5.2 Matrix Elements and RG Equations for Scalar Operators

First, we discuss the nucleon matrix elements for the quark/gluon scalar operators. The matrix elements for the quark scalar operators in O^q_S are given by

$$\langle N|m_q\bar{q}q|N\rangle \equiv f^{(N)}_{T_q} m_N, \tag{11.21}$$

with $f^{(N)}_{T_q}$ the mass fraction parameters of the quark q. Presently, the mass fraction parameters of quarks are evaluated with lattice QCD simulations. The following were derived by the ETM collaboration with $N_f = 2+1+1$ [28],

$$f^p_{T_u} = 0.0149(17)\left(^{21}_{16}\right), \; f^n_{T_u} = 0.0117(15)\left(^{18}_{12}\right),$$

$$f^p_{T_d} = 0.0234(23)\left(^{27}_{16}\right), \; f^n_{T_d} = 0.0298(23)\left(^{30}_{16}\right),$$

$$f^N_{T_s} = 0.0440(88)\left(^{72}_{15}\right),$$

$$f^N_{T_c} = 0.085(22)\left(^{11}_{7}\right), \tag{11.22}$$

$(N = p, n)$. The first and second parentheses are for statistical and systematical uncertainties, respectively. Heavier quarks have larger mass fraction parameters. In the simulations, the mass fraction parameter of the charm quark is also evaluated. We return to this later.

The nucleon matrix element of the gluon scalar operator is evaluated with the trace anomaly in QCD. The trace anomaly in QCD is [29]

$$\theta^\mu_\mu = \frac{\beta(\alpha_s)}{4\alpha_s} G^A_{\mu\nu} G^{A\mu\nu} + (1 - \gamma_m(\alpha_s)) \sum_q m_q \bar{q}q. \tag{11.23}$$

The beta function of α_s, $\beta(\alpha_s)$, and anomalous dimension of the quark mass, $\gamma_m(\alpha_s)$, are given by

$$\beta(\alpha_s) \equiv \mu\frac{\partial}{\partial\mu}\alpha_s \simeq 2b_1\frac{\alpha_s^2}{4\pi} + 2b_2\frac{\alpha_s^3}{(4\pi)^2},$$

$$\gamma_m(\alpha_s)m_q \equiv \mu\frac{\partial}{\partial\mu}m_q \simeq -6C_F\frac{\alpha_s}{4\pi}, \tag{11.24}$$

where $b_1 = -11N_c/3 + 2N_f/3$ and $b_2 = -34N_c^2/3 + 10N_cN_f/3 + 2C_FN_f$ ($N_c = 3$ and $C_F = 4/3$). The nucleon mass is given by the nucleon matrix element of the trace anomaly,

$$m_N \equiv \langle N|\theta_\mu^\mu|N\rangle.$$ (11.25)

Thus, the nucleon matrix element of the gluon scalar operator is given by

$$\langle N|\frac{\alpha_s}{\pi}G_{\mu\nu}^A G^{A\mu\nu}|N\rangle = m_N \frac{4\alpha_s^2}{\pi\beta(\alpha_s)}\left[1 - (1 - \gamma_m(\alpha_s))\sum_q f_{T_q}^{(N)}\right]$$

$$\simeq -\frac{8}{9}m_N(1 - \sum_q f_{T_q}^{(N)}) + O(\alpha_s).$$ (11.26)

We take $N_f = 3$ in the last of the above equations. When the gluon scalar operator is multiplied by α_s/π, the nucleon matrix element is O(1) in units of m_N.

In most UV models, the WIMPs do not directly couple with gluons, and the effective interactions of the WIMPs with the gluon are generated by the integration of the quarks or other coloured particles. This implies that the effective interaction for the gluon scalar operator is suppressed by α_s/π compared to those for the quark scalar operators. However, the nucleon matrix elements for the gluon scalar operators are O(1) even if the gluon scalar operator is multiplied by α_s/π. Thus, we have to evaluate the higher-order contributions to the gluon scalar operator by α_s/π, in contrast to the quark scalar operators.

Now we discuss the contribution of the heavy quark scalar operators $m_Q \bar{Q}Q\bar{X}X$ to the gluon scalar operator in order to see the counting of α_s. The nucleon matrix element of the trace anomaly is independent of the number of flavours N_f since it is a physical observable. This implies that

$$\langle N|m_Q\bar{Q}Q|N\rangle = \frac{\pi\Delta\beta(\alpha_s)}{4(1 - \gamma_m(\alpha_s))\alpha_s^2}\langle N|\frac{\alpha_s}{\pi}G_{\mu\nu}^A G^{A\mu\nu}|N\rangle$$

$$\simeq -\frac{1}{12}(1 + 11\frac{\alpha_s}{4\pi})\langle N|\frac{\alpha_s}{\pi}G_{\mu\nu}^A G^{A\mu\nu}|N\rangle$$

$$\simeq \frac{2}{27}m_N,$$ (11.27)

at $\mu \simeq m_Q$. Here, $\Delta\beta(\alpha_s) = \beta(\alpha_s)|_{N_f} - \beta(\alpha_s)|_{N_f-1}$. Thus, if the Wilson coefficients for the heavy and light quark scalar operators are common, the heavy quark contribution via the gluon scalar operator dominates over the light quark ones.

It is found that the numerical value in Eqn. 11.27 is consistent with the mass fraction of the charm quark in Eqn. 11.22. This coincidence is welcome, though a more precise determination of the mass fraction of the charm quark may reduce the uncertainty in the predicted direct detection rate. The charm quark mass is close to the hadronic scale so that the higher-dimensional operators might not be negligible after integrating out the charm quark. By integrating out the heavy quarks Q, the dimension-6 operators are generated as [30]

$$-\frac{\alpha_s}{12\pi} G^A_{\mu\nu} G^{A\mu\nu} + \frac{\alpha_s}{64\pi m_Q^2} (D^\nu G_{\nu\mu})^A (D_\rho G^{\rho\mu})^A - \frac{g_s \alpha_s}{720 m_Q^2} f_{ABC} G^A_{\mu\nu} G^{\mu\rho B} G^\nu_{\rho}{}^C.$$

(11.28)

The first term corresponds to the leading order term in the second line of Eqn. 11.27. When $\Lambda^2_{QCD}/m_c^2 \simeq O(10)\%$, the coefficients of the higher-dimensional operators are numerically suppressed so that they are expected to be a few per cent of the leading term. This should be justified in lattice QCD in order to derive reliable predictions about the DM direct detection.

The anomalous dimensions of the scalar operators are also derived from the RG invariance of the quark scalar operators and the trace anomaly,

$$\mu \frac{\partial}{\partial \mu} m_q \bar{q}q = 0, \quad \mu \frac{\partial}{\partial \mu} \theta^\mu_\mu = 0.$$

(11.29)

It is found that

$$\mu \frac{\partial}{\partial \mu} \left(C^q_S, C^g_S \right) = \left(C^q_S, C^g_S \right) \Gamma_S,$$

(11.30)

where Γ_S is an $(N_f + 1) \times (N_f + 1)$ matrix,

$$\Gamma_S = \begin{pmatrix} 0 & \cdots & 0 & 0 \\ \vdots & \ddots & \vdots & \vdots \\ 0 & \cdots & 0 & 0 \\ -\frac{4\alpha_s^2}{\pi} \frac{\partial \gamma_m(\alpha_s))}{\partial \alpha_s} & \cdots & -\frac{4\alpha_s^2}{\pi} \frac{\partial \gamma_m(\alpha_s)}{\partial \alpha_s} & \alpha_s^2 \frac{\partial}{\partial \alpha_s} \left(\frac{\beta(\alpha_s)}{\alpha_s^2} \right) \end{pmatrix}.$$

(11.31)

C^q_S and C^g_S are RG-invariant in the leading term of $O(\alpha_s)$. The solutions for these equations are

$$C^q_S(\mu) = C^q_S(\mu_0) - \frac{4}{\pi} C^g_S(\mu_0)(\gamma_m(\alpha_s(\mu)) - \gamma_m(\alpha_s(\mu_0))),$$

$$C^g_S(\mu) = \frac{\beta(\alpha_s(\mu))}{\alpha_s^2(\mu)} \frac{\alpha_s^2(\mu_0)}{\beta(\alpha_s(\mu_0))} C^g_S(\mu_0).$$

(11.32)

11.5.3   Matrix Elements and RG Equations for Twist-2 Operators

Next we discuss the matrix elements for the spin-2 twist-2 operators; they are given by the parton-distribution functions (PDFs) of the nucleons ($N = p, n$) as

$$\langle N(p)|O^q_{\mu\nu}(\mu)|N(p)\rangle = \frac{1}{m_N}\left(p_\mu p_\nu - \frac{1}{4}m_N^2 g_{\mu\nu}\right)(q^{(N)}(2;\mu) + \bar{q}^{(N)}(2;\mu)),$$

$$\langle N(p)|O^g_{\mu\nu}(\mu)|N(p)\rangle = -\frac{1}{m_N}\left(p_\mu p_\nu - \frac{1}{4}m_N^2 g_{\mu\nu}\right)g^{(N)}(2;\mu), \qquad (11.33)$$

where $q^{(N)}(2;\mu)$, $\bar{q}^{(N)}(2;\mu)$, and $g^{(N)}(2;\mu)$ are the second moments of the PDFs for the quark, antiquark, and gluon, respectively. The nth moments of the PDFs are defined as

$$q^{(N)}(n;\mu) \equiv \int_0^1 dx\, x^{n-1} q^{(N)}(x;\mu),$$

$$\bar{q}^{(N)}(n;\mu) \equiv \int_0^1 dx\, x^{n-1} \bar{q}^{(N)}(x;\mu),$$

$$g^{(N)}(n;\mu) \equiv \int_0^1 dx\, x^{n-1} g^{(N)}(x;\mu), \qquad (11.34)$$

where $q^{(N)}(x;\mu)$, $\bar{q}^{(N)}(x;\mu)$, and $g^{(N)}(x;\mu)$ are the PDFs of the quark, antiquark, and gluon, respectively, at the factorization scale μ.

The derivation of Eqn. 11.33 is given in standard textbooks of quantum field theory, e.g. Peskin and Schroeder [31] or Schwartz [32]. In the standard derivation, operator product expansions (OPEs) are applied to deeply inelastic scattering (DIS), $e^- + N \to e^- + X$. In the expansion, the twist-2 operators are dominant, and the higher-twist operators are suppressed by the momentum transfer. From the contour integral of the OPEs on the complex plane of $\omega = 1/x$ (x: the Bjorken x in PDFs), it can be shown that

$$\langle N(p)|O^q_{\mu_1\cdots\mu_n}(\mu)|N(p)\rangle = \frac{1}{m_N}\left\{p^{\mu_1}\cdots p^{\mu_n}\right\}_{\mathrm{TS}}(q^{(N)}(n;\mu) + (-1)^n \bar{q}^{(N)}(n;\mu)), \qquad (11.35)$$

where $O^q_{\mu_1\cdots\mu_n}(\mu)$ are twist-2 spin-n operators and 'TS' means traceless symmetric. This derivation is for only the leading order term of Eqn. 11.33. The μ dependence is introduced through the radiative correction. It is shown that the anomalous dimensions for twist-2 operators are consistent with the Altarelli–Parisi evaluation of PDFs.

There is another derivation of Eqn. 11.33. In the above derivation using DIS, it is assumed that the integral along $|\omega| \to \infty$ on the complex plane vanishes. If we define the PDFs using quantum fields, we may derive Eqn. 11.33 in a rigorous way without such an assumption of a specific process. It has been proposed by Collins and Soper [33] that the PDFs be defined in light-cone coordinates. Eqn. 11.33 is derived directly from the definition. Furthermore, Eqn. 11.35 is the Mellin transformation of the PDFs in the mathematical language. Thus, the inverse Mellin transformation of the matrix elements of twist-2 operators gives another definition of the PDFs. It gives a basis to evaluate of the PDFs in lattice QCD simulations [34].

Several groups evaluated the PDFs of partons by fitting with measurements in collider experiments, and they provide the PDFs at any factorization scale. The following are the second moments of the PDFs of proton derived by the CTEQ–Jefferson Lab. collaboration [35]

$$
\begin{aligned}
g^{(p)}(2,\mu) &= 0.464(2), \\
u^{(p)}(2,\mu) &= 0.223(3), \quad \bar{u}^{(p)}(2,\mu) = 0.036(2), \\
d^{(p)}(2,\mu) &= 0.118(3), \quad \bar{d}^{(p)}(2,\mu) = 0.037(3), \\
s^{(p)}(2,\mu) &= 0.0258(4), \quad \bar{s}^{(p)}(2,\mu) = s(2,\mu), \\
c^{(p)}(2,\mu) &= 0.0187(2), \quad \bar{c}^{(p)}(2,\mu) = c(2,\mu), \\
b^{(p)}(2,\mu) &= 0.0117(1), \quad \bar{b}^{(p)}(2,\mu) = b(2,\mu),
\end{aligned}
\tag{11.36}
$$

where $\mu = m_Z$ and $N_f = 5$. The second moments of valence quarks and the gluon are $O(1)$ and those of the sea quarks are sub-leading, as expected. Those of the neutron are to be obtained by exchanging the values of the up and down quarks.

When using PDFs at a fixed factorization scale, we may need to evaluate the radiative correction between this scale and the UV scale. The anomalous dimensions of spin-2 twist-2 operators at the two-loop level are given by [37]

$$
\mu \frac{d}{d\mu}(C_{T_i}^q, C_{T_i}^g) = (C_{T_i}^q, C_{T_i}^g)\, \Gamma_{\mathrm{T}},
\tag{11.37}
$$

with Γ_{T} an $(N_f + 1) \times (N_f + 1)$ matrix:

$$
\Gamma_{\mathrm{T}} = \begin{pmatrix}
\gamma_{qq} & 0 & \cdots & 0 & \gamma_{qg} \\
0 & \gamma_{qq} & & \vdots & \vdots \\
\vdots & & \ddots & 0 & \vdots \\
0 & \cdots & 0 & \gamma_{qq} & \gamma_{qg} \\
\gamma_{gq} & \cdots & \cdots & \gamma_{gq} & \gamma_{gg}
\end{pmatrix},
\tag{11.38}
$$

where

$$\gamma_{qq} = \frac{16}{3} C_F \cdot \frac{\alpha_s}{4\pi} + \left(-\frac{208}{27} C_F N_f - \frac{224}{27} C_F^2 + \frac{752}{27} C_F N_c \right) \left(\frac{\alpha_s}{4\pi} \right)^2,$$

$$\gamma_{qg} = \frac{4}{3} \cdot \frac{\alpha_s}{4\pi} + \left(\frac{148}{27} C_F + \frac{70}{27} N_c \right) \left(\frac{\alpha_s}{4\pi} \right)^2,$$

$$\gamma_{gq} = \frac{16}{3} C_F \cdot \frac{\alpha_s}{4\pi} + \left(-\frac{208}{27} C_F N_f - \frac{224}{27} C_F^2 + \frac{752}{27} C_F N_c \right) \left(\frac{\alpha_s}{4\pi} \right)^2,$$

$$\gamma_{gg} = \frac{4}{3} N_f \cdot \frac{\alpha_s}{4\pi} + \left(\frac{148}{27} C_F N_f + \frac{70}{27} N_c N_f \right) \left(\frac{\alpha_s}{4\pi} \right)^2. \tag{11.39}$$

11.5.4 Matrix Elements and RG Equations for Axial Vector Operators

Finally, we have the matrix elements of axial vector currents

$$\langle N | \bar{q} \gamma_\mu \gamma_5 q | N \rangle \equiv 2 S_\mu \Delta q_N, \tag{11.40}$$

where S_μ is for the nucleon spin and Δq_N is the spin fraction of quark q. The spin fractions of the quark are measured in DIS to be [36]

$$\Delta u_p = 0.77,$$
$$\Delta d_p = -0.47,$$
$$\Delta s_p = -0.15. \tag{11.41}$$

Those of the neutron are to be obtained by exchanging the values of the up and down quarks. The axial vector currents are RG invariant at the one-loop level.

11.5.5 Effective Interactions of WIMPs with Nucleons

The effective interactions of WIMPs with nucleons are evaluated using the nucleon matrix elements given in Section 1.5.4, as

$$f_N / m_N = \sum_{q=u,d,s} C_S^q(\mu_H) f_{T_q}^{(N)} + C_S^g(\mu_H) \frac{4\alpha_s^2(\mu_H)}{\pi \beta(\alpha_s(\mu_H))} \left(1 - (1 - \gamma_m(\mu_H)) \sum_{q=u,d,s} f_{T_q}^{(N)} \right)$$

$$+ \frac{3}{4} \sum_{i=1,2}^{N_f} \sum_q C_{Ti}^q(\mu)(q(2;\mu) + \bar{q}(2;\mu)) - \frac{3}{4} \sum_{i=1,2} C_{Ti}^g(\mu) g(2;\mu), \tag{11.42}$$

$$a_N = \sum_{q=u,d,s} C_{AV}^q(\mu_H) \Delta q. \tag{11.43}$$

As mentioned, C_S^q, C_S^g, and C_{AV}^q are RG-invariant at the one-loop level. When calculating f_N and a_N at the leading order of α_s,

$$C_S^q(\mu_H) = C_S^q(\mu_{UV}), \quad (q = u, d, s),$$

$$C_S^g(\mu_H) = C_S^g(\mu_{UV}) - \frac{1}{12} \sum_{q=c,b,t} C_S^q(\mu_{UV}),$$

$$C_{AV}^q(\mu_H) = C_{AV}^q(\mu_{UV}), \quad (q = u, d, s). \tag{11.44}$$

Thus, the scalar operator contribution to f_N is simplified as

$$f_N/m_N|\text{scalar op.} = \sum_{q=u,d,s} C_S^q(\mu_{UV}) f_{T_q}^{(N)} - \frac{8}{9} (C_S^g(\mu_{UV})$$

$$- \frac{1}{12} \sum_{q=c,b,t} C_S^q(\mu_{UV})) \left(1 - \sum_{q=u,d,s} f_{T_q}^{(N)} \right). \tag{11.45}$$

11.6 Examples (at Leading-Order of α_s)

We now evaluate the effective interaction of WIMPs with the nucleon at the leading order of α_s, using the formulae in Section 11.5. We consider three models: 1) gauge-singlet WIMPs coupled with the Higgs boson; 2) gauge-singlet WIMPs coupled with coloured scalars and quarks; and 3) $SU(2)_L$ non-singlet WIMPs. When the effective couplings of the WIMPs at the parton level are evaluated by integrating out heavy particles at the UV scale, we must consider the case of matching the Wilson coefficients between the UV and effective theories.

11.6.1 Gauge Singlet WIMPs Coupled with Higgs Boson

We now consider a case where the WIMPs are $SU(2)_L$ singlet fermions X, coupled with the SM Higgs boson h as

$$L_{\text{int}} = -f_X \bar{X} X h. \tag{11.46}$$

This interaction is not symmetric under $SU(2)_L \times U(1)_Y$. However, this interaction is introduced in models where an $SU(2)_L$ singlet Higgs boson, coupled with X, is introduced, and it is mixed with the SM Higgs boson. Alternatively, such as in the SUSY SM, $SU(2)_L$ singlet and doublet fermions couple with the SM Higgs boson so that Eqn. 11.46 is generated due to mixing of those fermions.

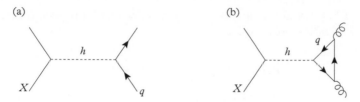

Figure 11.2 *Diagrams for the effective couplings of singlet WIMPs induced by Higgs boson exchange at the parton level.*

After integrating out the Higgs boson (Figure 11.2), the quark scalar operators are generated as

$$C_S^q(\mu_{UV}) = \frac{1}{v_h m_h^2} f_X, \tag{11.47}$$

where v_h is the vacuum expectation value of the Higgs field ($v_h \simeq 246$ GeV) and m_h is the SM Higgs mass ($m_h \simeq 125$ GeV). The gluon scalar operator is generated by the integration of the heavy quarks, and the other operators are not generated at the leading order of α_s. Thus, the SI coupling constants with nucleons are given at the leading order of α_s by

$$f_N/m_N = \frac{1}{v_h m_h^2} f_X \left(\bar{f}_{T_q}^{(N)} + 3 \times \frac{2}{27} (1 - \bar{f}_{T_q}^{(N)}) \right), \tag{11.48}$$

where $\bar{f}_{T_q}^{(N)} \equiv \sum_{q=u,d,s} f_{T_q}^{(N)}$. From this result, the SI cross section of proton is $\sigma_{SI}^p \simeq 2 \times 10^{-42}$ cm$^2 \times f_X^2$. The upperbound on σ_{SI}^p derived by XENON1T is about 10^{-46} cm^2 for $M_X \simeq 50$ GeV. Thus, $f_X \lesssim 10^{-2}$ for $M_X \simeq 50$ GeV.

11.6.2 Gauge Singlet WIMPs Coupled with Coloured Scalars and Quarks

Here, we consider the case where coloured scalars are introduced and the SU(2)$_L$ singlet WIMPs couple with the quarks and the coloured scalars. Binos in the SUSY SM have such couplings with scalar quarks. Thus, this example corresponds to the limit of a heavy Higgsino in the SUSY SM, where the Bino–Higgs coupling is suppressed.

The interactions of WIMPs with quarks and coloured scalars are given by

$$L_{\text{int}} = \sum_q \bar{q}(a_q + b_q \gamma_5) X \tilde{q} + \text{h.c.}, \tag{11.49}$$

where $\tilde{q}$ is the coloured scalar. The t-channel coloured scalar exchange diagrams (Figure 11.3) generate the quark scalar, twist-2, and axial vector operators, while the

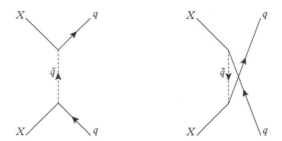

Figure 11.3 *Diagrams for the effective couplings of singlet WIMPs with quarks induced by coloured scalar exchange.*

gluon scalar and twist-2 operators come from one-loop diagrams. Now, consider the leading contribution of α_s to the effective couplings of the WIMPs with nucleons. Thus, the one-loop contribution to the gluon scalar operator (Figure 11.4) is included in the evaluation, while that to the gluon twist-2 operator is sub-leading, and is thus neglected.

From direct calculation, the Wilson coefficients for the quark operators are derived as

$$C_S^q(\mu_{UV}) = \frac{a_q^2 - b_q^2}{4m_q} \frac{1}{M_X^2 - M_{\tilde{q}}^2} + \frac{a_q^2 + b_q^2}{8} \frac{M_X}{(M_X^2 - M_{\tilde{q}}^2)^2},$$

$$C_{T1}^q(\mu_{UV}) = \frac{a_q^2 + b_q^2}{2} \frac{M_X}{(M_X^2 - M_{\tilde{q}}^2)},$$

$$C_{T2}^q(\mu_{UV}) = 0,$$

$$C_{AV}^q(\mu_{UV}) = -\frac{a_q^2 + b_q^2}{4} \frac{1}{M_X^2 - M_{\tilde{q}}^2}, \tag{11.50}$$

where $M_{\tilde{q}}$ is the coloured scalar mass and $\mu_{UV} \simeq M_{\tilde{q}}$. The above Wilson coefficients are for quarks q with $m_q < \mu_{UV}$.

On the other hand, we have to match the UV and effective theories in order to derive the Wilson coefficient for the gluon scalar operator. The evaluation of the Wilson coefficients for operators including gluon field strengths is always troublesome due to the tensor structure. However, when the momenta of the external gluons are negligibly small and the gluon fields may be included as background fields, the Fock–Schwinger gauge $(x^\mu A_\mu^A(x) = 0)$ is quite convenient for evaluating them, since we introduce propagators under the gluon field strength background with the gauge. Details of the technique are given in the Appendix.

Four one-loop diagrams in Figure 11.4 contribute to the gluon scalar operator. While diagrams A and C are automatically zero in the Fock–Schwinger gauge, diagrams B and D give contributions to the gluon scalar operator as

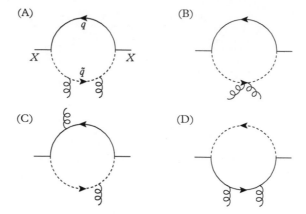

Figure 11.4 *Diagrams for the effective couplings of singlet WIMPs with gluons induced by coloured scalar exchange.*

$$C_S^g|_B = \sum_q \left[\frac{a_q^2 + b_q^2}{16} \frac{M_X}{6M_{\tilde{q}}^2(M_X^2 - M_{\tilde{q}}^2)} \right],$$

$$C_S^g|_D = \sum_q \left[-\frac{a_q^2 + b_q^2}{16} \frac{M_X}{6(M_X^2 - M_{\tilde{q}}^2)^2} - \frac{a_q^2 - b_q^2}{16} \frac{1}{3m_q M_{\tilde{q}}^2(M_X^2 - M_{\tilde{q}}^2)} \right]. \quad (11.51)$$

Here, we take the leading terms of m_q assuming $m_q \ll M_{\tilde{q}}$ to demonstrate the matching. We find that $C_S^g|_D = -1/12 \sum_q C_S^q$. The contribution of diagram D corresponds to the integration of heavy quarks in the effective theory. Thus,

$$C_S^g(\mu_{UV}) = \sum_{q=\text{all}} C_S^g|_D + \sum_{q(m_q > \mu_{UV})} C_S^g|_D. \quad (11.52)$$

In these equations, the leading term of m_q is shown, although we should use the exact formulae when m_q is not negligible compared to M_X and $M_{\tilde{q}}$.

We can now show some numerical results. We assume that the singlet WIMPs have interactions with only the top and bottom quarks ($a_q = b_q = 0$ for $q = u, d, c, s$, and $a_q = b_q = 1/2$ for $q = t, b$). This is a similar to the case where the binos are coupled with only the third-generation quarks and scalar quarks in the SUSY SM, assuming the other squarks are decoupled.

First, we discuss the renormalization scale-dependence of the contribution from the twist-2 operators to the SI coupling constants of WIMPs with nucleons (the last two terms in Eqn. 11.42). The Wilson coefficients for the quark/gluon scalar operators are RG invariant at the leading order of α_s, while those for the twist-2 operators are scale-dependent. As in Eqn. 11.42, they may be evaluated at any scale if the factorization scale of the PDFs is properly chosen.

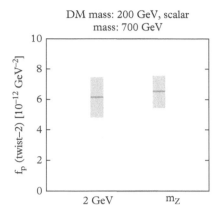

Figure 11.5 *Contribution from twist-2 operators to the SI coupling constant of WIMPs with protons, evaluated at μ = 2 GeV and m_Z. An explanation for the figure is given in the text. Figure courtesy of Hisano, Nagai and Nagata [3].*

Figure 11.5 shows the contribution from the twist-2 operators to the SI coupling constants of WIMPs with protons, which is evaluated at $\mu = 2$ GeV and m_Z. The PDFs for $\mu = 2$ GeV and m_Z are also used. Here, we take $M_X = 200$ GeV and $M_{\tilde{q}} = 700$ GeV. The dark and light gray bars denote the uncertainties coming from the PDF input and the perturbation in α_s, respectively. The method for evaluating the uncertainty from the PDFs' inputs is described in Owens, Accardi and Melnitchouk [35]. The uncertainty caused by neglecting the higher-order contributions in α_s is evaluated by varying the input and quark-mass threshold scales by a factor of two ($M_{\tilde{q}}/2 \leq \mu \leq 2M_{\tilde{q}}$, $m_t/2 \leq \mu \leq 2m_t$, and so on). The centre values of the two calculations are almost the same, and the uncertainty from the perturbation in α_s for the case $\mu = 2$ GeV is slightly larger. This is expected from the nature of asymptotic freedom in QCD. Thus, it is better to evaluate the contribution from the twist-2 operators at the weak scale.

Next, the SI coupling constant (left) and the SI cross section of WIMPs with protons (right) as functions of the coloured scalar mass are shown in Figure 11.6. Here, $M_X = 200$ GeV again. In the left-hand figure, the upper (lower) solid line shows the contribution of the scalar-type (twist-2-type) operators to the SI coupling constant of WIMPs with protons. For the twist-2 contribution, we use PDFs at $\mu = m_Z$ and show the calculations both with and without the RG effects between $M_{\tilde{q}}$ and m_Z in the solid and dashed lines, respectively. In the right-hand figure, the SI cross section of WIMPs with protons is shown with (solid) and without the RG effects (dashed line). The RG effects change the resulting value for the twist-2 contribution to the SI coupling constant by more than 50% when $M_{\tilde{q}} = 500$ GeV, and the scattering cross sections are modified by more than 20%. Thus, if the coloured mediator mass is much larger than the factorization scale in the PDFs adopted in the evaluation, it is important to include the RG effects.

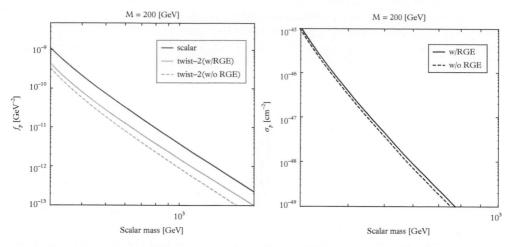

Figure 11.6 *SI coupling constant (left) and SI cross section of WIMPs with protons (right) as functions of coloured scalar mass. Here, $M_X = 200$ GeV. Explanation is given in text. Figures courtesy of Hisano, Nagai and Nagata [3].*

11.6.3   SU(2)$_L$ Non-Singlet WIMPs

Neutral components in the SU(2)$_L$ multiplets are candidates for WIMPs. If they are Dirac fermions or complex scalars with non-zero hypercharges, the DM direct detection experiments impose a strict bound on them. If the WIMP (X) is a fermion with hypercharge Y_X, the neutral current interaction induces SI interactions with nucleons ($N = n, p$) as,

$$L_{\text{eff}} = \sqrt{2} G_F Y_X \bar{X} \gamma_\mu X \left[(1 - 4\sin^2\theta_W) \bar{p}\gamma_\mu p - \bar{n}\gamma_\mu n \right], \qquad (11.53)$$

where G_F and θ_W are the Fermi constant and Weinberg angle, respectively. The SI cross section of WIMPs with neutrons is larger than that with protons due to the accidental cancellation in the latter. They are insensitive to the WIMP mass, and the SI cross section with neutrons is approximately given by $Y_X^2 \times 7 \times 10^{-40} \text{cm}^2$. If the WIMPs are the dominant component of the DM in our galaxy, the mass should be larger than $\sim 10^5$ TeV. This is much heavier than the unitarity bound on the WIMP mass in the assumption that the WIMPs are in thermal equilibrium in the early universe [38]. Even if the WIMPs are complex scalars with non-zero hypercharge, a similar bound is derived. On the other hand, this constraint is avoidable if the WIMPs are Majorana fermions or real scalars, since their vector coupling is forbidden automatically.

SU(2)$_L$ triplet Majorana fermions with zero hypercharge are one of the examples. They are called winos in the SUSY SM, and they are superpartners of the weak gauge bosons. The triplet fermions include a neutral Majorana fermion and a charged Dirac fermion. The radiative corrections due to the SU(2)$_L$ gauge interactions generate the mass splitting between the neutral and charged fermions as $\Delta M_X \simeq 165$ MeV when $M_X \gg m_Z$ [39], and it makes the neutral fermion lighter than the charged one. The radiative corrections may easily dominate over the contribution to the mass splitting from effective operators induced by the integration of heavier particles, since the operators have mass dimensions of seven and above.

Now, assume that the triplet fermions have only gauge interactions.[3] The WIMPs (the neutral component, X) only interact with the W boson and the charged component X^-:

$$L_{int} = -g_2(\bar{X}\gamma^\mu X^- W_\mu^\dagger + \text{h.c.}). \tag{11.54}$$

In this case, the WIMPs couple with quarks at the one-loop level (Figure 11.7), and with gluons at the two-loop level (Figure 11.8). The diagrams give finite contributions, and the loop momenta in the diagrams are typically around m_W. Thus, we consider the effective theory with $N_f = 5$ flavours at $\mu_{UV} = m_Z$.

The Higgs boson exchange diagrams in Figure 11.7 (a) and Figure 11.8 (a) generate the scalar operators

$$C_S^q(\mu_{UV})|_{\text{Higgs}} = \frac{\alpha_2^2}{4m_W m_h^2} g_H(\omega), \quad (q = u, d, s, c, b),$$

$$C_S^g(\mu_{UV})|_{\text{Higgs}} = -\frac{\alpha_2^2}{48 m_W m_h^2} g_H(\omega) \tag{11.55}$$

where $\omega = m_W^2/M_X^2$ and $\alpha_2 = g_2^2/4\pi$. The coefficient $C_S^g(\mu_{UV})$ comes from the integration of the top quark. Here and in the following calculations, the mass difference between the charged and neutral fermions is neglected. The mass functions including $g_H(\omega)$ in this section are reviewed in Hisano, Ishiwata and Nagata [4].

Next, we calculate the contributions to the scalar operators from the box diagrams in Figure 11.7 (b) and Figure 11.8(b) and Figure 11.8(c). In this calculation, it is convenient to first derive the OPEs of the charged current-charged current correlator,

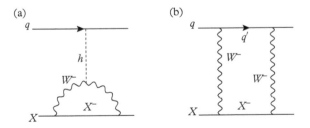

Figure 11.7 *Diagrams for effective couplings of SU(2)$_L$ triplet fermion WIMPs with quarks.*

[3] In the SUSY SM, this situation is realized when the superpartners, except for gauginos, are decoupled [40].

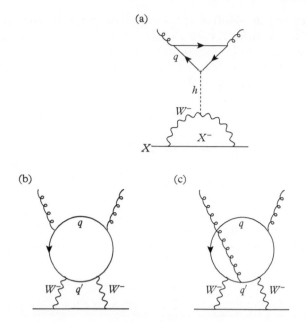

Figure 11.8 *Diagrams for the effective couplings of SU(2)$_L$ triplet fermion WIMPs with gluons.*

$$\Pi^W_{\mu\nu}(q) \equiv i \int d^4x \, e^{iqx} \, T \left\{ \mathcal{J}^W_\mu(x) \mathcal{J}^W_\nu(0)^\dagger \right\}, \tag{11.56}$$

where

$$\mathcal{J}^W_\mu(x) = \sum_{i=1,2,3} \frac{g_2}{\sqrt{2}} \bar{u}_i \gamma_\mu P_L d_i. \tag{11.57}$$

The correlator is decomposed into its transverse $\Pi^W_T(q^2)$ and longitudinal parts $\Pi^W_L(q^2)$,

$$\Pi^W_{\mu\nu}(q) = \left((-g_{\mu\nu} + \frac{q_\mu q_\nu}{q^2}) \Pi^W_T(q^2) + \frac{q_\mu q_\nu}{q^2} \Pi^W_L(q^2). \tag{11.58}$$

When connecting the correlator to the WIMP lines as in Figure 11.7(b) and Figure 11.8(b) and Figure 11.8(c), the longitudinal part does not contribute to the scalar operators due to the gauge invariance [42].

The quark and gluon scalar operators in the transverse part of the correlator are represented by

$$\Pi^W_T(q)|_{\text{scalar}} = \sum_q^{N_f=5} c^q_{W,S}(q^2) m_q \bar{q}q + c^g_{W,S}(q^2) \frac{\alpha_s}{\pi} G^A_{\mu\nu} G^{A\mu\nu}. \tag{11.59}$$

The tree-level diagrams contribute to $c^q_{W,S}(q^2)$, although the $c^q_{W,S}(q^2)$ are suppressed by the tiny quark masses, since $\mathcal{J}^W_\mu(x)$ is the *V-A* current. The exception is $c^b_{W,S}(q^2)$,

$$c^b_{W,S}(q^2) = \frac{g^2_2 m^2_t}{8(q^2 - m^2_t)^2}, \tag{11.60}$$

due to the large top quark mass. On the other hand, the one-loop diagrams induce $c^g_{W,S}(q^2)$ as

$$c^g_{W,S}(q^2) = \frac{g^2_2}{48q^2} \left[2 + \frac{q^2}{q^2 - m^2_t} \right], \tag{11.61}$$

where the first and second terms in the bracket come from loop diagrams including the first two and third generations, respectively. Using these results, the Wilson coefficients for the scalar operator are

$$C^b_S(\mu_{UV})|_{\text{box}} = \frac{\alpha^2_2}{m^3_W} \left[(-3)g_{\text{btm}}(\omega, \tau) \right],$$

$$C^g_S(\mu_{UV})|_{\text{box}} = \frac{\alpha^2_2}{4m^3_W} \left[2g_{\text{B1}}(\omega) + g_{\text{top}}(\omega, \tau) \right] \tag{11.62}$$

where $\tau = m^2_t / M^2_X$.

The contributions to the twist-2 operators also come from Figure 11.7(b) and Figure 11.8(b) and Figure 11.8(c), though those to the gluon twist-2 operators are at $O(\alpha_s)$ so that they are neglected here. The OPEs in the charged current-charged current correlator involve the quark twist-2 operators at the leading order of α_s:

$$\Pi^W_{\mu\nu} = \sum_{q=u,d,c,s} \frac{g^2_2}{2} \left[-\frac{g_{\mu\rho}g_{\nu\sigma}q^2 - g_{\mu\rho}q_\nu q_\sigma - q_\mu q_\rho g_{\nu\sigma} + g_{\mu\nu}q_\rho q_\sigma}{(q^2)^2} \right] O^{q\rho\sigma}$$

$$+ \frac{g^2_2}{2} \left[-\frac{g_{\mu\rho}g_{\nu\sigma}(q^2 - m^2_t) - g_{\mu\rho}q_\nu q_\sigma - q_\mu q_\rho g_{\nu\sigma} + g_{\mu\nu}q_\rho q_\sigma}{(q^2 - m^2_t)^2} \right] O^{b\rho\sigma}. \tag{11.63}$$

Then, the Wilson coefficients for the quark twist-2 operators are given by

$$C^q_{Ti}(\mu_{UV}) = \frac{\alpha^2_2}{m^3_W} g_{Ti}(\omega, 0),$$

$$C^b_{Ti}(\mu_{UV}) = \frac{\alpha^2_2}{m^3_W} g_{Ti}(\omega, \tau), \tag{11.64}$$

$(i = 1, 2)$.

The axial vector coupling is also evaluated in a similar way:

$$C_{AV}^q = \frac{\alpha_2^2}{8m_W^2} g_{AV}(\omega).$$ (11.65)

We have now shown the Wilson coefficients at the leading order of α_s. Some of the Wilson coefficients are not suppressed even if M_X is much heavier than m_W. The Wilson coefficients for the gluon and scalar operators at the hadronic scale are

$$C_S^q(\mu_H) \simeq \frac{\alpha_2^2}{4m_W m_h^2}(-2\pi),$$

$$C_S^g(\mu_H) \simeq -3\frac{\alpha_2^2}{48m_W m_h^2}(-2\pi) + 2\frac{\alpha_2^2}{4m_W^3}\frac{\pi}{12} + \frac{\alpha_2^2}{4m_W^3}\frac{\pi}{24}\frac{3x_{tw}+2}{(x_{tw}+1)^3}, \quad (11.66)$$

for $M_X \gg m_t, m_W$. Here, $x_{tw} = m_t/m_W$. The first term in $C_S^g(\mu_H)$ comes from the heavy quark contribution induced by the Higgs exchange, and the second and third are from box diagrams including the first two and third generations, respectively. The final is more suppressed by the top quark mass compared to the box diagrams including light quarks. The Wilson coefficients for the quark twist-2 operators at μ_{UV} are

$$C_{T1}^q(\mu_{UV}) = \frac{\alpha_2^2}{m_W^3}\frac{\pi}{3},$$

$$C_{T2}^q(\mu_{UV}) = O\left(\frac{m_W}{M_X}\right).$$ (11.67)

The axial vector coupling constants are also suppressed by $O(m_W/M_X)$. Thus, the SI coupling constants become independent of M_X in the limit of $M_X \gg m_W$ while the SD coupling vanishes in the limit [42, 43]. This implies that the SI cross sections are insensitive to the WIMP mass. This is positive in the search for the WIMPs.

On the other hand, the contributions are comparable to each other so that the accidental cancellation suppresses the SI cross sections. The SI coupling constants of WIMPs with nucleons are roughly estimated as

$$f_N/m_N \simeq \frac{\alpha_2^2}{m_W^3} \times 0.4 - \frac{\alpha_2^2}{m_W^3} \times 0.27 - \frac{\alpha_2^2}{m_W^3} \times 0.03,$$ (11.68)

where the first, second, and third terms come from the quark twist-2 ($C_{T1}^q(\mu_{UV})$), gluon scalar ($C_S^g(\mu_H)$), and quark scalar operators ($C_S^q(\mu_H)$), respectively. This cancellation reduces the SI cross sections by a factor of more than 10. In addition, it decreases the reliability of the calculation as the significance of the higher-order correction is relatively high [27]. The predicted SI cross sections are close to the neutrino SI scattering background, the neutrino floor. Thus, a more reliable evaluation method for the SI cross sections is needed.

The calculation can be extended to the cases of $SU(2)_L$ n-plets [44]. Even if the hypercharge is non-zero, the WIMPs may be Majorana fermions or real scalars by introducing the effective coupling with the Higgs boson [45]. Higgsinos in the SUSY SM are $SU(2)_L$ doublets with hypercharges $\pm 1/2$, while the effective operator with the Higgs boson induced by integrating out gauginos decomposes a neutral Dirac fermion into two Majorana fermions with mass splitting.

11.7 Towards the Next-Leading-Order Calculation of α_s

The previous sections showed the evaluation of the elastic scattering cross section of WIMPs with nucleons at the leading order of α_s. In many phenomenological studies, the leading-order evaluation is enough, since the DM abundance around the Earth still has large uncertainties. However, in some cases, evaluation at higher orders is required.

First, as in the case of $SU(2)_L$ triplet fermions, the accidental cancellation may suppress the leading order contribution. Many DM direct experiments are more sensitive to the SI cross section than the SD one. The SI cross section is induced from several contributions at the parton level that interfere with each other. In the case of $SU(2)_L$ triplet fermions, the contributions to the SI coupling at the parton level are comparable. Then, the destructive interference reduces the SI cross section significantly. The higher-order corrections have to be included in the evaluation of the SI cross section in order to be reliable.

Next is the uncertainty from the UV scale, μ_{UV}. The quark and gluon scalar operators are RG invariant at the leading order of α_s. Thus, their contribution is independent of μ_{UV} at the leading order. On the other hand, the SI cross section sensitive to the quark twist-2 operators at the leading order and their Wilson coefficients depend on μ_{UV}. We need to include the next-leading order contribution in order to reduce the uncertainty from μ_{UV}.

Third, some next-leading order corrections are known to be large. For example, the threshold correction to the gluon scalar operator at the quark mass scale at the next-leading order is large.

The next-leading order evaluation of the elastic scattering cross section of WIMPs with nucleons is not yet complete. Some results for the SI cross section are shown next.

11.7.1 Matching Condition at Quark Mass Threshold

The Wilson coefficients for the quark and gluon scalar operators are not RG invariant at $O(\alpha_s^2)$. Thus, the two-loop RG equations for the Wilson coefficients need to be solved in order to derive $C_S^q(\mu_H)$ and $C_S^g(\mu_H)$ with μ_H the hadronic scale at the next-leading order of α_s. When the factorization scale of the PDFs, used in the evaluating the matrix elements of twist-2 operators, is different from the UV scale, we also include the correction of their two-loop RG equations. The RG equations for the Wilson coefficients are shown in Section 11.5.

At the heavy quark mass thresholds, we also must include the threshold correction to the Wilson coefficients at $O(\alpha_s)$. When the bottom quark is decoupled and N_f is changed from 5 to 4, the matching conditions for the Wilson coefficients at the decoupling scale $\mu_b(\simeq m_b)$ are given by

$$C_S^q(\mu_b)|_{N_f=4} = C_S^q(\mu_b)|_{N_f=5}, \quad (q=u,d,s,c),$$

$$\alpha_s(\mu_b)C_S^g(\mu_b)|_{N_f=4} = \left[1 + \frac{\alpha_s(\mu_b)}{4\pi}\frac{2}{3}\log\frac{m_b^2}{\mu_b^2}\right]\alpha_s(\mu_b)C_S^g(\mu_b)|_{N_f=5}$$

$$-\frac{\alpha_s(\mu_b)}{12}\left[1 + \frac{\alpha_s(\mu_b)}{4\pi}\left(11 + \frac{2}{3}\log\frac{m_b^2}{\mu_b^2}\right)\right]\alpha_s(\mu_b)C_S^b(\mu_b)|_{N_f=5},$$

$$C_{Ti}^q(\mu_b)|_{N_f=4} = C_{Ti}^q(\mu_b)|_{N_f=5}, \quad (i=1,2),$$

$$C_{Ti}^g(\mu_b)|_{N_f=4} = \left[1 + \frac{\alpha_s(\mu_b)}{4\pi}\frac{2}{3}\log\frac{m_b^2}{\mu_b^2}\right]C_{Ti}^g(\mu_b)|_{N_f=5}$$

$$+\frac{\alpha_s(\mu_b)}{4\pi}\frac{2}{3}\log\frac{m_b^2}{\mu_b^2}\,C_{Ti}^b(\mu_b)|_{N_f=5}, \quad (i=1,2), \tag{11.69}$$

where

$$\frac{1}{\alpha_s(\mu_b)|_{N_f=4}} = \frac{1}{\alpha_s(\mu_b)|_{N_f=5}} + \frac{1}{(3\pi)}\log\frac{\mu_b}{m_b}. \tag{11.70}$$

These matching conditions are derived by comparing the two effective theories with $N_f = 5$ and 4. The logarithmic terms correspond to the RG equations. The large factor of 11 in the matching condition for C_S^g is found in the second line of Eqn. 11.27. Thus, the threshold correction C_S^g at α_s is not negligible even if it is of the next-leading order [48].

Similar matching conditions are applied for the other heavy quark threshold. For example, the matching conditions for the twist-2 operators at μ_b are irrelevant to the practical calculation when the factorization scale of the PDFs is higher than μ_b. However, if μ_{UV} is higher than the top quark mass and the factorization scale is below the top quark mass, we must include the radiative corrections between the two scales, taking account the top quark threshold. The matching conditions for C_{Ti}^g have to be included in the calculation.

11.7.2 Wilson Coefficients at UV Scale and SI Cross Sections at the Next-Leading Order of α_s

Here we evaluate the SI cross section at the next-leading order of α_s. The calculation is very involved. We have to evaluate the Wilson coefficient $C_S^g(\mu_{UV})$ for the gluon scalar operator at higher orders of α_s than for the other operators in the UV theory. In the case of the Higgs portal singlet WIMPs, the calculation is the straightforward. For singlet WIMPs coupled with coloured scalars and quarks, we must evaluate the following at the UV scale:

Operators	LO	NLO
Quark scalar	tree	1 loop
Quark twist-2	tree	1 loop
Gluon scalar	1 loop	2 loop
Gluon twist-2	–	1 loop

The coefficient for the gluon twist-2 operator from the one-loop diagrams is simultaneously calculated when calculating that of the gluon scalar operator at the leading order. However, the other contributions at the next-leading order have not yet been evaluated.

The last are the $SU(2)_L$ triplet fermions. When evaluating the SI cross sections for the $SU(2)_L$ triplet fermions at the next-leading order of α_s, we have to calculate the two- and three-loop diagrams. This calculation cannot be done by hand, if we include all the next-leading order contributions. Fortunately, the quark and gluon scalar operators in the charged current-charged current correlator $\Pi^W_{\mu\nu}$ are evaluated at the three-loop level [46], and the twist-2 operators at the two-loop level [47], assuming the quarks are massless. In the leading-order calculation, while the top quark mass is not negligible, the contributions of the third generation to the SI cross sections are suppressed by the top quark mass itself. Thus, we can neglect the next-leading order contributions of the third generation and evaluate the uncertainties from it.

Hisano, Ishiwata, and Nagata [4] include the following contributions in the evaluation of the Wilson coefficients at μ_{UV} with $N_f = 5$.

Paton	Orators	Higgs		Box	
		LO	NLO	LO	NLO
quark	scalar	1 loop	2 loop	–	2 loop
(1st and 2nd gen)	twist-2	–	–	1 loop	2 loop
quark	scalar	1 loop	2 loop	1 loop	2 loop (neglected)
(bottom)	twist-2	–	–	1 loop	2 loop (neglected)
gluon	scalar	2 loop	3 loop	2 loop	3 loop
(1st and 2nd gen)	twist-2	–	–	–	2 loop
gluon	scalar	2 loop	3 loop	2 loop	3 loop (neglected)
(3nd gen)	twist-2	–	–	–	2 loop (neglected)

Here, '$-$' means that the contributions vanish. The column 'Higgs' is for the Higgs exchange diagrams and 'box' is for box diagrams.

It is beyond the scope of this book to show the details of the calculation. The SI cross section in the limit of $M_X \gg m_W$ is [4]

$$S^p_{\mathrm{SI}} = 2.3^{+0.2+0.5}_{-0.3-0.4} \times 10^{-47} \mathrm{cm}^2 \qquad (11.71)$$

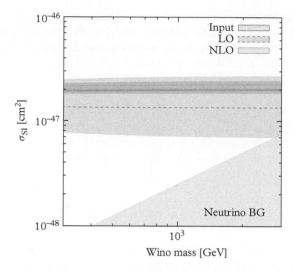

Figure 11.9 *SI cross section of SU(2)$_L$ triplet fermions with nucleons. The dashed and solid lines are for the leading-order and next-leading order results, respectively, with corresponding bands for perturbative uncertainties. The hatched band is for uncertainty from input errors. In the solid-shaded area the neutrino floor dominates the DM signals. Figure courtesy of Hisano, Ishiwata, and Nagata [4].*

where the first error comes from perturbative uncertainties, mainly from the choice of μ_{UV}. The second is from the input uncertainty, such as the PDFs. It is checked that neglecting the next-leading order contribution from the third-generation quarks does not lead to significant errors. The SI cross section with protons is shown as a function of M_X in Figure 11.9. The dashed and solid lines represent the leading-order and next-leading order results, respectively, with corresponding bands for perturbative uncertainties. The hatched band is for the uncertainty resulting from the input error. The yellow shaded area corresponds to the neutrino floor, where the neutrino background dominates the DM signal. Fortunately, the SI cross section in Eqn. 11.71 is larger than the neutrino floor even if the WIMP mass is around the 3 TeV predicted from thermal production.

The above calculation is applicable to other SU(2)$_L$ multiplets [4]. If the WIMPs come from SU(2)$_L$ doublet fermions (Higgsinos in the SUSY SM), the SI cross section with protons is around $10^{-49}\,\mathrm{cm}^2$, which is below the neutrino floor. On the other hand, larger SU(2)$_L$ multiplets predict larger SI cross sections. For example, SU(2)$_L$ quintuplet fermions with zero hypercharge predict $\sigma^p_{\mathrm{SI}} \simeq 2 \times 10^{-46}\,\mathrm{cm}^2$.

11.8   Summary

The DM direct detection experiments are important for tests of the WIMP DM, and they are a window to new physics at the TeV scale. The first-generation experiments, LUX, XENON1T, and PandaX-II, are giving the strict bounds on the SI cross sections of

WIMPs with nucleons. In a few years, the second-generation experiments will begin, LZ, XENONnT, and PandaX-nT with fiducial volumes $O(1)$ tonne. The third-generation experiments with fiducial volumes $O(10)$ tonne are also being planned in order to reach the neutrino floor. Thus, the theoretical predictions for the detection rates have to be more accurate and reliable in order to clarify their coverage to survey for WIMP models and their parameters.

In this chapter, the DM direct detection rates are evaluated with the effective theories. The effective theory approach is quite useful to evaluate them in a systematic way. We note the QCD aspects in the evaluation of the effective coupling constants of WIMPs with nucleons. It is found that they are well controllable, and we may evaluate them at the next-leading order of α_s in the effective theory approach.

11.9 Appendix: Fock–Schwinger gauge

The Fock–Schwinger gauge is convenient when evaluating the Wilson coefficients for the effective operators with $SU(3)_C$ field strengths. In the Feynman gauge, for example, it is tedious to maintain the tensor structure of the field strength. On the other hand, when using the Fock–Schwinger gauge, we may introduce the propagators of coloured particles in the background of the gluon field strength, or the derivatives. Thus, we may maintain the tensor structure automatically. This review is based on Novikov and colleagues [49].

The Fock–Schwinger gauge is given by

$$x^\mu A_\mu^A(x) = 0 \tag{11.72}$$

While this gauge fixing is not invariant under translation, the gauge field can be expanded at $x \simeq 0$ as

$$A_\mu^A(x) = \frac{1}{2 \cdot 0!} x^\rho G_{\rho\mu}^A(0) + \frac{1}{3 \cdot 1!} x^\alpha x^\rho (D_\alpha G_{\rho\mu}(0))^A$$
$$+ \frac{1}{4 \cdot 2!} x^\alpha x^\beta x^\rho (D_\alpha D_\beta G_{\rho\mu}(0))^A + \cdots . \tag{11.73}$$

The right-hand side in Eqn. 11.73 is given with the gauge covariant terms, such as the gluon field strength or its derivatives.

For the proof, we first need the following relation in the Fock–Schwinger gauge,

$$A_\mu^A(x) = \int_0^1 d\alpha \; G_{\rho\mu}^A(\alpha x) \, \alpha x^\rho . \tag{11.74}$$

It is derived as follows. Using the gauge fixing condition, it is found that

$$A_\mu^A(y) = -y^\rho \frac{\partial A_\rho^A(y)}{\partial y^\mu} = y^\rho G_{\rho\mu}^A(y) - y^\rho \frac{\partial A_\mu^A(y)}{\partial y^\rho} . \tag{11.75}$$

By moving the second term in the right-hand side to the left-hand side, and taking $y = \alpha x$, we get

$$\alpha x^{\rho} G^{A}_{\rho\mu}(\alpha x) = A^{A}_{\mu}(\alpha x) + x^{\rho}\frac{\partial A^{A}_{\mu}}{\partial x^{\rho}}$$

$$= \frac{d}{d\alpha}(\alpha A^{A}_{\mu}(\alpha x)). \tag{11.76}$$

Thus, Eqn. 11.74 is derived.

Next, by expanding the gauge fixing condition at $x \simeq 0$ as

$$x^{\mu}\left(A^{A}_{\mu}(0) + x_{\alpha_1}\partial^{\alpha_1}A^{A}_{\mu}(0) + \frac{1}{2}x_{\alpha_1}x_{\alpha_2}\partial^{\alpha_1}\partial^{\alpha_2}A^{A}_{\mu}(0) + \cdots\right) = 0, \tag{11.77}$$

we get

$$x^{\mu}A^{A}_{\mu}(0) = 0,$$
$$x^{\mu}x_{\alpha_1}\partial^{\alpha_1}A^{A}_{\mu}(0) = 0,$$
$$x^{\mu}x_{\alpha_1}x_{\alpha_2}\partial^{\alpha_1}\partial^{\alpha_2}A^{A}_{\mu}(0) = 0,$$
$$\cdots = 0. \tag{11.78}$$

Using these results, it is found that

$$x_{\alpha_1}\cdots x_{\alpha_n}\partial^{\alpha_1}\cdots\partial^{\alpha_n}G^{A}_{\rho\mu}(0) = x_{\alpha_1}\cdots x_{\alpha_n}(D^{\alpha_1}\cdots D^{\alpha_n}G_{\rho\mu}(0))^{A}. \tag{11.79}$$

By expanding Eqn. 11.74 around $x \simeq 0$ and applying Eqn. 11.79 to it, we get Eqn. 11.73. In addition, using Eqn. 11.78, the following equation for fermions (and also for scalars) is also derived,

$$\psi(x) = \psi(0) + x^{\alpha}D_{\alpha}\psi(0) + \frac{1}{2}x^{\alpha}x^{\beta}D_{\alpha}D_{\beta}\psi(0) + \cdots. \tag{11.80}$$

Next, we derive the quark propagator in the gluon field background, defined as

$$(i\partial_{\mu}\gamma^{\mu} - g_s A - m_q)iS(x,y) = i\delta^{(4)}(x-y), \tag{11.81}$$

where $A = A^{A}_{\mu}T_A\gamma_{\mu}$. The propagator is derived in a perturbative way as

$$iS(x,y) = iS^{(0)}(x-y) + \int d^4 z\, iS^{(0)}(x-z)(-ig_s A(z))iS^{(0)}(z-y)$$

$$+ \int d^4 z\, d^4 z'\, iS^{(0)}(x-z)(-ig_s A(z))iS^{(0)}(z-z')(-ig_s A(z'))iS^{(0)}(z'-y)$$

$$+ \cdots, \tag{11.82}$$

where $iS^{(0)}(x-y)$ is the propagator under no gluon background. The gluon background field A^{A}_{μ} may be replaced with the gluon field strength or its covariant derivatives using Eqn. 11.73 in the Fock–Schwinger gauge. In momentum space, A^{A}_{μ} is given by

$$A_\mu^A(k) = \int A_\mu^A(x) e^{ikx} d^4x$$

$$= \frac{i}{2} (2\pi)^4 G_{\mu\rho}^A(0) \frac{\partial}{\partial k_\rho} \delta^{(4)}(k) \cdots, \tag{11.83}$$

where the covariant derivatives of $G_{\mu\rho}^A(0)$ are omitted here since they are irrelevant to the calculation of the DM detection rates in the text.

Using the above results, we get

$$iS(p) = \int d^4x \, e^{ipx} \langle T\{\psi(x)\bar\psi(0)\}\rangle$$

$$= iS^{(0)}(p)$$

$$+ \int d^4k_1 \, iS^{(0)}(p) \, g_s\gamma^\alpha \left(\frac{1}{2} G_{\alpha\mu} \frac{\partial}{\partial k_{1\mu}} \delta^{(4)}(k_1)\right) iS^{(0)}(p-k_1)$$

$$+ \int d^4k_1 d^4k_2 \, iS^{(0)}(p) \, g_s\gamma^\alpha \left(\frac{1}{2} G_{\alpha\mu} \frac{\partial}{\partial k_{1\mu}} \delta^{(4)}(k_1)\right) iS^{(0)}(p-k_1) \, g_s\gamma^\beta$$

$$\times \left(\frac{1}{2} G_{\beta\nu} \frac{\partial}{\partial k_{2\nu}} \delta^{(4)}(k_2)\right) iS^{(0)}(p-k_1-k_2) + \cdots, \tag{11.84}$$

where $iS^{(0)}(p) = i/(\not p - m_q)$ and $G_{\mu\nu} \equiv G_{\mu\nu}^A T_A$.

Since translation invariance is broken due to the gauge fixing, $S(0,x)$ does not have the same form as $S(x,0)$,

$$i\tilde S(p) \equiv \int d^4x \, e^{-ipx} \langle T\{\psi(0)\bar\psi(x)\}\rangle$$

$$= iS^{(0)}(p)$$

$$+ \int d^4k_1 \, iS^{(0)}(p+k_1) \, g_s\gamma^\alpha \left(\frac{1}{2} G_{\alpha\mu} \frac{\partial}{\partial k_{1\mu}} \delta^{(4)}(p)\right) iS^{(0)}(p)$$

$$+ \int d^4k_1 d^4k_2 \, iS^{(0)}(p+k_1+k_2) \, g_s\gamma^\alpha \left(\frac{1}{2} G_{\alpha\mu} \frac{\partial}{\partial k_{2\mu}} \delta^{(4)}(k_2)\right)$$

$$\times iS^{(0)}(p+k_1) \, g_s\gamma^\beta \left(\frac{1}{2} g G_{\beta\nu} \frac{\partial}{\partial k_{1\nu}} \delta^{(4)}(k_1)\right) iS^{(0)}(p) + \cdots. \tag{11.85}$$

The coloured scalar propagator in the gluon background is also derived as

$$i\Delta(p) \equiv \int d^4x \, e^{ipx} \langle T\{\phi(x)\phi^\dagger(0)\}\rangle$$

$$= i\Delta^{(0)}(p)$$

$$+ \int d^4k_1 \, i\Delta^{(0)}(p) \, g_s(2p-k_1)^\alpha \left(\frac{1}{2} G_{\alpha\mu} \frac{\partial}{\partial k_{1\mu}} \delta^{(4)}(k_1)\right) i\Delta^{(0)}(p-k_1)$$

$$+ \int d^4 k_1 d^4 k_2 \, i\Delta^{(0)}(p) \, g_s(2p - k_1)^\alpha \left(\frac{1}{2} G_{\alpha\mu} \frac{\partial}{\partial k_{1\mu}} \delta^{(4)}(k_1) \right) i\Delta^{(0)}(p - k_1)$$

$$\times g_s(2p - 2k_1 - k_2)^\beta \left(\frac{1}{2} G_{\beta\nu} \frac{\partial}{\partial k_{2\nu}} \delta^{(4)}(k_2) \right) i\Delta^{(0)}(p - k_1 - k_2)$$

$$+ \int d^4 k_1 d^4 k_2 \, i\Delta^{(0)}(p)(-ig_s^2) \left(\frac{1}{2} G_{\alpha\mu} \frac{\partial}{\partial k_{1\mu}} \delta^{(4)}(k_1) \right) \left(\frac{1}{2} G^\alpha_{\ \nu} \frac{\partial}{\partial k_{2\nu}} \delta^{(4)}(k_2) \right)$$

$$\times i\Delta^{(0)}(p - k_1 - k_2) \,, \tag{11.86}$$

$$i\tilde{\Delta}(p) \equiv \int d^4 x \, e^{-ipx} \langle T\{\phi(0)\phi^\dagger(x)\} \rangle$$

$$= i\Delta^{(0)}(p)$$

$$+ \int d^4 k_1 \, i\Delta^{(0)}(p + k_1) \, g_s(2p + k_1)^\alpha \left(\frac{1}{2} G_{\alpha\mu} \frac{\partial}{\partial k_{1\mu}} \delta^{(4)}(k_1) \right) i\Delta^{(0)}(p)$$

$$+ \int d^4 k_1 d^4 k_2 \, i\Delta^{(0)}(p + k_1 + k_2) \, g_s(2p + 2k_1 + k_2)^\alpha \left(\frac{1}{2} G_{\alpha\mu} \frac{\partial}{\partial k_{1\mu}} \delta^{(4)}(k_2) \right)$$

$$\times i\Delta^{(0)}(p + k_1) \, g_s(2p + k_1)^\beta \left(\frac{1}{2} G_{\beta\nu} \frac{\partial}{\partial k_{2\nu}} \delta^{(4)}(k_1) \right) i\Delta^{(0)}(p)$$

$$+ \int d^4 k_1 d^4 k_2 \, i\Delta^{(0)}(p + k_1 + k_2)$$

$$\times (-ig_s^2) \left(\frac{1}{2} G_{\alpha\mu} \frac{\partial}{\partial k_{1\mu}} \delta^{(4)}(k_1) \right) \left(\frac{1}{2} G^\alpha_{\ \nu} \frac{\partial}{\partial k_{2\nu}} \delta^{(4)}(k_2) \right) i\Delta^{(0)}(p) \,, \tag{11.87}$$

where $i\Delta^{(0)}(p) = i/(p^2 - m^2)$. The scalar propagators are reduced to a more convenient form for practical usage as [50]

$$i\Delta(p) = i\Delta^{(0)}(p)$$

$$+ \int d^4 k_1 d^4 k_2 \, i\Delta^{(0)}(p)(-ig_s^2) \left(\frac{1}{2} G_{\alpha\mu} \frac{\partial}{\partial k_{1\mu}} \delta^{(4)}(k_1) \right) \left(\frac{1}{2} G^\alpha_{\ \nu} \frac{\partial}{\partial k_{2\nu}} \delta^{(4)}(k_2) \right)$$

$$\times i\Delta^{(0)}(p - k_1 - k_2)$$

$$+ \cdots,$$

$$i\tilde{\Delta}(p) = i\Delta^{(0)}(p)$$

$$+ \int d^4 k_1 d^4 k_2 \, i\Delta^{(0)}(p + k_1 + k_2)$$

$$\times (-ig_s^2) \left(\frac{1}{2} G_{\alpha\mu} \frac{\partial}{\partial k_{1\mu}} \delta^{(4)}(k_1) \right) \left(\frac{1}{2} G^\alpha_{\ \nu} \frac{\partial}{\partial k_{2\nu}} \delta^{(4)}(k_2) \right) i\Delta^{(0)}(p)$$

$$\cdots. \tag{11.88}$$

The scalar operator of the gluon $G^A_{\mu\nu} G^{A\mu\nu}$ is projected out from the bilinear term of the gluon field strength as

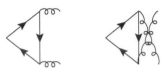

Figure 11.10 *Integrating out the heavy quark.*

$$G^A_{\alpha\mu} G^A_{\beta\nu} = \frac{1}{12} G^A_{\rho\sigma} G^{A\rho\sigma} \left(g_{\alpha\beta} g_{\mu\nu} - g_{\alpha\nu} g_{\beta\mu}\right)$$
$$- \frac{1}{2} g_{\alpha\beta} O^g_{\mu\nu} - \frac{1}{2} g_{\mu\nu} O^g_{\alpha\beta} - \frac{1}{2} g_{\alpha\nu} O^g_{\beta\mu} - \frac{1}{2} g_{\beta\mu} O^g_{\alpha\nu}$$
$$+ O^g_{\alpha\mu\beta\nu} \, , \tag{11.89}$$

where $O^g_{\mu\nu}$ is the twist-2 operator of gluon (Eqn. 11.20) and $O^g_{\alpha\mu\beta\nu}$ is given as

$$O^g_{\alpha\mu\beta\nu} \equiv G^A_{\alpha\mu} G^A_{\beta\nu}$$
$$- \frac{1}{2} g_{\alpha\beta} G^{A\rho}_\mu G^A_{\rho\nu} - \frac{1}{2} g_{\mu\nu} G^{A\rho}_\alpha G^A_{\rho\beta} + \frac{1}{2} g_{\alpha\nu} G^{A\rho}_\beta G^A_{\rho\mu} + \frac{1}{2} g_{\beta\mu} G^{A\rho}_\alpha G^A_{\rho\nu}$$
$$+ \frac{1}{6} G^A_{\rho\sigma} G^{A\rho\sigma} \left(g_{\alpha\beta} g_{\mu\nu} - g_{\alpha\nu} g_{\beta\mu}\right) . \tag{11.90}$$

Finally, we show Eqn. 11.27 by integrating out heavy quarks in the Fock–Schwinger gauge. In the text, we derived this using the trace anomaly in QCD. Here, we will show it in a diagrammatic way. As in Figure 11.10, two diagrams contribute there. However, in the Fock–Schwinger gauge, it is given by the trace of the quark propagator as

$$iM = -im_Q \int \frac{d^4 p}{(4\pi)^4} \mathrm{Tr}_{C+L}[iS(p)] = i\frac{\alpha_s}{12\pi} G^A_{\rho\sigma} G^{A\rho\sigma}. \tag{11.91}$$

While the Fock–Schwinger gauge is not invariant under translation, the invariance is recovered in the gauge-invariant results. The above calculation is quite easy compared with the other gauges, such as the Feynman gauge.

References

[1] For reviews of dark matter, G. Jungman, M. Kamionkowski, and K. Griest, (1996). *Phys. Rept.* 267: 195 [hep-ph/9506380]; K. A. Olive, TASI lectures on Dark Matter, Summary of Lectures Given at the Theoretical Advanced Study Institute in Elementary Particle Physics, Boulder, CO, 2-28 June, 2002. Available at: https://arxiv.org/abs/astro-ph/0301505; G. Bertone, D. Hooper, and J. Silk, (2005) *Phys. Rept.* 405: 279 [hep-ph/0404175]; J. L. Feng, (2010). *Ann. Rev. Astron. Astrophys.* 48: 495 [arXiv:1003.0904 [astro-ph.CO]].

[2] T. Marrodán Undagoitia and L. Rauch, (2016). *J. Phys. G* 43(1): 013001 [arXiv:150 9.08767 [physics.ins-det]].

[3] J. Hisano, R. Nagai, and N. Nagata, (2015). *JHEP* 1505: 037 [arXiv:1502.02244 [hep-ph]].

[4] J. Hisano, K. Ishiwata, and N. Nagata, (2015). *JHEP* 1506: 097 [arXiv:1504.00915 [hep-ph]].

[5] P. A. R. Ade, et al. [Planck Collaboration], (2014). *Astron. Astrophys.* [arXiv:1303.5076 [astro-ph.CO]].

[6] For reviews of SUSY SM and SUSY DM, G. Jungman, M. Kamionkowski, and K. Griest in [1]; S. P. Martin, (2010). *Adv. Ser. Direct. High Energy Phys.* 21: 1 [(1998). *Adv. Ser. Direct. High Energy Phys.* 18(1)] [hep-ph/9709356].

[7] T. Appelquist, H. C. Cheng, and B. A. Dobrescu, (2001). *Phys. Rev. D* 64: 035002 [hep-ph/0012100].

[8] H. C. Cheng, J. L. Feng, and K. T. Matchev, (2002). *Phys. Rev. Lett.* 89: 211301 [hep-ph/0207125]; H. C. Cheng, K. T. Matchev, and M. Schmaltz, (2002). *Phys. Rev. D* 66: 036005 [hep-ph/0204342].

[9] The Boltzmann equation for WIMPs is given in standard textbooks, e.g. E. W. Kolb and M. S. Turner, (1994). *The Early Universe*, Westview Press.

[10] J. Edsjo and P. Gondolo, (1997). *Phys. Rev. D* 56: 1879 [hep-ph/9704361].

[11] J. Hisano, S. Matsumoto, M. Nagai, O. Saito, and M. Senami, (2007). *Phys. Lett. B* 646: 34 [hep-ph/0610249]; M. Cirelli, A. Strumia, and M. Tamburini, (2007). *Nucl. Phys. B* 787: 152 [arXiv:0706.4071 [hep-ph]].

[12] M. Ajello, et al. [Fermi-LAT Collaboration], (2016). *Astrophys. J.* 819(1): 44 [arXiv:1511.02938 [astro-ph.HE]].

[13] M. Ackermann, et al. [Fermi-LAT Collaboration], (2015). *Phys. Rev. Lett.* 115(23): 231301 [arXiv:1503.02641 [astro-ph.HE]]; M. L. Ahnen, et al. [MAGIC and Fermi-LAT Collaborations], (2016). *JCAP* 1602(2): no.02, 039 [arXiv:1601.06590 [astro-ph.HE]].

[14] A. Abramowski, et al. [H.E.S.S. Collaboration], (2011). *Phys. Rev. Lett.* 106: 161301 [arXiv:1103.3266 [astro-ph.HE]]; M. L. Ahnen, et al. in [13].

[15] M. Aguilar, et al. [AMS Collaboration], (2016). *Phys. Rev. Lett.* 117(9): 091103.

[16] M. G. Aartsen, et al. [IceCube Collaboration], (2013). *Phys. Rev. Lett.* 110(13): 131302 [arXiv:1212.4097 [astro-ph.HE]].

[17] K. Choi, et al. [Super-Kamiokande Collaboration], (2015). *Phys. Rev. Lett.* 114(14): 141301 [arXiv:1503.04858 [hep-ex]].

[18] J. Hisano, S. Matsumoto, and M. M. Nojiri, (2004). *Phys. Rev. Lett.* 92: 031303 [hep-ph/0307216]; J. Hisano, S. Matsumoto, M. M. Nojiri, and O. Saito, (2005). *Phys. Rev. D* 71: 063528 [hep-ph/0412403]; J. Hisano, S. Matsumoto, O. Saito, and M. Senami, (2006). *Phys. Rev. D* 73: 055004 [hep-ph/0511118].

[19] E. Aprile, et al. [XENON Collaboration], (2017). *Phys. Rev. Lett.* 119(18): 181301 [arXiv:1705.06655 [astro-ph.CO]].

[20] X. Cui, et al. [PandaX-II Collaboration], (2017). *Phys. Rev. Lett.* 119(18): 181302 [arXiv:1708.06917 [astro-ph.CO]].

[21] J. Fan, M. Reece and L. T. Wang, (2010). *JCAP* 1011: 042 [arXiv:1008.1591 [hep-ph]].

[22] M. Drees and M. Nojiri, (1993). *Phys. Rev. D* 48: 3483 [hep-ph/9307208].

[23] J. D. Lewin and P. F. Smith, (1996). *Astropart. Phys.* 6: 87.

[24] J. I. Read, (2014). *J. Phys. G* 41: 063101 [arXiv:1404.1938 [astro-ph.GA]].

[25] D. S. Akerib, et al. [LUX Collaboration], (2017). *Phys. Rev. Lett.* 118(2): 021303 [arXiv: 1608.07648 [astro-ph.CO]].

[26] A. Gutlein, et al., (2010). *Astropart. Phys.* 34: 90 [arXiv:1003.5530 [hep-ph]];
J. Billard, L. Strigari, and E. Figueroa-Feliciano, (2014). *Phys. Rev. D* 89(2): 023524 [arXiv:1307.5458 [hep-ph]];
F. Ruppin, J. Billard, E. Figueroa-Feliciano, and L. Strigari, (2014). *Phys. Rev. D* 90(8): 083510 [arXiv:1408.3581 [hep-ph]].

[27] R. J. Hill and M. P. Solon, (2014). *Phys. Rev. Lett.* 112: 211602 [arXiv:1309.4092 [hep-ph]].

[28] A. Abdel-Rehim, et al. [ETM Collaboration], (2016). *Phys. Rev. Lett.* 116(25): 252001 [arXiv:1601.01624 [hep-lat]].

[29] M. A. Shifman, A. I. Vainshtein, and V. I. Zakharov, (1978). *Phys. Lett. B* 78: 443.

[30] L. Vecchi, [arXiv:1312.5695];
P. L. Cho and E. H. Simmons, (1995). *Phys. Rev. D* 51: 2360 [hep-ph/9408206].

[31] M. E. Peskin and D. V. Schroeder, (1995). An Introduction to *quantum field theory*, CRC Press.

[32] M. D. Schwartz, (2013). *Quantum Field Theory and the Standard Model*, Cambridge University Press.

[33] J. C. Collins and D. E. Soper, (1982). *Nucl. Phys. B* 194: 445.

[34] D. E. Soper, Nucl. *Phys. Proc. Suppl.* 53: 69 [hep-lat/9609018].

[35] J. F. Owens, A. Accardi, and W. Melnitchouk, (2013). *Phys. Rev. D* 87(9): 094012 [arXiv:1212.1702 [hep-ph]].

[36] D. Adams, et al. [Spin Muon Collaboration], (1995). *Phys. Lett. B* 357: 248.

[37] E. G. Floratos, D. A. Ross, and C. T. Sachrajda, (1979). *Nucl. Phys. B* 152: 493;
A. Gonzalez-Arroyo and C. Lopez, (1980). *Nucl. Phys. B* 166: 429.

[38] K. Griest and M. Kamionkowski, (1990). *Phys. Rev. Lett.* 64: 615.

[39] M. Ibe, S. Matsumoto, and R. Sato, (2013). *Phys. Lett. B* 721: 252 [arXiv:1212.5989 [hep-ph]].

[40] L. Randall and R. Sundrum, (1999). *Nucl. Phys. B* 557: 79 [hep-th/9810155];
G. F. Giudice, M. A. Luty, H. Murayama, and R. Rattazzi, (1998). *JHEP* 9812: 027 [hep-ph/9810442].

[41] J. Hisano, K. Ishiwata, and N. Nagata, (2013). *Phys. Rev. D* 87: 035020 [arXiv:1210.5985 [hep-ph]].

[42] J. Hisano, K. Ishiwata, and N. Nagata, (2010). *Phys. Lett. B* 690: 311 [arXiv:1004.4090 [hep-ph]].

[43] J. Hisano, S. Matsumoto, M. M. Nojiri, and O. Saito, (2005). *Phys. Rev. D* 71: 015007 [hep-ph/0407168].

[44] J. Hisano, K. Ishiwata, N. Nagata, and T. Takesako, (2011). *JHEP* 1107: 005 [arXiv: 1104.0228 [hep-ph]].

[45] J. Hisano, D. Kobayashi, N. Mori and E. Senaha, (2015). *Phys. Lett. B* 742: 80 [arXiv: 1410.3569 [hep-ph]].

[46] D. J. Broadhurst, P. A. Baikov, V. A. Ilyin, J. Fleischer, O. V. Tarasov, and V. A. Smirnov, (1994). *Phys. Lett. B* 329: 103 [hep-ph/9403274].

[47] W. A. Bardeen, A. J. Buras, D. W. Duke, and T. Muta, (1978). *Phys. Rev. D* 18: 3998.

[48] A. Djouadi and M. Drees, (2000). *Phys. Lett. B* 484: 183 [hep-ph/0004205].

[49] V. A. Novikov, M. A. Shifman, A. I. Vainshtein, and V. I. Zakharov, (1984). *Fortsch. Phys.* 32: 585.

[50] J. Hisano, K. Ishiwata, and N. Nagata, (2010). *Phys. Rev. D* 82: 115007 [arXiv:1007.2601 [hep-ph]].

12

Solutions to Problems at Les Houches Summer School on EFT

Marcel Balsiger[a], Marios Bounakis[b],
Mehdi Drissi[c,d], John Gargalionis[e],
Erik Gustafson[f], Greg Jackson[a], Matthew
Leak[g], Christopher Lepenik[h], Scott Melville[i,j],
Daniel Moreno[k], Michele Tammaro[l],
Selim Touati[m], Timothy Trott[n]

a Albert Einstein Center for Fundamental Physics, Institut für Theoretische Physik,
Universität Bern, Sidlerstrasse 5, CH-3012 Bern, Switzerland
b School of Mathematics, Statistics and Physics, Newcastle University,
Newcastle Upon Tyne, NE1 7RU, United Kingdom
c IRFU, CEA, Université Paris-Saclay, F-91191 Gif-sur-Yvette, France
d Department of Physics, University of Surrey, Guildford GU2 7XH,
United Kingdom
e ARC Centre of Excellence for Particle Physics at the Terascale, School of Physics,
The University of Melbourne, Victoria 3010, Australia
f Department of Physics and Astronomy, The University of Iowa, Iowa City, IA
52242, USA
g Theoretical Physics Division, Department of Mathematical Sciences,
University of Liverpool, Liverpool L69 3BX, United Kingdom
h University of Vienna, Faculty of Physics, Boltzmanngasse 5, A-1090 Wien, Austria
i DAMTP, University of Cambridge, Wilberforce Road, Cambridge,
CB3 0WA, United Kingdom
j Emmanuel College, University of Cambridge, St Andrew's Street, Cambridge,
CB2 3AP, United Kingdom

Balsiger, M., Bounakis, M., Drissi, M., Gargalionis, J., Gustafson, E., Jackson, G., Leak, M., Lepenik, C., Melville, S., Moreno, D.,
Tammaro, M., Touato, S., Trott, T., *Solutions to Problems at Les Houches Summer School on EFT* In: *Effective Field Theory in Particle
Physics and Cosmology.* Edited by: Sacha Davidson, Paolo Gambino, Mikko Laine, Matthias Neubert and Christophe Salomon,
Oxford University Press (2020). © Oxford University Press. DOI: 10.1093/oso/9780198855743.003.0012

k Grup de Física Teòrica, Dept. Física and IFAE-BIST, Universitat Autònoma de
Barcelona, E-08193 Bellaterra (Barcelona), Spain
l Department of Physics, University of Cincinnati, Cincinnati, Ohio 45221, USA
m Laboratoire de Physique Subatomique et de Cosmologie, Université
Grenoble-Alpes, CNRS/IN2P3, Grenoble INP, 38000 Grenoble, France
n Department of Physics, University of California, Santa Barbara, CA 93106, USA

Chapter Contents

This chapter features worked solutions to the various problems set by the lecturers during the course of the school. Throughout, equations from the lecture notes are referred to with a prefix L, e.g. eqn (L1.1). Further exercises that were added after the school are not solved here, and are left as a challenge for the enterprising reader.

12.1 Renormalisation and Effective Field Theories (Neubert): Solutions

The lectures [21] provide an overview of renormalization in quantum field theories, with particular attention paid to effective quantum field theories—in which the renormalization of composite operators, operator mixing under scale evolution, and the resummation of large logarithms are important.

This section contains three introductory problems, which cover propagators and 1PI diagrams, superficial degrees of divergence, and Renormalization Group Equations.

Exercise 1.1.6.4 Express the propagator of a spin-1 field in terms of 1PI self-energy diagrams.

SOLUTION:
From the Lagrangian, we can derive the classical equations of motion,

$$\mathcal{L} = -\tfrac{1}{4} F_{\mu\nu}^2 + \tfrac{1}{\xi}(\partial_\mu A^\mu)^2 + \mathcal{J}_\mu A^\mu \quad \Longrightarrow \quad \left[p^2 \eta_{\mu\nu} - \left(1 - \tfrac{1}{\xi}\right) p_\mu p_\nu \right] A_\nu = \mathcal{J}_\nu \quad (12.1)$$

Inverting this provides the Feynman propagator (classical Green's function),

$$P_{\mu\nu} = \tfrac{-i}{p^2} \left[\eta_{\mu\nu} - (1 - \xi) \tfrac{p_\mu p_\nu}{p^2} \right] \tag{12.2}$$

where the $+i\epsilon$ Feynman prescription for the poles is implicit.

The 1PI diagrams must satisfy the Ward identity by gauge invariance, and so we can write them as

$$p^2 \Pi(p) X^{\mu\nu} \quad \text{where} \quad X^{\mu\nu} = \eta^{\mu\nu} - p^\mu p^\nu / p^2 \tag{12.3}$$

where $X^{\mu\nu}$ is a projection onto the physical (transverse) polarization states. As a projection, it is idempotent

$$X^{\mu\nu} \eta_{\nu\alpha} X^{\alpha\beta} = X^{\mu\beta}$$

and it also annihilates the gauge-fixing term as $X^{\mu\nu} p_\mu = 0$. It is then straightforward to resum a geometric series of 1PI blobs,

$$\Pi^{\mu\nu} = P^{\mu\nu} + P^{\mu\alpha} X_{\alpha\beta} P^{\beta\nu} + PXPXP + \dots \tag{12.4}$$

$$= \frac{-i\xi p^{\mu} p^{\nu}}{p^4} - \frac{i}{p^2} \sum_{n=0}^{\infty} (\Pi(p^2))^n X^{\mu\nu} \tag{12.5}$$

$$= \frac{-i\xi p^{\mu} p^{\nu}}{p^4} - \frac{i}{p^2(1 - \Pi(p^2))} \left[\eta^{\mu\nu} - \frac{p^{\mu} p^{\nu}}{p^2} \right]. \tag{12.6}$$

Having extracted the factor of $k^2 X^{\mu\nu}$ from the 1PI self-energy diagrams, we find that the remaining $\Pi(k^2)$ function shifts the residue of the pole at $k^2 = 0$ in the resummed propagator to $1/(1 - \Pi(0))$. This is exactly analogous to the residue renormalization for scalars.

Exercise 1.2.3.1 Find the superficial degree of divergence for 1PI QCD Feynman graphs.

SOLUTION:
Counting powers of loop momenta, we define

$$D = 4L - P_q - 2P_g - P_c + V_{3g} \tag{12.7}$$

as the Grassmann-valued quarks and ghosts have propagators with only one power of momenta, while the bosonic gluons has two powers of momenta. The three-gluon vertex must also contain one power of momenta (by Lorentz invariance).

From the Euler characteristic,

$$L = P_q + P_g + P_c - (V_{qg} + V_{3g} + V_{4g} + V_{cg}) + 1$$

and from counting the total numbers of quark, ghost, and gluon legs,

$$2P_q + N_q = 2V_{qg} \tag{12.8}$$
$$2P_c + N_c = V_{cg} \tag{12.9}$$
$$2P_g + N_g = V_{qg} + V_{cg} + 3V_{3g} + 4V_{4g}. \tag{12.10}$$

Putting these together, we find that

$$D = 4 - N_g - \tfrac{3}{2}(N_q + N_c) \tag{12.11}$$

where N_X are the numbers of external particles. Although all physical diagrams have $N_c = 0$, it is useful to keep it explicitly for analysing subdiagrams.

Exercise 1.3.4.1 From the RGEs,

$$\frac{1}{\alpha_s}\beta = -2\epsilon - Z_{\alpha}^{-1} \frac{d}{d\ln\mu} Z_{\alpha} \tag{12.12}$$

$$\gamma_m = -\frac{1}{Z_m'} \frac{d}{d\ln\mu} Z_m' \tag{12.13}$$

in the MS scheme,

$$\frac{1}{\alpha_s}\beta = -2\epsilon - \beta Z_\alpha^{-1} \frac{d}{d\alpha_s} Z_\alpha \tag{12.14}$$

$$\gamma_m = -\frac{\beta}{Z_m'}\frac{d}{d\alpha_s} Z_m' \tag{12.15}$$

derive expressions for $\beta(\alpha_s,\epsilon)$, $Z_\alpha(\alpha_s,\epsilon)$, $\gamma(\alpha_s,\epsilon)$, $Z_\alpha(\alpha_s,\epsilon)$ as perturbative series in ϵ.

SOLUTION:

Assume that β is a smooth function of ϵ,

$$\beta = \sum_{k=0}^{\infty} \epsilon^k \beta^{(k)} \tag{12.16}$$

The renormalization Z_α receives pole contributions $1/\epsilon^k$ from k-loop diagrams, i.e.

$$Z = 1 + \sum_{k=1}^{\infty} \epsilon^{-k} Z_\alpha^{(k)} \tag{12.17}$$

where $Z_\alpha^{(k)} \sim \alpha_s^k$. The RGEs can be written

$$\beta(\alpha,\epsilon) = \frac{-2\epsilon\alpha_s Z_\alpha}{\left(1+\alpha_s\frac{d}{d\alpha_s}\right)Z_\alpha} \tag{12.18}$$

and hold for all ϵ. We can therefore equate the coefficients of each power of ϵ,

$$\sum_{k=0}^{\infty} \epsilon^k \beta^{(k)} = -2\epsilon\alpha_s + 2\alpha_s^2 \frac{d}{d\alpha_s} Z_\alpha^{(1)} + \mathcal{O}\left(\frac{1}{\epsilon}\right) \tag{12.19}$$

and conclude that

$$\beta(\alpha_s,\epsilon) = 2\alpha_s^2 \frac{d}{d\alpha_s} Z_\alpha^{(1)}(\alpha_s) - 2\alpha_s\epsilon. \tag{12.20}$$

Substituting this into the RGE, we find an infinite number of consistency conditions that fix all $Z_\alpha^{(k>1)}$ in terms of $Z_\alpha^{(1)}(\alpha_s)$,

$$\frac{dZ_\alpha^{(k+1)}}{d\alpha_s} = \frac{dZ_\alpha^{(1)}}{d\alpha_s}\left[1+\alpha_s\frac{d}{d\alpha_s}\right]Z_\alpha^{(k)} \quad \forall \quad k \geq 1 \tag{12.21}$$

Note that by taking repeated α_s derivatives, this gives every α_s^n coefficient of $Z_\alpha^{(k+1)}$ (we don't have $Z_{\alpha_s}^{(k+1)}|_{\alpha_s=0}$, but we know that this is zero).

Similarly, assume that γ_m is a smooth function of ϵ,

$$\gamma_m = \sum_{k=0}^{\infty} \epsilon^k \gamma_m^{(k)} \tag{12.22}$$

and then the positive powers of ϵ in the RGE give

$$\gamma_m = 2\alpha_s \frac{d}{d\alpha_s} Z_m'^{(1)}(\alpha_s) \tag{12.23}$$

and similarly we have consistency relations,

$$\frac{dZ_m'^{(k+1)}}{d\alpha_s} = \left[\frac{dZ_m'^{(1)}}{d\alpha_s} + \alpha_s \frac{dZ_\alpha^{(1)}}{d\alpha_s} \frac{d}{d\alpha_s} \right] Z_m'^{(k)}. \tag{12.24}$$

Physically, this is because the higher-order poles in Z come from multiple subdivergences, rather than than a genuine k-loop divergence.

12.2 Introduction to EFTs (Manohar): Solutions

The lectures [20] covered introductory material on EFTs as used in high-energy physics to compute experimentally observable quantities. The following exercise solutions treat power counting, loop corrections, field redefinitions and their relation to the equations of motion, decoupling of heavy particles, naive dimensional analysis, and the Standard Model Effective Field Theory (SMEFT) and many more basic concepts.

Useful references for solving the problems are

- EFT: [17, 19]
- Power counting: [7, 16]
- Matching in HQET and Field Redefinitions: [18]
- Invariants: [9, 10, 11, 12, 15]
- SMEFT: [1, 2, 13, 14].

Exercise 2.1 Show that for a *connected* graph, $V - I + L = 1$, where V is the number of vertices, I is the number of internal lines, and L is the number of loops. What is the formula if the graph has n connected components?

SOLUTION:

Consider a connected graph G where V is the number of vertices, I is the number of internal lines, and L is the number of loops. Since the identity does not depend on the external lines, erase them. Take an internal line (edge) that can be deleted without making the graph disconnected, and remove it. Then clearly $I \to I - 1$. If the edge joins vertices v_1 and v_2 and removing the edge leaves the graph connected, there must be a second path through the graph between v_1 and v_2, i.e. there is a loop which is removed when the edge is removed, so $L \to L - 1$. The operation leaves $V - I + L$ invariant. Proceed this way until removing an edge makes the graph disconnected, i.e. there are no loops, and we are left with a tree graph.

Now pick any vertex v in the graph, and move through the graph without retracing your path. There are no loops, so the path does not end back at v. Since the number of vertices is finite, eventually the path must end at a vertex (a 'leaf node' of the tree graph). Remove this last node and edge. Then $V \to V - 1$ and $I \to I - 1$

keeping $V - I + L$. Keep repeating this process until $I = 0$. Then we have a graph with one vertex $V = 1, I = 0, L = 0, V - I + L = 1$. Since our operators preserved $V - I + L = 1$, this completes the proof. If there are n connected components, the formula holds for each component, so the total is $V - I + L = n$ since the components have no vertices, edges, or loops in common.

Exercise 2.2 Work out the transformation of fermion bilinears $\overline{\psi}(\mathbf{x}, t) \Gamma \chi(\mathbf{x}, t)$ under C, P, T, where $\Gamma = P_L, P_R, \gamma^\mu P_L, \gamma^\mu P_R, \sigma^{\mu\nu} P_L, \sigma^{\mu\nu} P_R$. Use your results to find the transformations under CP, CT, PT, and CPT.

SOLUTION:
Under parity $\mathscr{P}$, charge conjugation $\mathscr{C}$, and time-reversal $\mathscr{T}$,

$$\mathscr{P}\psi(\mathbf{x}, t)\mathscr{P}^{-1} = \gamma^0 \psi(-\mathbf{x}, t),$$
$$\mathscr{C}\psi(\mathbf{x}, t)\mathscr{C}^{-1} = i\gamma^2 \psi(\mathbf{x}, t)^{\dagger T},$$
$$\mathscr{T}\psi(\mathbf{x}, t)\mathscr{T}^{-1} = i\gamma^1 \gamma^3 \psi(\mathbf{x}, -t). \tag{12.25}$$

Note that $\mathscr{C}$ is a *unitary* operator, and $\mathscr{T}$ is *anti-unitary* despite the $\dagger$ for the $\mathscr{C}$ transformation and not for the $\mathscr{T}$ transformation. $\mathscr{P}, \mathscr{C}$ are operators and commute with γ matrices; $\mathscr{T}$ complex conjugates the γ matrices. Transformations of $\overline{\psi}$ are given by taking conjugates of the above.

Under parity,

$$\mathscr{P}\overline{\psi}(\mathbf{x}, t)\Gamma\chi(\mathbf{x}, t)\mathscr{P}^{-1} = \mathscr{P}\overline{\psi}(\mathbf{x}, t)\mathscr{P}^{-1}\Gamma\mathscr{P}\chi(\mathbf{x}, t)\mathscr{P}^{-1} = \overline{\psi}(-\mathbf{x}, t)\gamma^0 \Gamma \gamma^0 \chi(-\mathbf{x}, t). \tag{12.26}$$

Under charge conjugation,

$$\mathscr{C}\overline{\psi}(\mathbf{x}, t)\Gamma\chi(\mathbf{x}, t)\mathscr{C}^{-1} = \mathscr{C}\overline{\psi}(\mathbf{x}, t)\mathscr{C}^{-1}\Gamma\mathscr{C}\chi(\mathbf{x}, t)\mathscr{C}^{-1} = \psi(\mathbf{x}, t)^T i\gamma^2 \gamma^0 \Gamma i\gamma^2 \chi(\mathbf{x}, t)^{\dagger T}. \tag{12.27}$$

Taking the transpose (and including the Fermi minus sign for the exchange of operators),

$$\mathscr{C}\overline{\psi}(\mathbf{x}, t)\Gamma\chi(\mathbf{x}, t)\mathscr{C}^{-1} = -\chi(\mathbf{x}, t)^\dagger (i\gamma^2)^T \Gamma^T (\gamma^0)^T (i\gamma^2)^T \psi(\mathbf{x}, t)$$
$$= -\overline{\chi}(\mathbf{x}, t)(i\gamma^0 \gamma^2)\Gamma^T (i\gamma^0 \gamma^2)\psi(\mathbf{x}, t)$$
$$= \overline{\chi}(\mathbf{x}, t)C\Gamma^T C^{-1} \psi(\mathbf{x}, t), \qquad C = i\gamma^0 \gamma^2. \tag{12.28}$$

Under time-reversal,

$$\mathscr{T}\overline{\psi}(\mathbf{x}, t)\Gamma\chi(\mathbf{x}, t)\mathscr{T}^{-1} = \mathscr{T}\overline{\psi}(\mathbf{x}, t)\mathscr{T}^{-1} \mathscr{T}\Gamma\mathscr{T}^{-1} \mathscr{T}\chi(\mathbf{x}, t)\mathscr{T}^{-1}$$
$$= \overline{\psi}(\mathbf{x}, -t)\gamma^0 i\gamma^1 \gamma^3 \gamma^0 \Gamma^* i\gamma^1 \gamma^3 \chi(\mathbf{x}, -t)$$
$$= \overline{\psi}(\mathbf{x}, -t)(i\gamma^1 \gamma^3)\Gamma^* (i\gamma^1 \gamma^3)\chi(\mathbf{x}, -t)$$
$$= \overline{\psi}(\mathbf{x}, -t)T\Gamma^* T^{-1}\chi(\mathbf{x}, -t), \qquad T = i\gamma^1 \gamma^3. \tag{12.29}$$

The transformations can then be determined by computing $\gamma^0 \Gamma \gamma^0$, $(i\gamma^0\gamma^2)$ $\Gamma^T(i\gamma^0\gamma^2)$ and $(i\gamma^1\gamma^3)\Gamma^*(i\gamma^1\gamma^3)$ to give the results in the table. The second line gives the coordinate arguments, and

$$\hat{\mu} = \begin{cases} \mu & \text{if } \mu = 0 \\ -\mu & \text{if } \mu = 1,2,3 \end{cases}$$

	$\mathscr{C}$	$\mathscr{P}$	$\mathscr{T}$
$(\mathbf{x},t)$	$(\mathbf{x},t)$	$(-\mathbf{x},t)$	$(\mathbf{x},-t)$
$\bar{\chi} P_L \psi$	$\bar{\psi} P_L \chi$	$\bar{\chi} P_R \psi$	$\bar{\chi} P_L \psi$
$\bar{\chi} P_R \psi$	$\bar{\psi} P_R \chi$	$\bar{\chi} P_L \psi$	$\bar{\chi} P_R \psi$
$\bar{\chi} \gamma^\mu P_L \psi$	$-\bar{\psi} \gamma^\mu P_R \chi$	$\bar{\chi} \gamma^{\hat{\mu}} P_R \psi$	$\bar{\chi} \gamma^{\hat{\mu}} P_L \psi$
$\bar{\chi} \gamma^\mu P_R \psi$	$-\bar{\psi} \gamma^\mu P_L \chi$	$\bar{\chi} \gamma^{\hat{\mu}} P_L \psi$	$\bar{\chi} \gamma^{\hat{\mu}} P_R \psi$
$\bar{\chi} \sigma^{\mu\nu} P_L \psi$	$-\bar{\psi} \sigma^{\mu\nu} P_L \chi$	$\bar{\chi} \sigma^{\hat{\mu}\hat{\nu}} P_R \psi$	$-\bar{\chi} \sigma^{\hat{\mu}\hat{\nu}} P_L \psi$
$\bar{\chi} \sigma^{\mu\nu} P_R \psi$	$-\bar{\psi} \sigma^{\mu\nu} P_R \chi$	$\bar{\chi} \sigma^{\hat{\mu}\hat{\nu}} P_L \psi$	$-\bar{\chi} \sigma^{\hat{\mu}\hat{\nu}} P_R \psi$

Combining the above gives

	$\mathscr{CP}$	$\mathscr{PT}$	$\mathscr{CT}$	$\mathscr{CPT}$
$(\mathbf{x},t)$	$(-\mathbf{x},t)$	$(-\mathbf{x},-t)$	$(\mathbf{x},-t)$	$(-\mathbf{x},-t)$
$\bar{\chi} P_L \psi$	$\bar{\psi} P_R \chi$	$\bar{\chi} P_R \psi$	$\bar{\psi} P_L \chi$	$\bar{\psi} P_R \chi$
$\bar{\chi} P_R \psi$	$\bar{\psi} P_L \chi$	$\bar{\chi} P_L \psi$	$\bar{\psi} P_R \chi$	$\bar{\psi} P_L \chi$
$\bar{\chi} \gamma^\mu P_L \psi$	$-\bar{\psi} \gamma^{\hat{\mu}} P_L \chi$	$\bar{\chi} \gamma^\mu P_R \psi$	$-\bar{\psi} \gamma^{\hat{\mu}} P_R \chi$	$-\bar{\psi} \gamma^\mu P_L \chi$
$\bar{\chi} \gamma^\mu P_R \psi$	$-\bar{\psi} \gamma^{\hat{\mu}} P_R \chi$	$\bar{\chi} \gamma^\mu P_L \psi$	$-\bar{\psi} \gamma^{\hat{\mu}} P_L \chi$	$-\bar{\psi} \gamma^\mu P_R \chi$
$\bar{\chi} \sigma^{\mu\nu} P_L \psi$	$-\bar{\psi} \sigma^{\hat{\mu}\hat{\nu}} P_R \chi$	$-\bar{\chi} \sigma^{\mu\nu} P_R \psi$	$\bar{\psi} \sigma^{\hat{\mu}\hat{\nu}} P_L \chi$	$\bar{\psi} \sigma^{\mu\nu} P_R \chi$
$\bar{\chi} \sigma^{\mu\nu} P_R \psi$	$-\bar{\psi} \sigma^{\hat{\mu}\hat{\nu}} P_L \chi$	$-\bar{\chi} \sigma^{\mu\nu} P_L \psi$	$\bar{\psi} \sigma^{\hat{\mu}\hat{\nu}} P_R$	$\chi\bar{\psi} \sigma^{\mu\nu} P_L \chi$

Note that for any operator $O(\mathbf{x},t)$, $\mathscr{CPT}$ transforms it to $(-1)^n O^\dagger(-\mathbf{x},-t)$ where n is the number of Lorentz indices. Thus the Lagrange density transforms as $\mathcal{L}(\mathbf{x},t) \to \mathcal{L}^\dagger(-\mathbf{x},-t)$, and a Hermitian action is $\mathscr{CPT}$ invariant.

Exercise 2.3 Show that for $SU(N)$,

$$[T^A]^\alpha_\beta [T^A]^\lambda_\sigma = \tfrac{1}{2} \delta^\alpha_\sigma \delta^\lambda_\beta - \tfrac{1}{2N} \delta^\alpha_\beta \delta^\lambda_\sigma, \tag{12.30}$$

where the $SU(N)$ generators are normalized to $\text{Tr}\, T^A T^B = \delta^{AB}/2$. From this, show that

$$\delta^\alpha_\beta \delta^\lambda_\sigma = \tfrac{1}{N} \delta^\alpha_\sigma \delta^\lambda_\beta + 2[T^A]^\alpha_\sigma [T^A]^\lambda_\beta,$$
$$[T^A]^\alpha_\beta [T^A]^\lambda_\sigma = \tfrac{N^2-1}{2N^2} \delta^\alpha_\sigma \delta^\lambda_\beta - \tfrac{1}{N}[T^A]^\alpha_\sigma [T^A]^\lambda_\beta. \tag{12.31}$$

SOLUTION:

The identity and the T^a $(a = 1, \ldots, N^2 - 1)$ provide a basis for $N \times N$ matrices, so any such matrix can be decomposed as

$$A = c_0 \mathbb{1} + c_a T^a. \tag{12.32}$$

If A is Hermitian, the coefficients are real. Now, assume that the Killing form is normalized as,

$$\text{Tr}\left(T^a T^b\right) = \tfrac{1}{2}\delta^{ab} \tag{12.33}$$

and therefore

$$\text{Tr}\,A = N c_0, \quad \text{Tr}\,T^a A = \tfrac{1}{2}c_a. \tag{12.34}$$

With explicit indices, this tells us that,

$$A_{ij} = \left(\tfrac{A_{k\ell}\delta_{k\ell}}{N}\right)\delta_{ij} + (2T^a_{k\ell}A_{\ell k})\,T^a_{ij}, \tag{12.35}$$

$$\Rightarrow \quad A_{\ell k}\left[\delta_{i\ell}\delta_{jk} - \tfrac{1}{N}\delta_{k\ell}\delta_{ij} - 2T^a_{ij}T^a_{k\ell}\right] = 0,$$

on picking out the coefficient of $A_{\ell k}$. As this identity holds for all elements A_{ij} of all Hermitian matrices, we conclude that the square bracket vanishes identically, and thus arrive at the desired Fierz identity,

$$T^a_{ij}T^a_{k\ell} = \tfrac{1}{2}\delta_{i\ell}\delta_{jk} - \tfrac{1}{2N}\delta_{ij}\delta_{k\ell}, \tag{12.36}$$

which is eqn (L2.2). Rewriting this equation as

$$\tfrac{1}{2N}\delta_{ij}\delta_{k\ell} + T^a_{ij}T^a_{k\ell} = \tfrac{1}{2}\delta_{i\ell}\delta_{jk}, \tag{12.37}$$

multiplying by 2, and renaming the indices gives the first of eqn (12.3). Adding eqn (12.1) and the corresponding equation with $\beta \leftrightarrow \sigma$ multiplied by $1/N$ gives the second of eqn (12.3).

Exercise 2.4 Spinor Fierz identities are relations of the form

$$(\overline{A}\,\Gamma_1 B)(\overline{C}\,\Gamma_2 D) = \sum_{ij} c_{ij}(\overline{C}\,\Gamma_i B)(\overline{A}\,\Gamma_j D)$$

where A, B, C, D are fermion fields, and c_{ij} are numbers. They are much simpler if written in terms of chiral fields using $\Gamma_i = P_L, P_R, \gamma^\mu P_L, \gamma^\mu P_R, \sigma^{\mu\nu}P_L, \sigma^{\mu\nu}P_R$, rather than Dirac fields. Work out the Fierz relations for

$$(\overline{A}P_L B)(\overline{C}P_L D), \quad (\overline{A}\gamma^\mu P_L B)(\overline{C}\gamma_\mu P_L D), \quad (\overline{A}\sigma^{\mu\nu}P_L B)(\overline{C}\sigma_{\mu\nu}P_L D),$$

$$(\overline{A}P_L B)(\overline{C}P_R D), \quad (\overline{A}\gamma^\mu P_L B)(\overline{C}\gamma_\mu P_R D), \quad (\overline{A}\sigma^{\mu\nu}P_L B)(\overline{C}\sigma_{\mu\nu}P_R D).$$

Do not forget the Fermi minus sign. The $P_R \otimes P_R$ identities are obtained from the $P_L \otimes P_L$ identities by using $L \leftrightarrow R$.

SOLUTION:

Define

$$\gamma^5 = \gamma_5 = i\gamma^0\gamma^1\gamma^2\gamma^3, \quad \sigma^{\mu\nu} = \tfrac{i}{2}[\gamma^\mu, \gamma^\nu] \tag{12.38}$$

and raise/lower spacetime indices with $\eta_{\mu\nu} = (+1, -1, -1, -1)$.

Use a chiral basis,

$$\left\{\Gamma^A\right\} = \{P_R, P_L, \gamma^\mu P_R, \gamma^\mu P_L, \sigma^{\mu\nu}\}.$$

$\gamma^\mu\gamma_5$ are four elements of the basis for $\mu = \{0, 1, 2, 3\}$, and $\sigma^{\mu\nu}$ are six elements of the basis for $\mu\nu = \{01, 02, 03, 12, 23, 31\}$. The sixteen elements of Γ^A are linearly independent and any 4×4 matrix can be written as a linear combination of Γ^A. Define a dual basis,

$$\{\Gamma_A\} = \left\{P_R, P_L, \gamma_\mu P_L, \gamma_\mu P_R, \tfrac{1}{2}\sigma_{\mu\nu}\right\}$$

such that,

$$\mathrm{Tr}[\Gamma_A\Gamma^B] = 2\delta_A^B.$$

Any 4×4 matrix can be written as

$$X = c_A\Gamma^A \qquad\qquad c_A = \tfrac{1}{2}\mathrm{Tr}[X\Gamma_A]. \tag{12.39}$$

Then it is straightforward to project the bilinears onto this basis,

$$\left(\Gamma^A\right)\left[\Gamma^B\right] = -\tfrac{1}{4}\mathrm{Tr}\left[\Gamma^A\Gamma_C\Gamma^B\Gamma_D\right]\left(\Gamma^D\right)\left[\Gamma^C\right] \tag{12.40}$$

where the overall minus sign accounts for the anti-commutativity of the fermion fields, and the type of parenthesis indicates whether the Γ matrix is contracted with $\overline{A}, B, \overline{C}$ or D.

Evaluating eqn (12.40) for the required cases:

$$(P_L)[P_L] = -\tfrac{1}{4}\mathrm{Tr}[P_L\Gamma_C P_L\Gamma_D]\left(\Gamma^D\right)\left[\Gamma^C\right]$$

$$= -\tfrac{1}{2}(P_L)[P_L] - \tfrac{1}{16}(\sigma^{\mu\nu}](\sigma_{\mu\nu}] - \tfrac{i}{16}\epsilon_{\mu\nu\alpha\beta}(\sigma^{\mu\nu}](\sigma_{\alpha\beta}] \tag{12.41}$$

[Note that $(\sigma^{\mu\nu}](\sigma_{\mu\nu}]$ sums over both $\mu < \nu$ and $\mu > \nu$ and so is twice the sum over $\sigma^{\mu\nu}$ in the basis set Γ^A.]

$$(P_L)[P_R] = -\tfrac{1}{2}(\gamma^\mu P_R](\gamma_\mu P_L] \tag{12.42}$$

$$(\gamma^\mu P_L)[\gamma_\mu P_L] = (\gamma^\mu P_L](\gamma_\mu P_L] \tag{12.43}$$

$$(\gamma^\mu P_L)[\gamma_\mu P_R] = -\tfrac{1}{4}\mathrm{Tr}\left[P_R\gamma^\mu\Gamma_C P_L\gamma_\mu\Gamma_D\right]\left(\Gamma^D\right)\left[\Gamma^C\right]$$

$$= -\tfrac{1}{4}\mathrm{Tr}\left[P_R\gamma^\mu P_L^2\gamma_\mu P_R\right](R][L)$$

$$= -2(R][L) \quad \text{as} \quad \gamma^\mu\gamma_\mu = 4\mathbb{1}, \quad \mathrm{Tr}[P_R^4] = \mathrm{Tr}[P_R] = 2 \tag{12.44}$$

$$(\sigma^{\mu\nu}P_L)[\sigma_{\mu\nu}P_L] = -6(P_L)[P_L] - \tfrac{1}{4}(\sigma^{\mu\nu})[\sigma_{\mu\nu}] - \tfrac{i}{4}\epsilon^{\mu\nu\alpha\beta}(\sigma_{\mu\nu})[\sigma_{\alpha\beta}] \qquad (12.45)$$

$$(\sigma^{\mu\nu}P_L)[\sigma_{\mu\nu}P_R] = 0. \qquad (12.46)$$

Note that

$$\tfrac{i}{2}\epsilon^{\mu\nu\alpha\beta}\sigma_{\alpha\beta} = -\gamma^5\sigma^{\mu\nu}$$

($\epsilon_{0123} = +1$) which can be used to replace the ϵ tensors by γ_5. eqn (12.41) becomes

$$(P_L)[P_L] = -\tfrac{1}{2}(P_L)[P_L] - \tfrac{1}{16}(\sigma^{\mu\nu})[\sigma_{\mu\nu}] + \tfrac{1}{16}(\sigma^{\mu\nu})[\gamma_5\sigma_{\mu\nu}]$$

$$= -\tfrac{1}{2}(P_L)[P_L] - \tfrac{1}{8}(\sigma^{\mu\nu}P_L)[\sigma_{\mu\nu}P_L] \qquad (12.47)$$

In the last line, we have put back the projectors on $\sigma^{\mu\nu}$, since they are contracted with left-handed fields, and used $\gamma_5 P_L = -P_L$. Similarly, eqn (12.45) becomes

$$(\sigma^{\mu\nu}P_L)[\sigma_{\mu\nu}P_L] = -6(P_L)[P_L] + \tfrac{1}{2}(\sigma^{\mu\nu}P_L)[\sigma_{\mu\nu}P_L] \qquad (12.48)$$

To summarize, the identities (including the Fermi minus sign) are:

$$(P_L)[P_L] = -\tfrac{1}{2}(P_L)[P_L] - \tfrac{1}{8}(\sigma^{\mu\nu}P_L)[\sigma_{\mu\nu}P_L]$$

$$(P_L)[P_R] = -\tfrac{1}{2}(\gamma^\mu P_R)[\gamma_\mu P_L]$$

$$(\gamma^\mu P_L)[\gamma_\mu P_L] = (\gamma^\mu P_L)[\gamma_\mu P_L]$$

$$(\gamma^\mu P_L)[\gamma_\mu P_R] = -2(R)[L]$$

$$(\sigma^{\mu\nu}P_L)[\sigma_{\mu\nu}P_L] = -6(P_L)[P_L] + \tfrac{1}{2}(\sigma^{\mu\nu}P_L)[\sigma_{\mu\nu}P_L]$$

$$(\sigma^{\mu\nu}P_L)[\sigma_{\mu\nu}P_R] = 0. \qquad (12.49)$$

Exercise 2.5 Compute the mass renormalization factor Z_m in QCD at one loop. Use this to determine the one-loop mass anomalous dimension γ_m,

$$\mu\frac{\mathrm{d}m}{\mathrm{d}\mu} = \gamma_m m, \qquad (12.50)$$

by differentiating $m_0 = Z_m m$, and noting that m_0 is μ-independent.

SOLUTION:

The Lagrangian for massless QCD is

$$\mathcal{L}_{m=0} = \bar{q}_0\,\slashed{D}q_0 - \tfrac{1}{4}G^A_{\mu\nu,0}\,G^{\mu\nu}_{A,0}, \qquad (12.51)$$

where q_0 is the bare quark field, $G_{\mu\nu,0}$ is the bare gluon field strength tensor, and $\slashed{D}$ is the covariant derivative (we are only interested in quark-gluon interactions, so other gauge bosons are neglected here)

$$D_\mu = \partial_\mu - ig_{0,s}t_A G^A_{\mu,0}, \qquad (12.52)$$

where t_A are the $SU(3)$ generators and $g_{0,s}$ the respective bare gauge coupling; $G^A_{\mu,0}$ is the bare gluon field. The insertion of the quark mass operator $\mathcal{O}_f = m_0\bar{q}_0 q_0$, where m_0 is the bare quark mass, modifies the Lagrangian

$$\mathcal{L} = \mathcal{L}_{m=0} - \mathcal{O}_f. \tag{12.53}$$

To renormalize the theory we define the renormalization constants

$$q_0 = Z_q^{1/2} q, \quad m_0 = Z_m m, \tag{12.54}$$

where the fields without subscripts are renormalized fields. These constants will absorb the divergences. The other renormalization constants (for the gauge field and coupling) are not needed for this problem. In the $\overline{\text{MS}}$ scheme we impose that only the $1/\epsilon$ pole and the dim-reg constants are absorbed, thus giving

$$Z_i = 1 + z_{i,1} \frac{g_s^2}{(4\pi)^2} \frac{1}{\epsilon} + \dots, \tag{12.55}$$

where the $z_{i,1}$ are μ-independent constants (all the μ dependence is in the coupling constant).

To solve this problem we are interested in the Lagrangian and operator counterterms

$$\mathcal{L}_{ct} = [(Z_q - 1)\slashed{p}], \qquad \mathcal{O}_{f,ct} = (Z_q Z_m - 1) m \bar{q} q \tag{12.56}$$

which make the one-loop corrections finite.

The loop integral with the insertion of the operator $\mathcal{O}_f$ is showed in Figure 12.1

$$I_{\bar{q}q} = m(ig_s)^2 \mu^{2\epsilon} \int \frac{d^d k}{(2\pi)^d} \gamma_\alpha t_A \frac{i\slashed{k}}{k^2} \frac{i\slashed{k}}{k^2} \gamma_\beta t_B \left(\frac{-ig^{\alpha\beta}\delta^{AB}}{(k-p)^2} \right). \tag{12.57}$$

Notice that we are only keeping the large k region, since we are only interested in the divergent part. We can simplify this expression noting that $\delta^{AB} t_A t_B = C_F = 4/3$ is the eigenvalue of the Casimir operator and that, in d dimensions, $g^{\alpha\beta} \gamma_\alpha \gamma_\beta = d$. Also $\slashed{k}\slashed{k} = k^2$, thus we have

$$I_{\bar{q}q} = -img_s^2 C_F \mu^{2\epsilon} \int \frac{d^d k}{(2\pi)^d} \frac{d}{k^2 (k-p)^2}. \tag{12.58}$$

Introducing the Feynman parameters we have

$$I_{\bar{q}q} = -ig_s^2 m C_F \mu^{2\epsilon} \int \frac{d^d k}{(2\pi)^d} \int_0^1 dx \frac{d}{[(1-x)k^2 + x(k-p)^2]^2}. \tag{12.59}$$

Figure 12.1 *One-loop correction to the mass operator.*

Define $\ell = k - xp$ and $\Delta = -x(1-x)p^2$, so we can write

$$I_{\bar{q}q} = -ig_s^2 mC_F \mu^{2\epsilon} \int \frac{d^d\ell}{(2\pi)^d} \int_0^1 dx \frac{d}{[\ell-\Delta]^2} = g_s^2 C_F \int_0^1 dx \frac{d}{(4\pi)^{\frac{d}{2}}} \frac{\Gamma(2-\frac{d}{2})}{\Gamma(2)} \left(\frac{\mu^2}{\Delta}\right)^{2-\frac{d}{2}}. \quad (12.60)$$

Expanding and integrating, we get

$$I_{\bar{q}q} = \frac{g_s^2 mC_F}{(4\pi)^2} 4 \left(\frac{1}{\epsilon} - \ln \frac{-p^2}{\mu^2} + \frac{3}{2}\right), \quad (12.61)$$

so that the divergent part is

$$I_{\bar{q}q,\text{div}} = \frac{4mg_s^2 C_F}{(4\pi)^2} \frac{1}{\epsilon}. \quad (12.62)$$

The counterterm $Z_q Z_m m$ cancels the divergence,

$$Z_q Z_m = 1 - \frac{4g_s^2 C_F}{(4\pi)^2} \frac{1}{\epsilon}. \quad (12.63)$$

We need now to calculate the one-loop contribution to the quark propagator, shown in Figure 12.2. This gives

$$I_2 = (ig_s)^2 \mu^{2\epsilon} \int \frac{d^d k}{(2\pi)^d} \gamma_\alpha t_A \frac{i(\not{p}-\not{k})}{(p-k)^2} \gamma_\beta t_B \left(\frac{-ig^{\alpha\beta}\delta^{AB}}{k^2}\right)$$
$$= -g_s^2 C_F \mu^{2\epsilon} \int \frac{d^d k}{(2\pi)^d} \frac{\gamma_\alpha(\not{p}-\not{k})\gamma^\alpha}{(p-k)^2 k^2}, \quad (12.64)$$

where we used again the Casimir operator eigenvalue. To simplify the Dirac structure in the numerator, we use the d-dimensional identity

$$\gamma_\mu \gamma^\nu \gamma^\mu = (2-d)\gamma^\nu, \quad (12.65)$$

thus

$$I_2 = -g_s^2 C_F \mu^{2\epsilon} \int \frac{d^d k}{(2\pi)^d} \frac{(2-d)(\not{p}-\not{k})}{(p-k)^2 k^2}. \quad (12.66)$$

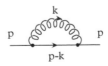

Figure 12.2 *One-loop diagram for the quark self energy.*

We introduce now the Feynman parameter and get

$$
\begin{aligned}
I_2 &= -g_s^2 C_F \mu^{2\epsilon} \int_0^1 dx \int \frac{d^d k}{(2\pi)^d} \frac{(2-d)(\not{p}-\not{k})}{[x(p-k)^2+(1-x)k^2]^2} \\
&= -g_s^2 C_F \mu^{2\epsilon} \int_0^1 dx \int \frac{d^d \ell}{(2\pi)^d} \frac{(2-d)((1-x)\not{p}-\not{\ell})}{[\ell^2-\Delta]^2},
\end{aligned}
\tag{12.67}
$$

where we defined $\ell = k - xp$ and $\Delta = -(1-x)xp^2$. The term linear in ℓ vanishes, so the only contribution is

$$
\begin{aligned}
I_2 &= -g_s^2 C_F \not{p} \int_0^1 dx(1-x) \frac{(2-d)}{(4\pi)^{\frac{d}{2}}} \frac{\Gamma(2-\frac{d}{2})}{\Gamma(2)} \left(\frac{\mu^2}{\Delta}\right)^{2-\frac{d}{2}} \\
&= g_s^2 C_F \not{p} \int_0^1 dx(1-x) \frac{2(1+\epsilon)}{(4\pi)^2} \left[\frac{1}{\epsilon} - \ln\left(\frac{-(x(1-x)p^2}{\mu^2}\right)\right],
\end{aligned}
\tag{12.68}
$$

where we used $d = 4 - 2\epsilon$, from which we have $2 - d = -2(1+\epsilon)$. The divergent part is then

$$
I_{2,\text{div}} = \frac{g_s^2 C_F \not{p}}{(4\pi)^2} \frac{1}{\epsilon}.
\tag{12.69}
$$

The counterterm Z_q cancels the divergence,

$$
Z_q = 1 - \frac{g_s^2 C_F}{(4\pi)^2} \frac{1}{\epsilon}.
\tag{12.70}
$$

We can now find the anomalous dimension. This can be computed from the equation

$$
\gamma_\mathcal{O} = \frac{1}{Z_m} \frac{dZ_m}{d\ln\mu} = -2g_s^2 \frac{dz_{m,1}}{dg_s^2},
\tag{12.71}
$$

where the last step can be deduced from eqn (12.55) (see Neubert's lectures). This means that $\gamma\mathcal{O}$ can be directly obtained from the $1/\hat{\epsilon}$ pole constant. From eqns (12.70) and (12.63),

$$
Z_m = \frac{1 - 4\frac{g_s^2 C_F}{(4\pi)^2} \frac{1}{\hat{\epsilon}}}{1 - \frac{g_s^2 C_F \not{p}}{(4\pi)^2} \frac{1}{\hat{\epsilon}}} = 1 - 3\frac{g_s^2 C_F}{(4\pi)^2} \frac{1}{\hat{\epsilon}} + \mathcal{O}(g_s^4).
\tag{12.72}
$$

Finally, we can now plug this result into (12.71) and get

$$
\gamma_\mathcal{O} = 6\frac{g_s^2 C_F}{(4\pi)^2} = C_F \frac{3\alpha_s}{2\pi} = \frac{2\alpha_s}{\pi}.
\tag{12.73}
$$

The anomalous dimension of m is the negative of that for O,

$$
\gamma_m = -C_F \frac{3\alpha_s}{2\pi} = -\frac{2\alpha_s}{\pi}.
\tag{12.74}
$$

The individual graphs, and thus Z_q and $Z_m Z_q$ depend on the choice of gauge, but Z_m and γ_m are gauge independent.

Exercise 2.6 Verify eqn (L2.23) and eqn (L2.24).

SOLUTION:

The one-loop QCD running,

$$\frac{1}{\alpha_s(\mu)} = \frac{1}{\alpha_s(Q)} + \frac{33-2n_f}{6\pi} \log\left(\frac{\mu}{Q}\right) \tag{12.75}$$

where n_f is the number of active flavours between μ and Q, allows us to relate a coupling fixed at 1 TeV to the hadronization scale Λ_{QCD} (at which $1/\alpha_s(\Lambda_{QCD}) = 0$) by running down through the heavy quark thresholds,

$$\frac{6\pi}{\alpha_s(1\,\text{TeV})} = 21\log\frac{1\,\text{TeV}}{m_t} + 23\log\frac{m_t}{m_b} + 25\log\frac{m_b}{m_c} + 27\log\frac{m_c}{\Lambda_{QCD}} \tag{12.76}$$

$$\implies \Lambda_{QCD} = e^{-\frac{6\pi}{27\alpha_s(1\,\text{TeV})}}(1\,\text{TeV})^{21/27}(m_t m_b m_c)^{2/27}$$

The proton mass is proportional to Λ_{QCD}, and so scales as $m_t^{2/27}$ when the coupling is fixed at an energy $\mu > m_t$ (providing other quark masses are held fixed). Applying $m_t\,d/dm_t$ to

$$\frac{6\pi}{\alpha_s(1\,\text{TeV})} = 21\log\frac{1\,\text{TeV}}{m_t} + 23\log\frac{m_t}{m_b} + 25\log\frac{m_b}{m_c} + 27\log\frac{m_c}{\mu_L} + \frac{6\pi}{\alpha_s(\mu_L)} \tag{12.77}$$

keeping $\alpha_s(1\,\text{TeV})$ fixed gives

$$0 = -21 + 23 + m_t\frac{d}{dm_t}\frac{6\pi}{\alpha_s(\mu_L)} \implies m_t\frac{d}{dm_t}\frac{1}{\alpha_s(\mu_L)} = -\frac{1}{3\pi}. \tag{12.78}$$

Exercise 2.7 In $d = 4$ spacetime dimensions, work out the field content of Lorentz-invariant operators with dimension $\mathscr{D}$ for $\mathscr{D} = 1,\ldots,6$. At this point, do not try and work out which operators are independent, just the possible structure of allowed operators. Use the notation ϕ for a scalar, ψ for a fermion, $X_{\mu\nu}$ for a field strength, and D for a derivative. For example, an operator of type $\phi^2 D$ such as $\phi D_\mu\phi$ is not allowed because it is not Lorentz-invariant. An operator of type $\phi^2 D^2$ could be either $D_\mu\phi D^\mu\phi$ or $\phi D^2\phi$, so a $\phi^2 D^2$ operator is allowed, and we will worry later about how many independent $\phi^2 D^2$ operators can be constructed.

SOLUTION:

We use the notation $[A] = n$ for the mass dimension of the operator A. The dimensionality of gauge boson, fermion and scalar fields in d spacetime dimensions can be determined from the kinetic terms in the Lagrangian imposing the condition

$$[\mathcal{L}] = d.$$

The three kinetic terms are (numerical constants are irrelevant for this discussion)

$$\partial_\mu\phi\partial^\mu\phi, \qquad \overline{\psi}i\partial\!\!\!/\psi, \qquad X_{\mu\nu}X^{\mu\nu}, \tag{12.79}$$

where ϕ, ψ, and X are the scalar field, the fermion field, and the gauge boson field strength, respectively. It follows immediately

$$[XX] = d \Rightarrow [X] = \tfrac{d}{2},$$ (12.80)

for the field strength. For the scalar field we have

$$[\partial_\mu \phi \partial^\mu \phi] = 2[\partial] + 2[\phi] = d \Rightarrow [\phi] = \tfrac{d-2}{2},$$ (12.81)

while for the fermion field we have

$$[\bar{\psi} i \partial\!\!\!/ \psi] = [\partial] + 2[\psi] = d \Rightarrow [\psi] = \tfrac{d-1}{2}.$$ (12.82)

In $d = 4$ we then have

$$[\phi] = 1, \quad [D] = 1, \quad [\psi] = \tfrac{3}{2}, \quad [X] = 2.$$ (12.83)

The Lorentz-invariant operator structures of mass dimension n are:

- $n = 1$: ϕ;
- $n = 2$: ϕ^2,
- $n = 3$: ψ^2, ϕ^3;
- $n = 4$: $\phi^2 D^2, \phi^4, X^2, \phi \psi^2, \psi^2 D$;
- $n = 5$: $\psi^2 D^2, \psi^2 \phi^2, \psi^2 X, \phi X^2, \phi^5, \phi^3 D^2, \psi^2 \phi D$;
- $n = 6$: $\phi^6, \phi^2 X^2, \phi^2 D^4, D^2 X^2, X^3, \psi^4, \psi^2 D^3, X \phi^2 D^2, \phi^4 D^2, \psi^2 \phi D^2, \psi^2 \phi^2 D$, $\psi^2 X \phi, \psi^2 \phi^3$.

D^n for $n = 2, 4, 6$ are not allowed, because there must be at least one field for D to act on. For $n = 0$, there is always the operator $\mathbb{1}$. $D^2 \phi$ for $n = 3$ is a total derivative and integrates to zero. $D^2 X$ for $n = 4$ is $D_\mu D_\nu X^{\mu\nu}$, and anti-symmetry of X converts this to $[D_\mu, D_\nu] X^{\mu\nu} \propto X^2$. Several operators, such as $\phi^2 D^4$ can be eliminated by field redefinitions. Similar simplifications occur in the next exercise.

Exercise 2.8 For $\mathbf{d} = 2, 3, 4, 5, 6$ dimensions, work out the field content of operators with dimension $\mathscr{D} \le \mathbf{d}$, i.e. the 'renormalizable' operators.

SOLUTION:

Using the general result for mass dimensionality in the previous exercise, we can now work out Lorentz-invariant renormalizable operators in different spacetime dimensions, i.e. those with dimension $n \le d$. Note that at $n = d$ there will be the kinetic term operators and these will always appear as $\phi^2 D^2$, $\psi^2 D$ and X^2 for scalar, fermion, and gauge boson, respectively. Except in the particular case of $n = 2$, we omit these operators in the lists. There is also the operator $\mathbb{1}$ (cosmological constant) which is also omitted in the lists.

- $d = 2$

$$[\phi] = 0, \quad [D] = 1, \quad [\psi] = \tfrac{1}{2}, \quad [X] = 1.$$ (12.84)

In this particular case, the scalar field can enter with an arbitrary power p by itself at $n = 0$ or in other operators at $n = 1, 2$ since it does not add any mass dimension to n and is Lorentz invariant. For each class of operator we then have an infinite set of operators when we change the value of p. So we have

★ $n = 0$: ϕ^p;
★ $n = 1$: $\phi^p \psi^2$;
★ $n = 2$: $\phi^p \psi^4, \phi^p D^2, \phi^p X^2, \phi^p \psi^2 D, \psi^2 X \phi^p$.

• $d = 3$

$$[\phi] = \tfrac{1}{2}, \quad [D] = 1, \quad [\psi] = 1, \quad [X] = \tfrac{3}{2}. \tag{12.85}$$

From now on there are no more non-trivial $n = 0$ operators.

★ $n = 1/2$: ϕ;
★ $n = 1$: ϕ^2;
★ $n = 3/2$: ϕ^3;
★ $n = 2$: ϕ^4, ψ^2;
★ $n = 5/2$: $\phi^5, \psi^2 \phi$;
★ $n = 3$: $\phi^6, \phi^2 \psi^2$;

• $d = 4$

$$[\phi] = 1, \quad [D] = 1, \quad [\psi] = \tfrac{3}{2}, \quad [X] = 2. \tag{12.86}$$

This case has already been analysed in the previous exercise, so we can skip it.

• $d = 5$

$$[\phi] = \tfrac{3}{2}, \quad [D] = 1, \quad [\psi] = 2, \quad [X] = \tfrac{5}{2}. \tag{12.87}$$

There are no more operator for $n = 1$ and from now on the only $n = 2$ operator is D^2, so we do not write it anymore.

★ $n = 3/2$: ϕ;
★ $n = 2$: none;
★ $n = 5/2$: none;
★ $n = 3$: ϕ^2;
★ $n = 3/2$: none;
★ $n = 4$: ψ^2.
★ $n = 9/2$: ϕ^3;

In this case, at $n = 5$ there are only the kinetic terms.

• $d = 6$

$$[\phi] = 2, \quad [D] = 1, \quad [\psi] = \tfrac{5}{2}, \quad [X] = 3. \tag{12.88}$$

★ $n = 1$: none;
★ $n = 2$: ϕ;
★ $n = 3$: none;
★ $n = 4$: ϕ^2;
★ $n = 5$: ψ^2;
★ $n = 6$: ϕ^3.

Exercise 2.9 Compute the decay rate $\Gamma(b \to c e^- \bar{\nu}_e)$ with the interaction Lagrangian

$$L = -\frac{4G_F}{\sqrt{2}} V_{cb}(\bar{c}\gamma^\mu P_L b)(\bar{\nu}_e \gamma_\mu P_L e)$$

with $m_e \to 0$, $m_\nu \to 0$, but retaining the dependence on $\rho = m_c^2/m_b^2$. It is convenient to write the three-body phase space in terms of the variables $x_1 = 2E_e/m_b$ and $x_2 = 2E_\nu/m_b$.

SOLUTION:

At tree level, the amplitude for the decay $b \to c e^- \bar{\nu}_e$ is given by,

$$i\mathcal{A} = \frac{4G_F}{\sqrt{2}} V_{cb}(\bar{c}\gamma^\mu b_L)(e_L \gamma_\mu \bar{\nu}_e) \tag{12.89}$$

where we use the same symbol for the spinor fields and their polarization spinors $u(p,s)$.

To compute the square of the amplitude, first consider

$$(\bar{c}_L \gamma^\mu b_L)(\bar{c}_L \gamma^\nu b_L)^\dagger = (\bar{c}_L \gamma^\mu b_L)(\bar{b}_L \gamma^\nu c_L) \tag{12.90}$$

$$= \tfrac{1}{4}\mathrm{Tr}\left[\gamma^\mu(1-\gamma_5)b\bar{b}\gamma^\nu(1-\gamma_5)c\bar{c}\right]. \tag{12.91}$$

Then performing a sum over the different available spin states,

$$\mathrm{Tr}\left[\gamma^\mu(1-\gamma_5)(\not{p}_b + m_b)\gamma^\nu(1-\gamma_5)(\not{p}_c + m_c)\right] = T^{\mu\nu}_{\alpha\beta} p_b^\alpha p_c^\beta \tag{12.92}$$

$$T^{\mu\nu}_{\alpha\beta} = 8\left(\eta^{\mu\alpha}\eta^{\nu\beta} - \eta^{\mu\nu}\eta^{\alpha\beta} + \eta^{\mu\beta}\eta^{\nu\alpha} + i\epsilon^{\mu\alpha\nu\beta}\right). \tag{12.93}$$

The probability is then,

$$\langle 0||\mathcal{A}|^2|0\rangle = \tfrac{1}{2}\sum_{\text{spins}} |i\mathcal{A}|^2 = \frac{G_F^2|V_{cb}|^2}{4} T^{\mu\nu}_{\alpha\beta} T^{\rho\sigma}_{\mu\nu} p_b^\alpha p_c^\beta p_{\nu,\rho} p_{e,\sigma} = 64 G_F^2|V_{cb}|^2 (p_b \cdot p_\nu)(p_c \cdot p_e) \tag{12.94}$$

summing over all spins, and including a 1/2 for spin-averaging over the initial b-quark spin.

The three body phase space is,

$$d\Pi_3 = \int \frac{d^3\mathbf{p}_c}{(2\pi)^3 2E_c} \int \frac{d^3\mathbf{p}_e}{(2\pi)^3 2|\mathbf{p}_e|} \int \frac{d^3\vec{p}_\nu}{(2\pi)^3 2|\vec{p}_\nu|} (2\pi)^4 \delta^{(4)}(p_c + p_e + p_\nu - p_b). \tag{12.95}$$

Note that when we integrate over the spin-summed probability, we are going to need the integral

$$I^{\alpha\beta}(q) = \int \frac{d^3\vec{p}_e}{(2\pi)^3 2|\vec{p}_e|} \int \frac{d^3\vec{p}_\nu}{(2\pi)^3 2|\vec{p}_\nu|} (2\pi)^4 \delta^{(4)}(p_e + p_\nu - q) p_e^\alpha p_\nu^\beta. \tag{12.96}$$

The simplest way to evaluate this integral is to exploit Lorentz invariance, to write it as,

$$I^{\alpha\beta}(q) = \eta^{\alpha\beta} I_1 + q^\alpha q^\beta I_2 \tag{12.97}$$

and then to calculate the two independent Lorentz scalars $I_{1,2}$ in the frame $q = 0$, and use

$$p_e \cdot p_v = \tfrac{1}{2}q^2 \qquad\qquad q \cdot p_v = \tfrac{1}{2}q^2 \qquad\qquad q \cdot p_e = \tfrac{1}{2}q^2 \qquad (12.98)$$

which follow from $q = p_e + p_v$, $p_e^2 = p_v^2 = 0$. Explicitly,

$$\eta_{\alpha\beta} I^{\alpha\beta} = \int \frac{d^3 p_e}{(2\pi)^3 2|p_e|} \int \frac{d^3 p_v}{(2\pi)^3 2|p_v|} (2\pi)^4 \delta^{(4)}(p_e + p_v - q)(p_e \cdot p_v) \qquad (12.99)$$

The space δ-function fixes $\mathbf{p}_v = -\mathbf{p}_e$, and $E_v = E_e = |\mathbf{p}_e|$, giving

$$= \frac{q^2}{2\pi} \int \frac{|\mathbf{p}_e|^2 d|\mathbf{p}_e|}{(2|\mathbf{p}_e|)(2|\mathbf{p}_e|)} \delta(q^0 - 2|\mathbf{p}_e|) = \frac{1}{16\pi} q^2 \qquad (12.100)$$

and

$$q_\alpha q_\beta I^{\alpha\beta} = \int \frac{d^3 p_e}{(2\pi)^3 2|p_e|} \int \frac{d^3 p_v}{(2\pi)^3 2|p_v|} (2\pi)^4 \delta^{(4)}(p_e + p_v - q)(q \cdot p_e)(q \cdot p_v)$$
$$= \frac{1}{32\pi} q^4 \qquad (12.101)$$

from eqn (12.98). These give

$$I^{\alpha\beta}(q) = \frac{1}{96\pi} \left(\eta^{\alpha\beta} q^2 + 2 q^\alpha q^\beta \right). \qquad (12.102)$$

The differential decay rate is

$$d\Gamma(b \to c e^- \bar{v}_e) = \frac{1}{2m_b} d\Pi_3 \langle 0||\mathcal{A}|^2|0\rangle \qquad (12.103)$$

and the δ-function in the three-body phase space eqn (12.95) can be written as

$$\delta^{(4)}(p_c + p_e + p_v - p_b) = \int d^4 q \, \delta^{(4)}(p_e + p_v - q) \delta^{(4)}(p_c + q - p_b). \qquad (12.104)$$

This gives

$$d\Gamma(b \to c e^- \bar{v}_e) = \frac{1}{2m_b} \frac{64 G_F^2 |V_{cb}|^2}{\pi} \int \frac{d^3 p_c}{(2\pi)^3 2E_c} \int d^4 q \, \delta^{(4)}(p_c + q - p_b) I^{\alpha\beta}(q) p_{c\alpha} p_{b\beta}$$
$$= \frac{G_F^2 |V_{cb}|^2}{3\pi^2} \int \frac{d^3 p_c}{(2\pi)^3 2E_c} \left[-4m_b E_c^2 + 3E_c(m_b^2 + m_c^2) - 2m_b m_c^2 \right]. \qquad (12.105)$$

The $\mathbf{p}_c$ integral is

$$\frac{d^3 p_c}{(2\pi)^3 2E_c} = \frac{1}{4\pi^2} |\mathbf{p}_c| dE_c \qquad (12.106)$$

Using the above results, we find,

$$\frac{d\Gamma(bce^- \bar{v})}{dE_c} = |V_{cb}|^2 \frac{G_F^2}{12\pi^3} \sqrt{E_c^2 - m_c^2} \left[-4m_b E_c^2 + 3E_c \left(m_b^2 + m_c^2 \right) - 2m_b m_c^2 \right]. \qquad (12.107)$$

The total decay rate is given by integrating between $E_c = m_c$ and $E_c = (m_c^2 + m_b^2)/2m_c$, the maximum energy kinematically allowed,

$$\Gamma(b \to c e^- \bar{\nu}_e) = |V_{cb}|^2 \frac{G_F^2 m_b^5}{192\pi^3}(1 - 8\rho + 8\rho^3 - \rho^4 - 12\rho^2 \log\rho) \qquad \rho = \frac{m_c^2}{m_b^2} \qquad (12.108)$$

Exercise 2.10 Compute the one-loop scalar graph Figure 12.3 with a scalar of mass m and interaction vertex $-\lambda\phi^4/4!$ in the $\overline{\text{MS}}$ scheme. Verify the answer is of the form eqn (L2.93). The overall normalization will be different, because this exercise uses a real scalar field, and H in the SM is a complex scalar field.

SOLUTION:
The Lagrangian is

$$\mathcal{L} = \tfrac{1}{2}\partial_\mu\phi\partial^\mu\phi - \tfrac{1}{2}m^2\phi^2 - \tfrac{\lambda}{4!}\phi^4. \qquad (12.109)$$

The Feynman vertex for the self-interaction is $-i\lambda$ (the 4! factor is cancelled by the 4! different ways of combining the four ϕ lines in the diagram).

Calling k the momentum running in the loop, we have (the graph has a symmetry factor of 1/2)

$$I = -i\tfrac{1}{2}\lambda\mu^{2\epsilon}\int \frac{d^d k}{(2\pi)^d} \frac{1}{[k^2 - m^2]} = \frac{\lambda m^2}{2(4\pi)^{\frac{d}{2}}} \frac{\Gamma\left(1 - \frac{d}{2}\right)}{\Gamma(1)} \left(\frac{\mu^2}{m^2}\right)^{2 - \frac{d}{2}}. \qquad (12.110)$$

Expanding it for $\epsilon \to 0$ we have

$$I = \frac{\lambda m^2}{32\pi^2}\left(\tfrac{1}{\epsilon} - \ln\frac{m^2}{\mu^2}\right). \qquad (12.111)$$

In the $\overline{\text{MS}}$ scheme the counterterm will cancel the $1/\hat\epsilon$ pole, leaving

$$I + I_{\text{c.t.}} = -\frac{\lambda m^2}{32\pi^2}\ln\frac{m^2}{\mu^2}. \qquad (12.112)$$

The Higgs mass correction has a slightly different prefactor because H is a complex field, and because of a different normalization convention for the $(H^\dagger H)^2$ vertex.

Figure 12.3 *One-loop correction to the scalar mass from the $-\lambda\phi^4/4!$ interaction.*

Exercise 2.11 Compute I_F and I_{EFT} given in eqns (L2.97, L2.99) in dimensional regularization in $d = 4 - 2\epsilon$ dimensions. Both integrals have UV divergences, and the $1/\epsilon$ pieces are canceled by counterterms. Determine the counterterm contributions $I_{F,\text{ct}}, I_{\text{EFT,ct}}$ to the two integrals.

SOLUTION:

The full integral is given by,

$$
\begin{aligned}
I_F &= -i\mu^{2\epsilon} \int_0^1 dx \int \frac{d^d k}{(2\pi)^d} \frac{1}{(k^2 - xm^2 - (1-x)M^2)^2} \\
&= \frac{1}{16\pi^2} \Gamma(\epsilon) \int_0^1 dx \left(\frac{4\pi\mu^2}{xm^2 + (1-x)M^2} \right)^\epsilon \\
&= \frac{1}{16\pi^2} \left[\frac{1}{\epsilon} + \int_0^1 dx \, \log\left(\frac{\bar{\mu}^2}{xm^2 + (1-x)M^2} \right) \right] \\
&= \frac{1}{16\pi^2} \left[\frac{1}{\epsilon} - \frac{1}{M^2 - m^2} \left(M^2 \log\frac{M^2}{\bar{\mu}^2} - m^2 \log\frac{m^2}{\bar{\mu}^2} \right) + 1 \right].
\end{aligned}
\tag{12.113}
$$

The full theory counterterm cancels the divergence,

$$
I_{F,\text{c.t.}} = -\frac{1}{16\pi^2} \frac{1}{\epsilon}
\tag{12.114}
$$

In an EFT expansion, we try to capture this answer by adding higher-dimension operators, each suppressed by the scale M. This looks like,

$$
\begin{aligned}
I_{\text{EFT}} &= i\mu^{2\epsilon} \int \frac{d^d k}{(2\pi)^d} \frac{1}{k^2 - m^2} \frac{1}{M^2} \left[1 + \frac{k^2}{M^2} + \frac{k^4}{M^4} + \cdots \right] \\
&= -\frac{1}{16\pi^2} \sum_{a=0}^{\infty} \left(\frac{m^2}{M^2} \right)^{a+1} (-1)^{a+1} \frac{\Gamma(\epsilon - a - 1)\Gamma(2 + a - \epsilon)}{\Gamma(2 - \epsilon)} \left(\frac{4\pi\mu^2}{m^2} \right)^\epsilon \\
&= -\frac{1}{16\pi^2} \sum_{a=0}^{\infty} \left(\frac{m^2}{M^2} \right)^{a+1} \left[\frac{1}{\epsilon} - \gamma_E + 1 + \mathcal{O}(\epsilon) \right] \left[1 + \epsilon \log\left(\frac{4\pi\mu^2}{m^2} \right) + \mathcal{O}(\epsilon^2) \right] \\
&= -\frac{1}{16\pi^2} \left[\frac{1}{\epsilon} + \log\frac{\bar{\mu}^2}{m^2} \right] \sum_{a=0}^{\infty} \left(\frac{m^2}{M^2} \right)^{a+1} \\
&= -\frac{1}{16\pi^2} \left[\frac{1}{\epsilon} - \log\frac{m^2}{\bar{\mu}^2} + 1 \right] \frac{m^2}{M^2 - m^2}.
\end{aligned}
\tag{12.115}
$$

Each term in the EFT expansion is UV divergent. Note that in the EFT, each of the operators has its own counterterm

$$
I_{\text{EFT,c.t.}}^{(n)} = \frac{1}{16\pi^2\epsilon} \left(\frac{m^2}{M^2} \right)^n
\tag{12.116}
$$

and so the anomalous dimensions in the two theories are different,

$$
\gamma_{\text{Full}} \neq \gamma_{\text{EFT}}
$$

which has important consequences for resumming large logarithms. The sum of all the counterterms (needed for Exercise 2.15) is

$$I_{\text{EFT,c.t.}} = \frac{1}{16\pi^2\epsilon} \frac{m^2}{M^2 - m^2} \tag{12.117}$$

The resummed answer captures the correct analyticity structure in the IR scale m (i.e. the branch cut from the logarithm), but not the UV scale M. Note that the series only formally converges for the soft part of the integral, $\int_0^M dk$, which is the regime of validity of this EFT. In order to successfully match the full integral, we need to add an additional term, given by the 'matching condition',

$$I_{\text{Match}} = I_{\text{Full}} - I_{\text{EFT}}. \tag{12.118}$$

Mathematically, this corresponds to the remainder of the integration $\int_M^\infty dk$, and physically reflects the high-energy modes that we have integrated out in going to the EFT description (see the next problem).

Exercise 2.12 Compute $I_M \equiv (I_F + I_{F,\text{ct}}) - (I_{\text{EFT}} + I_{\text{EFT,ct}})$ and show that it is analytic in m.

SOLUTION:
Using I_F and I_{EFT} from the previous solution, with the $1/\epsilon$ pieces dropped,

$$I_M = \frac{1}{16\pi^2}\left[-\frac{1}{M^2-m^2}\left(M^2\log\frac{M^2}{\mu^2} - m^2\log\frac{m^2}{\mu^2}\right) + 1\right] + \frac{1}{16\pi^2}\left[-\log\frac{m^2}{\mu^2} + 1\right]\frac{m^2}{M^2-m^2}$$

$$= \frac{1}{16\pi^2}\frac{M^2}{m^2-M^2}\left[\log\frac{M^2}{\mu^2} - 1\right]. \tag{12.119}$$

The non-analytic $\log m^2$ term has canceled, and I_M is analytic in m.

Exercise 2.13 Compute $I_F^{(\text{exp})}$, i.e. I_F with the IR m scale expanded out

$$I_F^{(\text{exp})} = -i\mu^{2\epsilon}\int\frac{d^dk}{(2\pi)^d}\frac{1}{(k^2-M^2)}\left[\frac{1}{k^2} + \frac{m^2}{k^4} + \dots\right].$$

Note that the first term in the expansion has a $1/\epsilon$ UV divergence, and the remaining terms have $1/\epsilon$ IR divergences.

SOLUTION:

$$I_F^{(\text{exp})} = -i\mu^{2\epsilon}\int\frac{d^dk}{(2\pi)^d}\frac{1}{k^2-M^2}\frac{1}{k^2}\left[1 + \frac{m^2}{k^2} + \frac{m^4}{k^4} + \dots\right] \tag{12.120}$$

$$= \frac{1}{16\pi^2}\left[\frac{1}{\epsilon} - \log\frac{M^2}{\mu^2} + 1\right]\frac{M^2}{M^2-m^2} \tag{12.121}$$

on integrating term-by-term and adding. The finite part agrees with eqn (12.119).

Exercise 2.14 Compute $I_F^{(\text{exp})} + I_{F,\text{ct}}$ using $I_{F,\text{ct}}$ determined in Exercise 2.11. Show that the UV divergence cancels, and the remaining $1/\epsilon$ IR divergence is the same as the UV counterterm $I_{\text{EFT},ct}$ in the EFT.

SOLUTION:

$$I_F^{(\text{exp})} = \frac{1}{16\pi^2}\left[\frac{1}{\epsilon} - \log\frac{M^2}{\bar{\mu}^2} + 1\right]\frac{M^2}{M^2 - m^2} \tag{12.122}$$

from Exercise 2.14 and

$$I_{F,\text{c.t.}} = -\frac{1}{16\pi^2}\frac{1}{\epsilon} \tag{12.123}$$

from Exercise 2.12, so

$$I_F^{(\text{exp})} + I_{F,\text{ct}} = \frac{1}{16\pi^2}\left\{\frac{1}{\epsilon}\frac{m^2}{M^2 - m^2} + \left[-\log\frac{M^2}{\bar{\mu}^2} + 1\right]\frac{M^2}{M^2 - m^2}\right\} \tag{12.124}$$

and the infinite part is the same as eqn (12.117).

Exercise 2.15 Compute $I_{\text{EFT}}^{(\text{exp})}$, i.e. I_{EFT} with the IR m scale expanded out. Show that it is a scaleless integral that vanishes. Using the known UV divergence from Exercise 2.11, write it in the form

$$I_{\text{EFT}}^{(\text{exp})} = -B\frac{1}{16\pi^2}\left[\frac{1}{\epsilon_{\text{UV}}} - \frac{1}{\epsilon_{\text{IR}}}\right],$$

and show that the IR divergence agrees with that in $I_F^{(\text{exp})} + I_{F,\text{ct}}$.

SOLUTION:

$$I_{\text{EFT}}^{(\text{exp})} = -i\mu^{2\epsilon}\int\frac{d^d k}{(2\pi)^d}\left[\frac{1}{k^2} + \frac{m^2}{k^4} + \cdots\right]\left[-\frac{1}{M^2} - \frac{k^2}{M^4} - \cdots\right]. \tag{12.125}$$

Multiplying out gives integrals of the form

$$-i\mu^{2\epsilon}\int\frac{d^d k}{(2\pi)^d}\frac{(m^2)^r}{(M^2)^s}(k^2)^{2+s-r} \tag{12.126}$$

which are scaleless and vanish. From Exercise 2.12,

$$B = \frac{m^2}{M^2 - m^2}. \tag{12.127}$$

and the $1/\epsilon_{\text{IR}}$ divergent term agrees with eqn (12.124). [Note the various signs.]

$$I_{\text{EFT}}^{(\text{exp})} + I_{\text{EFT,c.t.}}^{(\text{exp})} = \frac{1}{16\pi^2}\frac{1}{\epsilon_{\text{IR}}}B \tag{12.128}$$

We can also compute the integral directly. The log divergent terms are

$$I_{\text{EFT}}^{(\text{exp})} = -i\mu^{2\epsilon}\int\frac{d^d k}{(2\pi)^d}\left[-\frac{m^2}{M^2} - \frac{m^4}{M^4} - \cdots\right]\frac{1}{k^4}. \tag{12.129}$$

Using eqn (L2.112) gives

$$I_{\text{EFT}}^{(\text{exp})} = \left[-\tfrac{m^2}{M^2} - \tfrac{m^4}{M^4} - \cdots \right]\left[\tfrac{1}{\epsilon_{\text{UV}}} - \tfrac{1}{\epsilon_{\text{IR}}} \right] \tag{12.130}$$

so that

$$B = \left[\tfrac{m^2}{M^2} + \tfrac{m^4}{M^4} + \cdots \right] = \tfrac{m^2}{M^2 - m^2}. \tag{12.131}$$

The $1/\epsilon_{\text{UV}}$ terms are canceled by the counterterm eqn (12.117), and the remaining terms, which are IR divergent, agree with eqn (12.124).

Exercise 2.16 Compute $\left(I_F^{(\text{exp})} + I_{F,\text{ct}} \right) - \left(I_{\text{EFT}}^{(\text{exp})} + I_{\text{EFT,ct}} \right)$ and show that all the $1/\epsilon$ divergences (both UV and IR) cancel, and the result is equal to I_M found in Exercise 2.12.

SOLUTION:
This follows immediately using eqn (12.124) and eqn (12.128).

Exercise 2.17 Make sure you understand why you can compute I_M simply by taking $I_F^{(\text{exp})}$ and dropping all $1/\epsilon$ terms (both UV and IR).

SOLUTION:
No written solution needed.

Exercise 2.18 Compute the QED on-shell electron form factors $F_1(q^2)$ and $F_2(q^2)$ expanded to first order in q^2/m^2 using dimensional regularization to regulate the IR and UV divergences. This gives the one-loop matching to heavy-electron EFT. Note that it is much simpler to *first* expand and then do the Feynman parameter integrals. A more difficult version of the problem is to compute the on-shell quark form factors in QCD, which gives the one-loop matching to the HQET Lagrangian. For help with the computation, see Ref. [18]. Note that in the non-Abelian case, using background field gauge is helpful because the amplitude respects gauge invariance on the external gluon fields.

SOLUTION:
To calculate the QED electron on-shell form factors $F_1(q^2)$ and $F_2(q^2)$ up to one loop, we consider the vertex as pictured in Figure 12.4. The form factors $F_i(q^2)$ are defined as

$$\bar{u}(p_2)\Gamma^\mu(q^2)u(p_1) = \bar{u}(p_2)\left(F_1(q^2)\gamma^\mu + F_2(q^2)\tfrac{i\sigma^{\mu\nu}q_\nu}{2m} \right)u(p_1) \tag{12.132}$$

with $q = p_2 - p_1$, where we have obviously $F_1(0) = 1$ and $F_2(0) = 0$ at tree level. We define $\tilde{F}_i(q^2)$ as the one-loop contribution to the full form factors $F_i(q^2)$.

We will calculate the $\tilde{F}_i(q^2)$ by writing down the Feynman rules of Figure 12.4, identify the integrals that will give us $\tilde{F}_1(q^2)$ and $\tilde{F}_2(q^2)$, do the integrals over the loop momentum in d dimensions, expand in $\hat{q}^2 = q^2/m^2$ to first order, and evaluate the leftover Feynman parameter integrals.

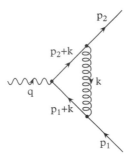

Figure 12.4 *Feynman diagram of 1-loop QED vertex correction, as calculated in Exercise 2.18.*

Applying the Feynman rules gives

$$-ie\Gamma^\mu(q^2) = \int \frac{d^4k}{(2\pi)^4}(-ie\gamma^\nu)\frac{i(\slashed{k}+\slashed{p}_2+m)}{(k+p_2)^2-m^2+i\epsilon}(-ie\gamma^\mu)\frac{i(\slashed{k}+\slashed{p}_1+m)}{(k+p_1)^2-m^2+i\epsilon}(-ie\gamma^\rho)\frac{-ig_{\nu\rho}}{k^2+i\epsilon}$$

$$= -e^3 \int \frac{d^4k}{(2\pi)^4}\frac{\gamma^\nu(\slashed{k}+\slashed{p}_2+m)\gamma^\mu(\slashed{k}+\slashed{p}_1+m)\gamma_\nu}{[(k+p_2)^2-m^2+i\epsilon][(k+p_1)^2-m^2+i\epsilon][k^2+i\epsilon]}. \tag{12.133}$$

First, we want to massage the denominator. By introducing Feynman parameters, we get

$$\frac{1}{[\ldots]} = \int_0^1 dx \int_0^1 dy \int_0^1 dz \frac{2\delta(1-x-y-z)}{(y[(k+p_2)^2-m^2]+x[(k+p_1)^2-m^2]+zk^2)^3}. \tag{12.134}$$

The denominator can be simplified further (using $x+y+z=1$, overall momentum conservation $q+p_1=p_2$ and on-shellness of the real electrons $p_1^2=p_2^2=m^2$) to

$$\begin{aligned}
(\ldots)^3 &= ([k+xp_1+yp_2]^2 - [xp_1+yp_2]^2 + x(p_1^2-m^2)+y(p_2^2-m^2))^3 \\
&= ([k+xp_1+yp_2]^2 - x^2p_1^2 + xy((p_2-p_1)^2 - p_1^2 - p_2^2) - y^2p_2^2)^3 \\
&= ([k+xp_1+yp_2]^2 + xyq^2 - x(x+y)p_1^2 - y(x+y)p_2^2)^3 \\
&= ([k+xp_1+yp_2]^2 + xyq^2 - (x+y)^2m^2)^3 \\
&= (l^2 + xyq^2 - (1-z)^2m^2)^3, \tag{12.135}
\end{aligned}$$

with shifted momenta $l = k+xp_1+yp_2$, which replaces the integrations over loop momentum k.

The numerator of (12.133) can be reduced using the following identities of the Dirac γ matrices in d dimensions

$$\gamma^\mu\gamma_\mu = d \tag{12.136}$$

$$\gamma^\mu\gamma^\alpha\gamma_\mu = (2-d)\gamma^\alpha \tag{12.137}$$

$$\gamma^\mu\gamma^\alpha\gamma^\beta\gamma_\mu = 4g^{\alpha\beta} - (4-d)\gamma^\alpha\gamma^\beta \tag{12.138}$$

$$\gamma^\mu\gamma^\alpha\gamma^\beta\gamma^\delta\gamma_\mu = -2\gamma^\delta\gamma^\beta\gamma^\alpha + (4-d)\gamma^\alpha\gamma^\beta\gamma^\delta. \tag{12.139}$$

We get

$$\gamma^\nu \dots \gamma_\nu = -2m^2\gamma^\mu + 4m(2k^\mu + p_1^\mu + p_2^\mu) - 2(\not k + \not p_1)\gamma^\mu(\not k + \not p_2) \tag{12.140}$$
$$+ (4-d)(\not k + \not p_2 - m)\gamma^\mu(\not k + \not p_1 - m).$$

By replacing k with parameter l as above ($k \to l - xp_1 - yp_2$), and erasing the terms linear in l (note that we will eventually integrate over l and everything else is even in l), we get

$$\gamma^\nu \dots \gamma_\nu = -2m^2\gamma^\mu + 4m(p_1^\mu(1-2x) + p_2^\mu(1-2y)) + (2-d)\not l\gamma^\mu\not l \tag{12.141}$$
$$- 2(-y\not p_2 + (1-x)\not p_1)\gamma^\mu(-x\not p_1 + (1-y)\not p_2)$$
$$+ (4-d)(-x\not p_1 + (1-y)\not p_2 - m)\gamma^\mu((1-x)\not p_1 - y\not p_2 - m)$$
$$+ \text{terms linear in } l.$$

Using relations $q + p_1 = p_2$ and $1 = x + y + z$, we get

$$\gamma^\nu \dots \gamma_\nu = -2m^2\gamma^\mu + 4mz(p_1^\mu + p_2^\mu) + 4m(x-y)q^\mu + (2-d)\not l\gamma^\mu\not l \tag{12.142}$$
$$- 2(z\not p_2 - (1-x)\not q)\gamma^\mu(z\not p_1 + (1-y)\not q)$$
$$+ (4-d)(x\not q + z\not p_2 - m)\gamma^\mu(z\not p_1 - y\not q - m).$$

Using on-shellness and completeness relations (we always have $\bar u(p_2)\Gamma^\mu(q^2)u(p_1)$ and therefore can write $\not p_2 A = mA$ as well as $A\not p_1 = Am$ for any A), we get for the numerator (after some rather uninspiring and tedious work)

$$\gamma^\nu \dots \gamma_\nu = (2-d)\not l\gamma^\mu\not l + 4mz(p_1^\mu + p_2^\mu) \tag{12.143}$$
$$+ mq^\mu(x-y)(4 - 2z - (4-d)(1-z))$$
$$+ m^2\gamma^\mu(-2 - 2z^2 + (4-d)(1-z)^2)$$
$$- q^2\gamma^\mu(2(z+xy) - (4-d)xy)$$
$$+ im\sigma^{\mu\nu}q_\nu(2z(1+z) - (4-d)(1-z)^2).$$

In the first line, we can apply the Gordon identity $p_1^\mu + p_2^\mu = 2m\gamma^\mu - i\sigma^{\mu\nu}q_\nu$ and make use of the relation $\not l\gamma^\mu\not l = \gamma^\alpha\gamma^\mu\gamma^\beta g_{\alpha\beta}\frac{l^2}{d} = (2-d)\gamma^\mu\frac{l^2}{d}$ to get

$$\gamma^\nu \dots \gamma_\nu = (2-d)^2\gamma^\mu\frac{l^2}{d} \tag{12.144}$$
$$+ mq^\mu(x-y)(4 - 2z - (4-d)(1-z))$$
$$+ m^2\gamma^\mu(8z - 2 - 2z^2 + (4-d)(1-z)^2)$$
$$- q^2\gamma^\mu(2(z+xy) - (4-d)xy)$$
$$- im\sigma^{\mu\nu}q_\nu(1-z)(2z + (4-d)(1-z)).$$

Obviously, the denominator (12.135) and the delta function is invariant under parameter exchange $x \leftrightarrow y$. Therefore, the second line of the numerator (12.144) gives zero contribution, as it changes sign under the same exchange.

Comparing to (12.132), the only contribution to $\tilde{F}_2(q^2)$ comes from the last line in (12.144), while lines 1, 3 and 4 give contributions to $\tilde{F}_1(q^2)$. We can finally write down the one-loop form factors as

$$\tilde{F}_i(q^2) = -2ie^2\mu^{4-d}\int_0^1 dx \int_0^1 dy \int_0^1 dz \int \frac{d^d l}{(2\pi)^d}\delta(1-x-y-z)\frac{A_i}{(l^2+xyq^2-(1-z)^2m^2)^3}$$
(12.145)

with

$$A_1 = \frac{(2-d)^2 l^2}{d} - (d-2)((1-z)^2 m^2 + xyq^2) + 2z(2m^2 - q^2)$$ (12.146)

$$A_2 = -2m^2(1-z)(2z+(4-d)(1-z)).$$ (12.147)

We encounter two types of phase space integrals in d dimensions, namely

$$\mathcal{I}_1 = \int \frac{d^d l}{(2\pi)^d}\frac{l^2}{(l^2-\Delta)^3}, \text{ and}$$ (12.148)

$$\mathcal{I}_2 = \int \frac{d^d l}{(2\pi)^d}\frac{1}{(l^2-\Delta)^3}.$$ (12.149)

It is worth to point out, that integrals of the form $\mathcal{I}_2$ are not divergent in $d=4$ dimensions. To be consistent, we will still keep the dependency on d explicit. This will help us to evaluate the Feynman parameter integrals, some of which diverge in $d=4$ dimensions. We will not go through the calculation of the integrals $\mathcal{I}_1$ and $\mathcal{I}_2$ step by step, as it is basically following the standard recipe of Wick rotating, parametrizing the loop momentum l in d dimensional polar coordinates, identifying the $d-1$ angular integrations as the surface of the d dimensional unit sphere (as we have only dependence on the radius of l) and computing the radial integration. We get the integrals to be

$$\mathcal{I}_1 = \frac{d}{4}\frac{i}{(4\pi)^{d/2}}\frac{1}{\Delta^{2-d/2}}\Gamma\left(\frac{4-d}{2}\right)$$ (12.150)

$$\mathcal{I}_2 = \frac{-i}{2(4\pi)^{d/2}}\frac{1}{\Delta^{3-d/2}}\Gamma\left(\frac{6-d}{2}\right).$$ (12.151)

Let us now calculate $\tilde{F}_1(q^2)$. From (12.150) and (12.151) as well as using the δ-function for the integration over z, we get (12.145) with (12.146) to become

$$\tilde{F}_1(q^2) = -2ie^2\mu^{4-d}\int_0^1 dx \int_0^{1-x} dy \frac{(2-d)^2 i}{4(4\pi)^{d/2}}\frac{1}{((x+y)^2 m^2 - xyq^2)^{2-d/2}}\Gamma\left(\frac{4-d}{2}\right)$$

$$- 2ie^2\mu^{4-d}\int_0^1 dx \int_0^{1-x} dy \frac{-i}{2(4\pi)^{d/2}}\frac{(2-d)((x+y)^2 m^2 + xyq^2)}{((x+y)^2 m^2 - xyq^2)^{3-d/2}}\Gamma\left(\frac{6-d}{2}\right)$$

$$- 2ie^2 \mu^{4-d} \int_0^1 dx \int_0^{1-x} dy \frac{-i}{2(4\pi)^{d/2}} \frac{2(1-x-y)(2m^2-q^2)}{((x+y)^2 m^2 - xyq^2)^{3-d/2}} \Gamma\left(\frac{6-d}{2}\right)$$

$$= 2e^2 \frac{(2-d)^2}{4(4\pi)^{d/2}} \frac{\mu^{4-d}}{m^{4-d}} \Gamma\left(\frac{4-d}{2}\right) \int_0^1 dx \int_0^{1-x} dy \frac{1}{((x+y)^2 - xy\hat{q}^2)^{2-d/2}}$$

$$- 2e^2 \frac{(2-d)}{2(4\pi)^{d/2}} \frac{\mu^{4-d}}{m^{4-d}} \Gamma\left(\frac{6-d}{2}\right) \int_0^1 dx \int_0^{1-x} dy \frac{((x+y)^2 + xy\hat{q}^2)}{((x+y)^2 - xy\hat{q}^2)^{3-d/2}}$$

$$- 2e^2 \frac{1}{(4\pi)^{d/2}} \frac{\mu^{4-d}}{m^{4-d}} \Gamma\left(\frac{6-d}{2}\right) \int_0^1 dx \int_0^{1-x} dy \frac{(1-x-y)(2-\hat{q}^2)}{((x+y)^2 - xy\hat{q}^2)^{3-d/2}}. \tag{12.152}$$

Along the way, we introduced $\hat{q}^2 = q^2/m^2$. Before we calculate the leftover integrals over x and y, we now want to expand the integrands in the heavy-electron-limit $\hat{q}^2 \approx 0$ to order $\hat{q}^2$. These expansions read

$$\frac{1}{(a-b\hat{q}^2)^{2-d/2}} = \frac{1}{a^{2-d/2}} - \frac{1}{2}\frac{b(d-4)}{a^{3-d/2}}\hat{q}^2 + \mathcal{O}(\hat{q}^4) \tag{12.153}$$

$$\frac{a+b\hat{q}^2}{(a-b\hat{q}^2)^{3-d/2}} = \frac{1}{a^{2-d/2}} - \frac{1}{2}\frac{b(d-8)}{a^{3-d/2}}\hat{q}^2 + \mathcal{O}(\hat{q}^4) \tag{12.154}$$

$$\frac{c-\hat{q}^2}{(a-b\hat{q}^2)^{3-d/2}} = \frac{c}{a^{3-d/2}} - \frac{1}{2}\frac{bc(d-6)+2a}{a^{4-d/2}}\hat{q}^2 + \mathcal{O}(\hat{q}^4). \tag{12.155}$$

Therefore, we can solve the Feynman parameter integrals of the first line in (12.152)

$$\int_0^1 dx \int_0^{1-x} dy \frac{1}{(...)^{2-d/2}} \approx \int_0^1 dx \int_0^{1-x} dy \frac{1}{(x+y)^{4-d}} - \frac{xy(d-4)}{2(x+y)^{6-d}}\hat{q}^2$$

$$= \frac{1}{d-2} - \frac{d-4}{12(d-2)}\hat{q}^2, \tag{12.156}$$

the one in the second line as

$$\int_0^1 dx \int_0^{1-x} dy \frac{(...)}{(...)^{3-d/2}} \approx \int_0^1 dx \int_0^{1-x} dy \frac{1}{(x+y)^{4-d}} - \frac{xy(d-8)}{2(x+y)^{6-d}}\hat{q}^2$$

$$= \frac{1}{d-2} - \frac{d-8}{12(d-2)}\hat{q}^2, \tag{12.157}$$

while the Feynman parameter integrals in the last line of (12.152) reads

$$\int_0^1 dx \int_0^{1-x} dy \frac{(...)(...)}{(...)^{3-d/2}} \approx \int_0^1 dx \int_0^{1-x} dy \left[\frac{2(1-x-y)}{(x+y)^{6-d}} \right.$$

$$\left. - \frac{(1-x-y)(xy(d-6)+(x+y)^2)}{(x+y)^{8-d}}\hat{q}^2 \right]$$

$$= \frac{2-\hat{q}^2}{(d-3)(d-4)} - \frac{d-6}{6(d-3)(d-4)}\hat{q}^2. \tag{12.158}$$

Note, that the last integral is divergent in $d = 4$ dimensions, which is the reason why we did the phase-space integral in d dimensions.

We will now put the solutions (12.156), (12.157) and (12.158) in (12.152). In the same step, we set the dimensions to $d = 4 - 2\epsilon$ and expand to order ϵ^0. The result (with $\bar{\mu}^2 = 4\pi e^{-\gamma_E} \mu^2$) turns out to be

$$\tilde{F}_1(\hat{q}^2) = \frac{e^2}{16\pi^2} \left(\frac{9-2\hat{q}^2}{3\epsilon} + \frac{8-\hat{q}^2}{2} - \frac{9-2\hat{q}^2}{3} \ln\left(\frac{m^2}{\bar{\mu}^2}\right) \right) + \mathcal{O}(\epsilon) + \mathcal{O}(\hat{q}^4). \tag{12.159}$$

The wave function graph Figure 12.2 gives

$$\delta Z = \frac{\alpha}{4\pi} \left(\frac{3}{\epsilon} + 4 - 3\log\frac{m^2}{\bar{\mu}^2} \right) \tag{12.160}$$

Adding the tree-level contribution, and subtracting δZ, we have

$$F_1(\hat{q}^2) = 1 + \frac{\alpha}{4\pi} \left(-\frac{2\hat{q}^2}{3\epsilon} - \frac{\hat{q}^2}{2} + \frac{2\hat{q}^2}{3} \ln\left(\frac{m^2}{\bar{\mu}^2}\right) \right) + \mathcal{O}(\hat{q}^4, \epsilon). \tag{12.161}$$

Now, we will calculate $\tilde{F}_2(q^2)$ in the same manner. This will be very straightforward as the procedure is exactly the same as in the calculation of $\tilde{F}_1(q^2)$. From (12.145) with (12.147), we have

$$\tilde{F}_2(q^2) = -2ie^2 \mu^{4-d} \int_0^1 dx \int_0^1 dy \int_0^1 dz \int \frac{d^d l}{(2\pi)^d} \tag{12.162}$$
$$\times \delta(1-x-y-z) \frac{-2m^2(1-z)(2z+(4-d)(1-z))}{(l^2+xyq^2-(1-z)^2 m^2)^3}.$$

We can use (12.151) for the phase-space integration and use the δ-function for the integration over z

$$\tilde{F}_2(q^2) = -2ie^2 \mu^{4-d} \int_0^1 dx \int_0^{1-x} dy \frac{-i}{2(4\pi)^{d/2}} \frac{-2m^2(x+y)(2(1+x+y)-d(x+y))}{((x+y)^2 m^2 - xyq^2)^{3-d/2}} \Gamma\left(\frac{6-d}{2}\right)$$
$$= 2e^2 \frac{1}{(4\pi)^{d/2}} \frac{\mu^{4-d}}{m^{4-d}} \Gamma\left(\frac{6-d}{2}\right) \int_0^1 dx \int_0^{1-x} dy \frac{2(x+y)+(2-d)(x+y)^2}{((x+y)^2 - xy\hat{q}^2)^{3-d/2}}, \tag{12.163}$$

where we again introduced $\hat{q}^2 = q^2/m^2$. The expansion of the integrand reads

$$\frac{1}{(a-b\hat{q}^2)^{3-d/2}} = \frac{1}{a^{3-d/2}} - \frac{1}{2} \frac{b(d-6)}{a^{4-d/2}} \hat{q}^2 + \mathcal{O}(\hat{q}^4), \tag{12.164}$$

and therefore we may write the leftover Feynman parameter integrations of $\tilde{F}_2(q^2)$ as follows

$$\int_0^1 dx \int_0^{1-x} dy \frac{\cdots}{(\ldots)^{3-d/2}} \approx \int_0^1 dx \int_0^{1-x} dy \left[\frac{2}{(x+y)^{5-d}} + \frac{(2-d)}{(x+y)^{4-d}} \right.$$
$$\left. - \left(\frac{xy(d-6)}{(x+y)^{7-d}} + \frac{(2-d)xy(d-6)}{2(x+y)^{6-d}} \right) \hat{q}^2 \right]$$
$$= \frac{2}{d-3} - 1 - \frac{d-6}{6(d-3)} \hat{q}^2 + \frac{d-6}{12} \hat{q}^2. \tag{12.165}$$

We end the calculation by injecting (12.165) into (12.163), set $d = 4 - 2\epsilon$ and expand to order ϵ^0:

$$F_2(\hat{q}^2) = \frac{e^2}{16\pi^2}\left(2 + \tfrac{1}{3}\hat{q}^2\right) + \mathcal{O}(\epsilon) + \mathcal{O}(\hat{q}^4)$$

$$= \frac{\alpha}{4\pi}\left(2 + \tfrac{1}{3}\hat{q}^2\right) + \mathcal{O}(\hat{q}^4, \epsilon) \tag{12.166}$$

which is finite.

Exercise 2.19 The SCET matching for the vector current $\overline{\psi}\gamma^\mu\psi$ for the Sudakov form factor is a variant of the previous problem. Compute $F_1(q^2)$ for on-shell massless quarks, in pure dimensional regularization with $Q^2 = -q^2 \neq 0$. Here Q^2 is the big scale, whereas in the previous problem q^2 was the small scale. The space-like calculation $Q^2 > 0$ avoids having to deal with the $+i0^+$ terms in the Feynman propagator which lead to imaginary parts. The time-like result can then be obtained by analytic continuation.

SOLUTION:

In SCET, the IR scales in which the full theory integral has to be expanded in are $m^2 \sim \lambda^2$ and $p^2 \sim \lambda^2$, where λ is the small power counting parameter in SCET. Consequently, expanding the integrand in the IR scales leads to the massless version of the full calculation.

Using standard QCD Feynman rules for the diagram in Figure 12.5, describing the one-loop correction of the vector current $\overline{\psi}\gamma^\mu\psi$, we get

$$\mathcal{A}^\mu = -i(4\pi)\alpha_s C_F \bar{u}(p_2, s_2)\tilde{\mu}^{2\varepsilon}\int \frac{d^d k}{(2\pi)^d}\frac{\gamma^\alpha(\slashed{p}_2 - \slashed{k})\gamma^\mu(\slashed{p}_1 - \slashed{k})\gamma_\alpha}{(k^2 - 2p_2\cdot k)(k^2 - 2p_1\cdot k)k^2}v(-p_1, s_2), \tag{12.167}$$

where $p_i^2 = 0$ was used when multiplying out the propagators, and the usual $+i0^+$ prescription is implied.

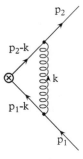

Figure 12.5 *The Feynman diagram encoding the virtual one-loop correction to the vector current $\overline{\psi}\gamma^\mu\psi$.*

After introducing standard Feynman parameters for the three propagators, i.e.

$$\frac{1}{(k^2-2p_2\cdot k)(k^2-2p_1\cdot k)k^2} \tag{12.168}$$

$$= 2\int_0^1 dx_1 \int_0^{1-x_1} dx_2 \frac{1}{(x_2k^2-2x_2p_2\cdot k+x_1k^2-2x_1p_1\cdot k+(1-x_1-x_2)k^2)^3} \tag{12.169}$$

$$= 2\int_0^1 dx_1 \int_0^{1-x_1} dx_2 \frac{1}{(k^2-2k\cdot(x_1p_1+x_2p_2))^3} \tag{12.170}$$

$$= 2\int_0^1 dx_1 \int_0^{1-x_1} dx_2 \frac{1}{((k-(x_1p_1+x_2p_2))^2-x_1x_2Q^2)^3}, \tag{12.171}$$

and using $p_1\cdot p_2 = 1/2Q^2$, we apply a shift in the loop momentum $k \equiv \ell + x_1p_1 + x_2p_2$. Consequently the denominator reduces to $(\ell^2 - x_1x_2Q^2)^3$, with ℓ the new loop momentum.

Next we simplify the Dirac structure in the numerator. Applying the standard identity $\gamma^\alpha\gamma^\beta\gamma^\mu\gamma^\delta\gamma_\alpha = -2\gamma^\delta\gamma^\mu\gamma^\beta + (4-d)\gamma^\beta\gamma^\mu\gamma^\delta$, and using that $\bar{u}(p_2,s_2)$ $\slashed{p}_2 = 0$ and $\slashed{p}_1 v(-p_1,s_1) = 0$ in the massless case, we get

$$\mathcal{N}^\mu \equiv \gamma^\alpha(\slashed{p}_2 - \slashed{k})\gamma^\mu(\slashed{p}_1 - \slashed{k})\gamma_\alpha \cong -2(\slashed{p}_1 - \slashed{k})\gamma^\mu(\slashed{p}_2 - \slashed{k}) + (4-d)\slashed{k}\gamma^\mu\slashed{k}, \tag{12.172}$$

where $\cong$ means that the equality is valid only when sandwiched between the spinors and under the loop integral. The loop momentum shift from above leads to (again using the spinors on the left and right)

$$\mathcal{N}^\mu \cong -2((1-x_1)\slashed{p}_1 - \slashed{\ell})\gamma^\mu((1-x_2)\slashed{p}_2 - \slashed{\ell}) + (4-d)(\slashed{\ell} + x_1\slashed{p}_1)\gamma^\mu(\slashed{\ell} + x_2\slashed{p}_2). \tag{12.173}$$

Reconsidering the denominator of the loop integral, we notice that the dependence is only quadratic in ℓ, such that in the numerator all terms linear in ℓ evaluate to zero when integrated over due to symmetry. Consequently

$$\mathcal{N}^\mu \cong (2-d)\slashed{\ell}\gamma^\mu\slashed{\ell} - 2(1-x_1)(1-x_2)\slashed{p}_1\gamma^\mu\slashed{p}_2 + (4-d)x_1x_2\slashed{p}_1\gamma^\mu\slashed{p}_2. \tag{12.174}$$

The two remaining Dirac structures can furthermore be rewritten as

$$\slashed{p}_1\gamma^\mu\slashed{p}_2 \cong -\gamma^\mu\slashed{p}_1\slashed{p}_2 \cong -2\gamma^\mu p_1\cdot p_2 = -\gamma^\mu Q^2, \tag{12.175}$$

and

$$\slashed{\ell}\gamma^\mu\slashed{\ell} = \gamma_\alpha\gamma^\mu\gamma_\beta\ell^\alpha\ell^\beta \cong \tfrac{1}{d}\gamma_\alpha\gamma^\mu\gamma_\beta\ell^2 g^{\alpha\beta} = \gamma^\mu\tfrac{2-d}{d}\ell^2, \tag{12.176}$$

where we used the identity $\gamma_\alpha\gamma^\mu\gamma^\alpha = (2-d)\gamma^\mu$ and the fact that due to the Lorentz structure of the loop integral we can rewrite $\ell^\alpha\ell^\beta \cong 1/d\ell^2 g^{\alpha\beta}$ under the integral sign.

Ultimately, we arrive at

$$\mathcal{N}^\mu \cong \gamma^\mu\left[\tfrac{(2-d)^2}{d}\ell^2 + Q^2(2(1-x_1)(1-x_2) - (4-d)x_1x_2)\right]. \tag{12.177}$$

The loop integral can now be solved by using the standard formulas

$$\tilde{\mu}^{2\varepsilon}\int\frac{d^d\ell}{(2\pi)^d}\frac{1}{(\ell^2-x_1x_2Q^2)^3}=-\frac{i\tilde{\mu}^{2\varepsilon}}{(4\pi)^{d/2}}\frac{\Gamma(3-d/2)}{\Gamma(3)}(x_1x_2Q^2)^{d/2-3},\tag{12.178}$$

$$\tilde{\mu}^{2\varepsilon}\int\frac{d^d\ell}{(2\pi)^d}\frac{\ell^2}{(\ell^2-x_1x_2Q^2)^3}=\frac{i\tilde{\mu}^{2\varepsilon}}{(4\pi)^{d/2}}\frac{\Gamma(1+d/2)\Gamma(2-d/2)}{\Gamma(d/2)\Gamma(3)}(x_1x_2Q^2)^{d/2-2},\tag{12.179}$$

and subsequently the Feynman parameter integrals by using

$$\int_0^1 dx_1\int_0^{1-x_1}dx_2\,x_1^a x_2^b=\frac{\Gamma(1+a)\Gamma(1+b)}{\Gamma(3+a+b)}.\tag{12.180}$$

After putting everything together and a few simplifications we can write the final result as

$$\mathcal{A}^\mu=\bar{u}(p_1,s_1)\gamma^\mu v(p_2,s_2)\frac{\alpha_s C_F}{4\pi}\frac{2^{-1+4\varepsilon}\pi^{1/2+\varepsilon}(\varepsilon(1-2\varepsilon)-2)\Gamma(1-\varepsilon)\Gamma(\varepsilon)}{\varepsilon\Gamma(3/2-\varepsilon)}\left(\frac{\tilde{\mu}^2}{Q^2}\right)^\varepsilon\tag{12.181}$$

$$=\bar{u}(p_1,s_1)\gamma^\mu v(p_2,s_2)\frac{\alpha_s C_F}{4\pi}$$
$$\left[-\frac{2}{\varepsilon^2}+\frac{1}{\varepsilon}\left(-2\log\frac{\mu^2}{Q^2}-3\right)-8+\frac{\pi^2}{6}-3\log\frac{\mu^2}{Q^2}-\log^2\frac{\mu^2}{Q^2}+\mathcal{O}(\varepsilon)\right],\tag{12.182}$$

where $d=4-2\varepsilon$ and $\tilde{\mu}^{2\varepsilon}=e^{\varepsilon\gamma_E}(4\pi)^{-\varepsilon}\mu^{2\varepsilon}$.

Dropping the $1/\varepsilon^n$ terms (and the spinor structure) we obtain the SCET matching coefficient for the vector current

$$C(\mu^2,Q^2)=1+\frac{\alpha_s C_F}{4\pi}\left[-8+\frac{\pi^2}{6}-3\log\frac{\mu^2}{Q^2}-\log^2\frac{\mu^2}{Q^2}\right]+\mathcal{O}(\alpha_s^2).\tag{12.183}$$

Exercise 2.20 Compute the SCET matching for time-like q^2, by analytically continuing the previous result. Be careful about the sign of the imaginary parts.

SOLUTION:

Remembering that $Q^2\equiv-q^2-i0^+$ and $\lim_{\epsilon\to0}\log(-1+i\epsilon)=\mathrm{sgn}\,(\epsilon)i\pi$ we can rewrite

$$\log\frac{\mu^2}{Q^2}=\log\frac{\mu^2}{-q^2-i0^+}=\log\frac{\mu^2}{q^2}+i\pi,\tag{12.184}$$

resulting in

$$C(\mu^2,q^2)=1+\frac{\alpha_s C_F}{4\pi}\left[-8+\frac{7\pi^2}{6}-3\log\frac{\mu^2}{q^2}-\log^2\frac{\mu^2}{q^2}-3i\pi-2i\pi\log\frac{\mu^2}{q^2}\right],\tag{12.185}$$

for the SCET matching coefficient and

$$H(\mu^2,q^2)=|C(\mu^2,q^2)|^2=1+\frac{\alpha_s C_F}{4\pi}\left[-16+\frac{7\pi^2}{3}-6\log\frac{\mu^2}{q^2}-2\log^2\frac{\mu^2}{q^2}\right],\tag{12.186}$$

for the 'hard function' appearing in SCET factorization theorems.

Exercise 2.21 Compute the anomalous dimension-mixing matrix in eqn (L2.21),

$$O_1 = (\bar{b}^\alpha \gamma^\mu P_L c_\alpha)(\bar{u}^\alpha \gamma^\mu P_L d_\alpha) \qquad O_2 = (\bar{b}^\alpha \gamma^\mu P_L c_\beta)(\bar{u}^\beta \gamma^\mu P_L d_\alpha)$$

$$\mu \frac{d}{d\mu} \begin{bmatrix} c_1 \\ c_2 \end{bmatrix} = \begin{bmatrix} \gamma_{11} & \gamma_{12} \\ \gamma_{21} & \gamma_{22} \end{bmatrix} \begin{bmatrix} c_1 \\ c_2 \end{bmatrix}.$$

Two other often-used bases are

$$Q_1 = (\bar{b}\gamma^\mu P_L c)(\bar{u}\gamma^\mu P_L d), \qquad Q_2 = (\bar{b}\gamma^\mu P_L T^A c)(\bar{u}\gamma^\mu P_L T^A d),$$

and

$$O_\pm = O_1 \pm O_2,$$

So let

$$\mathcal{L} = c_1 O_1 + c_2 O_2 = d_1 Q_1 + d_2 Q_2 = c_+ O + + c_- O_-$$

and work out the transformation between the anomalous dimensions for $d_{1,2}$ and $c_{+,-}$ in terms of those for $c_{1,2}$,

SOLUTION:

In the EFT, the following diagrams contribute to the renormalization of the four-fermion operators,

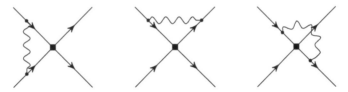

we compute these in the Feynman gauge, and set the external quark masses (IR scales) to zero.

We find,

$$\langle \mathcal{O}_1 \rangle_{1-\text{loop}} = \left[1 + \frac{\alpha_s}{4\pi}\left(2C_F + \frac{3}{N_c}\right)\left(\frac{1}{\epsilon} + \log\frac{\mu^2}{-p^2}\right)\right]\langle \mathcal{O}_1 \rangle_{\text{tree}} - \frac{3\alpha_s}{4\pi}\left(\frac{1}{\epsilon} + \log\frac{\mu^2}{-p^2}\right)\langle \mathcal{O}_2 \rangle_{\text{tree}}$$

$$(12.187)$$

where the angled brakcets denote the matrix element $\langle \bar{u}dc|\mathcal{O}|b\rangle$. The same expression holds for $\langle \mathcal{O}_2 \rangle_{1-\text{loop}}$, with $\mathcal{O}_1 \leftrightarrow \mathcal{O}_2$. If $\mathcal{L}$ has $c_1 O_1 + c_2 O_2$, one needs to add a counterterm

$$\mathcal{L}_{\text{ct}} = \frac{\alpha_s}{4\pi\epsilon} c_1 \left[-\left(2C_F + \frac{3}{N_c}\right)O_1 + 3O_2\right] + 1 \leftrightarrow 2. \qquad (12.188)$$

The wave function renormalization in $\overline{\text{MS}}$ is (see eqn 12.71)

$$Z_\psi = 1 - \frac{\alpha_s C_F}{4\pi\epsilon}, \qquad (12.189)$$

and hence

$$c_i^{\text{bare}} = Z_{ij} c_j \qquad\qquad Z_{ij} = \mathbb{1}_{ij} - \frac{\alpha_s}{4\pi\epsilon} \begin{pmatrix} 3/N_c & -3 \\ -3 & 3/N_c \end{pmatrix}. \tag{12.190}$$

Then using the fact that

$$\mu \tfrac{d}{d\mu} \alpha_s(\mu) = -2\epsilon\alpha_s + \beta_{\alpha_s} \tag{12.191}$$

where β_{α_s} is finite as $\epsilon \to 0$, we find the anomalous dimension,

$$\mu \tfrac{\text{d}}{\text{d}\mu} c_i = \gamma_{ij} c_j \qquad\qquad \gamma = -Z^{-1}\mu\tfrac{\text{d}}{\text{d}\mu} Z \tag{12.192}$$

so that

$$\gamma = \frac{\alpha_s}{4\pi} \begin{pmatrix} -6/N_c & 6 \\ 6 & -6/N_c \end{pmatrix}. \tag{12.193}$$

Transforming to the basis $\mathcal{O}_\pm = \mathcal{O}_1 \pm \mathcal{O}_2$ diagonalizes this matrix,

$$\gamma^{(+,-)} = \frac{\alpha_s}{4\pi} 6 \begin{pmatrix} 1 - 1/N_c & 0 \\ 0 & -1 - 1/N_c \end{pmatrix}. \tag{12.194}$$

For the operators Q_1, Q_2, they are related to the previous operators by colour and spin Fierz identities (see Exercise 2.3)

$$Q_1 = \mathcal{O}_1, \qquad Q_2 = \left(\bar{b}^\alpha \gamma^\mu P_L T^A_{\alpha\beta} c^\beta\right)\left(\bar{u}^\mu \gamma^\mu P_L T^A_{\mu\nu} d^\nu\right) = \tfrac{1}{2}\mathcal{O}_2 - \tfrac{1}{2N_c}\mathcal{O}_1 \tag{12.195}$$

and so we have,

$$\begin{pmatrix} Q_1 \\ Q_2 \end{pmatrix} = \begin{pmatrix} 1 & 0 \\ -1/2N_c & 1/2 \end{pmatrix} \begin{pmatrix} \mathcal{O}_1 \\ \mathcal{O}_2 \end{pmatrix} := M_{ij} O_j \tag{12.196}$$

and so the anomalous dimension matrix becomes,

$$\gamma^{(d)} = (M^T)^{-1} \gamma^{(c)} M^T$$

$$= \frac{\alpha_s}{4\pi} \begin{pmatrix} 0 & 3 - \frac{3}{N_c^2} \\ 12 & -\frac{12}{N_c} \end{pmatrix}. \tag{12.197}$$

Exercise 2.22 The classical equation of motion for $\lambda\phi^4$ theory,

$$L = \tfrac{1}{2}(\partial_\mu\phi)^2 - \tfrac{1}{2}m^2\phi^2 - \tfrac{\lambda}{4!}\phi^4,$$

is

$$E[\phi] = (-\partial^2 - m^2)\phi - \tfrac{\lambda}{3!}\phi^3.$$

The EOM Ward identity for $\theta = F[\phi]E$ is eqn (L2.175). Integrate both sides with

$$\int \text{d}x \, e^{-iq\cdot x} \prod_i \int \text{d}x_i \, e^{-ip_i\cdot x_i}$$

to get the momentum space version of the Ward identity

$$\langle 0|T\{\tilde{\phi}(p_1)\dots\tilde{\phi}(p_n)\tilde{\theta}(q)\}|0|0|T\{\tilde{\phi}(p_1)\dots\tilde{\phi}(p_n)\tilde{\theta}(q)\}|0\rangle \tag{12.198}$$

$$= i\sum_{r=1}^{n}\langle 0|T\{\tilde{\phi}(p_1)\dots\cancel{\tilde{\phi}(p_r)}\dots\tilde{\phi}(p_n)\tilde{F}(q+p_r)\}|0|0|T$$

$$\{\tilde{\phi}(p_1)\dots\cancel{\tilde{\phi}(p_r)}\dots\tilde{\phi}(p_n)\tilde{F}(q+p_r)\}|0\rangle.$$

(a) Consider the equation of motion operator

$$\theta_1 = \phi E[\phi] = \phi(-\partial^2 - m^2)\phi - \tfrac{\lambda}{3!}\phi^4,$$

and verify the Ward identity by explicit calculation at order λ (i.e. tree level) for $\phi\phi$ scattering, i.e. for $\phi\phi \to \phi\phi$.

(b) Take the on-shell limit $p_r^2 \to m^2$ at fixed $q \neq 0$ of

$$\prod_r (-i)(p_r^2 - m^2) \times \text{Ward identity},$$

and verify that both sides of the Ward identity vanish. Note that both sides do not vanish if we first take $q = 0$ and then takes the on-shell limit.

(c) Check the Ward identity to one loop for the equation of motion operator

$$\theta_2 = \phi^3 E[\phi] = \phi^3(-\partial^2 - m^2)\phi - \tfrac{\lambda}{3!}\phi^6.$$

SOLUTION:

(a) Here $F[\phi] = \phi$ and $\theta_1 = \phi E[\phi]$ in the Ward identity eqn (12.198). The left-hand side of the Ward identity is given by the matrix element

$$\langle 0|\tilde{\phi}(p_1)\tilde{\phi}(p_2)\tilde{\phi}(p_3)\tilde{\phi}(p_4)\theta_1(q)|0\rangle \tag{12.199}$$

with all momenta incoming so that $\sum_i p_i + q = 0$.

The graphs in Figure 12.6 give

$$\prod_i \Delta(p_i)\left[-4\lambda + \sum_k [p_k^2 - m^2 + (p_k+q)^2 - m^2]\frac{i}{(p_k+q)^2-m^2}(-i\lambda)\right]$$

$$= \prod_i \Delta(p_i)\lambda\left[\sum_k \frac{p_k^2-m^2}{(p_k+q)^2-m^2}\right] \tag{12.200}$$

since the $-\lambda\phi^3/3!$ and $\phi(-\partial^2 - m^2)\phi$ vertices with momentum q incoming have Feynman rules -4λ and $p^2 - m^2 + (p+q)^2 - m^2$. Here

$$\Delta(p) = \frac{i}{p^2-m^2} \tag{12.201}$$

Figure 12.6 *Tree-level matrix element of θ_1. The square box is the insertion of θ_1, and the grey circle is the $-\lambda\phi^4/4!$ vertex. One has to sum over all possible insertions of θ_1.*

Figure 12.7 *Tree-level matrix element of $\widetilde{\phi}(p_2)\widetilde{\phi}(p_3)\widetilde{\phi}(p_4)\widetilde{\phi}(p_1+q)$.*

Figure 12.8 *Tree-level matrix element of θ_2.*

is the scalar propagator. Note that part of the $\phi(-\partial^2 - m^2)\phi)$ insertion has cancelled the $-\lambda\phi^3/3!$ insertion.

The right-hand side of eqn (12.198) is given by Figure 12.7

$$i\langle 0|\widetilde{\phi}(p_2)\widetilde{\phi}(p_3)\widetilde{\phi}(p_4)\widetilde{\phi}(p_1+q)|0\rangle + \dots$$
$$= i\Delta(p_2)\Delta(p_3)\Delta(p_4)\frac{i}{(p_1+q)^2-m^2}(-i\lambda)$$
$$= \lambda\prod_i\Delta(p_i)\left[\frac{(p_1^2-m^2)}{(p_1+q)^2-m^2}+\dots\right] = \lambda\prod_i\Delta(p_i)\sum_k\left[\frac{(p_k^2-m^2)}{(p_k+q)^2-m^2}\right], \qquad (12.202)$$

which agrees with eqn (12.200), so the Ward identity is satisfied.

(b) Truncating the external legs, and taking the limit $p_i \to m^2$ at fixed q gives zero, since the numerators vanish. However, if we first set $q = 0$, then the numerator and denominator cancel, and one gets 4λ.

(c) Here $F[\phi] = \phi^3$ in the Ward identity eqn (12.198). The tree-level contribution to the $\phi\phi\phi\theta_2$ amplitude is show in Figure 12.8.

$$I = \sum_i 3!(p_i^2 - m^2) \tag{12.203}$$

omitting the four external propagators.

The right-hand side of the Ward identity is given by the matrix element in Figure 12.9,

$$\mathcal{J} = i \sum_i 3!(-i)(p_i^2 - m^2) \tag{12.204}$$

where the factor $(-i)(p_i^2 - m^2)$ arises to cancel the missing propagator in the diagram, again dropping the four external propagators.

Eqn. (12.203, 12.204) agree, so the Ward identity is satisfied. Multiplying by $p_i^2 \to m^2$ cancels the external propagators, and taking the limit $p_i^2 \to m^2$ gives zero on both sides.

The order λ^2 corrections to the left-hand side of the Ward identity are shown in Figure 12.10.

To evaluate the first graph, we need the one-loop graph from the bubble in the first graph of Figure 12.10,

$$I_m = \int \frac{i}{k^2 - m^2} + \text{c.t.} = -\frac{1}{16\pi^2} m^2 \left[1 - \log \frac{m^2}{\mu^2} \right] \tag{12.205}$$

which contributes to the one-loop mass shift

$$\delta m^2 = \tfrac{1}{2} \lambda I_m \tag{12.206}$$

of the scalar field. The first graph has an insertion of $\phi^3(-\partial^2 - m^2)\phi$. If the derivative acts on the internal line between the two vertices, it cancels the internal

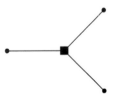

Figure 12.9 *Tree-level matrix element of the right-hand side of the Ward identity eqn (12.198) with $F = \phi^3$.*

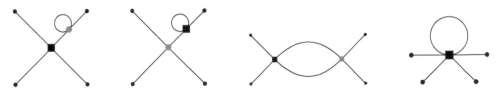

Figure 12.10 *One-loop matrix element of the left-hand side of the Ward identity eqn (12.198) with $F = \phi^3$. The square box is the insertion of θ_2, and the grey circle is the $-\lambda\phi^4/4!$ vertex.*

propagator. Otherwise, it produces a factor of $(p^2 - m^2)4$ on one of the external lines. The total is (again dropping the external propagators $\prod_i \Delta(p_i)$)

$$I_1 = \tfrac{3!}{2}\lambda I_m \times 4 + \tfrac{3!}{2}\lambda I_m \left\{ (p_1^2 - m^2) \left[\tfrac{1}{p_2^2 - m^2} + \tfrac{1}{p_3^2 - m^2} + \tfrac{1}{p_4^2 - m^2} \right] + \ldots \right\} \qquad (12.207)$$

where $\ldots$ is the sum over $1 \leftrightarrow 2$, $1 \leftrightarrow 3$, $1 \leftrightarrow 4$. The second graph has contributions where $p^2 - m^2$ cancels the loop propagator, which gives zero, and where $p^2 - m^2$ acts on one of the other lines. The λ from the $\lambda\phi^4$ vertex combines with the loop graph to give δm^2,

$$I_2 = \tfrac{3!}{2}\lambda I_m \left[1 + \tfrac{p_1^2 - m^2}{(p_1+q)^2 - m^2} \right] + \ldots = \tfrac{3!}{2}\lambda I_m \left[4 + \sum_i \tfrac{p_i^2 - m^2}{(p_i+q)^2 - m^2} \right]. \qquad (12.208)$$

The third graph can have $(-\partial^2 - m^2)$ acting on the internal lines, which do not satisfy $p^2 = m^2$. One has to sum over the s, t, and u channel diagrams. The basic loop integral for $\phi - \phi$ scattering is

$$G(p) = -i \int \tfrac{1}{[k^2 - m^2][(k+p)^2 - m^2]} + \text{c.t.} \qquad (12.209)$$

The s-channel graphs add to

$$I_{3s} = 3! \left\{ \tfrac{1}{2}\lambda G(p_3 + p_4)[p_1^2 - m^2 + p_2^2 - m^2] + \tfrac{1}{2}\lambda G(p_1 + p_2)[p_3^2 - m^2 + p_4^2 - m^2] + 2\lambda I_m \right\} \qquad (12.210)$$

The last term arises from $(-\partial^2 - m^2)$ acting on the internal lines, which converts the G integral eqn (12.209) into the mass integral eqn (12.205). The t and u channel results are given by $2 \leftrightarrow 3$ and $2 \leftrightarrow 4$. The fourth graph gives

$$I_4 = -\tfrac{6!}{3!}\tfrac{1}{2}\lambda I_m \qquad (12.211)$$

The right-hand side of the Ward identity is given by the $\phi\phi\phi^3$ correlator, with diagrams shown in Figure 12.11.

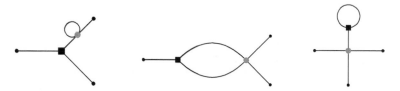

Figure 12.11 *One-loop matrix element of the right-hand side of the Ward identity eqn (12.198) with $F = \phi^3$. The square box is the insertion of ϕ^3, and the grey circle is the $-\lambda\phi^4/4!$ vertex.*

The graphs give

$$
\begin{aligned}
\mathcal{J}_1 &= \tfrac{3!}{2}\lambda I_m \left\{ \tfrac{1}{\Delta(p_1)}[\Delta(p_2)+\Delta(p_3)+\Delta(p_4)] + \dots \right\} \\
\mathcal{J}_{2s} &= \lambda \tfrac{3!}{2} \left\{ G(p_3+p_4)\left[\tfrac{1}{\Delta(p_1)}+\tfrac{1}{\Delta(p_2)}\right] + G(p_1+p_2)\left[\tfrac{1}{\Delta(p_3)}+\tfrac{1}{\Delta(p_4)}\right] \right\} \\
\mathcal{J}_3 &= \tfrac{3!}{2}\lambda I_m \left\{ \tfrac{\Delta(p_1+q)}{\Delta(p_1)} + \dots \right\}.
\end{aligned}
\tag{12.212}
$$

The Ward identity eqn (12.198) is satisfied, since $I_1 + I_2 + I_{3s} + I_{3t} + I_{3u} = \mathcal{J}_1 + \mathcal{J}_{2s} + \mathcal{J}_{2t} + \mathcal{J}_{2u} + \mathcal{J}_3$. Note that there is a non-trivial cancellation of the I_m terms between I_1, I_2, I_3.

Exercise 2.23 Write down all possible C-even dimension six terms in eqn (L2.42), and show how they can be eliminated by field redefinitions.

SOLUTION:

The C-even terms have an even number of field strength tensors, so the dimensions six terms have two field-strength tensors and two derivatives. The possible terms are

$$
\partial_\alpha F_{\mu\nu} \partial^\alpha F^{\mu\nu}, \partial^2 F_{\mu\nu} F^{\mu\nu}, \partial^\alpha F_{\alpha\mu} \partial_\beta F^{\beta\mu}.
\tag{12.213}
$$

The second term reduces to the first one on integration by parts. Also, using

$$
\partial_\alpha F_{\mu\nu} + \partial_\mu F_{\nu\alpha} + \partial_\nu F_{\alpha\mu} = 0
\tag{12.214}
$$

gives

$$
\partial_\alpha F_{\mu\nu} \partial^\alpha F^{\mu\nu} = (\partial_\mu F_{\nu\alpha} + \partial_\nu F_{\alpha\mu})^2 \implies \partial_\alpha F_{\mu\nu} \partial^\alpha F^{\mu\nu} = 2\partial^\alpha F_{\alpha\mu} \partial_\beta F^{\beta\mu}.
\tag{12.215}
$$

Thus the Lagrangian to dimension-six is

$$
\mathcal{L} = -\tfrac{1}{4} F_{\mu\nu} F^{\mu n u} + \tfrac{c}{m_e^2} \partial^\alpha F_{\alpha\mu} \partial_\beta F^{\beta\mu}.
\tag{12.216}
$$

Making the field redefinition

$$
A_\mu \to A_\mu + \tfrac{1}{m_e^2} X_\mu
\tag{12.217}
$$

gives the Lagrangian

$$
\mathcal{L} = -\tfrac{1}{4} F_{\mu\nu} F^{\mu n u} - \tfrac{1}{m_e^2} F_{\mu\nu} \partial_\mu X_\nu + \tfrac{c}{m_e^2} \partial^\alpha F_{\alpha\mu} \partial_\beta F^{\beta\mu} + \mathcal{O}\left(\tfrac{1}{m_e^4}\right).
\tag{12.218}
$$

Choosing $X_\mu = -c\partial^\alpha F_{\alpha\mu}$ and integrating by parts eliminates the $1/m_e^2$ terms.

Exercise 2.24 Take the heavy quark Lagrangian

$$\mathcal{L}_v = \bar{Q}_v \left\{ iv \cdot D + i\slashed{D}_\perp \frac{1}{2m + iv \cdot D} j\slashed{D}_\perp \right\} Q_v$$

$$= \bar{Q}_v \left\{ iv \cdot D - \frac{1}{2m} \slashed{D}_\perp \slashed{D}_\perp + \frac{1}{4m^2} \slashed{D}_\perp (iv \cdot D) \slashed{D}_\perp + \dots \right\} Q_v$$

and use a sequence of field redefinitions to eliminate the $1/m^2$ suppressed $v \cdot D$ term. The equation of motion for the heavy quark field is $(iv \cdot D)Q_v = 0$, so this example shows how to eliminate equation of motion operators in HQET. Here v^μ is the velocity vector of the heavy quark with $v \cdot v = 1$, and

$$D_\perp^\mu \equiv D^\mu - (v \cdot D)v^\mu.$$

If you prefer, you can work in the rest frame of the heavy quark, where $v^\mu = (1, 0, 0, 0)$, $v \cdot D = D^0$ and $D_\perp^\mu = (0, \mathbf{D})$. See [18] for help.

SOLUTION:
     The aim of the exercise is to show how operators can be eliminated by using the equations of motion/field redefinitions $(iv \cdot D)Q_v = 0$ and $\bar{Q}_v(iv \cdot D)Q_v = 0$. Indeed, these operators are not entirely eliminated, but sent to higher orders in the $1/m$ expansion. So, we must think about the field redefinitions as $(iv \cdot D)Q_v = \mathcal{O}(1/m)$ and $\bar{Q}_v(iv \cdot D)Q_v = \mathcal{O}(1/m)$. We will do the calculation in a general frame, even though we keep in mind that the field redefinition has to be applied in such a way that in the rest frame temporal derivatives are removed of the Lagrangian. That is important because blindly introducing the expression of $D_\perp$ in the $1/m$ term and using the equations of motion blindly, we could end up with a Lagrangian with time derivatives in the rest frame. In that case the equation of motion at order $1/m$ should be used. Let us start with the local HQET Lagrangian eqn (2) and introduce eqn (3) in it:

$$\mathcal{L}_v = \bar{Q}_v \left\{ iv \cdot D - \frac{1}{2m} \slashed{D}_\perp \slashed{D}_\perp + \frac{1}{4m^2} \slashed{D}_\perp (iv \cdot D) \slashed{D}_\perp + \mathcal{O}(1/m^3) \right\} Q_v$$

we will not write $\mathcal{O}(1/m^3)$ in the following, since it is understood that we are only interested in the Lagrangian up to $\mathcal{O}(1/m^2)$. Thus:

$$= \bar{Q}_v \left\{ iv \cdot D - \frac{1}{2m} \slashed{D}_\perp^2 + \frac{i}{4m^2} \slashed{D}_\perp (v \cdot D) \slashed{D}_\perp \right\} Q_v$$

$$= \bar{Q}_v \left\{ iv \cdot D - \frac{1}{4m}(\{\gamma^\mu, \gamma^\nu\} + [\gamma^\mu, \gamma^\nu]) D_{\perp,\mu} D_{\perp,\nu} + \frac{i}{8m^2}(\{\gamma^\mu, \gamma^\nu\} + [\gamma^\mu, \gamma^\nu]) D_{\perp\mu} (v \cdot D) D_{\perp\nu} \right\} Q_v$$

$$= \bar{Q}_v \left\{ iv \cdot D - \frac{1}{2m} D_\perp^2 + \frac{i}{2m} \sigma^{\mu\nu} D_{\perp,\mu} D_{\perp,\nu} + \frac{i}{4m^2} D_{\perp\mu} (v \cdot D) D_\perp^\mu + \frac{1}{4m^2} \sigma^{\mu\nu} D_{\perp\mu} (v \cdot D) D_{\perp\nu} \right\} Q_v \quad (12.219)$$

where we used $\{\gamma^\mu, \gamma^\nu\} = 2g^{\mu\nu} I_4$ and $[\gamma^\mu, \gamma^\nu] = -2i\sigma^{\mu\nu}$.

$$= \bar{Q}_v \left\{ iv \cdot D - \frac{1}{2m} D_\perp^2 + \frac{i}{2m} \sigma^{\mu\nu} D_{\perp,\mu} D_{\perp,\nu} + \frac{i}{4m^2} D_{\perp\mu} (v \cdot D) D_\perp^\mu + \frac{1}{4m^2} \sigma^{\mu\nu} D_{\perp\mu} (v \cdot D) D_{\perp\nu} \right\} Q_v \quad (12.220)$$

and

$$\mathcal{L}_v = \bar{Q}_v \left\{ iv \cdot D - \tfrac{1}{2m} D_\perp^2 + \tfrac{i}{4m} \sigma^{\mu\nu} [D_{\perp,\mu}, D_{\perp,\nu}] + \tfrac{i}{8m^2} [D_\perp^\mu, [v \cdot D, D_{\perp\mu}]] \right.$$
$$\left. + \tfrac{1}{8m^2} \sigma^{\mu\nu} \{D_{\perp\mu}, [v \cdot D, D_{\perp\nu}]\} \right\} Q_v \qquad (12.221)$$

where we added and subtracted $(v \cdot D) D_\perp^\mu$ or $D_\perp^\mu (v \cdot D)$ in order to make $v \cdot D$ to appear only inside commutators, and used the EOM conveniently. In this way, we ensure that, in the rest frame, temporal derivatives never act over heavy quark fields, and terms involving these time derivatives can be rewritten in terms of the chromoelectric and chromomagnetic fields.

In the rest frame, $v^\mu = (1, 0, 0, 0)$, $v \cdot D = D^0$ and $D_\perp^\mu = (0, -\mathbf{D})$, the Lagrangian takes the form:

$$\mathcal{L}_v = \bar{Q}_v \left\{ iD^0 + \tfrac{1}{2m} \mathbf{D}^2 + \tfrac{i}{4m} \sigma^{ij} [\mathbf{D}^i, \mathbf{D}^j] - \tfrac{i}{8m^2} [\mathbf{D}^i, [D^0, \mathbf{D}^i]] \right.$$
$$\left. + \tfrac{1}{8m^2} \sigma^{ij} \{\mathbf{D}^i, [D^0, \mathbf{D}^j]\} \right\} Q_v \qquad (12.222)$$

where $\mathbf{D}$ is the covariant derivative in Euclidean space with the metric $g = \mathrm{diag}(1, 1, 1)$, so we do not distinguish between up and down indices anymore. Recall that $G^{\mu\nu} = -\tfrac{i}{g} [D^\mu, D^\nu]$, so the chromoelectric and chromomagnetic fields are given by $\mathbf{E}^i = G^{i0} = \tfrac{i}{g} [\mathbf{D}^i, D^0]$ and $\mathbf{B}^i = -\tfrac{1}{2} \epsilon^{ijk} G^{jk} = \tfrac{i}{2g} \epsilon^{ijk} [\mathbf{D}^j, \mathbf{D}^k]$, respectively, whereas $\sigma^{ij} = \epsilon^{ijk} \sigma^k I_4$. Using all these relations the above Lagrangian can be written as:

$$\mathcal{L}_v = \bar{Q}_v \left\{ iD^0 + \tfrac{1}{2m} \mathbf{D}^2 + \tfrac{g}{2m} \sigma^i \mathbf{B}^i + \tfrac{g}{8m^2} [\mathbf{D}^i, \mathbf{E}^i] + \tfrac{ig}{8m^2} \epsilon^{ijk} \sigma^k \{\mathbf{D}^i, \mathbf{E}^j\} \right\} Q_v \qquad (12.223)$$

Exercise 2.25 Verify that the first term in eqn (L2.197) leads to the threshold correction in the gauge coupling given in eqn (L2.198). If one matches at $\bar{\mu} = m$, then $e_L(\bar{\mu}) = e_H(\bar{\mu})$, and the gauge coupling is continuous at the threshold. Continuity does not hold at higher loops, or when a heavy scalar is integrated out.

SOLUTION:

To simplify the algebra, absorb the coupling e into the gauge field, so that the covariant derivative is $D_\mu = \partial_\mu + ieA_\mu$. Then the gauge kinetic term above m is

$$\mathcal{L} = -\tfrac{1}{4e_H^2} F_{\mu\nu}^2. \qquad (12.224)$$

The one-loop fermion vacuum polarization graph is present in theory above m, and contributes as a correction to the Lagrangian below m. The matrix element of $(-1/4) F_{\mu\nu}^2$ in a photon state of momentum p is $-i(p^2 g_{\mu\nu} - p_\mu p_\nu)$, so eqn (L2.198) gives a shift in the gauge kinetic term, and the Lagrangian below m is

$$\mathcal{L} = -\tfrac{1}{4e_H^2} F_{\mu\nu}^2 - \tfrac{1}{2\pi^2} \tfrac{1}{6} \log \tfrac{m^2}{\bar{\mu}^2} \left(-\tfrac{1}{4} F_{\mu\nu}^2 \right) = -\tfrac{1}{4e_L^2} F_{\mu\nu}^2, \qquad (12.225)$$

where the vacuum polarization graph no longer has the factor of e^2 because of rescaling the gauge field. This gives

$$\frac{1}{e_L^2(\bar\mu)} = \frac{1}{e_H^2(\bar\mu)} - \frac{1}{12}\log\frac{m^2}{\bar\mu^2} \qquad (12.226)$$

Exercise 2.26 Assume the threshold correction is of the form

$$\frac{1}{e_L^2(\bar\mu)} = \frac{1}{e_H^2(\bar\mu)} + c\log\frac{m^2}{\bar\mu^2}.$$

Find the relation between c and the difference $\beta_H - \beta_L$ of the β-functions in the two theories, and check that this agrees with eqn (L2.198).

SOLUTION:
Differentiating both sides with regards to μ,

$$-\frac{2}{e_L^3}\beta_L(e_L) = -\frac{2}{e_H^3}\beta_H(e_H) - 2c \implies \beta_H - \beta_L = -ce^3 \qquad (12.227)$$

where $e = e_H = e_L$ to lowest order. In the example in the text, $\beta_H = e^3/(12\pi^2)$, $\beta_L = 0$ and $c = -1/12$.

Exercise 2.27 Show that the power counting formula eqn (L2.200) for an EFT Lagrangian is self-consistent, i.e. an arbitrary graph with insertions of vertices of this form generates an interaction that maintains the same form. (See [7] and [16]). Show that eqn (L2.200) is equivalent to

$$\hat{O} \sim \frac{\Lambda^4}{16\pi^2}\left[\frac{\partial}{\Lambda}\right]^{N_p}\left[\frac{4\pi\phi}{\Lambda}\right]^{N_\phi}\left[\frac{4\pi A}{\Lambda}\right]^{N_A}\left[\frac{4\pi\psi}{\Lambda^{3/2}}\right]^{N_\psi}\left[\frac{g}{4\pi}\right]^{N_g}\left[\frac{y}{4\pi}\right]^{N_y}\left[\frac{\lambda}{16\pi^2}\right]^{N_\lambda}.$$

SOLUTION:
For consistency, each field and derivative should be scaled by the same scale, and so the most general vertex could take the form,

$$L_{abcd}^{ABC} = \Lambda_x^4 \lambda^A g^B y^C \left(\frac{\partial}{\Lambda_\partial}\right)^d \left(\frac{\phi}{\Lambda_\phi}\right)^a \left(\frac{\psi}{\Lambda_\psi^{3/2}}\right)^b \left(\frac{A_\mu}{\Lambda_A}\right)^c. \qquad (12.228)$$

Consider a diagram made of V such vertices, with L loops,

$$\left[\int\frac{d^4k}{(2\pi)^4}\frac{1}{k^2}\right]^{I_\phi+I_A}\left[\int\frac{d^4k}{(2\pi)^4}\frac{1}{k}\right]^{I_\psi}[\Lambda_x^4(2\pi)^4\delta^4(p)]^{V-1}$$

$$\left(\frac{k}{\Lambda_\partial}\right)^{\Sigma_i d_i - d_E}\left(\frac{1}{\Lambda_\phi}\right)^{\Sigma_i a_i - a_E}\left(\frac{1}{\Lambda_\psi^{3/2}}\right)^{\Sigma_i b_i - b_E}\left(\frac{1}{\Lambda_A}\right)^{\Sigma_i c_i - c_E} L_{a_E b_E c_E d_E}^{\Sigma_i A_i, \Sigma_i B_i, \Sigma_i C_i}. \qquad (12.229)$$

We replace all of the momenta with a factor of Λ, associated with the cutoff of the EFT, and perform the loop integrals over the delta functions—this gives a factor of $\Lambda^4/16\pi^2$ per loop. By the conservation of ends, e.g. $\Sigma_i a_i - a_E = 2I_\phi$, and so we are left with,

$$\left(\frac{\Lambda^4}{16\pi^2}\right)^L (\Lambda^4)^{V-1+I_\phi+I_A+I_\psi} \left(\frac{\Lambda_x}{\Lambda}\right)^{4V-4} \left(\frac{\Lambda}{\Lambda_\partial}\right)^{\Sigma_i d_i - d_E} \left(\frac{\Lambda}{\Lambda_\phi}\right)^{2I_\phi} \left(\frac{\Lambda}{\Lambda_\psi}\right)^{3I_\psi} \left(\frac{\Lambda}{\Lambda_A}\right)^{2I_A}$$

$$\times L_{a_E b_E c_E d_E}^{\Sigma_i A_i, \Sigma_i B_i, \Sigma_i C_i}. \tag{12.230}$$

Using $V - I + L = 1$ for a connected graph, we find that,

$$(4\pi)^{-2L} \left(\frac{\Lambda_x}{\Lambda}\right)^{4V-4} \left(\frac{\Lambda}{\Lambda_\partial}\right)^{\Sigma_i d_i - d_E} \left(\frac{\Lambda}{\Lambda_\phi}\right)^{2I_\phi} \left(\frac{\Lambda}{\Lambda_\psi}\right)^{3I_\psi} \left(\frac{\Lambda}{\Lambda_A}\right)^{2I_A} L_{a_E b_E c_E d_E}^{\Sigma_i A_i, \Sigma_i B_i, \Sigma_i C_i}. \tag{12.231}$$

If every diagram is to give a correction to the tree-level L, which respects our various power counting rules, then we require

$$\left(\frac{4\pi\Lambda_x^2}{\Lambda^2}\right)^{2V-2} \left(\frac{\Lambda}{\Lambda_\partial}\right)^{\Sigma_i d_i - d_E} \left(\frac{\Lambda}{4\pi\Lambda_\phi}\right)^{2I_\phi} \left(\frac{\Lambda^{3/2}}{4\pi\Lambda_\psi^{3/2}}\right)^{2I_\psi} \left(\frac{\Lambda}{4\pi\Lambda_A}\right)^{2I_A} \leq 1 \tag{12.232}$$

for every $V, I_\phi, I_\psi, I_A, d_E$.

Further, note that this counting gives kinetic terms,

$$\frac{\Lambda_x^4}{\Lambda^2 \Lambda_\phi^2} (\partial_\mu \phi)^2, \quad \frac{\Lambda_x^4}{\Lambda \Lambda_\psi^3} \bar{\psi} \partial\!\!\!/ \psi, \quad \frac{\Lambda_x^4}{\Lambda^2 \Lambda_A^2} (F_{\mu\nu})^2 \tag{12.233}$$

and so maintaining a canonical normalization requires,

$$\frac{\Lambda_x^4}{\Lambda^2 \Lambda_\phi^2}, \quad \frac{\Lambda_x^4}{\Lambda \Lambda_\psi^3}, \quad \frac{\Lambda_x^4}{\Lambda^2 \Lambda_A^2} \leq 1,$$

which is guaranteed by the above diagram condition.

Finally, note that for the renormalizable operators, saturating these bounds gives

$$\frac{\Lambda_x^4}{\Lambda_\phi^4} \phi^4, \quad \frac{\Lambda_x^4}{\Lambda_\phi \Lambda_\psi^3} \phi \bar{\psi} \psi, \quad \frac{\Lambda_x^4}{\Lambda_A \Lambda_\psi^3} i \bar{\psi} A\!\!\!/ \psi,$$

and so to ensure order unity coefficients we must use the couplings $g' = g/4\pi, y' = y/4\pi$ and $\lambda' = \lambda/16\pi^2$. Altogether then, we have a consistent power counting of,

$$L_{abcd}^{ABC} = \frac{\mu^4}{16\pi^2} \left(\frac{\lambda}{16\pi^2}\right)^A \left(\frac{g}{4\pi}\right)^B \left(\frac{y}{4\pi}\right)^C \left(\frac{\partial}{\Lambda_\partial}\right)^d \left(\frac{4\pi\phi}{\Lambda_\phi}\right)^a \left(\frac{4\pi\psi}{\Lambda_\psi^{3/2}}\right)^b \left(\frac{4\pi A_\mu}{\Lambda_A}\right)^c \tag{12.234}$$

where all the Λ_i are greater than (or equal to) Λ, and the overall scale μ is less than (or equal to) Λ, where Λ is the EFT cut-off (the scale up to which we can safely integrate loop momenta). In practice, given an EFT we can determine the cut-off in terms of Λ_i by computing amplitudes and checking for the first breakdown of perturbative unitarity.

The most natural thing is to scale $\Lambda_x \Lambda_\partial = 1$, and $\Lambda_\partial = \Lambda_{\text{EFT}}$.

Exercise 2.28 Show (by explicit calculation) for a general 2×2 matrix A that

$$\tfrac{1}{6}\langle A \rangle^3 - \tfrac{1}{2}\langle A \rangle \langle A^2 \rangle + \tfrac{1}{3}\langle A^3 \rangle = 0, \quad \tfrac{1}{2}\langle A \rangle^2 - \tfrac{1}{2}\langle A^2 \rangle - \langle A \rangle A + A^2 = 0,$$

and for general 2×2 matrices A, B, C that

$$0 = \langle A \rangle \langle B \rangle \langle C \rangle - \langle A \rangle \langle BC \rangle - \langle B \rangle \langle AC \rangle - \langle C \rangle \langle AB \rangle + \langle ABC \rangle + \langle ACB \rangle$$

Identities analogous to this for 3×3 matrices are used to remove L_0 and replace it by $L_{1,2,3}$ in χPT, as discussed in Chapter 3.

SOLUTION:
For a 2×2 matrix,

$$A = \begin{pmatrix} a & b \\ c & d \end{pmatrix}$$

we have that,

$$\langle A \rangle^3 = (a+d)^3, \quad \langle A^2 \rangle = a^2 + d^2 + 2bc, \quad \langle A^3 \rangle = a^3 + d^3 + 3bc(a+d)$$
$$\implies 0 = \langle A \rangle^3 - 3\langle A \rangle \langle A^2 \rangle + 2\langle A^3 \rangle \; \checkmark \tag{12.235}$$

and that,

$$\langle A \rangle A = \begin{pmatrix} a(a+d) & b(a+d) \\ c(a+d) & d(a+d) \end{pmatrix}, \quad A^2 = \begin{pmatrix} a^2 + bc & b(a+d) \\ c(a+d) & d^2 + bc \end{pmatrix},$$
$$\tfrac{1}{2}(\langle A \rangle^2 - \langle A^2 \rangle)\mathbb{1} = \begin{pmatrix} ad - bc & 0 \\ 0 & ad - bc \end{pmatrix} \tag{12.236}$$
$$\implies 0 = \tfrac{1}{2}(\langle A \rangle^2 - \langle A^2 \rangle) - \langle A \rangle A + A^2 \; \checkmark$$

$$\langle A \rangle \langle B \rangle \langle C \rangle = (A_{11} + A_{22})(B_{11} + B_{22})(C_{11} + C_{22})$$
$$\langle AB \rangle \langle C \rangle = (C_{11} + C_{22})(A_{11}B_{11} + A_{21}B_{12} + A_{12}B_{21} + A_{22}B_{22})$$
$$\langle ABC \rangle = C_{12}(A_{21}B_{11} + A_{22}B_{21}) + A_{11}(B_{11}C_{11} + B_{12}C_{21})$$
$$+ A_{12}(B_{21}C_{11} + B_{22}C_{21}) + C_{22}(A_{21}B_{12} + A_{22}B_{22})$$

$$\implies 0 = \langle A \rangle \langle B \rangle \langle C \rangle - 3\langle A \rangle \langle BC \rangle + \langle ABC \rangle + \langle ACB \rangle + (A \leftrightarrow B) + (A \leftrightarrow C) \; \checkmark$$

Exercise 2.29 Show that the Jarlskog invariant

$$\mathcal{J} = \langle X_u^2 X_d^2 X_u X_d \rangle - \langle X_d^2 X_u^2 X_d X_u \rangle,$$

is the lowest order CP-odd invariant made of the quark mass matrices. Here,

$$X_u \equiv M_u M_u^\dagger, \quad X_d \equiv M_d M_d^\dagger, \quad M_u \to L M_u R_u^\dagger, \quad M_d \to L M_d R_d^\dagger$$

Show that $\mathcal{J}$ can also be written in the form

$$\mathcal{J} = \tfrac{1}{3}\langle [X_u, X_d]^3 \rangle,$$

and explicitly work out $\mathcal{J}$ in the SM using the CKM matrix convention of the PDG [22]. Verify eqns (L2.234, L2.235).

SOLUTION:

Under a change of basis in flavour space (bi-unitary transformation), M_u and M_d transform as follows:

$$M_u \rightarrow LM_uR_u^\dagger$$
$$M_d \rightarrow LM_dR_d^\dagger$$

where L, R_u and R_d are 3×3 unitary matrices. Let us try to construct a basis-invariant quantity from the quarks mass matrices. Because $R_u \neq R_d$, we must construct the Hermitian matrices $X_u \equiv M_uM_u^\dagger$ and $X_d \equiv M_dM_d^\dagger$, which transform as $U(3)$ octets:

$$X_u \rightarrow LX_uL^\dagger$$
$$X_d \rightarrow LX_dL^\dagger.$$

Any polynomial $p(X_u, X_d)$ also transforms as an octet:

$$p(X_u, X_d) \rightarrow Lp(X_u, X_d)L^\dagger.$$

In order to eliminate the remaining L and $L^\dagger$ to get an invariant, we take the trace of $p(X_u, X_d)$. Indeed, because the trace is cyclic $\langle AB \rangle = \langle BA \rangle$ and $L \in U(3)(LL^\dagger = \mathbb{I}_3)$, we have:

$$\langle p(X_u, X_d) \rangle \rightarrow \langle Lp(X_u, X_d)L^\dagger \rangle = \langle p(X_u, X_d)\mathbb{I}_3 \rangle = \langle p(X_u, X_d) \rangle.$$

Therefore, the trace of any polynomial of the Hermitian matrices X_u and X_d is a flavour invariant.

However, we want a CP-odd invariant. So, we must select only purely imaginary traces. In other words, we must find an invariant such that:

$$\mathrm{Re}\langle p(X_u, X_d) \rangle = 0$$
$$\mathrm{Im}\langle p(X_u, X_d) \rangle \neq 0.$$

Let us examine the simplest monomials, the powers of $X_{u,d}$ of the form X_u^n or X_d^n, the products of two powers of $X_{u,d}$ of the form $X_u^n X_d^m$, the products of three powers of $X_{u,d}$ of the form $X_u^n X_d^m X_u^p$, and so on... where n, m, and p are non-zero integers.

We have $(X_u^n)^\dagger = X_u^n$ and $(X_d^n)^\dagger = X_d^n$, so X_u^n and X_d^n are Hermitian matrices and then their diagonal entries are real and so is their traces. Thus, the corresponding invariants $\langle X_u^n \rangle$ and $\langle X_d^n \rangle$ are real and then CP-even.

Regarding the monomials which are products of two powers of $X_{u,d}$, of the form $X_{u,d}^n X_{d,u}^m$, let us calculate the imaginary part of the corresponding invariants, by using $\mathrm{Im}\langle M \rangle = \frac{1}{2i}\langle M - M^\dagger \rangle$:

$$2i\mathrm{Im}\langle X_u^n Xd\rangle = \left\langle X_u^n X_d^m - (X_u^n X_d^m)^\dagger\right\rangle$$
$$= \left\langle X_u^n X_d^m - X_d^m X_u^n\right\rangle$$
$$= \left\langle X_u^n X_d^m\right\rangle - \left\langle X_d^m X_u^n\right\rangle$$
$$= \left\langle X_u^n X_d^m\right\rangle - \left\langle X_u^n X_d^m\right\rangle = 0.$$

The same holds for $\langle X_d^n X_u^m\rangle$. Therefore, the invariants of the form $\langle X_u^n X_d^m\rangle$ and $\langle X_d^n X_u^m\rangle$ are real and then CP-even.

Let us now consider the monomials which are products of three powers of $X_{u,d}$, of the form $\langle X_{u,d}^n X_{d,u}^m X_{u,d}^p\rangle$:

$$2i\mathrm{Im}\langle X_u^n X_d^m X_u^p\rangle = \langle X_u^n X_d^m X_u^p - (X_u^n X_d^m X_u^p)^\dagger\rangle$$
$$= \langle X_u^n X_d^m X_u^p - X_u^p X_d^m X_u^n\rangle$$
$$= \langle X_u^n X_d^m X_u^p\rangle - \langle X_u^p X_d^m X_u^n\rangle$$
$$= \langle X_u^{n+p} X_d^m\rangle - \langle X_u^{p+n} X_d^m\rangle = 0.$$

The same holds for $\langle X_d^n X_u^m X_d^p\rangle$. So, these invariants of the form $\langle X_u^n X_d^m X_u^p\rangle$ and $\langle X_d^n X_u^m X_d^p\rangle$ are also real and then CP-even.

Next, we consider the monomials which are products of four powers of $X_{u,d}$, of the form $X_{u,d}^n X_{d,u}^m X_{u,d}^p X_{d,u}^q$:

$$2i\mathrm{Im}\langle X_u^n X_d^m X_u^p X_d^q\rangle = \langle X_u^n X_d^m X_u^p X_d^q - (X_u^n X_d^m X_u^p X_d^q)^\dagger\rangle$$
$$= \langle X_u^n X_d^m X_u^p X_d^q - X_d^q X_u^p X_d^m X_u^n\rangle$$
$$= \langle X_u^n X_d^m X_u^p X_d^q\rangle - \langle X_d^q X_u^p X_d^m X_u^n\rangle$$
$$= \langle X_u^n X_d^m X_u^p X_d^q\rangle - \langle X_u^n X_d^q X_u^p X_d^m\rangle = 0 \text{ if } m = q$$
$$= \langle X_d^q X_u^n X_d^m X_u^p\rangle - \langle X_d^q X_u^p X_d^m X_u^n\rangle = 0 \text{ if } n = p.$$

As we want a non-zero imaginary part, we must choose $m \neq q$ and $n \neq p$. The simplest choice (lowest order in quarks mass matrices) is then $(n,m,p,q) = (2,2,1,1)$ and corresponds to the invariant: $\langle X_u^2 X_d^2 X_u X_d\rangle$. This invariant is not CP-even (because it is complex) but is it CP-odd (purely imaginary)? Let us compute its real part using $\mathrm{Re}\langle M\rangle = \frac{1}{2}\langle M + M^\dagger\rangle$

$$2\mathrm{Re}\langle X_u^2 X_d^2 X_u X_d\rangle = \langle X_u^2 X_d^2 X_u X_d + (X_u^2 X_d^2 X_u X_d)^\dagger\rangle$$
$$= \langle X_u^2 X_d^2 X_u X_d + X_d X_u X_d^2 X_u^2\rangle$$
$$= \langle X_u^2 X_d^2 X_u X_d\rangle + \langle X_d X_u X_d^2 X_u^2\rangle$$
$$= \langle X_u^2 X_d^2 X_u X_d\rangle + \langle X_u^2 X_d X_u X_d^2\rangle \neq 0 \text{ (a priori)}$$

Thus, this invariant is not CP-odd. It is the sum of a CP-even and a CP-odd invariant:

$$\langle X_u^2 X_d^2 X_u X_d \rangle = \underbrace{\mathrm{Re}\langle X_u^2 X_d^2 X_u X_d \rangle}_{\text{CP-even}} + \underbrace{i\mathrm{Im}\langle X_u^2 X_d^2 X_u X_d \rangle}_{\text{CP-odd}},$$

Therefore, the lowest-order CP-odd invariant is:

$$
\begin{aligned}
i\mathrm{Im}\langle X_u^2 X_d^2 X_u X_d \rangle &= \tfrac{1}{2}\langle X_u^2 X_d^2 X_u X_d - (X_u^2 X_d^2 X_u X_d)^\dagger \rangle \\
&= \tfrac{1}{2}\langle X_u^2 X_d^2 X_u X_d - X_d X_u X_d^2 X_u^2 \rangle \\
&= \tfrac{1}{2}(\langle X_u^2 X_d^2 X_u X_d \rangle - \langle X_d^2 X_u^2 X_d X_u \rangle) \\
&= \tfrac{1}{2}\mathcal{J}.
\end{aligned}
$$

Let us calculate $\langle [X_u, X_d]^3 \rangle$:

$$
\begin{aligned}
\langle [X_u, X_d]^3 \rangle &= \langle (X_u X_d - X_d X_u)^3 \rangle \\
&= \langle X_u X_d X_u X_d X_u X_d - 3 X_u X_d X_u X_d^2 X_u + 3 X_u X_d^2 X_u X_d X_u - X_d X_u X_d X_u X_d X_u \rangle \\
&= \langle X_u X_d X_u X_d X_u X_d \rangle + 3\langle X_u X_d^2 X_u X_d X_u - X_u X_d X_u X_d^2 X_u \rangle - \langle X_d X_u X_d X_u X_d X_u \rangle \\
&= 3\langle X_u X_d^2 X_u X_d X_u - X_u X_d X_u X_d^2 X_u \rangle \\
&= 3\langle X_u^2 X_d^2 X_u X_d - X_d^2 X_u^2 X_d X_u \rangle \\
&= 3\mathcal{J}.
\end{aligned}
$$

Then, $\mathcal{J}$ can be written in the form $\mathcal{J} = \tfrac{1}{3}\langle [X_u, X_d]^3 \rangle$.

Let us work out explicitly $\mathcal{J}$. In the gauge basis in which all down-type quarks are mass eigenstates, we have: $M_u = V_{\mathrm{CKM}}^\dagger \begin{pmatrix} m_u & 0 & 0 \\ 0 & m_c & 0 \\ 0 & 0 & m_t \end{pmatrix}$ and $M_d = \begin{pmatrix} m_d & 0 & 0 \\ 0 & m_s & 0 \\ 0 & 0 & m_b \end{pmatrix}$.

- In the standard parametrization (PDG), the CKM matrix is written as:

$$
V_{\mathrm{CKM}} = \begin{pmatrix}
c_{12}c_{13} & s_{12}c_{13} & s_{13}e^{-i\delta} \\
-s_{12}c_{23} - c_{12}s_{23}s_{13}e^{i\delta} & c_{12}c_{23} - s_{12}s_{23}s_{13}e^{i\delta} & s_{23}c_{13} \\
s_{12}s_{23} - c_{12}c_{23}s_{13}e^{i\delta} & -c_{12}s_{23} - s_{12}c_{23}s_{13}e^{i\delta} & c_{23}c_{13}
\end{pmatrix},
$$

where $c_{ij} = \cos\theta_{ij}$ and $s_{ij} = \sin\theta_{ij}$. The explicit computation of $\mathcal{J} = \tfrac{1}{3}\langle [M_u M_u^\dagger, M_d M_d^\dagger]^3 \rangle$ yields:

$$
\begin{aligned}
\mathcal{J} = {}& \tfrac{i}{2}\sin(2\theta_{12})\sin(2\theta_{13})\sin(\theta_{13})\cos^2(\theta_{13})\sin(\delta) \\
& \times (m_b^2 - m_d^2)(m_b^2 - m_s^2)(m_s^2 - m_d^2)(m_t^2 - m_u^2)(m_t^2 - m_c^2)(m_c^2 - m_u^2).
\end{aligned}
$$

- In the Wolfenstein parametrization (PDG), we express the CKM matrix as:

$$V_{CKM} = \begin{pmatrix} 1 - \frac{\lambda^2}{2} & \lambda & A\lambda^3(\rho - i\eta) \\ -\lambda & 1 - \frac{\lambda^2}{2} & A\lambda^2 \\ A\lambda^3(1 - \rho - i\eta) & -A\lambda^2 & 1 \end{pmatrix} + \mathcal{O}(\lambda^4),$$

then we compute $\mathcal{J}$ and obtain:

$$\mathcal{J} = 2iA^2\eta\lambda^6(m_b^2 - m_d^2)(m_b^2 - m_s^2)(m_s^2 - m_d^2)(m_t^2 - m_u^2)(m_t^2 - m_c^2)(m_c^2 - m_u^2) + O(\lambda^7)$$

Exercise 2.30 Compute the Hilbert series for the ring of invariants generated by

(a) x, y (each of dimension 1), and invariant under the transformation $(x, y) \rightarrow (-x, -y)$.

(b) x, y, z (each of dimension 1), and invariant under the transformation $(x, y, z) \rightarrow (-x, -y, -z)$.

SOLUTION:

(a) In a first step we count the number of invariants N_n of degree n. Obviously, invariants can only be built for even n, namely of the form $x^a y^b$, with $a + b = n$ even. There are $n + 1$ non-equivalent possibilities to build such an invariant of degree n, such that

$$N_n = \begin{cases} n+1 & \text{for } n \text{ even} \\ 0 & \text{for } n \text{ odd} \end{cases}, \tag{12.237}$$

and consequently the Hilbert series is

$$H(q) = \sum_{n=0}^{\infty} N_n q^n = \sum_{n=0}^{\infty}(2n+1)q^{2n} = \frac{1+q^2}{(1-q^2)^2}. \tag{12.238}$$

This expressions can be easily interpreted: There are two invariant generators of degree $n = 2, x^2$ and y^2, (indicated by the power of the denominator and the power of q therein respectively) which are linearly independent regardless of to which power they are raised. Additionally, there is one invariant generator of degree $n = 2$, xy, which satisfies the relation $(xy)^2 = x^2y^2$ at degree $n = 4$ such that expressions of this form can be written as a linear combination of the other two generators.

We can rewrite the given expression for $H(q)$ as

$$H(q) = \frac{1-q^4}{(1-q^2)^3}, \tag{12.239}$$

allowing the same interpretation: there are three generators of degree $n = 2$, with one relation between them at degree $n = 4$.

(b) Analogous to exercise (a), we count the number of invariants N_n of degree n, which are now of the form $x^a y^b z^c$ with $a + b + c = n, n$ even. Writing these down in a systematic way we count

$$\underbrace{x^0 y^0 z^n, x^0 y^1 z^{n-1}, \ldots, x^0 y^{n-1} z^1, x^0 y^n z^0}_{n+1 \text{ terms}}, \underbrace{x^1 y^0 z^{n-1}, \ldots, x^1 y^{n-1} z^0}_{n \text{ terms}}, \ldots, \underbrace{x^n y^0 z^0}_{1 \text{ term}}, \quad (12.240)$$

such that

$$N_n = \begin{cases} \sum_{j=1}^{n+1} j = \frac{1}{2}(1+n)(2+n) & \text{for } n \text{ even} \\ 0 & \text{for } n \text{ odd} \end{cases}. \quad (12.241)$$

Therefore, the Hilbert series is

$$H(q) = \sum_{n=0}^{\infty} \frac{1}{2}(1+2n)(2+2n)q^{2n} = \frac{1+3q^2}{(1-q^2)^3}. \quad (12.242)$$

The result can be interpreted analogous to part (a): There are three invariant generators of degree $n = 2, x^2, y^2$ and z^2, where all powers of them remain linearly independent. This can be read off the power of the denominator and the power of q therein respectively. Additionally, there are three generators of degree $n = 2, xy, xz$ and yz, (indicated by the coefficient 3 in the numerator and the power of q respectively), which fulfill the relations $(xy)^2 = x^2 y^2, (xz)^2 = x^2 z^2$ and $(yz)^2 = y^2 z^2$ at degree $n = 4$, such that expressions of this form can, analogously to part (a), be written as a linear combination of the three generaltors identified at the beginning of this paragraph.

Exercise 2.31 Show that $(\psi_{Lr}^T C \psi_{Ls})$ is symmetric in rs and $(\psi_{Lr}^T C \sigma^{\mu\nu} \psi_{Ls})$ is anti-symmetric in rs.

SOLUTION:
  Solution:

$$(\psi_{Lr})^T C \psi_{Ls} = (\psi_{Lr\alpha}) C_{\alpha\beta} \psi_{Ls\beta} = -\psi_{Ls\beta} C_{\alpha\beta} (\psi_{Lr\alpha}) = -\psi_{Ls}^T C^T (\psi_{Lr}) \quad (12.243)$$

where the minus sign is from anti-commuting Fermi fields. Similarly

$$(\psi_{Lr}^T C \sigma^{\mu\nu} \psi_{Ls}) = -(\psi_{Ls}^T (C \sigma^{\mu\nu})^T \psi_{Lr}) \quad (12.244)$$

Using two-component left-handed fields, $C_{\alpha\beta} = \epsilon_{\alpha\beta} = i\sigma^2$ so $C^T = -C$, so the scalar operator is symmetric in rs. $\sigma^{\mu\nu}$ is proportional to the Pauli matrices $\sigma^k (\sigma^{ij} = \epsilon^{ijk} \sigma^k, \sigma^{0k} = -i\sigma^k)$ and

$$(i\sigma^2 \sigma^k)^T = (\sigma^k)^T (-i\sigma^2) = (\sigma^2)(-\sigma^k)(\sigma^2)(-i\sigma^2) = -i\sigma^2 \sigma^k \quad (12.245)$$

so the tensor operator is symmetric.

Exercise 2.32 Prove the duality relations eqns (L2.247, L2.248). The sign convention is $\gamma_5 = i\gamma^0 \gamma^1 \gamma^2 \gamma^3$ and $\epsilon_{0123} = +1$.

SOLUTION:
The identities are equivalent to showing that

$$\tfrac{i}{2}\epsilon^{\alpha\beta\mu\nu}\sigma_{\mu\nu}\gamma_5 = -\sigma^{\alpha\beta}. \tag{12.246}$$

The identity is true up to an overall normalization, since both sides are two-index anti-symmetric tensors of the same parity. The normalization is fixed by looking at one term, e.g.

$$\tfrac{i}{2}\epsilon^{01\mu\nu}\sigma_{\mu\nu}\gamma_5 = i\epsilon^{0123}\sigma_{23}\gamma_5 = i\epsilon^{0123}(i\gamma_2\gamma_3)(i\gamma^0\gamma^1\gamma^2\gamma^3) = -i\gamma^0\gamma^1 \overset{?}{=} -\sigma^{01}. \tag{12.247}$$

so the normalization is correct.

Exercise 2.33   Show that eqn (L2.249) is the unique dimension-five term in the SMEFT Lagrangian. How many independent operators are there for n_g generations?

SOLUTION:
This discussion is taken from Buchmüller and Wyler [4]. A dimension-five term in SMEFT cannot be built purely from fermions or scalars. Since fermions have mass-dimension 3/2 (in $d = 4$ dimensions), they must appear in pairs in a Lagrangian term to have integer mass dimension. Consequently, a maximum of two fermions can appear in an operator with mass dimension less than six. The only scalar in the SM is the Higgs doublet, and it is not possible to construct an SU(2) singlet from five SU(2) doublets. Therefore, for reasons of dimensionality and the requirement of building an SU(2) singlet, it is not possible to have an odd number of fermions or scalars. The only possibility is to have two fermions and two scalars.

If the scalars are taken to be H and H^*, then they have a total hypercharge of zero, and so the two fermions must also have total hypercharge zero so that the Lagrangian remains invariant under U(1)$_Y$. This is only possible by taking a multiplet and its charge conjugate, but their product cannot then form a Lorentz scalar. Alternatively, the two scalars can both be taken to be H, in which case it is possible to write the operators [27]

$$\left(\ell_i^T C \ell_j\right) H_k H_l \epsilon^{ik} \epsilon^{jl} \quad \text{and} \quad \left(\ell_i^T C \ell_j\right) H_k H_l \epsilon^{ij} \epsilon^{kl}. \tag{12.248}$$

However, the second term is forbidden if there is only a single Higgs doublet (i.e. no extended Higgs sector), since $H^T \epsilon H$ is identically zero. Therefore the only dimension-five term that can be written in SMEFT is

$$L_5 = c_5 \left(\ell_i^T C \ell_j\right) H_k H_l \epsilon^{ik} \epsilon^{jl}. \tag{12.249}$$

For n_g (fermion) generations, each lepton acquires a generation index that runs from $1, \ldots, n_g$, and the Wilson coefficient becomes a matrix in generation space (analogous to the Yukawa matrices). From Exercise 2.31, the operator is symmetric in generation indices, so there are $n_g(n_g+1)/2$ independent operators.

Exercise 2.34 Show that eqn (L2.249) generates a Majorana neutrino mass when H gets a vacuum expectation value, and find the neutrino mass matrix M_ν in terms of C_5 and v.

SOLUTION:
Solution: A Majorana mass term for a Dirac fermion ψ may be written as

$$-\tfrac{1}{2}M\psi_L^T C\psi_L, \tag{12.250}$$

where M is the Majorana mass.

Upon electroweak symmetry breaking, the Higgs acquires a VEV as

$$H \to \tfrac{1}{\sqrt{2}}\begin{pmatrix}0\\v\end{pmatrix}. \tag{12.251}$$

Then

$$c_5\left(\ell_i^T C\ell_j\right)H_kH_l\epsilon^{ik}\epsilon^{jl} \to \tfrac{c_5}{2}(\nu_L^T e_L^T)\begin{pmatrix}0&1\\-1&0\end{pmatrix}\begin{pmatrix}0\\v\end{pmatrix}(C_{\nu_L}Ce_L)\begin{pmatrix}0&1\\-1&0\end{pmatrix}\begin{pmatrix}0\\v\end{pmatrix} \tag{12.252}$$

$$= \tfrac{c_5v^2}{2}\nu_L^T C\nu_L. \tag{12.253}$$

The Majorana mass can then be identified as

$$M = -c_5v^2. \tag{12.254}$$

Exercise 2.35 Prove eqn (L2.261).

SOLUTION:
Done in eqn (12.49).

Exercise 2.36 In the SMEFT for n_g generations, how many operators are there of the following kind (in increasing order of difficulty): (a) Q_{He} (b) Q_{ledq} (c) $Q_{lq}^{(1)}$ (d) $Q_{qq}^{(1)}$ (e) Q_{ll} (f) Q_{uu} (g) Q_{ee} (h) Show that there are a total of 2,499 Hermitian dimension-six $\Delta B = \Delta L = 0$ operators.

SOLUTION:
The enumeration of operators of each combination of elementary field is given by the procedure in [10], summarized by their master formula for the Hilbert series of the dimension-six Standard Model in eqn (2.23) for one generation. As the number of operators for each elementary field combination is so small, this is sufficient to deduce a complete set of independent operators, such as those tabulated in the present section, by simply guessing. To extend the enumeration to n_g generations, as stated in [10], the Hilbert series may be derived by raising the

plethystic exponential to a power of n_g and then expanding and projecting onto the relevant components.

Alternatively, it is simple enough instead to inscribe flavour indices on the operators presented here and counting (see Appendix A of [2]). Giving the fermions in each operator a flavour index, the number of independent flavour components may be simply counted by decomposing into representation of the flavour group, subject to symmetry constraints of the elementary fields.

The operator class $Q_{ledq} \sim (L_{i,I}\bar{e}_j)(Q_k^I \bar{d}_l)$, where I denotes isospin indices (included for clarity) and i, j, k, l are flavour indices, consists of only distinct fields, so there are n_g^4 such operators, where each factor of n_g is simply the number of possible flavour identities that each field can take. Then, as the conjugate operator is distinct, exactly the same enumeration holds for its conjugates, giving a total of $2n_g^4$ operators.

Similarly, for $Q_{He} \sim \bar{e}^{\dagger i}\bar{\sigma}^\mu \bar{e}_j(H^\dagger iD_\mu H - iD_\mu H^\dagger H)$, the $\bar{e}^\dagger$ and $\bar{e}$ fields are distinguished as conjugates, so are distinct, implying that there are $n_g \times n_g$ possible operators. The adjoint of these flavoured operators just reverses the flavuor indices, so there are no additional contributions to these n_g^2 operators from conjugates. Identical arguments apply to operators with the replacement of $\bar{e}$ with any of the other fermion fields (and is unaffected if the bilinear has a triplet isospin structure that is contracted with the Higgs bilinear, which can occur if the fermion is isospinning). There is one exceptional example Q_{Hud}, which is not Hermitian (because the fermions are distinct), so has twice the number of flavour components.

The operator $Q_{lq}^{(1)} \sim (L^{\dagger i}\bar{\sigma}^\mu L_j)(Q^{\dagger k}\bar{\sigma}_\mu Q_l)$ has similar structure, where all gauge indices are contracted within each bilinear. Each operator is distinct, so there are n_g^4 possible flavour components, which transform into each other under conjugation. Identical counting applies to other operators consisting of two different bilinears of fermions of the same type with their conjugates, including those where the bilinears are not gauge singlets.

Other four-fermion operators involving identical fermions may be counted by giving them flavour indices and determining the number of non-zero entries. This may be done systematically beginning with the states of the most minimal internal structure.

Beginning generally with $(f^{\dagger ia}\bar{\sigma}^\mu f_{ja})(f^{\dagger kb}\bar{\sigma}_\mu f_{lb}) \sim (f^{\dagger ia}f^{\dagger kb})(f_{ja}f_{lb})$ by a Fierz identity. Here a and b denote internal gauge indices that are contracted within each original vector bilinear. The two fermion bilinears of identical gauge species may be decomposed into symmetric and anti-symmetric components in all indices. Fully anti-symmetric combinations are 0, because the fermion bilinear (ff) is symmetric. For $f = \bar{e}$ (operators of type Q_{ee}), there are no gauge indices and so $\bar{e}_{[j}\bar{e}_{l]} = 0$ by fermion statistics. This reduces the operators to $(\bar{e}^{\dagger(i}\bar{e}^{\dagger k)})(\bar{e}_{(j}\bar{e}_{l)})$, of which there are $\left(\frac{1}{2}n_g(n_g + 1)\right)^2 = \frac{1}{4}n_g^2(n_g + 1)^2$.

For $f = L(Q_{ll})$, the bilinear $L_j^I L_l^{\mathcal{J}}$ is only non-zero if both flavour and isospin indices are symmetrized with the same parity. This gives an isospin triplet bilinear with symmetric flavour indices and an isospin singlet with anti-symmetric flavour indices. These are then contracted in isospin indices with their conjugates. This requires that the isospin indices of each bilinear have the same parity in order to be non-zero, so results in $\left(\frac{1}{2}n_g(n_g+1)\right)^2$ terms $(\bar{L}^{\dagger(i}\bar{L}^{\dagger k)})(\bar{L}_{(j}\bar{L}_{l)})$ and $\left(\frac{1}{2}n_g(n_g-1)\right)^2$ terms $(\bar{L}^{\dagger[i}\bar{L}^{\dagger k]})(\bar{L}_{[j}\bar{L}_{l]})$, which gives a total of $\frac{1}{2}n_g^2(n_g^2+1)$.

For $f = \bar{u}(Q_{uu})$, isospin indices in the above case are replaced with colour indices, but the argument is otherwise identical. There are therefore $\frac{1}{2}n_g^2(n_g^2+1)$ of these terms as well. The same argument applies to Q_{dd}.

For $f = Q$ with the singlet colour structure $(Q_{qq}^{(1)})$, the fields have both colour and isospin. For overall symmetry, either all indices must be symmetrized over or exactly one set should be symmetrized and the other two anti-symmetrized. This gives four possible terms: $Q_{j,I,\alpha}Q_{l,\mathcal{J},\beta} = (3,6,S) + (1,\bar{3},S) + (1,6,A) + (3,\bar{3},A)$, where, to control indices, the representation labels of the form $(SU(2),SU(3),SU(N_f))$ have been written instead of the tensors (S and A denote symmetric and anti-symmetric rank-2 $SU(N_f)$ tensors, respectively). Exactly the same decomposition applies to the conjugate bilinear. Contracting the gauge indices of the decomposed bilinears, only the index contractions between pairs with the same symmetry parities are non-zero. This leaves four terms $(3\cdot3,\bar{6}\cdot6,S\otimes S) + (1\cdot1,3\cdot\bar{3},S\otimes S) + (1\cdot1,\bar{6}\cdot6,A\otimes A) + (3\cdot3,3\cdot\bar{3},A\otimes A)$, the first two and the last two add to give single tensors that are either symmetric in both $i \leftrightarrow k$ and $j \leftrightarrow l$ or anti-symmetric in both. These therefore have $(\frac{1}{2}n_g(n_g+1))^2$ and $(\frac{1}{2}n_g(n_g-1))^2$ components respectively, giving a total of $\frac{1}{2}n_g^2(n_g^2+1)$.

This argument also gives the counting for the operator involving isospin triplet bilinears instead. This is because, by the isospin Fierz identity (a.k.a. Clifford algebra), this may be decomposed into a linear combination of the above operator and an identical version with the isospin pairings of the quarks switched. Subtracting the former component, which is accounted for above, an identical tensor decomposition may be performed on the remaining term, with isospin indices swapped $I \leftrightarrow \mathcal{J}$ on the Q pair. This simply introduces a relative negative sign in the two terms, i.e. the decomposition is $(3\cdot3,\bar{6}\cdot6,S\otimes S) - (1\cdot1,3\cdot\bar{3},S\otimes S) - (1\cdot1,\bar{6}\cdot6,A\otimes A) + (3\cdot3,3\cdot\bar{3},A\otimes A)$. This operator has therefore been decomposed into two bisymmetric or biantisymmetric flavour tensors (independent to those from the isospin singlet operator above), so has the same number of independent components.

The remaining flavoured operators (neglecting baryon and lepton number violation) may be enumerated similarly. Those of the form $(\bar{L}R)H^3$ clearly have $2n_g^2$ flavour components each, because the two fermions are distinct and the operator is not self-adjoint (hence the factor of 2). An identical argument applies to the $(\bar{L}R)XH$ operators. Finally, in the $(\bar{L}R)(\bar{L}R)$ class, there are clearly $2n_g^4 Q_{lequ}$-type operators for each isospin configuration (by identical reasons as for the number of Q_{ledq} operators), while for Q_{quqd}-type operators, the isospin contraction ensures that the Q operators are never identical fermions, while the colour contractions with the different right-handed quarks distinguishes them, so there are also n_g^4 of each of these.

In the operator basis chosen here, the addition of fermion flavour indices does not modify the dimension-four equations of motion beyond the inscription of flavour indices on the fields.

There are also no examples of operators that cannot exist for $n_g = 1$ but can exist if the flavour provides an extra internal degree of freedom with which to distinguish what would otherwise be identical quanta (as would happen if e.g. right-handed neutrinos were included). This is simply because there are no such gauge-invariant combinations consistent with the Standard Model field content, which can be easily verified by counting operators with three fermions of the same type.

Having derived the number of independent flavoured operators in the baryon and lepton conserving dimension-six Standard Model, these may be added together to give the total number. Following the enumeration tabulated, for $n_g = 3$, there are 2,499.

12.3 EFT for Nuclear and (Some) Atomic Physics (van Kolck): Solutions

The chapter featuring these two lectures [25] introduced some applications of effective field theory EFT in the context of nuclear and atomic physics. It presented pionless EFT, a simple nuclear EFT containing contact interactions and discussed the prospects for including long-range forces.

Exercise 6.1 For a spherical well potential,

$$V(r) = -V_0 \Theta(R, r)$$

show that when the parameter $\alpha := \sqrt{mV_0}R$ is tuned close to the critical values $\alpha_c = (n + 1/2)\pi$ that the scattering length is given by,

$$a_2 \sim \frac{R}{\alpha_c(\alpha - \alpha_c)}$$

SOLUTION:
Eigenstates with energy $E = k^2/2m$ are described by,

$$\left[-\frac{\nabla^2}{2m} + V(r)\right]\psi = \frac{k^2}{2m}\psi \tag{12.255}$$

and can be separated for central potentials into,

$$\psi = \sum_{\ell=0}^{\infty} R_\ell(r) P_\ell(\cos\theta)$$

$$\left[\partial_r^2 + \frac{2}{r}\partial_r - \frac{\ell(\ell+1)}{r^2} - 2mV(r) + k^2\right]R_\ell(r) = 0$$

We define the *s*-wave scattering phase via the asymptotic behavior of the $\ell = 0$ mode,

$$R_0(r) \sim \tfrac{1}{kr} e^{i\delta_0(k)} \sin(kr + \delta_0(k)) \quad \text{as} \quad r \to \infty \tag{12.256}$$

and so for a spherical wave potential we can solve,

$$[\partial_r^2 + 2mV_0\Theta(R - r) + k^2] \, (rR_0) = 0 \tag{12.257}$$

using the boundary conditions $rR_0 \to 0$ as $r \to 0$ and (12.256),

$$rR_0(r) = \begin{cases} c\sin[r\sqrt{k^2 + 2mV_0}] & r < R, \\ \sin[kr + \delta_0] & r > R \end{cases} \tag{12.258}$$

where c is a constant of integration. Demanding that the wave function and its first derivative are continuous across the $r = R$ boundary, we find,

$$c = \frac{\sin[kR + \delta_0]}{\sin\left[R\sqrt{k^2 + 2mV_0}\right]}, \tag{12.259}$$

$$\sqrt{k^2 + 2mV_0}\cot\left[R\sqrt{k^2 + 2mV_0}\right] = k\cot[kR + \delta_0] \tag{12.260}$$

Taking the limit $k \to 0$, this gives,

$$k\cot(\delta_0) = \sqrt{2mV_0}\cot\left(R\sqrt{2mV_0}\right) + \mathcal{O}(k^2) \tag{12.261}$$

Now using, $\cot(\alpha) = -(\alpha - (n + 1/2)\pi) + \dots$ when $\alpha \approx (n + 1/2)\pi$, we find,

$$-\tfrac{1}{a_2} = \lim_{k \to 0} k\cot(\delta_0) = -\tfrac{\alpha_c}{R}(\alpha - (n + 1/2)\pi) + \dots \tag{12.262}$$

where ... are regular in the limit $\alpha \to \alpha_c$.

The divergence in the scattering length corresponds to a zero energy *s*-wave bound state accommodated by the potential when $\alpha \to \pi/2$. At $\alpha = n\pi$, the scattering cross section vanishes identically (the Ramsauer–Townsend effect).

Exercise 6.2 Solve the three-dimensional Schrodinger equation with,

$$V = \tfrac{4\pi c_0}{2m}\delta^3(r)$$

SOLUTION:

In momentum space, energy eigenstates obey

$$p^2\psi(p) - 4\pi c_0 \int \tfrac{d^3k}{(2\pi)^3}\psi(k) = 2mE\psi(p) \tag{12.263}$$

Note that the integral interaction is divergent. If we rewrite,

$$\int d^3k\left[(k^2 - 2mE)\delta^3(k - p) - \tfrac{c_0}{2\pi^2}\right]\psi(k) = 0 \tag{12.264}$$

then we see that the energy, E, is related to the momentum and the coupling c_0 by the condition,

$$1 = \frac{c_0}{2\pi^2} \int d^3k \frac{1}{k^2 - 2mE} \tag{12.265}$$

Introducing a UV cut-off, Λ, this can be written as,

$$1 = -\frac{c_0}{2\pi^2} 4\pi \left[\Lambda - i\frac{\pi}{2}\sqrt{2mE} + \mathcal{O}\left(\frac{\sqrt{mE}}{\Lambda}\right) \right] \tag{12.266}$$

Suppose we measure a bound state at energy $E < 0$. Then, we renormalization the coupling c_0 so that,

$$\sqrt{-2mE} := \frac{1}{c_0^R} = \frac{1}{c_0(\Lambda)} + \frac{2}{\pi}\Lambda \tag{12.267}$$

That is, as the cut-off Λ is taken to infinity, the coupling c_0 must run as above in order to maintain a bound state with energy E in the spectrum of the theory. The theory is only predictive once this renormalization has been performed. To analyse other bound states, we must look for other solutions to,

$$1 = \frac{c_0(\Lambda)}{2\pi^2} \int d^3k \frac{1}{k^2 - 2mE'} \tag{12.268}$$

$$1 = c_0(\Lambda) \left[\frac{2}{\pi}\Lambda + \sqrt{-2mE'} + \ldots \right] \tag{12.269}$$

which is satisfied iff $E' = E$, and so we conclude that there is only one bound state for the (renormalized) three-dimensional delta function.

The theory then predicts a wave function for this single bound state comprised of the spherical waves,

$$\left[\partial_r^2 - (c_0^R)^{-2} - \frac{\ell(\ell+1)}{r} \right](rR_\ell) = 0 \quad r \neq 0 \tag{12.270}$$

and so we have an asymptotic wave function dominated by the s-wave,

$$\psi \propto \frac{e^{-r/c_0^R}}{r} \quad \text{for} \quad r \to \infty$$

12.4   EFT with Nambu–Goldstone Modes (Pich): Solutions

These lectures [24] discussed EFTs that are useful for describing the dynamics of massless modes that emerge after spontaneous symmetry breaking. They emphasized chiral perturbation theory (χPT) and the electroweak sector of the Standard Model. For χPT, which is the focus of these exercises, [6] paved the way forward to carry out an effective low-energy expansion.

Exercise 3.2 The quadratic mass term of the $\mathcal{O}(p^2)\chi$PT Lagrangian generates a small mixing between the π_3 and η_* fields, proportional to the quark mass difference $\Delta m = m_d - m_u$.

a) Diagonalize the neutral meson mass matrix and find out the correct mass eigenstates and their masses.

b) When isospin is conserved, Bose symmetry forbids the decay $\eta \to \pi^0 \pi^+ \pi^-$ (why?). Compute the decay amplitude to first order in Δm.

SOLUTION:

a) Promoting the quark mass operator $\bar{q}\mathcal{M}q/2$, where $\mathcal{M} = \mathrm{diag}(m_u, m_d, m_s)$, to a spurion field, the low-energy chiral Lagrangian at order p^2 is,

$$\mathcal{L}_2 = \frac{f^2}{4}\langle \partial_\mu U \partial^\mu U^\dagger + B_0 \mathcal{M} U^\dagger + B_0 U \mathcal{M}^\dagger \rangle \qquad (12.271)$$

where f and B_0 are EFT parameters, and the angled brackets denote a trace over the $SU(3)$ valued matrix,

$$U = e^{i\sqrt{2}\Phi/f}, \quad \Phi = \begin{pmatrix} \frac{1}{\sqrt{2}}\pi_3 + \frac{1}{\sqrt{6}}\eta_8 & \pi^+ & K^+ \\ \pi^- & -\frac{1}{\sqrt{2}}\pi_3 + \frac{1}{\sqrt{6}}\eta_8 & K^0 \\ K^- & \bar{K}^0 & -\frac{2}{\sqrt{6}}\eta_8 \end{pmatrix}. \qquad (12.272)$$

The chiral Lagrangian contains the mass terms,

$$\mathcal{L}_2 \supset -\frac{1}{2}(\pi_3 \ \eta_8)\begin{pmatrix} M_{\pi_3}^2 & -\Delta/2 \\ -\Delta/2 & M_{\eta_8}^2 \end{pmatrix}\begin{pmatrix} \pi_3 \\ \eta_8 \end{pmatrix} \qquad (12.273)$$

$$\Delta = \frac{2B_0}{\sqrt{3}}(m_d - m_u), \quad M_{\pi_3}^2 = B_0(m_u + m_d), \quad M_{\eta_8}^2 = \frac{B_0}{3}(m_u + m_d + 4m_s).$$

For convenience, we also define,

$$M = M_{\eta_8}^2 - M_{\pi_3}^2, \quad a_\pm = M \pm \sqrt{M^2 + \Delta^2}.$$

The mass matrix is then diagonalized as,

$$\mathcal{L}_2 \supset -\frac{1}{2}(\pi^0 \ \eta)\begin{pmatrix} M_{\pi^0}^2 & 0 \\ 0 & M_\eta^2 \end{pmatrix}\begin{pmatrix} \pi^0 \\ \eta \end{pmatrix} \qquad (12.274)$$

$$M_{\pi^0}^2 = M_{\pi_3}^2 + \frac{1}{2}a_- \approx M_{\pi_3}^2 - \frac{\Delta^2}{4M}, \quad M_\eta^2 = M_{\pi_3}^2 + \frac{1}{2}a_+ \approx M_{\eta_8}^2 + \frac{\Delta^2}{4M}$$

$$\pi^0 = \frac{a_+ \pi_3 + \Delta \eta_8}{\sqrt{a_+^2 + \Delta^2}} \approx \pi_3 + \frac{\Delta}{2M}\eta_8, \quad \eta = \frac{a_- \pi_3 + \Delta \eta_8}{\sqrt{a_-^2 + \Delta^2}} \approx \eta_8 - \frac{\Delta}{2M}\pi_3.$$

b) The final state quark content,

$$\frac{1}{\sqrt{2}}(u\bar{u} - d\bar{d})(u\bar{d})(\bar{u}d)$$

is anti-symmetric under exchanging $u \leftrightarrow d$, i.e. under isospin symmetry. This is because π^0 is an anti-symmetric combination, and the $\pi^+\pi^-$ pair must be symmetric because Bose symmetry mandates a symmetric wave function. However, the original state $\eta = \frac{1}{\sqrt{6}}(u\bar{u} + d\bar{d} - 2s\bar{s})$ is symmetric under u, d isospin symmetry—and therefore the process $\eta \to \pi^0 \pi^+ \pi^-$ would violate isospin.

Explicitly, expanding $\mathcal{L}_2$ we find the operators,

$$\mathcal{L}_2 \supset \frac{B_0}{\sqrt{3}}(m_d - m_u)\pi^0\eta\left[1 - \frac{1}{3f^2}\pi^+\pi^-\right] \tag{12.275}$$

$$+ \frac{1}{3f^2}\left\{(\pi^0\overset{\leftrightarrow}{\partial}_\mu\pi^+)(\pi^-\overset{\leftrightarrow}{\partial}{}^\mu\pi^0) + \frac{1}{2}m_\pi^2\pi^+\pi^-(\pi^{02} + \eta^2)\right\} \tag{12.276}$$

from which we can construct three independent Feynman diagrams:

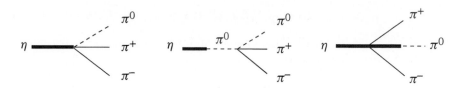

which give a total decay amplitude,

$$\mathcal{M}(\eta \to \pi^+\pi^-\pi^0) = \frac{B_0}{3\sqrt{3}f^2}(m_d - m_u)\left\{1 + \frac{3(p_2+p_3)^2 - 3m_\pi^2 - m_\eta^2}{m_\eta^2 - m_\pi^2}\right\}. \tag{12.277}$$

which indeed vanishes in the isospin limit $(m_d = m_u)$.

Exercise 3.3

a) Compute the axial current at $\mathcal{O}(p^2)$ in χPT and check that $f_\pi = f$ at this order.
b) Expand the $\mathcal{O}(p^2)$ axial current to $\mathcal{O}(\Phi^3)$ and compute the one-loop corrections to f_π. Remember to include the *pion* wave function renormalization.
c) Find the tree-level contribution of the $\mathcal{O}(p^4)\chi$PT Lagrangian to the axial current. Renormalize the UV loop divergences with the $\mathcal{O}(p^4)$ LECs.

SOLUTION:

Promoting an axial current $\bar{q}_L\slashed{\ell}q_L + \bar{q}_R\slashed{r}q_R$, where $\ell_\mu + r_\mu = 2a_\mu$, to a spurion field,

$$\ell_\mu \to g_L\ell_\mu g_L^\dagger + ig_L\partial_\mu g_L^\dagger \tag{12.278}$$

$$r_\mu \to g_R r_\mu g_R^\dagger + ig_R\partial_\mu g_R^\dagger \tag{12.279}$$

the low-energy chiral Lagrangian at order p^2 is

$$\mathcal{L}_2 = \frac{f^2}{4}\langle D_\mu U D^\mu U^\dagger\rangle, \quad D_\mu U = \partial_\mu U - ir_\mu U + iU\ell_\mu. \tag{12.300}$$

The currents are then,

$$\mathcal{J}_L^\mu = \frac{\partial}{\partial\ell_\mu}\mathcal{L}_2 = \frac{i}{2}f^2 D^\mu U^\dagger U = \frac{f}{\sqrt{2}}D^\mu\Phi + \frac{i}{2}[D^\mu\Phi,\Phi] - \frac{1}{3\sqrt{2}f}[[D^\mu\Phi,\Phi],\Phi] + \dots \tag{12.301}$$

$$\mathcal{J}_R^\mu = \frac{\partial}{\partial r_\mu} \mathcal{L}_2 = \tfrac{i}{2} f^2 D^\mu U U^\dagger = -\frac{f}{\sqrt{2}} D^\mu \Phi + \tfrac{i}{2}[D^\mu \Phi, \Phi] + \frac{1}{3\sqrt{2}f}[[D^\mu \Phi, \Phi], \Phi] + \dots$$

$$(12.302)$$

$$\implies \mathcal{J}_A^\mu = \mathcal{J}_L^\mu + \mathcal{J}_R^\mu = \sqrt{2}f\left(D^\mu \Phi - \tfrac{1}{3f^2}[[D^\mu \Phi, \Phi], \Phi] + \dots\right).$$

To leading order, the pion decay constant is then,

$$i\sqrt{2}f_\pi p^\mu = \langle 0|u\gamma^\mu \gamma_5 \bar{d}|\pi^+(p)\rangle = \langle 0|(\mathcal{J}_A^\mu)_{12}|\pi^-(p)\rangle = i\sqrt{2}f p^\mu + \dots \quad (12.303)$$

where we have used the charge basis (12.272) to write the order $D^\mu \pi^+$ part[1] of the current as,

$$(\mathcal{J}_A^\mu)_{12} \supset -\sqrt{2}f\left[1 - \tfrac{1}{3f^2}(2\pi^0 \pi^0 + 2\pi^- \pi^+ + \bar{K}^0 K^0 + K^+ K^-) + \dots\right]D^\mu \pi^+ \quad (12.304)$$

(we are working in the isospin limit where $\pi_3 \approx \pi^0$). The cubic vertices give rise to the following one-loop contributions to the matrix element

$$\tfrac{1}{f^2}\langle 0|\pi^0 \pi^0 D^\mu \pi^+|\pi^+(p)\rangle = i\frac{\mu^{2\epsilon}}{f^2}\int \frac{d^d k}{(2\pi)^d}\frac{-ip^\mu}{k^2 - M_\pi^2} = \tfrac{2}{f^2}\langle 0|\pi^+ \pi^- D^\mu \pi^+|\pi^+(p)\rangle \quad (12.305)$$

$$= -ip^\mu \frac{M_\pi^2}{(4\pi f)^2}\left(\frac{4\pi \mu^2}{M_\pi^2}\right)^\epsilon \Gamma(-1+\epsilon) \quad (12.306)$$

$$= -ip^\mu \frac{M_\pi^2}{(4\pi f)^2}\left[\tfrac{1}{\epsilon} - \gamma_E + 1 + \log\left(\frac{4\pi \mu^2}{M_\pi^2}\right) + \mathcal{O}(\epsilon)\right] \quad (12.307)$$

$$= ip^\mu \frac{M_\pi^2}{(4\pi f)^2}\log\frac{M_\pi^2}{\mu^2} \quad \text{after subtraction,} \quad (12.308)$$

where we have used (higher-order) counterterms to subtract the $1/\epsilon - \gamma_E + 1 + \log 4\pi$. Note that an overall factor of $1/2$ comes from the pion wave function normalization. The kaon elements are identical, with M_π replaced by M_K. This gives a pion decay constant,

$$f_\pi = f\left(1 - \frac{M_\pi^2}{(4\pi f)^2}\log\frac{M_\pi^2}{\mu^2} - \frac{M_K^2}{2(4\pi f)^2}\log\frac{M_K^2}{\mu^2}\right) + \text{Tree level } \mathcal{L}_4 \quad (12.309)$$

where to reliably include the loop corrections (which are order $1/(4\pi f)^2$), we must also include the tree-level corrections from the $\mathcal{O}(p^4)$ part of the Lagrangian.

---

[1] Note that the terms in π^+ do not contribute to the one-loop amplitude, because terms like $\pi^0 D^\mu \pi^0 \pi^+$ contain loop integrands $k^\mu/(k^2 - M_\pi^2)$ that vanish, and so we have omitted them from (12.304).

The terms in $\mathcal{L}_4$ which contribute at tree level are

$$\mathcal{L}_4 \supset L_4 \langle D_\mu U^\dagger D^\mu U \rangle \langle U^\dagger \mathcal{M} + \mathcal{M}^\dagger U \rangle + L_5 \langle D_\mu U^\dagger D^\mu U \left(U^\dagger \mathcal{M} + \mathcal{M}^\dagger U \right) \rangle \quad (12.310)$$

where we have absorbed a factor of B_0 into the conventional definitions of L_4 and L_5. The contribution to the current is

$$\mathcal{J}_L^\mu = \tfrac{\partial}{\partial \ell_\mu} \mathcal{L}_2 = i2L_4 D^\mu U^\dagger U \langle U^\dagger \mathcal{M} + \mathcal{M}^\dagger U \rangle + iL_5 \{D^\mu U^\dagger, U^\dagger \mathcal{M} + \mathcal{M}^\dagger U\} U \quad (12.311)$$

$$= \tfrac{\sqrt{2}}{f} [4L_4 \langle \mathcal{M} \rangle D^\mu \Phi + 2L_5 \mathcal{M} D^\mu \Phi + 2L_5 D^\mu \Phi \mathcal{M}] + \cdots \quad (12.312)$$

$$\mathcal{J}_R^\mu = \tfrac{\partial}{\partial r_\mu} \mathcal{L}_2 = -\tfrac{\sqrt{2}}{f} [4L_4 \langle \mathcal{M} \rangle D^\mu \Phi + 2L_5 \mathcal{M} D^\mu \Phi + 2L_5 D^\mu \Phi \mathcal{M}] + \cdots \quad (12.313)$$

and so restoring the overall factor of B_0,

$$(\mathcal{J}_A^\mu)_{12} = -\tfrac{8\sqrt{2}B_0}{f} [2L_4(m_u + m_d + m_s) + L_5(m_u + m_d)] D^\mu \pi^+$$

and so by using the L_4 and L_5 coefficients to absorb the one-loop UV divergence and running with μ, we find a pion decay constant,

$$f_\pi = f \Bigg(1 - \tfrac{M_\pi^2}{(4\pi f)^2} \log \tfrac{M_\pi^2}{\mu^2} - \tfrac{M_K^2}{2(4\pi f)^2} \log \tfrac{M_K^2}{\mu^2}$$

$$+ \tfrac{8M_K^2 + 4M_\pi^2}{f^2} \bar{L}_4(\mu) + \tfrac{4M_\pi^2}{f^2} \bar{L}_5(\mu) \Bigg). \quad (12.314)$$

Exercise 3.6 Assume the existence of a hypothetical light Higgs that couples to quarks with the Yukawa interaction

$$\mathcal{L}_{h^0 \bar{q}q} = -\tfrac{h^0}{v} \sum_q k_q m_q \bar{q}q.$$

a) Determine at lowest-order in the χPT expansion the effective Lagrangian describing the Higgs coupling to pseudoscalar mesons induced by the light-quark Yukawas.
b) Determine the effective $h^0 G_a^{\mu\nu} G_{\mu\nu}^a$ coupling induced by heavy quark loops.
c) The $G_a^{\mu\nu} G_{\mu\nu}^a$ operator can be related to the trace of the energy-momentum tensor, in the three-flavour QCD theory:

$$\Theta_\mu^\mu = \tfrac{\beta_1 \alpha_s}{4\pi} G_a^{\mu\nu} G_{\mu\nu}^a + \bar{q}\mathcal{M}q,$$

where $\beta_1 = -\tfrac{9}{2}$ is the first coefficient of the β function. Using this relation, determine the lowest-order χPT Lagrangian incorporating the Higgs coupling to pseudoscalar mesons induced by the heavy-quark Yukawas.
d) Compute the decay amplitudes $h^0 \to 2\pi$ and $\eta \to h^0 \pi^0$.

SOLUTION:

a) The light Higgs interaction term in the Lagrangian is, from the point of view of the quark fields, the same as the mass term. It is therefore convenient to (superficially) combine these terms,

$$\mathcal{L} \supset -\sum_q m_q \bar{q}q + \mathcal{L}_{h^0\bar{q}q} = -\sum_q \widetilde{m}_q \bar{q}q,$$

$$\widetilde{m}_q \equiv m_q \left(1 + \frac{h^0}{v} k_q\right). \qquad (12.315)$$

We can thus use the ordinary χPT expansion, with minor modifications to the light quark mass matrix. We shall consider $q = \{u,d,s\}$, which was discussed thoroughly in the lectures. To be explicit (angle brackets denote the trace),

$$\mathcal{L}_{h^0\bar{q}q}^{\text{eff}} = \frac{f^2}{4} \left\langle D_\mu U D^\mu U^\dagger + \chi U^\dagger + U\chi^\dagger \right\rangle, \qquad (12.316)$$

where U is a unitary SU(3) matrix, parametrized by the pseudoscalar octet Φ, and $f = f_\pi$ to this order (see problem 3.2). The (light) quark mass-matrix (which enters in $\chi = 2B_0\widetilde{\mathcal{M}}$) takes the form

$$\widetilde{\mathcal{M}} = \text{diag}(\widetilde{m}_u, \widetilde{m}_d, \widetilde{m}_s).$$

b) Coupling h^0 to the gluon-sector is provided via vertex corrections (in full QCD) arising from the heavy flavours. The lowest order contribution is the triangle diagram (Figure 12.12), where the heavy quark of mass m_Q, couples to the gluons. We consider here the frame in which the Higgs has zero four-momentum (and are hence neglecting the Higgs mass).

   Evaluating the diagram with standard techniques, we find it to be

$$-\frac{2\alpha_s}{3\pi} \frac{k_Q}{v} m_Q^2 \delta^{ab} \left(g_{\mu\nu} - \frac{p_\mu p_\nu}{p^2}\right) F\left(\frac{m_Q^2}{p^2}\right).$$

Here the function F is what remains of the usual Feynman parametrization:

$$F(x) \equiv \int_0^1 dy \frac{y}{y(1-y)-x} \simeq -\frac{1}{2x} + \dots; \quad x \to \infty. \qquad (12.317)$$

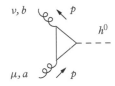

Figure 12.12 *Triangle diagram that couples the light Higgs to gluons. Here we consider the special case where the Higgs field has zero four-momentum. There is also a diagram with the gluons crossed (not shown).*

Taking the limit $m_Q^2 \gg p^2$ (i.e. $x \to \infty$) is needed for the low-energy theory. These heavy-quark loops are induced by the assumed Yukawa interaction with h^0. The Higgs-gluon interaction therefore follows from a term in the chiral Lagrangian,

$$\mathcal{L}_{h^0 gg} = \frac{\alpha_s}{4\pi} \frac{h^0}{v} \bar{k} G_a^{\mu\nu} G_{\mu\nu}^a, \tag{12.318}$$

where we have summed over the relevant heavy degrees of freedom and defined the 'average' Yukawa coupling $\bar{k} \equiv (k_c + k_b + k_t)/3$.

c) The trace of the energy-momentum tensor that follows from the effective chiral Lagrangian (12.316), but now omitting the Higgs-quark coupling ($\widetilde{\mathcal{M}} \to \mathcal{M}$),

$$\Theta_\mu^\mu = -\frac{f^2}{2} \Big\langle D_\mu U D^\mu U^\dagger + B_0(\mathcal{M}U^\dagger + U\mathcal{M}) \Big\rangle. \tag{12.319}$$

By identifying this expression with the corresponding Θ_μ^μ from full QCD, we may rewrite the $G_a^{\mu\nu} G_{\mu\nu}^a$ operator in terms of light quark and pseudoscalar fields. We also recall that the quark mass term $\bar{q}\mathcal{M}q$ is associated with factors proportional to B_0 in in the chiral Lagrangian. This is the matching procedure that reveals the low-energy representation of (12.318), namely

$$\mathcal{L}_{h^0 gg}^{\text{eff}} = -\frac{f^2}{2\beta_1} \frac{h^0}{v} \bar{k} \Big\langle D_\mu U D^\mu U^\dagger + 3B_0(\mathcal{M}U^\dagger + U\mathcal{M}) \Big\rangle, \tag{12.320}$$

where $\beta = -\frac{9}{2}$ is the first coefficient of the β-function.

d) Equations (12.316) and (12.319) together give the interaction of the light Higgs with the Goldstone bosons. Keeping then the terms proportional to h^0, we find that

$$\mathcal{L}_{h^0}^{\text{eff}} = -2\bar{k} \frac{h^0}{\beta_1 v} [\partial_\mu \pi^+ \partial^\mu \pi^- + \tfrac{1}{2} \partial_\mu \pi^0 \partial^\mu \pi^0 + \ldots] \tag{12.321}$$

$$- B_0 \frac{h^0}{v} \Big\{ (c_u m_u + c_d m_d)[\pi^+ \pi^- + \tfrac{1}{2}\pi^0 \pi^0]$$

$$+ \frac{1}{\sqrt{3}}(c_u m_u - c_d m_d)\pi^0 \eta + \ldots \Big\},$$

after expanding the matrix $U = \exp(i\sqrt{2}\Phi/2)$ and defining $c_q \equiv k_q - 3\bar{k}/\beta_1$. In (12.321) we have only kept the terms that are needed for decay amplitudes $h^0 \to 2\pi$ and $\eta \to h^0 \pi^0$; the Higgs also couples to strange mesons from the off-diagonal pieces in Φ.

In the isospin limit, $m_u = m_d \equiv \hat{m}$ and thus

$$B_0 = \frac{M_\pi^2}{m_u + m_d} \to \frac{M_\pi^2}{2\hat{m}},$$

where M_π is the pion mass. Then it is easy to read the decay amplitudes directly from the appropriate terms in (12.321),

$$T(h^0 \to 2\pi) = \tfrac{3}{\beta_1 v}\bar{k}(p_1 \cdot p_2) - B_0 \tfrac{3\hat{m}}{2v}(c_u + c_d)$$

$$= -\tfrac{M_\pi^2}{12v}(9(k_u + k_d) + 4\bar{k}),$$

where p_1 and p_2 are the outgoing four-momenta of the pions. We used the fact that $p_1 \cdot p_2 \approx -M_\pi^2$, neglecting the Higgs mass. The second decay amplitude is

$$T(\eta \to h^0 \pi^0) = -B_0 \tfrac{\hat{m}}{\sqrt{3}v}(c_u - c_d) = \tfrac{M_\pi^2}{2\sqrt{3}v}(k_u - k_d).$$

This channel is particularly interesting, since it is only possible if the up and down quark couple differently to h^0. In the $O(p^2)$ chiral Lagrangian, without the Yukawa coupling in (12.315), the $\eta - \pi^0$ term vanishes.

12.5   EFTs and Inflation (Burgess): Solutions

These three lectures [5] introduced inflationary cosmology, focusing on some uses of effective field theories in its analysis. This section contains some introductory problems relating to the field.

Exercise 12.5.1 Slow Growth of Fluctuations During Radiation Domination

The equation governing the growth of density fluctuations for non-relativistic matter in a spatially flat FRW geometry is

$$\ddot{\delta}_{\mathbf{k}} + 2H\dot{\delta}_{\mathbf{k}} + \left(\tfrac{c_s^2 k^2}{a^2} - 4\pi G\rho_{m0}\right)\delta_{\mathbf{k}} = 0, \tag{12.322}$$

where $\delta = \delta\rho_m/\rho_{m0}$ is the fractional fluctuation in the matter density, $\mathbf{k}$ is its Fourier label while a is the scale factor and $H = \dot{a}/a$ and so $H^2 = 8\pi G\rho_0/3$.

For a matter-dominated universe, for which $\rho_0 \simeq \rho_{m0} \propto 1/a^3$ and $a \propto t^{2/3}$, show that as $c_s \mathbf{k} \to 0$ eqn (12.322) gives power-law solutions of the form $\delta_0 \propto t^n$ with $n = \tfrac{2}{3}$ or $n = -1$. (The growing mode verifies the claim in class that $\delta_0 \propto a$ during matter domination.)

Consider now the transition between radiation and matter domination, for which $\rho_0 = \rho_{m0} + \rho_{r0}$ and so

$$H^2(a) = \tfrac{8\pi G\rho_0}{3} = \tfrac{H_{eq}^2}{2}\left[\left(\tfrac{a_{eq}}{a}\right)^3 + \left(\tfrac{a_{eq}}{a}\right)^4\right], \tag{12.323}$$

where radiation-matter equality occurs when $a = a_{eq}$, at which point $H(a = a_{eq}) = H_{eq}$. The matter part of this expansion comes from

$$H_m^2 := \tfrac{8\pi G\rho_{m0}}{3} = \tfrac{H_{eq}^2}{2}\left(\tfrac{a_{eq}}{a}\right)^3. \tag{12.324}$$

Verify that $\delta_0(x)$ satisfies

$$2x(1+x)\delta_0'' + (3x+2)\delta_0' - 3\delta_0 = 0, \tag{12.325}$$

where the scale factor, $x = a/a_{eq}$, is used as a proxy for time and primes denote differentiation with respect to x. Show that this is solved by $\delta_0 \propto (x + \frac{2}{3})$, and thereby show how the growing mode during matter domination does not grow during radiation domination. (*Bonus:* show that the linearly independent solutions to this one only grow logarithmically with x deep in the radiation-dominated era, for which $x \ll 1$.)

SOLUTION:

Slow growth of fluctuations during radiation domination

$$\ddot{\delta}_k + 2H\dot{\delta}_k + \left(\frac{c_s^2 k^2}{a^2} - 4\pi G\rho_{m0}\right)\delta_k = 0 \tag{12.326}$$

In a matter-dominated Universe, we can use the Friedmann equation, $3H^2 = 8\pi G\rho_0$, to write the evolution for long wavelength modes as

$$\ddot{\delta}_0 + 2H\dot{\delta}_0 - \frac{3}{2}H^2\delta_0 = 0 \tag{12.327}$$

Changing the dependent variable to $a(t)$, we can use, $da = aH\,dt$ and $H(t) = H_0 a^{-3/2}$ to write

$$0 = H_0 a^{-1/2}(H_0 a^{-1/2}\delta_0')' + 2H_0^2 a^{-2}\delta_0' - \frac{3}{2}H_0^2 a^{-3}\delta_0 \tag{12.328}$$

$$= H_0^2 a^{-3}\left(a^2 \partial_a^2 + \frac{3}{2}a\partial_a - \frac{3}{2}\right)\delta_0 \tag{12.329}$$

$$= H^2 \sum_n \left(n + \frac{3}{2}\right)(n - 1)\delta_0^{(n)} a^n \tag{12.330}$$

where we have written $\delta_0(a)$ as $\sum_n \delta_0^{(n)} a^n$, and can conclude that the general solution is,

$$\delta_0(t) = c_1 a + c_2 a^{-3/2} \propto c_1 t^{2/3} + c_2 t^{-1} \tag{12.331}$$

where c_1, c_2 are constants of integration (see method of Frobenius).

Now consider

$$H^2(a) = \frac{H_{eq}^2}{2}\left[x^{-3} + x^{-4}\right], \quad 4\pi G\rho_{m0} = \frac{3H_{eq}^2}{4}x^{-3} \tag{12.332}$$

where $x = a/a_{eq}$ is the dependent variable, and so we have

$$0 = \frac{H_{eq}^2}{2}\left[\frac{\sqrt{1+x}}{x}\left(\frac{\sqrt{1+x}}{x}\delta_0'\right)' + 2\frac{1+x}{x^2}\delta_0' - \frac{3}{2x^3}\delta_0\right] \tag{12.333}$$

$$= \frac{H_{eq}^2}{4x^3}[2x(1+x)\partial_x^2 + (3x+2)\partial_x - 3]\delta_0 \tag{12.334}$$

$$= 2H_m^2 \sum_n \left((2n+3)(n-1)\delta_0^{(n)} + 2n^2\delta_0^{(n-1)}\right). \tag{12.335}$$

One solution to this is, $\delta \propto (x + 2/3)$. During radiation domination, we have $x < 1$, and so the constant $2/3$ piece of the δ_0 fluctuations is important.

The other linearly independent solution is

$$\delta_0 = 3y - \left(1 + \tfrac{3x}{2}\right)\log\tfrac{1+y}{1-y}, \quad y = \sqrt{1+x},$$ (12.336)

which grows like $\log(x)$ for $x \ll 1$.

Exercise 12.5.2 Calculation of Vacuum Energy for a Scalar Field in a Static Space-time

There are a variety of ways commonly used to compute quantum corrections to the vacuum energy, and this tutorial is meant to show how they are related. For the purposes of the exercise a free real scalar field is used, with action $S = \int dt L = \int d^4x \mathcal{L}$ and Lagrangian $L = \int d^3x \mathcal{L}$. The Lagrangian density is

$$\mathcal{L} = -\sqrt{-g}\left[\tfrac{1}{2}\partial_\mu\phi\partial^\mu\phi + \tfrac{1}{2}m^2\phi^2\right],$$ (12.337)

and the resulting field equation is the Klein–Gordon equation

$$(-\Box + m^2)\phi = (-g^{\mu\nu}\nabla_\mu\nabla_\nu + m^2)\phi = 0.$$ (12.338)

But the relationship between the calculations described below is more general than just for this one example.

Canonical calculation The simplest approach to calculating the vacuum energy is the same calculation that identifies all of the energy eigenstates and eigenvalues. This starts by assuming a static background space-time with metric $ds^2 = -dt^2 + g_{ij}dx^i dx^j$, for which g_{ij} is time-independent and a conserved energy can be formulated. Using the above action the field's canonical momentum is

$$\pi(x) = \tfrac{\delta S}{\delta\dot\phi(x)} = \sqrt{-g}\dot\phi(x),$$ (12.339)

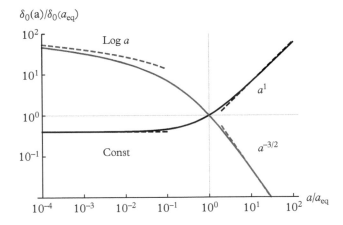

Figure 12.13 *The two independent solutions for growth of long wavelength fluctuations, $\delta_0(t)$.*

(where an over-dot as in $\dot\phi$ denotes ∂_t) and so the Hamiltonian density is

$$\mathcal{H} = \pi\dot\phi - \mathcal{L} = \frac{\pi^2}{2\sqrt{-g}} + \tfrac{1}{2}\sqrt{-g}\left[g^{ij}\nabla_i\phi\nabla_j\phi + m^2\phi^2\right]. \tag{12.340}$$

Background about quantization and mode functions Because this is quadratic in the fields it is essentially a fancy harmonic oscillator. To diagonalize it we expand the fields in terms of creation and annihilation operators

$$\phi(x) = \sum_n \left[a_n U_n(x) + a_n^\star U_n^*(x)\right], \tag{12.341}$$

where we choose the mode functions, $U_n(x)$, to be simultaneous eigenstates of $-g^{ij}\nabla_i\nabla_j$ and $i\partial_t$. That is they satisfy the Klein–Gordon equation, $(-\Box + m^2)$ $U_n = 0$, in a basis that also satisfies

$$-g^{ij}\nabla_i\nabla_j U_n(x) = \omega_n^2 U_n(x) \qquad \text{and} \qquad i\dot U_n = \varepsilon_n U_n(x), \tag{12.342}$$

for eigenvalues ω_n^2 and ε_n. The Klein–Gordon equation imposes a relation between these eigenvalues since $-\Box U_n = \ddot U_n - g^{ij}\nabla_i\nabla_j U_n = (-\varepsilon_n^2 - g^{ij}\nabla_i\nabla_j)U_n$ and so

$$\left(-g^{ij}\nabla_i\nabla_j + m^2\right)U_n = (\omega_n^2 + m^2)U_n = \varepsilon_n^2 U_n. \tag{12.343}$$

This shows how $\varepsilon_n^2 = \omega_n^2 + m^2$ gets determined by the spectrum of $g^{ij}\nabla_i\nabla_j$ for the spacetime of interest.

So we may write

$$U_n(\mathbf{x}, t) = \frac{1}{\sqrt{2\varepsilon_n}} u_n(\mathbf{x}) e^{-i\varepsilon_n t}, \tag{12.344}$$

where the prefactor is chosen for later convenience. Similarly

$$\pi(x) = \sqrt{-g}\dot\phi = -i\sqrt{-g}\sum_n \varepsilon_n\left[a_n U_n(x) - a_n^\star U_n^*(x)\right]. \tag{12.345}$$

The covariant normalization condition for the modes is defined using the Wronskian by

$$W_\Sigma(U_n, U_m) := -i\int_\Sigma d^3\mathbf{x}\sqrt{-g}\left[\dot U_n^*(x) U_m(x) - U_n^*(x)\dot U_m(x)\right] \tag{12.346}$$

$$= \int_\Sigma d^3\mathbf{x}\sqrt{-g}\left[\varepsilon_n U_n^* U_m + \varepsilon_m U_n^* U_m\right] = \delta_{mn}, \tag{12.347}$$

where Σ is a slice of fixed t. Similarly, because (12.342) tell us $i\dot U_n^* = -\varepsilon_n U_n^*$, we see that U_n^* and U_n are eigenstates for different energy eigenvalues (notice for $m \neq 0$ there are no zero eigenvalues), and so are also orthogonal

$$W_\Sigma(U_n^*, U_m) := -i \int_\Sigma d^3\mathbf{x}\sqrt{-g}\left[\dot{U}_n(x)U_m(x) - U_n(x)\dot{U}_m(x)\right] \tag{12.348}$$

$$= (\varepsilon_m - \varepsilon_n) \int_\Sigma d^3\mathbf{x}\sqrt{-g}\, U_n U_m = 0. \tag{12.349}$$

It does not matter which t we choose for W when evaluating these orthogonality conditions provided the falloff of U_n is sufficiently good at spatial infinity (if this exists), and this is the point of why W is defined the way it is. To see why notice $(-\Box + m^2)U_n = 0$ implies the following chain of equalities

$$0 = -i \int_\Sigma^{\Sigma'} d^4x\sqrt{-g}\left[[(-\Box + m^2)U_n]^* U_m - U_n^*[(-\Box + m^2)U_m]\right] \tag{12.350}$$

$$= i \int_\Sigma^{\Sigma'} d^4x\sqrt{-g}\nabla_\mu\left[(\nabla^\mu U_n)^* U_m - U_n^*(\nabla^\mu U_m)\right] \tag{12.351}$$

$$= i \oint_\Sigma^{\Sigma'} d^3x\sqrt{-g}n_\mu\left[(\nabla^\mu U_n)^* U_m - U_n^*(\nabla^\mu U_m)\right] \tag{12.352}$$

$$= W_\Sigma(U_n, U_m) - W_{\Sigma'}(U_n, U_m). \tag{12.353}$$

Here the integration in the first line is over a slab of spacetime lying between two constant-t slices, Σ and Σ'. The second line then integrates both terms by parts and the third line uses Gauss' theorem to write the result in terms of a surface integral over the boundaries of the spacetime region of interest, with n_μ being the outward-pointing normal. If there are no spatial boundaries (or if the boundary conditions are chosen at spatial infinity appropriately) then the only boundaries contributing to the integrals are Σ and Σ'. Then $n_\mu dx^\mu = \pm dt$ for the two surfaces, and the last line follows by recognizing that the surface integrals are precisely the Wronskians for each of the bounding constant-t surfaces. Comparing first and last lines shows that $W_\Sigma(U_n, U_m)$ does not depend on Σ.

Given the above conventions and normalization condition, completeness of the modes implies

$$\sum_n u_n(\mathbf{x})u_n^*(\mathbf{y}) = \frac{\delta^3(\mathbf{x}, \mathbf{y})}{[-g(\mathbf{x})]^{1/4}[-g(\mathbf{y})]^{1/4}}, \tag{12.354}$$

where the delta function transforms as a bi-density distribution that vanishes when $\mathbf{x} \neq \mathbf{y}$ and satisfies the defining condition

$$f(\mathbf{x}, t) = \int d^3\mathbf{y}\delta^3(\mathbf{x}, \mathbf{y})f(\mathbf{y}, t) \tag{12.355}$$

for all f without any metrics. The completeness condition is related to the normalization condition because multiplying (12.354) by $\sqrt{-g(\mathbf{y})}\, u_m(\mathbf{y})$ and integrating over $\mathbf{y}$ must give the tautology $u_m(\mathbf{x}) = u_m(\mathbf{x})$, which it does but only because the u_ms are orthogonal and (12.346) implies each mode satisfies the normalization condition

$$\int_{\Sigma} d^3x \sqrt{-g} u_n^*(\mathbf{x}) u_n(\mathbf{x}) = 2\varepsilon_n \int_{\Sigma} d^3x \sqrt{-g} U_n^*(\mathbf{x}) U_n(\mathbf{x}) = 1. \tag{12.356}$$

Finally, the harmonic oscillator (or creation and annihilation) operator algebra is equivalent to the canonical quantization conditions because

$$[a_n, a_m] = 0 \qquad \text{and} \qquad [a_n, a_m^\star] = \delta_{nm} \tag{12.357}$$

imply

$$[\phi(x), \phi(y)] = \sum_{nm} \{ U_n(x) U_m^*(y)[a_n, a_m^\star] + U_n^*(x) U_m(y)[a_n^\star, a_m] \} \tag{12.358}$$

$$= \sum_n \tfrac{1}{2\varepsilon_n} \{ u_n(\mathbf{x}) u_n^*(\mathbf{y}) - u_n^*(\mathbf{x}) u_n(\mathbf{y}) \} = 0, \tag{12.359}$$

and

$$[\pi(\mathbf{x}, t), \phi(\mathbf{y}, t)] = -i\sqrt{-g(\mathbf{x})} \sum_{nm} \varepsilon_n \{ U_n(x) U_m^*(y)[a_n, a_m^\star] - U_n^*(x) U_m(y)[a_n^\star, a_m] \}$$

$$\tag{12.360}$$

$$= -\tfrac{i}{2}\sqrt{-g(\mathbf{x})} \sum_n \{ u_n(\mathbf{x}) u_n^*(\mathbf{y}) + u_n^*(\mathbf{x}) u_n(\mathbf{y}) \} \tag{12.361}$$

$$= -i\frac{[-g(\mathbf{x})]^{1/4}}{[-g(\mathbf{y})]^{1/4}} \delta^3(\mathbf{x}, \mathbf{y}) = -i\delta^3(\mathbf{x}, \mathbf{y}). \tag{12.362}$$

Calculation of the energy eigenvalues and eigenstates

1. The point of the above is that the energy is diagonal when expressed in terms of the eigenstates of $a_n^\star a_n$, as we see by evaluating the Hamiltonian in terms of a_n and $a_n^\star$. To this end write

$$H = \int d^3x \,\mathcal{H} = \int d^3x \left\{ \frac{\pi^2}{2\sqrt{-g}} + \tfrac{1}{2}\sqrt{-g}\left[g^{ij}\nabla_i\phi\nabla_j\phi + m^2\phi^2 \right] \right\} \tag{12.363}$$

$$= \int d^3x \left\{ \frac{\pi^2}{2\sqrt{-g}} + \tfrac{1}{2}\sqrt{-g}\phi\left[-g^{ij}\nabla_i\nabla_j\phi + m^2\phi \right] \right\}, \tag{12.364}$$

and show that it can be written

$$H = \tfrac{1}{2}\sum_n \varepsilon_n \left(a_n^\star a_n + a_n a_n^\star \right). \tag{12.365}$$

2. The previous question shows that H is diagonal in the basis for which the operators $a_n^\star a_n$ are diagonal for all n. Show this by using the commutation relation $[a_n, a_m^\star] = \delta_{nm}$ to rewrite H as

$$H = E_0 + \sum_n \varepsilon_n a_n^\star a_n, \tag{12.366}$$

with the constant E_0 being formally written as

$$E_0 = \tfrac{1}{2} \sum_n \varepsilon_n. \tag{12.367}$$

This expression is 'formal' because the sum typically diverges. It can be regularized in many ways (and you might reasonably wonder whether or not physical results depend on which way is used). One such is zeta-function regularization, which defines

$$\zeta(s) := \sum_n \varepsilon^{-s} \tag{12.368}$$

for complex s. This often converges where the real part of s is sufficiently large and positive, and one tries to analytically extend this result down to the desired result $E_0 = \zeta(-1)$. Another way to proceed is instead to differentiate E_0 sufficiently many times with respect to m^2 that the sum converges, and then integrate the sum again to get E_0,

The energy eigenvalues for H are clearly given by $H|\{N_k\}\rangle = E|\{N_k\}\rangle$ with

$$E = E_0 + \sum_n N_n \varepsilon_n, \tag{12.369}$$

where the next exercise shows the allowed values for the N_n are $N_n = 0, 1, 2, \cdots$. The state $|0\rangle$ denotes the ground state (or vacuum) for which $N_n = 0$ for all n and has eigenvalue

$$H|0\rangle = E_0|0\rangle. \tag{12.370}$$

3. The basis diagonalizing $a_n^\star a_n$ for all n is called the 'occupation-number' basis and denoted $|\{N_k\}\rangle = |N_{n_1}, N_{n_2}, N_{n_3}, \cdots\rangle$ where the labels N_n are the eigenvalues for $a_n^\star a_n$, for all possible values taken by n. That is, they satisfy

$$a_n^\star a_n |\{N_k\}\rangle = N_n |\{N_k\}\rangle. \tag{12.371}$$

Prove that the $N_n = 0, 1, 2, \cdots$ are non-negative integers as follows. First prove $[a_n^\star a_n, a_m] = -\delta_{nm} a_n$ and $[a_n^\star a_n, a_m^\star] = +\delta_{nm} a_n^\star$. Show that these relations imply that if $|\{N_k\}\rangle$ is an eigenstate with eigenvector of $a_n^\star a_n$ with eigenvalue N_n then $a_n|\{N_k\}\rangle$ is also an eigenstate of $a_n^\star a_n$ but with eigenvalue $N_n - 1$ and $a_n^\star|\{N_k\}\rangle$ is an eigenvector with eigenvalue $N_n + 1$.

Next prove $N_n \geq 0$ by evaluating $\langle\{N_k\}|a_n^\star a_n|\{N_k\}\rangle = N_n\langle\{N_k\}|\{N_k\}\rangle = N_n$ and recognizing that the left-hand side is non-negative because it is the norm of the vector $a_n|\{N_k\}\rangle$. But this is inconsistent with the result that a_n always lowers the eigenvalue by one unit unless there exists an eigenstate for which $a_n|\Psi\rangle = 0$. Repeating this argument for all labels n shows there must be a state, $|0\rangle$, for which $a_n|0\rangle = 0$ for all n, and then all other eigenstates of $a_n^\star a_n$ are obtained by acting repeatedly on $|0\rangle$ with $a_n^\star$. (For example consider the particular state $|2_{n_5}, 6_{n_{20}}\rangle$, for which the particle state labelled by n_5 has eigenvalue $N_{n_5} = 2$ for $a_{n_5}^\star a_{n_5}$ and the state labeled by n_{20} has eigenvalue $N_{n_{20}} = 6$ for $a^{\star}_{n_{20}} a_{n_{20}}$. This is proportional to $(a^{\star}_{n_5})^2 (a^{\star}_{n_{20}})^6 |0\rangle$, and so on for any other choices for these eigenvalues.)

Path integral method of evaluating the vacuum energy

An alternate way to proceed instead uses the path integral formulation for the effective action

$$e^{i\Gamma[g]} = \int \mathcal{D}\phi \, e^{iS[\phi, g]}, \tag{12.372}$$

where the action $S[\phi,g]$ is given as the integral over (12.337), regarded as a function of the fields ϕ and $g_{\mu\nu}$. In this expression $\Gamma[g]$ is a contribution to the action for the metric, $g_{\mu\nu}$, obtained after integrating out the field ϕ. It is to be added to other terms (like the Einstein-Hilbert term), but our interest is in anything of the form $\Gamma = -\int d^4x \sqrt{-g} \, \rho_v$, because this gravitates like a cosmological constant (or vacuum energy). For time-translational invariant systems the integral over t diverges proportional to $\int_{-T}^{T} dt = T$ as $T \to \infty$ and it is the energy $E_0 = -\Gamma/T$ that should remain finite in this limit.

Because the functional integral is Gaussian it can be evaluated in terms of a functional determinant of the quadratic operator appearing in the action: $\Delta = (-\Box + m^2)\delta^4(x - y)$.

$$e^{i\Gamma} = \left[\det\left(-\Box + m^2 - i\epsilon\right)\right]^{-1/2}, \tag{12.373}$$

and so

$$\Gamma = \tfrac{i}{2}\ln\det\left(-\Box + m^2 - i\epsilon\right) = \tfrac{i}{2}\mathrm{Tr}\,\ln\left(-\Box + m^2 - i\epsilon\right). \tag{12.374}$$

Here ϵ is a positive quantity that is taken to zero at the end, and imposes (as usual for a Feynman propagator) the right boundary conditions to describe matrix elements in the vacuum. We suppress the $i\epsilon$ in what follows, but recall it when needed by regarding m^2 as having a small negative imaginary part.

To evaluate this again choose eigenfunctions that diagonalize $-g^{ij}\nabla_i\nabla_j$ and $i\partial_t$. That is choose a basis of functions, $V_n(x)$, for which

$$-g^{ij}\nabla_i\nabla_j V_n = \omega_n^2 V_n \quad \text{and} \quad i\partial_t V_n = \varepsilon V_n, \tag{12.375}$$

and so

$$(-\Box + m^2)V_n = (-\varepsilon^2 + \omega_n^2 + m^2)V_n \tag{12.376}$$

is diagonalized with eigenvalues $\lambda_n = -\varepsilon^2 + \omega_n^2 + m^2 = -\varepsilon^2 + \varepsilon_n^2$. Notice that unlike the previous section we do not also have $(-\Box + m^2)U_n = 0$ and so we *cannot* identify ε^2 with $\varepsilon_n^2 := \omega_n^2 + m^2$. Instead ε is the Fourier transform variable for time, arising generically for time-translationally invariant systems.

In terms of this our operator in this basis is

$$\langle n\varepsilon|\Delta|r\varepsilon'\rangle = (-\varepsilon^2 + \omega_n^2 + m^2)2\pi\delta(\varepsilon - \varepsilon')\delta_{nr}, \tag{12.377}$$

and so the trace may be given by taking diagonal elements and summing over their eigenvalues, with

$$\Gamma(m^2) = \tfrac{1}{2}\mathrm{Tr}\,\ln\left(-\Box + m^2\right) = \tfrac{i}{2}\sum_n \int_{-\infty}^{\infty}\frac{\mathrm{d}\varepsilon}{2\pi}\ln(-\varepsilon^2 + \omega_n^2 + m^2)2\pi\,\delta(0).$$

(12.378)

The factor of $\delta(0)$ arises due to time translation invariance, as may be seen by writing

$$2\pi\delta(E) = \lim_{T\to\infty}\int_{-T}^{T}\mathrm{d}t\,e^{-i\varepsilon t} \quad \text{and so} \quad 2\pi\delta(0) = \lim_{T\to\infty}T,$$

(12.379)

and so the well-behaved quantity is the energy

$$E_0 = -\lim_{T\to\infty}\frac{\Gamma}{T} = -\tfrac{i}{2}\sum_n \int_{-\infty}^{\infty}\frac{\mathrm{d}\varepsilon}{2\pi}\ln(-\varepsilon^2 + \omega_n^2 + m^2).$$

(12.380)

Again the remaining sums and integrals diverge. The integration over ε passes through singularities at $\pm\varepsilon_n$, which we should navigate by Wick rotating. That is, keeping in mind (as usual) that $m^2 \to m^2 - i\epsilon$ is required for the Feynman propagator, we can rotate our contour of integration counter-clockwise by ninety degrees in the complex ε plane by writing $\varepsilon \to i\varepsilon_E$ with ε_E also running from $-\infty$ to ∞. The integral converges if we first differentiate with respect to m^2, so show that

$$\frac{\partial E_0}{\partial m^2} = \tfrac{1}{2}\sum_n \int_{-\infty}^{\infty}\frac{\mathrm{d}\varepsilon_E}{2\pi}\left(\frac{1}{\varepsilon_E^2 + \omega_n^2 + m^2}\right)$$

(12.381)

$$= \tfrac{1}{4\pi}\sum_n\left[\frac{1}{\varepsilon_n}\arctan\left(\frac{\varepsilon}{\varepsilon_n}\right)\right]_{-\infty}^{\infty} = \tfrac{1}{4}\sum_n\frac{1}{\varepsilon_n},$$

(12.382)

where, as above, $\varepsilon_n = \sqrt{\omega_n^2 + m^2}$. Integrating again with respect to m^2 then gives

$$E_0(m^2) = \tfrac{1}{2}\sum_n \varepsilon_n,$$

(12.383)

up to an arbitrary m^2-independent constant. This is the same sum as was obtained in the canonical calculation earlier.

Flat space evaluation

As a particularly simple case consider the case of a flat geometry, for which $-g^{ij}\nabla_i\nabla_j = -\nabla^2$ can be diagonalized in Fourier space, with eigenfunctions $\exp(i\mathbf{p}\cdot\mathbf{x})$ and eigenvalues $\mathbf{p}^2$.

In terms of this the required operator in this basis is

$$\langle p|\Delta|q\rangle = (p_\mu p^\mu + m^2)(2\pi)^4\delta^4(p - q),$$

(12.384)

and so the trace may be given by taking diagonal elements and summing over their eigenvalues, with

$$\Gamma(m^2) = \tfrac{i}{2}\mathrm{Tr}\,\ln\left(-\Box + m^2\right) - \tfrac{i}{2}\int_{-\infty}^{\infty}\frac{\mathrm{d}^4p}{(2\pi)^4}\,\ln(p_\mu p^\mu + m^2)(2\pi)^4\delta^4(0). \quad (12.385)$$

The additional factor of $\delta^3(0)$ arises due to spatial translation invariance, as may be seen by writing (as we did before for time)

$$(2\pi)^3\delta^3(\mathbf{p}) = \lim_{L\to\infty}\int_{-L}^{L}\mathrm{d}^3\mathbf{x}\,e^{i\mathbf{p}\cdot\mathbf{x}} \quad \text{and so} \quad (2\pi)^3\delta^3(0) = \lim_{L\to\infty}L^3, \quad (12.386)$$

and so is proportional to the volume of space (as well as the previous proportionality to T). The well-behaved quantity for infinite translationally invariant systems is therefore the energy *density*,

$$\rho_v = \lim_{L\to\infty}\frac{E_0}{L^3} = -\lim_{L,T\to\infty}\frac{\Gamma}{TL^3} = -\tfrac{i}{2}\int_{-\infty}^{\infty}\frac{\mathrm{d}^4p}{(2\pi)^4}\,\ln(p_\mu p^\mu + m^2). \quad (12.387)$$

To avoid the singularities at $p^0 = \pm\sqrt{\mathbf{p}^2 + m^2}$, we again Wick rotate. In the resulting euclidean integral the angular integrals can be done once and for all, giving a factor of the volume of the unit three-sphere: $2\pi^2$. The remaining integral converges if we first differentiate with respect to m^2 thrice, so show that

$$\left(\frac{\partial}{\partial m^2}\right)^3\rho_v = \frac{2\pi^2}{2}\int_0^{\infty}\frac{p_E^3\,\mathrm{d}p_E}{(2\pi)^4}\frac{2}{(p_E^2+m^2)^3} = \frac{1}{32\pi^2 m^2}, \quad (12.388)$$

and so integrating three times with respect to m^2 then gives

$$\rho_v = \frac{m^4}{64\pi^2}\ln\left(\frac{m^2}{\mu^2}\right) + Am^4 + Bm^2 + C, \quad (12.389)$$

where μ, A, B and C are arbitrary m^2-independent constants. Although the values of A, B and C can depend on how the integrals were regulated, the logarithmic term cannot.

SOLUTION: Calculation of vacuum energy for a scalar field in a static spacetime.

1. The Hamiltonian can be written as,

$$H = \int\mathrm{d}^3x\left\{\frac{\pi^2}{2\sqrt{-g}} + \tfrac{1}{2}\sqrt{-g}\phi\left[-g^{ij}\nabla_i\nabla_j + m^2\right]\phi\right\} \quad (12.390)$$

where the canonical field and its conjugate momenta may be expanded in terms of annihilation and creation operators,

$$\phi(x) = \sum_n [a_n U_n(x) + a_n^* U_n^*(x)] \quad (12.391)$$

$$\pi(x) = -i\sqrt{-g}\sum_n \epsilon_n [a_n U_n(x) - a_n^* U_n^*(x)] \quad (12.392)$$

where the mode functions U_n are orthogonal and normalized with respect to the Klein-Gordon norm. This gives,

$$H = \tfrac{1}{2}\sum_n \epsilon_n(a_n^* a_n + a_n a_n^*), \quad (12.393)$$

which shows that number eigenstates (of $a_n^* a_n$) are also energy eigenstates.

2. Using $[a_n, a_n^*] = \delta_{nm}$, the Hamiltonian can be written

$$H = E_0 + \sum_n \epsilon_n a_n^* a_n, \quad (12.394)$$

where $E_0 = \tfrac{1}{2}\sum_n \epsilon_n$ is the zero point energy.

3. Using the canonical commutation relations, we have that

$$[a_n^* a_n, a_m] = -\delta_{nm} a_n \quad (12.395)$$

$$[a_n^* a_n, a_m^*] = +\delta_{nm} a_n^* \quad (12.396)$$

Then for a state with definite occupation numbers, $a_n^* a_n |\{N_k\}\rangle = N_n |\{N_n\}\rangle$, we have that

$$a_m^* a_m(a_n|\{N_k\}\rangle) = a_n a_m a_m^* |\{N_n\}\rangle - \delta_{nm} a_n |\{N_k\}\rangle = (N_m - \delta_{nm})a_n|\{N_k\}\rangle \quad (12.397)$$

and so a_n lowers the occupation number N_n by one, and similarly,

$$a_m^* a_m(a_n^*|\{N_k\}\rangle) = a_n^* a_m a_m^* |\{N_n\}\rangle + \delta_{nm} a_n^* |\{N_k\}\rangle = (N_m + \delta_{nm})a_n^*|\{N_k\}\rangle \quad (12.398)$$

and so a_n^* increases N_n by one.

In order for the Hilbert space to have positive-definite norm, there must be a null state, $a_n|0\rangle = 0$, which prevents any N_n from becoming negative (such states would have negative norm). From this single state, repeated action of a_n^* generates the entire Fock space of possible states.

Exercise 12.5.3 Quantum Fluctuations of a Scalar Field in a Class of Inflationary Space-times

For a change of pace we work in the Schrödinger picture, rather than the Heisenberg picture, and so compute the vacuum wavefunctional, $\Psi[\varphi, t]$, for a scalar field.

Action and Hamiltonian

Our starting point is the Lagrangian density for a spectator scalar

$$L = \int d^3x \; a(t)^3 \left[\tfrac{1}{2}\dot{\phi}^2 - \tfrac{1}{2u^2_{(t)}} (\nabla\phi)^2 - \tfrac{m^2(t)}{2} \phi^2 \right], \tag{12.399}$$

in an FRW spacetime with metric

$$ds^2 = -dt^2 + a^2(t)d\vec{x}^2 \tag{12.400}$$

and Hubble parameter $H(t) = \dot{a}/a$. Here $m(t)$ denotes the (possibly slowly time-dependent) mass.

Find the canonical momentum, π_k, for each Fourier mode, φ_k, of the scalar field. Given the quantization condition $\pi_k = -i\delta/\delta\varphi_k$, show that the Hamiltonian density in Schrödinger representation can be expressed in Fourier space as

$$\mathcal{H} = \mathcal{H}_0 + \sum_k \mathcal{H}_k, \tag{12.401}$$

with $\mathcal{H}_k$ for $k \neq 0$ given by

$$\mathcal{H}_k = -\frac{1}{a^3} \frac{\delta^2}{\delta\varphi_k\delta\varphi_{-k}} + a^3 \left[\frac{c_s^2 k^2}{a^2} + m^2 \right] \varphi_k \varphi_{-k} \tag{12.402}$$

where $\varphi_k^* = \varphi_{-k}$.

Ground state wave functional

Use this Hamiltonian to evolve the state wave-functional, $\Psi = \prod_k \Psi_k$, according to the Schrödinger equation,

$$i\frac{\partial \Psi_k}{\partial t} = \mathcal{H}_k \Psi_k, \tag{12.403}$$

and for free fields seek solutions subject to a Gaussian ansatz,

$$\Psi[\varphi] = \prod_k \Psi_k[\varphi] = \prod_k \mathcal{N}_k(t) \exp\left\{ -a^3(t) \left[\alpha_k(t)\varphi_k\varphi_{-k} \right] \right\} \tag{12.404}$$

and show that the variance of φ_k, $\langle |\varphi_k|^2 \rangle$, is given by $[a^3(\alpha_k + \alpha_k^*)]^{-1}$. Determine the evolution equations for the functions $\mathcal{N}_k(t)$, $\alpha_k(t)$ by substituting into (12.403). Show that they imply α_k must satisfy

$$0 = \dot{\alpha}_k + i\alpha_k^2 + 3H \; \alpha_k - i\left(\tfrac{k^2}{a^2} + m^2 \right) \text{ for } k \geq 0 \tag{12.405}$$

where all quantities (including the Hubble parameter) can be time dependent, and the dot denotes derivative with respect to time. The additional equation for $\mathcal{N}_k$ ensures it evolves in a way that is consistent with normalization, but is not needed in what follows.

The solution for α_k can be made very explicit if we assume power-law expansion, $a = a_0(t/t_0)^p$ (so that $H = p/t$ and $\epsilon = -\dot{H}/H^2 = 1/p$) and a time-independent ratio m/H. (Show that de Sitter space can be obtained as the special case where $p \to \infty$ and so $\epsilon \to 0$.)

Equations of the form of (12.405) are integrated by changing variables from α_k to u_k where

$$\alpha_k = -i\left(\frac{\dot{u}_k}{u_k}\right) = i\,a\,H\left[\frac{\partial_a u_k(a)}{u_k(a)}\right]. \tag{12.406}$$

Show that (12.405) is then satisfied if u_k solves the Klein–Gordon equation,

$$\ddot{u}_k + 3H\dot{u}_k + \left(\frac{k^2}{a^2} + m^2\right)u_k = 0. \tag{12.407}$$

For constant ϵ and m^2/H^2 show that this is solved by

$$u_k(a) = \tilde{C}_k\, y^q \sigma_k(y), \tag{12.408}$$

where $\tilde{C}_k$ is a-independent, provided q and y are chosen as

$$q = \frac{3-\epsilon}{2(1-\epsilon)}, \tag{12.409}$$

and

$$y(a,k) := \frac{1}{(1-\epsilon)}\left(\frac{k}{aH}\right) = \frac{1}{(1-\epsilon)}\left(\frac{k}{a_0 H_0}\right)\left(\frac{a_0}{a}\right)^{1-\epsilon}. \tag{12.410}$$

The point of these changes of variables is that they turn eqn (12.407) into the Bessel equation for σ_k:

$$y^2\sigma_k'' + y\sigma_k' + (y^2 - \nu^2)\sigma_k = 0, \tag{12.411}$$

where primes here denote derivatives with respect to y. Show that the order ν is given by

$$\nu^2 = \frac{1}{(1-\epsilon)^2}\left[\frac{(3-\epsilon)^2}{4} - \frac{m^2}{H^2}\right]. \tag{12.412}$$

The solutions for σ_k are (naturally) Bessel functions, and demanding agreement with the adiabatic vacuum before horizon exit tells us

$$u_k \propto \exp\left[\mp i \int dt\left(\frac{k}{a}\right)\right] \propto e^{\pm iy} \qquad \text{for } k/a \gg H, \tag{12.413}$$

of which we choose the lower sign since this turns out below to ensure the real part of α_k is positive (as required to ensure Ψ_k can be normalized). Show that this fixes the mode functions to be

$$u_k(a) = \tilde{C}_k y^q(a,k) H_\nu^{(2)}\left[y(a,k)\right] = \frac{C_k}{\sqrt{a^3 H}} H_\nu^{(2)}\left[y(a,k)\right] \tag{12.414}$$

where $C_k \propto k^q \tilde{C}_k$ relabels the integration constants and $H_\nu^{(2)}$ is the Hankel function of the second kind. The second equality in (12.414) follows from eqn (12.409), which implies $a^3 H y^{2q}$ is time-independent. Notice this reduces to the solution for a massive field in de Sitter space in the limit $\epsilon \to 0$.

Although $\mathcal{C}_k$ drops out of (12.406) and (so does not contribute directly to α_k), some later formulae are simpler if we choose $\mathcal{C}_k$ so that the Wronskian,

$$\mathcal{W}(u,v) := a^3(u^*\dot{v} - v^*\dot{u}), \tag{12.415}$$

satisfies $\mathcal{W}(u,u) = i$. Prove that in this case is the expression for the real and imaginary parts of α_k become

$$\alpha_k + \alpha_k^* = -i\left(\frac{u_k^*\dot{u}_k - u_k\dot{u}_k^*}{|u_k|^2}\right) = \frac{1}{a^3|u_k|^2} \tag{12.416}$$

$$\text{and} \quad \alpha_k - \alpha_k^* = -iaH\left[\frac{\partial_a(|u_k|^2)}{|u_k|^2}\right]. \tag{12.417}$$

What does the first of these imply for the variance of φ_k in terms of u_k?

Because $\mathcal{W}$ is independent of time (when evaluated with solutions to (12.407) it is convenient to compute the implications for $\mathcal{C}_k$ in the remote past, where $k \gg aH$, in which case the Hankel function has the asymptotic form

$$H_\nu^{(2)}(y) \to \sqrt{\frac{2}{\pi y}}e^{-iy + \frac{i\pi}{2}(\nu + \frac{1}{2})} \quad \text{for } y \to \infty. \tag{12.418}$$

Use this to show

$$|\mathcal{C}_k|^2 = \frac{\pi}{4(1-\epsilon)}, \tag{12.419}$$

for all k and ν.

Consequently the quantity relevant to fluctuations in the lectures is

$$|u_k|^2 = \frac{\pi}{4(1-\epsilon)a^3 H}|H_\nu^{(2)}(y)|^2. \tag{12.420}$$

Use the asymptotic expression

$$H_\nu^{(2)}(y) \to \frac{i\Gamma(\nu)}{\pi}\left(\frac{y}{2}\right)^{-\nu} \quad \text{for } y \to 0, \tag{12.421}$$

to derive the small-k limit

$$|u_k|^2 \to \frac{2^{2\nu-2}|\Gamma(\nu)|^2(1-\epsilon)^{2\nu-1}}{\pi a^3 H}\left(\frac{aH}{k}\right)^{2\nu}. \tag{12.422}$$

Evaluate this for the case $\nu = \frac{3}{2}$ (which for the de Sitter case $\epsilon = 0$ is a massless scalar field) and show that it agrees with the result obtained using the mode function directly, which in the case $\nu = \frac{3}{2}$ is very simple:

$$u_k = -(1-\epsilon)\frac{H}{\sqrt{2k^3}}(y-i)e^{-iy} \quad \text{for } \nu = \frac{3}{2} \tag{12.423}$$

up to an irrelevant phase. Prove that this does solve the Klein–Gordon equation in the case $\nu = \frac{3}{2}$.

The power spectrum $\Delta^2(k)$ is proportional to $k^3|u_k|^2$ evaluated for $k \ll aH$. For de Sitter space H is constant, and in this case what is the predicted k-dependence for $k^3|u_k|^2$ when $\nu = \frac{3}{2}$? When $\epsilon \neq 0$ H is time dependent and we are supposed to evaluate H at the moment where $aH = k$. If this were the whole story (and it is not quite), and if $\Delta^2(k) \propto A(k/k_0)^{n_s-1}$, what is the prediction for n_s as a function of ϵ?

SOLUTION:

Action and Hamiltonian:

$$\pi = \frac{\delta S}{\delta \phi} = a^3 \dot{\varphi} \tag{12.424}$$

Naïvely, we would define,

$$\pi_k = -i\omega_k a^3 \varphi_k$$

which is an explicitly time-dependent momentum. Unlike on flat space, where translation invariance guarantees constant momenta, on FLRW we can have time-dependence. The on-shell condition is,

$$\omega_k^2 = c_s^2 k^2 + m^2$$

where c_s is the sound speed for the scalar fluctuations on this background.

The Hamiltonian is given by,

$$H := \pi\varphi - \mathcal{L} = a^3 \left[\frac{1}{2a^3}\pi^2 + \phi \frac{-\nabla^2/a^2+m^2}{2}\phi \right] \tag{12.425}$$

$$= a^3 \left[\frac{1}{a^3}\pi_k\pi_{-k} + \left(\frac{c_s^2 k^2}{a^2} + m^2 \right)\varphi_k\varphi_{-k} \right] \tag{12.426}$$

Ground state wave functional:

Using the Gaussian ansatz,

$$\Psi[\varphi] = \prod_k \mathcal{N}_k(t) \exp\{-a^3(t)[\alpha_k(t)\varphi_k\varphi_{-k}]\} \tag{12.427}$$

we can solve the Schrödinger equation,

$$i\frac{\partial \Psi_k}{\partial t} = \mathcal{H}_k \Psi_k \tag{12.428}$$

to find

$$i\frac{\dot{\mathcal{N}}_k}{\mathcal{N}_k} - ia^3(3H\alpha_k + \dot{\alpha}_k)\varphi_k\varphi_{-k} = \alpha_k + a^3\left(-\alpha_k^2 + \frac{c_s^2 k^2}{a^2} + m^2 \right)\varphi_k\varphi_{-k} \tag{12.429}$$

As this holds for all φ_k, we must have that,

$$0 = \dot{\alpha}_k + i\alpha_k^2 + 3H\alpha_k - i\left(\frac{k^2}{a^2} + m^2 \right) \quad \text{for } k \geq 0 \tag{12.430}$$

along with the condition that $\dot{\mathcal{N}}_k = i\alpha_k\mathcal{N}_k$.

Power-law solutions: If $a = a_0(t/t_0)^p$, then $H = p/t$. We can take $p \to \infty$ in order to set $\dot{H}/H^2 \to 0$, and simultaneously take the limit $t = t_0 + \delta t \to t_0$ so that,

$$a(t_0) = \lim_{\substack{\delta t \to 0 \\ H \to \infty}} \exp\left(H(t_0 + \delta t)\log(1 + \tfrac{\delta t}{t_0})\right) = \begin{cases} a_0 & \text{if } H\delta t \to 0, \\ a_0 e^{H_0 t_0} & \text{if } H\delta t \to H_0 t_0, \\ 0 & \text{if } H\delta t \to \infty \end{cases} \qquad (12.431)$$

where H_0 is then the constant curvature of the dS space.

Using $\dot{\alpha}_k = -i\ddot{u}_k/u_k - i\alpha_k^2$, we get the Klein–Gordon equation for u_k. Then changing variables with $\dot{y} = -(1 - \epsilon)Hy$, we find,

$$\tilde{C}_k(1 - \epsilon)^2 H^2 y^q \left\{ y^2 \sigma_k'' + y\sigma_k'\left(1 + 2q + \tfrac{\epsilon - 3}{1 - \epsilon}\right) \right.$$
$$\left. + \sigma_k\left(y^2 + \tfrac{m^2}{H^2(1-\epsilon)^2} + q^2 + \tfrac{q(\epsilon - 3)}{1 - \epsilon}\right) \right\} = 0 \qquad (12.432)$$

and so choosing $q = (3 - \epsilon)/2(1 - \epsilon)$ we arrive at Bessel's equation for σ_k, with order

$$\nu^2 = \tfrac{1}{(1-\epsilon)^2}\left[\tfrac{(3-\epsilon)^2}{4} - \tfrac{m^2}{H^2}\right] \qquad (12.433)$$

The Bessel functions of first and second kind have the following asymptotic behaviour,

$$\lim_{z \to \infty} \mathcal{J}_n(z) \sim \sqrt{\tfrac{2}{\pi z}} \cos\left(z - \tfrac{m\pi}{2} - \tfrac{\pi}{4}\right) \qquad (12.434)$$

$$\lim_{z \to \infty} Y_n(z) \sim \sqrt{\tfrac{2}{\pi z}} \sin\left(z - \tfrac{m\pi}{2} - \tfrac{\pi}{4}\right) \qquad (12.435)$$

and therefore the desired asymptotic behaviour for σ is given by a Hankel functions of the second kind,

$$H_n^{(2)}(z) = \mathcal{J}_n(z) - iY_n(z),$$

which gives,

$$u_k = C_k/\sqrt{a^3 H}H_\nu^{(2)}\left(\tfrac{1}{1-\epsilon}\tfrac{k}{aH}\right).$$

The Wronskian,

$$\mathcal{W}(u, u) = a^3[u^*\dot{u} - u\dot{u}^*] = i. \qquad (12.436)$$

The corresponding α_k are,

$$\alpha_k = -i\tfrac{\dot{u}_k}{u_k} \implies \alpha_k + \alpha_k^* = -i\tfrac{\dot{u}_k u_k^* - \dot{u}_k^* u_k}{|u_k|^2} = \tfrac{1}{a^3|u_k|^2}$$

$$\alpha_k = -iaH\tfrac{\partial_a u_k}{u_k} \implies \alpha_k - \alpha_k^* = -iaH\tfrac{\partial_a u_k u_k^* + \partial_a u_k^* u_k}{|u_k|^2}.$$

The variance of φ_k is,

$$\langle|\varphi_k|^2\rangle = \tfrac{1}{a^3(\alpha_k + \alpha_k^*)} = |u_k|^2.$$

To find C_k, we use the asymptotics of the Hankel function to write,

$$u_k \sim \frac{C_k}{\sqrt{a^3 H}} \sqrt{\frac{2}{\pi y}} e^{-iy + \frac{i\pi}{2}(\nu + \frac{1}{2})} \tag{12.437}$$

$$\mathcal{W}(u, u) = a^3 \left(\frac{|C_k|^2}{a^3 H} \frac{2}{\pi y} 2i(1 - \epsilon) Hy \right) = i \tag{12.438}$$

$$\implies \quad |C_k|^2 = \frac{\pi}{4(1-\epsilon)}.$$

The variance is then given by,

$$\lim_{y \to 0} |u_k|^2 \to \frac{\pi}{4(1-\epsilon)a^3 H} \left| \frac{i\Gamma(\nu)}{\pi} \left(\frac{y}{2} \right)^{-\nu} \right|^2 \tag{12.439}$$

$$= \frac{2^{2\nu-2} |\Gamma(\nu)|^2 (1-\epsilon)^{2\nu-1}}{\pi a^3 H} \left(\frac{aH}{k} \right)^{2\nu}. \tag{12.440}$$

Massless modes: In the de Sitter limit, $\epsilon = 0$, we find that,

$$\nu^2 = \frac{9}{4} - \frac{m^2}{H^2}$$

and so massless fields on de Sitter correspond to $\nu = 3/2$.

Specializing to $\nu = 3/2$, and using $\Gamma(3/2) = \sqrt{\pi}/2$,

$$|u_k|^2 \to \frac{(1-\epsilon)^2 H^2}{2k^3}. \tag{12.441}$$

Note that this agrees with the exact mode function,

$$u_k = -(1 - \epsilon) \frac{H}{\sqrt{2k^3}} (y - i) e^{-iy} \implies |u_k|^2 = |1 - \epsilon|^2 \frac{H}{2k^3} |1 + iy|^2.$$

This corresponds to a σ_k,

$$\sigma_k = -\sqrt{\frac{2}{\pi}} y^{-3/2} (y - i) e^{-iy} \tag{12.442}$$

$$\implies \quad y^2 \sigma_k'' + y \sigma_k' = -(y^2 - \frac{3}{2}) \sigma_k$$

and therefore σ_k satisfies the Bessel equation with $\nu = 3/2$.

Considering,

$$k^3 |u_k|^2 |_{\nu=3/2} \to \frac{1}{2}(1 - \epsilon)^2 H^2 \quad \text{for } k \ll aH,$$

we conclude that $k^3 |u_k|^2$ is k-independent on de Sitter,

$$k^3 |u_k|^2 \Big|_{\nu=3/2} \to \frac{1}{2} H_0^2.$$

On FLRW, if we treat ϵ as a constant then we can solve,

$$a(t) = k_0 (\epsilon(t - t_0))^{1/\epsilon}$$

$$aH = k \implies H = (k/k_0)^{-\frac{\epsilon}{1-\epsilon}}$$

and so we find that,

$$n_s - 1 = \frac{-2\epsilon}{1-\epsilon} + \mathcal{O}(\dot{\epsilon}). \tag{12.443}$$

This is the effect that a spectator field has on the spectral tilt—note that there must also be a inflation field to drive the expansion.

12.6 EFT of Large-Scale Structure (Baldauf): Solutions

The series of four lectures [3] in Chapter 7 provides an introduction to the topic of Large-Scale Structure (LSS) with an emphasis on the EFT approach. It begins with a general introduction to LSS phenomenology and relevant statistical tools, and then develops the standard cosmological perturbation theory, which allows us to compute relevant observables. Building on the shortcomings of standard perturbation theory, the EFT approach is introduced and emphasizes different problems that it successfully addresses. Finally, it discusses the relation to observations of biased tracers, which are discussed together with the need to account for redshift-space distortion.

This section contains five introductory problems useful for a familiarization with the different aspects of LSS theory. The first two problems deal with standard phenomenology in cosmology as well as some statistical aspects of Gaussian fields. Problems three and four cover fluid mechanic aspects in an expanding Universe and problem five is an application of perturbation theory. Particular aspects of EFT are briefly touched via questions about counterterms and UV-sensitivity in problem one and five.

Exercise 12.6.1 Clustering of Fixed Height Subsamples Consider a Gaussian random field δ described by a power spectrum P. The corresponding real-space correlation function is ξ and the variance $\sigma^2 = \xi(0)$. Consider the PDF of fluctuations

$$\mathbb{P}_{2\mathrm{pt}}(\delta_1, \delta_2 | r) = \frac{1}{(2\pi)^2 |C(r)|} \exp\left[-\frac{1}{2} Y^T C^{-1}(r) Y\right] \tag{12.444}$$

where

$$\begin{cases} Y = \begin{pmatrix} \delta_1 \\ \delta_2 \end{pmatrix} = \sigma \begin{pmatrix} \nu_1 \\ \nu_2 \end{pmatrix} \\ C_{ij}(r) = \langle \delta_i \delta_j \rangle = \sigma^2 \delta^{(K)} + (1 - \delta^{(K)}) \xi(r) \end{cases} \tag{12.445}$$

are respectively the state vector and the covariance matrix. The correlation matrix of the field can be recovered as

$$\xi(r) = \langle \delta(\mathbf{x}) \delta(\mathbf{x} + \mathbf{r}) \rangle = \sigma^4 \int d\nu_1 \int d\nu_2 \nu_1 \nu_2 \, \mathbb{P}_{2\mathrm{pt}}(\sigma \nu_1, \sigma \nu_2 | r) \tag{12.446}$$

Consider the subset of fluctuations of fixed amplitude ν_c and calculate their correlation function

$$\xi_c(r) = \frac{\mathbb{P}_{2\mathrm{pt}}(\sigma \nu_1, \sigma \nu_2 | r)}{\mathbb{P}_{1\mathrm{pt}}(\sigma \nu_c) \mathbb{P}_{1\mathrm{pt}}(\sigma \nu_c)} - 1 \tag{12.447}$$

For large separations this allows an expansion in the small quantity ξ/σ^2. Write down this expansion to second order. The prefactors of this expansion are called bias parameters. Fourier transform the expression to k-space and consider the low-k limit. Can you identify contributions that would require a counterterm?

Remark: The above model can be used to model regions that will eventually form dark matter halos, i.e. formation sites of galaxies. In this context the field is smoothed on a scale R to consider fluctuations of a given mass $M \propto \bar{\rho} R^3$

$$\delta_R(\mathbf{x}) = \int d^3 x \, W_R(|\mathbf{x} - \mathbf{x}'|) \delta(\mathbf{x}').$$

(12.448)

where $W_R(r)$ is a Gaussian or top hat filter.

SOLUTION:

Computation of $\xi_c(r)$

As δ is a Gaussian random field of variance $\xi(\mathbf{r})$, $\delta_1 = \delta(\mathbf{x})$ and $\delta_2 = \delta(\mathbf{x}+\mathbf{r})$ are, by definition, Gaussian random variables of variance $\xi(0) = \sigma^2$. Thus,

$$\mathbb{P}_{1pt}(\sigma \nu_1) = \frac{1}{\sigma\sqrt{2\pi}} \exp\left[-\frac{\nu_1^2}{2}\right].$$

(12.449)

Moreover,

$$C(r) = \begin{pmatrix} \sigma^2 & \xi(r) \\ \xi(r) & \sigma^2 \end{pmatrix}$$

(12.450)

so that,

$$\begin{cases} |C(r)| = \sigma^4 - \xi(r)^2 & (12.451a) \\ C^{-1}(r) = \dfrac{1}{|C(r)|} \begin{pmatrix} \sigma^2 & -\xi(r) \\ -\xi(r) & \sigma^2 \end{pmatrix}. & (12.451b) \end{cases}$$

Using (12.451) to compute $\mathbb{P}_{2pt}$ and plugging it into (12.447) (together with the explicit form of $\mathbb{P}_{1pt}$,) we get

$$\xi_c(r) = \sqrt{\frac{1}{1 - \left(\frac{\xi(r)}{\sigma^2}\right)^2}} \exp\left[\nu_c^2 \frac{\frac{\xi(r)}{\sigma^2}}{1 + \frac{\xi(r)}{\sigma^2}}\right] - 1.$$

(12.452)

In the following, the explicit r dependence in ξ is dropped off.

Expansion in $\frac{\xi}{\sigma^2}$

Let us assume $\xi \ll \sigma^2$. Expanding in this case,

$$\xi_c = \left(1 - \left(\tfrac{\xi}{\sigma^2}\right)^2\right)^{-\frac{1}{2}} \exp\left[v_c^2 \frac{\frac{\xi}{\sigma^2}}{1 + \frac{\xi}{\sigma^2}}\right] - 1 \tag{12.453}$$

$$= \left\{1 + \tfrac{1}{2}\left(\tfrac{\xi}{\sigma^2}\right)^2 + O\left(\tfrac{\xi}{\sigma^2}\right)^3\right\}$$

$$\times \exp\left[v_c^2 \left\{\tfrac{\xi}{\sigma^2} - \left(\tfrac{\xi}{\sigma^2}\right)^2 + O\left(\tfrac{\xi}{\sigma^2}\right)^3\right\}\right] - 1 \tag{12.454}$$

$$= \left\{1 + \tfrac{1}{2}\left(\tfrac{\xi}{\sigma^2}\right)^2 + O\left(\tfrac{\xi}{\sigma^2}\right)^3\right\}$$

$$\times \left\{1 + v_c^2 \tfrac{\xi}{\sigma^2} + \left(v_c^2 - \tfrac{1}{2}v_c^4\right)\left(\tfrac{\xi}{\sigma^2}\right)^2 + O\left(\tfrac{\xi}{\sigma^2}\right)^3\right\} - 1 \tag{12.455}$$

and finally,

$$\xi_c = v_c^2 \tfrac{\xi}{\sigma^2} + \left(\tfrac{1}{2} + v_c^2 - \tfrac{1}{2}v_c^4\right)\left(\tfrac{\xi}{\sigma^2}\right)^2 + O\left(\tfrac{\xi}{\sigma^2}\right)^3. \tag{12.456}$$

Low-k limit

As $P(k) = \mathrm{FT}[\xi](k)$, the Fourier transformed of (12.456) reads

$$P_c(k) = v_c^2 \frac{P(k)}{\sigma^2} + \left(\tfrac{1}{2} + v_c^2 - \tfrac{1}{2}v_c^4\right)\frac{(P * P)(k)}{\sigma^4}. \tag{12.457}$$

where P_c is the Fourier transformed of ξ_c. Taking the low-k limit, we have

$$P(k) \propto k^{n_S} \tag{12.458}$$

and

$$(P * P)(k) = \int \frac{d^3q}{(2\pi)^3} P(q) P(k - q) \tag{12.459}$$

$$\simeq \int \frac{d^3q}{(2\pi)^3} P(q) P(-q) \tag{12.460}$$

$$\propto k^0. \tag{12.461}$$

Therefore, considering $n_S > 0$, the first term in (12.457) goes to 0 in the low-k limit, while the second term goes to a constant, provided that (12.460) is converging to a non-zero constant.

The necessity (or not) of a counter-term is analysed by looking at the UV sensitivity of each term. Consider a certain regularization/smoothing of the δ field. The power spectrum $P = P_R$ picks up a dependence on the smoothing scale R

which, when integrated in a UV-divergent integral, creates a dependence of P_c on R. Thus the $P * P$ term will require a counter-term to cancel this dependence (while the P term does not)[2]. More explicitly, let us consider a family of regulator $W_R(\mathbf{x})$ such that $\lim_{R\to 0} W_R(\mathbf{x}) = \delta^{(D)}(\mathbf{x})$ or equivalently $\lim_{R\to 0} W_R(\mathbf{q}) = 1$. By definition the regularized fluctuation field is,

$$\delta_R(\mathbf{x}) = \int d^3\mathbf{y}\, W_R(\mathbf{x} - \mathbf{y})\delta(\mathbf{y}). \tag{12.462}$$

Thus

$$\xi_R(r) = \langle \delta_R(\mathbf{x})\delta_R(\mathbf{x} + \mathbf{r}) \rangle \tag{12.463}$$

$$= \int d^3\mathbf{y}\, d^3\mathbf{y}'\, W_R(\mathbf{x} - \mathbf{y})\, W_R(\mathbf{x} + \mathbf{r} - \mathbf{y}')\langle \delta(\mathbf{y})\delta(\mathbf{y}')\rangle$$

$$= \int d^3\mathbf{y}\, d^3\mathbf{r}'\, W_R(\mathbf{x} - \mathbf{y})\, W_R(\mathbf{x} - \mathbf{y} + \mathbf{r} - \mathbf{r}')\langle \delta(\mathbf{y})\delta(\mathbf{y} + \mathbf{r}')\rangle$$

$$= \int d^3\mathbf{y}\, d^3\mathbf{r}'\, W_R(\mathbf{x} - \mathbf{y})\, W_R(\mathbf{x} - \mathbf{y} + \mathbf{r} - \mathbf{r}')\xi(\mathbf{r}')$$

$$= \int d^3\mathbf{y}\, W_R(\mathbf{x} - \mathbf{y})(W_R * \xi)(\mathbf{x} - \mathbf{y} + \mathbf{r})$$

$$= \int d^3\mathbf{y}\, W_R(\mathbf{y})(W_R * \xi)(\mathbf{y} + \mathbf{r})$$

$$= \left(\tilde{W}_R * W_R * \xi \right)(\mathbf{r}) \tag{12.464}$$

where for the last line we defined

$$\tilde{W}_R(\mathbf{x}) \equiv W_R(-\mathbf{x}). \tag{12.465}$$

Assuming $W_R(\mathbf{x})$ to be real and even, the Fourier transformed of (12.464) reads

$$P_R(\mathbf{q}) = W_R^2(\mathbf{q})P(\mathbf{q}). \tag{12.466}$$

Now getting back to (12.457), the first term proportional to P_R converges to P so that the R dependence vanishes (regardless of any assumption on P). However, in the second term, we cannot simply take the $R \to 0$ limit inside the integral

$$P_R * P_R(\mathbf{k}) = \int \frac{d^3\mathbf{q}}{(2\pi)^3} W_R^4(\mathbf{q})P(\mathbf{q})P(\mathbf{k} - \mathbf{q}) \tag{12.467}$$

as it diverges when assuming boldly the low-k form of P to hold for arbitrarily high momentum q^3. Therefore, either we have to take into account the full k dependence of P (but its UV part is not known) or we can employ an EFT approach. In the latter

---

[2] Even if the integral would be convergent, the fact that it runs over non-perturbative wave number requires potential counterterms to capture effectively the physics out of the domain of validity, say for $k > k_{NL}$.
[3] The conclusion is the same even when taking into account the transfer function $T(\mathbf{q})$.

case one assumes the decoupling between the UV physics and checks it a posteriori (unless it can be proven from a known underlying theory.) Such decoupling has for consequence that observables are independent of the UV part of P. As such, we can use an arbitrary regularized P_R to the price of adding counter-terms in the theory. The R dependence of these counterterms are such that they cancel the arbitrariness of the regularization and allow recovery of the independence of observables regarding the UV physics. In particular, they cancel the divergence in $P_R * P_R$ when taking $R \to 0$. Eventually, the remaining R-independent terms can be matched to experimental data at a given k momentum. In this case the counter-term should have a k^0 dependence. Such a counter-term corresponds to a stochasticity correction $\epsilon_{2,R}$ in the power spectrum of fixed amplitude fluctuations such that

$$\lim_{R \to 0} \epsilon_{2,R} + \left(\tfrac{1}{2} + v_c^2 - \tfrac{1}{2}v_c^4\right) \frac{(P_R * P_R)(\mathbf{k})}{\sigma^4} \tag{12.468}$$

is finite and denoted as

$$\epsilon_2 + \left(\tfrac{1}{2} + v_c^2 - \tfrac{1}{2}v_c^4\right) \frac{(P*P)(\mathbf{k})}{\sigma^4}. \tag{12.469}$$

Consequently, the model is modified by including the stochasticicy counterterm and the power spectrum of fixed-amplitude fluctuations now reads as

$$P_c(\mathbf{k}) = \epsilon_2 + v_c^2 \frac{P(\mathbf{k})}{\sigma^2} + \left(\tfrac{1}{2} + v_c^2 - \tfrac{1}{2}v_c^4\right) \frac{(P*P)(\mathbf{k})}{\sigma^4}. \tag{12.470}$$

Exercise 12.6.2 Equality Scale Integrating the Bose–Einstein distribution, we get that the radiation energy density is related to the temperature T_{CMB} of the CMB photons by $\rho_{rad} = \frac{\pi^2}{15} T_{CMB}^4$. Use the measured values of the CMB temperature of $T_{CMB} = 2.725$K and matter density $\Omega_{m,0} = 0.28$ to calculate the scale factor of matter-radiation equality a_{eq}. Calculate the size of the horizon at a_{eq} and the wave number k_{eq} of fluctuations entering at matter-radiation equality. This is the characteristic scale, at which the transfer function transitions from the large-scale k^0 behaviour to the small-scale $\ln(k)/k^2$ behaviour.

SOLUTION:

Scale factor a_{eq} at matter-radiation equality

At matter-radiation equality we have

$$\Omega_r(a_{eq}) = \Omega_m(a_{eq}) \tag{12.471}$$

$$\Omega_{r,0} a_{eq}^{-4} = \Omega_{m,0} a_{eq}^{-3} \tag{12.472}$$

$$a_{eq} = \frac{\Omega_{r,0}}{\Omega_{m,0}} \tag{12.473}$$

$$a_{eq} = \frac{\rho_{r,0}}{\rho_c \Omega_{m,0}} \tag{12.474}$$

$$\tag{12.475}$$

Assuming only photons contribute to radiation energy-density, we get $\rho_{r,}$ $0 = \frac{\pi^2}{15} T_{\text{CMB}}^4$ and

$$a_{\text{eq}} = \frac{\pi^2}{15} \frac{T_{\text{CMB}}^4}{\rho_c \Omega_{m,0}} \qquad (12.476)$$

$$a_{\text{eq}} = 8.8 \times 10^{-5} h^{-2}. \qquad (12.477)$$

Contribution of Neutrinos

To be more precise in the computation of a_{eq}, we should take neutrinos into account. Before a_{eq}, the temperature is sufficiently large so that we can neglect the contribution from their masses. As for photons, neutrino energy density is obtained by integrating its distribution function (Fermi–Dirac this time)

$$\rho_\nu = 2 \int \frac{d^3\mathbf{p}}{(2\pi)^3} \frac{\sqrt{\mathbf{p}^2 + m_\nu^2}}{e^{\frac{p}{T_\nu}} + 1}. \qquad (12.478)$$

To see why we can disregard its mass, we can make the change of variable $x = \frac{p}{T_\nu}$

$$\rho_\nu = \frac{1}{\pi^2} T_\nu^4 \int_0^{+\infty} dx \; x^2 \frac{\sqrt{x^2 + \left(\frac{m_\nu}{T_\nu}\right)^2}}{e^x + 1} \qquad (12.479)$$

so that for $T_\nu \gg m_\nu$

$$\rho_\nu = \frac{1}{\pi^2} T_\nu^4 \int_0^{+\infty} dx \frac{x^3}{e^x + 1}. \qquad (12.480)$$

After integration, taking into account three families of neutrinos leads to

$$\rho_\nu = 3 \frac{7}{8} \frac{\pi^2}{15} T_\nu^4. \qquad (12.481)$$

To get the energy density contribution of neutrinos, we still need to know their temperature T_ν. It can be related to the temperature of photons T_γ from entropy conservation before and after electrons and positrons annihilate. This is a standard result in cosmology which gives $T_\nu = \left(\frac{4}{11}\right)^{\frac{1}{3}} T_\gamma$. Therefore,

$$\rho_\nu = 3 \frac{7}{8} \left(\frac{4}{11}\right)^{\frac{4}{3}} \rho_\gamma \qquad (12.482)$$

$$\rho_r = \left(1 + 3 \frac{7}{8} \left(\frac{4}{11}\right)^{\frac{4}{3}}\right) \rho_r^\gamma \qquad (12.483)$$

and the correction to a_{eq} reads as

$$a_{eq} = \left(1 + 3\frac{7}{8}\left(\frac{4}{11}\right)^{\frac{4}{3}}\right) a_{eq}^{v} \tag{12.484}$$

$$a_{eq} = 1.5 \times 10^{-4} h^{-2}. \tag{12.485}$$

Horizon $\chi(a_{eq})$

The size of the horizon at a_{eq} is

$$\chi(a_{eq}) = c \int_0^{t_{eq}} \frac{dt}{a(t)} \tag{12.486}$$

$$= c \int_0^{a_{eq}} \frac{da}{a^2 H(a)}. \tag{12.487}$$

In radiative-dominated era,

$$H(a) \simeq H_0 \sqrt{\Omega_{r,0} a^{-4}} \tag{12.488}$$

so that,

$$\chi(a_{eq}) = \frac{c \; a_{eq}}{H_0 \sqrt{\Omega_{r,0}}} \tag{12.489}$$

$$= \frac{c}{H_0 \Omega_{m,0}} \sqrt{\Omega_{r,0}} \tag{12.490}$$

$$= \sqrt{1 + 3\frac{7}{8}\left(\frac{4}{11}\right)^{\frac{4}{3}}} \frac{c}{H_0 \Omega_{m,0}} \frac{\pi T_{CMB}^2}{\sqrt{15\rho_c}} \tag{12.491}$$

$$\chi(a_{eq}) = 5.1 \times 10^2 h^{-2} \text{Mpc} \tag{12.492}$$

and the corresponding wave number k_{eq} gives

$$k_{eq} = \frac{2\pi}{\chi(a_{eq})} \tag{12.493}$$

$$k_{eq} = 1.2 \times 10^{-2} h^2 \text{Mpc}^{-1}. \tag{12.494}$$

Influence of Matter Around a_{eq}

The approximation made in (12.488) might be too crude, especially near matter-radiation equality. To be more precise, let us do the calculation taking into account matter energy density. Thus (12.488) becomes

$$H(a) \simeq H_0 \sqrt{\Omega_{r,0} a^{-4} + \Omega_{m0} a^{-3}} \tag{12.495}$$

so that

$$\chi(a_{eq}) = c \int_0^{a_{eq}} \frac{da}{a^2 H(a)} \tag{12.496}$$

$$= \frac{c}{H_0\sqrt{\Omega_{r,0}}} \int_0^{a_{eq}} \frac{da}{\sqrt{1+\frac{\Omega_{m,0}}{\Omega_{r,0}}a}} \tag{12.497}$$

$$= \frac{c}{H_0\sqrt{\Omega_{r,0}}} \int_0^{a_{eq}} \frac{da}{\sqrt{1+\frac{a}{a_{eq}}}} \tag{12.498}$$

$$= \frac{c\ a_{eq}}{H_0\sqrt{\Omega_{r,0}}} \int_0^1 \frac{dx}{\sqrt{1+x}} \tag{12.499}$$

$$= \frac{c\ a_{eq}}{H_0\sqrt{\Omega_{r,0}}} \left(2\sqrt{2}-2\right) \tag{12.500}$$

$$\chi(a_{eq}) = 4.2 \times 10^2 h^{-2} \text{Mpc} \tag{12.501}$$

and the wavenumber at equivalence becomes

$$k_{eq} = 1.4 \times 10^{-2} h^2 \text{Mpc}^{-1}. \tag{12.502}$$

Exercise 12.6.3 Fluid Equations Using the definitions of density, mean streaming velocity, and velocity dispersion in terms of the distribution function $f(\mathbf{x},\mathbf{p},\tau)$ and the conservation of phase space density $df/d\tau = 0$, derive the continuity and Euler equations for collisionless dark matter. You will need to use the energy momentum conservation of the homogeneous background Universe $\bar{\rho}' + 3\mathcal{H}\bar{\rho} = 0$.

SOLUTION:
Vlasov's Equation

Using Liouville's theorem for the dark matter distribution function $f(\mathbf{x},\mathbf{p},\tau)$,

$$\frac{df}{d\tau} = \frac{\partial f}{\partial \tau} + \frac{dx_i}{d\tau}\frac{\partial f}{\partial x_i} + \frac{dp_i}{d\tau}\frac{\partial f}{\partial p_i} = 0, \tag{12.503}$$

where summation on repeated indices are implied. Besides, the equations of motion for collisionless dark matter reads

$$\begin{cases} x_i' = \frac{p_i}{am} & (12.504a) \\ p_i' = -am\nabla_i\phi & (12.504b) \end{cases}$$

with ϕ the peculiar potential. Hence the collisionless Boltzmann's equation (or Vlasov's equation)

$$\underbrace{\left[\frac{\partial}{\partial \tau} + \frac{p_i}{am}\frac{\partial}{\partial x_i} - am\nabla_i\phi\frac{\partial}{\partial p_i}\right]}_{\mathcal{L}}f = 0 \tag{12.505}$$

where we have defined the Liouville operator $\mathcal{L}$ for notation convenience.

Continuity Equation

Now taking the 0th order comoving velocity moment of (12.505)

$$\int d^3\mathbf{p}\,\mathcal{L}f = 0 \tag{12.506}$$

we get

$$\frac{\partial}{\partial\tau}\left(\int d^3\mathbf{p}f\right) + \int d^3\mathbf{p}\left(\frac{p_i}{am}\frac{\partial f}{\partial x_i}\right) - \int d^3\mathbf{p}\left(am\nabla_i\phi\frac{\partial f}{\partial p_i}\right) = 0. \tag{12.507}$$

Using vanishing boundary conditions on f one can discards the third term and we obtain using definitions of ρ and v_i,

$$\frac{\partial}{\partial\tau}\left(\frac{a^3}{m}\rho\right) + \frac{\partial}{\partial x_i}\left(v_i\frac{a^3}{m}\rho\right) = 0. \tag{12.508}$$

Developing $\rho = \bar\rho(1+\delta)$ and using the energy-momentum conservation of the homogeneous background universe, i.e. $\bar\rho' + 3\mathcal{H}\bar\rho = 0$,

$$\frac{a^3}{m}\bar\rho'(1+\delta) + \frac{a^3}{m}\bar\rho\delta' + 3\frac{a^2a'}{m}\bar\rho(1+\delta) + \bar\rho\frac{a^3}{m}\frac{\partial}{\partial x_i}(v_i(1+\delta)) = 0 \tag{12.509}$$

$$\frac{a^3}{m}\bar\rho\left\{-3\mathcal{H} + 3\mathcal{H} + \frac{\partial}{\partial x_i}(v_i(1+\delta)) + \delta' + 3\mathcal{H}\delta - 3\mathcal{H}\delta\right\} = 0 \tag{12.510}$$

and finally we recover the continuity equation

$$\delta' + \frac{\partial}{\partial x_i}((1+\delta)v_i) = 0. \tag{12.511}$$

Euler Equation

Similarly, for the Euler equation, we need to take the first comoving velocity moments

$$\int d^3\mathbf{p}\frac{p_i}{am}\mathcal{L}f = 0 \tag{12.512}$$

which reads

$$\left(\int d^3\mathbf{p}\frac{p_i}{am}\frac{\partial f}{\partial\tau}\right) + \int d^3\mathbf{p}\left(\frac{p_i}{am}\frac{p_j}{am}\frac{\partial f}{\partial x_j}\right) - \int d^3\mathbf{p}\left(\frac{p_i}{am}am\nabla_j\phi\frac{\partial f}{\partial p_j}\right) = 0. \tag{12.513}$$

Integrating by part, the 1st term of (12.513) reads as

$$\frac{\partial}{\partial\tau}\left(\frac{a^3}{m}\rho v_i\right) + \frac{a^3\rho}{m}\mathcal{H}v_i \tag{12.514}$$

while the second term reads as

$$\frac{a^3}{m}\frac{\partial}{\partial x_j}(\sigma_{ij}\rho) + \frac{a^3}{m}\frac{\partial}{\partial x_j}(v_iv_j\rho) \tag{12.515}$$

and the last term as

$$-\nabla_j \phi \int d^3 \mathbf{p} \left[\tfrac{\partial}{\partial p_j}(p_i f) - \tfrac{\partial p_i}{\partial p_j} f \right] = \nabla_j(\phi) \tfrac{a^3}{m} \rho, \tag{12.516}$$

again assuming vanishing boundary conditions. Dividing (12.513) by $\tfrac{a^3}{m}\rho$ we get

$$\tfrac{1}{a^3 \rho} \tfrac{\partial}{\partial \tau}(a^3 \rho v_i) + \tfrac{1}{\rho} \tfrac{\partial}{\partial x_j}(v_i v_j \rho) + \mathcal{H} v_i = -\nabla_i(\phi) - \tfrac{1}{\rho} \tfrac{\partial}{\partial x_j}(\sigma_{ij}\rho). \tag{12.517}$$

Developing the first two terms and cancelling the background contributions using again $\bar{\rho}' + 3\mathcal{H}\bar{\rho} = 0$, we get

$$\tfrac{1}{a^3 \rho} \tfrac{\partial}{\partial \tau}(a^3 \rho v_i) + \tfrac{1}{\rho} \tfrac{\partial}{\partial x_j}(v_i v_j \rho)$$
$$= \tfrac{\partial v_i}{\partial \tau} + \tfrac{v_i}{1+\delta}\left[\delta' + \tfrac{\partial \delta}{\partial x_j}v_j\right] + v_i\left(\tfrac{\partial}{\partial x_j}v_j\right) + \left(v_j \tfrac{\partial}{\partial x_j}\right)v_i. \tag{12.518}$$

Eventually, using the continuity relation (12.511) which reads

$$\delta' + \tfrac{\partial \delta}{\partial x_j}v_j = -\left(\tfrac{\partial}{\partial x_j}v_j\right) \times (1+\delta) \tag{12.519}$$

we get the Euler equation

$$\tfrac{\partial v_i}{\partial \tau} + \left(v_j \tfrac{\partial}{\partial x_j}\right)v_i + \mathcal{H}v_i = -\nabla_i(\phi) - \tfrac{1}{\rho}\tfrac{\partial}{\partial x_j}(\sigma_{ij}\rho). \tag{12.520}$$

Exercise 12.6.4 Recursion Relations Starting from the k-space version of the continuity and Euler equations in a matter-only EdS Universe, and the ansatz

$$\delta(\mathbf{k},\tau) = \sum_{i=1}^{\infty} a^i(\tau)\delta^{(i)}(\mathbf{k}), \qquad \theta(\mathbf{k},\tau) = -\mathcal{H}(\tau)\sum_{i=1}^{\infty} a^i(\tau)\theta^{(i)}(\mathbf{k})$$

derive the recursion relations for the gravitational coupling kernels F_n and G_n relating the nth order fields to the linear density fields

$$\delta^{(n)}(\mathbf{k}) = \prod_{m=1}^{n}\left\{\int \tfrac{d^3 q_m}{(2\pi)^3}\delta^{(1)}(\mathbf{q}_m)\right\} F_n(\mathbf{q}_1,\ldots,\mathbf{q}_n)\delta^{(D)}(\mathbf{k}-\mathbf{q}|_1^n),$$

$$\theta^{(n)}(\mathbf{k}) = \prod_{m=1}^{n}\left\{\int \tfrac{d^3 q_m}{(2\pi)^3}\delta^{(1)}(\mathbf{q}_m)\right\} G_n(\mathbf{q}_1,\ldots,\mathbf{q}_n)\delta^{(D)}(\mathbf{k}-\mathbf{q}|_1^n).$$

SOLUTION:
Let us consider a matter-only EdS Universe, neglecting vorticity, the decaying mode, and taking the ansatz

$$
\begin{cases}
\delta(\mathbf{k}, \tau) = \displaystyle\sum_{n=1}^{\infty} a^n(\tau)\delta^{(n)}(\mathbf{k}) & \text{(12.521a)} \\[3ex]
\theta(\mathbf{k}, \tau) = -\mathcal{H}(\tau)\displaystyle\sum_{n=1}^{\infty} a^n(\tau)\theta^{(n)}(\mathbf{k}) & \text{(12.521b)}
\end{cases}
$$

as solution of Euler and continuity equations

$$
\delta'(\mathbf{k}) + \theta(\mathbf{k}) = -\int \frac{d^3q}{2\pi}\frac{d^3q'}{2\pi}\delta^{(D)}(\mathbf{k}-\mathbf{q}-\mathbf{q}') \tag{12.522a}
$$
$$
\times \alpha(\mathbf{q},\mathbf{q}')\theta(\mathbf{q})\delta(\mathbf{q}')
$$
$$
\theta'(\mathbf{k}) + \mathcal{H}\theta(\mathbf{k}) + \tfrac{3}{2}\Omega_m \mathcal{H}\delta(\mathbf{k}) = -\int \frac{d^3q}{2\pi}\frac{d^3q'}{2\pi}\delta^{(D)}(\mathbf{k}-\mathbf{q}-\mathbf{q}')
$$
$$
\times \beta(\mathbf{q},\mathbf{q}')\theta(\mathbf{q})\theta(\mathbf{q}'). \tag{12.522b}
$$

In this case $\Omega_m = 1$ and Friedman's equation gives us

$$
\mathcal{H}'(\tau) = -\tfrac{1}{2}\mathcal{H}^2(\tau). \tag{12.523}
$$

Therefore,

$$
\delta'(\mathbf{k}, \tau) = \sum_{n=1}^{\infty} a^n(\tau)\left[n\mathcal{H}(\tau)\delta^{(n)}(\mathbf{k}) \right] \tag{12.524a}
$$

and

$$
\theta'(\mathbf{k}, \tau) = \sum_{n=1}^{\infty} a^n(\tau)\left[\left(\tfrac{1}{2} - n\right)\mathcal{H}^2(\tau)\theta^{(n)}(\mathbf{k}) \right]. \tag{12.524b}
$$

Since we are only interested in the finite perturbation expansion[4], $\theta \times \delta$ at order n is just the Cauchy product, i.e.

$$
\theta(\mathbf{q}, \tau)\delta(\mathbf{q}', \tau) = \sum_{n=1}^{\infty} a^n(\tau)\left[-\mathcal{H}(\tau)\sum_{p=1}^{n-1}\theta^{(p)}(\mathbf{q})\delta^{(n-p)}(\mathbf{q}') \right] \tag{12.524c}
$$

$$
\theta(\mathbf{q}, \tau)\theta(\mathbf{q}', \tau) = \sum_{n=1}^{\infty} a^n(\tau)\left[\mathcal{H}^2(\tau)\sum_{p=1}^{n-1}\theta^{(p)}(\mathbf{q})\theta^{(n-p)}(\mathbf{q}') \right] \tag{12.524d}
$$

---

[4] The derivations can be extended to several resummation schemes. For example, using the usual limit of partial sums or a Borel resummation would lead to similar results as they are both compatible with product, sum, and derivative of series. The only care to be taken in these cases would be to consider the domain of convergence.

Inserting (12.524) in the continuity equation we get that for all $n \geq 1$,

$$
n\delta^{(n)}(\mathbf{k}) - \theta^{(n)}(\mathbf{k}) = \int \frac{d^3 q}{2\pi} \frac{d^3 q'}{2\pi} \delta^{(D)}(\mathbf{k} - \mathbf{q} - \mathbf{q}') \alpha(\mathbf{q}, \mathbf{q}')
$$
$$
\times \left[\sum_{p=1}^{n-1} \theta^{(p)}(\mathbf{q}) \delta^{(n-p)}(\mathbf{q}') \right].
$$
(12.525)

Now inserting the explicit dependence of $\delta^{(n)}$ and $\theta^{(n)}$ in terms of $\delta^{(1)}$ and the kernels F_n and G_n, we get the equality for the integrand of $\mathbf{q}_1, \ldots, \mathbf{q}_n$

$$
[n F_n(\mathbf{q}_1, \ldots, \mathbf{q}_n) - G_n(\mathbf{q}_1, \ldots, \mathbf{q}_n)] \delta^{(D)}(\mathbf{k} - \mathbf{q}|_1^n)
$$
$$
= \int \frac{d^3 q}{2\pi} \frac{d^3 q'}{2\pi} \delta^{(D)}(\mathbf{k} - \mathbf{q} - \mathbf{q}') \alpha(\mathbf{q}, \mathbf{q}') \times \left[\sum_{p=1}^{n-1} G_p(\mathbf{q}_1, \ldots, \mathbf{q}_p) \delta^{(D)}(\mathbf{q} - \mathbf{q}|_1^p) \right.
$$
$$
\left. F_{n-p}(\mathbf{q}_{p+1}, \ldots, \mathbf{q}_n) \delta^{(D)}(\mathbf{q} - \mathbf{q}|_{p+1}^n) \right].
$$
(12.526)

Killing the integral on $\mathbf{q}$ and $\mathbf{q}'$, we get the first recursion relation on the kernels

$$
n F_n(\mathbf{q}_1, \ldots, \mathbf{q}_n) - G_n(\mathbf{q}_1, \ldots, \mathbf{q}_n) = \sum_{p=1}^{n-1} \alpha(\mathbf{q}|_1^p, \mathbf{q}|_p^n + 1) G_p(\mathbf{q}_1, \ldots, \mathbf{q}_p)
$$
$$
\times F_{n-p}(\mathbf{q}_{p+1}, \ldots, \mathbf{q}_n)
$$
(12.527a)

Similarly, by inserting (12.524) into the Euler equation, we get the second recursion relation on the kernels which reads,

$$
(2n+1) G_n(\mathbf{q}_1, \ldots, \mathbf{q}_n) - 3 F_n(\mathbf{q}_1, \ldots, \mathbf{q}_n) = \sum_{p=1}^{n-1} 2\beta(\mathbf{q}|_1^p, \mathbf{q}|_p^n + 1) G_p(\mathbf{q}_1, \ldots, \mathbf{q}_p)
$$
$$
\times G_{n-p}(\mathbf{q}_{p+1}, \ldots, \mathbf{q}_n).
$$
(12.527b)

Finally, by combining (12.527a) and (12.527b) we obtain the recursion relations for the gravitational kernels

$$
F_n(\mathbf{q}_1, \ldots, \mathbf{q}_n) = \sum_{p=1}^{n-1} \frac{G_p(\mathbf{q}_1, \ldots, \mathbf{q}_p)}{(2n+3)(n-1)} \{ (2n+1)\alpha(\mathbf{q}|_1^p, \mathbf{q}|_{p+1}^n) F_{n-p}(\mathbf{q}_{p+1}, \ldots, \mathbf{q}_n)
$$
$$
+ 2\beta(\mathbf{q}|_1^p, \mathbf{q}|_{p+1}^n) G_{n-p}(\mathbf{q}_{p+1}, \ldots, \mathbf{q}_n) \} \quad (12.528)
$$

$$G_n(\mathbf{q}_1,\ldots,\mathbf{q}_n) = \sum_{p=1}^{n-1} \frac{G_p(\mathbf{q}_1,\ldots,\mathbf{q}_p)}{(2n+3)(n-1)} \{3\alpha(\mathbf{q}|_1^p,\mathbf{q}|_{p+1}^n)F_{n-p}(\mathbf{q}_{p+1},\ldots,\mathbf{q}_n)$$

$$+ 2n\beta\ (\mathbf{q}|_1^p,\mathbf{q}|_{p+1}^n)G_{n-p}(\mathbf{q}_{p+1},\ldots,\mathbf{q}_n)\} \quad (12.529)$$

Exercise 12.6.5 Two-Loop Power Spectrum Write down the the diagrams and integrals contributing to the two-loop matter power spectrum in terms of the gravitational coupling kernels F_n. Try to identify the diagrams with the strongest UV-sensitivity.

SOLUTION:
Diagrams and Integral at Two-Loop Order

In perturbation theory, connected diagrams contributing to the power spectrum $P(\mathbf{k}) \equiv \langle \delta(\mathbf{k})\delta(-\mathbf{k})\rangle$ contains only two vertices F_{n_1} and F_{n_2} arising from the expansion of δ functions using the ansatz (12.521b). Using the topological identity for connected planar graphs $L = I - V + 1$ where L is the number of loops, I the number of internal lines and V the number of vertices, leads to

$$n_1 + n_2 = 6 \quad (12.530)$$

for two-loop diagrams. Two-loop contributions to $P(\mathbf{k})$ corresponds to any contribution verifying (12.530).
Therefore[5],

$$P_{2-\text{loop}}(\mathbf{k}) = 2\langle\delta^{(1)}(\mathbf{k})\delta^{(5)}(-\mathbf{k})\rangle + 2\langle\delta^{(2)}(\mathbf{k})\delta^{(4)}(-\mathbf{k})\rangle + \langle\delta^{(3)}(\mathbf{k})\delta^{(3)}(-\mathbf{k})\rangle, \quad (12.531)$$

$$\equiv P_{15}(\mathbf{k}) + P_{24}(\mathbf{k}) + P_{33}(\mathbf{k}), \quad (12.532)$$

where $P_{n_1 n_2}(\mathbf{k})$ corresponds to the sum of all diagrams with vertices F_{n_1} and F_{n_2}. Assuming Gaussianity of $\delta^{(1)}(\mathbf{k})$ one can apply the Wick theorem leading to four distinct diagrams depicted in Figure 12.14. Defining $F_n^{(s)}$ as the symmetrized kernel F_n, associated integrals read as

$$P_{15}(\mathbf{k}) = 30 \int \frac{d^3 q_1}{(2\pi)^3} \frac{d^3 q_2}{(2\pi)^3} F_5^{(s)}(\mathbf{q}_1,-\mathbf{q}_1,\mathbf{q}_2,-\mathbf{q}_2,\mathbf{k})$$

$$\times P_{11}(\mathbf{q}_1)P_{11}(\mathbf{q}_2)P_{11}(\mathbf{k}), \quad (12.533a)$$

$$P_{24}(\mathbf{k}) = 24 \int \frac{d^3 q_1}{(2\pi)^3} \frac{d^3 q_2}{(2\pi)^3} F_4^{(s)}(\mathbf{q}_1,\mathbf{k}-\mathbf{q}_1,\mathbf{q}_2,-\mathbf{q}_2)F_2^{(s)}(-\mathbf{q}_1,-(\mathbf{k}-\mathbf{q}_1))$$

$$\times P_{11}(\mathbf{q}_1)P_{11}(\mathbf{k}-\mathbf{q}_1)P_{11}(\mathbf{q}_2), \quad (12.533b)$$

---

[5] The time-dependence is omitted for clarity. Assuming the time-evolution factorizes at linear order leads to a simple power of linear growth factor.

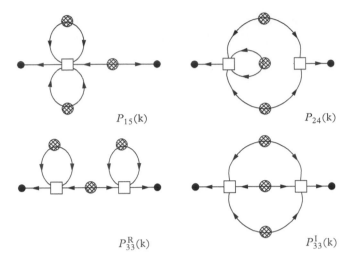

$P_{15}(k)$

$P_{24}(k)$

$P_{33}^{R}(k)$

$P_{33}^{I}(k)$

Figure 12.14 *Two-loop diagrams contributing to the power spectrum.*

$$P_{33}^{R}(\mathbf{k}) = 9 \int \frac{d^3 q_1}{(2\pi)^3} \frac{d^3 q_2}{(2\pi)^3} F_3^{(s)}(\mathbf{q}_1, -\mathbf{q}_1, \mathbf{k}) F_3^{(s)}(\mathbf{q}_2, -\mathbf{q}_2, -\mathbf{k})$$
$$\times P_{11}(\mathbf{q}_1) P_{11}(\mathbf{q}_2) P_{11}(\mathbf{k}), \qquad (12.533c)$$

$$P_{33}^{I}(\mathbf{k}) = 6 \int \frac{d^3 q_1}{(2\pi)^3} \frac{d^3 q_2}{(2\pi)^3} F_3^{(s)}(\mathbf{q}_1, \mathbf{q}_2, \mathbf{k} - \mathbf{q}_1 - \mathbf{q}_2) F_3^{(s)}(-\mathbf{q}_1, -\mathbf{q}_2, -(\mathbf{k} - \mathbf{q}_1 - \mathbf{q}_2))$$
$$\times P_{11}(\mathbf{q}_1) P_{11}(\mathbf{q}_2) P_{11}(\mathbf{k} - \mathbf{q}_1 - \mathbf{q}_2), \qquad (12.533d)$$

where $P_{33}^{R}(\mathbf{k})$ and $P_{33}^{I}(\mathbf{k})$ correspond respectively to the 1PR and 1PI diagram with two F_3 vertices.

Ultraviolet Sensitivity

To analyse systematically the ultraviolet sensitivity of integrals in (12.533) we can in principle simply apply Weinberg's asymptotic theorem [26] to the integrands. However, it quickly becomes cumbersome to analyse all possible subspaces of integration and the associated asymptotic coefficients of the gravitational coupling kernels $F_n^{(s)}$ in these subspaces. Instead we can derive an upper bound on the superficial degree of divergence D_{UV} for any diagram.

The linear power spectrum P_{11} scales as a power n of the momentum in the UV limit. Combined with a factor 3 for each loop integration leads to

$$D_{UV} \leq 3L - nI + \sum_i D\left(F_{n_i}^{(s)}\right) \qquad (12.534)$$

where the sum is on the set of vertices (inside a loop) and $D(F_{n_i}^{(s)})$ is an upper bound on the asymptotic coefficients of $F_{n_i}^{(s)}$. Since for $p \gg k$ [3]

$$F_{n_i}^{(s)}(\mathbf{q}_1, \ldots, \mathbf{q}_{n_i} \ 2\mathbf{p}, -\mathbf{p}) \propto \frac{k^2}{p^2}, \tag{12.535}$$

it is clear that $D(F_{n_i}^{(s)}) \geq -2$. However, the asymptotic behaviour of the kernel could be different when taking more than two momenta in the UV limit. To easily obtain an upper bound on $D(F_{n_i}^{(s)})$ we typically use a decoupling argument from the UV physics so that $D(F_{n_i}^{(s)}) < 0$ (otherwise the mass fluctuation derived in perturbation theory would strongly depend on the UV initial conditions.) See [8] where it is claimed that $D(F_{n_i}^{(s)}) < 0$ has been checked by explicit computation for $n_i \leq 6$ (which is enough in our case.) Then, because $F_{n_i}^{(s)}$ is a rational function in the momenta $D(F_{n_i}^{(s)})$ can only be an integer so that $D(F_{n_i}^{(s)}) \leq -1$. Thus,

$$D_{\mathrm{UV}} \leq 3L - nI - V, \tag{12.536}$$

where V is the number of vertices (inside a loop) of the diagram. Eventually, using (12.535), (12.536) can be refined to

$$D_{\mathrm{UV}} \leq 3L - nI - V - T, \tag{12.537}$$

where T denotes the number of tadpoles of the diagram.

Applying (12.537) to two-loop diagrams leads to

$$D_{\mathrm{UV}}(P_{15}) \leq 3 + 2n, \tag{12.538a}$$

$$D_{\mathrm{UV}}(P_{24}) \leq 3 + 3n, \tag{12.538b}$$

$$D_{\mathrm{UV}}(P_{33}^{\mathrm{R}}) \leq 2 + 2n, \tag{12.538c}$$

$$D_{\mathrm{UV}}(P_{33}^{\mathrm{I}}) \leq 4 + 3n. \tag{12.538d}$$

As seen in (12.538) the UV-sensitivity of a diagram depends on n. From now on we assume the inequalities (12.538) to be saturated. Regardless of n

$$D_{\mathrm{UV}}(P_{33}^{\mathrm{R}}) \leq D_{\mathrm{UV}}(P_{15}), \tag{12.539a}$$

$$D_{\mathrm{UV}}(P_{24}) \leq D_{\mathrm{UV}}(P_{33}^{\mathrm{I}}), \tag{12.539b}$$

so that the most UV-sensitive two-loop diagram is either P_{15} or P_{33}^{I}. We conclude that the most UV-sensitive diagram depends on the model. *If $n > -1$ then the most UV-sensitive diagram is P_{33}^{I} else it is P_{15}.* Note that to be complete, we should also check the superficial degree of divergence of any subdiagram.

References

[1] R. Alonso, E. E. Jenkins, and A. V. Manohar (2014). Holomorphy without Supersymmetry in the Standard Model Effective Field Theory. *Phys. Lett. B* 739: 95–8.

[2] R. Alonso, E. E. Jenkins, and A. V. Manohar, and M. Trott (2014). Renormalization Group Evolution of the Standard Model Dimension Six Operators III: Gauge Coupling Dependence and Phenomenology. *JHEP* 4: 159.

[3] T. Baldauf (2019). Effective Field Theory of Large Scale Structure. In *Les Houches Summer School on EFT in Particle Physics and Cosmology*. July 3–28, 2017, Les Houches, France.

[4] W. Buchmuller and D. Wyler (1986). Effective Lagrangian Analysis of New Interactions and Flavor Conservation. *Nucl. Phys. B* 268: 621.

[5] C. P. Burgess (2019). Introduction to Effective Field Theories and Inflation. Chapter 4, this volume.

[6] J. Gasser and H. Leutwyler (1984). Chiral Perturbation Theory to One Loop. *Annals Phys.* 158: 142.

[7] B. M. Gavela, E. E. Jenkins, A. V. Manohar, and L. Merlo (2016). Analysis of General Power Counting Rules in Effective Field Theory. *Eur. Phys. J. C* 76(9): 485.

[8] M. H. Goroff, B. Grinstein, S. J. Rey, and M. B. Wise (1986). Coupling of Modes of Cosmological Mass Density Fluctuations. *Astrophys. J.* 311: 6–14.

[9] A. Hanany, E. E. Jenkins, A. V. Manohar, and G. Torri (2011). Hilbert Series for Flavor Invariants of the Standard Model. *JHEP* 3: 096.

[10] B. Henning, X. Lu, T. Melia, and H. Murayama (2017). 2, 84, 30, 993, 560, 15456, 11962, 261485, . . . : Higher-Dimension Operators in the SMEFT. *JHEP* 08: 016.

[11] B. Henning, X. Lu, T. Melia, and H. Murayama (2017). Operator Bases, S-matrices, and their Partition Functions. *JHEP* 10: 199.

[12] E. E. Jenkins and A. V. Manohar (2009). Algebraic Structure of Lepton and Quark Flavor Invariants and CP Violation. *JHEP* 10: 094.

[13] E. E. Jenkins, A. V. Manohar, and M. Trott (2013). Renormalization Group Evolution of the Standard Model Dimension Six Operators I: Formalism and λ Dependence. *JHEP* 10: 087.

[14] E. E. Jenkins, A. V. Manohar, and M. Trott (2014). Renormalization Group Evolution of the Standard Model Dimension Six Operators II: Yukawa Dependence. *JHEP* 01: 035.

[15] L. Lehman and A. Martin (2015). Hilbert Series for Constructing Lagrangians: Expanding the Phenomenologist's Toolbox. *Phys. Rev. D* 91: 105014.

[16] A. V. Manohar and H. Georgi (1984). Chiral Quarks and the Nonrelativistic Quark Model. *Nucl. Phys. B* 234: 189.

[17] A. V. Manohar (1997). Effective Field Theories. In H. Latal and W. Schweiger (Eds), *Perturbative and Nonperturbative Aspects of Quantum Field Theory*, 311–62, Springer.

[18] A. V. Manohar (1997). The HQET/NRQCD Lagrangian to Order α s/m3. *Phys. Rev. D* 56: 230–7.

[19] A. V. Manohar (2003). Deep Inelastic Scattering as x $\rightarrow 1$ Using Soft-Collinear Effective Theory. *Phys. Rev. D* 68: 114019.

[20] A. V. Manohar (2018). Introduction to Effective Field Theories. In Les Houches Summer School on EFT in Particle Physics and Cosmology. July 3–28, 2017, Les Houches, France.

[21] Neubert, Matthias (2019). Introduction to Renormalisation and the Renormalisation Group. In *Les Houches Summer School on EFT in Particle Physics and Cosmology*. July 3–28, 2017, Les Houches, France.

[22] C. Patrignani, et al. (2016). Review of Particle Physics. *Chin. Phys. C* 40(10): 100001.

[23] A. Pich (2019) Effective Field Theory with Nambu–Goldstone Modes. Chapter 3, this volume.

[24] A. Pich (2018). Effective Field Theory with Nambu–Goldstone Modes. In *Les Houches Summer School on EFT in Particle Physics and Cosmology*. July 3–28, 2017, Les Houches, France.

[25] U. van Kolck (2019). Effective Field Theories for Nuclear and (Some) Atomic Physics. In *Les Houches Summer School on EFT in Particle Physics and Cosmology*. July 3–28, 2017, Les Houches, France.

[26] S. Weinberg (1960). High-Energy Behavior in Quantum Field Theory. *Phys. Rev.* 118: 838–49.

[27] S. Weinberg (1979). Baryon and Lepton Nonconserving Processes. *Phys. Rev. Lett.* 43: 1566–70.